Robynn Vargas

							8A
							2 He 4.0026

		3A	4A	5A	6A	7A	
		5 B 10.81	6 C 12.011	7 N 14.0067	8 O 15.9994	9 F 18.998	10 Ne 20.179

	1B	2B	13 Al 26.9815	14 Si 28.085	15 P 30.974	16 S 32.06	17 Cl 35.453	18 Ar 39.948
28 Ni 8.69	29 Cu 63.546	30 Zn 65.39	31 Ga 69.72	32 Ge 72.59	33 As 74.922	34 Se 78.96	35 Br 79.904	36 Kr 83.80
46 Pd 6.42	47 Ag 107.868	48 Cd 112.41	49 In 114.82	50 Sn 118.71	51 Sb 121.75	52 Te 127.60	53 I 126.905	54 Xe 131.29
78 Pt 95.08	79 Au 196.967	80 Hg 200.59	81 Tl 204.383	82 Pb 207.2	83 Bi 208.98	84 Po 210.0	85 At 210.0	86 Rn 222.0

64* Gd 7.25	65* Tb 158.925	66* Dy 162.50	67* Ho 164.930	68* Er 167.26	69* Tm 168.934	70* Yb 173.04	71* Lu 174.967
96† Cm 2.1	97† Bk 247.1	98† Cf 252.1	99† Es 252.1	100† Fm 257.1	101† Md 256.1	102† No 259.1	103† Lr 260.1

ORGANIC
CHEMISTRY

THIRD EDITION

ORGANIC CHEMISTRY

THIRD EDITION

G. MARC LOUDON

PURDUE UNIVERSITY

The Benjamin/Cummings Publishing Company, Inc.

A Division of Addison-Wesley Publishing Company

Redwood City, California § Menlo Park, California
Reading, Massachusetts § New York § Don Mills, Ontario § Wokingham, U.K.
Amsterdam § Bonn § Paris § Milan § Sydney § Singapore § Tokyo
Seoul § Taipei § Mexico City § San Juan

SPONSORING EDITOR: *Anne Scanlan-Rohrer*
ASSISTANT EDITOR: *Leslie With*
EDITORIAL ASSISTANT: *Karmen Butterer*
PRODUCTION SUPERVISOR: *Rani Cochran*
PROJECT MANAGER AND ART COORDINATOR: *Joan Keyes, Jonathan Peck Typographers*
BOOK DESIGN: *Patrick Devine, Jonathan Peck Typographers*
DESIGN MANAGER: *Michele Carter*
COVER DESIGN: *Yvo Riezebos*
COVER ILLUSTRATION: *Joseph Maas*
COPY EDITOR: *John A. Servideo*
FIGURATIVE ART: *Alexander Teshin Associates and Jonathan Peck Typographers*
COMPOSITION AND FILM COORDINATOR: *Vivian McDougal*
COMPOSITION AND FILM: *Jonathan Peck Typographers*
EXECUTIVE MARKETING MANAGER: *Anne Emerson*
MARKETING COMMUNICATIONS MANAGER: *Nathalie Mainland*
SENIOR MANUFACTURING COORDINATOR: *Merry Free Osborn*
PRINTING: *Von Hoffmann*

Library of Congress Cataloging-in-Publication Data

Loudon, G. Marc.
    Organic Chemistry/ G. Marc Loudon.—3rd ed.
      p.   cm.
    Includes index.
    ISBN 0-8053-6650-4
    1. Chemistry, Organic.   I. Title
OD251.2.L68   1995
547—dc20   94-42763
                                                         CIP

2 3 4 5 6 7 8 9 10 — VH — 98 97 96 95

The Benjamin/Cummings Publishing Company, Inc.
390 Bridge Parkway
Redwood City, CA 94065

To my teachers, especially
Don Noyce and Dan Koshland

"The beginning is the most important part of the work."
—Plato

Preface

This book was written because I felt the need for a text that corresponds closely to the course I teach. Although it is organized along tried-and-true functional-group lines, the book contains some unique features that have served me well in both my teaching and my learning of organic chemistry.

I have had four major concerns in both the initial writing and the revisions of this text: *readability, presentation, organization,* and *scientific accuracy.*

Readability

A number of techniques were employed to enhance readability.

1. **Common everyday analogies** bring chemical phenomena within the student's own frame of reference. Two of the many examples can be found on pp. 249 and 705.

2. **Numerous figures and diagrams** illustrate important concepts. This edition features an increased use of molecular models drawn to scale in both ball-and-stick and space-filling representations.

3. **Frequent cross-referencing** within the text assists students in finding initial discussions of seminal topics.

4. **Guidance on problem solving** helps students see learning as a process rather than as a mere assimilation of facts. This guidance is provided in two major ways. First are the solved Study Problems, new to this edition, that can be found in each chapter; an example is Study Problem 9.2, p. 425. In addition, some of the *Study Guide Links* (discussed in more detail below) deal in very detailed ways with approaches to solving certain important and frequently occurring types of problems; see, for example, the callout to Study Guide Link 4.7, p. 172. In addition, hints in many problems help put students on the right track without forcing them to consult the solutions; Problem 9.42, p. 436, illustrates this technique.

5. **Framing problems in humorous or "real-life" contexts,** illustrated by Problem 14.38, p. 674, helps reduce the formality of the presentation.

6. **Frequent summaries and recapitulations of key ideas** help students see the forest while amidst the trees. A **Key Ideas** summary at the end of each chapter brings together important concepts from within the chapter, often from a somewhat different perspective than is found in the text itself. In a number of sections, important points are collected into numbered lists, as in Sec. 9.4F. A few reactions that seem to cause students particular difficulty are summarized in tables (for example, Table 5.1, p. 191, and Table 9.4, p. 424) or in retrosynthetic diagrams within the text (for example, Eq. 22.59, p. 1071). Appendices at the end of the text summarize rules of nomenclature, key spectroscopic information, acidity and basicity data, and synthetic methods. Because I try to discourage students from memorizing tables of reactions, I have placed reaction summaries within the *Study Guide and Solutions Manual* supplement rather than within the text. I encourage students to make their own summaries and check them against those in the supplement.

7. **Use of a second color within chemical equations,** a device originated in organic chemistry texts with the first edition of this book, helps students follow the changes involved; Eq. 5.9, p. 179, is one of literally hundreds of examples that employ this technique. Yet this is not done in every equation, because eventually students must learn to parse equations for themselves. I believe that the decoding of color in equations must be completely intuitive and must not require detailed explanation. The use of four colors was intentionally avoided for this reason; despite its cosmetic appeal, it can actually be a source of confusion in many cases.

8. **Historical sketches presented in anecdotal style** stress that chemistry is a human endeavor, and that the road to discovery is paved with serendipity and humor. Examples are found in Sec. 6.12 and in the boxed sidebar on pp. 893–894.

Presentation

A number of elements of the presentation used in this text are worth special emphasis.

1. I have chosen a mechanistic approach within the overall functional-group framework of the text. This emphasis has as its basis the philosophy that the only way for students truly to understand organic chemistry is for them to see the unifying elements that connect what might at first appear to be unrelated phenomena. Thus, a student will learn a given reaction more easily when its mechanistic connection to an earlier reaction is apparent. Eqs. 11.25 and 11.26, p. 508, and Eq. 22.52, p. 1067, illustrate this approach. I have set the stage for all mechanistic discussions of heterolytic reactions with an introductory chapter on acid-base reactions (Chapter 3), and have subdivided Lewis acid-base reactions into two types, Lewis acid-base associations and electron-pair displacement reactions. Out of this introduction flows logically the "curved-arrow" formalism, which is also introduced and thoroughly explained in Chapter 3, and then used consistently throughout the text.

2. Acid-base chemistry—both Lewis and Brønsted—has always been a focus of this text, but with the introduction of a new chapter (Chapter 3) devoted to this subject, this edition gives even more emphasis to acid-base chemistry as the foundation for understanding a great portion of organic chemistry.

3. One of the key elements in the presentation of organic chemistry by any textbook is the *mechanistic centerpiece*—the reaction used to introduce the notions of mechanism such as multistep reactions, reactive intermediates, reaction free-energy diagrams,

rate-limiting step, and the like. Traditionally, free-radical halogenation has been used for this purpose, but an increasingly large number of texts have followed the practice, introduced in the first edition of this text, of using a heterolytic reaction as the mechanistic centerpiece, because it is this type of reaction that will occupy the student's attention for most of the course. Two types of heterolytic reactions have been most popular for use as the mechanistic centerpiece: additions to alkenes and nucleophilic substitution reactions. I have chosen the former, because an unsymmetrical alkene can in principle undergo two competing addition reactions. The relative merits of the two reaction pathways can be evaluated by a direct comparison of transition-state stabilities (and through Hammond's postulate, the relative stabilities of the reactive intermediates involved); the issue is not complicated by the relative stabilities of starting materials, because the starting materials are the same for both pathways.

Free-radical reactions are not ignored, but are merely postponed until students have had a chance to master the essentials of heterolytic reactions. At that time (Sec. 5.6), free-radical reactions (and the corresponding "fishhook" formalism) are introduced and thoroughly discussed.

4. This text contains 1506 problems, many with multiple parts, some within the text and others at the ends of the chapters. Problems utilize in many cases a paired-problem format that is new to this edition (discussed below). Problems range in difficulty from drill-type problems to problems that will challenge the best students.

5. A thorough discussion of stereochemistry continues to be a hallmark of this text, beginning with alkene stereochemistry in Sec. 4.1B. Unique to this text is correct use of the term *stereocenter* (Secs. 4.1B and 6.1B), a discussion of reaction stereochemistry (Secs. 7.7–7.9), and a discussion of chemical equivalence (Sec. 10.8A) that provides the basis for understanding both NMR spectra and enzyme-catalyzed reactions.

6. The first edition of this text brought to introductory organic chemistry the important topic of solvents in organic chemistry, and the relationship of gas-phase chemistry to solvent effects. This topic is important in this edition as well, including Sec. 9.4D, which discusses the important concept of nucleophilicity in terms of solvent effects.

7. A systematic method for identifying oxidations and reductions is continued in this edition.

8. A rational approach to the construction of organic syntheses is also continued in this edition.

9. In recognition of the fact that a large number of students who take organic chemistry have special interests in the life sciences, biological examples of organic chemistry are sprinkled throughout the text. These are not set apart in "special topics" chapters, but instead are included in sections adjacent to the relevant laboratory chemistry. In providing these examples, I have not forgotten that this is a chemistry text and not a biochemistry text; the underlying theme of these examples is not the details of the biology involved, but rather the close analogy of biological chemistry to laboratory reactions.

10. Nomenclature is treated thoroughly because I believe that after a first course in organic chemistry, a student should be able to construct a systematic name for any simple organic compound.

11. Finally, I have provided discussions of industrial processes (including polymer chemistry) as some indication of the important role that organic chemistry plays in today's economy. Because we teach so many students who will not become practicing chemists, some of whom may become influential in shaping public policy, I believe that these students must leave our classes convinced that organic chemistry is a potent economic force, which can and does provide significant social and economic benefits.

Organization

Chapters 1–3 of the text deal respectively with structure and bonding, alkanes and nomenclature, and acids and bases. Alkenes, alkyl halides, alcohols, ethers, epoxides, thiols, and sulfides have been grouped together in the next chapters (Chapters 4–11) for two reasons: first, because the chemistry of these functional types is strongly interrelated; and second, because a substantial amount of nonhydrocarbon chemistry can thereby be covered in the early part of the course. Following an interlude (Chapters 12–13) dealing with NMR, IR, and mass spectroscopy (which could be placed anywhere with little adjustment in a course that uses this text), are chapters on alkynes, dienes, resonance, and aromatic chemistry (Chapters 14–18), including a chapter that deals with allylic and benzylic reactivity (Chapter 17). Then comes carbonyl chemistry (Chapters 19–22), the last chapter of which is devoted to a discussion of enols, enolate ions, and condensation reactions. This is followed by amine and heterocyclic chemistry (Chapters 23–24), pericyclic reactions (Chapter 25), and finally, introductory treatments of amino acid, peptide and protein chemistry (Chapter 26) as well as carbohydrate and nucleic acid chemistry (Chapter 27).

Scientific Accuracy

Each topic in this book was researched back to the original or review literature. I am indebted to the many students, reviewers, and faculty who have sent me concrete suggestions for improving both the accuracy and the presentation of the text. We have also attempted to ensure typographic accuracy to the greatest extent humanly possible through a five-stage revision process, and through the involvement of accuracy checkers during both the manuscript and page-proof stages of production. The *Study Guide and Solutions Manual* has been through a similar process.

Changes within the Third Edition

The most obvious change within the third edition is Chapter 3 (Acids and Bases; the Curved-Arrow Formalism). In addition, Chapters 1 and 2 of the second edition have been consolidated into Chapter 1 of this edition; a section on the Simmons-Smith reaction (Sec. 9.8B) has been provided in response to user requests; and a section on ether synthesis (Sec. 11.1C) has been added for the mechanistic insight it affords. Carbon NMR has received increased emphasis in the spectroscopy sections, including a short discussion of the DEPT technique. Most importantly, the entire text has been completely rewritten; many organizational changes will be found in the way that topics have been presented.

A major facet of this edition is the introduction of several important new pedagogical devices. First, I have included numerous *Study Problems* that illustrate not only the answers to problems, but also the *solutions* to them; that is, the intellectual processes by which these answers were obtained; an example is Study Problem 10.6 on p. 487.

Second, I have introduced a unique "paired problems" approach. Problems (or problem parts) marked with an asterisk are solved in the *Study Guide and Solutions Manual*. In many cases, such a problem (or problem part) is followed by an unmarked problem (or part) for which the answer is *not* provided in the supplement. Each unmarked problem is designed to be *of the same type as the problem that precedes it and of equal or lesser difficulty.* The purpose of this approach is to decrease students' dependence on the *Study Guide and Solutions Manual* for answers, especially those students who tend to consult solutions before they have wrestled adequately with the problems. Because instructors may differ in their philosophies about this issue, I have also provided the answers to the unmarked problems in an *Instructor's Solutions Manual to Accompany Organic Chemistry*. Instructors of adopting institutions may, without further permission, duplicate or post these additional solutions or otherwise make them available to students if they desire. Alternatively, instructors may also find the unmarked problems a convenient source of examination questions. Students may wish to use the marked problems and their solutions as additional study problems, and then test themselves on the unmarked problems. Not every problem is paired; some problems, particularly the more difficult ones, do not have an unmarked partner. The scheme is simple: if a problem (or part) has no asterisk, the previous asterisked problem (or part) is its paired partner.

The third pedagogical change is the introduction of a new and unique feature, the *Study Guide Links*. These are discussions within the *Study Guide and Solutions Manual* that provide additional instruction or details on key topics. They are called out in the left margin of the text with a "pointed finger" icon. The callout for Study Guide Link 3.2, p. 93, is an example of such a callout. There are two types of Study Guide Links. Those marked with a check (✓) provide extra assistance with key topics that I have found through experience are either especially important or difficult for many students. Those not marked with a check provide more in-depth information about a topic. Although a case could be made that such elements should be included within the text, their incorporation would have made the text unacceptably long. Furthermore, I believe that the digressive nature of these elements would have distracted from the flow of the text.

Users of the second edition will notice that this edition is somewhat longer. The major sources of this additional length are the Study Problems and the paired problems. Despite the tendency toward shorter textbooks, I believe that the increased length can be justified by the potential benefits of these additions. Other aspects of the text that account for its greater length are more complete explanations of traditionally difficult topics, a more open design, and a somewhat more spacious typeface. Thus, students using this edition will not be contending with a significantly different amount of information; rather, they will have more assistance in *understanding* the information that is there.

Supplements

Several supplements augment this text. The most important of these is the *Study Guide and Solutions Manual* by G. Marc Loudon and Joseph G. Stowell, which is available for student purchase. Composed in an attractive format much like that of the text, this supplement contains terms, conceptual outlines, reaction summaries, Study Guide Links, and the solutions to asterisked problems. The *Instructor's Solutions Manual to Accompany Organic Chemistry*, by G. Marc Loudon, which contains answers to all nonasterisked problems, is available free of charge to instructors at qualified adopting institutions. A set of seventy-five two-color *Transparency Acetates* for use with overhead projectors is also available free of charge to each qualified adopting institution. Perhaps most exciting is an original software supplement, *Computer Animation Modules for Organic Chemistry*, by David Allen and G. Marc Loudon, which consists of professionally produced, full-color animations of organic chemistry concepts that have spatial and/or dynamic components which are difficult to illustrate with static media such as a chalkboard or printed page. These modules are suitable for use in lecture demonstrations or computer laboratories, and require a Macintosh II computer (or later) with 8 megabytes of RAM. They were produced as part of a Purdue University-sponsored initiative on computer-assisted instruction, and have been extensively class-tested by the authors. These modules are also provided free of charge to qualified adopters. Rounding out the supplement package is a *ChemDraw/Chem3D Viewer* demonstration software package from Cambridge Scientific Computing, Inc., free to students who purchase the text; and the Benjamin/Maruzen *HGS Molecular Structure Models* set, which is available for student purchase.

My colleague Professor Phil Fuchs in the Department of Chemistry at Purdue has also developed a complete set of lecture notes based on this text which are available through the Stipes Publishing Company, 10–12 Chester Street, Champaign, Illinois. Institutions that adopt Professor Fuch's notes are given a complimentary set of transparencies created by Professor Fuchs to accompany his notes. (Please note that these lecture notes are *not* available through Benjamin/Cummings.)

Acknowledgements

I am particularly indebted to my Department Head, Professor Stephen R. Byrn, and to my Dean, Charles O. Rutledge, for providing encouragement for this project, an atmosphere in which it could be completed, and a sabbatical semester that made timely completion possible. I gratefully acknowledge the able assistance of Ms. Vicki Killion and Mr. John Pinzelik of the Purdue Libraries for their constant help and assistance in the library phase of this project. Thanks go to my colleagues, John Schwab, V. Jo Davisson, Mark Cushman, Don Bergstrom, Joe Stowell, Phil Fuchs, John Grutzner, and George Bodner for advice, suggestions, and reactions to this text and the teaching that sprang from it. Special thanks go to the many reviewers and accuracy checkers, listed on subsequent pages, for their critical reading of the manuscript; but thanks particularly go to Professor Ron Magid, an award-winning teacher at the University of Tennessee, Knoxville, for his unflagging attention to detail and good humor. Ron, I hope the Mets win it all again some day! Most of all, thanks to the students, both those from Purdue and those from other universities, who took the time to drop by, call, or write with comments about the text. Many of the improvements in this edition came from their suggestions. I welcome questions and comments about this edition from both instructors and students; you can reach me by mail, by telephone (317/494-1462), or over the Internet (*loudonm@sage.cc.purdue.edu*).

I gratefully acknowledge the collaboration of the professionals at Benjamin/Cummings Publishing Company: Anne Scanlan-Rohrer, Senior Editor, and a friend as much as an editor; Leslie With, assistant editor; Cathy Greulich, developmental editor; Nathalie Mainland and Rob Hughes in marketing; and Rani Cochran, surely one of the best production managers in the industry. Working with the pros at Jonathan Peck Typographers, especially their project manager, Joan Keyes, has been a genuine pleasure.

I would like to thank the authors and publishers acknowledged separately in the credits section for permission to reproduce copyrighted materials.

Finally, I acknowledge the love and support of Judy, Kyle, and Chris, without which this project could not have been completed.

My wish is that students and professors will enjoy using of this text and will benefit from it as much as I have enjoyed writing it!

Marc Loudon
West Lafayette, Indiana
September 1994

About the Author

Marc Loudon (on the right in the photo below) received his B.S. (Magna cum Laude) in chemistry from Louisiana State University in 1964 and his Ph.D. in organic chemistry from the University of California, Berkeley, where he worked with Professor Donald S. Noyce. After two years of postdoctoral study with Professor Daniel E. Koshland in the Biochemistry Department at Berkeley, Dr. Loudon joined the Chemistry Department at Cornell University, where he taught organic chemistry to both premedical students and science majors. He received the Clark Award for Distinguished Teaching in 1976. Since 1977, Dr. Loudon has been Professor of Medicinal Chemistry at Purdue University and, since 1988, Associate Dean for Research and Graduate Programs of the School of Pharmacy and Pharmacal Sciences. Dr. Loudon teaches organic chemistry to pharmacy and prepharmacy students at Purdue, where he has twice won the Henry Heine Outstanding Teacher Award and the Class of 1922 Helping Students Learn Award.

Dr. Loudon's research interests are in the application of organic chemistry to biological problems, particularly in the peptide and protein field, and in the mechanisms of organic reactions in aqueous solution. He is also involved in several projects that are exploring innovative ways of teaching organic chemistry.

Dr. Loudon is an accomplished pianist and organist and has performed professionally in the San Francisco Bay Area, at Cornell University, and in Indiana. He also enjoys playing competitive tennis and, despite a steadily improving backhand, continues to participate in tournaments with no success whatsoever.

Reviewers and Accuracy Checkers

REVIEWERS

Gloria Anderson
MORRIS BROWN COLLEGE

Elizabeth Armstrong
SKYLINE COLLEGE

William F. Bailey
UNIVERSITY OF CONNECTICUT

R. G. Bass
VIRGINIA COMMONWEALTH UNIVERSITY

Wesley G. Bentrude
UNIVERSITY OF UTAH

David Boyles
SOUTH DAKOTA SCHOOL OF MINES AND
TECHNOLOGY

Barry Carpenter
CORNELL UNIVERSITY

Barry Coddens
NORTH CENTRAL UNIVERSITY

Arnold Craig
MONTANA STATE UNIVERSITY

William P. Dailey
UNIVERSITY OF PENNSYLVANIA

Dennis D. Davis
NEW MEXICO STATE UNIVERSITY

James Deyrup
UNIVERSITY OF FLORIDA, GAINSVILLE

Gideon Fraenkel
OHIO STATE UNIVERSITY

Rainer Glaser
UNIVERSITY OF MISSOURI, COLUMBIA

Steven Hardinger
CALIFORNIA STATE UNIVERSITY,
FULLERTON

Norbert R. Hepfinger
RENSSELAER POLYTECHNIC INSTITUTE

William H. Hersh
QUEENS COLLEGE, CITY UNIVERSITY OF
NEW YORK

Richard Hill
UNIVERSITY OF GEORGIA

Earl Huyser
UNIVERSITY OF KANSAS

James Keefe
San Francisco State University

Robert C. Kerber
SUNY AT STONY BROOK

Michael O. Killpack
UNITED STATES AIR FORCE ACADEMY

Charles Kingsbury
UNIVERSITY OF NEBRASKA, LINCOLN

D. B. Knight
UNIVERSITY OF NORTH CAROLINA,
GREENSBORO

Ronald Magid
UNIVERSITY OF TENNESSEE, KNOXVILLE

Kenneth Marsi
CALIFORNIA STATE UNIVERSITY,
LONG BEACH

Donald Matteson
WASHINGTON STATE UNIVERSITY

David A. Nelson
UNIVERSITY OF WYOMING

Martin Newcomb
WAYNE STATE UNIVERSITY

Bruce Phillips
ST. LOUIS COLLEGE OF PHARMACY

Bryan Roberts
UNIVERSITY OF PENNSYLVANIA

Harold Rogers
CALIFORNIA STATE UNIVERSITY,
FULLERTON

William Scott
MILES INC., PHARMACEUTICAL DIVISION

George H. Wahl, Jr.
NORTH CAROLINA STATE UNIVERSITY

Richard H. Wallace
THE UNIVERSITY OF ALABAMA

ACCURACY CHECKERS

Morton A. Golub

John Pendry

James W. Long
UNIVERSITY OF OREGON

Henk Juve

James Keefe
SAN FRANCISCO STATE UNIVERSITY

Contents

 1 Chemical Bonding and Chemical Structure 1

1.1 **Introduction** 1
   A. What is Organic Chemistry? 1
   B. Emergence of Organic Chemistry 1
   C. Why Study Organic Chemistry? 2

1.2 **Classical Theories of Chemical Bonding** 3
   A. Electrons in Atoms 3
   B. The Ionic Bond 4
   C. The Covalent Bond 5
   D. The Polar Covalent Bond 10

1.3 **Structures of Covalent Compounds** 12
   A. Methods for Determining Molecular Structure 12
   B. Prediction of Molecular Geometry 13

1.4 **Resonance Structures** 19

1.5 **Wave Nature of the Electron** 21

1.6 **Electronic Structure of the Hydrogen Atom** 22
   A. Orbitals, Quantum Numbers, and Energy 22
   B. Spatial Characteristics of Orbitals 24
   C. Summary: Atomic Orbitals of Hydrogen 27

1.7 **Electronic Structures of More Complex Atoms** 28

1.8 **Another Look at the Covalent Bond: Molecular Orbitals** 30
   A. Molecular Orbital Theory 30
   B. Molecular Orbital Theory and Lewis Structures 34

1.9 **Hybrid Orbitals** 34
   A. Bonding in Methane 34
   B. Bonding in Ammonia 36

Key Ideas in Chapter 1 37

Additional Problems 39

2 Alkanes 45

2.1 **Hydrocarbons** 45

2.2 **Unbranched Alkanes** 47

2.3 **Conformations of Alkanes** 49
   A. Conformation of Ethane 49
   B. Conformations of Butane 52

2.4 **Constitutional Isomers and Nomenclature** 55
   A. Isomers 55
   B. Organic Nomenclature 56
   C. Substitutive Nomenclature of Alkanes 56
   D. Highly Condensed Structures 61
   E. Classification of Carbon Substitution 63

2.5 **Cycloalkanes and Skeletal Structures** 64

2.6 **Physical Properties of Alkanes** 67
   A. Boiling Points 67
   B. Melting Points 70
   C. Other Physical Properties 71

2.7   **Combustion and Elemental Analysis** 72

   A.   Reactivity of Alkanes; Combustion 72

   B.   Elemental Analysis by Combustion 72

2.8   **Occurrence and Use of Alkanes** 75

2.9   **Functional Groups and the "R" Notation** 78

Key Ideas In Chapter 2 80

Additional Problems 81

3   Acids and Bases; the Curved Arrow Formalism   87

3.1   **Lewis Acid-Base Association Reactions** 87

   A.   Electron-Deficient Compounds 87

   B.   Reactions of Electron-Deficient Compounds with Lewis Bases 88

   C   The Curved-Arrow Formalism 89

3.2   **Electron-Pair Displacement Reactions** 90

   A.   Electron-Pair Dispacement Reactions as Lewis Acid-Base Reactions 90

   B.   The Curved-Arrow Formalism 91

3.3   **Review of the Curved-Arrow Formalism** 94

   A.   Use of the Curved-Arrow Formalism to Represent Lewis Acid-Base Reactions 94

   B.   Use of the Curved-Arrow Formalism to Derive Resonance Structures 95

3.4   **Brønsted-Lowry Acids and Bases** 96

   A.   Definition of Brønsted Acids and Bases 96

   B.   Strengths of Brønsted Acids 98

   C.   Strengths of Brønsted Bases 100

   D.   Equilibria among Different Acids and Bases 101

3.5   **Free Energy and Chemical Equilibrium** 102

3.6   **Relationship of Structure to Acidity** 105

   A.   The Element Effect 105

   B.   The Polar Effect 108

Key Ideas In Chapter 3 112

Additional Problems 113

4   Introduction to Alkenes; Reaction Rates   121

4.1   **Structure and Bonding in Alkenes** 121

   A.   Carbon Hybridization in Alkenes 122

   B.   *Cis-Trans* Isomerism 126

4.2   **Nomenclature of Alkenes** 129

   A.   IUPAC Substitutive Nomenclature 129

   B.   Nomenclature of Stereoisomers: the *E,Z* System 132

4.3   **Unsaturation Number** 136

4.4   **Physical Properties of Alkenes** 138

4.5   **Relative Stabilities of Alkene Isomers** 139

   A.   Heats of Formation 139

   B.   Relative Stabilities of Alkene Isomers 141

4.6   **Addition Reactions of Alkenes** 144

4.7   **Addition of Hydrogen Halides to Alkenes** 145

   A.   Regioselectivity of Hydrogen Halide Addition 145

   B.   Carbocation Intermediates in Hydrogen Halide Addition 146

   C.   Structure and Stability of Carbocations 148

   D.   Carbocation Rearrangement in Hydrogen Halide Addition 151

4.8   **Reaction Rates** 154

   A.   The Transition State 154

B. Multistep Reactions and the Rate-Limiting Step 156

C. Hammond's Postulate 158

4.9 Catalysis 161

A. Catalytic Hydrogenation of Alkenes 161

B. Hydration of Alkenes 163

C. Enzyme Catalysis 165

Key Ideas in Chapter 4 166

Additional Problems 167

5 Addition Reaction of Alkenes 175

5.1 Reactions of Alkenes with Halogens 175

A. Addition of Chlorine and Bromine 175

B. Halohydrins 177

5.2 Writing Organic Reactions 179

5.3 Conversion of Alkenes into Alcohols 180

A. Oxymercuration-Reduction of Alkenes 180

B. Hydroboration-Oxidation of Alkenes 183

C. Comparison of Methods for the Synthesis of Alcohols from Alkenes 187

5.4 Ozonolysis of Alkenes 188

5.5 Conversion of Alkenes into Glycols 192

5.6 Free-Radical Addition of Hydrogen Bromide to Alkenes 194

A. The Peroxide Effect 194

B. Free Radicals 196

C. Free-Radical Chain Reactions 197

D. Explanation of the Peroxide Effect 200

E. Bond Dissociation Energies 204

5.7 Industrial Use and Preparation of Alkenes 207

A. Free-Radical Polymerization of Alkenes 207

B. Thermal Cracking of Alkanes 210

C. Commercial Importance of Ethylene and Propene 212

Key Ideas in Chapter 5 214

Additional Problems 215

6 Introduction to Stereochemistry 225

6.1 Enantiomers, Chirality, and Symmetry 225

A. Enantiomers and Chirality 225

B. Asymmetric Carbon and Stereocenters 228

C. Chirality and Symmetry 229

6.2 Nomenclature of Enantiomers: The R,S System 231

6.3 Physical Properties of Enantiomers: Optical Activity 234

A. Polarized Light 234

B. Optical Activity 235

C. Optical Activities of Enantiomers 238

6.4 Racemates 239

6.5 Stereochemical Correlation 240

6.6 Diastereomers 241

6.7 Meso Compounds 244

6.8 Enantiomeric Resolution 249

6.9 Chiral Molecules without Asymmetric Atoms 251

6.10 Conformational Stereoisomers 253

A. Stereoisomers Interconverted by Internal Rotations 253

B. Asymmetric Nitrogen: Amine Inversion 254

6.11 Fischer Projections 256

6.12 The Postulation of Tetrahedral Carbon 260

Key Ideas In Chapter 6 264

Additional Problems 265

7 **Cyclic Compounds. Stereochemistry of Reactions   271**

7.1   **Relative Stabilities of the Monocyclic Alkanes   271**

7.2   **Conformations of Cyclohexane   272**
   A.   The Chair Conformation   272
   B.   Interconversion of Chair Conformations   276
   C.   Boat and Twist-Boat Conformations   277

7.3   **Monosubstituted Cyclohexanes and Conformational Analysis   279**

7.4   **Disubstituted Cyclohexanes   283**
   A.   *Cis-Trans* Isomerism   283
   B.   Conformational Analysis   286
   C.   Use of Planar Structures for Cyclic Compounds   287
   D.   Stereochemical Consequences of the Chair Flip   289

7.5   **Cyclopentane, Cyclobutane, and Cyclopropane   292**
   A.   Cylopentane   292
   B.   Cyclobutane and Cyclopropane   293

7.6   **Bicyclic and Polycyclic Compounds   294**
   A.   Classification and Nomenclature   294
   B.   *Cis* and *Trans* Ring Fusion   297
   C.   *Trans*-Cycloalkenes and Bredt's Rule   299
   D.   Steroids   300

7.7   **Relative Reactivities of Stereoisomers   302**
   A.   Relative Reactivities of Enantiomers   302
   B.   Relative Reactivities of Diastereomers   304

7.8   **Reactions That Form Stereoisomers   305**
   A.   Reactions of Achiral Compounds That Give Enantiomeric Products   305

   B.   Reactions That Give Diastereomeric Products   307

7.9   **Stereochemistry of Chemical Reactions   310**
   A.   Stereochemistry of Addition Reactions   310
   B.   Stereochemistry of Substitution Reactions   310
   C.   Stereochemistry of Bromine Addition   313
   D.   Stereochemistry of Hydroboration-Oxidation   317
   E.   Stereochemistry of Other Addition Reactions   319

Key Ideas in Chapter 7   321

Additional Problems   322

8 **Introduction to Alkyl Halides, Alcohols, Ethers, Thiols, and Sulfides   335**

8.1   **Nomenclature   336**
   A.   Nomenclature of Alkyl Halides   336
   B.   Nomenclature of Alcohols and Thiols   338
   C.   Nomenclature of Ethers and Sulfides   343

8.2   **Structures   345**

8.3   **Effect of Molecular Polarity and Hydrogen Bonding on Physical Properties   346**
   A.   Boiling Points of Ethers and Alkyl Halides   346
   B.   Boiling Points of Alcohols   348
   C.   Hydrogen Bonding   348

8.4   **Solvents in Organic Chemistry   351**
   A.   Classification of Solvents   352
   B.   Solubility   353
   C.   Cation-Binding Molecules   358

8.5   **Acidity of Alcohols and Thiols   360**
   A.   Formation of Alkoxides and Mercaptides   362

B. Polar Effects on Alcohol
    Acidity 364
C. Role of the Solvent in
    Acidity 365

8.6 **Basicity of Alcohols and
     Ethers** 366

8.7 **Grignard and Organolithium
     Reagents** 367
A. Formation of Grignard and
    Organolithium Reagents 368
B. Protonolysis of Grignard and
    Organolithium Reagents 370

8.8 **Industrial Preparation and Use
     of Alkyl Halides, Alcohols, and
     Ethers** 371
A. Free-Radical Halogenation
    of Alkanes 371
B. Uses of Halogen-Containing
    Compounds 373
C. Production and Use of Alcohols
    and Ethers 374
D. Safety Hazards of Ethers 377

Key Ideas in Chapter 8 378

Additional Problems 379

9   **Chemistry of Alkyl
    Halides** 385

9.1 **An Overview of Nucleophilic
     Substitution and β-Elimination
     Reactions** 385
A. Nucleophilic Substitution
    Reactions 385
B. β-Elimination Reactions 387
C. Competition between Nucleophilic
    Substitution and β-Elimination
    Reactions 388

9.2 **Equilibrium in Nucleophilic
     Substitution Reactions** 389

9.3 **Reaction Rates** 390
A. Definition of Reaction Rate 391
B. The Rate Law 391
C. Reaction Rate and the Standard
    Free Energy of Activation 392

9.4 **The S_N2 Reaction** 394
A. Rate Law and Mechanism of the
    S_N2 Reaction 394
B. Stereochemistry of the S_N2
    Reaction 394
C. Effect of Alkyl Halide Structure
    on the S_N2 Reaction 397
D. Solvent Effects on Nucleophilicity
    in the S_N2 Reaction 399
E. Leaving-Group Effects in the
    S_N2 Reaction 403
F. Summary of the S_N2 Reaction 403

9.5 **The E2 Reaction** 404
A. Rate Law and Mechanism of the
    E2 Reaction 404
B. Leaving-Group Effects on the
    E2 Reaction 404
C. Deuterium Isotope Effects in the
    E2 Reaction 405
D. Stereochemistry of the
    E2 Reaction 407
E. Regioselectivity of the
    E2 Reaction 410
F. Competition between the E2 and
    S_N2 Reactions: a Closer Look 411
G. Summary of the E2 Reaction 416

9.6 **The S_N1 and E1 Reactions** 416
A. Rate Law and Mechanism of the
    S_N1 and E1 Reactions 417
B. Rate-Limiting and
    Product-Determining Steps 419
C. Reactivity and Product
    Distributions in S_N1-E1
    Reactions 421
D. Summary of the S_N1 and
    E1 Reactions 423

9.7 **Summary of Substitution and
     Elimination Reactions of Alkyl
     Halides** 423

9.8 **Carbenes and Carbenoids** 427
A. α-Elimination Reactions 427
B. The Simmons–Smith
    Reaction 430

Key Ideas in Chapter 9 432

Additional Problems 433

Chemistry of Alcohols, Glycols, and Thiols 443

10.1   **Dehydration of Alcohols 443**

10.2   **Reactions of Alcohols with Hydrogen Halides 447**

10.3   **Sulfonate and Inorganic Ester Derivatives of Alcohols 450**
   A. Sulfonate Ester Derivatives of Alcohols 450
   B. Alkylating Agents 454
   C. Ester Derivatives of Strong Inorganic Acids 455
   D. Reactions of Alcohols with Thionyl Chloride 456

10.4   **Conversion of Alcohols into Alkyl Halides: Summary 457**

10.5   **Oxidation and Reduction in Organic Chemistry 459**
   A. Oxidation Numbers 459
   B. Oxidizing and Reducing Agents 463

10.6   **Oxidation of Alcohols 467**
   A. Oxidation to Aldehydes and Ketones 467
   B. Oxidation to Carboxylic Acids 469
   C. Oxidative Cleavage of Glycols 470

10.7   **Biological Oxidation of Ethanol 472**

10.8   **Chemical and Stereochemical Group Equivalence 475**
   A. Chemical Equivalence and Nonequivalence 476
   B. Stereochemistry of the Alcohol Dehydrogenase Reaction 480

10.9   **Oxidation of Thiols 482**

10.10 **Synthesis of Alcohols and Glycols 485**

10.11 **Design of Organic Synthesis 486**

Key Ideas in Chapter 10 488

Additional Problems 489

 Chemistry of Ethers, Epoxides, and Sulfides 497

11.1   **Synthesis of Ethers and Sulfides 497**
   A. Williamson Ether Synthesis 497
   B. Alkoxymercuration-Reduction of Alkenes 499
   C. Ethers from Alcohol Dehydration and Alkene Addition 500

11.2   **Synthesis of Epoxides 503**
   A. Oxidation of Alkenes with Peroxycarboxylic Acids 503
   B. Cyclization of Halohydrins 505

11.3   **Cleavage of Ethers 507**

11.4   **Nucleophilic Substitution Reactions of Epoxides 509**
   A. Ring-Opening Reactions under Basic Conditions 509
   B. Ring-Opening Reactions under Acidic Conditions 511
   C. Reaction of Ethylene Oxide with Grignard Reagents 515

11.5   **Oxonium and Sulfonium Salts 516**
   A. Reactions of Oxonium and Sulfonium Salts 516
   B. S-Adenosylmethionine: Nature's Methylating Agent 517

11.6   **Neighboring-Group Participation 518**

11.7   **Oxidation of Ethers and Sulfides 522**

11.8   **The Three Fundamental Operations of Organic Synthesis 523**

Key Ideas in Chapter 11 525

Additonal Problems 526

 Infrared Spectroscopy and Mass Spectrometry 535

12.1   **Introduction to Spectroscopy 535**
   A. Electromagnetic Radiation 535
   B. Absorption Spectroscopy 538

12.2 **Infrared Spectroscopy** 540
  A. The Infrared Spectrum 540
  B. Physical Basis of IR
      Spectroscopy 541

12.3 **Infrared Absorption and Chemical
      Structure** 543
  A. Factors that Determine IR
      Absorption Position 544
  B. Factors that Determine IR
      Absorption Intensity 547

12.4 **Functional-Group Infrared
      Absorptions** 549
  A. IR Spectra of Alkanes 549
  B. IR Spectra of Alkyl Halides 550
  C. IR Spectra of Alkenes 550
  D. IR Spectra of Alcohols and
      Ethers 553

12.5 **The Infrared Spectrometer** 555

12.6 **Introduction to Mass
      Spectrometry** 556
  A. Production of the Mass
      Spectrum 556
  B. Isotopic Peaks 558
  C. Fragmentation Mechanisms 561
  D. Odd-Electron Ions and
      Even-Electron Ions 565
  E. Identifying the Molecular Ion 566
  F. The Mass Spectrometer 567

Key Ideas in Chapter 12 569
Additional Problems 570

**13 Nuclear Magnetic
      Resonance
      Spectroscopy** 579

13.1 **Introduction to NMR
      Spectroscopy** 579

13.2 **The NMR Spectrum: Chemical Shift
      and Integral** 582
  A. Obtaining the NMR Spectrum 582
  B. Chemical Shift Scales 582
  C. The Number of Absorptions in an
      NMR Spectrum 584

  D. Relationship of Chemical Shift to
      Structure 587
  E. Counting Protons with the
      Integral 591
  F. Using the Chemical Shift and
      Integral to Determine Unknown
      Structures 592

13.3 **The NMR Spectrum: Spin–Spin
      Splitting** 594
  A. The $n + 1$ Splitting Rule 595
  B. Why Splitting Occurs 599
  C. Solving Unknown Structures
      with NMR Spectra Involving
      Splitting 600

13.4 **Complex NMR Spectra** 603
  A. Multiplicative Splitting 603
  B. Breakdown of the $n + 1$ Rule 607

13.5 **Use of Deuterium in Proton
      NMR** 611

13.6 **Characteristic Functional-Group
      NMR Absorptions** 611
  A. NMR Spectra of Alkenes 612
  B. NMR Spectra of Alkanes and
      Cycloalkanes 615
  C. NMR Spectra of Alkyl Halides and
      Ethers 616
  D. NMR Spectra of Alcohols 616

13.7 **NMR Spectroscopy of Dynamic
      Systems** 619

13.8 **Carbon NMR** 622

13.9 **Solving Structure Problems
      with Spectroscopy** 628

13.10 **The NMR Spectrometer** 632

13.11 **Other Uses of NMR** 633

Key Ideas in Chapter 13 634
Additional Problems 635

14 Chemistry of Alkynes   645

14.1 **Nomenclature of Alkynes** 645

14.2 **Structure And Bonding
      in Alkynes** 647

14.3   **Physical Properties of Alkynes** 649
    A.   Boiling Points and Solubilities 649
    B.   IR Spectroscopy of Alkynes 650
    C.   NMR Spectroscopy of Alkynes 650

14.4   **Introduction to Addition Reaction of the Triple Bond** 653

14.5   **Conversion of Alkynes into Aldehydes and Ketones** 654
    A.   Hydration of Alkynes 654
    B.   Hydroboration-Oxidation of Alkynes 657

14.6   **Reduction of Alkynes** 659
    A.   Catalytic Hydrogenation of Alkynes 659
    B.   Reduction of Alkynes with Sodium in Liquid Ammonia 660

14.7   **Acidity of 1-Alkynes** 662
    A.   Acetylenic Anions 662
    B.   Acetylenic Anions as Nucleophiles 665

14.8   **Organic Synthesis Using Alkynes** 666

14.9   **Pheromones** 668

14.10   **Occurrence and Use of Alkynes** 670

Key Ideas in Chapter 14 671

Additional Problems 672

15   **Dienes, Resonance, and Aromaticity** 679

15.1   **Structure and Stability of Dienes** 680
    A.   Structure and Stability of Conjugated Dienes 680
    B.   Structure and Stability of Cumulated Dienes 682

15.2   **Ultraviolet Spectroscopy** 684
    A.   The UV Spectrum 684
    B.   Physical Basis of UV Spectroscopy 686
    C.   UV Spectroscopy of Conjugated Alkenes 688

15.3   **The Diels-Alder Reaction** 690
    A.   Reactions of Conjugated Dienes with Alkenes 690
    B.   Effect of Diene Conformation on the Diels-Alder Reaction 694
    C.   Stereochemistry of the Diels-Alder Reaction 696

15.4   **Addition of Hydrogen Halides to Conjugated Dienes** 699
    A.   1,2- and 1,4-Additions 699
    B.   Allylic Carbocations 701
    C.   Kinetic and Thermodynamic Control 703

15.5   **Diene Polymers** 707

15.6   **Resonance** 708
    A.   Writing Resonance Structures 709
    B.   Relative Importance of Resonance Structures 710
    C.   Use of Resonance Structures 712

15.7   **Introduction to Aromatic Compounds** 715
    A.   Benzene, a Puzzling "Alkene" 715
    B.   Structure of Benzene 717
    C.   Stability of Benzene 719
    D.   Aromaticity and the Hückel $4n + 2$ Rule 719
    E.   Antiaromatic Compounds 725

Key Ideas in Chapter 15 726

Additional Problems 727

16   **Chemistry of Benzene and Its Derivatives** 737

16.1   **Nomenclature of Benzene Derivatives** 737

16.2   **Physical Properties of Benzene Derivatives** 740

16.3   **Spectroscopy of Benzene Derivatives** 741
    A.   IR Spectroscopy 741
    B.   NMR Spectroscopy 742
    C.   CMR Spectroscopy 745
    D.   UV Spectroscopy 746

16.4   **Electrophilic Aromatic Substitution
Reactions of Benzene** 748
   A. Halogenation of Benzene 748
   B. Electrophilic Aromatic
      Substitution 750
   C. Nitration of Benzene 752
   D. Sulfonation of Benzene 753
   E. Friedel-Crafts Acylation
      of Benzene 754
   F. Friedel-Crafts Alkylation
      of Benzene 757

16.5   **Electrophilic Aromatic Substitution
Reactions of Substituted
Benzenes** 760
   A. Directing Effects
      of Substituents 760
   B. Activating and Deactivating Effects
      of Substituents 767
   C. Use of Electrophilic Aromatic
      Substitution in Organic
      Synthesis 771

16.6   **Hydrogenation of Benzene
Derivatives** 775

16.7   **Source and Industrial Use
of Aromatic Hydrocarbons** 776

Key Ideas in Chapter 16 779

Additional Problems 780

17   **Allylic and Benzylic
Reactivity** 789

17.1   **Reactions Involving Allylic and
Benzylic Carbocations** 789

17.2   **Reactions Involving Allylic and
Benzylic Radicals** 794

17.3   **Reactions Involving Allylic and
Benzylic Anions** 799
   A. Allylic Grignard Reagents 800
   B. E2 Eliminations Involving Allylic
      or Benzylic Hydrogens 802

17.4   **Allylic and Benzylic
S_N2 Reactions** 803

17.5   **Benzylic Oxidation
of Alkylbenzenes** 804

17.6   **Terpenes** 806
   A. The Isoprene Rule 806
   B. Biosynthesis of Terpenes 808

Key Ideas in Chapter 17 812

Additional Problems 813

18   **Chemistry of Aryl Halides,
Vinylic Halides, and
Phenols** 823

18.1   **Lack of Reactivity of Vinylic
and Aryl Halides Under S_N2
Conditions** 824

18.2   **Elimination Reactions of Vinylic
Halides** 826

18.3   **Lack of Reactivity of Vinylic
and Aryl Halides Under S_N1
Conditions** 826

18.4   **Nucleophilic Substitution Reactions
of Aryl Halides** 829
   A. Nucleophilic Aromatic
      Substitution 829
   B. Substitution
      by Elimination-Addition:
      Benzyne 832
   C. Summary: Nucleophilic
      Substitution Reactions of Aryl
      Halides 835

18.5   **Aryl and Vinylic Grignard
Reagents** 836

18.6   **Acidity of Phenols** 837
   A. Resonance Effects on
      Acidity 837
   B. Formation and Use
      of Phenoxides 840

18.7   **Oxidation of Phenols
to Quinones** 841

18.8   **Electrophilic Aromatic Substitution
Reactions of Phenols** 846

18.9   **Lack of Reactivity of the
Aryl-Oxygen Bond** 849

18.10   **Industrial Preparation and Use
of Phenol** 851

Key Ideas in Chapter 18 852

Additional Problems 853

19 Chemistry of Aldehydes and Ketones; Carbonyl-Addition Reactions 865

19.1 Nomenclature of Aldehydes and Ketones 867
    A. Common Nomenclature 867
    B. Substitutive Nomenclature 869

19.2 Physical Properties of Aldehydes and Ketones 871

19.3 Spectroscopy of Aldehydes and Ketones 872
    A. IR Spectroscopy 872
    B. Proton NMR Spectroscopy 874
    C. Carbon NMR Spectroscopy 875
    D. UV Spectroscopy 876
    E. Mass Spectrometry 878

19.4 Synthesis of Aldehydes and Ketones 880

19.5 Introduction to Aldehyde and Ketone Reactions 880

19.6 Basicity of Aldehydes and Ketones 881

19.7 Reversible Addition Reactions of Aldehydes and Ketones 884
    A. Mechanisms of Carbonyl-Addition Reactions 884
    B. Equilibria and Rates in Carbonyl-Addition Reactions 887

19.8 Reduction of Aldehydes and Ketones to Alcohols 890

19.9 Reactions of Aldehydes and Ketones with Grignard and Related Reagents 895

19.10 Acetals and Their Use as Protecting Groups 898
    A. Preparation and Hydrolysis of Acetals 898
    B. Protecting Groups 902

19.11 Reactions of Aldehydes and Ketones with Amines 904
    A. Reaction with Primary Amines and Other Monosubstituted Derivatives of Ammonia 904
    B. Reaction with Secondary Amines 907

19.12 Reduction of Carbonyl Groups to Methylene Groups 909

19.13 The Wittig Alkene Synthesis 911

19.14 Oxidation of Aldehydes to Carboxylic Acids 915

19.15 Manufacture and Use of Aldehydes and Ketones 917

Key Ideas in Chapter 19 918

Additional Problems 919

20 Chemistry of Carboxylic Acids 929

20.1 Nomenclature of Carboxylic Acids 929
    A. Common Nomenclature 929
    B. Substitutive Nomenclature 931

20.2 Structure and Physical Properties of Carboxylic Acids 934

20.3 Spectroscopy of Carboxylic Acids 935
    A. IR Spectroscopy 935
    B. NMR Spectroscopy 936

20.4 Acid-Base Properties of Carboxylic Acids 937
    A. Acidity of Carboxylic and Sulfonic Acids 937
    B. Basicity of Carboxylic Acids 941

20.5 Fatty Acids, Soaps, and Detergents 942

20.6 Synthesis of Carboxylic Acids 944

20.7 Introduction to Carboxylic Acid Reactions 945

20.8 Conversion of Carboxylic Acids into Esters 946

A. Acid-Catalyzed Esterification 946

B. Esterification by Alkylation 950

20.9 Conversion of Carboxylic Acids into Acid Chlorides and Anhydrides 952

A. Synthesis of Acid Chlorides 952

B. Synthesis of Anhydrides 954

20.10 Reduction of Carboxylic Acids to Primary Alcohols 956

20.11 Decarboxylation of Carboxylic Acids 958

Key Ideas in Chapter 20 960

Additional Problems 961

 21 Chemistry of Carboxylic Acid Derivatives 971

21.1 Nomenclature and Classification of Carboxylic Acid Derivatives 971

A. Esters and Lactones 971

B. Acid Halides 973

C. Anhydrides 974

D. Nitriles 974

E. Amides, Lactams, and Imides 975

F. Nomenclature of Substituent Groups 976

G. Carbonic Acid Derivatives 977

21.2 Structures of Carboxylic Acid Derivatives 978

21.3 Physical Properties of Carboxylic Acid Derivatives 980

A. Esters 980

B. Anhydrides and Acid Chlorides 980

C. Nitriles 981

D. Amides 981

21.4 Spectroscopy of Carboxylic Acid Derivatives 982

A. IR Spectroscopy 982

B. NMR Spectroscopy 982

21.5 Basicity of Carboxylic Acid Derivatives 987

21.6 Introduction to Reactions of Carboxylic Acid Derivatives 988

21.7 Hydrolysis of Carboxylic Acid Derivatives 989

A. Hydrolysis of Esters 989

B. Hydrolysis of Amides 992

C. Hydrolysis of Nitriles 994

D. Hydrolysis of Acid Chlorides and Anhydrides 995

E. Mechanisms and Reactivity in Nucleophilic Acyl Substitution Reactions 996

21.8 Reactions of Carboxylic Acid Derivatives with Nucleophiles 1000

A. Reactions of Acid Chlorides with Nucleophiles 1000

B. Reactions of Anhydrides with Nucleophiles 1004

C. Reactions of Esters with Nucleophiles 1005

21.9 Reactions of Carboxylic Acid Derivatives 1007

A. Reduction of Esters to Primary Alcohols 1007

B. Reduction of Amides to Amines 1008

C. Reduction of Nitriles to Primary Amines 1010

D. Reduction of Acid Chlorides to Aldehydes 1012

E. Relative Reactivities of Carbonyl Compounds 1014

21.10 Reactions of Carboxylic Acid Derivatives with Organometallic Reagents 1015

A. Reaction of Esters with Grignard Reagents 1015

B. Reaction of Acid Chlorides with Lithium Dialkylcuprates 1017

21.11 Synthesis of Carboxylic Acid Derivatives 1018

21.12 Use and Occurrence of Carboxylic Acids and Their Derivatives 1020

A. Nylon and Polyesters 1020

B. Waxes, Fats, and Phospholipids 1023

Key Ideas in Chapter 21 1026

Additional Problems 1027

 22 Chemistry of Enolate Ions, Enols, and α,β-Unsaturated Carbonyl Compounds 1039

22.1 **Acidity of Carbonyl Compounds 1039**
   A. Formation of Enolate Anions 1039
   B. Introduction to Reactions of Enolate Ions 1042

22.2 **Enolization of Carbonyl Compounds 1044**

22.3 **α-Halogenation of Carbonyl Compounds 1048**
   A. Acid-Catalyzed α-Halogenation of Aldehydes and Ketones 1048
   B. Halogenation of Aldehydes and Ketones in Base: the Haloform Reaction 1050
   C. α-Bromination of Carboxylic Acids 1052
   D. Reactions of α-Halo Carbonyl Compounds 1053

22.4 **Aldol Addition and Aldol Condensation 1055**
   A. Base-Catalyzed Aldol Reactions 1055
   B. Acid-Catalyzed Aldol Condensation 1058
   C. Special Types of Aldol Reactions 1060
   D. Synthesis with the Aldol Condensation 1062

22.5 **Condensation Reactions Involving Ester Enolate Ions 1064**
   A. Claisen Condensation 1065
   B. Dieckmann Condensation 1068
   C. Crossed Claisen Condensation 1069
   D. Synthesis with the Claisen Condensation 1071

22.6 **Alkylation of Ester Enolate Ions 1074**
   A. Malonic Ester Synthesis 1074
   B. Direct Alkylation of Enolate Ions Derived from Simple Monoesters 1076
   C. Acetoacetic Ester Synthesis 1079

22.7 **Biosynthesis of Fatty Acids 1082**

22.8 **Conjugate-Addition Reactions 1085**
   A. Conjugate Addition to α,β-Unsaturated Carbonyl Compounds 1085
   B. Conjugate Addition vs. Carbonyl-Group Reactions 1088
   C. Conjugate Addition of Enolate Ions 1090

22.9 **Reduction of α,β-Unsaturated Carbonyl Compounds 1093**

22.10 **Reactions of α,β-Unsaturated Carbonyl Compounds with Organometallic Reagents 1094**
   A. Addition of Organolithium Reagents to the Carbonyl Group 1094
   B. Conjugate Addition of Lithium Dialkylcuprate Reagents 1095

22.11 **Organic Synthesis with Conjugate-Addition Reactions 1097**

Key Ideas in Chapter 22 1098

Additional Problems 1099

23 Chemistry of Amines 1113

23.1 **Nomenclature of Amines 1114**
   A. Common Nomenclature 1114
   B. Substitutive Nomenclature 1114

23.2 **Structure of Amines 1116**

23.3 **Physical Properties of Amines 1117**

23.4 **Spectroscopy of Amines 1118**
   A. IR Spectroscopy 1118

B. NMR Spectroscopy 1119

C. Mass Spectrometry 1119

23.5 **Basicity and Acidity of Amines** 1120

A. Basicity of Amines 1120

B. Substituent Effects on Amine Basicity 1121

C. Separations Using Amine Basicity 1125

D. Acidity of Amines 1127

E. Summary of Acidity and Basicity 1127

23.6 **Quaternary Ammonium Salts** 1127

23.7 **Alkylation and Acylation Reactions of Amines** 1128

A. Direct Alkylation of Amines 1128

B. Reductive Amination 1130

C. Acylation of Amines 1132

23.8 **Hofmann Elimination of Quaternary Ammonium Hydroxides** 1133

23.9 **Aromatic Substitution Reactions of Aniline Derivatives** 1137

23.10 **Diazotization; Reactions of Diazonium Ions** 1139

A. Formation and Substitution Reactions of Diazonium Salts 1139

B. Aromatic Substitution with Diazonium Ions 1143

C. Reactions of Secondary and Tertiary Amines with Nitrous Acid 1144

23.11 **Synthesis of Amines** 1146

A. Gabriel Synthesis of Primary Amines 1146

B. Reduction of Nitro Compounds 1147

C. Curtius and Hofmann Rearrangements 1148

D. Synthesis of Amines: Summary 1153

23.12 **Use and Occurrence of Amines** 1153

A. Industrial Use of Amines and Ammonia 1153

B. Naturally Occurring Amines 1154

Key Ideas in Chapter 23 1156

Additional Problems 1157

24   Chemistry of Naphthalene and the Aromatic Heterocycles 1169

24.1 **Chemistry of Naphthalene** 1171

A. Physical Properties and Structure 1171

B. Nomenclature 1171

C. Electrophilic Substitution Reactions 1173

24.2 **Introduction to the Aromatic Heterocycles** 1177

A. Nomenclature 1178

B. Structure and Aromaticity 1179

C. Basicity and Acidity of the Nitrogen Heterocycles 1182

24.3 **Chemistry of Furan, Pyrrole, and Thiophene** 1184

A. Electrophilic Aromatic Substitution 1184

B. Addition Reactions of Furan 1187

C. Side-Chain Reactions 1188

24.4 **Synthesis of Indoles** 1189

A. Fischer Indole Synthesis 1190

B. Reissert Indole Synthesis 1192

24.5 **Chemistry of Pyridine and Quinoline** 1194

A. Electrophilic Aromatic Substitution 1194

B. Nucleophilic Aromatic Substitution 1197

C. Pyridinium Salts and Their Reactions 1201

D. Side-Chain Reactions of Pyridine Derivatives 1202

E. Skraup Synthesis of Quinolines 1204

24.6 **Occurrence of Heterocyclic Compounds** 1206

Key Ideas in Chapter 24 1208

Additional Problems 1209

25 Pericyclic Reactions 1219

25.1 **Molecular Orbitals of Conjugated π-Electron Systems** 1222
   A. Molecular Orbitals of Conjugated Alkenes 1222
   B. Molecular Orbitals of Conjugated Ions and Radicals 1226
   C. Excited States 1227

25.2 **Electrocyclic Reactions** 1229
   A. Thermal Electrocyclic Reactions 1229
   B. Excited-State (Photochemical) Electrocyclic Reactions 1231
   C. Selection Rules and Microscopic Reversibility 1232

25.3 **Cycloaddition Reactions** 1235

25.4 **Sigmatropic Reactions** 1239
   A. Classification and Stereochemistry 1239
   B. [3,3] Sigmatropic Rearrangements 1246
   C. Summary: Selection Rules for Sigmatropic Reactions 1248

25.5 **Summary of The Pericyclic Selection Rules** 1250

25.6 **Fluxional Molecules** 1251

25.7 **Formation of Vitamin D** 1252

Key Ideas in Chapter 25 1254

Additional Problems 1255

 26 Amino Acids, Peptides, and Proteins 1263

26.1 **Nomenclature of Amino Acids and Peptides** 1264
   A. Nomenclature of Amino Acids 1264
   B. Nomenclature of Peptides 1265

26.2 **Stereochemistry of the α-Amino Acids** 1268

26.3 **Acid-Base Properties of Amino Acids and Peptides** 1269

   A. Zwitterionic Structures of Amino Acids and Peptides 1269
   B. Isoelectric Points of Amino Acids and Peptides 1271
   C. Separations of Amino Acids and Peptides Using Acid-Base Properties 1275

26.4 **Synthesis and Enantiomeric Resolution of α-Amino Acids** 1278
   A. Alkylation of Ammonia 1278
   B. Alkylation of Aminomalonate Derivatives 1278
   C. Strecker Synthesis 1279
   D. Enantiomeric Resolution of α-Amino Acids 1280

26.5 **Acylation and Esterfication Reactions of α-Amino Acids** 1281

26.6 **Determination of Peptide Structure** 1282
   A. Hydrolysis of Peptides; Amino Acid Analysis 1282
   B. Sequential Degradation of Peptides 1284
   C. Specific Cleavage of Peptides 1286

26.7 **Solid-Phase Peptide Synthesis** 1291

26.8 **Structures of Peptides and Proteins** 1299
   A. Primary Structure 1299
   B. Secondary Structure 1301
   C. Tertiary and Quaternary Structures 1303

26.9 **Enzymes: Biological Catalysts** 1308

26.10 **Occurrence of Peptides and Proteins** 1311

Key Ideas in Chapter 26 1313

Additional Problems 1315

 27 Carbohydrates and Nucleic Acids 1325

27.1 **Classification and Properties of Carbohydrates** 1326

27.2 **Structures of the Monosaccharides** 1327

    A. Stereochemistry and Configuration 1327

    B. Cyclic Structures of the Monosaccharides 1331

27.3 **Mutarotation of Carbohydrates** 1337

27.4 **Base-Catalyzed Isomerization of Aldoses and Ketoses** 1341

27.5 **Glycosides** 1343

27.6 **Ether and Ester Derivatives of Carbohydrates** 1346

27.7 **Oxidation and Reduction Reactions of Carbohydrates** 1348

    A. Oxidation to Aldonic Acids 1349

    B. Oxidation to Aldaric Acids 1350

    C. Periodate Oxidation 1352

    D. Reduction to Alditols 1353

27.8 **Kiliani-Fischer Synthesis** 1353

27.9 **Proof of Glucose Stereochemistry** 1355

    A. Which Diastereomer? The Fischer Proof 1355

    B. Which Enantiomer? The Absolute Configuration of D-(+)-Glucose 1359

27.10 **Disaccharides and Polysaccharides** 1361

    A. Disaccharides 1361

    B. Polysaccharides 1364

27.11 **Nucleosides, Nucleotides, and Nucleic Acids** 1367

    A. Nucleosides and Nucleotides 1367

    B. Structures of DNA and RNA 1370

27.12 **DNA, RNA, and the Genetic Code** 1375

    A. Role of DNA and RNA in Protein Synthesis 1375

    B. DNA Modification and Chemical Carcinogenesis 1378

Key Ideas in Chapter 27 1380

Additional Problems 1382

Appendices A-1

I.   **Substitutive Nomenclature of Organic Compounds** A-1

II.   **Important Infrared Absorptions** A-2

III. **Important Proton NMR Chemical Shifts** A-4

    A. Protons within Functional Groups A-4

    B. Alkyl Protons Adjacent to Functional Groups A-4

IV. **Summary of Synthetic Methods** A-5

V.   **Reactions Used to Form Carbon-Carbon Bonds** A-11

VI. **Typical Acidities and Basicities of Organic Functional Groups** A-12

    A. Acidities of Groups That Ionize to Give Anionic Conjugate Bases A-12

    B. Basicities of Groups That Ionize to Give Cationic Conjugate Bases A-13

Credits C-1

Index I-13

1

Chemical Bonding and Chemical Structure

1.1 Introduction

A. What Is Organic Chemistry?

Organic chemistry is the branch of science that deals generally with compounds of carbon. Yet the name *organic* seems to imply a connection with living things. Let's explore this connection, for the emergence of organic chemistry as a science is linked to the early evolution of the life sciences.

B. Emergence of Organic Chemistry

As early as the sixteenth century, scholars seem to have had some realization that the phenomenon of life has chemical attributes. Theophrastus Bombastus von Hohenheim, a Swiss physician and alchemist (ca. 1493–1541) also known as Paracelsus, sought to deal with medicine in terms of its "elements" mercury, sulfur, and salt. An ailing person was thought to be deficient in one of these elements and therefore in need of supplementation with the missing substance. Paracelsus was said to have effected some dramatic "cures" based on this idea.

By the eighteenth century, chemists were beginning to recognize the chemical aspects of life processes in a modern sense. Antoine Laurent Lavoisier (1743–1794) recognized the similarity of respiration to combustion in the uptake of oxygen and expiration of carbon dioxide.

At about the same time, it was found that certain compounds are associated with living systems; it was observed that these compounds generally contain carbon. They were thought to have arisen from, or to be a consequence of, a "vital force" responsible for the life process. The term *organic* was applied to substances isolated from living things by Jöns Jacob Berzelius (1779–1848). Somehow, the fact that these chemical substances were organic in nature was thought to put them beyond the scope of the experimentalist.

1

The logic of the time seems to have been that life is not understandable; organic compounds spring from life; therefore, organic compounds are not understandable.

The supposed barrier between organic (living) and inorganic (nonliving) chemistry fell in 1828 because of a serendipitous (accidental) discovery of Friedrich Wöhler (1800–1882), a German analyst originally trained in medicine. When Wöhler heated ammonium cyanate, an inorganic compound, he isolated urea, a known urinary excretion product of mammals.

$$\overset{+}{N}H_4NCO^- \xrightarrow{\text{heat}} H_2N-CO-NH_2 \qquad (1.1)$$

ammonium   **urea**
cyanate

Wöhler recognized that he had synthesized this biological material "without benefit of a kidney, a bladder, or a dog." Not long thereafter followed the synthesis of acetic acid by Kolbe in 1845 and the preparation of acetylene and methane by Berthelot in the period 1856–1863. Although "vitalism" was not so much a textbook theory as an intuitive idea that something might be special and beyond human grasp about the chemistry of living things, Wöhler did not identify his urea synthesis with the demise of the vitalistic idea; rather, his work signaled the start of a period in which the synthesis of so-called organic compounds was no longer regarded as something outside the province of laboratory investigation. Organic chemists now investigate not only molecules of biological importance, but also intriguing molecules of bizarre structure and purely theoretical interest. Thus, organic chemistry deals with compounds of carbon regardless of their origin. Wöhler seems to have anticipated these developments when, a little more than a century ago, he wrote to his mentor Berzelius, "Organic chemistry appears to be like a primeval tropical forest, full of the most remarkable things."

C. Why Study Organic Chemistry?

The study of organic chemistry is important for several reasons. First, the field has an independent vitality as a branch of science. Organic chemistry is characterized by continuing development of new knowledge, a fact evidenced by the large number of journals devoted exclusively or in large part to the subject. Second, organic chemistry lies at the heart of a substantial fraction of modern chemical industry and therefore contributes to the economies of many nations.

Many students who take organic chemistry nowadays are planning careers in the biological sciences or in the allied health disciplines such as pharmacy or medicine. Organic chemistry is immensely important as a foundation to these fields, and its importance is sure to increase. One need only open a modern textbook or journal of biochemistry or biology to appreciate the sophisticated organic chemistry that is central to these areas.

Even for those who do not plan a career in any of the sciences, a study of organic chemistry is important. We live in a technological age that is made possible in large part by applications of organic chemistry to industries as diverse as plastics, textiles, communications, transportation, food, and clothing. In addition, problems of energy, pollution, and depletion of resources are all around us. If organic chemistry has played a part in creating these problems, it will almost surely have a role in their solution.

As a science, organic chemistry lies at the interface of the physical and biological sciences. Research in organic chemistry is a mixture of sophisticated logic and empirical observation. At its best, it can assume artistic proportions. You can use the study of organic chemistry to develop and apply basic skills in problem solving and at the same time learn a subject of immense practical value. Thus, to develop as a chemist, to remain in the mainstream of a health profession, or to be a well informed citizen in a technological age, you will find value in the study of organic chemistry.

In this text we have several objectives. Of course, we'll present the "nuts and bolts"—the nomenclature, classification, structure, and properties of organic compounds. We'll cover the principal reactions and the syntheses of organic molecules. But more than this, we'll develop underlying principles that allow us to understand, and sometimes predict, reactions rather than simply memorizing them. We'll consider some of the organic chemistry that is industrially important. Finally, we'll examine some of the beautiful applications of organic chemistry in biology, for example, how organic chemistry is executed in nature and how the biological world has provided the impetus for much of the research in organic chemistry.

1.2 Classical Theories of Chemical Bonding

To understand organic chemistry, it is necessary to have some understanding of the **chemical bond**—the forces that hold atoms together within molecules. First, we'll review some of the older or "classical" ideas of chemical bonding—ideas that, despite their age, remain useful today. Then, in the last part of this chapter, we'll consider more modern ways of describing the chemical bond.

A. Electrons in Atoms

Chemistry happens because of the behavior of electrons in molecules. At the basis of this behavior is the way electrons are arranged within atoms. The arrangements of electrons in atoms, in turn, is suggested by the periodic table. Consequently, let's first review the organization of the periodic table (inside front cover). The shaded elements are of greatest importance in organic chemistry; knowing their atomic numbers and relative positions will be valuable later on. For the moment, however, the following details of the periodic table should be considered, for these were important in the development of concepts of bonding.

A neutral atom of each element contains a number of both protons and electrons equal to its atomic number. The organization of the periodic table lends itself to the idea that electrons reside in layers, or *shells*, about the nucleus. The outermost shell of electrons in an atom is called its **valence shell**, and the electrons in this shell are called **valence electrons**. *The number of valence electrons for any neutral atom in an A group of the periodic table* (except helium) *equals its group number*. Thus lithium, sodium, and potassium (Group 1A) have one valence electron; carbon (Group 4A) has four valence electrons, the halogens (Group 7A) have seven valence electrons, and the noble gases (except helium) have eight valence electrons; helium has two valence electrons.

Walther Kossel noted in 1916 that *stable ions are formed when atoms gain or lose valence electrons in order to have the same number of electrons as the noble gas of closest atomic number*. Thus, potassium, with one valence electron (and 19 total electrons), tends to lose an electron to become K^+, the potassium ion, which has the same number of electrons (18) as the nearest noble gas, argon. Chlorine, with seven valence electrons (and 17 total electrons) tends to accept an electron to become the 18-electron chloride ion,

Cl^-, which also has the same number of electrons as argon. Because the noble gases have an octet of electrons (that is, eight electrons) in their valence shells, the tendency of atoms to gain or lose valence electrons to form ions with the noble-gas configuration has been known as the **octet rule**. According to the octet rule, *an atomic species tends to be especially stable when its valence shell contains eight electrons.* (The corresponding rule for elements near helium, of course, is a "duet rule.")

It is not equally easy for all atoms to achieve the valence-electron configuration of a noble gas. *The ease with which neutral atoms lose electrons to form positive ions increases to the left and toward the bottom of the periodic table.* Thus, the alkali metals, on the extreme left of the periodic table, have the greatest tendency to exist as positive ions; and within the alkali metals, cesium has the greatest tendency to form a positive ion. *The ease with which neutral atoms gain electrons to form negative ions increases to the right and toward the top of the periodic table.* The halogens, which are on the extreme right of the periodic table, have the greatest tendency to form negative ions; and within the halogens, fluorine has the greatest tendency to form a negative ion. (The noble gases are written on the right of the periodic table for convenience, but as we know, these atoms do not easily form ions, and are not considered in the above trends.)

B. The Ionic Bond

A common type of chemical compound is one in which the component atoms exist as ions. Such a compound is called an **ionic compound**. Potassium chloride, KCl, is a common ionic compound. Because potassium and chlorine come from the extreme left and right, respectively, of the periodic table, it is not surprising that potassium readily exists as the positive ion K^+ and chlorine as the negative ion Cl^-. The electronic configurations of these two ions obey the octet rule.

The structure of crystalline KCl is shown in Fig. 1.1. The KCl structure is typical of many ionic compounds. In the KCl structure, each positive ion is surrounded by negative ions, and each negative ion by positive ions. A force of attraction exists between the ions of opposite charge. Such an attractive force between opposite charges is called an **electrostatic attraction**. The electrostatic attraction between positive and negative ions in crystalline KCl (and in other ionic compounds) is a *stabilizing* force; it holds the ions together within the crystal. For this reason, this attraction is sometimes called the **ionic bond**. The ionic bond is the same in all directions; that is, a positive ion has the same attraction for each of its neighboring negative ions, and a negative ion has the same attraction for each of its neighboring positive ions.

When an ionic compound such as KCl dissolves in water, it dissociates into free ions (each surrounded by water). Each potassium ion moves around in solution more or less independent of each chloride ion. The conduction of electricity by KCl solutions shows that the ions are present. Thus, the ionic bond is broken when KCl dissolves in water.

To summarize, the ionic bond

1. is an electrostatic attraction between ions;
2. is the same in all directions, that is, it has no preferred orientation in space;
3. is broken when an ionic compound dissolves in water;

tighten the lattice structure

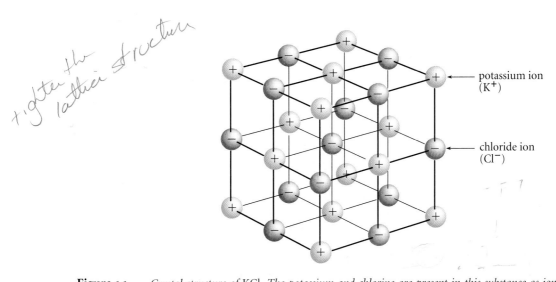

potassium ion (K$^+$)

chloride ion (Cl$^-$)

Figure 1.1   *Crystal structure of KCl. The potassium and chlorine are present in this substance as ions. The attractive forces between the potassium ions and chloride ions are purely electrostatic. Note that each positive ion is surrounded by negative ions, and each negative ion by positive ions. Thus, the attraction between ions—the ionic bond—is the same in all directions.*

4. is most likely to form between atoms at opposite ends of the periodic table.

PROBLEMS

1.1   How many valence electrons are found in each of the following species?
*(a) Na   (b) Ca   *(c) O^{2-} (d) Ca$^+$

1.2   When two different species have the same number of valence electrons they are said to be *isoelectronic*. Name the species that satisfies each of the following criteria:
*(a)   the singly charged negative ion isoelectronic with neon
(b)   the dipositive ion isoelectronic with argon
*(c)   the neon species that is isoelectronic with neutral fluorine
(d)   the singly charged negative ion isoelectronic with helium

C. The Covalent Bond

Many compounds contain bonds that are not at all like the ionic bond in KCl. Neither these compounds nor their solutions conduct electricity. This fact indicates that these compounds are *not* ionic. How are the bonding forces that hold atoms together in such compounds different from those in KCl? In 1916, G. N. Lewis, an American physical

Asterisks denote problems that have worked-out solutions in the Solutions Guide. The solutions to the remaining problems are provided in the instructor's supplement.

chemist, proposed an electronic model for bonding in nonionic compounds. According to this model, the chemical bond in a nonionic compound is a **covalent bond**, which consists of an electron pair that is *shared* between bonded atoms. Let's examine some of the ideas associated with the covalent bond.

Lewis Structures   One of the simplest examples of a covalent bond is the bond between the two hydrogen atoms in the hydrogen molecule.

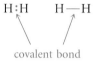

$$\text{H:H} \qquad \text{H——H}$$

covalent bond

The symbols **:** and ——— are both used to denote an electron pair; a shared electron pair is the essence of the covalent bond. Molecular structures that use this notation for the electron-pair bond are called **Lewis structures.** In the hydrogen molecule, an electron-pair bond holds the two hydrogen atoms together. Conceptually, the bond can be envisioned to come from the pairing of the valence electrons of two hydrogen atoms:

$$\text{H} \cdot + \text{H} \cdot \longrightarrow \text{H:H} \tag{1.2}$$

Both electrons in the covalent bond are shared equally between the hydrogen atoms. Even though electrons are mutually repulsive, bonding occurs because the electron of each hydrogen atom is attracted to both hydrogen nuclei (protons) simultaneously.

Another example of a covalent bond is provided by the simplest stable organic molecule, methane, CH_4. Conceptually, methane results by pairing each of the four carbon valence electrons with a hydrogen valence electron to make four C—H electron-pair bonds.

$$\cdot \overset{\cdot}{C} \cdot + 4\text{H} \cdot \longrightarrow \overset{\overset{..}{H}}{\underset{\overset{..}{H}}{\text{H:}\overset{..}{C}\text{:H}}} \quad \text{or} \quad \overset{\overset{\textstyle H}{|}}{\underset{\underset{\textstyle H}{|}}{\text{H——C——H}}} \quad \text{or} \quad CH_4 \tag{1.3}$$

Water, H_2O, represents another example of a covalent compound. Oxygen has six valence electrons. Two of these combine with hydrogens to make two O—H covalent bonds; four of the oxygen valence electrons are left over. These are represented in the Lewis structure of water as electron pairs on the oxygen. In general, unshared valence electrons in Lewis structures are depicted as paired dots and referred to as **unshared pairs** of electrons.

$$\text{unshared pairs} \qquad \text{H——}\overset{..}{\underset{..}{O}}\text{——H}$$

Although we often write water as H—O—H, or even H_2O, it is a good habit to indicate all unshared valence electrons with paired dots until we remember intuitively that they are there.

These examples, H_2, CH_4, and $H_2\overset{..}{O}$, illustrate an important point: if *all* shared and unshared electrons around a given atom are counted, *the octet rule is often obeyed in covalent bonding.* For example, consider the structure of methane shown in Eq. 1.3. Four shared pairs surround the carbon atom—that is, eight shared electrons, an octet. Two shared electrons surround each hydrogen; as we've already observed, the "octet" for hydrogen is two electrons. Similarly, the unshared electrons and shared electrons around the oxygen of water total eight electrons.

Two atoms in covalent compounds may be connected by more than one covalent bond. The following compounds are well-known examples:

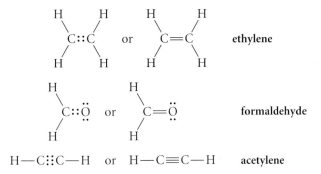

Ethylene and formaldehyde each contain a **double bond**—a bond consisting of two electron pairs. Acetylene contains a **triple bond**—a bond with three electron pairs.

Covalent bonds are especially important in organic chemistry because *all organic molecules contain covalent bonds.*

Formal Charge   The Lewis structures considered in the previous discussion are those of neutral molecules. However, many familiar ionic species such as SO_4^{2-}, NH_4^+, and BF_4^- also contain covalent bonds. Consider the tetrafluoroborate anion, which contains covalent B—F bonds:

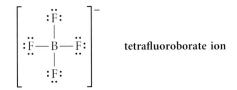

Since the ion bears a negative charge, one or more of the atoms within the ion must be charged—but which one(s)? The rigorous answer is that the charge is shared by all of the atoms. However, there is a useful procedure for electronic bookkeeping that assigns a charge to specific atoms. The charge on each atom assigned by this procedure is called its **formal charge**. The sum of the formal charges on the individual atoms must, of course, equal the total charge on the ion.

To assign a formal charge to an atom, first assign a *valence electron count* to that atom by adding *all* unshared valence electrons on the atom and *one* electron from every covalent bond to the atom. Subtract this electron count from the group number of the atom in the periodic table, which is equal to the number of valence electrons in the neutral atom. The resulting difference is the formal charge. This procedure is illustrated in the following study problem.

STUDY PROBLEM 1.1

Assign a formal charge to each of the atoms in the tetrafluoroborate ion $[BF_4]^-$.

Solution Let's first apply the procedure outlined above to fluorine:

Group number of fluorine: 7
Valence electron count: 7
(Unshared pairs contribute 6 electrons; covalent bond contributes 1 electron)
Formal charge on fluorine:
Group number − valence electron count = 7 − 7 = 0

STUDY GUIDE LINK: ✓1.1 *Formal Charge*

Because all fluorine atoms of $[BF_4]^-$ are equivalent, they all must have the same formal charge—zero. Thus, it follows that the boron must bear the formal negative charge. Let's compute it to be sure.

Group number of boron: 3
Valence electron count: 4
(Four covalent bonds contribute 1 electron each)
Formal charge on boron: 3 − 4 = −1

Because the formal charge of boron is −1, the structure of $[BF_4]^-$ is written with the minus charge assigned to boron:

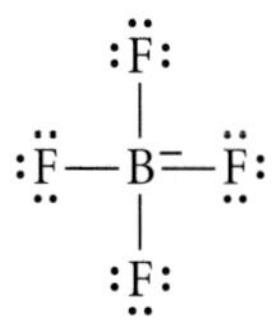

Rules for Writing Lewis Structures The previous two sections can be summarized in the following guidelines for writing Lewis structures.

1. Hydrogen can share no more than two electrons.
2. The sum of all bonding electrons and unshared pairs for atoms in the second period of the periodic table—the row beginning with lithium—will under no circumstances be greater than eight (octet rule). These atoms may, however, have fewer than eight electrons.
3. In some cases, atoms below the second period of the periodic table may have more than eight electrons. However, these cases occur so infrequently that Rule 2 should also be followed until exceptions are discussed later in the text.

✓Study Guide Links *flagged with a checkmark refer to material in the* Study Guide/Solutions Manual *that provides extra suggestions or shortcuts that should help you master the subject more easily.*

4 Nonvalence electrons are not shown in Lewis structures.

5. The formal charge on each atom is computed by the formalism illustrated in Study Problem 1.1 and, if not equal to zero, is indicated with a plus or minus sign.

Notice that we've used two types of electron counting. When we want to know whether an atom has a complete octet, we count all unshared valence electrons and *all bonding electrons* (Rule 2 above). When we count to determine formal charge, we count all unshared valence electrons and *half of the bonding electrons.*

STUDY PROBLEM 1.2

Draw a Lewis structure for the covalent compound methanol, CH_4O. Assume that the octet rule is obeyed, and that none of the atoms have formal charges.

Solution   In order for carbon to be *both* neutral *and* consistent with the octet rule at the same time, it must have four covalent bonds:

There is also only one way each for oxygen and hydrogen to have zero formal charge and simultaneously not violate the octet rule:

If we connect the carbon and the oxygen, and fill in the remaining bonds with hydrogens, we obtain a structure that meets all the criteria in the problem:

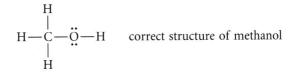

PROBLEMS

1.3   Draw a Lewis structure for each of the following species. Show all unshared pairs and the formal charges, if any. Assume that bonding follows the octet rule in all cases.
   *(a) $CHCl_3$ (b) HCl   *(c) ammonia, NH_3
   (d) ammonium ion, $[NH_4]^+$ *(e) $[H_3O]^+$ (f) $[AlCl_4]^-$

*1.4   Write two reasonable structures corresponding to the formula C_2H_6O. Assume that all bonding adheres to the octet rule, and that no atom bears a formal charge.

1.5   Draw a Lewis structure for each of the following compounds, assuming that all bonding obeys the octet rule, and that no atom bears a formal charge.
   *(a) allene, C_3H_4, which contains only double bonds between carbons
   (b) acetonitrile, C_2H_3N, which contains a carbon-nitrogen triple bond

*1.6   Compute the formal charges on each atom of the following structure. What is the charge on the entire structure?

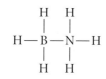

D. The Polar Covalent Bond

In many covalent bonds the electrons are not shared equally between two bonded atoms. Consider, for example, the covalent compound hydrogen chloride, HCl. (Although HCl dissolves in water to form H_3O^+ and Cl^- ions, in the pure gaseous state HCl is a covalent compound.) The electrons in the H—Cl covalent bond are unevenly distributed between the two atoms; they are polarized, or "pulled," toward the chlorine and away from the hydrogen. A bond in which electrons are shared unevenly is called a **polar bond**. The H—Cl bond is thus an example of a polar bond.

How can we determine whether a bond is polar? Think of the two atoms at each end of the bond as if they were engaging in a tug-of-war for the bonding electrons. *The tendency of an atom to attract electrons to itself in a covalent bond is indicated by its electronegativity*. The electronegativities of several elements are given in Table 1.1. If two atoms have equal electronegativities, then the bonding electrons are shared equally. But if one atom is more electronegative, then the electrons are unequally shared, and the bond is polar. (We might think of a polar covalent bond as a covalent bond that is trying

Table 1.1   Average Pauling Electronegativities of Some Main Group Elements

			H			
			2.20			
Li	Be	B	C	N	O	F
0.98	1.57	2.04	2.55	3.04	3.44	3.98
Na	Mg	Al	Si	P	S	Cl
0.93	1.31	1.61	1.90	2.19	2.58	3.16
K	Ca	Ga	Ge	As	Se	Br
0.82	1.00	1.81	2.01	2.18	2.55	2.96
Rb	Sr	In	Sn	Sb	Te	I
0.82	0.95	1.78	1.80	2.05	2.10	2.66
Cs	Ba	Tl	Pb	Bi		
0.79	0.89	1.62	1.87	2.02		

to become ionic!) It should now be clear how to recognize whether a bond is polar: *a polar bond is a bond between atoms of significantly different electronegativity.*

Sometimes we'll indicate the polarity of a bond in the following way:

$$\overset{\delta^+}{H} — \overset{\delta^-}{Cl}$$

In this notation, the delta (δ) is read "partially," or "somewhat." Thus, this notation indicates that the hydrogen atom of HCl is "partially positive," and the chlorine atom is "partially negative."

How can bond polarity be measured experimentally? The uneven electron distribution in a compound containing covalent bonds is measured by a quantity called the **dipole moment**, which is abbreviated with the Greek letter μ. The dipole moment is commonly given in derived units called *debyes*, abbreviated D, and named for the physical chemist Peter Debye. For example, the HCl molecule has a dipole moment of 1.08 D. If the H—Cl bond in HCl were not polar, the dipole moment of HCl would be zero.

An important aspect of the dipole moment is that it is a *vector quantity.* That is, it has not only magnitude, but direction. The direction of a dipole moment vector is taken by convention to be *from* the partial positive charge *to* the partial negative charge. Thus, the dipole moment vector for the HCl molecule is oriented along the H—Cl bond from the H to the Cl:

Study Guide Link:†
1.2
Dipole Moments

Molecules that have permanent dipole moments are called **polar molecules.** Thus, HCl is a polar molecule. Some molecules contain several polar bonds. Each polar bond has associated with it a dipole moment contribution, called a **bond dipole.** The net dipole moment of such a polar molecule is the vector sum of its bond dipoles. (Because HCl has only one bond, its dipole moment is equal to the H—Cl bond dipole.) Dipole moments of typical polar organic molecules are in the 1–3 D range.

The vectorial aspect of bond dipoles can be illustrated in a relatively simple way with the carbon dioxide molecule, CO_2.

$$\text{C—O bond dipoles}$$
$$\overset{\leftarrow\ \ \rightarrow}{O=C=O}$$

Because the CO_2 molecule is *linear*, the C—O bond dipoles are oriented in opposite directions. Because they have equal magnitudes, they *exactly cancel.* (Two vectors of equal magnitude oriented in opposite directions always cancel.) Consequently, CO_2 is *not* a polar molecule, *even though it has polar bonds.* In contrast, if a molecule contains several bond dipoles that do not cancel, the various bond dipoles add vectorially to give the overall resultant dipole moment.

The polarity of a molecule can significantly affect its chemical and physical properties. For example, its polarity may give some indication of how a molecule reacts chemically.

†Study Guide Links *not flagged with a checkmark (✓) contain additional material that covers the subject in greater depth.*

Returning to the HCl molecule, we know that HCl in water dissociates to its ions in a manner suggested by its bond polarity.

$$H_2O + \overset{\delta+}{H} \overset{\delta-}{-} Cl \xrightarrow{H_2O} H_3O^+ + Cl^- \tag{1.4}$$

We'll find many similar examples in organic chemistry in which bond polarity provides a clue to chemical reactivity.

PROBLEM

*1.7   Analyze the polarity of each bond in the following organic compound. Which bond, other than the C—C bond, is the least polar one in the molecule? Which carbon has the most partial positive character?

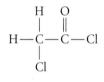

1.3   Structures of Covalent Compounds

You probably realize from your earlier chemical training that each covalent chemical compound has a **structure**—that is, a definite arrangement of its constituent atoms in space. The concept of covalent compounds as three-dimensional objects was developed in the latter part of the last century. Chemists who lived before that time regarded covalent compounds as shapeless groups of atoms held together in a rather undefined way by poorly understood electrical forces. Although the currently accepted structural characteristics of organic compounds were first suggested in 1874, these postulates were based on indirect chemical and physical evidence. Until the early twentieth century no one knew whether they had any physical reality, because scientists had no techniques for viewing molecules at the atomic level. Thus, as recently as the second decade of the twentieth century investigators could ask two questions: (1) Do organic molecules have specific geometries and, if so, what are they? (2) Are there simple principles that can predict molecular geometry?

A. Methods for Determining Molecular Structure

Among the greatest developments of chemical physics in the early twentieth century were the discoveries of ways to peer into molecules and deduce the arrangement in space of their constituent atoms. Most information of this type today comes from three sources: X-ray crystallography, electron diffraction, and microwave spectroscopy. The arrangement of atoms in the crystalline solid state can be determined by *X-ray crystallography*. This technique, invented in 1915 and revolutionized by the availability of high-speed computers, uses the fact that X-rays are diffracted from the atoms of a crystal in precise patterns that can be translated into a molecular structure. In 1930 another technique, *electron diffraction*, was developed from the observation that electrons are scattered by the atoms in molecules of gaseous substances. The diffraction patterns resulting from this scattering can also be used to deduce the arrangements of atoms in molecules. Following the development of radar in World War II came *microwave spectroscopy*, in which the absorption of microwave radiation by gaseous molecules provides structural information.

Most of the details of molecular structure in this book are derived from gas-phase methods—electron diffraction and microwave spectroscopy. For molecules that are not readily studied in the gas phase, X-ray crystallography is the most important source of structural information. No comparable method allows the study of structures in solution, a fact that is unfortunate because most chemical reactions take place in solution. The consistency of gas-phase and crystal structures suggests, however, that molecular structures in solution probably differ little from those of molecules in the solid or gaseous state.

B. Prediction of Molecular Geometry

Molecular structure is important because the way that a molecule reacts chemically is closely linked to its structure. Consequently, it will be important for us to predict structures of covalently bonded molecules in a general way.

What do we need to describe the structure of a covalent molecule? The structure of a simple diatomic molecule such as HCl is completely defined by the **bond length**, the distance between the centers of the bonded nuclei. Bond length is usually given in angstroms; 1 Å = 10^{-10} meter = 10^{-8} cm = 100 picometers (pm). Thus, the structure of HCl is completely specified by the separation of the H and Cl nuclei, 1.274 Å. When a molecule has more than two atoms, we need to specify not only each bond length, but also each **bond angle**, the angle between each pair of bonds to the same atom. Consider, for example, the molecule methane, CH_4. When we know the C—H bond lengths and the H—C—H bond angles, the structure of methane is completely determined.

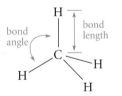

Bond Length   The following generalizations can be made about bond length, *in order of importance.*

1. *Bond lengths increase significantly toward higher periods (rows) of the periodic table.* This trend is illustrated in Fig. 1.2. For example, the H—S bond in hydrogen sulfide is longer than the other bonds to hydrogen in Fig. 1.2; sulfur is in the third period of the periodic table, and carbon, nitrogen, and oxygen are in the second period. Similarly, a C—H bond is shorter than a C—F bond, which is shorter than a C—Cl bond. These effects all reflect atomic size. Because bond length is the distance between centers of bonded atoms, larger atoms form longer bonds.

2. *Bond lengths decrease with increasing bond order.* **Bond order** describes the number of covalent bonds shared by two atoms. For example, a C—C

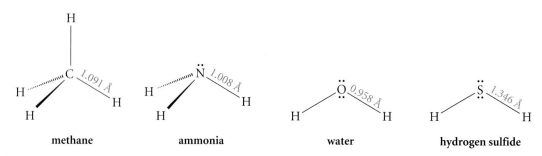

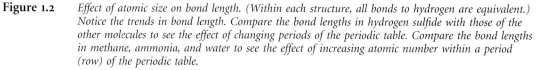

methane   ammonia   water   hydrogen sulfide

Figure 1.2   *Effect of atomic size on bond length. (Within each structure, all bonds to hydrogen are equivalent.) Notice the trends in bond length. Compare the bond lengths in hydrogen sulfide with those of the other molecules to see the effect of changing periods of the periodic table. Compare the bond lengths in methane, ammonia, and water to see the effect of increasing atomic number within a period (row) of the periodic table.*

single bond has a bond order of 1, a C=C double bond a bond order of 2, and a C≡C triple bond a bond order of 3. The decrease of bond length with increasing bond order is illustrated in Fig. 1.3. Notice that the bond lengths for carbon-carbon bonds are in the order C—C > C=C > C≡C.

3. *Bonds of a given order decrease in length toward higher atomic number (to the right) along a given row (period) of the periodic table.* Compare, for example, the H—C, H—N, and H—O bond lengths in Fig. 1.2. Likewise, the C—F bond in CH_3—F is shorter than the C—C bond in CH_3—CH_3. Because atoms on the right of the periodic table in a given row are smaller, this trend is also an effect of atomic size. However, this effect is much less drastic than the trend in atomic size observed between periods discussed above in Point 1.

Bond Angle   The bond angles within a molecule determine its *shape*—whether it is bent or linear. Two generalizations allow us to predict the approximate bond angles, and therefore the general shapes, of many simple molecules. The first is that *the groups bound to a central atom are arranged so that they are as far from one another as possible.* For

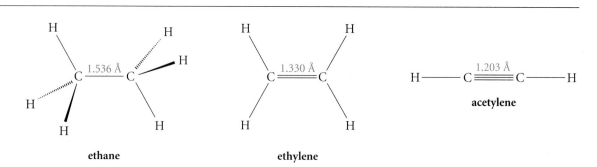

ethane   ethylene

Figure 1.3   *Effect of bond order on bond length. Notice that as the carbon-carbon bond order increases, the bond length decreases.*

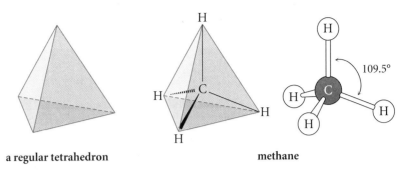

a regular tetrahedron   methane

Figure 1.4   *Structure of methane. The hydrogens are equidistant from each other and from the central carbon, and all H—C—H bond angles are 109.5°.*

methane, CH_4, the central atom is carbon and the groups are the four hydrogens. The hydrogens of methane are farthest apart when they occupy the vertices of a *tetrahedron* centered on the carbon atom (Fig. 1.4). Because the four C—H bonds of methane are identical, the hydrogen atoms of methane lie at the vertices of a *regular tetrahedron* (a tetrahedron with equal sides). The tetrahedral shape of methane requires an H—C—H bond angle of 109.5°.

In applying this rule for the purpose of predicting bond angles, we regard all groups as identical. Thus the groups that surround carbon in H_3CCl (methyl chloride) are treated as if they were identical, even though in reality the C—Cl bond is considerably longer than the C—H bonds. Although the bond angles show minor deviations from the exact tetrahedral bond angle of 109.5°, methyl chloride in fact has the general tetrahedral shape.

The tetrahedral structure, then, is assumed by molecules when four groups are arranged about a central atom. Because we'll see this geometry repeatedly, it is worth the effort to become familiar with it. Tetrahedral carbon is often represented, as shown below in the structure of methylene chloride, CH_2Cl_2, with two of its bound groups in the plane of the page. One of the remaining groups, indicated with a dashed line, is behind the page, and the other, indicated with a wedge-shaped heavy line, is in front of the page.

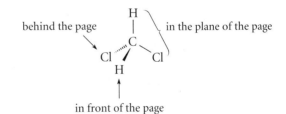

behind the page   in the plane of the page

in front of the page

☞
STUDY GUIDE LINK:
✓**1.3**
*Structure Drawing
Conventions*

A good way to become familiar with the tetrahedral shape is to use **molecular models**, commercially available scale models from which you can construct simple organic molecules. Perhaps your instructor has required that you purchase a set of models or can recommend a set to you. Professional organic chemists often use such models to study questions related to molecular geometry, and almost all beginning students require models, at least initially, to visualize the three-dimensional aspects of organic chemistry. Some of the types of models available are shown in Fig. 1.5. In this text, we'll use ball-and-stick models (Fig. 1.5a) to visualize the directionality of chemical bonds, and we'll use space-filling models to see the consequences of atomic and molecular volumes. You

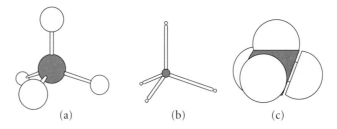

(a)   (b)   (c)

Figure 1.5   *Molecular models of methane. (a) Ball-and-stick models show the atoms as balls and the bonds as connecting sticks. Most inexpensive sets of student models are of this type. (b) A wire-frame model shows a nucleus (in this case, carbon) and its attached bonds. (c) Space-filling models depict atoms as spheres with radii proportional to their covalent or atomic radii. Space-filling models are particularly effective at showing the volume occupied by atoms or molecules.*

should obtain an inexpensive set of molecular models and use them frequently. Begin using them by building a model of the methylene chloride molecule discussed above and relating it to the line-wedge formula.

MOLECULAR MODELING BY COMPUTER

In recent years, scientists have come to use computers to depict molecular models. Computerized molecular modeling is particularly useful for very large molecules, because building real molecular models in these cases can be prohibitively expensive in both time and money. The decreasing cost of computing power has made computerized molecular modeling increasingly more practical. Most of the models shown in this text were drawn to scale from the output of a molecular modeling program on a desktop computer.

When *three* groups surround a central atom, the groups are as far apart as possible when all bonds lie in the same plane with bond angles of 120°. This is, for example, the geometry of boron trifluoride:

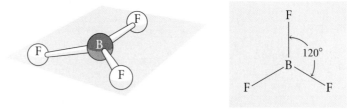

In such a situation the central atom (in this case boron) is said to have **trigonal planar** geometry.

When a central atom is surrounded by *two* groups, maximum separation of the groups demands a bond angle of 180°. This is the situation with each carbon in acetylene, $H—C{\equiv}C—H$. Each carbon is surrounded by two groups, a hydrogen and another carbon. (It makes no difference that the carbon has a triple bond.) Because of the 180° bond angle at each carbon, acetylene is a **linear** molecule.

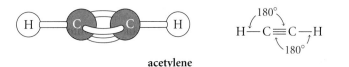

acetylene

The second generalization about bond angles applies to molecules with unshared electron pairs. In predicting the geometry of a molecule, *an unshared electron pair can be considered as a bond without a nucleus at one end.* This rule predicts, for example, the geometry of ammonia. In view of this rule, ammonia, $:NH_3$, has four groups about the central nitrogen: three hydrogens and an electron pair. To a first approximation, these groups adopt the tetrahedral geometry with the electron pair occupying one corner of the tetrahedron. (This geometry is sometimes called **trigonal pyramidal**, or pyramid-like geometry—*trigonal* because there are three hydrogens, and *pyramidal* because the N—H bonds lie along the edges of a pyramid.) We can refine our prediction of geometry even more if we recognize that an electron pair without a nucleus at one end has an especially repulsive interaction with electrons in adjacent bonds. As a result, the bond angle between the electron pair and the N—H bond is a little larger than tetrahedral, leaving the H—N—H angle a little smaller than tetrahedral; in fact, the H—N—H angle is 107.3°.

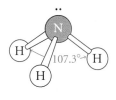

ammonia

STUDY PROBLEM 1.3

Estimate each bond angle in the following molecule, and order the bonds according to length, shortest first.

$$
\begin{array}{c}
\overset{5}{O} \\
\parallel {\scriptstyle (e)} \\
\overset{1}{H} \!-\! \overset{2}{C} \!\equiv\! \overset{3}{C} \!-\! \overset{4}{C} \!-\! \overset{6}{Cl} \\
{\scriptstyle (a)} \quad {\scriptstyle (b)} \quad {\scriptstyle (c)} \quad {\scriptstyle (d)}
\end{array}
$$

Solution   Because carbon-2 is bound to two groups (H and C), its geometry is linear. Similarly, carbon-3 also has linear geometry. The remaining carbon (carbon-4) is bound to three groups (C, O, and Cl), and therefore has approximately trigonal planar geometry. To arrange the bonds in order of length, recall the *order of importance* of the bond-length generalizations. The major effect on length is the row in the periodic table from which the bonded atoms are taken. Hence, the H—C bond is shorter than *all* carbon-carbon or carbon-oxygen bonds, which are shorter than the C—Cl bond. The next major effect is the bond order. Hence, the C≡C bond is shorter than the C=O bond, which is shorter than the C—C bond. Putting these conclusions together, the required order of bond lengths is

$$(a) < (b) < (e) < (c) < (d)$$

PROBLEMS

1.8   Predict the approximate geometry in each of the following molecules.

*(a)   water   (b)   the BF_4^- anion

*(c) $CH_2\!=\!\ddot{O}$ (formaldehyde)   (d) $CH_3\!-\!C\!\equiv\!N\!:$ (acetonitrile)

1.9   Estimate each of the bond angles and order the bond lengths, smallest first, for the following molecule.

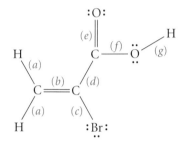

Conformation   For completely describing the shapes of molecules that are more complex than the ones we've just discussed, we need to specify not only the bond lengths and bond angles, but also *the spatial relationship of the bonds on adjacent atoms.*

To illustrate this problem, consider the molecule ethylene, $H_2C\!=\!CH_2$. The knowledge that both carbons have trigonal-planar geometry is not sufficient to describe completely the shape of ethylene. To see this point, imagine two planes, each containing one of the CH_2 groups of ethylene (Fig. 1.6). Nothing in what you've learned so far specifies the relationship of these two planes. The limiting possibilities, shown in Fig. 1.6, are that these planes lie at an angle of either 0° or 90°. The angular relationship of these planes is called the **dihedral angle**. Chemists frequently use another method of viewing the dihedral angle, also shown in Fig. 1.6, which will prove to be more generally useful. In this method, we sight along a bond of interest (the carbon-carbon double bond in the case of ethylene) from one end of the structure and project all atoms into the plane of the page. Such a projection is called a **Newman projection**. When we view the ethylene molecule in a Newman projection, the dihedral angle is the angle between the $C\!-\!H$ bonds on *adjacent* carbons; again the angle between closer $C\!-\!H$ bonds is 0° in the conformation at the top of Fig. 1.6, and 90° in the conformation at the bottom of the figure. A description of the dihedral angles within a molecule is called the **conformation** of the molecule. The conformation of ethylene is planar (dihedral angle = 0°), for reasons that are discussed in Chapter 4.

Molecules containing many bonds typically contain many dihedral angles to be specified. We'll examine molecular conformations further, and begin to learn some of the principles that allow us to predict conformations, in Chapter 2.

In summary, then, the structure of a molecule is completely determined by three elements: its bond lengths, its bond angles, and its dihedral angles, or conformation. The structures of some molecules are completely determined only by bond lengths and/or bond angles. Conformation enters the picture for more complex molecules.

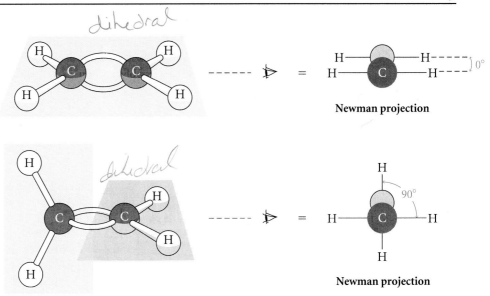

Figure 1.6   *The possible conformations and dihedral angles in ethylene. The H—C—C—H dihedral angle is the angle between the C—H bonds on adjacent carbons when one sights along the C=C bond from one end of the molecule and projects all C—H bonds into the plane of the page to give a Newman projection (shown at right). In planar ethylene the dihedral angle between nearest hydrogens is 0°. In the nonplanar structure, the dihedral angle is 90°.*

PROBLEMS

*1.10   Using molecular models, examine the conformations of *ethane*, H_3C—CH_3. What are the dihedral angles between closest C—H bonds (a) when they are as far apart as possible? (b) when they are as close as possible? Draw a diagram like the one in Fig. 1.6 to illustrate these two conformations.

1.11   Using molecular models to assist you, draw three conformations for the hydrogen peroxide (H_2O_2) molecule, in which the dihedral angles between O—H bonds are, respectively, (a) 0°, (b) 90°, and (c) 180°. What type of motion is required for these conformations to interconvert? (See also Problem 1.35.)

1.4   Resonance Structures

Some compounds are not accurately described by a single Lewis structure. Consider, for example, nitromethane, H_3C—NO_2.

This Lewis structure shows an N—O single bond and an N=O double bond. On the basis of the last section, we expect double bonds to be shorter than single bonds. However, it is found experimentally that the two nitrogen-oxygen bonds of nitromethane have the same length, intermediate between the lengths expected for single and double bonds. In order to convey this idea, nitromethane can be written as follows:

$$
\begin{bmatrix}
H_3C-\overset{+}{N}\begin{matrix}\ddots\ddot{O}:^-\\[2pt]\ddots\\[-2pt]:\ddot{O}:\end{matrix}
\quad\longleftrightarrow\quad
H_3C-\overset{+}{N}\begin{matrix}:\ddot{O}:\\[2pt]\\[-2pt]\ddot{O}:^-\end{matrix}
\end{bmatrix}
\qquad (1.5)
$$

The double-headed arrow means that nitromethane is a single compound that has characteristics of both structures; nitromethane is said to be a **resonance hybrid** of these two structures. The double-headed arrow does *not* mean that the two structures are in rapid equilibrium; rather, nitromethane is *one* compound with *one* structure.

It is important to understand that the two structures in Eq. 1.5 are *fictitious*, but nitromethane is a real molecule. Because there is no way to describe nitromethane with a single real structure, we are forced to describe it as the *hybrid* of two fictitious structures. An analogy to this situation is a description of Fred Flatfoot, a *real* detective. Lacking words to describe Fred, we picture him as a resonance hybrid of two *fictional* characters:

Fred Flatfoot = [Sherlock Holmes ⟷ James Bond]

This suggests that Fred is a dashing, violin-playing, pipe-smoking, highly intelligent British agent with an assistant named Watson, and that Fred likes his martinis shaken, not stirred.

When two resonance structures are identical, as they are for nitromethane, they are equally important in describing the molecule. We can think of nitromethane as a 1:1 average of the structures in Eq. 1.5. For example, each oxygen bears half a negative charge, and each nitrogen-oxygen bond is neither a single bond nor a double bond, but a bond halfway in between. The hybrid character of nitromethane is sometimes conveyed in a single structure with dashed lines to represent partial bonds. In this notation, the minus charge is understood to be equally shared by the atoms at the ends of the dashed semicircle—the two oxygens.

When two resonance structures are not identical, then the molecule they represent is a *weighted average* of the two. That is, one of the structures is more important than the other in describing the molecule. Such is the case, for example, with the methoxymethyl cation:

$$
\left[H_2\overset{+}{C}-\ddot{\ddot{O}}-CH_3 \quad\longleftrightarrow\quad H_2C=\underset{+}{\ddot{O}}-CH_3\right]
\qquad (1.6)
$$

methoxymethyl cation

It turns out that the structure on the right is a better description of this cation because all atoms have complete octets. Hence, there is significant double-bond character in the C—O bond, and most of the positive charge resides on the oxygen.

Another very important aspect of resonance structures is that they have implications for the stability of the molecule they represent. *A molecule represented by resonance structures is more stable than its fictional resonance contributors.* For example, the actual molecule nitromethane is more stable than either one of the fictional molecules described by the contributing resonance structures in Eq. 1.5. Nitromethane is thus said to be a *resonance-stabilized molecule*, as is the methoxymethyl cation.

How do you know when to use resonance structures, how to draw them, or how to assess their relative importance? In Chapter 3, you'll learn a technique for deriving resonance structures, and in Chapter 15, we'll return to a more detailed study of the other aspects of resonance. In the meantime, we'll draw resonance structures for you and tell you when they're important. Just try to remember the following points:

1. Resonance structures are used for compounds that are not adequately described by a single Lewis structure.

2. Resonance structures are *not* in equilibrium; that is, the compound they describe is not one structure part of the time and another structure part of the time, but rather *one structure*.

3. The structure of a molecule is the *weighted average* of its resonance structures. When resonance structures are identical, they are equally important descriptions of the molecule.

4. Resonance hybrids are more stable than any of the fictional structures used to describe them. Molecules described by resonance structures are said to be *resonance stabilized*.

PROBLEM

*1.12   The compound benzene has only one type of carbon-carbon bond, and this bond is found to have a length intermediate between that of a single bond and a double bond. Draw a resonance structure of benzene that, taken with the structure below, accounts for the carbon-carbon bond length.

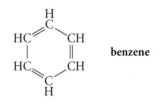

benzene

1.5   Wave Nature of the Electron

You've learned that the covalent chemical bond can be viewed as the sharing of one or more electron pairs between two atoms. Although this simple model of the chemical bond is very useful, in some situations it will not be adequate. A deeper insight into the nature of the chemical bond can be obtained from an area of science called *quantum mechanics*. Quantum mechanics deals in detail with, among other things, the behavior of electrons in atoms and molecules. Although the theory involves some sophisticated mathematics, we need not explore the mathematical detail in order to appreciate some general conclusions of the theory. The starting point for quantum mechanics is the idea that small particles such as electrons have the character of waves. How did this idea evolve?

As the twentieth century opened, it became clear that certain things about the behavior of electrons could not be explained by conventional theories. There seemed to be no doubt that the electron was a particle; after all, both its charge and mass had been measured. However, electrons could also be diffracted like light, and diffraction phenomena were associated with waves, not particles. The traditional views of the physical world treated particles and waves as unrelated phenomena. In the mid-1920s, this mode of thinking was changed by the advent of quantum mechanics. This theory holds that in the submicroscopic world of the electron and other small particles, there is no real distinction between particles and waves. The behavior of small particles such as the electron can be described by the physics of waves. In other words, matter can be regarded as a *wave–particle duality*.

How does this wave–particle duality require us to alter our thinking about the electron? In our everyday life we are accustomed to a deterministic world. That is, the position of any familiar object can be measured precisely, and its velocity can be determined for all practical purposes to any desired degree of accuracy. For example, we can point to a ball resting on a table and state with confidence, "That ball is at rest (its velocity is zero), and it is located exactly one foot from the edge of the table." Nothing in our experience indicates that we could not make similar measurements for an electron. The problem is that humans, chemistry books, and baseballs are of a certain scale. Electrons and other tiny objects are of a much smaller scale. A central principle of quantum mechanics, the **Heisenberg uncertainty principle**, tells us that *the precision with which we can determine the position and velocity of a particle depends on its scale.* According to this principle, it is impossible to define exactly both the position and velocity of an electron. Rather, we are limited to stating the *probability* with which we might expect to find an electron in any given region of space.

In summary:

1. Electrons have wavelike properties.
2. The exact position of an electron cannot be specified; only the probability that it occupies a certain region of space can be specified.

1.6   Electronic Structure of the Hydrogen Atom

In order to understand the implications of quantum theory for covalent bonding, you must first understand what the theory has to say about the electronic structure of atoms. This section presents the applications of quantum theory to the simplest atom, hydrogen. In the next section, we'll consider the electronic structures of more complex atoms.

A. Orbitals, Quantum Numbers, and Energy

In an earlier model of the hydrogen atom, the electron was thought to circle the nucleus in a well-defined orbit, much as the earth circles the sun. Quantum theory replaced the orbit with the *orbital*, which, despite the similar name, is something quite different. An **atomic orbital** is a description of the wave properties of the electron in an atom. We can think of an atomic orbital of hydrogen as an *allowed state—that is, an allowed wave motion—of an electron in the hydrogen atom.*

Table 1.2   Relationship Among All Four Quantum Numbers

n	l	m	s	n	l	m	s	n	l	m	s
1	0 (1s)	0	$\pm\frac{1}{2}$[a]	2	0 (2s)	0	$\pm\frac{1}{2}$	3	0 (3s)	0	$\pm\frac{1}{2}$
					1 (2p)	-1	$\pm\frac{1}{2}$		1 (3p)	-1	$\pm\frac{1}{2}$
						0	$\pm\frac{1}{2}$			0	$\pm\frac{1}{2}$
						$+1$	$\pm\frac{1}{2}$			$+1$	$\pm\frac{1}{2}$
									2 (3d)	-2	$\pm\frac{1}{2}$
										-1	$\pm\frac{1}{2}$
										0	$\pm\frac{1}{2}$
										$+1$	$\pm\frac{1}{2}$
										$+2$	$\pm\frac{1}{2}$

[a] $\pm\frac{1}{2}$ means that the spin quantum number s may assume either the value $+\frac{1}{2}$ or $-\frac{1}{2}$

Many possible orbitals, or states, are available to the electron in the hydrogen atom. In the mathematics of "electron waves," each orbital is described by a series of three **quantum numbers**. Quantum numbers serve as labels, or designators, for the various orbitals, or wave motions, available to the electron.

The **principal quantum number**, abbreviated n, can assume any integral value greater than zero—that is, $n = 1, 2, 3, \ldots$.

The **angular momentum quantum number**, abbreviated l, depends on the value of n. The l quantum number can assume any integral value from zero through $n - 1$, that is, $l = 0, 1, 2, \ldots, n - 1$. So that they are not confused with the principal quantum number, the values of l are encoded as letters. To $l = 0$ is assigned the letter s; to $l = 1$ the letter p; to $l = 2$ the letter d; and to $l = 3$ the letter f. The values of l are summarized in Table 1.2. It follows that there can be only one orbital with $n = 1$, an orbital with $l = 0$—a $1s$ orbital. However, two values of l, 0 and 1, are allowed for $n = 2$. Consequently, it is possible to have both $2s$ and $2p$ orbitals.

The **magnetic quantum number**, abbreviated m, is the last orbital quantum number. Its values depend on the value of l. The m quantum number can be zero as well as both positive and negative integers up to $\pm l$—that is, $0, \pm 1, \pm 2, \ldots, \pm l$. Thus, for $l = 0$ (an s orbital), m can only be 0. For $l = 1$ (a p orbital), m can have the values -1, 0, and $+1$. In other words, there is one s orbital with a given principal quantum number, but (for $n > 1$) there are three p orbitals with a given principal quantum number, one corresponding to each value of m. Because of the multiple possibilities for l and m, there are an increasingly large number of orbitals as n increases. This point is illustrated in Table 1.2 up to $n = 3$.

An electron is characterized by a fourth quantum number, called the **spin quantum number**, abbreviated s. This quantum number can assume either of two values: $+\frac{1}{2}$ or $-\frac{1}{2}$. The significance of the spin quantum number for atomic structure will become clear when we deal with atoms other than hydrogen (Sec. 1.7).

Just as an electron in the hydrogen atom can exist only in certain states, or orbitals, it can also have only certain allowed energies. Each orbital is associated with a characteristic electron energy. *The energy of an electron in a hydrogen atom is determined by the principal quantum number n of its orbital.* This is one of the central ideas of quantum theory. The energy of the electron is said to be *quantized*, or limited to certain values. This feature of the atomic electron is a direct consequence of its wave properties. An electron in an orbital with $n = 1$ remains in that state unless the atom is subjected to the exact amount of energy (say, from light) required to increase the energy of the electron to a state with a higher n, say $n = 2$. If that happens, the electron absorbs energy and instantaneously assumes the new, more energetic, wave motion characteristic of the orbital with $n = 2$. (Such energy-absorption experiments gave the first clues to the quantized nature of the atom.) An analogy to this may be familiar. If you have ever blown across the opening of a soda-pop bottle (or a flute, which is a more sophisticated example of the same thing), you know that only a certain pitch can be produced by a bottle of a given size. If you blow harder, the pitch does not rise, but only becomes louder. However, if you blow hard enough, the sound suddenly jumps to a note of higher pitch. The pitch is quantized; only certain sound frequencies (pitches) are allowed. Such phenomena are observed because sound is a wave motion of the air in the bottle, and only certain pitches can exist in a cavity of a given size without cancelling themselves out. The progressively higher pitches you hear as you blow harder (called *overtones* of the lowest pitch) are analogous to the progressively higher energy states (orbitals) of the electron in the atom.

B. Spatial Characteristics of Orbitals

One of the most important aspects of atomic structure for organic chemistry is that *each orbital is characterized by a three-dimensional region of space in which the electron is most likely to exist.* That is, orbitals have *spatial* characteristics. The *size* of an orbital is governed by its principal quantum number n: the larger n is, the greater the region of space occupied by the corresponding orbital. The *shape* of an orbital is governed by its angular momentum quantum number l. The *directionality* of an orbital is governed by its magnetic quantum number m. These points are best illustrated by example.

When an electron occupies a 1s orbital, it is most likely to be found in a sphere surrounding the atomic nucleus (Fig. 1.7). *We cannot say exactly where in that sphere the electron is* by the uncertainty principle; locating the electron is a matter of probability. The mathematics of quantum theory indicate that there is about a 90% probability that an electron in a 1s orbital will be found within a sphere of radius 1 Å about the nucleus. This "90% probability level" is taken as the approximate size of an orbital. Thus, we can

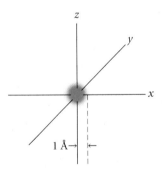

Figure 1.7   *The 1s orbital. Most of the electron density lies in a sphere within 1 Å of the nucleus.*

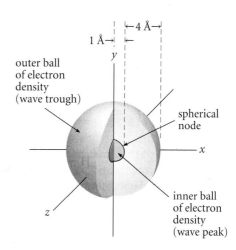

Figure 1.8   *A 2s orbital in a cutaway view, showing the peak (color) and trough (gray) of the electron wave. This orbital can be described as two concentric spheres of electron density. Note that a 2s orbital is considerably larger than a 1s orbital.*

depict an electron in a 1s orbital as *a smear of electron density*, most of which is within 1 Å of the nucleus.

When an electron occupies a 2s orbital, it also lies in a sphere, but the sphere is considerably larger—about four times the radius of the 1s orbital (Fig. 1.8). A 3s orbital is even larger still. The size of the orbital reflects the fact that the electron has greater energy; a more energetic electron can escape the attraction of the positive nucleus to a greater extent.

The 2s orbital illustrates another spatial aspect of orbitals, a *node*. You may be familiar with a simple wave motion such as the wave in a vibrating string, or waves in a pool of water. If so, you know that waves have *peaks* and *troughs*, regions where the waves are, respectively, at their maximum and minimum heights (Fig. 1.9). Suppose we define a wave to be positive at a wave peak and negative at a wave trough. Then, somewhere in between the peak and the trough the wave is zero. A **node** is a point or, in a three-dimensional wave, a surface at which the wave is zero.

As shown in Fig. 1.8, the 2s orbital has a node. This node separates a wave peak near the nucleus from a wave trough further out. Because the 2s orbital is a three-dimensional wave, its node is a *surface*. The nodal surface in the 2s orbital is an infinitely thin sphere. Thus, the 2s orbital has the characteristics of two concentric balls of electron density.

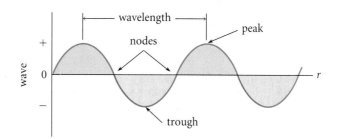

Figure 1.9   *A planar wave showing peaks, troughs, and nodes.*

Physically, the wave peak corresponds to a positive value in the mathematical function that describes the $2s$ wave motion, and a wave trough corresponds to a negative value. (This has *nothing* to do with the charge on the electron.) The node—the spherical shell of zero electron density—lies between the peak and the trough. Some students ask, "If the electron cannot exist at the node, how does it cross the node?" The answer is that the electron *is* a wave, and the node is part of its wave motion, just as the node is part of the wave in a vibrating string. The electron is not analogous to the string; it is analogous to the *wave* in the string.

The $2p$ orbital (Fig. 1.10a) illustrates how the l quantum number governs the *shape* of an orbital. All s orbitals are spheres. In contrast, all p orbitals have dumbbell shapes and are directed in space (that is, they lie along a particular axis). One lobe of the $2p$ orbital corresponds to a wave peak, and the other to a wave trough; the electron density is identical in corresponding parts of each lobe. Note that the two lobes are *parts of the same orbital*. The node in the $2p$ orbital, which passes through the nucleus and separates the two lobes, is a plane. The size of the $2p$ orbital, like that of other orbitals, is governed by its principal quantum number; it extends about the same distance from the nucleus as a $2s$ orbital.

The three $2p$ orbitals illustrate how the m quantum number governs the *directionality* of orbitals. The three equivalent $2p$ orbitals, which differ only in their values of m, are oriented in different directions: they are mutually perpendicular (Fig. 1.10b).

A $3p$ orbital has the typical dumbbell shape characteristic of p orbitals (Fig. 1.11). However, a $3p$ orbital contains *two* nodes. One node is a plane through the nucleus, much like the node of a $2p$ orbital. The other node is a spherical node that separates the inner part of each lobe from the larger outer part. *An orbital with principal quantum number n has $n - 1$ nodes.* Because the $3p$ orbital has $n = 3$, it has $(3 - 1) = 2$ nodes. The greater number of nodes in orbitals with higher n is a reflection of their higher energies. Again, the analogy to sound waves is very striking: overtones of higher pitch have larger numbers

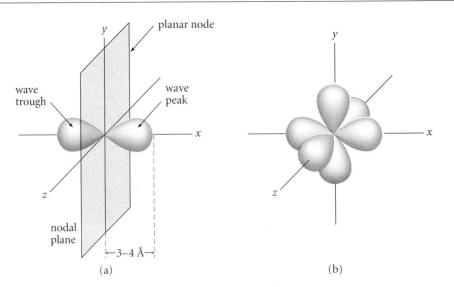

(a)   (b)

Figure 1.10   *(a) A 2p orbital. Notice the planar node that separates the orbital into two lobes. (b) The three 2p orbitals shown together. Each orbital has a different value of the quantum number m.*

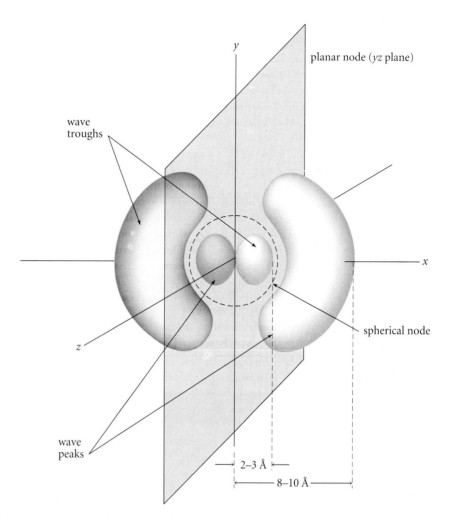

planar node (*yz* plane)

wave
troughs

spherical node

wave
peaks

2–3 Å

8–10 Å

Figure 1.11   *A 3p orbital. There are three such orbitals, each mutually perpendicular. Notice that the 3p orbital is much larger than the 2p orbital in Fig 1.10.*

of nodes. The size of a *3p* orbital also follows the trend of increased size that accompanies a higher principal quantum number: the 90% probability surface encompasses a larger region of space; this orbital extends to about 8–10 Å from the nucleus.

C. Summary: Atomic Orbitals of Hydrogen

Let's summarize the important points about orbitals in the hydrogen atom.

1. An orbital is an allowed state for the electron. It is a description of the wave motion of the electron.

2. Electron density within an orbital is a matter of probability, by the Heisenberg uncertainty principle. We can think of an orbital as a "smear" of electron density.

3. Orbitals are described by three quantum numbers:
   a. The principal quantum number n governs the energy of an orbital; orbitals of higher n have higher energy.
   b. The angular momentum quantum number l governs the shape of an orbital. Orbitals with $l = 0$ (s orbitals) are spheres; orbitals with $l = 1$ (p orbitals) are dumbbells.
   c. The magnetic quantum number m governs the orientation of an orbital.

4. An electron also has a property called spin which is described by a fourth quantum number s, which can have a value of $+\frac{1}{2}$ or $-\frac{1}{2}$.

5. Orbitals with $n > 1$ have nodes, which are surfaces of zero electron density. The nodes separate peaks of electron density from troughs. Orbitals with principal quantum number n have $n - 1$ nodes.

6. Orbital size increases with increasing n.

PROBLEM

1.13   Use the trends in orbital shapes you've just learned to describe the general features of the following orbitals.
*(a)   a 3s orbital   (b)   a 4s orbital

1.7   Electronic Structures of More Complex Atoms

The orbitals available to electrons in atoms with atomic number greater than 1 are, to a useful approximation, essentially like those of the hydrogen atom. There is, however, one important difference: In atoms other than hydrogen, electrons with the same principal quantum number n but with different values of l have different energies. Thus helium, carbon, and oxygen, like hydrogen, have 2s and 2p orbitals, but, unlike hydrogen, electrons in these orbitals differ in energy. The ordering of energy levels for atoms with more than one electron is illustrated schematically in Fig. 1.12. As this figure shows, the gaps between energy levels become progressively smaller as the principal quantum number increases. Furthermore, the energy gap between orbitals that differ in principal quantum number is greater than the gap between two orbitals within the same principal quantum level. Thus, the difference in energy between 2s and 3s orbitals is greater than the difference in energy between 3s and 3p orbitals.

Atoms beyond hydrogen, of course, have more than one electron. Let's now consider the **electronic configurations** of these atoms, that is, the way their electrons are distributed among their atomic orbitals. The **Aufbau principle** (literally, "buildup principle") tells us how to determine electronic configurations. This principle says to place electrons one by one into orbitals of the lowest possible energy in a manner consistent with the Pauli exclusion principle and Hund's rules. The **Pauli exclusion principle** states that no two electrons may have all four quantum numbers the same. As a consequence of this principle, *a maximum of two electrons may be placed in any one orbital, and these electrons must have different spins.* To illustrate, consider the electronic configuration of the helium atom, which contains two electrons. Both electrons can be placed into the 1s orbital as long as they have differing spin. Consequently, we can write the electronic configuration

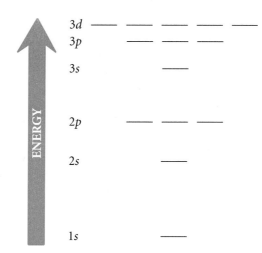

Figure 1.12   *A schematic representation of the relative energies of different orbitals in a many-electron atom. The exact scale varies from atom to atom, but the energy levels tend to be closer together as the principal quantum number increases.*

of helium as follows:

helium, He: $1s^2$

This notation means that there are two electrons of differing spin in the $1s$ orbital.

To illustrate Hund's rules, consider the electronic configuration of carbon, obviously a very important element in organic chemistry. A carbon atom has six electrons. The first two electrons (with opposite spins) go into the $1s$ orbital; the next two (also with opposite spins) go into the $2s$ orbital. Hund's rules tell us how to distribute the remaining two electrons among the *three equivalent* $2p$ orbitals. **Hund's rules** state, first, that to distribute electrons among identical orbitals of equal energy, single electrons are placed into separate orbitals before the orbitals are filled; and second, that the spins of these unpaired electrons are the same. Representing electrons as colored arrows, and letting their relative directions correspond to their relative spins, the electronic configuration of carbon can be indicated as follows:

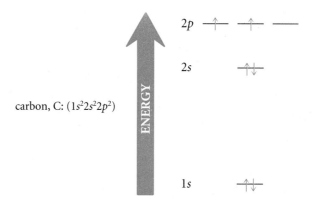

carbon, C: $(1s^2 2s^2 2p^2)$

Notice, in accordance with Hund's rules, that the electrons in the carbon $2p$ orbitals are unpaired with identical spin. This configuration ensures that repulsions between electrons are minimized, because electrons in different p orbitals occupy different regions of space. (Recall that the three $2p$ orbitals are mutually perpendicular; Fig. 1.10b.) As shown above, we can also write the electronic configuration of carbon more concisely as $1s^2 2s^2 2p^2$; when this notation is used, it is understood that the $2p$ electrons are distributed in accordance with Hund's rules.

Describe the electronic configuration of the oxygen atom.

Solution   Because oxygen has atomic number $= 8$, it has eight electrons. Following the Aufbau principle, the first two electrons occupy the $1s$ orbital with opposite spins. The next two, again with opposite spins, occupy the $2s$ orbital. Taking Hund's rules into account, the next three electrons are placed, unpaired and with identical spin, into the three equivalent $2p$ orbitals. The one remaining electron is then placed, with opposite spin, into a $2p$ orbital. To summarize:

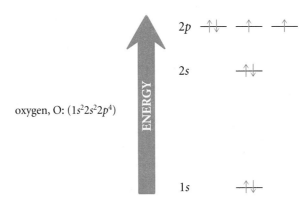

oxygen, O: $(1s^2 2s^2 2p^4)$

1.14   Give the electronic configurations of each of the following atoms and ions.
  *(a)   the nitrogen atom   (b)   the lithium atom
  *(c)   the sodium atom   *(d)   the magnesium atom
  (e)   the fluoride ion, F^- (f)   the potassium ion, K^+

1.8   Another Look at the Covalent Bond: Molecular Orbitals

A. Molecular Orbital Theory

When atoms combine into a molecule, the electrons contributed to the chemical bonds by each atom are no longer localized on individual atoms, but "belong" to the entire molecule. Consequently, atomic orbitals are no longer appropriate descriptions for the state of electrons in molecules. Instead, **molecular orbitals**, which are orbitals for the *entire molecule*, are used. The electronic configuration of a molecule is derived just like that of an atom: we arrange the molecular orbitals in order of increasing energy, and then add the available electrons to them in a manner consistent with the Pauli principle and Hund's rules. The question then becomes: How do we find out about the molecular orbitals?

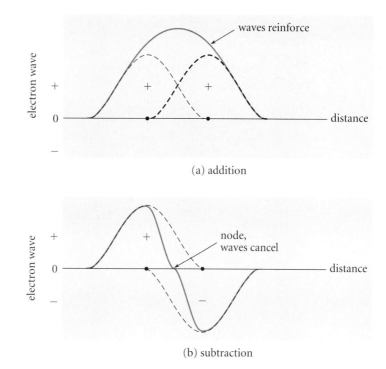

Figure 1.13 *The addition and subtraction of two waves such as the 1s orbitals of two hydrogen atoms. The dots represent the hydrogen nuclei; the dashed lines are the orbitals of the individual nuclei; and the colored line is the wave resulting from the combination. The plus and minus signs indicate wave peaks and troughs, respectively.*

Quantum mechanics specifies that we can derive molecular orbitals by combining atomic orbitals of the constituent atoms in a certain way. Conceptually, this is reasonable: molecules come from a combination of atoms, and molecular orbitals come from a combination of atomic orbitals. Moreover, quantum mechanics also specifies that if we combine *n* atomic orbitals, we get *n* molecular orbitals. This means, for example, that two molecular orbitals of the hydrogen molecule H—H can be derived by a combination of the 1s atomic orbitals of two hydrogen atoms.

To form the molecular orbitals of the hydrogen molecule from the atomic orbitals of its constituent atoms, let's imagine that we bring together two isolated hydrogen atoms until their nuclei are separated by the bond length in the hydrogen molecule. As a result, their 1s orbitals overlap. This overlap causes the 1s orbitals to interact. This interaction results in the formation of the two molecular orbitals of the hydrogen molecule. One molecular orbital is formed by the *addition* (additive overlap) of the individual atomic 1s orbitals. The second molecular orbital is generated by the *subtraction* (subtractive overlap) of the individual atomic 1s orbitals.

Let's first examine the molecular orbital formed by the addition of the two 1s orbitals. In the region in which these two orbitals overlap, they reinforce, because a wave peak is being added to a wave peak (Fig. 1.13a). (Remember, the orbitals really are "electron waves.") The result of the addition is the **bonding molecular orbital** of the hydrogen molecule (Fig. 1.14).

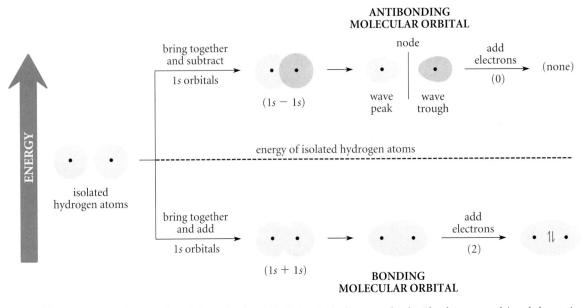

Figure 1.14 *Construction of the molecular orbitals for the hydrogen molecule. The dots are nuclei and the vertical arrows are electrons.*

The second molecular orbital is formed by subtraction of the two atomic orbitals in the overlap region. Subtraction of two atomic $1s$ orbitals is the same as overlap of a wave peak with a wave trough (Fig. 1.13b). In the region of subtractive overlap, the orbitals (or electron waves) cancel each other out, and the resulting molecular orbital contains a node between the two nuclei. This is the second molecular orbital of the hydrogen molecule, called the **antibonding molecular orbital** (Fig. 1.14).

The energy of an electron in the bonding molecular orbital is lower than that of an electron in an isolated hydrogen atom. On the other hand, the energy of an electron in the antibonding molecular orbital is higher than its energy in the hydrogen atom. Notice that the antibonding molecular orbital, unlike the bonding molecular orbital, has a node; like atomic orbitals, molecular orbitals of higher energy have more nodes.

The Aufbau principle operates for molecular orbitals just as it does for atomic orbitals. Hence, two electrons occupy the bonding molecular orbital of the hydrogen molecule; the antibonding molecular orbital is unoccupied. Because electrons in the bonding orbital of H_2 have lower energy than electrons in two isolated hydrogen atoms, it is evident that *chemical bonding is an energetically favorable process.* In fact, formation of a mole of hydrogen molecules from two moles of hydrogen atoms releases 435 kJ/mol (104 kcal/mol) of energy. This is a large amount of energy on a chemical scale—more than enough to raise the temperature of one kilogram of water from freezing to boiling.

According to the picture just developed, the chemical bond in a hydrogen molecule results from the occupation of a bonding molecular orbital by two electrons. You may wonder why we concern ourselves with the antibonding molecular orbital if it is not occupied. The reason is that it *can* be occupied! Introduction of a third electron into the hydrogen molecule requires occupation of the antibonding molecular orbital. The resulting three-electron species is the hydrogen molecule anion, H_2^- (Problem 1.15). Could such a species exist? It does, and this is why: Each electron in the bonding molecular

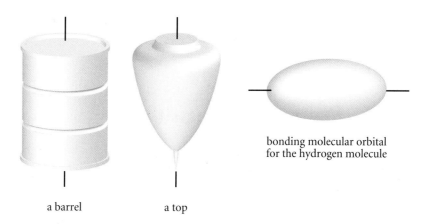

bonding molecular orbital
for the hydrogen molecule

a barrel   a top

Figure 1.15   *Some cylindrically symmetric objects. Objects are cylindrically symmetric when they appear the same no matter how they are rotated about their cylindrical axis (black line).*

orbital of the hydrogen molecule contributes half the stability of the molecule. The third electron in H_2^-, the one in the antibonding molecular orbital, has a high energy that offsets the stabilization afforded by *one* of the bonding electrons. However, the stabilization due to the second bonding electron remains. Thus, H_2^- is a stable species, but only about half as stable as the hydrogen molecule.

The importance of the antibonding orbital is also evident from an attempt to construct the molecule He_2, diatomic helium. The molecular orbitals for this molecule are conceptually identical to those for H_2 (Fig. 1.14), that is, one bonding and one antibonding molecular orbital, formed by addition and subtraction of the 1s orbitals of the isolated helium atoms. However, diatomic helium would have *four* electrons (two from each helium atom) to distribute between the molecular orbitals. The Aufbau principle dictates that two electrons occupy the bonding molecular orbital, and the remaining two occupy the antibonding molecular orbital. The lower energy of the electrons in the bonding molecular orbital is offset by the higher energy of the electrons in the antibonding molecular orbital. Because bonding in the He_2 molecule has no energetic advantage, helium is monatomic.

In the bonding molecular orbital of the hydrogen molecule the electrons occupy an ellipsoidal region of space. No matter how we turn the hydrogen molecule about a line joining the two nuclei, its electron density looks the same. This is another way of saying that the bond in the hydrogen atom has **cylindrical symmetry**. Other cylindrically symmetric objects are shown in Fig. 1.15. Bonds that are cylindrically symmetric about the internuclear axis are called **sigma bonds** (abbreviated σ bonds). The bond in the hydrogen molecule is thus a σ bond. The Greek letter sigma was chosen to describe the bonding molecular orbital of hydrogen because it is the Greek letter equivalent of *s*, the letter used to describe the lowest-energy atomic orbital.

PROBLEM

1.15   (a) Draw an orbital diagram corresponding to Figure 1.14 for *(1) the He_2^+ ion; (2) the H_2^- ion; *(3) the H_2^{2-} ion; (4) the H_2^+ ion. Which of these species are likely to exist as diatomic species, and which would dissociate into monatomic fragments? Explain.

*(b) The bond dissociation of energy of H_2 is 435 kJ/mol (104 kcal/mol); that is, it takes this amount of energy to dissociate H_2 into its atoms. Estimate the bond dissociation energy of H_2^+ and explain your answer.

B. Molecular Orbital Theory and Lewis Structures

Let's now relate the quantum mechanical view of the chemical bond to the concept of the Lewis electron-pair bond. In the Lewis picture, each covalent bond is represented by at least one electron pair shared between two nuclei. In the quantum-mechanical description, the σ bond exists because of the presence of electrons in a bonding molecular orbital and the resulting electron density between the two nuclei. Both electrons are attracted to each nucleus and therefore act as the cement that holds the nuclei together. Since a bonding molecular orbital can hold two electrons, *the Lewis view of the electron-pair bond is approximately equivalent to the quantum-mechanical idea of a bonding molecular orbital occupied by a pair of electrons.* The Lewis picture places the electrons squarely between the nuclei. Quantum theory says that although the electrons have a high probability of being between the bound nuclei, they can also occupy other regions of space.

Molecular orbital theory shows, however, that a chemical bond need not be an electron *pair*. For example, H_2^+ (the hydrogen molecule cation, which we might represent in the Lewis sense as H$\overset{+}{\cdot}$H) is a stable species in the gas phase (Problem 1.15). It is not so stable as the hydrogen molecule itself because the ion has only one electron in the bonding molecular orbital, rather than the two found in a neutral hydrogen molecule. The hydrogen molecule anion, H_2^-, discussed in the previous section, might be considered to have a three-electron bond consisting of two bonding electrons and one antibonding electron. The electron in the antibonding orbital is also shared by the two nuclei, but in a way that reduces the energetic advantage of bonding. (H_2^- is not as stable as H_2; Sec. 1.8A.) This example illustrates the point that the sharing of electrons between nuclei need not contribute to bonding. It is not surprising, then, that *the most common bonding situations occur when bonding molecular orbitals contain electron pairs and antibonding molecular orbitals are empty.* This is why ordinary chemical bonds can be represented as electron pairs.

1.9 Hybrid Orbitals

A. Bonding in Methane

When quantum theory is applied to methane, CH_4, the result is that the bonding to four hydrogen atoms alters the simple atomic-orbital picture derived for carbon in Sec. 1.7. That is, *the carbon in methane has an arrangement of orbitals that is different from the orbitals in atomic carbon.* The orbital arrangement for carbon in methane can, however, be derived from that of free carbon. For carbon in methane, we imagine that the $2s$ orbital and the three $2p$ orbitals are mixed to give four *equivalent* orbitals, each with a character intermediate between pure s and pure p. Since each carbon orbital in methane is one part s and three parts p, it is called an sp^3 orbital (pronounced "s-p-three," not "s-p-cubed"). This means that the six carbon electrons are distributed between one $1s$

orbital and four equivalent sp^3 orbitals. This mental transformation can be summarized as follows:

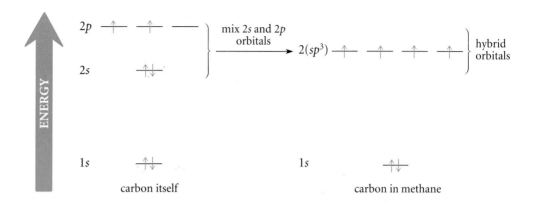

The sp^3 orbitals are examples of **hybrid orbitals**, because they come from the mixing of pure orbitals, just as a hybrid rose is formed by the mixing of pure strains. The shape of an sp^3 orbital is shown in Fig. 1.16a. The orbital consists of two lobes separated by a node, much like a p orbital. However, one of the lobes is very small, and the other is very large. In other words, the electron density in an sp^3 hybrid orbital is highly *directed* in space.

The number of hybrid orbitals (four in this case) is the same as the number of orbitals that are mixed to obtain them. (One s orbital + three p orbitals = four sp^3 orbitals.) It turns out that the large lobes of the four carbon sp^3 orbitals are directed to the corners of regular tetrahedron as shown in Figure 1.16b. Four hydrogen atoms, each with a single 1s electron, overlap with the four carbon sp^3 orbitals, each also with a single electron, to give the four bonding C—H molecular orbitals containing two electrons apiece (Fig. 1.16c). (As a result of this overlap four unoccupied antibonding orbitals also arise, which are not important here.) The four C—H bonds thus produced are σ bonds.

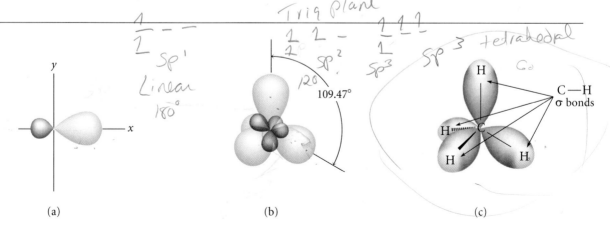

Figure 1.16   (a) A carbon sp^3 hybrid orbital. (b) The four sp^3 orbitals of carbon shown together. (c) An orbital picture of tetrahedral methane showing the four equivalent σ bonds formed from overlap of carbon sp^3 and hydrogen 1s orbitals. The rear lobes of the orbitals shown in (b) are omitted for clarity.

Why are hybrid orbitals formed? First, hybridization gives bonds that are as far apart as possible. The pure *s* and *p* orbitals available on nonhybridized carbon, in contrast, are not directed tetrahedrally. Second, rehybridization provides orbitals that have the bulk of their electron density directed toward the hydrogen nuclei. This directional character provides more electron "cement" between the nuclei and gives stronger (that is, more stable) bonds.

B. Bonding in Ammonia

Ammonia, :NH₃, is an example of a compound with an unshared electron pair. The electronic configuration of nitrogen in ammonia is, like carbon in methane, hybridized to yield four sp^3 hybrid orbitals; however, unlike the corresponding carbon orbitals, one of these orbitals contains an unshared electron pair.

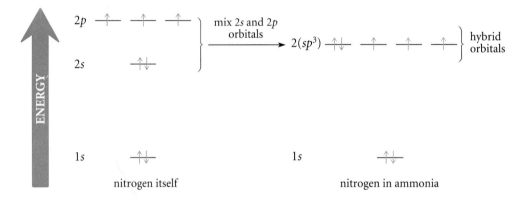

Each sp^3 orbital on nitrogen containing one electron can overlap with the 1*s* orbital of a hydrogen atom, also containing one electron, to give the three N—H σ bonds of ammonia. The sp^3 orbital containing the pair of electrons is filled. The electrons in this orbital become the nitrogen unshared pair. The unshared pair and the three N—H bonds, because they are made up of sp^3 hybrid orbitals, are directed to the corners of a regular tetrahedron (Fig. 1.17). The advantage of orbital hybridization in ammonia is the same as in carbon: hybridization accommodates the maximum separation of the unshared pair and the three hydrogens and, at the same time, provides strong, directed N—H bonds.

Notice the connection between hybridization of an atom and the arrangement in space of the bonds around that atom. Atoms surrounded by four groups (including

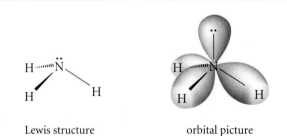

Lewis structure   orbital picture

Figure 1.17   *Bonding molecular orbital picture for ammonia, NH₃.*

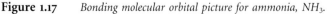

unshared pairs) in a tetrahedral arrangement are inevitably sp^3 hybridized. A trigonal-planar bonding arrangement is associated with a different hybridization, and a linear bonding arrangement with yet a third type of hybridization. In other words, *hybridization and molecular geometry are closely correlated.* The picture of covalent bonding derived from molecular orbital theory also drives home one of the most important differences between the ionic and covalent bond: *the covalent bond has a definite direction in space,* whereas the ionic bond is the same in all directions. The directionality of covalent bonding is responsible for molecular shape; and, as we shall see, molecular shape can have some very important chemical consequences.

Finally, you are now in a position to appreciate the quantum-mechanical reason for the wide adherence to the octet rule in covalent compounds. The atoms in the second period of the periodic table have partially filled orbitals in quantum level 2. There are four of these valence orbitals in all, one $2s$ orbital and three $2p$ orbitals. (Other atoms in the A groups of the periodic table have a similar valence orbital arrangement in higher principal quantum levels.) When these four orbitals combine with orbitals from other atoms, four bonding molecular orbitals are formed. Maximum stability results when these bonding molecular orbitals are completely filled with electrons. Filling these bonding molecular orbitals takes eight electrons—an octet.

PROBLEM

*1.16   Construct a hybrid orbital picture for the water molecule using oxygen sp^3 hybrid orbitals.

KEY IDEAS IN CHAPTER 1

⚗ Chemical compounds contain two types of bonds: ionic and covalent. In ionic compounds, ions are held together by electrostatic attraction (the attraction of opposite charges). In covalent compounds, atoms are held together by the sharing of electrons.

⚗ Both the formation of ions and bonding in covalent compounds tend to follow the octet rule: each atom is surrounded by eight valence electrons (two electrons for hydrogen).

⚗ The formal-charge convention assigns charges within a given species to its constituent atoms.

⚗ In covalent compounds, electrons may be shared unequally between bonded atoms. This unequal sharing results in a bond dipole. The vector sum of all bond dipoles in a molecule is its dipole moment.

(continues)

The structure of a molecule is determined by its bond lengths, bond angles, and conformation. Bond lengths are governed, in order of importance, by the period of the periodic table from which the bonded atoms are derived; by the bond order (whether the bond is single, double, or triple); and by the column (group) of the periodic table from which the atoms in the bond are derived. Approximate bond angles can be predicted by assuming that the groups bound to a central atom are as far apart as possible. Conformation can be described by the dihedral angles between bonds on adjacent atoms.

Molecules that are not adequately described by a single Lewis structure are represented as resonance hybrids, which are weighted averages of two or more fictitious Lewis structures. Resonance hybrids are more stable than any of their contributing resonance structures.

As a consequence of their wave properties, electrons in atoms and molecules can exist only in certain allowed energy states, called orbitals. Orbitals are descriptions of the wave properties of electrons in atoms and molecules, including their spatial distribution.

Electrons in orbitals are characterized by quantum numbers which, for atoms, are designated n, l, and m. Electron spin is described by a fourth quantum number s. The higher the principal quantum number n of an electron, the higher is its energy. In atoms other than hydrogen, the energy is also a function of the l quantum number.

Some orbitals contain nodes, which separate the wave-peak parts of the orbitals from the wave-trough parts. The number of nodes in an atomic orbital increases with its principal quantum number.

The distribution of electron density in a given type of orbital has a characteristic arrangement in space: all s orbitals are spheres, all p orbitals have two equal-sized lobes, etc.

Both atomic orbitals and molecular orbitals are populated with electrons according to the Aufbau principle.

Covalent bonds are formed when the orbitals of different atoms overlap. Covalent bonding can be understood to arise from the filling of bonding molecular orbitals by electron pairs.

The directional properties of bonds can be understood by the use of hybrid orbitals. The hybridization of an atom and the geometry of the atoms attached to it are closely related. All sp^3-hybridized atoms have tetrahedral geometry.

ADDITIONAL
PROBLEMS

1.17   In each of the following sets, specify the *one* compound that is most likely to exist as free ions in its liquid (molten) state.
*(a) (1) CCl_4 (2) HCl   (3) NaAt   (4) K_2
(b) (1) CS_2 (2) CsF   (3) HF   (4) XeF_4

*1.18   Which of the atoms in the following species have a complete octet? What is the formal charge on each? Assume all valence electrons are shown.
(a) CH_3 (b) $:NH_3$ (c) $:CH_3$ (d) BH_3 (e) $:\ddot{I}:$

1.19   Draw one Lewis structure for each of the following compounds; show all unshared electron pairs. None of the compounds bears a net charge, and all atoms except hydrogen have octets.
*(a) C_2H_3Cl (b) C_3H_6O
*(c) ketene, C_2H_2O, which has a carbon–carbon double bond
(d) a compound other than allene (Problem 1.5a), C_3H_4

1.20   Give the formal charge on each atom and the net charge on each species in each of the following structures. All valence electrons are shown.

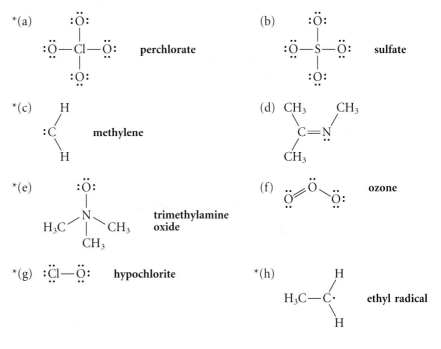

*(a)   perchlorate

(b)   sulfate

*(c)   methylene

(d)

*(e)   trimethylamine oxide

(f)   ozone

*(g) $:\ddot{C}l—\ddot{O}:$ hypochlorite

*(h)   ethyl radical

1.21   Give the electronic configuration of *(a) the chlorine atom; (b) the chloride ion; *(c) the argon atom; (d) the magnesium atom.

*1.22   Which of the following orbitals is/are *not* permitted by the quantum mechanics of the hydrogen atom? Explain.
(a) *2s*   (b) *6s*   (c) *5d*   (d) *2d*   (e) *3p*

1.23   Predict the approximate bond angles in each of the following molecules.

*(a) $:CH_2$

(b) BeH_2

*(c) $\overset{+}{C}H_3$

(d) $:\overset{..}{\underset{..}{Cl}}-Si(CH_3)_3$ (Give Cl—Si—C angle.)

*(e) $\overset{..}{O}=\overset{+}{O}-\overset{\overline{..}}{O}:$ **ozone**

(f) $H_2C=C=CH_2$ **allene**   (Give H—C—C and C—C—C angles.)

*(g)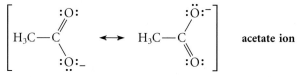

(h) $H_3\overset{+}{O}:$

*(i)   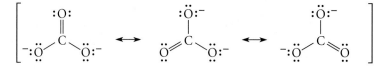   **carbonate ion**

*1.24   The *acetate ion* is a resonance hybrid of two structures.

$$\left[H_3C-C\overset{:O:}{\underset{:\overset{..}{O}:_-}{\Bigg\langle}} \longleftrightarrow H_3C-C\overset{:\overset{..}{O}:^-}{\underset{:O:}{\Bigg\langle}} \right]$$ **acetate ion**

In this ion, identify

(a)   the longest bond(s)   (c)   the bonds that are equivalent by resonance

(b)   the shortest bond(s)   (d)   the most polar bonds

*1.25   Consider the resonance structures for the *carbonate ion*.

$$\left[\underset{-:\overset{..}{O}}{\overset{:O:}{}}\overset{\|}{\underset{}{C}}\overset{}{\underset{\overset{..}{O}:^-}{}} \longleftrightarrow \underset{\overset{..}{O}}{\overset{:\overset{..}{O}:^-}{}}\overset{|}{\underset{}{C}}\overset{}{\underset{\overset{..}{O}:^-}{}} \longleftrightarrow \underset{-:\overset{..}{O}}{\overset{:\overset{..}{O}:^-}{}}\overset{|}{\underset{}{C}}\overset{}{\underset{\overset{}{O}}{}} \right]$$

(a)   How much negative charge is on each oxygen of the carbonate ion?

(b)   What is the bond order of each carbon-oxygen bond in the carbonate ion?

1.26   The *allyl cation* can be represented by the following resonance structures.

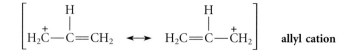   **allyl cation**

(a)   What is the *bond order* of each carbon-carbon bond in the allyl cation?

(b)   How much positive charge resides on each carbon of the allyl cation?

*(c) Although the structures above are reasonable descriptions of the allyl cation, the following cation *cannot* be described by analogous resonance structures. Explain why the structure on the right is not a reasonable resonance structure.

$$\left[\underset{H_2C=C-\overset{+}{N}H_3}{\overset{\overset{\displaystyle H}{|}}{}} \quad \bowtie \quad \underset{H_2\overset{+}{C}-C=NH_3}{\overset{\overset{\displaystyle H}{|}}{}} \right]$$

*1.27   (a) Two types of nodes occur in atomic orbitals: spherical surfaces and planes. Examine the nodes in $2s$, $2p$, and $3p$ orbitals and show that they agree with the following statements:
1. an orbital of principal quantum number n has $n - 1$ nodes;
2. the value of l gives the number of planar nodes.
   (b) How many spherical nodes are there in a $5s$ orbital? In a $3d$ orbital? How many nodes of all types are there in a $3d$ orbital?

*1.28   The shape of one of the five equivalent $3d$ orbitals is shown below. From your answer to the previous question, sketch the nodes of this $3d$ orbital, and associate a wave peak or a wave trough with each lobe of the orbital.

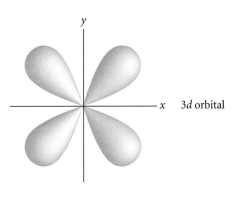

1.29   Orbitals with $l = 3$ are called f orbitals. (a) How many equivalent f orbitals are there? (b) In what principal quantum level do f orbitals first appear? (c) How many nodes are there in a $5f$ orbital?

*1.30   Sketch a $4p$ orbital. Show the nodes and the regions of wave peaks and wave troughs.

*1.31   Account for the fact that $H_3C—Cl$ (dipole moment 1.90 D) and $H_3C—F$ (dipole moment 1.85 D) have almost identical dipole moments, even though fluorine is considerably more electronegative than chlorine. (See Study Guide Link 1.2.)

*1.32   In Sec. 1.3B we noted that the principles for predicting bond angles do not permit a distinction between the following two conceivable forms of ethylene.

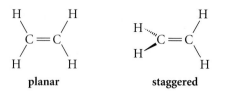

planar   staggered

The dipole moment of ethylene is zero. Does this experimental fact provide a clue to the conformation of ethylene? Why or why not?

*1.33   The difference in energy between an electron in the first quantum level of the hydrogen atom and one in the second is 1.635×10^{-18} J. The energy available from a photon is

$$E = hc/\lambda$$

where h is Planck's constant, 6.626×10^{-34} J-sec, c is the velocity of light (3×10^8 m/sec), and λ is the wavelength of the light. Calculate the wavelength of light required to bring about the jump of an electron in a hydrogen atom from the $n = 1$ to $n = 2$ level. (The wavelength is defined in Fig. 1.9.) The hydrogen atom is found to absorb light at precisely this wavelength, which is in the far ultraviolet region of electromagnetic radiation.

*1.34   A well-known chemist, Havno Szents, has heard you apply the rules for predicting molecular geometry to water; you have proposed (Problem 1.8a) a bent geometry for this compound. Dr. Szents is unconvinced by your arguments and continues to propose that water is a linear molecule. He demands that you debate the issue with him before a learned academy. You must therefore come up with *experimental* data that will prove to an objective body of scientists that water indeed has bent geometry. Explain why the dipole moment of water, 1.84 D, could be used to support your case.

*1.35   Consider again three possible conformations of the hydrogen peroxide molecule, H_2O_2, discussed in Problem 1.11, in which the dihedral angles between O—H bonds are, respectively, 0°, 180°, and 90°. Assuming that the H_2O_2 molecule exists predominantly in one conformation, which of these conformations can be *ruled out* by the fact that H_2O_2 has a large dipole moment (2.13 D)? Explain.

*1.36   Consider two $2p$ orbitals, one on each of two different atoms, oriented side to side, as follows:

nuclei

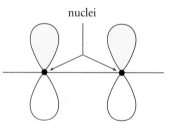

Imagine bringing these nuclei together so that overlap occurs between the wave peaks and between the wave troughs. This overlap results in a system of molecular orbitals.

(a) Sketch the shape of the resulting bonding and antibonding molecular orbitals.

(b) Identify the node(s) in each.

(c) When two electrons occupy the bonding molecular orbital, is the resulting bond a σ bond? Explain.

1.37 Consider two $2p$ orbitals, one on each of two atoms, oriented head to head as follows:

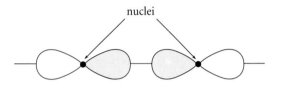

nuclei

Imagine gradually bringing the nuclei closer together until the two wave peaks (the colored lobes) of the orbitals just overlap. A new system of molecular orbitals is formed by this overlap.

(a) Sketch the shape of the resulting bonding and antibonding molecular orbitals.

(b) Identify the nodes in each molecular orbital.

(c) If two electrons occupy the bonding molecular orbital, is the resulting bond a σ bond? Explain.

*1.38 When a hydrogen molecule absorbs light, an electron jumps from the bonding molecular orbital to the antibonding molecular orbital. Explain why this light absorption can lead to the dissociation of the hydrogen molecule into two hydrogen atoms. (This process, called *photodissociation*, can sometimes be used to initiate chemical reactions.)

*1.39 Suppose you take a trip to a distant universe and find that the periodic table there is derived from an arrangement of quantum numbers different from the one on Earth. The rules there are:

1. principal quantum number $n = 1, 2, \ldots$ (as on Earth)
2. angular momentum quantum number $l = 0, 1, 2, \ldots, n - 1$ (as on Earth)
3. magnetic quantum number $m = 0, 1, 2, \ldots, l$ (that is, only *positive* integers up to and including l are allowed)
4. spin quantum number $s = -1, 0, +1$ (that is, *three* allowed values of spin)

(a) Write the electronic configuration of the element with atomic number 8 in the periodic table.

(b) What is the atomic number of the second noble gas?

(c) Assuming that the Pauli principle remains valid, what is the maximum number of electrons that can populate a given orbital?

(d) What rule would replace the octet rule?

2

Alkanes

2.1 Hydrocarbons

Our study of the various classes of organic compounds begins with the **hydrocarbons**, compounds that contain only the elements carbon and hydrogen. Methane, CH_4, is the simplest hydrocarbon. As you learned in Chapter 1, all the hydrogen atoms of methane are equivalent, occupying the corners of a regular tetrahedron. Imagine now, that instead of being bound only to hydrogens, a carbon atom could be bound to a second carbon with enough hydrogens to fulfill the octet rule. The resulting compound is *ethane*.

Lewis structures of ethane:

$$H:\overset{\overset{\displaystyle H}{\cdot\cdot}}{\underset{\underset{\displaystyle H}{\cdot\cdot}}{C}}:\overset{\overset{\displaystyle H}{\cdot\cdot}}{\underset{\underset{\displaystyle H}{\cdot\cdot}}{C}}:H \qquad H-\overset{\overset{\displaystyle H}{|}}{\underset{\underset{\displaystyle H}{|}}{C}}-\overset{\overset{\displaystyle H}{|}}{\underset{\underset{\displaystyle H}{|}}{C}}-H \qquad H_3C-CH_3$$

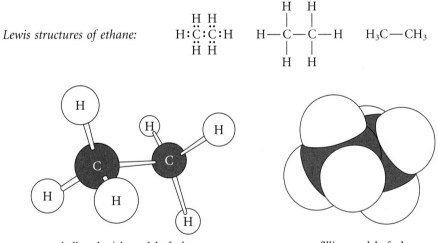

ball-and-stick model of ethane   space-filling model of ethane

In ethane, the bond between the two carbon atoms is longer than a C—H bond, but it is an electron-pair bond in the Lewis sense much like the C—H bond. In terms of orbitals, the carbon-carbon bond in ethane consists of two electrons in a bonding molecular orbital formed by the overlap of two sp^3 hybrid orbitals, one from each carbon. Thus, the carbon-carbon bond in ethane is an sp^3-sp^3 sigma bond (Fig. 2.1). The C—H bonds in ethane are like those of methane. They consist of electron pairs in bonding molecular orbitals, each of which is formed by the overlap of a carbon sp^3 orbital with a hydrogen $1s$ orbital. Both the H—C—C and H—C—H bond angles in ethane are approximately tetrahedral, because each carbon bears four groups.

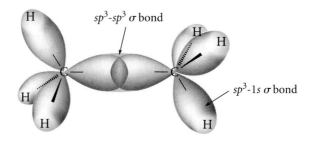

Figure 2.1   *Bonding molecular orbitals in ethane.*

We can go on to envision other hydrocarbons in which any number of carbons are bonded in this way to form chains of carbons bearing their associated hydrogen atoms. Indeed, the ability of a carbon to form stable bonds to other carbons is what gives rise to the tremendous number of known organic compounds. The idea of carbon chains, a revolutionary one in the early days of chemistry, was developed independently by the German chemist August Kekulé and the Scotsman Archibald Scott Couper in about 1858. Kekulé's account of his inspiration for this idea is amusing.

> During my stay in London I resided for a considerable time in Clapham Road in the neighborhood of Clapham Common. . . . One fine summer evening I was returning by the last bus, "outside" as usual, through the deserted streets of the city that are at other times so full of life. I fell into a reverie, and lo, the atoms were gamboling before my eyes. Whenever, hitherto, these diminutive beings had appeared to me they had always been in motion. Now, however, I saw how, frequently, two smaller atoms united to form a pair. . . . *I saw how the larger ones formed a chain*, dragging the smaller ones after them but only at the ends of the chain. . . . The cry of the conductor, "Clapham Road," awakened me from my dreaming, but I spent a part of the night putting on paper at least sketches of these dream forms. This was the origin of the "Structure Theory."

Hydrocarbons are divided into two broad classes: **aliphatic hydrocarbons** and **aromatic hydrocarbons**. There are three types of aliphatic hydrocarbons: *alkanes*, *alkenes*, and *alkynes*. We'll begin our study of aliphatic hydrocarbons with the **alkanes**, also known as **paraffins**. These are hydrocarbons that contain only single bonds. (Methane and ethane are the simplest alkanes.) Later we'll consider the **alkenes**, or **olefins**, hydrocarbons that contain carbon-carbon double bonds; and the **alkynes**, or **acetylenes**, hydrocarbons that contain carbon-carbon triple bonds. The class of **aromatic hydrocarbons** consists of benzene and its substituted derivatives.

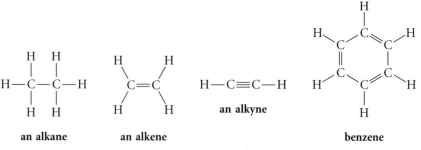

2.2   Unbranched Alkanes

Carbon chains take many forms in the alkanes; they may be branched or unbranched, and they can even exist as rings (cyclic alkanes). Alkanes with unbranched carbon chains are sometimes called **normal alkanes**, or **n-alkanes**. A few of the unbranched alkanes are shown in Table 2.1, along with some of their physical properties. You should learn the names of the first twelve unbranched alkanes because these names are the basis for naming many other organic compounds. The names *ethane*, *propane*, and *butane* have their origins in the early history of organic chemistry, but the names of the higher alkanes are derived from the corresponding Greek numerical names: pentane (*pent* = five); hexane (*hex* = six); etc.

Organic molecules are represented in different ways which we'll illustrate using the alkane hexane. The **molecular formula** of a compound (C_6H_{14} for hexane) gives its atomic composition. All noncyclic alkanes (alkanes without rings) have the general formula C_nH_{2n+2}, in which n is the number of carbon atoms. The **structural formula** of a molecule is its Lewis structure, which shows the **connectivity** of its atoms, that is, the order in which its atoms are connected. For example, a structural formula for hexane is the following:

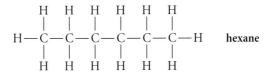

hexane

Table 2.1   The Unbranched Alkanes

Compound name	Molecular formula	Condensed structural formula	Melting point, °C	Boiling point, °C	Density,[a] g/mL
methane	CH_4	CH_4	−182.5	−161.7	—
ethane	C_2H_6	CH_3CH_3	−183.3	−88.6	—
propane	C_3H_8	$CH_3CH_2CH_3$	−187.7	−42.1	0.5005
butane	C_4H_{10}	$CH_3(CH_2)_2CH_3$	−138.3	−0.5	0.5788
pentane	C_5H_{12}	$CH_3(CH_2)_3CH_3$	−129.8	36.1	0.6262
hexane	C_6H_{14}	$CH_3(CH_2)_4CH_3$	−95.3	68.7	0.6603
heptane	C_7H_{16}	$CH_3(CH_2)_5CH_3$	−90.6	98.4	0.6837
octane	C_8H_{18}	$CH_3(CH_2)_6CH_3$	−56.8	125.7	0.7026
nonane	C_9H_{20}	$CH_3(CH_2)_7CH_3$	−53.5	150.8	0.7177
decane	$C_{10}H_{22}$	$CH_3(CH_2)_8CH_3$	−29.7	174.0	0.7299
undecane	$C_{11}H_{24}$	$CH_3(CH_2)_9CH_3$	−25.6	195.8	0.7402
dodecane	$C_{12}H_{26}$	$CH_3(CH_2)_{10}CH_3$	−9.6	216.3	0.7487
eicosane	$C_{20}H_{42}$	$CH_3(CH_2)_{18}CH_3$	+36.8	343.0	0.7886

[a] The densities tabulated in this text are at 20° unless otherwise noted.

Writing each hydrogen atom in this way is very time-consuming, and a simpler representation of this molecule, called a **condensed structural formula**, conveys the same information.

$$CH_3—CH_2—CH_2—CH_2—CH_2—CH_3 \qquad \textbf{hexane}$$

In such a structure the hydrogen atoms are understood to be connected to carbon atoms with single bonds, and the bonds shown explicitly are *bonds between carbon atoms.* Sometimes even these bonds are omitted, so that hexane can also be written $CH_3CH_2CH_2CH_2CH_2CH_3$. The structural formula may be further abbreviated as shown in the third column of Table 2.1. In this type of formula, for example, $(CH_2)_4$ means $—CH_2—CH_2—CH_2—CH_2—$, and hexane can thus be written $CH_3(CH_2)_4CH_3$.

The family of unbranched alkanes form a series in which successive members differ from each other by one $—CH_2—$ group (**methylene group**) in the carbon chain. Generally, physical properties within such a series vary in a regular way. An examination of Table 2.1, for example, reveals that the boiling points, melting points, and densities of the unbranched alkanes vary regularly with increasing number of carbon atoms. This variation can be useful for quickly estimating the properties of a member of the series whose properties are not known.

Even more important, the members of a series usually undergo the same chemical reactions. This observation greatly simplifies the learning of organic chemistry. For example, we can study the chemical reactions of propane with the confidence that ethane, butane, or dodecane will undergo analogous reactions.

> The French chemist Charles Gerhardt (1816–1856) recognized in 1845 that many related compounds differ in their molecular formulas by multiples of (CH_2). He wrote, "These (related) substances undergo reactions according to the same equations, and it is only necessary to know the reactions of one in order to predict the reactions of the others."

PROBLEMS

2.1   *(a)   How many hydrogen atoms are in the unbranched alkane with 18 carbon atoms?

(b)   How many carbon atoms are in the unbranched alkane with 30 hydrogen atoms?

*(c)   Is there an unbranched alkane containing 23 hydrogen atoms? If so, give its formula; if not, explain why not.

2.2   Estimate the boiling points of *(a) tridecane, $C_{13}H_{28}$, and (b) tetradecane, $C_{14}H_{30}$. Give a condensed structural formula for each compound.

2.3   Conformations of Alkanes

In Section 1.3B you learned that understanding the structures of many molecules requires that we specify not only their bond lengths and bond angles, but also their *conformations.*

In this section, we'll use the simple alkanes ethane and butane to develop some simple principles that will allow us to predict the conformations of more complex molecules.

A. Conformation of Ethane

In order to specify the conformation of ethane, we must define the relationship of the C—H bonds on one carbon to those on the other. To do this, view the molecule in a *Newman projection* (Sec. 1.3B). A Newman projection for ethane is shown in Fig. 2.2. In a typical Newman projection, we represent the near carbon of the bond as an open circle. The remaining carbon is hidden from view behind the circle. The bonds drawn to the center of the circle are attached to the near carbon; the bonds drawn to the periphery of the circle are attached to the hidden carbon. The carbon-carbon bond is, of course, also

ball-and-stick models:

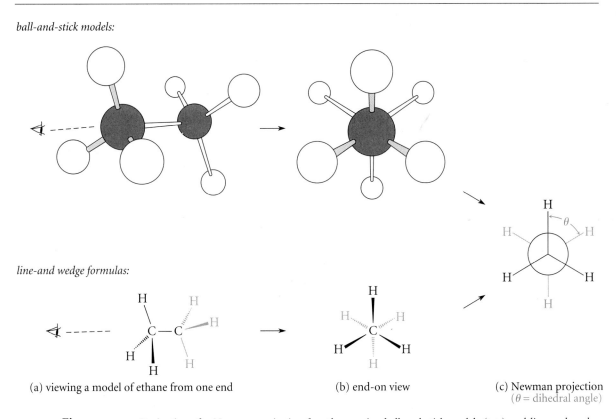

line-and wedge formulas:

(a) viewing a model of ethane from one end   (b) end-on view   (c) Newman projection
(θ = dihedral angle)

Figure 2.2   *Derivation of a Newman projection for ethane using ball-and-stick models (top) and line-and-wedge formulas (bottom). First view the ethane molecule from the end of the bond you wish to project, as in (a). The resulting end-on view is shown in (b). This is represented as a Newman projection (c) in the plane of the page, in which the carbon closer to the observer is represented by a circle. The bonds drawn to the center of the circle are attached to the carbon closer to the observer; the bonds drawn to the periphery of the circle (grey) are attached to the other (hidden) carbon.*

hidden. In the Newman projection of ethane, the *dihedral angle* θ between the carbon-hydrogen bonds on the different carbons defines the conformation of ethane.

Two limiting possibilities for the conformation of ethane can be seen from its Newman projections. In one, a C—H bond of one carbon bisects the angle between two C—H bonds of the other; this conformation of ethane is called the **staggered conformation** ($\theta = 60°$). In the other conformation, the **eclipsed conformation**, the C—H bonds on the respective carbons are superimposed ($\theta = 0°$).

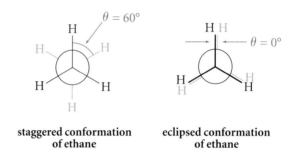

staggered conformation **of ethane**	**eclipsed conformation** **of ethane**

Of course, dihedral angles other than 60° and 0° are possible, but these two will prove to be of central importance. Which is the preferred conformation of ethane?

The energies of the ethane conformations can be described by a plot of relative energy *vs.* dihedral angle, which is shown in Fig. 2.3. In this figure, a dihedral angle of 0° is arbitrarily taken to be the eclipsed conformation, and the dihedral angle is the angle between the bonds to the colored hydrogens on the different carbons. To see the relationships in Fig. 2.3, build a model of ethane and use it in the following way. Hold either carbon fixed and turn the other about the C—C bond; as the angle of rotation changes, the model passes alternately through three identical staggered and three identical eclipsed conformations. As shown by Fig. 2.3, identical conformations have identical energies, as intuition dictates. The graph also shows that the eclipsed conformation is characterized by an energy *maximum*, and the staggered conformation is characterized by an energy *minimum*. The staggered conformation is thus the *more stable conformation* of ethane. The graph shows that the staggered conformation is more stable than the eclipsed conformation by about 12 kJ/mol (about 2.9 kcal/mol). This means that it would take about 12 kJ of energy to convert one mole of staggered ethane into one mole of eclipsed ethane, if such a conversion were practical.

Why is the staggered conformation of ethane more stable? According to one argument that is based on molecular-orbital theory, the molecular orbitals in the C—H bonds have an unfavorable interaction in the eclipsed conformation that is not present in the staggered conformation. This interaction is manifested as a greater energy for eclipsed ethane.

One staggered conformation of ethane can convert into another by rotation about the carbon-carbon bond. This rotation is called an **internal rotation** (to differentiate it from a rotation of the entire molecule). When an internal rotation occurs, an ethane molecule must briefly pass through the eclipsed conformation; thus, it must acquire the additional energy of the eclipsed conformation and then lose it again. What is the source of this energy?

At temperatures above absolute zero, molecules are in constant motion and therefore have kinetic energy. Heat is a manifestation of this energy. In a sample of ethane the molecules move about in a random manner, much as thousands of people might mill

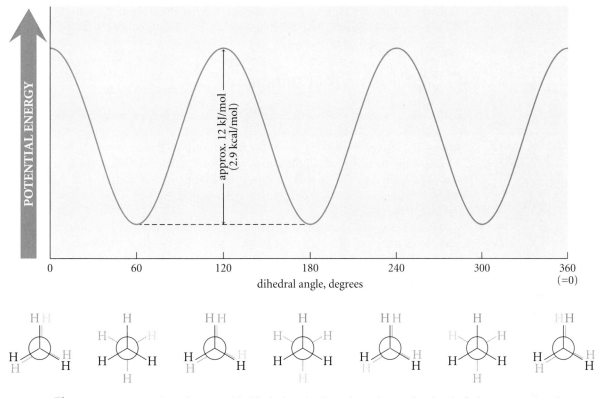

Figure 2.3   *Variation of energy with dihedral angle about the carbon-carbon bond of ethane. Note that the staggered conformations are at the energy minima, and the eclipsed conformations are at the energy maxima.*

about in a crowd. These moving molecules frequently collide, and molecules can gain or lose energy in such collisions. (An analogy is the collision of a bat with a ball; some of the kinetic energy of the bat is lost to the ball.) When an ethane molecule gains sufficient energy from a collision, it can pass through the higher-energy eclipsed conformation into another staggered conformation, that is, it can undergo internal rotation. Whether a given ethane molecule acquires sufficient energy to undergo an internal rotation is strictly a matter of *probability* (random chance). However, an internal rotation is more likely at higher temperature, because molecules have greater kinetic energy at higher temperature.

The likelihood that ethane undergoes internal rotation is reflected as its *rate* of rotation: how many times per second the molecule converts from one staggered conformation into another. This rate is determined by how much energy must be acquired in order for the rotation to occur, 12 kJ/mol (2.9 kcal/mol) in the case of ethane. This amount of energy is small enough that the internal rotation of ethane is very rapid even at very low temperatures. At 25 °C a typical ethane molecule undergoes a rotation from one staggered conformation to another at a rate of about 10^{11} times per second! This means that the interconversion between staggered conformations takes place about once every 10^{-11} second. Despite this short lifetime for any one staggered conformation, an ethane molecule spends most of its time in its staggered conformations, passing only

transiently through its eclipsed conformations. Thus, an internal rotation is best characterized not as a continuous spinning, but a constant succession of abrupt jumps from one
staggered conformation to another.

B. Conformations of Butane

Internal rotation about the central carbon-carbon bond of butane represents an important
and somewhat more complex conformational situation. The Newman projections for this
rotation are derived by looking down the *central* carbon-carbon bond as shown in Fig.
2.4. The graph of energy as a function of angle of internal rotation is given in Fig. 2.5.
Note once again that the various rotational possibilities are generated with a model by
holding either carbon fixed (the carbon away from the observer in Fig. 2.5) and rotating
the other one.

   Figure 2.5 shows that the staggered conformations of butane, like those of ethane,
are at energy minima, and are thus the stable conformations of butane. However, there are
important differences between the butane and ethane situations. Not all of the staggered
conformations (nor the eclipsed conformations) are alike. The different staggered conformations have been given special names. The conformations with a dihedral angle of $\pm60°$

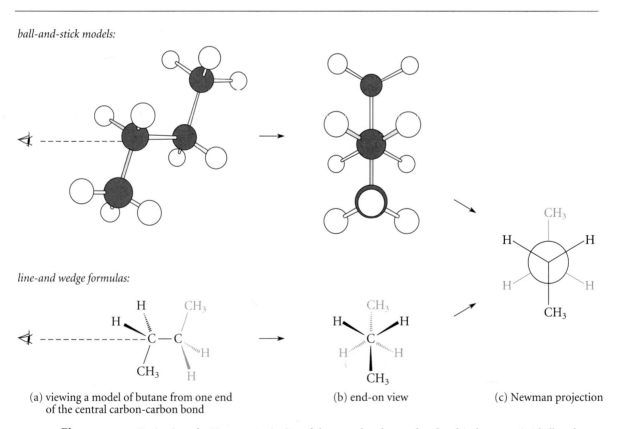

Figure 2.4   *Derivation of a Newman projection of the central carbon-carbon bond in butane using ball-and-
stick models (top) and line-and-wedge formulas (bottom). The drawing conventions are the same
as in Fig. 2.2. (Only one of the butane conformations is shown.)*

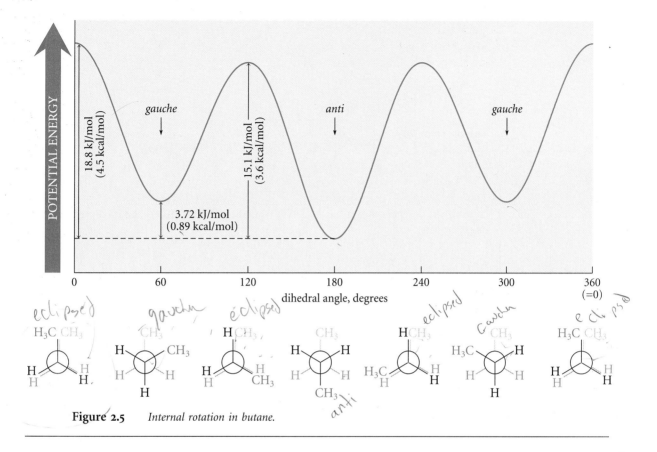

Figure 2.5 *Internal rotation in butane.*

STUDY GUIDE LINK:
✓2.1
Newman Projections

between the two C—CH$_3$ bonds are called *gauche* conformations; the form in which the dihedral angle is 180° is called the *anti* conformation.

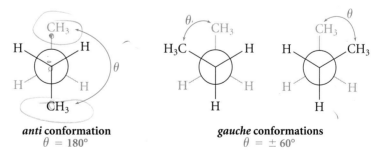

anti conformation
$\theta = 180°$

gauche conformations
$\theta = \pm 60°$

Figure 2.5 shows that the *gauche* and *anti* conformations of butane have different energies. The *anti* conformation is the more stable of the two by 3.72 kJ/mol (0.89 kcal/mol). The *gauche* conformation is the less stable of the two because the CH$_3$ groups are very close together—so close that the hydrogens on the two groups occupy each other's space. You can see this with the aid of the space-filling model in Fig. 2.6a. One measure of an atom's size is its **van der Waals radius**. Energy is required to force two *nonbonded* atoms together more closely than the sum of their van der Waals radii. Since the van der Waals radius of a hydrogen atom is about 1.2 Å, forcing the centers of two nonbonded hydrogens to be closer than twice this distance requires energy. Furthermore, the more

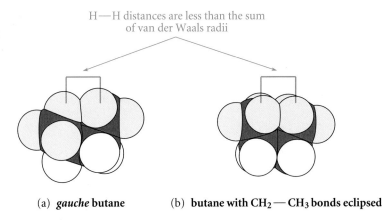

H—H distances are less than the sum
of van der Waals radii

(a) *gauche* butane   (b) butane with CH_2—CH_3 bonds eclipsed

Figure 2.6   *(a) Space-filling models of* gauche *butane with the methyl hydrogens shown in color. A hydrogen atom from one* CH_3 *group is so close to a hydrogen atom of the other* CH_3 *group that they violate each other's van der Waals radii. This unfavorable interaction causes* gauche *butane to have a higher energy than* anti *butane, in which this interaction is absent. (b) When the two* CH_3 *groups of butane are eclipsed, van der Waals repulsions are even greater than they are in* gauche *butane.*

the two hydrogens are pushed together, the more energy is required. This extra energy required to force two nonbonded atoms within the sum of their van der Waals radii is called a **van der Waals repulsion**. Thus, in order to attain the *gauche* conformation, butane must acquire more energy. In other words, *gauche butane is destabilized by van der Waals repulsions between nonbonded hydrogens on the two* CH_3 *groups.*

The eclipsed conformations of butane are unstable for the same reason that the eclipsed conformations of ethane are unstable. Moreover, in the conformation of butane in which the two C—CH_3 bonds are eclipsed, the hydrogens in the two CH_3 groups are even closer than they are in the *gauche* conformation (Fig. 2.6b). The van der Waals repulsions and the resultant energy cost are correspondingly greater. Notice that of all the eclipsed conformations, this one is the most unstable ($\theta = 0°$ in Fig. 2.5).

Why should we care about the relative energies of the butane conformations? The reason is that when conformations are in equilibrium, as the various conformations of butane are, *the most stable conformation—the conformation of lowest energy—is present in greatest amount.* Thus, the *anti* conformation of butane is the predominant conformation of butane. At room temperature, there are about twice as many molecules of butane in the *anti* conformation as there are in the *gauche* conformation.

The *gauche* and *anti* conformations of butane interconvert rapidly at room temperature—almost as rapidly as the staggered forms of ethane. Because the eclipsed conformations are unstable, they do not exist to any measurable extent.

Several of the things you've learned from studying the conformations of ethane and butane can be generalized to other alkanes.

1. Staggered conformations about single bonds are favored.

2. Conformations in which larger groups are brought closer together are less stable than conformations in which these groups are farther apart.

3. Van der Waals repulsions (repulsions between nonbonded atoms) are a source of instability.

*2.3   (a) Draw a Newman projection for each conformation about the C2–C3 bond of isopentane, a compound containing a branched carbon chain.

$$CH_3 \overset{2}{-} \overset{|}{CH} \overset{3}{-} CH_2 - CH_3 \qquad \textbf{isopentane}$$
$$\underset{CH_3}{|}$$

Show both staggered and eclipsed conformations.

(b) Sketch a curve of potential energy *vs.* dihedral angle for isopentane similar to that of butane in Fig. 2.5. Your curve should show the relative stabilities of the different conformations. (Do not try to give numerical values for relative energies.) Label each energy maximum and minimum with one of the conformations you drew in (a).

(c) Which one conformation is likely to be present in greatest amount in a sample of isopentane? Explain.

2.4   Repeat Problem 2.3 for the C1–C2 bond of butane.

$$\overset{1}{CH_3} - \overset{2}{CH_2} - CH_2 - CH_3 \qquad \textbf{butane}$$

Constitutional Isomers and Nomenclature

A. Isomers

When a carbon atom in an alkane is bound to more than two other carbon atoms, a branch in the carbon chain occurs at that position. The smallest branched alkane has four carbon atoms. As a result, there are two four-carbon alkanes; one is *butane*, and the other is *isobutane*.

$$CH_3 - CH_2 - CH_2 - CH_3 \qquad \qquad \overset{CH_3}{\underset{CH_3}{\overset{\diagdown}{\underset{\diagup}{CH}}} - CH_3}$$

butane
bp −0.5°

isobutane
bp −11.7°

These are different compounds with different properties. For example, the boiling point of butane is −0.5 °C, whereas that of isobutane is −11.7 °C. Yet both have the same molecular formula, C_4H_{10}. Different compounds that have the same molecular formula are said to be **isomers**.

There are different types of isomers. Isomers such as butane and isobutane that differ in the *connectivity* of their atoms are termed **constitutional isomers** (in earlier literature called **structural isomers**). Recall (Sec. 2.2) that *connectivity* is the order in which the atoms of the molecule are connected. Notice that the atomic connectivities of butane and isobutane differ because, in isobutane, a carbon is attached to three other carbons, whereas in butane, no carbon is attached to more than two other carbons.

Butane and isobutane are the only constitutional isomers with the formula C_4H_{10}. However, more constitutional isomers are possible for alkanes with more carbon atoms. There are nine isomers of the heptanes (C_7H_{16}); 75 isomers of the decanes ($C_{10}H_{22}$); and 366,319 isomers of the eicosanes ($C_{20}H_{42}$)! The large number of isomers that are possible for organic compounds of even modest size creates a problem of nomenclature. How can one unambiguously designate each one of many isomeric compounds with a name that can be easily constructed and remembered?

B. Organic Nomenclature

An organized effort to standardize organic nomenclature dates from proposals made at Geneva in 1892. From these proposals have evolved several accepted systems of nomenclature developed and sanctioned by the International Union on Pure and Applied Chemistry (IUPAC), a professional association of chemists. The most widely applicable system of nomenclature in use today is called **substitutive nomenclature**.

The IUPAC rules for nomenclature of alkanes form the basis for the substitutive nomenclature of most other compound classes. Hence, it is important to learn these rules and be able to apply them.

C. Substitutive Nomenclature of Alkanes

Alkanes are named by applying the following rules *in order*. This means that if one rule doesn't unambiguously determine the name of a compound of interest, we proceed down the list *in order* until we find a rule that does.

1. *The unbranched alkanes are named according to the number of carbons as shown in Table 2.1.*
2. *For alkanes containing branched carbon chains, determine the principal chain.*

The **principal chain** is the longest continuous carbon chain in the molecule. To illustrate:

$$CH_3—CH_2—CH_2—CH—CH_2—CH_3 \qquad \text{principal chain}$$
$$| $$
$$CH_3$$

When identifying the principal chain, take into account that *the condensed structure of a given molecule may be drawn in several different ways.* Thus, the following structures represent the *same molecule*, with the principal chain shown in color:

$$CH_3—CH_2—CH_2—CH—CH_2—CH_3 \qquad CH_3—CH—CH_2—CH_3$$
$$| \qquad\qquad\qquad\qquad\qquad | $$
$$CH_3 \qquad\qquad\qquad\qquad\qquad CH_2—CH_2—CH_3$$

Although these structures don't look the same, they are the same! How can you know? The key point to understand is that condensed structures like the ones used here are *not* meant to depict orientations in space. In order for two condensed structures to represent the same molecule, they need only have the same *connectivities*. In both the structures above, for example, the connectivity sequence is the same: CH_3, CH_2, CH_2, (CH connected to CH_3), CH_2, CH_3. Hence, both structures represent the same molecule. Thus, identification of the principal chain, as well as all other aspects of substitutive nomenclature, are based on connectivity—not on the way a molecule happens to be drawn.

3. *If two or more chains within a structure have the same length, choose as the principal chain the one with the greater number of branches.*

The following structure is an example of such a situation:

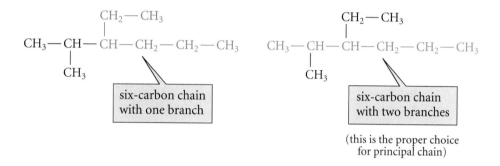

(this is the proper choice for principal chain)

The correct choice of principal chain is the one on the right, because it has two branches; the choice on the left has only one. (It makes no difference that the branch on the left is larger, or that it has additional branching within itself.)

4. *Number the carbons of the principal chain consecutively from one end to the other in the direction that gives the branch the lower number.*

In the following structure, the carbons of the principal chain are numbered to give the lower number to the $—CH_3$ branch.

$$\overset{6}{CH_3}—\overset{5}{CH_2}—\overset{4}{CH_2}—\overset{3}{CH}—\overset{2}{CH_2}—\overset{1}{CH_3} \quad \longleftarrow \text{proper numbering}$$
$$\underset{1}{}\quad\underset{2}{}\quad\underset{3}{}\quad\underset{4}{}\Big|\quad\underset{5}{}\quad\underset{6}{} \quad \longleftarrow \text{improper numbering}$$
$$CH_3$$

5. *Name each branch and identify the carbon number of the principal chain at which it occurs.*

In the previous example, the branching group is a $—CH_3$ group. This group is called a *methyl group*, and it is located at carbon-3 of the principal chain.

Branching groups are in general termed **substituents**, and substituents derived from alkanes are called **alkyl groups**. An alkyl group may contain more than one carbon. The name of an unbranched alkyl group is derived from the unbranched alkane with the same number of carbons by dropping the final *ane* and adding *yl*.

$—CH_3$	methyl (= meth~~ane~~ + yl)
$—CH_2CH_3$ or $—C_2H_5$	ethyl (= eth~~ane~~ + yl)
$—CH_2CH_2CH_3$	propyl

Table 2.2 Nomenclature of Some Short Branched-Chain Alkyl Groups

Group structure	Condensed structure	Written name	Pronounced name
CH₃\CH—/CH₃	$(CH_3)_2CH—$	*isopropyl*	isopropyl
CH₃\CHCH₂—/CH₃	$(CH_3)_2CHCH_2—$	*isobutyl*	isobutyl
CH₃CH₂CH— | CH₃	—	*sec*-butyl	secondary butyl
CH₃ | CH₃—C— | CH₃	$(CH_3)_3C—$	*tert*-butyl (or *t*-butyl)	tertiary butyl
CH₃ | CH₃—C—CH₂— | CH₃	$(CH_3)_3CCH_2—$	neopentyl	neopentyl

Alkyl substituents themselves may be branched. The most common branched alkyl groups have special names, given in Table 2.2. These should be learned, because they will be encountered frequently. Notice that the "iso" prefix is used for substituents containing two methyl groups at the end of a carbon chain. Also notice carefully the difference between an isobutyl group and a *sec*-butyl group; these two groups are frequently confused by beginning students.

6. *Construct the name by writing the carbon number of the principal chain at which the substituent occurs, a hyphen, the name of the branch, and the name of the alkane corresponding to the principal chain.*

$$CH_3—CH_2—CH_2—CH—CH_2—CH_3$$
$$| \atop CH_3$$

name: **3-methylhexane**

name of principal chain

number and name of alkyl substituent

STUDY GUIDE LINK:
✓2.2
*Nomenclature of
Simple Branched
Compounds*

Notice that the name of the branch and the name of the principal chain are written together as one word. Notice also that the name itself does *not* necessarily convey any isomeric relationships; that is, the compound above is a *constitutional isomer of heptane* because it has seven carbon atoms, but *it is named as a derivative of hexane,* because its principal chain contains six carbon atoms.

**STUDY
PROBLEM
2.1**

Name the following compound, and give the name of the unbranched alkane of which it is a constitutional isomer.

$$CH_3-CH_2-CH_2-CH-CH_2-CH_2-CH_3$$
$$|$$
$$CH-CH_3$$
$$|$$
$$CH_3$$

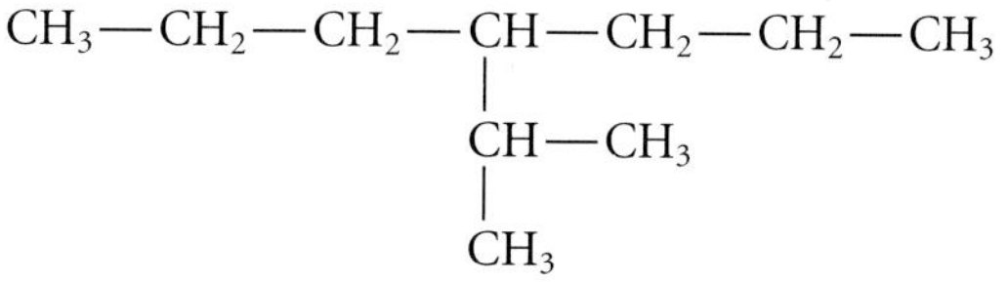

Solution   Because the principal chain has seven carbons, the compound is named as a substituted heptane. The branch is at carbon-4, and the substituent group at this branch is

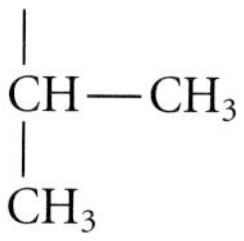

$$|$$
$$CH-CH_3$$
$$|$$
$$CH_3$$

Table 2.2 shows that this group is an *isopropyl group*. Thus, the name of the compound is 4-isopropylheptane:

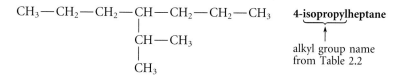

$$CH_3-CH_2-CH_2-CH-CH_2-CH_2-CH_3$$ **4-isopropylheptane**
$$|$$
$$CH-CH_3$$ alkyl group name
$$|$$ from Table 2.2
$$CH_3$$

Because this compound has the molecular formula $C_{10}H_{22}$, it is a constitutional isomer of the unbranched alkane *decane.*

7. *When there are multiple substituent groups on the principal chain, each sub-stituent receives its own number. The prefixes di, tri, tetra, etc., are used to indicate the number of identical substituents.*

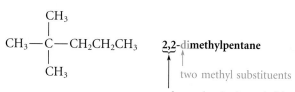

$$CH_3$$
$$|$$
$$CH_3-C-CH_2CH_2CH_3$$ **2,2-dimethylpentane**
$$|$$
$$CH_3$$ two methyl substituents

shows that both methyl branches are
at carbon-2 of the principal chain

....

Which two of the following structures represent the same compound? Name the compound.

$$CH_3—CH_2$$
$$|$$
$$CH—CH_3$$
$$|$$
$$CH_3—CH_2—CH$$
$$|$$
$$CH_3 \quad A$$

$$CH_3—CH—CH_2—CH—CH_2—CH_3$$
$$| \qquad\qquad |$$
$$CH_3 \qquad\quad CH_3$$
$$B$$

$$CH_3—CH—CH—CH_2—CH_3$$
$$| \qquad |$$
$$CH_2 \quad CH_3$$
$$|$$
$$CH_3$$
$$C$$

Solution   The connectivities of both *A* and *C* are the same: (CH_3, CH_2, [CH connected to CH_3], [CH connected to CH_3], CH_2, CH_3). The compound represented by these structures has six carbons in its principal chain, and is therefore named as a hexane. There are methyl branches at carbons 3 and 4. Hence the name is 3,4-dimethylhexane. (You should name compound *B* after you study the next rule.)

....

8. *When there are substituent groups at more than one carbon of the principal chain, alternative numbering schemes are compared number by number, and the one is chosen that gives the lower number at the first point of difference.*

To apply this rule, write the two possible numbering schemes derived by numbering from either end of the chain. In the next example, the two schemes are 2,5,5- and 3,3,6-.

$$CH_3$$
$$|$$
$$CH_3—CH_2—C—CH_2—CH_2—CH—CH_3$$
$$| \qquad\qquad\qquad\qquad |$$
$$CH_3 \qquad\qquad\qquad CH_3$$

possible names:
2,5,5-trimethylheptane (correct)
3,3,6-trimethylheptane (incorrect)

A decision between the two numbering schemes is made by a pairwise comparison of the number sets (2,5,5) and (3,3,6). Because the *first point of difference* in these sets occurs at the first pair—2 *vs.* 3—the decision is made at this point and the first scheme is chosen, because 2 is lower than 3. The second point of difference, 5 *vs.* 3, does not enter the choice. Also, it makes no difference whether the names of the substituents are the same or different; only their numerical locations are used. (You should now be able to name compound *B* in Study Problem 2.2.)

9. *Substituent groups are cited in alphabetical order regardless of their location in the principal chain.* The numerical prefixes di, tri, etc., as well as the prefixes *tert-* and *sec-* are ignored in alphabetizing, but the prefixes iso, neo, and cyclo are considered in alphabetizing substituent groups.

The following compounds illustrate the application of this rule:

$$\overset{7}{CH_3}—\overset{6}{CH_2}—\overset{5}{CH}—\overset{4}{CH_2}—\overset{3}{CH_2}—\overset{2}{CH}—\overset{1}{CH_3}$$
$$\qquad\qquad | \qquad\qquad\qquad\qquad |$$
$$\qquad\qquad CH_2 \qquad\qquad\qquad CH_3$$
$$\qquad\qquad |$$
$$\qquad\qquad CH_3$$

5-ethyl-2-methylheptane
(*ethyl* is cited before *methyl* even though it has a higher number)

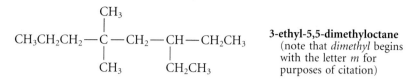

3-ethyl-5,5-dimethyloctane
(note that *dimethyl* begins
with the letter *m* for
purposes of citation)

10. *When the numbering of different groups is not resolved by the other rules,
    the first-cited group receives the lowest number.*

In the following compound, rules 1–9 do not dictate a choice between the names
3-ethyl-5-methylheptane and 5-ethyl-3-methylheptane. Because the ethyl group is cited
first in the name, it receives the lower number, by rule 10.

$$CH_3—CH_2—CH—CH_2—CH—CH_2—CH_3$$
$$\quad\qquad\;\; | \qquad\qquad\; |$$
$$\quad\qquad CH_3 \qquad\quad C_2H_5$$

3-ethyl-5-methylheptane

There are situations of greater complexity that are not covered by these ten rules;
however, these rules will suffice for most cases.

D. Highly Condensed Structures

When space is at a premium, parentheses are sometimes used to form highly condensed
structures that can be written on one line, as in the following example.

$$(CH_3)_4C \quad \text{or} \quad C(CH_3)_4 \quad \text{means} \quad CH_3—\overset{\displaystyle CH_3}{\underset{\displaystyle CH_3}{\overset{|}{\underset{|}{C}}}}—CH_3$$

When such structures are complex, it is sometimes not immediately obvious, particularly to the beginner, which atom inside the parentheses is connected to the atom outside
the parentheses, but a little analysis will generally solve the problem. Usually the structure
is drawn so that one of the parentheses intervenes between the atoms that are connected
(except for attached hydrogens). However, if in doubt, look for the atom within the
parentheses that is missing its usual number of bonds. When the group inside the
parentheses is CH_3, for example, the carbon has only three bonds (to the Hs). Hence, it
must be bound to the group outside the parentheses. Consider as another example the
CH_2OH groups in the following structure.

$$(CH_3)_2CH—CH(CH_2OH)_2 \quad \text{means} \quad \overset{\displaystyle CH_3 \qquad\quad CH_2OH}{\underset{\displaystyle CH_3 \qquad\quad CH_2OH}{CH—CH}}$$

Because the oxygen is bound to a carbon and to a hydrogen, it has its full complement
of two bonds (the two unshared pairs are understood). The carbon, however, is bound
to only three groups (two Hs and the oxygen); hence, it is the atom that is connected to
the carbon outside the parentheses.

If the meaning of a condensed structure is not immediately clear, *do not hesitate to
write it out in less condensed form.* If you will take the time to do this in a few cases, it
should not be long before interpretation of condensed structures becomes more routine.

NOMENCLATURE AND CHEMICAL INDEXING

From the large number of alkane isomers alone, perhaps you can appreciate that vast numbers of organic compounds could (and do) exist. Many have never been prepared—not because their preparation would be unreasonably difficult, but simply because there has never been a need for them. How can one determine whether a given compound has ever been prepared, and, if so, what its properties are? The only way is to search exhaustively the existing published body of chemical knowledge, called the **chemical literature**, for the information of interest. This would be an impossible task were it not for a well-organized index and summary of the chemical literature, called *Chemical Abstracts*. However, the explosion of chemical knowledge has made even the searching of *Chemical Abstracts* indexes a very time-consuming and tedious task in some cases. As more organic compounds are added each year to the list of known compounds, the importance of an automated information storage and retrieval system for carrying out chemical literature searches increases. Nomenclature is of central importance in such a system. Computer-based information technology has been developed that allows chemists to search for compounds using name fragments as well as complete names. Moreover, it is also possible, although currently somewhat less reliable, to search the chemical literature for compounds of interest solely by their structures. Through the on-line information-retrieval system of *Chemical Abstracts*, a scientist can now transmit a structure drawing and receive in return an indexing number, called the *registry number*, and the "official" name of the compound, if it is a known entity. From this information an efficient computer-based search for all known information about the compound can be conducted.

STUDY PROBLEM 2.3

Write the Lewis structure of 4-*sec*-butyl-5-ethyl-3-methyloctane. Then write the structure in a condensed form.

Solution To this point we've been giving names to structures. This problem now requires that we work "in reverse" and construct a structure from a name. As in most nomenclature problems, a systematic approach is required. First, write the principal chain. Because the name ends in *octane*, the principal chain contains eight carbons. You'll find it helpful at first to draw the principal chain without its hydrogen atoms:

$$C—C—C—C—C—C—C—C$$

Next, attach the branches indicated in the name at the appropriate positions: a *sec*-butyl group at carbon-4, an ethyl group at carbon-5, and a methyl group at carbon-3. (Use Table 2.2 if necessary.)

$$CH_3—CH—CH_2—CH_3 \longleftarrow \textit{sec-butyl group}$$

$$C—C—C—\overset{|}{C}—\overset{|}{C}—C—C—C$$

$$\overset{|}{CH_3} \qquad \overset{|}{CH_2—CH_3}$$

Finally, fill in the proper number of hydrogens at each carbon of the principal chain so that each carbon has a total of four bonds:

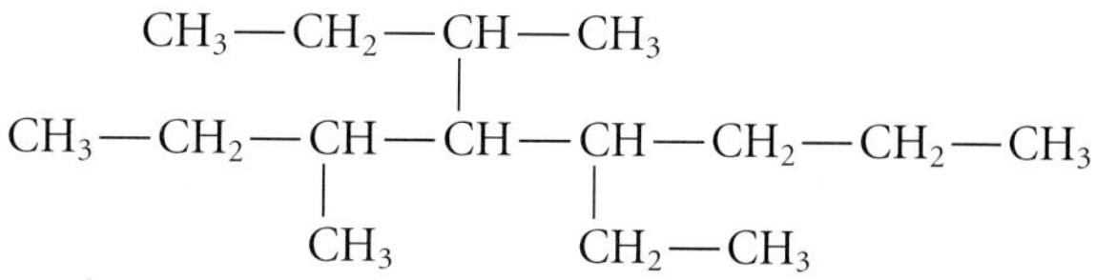

$$CH_3-CH_2-CH-CH_3$$
$$CH_3-CH_2-CH-CH-CH-CH_2-CH_2-CH_3$$
$$| \quad\quad\quad |$$
$$CH_3 \quad\quad CH_2-CH_3$$

4-*sec*-butyl-5-ethyl-3-methyloctane

To write the structure in condensed form, put like groups attached to the same carbon within parentheses. Notice that the structure contains within it two *sec*-butyl groups (color in the structure below), even though only one is mentioned in the name; the other consists of a methyl branch and part of the principal chain.

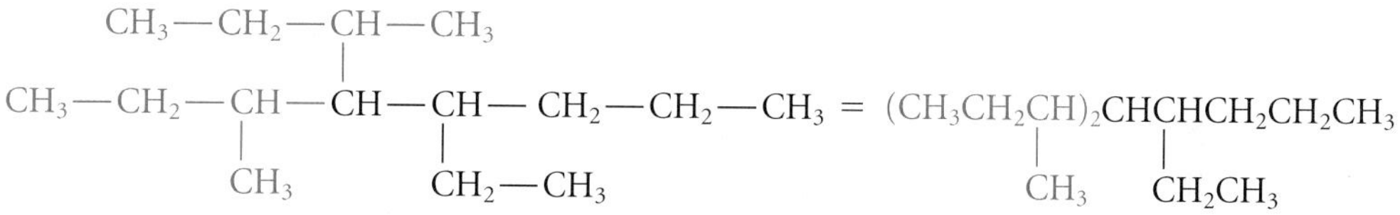

$$CH_3-CH_2-CH-CH_3$$
$$CH_3-CH_2-CH-CH-CH-CH_2-CH_2-CH_3 = (CH_3CH_2CH)_2CHCHCH_2CH_2CH_3$$
$$| \quad\quad\quad | \quad\quad\quad\quad\quad\quad\quad\quad | \quad |$$
$$CH_3 \quad\quad CH_2-CH_3 \quad\quad\quad\quad\quad\quad CH_3 \quad CH_2CH_3$$

PROBLEMS

2.5 Name the following compounds. Be sure to designate the principal chain properly before constructing the name.

*(a) $CH_3CH_2-CH-\!-\!-CH-CH_2CH_2CH_3$
$\quad\quad\quad\quad\quad\quad | \quad\quad\quad |$
$\quad\quad\quad\quad\quad CH_2 \quad\quad CH_3$
$\quad\quad\quad\quad\quad\quad |$
$\quad\quad\quad\quad\quad CH_2-CH_2-CH_3$

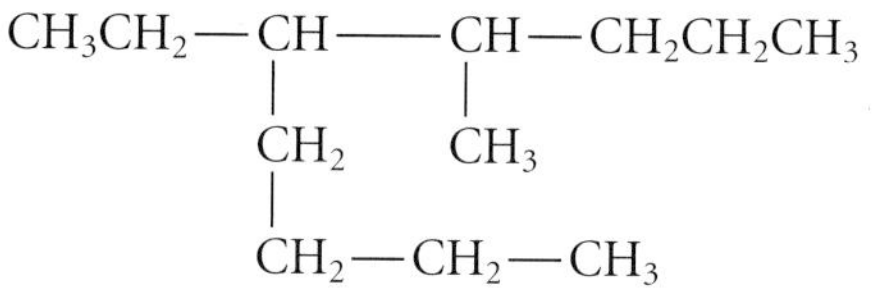

(b)
$\quad\quad\quad\quad\quad\quad\quad\quad\quad\quad CH_3 \quad C_2H_5$
$\quad\quad\quad\quad\quad\quad\quad\quad\quad\quad | \quad\quad |$
$\quad CH_3CH_2CH_2CH-C-\!-\!-CH-CH_2CH_2CH_3$
$\quad\quad\quad\quad\quad\quad\quad\quad | \quad\quad |$
$\quad\quad\quad\quad\quad\quad\quad\quad CH_3 \quad CH_3$

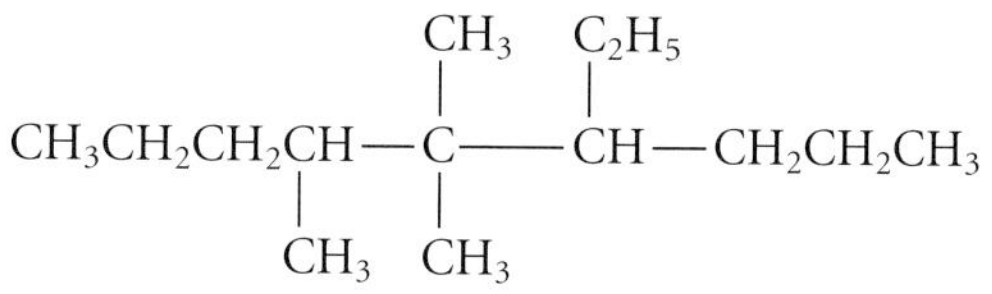

*(c) $(CH_3CH_2CH_2)_2CHCH(CH_2CH_3)_2$

*2.6 Draw condensed structures for all isomers of heptane and give their systematic names.

2.7 Draw the structures of
*(a) 4-isopropyl-2,4,5-trimethylheptane
(b) 6-*tert*-butyl-2-methyl-5-propylnonane

E. Classification of Carbon Substitution

It is important to recognize different types of carbon substitution in branched compounds, because chemical reactivity often varies with the degree of substitution. A carbon is said to be **primary**, **secondary**, **tertiary**, or **quaternary** when it is bonded to one, two, three, or four other carbons, respectively.

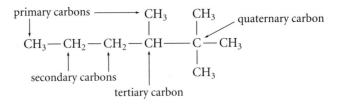

Likewise, the hydrogens bonded to each type of carbon are called primary, secondary, or tertiary hydrogens, respectively.

PROBLEMS

*2.8   In the structure of 4-isopropyl-2,4,5-trimethylheptane (Problem 2.7a):
(a) Identify the primary, secondary, tertiary, and quaternary carbon atoms.
(b) Circle one example of each of the following groups: a methyl group; an ethyl group; an isopropyl group; a *sec*-butyl group; an isobutyl group.

2.9   How many ethyl groups and how many methyl groups are there in the structure of 4-*sec*-butyl-5-ethyl-3-methyloctane, the compound discussed in Study Problem 2.3?

2.5   Cycloalkanes and Skeletal Structures

Alkanes that contain carbon chains in closed loops, or rings, are called **cycloalkanes**. These compounds are named by adding the prefix *cyclo* to the name of the alkane. Thus, the six-membered cycloalkane is called *cyclohexane*.

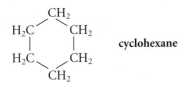

cyclohexane

The names and some physical properties of the simple cycloalkanes are given in Table 2.3. Notice that the general formula for an alkane containing a single ring has two fewer hydrogens than that of the open-chain alkane with the same number of carbon atoms. For example, cyclohexane has the formula C_6H_{12}, but hexane has the formula C_6H_{14}. The general formula for the cycloalkanes with one ring is C_nH_{2n}.

Because of the tetrahedral configuration of carbon in the cycloalkanes, the carbon skeletons of the cycloalkanes (except for cyclopropane) are not planar, but are puckered. We'll consider the conformations of cycloalkanes in Chapter 7. For now, remember only that planar condensed structures for the cycloalkanes convey no information about their conformations.

Table 2.3   Physical Properties of Some Cycloalkanes

Compound	Boiling point, °C	Melting point, °C	Density, g/mL
cyclopropane	−32.7	−127.6	
cyclobutane	12.5	−50.0	
cyclopentane	49.3	−93.9	0.7457
cyclohexane	80.7	6.6	0.7786
cycloheptane	118.5	−12.0	0.8098
cyclooctane	150.0	14.3	0.8340

Skeletal Structures   An important structure-drawing convention that is almost universally used for cyclic molecules is to draw them using **skeletal structures**—structures that show only the carbon-carbon bonds. In this notation a cycloalkane is drawn as a closed geometrical figure. In a skeletal structure, it is understood that a carbon is at each vertex of the figure, and that enough hydrogens are present on each carbon to fulfill its tetravalence. Thus, the skeletal structure of cyclohexane is drawn as follows:

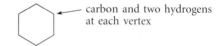

carbon and two hydrogens
at each vertex

Skeletal structures may also be drawn for open-chain alkanes. For example, hexane can be indicated this way:

for   CH₃ CH₂ CH₂ CH₂ CH₂ CH₃

When drawing a skeletal structure for an open-chain compound, don't forget that carbons are not only at each vertex, but also *at the ends of the structure.* Thus, the six carbons of hexane above are indicated by the four vertices and two ends of the skeletal structure. Here are a few other examples of skeletal structures:

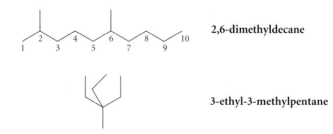

2,6-dimethyldecane

3-ethyl-3-methylpentane

Nomenclature of Cycloalkanes   The nomenclature of cycloalkanes follows essentially the same rules used for open-chain alkanes, as the following examples show:

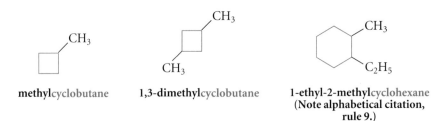

methylcyclobutane   **1,3-dimethyl**cyclobutane   **1-ethyl-2-methyl**cyclohexane
(**Note alphabetical citation,**
rule 9.)

Notice that the numerical prefix 1- is not necessary for monosubstituted cycloalkanes. Thus, the first compound is methylcyclobutane, not 1-methylcyclobutane. Two or more substituents, however, must be numbered to indicate their relative position. The lowest number is assigned in accordance with the usual rules.

Most of the cyclic compounds in this text, like the examples above, involve rings with small alkyl branches. In such cases, the ring is treated as the principal chain. However, when a noncyclic carbon chain contains more carbons than an attached ring, the ring is treated as the substituent.

$CH_3CH_2CH_2CH_2CH_2$ **1-cyclopropylpentane**

Study Problem 2.4

Name the following compound.

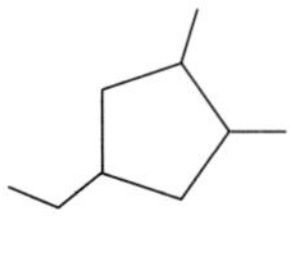

Solution   This compound is a cyclopentane with two methyl substituents and one ethyl substituent. If we number ring carbons consecutively, the following numbering schemes (and corresponding names) are possible, depending on which carbon is designated as carbon-1:

1,2,4-	4-ethyl-1,2-dimethylcyclopentane
1,3,4-	1-ethyl-3,4-dimethylcyclopentane
1,3,5-	3-ethyl-1,5-dimethylcyclopentane

The correct name is decided by nomenclature rule 8, the "first point of difference" rule, using the numbering schemes (*not* the names themselves). The first point of difference in the three schemes is at the second number (2 *vs.* 3 *vs.* 3); the scheme 1,2,4- has the lowest number at this point. Consequently, the correct name is 4-ethyl-1,2-dimethylcyclopentane.

Draw a skeletal structure of *tert*-butylcyclohexane.

Solution   The real question in this problem is how to represent a *tert*-butyl group with a skeletal structure. The branched carbon in this group has four other bonds, three of which go to CH_3 groups. Hence:

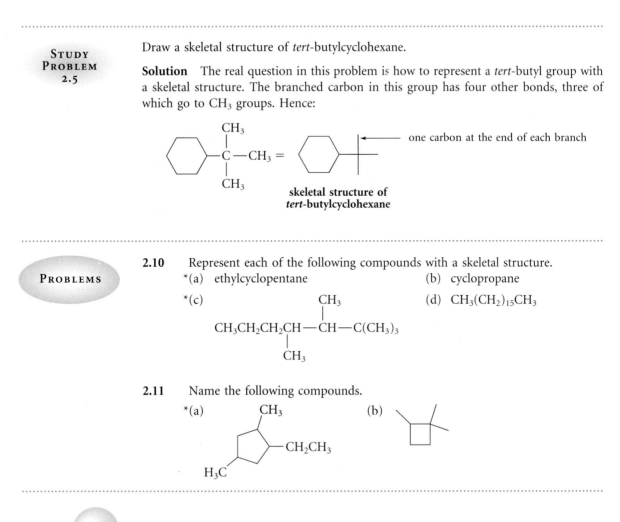

skeletal structure of
***tert*-butylcyclohexane**

PROBLEMS

2.10   Represent each of the following compounds with a skeletal structure.
*(a)   ethylcyclopentane   (b)   cyclopropane
*(c) CH_3 (d) $CH_3(CH_2)_{15}CH_3$

$$CH_3CH_2CH_2CH—CH—C(CH_3)_3$$
$$| \qquad\qquad\qquad |$$
$$CH_3 \qquad\qquad\qquad CH_3$$

2.11   Name the following compounds.
*(a) CH_3 (b)

2.6   Physical Properties of Alkanes

Each time we come to a new family of organic compounds, we'll consider the trends in their melting points, boiling points, densities, and solubilities, collectively referred to as their *physical properties*. The physical properties of an organic compound are important because they determine the conditions under which the compound is handled and used. For example, the form in which a drug is manufactured and dispensed is affected by its physical properties. In commercial agriculture, ammonia (a gas at ordinary temperatures) and urea (a crystalline solid) are both very important sources of nitrogen, but their physical properties dictate that they are handled and dispensed in very different ways.

Your goal should *not* be to memorize physical properties of individual compounds, but rather to learn to predict trends in how physical properties vary with structure.

A. Boiling Points

The **boiling point** is the temperature above which a substance is transformed spontaneously and completely from the liquid to the gaseous state. Table 2.1 shows that there is a regular change in the boiling points of the unbranched alkanes with increasing number of carbons. This trend of boiling point within the series of unbranched alkanes is particularly apparent in a plot of boiling point against carbon number (Fig. 2.7). *The regular increase in boiling point of 20–30° per carbon atom within a series is a general trend observed for many types of organic compounds.*

What is the reason for this increase? The boiling point is a crude measure of the attractive forces among molecules in the liquid state compared to those in the gaseous state. Because, to a useful approximation, the attractive forces among molecules in the gaseous state are negligible at atmospheric pressure, the boiling point measures the relative strengths of cohesive interactions among the molecules of a liquid. Yet no chemical bond is formed between separate molecules. What, then, is the origin of this attraction of one molecule for another?

In the previous chapter you learned that electrons in bonds are not confined between the nuclei, but rather reside in bonding molecular orbitals that surround the nuclei; we can think of these occupied molecular orbitals as "electron clouds." Imagine an organic molecule such as an alkane as something akin to a cotton ball (the electron clouds) with embedded cotton seeds (the nuclei). The shapes of these electron clouds can be altered by external forces. One such external force is the electric field of the electrons in nearby molecules. When two molecules approach each other closely, as in a liquid, the electron clouds of one molecule repel the electron clouds of the other. As a result, both molecules *temporarily* acquire small localized separations of charge (Fig. 2.8) called *induced dipoles*.

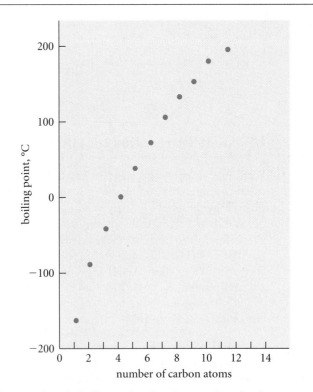

Figure 2.7   *Boiling points of some unbranched alkanes plotted against number of carbon atoms. Notice the steady increase with the size of the alkane, which is in the range of 20–30° per carbon atom.*

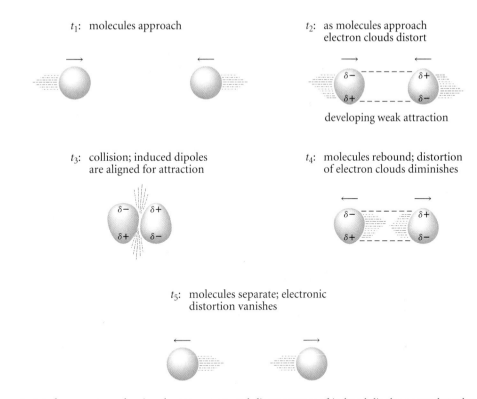

t_1: molecules approach

t_2: as molecules approach electron clouds distort

developing weak attraction

t_3: collision; induced dipoles are aligned for attraction

t_4: molecules rebound; distortion of electron clouds diminishes

t_5: molecules separate; electronic distortion vanishes

Figure 2.8 *A stop-frame cartoon showing the appearance and disappearance of induced dipoles as two hypothetical molecules collide and rebound. Notice that the distortion of the electron clouds is not permanent but varies with time. The electronic distortion in one species is induced by the proximity to the other. The frames are labeled t_1, t_2, etc., for successive times.*

That is, the molecules take on a *temporary* dipole moment. The deficiency of electrons (positive charge) in part of one molecule is attracted by the excess of electrons (negative charge) in part of the other. This attraction, an example of an **attractive van der Waals force** or **dispersion force**, is the cohesive force that must be overcome in order to vaporize a liquid hydrocarbon. Note that alkanes do *not* have appreciable permanent dipole moments. The dipole moment that causes the attraction between alkane molecules is induced *temporarily* in one molecule by the proximity of another. We might say that "nearness makes the molecules grow fonder."

You are now in a position to understand why larger molecules have higher boiling points. A larger molecule has a greater surface of electron clouds available from van der Waals interactions with other molecules. Because van der Waals attractions are greater between large molecules, large molecules have higher boiling points.

The *shape* of a molecule is also important in determining its boiling point. For example, a comparison of the boiling point of the highly branched alkane neopentane (9.4°) and its unbranched isomer pentane (36.1°) is particularly striking. Neopentane has four methyl groups disposed in a tetrahedral fashion about a central carbon. As the following space-filling models show, the molecule almost resembles a compact ball, and could fit readily within a sphere. On the other hand, pentane is rather extended, is ellipsoidal in shape, and would not fit within the same sphere.

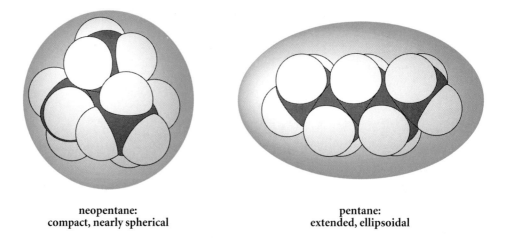

neopentane:
compact, nearly spherical

pentane:
extended, ellipsoidal

The more a molecule approaches spherical proportions, the less surface area it presents to other molecules, because a sphere is the three-dimensional object with the minimum surface-to-volume ratio. Since neopentane has less surface area at which van der Waals interactions with other neopentane molecules can occur, it has fewer cohesive interactions than pentane, and thus, a lower boiling point.

In summary, two trends in the variation of boiling point with structure are:

1. Boiling points increase with increasing molecular weight within a homologous series—typically 20–30° per carbon atom. This increase is due to the greater van der Waals attractions among larger molecules.

2. Boiling points tend to be lower for highly branched molecules that approach spherical proportions because they have less molecular surface available for van der Waals attractions.

B. Melting Points

The **melting point** of a substance is the temperature above which it is transformed spontaneously and completely from the solid to the liquid state. The melting point is an especially important physical property in organic chemistry because it is used both to identify organic compounds and to make a general assessment of their purity. Melting points are usually depressed, or lowered, by impurities, and the melting range (the range of temperature over which a substance melts), usually quite narrow for a pure substance, is substantially broadened by impurities. The melting point is a measure of the forces stabilizing the solid state weighed against those stabilizing the liquid state. Figure 2.9, which is a plot of melting point against carbon number for the unbranched alkanes, shows that melting points tend to increase with number of carbons. This is a general trend within many series of organic compounds.

Figure 2.9 also shows that the melting points of unbranched alkanes with an even number of carbon atoms lie on a separate, higher curve from those of the alkanes with an odd number of carbons. This reflects a more effective packing of the even-carbon alkanes in the crystalline solid state. In other words, the odd-carbon alkane molecules do not "fit together" as well in the crystal as the even-carbon alkanes. Similar alternation

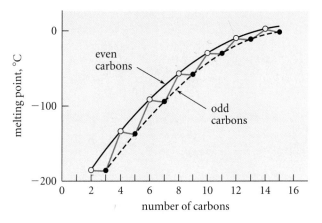

Figure 2.9   *Plot of melting points of the unbranched alkanes against number of carbon atoms. Notice that the alkanes with an even number of carbons lie on a different curve from the alkanes with an odd number of carbons.*

of melting points is observed in other series of compounds. For example, a similar trend is apparent in the melting points of cycloalkanes in Table 2.3.

Branched-chain hydrocarbons tend to have lower melting points than linear ones because the branching interferes with regular packing in the crystal. When a branched molecule has a substantial symmetry, however, its melting point is typically relatively high because of the ease with which symmetrical molecules fit together within the crystal. For example, the melting point of the very symmetrical molecule neopentane, −16.8°, is considerably higher than that of the less symmetrical pentane, −129.8°, which is higher still than that of isopentane, −159.9°.

In summary, melting points show the following general trends.

1. Melting points tend to increase with increasing molecular weight within a series.

2. Many highly symmetrical molecules have unusually high melting points.

3. A sawtooth pattern of melting point behavior (see Fig. 2.9) is often observed within a series.

PROBLEM

*2.12   Match each of the following compounds with the correct boiling points and melting points. Explain your choices.
Compounds: 2,2,3,3-tetramethylbutane and octane
Boiling points: 106.5°, 125.7°
Melting points: −56.8°, +100.7°

C. Other Physical Properties

Alkanes have negligible dipole moments, and thus are nonpolar molecules. The alkanes are, for all practical purposes, insoluble in water—thus the saying, "Oil and water don't mix." (Alkanes are a major constituent of crude oil.) Alkanes also have considerably lower densities than water. For these reasons a mixture of an alkane and water will separate into two distinct layers with the less dense alkane layer on top. An oil slick is an example of this behavior.

PROBLEM

*2.13   Gasoline consists mostly of alkanes. Explain why water is not usually very effective in extinguishing a gasoline fire.

2.7   Combustion and Elemental Analysis

A.   Reactivity of Alkanes; Combustion

Alkanes are among the least reactive types of organic compounds. They do not react with common acids or bases, nor do they react with common oxidizing or reducing agents.

Alkanes do, however, share one type of reactivity with many other types of organic compounds: they are flammable; that is, they react rapidly with oxygen to give carbon dioxide and water, provided that the reaction is initiated by a suitable heat source such as a flame or the spark from a spark plug. This reaction is called **combustion**. For example, the combustion of methane, the major alkane in natural gas, is written as follows.

$$CH_4 + 2O_2 \longrightarrow CO_2 + 2H_2O \tag{2.1}$$

The general combustion reaction for a noncyclic alkane can be written:

$$C_nH_{2n+2} + \frac{3n+1}{2}O_2 \longrightarrow (n+1)H_2O + nCO_2 \tag{2.2}$$

These reactions are examples of *complete combustion*: combustion in which carbon dioxide and water are the only combustion products. Under conditions of oxygen deficiency, incomplete combustion may also occur with the formation of such by-products as carbon monoxide, CO.

The fact that one can carry a container of gasoline in the open air shows that simple mixing of alkanes and oxygen does not initiate combustion. However, once a small amount of heat is applied (in the form of a flame or a spark from a spark plug), the combustion reaction proceeds vigorously and spontaneously with the liberation of large amounts of energy.

Combustion is of tremendous commercial importance, because it liberates energy that can be used to keep us warm, generate electricity, or move motor vehicles.

B.   Elemental Analysis by Combustion

Combustion is used in the quantitative determination of elemental compositions, called **elemental analysis**. Combustion has been used for this purpose since the beginning of the modern era of chemistry. The results of combustion analyses led early chemists to the realization that most compounds contain their constituent elements in definite whole-number ratios.

Because most organic compounds contain both carbon and hydrogen, the analysis for these elements is particularly important in organic chemistry. The proportions of both carbon and hydrogen in a compound can be determined simultaneously by a technique that has changed little since its inception. A small sample (typically 5–10 mg) of the substance to be analyzed is completely burned, and the CO_2 and H_2O produced in the combustion are collected and weighed. From the mass of the CO_2 produced, the mass of carbon in the sample can be determined. Similarly, from the mass of the H_2O produced, the amount of hydrogen in the sample can be determined. Oxygen, if present, is usually not determined directly, but by difference. The result of such an elemental analysis is expressed as the mass percent of each element in the compound. The following study problem illustrates how elemental analysis works.

STUDY **P**ROBLEM **2.6**

Suppose 7.00 mg of a liquid hydrocarbon are burned and found to yield 21.58 mg of CO_2 and 9.94 mg of H_2O. Find the mass percent of carbon and hydrogen in the original sample.

Solution   First find the mass of carbon that is present in the CO_2; this same mass of carbon must have been present in the sample.

$$\text{mass of carbon} = (\text{fraction of carbon in } CO_2)(\text{mass of } CO_2) \qquad (2.3)$$

The fraction of carbon in CO_2 is the atomic mass of carbon over the molecular mass of CO_2. Thus, Eq. 2.3 becomes

$$\text{mass of carbon} = (12.01 \text{ mg of C}/44.01 \text{ mg of } CO_2)(21.58 \text{ mg of } CO_2)$$
$$= 5.89 \text{ mg} \qquad (2.4)$$

Next, calculate the mass of hydrogen in the water, noting that there are two hydrogens in every water molecule:

$$\text{mass of hydrogen} = (2 \text{ hydrogens}/H_2O)(1.008 \text{ mg of H}/18.02 \text{ mg of } H_2O)(9.94 \text{ mg of } H_2O)$$
$$= 1.11 \text{ mg} \qquad (2.5)$$

Because the mass of hydrogen plus the mass of carbon equals the mass of the sample, 7.00 mg, the sample can contain no other elements. The mass percent of carbon is $(5.89/7.00) \times 100 = 84.14\%$, and the mass percent of hydrogen is $(1.11/7.00) \times 100 = 15.86\%$.

The mass percents of the elements in a compound can be used to determine directly the **empirical formula**. This is the formula that gives the smallest whole-number molar proportions of the elements. This process can be summarized in the following steps.

1. *Convert the relative masses of the elements into molar proportions by dividing the mass percent of each element by its atomic mass.*

2. *Divide the molar proportion of each element by that of the element present in the smallest proportion.*

3. *Multiply the resulting proportions by successive integers (2, 3, 4, ...) until whole-number proportions for all elements are obtained. The result is the empirical formula.*

This process is illustrated in the following study problem.

STUDY PROBLEM 2.7

Calculate the empirical formula of the compound used in Study Problem 2.6.

Solution Because the formula of a compound expresses the molar proportions of its elements, first convert the mass percent data to mole ratios. The mass percentages of carbon and hydrogen calculated in Study Problem 2.6 mean that there are 84.14 g of carbon and 15.86 g of hydrogen in 100 g of the sample. Hence there are (84.14/12.01) = 7.00 moles of carbon and (15.83/1.008) = 15.73 moles of hydrogen in 100 g of the sample. A formula that expresses the relative molar proportions of carbon and hydrogen is therefore $C_{7.00}H_{15.73}$. Because organic compounds have whole numbers of elements, this must be converted into a formula in which both elements are present as whole numbers. The easiest way to do this is first to divide through the formula above by the element present in least molar proportion, in this case, carbon. This yields the formula $C_{1.00}H_{2.25}$. Then multiply the formula by successive integers (2, 3, 4, ...) until whole numbers for both elements are obtained. Multiplication by 4 yields the smallest whole-number formula, C_4H_9. This is the empirical formula.

The **molecular formula** of a compound is the formula that gives the actual number of each type of atom in one molecule of the compound. The molecular formula is always some integral multiple of the empirical formula. In some cases the molecular formula is the same as the empirical formula; in other cases it isn't. The empirical formula C_4H_9 of the compound in the last study problem, for example, *cannot* be the molecular formula, because *all hydrocarbons have even numbers of hydrogens.* The smallest molecular formula that has a carbon-hydrogen ratio of 4:9 is C_8H_{18}. This is an acceptable candidate for the molecular formula. Hence, the compound could be octane or one of its isomers.

PROBLEM

*2.14   Multiplication of the formula C_4H_9 by 4 gives the formula $C_{16}H_{36}$. Would this be an acceptable candidate for the molecular formula of a hydrocarbon? Explain.

If more than one molecular formula is possible for a given empirical formula, determination of the correct molecular formula requires an additional piece of information: the *molecular mass*. Suppose, for example, that the empirical formula of a compound is determined by combustion to be CH_2. This empirical formula could correspond to an infinite number of molecular formulas: C_2H_4, C_3H_6, C_4H_8, and so on. The molecular mass determines which of these alternatives is correct. For example, if the molecular mass were determined to be 84, then the molecular formula would have to be C_6H_{12}. Molecular masses are determined most often today by *mass spectrometry*, a physical technique discussed in Chapter 12.

PROBLEMS

2.15   Calculate *(a) the mass percent of carbon, hydrogen, and oxygen in a sample with the molecular formula $C_3H_6O_2$; (b) the mass percent carbon and hydrogen in a sample with the molecular formula C_6H_{14}.

2.16   Give the empirical formulas for the compounds described in *(a) and (b) of Problem 2.15.

2.17   *(a)   In a laboratory you have found a bottle labeled "alkane *X*." In an attempt to determine its structure, you carry out an elemental analysis. Combustion of 10.00 mg of *X* in a stream of O_2 yields 31.95 mg of CO_2 and 11.44 mg of H_2O. The molecular mass of *X* is found to be 110. Determine the molecular formula of compound *X*. Assuming that the label on the bottle can be believed, what can be said about the structure of *X* on the basis of its molecular formula?

(b)   How many mg of O_2 are consumed in the combustion analysis in part (a)?

2.8   Occurrence and Use of Alkanes

Most alkanes come from **petroleum**, or crude oil. (The word *petroleum* comes from the Latin words for "rock" and "oil": thus, "oil from rocks.") Petroleum is a dark, viscous mixture composed mostly of alkanes and aromatic hydrocarbons that are separated by a technique called **fractional distillation**. In fractional distillation, a mixture of compounds is slowly boiled; the vapor is then collected, cooled, and recondensed to a liquid. Because the compounds with the lowest boiling points vaporize most readily, the condensate from a fractional distillation is richest in the more volatile components of the mixture. As distillation continues, components of progressively higher boiling point appear in the condensate. A student who takes an organic laboratory course will almost certainly become acquainted with this technique on a laboratory scale. Industrial fractional distillations are carried out on a large scale in fractionating towers that are several stories tall (Fig. 2.10). The typical fractions obtained from distillation of petroleum are given in Table 2.4.

Table 2.4   Components of Petroleum

Fraction name	Boiling range, °C	Number of carbon atoms
gas	<20	1–4
petroleum ether	30–60	5–6
ligroin (light naphtha)	60–90	6–7
gasoline (naphtha)	85–200	6–12
kerosene	200–300	12–15
heating oil	300–400	15–18
lubricating oil, asphalt	>400	16–24

Figure 2.10   *Industrial fractionating towers used to separate mixtures of compounds on the basis of their boiling points.*

Another important alkane source is natural gas, which is mostly methane. Natural gas comes from gas wells of various types. There are also significant biological sources of methane that could someday be exploited commercially. Methane is produced by the action of certain anaerobic bacteria (bacteria that function without oxygen) on decaying organic matter (Fig. 2.11). It is this type of process, for example, that produces "marsh gas," as methane was known before it was characterized by organic chemists. Biological production of methane could become a source of natural gas in the future.

Alkanes of low molecular mass are in great demand for a variety of purposes—especially motor fuels—and alkanes available directly from wells do not satisfy the demand. The petroleum industry has developed methods for *cracking* high molecular mass alkanes into lower molecular mass alkanes and alkenes (Sec. 5.7B), and has also developed processes for *reforming* unbranched alkanes into branched-chain ones, which have superior ignition properties as motor fuels.

At present the greatest use of alkanes is for fuel. Typically, motor fuels, fuel oils, and aviation fuels account for about 80% of all hydrocarbon consumption. An Arabian oil minister was once heard to remark, "Oil is too precious to burn." He was undoubtedly referring to the fact that there are important uses for petroleum besides fuels. Petroleum will remain for the foreseeable future the principal source of *carbon*, from which organic starting materials are made for such diverse products as plastics and pharmaceuticals. Petroleum is thus the basis for organic chemical *feedstocks*—the basic organic compounds from which more complex chemical substances are fabricated.

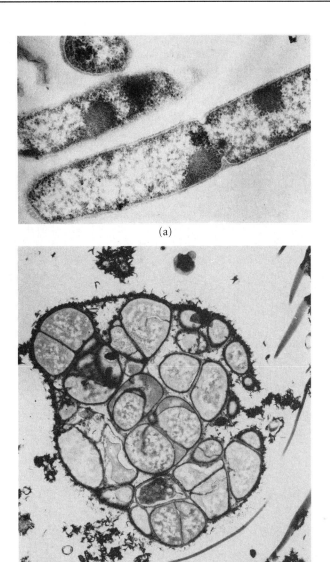

(a)

(b)

Figure 2.11   *Methanogens (methane-producing bacteria). (a) A methanobacterium species and (b) a methanosarcina species are found in anaerobic sewage digesters and in anaerobic sediments of natural waters.*

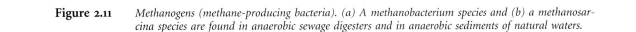

ALKANES AS MOTOR FUELS

Alkanes vary significantly in their quality as motor fuels. Branched-chain alkanes are better motor fuels than unbranched ones. The quality of a motor fuel relates to its rate of ignition in an internal combustion engine. Premature

(*continues*)

ignition results in engine knock, a condition that indicates poor engine performance. Severe engine knock can result in significant engine damage. The *octane number* is a measure of the quality of a motor fuel: the higher the octane number, the better the fuel. Octane numbers of 100 and 0 are assigned to 2,2,4-trimethylpentane and heptane, respectively. Mixtures of the two compounds are used to define octane numbers between 0 and 100. For example, a fuel that performs as well as a 1:1 mixture of 2,2,4-trimethylpentane and heptane has an octane number of 50. Good quality motor fuels used in modern automobiles have octane numbers in the 87–95 range.

Various additives can be used to improve the octane number of motor fuels. In the past, tetraethyllead, $(C_2H_5)_4Pb$, was used extensively for this purpose, but concerns over atmospheric lead pollution and the use of catalytic converters (which are adversely affected by lead) have made leaded gasoline, for all practical purposes, a thing of the past. Today, *tert*-butyl methyl ether (MTBE, $(CH_3)_3C—O—CH_3$) is the major additive used for improving octane number. Accordingly, production of this compound on an industrial scale has enjoyed a meteoric rise in the last fifteen years.

Recently, periods of relative scarcity of petroleum products have alternated with periods of relative abundance. There is no doubt, however, that eventually the world will exhaust its natural petroleum reserves. It is thus important that scientists develop new sources of energy, which may include new ways of producing petroleum.

2.9 Functional Groups and the "R" Notation

Alkanes are the conceptual "rootstock" of organic chemistry. Replacing C—H bonds of alkanes gives the many functional groups of organic chemistry. A **functional group** is a characteristically bonded group of atoms that has about the same chemical reactivity whenever it occurs in a variety of compounds. Some examples of simple organic compounds, with their functional groups shown in color, are the following:

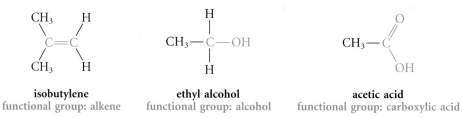

| isobutylene | ethyl alcohol | acetic acid |
| functional group: alkene | functional group: alcohol | functional group: carboxylic acid |

For example, the functional group characteristic of alkenes is the carbon–carbon double bond. Most alkenes undergo the same types of reactions, and these reactions occur at or near the double bond. Similarly, all alcohols contain an —OH group bound to a saturated carbon atom (that is, a carbon bearing only single bonds). The characteristic reactions of alcohols occur at the —OH group or the adjacent carbon, and this functional group undergoes the same general chemical transformations regardless of the structure of the remainder of the molecule. The organization of this text is centered for the most part around the common functional groups. Although you will study in detail each major functional group in subsequent chapters, you should learn to recognize the common functional groups now. These are shown inside the back cover.

Because the chemical properties of functional groups are general, it is often convenient to employ a general notation for organic compounds. When organic chemists wish to indicate a general structure, they use an R, much as mathematicians use x to indicate a general number. The R, unless otherwise indicated, stands for an alkyl group. Thus, the general formula of an alkyl chloride, R—Cl, might stand for any of the following structures, or a host of others.

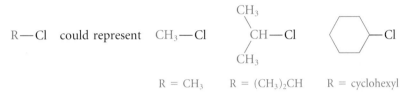

Just as alkyl groups such as methyl, ethyl, and isopropyl are substituent groups derived from alkanes, **aryl groups** are substituent groups derived from benzene and its derivatives. The simplest aryl group is the **phenyl group**, abbreviated Ph—, which is derived from the hydrocarbon benzene.

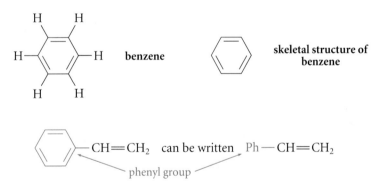

Other aryl groups are designated by Ar—. Thus Ar—OH could refer to any one of the following compounds, or many others.

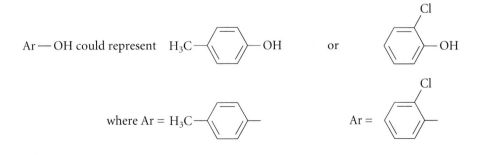

Although you will not study benzene and its derivatives until Chapter 15, you will see many examples in the meantime in which phenyl and aryl groups are used as substituent groups.

PROBLEMS

2.18   Draw a structural formula for each of the following compounds. (Several formulas may be possible.)

*(a)   an amine with molecular formula $C_4H_{11}N$

(b)   an alcohol with molecular formula $C_5H_{10}O$

*(c)   a carboxylic acid with elemental analysis 40.00% carbon, 6.71% hydrogen, and 53.29% oxygen

(d)   an alkyne with elemental analysis 88.82% carbon, 11.18% hydrogen, and molecular mass 54

*2.19   A certain compound was found to have the molecular formula $C_5H_{12}O_2$. Which of the following functional groups could be present in the compound? Give one example for each positive answer, and explain any negative responses.

(a)   an amide   (b)   an ether   (c)   a carboxylic acid

(d)   a phenol   (e)   an alcohol   (f)   an ester

KEY IDEAS IN CHAPTER 2

Alkanes are hydrocarbons that contain only carbon-carbon single bonds; alkanes may contain branched chains, unbranched chains, or rings.

Alkanes have sp^3-hybridized carbon atoms with tetrahedral geometry. They exist in various staggered conformations that are rapidly interconverted at room temperature, and the conformation that minimizes van der Waals repulsions has the lowest energy and is the predominant one. In butane, the major conformation is the *anti* conformation; the *gauche* conformations exist to a lesser extent.

Isomers are different compounds with the same molecular formula. Compounds that have the same molecular formula but differ in their atomic connectivities are called *constitutional isomers*.

Alkanes are named systematically according to the substitutive nomenclature rules of the IUPAC. The name of a compound is based on its principal chain, which, for an alkane, is the longest continuous carbon chain in the molecule.

- The boiling point of an alkane is determined by van der Waals attractions between molecules, which in turn depend on molecular size and shape. Large molecules have relatively high boiling points; highly branched molecules have relatively low boiling points. The boiling points of compounds within a series increase 20–30° per carbon atom.
- Melting points of alkanes increase with molecular mass. Highly symmetrical molecules have particularly high melting points.
- Combustion is the most important reaction of alkanes. It finds practical application in the generation of much of the world's energy. Combustion is also used analytically to determine the empirical formula of organic compounds by elemental analysis.
- Alkanes are derived from petroleum and are used mostly as fuels; however, they are also important as raw materials for the preparation of other organic compounds.
- Organic compounds are classified by their functional groups. Different compounds containing the same functional groups undergo the same types of reactions.
- The "R" notation is used as a general abbreviation for alkyl groups; Ph is the abbreviation for a phenyl group, and Ar is the abbreviation for an aryl (substituted phenyl) group.

Additional Problems

2.20 Given the boiling point of the first compound in each set, estimate the boiling point of the second.

*(a) $CH_3\overset{\overset{\large O}{\|}}{C}CH_2CH_2CH_2CH_2CH_3$ (bp 152°) $\quad CH_3CH_2\overset{\overset{\large O}{\|}}{C}CH_2CH_2CH_2CH_3$

(b) $CH_3\overset{\overset{\large O}{\|}}{C}CH_2CH_2CH_2CH_2CH_3$ (bp 152°) $\quad CH_3\overset{\overset{\large O}{\|}}{C}CH_2CH_2CH_2CH_3$

*(c) $CH_3CH_2CH_2CH_2CH_2CH_2Br$ (bp 155°) $\quad CH_3CH_2CH_2CH_2CH_2CH_2CH_2CH_2Br$

(d) $CH_3CH_2CH_2CH{=}CH_2$ (bp 30°) $\quad CH_3CH_2CH_2CH_2CH{=}CH_2$

2.21 *(a) Draw the structures and give the names of all isomers of octane with five carbons in their principal chains.

(b) Draw the structures and give the names of all isomers of octane with six carbons in their principal chains.

2.22   Label each carbon in the following molecules as primary, secondary, tertiary, or quaternary.

*(a)   CH₃   (b)

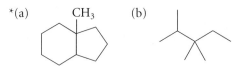

2.23   Draw the structure and give the name of an alkane
*(a)   that has more than three carbons and only primary hydrogens,
(b)   that has five carbons and only secondary hydrogens,
*(c)   that has only tertiary hydrogens,
(d)   that has a molecular mass of 84.2.

2.24   Name each of the following compounds using IUPAC substitutive nomenclature.

*(a)
$$CH_3-C-CH-CH_2-CH_3$$
with CH_3 and $CH_2-CH_2-CH_3$ above the C and CH, and $CH_2-CH_2-CH_3$ below the C

(b)
$$CH_3-CH-CH-CH_2-CH_3$$
with $CH_2-CH_2-CH_3$ above the second CH, and $CH_2-CH_2-CH_3$ below the first CH

*(c) $CH_3-CH-CH_2-CH_2-CH_3$ (d)
with CH below, and H_3C CH_3

*(e) H_3C CH_3 (f)
CH_3-CH_2 [cyclopentane ring] CH_2-CH_3

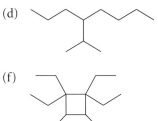

2.25   Draw structures that correspond to the following names.
*(a)   4-isobutyl-2,5-dimethylheptane
(b)   2,2-dimethylpentane
*(c)   5-*sec*-butyl-6-*tert*-butyl-2,2-dimethylnonane
(d)   2,3,5-trimethyl-4-propylheptane

2.26   The following labels were found on bottles of liquid hydrocarbons in the laboratory of Dr. Ima Turkey following his disappearance under mysterious circumstances. Although each name defines a structure unambiguously, some are not correct IUPAC substitutive names. Give the correct name for any compounds that are not named correctly.

*(a) 5-neopentyldecane

(b) 2-ethyl-2,4,6-trimethylheptane

*(c) 1-cyclopropyl-3,4-dimethylcyclohexane

(d) 3-butyl-2,2-dimethylhexane

*2.27   Within each set, which two structures represent the same compound?

(a)

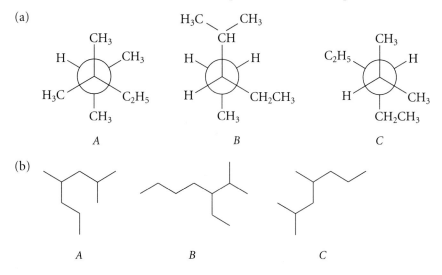

(b)

*2.28   Sketch a diagram of potential energy *vs.* angle of rotation about the carbon-carbon bond of chloroethane, CH_3—CH_2—Cl. The magnitude of the energy barrier to internal rotation is 15.5 kJ/mol (3.7 kcal/mol). Label this barrier on your diagram.

2.29   Explain how you would expect the diagram of potential energy *vs.* angle of internal rotation about the C2—C3 (central) carbon-carbon bond of 2,2,3,3-tetramethylbutane to differ from that for ethane.

*2.30   The *anti* conformation of 1,2-dichloroethane, Cl—CH_2—CH_2—Cl, is 4.81 kJ/mol (1.15 kcal/mol) more stable than the *gauche* conformation. The two energy barriers (measured relative to the energy of the *gauche* conformation) for carbon-carbon bond rotation are 21.5 kJ/mol (5.15 kcal/mol) and 38.9 kJ/mol (9.3 kcal/mol).

(a) Sketch a graph of potential energy *vs.* angle of rotation about the carbon-carbon bond. Show the energy differences on your graph and label each minimum and maximum with the appropriate conformation of 1,2-dichloroethane.

(b) Which conformation of this compound is present in greatest amount? Explain.

*2.31   When the structure of compound A on the left was determined in 1972, it was found to have an unusually long C—C bond and unusually large C—C—C bond angles, compared to the similar parameters for compound B (isobutane).

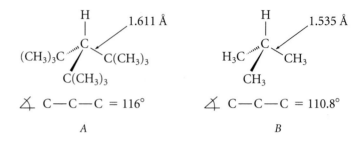

A   B

Explain why the indicated bond length and bond angle are larger for compound A.

*2.32   Predict the most stable conformation of hexane. (*Hint*: Consider the conformation about each carbon-carbon bond separately.) Make a model and draw a line-wedge formula for this conformation.

*2.33   Predict which of the following compounds should have the larger energy barrier to internal rotation about the central bond. Explain.
(a) $(CH_3)_3C—C(CH_3)_3$ (b) $(CH_3)_3Si—Si(CH_3)_3$

2.34   From what you learned in Sec. 1.3B about the relative lengths of C—C and C—O bonds, predict which of the following compounds should have the largest energy difference between *gauche* and *anti* conformations. Explain.
(a) $CH_3—O—CH_2—CH_3$ (b) $CH_3—CH_2—CH_2—CH_3$

*2.35   (a) What value is expected for the dipole moment of the *anti* conformation of 1,2-dibromoethane, $Br—CH_2—CH_2—Br$? Explain.
(b) The dipole moment μ of any compound that undergoes internal rotation can be expressed as a weighted average of the dipole moments of each of its conformations by the following equation:

$$\mu = \mu_1 N_1 + \mu_2 N_2 + \mu_3 N_3$$

in which μ_i is the dipole moment of conformation i, and N_i is the mole fraction of conformation i. (The mole fraction of any conformation i is the number of moles of i divided by the total moles of all conformations.) There are about 82 mole percent of *anti* conformation and about 9 mole percent of each *gauche* conformation present at equilibrium in 1,2-dibromoethane, and the observed dipole moment of 1,2-dibromoethane is 1.0 D. Using the above equation, and the answer to part (a), calculate the dipole moment of the *gauche* conformation of 1,2-dibromoethane.

*2.36   Write a balanced equation for the combustion of a general cycloalkane C_nH_{2n}.

2.37   What is the mass percent carbon in each of the following compounds?

*(a)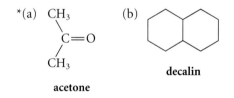

acetone

(b)

decalin

2.38   *(a) A hydrocarbon is found to contain 87.17% carbon and 12.83% hydrogen by mass. Calculate the minimum molecular formula for this compound.

  (b) Draw the structure of an alkane (which may contain one or more rings) consistent with the analysis given in part (a) that has two tertiary carbons and all other carbons secondary. (More than one correct answer is possible.)

  *(c) Draw the structure of an alkane (which may contain one or more rings) consistent with the analysis given in part (a) that has no primary hydrogens, no tertiary carbon atoms, and one quaternary carbon atom. (More than one correct answer is possible.)

*2.39   A 7.00-mg sample of a hydrocarbon with a molecular mass of 140.3 is burned in a stream of oxygen to yield 21.96 mg of CO_2 and 8.99 mg of water.

  (a) How many milliliters of oxygen (assume 25 °C, 1 atm pressure) are consumed in this experiment?

  (b) What is the molecular formula of the hydrocarbon?

2.40   *(a) In a laboratory you have found a bottle marked "amide X, elemental analysis 55.14% carbon, 10.41% hydrogen, and 16.08% nitrogen, molecular mass 87." Evidence gathered by certain physical methods leads you to believe that the compound contains an isopropyl group. Suggest two structures for X that are consistent with these data.

  (b) In the same laboratory you have found another amide Y that has the same molecular mass and elemental analysis. This amide contains neither an isopropyl nor a propyl group. Suggest two structures for Y consistent with these facts. What is the relationship between X and Y?

*2.41   Identify the different functional groups (aside from the alkane carbons) present in acebutolol, a drug that blocks a certain part of the nervous system.

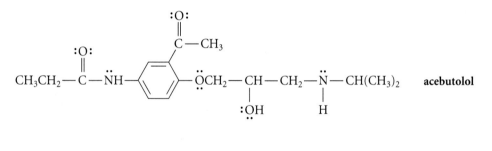

acebutolol

2.42   Name the functional group in each compound.

(a)   CH$_3$CH$_2$

(b)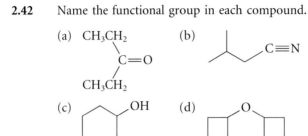

$$\underset{\underset{\displaystyle CH_3CH_2}{}}{C}=O$$

(c)   OH

(d)

3

Acids and Bases; the Curved-Arrow Formalism

This chapter concentrates on acid-base reactions, a topic that you have probably studied in earlier chemistry courses. Why are acid-base reactions worth special attention in an organic chemistry course? First, many organic reactions are themselves acid-base reactions, or are close analogs of common inorganic acid-base reactions. This means that if you understand the principles behind simple acid-base reactions, you also understand the principles behind the analogous organic reactions. Second, many organic reactions occur as a sequence of several individual steps, each of which is an acid-base reaction. This means that if you understand acid-base reactions, you can understand the details of many organic reactions. Third, acid-base reactions provide simple examples that can be used to illustrate some ideas that will prove useful in more complicated situations. In particular, you'll learn in this chapter about the *curved-arrow formalism*, a powerful piece of symbolic logic that will help you to follow, understand, and even predict organic reactions. Finally, acid-base reactions provide some examples for discussion of some principles of chemical equilibrium.

3.1 Lewis Acid-Base Association Reactions

A. Electron-Deficient Compounds

In Section 1.2C, you learned that covalent bonding in many cases conforms to the *octet rule*. That is, the sum of the bonding and unshared valence electrons surrounding a given atom equals eight (two for hydrogen). Although you can safely assume that none of the atoms in the compounds you'll encounter in your early studies possesses more than an octet of electrons, some compounds may contain atoms with fewer than an octet of electrons. In particular, some compounds contain atoms that are *short of an octet by one or more electron pairs.* Such species are termed **electron-deficient compounds**. One example of an electron-deficient compound is boron trifluoride:

boron trifluoride

BF_3 is electron-deficient because the boron, with six electrons in its valence shell, is two electrons, or one electron pair, short of an octet.

B. Reactions of Electron-Deficient Compounds with Lewis Bases

Electron-deficient compounds have a tendency to undergo chemical reactions that complete their valence-shell octets. In such reactions, an electron-deficient compound reacts with a species that has one or more unshared valence electron pairs. An example of such a reaction is the association of boron trifluoride and fluoride ion:

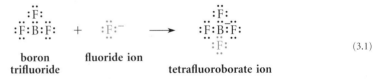

(3.1)

In such reactions, the electron-deficient compound acts as a **Lewis acid**: *an electron-pair acceptor.* Boron trifluoride is the Lewis acid in Eq. 3.1. The species that provides the electron pair acts as a **Lewis base**: *an electron-pair donor.* Fluoride ion is the Lewis base in Eq. 3.1. When a Lewis acid and a Lewis base combine to give a single product, as in this example, the reaction is termed a **Lewis acid-base association reaction**. Notice that as a result of this association reaction, each atom in the product tetrafluoroborate ion has a complete octet.

A term used almost interchangeably with "Lewis acid" in organic chemistry is **electrophile** (*phile* = loving; *electrophile* = electron-loving). *Electron-deficient compounds constitute an important class of electrophiles.* A term used synonymously with "Lewis base" in organic chemistry is **nucleophile** ("nucleus-loving"). The reactions of *Lewis acids with Lewis bases* are the same as the reactions of *electrophiles with nucleophiles.*

> A peculiarity in the octet-counting formalism is evident in Eq. 3.1. The fluoride ion has an octet. After it shares an electron pair with BF_3, the fluorine still has an octet in the product BF_4^-. You might ask, "How can fluorine have an octet both before and after it shares electrons?" The answer is that we count unshared pairs of electrons in the fluoride ion, but in BF_4^-, we assign to the fluorine the electrons in its unshared pairs as well as *both* electrons in the newly formed chemical bond. There is an apt analogy to this situation when a poor person *P* marries a wealthy person *W*. Before the marriage, one person is poor and the other wealthy; after the marriage, *W* is still wealthy, and *P*, like the boron in BF_4^-, has become wealthy by marriage! The justification for this practice of counting electrons twice is that it provides an extremely useful framework for predicting chemical reactivity. Note

again that the procedure used in counting electrons for the octet differs from the one used in calculating formal charge (Sec. 1.2C).

STUDY
PROBLEM
3.1

Which of the following compounds can react with the Lewis base Cl^- in a Lewis acid-base association reaction?

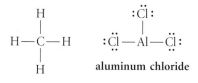

methane   aluminum chloride

Solution In order for a compound to react as a Lewis acid in an association reaction, it must be able to accept an electron pair from the Lewis base Cl^-. In aluminum chloride, the aluminum is short of an octet by one pair. Hence, aluminum chloride is an electron-deficient compound, and can readily accept an electron pair from chloride ion in an association reaction, as follows:

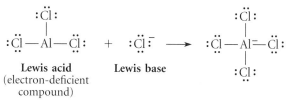

Lewis acid   Lewis base
(electron-deficient
compound)

In contrast, every atom in methane has the noble-gas number of electrons (carbon has eight, hydrogen has two). Hence, methane is not electron-deficient, and cannot undergo a Lewis acid-base association reaction.

C. The Curved-Arrow Formalism

Organic chemists have developed a symbolic device for keeping track of electron pairs in chemical reactions; this device is called the **curved-arrow formalism**. As this formalism is applied to the reactions of Lewis bases with electron-deficient Lewis acids, the formation of a chemical bond is described by a "flow" of electrons *from the electron donor* (Lewis base) *to the electron acceptor* (Lewis acid). This "electron flow" is indicated by a curved arrow drawn *from the electron source to the electron acceptor*. This formalism is applied to the reaction of Eq. 3.1 in the following way:

electron source   newly formed bond

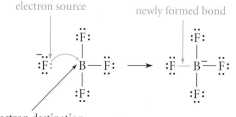

electron destination

(3.2)

The colored arrow indicates that the electron pair on the fluoride ion becomes the shared electron pair in the newly formed bond of BF_4^-. In this formalism, the fluoride ion is said to **attack** the boron atom of the BF_3. The word *attack* when used this way in organic chemistry means "donate electrons to."

The reverse of a Lewis acid-base reaction can be termed a **Lewis acid-base dissociation**. The ion BF_4^- can dissociate to give BF_3 and fluoride ion, in the reverse of Eq. 3.2. This dissociation is shown with the curved-arrow formalism as follows:

$$\begin{array}{ccc}
:\ddot{F}: & & :\ddot{F}: \\
| & & | \\
:\ddot{F}-\overset{\frown}{B}-\ddot{F}: & \longrightarrow & :\ddot{F}-B \;+\; :\ddot{F}:^- \\
| & & | \\
:\ddot{F}: & & :\ddot{F}:
\end{array}$$

(3.3)

Because the B—F bond is breaking, it is the source of the electron pair that is transferred to a fluorine to give fluoride ion.

PROBLEM

3.1   (a)   Suggest a structure for the product of each of the following Lewis acid-base association reactions; be sure to assign formal charges. Label the Lewis acid and the Lewis base, and identify the attacking atom in each case.

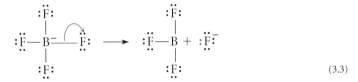

(b)   Draw the curved-arrow formalism for the forward and reverse of each of the reactions in part (a).

3.2 Electron-Pair Displacement Reactions

A. Electron-Pair Displacement Reactions as Lewis Acid-Base Reactions

In some reactions an electron pair is donated to an atom that is *not* electron deficient. When this happens, another electron pair must simultaneously depart from the receiving atom so that the octet rule is not violated. The following reaction is an example of such a process.

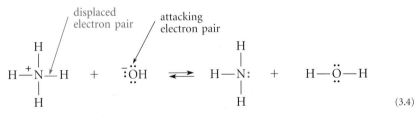

In this reaction, a hydrogen of the ammonium ion is attacked by an electron pair of the hydroxide ion. As a result, this hydrogen becomes bonded to the oxygen to give water, and the electron pair in the N—H bond of the ammonium ion becomes the unshared pair in the product ammonia. If the latter electron pair had not departed, hydrogen would have more electrons than allowed by the octet rule.

This type of reaction is termed an **electron-pair displacement reaction** because one electron pair is displaced from an atom (in this case, from a hydrogen) by the attack of another electron pair. In many such reactions, an atom is transferred between two other atoms. In this example, a proton is transferred from the nitrogen of the ammonium ion to the oxygen of the hydroxide ion.

Electron-pair displacement reactions can also be classified as Lewis acid-base reactions. In this view, the Lewis acid in Eq. 3.4 is the proton of the ammonium ion, which accepts an electron pair from hydroxide ion, and the Lewis base is the hydroxide ion, which donates an electron pair to the proton. However, the Lewis acids in electron-pair displacement reactions are different from the Lewis acids discussed in the previous section, because in electron-pair displacement reactions *the Lewis acids are not electron-deficient*. In order to accept an electron pair, they must at the same time give up an electron pair so as not to violate the octet rule.

B. The Curved-Arrow Formalism

The curved-arrow formalism is also useful for electron-pair displacement reactions. This can be illustrated with the reaction of Eq. 3.4. In this case, *two* arrows are required, one for the attacking electron pair and one for the displaced electron pair:

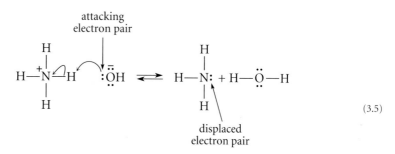

(3.5)

STUDY GUIDE LINK:
✓3.1
*The Curved-Arrow
Formalism*

Notice that in all uses of the curved-arrow formalism each curved arrow originates at the *source* of electrons—an unshared pair or a bond—and terminates at the *destination* of the electron pair.

Attacking electron pairs can originate from bonds as well as unshared pairs:

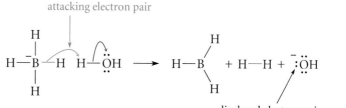

(3.6)

The acceptor atom can of course be an atom other than hydrogen, a point illustrated by the following study problem.

STUDY PROBLEM 3.2

Give the curved-arrow formalism for the following reaction.

$$\text{H} - \overset{\overset{\displaystyle H}{|}}{\underset{\underset{\displaystyle H}{|}}{\text{C}}} - \overset{..}{\underset{..}{\text{Cl}}}: \;+\; :\overset{..}{\underset{..}{\text{O}}}\text{H} \;\longrightarrow\; \text{H} - \overset{\overset{\displaystyle H}{|}}{\underset{\underset{\displaystyle H}{|}}{\text{C}}} - \overset{..}{\underset{..}{\text{O}}}\text{H} \;+\; :\overset{..}{\underset{..}{\text{Cl}}}:$$

Solution   In this reaction an unshared electron pair from the oxygen of ⁻OH displaces the electron pair from the C—Cl bond onto the chlorine; the carbon atom is transferred from the Cl to the oxygen. Because this is an electron pair displacement reaction, two arrows are required. Remember that a curved arrow is drawn from the *source* of an electron pair to its *destination*. The *source* of the attacking electron pair is the ⁻OH ion. The *destination* of the attacking electron pair is the carbon atom. Hence, one curved arrow goes from an electron pair of the ⁻OH (any one of the three pairs) to the carbon atom. Because carbon can have only eight electrons, it must lose a pair of electrons to the chloride ion, which is formed in the reaction. Hence, the source of the second electron pair is the electron pair in the C—Cl bond; its destination is the chlorine. The curved-arrow formalism for this reaction is as follows:

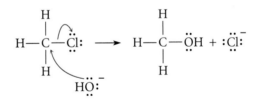

(Be sure to read Study Guide Link 3.1 about the appearance of curved arrows.)

The displaced electron pair in the curved-arrow formalism may come from a double or triple bond, as in the following study problem.

STUDY PROBLEM 3.3

Given the following two reactants and the curved-arrow formalism for their reaction, give the structure of the product.

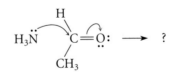

Solution   In this problem you are asked to complete a transformation using the given curved-arrow formalism as a guide. The bonds or unshared electron pairs at the tails of the arrows are the ones that will not be in the same place in the product. The heads of the arrows point to the places at which new bonds or unshared pairs exist in the product. Use the following steps to draw the product.

Step 1   Redraw all atoms just as they were in the reactants:

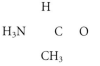

Step 2   Put in the bonds and electron pairs that do not change:

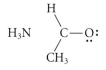

Step 3   Draw the new bonds or electron pairs indicated by the curved-arrow formalism:

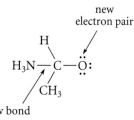

Step 4   Complete the formal charges to give the product. Note that the *sum* of the formal charges in the reactants and products must be the same—zero in this case.

Always keep in mind that curved arrows show the flow of *electron pairs*, not the movement of nuclei. Beginning students sometimes forget this point. For example, the proton transfer from HCl to ⁻OH might be *incorrectly* written as follows:

INCORRECT CURVED-ARROW FORMALISM!

$$\text{H}\ddot{\text{O}}\text{:}^{-} \quad \overset{\frown}{\text{H}}\text{—}\ddot{\text{C}}\text{l:} \quad \longrightarrow \quad \text{H}\ddot{\text{O}}\text{—H} + \text{:}\ddot{\text{C}}\text{l:}^{-} \qquad (3.7)$$

Study Guide Link:
✓3.2
Rules for Use of the Curved-Arrow Formalism

This is not correct because it shows the movement of the proton rather than the flow of electron pairs. Someone accustomed to using the formalism correctly would take this to imply the transfer of H⁻ to ⁻OH, an impossible reaction! The *correct* use of the curved-arrow formalism shows the flow of electron pairs, as follows:

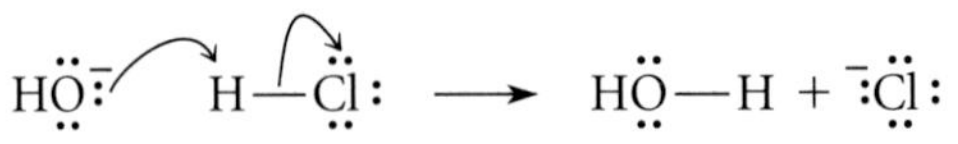

$$\ddot{\text{H}\ddot{\text{O}}}\text{:}^{-} \quad \text{H}-\ddot{\text{C}}\text{l:} \quad \longrightarrow \quad \text{H}\ddot{\text{O}}-\text{H} + \text{:}\ddot{\text{C}}\text{l:}^{-} \qquad \textit{CORRECT!} \qquad (3.8)$$

PROBLEMS

3.2   For each of the following cases, give the product(s) of the transformation indicated by the curved-arrow formalism.

*(a)

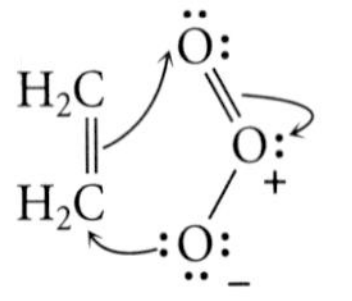

*(b)

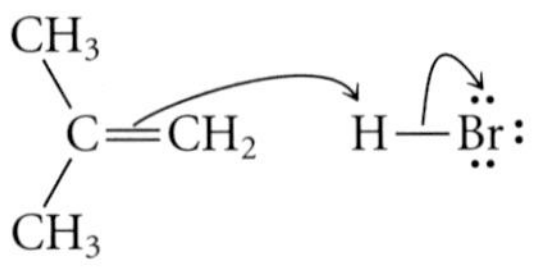

(c)

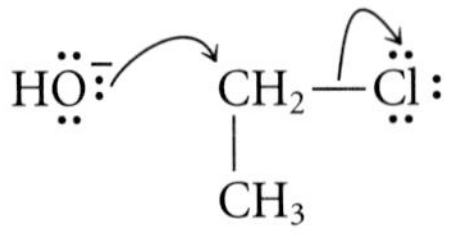

(d)

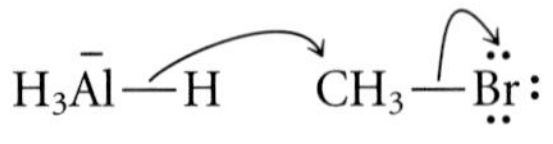

3.3   Suggest a curved-arrow formalism for each of the following reactions.

*(a) 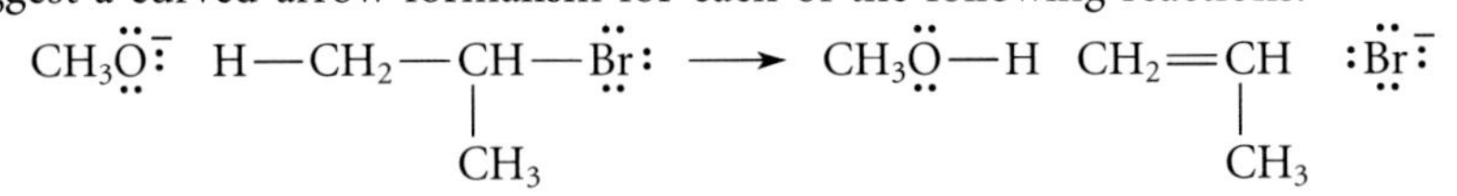 $\text{CH}_3\ddot{\text{O}}\text{:}^{-}$ $\text{H}-\text{CH}_2-\text{CH}-\ddot{\text{Br}}\text{:}$ $\longrightarrow$ $\text{CH}_3\ddot{\text{O}}-\text{H}$ $\text{CH}_2{=}\text{CH}$ $\text{:}\ddot{\text{Br}}\text{:}^{-}$
          $\qquad\qquad\qquad\qquad\quad |$ $\qquad\qquad\qquad\qquad\quad |$
          $\qquad\qquad\qquad\qquad\ \text{CH}_3$ $\qquad\qquad\qquad\qquad\ \text{CH}_3$

(b) 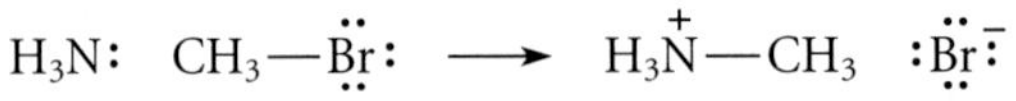 $\text{H}_3\text{N:}$ $\text{CH}_3-\ddot{\text{Br}}\text{:}$ $\longrightarrow$ $\text{H}_3\overset{+}{\text{N}}-\text{CH}_3$ $\text{:}\ddot{\text{Br}}\text{:}^{-}$

3.3   **Review of the Curved-Arrow Formalism**

A.   Use of the Curved-Arrow Formalism to Represent Lewis Acid-Base Reactions

In this chapter you've learned that two classes of Lewis acid-base reactions are

1. the association reactions of Lewis bases with electron-deficient compounds (and their reverse dissociation reactions); and
2. electron-pair displacement reactions.

You'll find that *every reaction you study involving electron pairs can be analyzed as one of these two reaction types, or as a combination of them.* In other words, all reactions involving electron pairs can be dissected ultimately into only two fundamental Lewis acid-base processes! Since both of these fundamental processes can be described with curved arrows, then it follows that *any reaction involving electron pairs can be described with the curved-arrow formalism.* Because of its fundamental importance, the curved-arrow formalism can be, if properly used, a tool of great power for following, understanding, simplifying, and even predicting the reactions of organic chemistry.

B. Use of the Curved-Arrow Formalism to Derive Resonance Structures

In Sec. 1.4 you learned that resonance structures are used when the structure of a compound is not adequately represented by a single Lewis structure. In many cases *resonance structures differ by the movement of electron pairs.* Because the curved-arrow formalism can represent all transformations involving electron pairs, this formalism can also be used to derive correct resonance structures, that is, to show how one resonance structure can be converted into another. The following study problem illustrates this point with two resonance-stabilized molecules that were discussed in Sec. 1.4.

STUDY PROBLEM 3.4

In each set below, show how the second resonance structure can be derived from the first by the curved-arrow formalism.

(a)

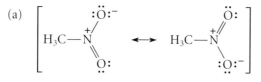

nitromethane

(b) $\left[H_3C\ddot{O}—\overset{+}{C}H_2 \quad \longleftrightarrow \quad H_3C\overset{+}{O}=CH_2 \right]$

methoxymethyl cation

Solution

(a) To convert the structure on the left into the one on the right, an unshared electron pair on the upper oxygen must be used to form a bond to the nitrogen, and a bond to the lower oxygen must be used to form an unshared electron pair, as follows:

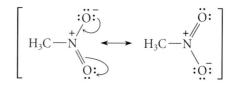

Two arrows are required because the formation of the new bond requires the *displacement* of another. Thus, the appropriate curved-arrow formalism is the electron-pair displacement formalism. Remember: even though this use of curved arrows is identical to that used for describing a reaction, the interconversion of resonance structures is *not* a reaction. The two structures are, taken together, a representation of a *single molecule.*

(b) In the structure on the left, the positively charged carbon is electron-deficient. The structure on the right is derived by the donation of an oxygen unshared pair to this carbon.

$$\left[H_3C\ddot{O}\overset{\frown}{—}\overset{+}{C}H_2 \quad \longleftrightarrow \quad H_3C\overset{+}{O}=CH_2 \right]$$

This transformation resembles a Lewis acid-base association reaction, and the same curved-arrow formalism is used: a single curved arrow showing the donation of electrons to the electron-deficient carbon.

PROBLEM

3.4   *(a)   Draw a resonance structure for the allyl anion (below) which shows that the two carbon-carbon bonds have an identical bond order of 1.5 and that the unshared electron pair (and negative charge) is shared equally by the two terminal carbons. Use the curved-arrow formalism to derive your structure.

$$\left[:\bar{C}H_2 - CH{=}CH_2 \quad \longleftrightarrow \quad ? \quad \right]$$

allyl anion

(b)   Using the curved-arrow formalism, derive a resonance structure for the allyl cation (below) which shows that each carbon-carbon bond has a bond order of 1.5 and that the positive charge is shared equally by both terminal carbon atoms.

$$\left[\overset{+}{C}H_2 - CH{=}CH_2 \quad \longleftrightarrow \quad ? \quad \right]$$

allyl cation

*(c)   Using the curved-arrow formalism, derive a resonance structure for benzene (below) which shows that all carbon-carbon bonds are identical and have a bond order of 1.5.

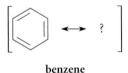

benzene

3.4   Brønsted-Lowry Acids and Bases

A.   Definition of Brønsted Acids and Bases

Although less general than the Lewis concept, the *Brønsted-Lowry acid-base concept* is another view of acids and bases that is extremely important and useful in organic chemistry. The Brønsted-Lowry definition of acids and bases was published in 1923, the same year that Lewis formulated his ideas of acidity and basicity. A **Brønsted acid** is defined as a species that reacts by donating a proton; a **Brønsted base** is a species that reacts by accepting a proton.

The reaction of ammonium ion with hydroxide ion is an example of a Brønsted acid-base reaction.

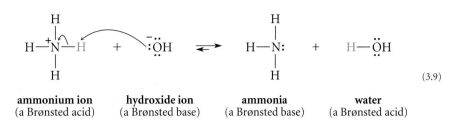

ammonium ion	hydroxide ion	ammonia	water
(a Brønsted acid)	(a Brønsted base)	(a Brønsted base)	(a Brønsted acid)

$$(3.9)$$

On the left side of this equation, the ammonium ion is acting as a Brønsted acid and the hydroxide ion is acting as a Brønsted base; looking at the equation from right to left, water is acting as a Brønsted acid, and ammonia as a Brønsted base.

A Brønsted acid-base reaction is an electron-pair displacement reaction in which a proton is transferred from one atom to another. In terms of the Lewis acid-base concept, a proton of the ammonium ion is the Lewis acid; it accepts an electron pair when it undergoes attack by hydroxide ion, a Lewis base. Notice that the curved-arrow formalism is the two-arrow formalism used for all electron-pair displacements. Notice carefully that even though the Brønsted acid-base definition focuses on the transfer of *protons*, the curved-arrow formalism, as always, shows the movement of *electron pairs*.

When a Brønsted acid loses a proton, its **conjugate base** is formed; when a base gains a proton, its **conjugate acid** is formed. Hence, in Eq. 3.9, NH_4^+ and NH_3 are a **conjugate acid-base pair**; and H_2O and OH^- are also a conjugate acid-base pair:

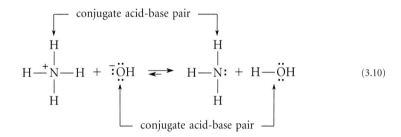

$$(3.10)$$

Notice that the conjugate acid-base relationship is *across* the equilibrium arrows. For example, NH_4^+ and NH_3 are a conjugate acid-base pair, but NH_4^+ and OH^- are *not* a conjugate acid-base pair.

Study Guide Link:
✓3.3
*Identification of Acids
and Bases*

Some compounds, called **amphoteric compounds**, can act as either acids or bases. Water is the archetypical example of an amphoteric compound. For example, in Eq. 3.10, water is the conjugate *acid* of the acid-base pair H_2O/OH^-; in the following reaction, water is the conjugate *base* of the acid-base pair H_3O^+/H_2O:

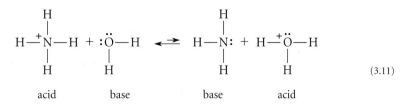

$$(3.11)$$

The Brønsted-Lowry acid-base concept is important for organic chemistry because many organic reactions *are* Brønsted acid-base reactions, and many others have close analogy to Brønsted acid-base reactions.

3.5 In the following reactions, label the conjugate acid-base pairs. Then draw the curved-arrow formalism for these reactions in the left-to-right direction.

*(a) $\ddot{N}H_3 + \ddot{N}H_3 \;\rightleftharpoons\; {}^-\colon\!\ddot{N}H_2 + \overset{+}{N}H_4$

(b) $\ddot{N}H_3 + {}^-\colon\!\ddot{O}H \;\rightleftharpoons\; {}^-\colon\!\ddot{N}H_2 + H_2\ddot{O}\colon$

3.6 Give the curved-arrow formalism for the reactions in Problem 3.5 in the right-to-left direction.

*3.7 Write a Brønsted acid-base reaction in which $H_2\ddot{O}/\colon\!\ddot{O}H$ and $CH_3\ddot{O}H/CH_3\ddot{O}\colon^-$ act as conjugate acid-base pairs.

B. Strengths of Brønsted Acids

Many reactions of organic chemistry can be understood and even predicted by a knowledge of the relative strengths of the acids and bases involved. The relative strengths of Brønsted acids are determined by how well they transfer a proton to a standard Brønsted base. The standard base traditionally used for comparison is water. The transfer of a proton from a general acid HA to water is indicated by the following equilibrium:

$$HA + H_2O \;\rightleftharpoons\; A\colon^- + H_3O^+ \tag{3.12}$$

The equilibrium constant for this reaction is given by

$$K_{eq} = \frac{[A\colon^-][H_3O^+]}{[HA][H_2O]} \tag{3.13}$$

(The quantities in brackets are molar concentrations at equilibrium.) Because water is the solvent, and its concentration remains effectively constant regardless of the concentrations of the other species in the equilibrium, we multiply Eq. 3.13 through by $[H_2O]$ and thus define another constant K_a, called the **dissociation constant**:

$$K_a = K_{eq}[H_2O] = \frac{[A\colon^-][H_3O^+]}{[HA]} \tag{3.14}$$

Each acid has its own unique dissociation constant. The larger the dissociation constant of an acid, the more H_3O^+ ions are formed when the acid is dissolved in water at a given concentration. Thus, *the strength of an acid is measured by the magnitude of its dissociation constant.*

Because the dissociation constants of different Brønsted acids cover a range of many powers of ten, it is useful to express acid strength in a logarithmic manner. Using p as an abbreviation for negative logarithm, we can write the following definitions:

$$pK_a = -\log K_a \tag{3.15}$$

$$pH = -\log [H_3O^+] \tag{3.16}$$

The approximate pK_a values of several Brønsted acids are given in Table 3.1 in order of decreasing pK_a. Since stronger acids have larger K_a values, it follows from Eq. 3.15

Table 3.1 Relative Strengths of Some Acids and Bases

Conjugate acid	Conjugate base	pK_a
$\overset{..}{N}H_3$ (ammonia)	$\overset{..}{:}NH_2^-$ (amide)	~35
$R\overset{..}{O}H$ (alcohol)	$R\overset{..}{O}:^-$ (alkoxide)	15–19[a]
$H\overset{..}{O}H$ (water)	$H\overset{..}{O}:^-$ (hydroxide)	15.7
$R\overset{..}{S}H$ (thiol)	$R\overset{..}{S}:^-$ (thiolate)	10–12[a]
R_3NH^+ (trialkylammonium ion)	$R_3N:$ (trialkylamine)	9–11[a]
NH_4^+ (ammonium ion)	$H_3N:$ (ammonia)	9.25
HCN (hydrocyanic acid)	$^-:CN$ (cyanide)	9.40
$H_2\overset{..}{S}$ (hydrosulfuric acid)	$H\overset{..}{S}:^-$ (hydrosulfide)	7.0
$R\!-\!\overset{\overset{\displaystyle O}{\|}}{C}\!-\!\overset{..}{O}H$ (carboxylic acid)	$R\!-\!\overset{\overset{\displaystyle O}{\|}}{C}\!-\!\overset{..}{O}:^-$ (carboxylate)	4–5[a]
$H\overset{..}{F}:$ (hydrofluoric acid)	$:\overset{..}{F}:^-$ (fluoride)	3.2
$H_3C\!-\!\langle\rangle\!-\!SO_3H$ (p-toluene-sulfonic acid)	$H_3C\!-\!\langle\rangle\!-\!SO_3^-$ (p-toluene-sulfonate, or "tosylate")	−1
$H_3\overset{..}{O}^+$ (hydronium ion)	$H_2\overset{..}{O}$ (water)	−1.7
H_2SO_4 (sulfuric acid)	HSO_4^- (bisulfate)	−3[b]
$H\overset{..}{C}l:$ (hydrochloric acid)	$:\overset{..}{C}l:^-$ (chloride)	−6 to −7[b]
$H\overset{..}{B}r:$ (hydrobromic acid)	$:\overset{..}{B}r:^-$ (bromide)	−8 to −9.5[b]
$H\overset{..}{I}:$ (hydroiodic acid)	$:\overset{..}{I}:^-$ (iodide)	−9.5 to −10[b]
$HClO_4$ (perchloric acid)	ClO_4^- (perchlorate)	−10 (?)[b]

[a] Precise value varies with the structure of R.
[b] Estimates; exact measurement is not possible.

that *stronger acids have smaller* pK_a *values.* Thus, HCN (pK_a = 9.4) is a stronger acid than water (pK_a = 15.7). In other words, the strengths of acids in the first column of Table 3.1 increase from the top to the bottom of the table.

PROBLEMS

3.8 What is the pK_a of an acid that has a dissociation constant of
*(a) 5.8×10^{-6} (b) 10^{-3} *(c) 50

3.9   What is the dissociation constant of an acid that has a pK_a of
*(a) 7.8   (b) 4   *(c) −2

3.10   Which acid is the strongest
*(a) in Problem 3.8?   (b) in Problem 3.9?

C. Strengths of Brønsted Bases

The strength of a Brønsted base is conveniently expressed in terms of the pK_a of its conjugate acid. Thus, the base strength of fluoride ion is measured by the pK_a of its conjugate acid HF; the base strength of ammonia is measured by the pK_a of its conjugate acid, the ammonium ion, $^+NH_4$. That is, when we say that a base is weak, we are also saying that its conjugate acid is strong; or, if a base is strong, its conjugate acid is weak. Thus, it is easy to tell which of two bases is stronger by looking at the pK_a values of their conjugate acids: *the stronger base has the conjugate acid with the greater (or less negative) pK_a.* For example, ^-CN, the conjugate base of HCN, is a weaker base than ^-OH, the conjugate base of water, because the pK_a of HCN is less than that of water. Thus, the strengths of the bases in the second column of Table 3.1 increase from the bottom to the top of the table.

BASICITY CONSTANTS

In the older literature, a quantity called the **basicity constant**, K_b, was used as a measure of base strength. The basicity constant is essentially a measure of the ability of a base to accept a proton from water rather than from H_3O^+. For example, the reaction in which ammonia accepts a proton from water is as follows:

$$NH_3 + H_2O \; \xleftrightarrow{\quad} \; \overset{+}{N}H_4 + \overset{-}{O}H \qquad (3.17)$$

The equilibrium constant for this reaction is

$$K_{eq} = \frac{[NH_4^+][^-OH]}{[NH_3][H_2O]} \qquad (3.18)$$

Since $[H_2O]$ is effectively a constant, it is convenient to incorporate it into the equilibrium constant. Multiplying through Eq. 3.18 by $[H_2O]$ gives the *basicity constant*, K_b:

$$K_b = K_{eq}[H_2O] = \frac{[NH_4^+][^-OH]}{[NH_3]} \qquad (3.19)$$

The K_b for a base is simply related to the dissociation constant K_a of its conjugate acid. By multiplying the numerator and denominator of this equation by $[H_3O^+]$, and recognizing that the ion-product constant of water K_w equals $[H_3O^+][^-OH]$, we find that

$$K_b = \frac{K_w}{K_a} \qquad (3.20)$$

(The value of K_w is 10^{-14} mol^2/L^2.) You should verify Eq. 3.20. Taking negative logarithms, and recognizing that pK_w = 14,

$$pK_b = 14 - pK_a \qquad (3.21)$$

Although most people today use K_a (or pK_a) values of conjugate acids as quantitative measures of base strength, calculation of pK_b values is a simple matter with Eq. 3.21.

D. Equilibria among Different Acids and Bases

When a Brønsted acid and base react, we can tell immediately whether the equilibrium lies to the right or left by comparing the pK_a values of the two acids involved. *The equilibrium in the reaction of an acid and a base always favors the side with the weaker acid and weaker base.* For example, in the following acid-base reaction, the equilibrium lies well to the right, because H_2O is the weaker acid and ^-CN the weaker base.

$$
\begin{array}{ccccccc}
 & & \text{(stronger} & & & \text{(weaker} & \\
 & & \text{base)} & & & \text{base)} & \\
\text{HCN} & + & \text{OH}^- & \rightleftharpoons & {}^-\text{CN} & + & \text{H}_2\text{O} \\
\text{p}K_a = 9.4 & & & & & \text{p}K_a = 15.7 & \\
\text{(stronger acid)} & & & & & \text{(weaker acid)} &
\end{array}
\qquad (3.22)
$$

We'll frequently find it useful to estimate the equilibrium constants of acid-base reactions. The equilibrium constant for an acid-base reaction can be calculated in a straightforward way from the pK_a values of the two acids involved. To do this calculation, subtract the pK_a of the acid on the left side of the equation from the pK_a of the acid on the right and take the antilog of the resulting number. This procedure is illustrated for the reaction in Eq. 3.22 in the following study problem, and is justified in Problem 3.39 at the end of the chapter.

STUDY PROBLEM 3.5

Calculate the equilibrium constant for the reaction of HCN with hydroxide ion (Eq. 3.22).

Solution   Subtracting the pK_a of the acid on the left of Eq. 3.22 (HCN) from the one on the right (H_2O) gives the logarithm of the desired equilibrium constant K_{eq}.

$$\log K_{eq} = 15.7 - 9.4 = 6.3$$

The equilibrium constant for this reaction is the antilog of this number, or

$$K_{eq} = 10^{6.3} = 2 \times 10^6$$

This large number means that the equilibrium of Eq. 3.22 lies *far to the right*. That is, if we dissolve HCN in an equimolar solution of NaOH, a reaction occurs to give a solution in which there is *much* more ^-CN than either ^-OH or HCN. On the other hand, in an aqueous solution of NaCN, only a very small amount of ^-CN reacts with the H_2O to give ^-OH and HCN.

3.11   What is *(a) the strongest base and (b) the weakest base listed in the second column of Table 3.1?

3.12   Using the pK_a values in Table 3.1, calculate the equilibrium constant for the following reactions.
*(a) NH_3 acting as a base toward the acid HCN
(b) F^- acting as a base toward the acid HCN

Sometimes students confuse acid strength and base strength when they encounter an *amphoteric compound*—a compound that can act as *both* an acid and a base. Water presents this sort of problem. According to the definitions just developed, the *base strength* of water is measured by the pK_a of its *conjugate acid*, H_3O^+, whereas the *acid strength of water* (or the base strength of its conjugate base hydroxide) is measured by the pK_a of water itself. These two quantities refer to very different reactions of water:

Water acting as a base:

$$H_2O \;+\; AH \;\;\rightleftharpoons\;\; H_3O^+ \;+\; A{:}^- \tag{3.23a}$$

$$pK_a = -1.7$$

Water acting as an acid:

$$B{:}^- \;+\; H_2O \;\;\rightleftharpoons\;\; BH \;+\; \overline{O}H \tag{3.23b}$$

$$pK_a = 15.7$$

*3.13   Write an equation for each of the following equilibria, and use Table 3.1 to identify the pK_a value associated with the acidic species in each equilibrium.
(a)   ammonia acting as a base toward the acid water
(b)   ammonia acting as an acid toward the base water

3.5   Free Energy and Chemical Equilibrium

As you learned in the previous section, the equilibrium constant for a reaction tells us which species in a chemical equilibrium are present in highest concentration. The equilibrium constant also tells us something more fundamental about the compounds in a chemical equilibrium: their *relative standard free energies*.

   Let's consider a specific example—the dissociation equilibrium of hydrofluoric acid, a relatively weak acid:

$$H{-}F + H_2O \;\;\rightleftharpoons\;\; F^- + H_3O^+ \tag{3.24}$$

From Table 3.1, the pK_a of HF is 3.2. Hence, the dissociation constant K_a of HF is $10^{-3.2}$, or 6.3×10^{-4}. The small magnitude of this equilibrium constant means that HF is dissociated to only a small extent in aqueous solution. For example, in an aqueous solution containing 0.1 M HF, we can calculate using methods learned in general chemistry that only about 8% of the acid is dissociated to fluoride ions and hydrated protons.

The dissociation constant is related to the *standard free energy difference* between products and reactants in the following way. If K_a is the dissociation constant as defined in Eq. 3.14, then the standard free energy of dissociation is given by the following equation:

$$\Delta G_a^\circ = -RT \ln K_a \qquad (3.25)$$

The **standard free energy of dissociation** ΔG_a° is equal to the standard free energies of the ionization products (H_3O^+ and F^-) minus the standard free energy of the un-ionized acid (H—F). The standard free energy of the solvent water, because it is the same for all acids, is arbitrarily set to zero (that is, ignored). R is the molar gas constant (8.314 $\times$ 10^{-3} kJ/deg-mol or 1.987 $\times$ 10^{-3} kcal/deg-mol) and T is the absolute temperature in kelvins (degrees K). In common logarithms,

$$\Delta G_a^\circ = -2.3RT \log K_a \qquad (3.26)$$

Because $-\log K_a$ is by definition the pK_a (Eq. 3.15), then

$$\Delta G_a^\circ = 2.3RT \, (pK_a) \qquad (3.27)$$

Introducing the pK_a of HF (3.2) into this equation, we find, at 25 °C (298 K)

$$\Delta G_a^\circ = 18.2 \text{ kJ/mol (4.36 kcal/mol)} \qquad (3.28)$$

What is the meaning of this standard free energy change? It means that the products of the dissociation equilibrium, H_3O^+ and F^-, have 18.2 kJ/mol (4.36 kcal/mol) more free energy than the undissociated acid HF—that is, the products are *less stable* than the reactants by 18.2 kJ/mol (4.36 kcal/mol) under standard conditions, usually taken as 1 atm pressure (for gases) or 1 mole per liter for liquid solutions. Physically, this means that if we could somehow connect a free-energy source such as a battery to an aqueous solution containing one mole per liter of HF, it would take 18.2 kJ (4.36 kcal) of energy to drive the reaction completely to one mole per liter of hydrated protons and one mole per liter of fluoride ions. Or, we can turn the idea around: if we could somehow generate a solution containing one mole per liter of hydrated protons and one mole per liter of fluoride ions, this solution would release 18.2 kJ/mol (4.36 kcal/mol) of free energy if the two reacted completely to give water and one mole per liter of HF.

Let's now generalize this result for a reaction in which the reactant is R and the product is P. The equilibrium constant K_{eq} for the interconversion of R and P is related to the standard free-energy difference ($G_P^\circ - G_R^\circ$) between P and R as follows:

$$\Delta G^\circ = (G_P^\circ - G_R^\circ) = -2.3RT \log K_{eq} \qquad (3.29)$$

Suppose ΔG° is positive (Fig. 3.1a). This means that P has a greater free energy than R, or P is less stable than R. If R and P are brought to equilibrium, R will be present in greater amount. This is shown by rearranging Eq. 3.29:

$$\log K_{eq} = \frac{-\Delta G^\circ}{2.3RT} \qquad (3.30)$$

When ΔG° is positive, $\log K_{eq}$ is a negative number and consequently $K_{eq} = [P]/[R]$ is less than unity—in other words, the concentration of R is higher than that of P at equilibrium. On the other hand, suppose that the standard free-energy difference is negative, as shown in Fig. 3.1b; that is, products have lower free energy than reactants.

Table 3.2   **Relationship between Equilibrium Constant and Standard Free-Energy Changes for Reactions at 25 °C**

$$\Delta G° = -2.3RT \log K_{eq} \text{ or } K_{eq} = 10^{-\Delta G°/2.3RT}$$

$\Delta G°$		K_{eq}	K_{eq}	$\Delta G°$	
kJ/mol	kcal/mol			kJ/mol	kcal/mol
+40	9.56	9.8×10^{-8}	10^6	−34.2	−8.18
+20	4.78	3.1×10^{-4}	10^4	−22.8	−5.46
+10	2.39	0.018	10^2	−11.4	−2.73
+5	1.20	0.13	50	−9.71	−2.32
0	0	1.0	30	−8.41	−2.01
−5	−1.20	7.5	10	−5.69	−1.36
−10	−2.39	56.6	5	−3.98	−0.95
−20	−4.78	3.2×10^3	3	−2.72	−0.65
−40	−9.56	1.0×10^7	1	0	0

In this case log K_{eq} is positive and K_{eq} is greater than unity; in other words, the concentration of *P* is higher than that of *R* at equilibrium. These examples show that *chemical equilibrium favors the species of lower free energy.*

The more two compounds differ in free energy, the greater is the difference in their concentrations at equilibrium. Because of the *logarithmic* relationship of free energy and equilibrium constant, relatively small changes in $\Delta G°$ bring about relatively large changes in equilibrium constant or equilibrium concentration (Table 3.2).

The preceding discussion shows the close relationship between the relative stabilities of compounds and their concentrations at equilibrium. This relationship means that if

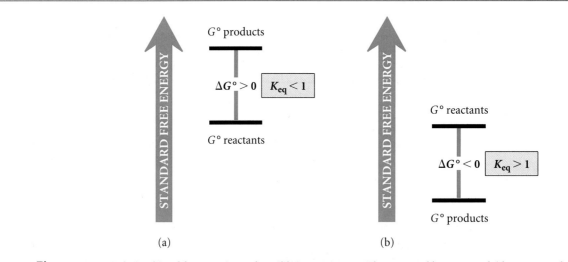

Figure 3.1   *Relationship of free energies and equilibrium constants. The more stable compound (the compound of lower free energy) is favored in a chemical equilibrium. In (a) the reactants are more stable and are favored at equilibrium; in (b) the products are more stable and are favored at equilibrium.*

we can predict relative stabilities, we can then predict chemical effects such as equilibrium constants. This will be our task in the next section.

PROBLEMS

*3.14   A reaction has a standard free-energy change of -14.6 kJ/mol (-3.5 kcal/mol). Calculate the equilibrium constant for the reaction at 25 °C.

3.15   Calculate the standard free-energy change of a reaction that has an equilibrium constant of 305.

3.16   *(a)   A reaction $A + B \rightleftarrows C$ has a standard free-energy change of -2.93 kJ/mol (-0.7 kcal/mol) at 25 °C. What are the concentrations of A, B, and C at *equilibrium* if, at the beginning of the reaction, their concentrations are 0.1 M, 0.2 M, and 0 M, respectively?
        (b)   In a qualitative sense, how would you expect your answer for part (a) to change if the reaction has a standard free-energy change of $+2.93$ kJ/mol ($+0.7$ kcal/mol)? Verify your prediction with a calculation.

3.6   Relationship of Structure to Acidity

The goal of this section is to help you learn to do something very powerful indeed—to use the *structures* of compounds to predict trends in their chemical properties. The chemical property we are going to deal with here is Brønsted acidity, but what you learn can be brought to bear on other chemical properties. This section will deal with the following question: How can we predict the relative strengths of Brønsted acids within a series? Your ability to deal with questions like this will require that you use all that you have learned in the previous sections.

A. The Element Effect

One of the most important things that determines the acidity of a Brønsted acid is *the identity of the atom to which the acidic hydrogen is attached*. For example, consider the acidities of the following two compounds:

$$CH_3CH_2-O-H \qquad\qquad CH_3CH_2-S-H$$

ethanol (an alcohol)   **ethanethiol (a thiol, or mercaptan)**
$pK_a = 15.9$ $pK_a = 10.5$

These two compounds are structurally similar, the sole difference between them being the element (color) to which the acidic proton is attached. The elements come from the same group in the periodic table; yet the thiol is almost a *million times as acidic* as the alcohol (which is about as acidic as water). Another important example of the same trend is the relative acidities of the hydrogen halides. HI is the strongest of these acids; HF is the weakest. (The relevant pK_a data are found in Table 3.1.) These data illustrate a noteworthy trend: *Brønsted acidity increases as the atom to which the acidic hydrogen is attached has a greater atomic number within a column (group) of the periodic table.*

Now let's see how acidities vary across the periodic table within the same row, or period:

$$H—CH_3 \quad H—NH_2 \quad H—OH \quad H—F \qquad (3.31)$$

$$pK_a \qquad \sim 55 \qquad \sim 35 \qquad 15.7 \qquad 3.2$$

(The pK_a values of methane and ammonia are so high that they are not known with certainty.) These data demonstrate another important trend: *Brønsted acidity increases as the atom to which the acidic hydrogen is attached is farther to the right within a row (period) of the periodic table.*

How can these trends be explained? Let's divide the ionization process of a typical acid H—A into three steps, shown in Eqs. 3.32a–c. Of course, ionization occurs in one step, but we can *think* of the actual process as the sum of fictitious processes to help us understand the observed trends in acidity.

Bond breaking	$H{\overset{\text{\small$\xi$}}{—}}A \longrightarrow H\cdot + A\cdot$	(3.32a)
Electron transfer to A·	$e^- + A\cdot \longrightarrow A{:}^-$	(3.32b)
Loss of an electron from H·	$H\cdot \longrightarrow H^+ + e^-$	(3.32c)
Sum	$H—A \longrightarrow H^+ + A{:}^-$	(3.32d)

Notice that if we cancel identical items from opposite sides, the sum of these steps (Eq. 3.32d) is the overall dissociation reaction. *Anything that makes any of these steps more favorable tends to increase acidity.* Let us consider the energetics of each of these steps in turn.

The first step (Eq. 3.32a) is the breaking of the H—A bond "in half," with one bonding electron going to one atom and the other bonding electron going to the other atom. The energy required for this step is called the **bond dissociation energy**. Trends in bond dissociation energy are indicated by the following data:

Within Group 7A of the periodic table:

Bond:	H—F	H—Cl	H—Br	H—I	
Bond dissociation energy:					
(kJ/mol)	569	431	368	297	
(kcal/mol)	136	103	88	71	(3.33a)

Within the second period of the periodic table:

Bond:	H—CH$_3$	H—NH$_2$	H—OH	H—F	
Bond dissociation energy:					
(kJ/mol)	439	448	498	569	
(kcal/mol)	105	107	119	136	(3.33b)

Because the bond dissociation energy is the energy *required* for dissociation to occur, smaller numbers represent more favorable reactions.

The second step of acid dissociation (Eq. 3.32b) shows an atom or group A· accepting an electron to form the corresponding anion. The energy released in this step is the **electron affinity** of A. Trends in electron affinity are indicated by the following data:

Within Group 7A of the periodic table:

Atom:	$:\ddot{I}\cdot$	$:\ddot{Br}\cdot$	$:\ddot{Cl}\cdot$	$:\ddot{F}\cdot$	
Electron affinity:					
(kJ/mol)	295	324	349	328	
(kcal/mol)	70	78	83	78	(3.34a)

Within the second period of the periodic table:

Group:	$\cdot CH_3$	$\cdot \ddot{N}H_2$	$\cdot \ddot{O}H$	$\cdot \ddot{F}:$	
Electron affinity:					
(kJ/mol)	7.7	74	177	328	
(kcal/mol)	1.8	18	42	78	(3.34b)

Because the electron affinity is the energy *released* when a group combines with an electron, larger numbers represent more favorable reactions.

☞
STUDY GUIDE LINK:
3.4
*Factors that Affect
Acidity*

The remaining step of acid dissociation (Eq. 3.32c) is loss of an electron from a hydrogen atom. The energy required for this step is the **ionization potential** of the hydrogen atom. Because this is the same for all Brønsted acids, it does not enter into a comparison of different acids.

The data in Eqs. 3.33a and 3.34a show that within a column (group) of the periodic table, electron affinities do not change as much as bond dissociation energies. For example, the electron affinities of Cl and I differ by only 54 kJ/mol (13 kcal/mol), whereas the bond dissociation energies of H—Cl and H—I differ by 134 kJ/mol (32 kcal/mol). Hence, *the greater strength of Brønsted acids H—A toward high atomic number within a column (group) of the periodic table is due to weaker H—A bonds.*

Across a row (period) of the periodic table, electron affinities change much more than bond dissociation energies (Eqs. 3.33b and 3.34b). The increase in electron affinities from $\cdot CH_3$ to $\cdot F$ is 320 kJ/mol (76 kcal/mol), but the increase in bond energies from H—CH_3 to H—F is 130 kJ/mol (31 kcal/mol). Hence, *the trend toward higher acidities of Brønsted acids H—A along a row (period) of the periodic table can be attributed to the ability of the atoms or groups A to attract electrons.* You may have noticed that this trend is similar to the trend in electronegativities (Table 1.1), and this should not be surprising: an atom's electronegativity measures its ability to attract electrons—not isolated electrons, but rather electrons within a chemical bond.

Because fluorine is much more electronegative than iodine, some students are surprised to learn that H—F is a much weaker acid than H—I. However, as you have just seen, it is the weaker H—I bond, not the electron-attracting ability of iodine or fluorine, that accounts for the greater acidity of H—I. Likewise, H—SH is a stronger acid than H—OH, and thiols (H—SR) are stronger acids than alcohols (H—OR), for the same reason.

The effect of the attached atom A on the acidity of a Brønsted acid H—A is termed the **element effect**. To summarize:

1. The element effect on the Brønsted acidities of a series of acids H—A within a column (group) of the periodic table is dominated by the bond dissociation energies of H—A; stronger acids have weaker H—A bonds.

2. The element effect on the Brønsted acidities of a series of acids H—A within a row (period) of the periodic table is dominated by electron-attracting abilities of the elements A; stronger acids have more electron-attracting A groups.

B. The Polar Effect

Another important effect on acidities can be illustrated using as examples the acidities of certain *carboxylic acids*. Carboxylic acids are in most cases weak acids that ionize somewhat in water to give their conjugate bases, called *carboxylate ions*.

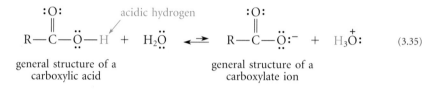

$$(3.35)$$

general structure of a
carboxylic acid

general structure of a
carboxylate ion

Carboxylic acids are among the most common and important of the acidic organic compounds.

Consider the trend in acidity indicated by the following data for the acidity of acetic acid and some of its substituted derivatives:

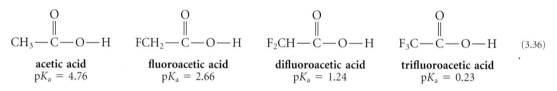

$$(3.36)$$

acetic acid
pK_a = 4.76

fluoroacetic acid
pK_a = 2.66

difluoroacetic acid
pK_a = 1.24

trifluoroacetic acid
pK_a = 0.23

Within this series, the only structural difference from compound to compound is that hydrogens have been substituted by fluorines several atoms away from the acidic hydrogen. The more fluorines there are, the stronger is the acid. A similar effect is observed when other electronegative atoms or groups are substituted into a carboxylic acid molecule. The following data illustrate the same type of effect:

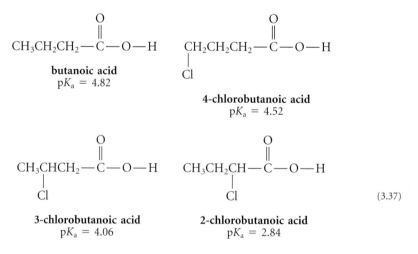

butanoic acid
pK_a = 4.82

4-chlorobutanoic acid
pK_a = 4.52

3-chlorobutanoic acid
pK_a = 4.06

2-chlorobutanoic acid
pK_a = 2.84

$$(3.37)$$

Notice from these data that the closer the electronegative group is to the acidic hydrogen, the greater is its effect on acidity.

What is the reason for these effects? To begin with, consider the standard free energy of the ionization process. Recall (Sec. 3.5, Eq. 3.27) that the standard free energy of ionization ΔG_a° is related to the dissociation constant K_a of an acid by the equation

$$\Delta G_a^\circ = 2.3RT\,(pK_a) \tag{3.38}$$

That is, for the dissociation of a carboxylic acid, the pK_a is directly proportional to ΔG_a°, the standard free-energy difference between the reactant acid and the products of the dissociation reaction. This idea is shown diagrammatically in Fig. 3.2. Notice in this diagram that *when the free energy of the conjugate-base carboxylate ion is lower (that is, when the ion is more stable), the pK_a of the acid is smaller (that is, the acid is more acidic).*

Electronegative substituent groups such as halogens increase the acidities of carboxylic acids by stabilizing their conjugate-base carboxylate ions. This stabilization originates in the polarity of the carbon-halogen bond. To visualize this idea, consider the *electrostatic interaction* (interaction between charges) of the negatively charged carboxylate oxygen with the nearby carbon-halogen bond dipole:

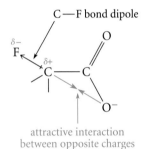

$$\tag{3.39}$$

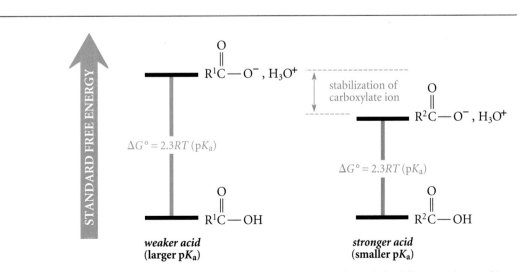

Figure 3.2   *Carboxylic acids of greater acidity have more stable conjugate bases. (The different acids are arbitrarily placed at the same free energy for comparison purposes.)*

This interaction is governed by the following equation, called the **electrostatic law**:

$$E = k\frac{q_1 q_2}{\epsilon r} \tag{3.40}$$

In this equation q_1 and q_2 are charges, ϵ is a constant, k is a proportionality constant, and r is the distance between the charges. According to this law, charges of the opposite sign interact to give a negative (stabilizing) contribution to total energy. This means that the negative charge on the carboxylate oxygen and the positive end of the C—F bond dipole interact favorably. Although the electrostatic law also indicates that the negative charge on the oxygen and the negative end of the bond dipole interact unfavorably, *this interaction occurs across a larger distance* (r in the electrostatic law is larger), and therefore the resulting contribution to total energy E is smaller (less important). Hence, the *net* interaction of the carboxylate oxygen and the nearby C—F bond dipole is an attractive, stabilizing one. As you can see from the right side of Fig. 3.2, this stabilization lowers the pK_a of an acid, or strengthens the acid. Because the carboxylic acid itself is uncharged, the polar effect on its stability is much less important and can be ignored.

The effect on chemical properties caused by the interactions between charges and/or dipoles is called a **polar effect**. (It is also known as an **inductive effect**.) Thus, in the present examples halogens (or other electronegative substituents) have an *acid-strengthening polar effect* on the acidity of carboxylic acids. As the series in Eq. 3.36 shows, the more halogens there are, the greater is the effect on acidity. In fact, trifluoroacetic acid borders on being a strong acid.

Another way to describe the polar effect of halogens and other electronegative groups is to say that they exert an **electron-withdrawing polar effect**, because they pull electrons toward themselves and away from the carbon to which they are attached. As we might imagine, there are other groups that exert an opposite polar effect, called an **electron-donating polar effect** (see Problem 3.36), and such groups raise the pK_a, or reduce the acidity, of nearby carboxylic acids.

The inverse relationship between the interaction energy E and distance r in Eq. 3.40 tells us that polar effects should diminish as the distance between the interacting groups increases. Indeed, within the series of Eq. 3.37, the influence of a chlorine on the pK_a decreases significantly as the chlorine is more remote from the carboxylate oxygen.

From this section, you've learned about two effects of *structure* on *chemical properties*, in this case, the chemical properties of acidity and basicity. The *element effect*, the larger of the two effects, is how the atom attached to the acidic proton affects acidity. The element effect has its origins in bond energies and electron affinities (or electronegativities). Trends in acidity based on the element effect can be predicted by noting the relationship of the elements in the periodic table. The *polar effect* is how remote groups affect acidity through interactions between charges and/or dipoles. Trends in acidity based on the polar effect can be predicted by analyzing the effect of charge-charge or charge-dipole interactions on the stability of the charged species in acid-base equilibria.

STUDY PROBLEM 3.6

Rank the following compounds in order of increasing basicity.

$$\underset{\substack{\text{acetate ion} \\ A}}{CH_3-\overset{\overset{\displaystyle O}{\|}}{C}-O^-}
\qquad
\underset{\substack{\text{acetamide anion} \\ B}}{CH_3-\overset{\overset{\displaystyle O}{\|}}{C}-\overset{-}{N}H}
\qquad
\underset{\substack{\text{glycine (an amino acid)} \\ C}}{\overset{+}{H_3}N-CH_2-\overset{\overset{\displaystyle O}{\|}}{C}-O^-}$$

Solution   First recognize that a problem in relative basicity is equivalent to a problem in relative acidity. If you can rank the acidities of the conjugate acids, you've solved the problem. The relevant conjugate acids are

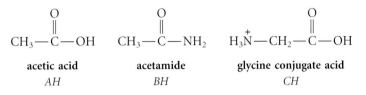

<table>
<tr><td>acetic acid</td><td>acetamide</td><td>glycine conjugate acid</td></tr>
<tr><td>AH</td><td>BH</td><td>CH</td></tr>
</table>

Both *AH* and *CH* are carboxylic acids; in both cases the acidic hydrogen is bound to an oxygen. The acidic hydrogen in compound *BH* is bound to a nitrogen. The difference in acidities of *BH* and the other two compounds is therefore due primarily to the *element effect* along the first row of the periodic table. This effect predicts that the O—H group should be more acidic than a comparably substituted N—H group because oxygen is more electron-attracting than nitrogen. Thus, the acidities of both *AH* and *CH* are greater than the acidity of *BH*. The difference in the acidities of *AH* and *CH* is due to the *polar effect* of the $\overset{+}{H_3N}$— group in compound *CH* on the acidity of its carboxylic acid group. The full positive charge on the nitrogen has a favorable interaction with the negatively charged carboxylate oxygen that should, by Fig. 3.2, enhance the acidity of *CH*. Hence, the final order of acidity is *CH* > *AH* > *BH*. Since stronger acids have weaker conjugate bases, the basicity order of the conjugate bases is *C* < *A* < *B*.

Element effects and polar effects are but two of many known effects on acidity. And acid-base reactions represent but one type of chemical reaction (albeit a very important one). However, the *process* of analyzing these effects that has been used here—going from *structures* to *energies* to *chemical behavior*—is used throughout organic chemistry. Hence, it is very important that this process be thoroughly understood.

PROBLEMS

3.17   In each of the following sets, arrange the compounds in order of decreasing pK_a, and explain your reasoning.

*(a) $ClCH_2CH_2SH$, $ClCH_2CH_2OH$, CH_3CH_2OH

(b)

$$CH_3O—CH_2—\overset{\displaystyle O}{\overset{\|}{C}}—OH, \qquad CH_3—\overset{\displaystyle O}{\overset{\|}{C}}—OH, \qquad \underset{\underset{\displaystyle OCH_3}{|}}{CH_2}—\underset{\underset{\displaystyle OCH_3}{|}}{CH}—\overset{\displaystyle O}{\overset{\|}{C}}—OH$$

3.18   Calculate the standard free energy for dissociation of
*(a) fluoroacetic acid (pK_a = 2.66)
(b) acetic acid (pK_a = 4.76)

***3.19**   Rationalize your answer to the previous problem by explaining why more energy is required to ionize acetic acid than fluoroacetic acid.

KEY IDEAS IN CHAPTER 3

A compound is a Lewis acid (or electrophile) when it reacts by accepting an electron pair; a compound is a Lewis base (or nucleophile) when it reacts by donating an electron pair.

Electron-deficient compounds contain an atom that is short of an octet by one or more electron pairs.

Electron-deficient compounds react as Lewis acids with Lewis bases in Lewis acid-base association reactions; the reverse of a Lewis acid-base association reaction is a Lewis acid-base dissociation.

When a Lewis base attacks an atom that is not electron-deficient, an electron pair must also depart from the atom undergoing attack. The resulting reaction is an electron-pair displacement reaction.

All Lewis acid-base reactions involve either the reactions of Lewis bases with electron-deficient compounds, or electron-pair displacements.

The curved-arrow formalism is an important logical symbolism for depicting the flow of electron pairs in chemical reactions. The reaction of a Lewis base with an electron-deficient compound requires one curved arrow; an electron-pair displacement reaction requires two.

The curved-arrow formalism can also be used to derive resonance structures that are related by the movement of one or more electron pairs.

A compound is a Brønsted acid when it reacts by donating a proton. A compound is a Brønsted base when it reacts by accepting a proton. Brønsted acid-base reactions are electron-pair displacement reactions that occur by attack of an electron pair on a proton.

The strength of a Brønsted acid is indicated by the magnitude of its dissociation constant K_a. Because dissociation constants for various acids can differ by many orders of magnitude, a logarithmic pK_a scale is used, in which $pK_a = -\log K_a$. The strength of a Brønsted base is inferred from the K_a (or pK_a) of its conjugate acid.

The equilibrium constant K_{eq} for a reaction is related to the standard free-energy difference $\Delta G°$ between products and reactants by the relationship $\Delta G° = -2.3RT \log K_{eq}$. Reactions with $K_{eq} < 1$ have positive $\Delta G°$ values, and favor reactants at equilibrium. Reactions with $K_{eq} > 1$ have negative $\Delta G°$ values and favor products at equilibrium.

Acidity, basicity, and other chemical properties vary with structure. Two structural effects on Brønsted acidity are the *element effect* and the *polar effect*. The element effect is dominated by the change in bond dissociation energies within a column (group) of the periodic table and by the change in electron affinities (or electronegativities) within a row (period) of the periodic table. The polar effect is caused largely by the interaction of charges formed in the acid-base reaction with polar bonds or other charged groups in the acid or base molecule.

The process for analyzing the effect of structure on acidity is to assess the effect of structure on energy, and then to consider how the resulting energy (viewed as a free energy ΔG_a°) affects the pK_a.

ADDITIONAL PROBLEMS

*3.20   Which of the following are electron-deficient compounds? Explain.

(a)

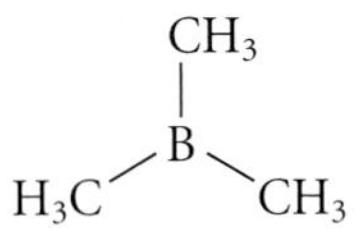

$$H_3C \overset{\overset{\textstyle CH_3}{|}}{\underset{}{B}} \diagdown CH_3$$

(b)

$$CH_3 = \overset{\overset{\textstyle CH_3}{|}}{\underset{\underset{\textstyle CH_3}{|}}{B}} - CH_3$$

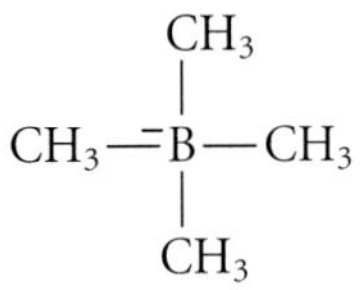

(c) :CH$_2$

3.21   Which of the following are electron-deficient compounds? Explain.

(a) CH$_3$—$\overset{+}{\ddot{N}}$H$_3$ (b) 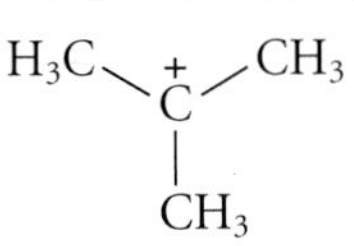 $H_3C \diagdown \overset{+}{\underset{\underset{\textstyle CH_3}{|}}{C}} \diagup CH_3$ (c) CH$_3$—$\underset{+}{\overset{\cdot\cdot}{\ddot{N}}H}$

3.22   Give the curved-arrow formalism for, and predict the immediate product of, each of the following reactions. Each involves an electron-deficient Lewis acid and a Lewis base.

*(a) CH$_3$—$\ddot{\underset{\cdot\cdot}{O}}$—CH$_3$ + BF$_3$ $\longrightarrow$ (b) 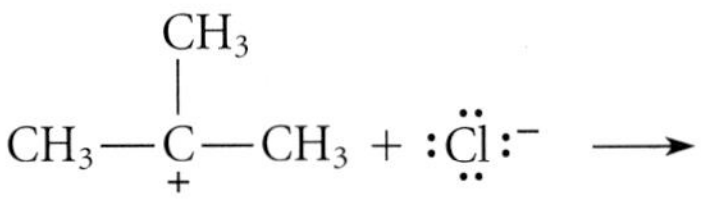 $CH_3 - \overset{\overset{\textstyle CH_3}{|}}{\underset{\underset{\textstyle +}{|}}{C}} - CH_3 + :\ddot{\underset{\cdot\cdot}{Cl}}:^- \longrightarrow$

*(c)

$$CH_3 - Mg - \ddot{\underset{\cdot\cdot}{Cl}}: \; + \; 2 \quad \underset{H_2C \diagdown}{\overset{H_2C \diagup}{\underset{\diagdown CH_2}{\overset{CH_2 \diagdown}{|}}}} \ddot{O}: \longrightarrow$$

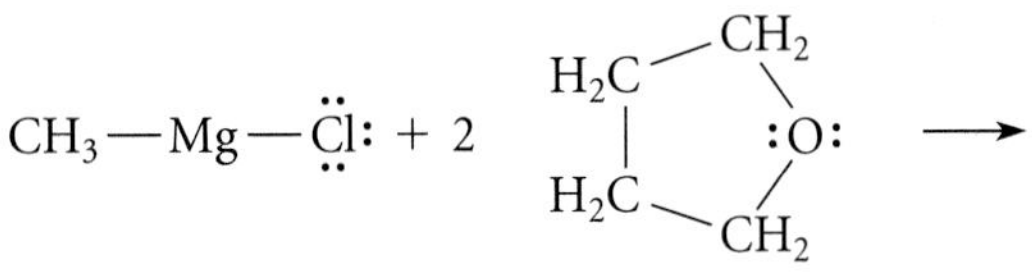

(d)

$$CH_3 - \overset{\overset{\textstyle :\ddot{O}H}{|}}{\underset{+}{C}} - CH_3 + CH_3 - \ddot{\underset{\cdot\cdot}{N}}H_2 \longrightarrow$$

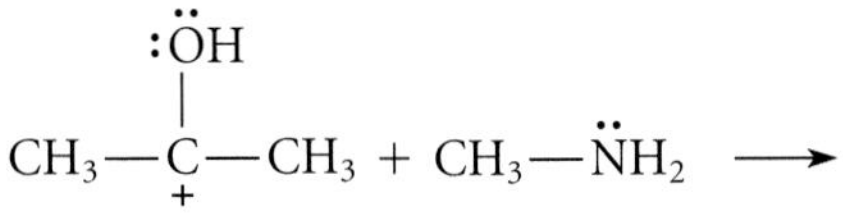

(Problem 3.22 continues)

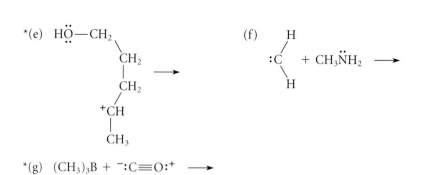

*(e) HÖ—CH₂
              \
               CH₂
               |
               CH₂
              /
            ⁺CH
              |
             CH₃   ⟶

(f)   H
             /
          :C   + CH₃ṄH₂   ⟶
             \
              H

*(g) (CH₃)₃B + ⁻:C≡O:⁺   ⟶

3.23   In each of the following Brønsted acid-base reactions, label the conjugate acid-base pairs. Then give the curved-arrow formalism for each reaction.

*(a)   :O: :O:
              ‖   ‖
      CH₃—C—Ö—H + ⁻:ÖH   ⟶   CH₃—C—Ö:⁻ + H₂Ö

(b)   O   :O:
              ‖   ‖
      CH₃—C—Ö—H + H₂Ö   ⟶   CH₃—C—Ö:⁻ + H₃Ö⁺

*(c)   :Ö—H   ⁻:Ö: :Ö⁻   H—Ö:
            /   \   /   \
      O=C   C=O   ⟶   O=C   C=O
            \   /   \   /
      CH₃—CH—CH₂—CH₂   CH₃—CH—CH₂—CH₂

(d) (CH₃)₃N—H + CH₃Ö:⁻   ⟶   (CH₃)₃N: + CH₃Ö—H
            ⁺

*(e) CH₃   CH₃
        \   \⁺
         C=CH₂ + H₃Ö⁺   ⟶   C—CH₃ + H₂Ö:
        /   /
      CH₃   CH₃

(f)

      :O: :O:
      ‖   ‖
  CH₃—C—Ö—H + H₂C̈—N̈≡N: ⟶   CH₃—C—Ö:⁻ + CH₃—N̈≡N:
              ⁻ ⁺   ⁺

*3.24   The conversion of alcohols into alkenes, a process called *dehydration*, takes place as a succession of three simple acid-base reactions, shown on the next page.
    (a) Classify each reaction step with one or more of the following terms:
        (1) a Lewis acid-base reaction
        (2) an association reaction of a Lewis base with an electron-deficient Lewis acid
        (3) a Lewis acid-base dissociation reaction
        (4) an electron-pair displacement reaction
        (5) a Brønsted acid-base reaction

(b) If the step is a Brønsted acid-base reaction, indicate the conjugate acid-base pairs.

(c) Draw the curved-arrow formalism for each step in the left-to-right direction.

Step 1

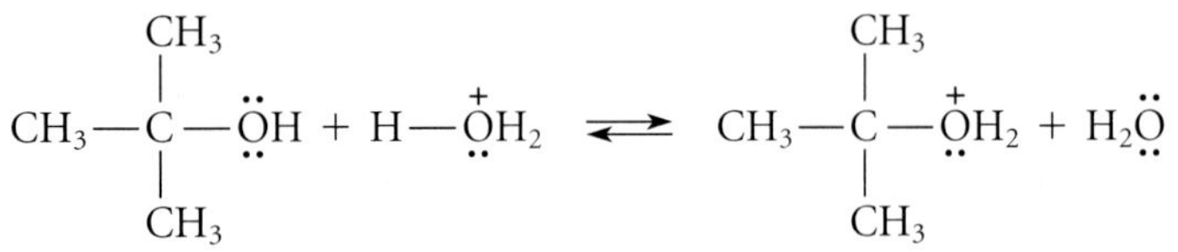

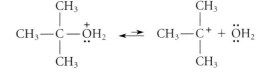

Step 2

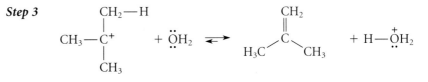

Step 3

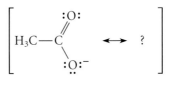

3.25 *(a) The two C—O bonds in the acetate ion (the conjugate base of acetic acid) are equivalent, and the negative charge is shared equally on each oxygen. Use the curved-arrow formalism to draw a resonance structure which, taken together with the structure below, conveys this idea.

$$\left[H_3C-C\overset{\displaystyle :O:}{\underset{\displaystyle :O:^-}{}} \quad \longleftrightarrow \quad ? \right]$$

acetate ion

(b) The resonance structures of carbon monoxide are shown below. Show how each structure can be converted into the other using the arrow formalism.

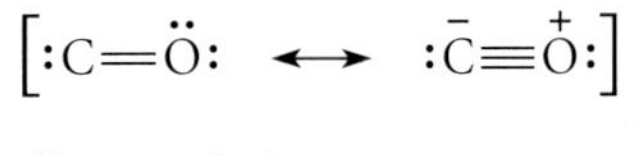

$$\left[:C\!\!=\!\!\ddot{O}: \quad \longleftrightarrow \quad :\bar{C}\!\!\equiv\!\!\overset{+}{O}: \right]$$

3.26 Use the curved-arrow formalism to derive resonance structures that convey the following ideas.

*(a) The outer oxygens of ozone, $:\!\ddot{O}\!\!=\!\!\overset{+}{\underset{..}{O}}\!\!-\!\!\ddot{\underset{..}{O}}\!:^-$, have an equal amount of negative charge.

(b) All C—O bonds in the carbonate ion are of equal length.

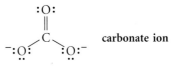

carbonate ion

*(c) The conjugate acid of formaldehyde, $CH_2\!\!=\!\!\overset{+}{\underset{..}{O}}\!\!-\!\!H$, has substantial positive charge on carbon.

(Problem 3.26 continues)

(d) The inner carbon of acetonitrile oxide, $CH_3—C{\equiv}\overset{+}{N}—\overset{..}{\underset{..}{O}}{:}^-$, has some positive charge.

3.27 Draw the products of each of the following reactions indicated by the curved-arrow formalism.

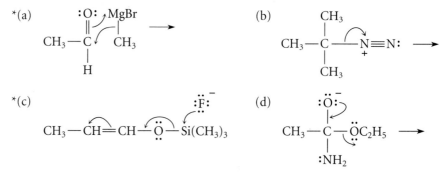

*(a)   (b)

*(c)   (d)

3.28 Draw a curved-arrow formalism that indicates the flow of electrons in each of the following transformations.

*(a)
$$:\overset{..}{\underset{..}{Br}}—CH_2—CH_2—\overset{\overset{\textstyle :O:}{\|}}{C}—\overset{..}{\underset{..}{O}}{:}^- \longrightarrow :\overset{..}{\underset{..}{Br}}{:}^- + CH_2{=}CH_2 + :\overset{..}{\underset{..}{O}}{=}C{=}\overset{..}{\underset{..}{O}}{:}$$

(b)
$$(CH_3)_2\overset{..}{N}H + CH_2{=}CH—\overset{\overset{\textstyle :O:}{\|}}{C}—\overset{..}{\underset{..}{O}}C_2H_5 \longrightarrow (CH_3)_2\overset{+}{N}H—CH_2—CH{=}\overset{\overset{\textstyle :\overset{..}{\underset{..}{O}}{:}^-}{|}}{C}—\overset{..}{\underset{..}{O}}C_2H_5$$

*(c)
$$\underset{\underset{\textstyle :\overset{..}{\underset{..}{S}}{:}^-}{|}}{CH_2}—CH_2—CH_2—CH_2—\overset{..}{\underset{..}{C}}l{:} \longrightarrow \begin{matrix} H_2C—CH_2 \\ / \qquad \backslash \\ H_2C \qquad\quad CH_2 \\ \backslash \qquad / \\ :\overset{..}{S}{:} \end{matrix} + :\overset{..}{\underset{..}{C}}l{:}^-$$

(d)
$$\begin{matrix} :O: \\ / \quad \backslash \\ H_2C{—}CH_2 \end{matrix} + :\overset{-}{C}N \longrightarrow \overset{\overset{\textstyle {}^-:\overset{..}{\underset{..}{O}}{:}}{|}}{CH_2}—CH_2—CN$$

*(e)
$$CH_3\overset{..}{N}H_2 + CH_2{=}CH—\underset{\underset{\textstyle CH_3}{|}}{CH}—\overset{..}{\underset{..}{C}}l{:} \longrightarrow CH_3\overset{+}{N}H_2—CH_2—CH{=}\underset{\underset{\textstyle CH_3}{|}}{CH} \quad :\overset{..}{\underset{..}{C}}l{:}^-$$

(f)
$$\underset{\underset{\textstyle :\overset{..}{\underset{..}{Br}}{:}}{|}}{CH_2}—CH_2—Ph + {}^-:\overset{..}{\underset{..}{O}}CH_3 \longrightarrow CH_2{=}CH—Ph + :\overset{..}{\underset{..}{Br}}{:}^- + H—\overset{..}{\underset{..}{O}}CH_3$$

3.29 The following examples of *incorrect* curved-arrow formalism were found in the notebooks of Barney Bottlebrusher, a student who was known to have difficulty with organic chemistry. Explain what is wrong in each case.

*(a)

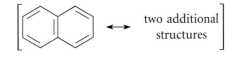

(b) $CH_3\ddot{O}:$ $CH_3—\ddot{B}r:$ $\longrightarrow$ $CH_3—\ddot{O}—CH_3 + :\ddot{B}r\overline{:}$

3.30   *(a) Naphthalene can be described by two resonance structures in addition to the structure below. Derive these structures with the curved-arrow formalism.

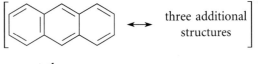

naphthalene

(b) Derive three additional resonance structures for anthracene using the curved-arrow formalism.

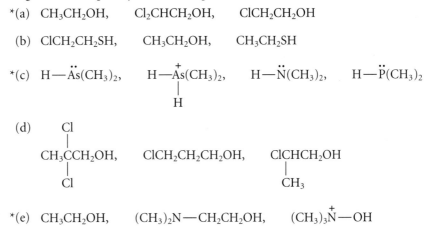

anthracene

3.31   *(a) If the standard free-energy difference between *anti* butane and either one of the *gauche* conformations is 3.8 kJ/mol (0.9 kcal/mol), calculate the amounts of *gauche* and *anti* butane present in equilibrium in one mole of butane at 25 °C. (Remember that there are two *gauche* conformations.)

(b) The standard free-energy difference between 2,2-dimethylpropane and pentane is 6.86 kJ/mol (1.64 kcal/mol); 2,2-dimethylpropane is the more stable compound. If the two were present in an equilibrium mixture, what would be the percentage of each in the mixture at 25 °C?

3.32   Arrange the compounds in each of the following sets in order of decreasing pK_a, highest first. Explain your reasoning.

*(a) CH_3CH_2OH, Cl_2CHCH_2OH, $ClCH_2CH_2OH$

(b) $ClCH_2CH_2SH$, CH_3CH_2OH, CH_3CH_2SH

*(c) $H—\ddot{A}s(CH_3)_2$, $H—\overset{+}{A}s(CH_3)_2$, $H—\ddot{N}(CH_3)_2$, $H—\ddot{P}(CH_3)_2$
                             |
                             H

(d)   Cl
           |
    CH_3CCH_2OH, $ClCH_2CH_2CH_2OH$, $ClCHCH_2OH$
           | |
           Cl CH_3

*(e) CH_3CH_2OH, $(CH_3)_2N—CH_2CH_2OH$, $(CH_3)_3\overset{+}{N}—OH$

3.33   Using data in Table 3.1 as well as the data below, estimate the equilibrium constants for each of the following reactions at 25 °C.

*(a) $(CH_3)_3N + H{-}CN \rightleftharpoons (CH_3)_3\overset{+}{N}H + {}^-CN$

$$pK_a = 9.76$$

(b) $CH_3CH_2S{-}H + {}^-OH \rightleftharpoons CH_3CH_2S^- + H_2O$

$$pK_a = 10.5$$

3.34   *(a)   What is the standard free-energy change at 25 °C for reaction (a) in Problem 3.33?

(b)   What is the standard free-energy change at 25 °C for reaction (b) in Problem 3.33?

*(c)   In reaction (a) of Problem 3.33, how much of each species will be present at equilibrium if the initial concentrations of $(CH_3)_3N$ and HCN are both zero, and $(CH_3)_3\overset{+}{N}H$ and ^-CN are present initially at concentrations of 0.1 *M*?

3.35   Phenylacetic acid has a pK_a of 4.31; acetic acid has a pK_a of 4.76.

O   O
‖   ‖
Ph—CH₂—C—O—H CH₃—C—O—H

phenylacetic acid   **acetic acid**

*(a)   Which acid has the more favorable (smaller) standard free energy of dissociation?

(b)   What free energy would be expended to dissociate a 1 *M* solution of phenylacetic acid completely into 1 *M* H_3O^+ and 1 *M* phenylacetate ion?

*(c)   According to the pK_a data, which type of polar effect is characteristic of the phenyl group: an electron-withdrawing polar effect, or an electron-donating polar effect? Explain your reasoning.

*3.36**   Malonic acid has two carboxylic acid groups, and consequently undergoes two ionization reactions. The pK_a for the first ionization of malonic acid is 2.86; the pK_a for the second is 5.70. The pK_a of acetic acid is 4.76.

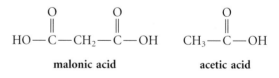

malonic acid   **acetic acid**

(a)   Write out the equations for the first and second ionizations of malonic acid and label each with the appropriate pK_a value.

(b)   Why is it that the first pK_a of malonic acid is much *lower* than the pK_a of acetic acid, but the second pK_a of malonic acid is much *higher* than the pK_a of acetic acid?

(c) Consider a series of dicarboxylic acids given by the following general structure:

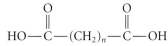

How would you expect the *difference* between the first and second pK_a values to change as n increases? Explain. (*Hint:* Look at the denominator of the electrostatic law, Eq. 3.40.)

*3.37   Which of the following two reactions would have an equilibrium constant most favorable to the right? Explain your answer.

(1)

$$CH_3-\underset{\underset{CH_3}{|}}{\overset{\overset{CH_3}{|}}{C}}-OH + H_3O^+ \;\rightleftharpoons\; CH_3-\underset{\underset{CH_3}{|}}{\overset{\overset{CH_3}{|}}{C}}{}^+ + 2\,H_2O$$

(2)

$$CF_3-\underset{\underset{CH_3}{|}}{\overset{\overset{CH_3}{|}}{C}}-OH + H_3O^+ \;\rightleftharpoons\; CF_3-\underset{\underset{CH_3}{|}}{\overset{\overset{CH_3}{|}}{C}}{}^+ + 2\,H_2O$$

3.38   From Figure 3.2, how would a structural effect that destabilizes the *acid* form of a conjugate acid-base pair affect its acidity? Use your analysis to predict which of the following two compounds is more basic.

$$Cl-CH_2CH_2-\overset{..}{N}H_2 \quad \text{and} \quad CH_3CH_2-\overset{..}{N}H_2$$

*3.39   Let the equilibrium constant for the following acid-base reaction be K_{eq}.

$$HA + B^- \;\rightleftharpoons\; A^- + HB$$

Let K_{HA} be the dissociation constant of HA, and K_{HB} be the dissociation constant of HB. Justify the procedure used in Study Problem 3.5 to calculate the equilibrium constant for Eq. 3.22 by showing that

$$K_{eq} = 10^{(pK_{HB} - pK_{HA})}$$

(*Hint:* Show first that $K_{eq} = K_{HA}/K_{HB}$.)

3.40   *(a) The acid HI is considerably stronger than HCl (see Table 3.1). Why, then, does a $10^{-3}\ M$ aqueous solution of either acid in water give the same pH reading of about 3?

(b) Potassium amide, $K^+\ \overset{..}{:}NH_2$, whose conjugate acid has a pK_a of 35, is a much stronger base than potassium *tert*-butoxide, $K^+\ (CH_3)_3C-\overset{..}{\underset{..}{O}}{:}^-$, whose conjugate acid has a pK_a of 19. Yet a $10^{-3}\ M$ solution of either compound in water has an identical pH reading of about 11. Explain.

*3.41   The borohydride anion reacts with water in the following way:

$$\overset{-}{B}H_4 \;\; + \;\; H_2\overset{..}{\underset{..}{O}} \longrightarrow H\overset{..}{\underset{..}{O}}\!\!-\!\!\overset{-}{B}H_3 + H_2$$

**borohydride
anion**

This reaction occurs in a sequence of two steps, the first of which is shown in Eq. 3.6. Write out both steps and give the curved-arrow formalism for each.

*3.42   Astatine (At) is the radioactive halogen that lies below iodine in the periodic table. How would you expect the following properties to compare (greater or less)?
(a) bond dissociation energy of H—At *vs.* that of HI
(b) electron affinity of At *vs.* that of I
(c) dissociation constant of H—At *vs.* that of HI

4

Introduction to Alkenes; Reaction Rates

Alkenes are hydrocarbons that contain carbon-carbon double bonds. Alkenes are sometimes called **olefins**, particularly in older literature. *Ethylene* is the simplest alkene.

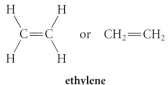

ethylene

Because compounds containing double or triple bonds have fewer hydrogens than the corresponding alkanes, they are classified as **unsaturated hydrocarbons**, in contrast to alkanes, which are classified as **saturated hydrocarbons**.

This chapter covers the structure, bonding, nomenclature, and physical properties of alkenes. Then, using a few alkene reactions, some of the physical principles are discussed that are important in understanding the reactions of organic compounds in general.

4.1 Structure and Bonding in Alkenes

The double-bond geometry of ethylene is typical of that found in other alkenes. Ethylene follows the rules for predicting molecular geometry (Sec. 1.3B), which suggest that each carbon of ethylene should have trigonal-planar geometry; that is, all the atoms surrounding each carbon lie in one plane with bond angles approximating 120°. The experimentally determined structure of ethylene agrees with these expectations, and shows further that ethylene is a planar molecule. For alkenes in general, the carbons of a double bond and the atoms attached to them all lie in the same plane.

Models of ethylene are shown in Fig. 4.1, and a comparison of the geometries of ethylene and propene with those of ethane and propane is given in Fig. 4.2. Notice that the carbon-carbon double bonds of ethylene and propene are shorter than the carbon-

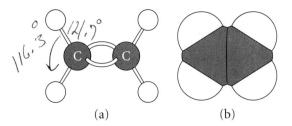

Figure 4.1   *Models of ethylene. (a) A ball-and-stick model. (b) A space-filling model. Notice that ethylene is a planar molecule.*

carbon single bonds of ethane and propane. This illustrates the relationship of bond length and bond order (Sec. 1.3B): double bonds are shorter than single bonds between the same atoms.

Another feature of alkene structure comes from comparing the structures of propene and propane in Fig. 4.2. Notice that the carbon-carbon *single* bond of propene is shorter than the carbon-carbon *single* bonds of propane. The shortening of all these bonds is a consequence of the particular way that carbon atoms are hybridized in alkenes.

A. Carbon Hybridization in Alkenes

The carbons of an alkene double bond are hybridized differently from those of an alkane. In this hybridization (Fig. 4.3), the carbon 2s orbital is mixed, or hybridized, with only

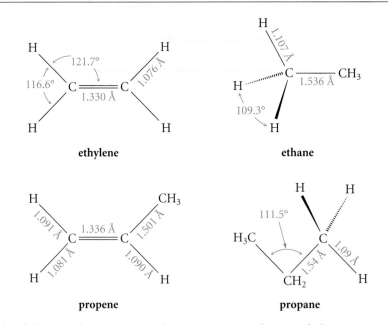

Figure 4.2   *Structures of ethylene, ethane, propene, and propane. Compare the trigonal-planar geometry of ethylene (bond angles near 120°) with the tetrahedral geometry of ethane (bond angles near 109.5°). Notice that all carbon-carbon double bonds are shorter than carbon-carbon single bonds. Notice also that the carbon-carbon single bond in propene is somewhat shorter than the carbon-carbon bonds of propane.*

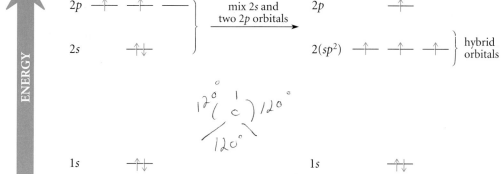

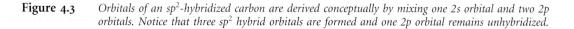

Figure 4.3   *Orbitals of an sp²-hybridized carbon are derived conceptually by mixing one 2s orbital and two 2p orbitals. Notice that three sp² hybrid orbitals are formed and one 2p orbital remains unhybridized.*

two of the three available $2p$ orbitals. Because three orbitals are mixed, the result is three hybrid orbitals. Each has one part s character and two parts p character. These hybrid orbitals are called sp^2 (pronounced s-p-two) orbitals, and the carbon is said to be sp^2-hybridized. The shape of an individual sp^2 orbital (Fig. 4.4a) is much like that of an sp^3 orbital. A diagram of the orbitals of an sp^2-hybridized carbon (Fig. 4.4b) shows that the axes of the three sp^2 orbitals are disposed in the same plane at mutual angles of 120°. Thus, an sp^2-hybridized carbon atom bears three sp^2-hybrid orbitals and one p orbital. The axis of the p orbital is perpendicular to the plane containing the sp^2 orbitals.

The carbon hybridization in alkenes is another example of the idea that *hybridization and geometry are related* (Sec. 1.9). From the previous discussion you've learned that the bonding geometry of an sp^2-hybridized carbon atom is trigonal planar. Whenever a main-group atom has trigonal-planar geometry, its hybridization is sp^2. Whenever such an atom has tetrahedral geometry, its hybridization is sp^3.

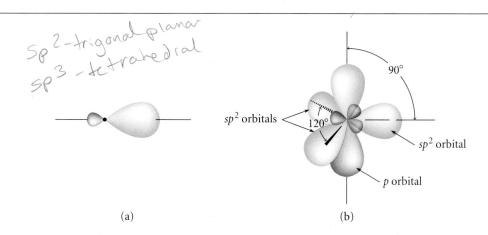

(a)   (b)

Figure 4.4   *(a) The general shape of an sp² hybrid orbital is very similar to that of an sp³ hybrid orbital, with a large and small lobe of electron density separated by a node. (b) Spatial distribution of orbitals on an sp²-hybridized carbon atom. Notice that the axes of the three sp² orbitals lie in the same plane at angles of 120°, and the axis of the p orbital is perpendicular to this plane.*

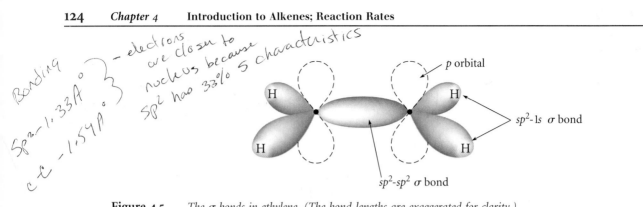

[Handwritten margin notes: Bonding / Sp² ~ 1.33Å / c-c ~ 1.54Å / electrons are closer to nucleus because sp² has 33% s characteristics]

[Handwritten margin notes (left side): C-H sp² - 1s 6 / -two remaining sp² / orbitals on each / Carbon overlap with / a hydrogen 1s orbital / π molecular orbital - additive / overlap of two carbon 2p / orbitals (low σ energy) / π molecular orbital - subtractive / overlap of the two carbon 2p / orbitals (high σ energy)]*

Figure 4.5   *The σ bonds in ethylene. (The bond lengths are exaggerated for clarity.)*

Like an sp^3 orbital, an sp^2 orbital is highly directed in space. The difference between these two types of hybrid orbitals is that the electron density within an sp^2 orbital is somewhat closer to the nucleus. This difference is a consequence of the larger amount of s character in an sp^2 orbital. An sp^3 orbital has 25% s character (one s orbital mixed with three p orbitals), but an sp^2 orbital has 33% s character. Since s electrons, on the average, are closer to the nucleus than p electrons, electrons in hybrid orbitals with more s character also have a greater probability of being found closer to the nucleus.

Conceptually ethylene can be formed by the bonding of two sp^2-hybridized carbon atoms and four hydrogen atoms. An sp^2 orbital on one carbon overlaps with an sp^2 orbital on another to form a bonding molecular orbital containing two electrons—a carbon-carbon σ bond. (There are also corresponding antibonding orbitals that are not important here.) Each of the two remaining sp^2 orbitals on each carbon overlaps with a hydrogen 1s orbital to form a C—H sp^2-1s σ bond containing two electrons. These orbitals account for the four carbon-hydrogen bonds and *one* of the two carbon-carbon bonds of ethylene (Fig. 4.5).

The second carbon-carbon bond of ethylene comes from the overlap of the p orbitals on each carbon atom—the ones "left over" after hybridization. This overlap results in the formation of both bonding and antibonding molecular orbitals (Fig. 4.6; see also Problem 1.36, Chapter 1). These molecular orbitals are formed by additive and subtractive combinations of carbon 2p orbitals in much the same way that the molecular orbitals of the hydrogen molecule are formed by additive and subtractive combinations of hydrogen 1s orbitals (Sec. 1.8A). However, the overlap of p orbitals is not "head-to-head," as in a σ bond, but rather is "side-to-side." This type of bonding molecular orbital, which results from additive overlap of the two carbon 2p orbitals, is called a **π molecular orbital**. (Note that π is the Greek equivalent of p.) This molecular orbital, like the p orbitals from which it is formed, has a nodal plane (dashed lines in Fig. 4.6); this plane coincides with the plane of the ethylene molecule. The antibonding molecular orbital, which results from subtractive overlap of the two carbon 2p orbitals, is called a **π* molecular orbital**. It has two nodes. One of these nodes is a plane coinciding with the plane of the molecule, and the other is a plane between the two carbons, perpendicular to the plane of the molecule. The bonding (π) molecular orbital lies at lower energy than the isolated p orbitals, whereas the antibonding (π*) molecular orbital lies at higher energy. The two 2p electrons (one from each carbon), by the Aufbau principle, occupy the molecular orbital of lowest energy—the π molecular orbital. The result of this occupation is the second of the carbon-carbon bonds, called a **π bond**. The antibonding molecular orbital is unoccupied.

Surbital
P orbital — ○ — closer to nucleus

Unlike a σ bond, a π bond is not cylindrically symmetric about the line connecting the two nuclei. The π bond has electron density both above and below the plane of the ethylene molecule, with a wave peak on one side of the molecule, a wave trough on the other, and a node in the plane of the molecule. The π bond is *one bond* with two lobes, just as a *p* orbital is *one orbital* with two lobes. In this bonding picture, then, there are two types of carbon-carbon bonds: a σ bond, with most of its electron density relatively concentrated between the carbon atoms, and a π bond, with most of its electron density concentrated above and below the plane of the ethylene molecule.

This bonding picture accounts nicely for the planar conformation of ethylene. If the two CH_2 groups were twisted away from coplanarity, the *p* orbitals could not overlap to form the π bond. Thus, the overlap of the *p* orbitals and, consequently, the existence of the π bond *require* the planarity of the ethylene molecule.

This bonding picture also shows why the single bonds at an sp^2-hybridized carbon are somewhat shorter than those at an sp^3-hybridized carbon. The carbon-carbon single bond of propene, for example, is derived from the overlap of a carbon sp^3 orbital of the CH_3 group with a carbon sp^2 orbital of the alkene carbon. A carbon-carbon bond of propane is derived from the overlap of two carbon sp^3 orbitals. Because the electron density of sp^2 orbitals is somewhat closer to the nucleus, it follows that the sp^2-sp^3

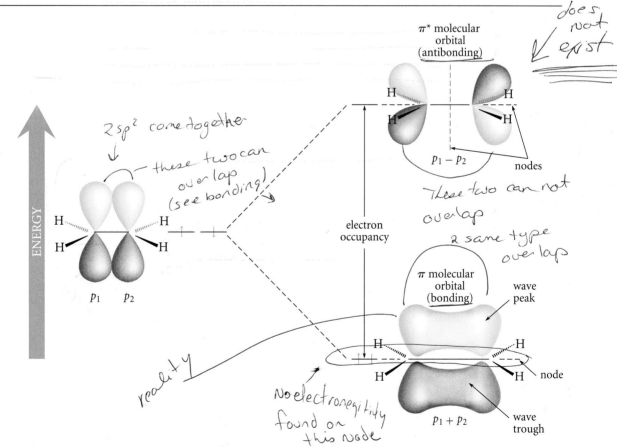

Figure 4.6   *Overlap of p orbitals to form bonding and antiboding π molecular orbitals of ethylene. The π bond is formed when two electrons occupy the bonding π molecular orbital. The nodes are planes perpendicular to the page and are indicated with dashed lines.*

σ bond of propene, with a length of 1.50 Å, is somewhat shorter than the sp^3-sp^3 σ bond of propane, with a length of 1.54 Å (Fig. 4.2).

B. *Cis-Trans* Isomerism

The alkenes with four carbons (the butenes) exist in isomeric forms. In the butenes with unbranched carbon chains, the double bond may be located either at the end or in the middle of the carbon chain.

$$CH_2{=}CH{-}CH_2{-}CH_3 \qquad CH_3{-}CH{=}CH{-}CH_3$$

1-butene **2-butene**

Isomeric alkenes such as these that differ in the position of their double bonds are examples of *constitutional isomers* (Sec. 2.4A).

The structure of 2-butene illustrates another important type of isomerism. It turns out that *there are two separable, distinct 2-butenes*. One has a boiling point of 3.7 °C; the other has a boiling point of 0.88 °C. In the compound with the higher boiling point, called *cis*-2-butene or (*Z*)-2-butene, the methyl groups are on the same side of the double bond. In the other 2-butene, called *trans*-2-butene, or (*E*)-2-butene, the methyl groups are on opposite sides of the double bond.

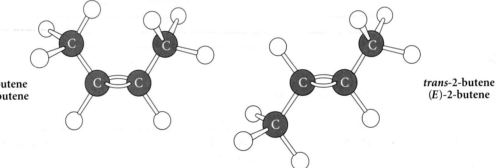

cis-2-butene
(*Z*)-2-butene

trans-2-butene
(*E*)-2-butene

These isomers have identical atomic *connectivities* (CH_3 connected to CH, CH doubly bonded to CH, CH connected to CH_3). Despite their identical connectivities, *the two compounds differ in the way their constituent atoms are arranged in space*. Compounds with identical connectivities that differ in the spatial arrangement of their atoms are called **stereoisomers**. *Cis-trans* (or *E,Z*) isomerism is only one type of stereoisomerism; other types will be considered in Chapter 6. The (*E*) and (*Z*) notation has been adopted by the IUPAC as a general way of naming *cis* and *trans* isomers, and is discussed in Sec. 4.2B.

The interconversion of *cis*- and *trans*-2-butene requires a 180° internal rotation about the double bond.

cis-2-butene $\xrightarrow[\text{rotation}]{180°}$ *trans*-2-butene   (4.1)

(interconversion does not occur at ordinary temperatures)

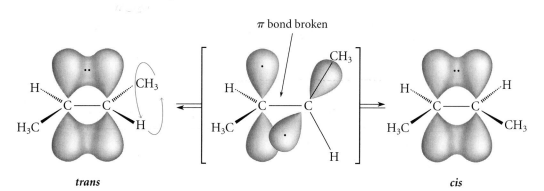

π bond broken

trans *cis*

Figure 4.7   *Internal rotation about the carbon-carbon double bond in an alkene requires breaking the π bond. This does not occur at ordinary temperatures because too much energy is required.*

Because *cis-* and *trans*-2-butene do not interconvert at ordinary temperatures, it follows that this rotation must be very slow. Why is this so? In order for such an internal rotation to occur, the *p* orbitals on each carbon must be twisted away from coplanarity; that is, *the π bond must be broken* (Fig. 4.7). Since bonding is energetically favorable, lack of it is energetically costly. It takes more energy to break the π bond than is available under normal conditions; thus, the π bond in alkenes remains intact and internal rotation about the double bond does not occur. In contrast, internal rotation about the carbon-carbon *single* bonds of ethane or butane is very rapid (Sec. 2.3) because no chemical bond is broken in the process.

How can you know whether an alkene can exist as *cis* and *trans* (or *E* and *Z*) stereoisomers? One way to tell is to mentally exchange the two groups attached to either carbon of the double bond. This process will give you one of two results: either the resulting molecule will be identical to the original—superimposable on the original atom-for-atom—or it will be different. If it is different, it can *only* be a stereoisomer (because its connectivity is the same). Let's illustrate with two cases.

In the following case, interchange of two groups at either carbon of the double bond gives different molecules, hence, stereoisomers.

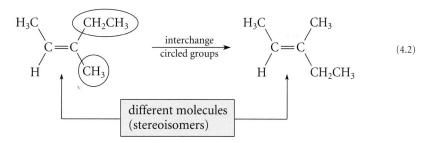

$$\text{(4.2)}$$

(Verify that interchange of the two groups at the other carbon of the double bond would give the same result.) In contrast, interchange of two groups attached to a carbon of the double bond in the following structure gives back the same molecule:

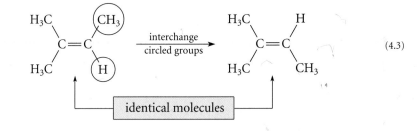

(4.3)

*Stereocenter -
an atom at which the two
the interchange of a stereoisomer
groups gives a stereoisomer*

STUDY GUIDE LINK:
✓ 4.1
*Different Ways to
Draw the Same
Structure*

The molecule on the right at first may not *look* identical to the one on the left, but it is; the two are just drawn differently. *If this is not clear, build a model of each molecule and show that the two can be superimposed atom-for-atom!*

When an alkene can exist as *cis* and *trans* (or *E* and *Z*) stereoisomers, both carbons of the double bond are stereocenters. A **stereocenter** is an atom at which the interchange of two groups gives a stereoisomer. (Another term that means the same thing is **stereogenic atom**.)

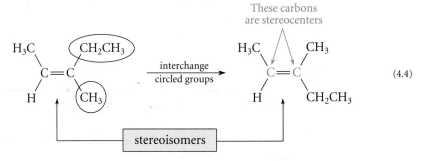

(4.4)

(Interchange of the CH_3 and the H at the other carbon of the double bond gives the same result; that is why both carbons of the double bond are stereocenters.) In contrast, the carbons of the double bond in the alkene of Eq. 4.3 are *not* stereocenters:

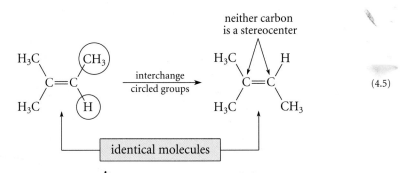

(4.5)

We'll show in Chapter 6 that *cis-trans* isomers are not the only type of stereoisomers. In every set of stereoisomers we encounter, we'll be able to identify one or more stereocenters that can be used to generate the set of stereoisomers by the interchange of attached groups.

PROBLEM

4.1   Which of the following alkenes can exist as *cis-trans* (or *E,Z*) isomers? Identify the stereocenters in each.

*(a) $CH_2{=}CHCH_2CH_2CH_3$ (**1-pentene**)

(b) $CH_3CH_2CH{=}CHCH_2CH_3$ (**3-hexene**)

*(c) $CH_2{=}CH{-}CH{=}CH{-}CH_3$ (**1,3-pentadiene**)

(d) $CH_3CH_2CH{=}CCH_3$ (**2-methyl-2-pentene**)
$$\qquad\qquad\quad |$$
$$\qquad\qquad\ CH_3$$

*(e)   ⬜   (**cyclobutene**)

(*Hint*: Try to build a model of both stereoisomers, but don't break your models!)

4.2   Nomenclature of Alkenes

A.   IUPAC Substitutive Nomenclature

The IUPAC substitutive nomenclature of alkenes is derived by modifying alkane nomenclature in a simple way. An unbranched alkene is named by replacing the *ane* suffix in the name of the corresponding alkane with the ending *ene*. The carbons are numbered from one end to the other so that the double bond receives the lowest number.

$$\overset{1}{C}H_2{=}\overset{2}{C}H{-}\overset{3}{C}H_2\overset{4}{C}H_2\overset{5}{C}H_2\overset{6}{C}H_3 \qquad \textbf{1-hexene}$$

hexane + ene = hexene   ↑ position of double bond

The IUPAC recognizes an exception to this rule for the name of the simplest alkene, $CH_2{=}CH_2$, which is usually called ethylene rather than ethene.

The names of alkenes with branched chains are, like those of alkanes, derived from their *principal chains*. In an alkene, the principal chain is defined as *the carbon chain containing the greatest number of double bonds*, even if this is not the longest chain. If more than one candidate for the principal chain has equal numbers of double bonds, the principal chain is the longest of these. The principal chain is numbered from the end so as to give the lowest numbers to double bonds at the first point of difference. (The meaning of "the first point of difference" was discussed in Sec. 2.4C.)

When the alkene contains an alkyl substituent in addition to the principal chain, it is the position of the double bond, not the position of the branch, that determines the numbering of the chain. The position of the double bond is cited in the name *after* the name of the alkyl group.

STUDY PROBLEM 4.1

Name the following compound using IUPAC substitutive nomenclature.

$$CH_2{=}C{-}CH_2CH_2CH_3$$
$$\qquad\ \ |$$
$$\quad CH_2CH_2CH_2CH_2CH_3$$

Solution   The principal chain is the longest continuous carbon chain containing *both carbons* of the double bond, as shown in color below. Note that this is *not* the longest carbon chain in the molecule. The substituent group is a propyl group. Hence, the name of the compound is 2-propyl-1-heptene:

position of substituent group

position of double bond

$$CH_2{=}\overset{1}{C}{-}\overset{2}{C}H_2CH_2CH_3$$

$$\underset{3\quad 4\quad 5\quad 6\quad 7}{CH_2CH_2CH_2CH_2CH_3}$$

2-propyl-1-heptene

⟵ principal chain (longest chain containing double bond; double bond receives lowest number)

✎ If a compound contains more than one double bond, the *ane* ending of the corresponding alkane is replaced by *adiene* (if there are two double bonds), *atriene* (if there are three double bonds), etc.

$$CH_2{=}CHCH_2CH_2CH{=}CH_2 \qquad \textbf{1,5-hexadiene}$$

STUDY PROBLEM 4.2

Name the following compound:

$$\begin{array}{c} CH_2{-}CH{=}CH_2 \\ | \\ CH_3CH_2CH_2CH_2{-}C{=}CH{-}CH_3 \end{array}$$

(handwritten: 1,4 hexadiene · 4 butyl)

Solution   The principal chain (color in the structure below) is the chain containing the greatest number of double bonds. The compound is a 1,4-hexadiene, with a butyl branch at carbon-4:

(handwritten: 4butyl-1,4 hexadiene)

$$\begin{array}{c} \overset{3}{C}H_2{-}\overset{2}{C}H{=}\overset{1}{C}H_2 \\ | \\ CH_3CH_2CH_2CH_2{-}\underset{4}{C}{=}\underset{5}{C}H{-}\underset{6}{C}H_3 \end{array}$$

⟵ principal chain

4-butyl-1,4-hexadiene

number of double bonds

positions of double bonds

position of substituent

If the name remains ambiguous after determining the correct numbers for the double bonds, then the principal chain is numbered so that the lowest numbers are given to the branches at the first point of difference.

STUDY PROBLEM 4.3

Name the following compound:

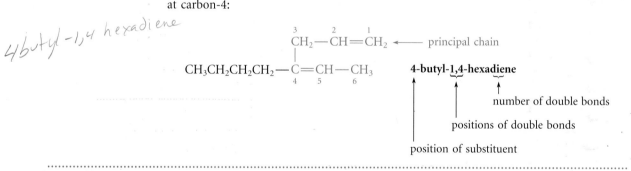

(handwritten annotations: 1,6dimethyl cyclohexene · 2,3dimethyl cyclohexene · 2,3 dimethyl 1 cyclohexene · 1,2dimethyl cyclohexene)

Solution Two ways of numbering this compound give the double bond the number 1.

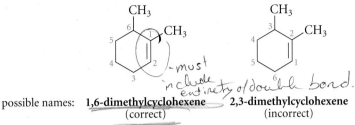

possible names: **1,6-dimethylcyclohexene** **2,3-dimethylcyclohexene**
(correct) (incorrect)

In this situation choose the numbering scheme that gives the lowest number for the methyl substituents at the first point of difference. In comparing the substituent numbering schemes (1,6) with (2,3), the first point of difference occurs at the first number (1 *vs.* 2). The (1,6) numbering scheme is correct because 1 is lower than 2. Notice that the number 1 for the double bond is not given explicitly in the name, because this is the only possible number. That is, when the double bond in a ring receives numerical priority, its carbons *must* receive the numbers 1 and 2.

Substituent groups may also contain double bonds. Some widely occurring groups of this type have special names that must be learned:

CH_2=CH— **vinyl** CH_2=CH—CH_2— **allyl** CH_2=C— **isopropenyl**
CH_3

Other substituent groups are numbered *from the point of attachment to the principal chain.*

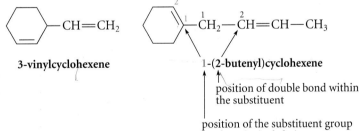

3-vinylcyclohexene **1-(2-butenyl)cyclohexene**

position of double bond within the substituent

position of the substituent group on the principal chain

The names of these groups, like the names of ordinary alkyl groups, are constructed from the name of the parent hydrocarbon by dropping the final *e* and replacing it with *yl.* Thus, the substituent in the second example above is buten*e* + yl = butenyl. Notice the use of parentheses to set off the names of substituents with internal numbering.

Finally, some alkenes have nonsystematic traditional names that are recognized by the IUPAC. These can be learned as they are encountered. Two examples are styrene and isoprene:

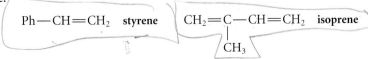

Ph—CH=CH_2 **styrene** CH_2=C—CH=CH_2 **isoprene**
CH_3

(Recall from Sec. 2.9 that Ph— refers to the phenyl group, a singly substituted benzene ring.)

PROBLEMS

4.2   Give the structure of each of the following:
*(a)   1-isopropenylcyclopentene   (b)   4-methyl-1,3-hexadiene
*(c)   5-(3-pentenyl)-1,3,6,8-decatetraene   (d)   3-ethyl-1,4-cyclohexadiene

4.3   Name the following compounds:

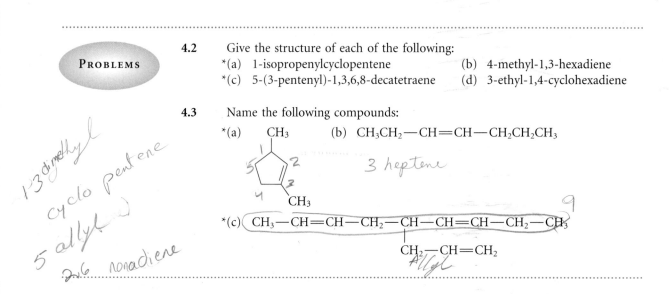

*(a)   CH₃   (b) $CH_3CH_2-CH=CH-CH_2CH_2CH_3$

*1,3 dimethyl
cyclo pentene*

5 allyl

2,6 nonadiene

3 heptene

(c) $CH_3-CH=CH-CH_2-CH-CH=CH-CH_2-CH_3$

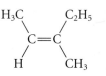

$CH_2-CH=CH_2$

Allyl

B. Nomenclature of Stereoisomers: the *E,Z* System

The *cis* and *trans* designations for stereoisomers, although no longer officially sanctioned by the IUPAC, continue to be widely used. They are unambiguous when each carbon of a double bond has a single hydrogen, as in *cis*- and *trans*-2-butene. However, in some important situations, the use of the terms *cis* and *trans* is ambiguous. For example, is the following compound, a stereoisomer of 3-methyl-2-pentene, the *cis*- or the *trans*-isomer?

H₃C   C₂H₅

C=C

H   CH₃

One person might label this compound *trans*, because the two identical groups are on opposite sides of the double bond. Another might label it *cis*, because the larger groups are on the same side of the double bond. Exactly this sort of ambiguity—and the use of both conventions simultaneously in the chemical literature—brought about the adoption of an unambiguous system for the nomenclature of stereoisomers, called the **Cahn–Ingold–Prelog system** after its inventors. As this system is applied to alkenes, we assign *relative priorities* to the two groups on each carbon of the double bond according to a set of rules given in the list that follows. We then compare the relative locations of these groups on each alkene carbon. If the groups of higher priority are on the same side of the double bond, the compound is said to have the *Z* configuration (*Z = zusammen*, German, "together"). If the groups of higher priority are on opposite sides of the double bond, the compound is said to have the *E* configuration (*E = entgegen*, German, "across").

groups of higher priority on same side of double bond
Z configuration

groups of higher priority on opposite sides of double bond
E configuration

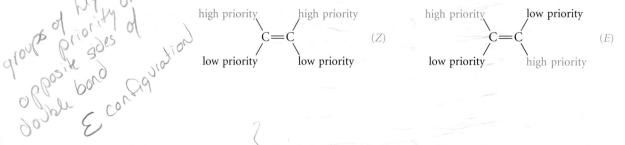

For a compound with more than one double bond, the configuration of each double bond is specified independently.

To assign relative priorities, follow each of the steps below in order until a decision is reached. Study Problems 4.4 and 4.5 illustrate the use of these steps.

Step 1 Examine the atoms directly attached to a given carbon of the double bond, and then follow the first rule that applies.

> *Rule 1a* Assign higher priority to the group containing the atom of higher atomic number.

> *Rule 1b* Assign higher priority to the group containing the isotope of higher atomic mass.

Step 2 If the atoms directly attached to the double bond are the same, then, working outward from the double bond, consider within each group the set of attached atoms. You'll have two sets—one for each group on the double bond.

> *Rule 2* Arrange the attached atoms within each set in descending priority order, and make a pairwise comparison of the atoms in the two sets. The higher priority is assigned to the atom of higher atomic number (or atomic mass in the case of isotopes) at the first point of difference.

Step 3 If the sets of attached atoms are identical, move away from the double bond within each group to the next atom following the *path of highest priority* and identify new sets of attached atoms. Then apply Rule 2 to these new sets. Keep following this step until a decision is reached. Remember that a priority decision *must* be made at the first point of difference.

STUDY
PROBLEM
4.4

What is the configuration of the following stereoisomer of 3-methyl-2-pentene? (The numbers and letters are for reference in the solution.)

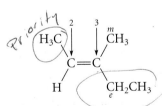

Solution   First, consider the relative priorities of the groups attached to carbon-2. Applying Rule 1a, the two atoms directly attached to carbon-2 are C and H. Because C has a higher atomic number (6) than H (1), the CH_3 group is assigned the higher priority. Now consider the groups attached to carbon-3. Step 1 leads to no decision, because in both groups the atom directly attached to the double bond is the same—a carbon *m* in the case of the methyl group, and a carbon *e* in the case of the ethyl group. Following Step 2, represent the atoms attached to these carbons as a set in descending priority order. For carbon *e*, the set is (C,H,H); notice that *the carbon of the double bond is not*

included in the set. For carbon *m*, the set is (H,H,H). Now make a pairwise comparison of (C,H,H) with (H,H,H). The first point of difference occurs at the comparison of the first atoms of each set, C and H. Because C has higher priority, the group containing this atom—the ethyl group—also has higher priority. The priority pattern is therefore

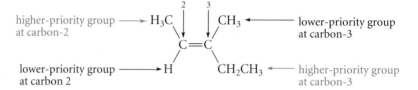

Because groups of like priority are on opposite sides of the double bond, this alkene is the *E* isomer, and is named (*E*)-3-methyl-2-pentene.

Study Problem 4.5

Name the following alkene. (The numbers and letters are for reference in the solution.)

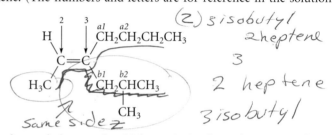

Solution   At carbon-2, the methyl group has higher priority, by Rule 1a. At carbon-3, Rule 1a allows no decision, because atoms *a1* and *b1* are identical—both are carbons. Proceeding to Step 2, the set of atoms attached to either carbons *a1* or *b1* can be represented as (C,H,H); again, no decision is possible. Step 3 says that we must now consider the next atoms in each chain along the path of highest priority. We therefore move to the next carbon atom (*a2* and *b2*) rather than the hydrogen in each chain, because carbon has higher priority than hydrogen. The set of atoms attached to *a2* is (C,H,H); the set attached to *b2* is (C,C,H). Notice that *carbons a1 and b1 considered in the previous step are not considered as members of these sets,* because we always work outward, away from the double bond, by Step 2. The difference in the second atoms of each set—C *vs.* H—dictates a decision. Because the set of atoms at carbon *b2* has higher priority, the group containing carbon *b2* (the isobutyl group) also has the higher priority. The process used can be summarized as follows:

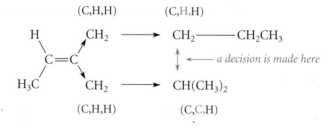

Because the groups of like priority are on the same side of the double bond, this alkene has the *Z* configuration.

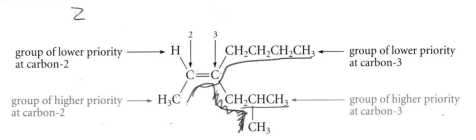

group of lower priority at carbon-2 ⟶ H

CH₂CH₂CH₂CH₃ ⟵ *group of lower priority* at carbon-3

group of higher priority at carbon-2 ⟶ H₃C

CH₂CHCH₃ ⟵ *group of higher priority* at carbon-3

Application of the nomenclature rules completes the name: (Z)-3-isobutyl-2-heptene.

Sometimes the groups to which we must assign priorities themselves contain double bonds. Double bonds are treated by a special convention, in which the double bond is rewritten as a single bond and the atoms at each end of the double bond are duplicated:

—CH=CH₂ is treated as —CH—CH₂ and —CH=O is treated as —CH—O

Notice that the duplicated atoms bear only one bond. (The developers of this scheme preferred to say that each of these duplicated carbons "bears three phantom (that is, imaginary) atoms of priority zero.") The treatment of triple bonds requires triplicating the atoms involved:

—C≡CH is treated as —C—CH and —C≡N is treated as —C—N

This convention allows us to establish, for example, the relative priorities of the vinyl and isopropyl groups:

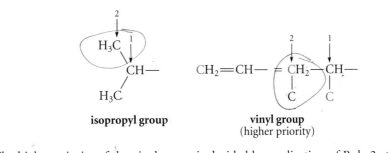

isopropyl group

vinyl group
(higher priority)

The higher priority of the vinyl group is decided by application of Rule 2 at carbon-2 of each group.

The following examples illustrate the *E,Z* nomenclature of compounds with more than one double bond:

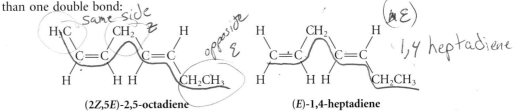

(2Z,5E)-2,5-octadiene

(E)-1,4-heptadiene

In the second example, no number before the *E* is required, since the *E,Z* designation is only relevant to one of the double bonds.

Additional rules have been developed to deal with more complex situations, but we don't need to consider these here.

PROBLEMS

4.4   Name each of the following compounds, including the proper designation of double-bond stereochemistry:

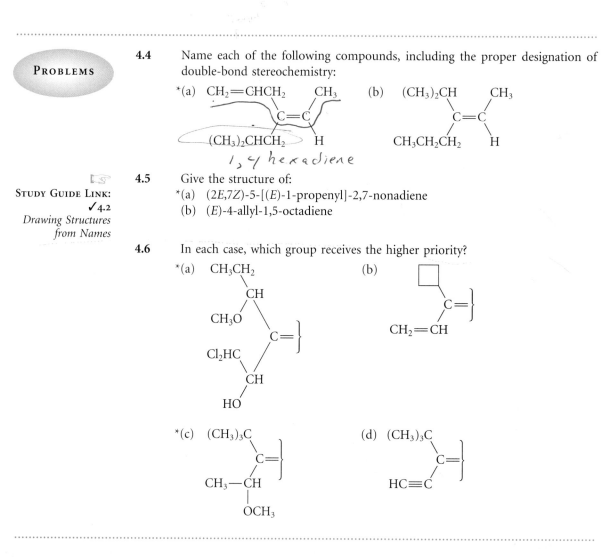

*(a)

CH_2=$CHCH_2$ ... CH_3

$(CH_3)_2CHCH_2$ ... H

C=C

1,4 hexadiene

(b) $(CH_3)_2CH$ ... CH_3

$CH_3CH_2CH_2$ ... H

C=C

STUDY GUIDE LINK:
✓ **4.2**
Drawing Structures from Names

4.5   Give the structure of:
*(a)   (2E,7Z)-5-[(E)-1-propenyl]-2,7-nonadiene
(b)   (E)-4-allyl-1,5-octadiene

4.6   In each case, which group receives the higher priority?

*(a) CH_3CH_2
        CH
    CH_3O
                C=}
    Cl_2HC
        CH
    HO

(b)   [cyclobutane]
                C=}
    CH_2=CH

*(c) $(CH_3)_3C$
                C=}
    CH_3—CH
        OCH_3

(d) $(CH_3)_3C$
                C=}
    HC≡C

4.3   **Unsaturation Number**

The empirical formula of an alkene, like that of an alkane, can be determined by combustion analysis (Sec. 2.7). An alkene with one double bond has two fewer hydrogens than the alkane with the same carbon skeleton. Likewise, a compound containing a ring also has two fewer hydrogens in its molecular formula than the corresponding noncyclic compound. (Compare cyclohexane, C_6H_{12}, with hexane, C_6H_{14}.) As this simple example shows, *the molecular formula of an organic compound contains information about the number of rings and double (or triple) bonds in the compound.*

The presence of rings or double bonds within a molecule is indicated by a quantity called the **unsaturation number**, or **degree of unsaturation**, *U*. *The unsaturation number of a molecule is equal to the total number of its rings and multiple bonds.* The unsaturation number of a hydrocarbon is readily calculated from the molecular formula as follows. The maximum number of hydrogens possible in a hydrocarbon with *C* carbon atoms is $2C + 2$. Since *every ring or double bond reduces the number of hydrogens from this maximum by 2*, the unsaturation number is equal to half the difference between the maximum number of hydrogens and the actual number *H*:

$$U = \frac{2C + 2 - H}{2} = \text{number of rings} + \text{multiple bonds} \qquad (4.6)$$

For example, cyclohexene, C_6H_{10}, has $U = [2(6) + 2 - 10]/2 = 2$. Cyclohexene has two degrees of unsaturation: one ring and one double bond.

How does the presence of other elements affect the calculation of the unsaturation number? You can readily convince yourself from common examples (for instance, ethanol, C_2H_5OH) that Eq. 4.6 remains valid when *oxygen* is present in an organic compound. Because a *halogen* is monovalent, each halogen atom in an organic compound always reduces the maximum possible number of hydrogens by one. Thus, if the number of carbons is *C*, the maximum number of *halogens plus hydrogens* is $2C + 2$. If *X* equals the actual number of halogens present, the formula for unsaturation number therefore becomes

$$U = \frac{2C + 2 - (X + H)}{2} = \frac{2C + 2 - X - H}{2} \qquad (4.7)$$

Finally, when *nitrogen* is present, the number of hydrogens in a saturated compound increases by one for each nitrogen. (For example, the saturated compound methylamine, $CH_3 - NH_2$, has $2C + 3$ hydrogens.) Therefore, if *N* is the number of nitrogens, the formula for unsaturation number becomes

$$U = \frac{2C + 2 + N - (H + X)}{2} = \frac{2C + 2 + N - H - X}{2} \qquad (4.8)$$

The utility of the unsaturation number is that it gives us structural information about an unknown compound from the molecular formula alone. This idea is illustrated in Problem 4.9.

PROBLEMS

4.7   Calculate the unsaturation number for each of the following compounds:
  *(a) $C_5H_8N_2$ (b) $C_3H_4Cl_4$

4.8   Without writing a formula or structure, give the unsaturation number of each of the following compounds:
  *(a) 2,4,6-octatriene   (b) methylcyclohexane

***4.9**   An unknown compound contains 85.60% carbon and 14.40% hydrogen. How many rings or double bonds does it have?

4.10   What is the contribution of a triple bond to the unsaturation number?

<table>
</table>

4·4 Physical Properties of Alkenes

Except for their melting points and dipole moments, many alkenes differ little in their physical properties from the corresponding alkane.

	$CH_2{=}CH(CH_2)_3CH_3$	$CH_3(CH_2)_4CH_3$
	1-hexene	hexane
boiling point	63.4 °C	68.7 °C
melting point	−139.8 °C	−95.3 °C
density	0.673 g/mL	0.660 g/mL
water solubility	negligible	negligible
dipole moment	0.46 D	0.085 D

Like alkanes, alkenes are flammable, nonpolar compounds that are less dense than, and insoluble in, water. The lower molecular weight alkenes are gases.

The dipole moments of some alkenes, though small, are greater than those of the corresponding alkanes.

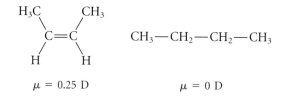

$$\mu = 0.25 \text{ D} \qquad\qquad \mu = 0 \text{ D}$$

How can we account for the dipole moments of alkenes? Remember that the electron density in sp^2 orbitals lies closer to the nucleus than it does in sp^3 orbitals (Sec. 4.1A). As a result, in alkenes with alkyl groups attached to double bonds, electrons are polarized, or pulled, *away* from the alkyl group, toward the trigonal carbon atom.

This polarization results in a bond dipole. The dipole moment of *cis*-2-butene is the vector sum of the $CH_3{-}C{=}$ and $H{-}C{=}$ bond dipoles. Although both types of bond dipoles are probably oriented toward the alkene carbon, there is good evidence (Problem 4.60) that the polarization of the $CH_3{-}C$ bond is greater. This is why *cis*-2-butene has a net dipole moment.

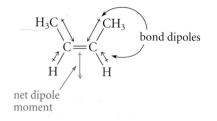

In summary: Bonds from alkyl groups to trigonal carbon are polarized so that *electrons are drawn away from alkyl groups toward trigonal carbon.* An equivalent statement is that *trigonal carbon withdraws electrons from alkyl groups.*

4.11   Which compound in each set should have the larger dipole moment? Explain.
*(a)   propene or 2-methylpropene   (b)   *cis*-2-butene or *trans*-2-butene

4.5   Relative Stabilities of Alkene Isomers

When we ask which of two compounds is more stable, we are asking which compound has lower energy. However, energy can take different forms, and the energy we use to measure relative "stability" depends on the purpose we have in mind. If we are interested in the position of chemical equilibrium, then the *standard free energy change* ($\Delta G°$) for the equilibrium is the energy measurement of interest (Sec. 3.5). However, the free energy change for a reaction is not the same as the total energy change. If we want to inquire about the relative stabilities of the bonding arrangements in two different molecules, we need to know their *relative total energies*. For our purposes, the relative total energies of two compounds are essentially the same as their *relative standard enthalpies*, or *heat contents*, abbreviated $\Delta H°$. Section 4.5A will show how the standard enthalpies of organic compounds are expressed. Then, in Section 4.5B, you'll learn what the enthalpies of alkenes can reveal about the relative stabilities of different bonding arrangements in alkenes.

A. Heats of Formation

The relative enthalpies of many organic compounds are available in standard tables as *heats of formation*. The standard **heat of formation** of a compound, abbreviated $\Delta H_f°$, is the heat of the reaction in which the compound is formed from its elements in their natural state at 1 atm pressure and 25 °C. Thus, the heat of formation of *trans*-2-butene is the heat absorbed in the following reaction:

$$4H_2 + 4C \longrightarrow \textit{trans-2-butene} \tag{4.9}$$

The conventions used in dealing with heats of reaction are the same as with free energies: the heat of any reaction is the *difference* in the enthalpies of products and reactants.

$$\Delta H°(\text{reaction}) = H°(\text{products}) - H°(\text{reactants}) \tag{4.10}$$

A reaction in which heat is liberated is said to be an **exothermic reaction**, and one in which heat is absorbed is said to be an **endothermic reaction**. The $\Delta H°$ of an exothermic reaction, by Eq. 4.10, has a negative sign; the $\Delta H°$ of an endothermic reaction has a positive sign. The heat of formation of *trans*-2-butene (Eq. 4.9) is -11.2 kJ/mol (-2.67 kcal/mol); this means that heat is liberated in the formation of *trans*-2-butene from carbon and hydrogen, and that the alkene has lower energy than the four moles each of C and H_2 from which it is formed.

Heats of formation are used to determine the relative enthalpies of molecules, that is, which of two molecules has lower energy. How this is done is illustrated in the following study problem.

Calculate the standard enthalpy difference between the *cis* and *trans* isomer of 2-butene. Specify which stereoisomer is more stable. The heats of formation are, for the *cis*-isomer, -6.99 kJ/mol, and for the *trans*-isomer, -11.2 kJ/mol (-1.67 and -2.67 kcal/mol, respectively).

Solution   The enthalpy difference requested in the problem corresponds to the standard ΔH° of the following reaction:

$$\begin{array}{cccc} & cis\text{-2-butene} & \longrightarrow & trans\text{-2-butene} \\ \Delta H_f^\circ & -6.99 & & -11.2 \quad \text{kJ/mol} \\ & -1.67 & & -2.67 \quad \text{kcal/mol} \end{array} \qquad (4.11)$$

To obtain the standard enthalpy difference, apply Eq. 4.10 *using the corresponding heats of formation in place of the H° values*. Thus, ΔH_f° for the reactant, *cis*-2-butene, is subtracted from that of the product, *trans*-2-butene. The ΔH° for this reaction, then, is $-11.2 - (-6.99)$ or -4.2 kJ/mol (-1.0 kcal/mol). This means that *trans*-2-butene is more stable than *cis*-2-butene by 4.2 kJ/mol (1.0 kcal/mol).

The procedure used in Study Problem 4.6 utilizes the fact that *chemical reactions and their associated energies can be added algebraically*. (This principle is known as **Hess's law of constant heat summation**.) What we have really done in the study problem is to subtract the two formation reactions and their associated energies:

Equations:			ΔH° (kJ/mol):	ΔH° (kcal/mol):	
$4\cancel{C} + 4\cancel{H_2}$	$\longrightarrow$	*trans*-2-butene	-11.2	-2.67	(4.12a)
cis-2-butene	$\longrightarrow$	$4\cancel{C} + 4\cancel{H_2}$	$+6.99$	$+1.67$	(4.12b)
Sum: *cis*-2-butene	$\longrightarrow$	*trans*-2-butene	-4.2	-1.0	(4.12c)

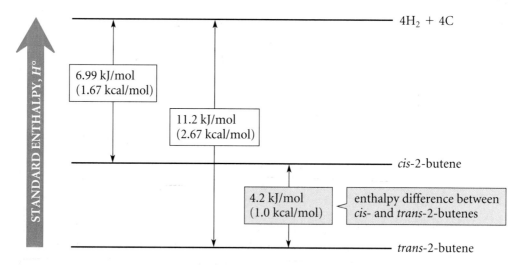

Figure 4.8   *Use of heats of formation to derive relative enthalpies of two isomeric compounds. The enthalpies of both compounds are measured relative to a common reference, the elements from which they are formed. The difference between the enthalpies of formation is equal to the enthalpy difference between the two isomers.*

Because *cis-* and *trans-*2-butene are isomers, the elements from which they are formed are the same and *cancel in the comparison.* This is shown by the diagram in Fig. 4.8. Were we to compare the enthalpies of compounds that are not isomers, the two formation equations would have different quantities of carbon and hydrogen, and the sum would contain leftover C and H_2. This sum would not correspond to the direct comparison desired.

STUDY GUIDE LINK:
4.3
Source of Heats of Formation

Using heats of formation to calculate the standard enthalpy difference between two compounds (Study Problem 4.6) is analogous to measuring the relative heights of two objects by comparing their distances from a common reference, say, the ceiling. If a table top is five feet below the ceiling, and an electrical outlet is seven feet below the ceiling, then the table top is two feet above the outlet. Notice the height of the ceiling can be taken arbitrarily as zero; its absolute height is irrelevant. When heats of formation are compared, the enthalpy reference point is the enthalpy of the elements in their "standard states," their normal states at 25 °C and 1 atm pressure; the enthalpies of formation of the elements in their standard states are arbitrarily taken to be zero.

STUDY GUIDE LINK:
4.4
Free Energy and Enthalpy

PROBLEM

*4.12   (a) Calculate the enthalpy change for the reaction 1-butene → 2-methylpropene. The heats of formation are 1-butene, −0.13 kJ/mol (−0.03 kcal/mol); 2-methylpropene, −16.90 kJ/mol (−4.04 kcal/mol).
        (b) Which butene isomer in (a) is more stable?

4.13   (a) If the standard enthalpy change for the reaction 2-ethyl-1-butene → 2-methyl-1-pentene is +7.45 kJ/mol (+1.78 kcal/mol), and if ΔH_f° for 2-ethyl-1-butene is −51.55 kJ/mol (−12.32 kcal/mol), what is ΔH_f° for 2-methyl-1-pentene?
        (b) Which isomer is more stable?

*4.14   Calculate ΔH_f° for 1-hexene from its heat of combustion, −3770.4 kJ/mol (−901.14 kcal/mol), using the method and the combustion data for carbon and hydrogen given in Study Guide Link 4.3.

B. Relative Stabilities of Alkene Isomers

The heats of formation of alkenes can be used to determine how various structural features of alkenes affect their stabilities. We'll answer two questions using heats of formation. First, which is more stable: a *cis* alkene or its *trans* isomer? Second, how does the amount of branching at the double bond affect the stability of an alkene?

Study Problem 4.6 showed that *trans-*2-butene has a lower enthalpy of formation than *cis-*2-butene by 4.2 kJ/mol (1.0 kcal/mol) (Eq. 4.12c). In fact, most *trans* alkenes are more stable than their *cis* isomers. Why is this so? The methyl groups in *cis-*2-butene are forced to occupy the same plane on the same side of the double bond. A space-filling model of *cis-*2-butene (Fig. 4.9) shows that one hydrogen in each of the *cis* methyl groups is within a van der Waals radius of the other. Hence, van der Waals repulsions occur

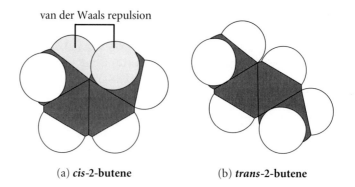

(a) *cis*-**2-butene**   (b) *trans*-**2-butene**

Figure 4.9   *Space-filling models of (a) cis-2-butene and (b) trans-2-butene. In cis-2-butene, van der Waals repulsions exist between hydrogen atoms of the two methyl groups (color). In trans-2-butene, these van der Waals repulsions are not present.*

between the methyl groups much like those in *gauche* butane (Sec. 2.3B). In contrast, no such repulsions occur in the *trans* isomer, in which the methyl groups are far apart. Not only do the heats of formation suggest the presence of van der Waals repulsions in *cis* alkenes, but they give us quantitative information on the magnitude of such repulsions.

How does branching at the double bond affect the relative energies of alkenes? First, let's compare the heats of formation of the following two isomers. The first has a single alkyl group directly attached to the double bond. The second has two alkyl groups attached to the double bond, in other words, has a single carbon branch at the double bond.

$$CH_2\!\!=\!\!CH\!-\!CH(CH_3)_2 \qquad \Delta H_f^\circ = -28.95 \text{ kJ/mol} \qquad\qquad (4.13a)$$
$$-6.92 \text{ kcal/mol}$$

$$\overset{\overset{\textstyle CH_3}{\textstyle |}}{CH_2\!\!=\!\!C\!-\!C_2H_5} \qquad \Delta H_f^\circ = -36.32 \text{ kJ/mol} \qquad\qquad (4.13b)$$
$$-8.68 \text{ kcal/mol}$$

Notice that both compounds have a single branch; they differ only in the *position* of the branch. In this case, the isomer with the branch at a carbon of the double bond has the smaller (more negative) heat of formation, and is therefore more stable. The data in Table 4.1 for some isomeric hexenes show that this trend continues for increasing numbers of alkyl groups directly attached to the double bond. These data show that *an alkene is stabilized by alkyl substituents on the double bond.* When we compare the stability of alkene isomers we find that *the alkene with the greatest number of alkyl substituents on the double bond is usually the most stable one.*

Notice that, to a useful approximation, it is the *number* of alkyl groups on the double bond more than their *identities* that govens the stability of an alkene. In other words, a molecule with two small alkyl groups on the double bond is more stable than its isomer with one large group on the double bond. The first two entries in Table 4.1 demonstrate this point. The second entry, 2-ethyl-1-butene, with two ethyl groups on the double bond, is more stable than the first entry, 2-methyl-1-pentene, which has a single isobutyl group on the double bond.

Table 4.1   Effect of Branching on the Stabilities of Hexene Isomers

Hexene isomer	Number of alkyl groups directly attached to double bond	ΔH_f°	Enthalpy difference
$CH_2{=}CH{-}CH_2CH(CH_3)_2$	1	-44.10 kJ/mol -10.54 kcal/mol	-7.45 kJ/mol -1.78 kcal/mol
$CH_2{=}C$ (C$_2$H$_5$, C$_2$H$_5$)	2	-51.55 kJ/mol -12.32 kcal/mol	
H$_3$C, CH$_3$ C=C H, C$_2$H$_5$	3	-58.66 kJ/mol -14.02 kcal/mol	-7.11 kJ/mol -1.70 kcal/mol
H$_3$C, CH$_3$ C=C H$_3$C, CH$_3$	4	-59.20 kJ/mol -14.15 kcal/mol	-0.54 kJ/mol -0.13 kcal/mol

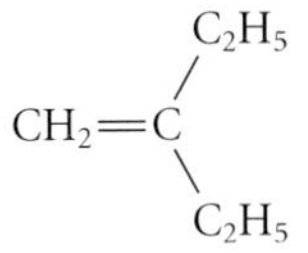

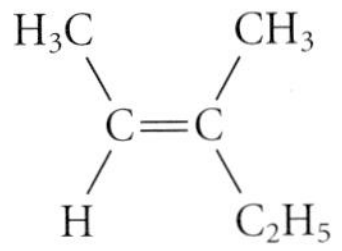

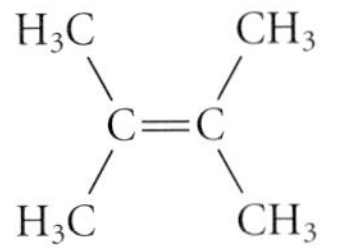

Heats of formation have given us considerable information about how alkene stabilities vary with structure. To summarize:

Increasing stability:

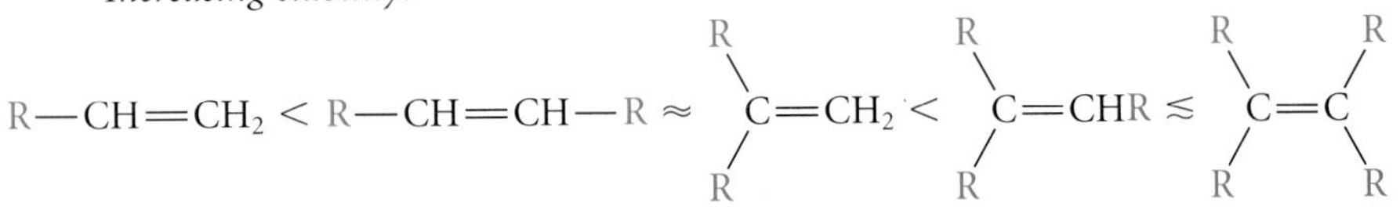

$$R{-}CH{=}CH_2 < R{-}CH{=}CH{-}R \approx \underset{R}{\overset{R}{C}}{=}CH_2 < \underset{R}{\overset{R}{C}}{=}CHR \lesssim \underset{R}{\overset{R}{C}}{=}\underset{R}{\overset{R}{C}} \qquad (4.14a)$$

and

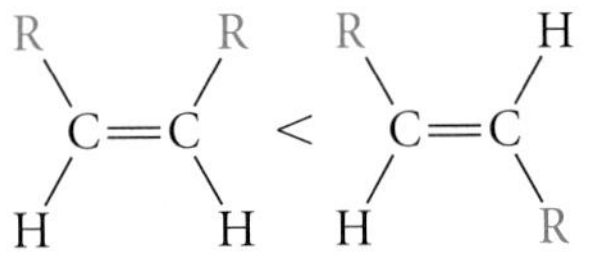

$$\underset{H}{\overset{R}{C}}{=}\underset{H}{\overset{R}{C}} < \underset{H}{\overset{R}{C}}{=}\underset{R}{\overset{H}{C}} \qquad (4.14b)$$

PROBLEMS

4.15   Within each series arrange the compounds in order of increasing stability:

*(a)

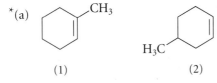

(1)   (2)

(Problem 4.15 continues)

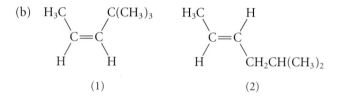

(1)   (2)

*4.16   As shown by the third and fourth entries of Table 4.1, the heats of forma-
tion of (E)-3-methyl-2-pentene and 2,3-dimethyl-2-butene differ very little
(−0.54 kJ/mol, −0.13 kcal/mol). This contrasts with the trend shown in the
other examples, that each branch at the double bond contributes about 7.1–7.3
kJ/mol (1.7–1.8 kcal/mol) to additional alkene stability. Suggest one reason why
the latter compound, despite its greater number of branches at the double bond,
is not much more stable than the former.

4.6   Addition Reactions of Alkenes

The remainder of this chapter considers three reactions of alkenes: the reaction with
hydrogen halides; the reaction with hydrogen, called *catalytic hydrogenation*; and the
reaction with water, called *hydration*. These reactions will be used to establish some
important principles of chemical reactivity that are very useful in organic chemistry.
Other alkene reactions are presented in Chapter 5.

The most characteristic type of alkene reaction is **addition** at the carbon-carbon
double bond. The addition reaction can be represented generally as follows:

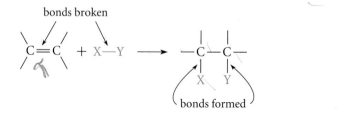

(4.15)

In an addition reaction, the carbon-carbon π bond of the alkene and the X—Y bond
of the reagent are broken, and new C—X and C—Y bonds are formed.

**STUDY
PROBLEM
4·7**

H—Br undergoes an addition reaction with most alkenes. Write the product of addition
of H—Br with ethylene.

Solution   Follow the pattern in Eq. 4.15, with X—Y = H—Br. In the product of
addition, the carbon-carbon π bond and the H—Br bond are broken, and new C—H
and C—Br bonds are formed, as follows:

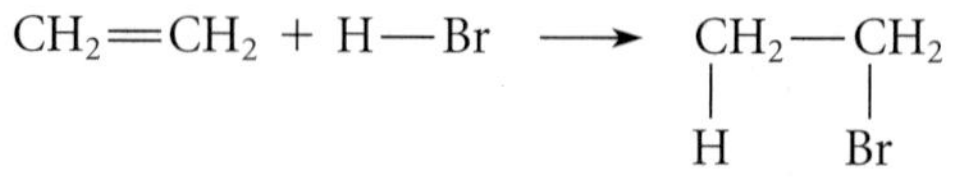

$$CH_2{=}CH_2 + H{-}Br \longrightarrow \underset{\displaystyle \substack{| \quad\; |\\ H \quad Br}}{CH_2{-}CH_2}$$

PROBLEM

4.17   Give the structure of the addition product formed when ethylene reacts with each of the following reagents:

*(a) HO—Br   (b) H—I

*(c) BH_3 (*Hint:* Each of the B—H bonds undergoes an addition to one molecule of ethylene. That is, three moles of ethylene react with one mole of BH_3.)

(d) Br_2

4.7   Addition of Hydrogen Halides to Alkenes

The hydrogen halides H—Cl, H—Br, and H—I undergo addition to carbon-carbon double bonds to give products called *alkyl halides*, compounds in which a halogen is bound to a saturated carbon atom:

$$CH_3—CH{=}CH—CH_3 + H—Br \longrightarrow CH_3—CH—CH—CH_3 \qquad (4.16)$$
$$\qquad\qquad\qquad\qquad\qquad\qquad\qquad\qquad\qquad\qquad |\quad\;\; |$$
$$\qquad\qquad\qquad\qquad\qquad\qquad\qquad\qquad\qquad\qquad H\quad Br$$

2-butene
(Z or E)

2-bromobutane
(an alkyl halide)

A. Regioselectivity of Hydrogen Halide Addition

When the alkene has an unsymmetrically located double bond (one that is not in the center of the molecule), two isomeric products are possible.

$$CH_2{=}CH—(CH_2)_3CH_3 + HI \longrightarrow$$

1-hexene

$$CH_3—CH—(CH_2)_3CH_3 \quad or \quad I—CH_2—CH_2—(CH_2)_3CH_3 \qquad (4.17)$$
$$\qquad\qquad |$$
$$\qquad\qquad I$$

1-iodohexane
(not observed)

2-iodohexane
(observed)

As shown in Eq. 4.17, *only one of the two possible products is formed* from a 1-alkene in significant amount. Generally, *the main product is that isomer in which the halogen is bonded to the carbon of the double bond with the greater number of alkyl substituents, and the hydrogen is bonded to the carbon with the smaller number of alkyl substituents.*

$$CH_2{=}CH—(CH_2)_3CH_3$$

H goes here   I goes here

A reaction that gives only one of several possible constitutional isomers is said to be a **regioselective reaction**. Hydrogen halide addition to alkenes is a regioselective reaction because addition of the hydrogen halide across the double bond gives only one of the two possible constitutionally isomeric addition products.

When the two carbons of the alkene double bond have equal numbers of alkyl substituents, little or no regioselectivity is observed in hydrogen halide addition, even if the alkyl groups are of different size.

$$HBr + CH_3—CH=CH—C_2H_5 \longrightarrow$$

2-pentene

$$CH_3—\underset{\underset{Br}{|}}{CH}—CH_2—C_2H_5 + CH_3—CH_2—\underset{\underset{Br}{|}}{CH}—C_2H_5 \qquad (4.18)$$

2-bromopentane   **3-bromopentane**

(nearly equal amounts)

MARKOWNIKOFF'S RULE

The regioselective addition of hydrogen halides to alkenes was first reported in 1870 by Vladimir Markownikoff (1838–1904), director of the Chemical Institute of Moscow. The regioselectivity of hydrogen halide addition to alkenes was generalized as **Markownikoff's rule**, which was stated as follows: "The halogen of a hydrogen halide attaches itself to the carbon of the alkene bearing the least number of hydrogens and the greater number of carbons."

PROBLEM

4.18   Using the known regioselectivity of hydrogen halide addition to alkenes, predict the addition product that results from the reaction of:
*(a)   H—Br with 1-methylcyclohexene
(b)   H—Cl with 2-methylpropene

B. Carbocation Intermediates in Hydrogen Halide Addition

For many years the regioselectivity of hydrogen halide addition had only an *empirical* (experimental) basis. The modern understanding of this regioselectivity begins with the fact that the overall addition reaction actually occurs as *two successive reactions*. Let's consider each of these in turn.

In the first reaction, the electron pair in the π bond of an alkene attacks the proton of the hydrogen halide. The electrons of the π bond react rather than the electrons of σ bonds because π electrons are farther from the nucleus, less strongly attracted to their parent carbons, and therefore more easily shared with other atoms. As a result, the carbon-carbon double bond is **protonated** on a carbon atom; that is, one of the carbon atoms accepts a proton. The other carbon becomes positively charged and electron deficient:

$$\text{R—CH}=\text{CH—R} \longrightarrow \text{R—}\overset{+}{\text{CH}}\text{—CH—R} \quad :\ddot{\text{Br}}: \qquad (4.19a)$$

a carbocation

The species with a positively charged, electron-deficient carbon is called a **carbocation**, pronounced CAR-bo-CAT-ion. (The term **carbonium ion** was used in earlier literature.) Notice that the formation of the carbocation from the alkene is an *electron-pair displacement reaction* (Sec. 3.2A) in which the π bond acts as a *Brønsted base* (Sec. 3.4A) toward the *Brønsted acid* H—Br. The π bond is not an ordinary base in the sense that we consider ammonia, hydroxide ion, or even water as bases. Rather, it is a *very* weak base. Nevertheless, it can be protonated to a small extent by the strong acid HBr.

The resulting carbocation is a powerful electron-deficient Lewis acid, and is thus a potent electrophile. In the second reaction of hydrogen halide addition, the carbocation is attacked at its electron-deficient carbon by the halide ion, which is a Lewis base, or nucleophile:

$$\text{R—}\underset{+}{\text{CH}}\text{—CH}_2\text{—R} \longrightarrow \text{R—CH—CH}_2\text{—R} \qquad (4.19b)$$

The carbocations involved in hydrogen halide addition to alkenes are examples of **reactive intermediates** or **unstable intermediates**: species that react so rapidly that they never accumulate in more than very low concentration. Most carbocations are too reactive to be isolated except under special circumstances. Thus, carbocations cannot be isolated from the reactions of hydrogen halides and alkenes because they react instantaneously with halide ions.

The complete description of a reaction pathway, including any reactive intermediates such as carbocations, is called the **mechanism** of the reaction. To summarize the two steps in the mechanism of hydrogen halide addition to alkenes:

1. A carbon of the π bond is protonated.

2. A halide ion attacks the resulting carbocation.

When the double bond of an alkene is not located symmetrically within the molecule, then protonation of the double bond can occur in two ways to give two different carbocations. For example, protonation of 2-methylpropene can give either the *tert*-butyl cation (Eq. 4.20a) or the isobutyl cation (Eq. 4.20b):

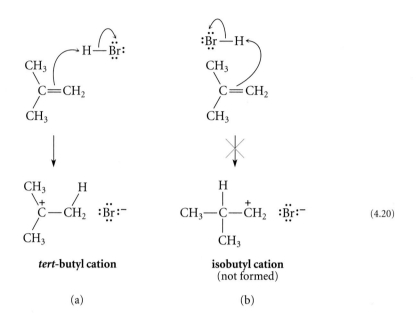

tert-butyl cation   **isobutyl cation**
                               (not formed)

(a)   (b)   (4.20)

These two reactions are in *competition*—that is, one can only happen at the expense of the other because they compete for the same starting material. Only the *tert*-butyl cation is formed in this reaction. The *tert*-butyl cation is formed exclusively because reaction 4.20a is *much faster* than reaction 4.20b. Because the *tert*-butyl cation is the only carbocation formed, it is the only carbocation available to react with the bromide ion. Hence, the only product of HBr addition to 2-methylpropene is *tert*-butyl bromide.

$$\underset{\underset{H_3C}{\diagup}}{\overset{H_3C}{\diagdown}}\overset{+}{C}-CH_3 \quad\longrightarrow\quad CH_3-\underset{\underset{CH_3}{|}}{\overset{\overset{:\!\ddot{B}r\!:}{|}}{C}}-CH_3 \qquad\qquad (4.21)$$

tert-butyl bromide

Notice that the bromide ion has become attached to the carbon of 2-methylpropene bearing the greater number of alkyl groups. In other words, *the regioselectivity of hydrogen halide addition is due to the formation of only one of two possible carbocations.*

The reason that the *tert*-butyl cation is formed is that the *tert*-butyl cation is *more stable* than the isobutyl cation. Thus, *the regioselectivity of hydrogen halide addition is due to the formation of the more stable carbocation intermediate.* In order to complete our understanding of hydrogen halide addition, then, we need to understand the factors that govern the relative stabilities of carbocations.

C. Structure and Stability of Carbocations

Carbocations are classified by the degree of alkyl substitution at their electron-deficient carbon atoms.

Table 4.2   Heats of Formation of the Isomeric Butyl Cations (Gas Phase, 25 °C)

Cation structure	Name	Heat of formation		Relative energy[a]	
		kJ/mol	kcal/mol	kJ/mol	kcal/mol
$CH_3CH_2CH_2\overset{+}{C}H_2$	butyl cation	845	202	155	37
$(CH_3)_2CH\overset{+}{C}H_2$	isobutyl cation	828	198	138	33
$CH_3\overset{+}{C}HCH_2CH_3$	sec-butyl cation	757	181	67	16
$(CH_3)_3\overset{+}{C}$	tert-butyl cation	690	165	(0)	(0)

[a] Energy difference between each carbocation and the tert-butyl cation

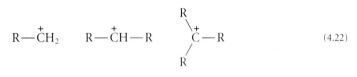

$$R\!-\!\overset{+}{C}H_2 \qquad R\!-\!\overset{+}{C}H\!-\!R \qquad \overset{\displaystyle R}{\underset{\displaystyle R}{\overset{\diagdown}{\underset{\diagup}{C}}}}\!\overset{+}{-}\!R \qquad (4.22)$$

primary   secondary   tertiary

That is, primary carbocations have one alkyl group bound to the electron-deficient carbon, secondary carbocations have two, and tertiary carbocations have three. For example, the isobutyl cation in Eq. 4.20b is a primary carbocation, and the tert-butyl cation in Eq. 4.20a is a tertiary carbocation.

The gas-phase heats of formation of the isomeric butyl carbocations are given in Table 4.2. The data in this table show that branching at the electron-deficient carbon strongly stabilizes carbocations. (A comparison of the first two entries shows that branching at other carbons is much less significant.) The relative stability of isomeric carbocations is therefore as follows:

Stability of carbocations: tertiary > secondary > primary   (4.23)

The reason for this stability order can be found in the geometry and electronic structure of carbocations, shown in Fig. 4.10 for the tert-butyl cation. The electron-deficient carbon of the carbocation has *trigonal planar* geometry (Sec. 1.3B) and is therefore sp^2-hybridized (Sec. 4.1A); the p orbital on this carbon contains no electrons.

Because alkenes are stabilized by alkyl groups on their trigonal, sp^2-hybridized carbons, it is perhaps not surprising that carbocations are also stabilized by alkyl substituents. However, if you compare the data in Tables 4.1 and 4.2, you'll notice that each alkyl branch stabilizes an alkene by about 7 kJ/mol, but each branch stabilizes a carbocation by about 70 kJ/mol. In other words, the stabilization of carbocations by alkyl substituents is considerably greater than the stabilization of alkenes.

The explanation for the stabilization of carbocations by alkyl branching is a phenomenon called **hyperconjugation**, which is the overlap of bonding electrons from the adjacent σ bonds with the unoccupied p orbital of the carbocation.

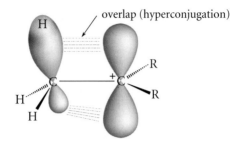

In this diagram, the σ bond that provides the bonding electrons is a C—H bond. Hyperconjugation can be shown with resonance structures as follows:

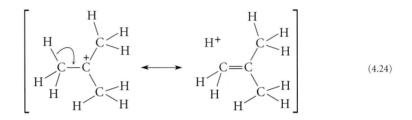

(4.24)

The energetic advantage of hyperconjugation comes from the additional bonding symbolized by the double bond in the resonance structure on the right. Additional bonding is a stabilizing effect. Another way to look at hyperconjugation is that it allows the electrons in the C—H bonds to occupy a larger region of space, thus reducing their repulsive interactions with one another. However we think of the phenomenon, the number of adjacent σ bonds available for hyperconjugation is greater when there are more alkyl branches at the electron-deficient carbon. Consequently, alkyl substitution at the electron-deficient carbon stabilizes carbocations.

Let's now bring together what you've learned about carbocation stability and the mechanism of hydrogen halide addition to alkenes. The addition occurs in two steps. In the first step, protonation of the alkene double bond occurs *at the carbon with the fewer alkyl branches* so that the more stable carbocation is formed—that is, *the one with the greater number of alkyl branches at the electron-deficient carbon*. The reaction is completed when the halide ion attacks the electron-deficient carbon.

The understanding of many organic reactions hinges on an understanding of the reactive intermediates involved. Carbocations are important reactive intermediates that

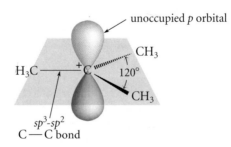

Figure 4.10   *Hybridization and geometry of the tert-butyl cation. Notice the trigonal-planar geometry and the unoccupied p orbital perpendicular to the plane of the cation.*

occur not only in the mechanism of hydrogen halide addition, but in the mechanisms of many other reactions as well. Hence, your knowledge of carbocations will be put to use often.

PROBLEMS

*4.19   Predict the product of the addition reaction between 2-methyl-1-butene and I—N$_3$ using the fact that the reaction begins by attack of the π electrons on the iodine and breaking of the I—N bond.

4.20   Predict the product of the reaction of I—Cl with 2-methylpropene.

4.21   In each case, give *two different* alkene starting materials that would react with H—Br to give the compound shown as the major addition product.

STUDY GUIDE LINK:
✓4.5
Solving Reaction Problems

*(a)
$$CH_3\overset{\overset{\displaystyle CH_3}{|}}{\underset{\underset{\displaystyle Br}{|}}{C}}\overset{\overset{\displaystyle CH_3}{|}}{CH}CH_3$$

(b)
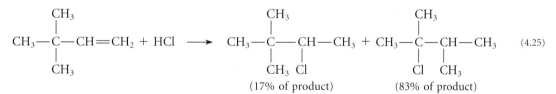

D. Carbocation Rearrangement in Hydrogen Halide Addition

In some cases the addition of a hydrogen halide to an alkene gives an unusual product, as in the following example.

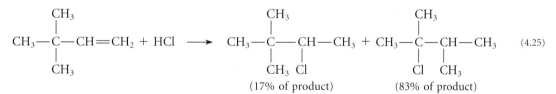

$$CH_3\overset{\overset{\displaystyle CH_3}{|}}{\underset{\underset{\displaystyle CH_3}{|}}{C}}CH{=}CH_2 + HCl \longrightarrow CH_3\overset{\overset{\displaystyle CH_3}{|}}{\underset{\underset{\displaystyle CH_3\ Cl}{|\ \ |}}{C}}CH{-}CH_3 + CH_3\overset{\overset{\displaystyle CH_3}{|}}{\underset{\underset{\displaystyle Cl\ \ CH_3}{|\ \ |}}{C}}CH{-}CH_3 \qquad (4.25)$$
$$\text{(17\% of product)} \qquad \text{(83\% of product)}$$

The minor product is the result of ordinary regioselective addition of HCl across the double bond. The origin of the major product, however, is not obvious. Examination of the carbon skeleton of the major product shows that a **rearrangement** has occurred. In a rearrangement, a group from the starting material has moved to a different position in the product. In this case, a methyl group of the alkene has changed positions. As a result, the carbons of the alkyl halide product are connected differently from the carbons of the alkene starting material. Although the rearrangement leading to the second product may seem strange at first sight, it is readily understood by considering the fate of the carbocation intermediate in the reaction.

The reaction begins like a normal addition of HCl, that is, by protonation of the double bond to yield the carbocation with the greater number of alkyl branches at the electron-deficient carbon.

$$CH_3-\underset{\underset{CH_3}{|}}{\overset{\overset{CH_3}{|}}{C}}-CH{=}CH_2 \quad H-\ddot{\underset{\cdot\cdot}{Cl}}: \longrightarrow CH_3-\underset{\underset{CH_3}{|}}{\overset{\overset{CH_3}{|}}{C}}-\overset{+}{C}H-CH_3 + :\ddot{\underset{\cdot\cdot}{Cl}}:^- \qquad (4.26)$$

Reaction of this carbocation with Cl^- occurs as expected to yield the minor product of Eq. 4.25. However, the carbocation can also undergo a second type of reaction: it can *rearrange*.

$$CH_3-\underset{\underset{CH_3}{|}}{\overset{\overset{CH_3}{|}}{C}}-\overset{+}{C}H-CH_3 \longrightarrow CH_3-\underset{\overset{+}{}}{\overset{\overset{CH_3}{|}}{C}}-CH-CH_3 \qquad (4.27a)$$

In this reaction, the methyl group moves *with its pair of bonding electrons* from the carbon adjacent to the electron-deficient carbon. The carbon from which this group departs, as a result, becomes electron-deficient and positively charged. That is, the rearrangement converts one carbocation into another. The major product of Eq. 4.25 is formed by the attack of Cl^- on the new carbocation.

$$CH_3-\underset{\overset{+}{}}{\overset{\overset{CH_3}{|}}{C}}-CH(CH_3)_2 + :\ddot{\underset{\cdot\cdot}{Cl}}:^- \longrightarrow CH_3-\underset{\underset{:\ddot{Cl}:}{|}}{\overset{\overset{CH_3}{|}}{C}}-CH(CH_3)_2 \qquad (4.27b)$$

Why does rearrangement of the carbocation occur? In the case of reaction 4.27a, a more stable tertiary carbocation is formed from a less stable secondary one. Therefore, *rearrangement is favored by the increased stability of the rearranged ion.*

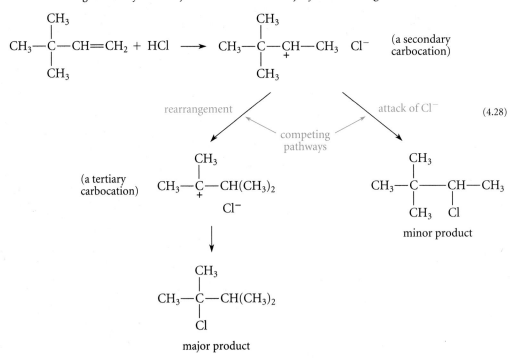

You've now learned two pathways by which carbocations can react. They can (1) react with a nucleophile and (2) rearrange to more stable carbocations. The outcome of Eq. 4.25 represents a competition between these two pathways. In any particular case, one cannot predict exactly how much of each different product will be obtained. Nevertheless, the reactions of carbocation intermediates show why both products are reasonable. Rearrangements do not occur in the reactions of many alkenes (for example, Eqs. 4.17 and 4.18) because rearrangement would not give a more stable carbocation.

Carbocation rearrangements are not limited to the migrations of alkyl groups. In the following reaction, the major product is also derived from the rearrangement of a carbocation intermediate. This rearrangement involves a **hydride shift**, the migration of a hydrogen with its two bonding electrons.

$$\underset{\underset{\overset{|}{H}}{\overset{|}{\underset{}{C}}}{CH_3-\overset{\overset{CH_3}{|}}{C}-CH=CH_2}} + HBr \longrightarrow \underset{\underset{Br}{\overset{|}{}}}{CH_3-\overset{\overset{CH_3}{|}}{CH}-CH-CH_3} + \underset{\underset{Br}{\overset{|}{}}}{CH_3-\overset{\overset{CH_3}{|}}{C}-CH_2CH_3} \quad (4.29)$$

(about 45% of product)   (about 55% of product)

THE FIRST DESCRIPTION OF CARBOCATION REARRANGEMENTS

The first clear formulation of the involvement of carbocations in molecular rearrangements was proposed by Frank C. Whitmore (1887–1947) of Pennsylvania State University. Whitmore said that carbocation rearrangements result when "an atom in an electron-hungry condition seeks its missing electron pair from the next atom in the molecule." Whitmore's description shows that a carbocation rearrangement is simply another Lewis acid-base reaction. The Lewis acid is the electron-deficient carbon of the carbocation, and the Lewis base is the electron pair in the bond to the group that rearranges.

PROBLEMS

*4.22   Give a curved-arrow mechanism for the reaction in Eq. 4.29 that accounts for the formation of both products.

*4.23   Only one of the following three alkyl halides can be prepared as the *major* product of the addition of HBr to an alkene. Which compound can be prepared in this way? Explain why the other two *cannot* be prepared in this way.

$$CH_3CH_2CH_2CH_2CH_2Br \qquad \underset{\overset{|}{Br}}{CH_3CHCH_2CH_2CH_3} \qquad \underset{\overset{|}{CH_3}}{CH_3-\underset{\overset{|}{Br}}{CH}-\overset{\overset{CH_3}{|}}{C}-C_2H_5}$$

$$A \qquad\qquad\qquad B \qquad\qquad\qquad\qquad C$$

4.8   Reaction Rates

Whenever a reaction can give more than one possible product, two or more reactions are in competition. (You've already seen examples of competing reactions in hydrogen halide addition to alkenes.) One reaction predominates when it occurs *more rapidly* than the other competing reactions. Understanding the factors that control the rates at which reactions take place is the subject of this section.

A. The Transition State

The **rate** of a chemical reaction can be defined for our purposes as the number of reactant molecules converted into product in a given time. The theory of reaction rates used by many organic chemists assumes that as the reactants change into products, they pass through an unstable state of maximum free energy, called the **transition state**. The transition state has a higher energy than either the reactants or products and therefore represents an **energy barrier** to their interconversion. This energy barrier is shown graphically in a **reaction free-energy diagram** (Fig. 4.11). This is a diagram of the standard free energy of a reacting system as old bonds break and new ones form along the reaction pathway. In this diagram the pathway of the reaction from reactants to products is called the *reaction coordinate*. The energy barrier, $\Delta G^{\circ\ddagger}$, called the **standard free energy of activation**, is equal to the difference between the standard free energies of the transition state and reactants. (The double dagger, ‡, is the symbol used for transition states.) The size of the energy barrier $\Delta G^{\circ\ddagger}$ determines the rate of a reaction: *the higher the barrier, the lower the rate.* Thus, the reaction shown in Fig. 4.11a is slower than the one in Fig. 4.11b because it has a larger energy barrier. In the same sense that relative free energies of reactants and products determine the equilibrium constant, the relative free energies of transition state and reactants determine the reaction rate.

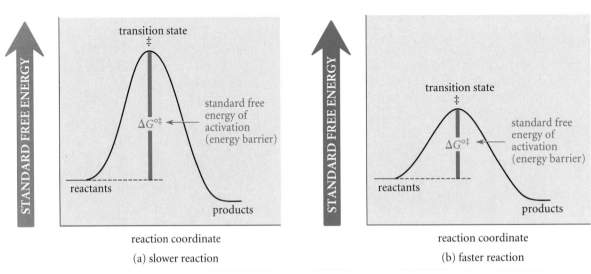

(a) slower reaction   (b) faster reaction

Figure 4.11   *Reaction free-energy diagrams for two hypothetical reactions. The standard free energy of activation ($\Delta G^{\circ\ddagger}$) is the energy barrier that must be overcome in order for the reaction to occur. The reaction in (a) is intrinsically slower than the one in (b) because it has a larger $\Delta G^{\circ\ddagger}$.*

Where do reactant molecules get the energy to cross the energy barrier and form products? In general, molecules obtain the energy required to react from their thermal motions. At a given temperature, a collection of molecules can be characterized by an average energy, just as a chemistry class might be characterized by an average student. However, such a collection of molecules contains a distribution of energies, just as a chemistry class has a distribution of abilities. The rate of a reaction is directly related to the number of molecules that have enough energy to cross the energy barrier in a given amount of time. If the energy barrier ($\Delta G^{\circ\ddagger}$) is low, then more molecules possess enough energy to cross the barrier, and the rate of the reaction is greater. If $\Delta G^{\circ\ddagger}$ is high, then fewer molecules have sufficient energy to cross the barrier, and the rate of the reaction is smaller.

For a given reaction under a given set of conditions, we cannot control the size of the energy barrier; it is a natural property of the reaction. Some reactions are intrinsically slow, and some are intrinsically fast. What we can sometimes control is the fraction of molecules with enough energy to cross the energy barrier. That is, we can increase the energy of a chemical system by *raising the temperature*. The rates of reactions increase when the temperature is raised.

Let's summarize. Two factors that govern the intrinsic reaction rate are

1. the size of the energy barrier, or standard free energy of activation $\Delta G^{\circ\ddagger}$: reactions with larger $\Delta G^{\circ\ddagger}$ are slower.

2. the temperature: reactions are faster at higher temperatures.

An Analogy for Energy Barriers

An analogy that can help in visualizing these concepts is shown in Fig. 4.12. Water in the upper cup would flow into the pan below if it could somehow gain enough kinetic energy to surmount the wall of the cup. The wall of the cup is a potential-energy barrier to the downhill flow of water. Likewise, molecules have to achieve a transient state of high energy—the *transition state*—in order to break stable chemical bonds and undergo reaction. An analogy to thermal motion is what happens if we shake the cup. If the cup is shallow (low energy barrier), there is a good likelihood that the shaking will cause water to slosh over the sides of the cup and drop into the pan. This will occur at some characteristic *rate*—some number of milliliters per second. If the cup is very deep (high barrier), it is less likely that water will flow from cup to pan. Consequently, the rate at which water collects in the pan is smaller. Shaking the cup more vigorously provides an analogy to the effect of increasing temperature. As the "sloshing" becomes more violent, the water acquires more kinetic energy, and water accumulates in the pan at a higher rate. Likewise, high temperature increases the rate of a chemical reaction by increasing the energy of the reacting molecules.

It is very important to understand that the equilibrium constant for a reaction tells us *absolutely nothing* about its rate. Some reactions with very large equilibrium constants are slow. For example, the equilibrium constant for combustion of alkanes is very large; yet a container of gasoline (alkanes) can be handled in the open air because the reaction

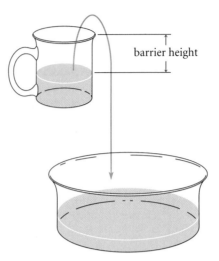

Figure 4.12 *A difference of potential energy is not enough to cause the water in the cup to drop to the bowl below. The water must first overcome the barrier imposed by the walls of the cup.*

of gasoline with oxygen, in the absence of heat, is immeasurably slow. On the other hand, some unfavorable reactions come to equilibrium almost instantaneously. For example, the reaction of ammonia with water to give ammonium hydroxide has a very unfavorable equilibrium constant; but the small extent of reaction that does occur takes place very rapidly.

B. Multistep Reactions and the Rate-Limiting Step

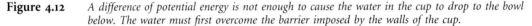

Many chemical reactions take place with the formation of reactive intermediates. When intermediates exist in a chemical reaction, then what we commonly express as one reaction, for example, addition of HBr to 2-methylpropene

$$(CH_3)_2C{=}CH_2 + HBr \longrightarrow (CH_3)_3C{-}Br \qquad (4.30)$$

is really a sequence of two reactions:

$$(CH_3)_2C{=}CH_2 + HBr \longrightarrow (CH_3)_3C^+ + Br^- \qquad (4.31a)$$

$$(CH_3)_3C^+ + Br^- \longrightarrow (CH_3)_3C{-}Br \qquad (4.31b)$$

Each step of a multistep reaction has its own characteristic rate, and therefore its own transition state. The energy changes in such a reaction can also be depicted in a reaction free-energy diagram. Such a diagram for the addition of HBr to 2-methylpropene is shown in Fig. 4.13. Each free-energy maximum between reactants and products represents a transition state, and the minimum represents the carbocation intermediate.

The rate of the overall reaction depends in detail on the rates of its various steps. However, it often happens that one step of a multistep reaction is considerably slower than any of the others. This slowest step in a multistep chemical reaction is called the

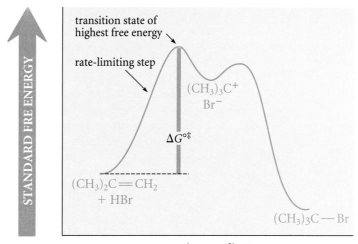

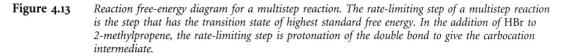

reaction coordinate

Figure 4.13   *Reaction free-energy diagram for a multistep reaction. The rate-limiting step of a multistep reaction is the step that has the transition state of highest standard free energy. In the addition of HBr to 2-methylpropene, the rate-limiting step is protonation of the double bond to give the carbocation intermediate.*

rate-limiting step or **rate-determining step** of the reaction. In such a case *the rate of the overall reaction is equal to the rate of the rate-limiting step.* In terms of the reaction free-energy diagram in Fig. 4.13, *the rate-limiting step is the step with the transition state of highest free energy.* This diagram indicates that in the addition of HBr to 2-methylpropene, the rate-limiting step is the first step of the reaction—the protonation of the alkene to give the carbocation. The overall rate of addition of HBr to 2-methylpropene is equal simply to the rate of this first step.

The rate-limiting step of a reaction has a special importance. Anything that increases the rate of this step increases the overall reaction rate. Conversely, if a change in the reaction conditions (for example, a change in temperature) affects the rate of the reaction, it is the effect on the rate-limiting step that is being observed. Because the rate-limiting step of a reaction has special importance, its identification receives particular emphasis when we attempt to understand the mechanism of a reaction.

AN ANALOGY FOR RATE-LIMITING STEP

The idea of a rate-limiting step is analogous to a toll station on a modern freeway at rush hour. We can divide the rate of passage of cars through a toll plaza into three steps: (1) entry of the cars into the toll area; (2) taking of the toll by the collector; and (3) exit of cars from the toll area. Suppose that our toll station has a very slow, lackadaisical toll collector. He takes the toll so slowly that the rate of passage of cars through the plaza is determined strictly by how fast he works. Toll-taking—the second step—is the rate-limiting process for passage of cars through the toll booth. Cars can line up

(continues)

more or less frequently, but as long as there is a line of cars, the rate of passage through the toll plaza is the same. If the collector takes one toll per minute, then cars exit at one per minute.

Imagine now a different situation: a super-fast toll collector. He is so fast, in fact, that he can keep pace with any number of cars likely to pass through the toll booth in a given time. The rate as which cars exit from the toll plaza is determined strictly by how fast they arrive. In this case, step (1), the entry of cars into the toll plaza, is the rate-limiting step.

Now imagine that an efficiency expert has been hired to increase the rate of passage of cars through the toll plaza with the slow toll collector. Her first job must be to locate the bottleneck. Only if she affects the rate-limiting step by replacing the slow toll collector will she improve the rate of passage through the toll plaza. Likewise, if we want to increase the rate of reaction, we must do something to increase the rate of its slowest step.

PROBLEMS

*4.24   Draw a reaction free-energy diagram for a reaction $A \rightleftarrows B \rightleftarrows C$ that meets the following criteria: The standard free energies are in the order $C < A < B$, and the rate-limiting step of the reaction is $B \rightleftarrows C$.

4.25   Repeat Problem 4.24 for a case in which the standard free energies are in the order $A < C < B$, and the rate-limiting step of the reaction is $A \rightleftarrows B$.

C. Hammond's Postulate

Transition states possess a maximum of free energy with respect to reactants and products, and therefore are not stable and cannot be isolated. Nevertheless, the transition-state concept is useful because *transition states can be visualized as structures.* Typically, we think of a transition state as a structure somewhere in between the structures of reactants and products. For example, in the addition of HBr to an alkene, the transition state of the first step is visualized as a structure along the reaction pathway somewhere between the structures of the starting materials, the alkene and HBr, and the products of this step, the carbocation and a bromide ion:

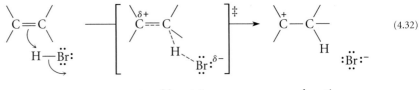

(4.32)

                           transition state   carbocation

In a transition state various bonds are in the process of breaking and forming; these are represented as dashed lines. Thus, in Eq. 4.32, the π bond of the alkene and the H—Br bond are breaking, and a C—H bond is forming; all these bonds are represented as dashed lines. The full charges present in the carbocation and bromide ion products are partially formed in the transition state, and are indicated as partial charges. (Recall that $\delta+$ and $\delta-$ mean "somewhat positive" and "somewhat negative," respectively; Sec. 1.2D.)

The major factors contributing to the instability of the transition state in Eq. 4.32 are the same ones that make the carbocation unstable: the separation of positive and negative charge and the development of an electron-deficient site. Recognizing the strong resemblance of the transition state to the carbocation, let's make an approximation: *assume that the structure of the transition state closely resembles the structure of the carbocation intermediate.* This approximation can be generalized: *assume that the transition states for reactions involving unstable intermediates can be closely approximated by the intermediates themselves.* This assumption is called **Hammond's postulate**, and was first applied to organic reactions in 1955 by George S. Hammond, then a Professor of Chemistry at Iowa State University. If the *structure* of a transition state resembles that of an unstable intermediate, then it stands to reason that the *free energy* of a transition state also resembles the free energy of the unstable intermediate. For example, for the transformation in Eq. 4.32, we assume that the standard free energies of the carbocation and the transition state for its formation are almost the same.

The utility of Hammond's postulate in dealing with reaction rates can be demonstrated by showing how we could have used it along with a knowledge of carbocation stability to predict the regioselectivity of HBr addition to 2-methylpropene. Recall (Sec. 4.8B) that the rate-limiting step in this reaction is the first step: protonation of the alkene by HBr to give a carbocation. As shown in Eqs. 4.20a and 4.20b, this protonation could occur in two different and competing ways. Protonation of the double bond at one carbon gives the *tert*-butyl cation as the unstable intermediate; protonation of the double bond at the other carbon gives the isobutyl cation. We apply Hammond's postulate by assuming that *the structures and energies of the transition states are approximated by the structures and energies of the unstable intermediates—the carbocations—themselves.*

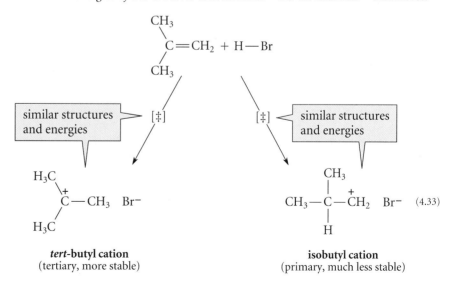

Because the tertiary carbocation is more stable, *the transition state leading to the tertiary carbocation should also be the one of lower energy.* As a result, protonation of 2-methylpropene to give the tertiary carbocation has the transition state with the smaller free energy,

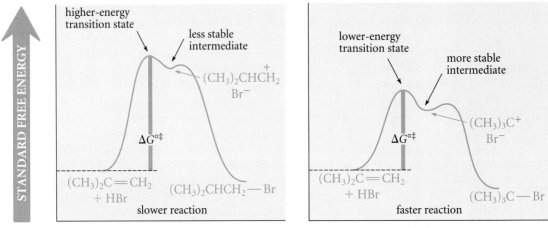

reaction coordinate

Figure 4.14 *A reaction free-energy diagram for the two possible modes of HBr addition to 2-methylpropene. The formation of tert-butyl bromide (right panel) is faster because it involves the more stable carbocation intermediate, and therefore the transition state of lower energy.*

and is thus the faster of the two competing reactions (Fig. 4.14). Addition of HBr to alkenes is regioselective because protonation of a double bond to give a more branched carbocation has a transition state of lower energy than the transition state for protonation to give a less branched carbocation. *It is not the stabilities of the carbocations themselves* that determine which reaction is faster; *it is the relative free energies of the transition states for carbocation formation* that determine the relative rates of the two processes. Only the validity of Hammond's postulate allows us to assume an approximate equivalence between the two possible transition states and the respective carbocation intermediates.

In this text we'll frequently analyze or predict reaction rates by considering the stabilities of reactive intermediates such as carbocations. When we do this, we are invoking Hammond's postulate.

PROBLEMS

4.26 Using dashed lines and partial charges where appropriate, suggest structures for the transition states of the following reactions.

*(a)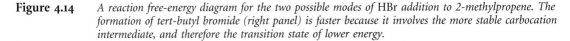

(b) The attack of bromide ion on the *tert*-butyl cation to give $(CH_3)_3CBr$ (*tert*-butyl bromide).

*4.27 Apply Hammond's postulate to decide which reaction is faster: addition of HBr to 2-methylpropene, or addition of HBr to *trans*-2-butene. Assume that the energy difference between the starting alkenes can be ignored. Why is this assumption necessary?

4.9   Catalysis

Some reactions take place much more rapidly in the presence of certain substances that are themselves left unchanged by the reaction. A substance that increases the rate of a reaction without being consumed is called a **catalyst**. A practical example is the catalytic converter on the modern automobile. The platinum catalyst in the converter brings about the rapid oxidation (combustion) of hydrocarbon exhaust emissions. This reaction would not occur were it not for the catalyst; yet the catalyst is left unchanged by the combustion reaction. The catalyst increases the rate of the combustion reaction by many orders of magnitude.

When a catalyst and the reactants exist in separate phases, the catalyst is termed a **heterogeneous catalyst**. The catalyst in a catalytic converter is a heterogeneous catalyst because it is a solid and the reactants are gases. In other cases, a reaction in solution may be catalyzed by a soluble catalyst. A catalyst that is soluble in a reaction solution is called a **homogeneous catalyst**.

In this section, we'll consider three examples of catalyzed alkene reactions. The first example, *catalytic hydrogenation*, is an example of heterogeneous catalysis. The second example, *hydration*, is an example of homogeneous catalysis. The last example involves catalysis of a biological reaction.

CATALYST POISONS

Although catalysts should in theory function indefinitely, in practice many catalysts, particularly heterogeneous catalysts, slowly become less effective. It is as if they "wear out." One reason for this behavior is that they slowly absorb impurities called *catalyst poisons* from the surroundings, and these impurities impede the functioning of the catalyst. An example of this phenomenon also occurs with the catalytic converter. The lead in leaded gasoline is a potent poison of the catalyst in a catalytic converter. This fact, as well as atmospheric lead pollution, are the major reasons why leaded gasoline is no longer used in most automotive engines.

A.   Catalytic Hydrogenation of Alkenes

When a solution of an alkene is stirred under an atmosphere of hydrogen, nothing happens. But if the same solution is stirred under hydrogen in the presence of certain catalysts, the hydrogen is rapidly absorbed by the solution. The hydrogen is consumed because it undergoes an *addition* to the alkene double bond.

$$(4.34)$$

$$CH_3(CH_2)_5CH{=}CH_2 + H_2 \xrightarrow{\ Pt/C\ } CH_3(CH_2)_5CH_2{-}CH_3 \qquad (4.35)$$

These reactions are examples of **catalytic hydrogenation**, an addition of hydrogen in the presence of a catalyst. Catalytic hydrogenation is one of the best ways to convert alkenes into alkanes. Catalytic hydrogenation is an important reaction in both industry and the laboratory. The inconvenience of using a special apparatus for the handling of a flammable gas (hydrogen) is more than offset by the great utility of the reaction.

In the reactions above, the catalyst is written over the reaction arrows. Pt/C is read, "Platinum supported on carbon," or simply, "Platinum on carbon." This catalyst is a finely divided platinum metal that has been precipitated, or "supported," on activated charcoal. A number of noble metals, such as platinum, palladium, and nickel, are useful as hydrogenation catalysts, and they are often used in conjunction with solid support materials such as alumina (Al_2O_3), barium sulfate ($BaSO_4$), or, as in the examples above, activated carbon.

Because hydrogenation catalysts are insoluble in the reaction solution, they are examples of *heterogeneous catalysts*. Even though they involve relatively expensive noble metals, they are very practical because they can be filtered off and reused. Furthermore, because they are exceedingly effective, they can be used in very small amounts. For example, typical catalytic hydrogenation reactions can be run with reactant:catalyst ratios of 100 or more.

How do hydrogenation catalysts work? Research has shown that both the hydrogen and the alkene must be adsorbed on the surface of the catalyst in order for a reaction to occur. The catalyst is believed to form reactive metal-carbon and metal-hydrogen bonds that ultimately are broken to form the products and to regenerate the catalyst sites. Beyond this, the chemical details of catalytic hydrogenation are poorly understood. This is not a reaction for which a curved-arrow mechanism can be written. The mechanism of noble-metal catalysis is an active area of research in many branches of chemistry.

Notice that the benzene ring is inert to conditions under which normal double bonds react readily:

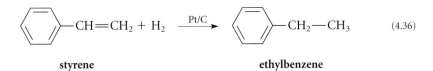

styrene   **ethylbenzene**

(4.36)

(Benzene rings can be hydrogenated, however, with certain catalysts under conditions of high temperature and pressure.) You will see that many other alkene reactions do not affect the "double bonds" of a benzene ring. The relative inertness of benzene rings to alkene reactions was one of the great puzzles of organic chemistry that was ultimately explained by the theory of aromaticity, which is introduced in Chapter 15.

PROBLEMS

4.28   Give the product formed when each of the following alkenes reacts with a large excess of hydrogen in the presence of Pd/C.
     *(a) (E)-1,3-hexadiene   (b) 1-pentene

*4.29   Give the structures of five alkenes, each with the formula C_6H_{12}, that would give hexane as the product of catalytic hydrogenation.

B. Hydration of Alkenes

The alkene double bond undergoes addition of water in the presence of moderately concentrated strong acids such as H_2SO_4, $HClO_4$, and HNO_3.

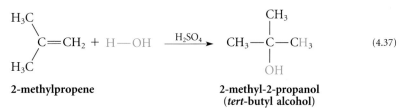

(4.37)

2-methylpropene   **2-methyl-2-propanol**
                                    (***tert*-butyl alcohol**)

The addition of the elements of water is in general called **hydration**. Hence, the addition of water to the alkene double bond is called **alkene hydration**.

Hydration does not occur at a measurable rate in the absence of an acid, and the acid is not consumed in the reaction. Hence, alkene hydration is an *acid-catalyzed reaction*. Because the catalyzing acid is soluble in the reaction solution, it is a *homogeneous catalyst*.

Notice that this reaction, like the addition of HBr, is *regioselective*. As in the addition of HBr, the hydrogen adds to the carbon of the double bond with the smaller number of alkyl substituents. The more electronegative partner of the H—OH bond, the OH group, like the Br in HBr addition, adds to the carbon of the double bond with the greater number of alkyl substituents.

In this reaction, the manner in which the catalyst functions can be understood by considering the mechanism of the reaction, which is very similar to that of HBr addition. In the first step of the reaction, which is the rate-limiting step, the double bond is protonated to give a carbocation. Since water is present, the actual acid is the hydrated proton (H_3O^+).

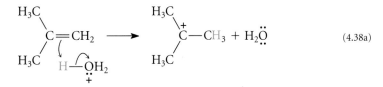

(4.38a)

Notice that this is a Brønsted acid-base reaction. Because this is the rate-limiting step, the rate of the hydration reaction is increased when the rate of this step is increased. The strong acid H_3O^+ is more effective than the considerably weaker acid water in protonating a weak base (the alkene). If a good source of protons is not present, the reaction does not occur because water alone is too weak an acid to protonate the alkene.

In the next step of the hydration reaction, the carbocation is attacked by the Lewis base water in a Lewis acid-base association reaction:

$$(CH_3)_3C^+ \quad :\ddot{O}H_2 \quad \longrightarrow \quad (CH_3)_3C—\overset{+}{\ddot{O}}H_2$$

(4.38b)

Finally, a proton is lost to solvent in another Brønsted acid-base reaction to give the alcohol product and regenerate the catalyzing acid H_3O^+:

$$(CH_3)_3C—\overset{+}{\ddot{O}}H \quad \rightleftharpoons \quad (CH_3)_3C—\ddot{O}H + H_3O^+$$

$$H_2O:$$

(4.38c)

Notice in this mechanism the importance of both Brønsted acid-base and Lewis acid-base reactions. Notice also that the proton consumed in Eq. 4.38a is not the same one that is produced in Eq. 4.38c. Nevertheless, the overall reaction is acid-catalyzed, because there is no *net* consumption of protons.

Because the hydration reaction involves carbocation intermediates, some alkenes give rearranged hydration products.

$$
\underset{\underset{CH_3}{|}}{\overset{\overset{H}{|}}{CH_3-C-CH=CH_2}} + H_2O \xrightarrow{\text{H}_3\text{O}^+} \underset{\underset{CH_3}{|}}{\overset{\overset{OH}{|}}{CH_3-C-CH_2-CH_3}} \qquad (4.39)
$$

PROBLEMS

*4.30   Give the mechanism for the reaction in Eq. 4.39. Show each step of the mechanism separately with careful use of the curved-arrow formalism. Explain why the rearrangement takes place.

4.31   The alkene 3,3-dimethyl-1-butene undergoes acid-catalyzed hydration with rearrangement. Use the mechanism of hydration and rearrangement to predict the structure of the hydration product of this alkene.

Alkene hydration in most cases is *not* a useful laboratory method for the preparation of most alcohols because of rearrangements and other side reactions that can occur. Nevertheless, it is important in organic chemistry because it is a particularly well-understood example of homogeneous catalysis.

Alkene hydration is also important because of its use in the industrial synthesis of ethanol (ethyl alcohol). About 500,000,000 pounds of ethanol are produced annually in the United States by the hydration of ethylene:

$$
CH_2=CH_2 + H_2O \xrightarrow{\text{H}_2\text{SO}_4,\ 100°} CH_3CH_2OH \qquad (4.40)
$$

$$
CH_2=CH_2 + H_2O \xrightarrow[\text{H}_3\text{PO}_4]{300°,\ \text{vapor phase}} CH_3CH_2OH \qquad (4.41)
$$

Actually, the reaction with sulfuric acid in solution occurs in two distinct stages. In the first stage, sulfuric acid undergoes an addition to ethylene to give an addition product, ethyl sulfate:

$$
CH_2=CH_2 + H-OSO_3H \longrightarrow CH_3-CH_2-OSO_3H \qquad (4.42a)
$$

                 sulfuric acid   **ethyl sulfate**

(You should be able to write a mechanism for this reaction.) In the second step, the ethyl sulfate reacts with water to give ethanol and regenerate sulfuric acid:

$$
CH_3-CH_2-OSO_3H + H-OH \longrightarrow CH_3-CH_2-OH + H-OSO_3H \qquad (4.42b)
$$

      ethyl sulfate   **ethanol**
                                     (ethyl alcohol)

The overall result of these two steps is a net hydration.

PROBLEMS

*4.32   *Isopropyl alcohol* is produced commercially by the hydration of propene. Show the steps in this process analogous to Eqs. 4.42a–b. If you do not know the structure of isopropyl alcohol, try to deduce it by analogy from the structure of propene and the mechanism of alkene hydration.

4.33   What alkene undergoes acid-catalyzed hydration to give the following alcohol?

*4.34   What alcohol is formed when methylenecyclobutane undergoes acid-catalyzed hydration?

C. Enzyme Catalysis

Catalysis is not limited to the laboratory or chemical industry. The biological processes of nature involve thousands of chemical reactions, most of which have their own unique naturally occurring catalysts. These biological catalysts are called **enzymes**. Under physiological conditions, most important biological reactions would be too slow to be useful in the absence of their enzyme catalysts. Enzyme catalysts are important not only in nature; they are finding increasing use both in industry and the laboratory.

Many of the best-characterized enzymes are soluble in aqueous solution and hence are homogeneous catalysts. However, other enzymes are known that are immobilized within biological substructures such as membranes, and can be viewed as heterogeneous catalysts.

An example of an important enzyme-catalyzed addition to an alkene is the hydration of fumarate ion to malate ion.

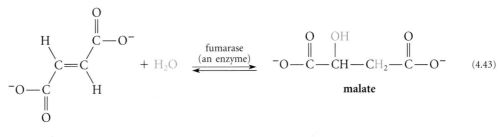

(4.43)

This reaction is catalyzed by the enzyme *fumarase*. It is one reaction in the Krebs cycle, or citric acid cycle, a series of reactions that plays a central role in the generation of energy in biological systems. The effectiveness of fumarase catalysis can be appreciated by the following comparison: At physiological pH and temperature (pH = 7, 37°), the enzyme-catalyzed reaction is 10^4 times as fast as the same reaction in the absence of enzyme at 175°! (At 37° the reaction rate in the absence of enzyme is too slow to measure.)

Compared at a common temperature, the enzyme-catalyzed reaction is many orders of magnitude faster.

Fumarase catalysis illustrates a very important point about catalysis in general. The reaction in Eq. 4.43 is one in which appreciable amounts of both fumarate and malate exist at equilibrium. *The presence of a catalyst cannot change the value of the equilibrium constant* because the equilibrium constant depends only on the relative free energies of reactants and products. Hence, because the enzyme catalyst fumarase catalyzes the forward reaction, it must also catalyze the reverse reaction. (If this were not so, the reaction would go completely to the right, and the equilibrium constant would change.) Consequently, *a catalyst has an equal effect on both forward and reverse reactions of an equilibrium.* This means that if the forward reaction of an equilibrium is accelerated a million-fold by a catalyst, then the reverse reaction is also accelerated by the same amount.

To summarize: a catalyst cannot affect the position of an equilibrium, but it does affect the rate at which a reaction comes to equilibrium.

KEY IDEAS IN CHAPTER 4

- Alkenes are compounds containing carbon-carbon double bonds. Alkene carbon atoms, as well as other trigonal-planar atoms, are sp^2 hybridized.

- The carbon-carbon double bond consists of a σ bond and a π bond. The π electrons can react with Brønsted or Lewis acids.

- In the IUPAC substitutive nomenclature of alkenes, the principal chain, which is the carbon chain containing the greatest number of double bonds, is numbered so that the double bonds receive the lowest numbers.

- Because rotation about the alkene double bond does not occur under normal conditions, some alkenes exist as *cis* and *trans* isomers. These are named using the *E,Z* priority system.

- The unsaturation number of a compound, which is equal to the number of rings plus multiple bonds in the compound, can be calculated from the molecular formula.

- Heats of formation (enthalpies of formation) can be used to determine the relative stabilities of various bonding arrangements. Heats of formation reveal that alkenes with more alkyl branches at their double bonds are more stable than isomers with fewer branches; and *trans* alkenes are more stable than their *cis* isomers.

🧪 Reactants are converted into products through unstable species called transition states. A reaction rate is determined by the standard free energy of activation $\Delta G^{\circ\ddagger}$, the standard free energy difference between the transition state and the reactants. Reactions with smaller standard free energies of activation are faster.

🧪 The rates of multistep reactions are determined in many cases by the rate of the slowest step, called the rate-limiting step. This step is the one with the transition state of highest standard free energy.

🧪 Dipolar molecules such as H—Br, H—OH, and H—OSO$_3$H add to alkenes in a regioselective manner so that the hydrogen adds to the less branched carbon, and the electronegative group to the more branched carbon of the double bond.

🧪 The regioselectivity observed in the addition reactions of hydrogen halides or water to alkenes is a consequence of several facts: (a) the rate-limiting transition state of each reaction resembles a carbocation; (b) the relative stability of cations is in the order tertiary > secondary > primary; and (c) the structures and energies of transition states for reactions involving unstable intermediates (such as carbocations) resemble the structures and energies of the unstable intermediates themselves (Hammond's postulate).

🧪 Reactions involving carbocation intermediates show rearrangements in some cases.

🧪 A catalyst increases the rate of a reaction without being consumed in the reaction. A catalyst does not affect the position of a chemical equilibrium. Catalysts catalyze the forward and reverse reactions of an equilibrium equally.

🧪 Catalysts are of two types: heterogeneous and homogeneous. Catalytic hydrogenation of alkenes involves heterogeneous catalysis; acid-catalyzed hydration of alkenes involves homogeneous catalysis.

🧪 Enzymes are biological catalysts.

ADDITIONAL PROBLEMS

*4.35   Give the structures and IUPAC substitutive names of the isomeric alkenes with molecular formula C$_6$H$_{12}$ containing five carbons in their principal chains.

4.36   Give the structures and IUPAC substitutive names of the isomeric alkenes with molecular formula C$_6$H$_{12}$ containing four carbons in their principal chains.

4.37   Which alkenes within each set give predominantly a single constitutional isomer when treated with HBr, and which give a mixture of isomers? Explain.
    *(a) the alkenes in Problem 4.35   (b) the alkenes in Problem 4.36

4.38   Arrange the compounds within each set in order of increasing heats of formation. (Some may be classified as "about the same.")
*(a)   the alkenes in Problem 4.35   (b)   the alkenes in Problem 4.36

4.39   Give a structural formula for each of the following compounds.
*(a)   (*Z*)-3-methyl-2-octene   (b)   cyclobutene
*(c)   styrene   (d)   isoprene
*(e)   5,5-dimethyl-1,3-cycloheptadiene   (f)   3-methyl-1-octene
*(g)   1-vinylcyclohexene   (h)   3-allylcyclopentene

4.40   Give an IUPAC substitutive name for each of the following compounds. Include the *E,Z* designations where appropriate.

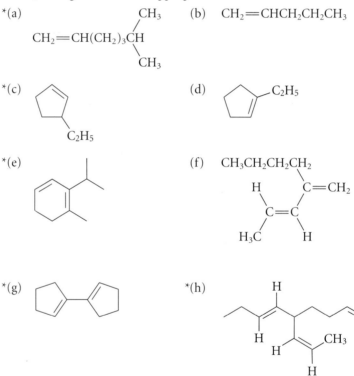

*(a) CH_3 (b) $CH_2{=}CHCH_2CH_2CH_3$

$CH_2{=}CH(CH_2)_3CH$

CH_3

*(c)   (d) C_2H_5

C_2H_5

*(e)   (f) $CH_3CH_2CH_2CH_2$

H $C{=}CH_2$

$C{=}C$

H_3C H

*(g)   *(h) H

CH_3

4.41   A confused chemist Al Keane used the following names in a paper about alkenes. Although each name specifies a structure, in some cases the name is not correct. Correct the names that are wrong.
*(a)   *trans*-1-*tert*-butylpropene   (b)   3-butene
*(c)   2-methylcyclopropene   (d)   (*Z*)-2-hexene
*(e)   2-methyl-1,3-butadiene   (f)   6-methylcycloheptene

4.42   Give the configuration (*E* or *Z*) of each of the following alkenes. Note that D is deuterium, or ^2H, the isotope of hydrogen with atomic mass = 2.

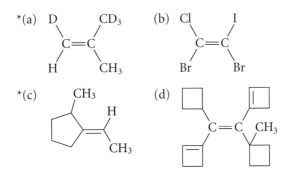

4.43 *(a) The following compound can be prepared by the addition of HBr to either of two alkenes; give their structures.

(b) Starting with the same two alkenes, would the products be different if DBr were used? Explain. (See note about deuterium in Problem 4.42.)

4.44 Classify the compounds within each of the following pairs as either identical molecules (I), constitutional isomers (C), stereoisomers (S), or none of the above (N).

*(a) cyclohexane and 1-hexene

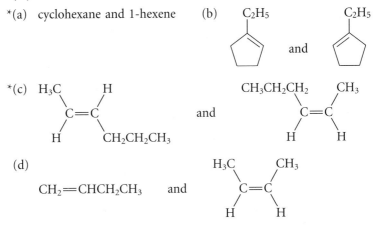

*(e) cyclopentane and cyclopentene

*4.45 Predict the geometry of BF₃. What hybridization of boron is suggested by this geometry? Draw an orbital diagram for boron similar to that for the carbons in ethylene shown in Fig. 4.3.

4.46 Classify each of the labeled bonds in the structure on the next page in terms of the bond type (σ or π) and the component orbitals that overlap to form the bond. (For example, the carbon-carbon bond in ethane is an sp^3-sp^3 σ bond.)

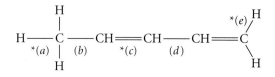

☞
Study Guide Link:
✓ 4.6
*Solving Structure
Problems*

*4.47   An alkene X with molecular formula C_7H_{12} adds HBr to give a *single* alkyl halide Y with molecular formula $C_7H_{13}Br$ and undergoes catalytic hydrogenation to give 1,1-dimethylcyclopentane. Draw the structures of X and Y.

4.48   Give the structures of the two stereoisomeric alkenes with molecular formula C_6H_{12} that will react with HI to give the same *single* product, and will undergo catalytic hydrogenation to give hexane.

*4.49   You have been called in as a consultant for the firm Alcohols Unlimited, which wants to build a plant to produce 1-propanol, $CH_3CH_2CH_2OH$. The research director, Al Keyhall, has proposed that acid-catalyzed hydration of propene be used to prepare this compound. The company president, O. H. Gruppa, has asked you to evaluate this suggestion. Millions of dollars are on the line. What is your answer? Can 1-propanol be prepared in this way?

*4.50   A certain compound A is converted into a compound B in a reaction without intermediates. The reaction has an equilibrium constant $K_{eq} = [B]/[A] = 150$ and, with the free energy of A as a reference point, a standard free energy of activation of 96 kJ/mol (23 kcal/mol).
(a)   Draw a reaction free-energy diagram for this process, showing the relative free energies of A, B, and the transition state for the reaction.
(b)   What is the standard free energy of activation for the reverse reaction B → A? How do you know?

4.51   A reaction $A \rightleftharpoons B \rightleftharpoons C \rightleftharpoons D$ has the reaction free-energy diagram shown in Fig. 4.15.
(a)   Which compound is present in greatest amount when the reaction comes to equilibrium? In least amount?
(b)   What is the rate-limiting step of the reaction?
(c)   Using a vertical arrow, label the standard free energy of activation for the overall $A \rightleftharpoons D$ reaction.
(d)   Which reaction of compound C is faster: $C \rightarrow B$ or $C \rightarrow D$? How do you know?

*4.52   Draw the structure of the reactive intermediate which, according to Hammond's postulate, should most closely resemble the transition state of the rate-limiting step for the hydration of 1-methylcyclohexene.

4.53   For the addition of HBr to 2-ethyl-1-butene, draw:
(a)   a structural representation of the transition state for the initial (rate-limiting) step.
(b)   a structural representation of the transition state for attack of bromide ion on the carbocation intermediate.

*4.54   When 3-methyl-1-butene is burned to CO_2 and H_2O, 3147.75 kJ/mol (752.33 kcal/mol) of heat is produced. How much heat is liberated when 2-methyl-1-butene is burned? Heats of formation are: 3-methyl-1-butene, -28.95 kJ/mol (-6.92 kcal/mol); 2-methyl-1-butene, -36.32 kJ/mol (-8.68 kcal/mol). The heats of combustion of carbon and hydrogen are *not* necessary to work this problem. (*Hint:* Draw a diagram of the energy relationships among each hydrocarbon, $5C + 5H_2$, and $5CO_2 + 5H_2O$.)

4.55   The heat of formation of 1,3-pentadiene is 77.82 kJ/mol (18.60 kcal/mol), and that of 1,4-pentadiene is 105.44 kJ/mol (25.20 kcal/mol).
(a) In which alkene is the bonding arrangement more stable?
(b) Calculate the heat of combustion of 1,3-pentadiene. The heat of combustion of carbon is -393.5 kJ/mol (-94.05 kcal/mol), and that of H_2 is -241.8 kJ/mol (-57.80 kcal/mol).

*4.56   Enthalpies of hydrogenation were at one time an important source of enthalpy data. If the heat liberated on catalytic hydrogenation of *cis*-2-butene is 119.66 kJ/mol (28.6 kcal/mol), and if *trans*-2-butene is 4.18 kJ/mol (1.00 kcal/mol) more stable than *cis*-2-butene, how much heat is liberated when *trans*-2-butene undergoes catalytic hydrogenation?

4.57   You are the purchasing officer for a well-known company and you are looking for savings on heat. You have been offered a ton of liquefied cyclobutane at a certain price by Murky Fuels, Inc., and a ton of liquefied 1-butene at the same price by Bottled Butyl, Inc. Assuming that you could burn each with equal efficiency, which do you choose? The heats of formation are: cyclobutane, 26.65 kJ/mol (6.37 kcal/mol); and 1-butene, -0.13 kJ/mol (-0.03 kcal/mol).

*4.58   Make a model of cycloheptene with the *trans* (or *E*) configuration at the double bond. Now make a model of *cis*-cycloheptene. By examining your models, determine which compound should have the greater heat of formation. Explain.

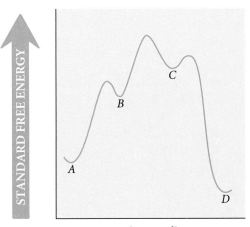

Figure 4.15   *Reaction free-energy diagram for Problem 4.51*

4.59 Which of the following two reactions should have the greatest $\Delta H°$ change? Why? (*Hint:* Examine a model of the two *cis*-alkenes.)

(1)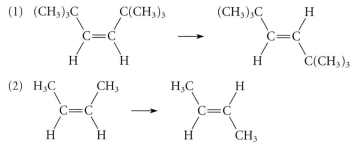

(2)

*4.60 Consider the following compounds and their dipole moments:

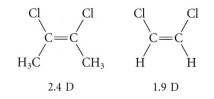

2.4 D 1.9 D

Assume that the C—Cl bond dipole is oriented as follows in each of these compounds.

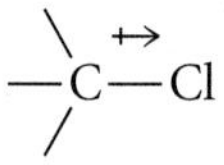

(a) According to the dipole moments above, which is *more* electron-donating toward a double bond, methyl or hydrogen? Explain.

(b) Which of the compounds below should have the greater dipole moment? Explain.

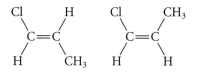

STUDY GUIDE LINK:
✓4.7
*Solving Mechanistic
Problems*

*4.61 Provide a mechanism, including the curved-arrow formalism, for the following reaction.

dilute aqueous H_2SO_4

4.62 The industrial synthesis of *tert*-butyl methyl ether involves treatment of 2-methylpropene with methanol (CH_3OH) in the presence of an acid catalyst, as shown in the following equation.

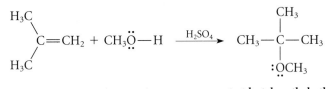

tert-**butyl methyl ether**

This ether is used commercially as an antiknock gasoline additive and as such has become one of the most important organic compounds manufactured on a large industrial scale. Using the curved-arrow formalism, propose a mechanism for this reaction.

*4.63   Using the curved-arrow formalism, suggest a mechanism for the following reaction.

$$H_3C \underset{H_3C}{\overset{H_3C}{\diagdown}}C=CH-CH_2-CH_2-CH=CH_2 + H_2\ddot{O}\colon \xrightarrow{H_2SO_4} \quad$$

(*Hint:* Use Hammond's postulate to decide which double bond should protonate first.)

4.64   Supply the correct curved-arrow formalism for the following acid-catalyzed isomerization.

$$H-\ddot{O}SO_3H$$

*4.65   The standard free energy of formation, ΔG_f°, is the free-energy change for the formation of a substance at 25 °C and 1 atm pressure from its elements in their natural states under the same conditions.

(a) Calculate the equilibrium constant for the interconversion of the alkenes below, given the standard free energy of formation of each. Indicate which compound is favored at equilibrium.

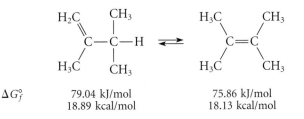

ΔG_f°	79.04 kJ/mol	75.86 kJ/mol
	18.89 kcal/mol	18.13 kcal/mol

(b) What does the equilibrium constant tell us about the rate at which this interconversion takes place?

*4.66   The difference in the standard free energies of formation (see Problem 4.65) for 1-butene and 2-methylpropene is 13.39 kJ/mol (3.2 kcal/mol).

(a) Which compound is more stable? Why?

The standard free energy of activation for the hydration of 2-methylpropene is 22.84 kJ/mol (5.46 kcal/mol) less than that for the hydration of 1-butene.

(b) Which reaction is faster?

(c) Draw reaction free-energy diagrams on the same scale for the hydration reactions of these two alkenes, showing the relative free energies of both starting materials and rate-determining transition states.

(d) By how much free energy is the transition state for the hydration of 2-methylpropene more stable than that for hydration of 1-butene? Using the mechanism of the reaction, suggest a reason why it is more stable.

5

Addition Reactions of Alkenes

The most common reactions of alkenes are *addition reactions*. Chapter 4 covered hydrogen halide addition, catalytic hydrogenation, and hydration, and showed how the curved-arrow formalism and the properties of reactive intermediates can be used to understand the regioselectivity of these additions. This chapter surveys some other addition reactions of alkenes using the same approach, and also presents a different arrow formalism used for describing reactions involving unpaired electrons rather than electron pairs.

5.1 Reaction of Alkenes with Halogens

A. Addition of Chlorine and Bromine

Halogens undergo addition to alkenes.

$$CH_3-CH=CH-CH_3 + Br_2 \xrightarrow[\text{(solvent)}]{CH_2Cl_2} CH_3-CH-CH-CH_3 \quad (5.1)$$

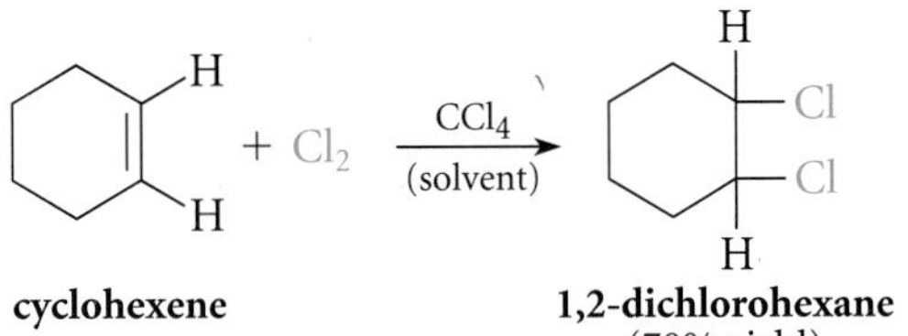

$$\underset{\substack{\textbf{2-butene} \\ (\textit{cis or trans})}}{} \qquad \underset{\substack{Br \quad Br \\ \textbf{2,3-dibromobutane}}}{}$$

cyclohexene + Cl₂ → (CCl₄ solvent) → 1,2-dichlorohexane (70% yield) (5.2)

The products of these reactions are **vicinal dihalides**. "Vicinal" (*vicinus*, Latin "neighborhood") means "on adjacent sites." Thus, vicinal dihalides are compounds with halogens on adjacent carbons.

Bromine and chlorine are the two halogens used most frequently in halogen addition. Fluorine is so reactive that it not only adds to the double bond but also rapidly replaces all the hydrogens with fluorines, often with considerable violence. Iodine adds to alkenes at low temperature, but most diiodides are unstable and decompose to the corresponding alkenes and I_2 at room temperature. Because bromine is a liquid that is more easily handled than chlorine gas, many halogen additions are carried out with bromine. Inert solvents such as carbon tetrachloride (CCl_4) or methylene chloride (CH_2Cl_2) are typically used for halogen additions because these solvents dissolve both halogens and alkenes. The addition of bromine to most alkenes is so fast that when bromine is added dropwise to a solution of the alkene the red bromine color disappears almost immediately. In fact *this discharge of color is a useful test for alkenes.*

Bromine addition can occur by a variety of mechanisms, depending on the solvent, the alkene, and the reaction conditions. One of the most common mechanisms involves a reactive intermediate called a **bromonium ion**.

$$CH_3—CH=CH—CH_3 + Br_2 \longrightarrow CH_3—CH—CH—CH_3 \quad :\overset{..}{\underset{..}{Br}}:^-$$

a bromonium ion   (5.3)

In a bromonium ion, a positively charged bromine with an octet of electrons is bonded to two atoms. Although the bromonium ion forms in one step, it is helpful to envision its formation as follows:

$$CH_3—CH=CH—CH_3 \longrightarrow CH_3—\overset{+}{CH}—CH—CH_3 \longrightarrow CH_3—CH—CH—CH_3 \quad (5.4a)$$

The bromonium ion rather than a carbocation is formed because, in the bromonium ion, every atom has an octet of electrons. One piece of experimental evidence that a carbocation is not involved is that rearrangements are not observed in halogen addition reactions. Other evidence supporting the bromonium ion intermediate has to do with the stereochemical aspects of the reaction, which are discussed in Sec. 7.9C. Analogous halonium ions form when chlorine or iodine reacts with alkenes.

Attack of the bromide ion at either of the carbons bound to bromine in the bromonium ion completes the addition of bromine.

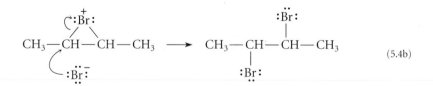

$$CH_3—CH—CH—CH_3 \longrightarrow CH_3—CH—CH—CH_3 \quad (5.4b)$$

This attack occurs because the positively charged bromine is very electronegative and readily accepts an electron pair.

(handwritten margin note: most used solvents (CCl4) + (CH2Cl2))

$CH_3-CH=CH-CH_3 + Br \cdot Br \rightarrow CH_3-CH-CH-CH_3$ *(handwritten annotation)*

B. Halohydrins

$\rightarrow CH_3-CH-CH-CH_3$ *(handwritten annotation)*

In the addition of bromine, the only nucleophile available to attack the bromonium ion is the bromide ion (Eq. 5.4b). When other nucleophiles (Lewis bases) are present, they, too, can attack the bromonium ion to form products other than dibromides. A common situation of this type occurs when the solvent itself is a Lewis base. For example, when an alkene is brominated in a solvent containing a large amount of water, water attacks the bromonium ion.

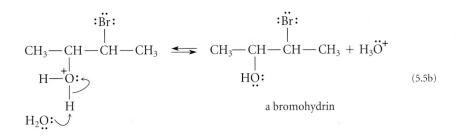

$$(5.5a)$$

Loss of a proton from the oxygen gives the product, which is an example of a **bromo-hydrin**: a compound containing both an —OH and a —Br group.

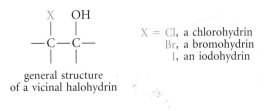

$$(5.5b)$$

a bromohydrin

Bromohydrins are members of the general class of compounds called **halohydrins**, which are compounds containing both a halogen and an —OH group. In the most common type of halohydrin, the two groups occupy adjacent, or *vicinal*, positions.

X   OH
 | |
—C—C—   X = Cl, a chlorohydrin
 | | Br, a bromohydrin
                      I, an iodohydrin

general structure
of a vicinal halohydrin

Halohydrin formation involves the net addition to the double bond of the elements of hypohalous acid, for example, hypobromous acid, HOBr, or hypochlorous acid, HOCl. In fact, aqueous solutions of the halogens contain appreciable amounts of the corresponding hypohalous acids. Household chlorine bleach, a solution of the sodium salt of hypochlorous acid, can be used as a reagent for the preparation of chlorohydrins. Notice that although the products of I_2 addition are unstable, iodohydrins can be prepared.

When the alkene is unsymmetrical, attack of water on the bromonium ion can give two possible products, each resulting from breaking of a different carbon-bromine bond. The reaction is completely regioselective, however, when one carbon of the alkene contains *two* alkyl substituents.

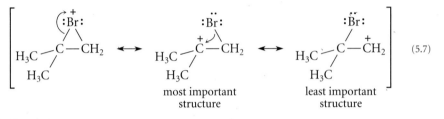

$$CH_3 \diagdown C=CH_2 + Br_2 + H_2O \text{ (solvent)} \longrightarrow CH_3-C-CH_2 + HBr \qquad (5.6)$$

(77% yield)

The reason for this regioselectivity is found in the structure of the bromonium ion intermediate, shown as a hybrid of three resonance structures:

$$\begin{bmatrix} \overset{+}{:}\!\ddot{B}r: & & :\ddot{B}r: & & :\ddot{B}r: \\ H_3C-C-CH_2 & \longleftrightarrow & H_3C-\overset{+}{C}-CH_2 & \longleftrightarrow & H_3C-C-\overset{+}{C}H_2 \\ H_3C & & H_3C & & H_3C \end{bmatrix} \qquad (5.7)$$

most important   least important
structure   structure

When one carbon has two alkyl substituents, the center structure, which resembles a tertiary carbocation, is particularly important. (In fact, it may be in this case that the reactive intermediate *is* a tertiary carbocation rather than a bromonium ion.) This means that in the bromonium ion, if a bond does exist between the bromine and the carbon bearing the two alkyl groups, this bond is very weak. When water attacks the bromonium ion, it is this bond that opens, and the —OH group thus becomes bound to the more branched carbon.

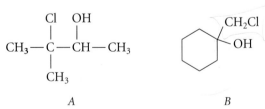

(5.8)

....

**STUDY
PROBLEM
5.1**

Which of the following chlorohydrins could be formed by addition of Cl_2 in water to an alkene? Explain.

$$\begin{matrix} Cl & OH \\ | & | \\ CH_3-C-CH-CH_3 \\ | \\ CH_3 \end{matrix}$$

A

B

Solution   The mechanistic reasoning used in this section shows that the nucleophile (water) attacks the *more branched* carbon of the double bond. In compound *A*, the —OH group is at a carbon with *fewer* alkyl branches, and the Cl at a carbon with *more* alkyl branches. Hence, this compound could *not* be formed in the reaction of Cl_2 and water

with an alkene. Compound *B* could be formed by such a reaction, however, because the
—OH group is at a more branched carbon than the Cl. Don't forget that the carbons
of the ring are branches even though they are part of the ring structure.

PROBLEMS

5.1 Give the products, and the mechanisms for their formation, when 2-methyl-1-hexene reacts with each of the following reagents.

*(a) iodine azide (I—N_3) (*Hint:* The azide ion, N_3^-, behaves much like a halide ion.)

(b) Br_2 in H_2O

*5.2 Give the structure of the alkene that could be used as a starting material to form chlorohydrin *B* in Study Problem 5.1.

5.2 Writing Organic Reactions

Organic chemists use a variety of conventions in writing reactions. Of course, the most
thorough way to write a reaction is to use a complete, balanced equation. Equations 5.1 and
5.6 are examples of such equations. Other information, such as the reaction conditions, is
sometimes included in equations. For example, in Eq. 5.1 the solvent is written over the
arrow, even though the solvent is not an actual reactant. In the following equation, the
H_3O^+ written over the arrow indicates that an acid catalyst (Sec. 4.9B) is required.

$$\begin{array}{c} CH_3 \\ \diagdown \\ C{=}CH_2 + H_2O \xrightarrow{H_3O^+} CH_3{-}\overset{\displaystyle CH_3}{\underset{\displaystyle OH}{\overset{|}{\underset{|}{C}}}}{-}CH_3 \\ \diagup \\ CH_3 \end{array} \qquad (5.9)$$

Equation 5.6 includes a *percentage yield* figure. A **percentage yield** is the percentage
of the theoretical amount of product formed that has actually been isolated from the
reaction mixture by a chemist in the laboratory. Although different chemists might obtain
different yields in the same reaction, the percentage yield gives some idea of how free
the reaction is from contaminating by-products and/or how easy it is to isolate the
product from the reaction mixture. Thus, a reaction $2A + B \rightarrow 3C + D$ should give
three moles of *C* for every one of *B* and two of *A* used (assuming that one of these
reactants is not present in excess). A 90% yield of *C* means that 2.7 moles of *C* were
actually isolated under these conditions. The 10% loss may have been due to handling
problems, small amounts of by-products, or other reasons. Virtually all of the reactions
given in this book are actual laboratory examples; the percentage yield figures included
in many of these reactions are not meant to be learned, but are given simply to indicate
how successful a reaction actually is in practice.

In many cases it is convenient to abbreviate reactions by showing only the *organic
starting materials* and the *major organic product(s)*. The other reactant and conditions are
written over the arrow. Thus, Eq. 5.1 might have been written

$$CH_3{-}CH{=}CH{-}CH_3 \xrightarrow[CCl_4]{Br_2} CH_3{-}\underset{\displaystyle Br}{\overset{|}{CH}}{-}\underset{\displaystyle Br}{\overset{|}{CH}}{-}CH_3 \qquad (5.10)$$

When equations are written this way, by-products are not given and, in many cases, the equation is not balanced. We can tell reactants from catalysts because reactants are consumed and catalysts are not. Thus in Eq. 5.10 we know that Br_2 is a reactant rather than a catalyst because it is consumed in the reaction. This "shorthand" way of writing organic reactions is frequently used because it saves space and time.

5.3   Conversion of Alkenes into Alcohols

Although hydration of alkenes (Sec. 4.9B) is used industrially for the preparation of particular alcohols, it is rarely used for the laboratory preparation of alcohols. This section presents two reaction sequences that are especially useful in the laboratory for the conversion of alkenes into alcohols. These two sequences are complementary because they occur with opposite regiospecificities.

A. Oxymercuration-Reduction of Alkenes

Oxymercuration of Alkenes   In a reaction called **oxymercuration**, alkenes react with mercuric acetate in aqueous solution to give addition products in which an —HgOAc (acetoxymercuri) group and an —OH (hydroxy) group derived from water have added to the double bond.

$$CH_3CH_2CH_2CH_2 - CH = CH_2 + AcO - Hg - OAc + H_2O \xrightarrow{\text{THF, H}_2\text{O}} \qquad (5.11)$$

$$\underset{\text{1-hexene}}{} \qquad \underset{\text{mercuric acetate}}{}$$

$$CH_3CH_2CH_2CH_2 - \underset{\underset{\text{OH}}{|}}{CH} - \underset{\underset{\text{HgOAc}}{|}}{CH_2} \quad + \quad HOAc$$

$$\underset{\text{(95\% yield)}}{} \qquad \underset{\textbf{acetic acid}}{}$$

In this equation, the abbreviation —OAc or AcO— stands for the *acetate* group:

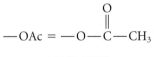

$$-OAc = -O - \overset{\overset{\displaystyle O}{\|}}{C} - CH_3$$

acetate group

The solvent (written over the arrow) is a mixture of water and THF (tetrahydrofuran), a widely used ether.

tetrahydrofuran, or THF

Water is required both as a reactant and to dissolve the mercuric acetate. THF is added to dissolve the alkene, which is not soluble in water alone. The oxymercuration reaction bears a close resemblance to halohydrin formation, which was discussed in the previous section. The first step of the reaction mechanism involves the formation of a cyclic ion called a *mercurinium ion*:

$$R—CH{=}CH_2 \longrightarrow R—\overset{+}{C}H—CH_2 \quad \overset{\text{..}}{\underset{\text{..}}{:}O} Ac^- \longrightarrow R—CH—CH_2 \quad \overset{\text{..}}{\underset{\text{..}}{:}O} Ac^- \tag{5.12a}$$

a mercurinium ion

(Contrast this equation with Eq. 5.4a for the formation of a bromonium ion.) Although the formation of a mercurinium ion in Eq. 5.12a is shown in two steps, this process, like formation of a bromonium ion, actually takes place in a single step.

Just as the bromonium ion in Eq. 5.8 is attacked by the solvent water, which is present in large excess, the mercurinium ion is also attacked by the solvent water:

$$\tag{5.12b}$$

Notice that, of the two carbons in the ring, attack of solvent occurs at the more branched carbon, just as in solvent attack on a bromonium ion (Eq. 5.8). A difference between oxymercuration and halohydrin formation, however, is the degree of regioselectivity. In oxymercuration, attack of water occurs exclusively at the more branched carbon, even if that carbon has only *one* alkyl substituent (as in Eq. 5.12b). (Recall that in halohydrin formation, the reaction is completely regioselective only if one of the alkene carbons has *two* alkyl branches.)

The addition is completed by transfer of a proton to the acetate ion that was formed in Eq. 5.12a.

$$\tag{5.12c}$$

acetate ion

Conversion of Oxymercuration Adducts into Alcohols The utility of oxymercuration lies in the fact that oxymercuration adducts are easily converted into alcohols by treatment with the reducing agent sodium borohydride ($NaBH_4$) in base.

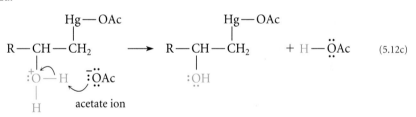

$$4CH_3CH_2CH_2CH_2—\underset{\underset{\displaystyle OH}{|}}{CH}—CH_2—Hg—OAc + 4OH^- + NaBH_4 \longrightarrow$$

sodium
borohydride

$$\tag{5.13}$$

$$4CH_3CH_2CH_2CH_2—\underset{\underset{\displaystyle OH}{|}}{CH}—CH_3 + Na^+ \, B(OH)_4^- + 4Hg^0\!\downarrow + 4AcO^-$$

The key thing to notice about this reaction is that the carbon-mercury bond is replaced by a carbon-hydrogen bond (color in Eq. 5.13). The oxymercuration adducts are usually not isolated, but are treated directly with a basic solution of $NaBH_4$ in the same reaction vessel.

The oxymercuration and $NaBH_4$ reactions, when used sequentially, are referred to as **oxymercuration-reduction** of an alkene. The overall result of oxymercuration-reduction is the *net* addition of the elements of water (H and OH) to an alkene double bond in a *regioselective* manner: the —OH group is added to the *more branched carbon of the double bond*.

$$CH_3CH_2CH_2CH_2—CH{=}CH_2 \xrightarrow[\text{2) } NaBH_4/OH^-]{\text{1) } Hg(OAc)_2/H_2O} CH_3CH_2CH_2CH_2—\underset{\underset{OH}{|}}{CH}—CH_3 \quad \text{(96\% yield)} \quad (5.14)$$

Notice that this is the same overall transformation provided by the hydration reaction (Sec. 4.9B). However, oxymercuration-reduction is much more convenient to run on a laboratory scale than alkene hydration, and is free of rearrangements and other side reactions that are encountered in hydration. For example, the alkene used in the following equation gives products derived from carbocation rearrangement when it undergoes hydration (Problem 4.31, Chapter 4). However, no rearrangements are observed in oxymercuration-reduction:

$$(CH_3)_3C—CH{=}CH_2 \xrightarrow[\text{2) } NaBH_4/OH^-]{\text{1) } Hg(OAc)_2/H_2O} (CH_3)_3C—\underset{\underset{OH}{|}}{CH}—CH_3 \quad (5.15)$$

$$\text{(94\% yield; note lack of rearrangement)}$$

The absence of rearrangements is one reason that mercurinium ions, rather than carbocations, are thought to be the reactive intermediates in oxymercuration.

PROBLEMS

5.3 Give the products expected when each of the following alkenes is subjected to oxymercuration-reduction.
*(a) cyclohexene   (b) ethylene
*(c) *trans*-4-methyl-2-pentene   (d) 2-methyl-2-pentene

5.4 Give the structure of an alkene that would give each of the following alcohols as the major (or only) product as a result of oxymercuration-reduction.

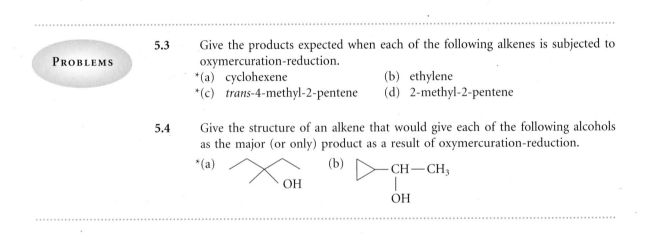

B. Hydroboration-Oxidation of Alkenes

Conversion of Alkenes into Organoboranes   Borane (BH_3) adds regioselectively to alkenes so that the boron becomes bonded to the less branched carbon of the double bond, and the hydrogen becomes bonded to the more branched carbon:

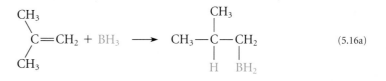

$$(5.16a)$$

Because borane has *three* B—H bonds, one borane molecule can add to three alkene molecules. The first of these additions to 2-methylpropene is shown in Eq. 5.16a. The second and third additions are as follows:

Second addition:

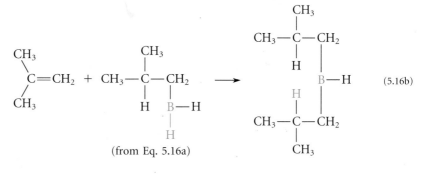

$$(5.16b)$$

(from Eq. 5.16a)

Third addition:

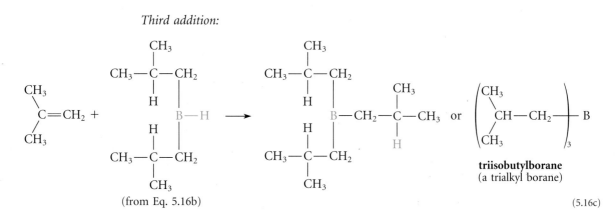

(from Eq. 5.16b)

triisobutylborane
(a trialkyl borane)

$$(5.16c)$$

The addition of BH_3 to alkenes is called **hydroboration**. The product of hydroboration is a trialkylborane such as the triisobutylborane shown in Eq. 5.16c.

BORANE AND DIBORANE

Borane actually exists as a toxic, colorless gas called *diborane*, which has the formula B_2H_6. Because borane is a Lewis acid, the boron has a strong tendency to acquire an additional electron pair. This tendency is satisfied by the formation of diborane, in which two hydrogens are shared between the two borons in remarkable "half bonds." This bonding can be depicted with resonance structures:

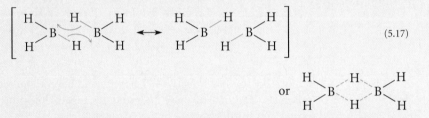

(5.17)

When dissolved in an ether solvent, diborane dissociates to form a borane-ether complex. Because ethers are Lewis bases, they can satisfy the electron deficiency at boron:

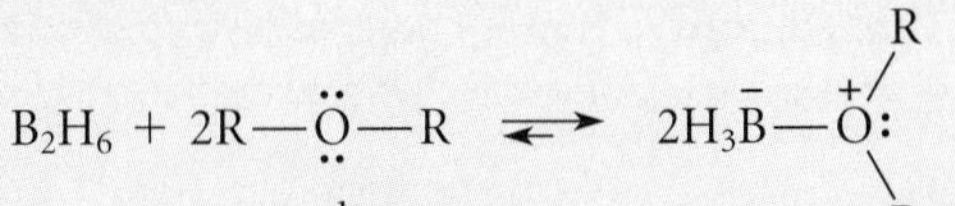

$$B_2H_6 + 2R\overset{..}{\underset{..}{O}}R \;\rightleftharpoons\; 2H_3\bar{B}\overset{R}{\underset{R}{\overset{+}{O}:}}$$

(5.18)

an ether

borane-ether complex

The following ethers are commonly used as solvents in the hydroboration reaction:

$$C_2H_5\overset{..}{\underset{..}{O}}C_2H_5 \qquad \qquad$$

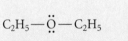

diethyl ether   **tetrahydrofuran (THF)**

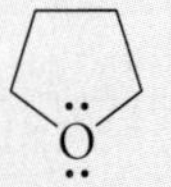

$$CH_3\overset{..}{\underset{..}{O}}\!-CH_2CH_2\!-\overset{..}{\underset{..}{O}}\!-CH_2CH_2\!-\overset{..}{\underset{..}{O}}CH_3$$

diethylene glycol dimethyl ether (diglyme)

Borane-ether complexes are the actual reagents involved in hydroboration reactions. For simplicity, the simple formula BH_3 will be used for borane.

Each of the three additions in the hydroboration reaction is believed to occur as a single step, which can be represented with the curved-arrow formalism as follows:

$$\begin{array}{c}\text{(H—BH}_2 \text{ adding across } H_3C_2C=CH_2)\end{array} \longrightarrow \begin{array}{c}\text{H}_3C \\ \text{H}_3C\end{array}C\text{—CH}_2 \longrightarrow \text{ further addition} \qquad (5.19)$$

A reaction that occurs, like this one, in a single step without intermediates is said to occur by a **concerted mechanism**, because everything happens "in concert"—at the same time. No intermediates are involved in a concerted reaction. Because carbocation intermediates are not involved, rearrangements do not occur in hydroboration.

If ionic intermediates are not involved in hydroboration, how can the regioselectivity of the reaction be explained? The regioselectivity arises from the charge distribution in the transition state. Because BH_3 is an electron-deficient Lewis acid, it is readily attacked by alkene π electrons. Such an attack leaves a carbon of the alkene somewhat electron-deficient. If boron becomes bound to the terminal carbon, the carbon bearing the R-group—the more branched carbon—is somewhat electron-deficient in the transition state:

$$\begin{array}{c}\text{H—BH}_2 \\ \text{R—CH=CH}_2\end{array} \longrightarrow \left[\begin{array}{c}\overset{\delta-}{\text{H-----BH}_2} \\ \text{R—CH—CH}_2 \\ \quad\quad\quad_{\delta+}\end{array}\right]^{\ddagger} \longrightarrow \begin{array}{c}\text{H} \quad \text{BH}_2 \\ \text{R—CH—CH}_2\end{array} \qquad (5.20a)$$

In contrast, if boron becomes bound to the carbon bearing the R-group, then the terminal carbon is somewhat electron deficient in the transition state:

$$\begin{array}{c}\text{H}_2\text{B—H} \\ \text{R—CH=CH}_2\end{array} \longrightarrow \left[\begin{array}{c}\overset{\delta-}{\text{H}_2\text{B-----H}} \\ \text{R—CH—CH}_2 \\ \quad\quad\quad_{\delta+}\end{array}\right]^{\ddagger} \longrightarrow \begin{array}{c}\text{H}_2\text{B} \quad \text{H} \\ \text{R—CH—CH}_2\end{array} \qquad (5.20b)$$

As these structures show, the two possible transition states have some *carbocation character* even though carbocations themselves are not intermediates in the reaction. Because alkyl substitution at electron-deficient carbons is a stabilizing effect (Sec. 4.7C), the transition state in Eq. 5.20a has a lower energy than the transition state in Eq. 5.20b. A reaction with a transition state of lower energy has a larger rate. Consequently, the reaction in Eq. 5.20a is faster than the one in Eq. 5.20b, and, as a result, hydroboration is regioselective.

Conversion of Organoboranes into Alcohols   The utility of hydroboration lies in the many reactions of organoboranes themselves. One of the most important reactions of organoboranes is their conversion into alcohols with hydrogen peroxide (H_2O_2) and aqueous base.

$$\left(\begin{array}{c}\text{CH}_3 \\ \text{CH—CH}_2 \\ \text{CH}_3\end{array}\right)_3\text{B} + 3H_2O_2 + {}^-\text{OH} \longrightarrow 3 \begin{array}{c}\text{CH}_3 \\ \text{CH—CH}_2\text{—OH} \\ \text{CH}_3\end{array} + {}^-\text{B(OH)}_4 \qquad (5.21)$$

triisobutylborane   hydrogen peroxide

2-methyl-1-propanol
(isobutyl alcohol)
(95% yield)

It is important to notice that the net result of this transformation is *replacement of the boron by an* —OH *in each alkyl group.* The oxygen of the —OH group comes from the H_2O_2.

Typically, the organoborane product of hydroboration is not isolated, but is treated directly with alkaline hydrogen peroxide to give an alcohol. The addition of borane and subsequent reaction with H_2O_2, taken together, are referred to as **hydroboration-oxidation**.

Tracing the fate of an alkene through the entire hydroboration-oxidation sequence reveals that the *net* result is addition of the elements of water (H, OH) to the double bond in a regioselective manner so that the —OH ends up at the *less branched carbon atom of the double bond* (color in Eq. 5.22):

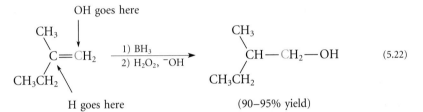

$$(5.22)$$

Hydroboration-oxidation is therefore a useful way to synthesize certain alcohols from alkenes. It is particularly useful to prepare alcohols of the general structure R_2CH—CH_2—OH or R—CH_2—CH_2—OH, as in Eqs. 5.22 and 5.23. Because carbocations are not involved in either reaction, the alcohol products are not contaminated by constitutional isomers arising from rearrangements.

The following example shows that the benzene ring does not react with BH_3 even though the ring apparently contains double bonds:

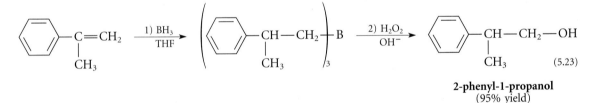

$$(5.23)$$

This calls to mind a similar resistance of benzene rings to other addition reactions, such as catalytic hydrogenation (Sec. 4.9A).

H. C. BROWN AND HYDROBORATION

Hydroboration was discovered accidentally in 1955 by Professor H. C. Brown (1912–   ) and his colleagues of Purdue University. Brown quickly realized its significance and in subsequent years has carried out research demonstrating the versatility of organoboranes as intermediates in organic synthesis. Brown, now an Emeritus Professor at Purdue, still conducts research in this area, which he calls "a vast unexplored continent." In 1979 his research was recognized with the Nobel Prize in Chemistry.

5.5   Give the product expected from the hydroboration-oxidation of each of the following alkenes.

*(a)   cyclohexene   (b)   ethylene

*(c)   *trans*-4-methyl-2-pentene   (d)   2-methyl-2-pentene

*5.6   Contrast the answers for Problem 5.5 with the answers for the corresponding parts of Problem 5.3. For which alkenes are the alcohol products the same? For which are they different? Explain why the same alcohols are obtained in some cases and why different alcohols are obtained in others.

5.7   For each case below, give the structure of an alkene that would give the alcohol as the major (or only) product of hydroboration-oxidation.

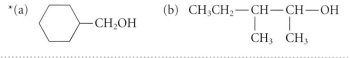

*(a) ⬡—CH₂OH   (b)   CH₃CH₂—CH—CH—OH
                                       | |
                                      CH₃   CH₃

C. Comparison of Methods for the Synthesis of Alcohols from Alkenes

Let's now compare the different ways of preparing alcohols from alkenes. *Hydration of alkenes* is a useful industrial method for the preparation of a few alcohols, but it is not a good laboratory method (Sec. 4.9B). Indeed, many industrial methods for the preparation of organic compounds are not *general*. That is, an industrial method typically works well in the specific case for which it was designed, but cannot be applied to other related cases. The reason is that the chemical industry has gone to great effort to work out conditions that are optimal for the preparation of *particular compounds* of great commercial importance (such as ethanol) using reagents that are readily available and cheap (such as water and common inorganic acids). Although the industrial ethanol synthesis could be duplicated in the laboratory, there is no need to do so because ethanol *is* cheap and readily available from industrial sources. For laboratory work it is not practical for chemists to design a specific procedure for each new compound. Thus the development of general methods that work with a wide variety of compounds is important. Because laboratory synthesis is generally carried out on a relatively small scale, the expense of reagents is less of a concern.

Hydroboration-oxidation and oxymercuration-reduction are both *general laboratory methods* for the preparation of alcohols from alkenes. That is, they can be applied successfully to a large variety of alkene starting materials. Which method is better for preparation of a given alcohol? A choice between the two methods usually hinges on the difference in their regioselectivities. As shown in the following equation, hydroboration-oxidation gives an alcohol in which the —OH group has been added to the *less branched carbon* of the double bond. Oxymercuration-reduction gives an alcohol in which the —OH group has been added to the *more branched carbon* of the double bond.

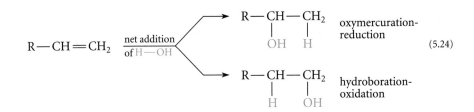

(5.24)

Study Guide Link:
✓**5.1**
*How to Study
Organic Reactions*

For alkenes that yield the same alcohol by either method, the choice between the two is in principle arbitrary.

Problems

5.8   Which of the following alkenes would give the same alcohol on either oxymercuration-reduction or hydroboration-oxidation, and which would give different alcohols? Explain.

*(a) *cis*-2-butene   (b)   cyclobutene
*(c)   3-methyl-1-pentene   (d)   1-methylcyclohexene

5.9   From what alkene and by which method would you prepare each of the following alcohols essentially free of constitutional isomers?

*(a) $(C_2H_5)_3C$—OH   (b) 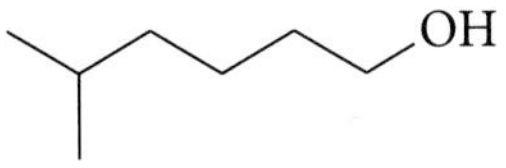 ![branched structure with OH]   *(c) 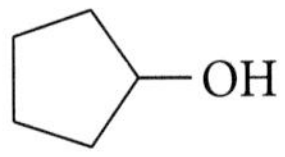 ![cyclopentane with OH]

5.4   Ozonolysis of Alkenes

Ozone, O_3, adds to alkenes at low temperature to yield an unstable compound called a *molozonide*. The molozonide is rapidly transformed into a second adduct, called simply an *ozonide*.

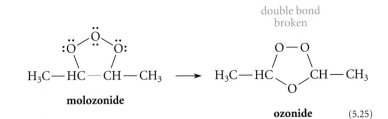

(5.25)

Notice that both bonds of the double bond are broken in the final ozonide.

The reaction of an alkene with ozone to yield products of double-bond cleavage is called *ozonolysis*. (The suffix *-lysis* is used for describing bond-breaking processes. Examples are *hydrolysis*, "bond-breaking by water," *thermolysis*, "bond-breaking by heat," and, of course, *ozonolysis*, "bond-breaking by ozone.")

PREPARATION OF OZONE

Ozone is generated by passing oxygen through an electrical discharge in an apparatus called an *ozonator*. The following reaction takes place:

$$3O_2 \xrightarrow{\text{electrical discharge}} 2O_3$$

This reaction is not unlike what happens in the atmosphere when ozone is produced under the influence of lightning.

The first step in ozonolysis, formation of the molozonide, is another addition reaction of the alkene double bond. The central oxygen of ozone is a positively charged electronegative atom and therefore strongly attracts electrons. Use of the curved-arrow formalism shows that this oxygen can accept an electron pair when the other oxygen of the $O=O$ bond is attacked by the π electrons of the alkene.

$$\ddot{O} \overset{+}{\underset{}{\diagup}} \overset{\ddot{O}}{\diagdown} \ddot{O}:^- \qquad :\ddot{O} \diagup \overset{\ddot{O}}{\diagdown} \ddot{O}:$$

$$H_3C\!-\!HC\!=\!CH\!-\!CH_3 \longrightarrow H_3C\!-\!HC\!-\!CH\!-\!CH_3 \qquad (5.26)$$

Notice that this reaction results in the formation of a ring, because the three oxygens of the ozone molecule remain intact. Additions that give rings are called **cycloadditions**. Furthermore, the cycloaddition of ozone occurs in a single step. Hence, this is another example, like hydroboration, of a *concerted mechanism*.

The molozonide cycloaddition product is unstable, and spontaneously forms the ozonide. In this reaction, the remaining carbon-carbon bond of the alkene is broken.

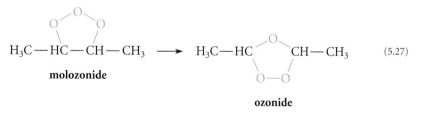

$$H_3C\!-\!HC\!-\!CH\!-\!CH_3 \longrightarrow H_3C\!-\!HC \diagup\!\!\overset{O}{\diagdown}\!\!CH\!-\!CH_3 \qquad (5.27)$$

molozonide

ozonide

STUDY GUIDE LINK:
✓5.2
*Mechanism of
Ozonolysis*

Ozonolysis is useful because of the reactions that ozonides can undergo. Ozonides can be converted into aldehydes, ketones, or carboxylic acids, depending on the structure of the alkene starting material and the reaction conditions. When the ozonide is treated with dimethyl sulfide, $(CH_3)_2S$, the ozonide is split:

$$CH_3-HC \underset{O-O}{\overset{O}{\diagup\diagdown}} CH-CH_3 + CH_3-\overset{..}{\underset{..}{S}}-CH_3 \longrightarrow 2\ CH_3-CH{=}O + \overset{CH_3}{\underset{CH_3}{\overset{\diagup}{\underset{\diagdown}{\overset{+}{S}}}}}-\overset{..}{\underset{..}{\overset{..}{O}}}\!\!\overset{-}{:} \quad (5.28)$$

dimethyl sulfide   an aldehyde

Notice that the net transformation resulting from ozonolysis of an alkene followed by dimethyl sulfide treatment is the replacement of a $\diagup C{=}C\diagdown$ group by two $\diagdown C{=}O$ groups:

double bond is split

$$CH_3-CH{\overset{\Vert}{\Vert}}CH-CH_3 \xrightarrow[\text{2) } (CH_3)_2S]{\text{1) } O_3} CH_3-CH{=}O + O{=}CH-CH_3 \quad (5.29)$$

$={=}O$ here $O{=}$ here

If the two ends of the double bond are identical, as in Eq. 5.29, then two equivalents of the same product are formed. If the two ends of the alkene are different, then a mixture of two different products is obtained:

$$CH_3(CH_2)_5CH{=}CH_2 \xrightarrow[CH_2Cl_2]{O_3} \xrightarrow{(CH_3)_2S} CH_3(CH_2)_5CH{=}O + O{=}CH_2 \quad (5.30)$$

heptanal   formaldehyde
(75% yield)

If a carbon of the double bond in the starting alkene bears a hydrogen, then an aldehyde is formed, as in Eq. 5.29 and Eq. 5.30. In contrast, if a carbon of the double bond bears no hydrogens, then a ketone is formed instead:

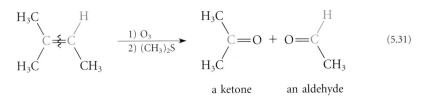

a ketone   an aldehyde

If the ozonide is simply treated with water, hydrogen peroxide (H_2O_2) is formed as a by-product. (The purpose of the dimethyl sulfide in Eq. 5.28 is to react with the hydrogen peroxide.) Under these conditions (or if hydrogen peroxide is added specifically), aldehydes are converted into carboxylic acids, but ketones are unaffected. Hence, the alkene in Eq. 5.31 would react as follows:

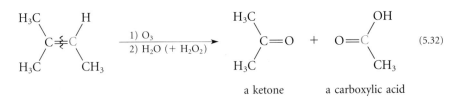

a ketone   a carboxylic acid

The different results obtained in ozonolysis are summarized in Table 5.1.

If the structures of its ozonolysis products are known, then the structure of an unknown alkene can be deduced. This idea is illustrated in the following study problem.

Table 5.1 Summary of Ozonolysis Results under Different Conditions

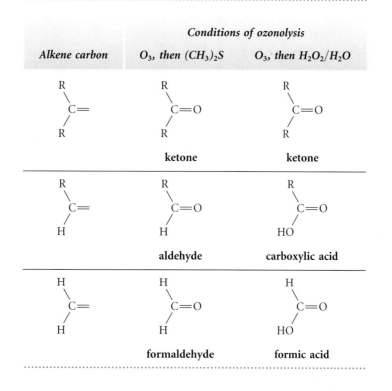

Alkene carbon	*Conditions of ozonolysis*	
	O_3, then $(CH_3)_2S$	*O_3, then H_2O_2/H_2O*
$R_2C=$	$R_2C=O$ **ketone**	$R_2C=O$ **ketone**
$R(H)C=$	$R(H)C=O$ **aldehyde**	$R(HO)C=O$ **carboxylic acid**
$H_2C=$	$H_2C=O$ **formaldehyde**	$H(HO)C=O$ **formic acid**

STUDY PROBLEM 5.2

Alkene *X* of unknown structure gives the following products after treatment with ozone followed by aqueous H_2O_2:

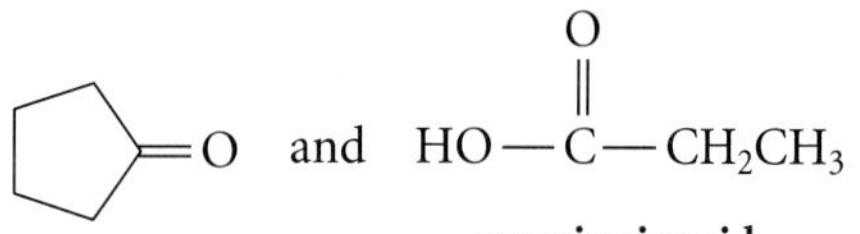

propionic acid

cyclopentanone

What is the structure of *X*?

Solution The structure of the alkene can be deduced by *mentally reversing* the ozonolysis reaction. To do this, rewrite the C=O double bonds as "dangling" double bonds by dropping the oxygen:

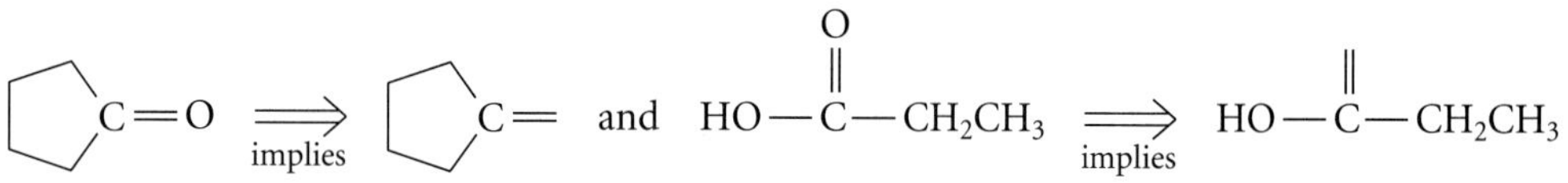

Next, replace the HO— group of any carboxylic acid fragments with H—. This is done because a carboxylic acid is formed only when there is a hydrogen on the carbon of the double bond (Table 5.1).

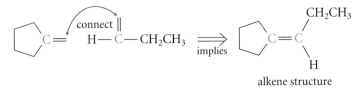

Finally, connect the "dangling ends" of the double bonds in the two partial structures to generate the structure of the alkene:

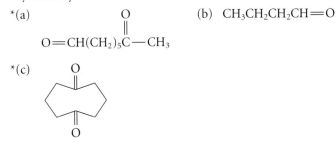

alkene structure

PROBLEMS

5.10 Give the products (if any) expected from treatment of each of the following compounds with ozone followed by dimethyl sulfide.

*(a) [structure: cyclohexane with =CH₂]   (b) 3-methyl-2-pentene

*(c) cyclooctene   *(d) 2-methylpentane

*5.11 Give the products (if any) expected when the compounds in Problem 5.10 are treated with ozone followed by aqueous hydrogen peroxide.

5.12 In each case, give the structure of an eight-carbon alkene that would yield each of the following compounds (and no others) after treatment with ozone followed by dimethyl sulfide.

*(a)
$$O=CH(CH_2)_5\overset{\displaystyle O}{\overset{\|}{C}}-CH_3$$

(b) $CH_3CH_2CH_2CH=O$

*(c) [structure: bicyclic compound with two C=O groups]

*5.13   What aspect of alkene structure cannot be determined by ozonolysis?

5.5

Conversion of Alkenes into Glycols

Alkenes are readily converted into *vicinal glycols* using either osmium tetroxide (OsO_4) or alkaline potassium permanganate ($KMnO_4$). A **glycol**, or **diol**, is an alcohol containing two —OH groups; in a **vicinal glycol** these —OH groups are on adjacent carbons.

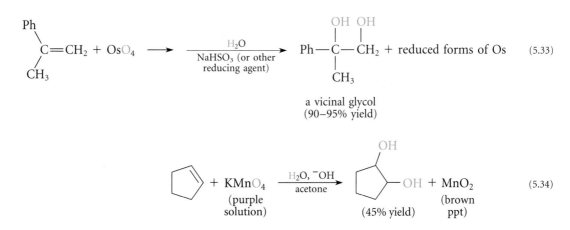

a vicinal glycol
(90–95% yield)

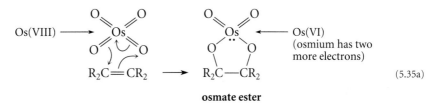

The reaction with KMnO$_4$ is the basis for a well known qualitative test for alkenes called the *Baeyer test for unsaturation.* When an alkene is added to a solution containing aqueous permanganate, the brilliant purple color of the permanganate is replaced by a murky brown precipitate of MnO$_2$ (manganese dioxide). (Other compounds such as aldehydes and some types of alcohols give a positive test as well.)

The mechanisms of these two reactions are similar. The OsO$_4$ reaction, for example, involves a cyclic intermediate called an *osmate ester.* Although the mechanism by which it is formed is still debated, it can be represented in the following way:

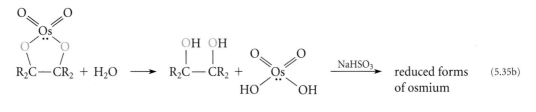

Note that this reaction is a *cycloaddition* much like ozonolysis. The osmium in OsO$_4$ is relatively electronegative because it is in a +8 oxidation state. The attraction of osmium for electrons in its bonds to oxygen makes the oxygens rather electrophilic, much like the outer oxygens of ozone. Consequently, one of these oxygens can serve as a point of attack of the alkene π electrons, as shown in Eq. 5.35a. The mechanism of the KMnO$_4$ reaction is probably similar.

A glycol is formed when the cyclic osmate ester is treated with water and a mild reducing agent such as sodium bisulfite, NaHSO$_3$. The water converts the osmate ester into a glycol, and the sodium bisulfite converts the osmium-containing by-products into reduced forms of osmium that are easy to separate by filtration.

In the KMnO$_4$ reaction, the glycol product forms spontaneously, and the MnO$_2$ by-product precipitates; no other reagents are necessary.

Osmium tetroxide is very toxic and quite expensive, but relatively simple procedures have been developed that permit the osmium by-products to be re-oxidized to OsO_4. Because the osmium can be recycled in this way, the expense of OsO_4 is not a serious problem for small-scale reactions. For larger scale reactions, the less expensive $KMnO_4$ is the preferred reagent, although yields are usually lower with this reagent because of side reactions that occur. (The alkaline conditions of the permanganate reaction minimize these side reactions.)

PROBLEMS

5.14   Give the products expected when each of the following alkenes reacts with either OsO_4 followed by aqueous $NaHSO_3$, or with alkaline $KMnO_4$.
   *(a)   2-methylpropene   (b)   1-methylcyclopentene

   *(c)

5.15   From what alkene could each of the following glycols be prepared by the methods described in this section?

   *(a) $HO-CH_2-\underset{\underset{OH}{|}}{CH}-\underset{\underset{OH}{|}}{CH}-CH_2-OH$ (b) $HO-\underset{\underset{CH_3}{|}}{CH}-\underset{\underset{OH}{|}}{CH}-CH_3$

*5.16   Suggest a structure for the cyclic intermediate in the reaction of alkenes with $KMnO_4$. Show the formation of this intermediate with the curved-arrow formalism. (*Hint:* The Lewis structure of MnO_4^- can be written as follows.)

Free-Radical Addition of Hydrogen Bromide to Alkenes

5.6

A. The Peroxide Effect

Recall that addition of HBr to alkenes is a regioselective reaction in which the bromine is directed to the more branched carbon of a double bond (Sec. 4.7). For example, 1-pentene reacts with HBr to give exclusively 2-bromopentane:

$$CH_3CH_2CH_2CH=CH_2 + H-Br \longrightarrow CH_3CH_2CH_2\underset{\underset{Br}{|}}{CH}CH_3 \qquad (5.36)$$

1-pentene

2-bromopentane
(79% yield)

For many years, results such as this were at times difficult to reproduce. Some investigators found that the addition of HBr was a regioselective reaction. Others, however, obtained mixtures of constitutional isomers in which the second isomer had bromine bound at the *less branched* carbon of the double bond. In the late 1920s, Morris Kharasch of the University of Chicago began investigations that led to a solution of this puzzle. He found that when traces of *peroxides* (compounds of the general structure R—O—O—R) are added to the reaction mixture, *the regioselectivity of HBr addition is reversed!* In other words, 1-pentene was found to react in the presence of peroxides so that the bromine adds to the less branched carbon of the double bond:

$$
\underset{\textbf{1-pentene}}{CH_3CH_2CH_2CH=CH_2} + H-Br \xrightarrow[\substack{\text{(benzoyl peroxide;} \\ \text{small amount used)}}]{\substack{O \quad\quad O \\ \| \quad\quad \| \\ Ph-C-O-O-C-Ph}} \underset{\substack{\textbf{1-bromopentane} \\ \text{(96\% yield)}}}{CH_3CH_2CH_2CH_2CH_2-Br} \quad (5.37)
$$

(Contrast this result with that in Eq. 5.36.) This reversal of regioselectivity in HBr addition is termed the **peroxide effect**.

> Because the regioselectivity of ordinary HBr addition is described by *Markownikoff's rule* (Sec. 4.7A), the peroxide-promoted addition is sometimes said to have *anti-Markownikoff* regioselectivity. This means simply that the bromine is directed to the carbon of the alkene double bond with fewer carbon branches.

It was also found that light further promotes the peroxide effect. When Kharasch and his colleagues scrupulously excluded peroxides and light from the reaction, they found that HBr addition has the "normal" regioselectivity shown in Eq. 5.36.

The "peroxide effect" is quite general. In other words, *in the presence of peroxides, the addition of HBr to alkenes occurs such that the hydrogen is bound to the carbon of the double bond bearing the greater number of alkyl branches.* Very small amounts of peroxides are required in order to bring about this effect.

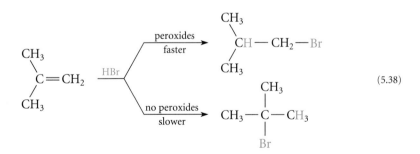

Furthermore, the regioselectivity of HI or HCl addition to alkenes is not affected by the presence of peroxides. For these hydrogen halides, the normal regioselectivity of addition predominates whether peroxides are present or not!

The addition of HBr to alkenes in the presence of peroxides occurs by a mechanism that is completely different from that for normal addition. This mechanism involves reactive intermediates known as *free radicals*. In order to appreciate the reasons for the peroxide effect, then, let's digress and learn some basic facts about free radicals.

B. Free Radicals

In all reactions considered previously we used an arrow formalism that indicates the movement of electrons in pairs. The dissociation of HBr, for example, is written

$$H \overset{\frown}{\longrightarrow} \ddot{B}r\colon\ \longrightarrow\ H^+ + \colon\ddot{\underset{\cdot\cdot}{B}}r\colon^- \tag{5.39}$$

In this reaction, both electrons of the H—Br covalent bond move to the bromine to give a bromide ion, and the hydrogen is left as the electron-deficient proton. This type of bond breaking is called **heterolytic cleavage**, or **heterolysis** (*hetero* = different; *lysis* = bond breaking).

However, bond rupture can occur in another way. An electron-pair bond may also break so that each bonding partner retains one electron of the chemical bond.

$$H \longrightarrow \ddot{B}r\colon\ \longrightarrow\ H\cdot + \cdot\ddot{B}r\colon \tag{5.40}$$

In this process, a hydrogen *atom* and a bromine *atom* are produced. As you should verify, these atoms are uncharged. This type of bond breaking is called **homolytic cleavage**, or **homolysis** (*homo* = the same; *lysis* = bond breaking).

A different curved-arrow formalism is used for homolysis. In this formalism, called the **fishhook formalism**, *electrons are moved individually rather than in pairs.* This type of electron flow is represented with singly barbed arrows, or fishhooks; one fishhook is used for each electron:

$$H \overset{\frown\frown}{\longrightarrow} \ddot{B}r\colon\ \longrightarrow\ H\cdot + \cdot\ddot{B}r\colon \tag{5.41}$$

Homolytic bond cleavage is not restricted to diatomic molecules. For example, peroxides undergo homolytic cleavage at the O—O bond.

$$(CH_3)_3C \longrightarrow \overset{\frown\frown}{\underset{\cdot\cdot}{\ddot{O}}} \ \underset{\cdot\cdot}{\ddot{O}} \longrightarrow C(CH_3)_3\ \longrightarrow\ 2(CH_3)_3C \longrightarrow \underset{\cdot\cdot}{\ddot{O}}\cdot \tag{5.42}$$

di-*tert*-butyl peroxide   ***tert*-butoxy radical**

The fragments on the right side of this equation possess unpaired electrons. Any species with at least one unpaired electron is called a **free radical**. The hydrogen atom and the bromine atom on the right side of Eq. 5.40, and the *tert*-butoxy radical on the right side of Eq. 5.42, are all examples of free radicals.

PROBLEMS

5.17   Draw the products of each of the following transformations shown by the fishhook formalism.

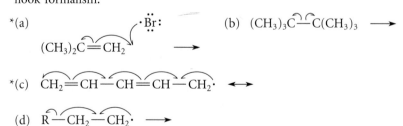

5.18   Indicate whether each of the following reactions is homolytic or heterolytic. Write the appropriate fishhook or curved-arrow formalism for each.

*(a) $H_2\ddot{O}$ ⟶ $H\cdot + \cdot\ddot{O}H$ (b) $H_2\ddot{O}$ ⟶ $H^+ + :\ddot{O}H^-$

*(c) $CH_2=CH_2 + HBr$ ⟶ $^+CH_2—CH_3 + Br^-$

(d) H_2 ⟶ $2H\cdot$

C. Free-Radical Chain Reactions

Although a few stable free radicals are known, most free radicals are very reactive. When they are generated in chemical reactions, they generally behave as *reactive intermediates*, that is, they react before they can accumulate in significant amounts. This section shows how free radicals are involved as reactive intermediates in the peroxide-promoted addition of HBr to alkenes. This discussion will provide the basis for an understanding of the peroxide effect, which is the subject of the next section.

Free-radical reactions typically involve three types of steps:

1. *initiation* steps,

2. *propagation* steps, and

3. *termination* steps.

Let's examine each of these steps using the peroxide-promoted addition of HBr to alkenes as an example of a typical free-radical reaction.

Initiation   In the **initiation** steps, the free radicals that take part in subsequent steps of the reaction are formed from a molecule that readily undergoes homolysis, called a **free-radical initiator**. The initiator is in effect the source of free radicals. Peroxides such as di-*tert*-butyl peroxide are frequently used as free-radical initiators. The first initiation step in the free-radical addition of HBr to an alkene is the homolysis of the peroxide.

$$(CH_3)_3C—\ddot{O}—\ddot{O}—C(CH_3)_3 \longrightarrow 2(CH_3)_3C—\ddot{O}\cdot \qquad (5.43)$$

di-*tert*-butyl peroxide   ***tert*-butoxy radical**

> Although most peroxides can serve as free-radical initiators, a notable exception is hydrogen peroxide (H_2O_2), which is *not* commonly used as a source of initiating free radicals. The reason is that the O—O bond in hydrogen peroxide is considerably stronger than the O—O bonds in most other peroxides, and is therefore harder to break homolytically. The cleavage of organoboranes by hydrogen peroxide (the oxidation part of hydroboration-oxidation; Eq. 5.21, Sec. 5.3B) is *not* a free-radical reaction.

Peroxides are not the only source of free-radical initiators. Another widely used initiator is azoisobutyronitrile, known by the acronym AIBN. This substance readily forms free radicals because the very stable nitrogen molecule is liberated as a result of homolytic cleavage:

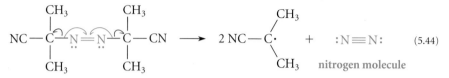

$$\qquad (5.44)$$

azoisobutyronitrile (AIBN)

Sometimes heat or light initiates a free-radical reaction. This usually happens because the additional energy promotes homolysis of the free-radical initiator, or less often, the reactants themselves, into free radicals.

The effects of initiators provide some of the best clues that a reaction occurs by a free-radical mechanism. If a reaction occurs in the presence of a known free-radical initiator, but does not occur in its absence, we can be fairly certain that the reaction involves free-radical intermediates.

A second initiation step is required in the free-radical addition of HBr to alkenes: the removal of a hydrogen atom from HBr by the *tert*-butoxy free radical that was formed in the first initiation step (Eq. 5.43).

$$(CH_3)_3C—\ddot{O}\cdot \quad H—\ddot{B}r: \quad \longrightarrow \quad (CH_3)_3C—\ddot{O}—H + \cdot\ddot{B}r: \qquad (5.45)$$

(from Eq. 5.43)

This is an example of another common type of free-radical process, called *atom abstraction*. In an atom abstraction reaction, a free radical removes an atom from another molecule, and a new free radical is formed (Br· in Eq. 5.45). The bromine atom is involved in the next phase of the reaction: the propagation steps.

Propagation   In the **propagation** steps of a free-radical reaction, products are formed and no *net* consumption or destruction of free radicals occurs. The free-radical by-product of one propagation step serves as the starting material for another.

The first propagation step of free-radical addition of HBr to an alkene is the *addition* of the bromine atom generated in Eq. 5.45 to the double bond.

$$R—CH{=}CH—R \quad \longrightarrow \quad R—\overset{\cdot}{C}H—\underset{:\ddot{B}r:}{CH}—R \qquad (5.46a)$$

$$\cdot\ddot{B}r:$$

Addition of a free radical to a double or triple bond is another common process encountered in free-radical chemistry.

The second propagation step is another *atom abstraction reaction*: removal of a hydrogen atom from HBr by the free-radical product of Eq. 5.46a to give the addition product and a new bromine atom.

$$R—\overset{\cdot}{C}H—\underset{Br}{CH}—R \quad \longrightarrow \quad R—\underset{H}{CH}—\underset{Br}{CH}—R + :\ddot{B}r\cdot \qquad (5.46b)$$

$$:\ddot{B}r—H$$

The bromine atom, in turn, can react with another molecule of alkene (Eq. 5.46a), and this can be followed by the generation of another molecule of product along with another bromine atom (Eq. 5.46b). These two processes continue in a chainlike fashion until the reactants are consumed. For this reason, free-radical reactions that involve propagation steps and their associated initiation and termination steps are said to be **free-radical chain reactions**. For each "link in the chain," or cycle of the two propagation steps, one molecule of the product is formed and one molecule of alkene starting material is consumed. Notice that for each free radical consumed in the propagation steps, one is produced. Because there is no net destruction of free radicals, the initial concentration of free radicals provided by the initiator, and thus the concentration of the initiator itself, can be small. Typically, the initiator concentration is only 1–2% of the alkene concentration.

The free radicals involved in the propagation steps of a chain reaction are said to *propagate the chain*. The free radicals in Eq. 5.46a–b are the chain-propagating radicals.

> An analogy for a chain reaction can be found in the world of business. A businessperson uses a little seed money, or capital, to purchase a small business. In time, this business produces profit that is used to buy another business. This business produces profit that can be used to buy yet another business, and so on. All this time, the businessperson is accumulating business property (instead of alkyl bromides) although the total amount of cash on hand is, by analogy to the chain-propagating free radicals in the reaction sequence above, small compared with the total amount of the investments.

Termination   In the **termination** steps of a free-radical chain reaction, free radicals are destroyed by *recombination reactions*. In a recombination reaction, two free radicals come together to form a covalent bond. In other words, a recombination reaction is the reverse of a homolysis reaction. The following reactions are two examples of termination reactions that take place in free-radical addition of HBr to alkenes. In these reactions, the chain-propagating radicals of Eqs. 5.46a and 5.46b, respectively, recombine to form by-products.

$$:\!\overset{..}{\underset{..}{Br}}\!\cdot \; + \; :\!\overset{..}{\underset{..}{Br}}\!\cdot \; \longrightarrow \; Br_2 \qquad (5.47)$$

$$2R\!-\!\underset{\underset{Br}{|}}{CH}\!-\!CH\!-\!R \; \longrightarrow \; R\!-\!\underset{\underset{R-CH-CH-R}{|}}{CH}\!-\!\overset{\overset{Br}{|}}{CH}\!-\!R \qquad (5.48)$$

Because such recombination reactions destroy free radicals, they "break" the free-radical chains and terminate the propagation reactions.

The recombination reactions of free radicals are in general highly exothermic; that is, they have very favorable, or negative, $\Delta H°$ values. They typically occur on every encounter of two free radicals. In view of this fact, you might ask why free radicals do not simply recombine before they propagate any chains. The answer is simply a matter of the relative concentrations of the various species involved. Free-radical intermediates are present in *very low concentration*, but the other reactants are present in much higher concentration. Consequently, it is much more probable for a bromine atom to be involved in the propagation reaction of Eq. 5.46a than in a recombination reaction with another bromine atom:

$$\ddot{Br}\cdot \begin{cases} \xrightarrow[\text{common occurrence}]{\text{[RCH=CHR] large;}} & :\ddot{Br}-\underset{\underset{R}{|}}{CH}-\overset{\cdot}{CH}-R \\[2em] \xrightarrow[\text{rare occurrence}]{[\,:\ddot{Br}\cdot\,]\text{ small;}} & :\ddot{Br}-\ddot{Br}: \end{cases} \tag{5.49}$$

In a typical free-radical chain reaction, a termination reaction occurs once for every 10,000 propagation reactions. As the reactants are depleted, however, the probability becomes significantly greater that one free radical will survive long enough to find another with which it can recombine. Small amounts of by-products resulting from termination reactions are typically observed in free-radical chain reactions.

In most cases, only exothermic propagation steps—steps with favorable, or negative, $\Delta H°$ values—occur rapidly enough to compete with the recombination reactions that terminate free-radical chain processes. Both of the propagation steps in the free-radical HBr addition to alkenes are exothermic, and they both occur readily. However, the first propagation step of free-radical HI addition, and the second propagation step of free-radical HCl addition, are quite *endothermic*. For this reason, these processes occur to such a small extent that they cannot compete with the recombination processes that terminate these chain reactions. Consequently, the free-radical addition of neither HCl nor HI to alkenes is observed.

This section has discussed the characteristics of free-radical chain reactions using the free-radical addition of HBr to alkenes as an example. A number of other useful laboratory reactions involve free-radical chain mechanisms. Many very important industrial processes are also free-radical chain reactions (Sec. 5.7). Free-radical chain reactions of great environmental importance occur in the upper atmosphere when ozone is destroyed by chlorofluorocarbons (Freons), the compounds that until recently have been exclusively used as coolants in air conditioners and refrigerators. A number of free-radical processes have also been characterized in biological systems.

PROBLEMS

*5.19   In the free-radical addition of HBr to 1-hexene, suggest a structure for three recombination products that might be found in small amounts in the reaction mixture.

5.20   Suggest a reason why small amounts of a dihalide R—CHBr—CHBr—R are formed during the peroxide-initiated addition of HBr to an alkene R—CH=CH—R.

D. Explanation of the Peroxide Effect

The free-radical mechanism is the basis for understanding the "peroxide effect" on HBr addition to alkenes, that is, why the presence of peroxides reverses the normal regioselectivity of HBr addition. The following example will serve as the basis for our discussion (see also Eq. 5.38).

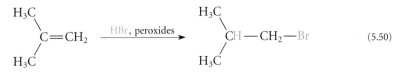

$$\text{(5.50)}$$

Recall that the reaction is initiated by the formation of a bromine atom from HBr (Eq. 5.45). When the bromine atom adds to the π bond of an alkene, two reactions are in competition: the bromine atom can react at either of the two carbons of the double bond to give different free-radical intermediates.

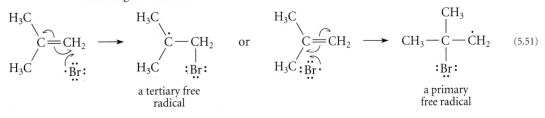

$$\text{(5.51)}$$

Free radicals, like carbocations, can be classified as primary, secondary, or tertiary.

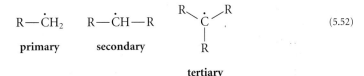

$$\text{(5.52)}$$

Equation 5.51 thus involves a competition between the formation of a primary and the formation of a tertiary free radical. *The formation of the tertiary free radical is faster.*

The tertiary free radical is formed more rapidly for two reasons. The first reason is that when the rather large bromine atom reacts at the more branched carbon of the double bond, it experiences van der Waals repulsions with the hydrogens in the branches (Fig. 5.1a). These repulsions increase the energy of this transition state. When the bromine atom reacts at the less branched carbon of the double bond, these van der Waals repulsions are absent (Fig. 5.1b). Because the reaction with the transition state of lower energy is the faster reaction, attack of the bromine atom on *the less branched carbon of the alkene double bond* to give *the more branched free radical* is faster. Subsequent reaction of this free radical with HBr leads to the observed products. To summarize:

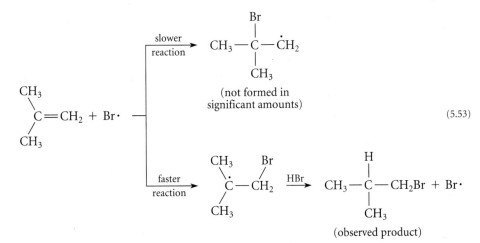

$$\text{(5.53)}$$

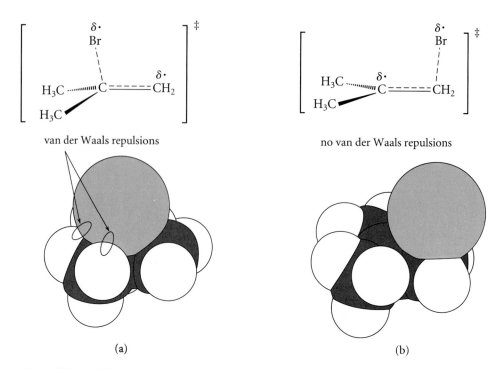

(a)   (b)

Figure 5.1   *Space-filling models of the alternative transition states for addition of a bromine atom to 2-methyl-propene. The bromine atom is shown in color. In (a), the bromine is adding to the more branched carbon of the double bond. This transition state contains van der Waals repulsions between the bromine and four of the six methyl hydrogens. (Two of these are shown; two are hidden from view.) In (b), the bromine is adding to the less branched carbon of the double bond, and the van der Waals repulsions shown in (a) are absent. The transition state in (b) has lower energy, and therefore leads to the observed product.*

Any effect on a reaction that can be attributed to van der Waals repulsions is termed a **steric effect** (Greek *stereos*, "solid"). Thus, the regioselectivity of free-radical HBr addition to alkenes is due in part to a steric effect. Other examples of steric effects are the preference of butane for the *anti* rather than the *gauche* conformation (Sec. 2.3B), and the greater stability of *trans*-2-butene over *cis*-2-butene (Sec. 4.5B).

The second reason that the tertiary radical is formed in Eq. 5.53 has to do with its relative stability. The heats of formation of several free radicals are given in Table 5.2. Comparing the heats of formation for propyl and isopropyl radicals shows that the secondary radical is more stable than the primary one by 12.5 kJ/mol (3.0 kcal/mol). Similarly, the *tert*-butyl radical is more stable than the *sec*-butyl radical by 18.4 kJ/mol (4.4 kcal/mol). Therefore:

Stability of free radicals:

$$\text{tertiary} > \text{secondary} > \text{primary} \tag{5.54}$$

Notice that free radicals have the same stability order as carbocations. However, the energy differences between isomeric free radicals are only about one-fifth the magnitude of the differences between corresponding carbocations. (Compare Tables 4.2 and 5.2.)

Table 5.2   Heats of Formation of Some Free Radicals (25°)

Radical	Structure	ΔH_f° (kJ/mol)	ΔH_f° (kcal/mol)
methyl	$\cdot CH_3$	145.6	34.8
ethyl	$\cdot CH_2CH_3$	117.2	28.0
propyl	$\cdot CH_2CH_2CH_3$	100.4	24.0
isopropyl	$CH_3\dot{C}HCH_3$	87.9	21.0
sec-butyl	$CH_3\dot{C}HCH_2CH_3$	66.9	16.0
tert-butyl	$(CH_3)_3C\cdot$	48.5	11.6

This free-radical stability order can be understood from the geometry and hybridization of a typical carbon radical (Fig. 5.2). Although the subject continues to be debated, it appears that alkyl radicals have trigonal-planar, or nearly planar, geometries. Recall that trigonal-planar carbons are sp^2-hybridized (Sec. 4.1A). Hence, alkyl radicals are sp^2-hybridized; the unpaired electron is in a carbon $2p$ orbital. The stability order in Eq. 5.54 implies, then, that *free radicals, like alkenes and carbocations, are stabilized by alkyl-group substitution at sp^2-hybridized carbons.*

By Hammond's postulate (Sec. 4.8C), a more stable free radical should be formed more rapidly than a less stable one. Thus, when a bromine atom adds to the π bond of an alkene, it adds to the less branched carbon of the double bond because the *more branched*, and hence *more stable*, free radical is formed as a result. The product of HBr addition is formed by the subsequent reaction of this free radical with HBr (Eq. 5.53). Notice that whether we consider steric effects in the transition state or the relative stabilities of free radicals, the same outcome of free-radical HBr addition is predicted.

Understanding of the regioselectivity of free-radical HBr addition to alkenes provides an understanding of the "peroxide effect"—why the regioselectivity of HBr addition to alkenes differs in the presence and absence of peroxides. Both reactions begin by attachment of an atom to the *less branched carbon* of the alkene double bond. In the absence of peroxides, the *proton* adds first, and bromine ends up on the more branched carbon. In the presence of peroxides, the free-radical mechanism takes over; a *bromine atom* adds

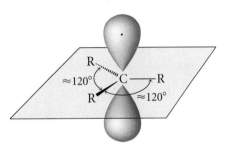

Figure 5.2   *Carbon radicals are sp^2-hybridized and have trigonal-planar geometry. The unpaired electron occupies the p orbital.*

first, and hydrogen ends up on the more branched carbon. The peroxide effect is not observed in the addition of HCl and HI to alkenes because free-radical addition of these hydrogen halides does not take place (Sec. 5.6C). Finally, it should be clear that the "peroxide effect" is not limited to peroxides; any good free-radical initiator will bring about the same effect.

5.21   Predict the major organic product(s) formed when HBr reacts with each of the following alkenes in the presence of peroxides.
*(a) cis-2-butene   (b) trans-2-hexene
*(c) 1-methylcyclopentene   *(d) (E)-3-methyl-2-hexene

5.22   Write the free-radical chain mechanism for the reactions of *(c) and (a) in Problem 5.21. Identify the initiation, propagation, and termination steps.

E. Bond Dissociation Energies

How easily does a chemical bond break homolytically to form free radicals? The question can be answered by examining the **bond dissociation energy**. For a bond between two atoms A—B, the bond dissociation energy is defined as the enthalpy $\Delta H°$ of the reaction

$$A\overset{\frown}{}\overset{\frown}{}B \longrightarrow A\cdot + B\cdot \tag{5.55}$$

Notice that a bond dissociation energy always corresponds to the enthalpy required to break a bond *homolytically*. Thus, the bond energy of H—Br refers to the process

$$H\overset{\frown}{}\overset{..}{\underset{..}{Br}}: \longrightarrow H\cdot + \cdot\overset{..}{\underset{..}{Br}}: \tag{5.56}$$

and *not* to the heterolytic process

$$H\overset{\frown}{}\overset{..}{\underset{..}{Br}}: \longrightarrow H^+ + \overset{-}{:}\overset{..}{\underset{..}{Br}}: \tag{5.57}$$

Some bond dissociation energies are collected in Table 5.3. *A bond dissociation energy measures the intrinsic strength of a chemical bond.* For example, breaking the H—H bond requires 435 kJ/mol (104 kcal/mol). It then follows that forming the hydrogen molecule from two hydrogen atoms liberates 435 kJ/mol (104 kcal/mol) of energy. Table 5.3 shows that different bonds exhibit significant differences in bond strength; even bonds of the same general type, for example, the various C—H bonds, can differ in bond strength by many kilojoules per mole.

Bond dissociation energies like those in Table 5.3 can be used in a number of ways. For example, these data show why di-*tert*-butyl peroxide (the second entry of the "other" classification in the table) is an excellent free-radical initiator. The lower a bond dissociation energy is, the lower the temperature required to rupture the bond in question and form free radicals at a reasonable rate. The homolysis of the O—O bond in di-*tert*-butyl peroxide requires only 159 kJ/mol (38 kcal/mol); this is one of the smallest bond dissociation energies in Table 5.3. With such a small bond dissociation energy, this peroxide readily forms small amounts of free radicals on gentle heating.

Table 5.3 Bond Dissociation Energies (25 °C)

Bond	$\Delta H°$ (kJ/mol)	$\Delta H°$ (kcal/mol)	Bond	$\Delta H°$ (kJ/mol)	$\Delta H°$ (kcal/mol)
C—H bonds			**C—O bonds**		
CH_3—H	439	105	CH_3—OH	385	92
CH_3CH_2—H	418	100	CH_3—OCH_3	347	83
$(CH_3)_2CH$—H	410	98	Ph—OH	464	111
$(CH_3)_3C$—H	402	96	CH_2=O (both)	732	175
$PhCH_2$—H	368	88	CH_2=O (π bond)	377	90
CH_2=$CHCH_2$—H	360	86	**C—N bonds**		
RCH=CH—H	460	110	CH_3—NH_2	356	85
Ph—H	464	111	Ph—NH_2	427	102
$RC\equiv C$—H	548	131	CH_2=NH	$\approx$644	$\approx$154
H—CN	519	124	$HC\equiv N$	937	224
C—Halogen bonds			**H—X bonds**		
CH_3—F	459	110	H—OH	498	119
CH_3CH_2—F	454	109	H—OCH_3	435	104
CH_3—Cl	356	85	H—O_2CCH_3	444	106
CH_3CH_2—Cl	335	80	H—OPh	360	86
CH_3—Br	293	70	H—F	569	136
CH_3CH_2—Br	284	68	H—Cl	431	103
CH_3—I	238	57	H—Br	368	88
CH_3CH_2—I	222	53	H—I	297	71
Ph—F	527	126	H—NH_2	448	107
Ph—Cl	402	96	H—SH	381	91
Ph—Br	339	81	**X—X bonds**		
Ph—I	272	65	H—H	435	104
C—C bonds			F—F	159	38
CH_3—CH_3	377	90	Cl—Cl	247	59
Ph—CH_3	423	101	Br—Br	192	46
$PhCH_2$—CH_3	314	75	I—I	151	36
CH_3—CN	510	122	**Other**		
CH_2=CH_2 (both)	715	171	HO—OH	213	51
CH_2=CH_2 (π bond)	264–276	63–66	$(CH_3)_3CO$—$OC(CH_3)_3$	159	38
$HC\equiv CH$	$\approx$962	$\approx$230	HO—Br	234	56
			$(CH_3)_3CO$—Br	205	49

An important use of bond dissociation energies is to calculate or estimate the $\Delta H°$ of reactions. As an illustration, consider the second initiation step of free-radical bromination, in which a *tert*-butoxy radical reacts with H—Br.

$$(CH_3)_3C—\overset{..}{\underset{..}{O}}\cdot \; + \; H—\overset{..}{\underset{..}{Br}}: \quad \longrightarrow \quad (CH_3)_3C—\overset{..}{\underset{..}{O}}—H \; + \; \cdot\overset{..}{\underset{..}{Br}}: \tag{5.58}$$

***tert*-butoxy radical**

In this reaction, a hydrogen is abstracted from HBr by the *tert*-butoxy radical. This is not the only reaction that might occur. Instead, the *tert*-butoxy radical might abstract a bromine atom from HBr:

$$(CH_3)_3C\text{—}\overset{..}{\underset{..}{O}}\cdot \; + \; H\text{—}\overset{..}{\underset{..}{Br}}: \; \longrightarrow \; (CH_3)_3C\text{—}\overset{..}{\underset{..}{O}}\text{—}\overset{..}{\underset{..}{Br}}: \; + \; \cdot H \qquad (5.59)$$

tert-butoxy radical

However, hydrogen and not bromine is abstracted. The reason lies in the relative enthalpies of the two reactions. These enthalpies are not known by direct measurement, but can be calculated using bond dissociation energies. To calculate the $\Delta H°$ for a reaction, *subtract the bond dissociation energies of the bonds formed from the bond dissociation energies of the bonds broken.* This procedure is illustrated in the following study problem.

STUDY PROBLEM 5.3

STUDY GUIDE LINK:
✓5.3
Bond Dissociation Energies and Heats of Reaction

Estimate the standard enthalpies of the reactions shown in Eqs. 5.58 and 5.59.

Solution   To obtain the required estimates, subtract the bond dissociation energy of the bond formed from that of the bond broken in each reaction. In both equations, the bond broken is the H—Br bond. From Table 5.3, the bond dissociation energy of this bond is 368 kJ/mol (88 kcal/mol). The bond formed in Eq. 5.58 is the O—H bond in $(CH_3)_3CO$—H (*tert*-butyl alcohol). This exact compound is not found in Table 5.3! What do we do? Look for the same type of bond in as similar a compound as possible. For example, the table includes an entry for the alcohol CH_3O—H (methyl alcohol). (The O—H bond dissociation energies for methyl alcohol and *tert*-butyl alcohol differ very little.) Subtracting the enthalpy of the bond formed (the O—H bond) from that of the bond broken (the H—Br bond), we obtain $368 - 435 = -67$ kJ/mol or $88 - 104 = -16$ kcal/mol. This is the enthalpy of the reaction in Eq. 5.58. Following the same procedure for Eq. 5.59, use the bond dissociation energy of the O—Br bond in $(CH_3)_3CO$—Br, which is given in Table 5.3 as 205 kJ/mol (49 kcal/mol). The calculated enthalpy for Eq. 5.59 is then $368 - 205 = 163$ kJ/mol or $88 - 49 = 39$ kcal/mol. These $\Delta H°$ estimates are the required solution to the problem.

The calculation in Study Problem 5.3 shows that reaction 5.58 is highly exothermic (favorable) and reaction 5.59 is highly endothermic (unfavorable). Thus, *the observed reaction* (abstraction of H, Eq. 5.58) *is the only one that can readily occur.* Abstraction of Br (Eq. 5.59) is so unfavorable energetically that it is not likely to occur to any appreciable extent.

> Strictly speaking, we need $\Delta G°$ values to determine whether a reaction will occur spontaneously. However, for the reactions discussed here, $\Delta H°$ is a reasonable approximation of $\Delta G°$.

Bond dissociation energies can be used to calculate the enthalpy of any reaction (not just free-radical reactions) provided the bond dissociation energies of the appropriate bonds are known or can be closely estimated. When bond dissociation energies (or any other gas-phase enthalpies) are used in this way, remember that they apply to the gas phase. Bond dissociation energies can be used, however, to compare the enthalpies of two reactions in solution provided that the effect of solvent is either negligible or is the same for both of the reactions being compared (and thus cancels in the comparison). This assumption is valid for many free-radical reactions, including the ones in Study Problem 5.3.

PROBLEMS

5.23   Estimate the $\Delta H°$ values for each of the following gas-phase reactions using bond dissociation energies.

*(a) $CH_2{=}CH_2 + Cl_2 \longrightarrow Cl{-}CH_2{-}CH_2{-}Cl$

(b) $CH_4 + Cl_2 \longrightarrow CH_3{-}Cl + H{-}Cl$

*5.24   Consider the second propagation step of peroxide-promoted HBr addition to alkenes (Eq. 5.46b). Explain why hydrogen, and not bromine, is abstracted from HBr by the free-radical reactant.

*5.25   Use the information from Table 5.3 to calculate the heat of formation of (a) the hydrogen atom and (b) the chlorine atom. (*Hint:* Remember the definition of the heat of formation; Sec. 4.5A.)

5.26   Calculate the bond dissociation energy of a $C{-}H$ bond in ethane using the following heats of formation: ethane, -84.5 kJ/mol (-20.2 kcal/mol); the hydrogen atom (Problem 5.23); and the ethyl radical (Table 5.2). Check your answer in Table 5.3.

5.7   # Industrial Use and Preparation of Alkenes

The chemical industry is a major force in the world's economy, and the production and use of alkenes are foundations of this industry. This section presents briefly two industrial processes, *free-radical polymerization* and *thermal cracking of alkanes*. Free-radical polymerization accounts for a large fraction of the alkenes used commercially. Thermal cracking of alkanes, which also involves free-radical intermediates, is a major process by which alkenes are produced industrially. This section concludes with a discussion of ethylene, commercially the most important organic compound.

A. Free-Radical Polymerization of Alkenes

In the presence of free-radical initiators such as peroxides or AIBN, many alkenes react to form **polymers**, which are very large molecules composed of repeating units. Polymers are derived from small molecules in the same sense that a wall is composed of bricks, or a train is composed of boxcars. In a **polymerization reaction**, small molecules known as **monomers** react to form a polymer. For example, ethylene can be used as a monomer and polymerized under free-radical conditions to yield an industrially important polymer called **polyethylene**.

$$n \; H_2C\!\!=\!\!CH_2 \xrightarrow{\text{initiator}} \quad \left[\!CH_2\!-\!CH_2\!\right]_n \tag{5.60}$$

ethylene   **polyethylene**
(the polymer of ethylene)

In this formula for polyethylene, the subscript n means that each polyethylene molecule contains a very large and not very precisely defined number of repeating units, $-CH_2-CH_2-$. Typically n might be in the range of 3,000 to 40,000. Polyethylene is an example of an **addition polymer**, that is, a polymer in which no atoms of the monomer unit have been lost as a result of the polymerization reaction. (Other types of polymers are discussed later in the text.)

Because the polymerization of ethylene shown in Eq. 5.60 occurs by a free-radical mechanism, it is an example of **free-radical polymerization**. The reaction is initiated when a radical R·, derived from peroxides or other initiators, adds to the double bond of ethylene to form a new radical.

$$R\cdot \overset{\frown}{CH_2}\!\!=\!\!\overset{\frown}{CH_2} \quad \longrightarrow \quad R-CH_2-\dot{C}H_2 \tag{5.61a}$$

initiating
radical

The propagation phase of the reaction begins when the new radical adds to another molecule of ethylene.

$$R-CH_2-\dot{C}H_2 \; \overset{\frown}{CH_2}\!\!=\!\!\overset{\frown}{CH_2} \quad \longrightarrow \quad R-CH_2-CH_2-CH_2-\dot{C}H_2 \tag{5.61b}$$

Notice that Eqs. 5.61a and b are further examples of a typical free-radical reaction: addition to a double bond. (Compare with Eq. 5.46a.) This process continues indefinitely until the ethylene supply is exhausted.

$$R\left[\!CH_2\!-\!CH_2\!\right]_n CH_2-\dot{C}H_2 + CH_2\!\!=\!\!CH_2 \quad \longrightarrow \quad R\left[\!CH_2\!-\!CH_2\!\right]_{n+1} CH_2-\dot{C}H_2 \tag{5.61c}$$

The reaction terminates when the radicals present in the reaction mixture are eventually consumed by any one of several termination reactions. Typically, polymers with molecular weights of 10^5 to 10^7 daltons are formed in free-radical polymerizations. The polymer chain is so long that the groups at its ends represent an insignificant part of the total structure. Hence, these terminating groups are ignored, and the polymer structure is written as $\left[\!CH_2\!-\!CH_2\!\right]_n$.

Free-radical polymerization of alkenes is very important commercially. About 21 billion pounds of polyethylene is manufactured annually in the United States, of which more than half is made by free-radical polymerization. The free-radical process yields a very transparent polymer, called *low-density polyethylene*, used in films and packaging. (Freezer bags and sandwich bags are usually made of low-density polyethylene.) Another method of polyethylene manufacture, called the *Ziegler process*, employs a titanium catalyst and does not involve free-radical intermediates. This process yields a *high-density polyethylene* used in blow-molded plastic containers.

Many other commercially important polymers are produced from alkene monomers by free-radical polymerization. Some of these are listed in Table 5.4. Alkene polymers

Table 5.4   Some Addition Polymers Produced by Free-Radical Polymerization

Polymer name (Trade name)	Structure of monomer	Properties of the polymer	Uses
Polyethylene	$CH_2{=}CH_2$	Flexible, semiopaque, generally inert	Containers, film
Polystyrene	$Ph{-}CH{=}CH_2$	Clear, rigid; can be foamed with air	Containers, toys, packing material and insulation
Poly(vinyl chloride) (PVC)	$CH_2{=}CH{-}Cl$	Rigid, but can be plasticized with certain additives	Plumbing, leatherette, hoses. Monomer has been implicated as a carcinogen.
Polychlorotrifluoroethylene (Kel-F)	$CF_2{=}CF{-}Cl$	Inert	Chemically inert apparatus, fittings, and gaskets
Polytetrafluoroethylene (Teflon)	$CF_2{=}CF_2$	Very high melting point; chemically inert	Gaskets; chemically resistant apparatus and parts
Poly(methyl methacrylate) (Plexiglas, Lucite)	$CH_2{=}\underset{\underset{\textstyle CH_3}{\mid}}{C}{-}CO_2CH_3$	Clear and semiflexible	Lenses and windows; fiber optics
Polyacrylonitrile (Orlon, Acrilan)	$CH_2{=}CH{-}CN$	Crystalline, strong, high luster	Fibers

surround us in many of the articles we use every day. Telephones, computers, automobiles, sports equipment, stereo systems, food packaging, and many other items have important components fabricated from alkene polymers.

DISCOVERY OF TEFLON

In April 1938 Roy J. Plunkett, who had obtained his Ph. D. only two years earlier from Ohio State University, was working in the laboratories of the DuPont company. He decided to use some tetrafluoroethylene (a gas) in the preparation of a refrigerant. When he opened the valve on the cylinder of tetrafluoroethylene, no gas escaped. Because the weight of the empty cylinder was known, Plunkett was able to determine that the cylinder had the weight expected for a full cylinder of the gas. It was at this point that Plunkett's

(*continues*)

scientific curiosity paid a handsome dividend. Rather than discard the cylinder, he checked to be sure the valve was not faulty, and then cut the cylinder open. Inside he found a polymeric material that felt slippery to the touch, could not be melted with extreme heat, and was chemically inert to almost everything. Thus Plunkett accidentally discovered the polymer we know today as Teflon. At that time, no one imagined the commercial value of Teflon. Only with the advent of the atomic bomb project during World War II did it find a use: to form gaskets that were inert to the highly corrosive gas UF_6 used to purify the isotopes of uranium. In the 1960s Teflon was introduced to consumers as the nonstick coating on cookware.

PROBLEMS

5.27   Using the monomer structures of the alkenes in Table 5.4, draw the structures of *(a) polypropylene, the polymer of propene; and (b) poly(vinyl chloride) (PVC), the polymer used in household drainpipes.

5.28   Give the mechanism for free-radical polymerization of *(a) propene and (b) tetrafluoroethylene.

B. Thermal Cracking of Alkanes

Simple alkenes are considered to be petroleum products, but they are not obtained directly from oil wells. Rather, they are produced industrially from alkanes in a process called **cracking**. Cracking breaks larger alkanes into a mixture of smaller hydrocarbons, some of which are alkenes. The chemistry of the cracking process is of interest, not only because of its commercial significance, but also because it illustrates some additional chemistry of free radicals.

Ethylene, the alkene of greatest commercial importance, is produced by a process called *thermal cracking*. In this process, a mixture of alkanes from the distillation of petroleum (Sec. 2.8) is mixed with steam and heated in a furnace at 750–900 °C for a fraction of a second, and is then quenched (rapidly cooled) to prevent secondary reactions. The products of cracking are then separated.

In the United States, the hydrocarbon most often used to produce ethylene is ethane, a component of natural gas. In the cracking of ethane, ethylene and hydrogen are formed at very high temperature. In cracking, the temperatures used are so high that initiating radicals are formed by spontaneous bond rupture. Only two types of bonds in ethane can break: the C—C bond and the C—H bonds. The bond dissociation energies in Table 5.3 show that rupture of the C—C bond requires somewhat less energy. Hence, fragmentation of a few ethane molecules into two methyl radicals takes place.

First initiation step:

$$CH_3 \overset{\frown}{} CH_3 \xrightarrow{\text{heat}} 2 \cdot CH_3 \qquad (5.62a)$$

In a second initiation step, a methyl radical abstracts a hydrogen atom from another ethane molecule:

Second initiation step:

$$H_3C \cdot \overset{\frown}{\quad} H \overset{\frown}{\quad} CH_2 - CH_3 \quad \longrightarrow \quad CH_4 \; + \; \cdot CH_2 - CH_3 \qquad (5.62b)$$

methane   ethyl radical

This step accounts for small amounts of methane formed in the cracking process. The ethyl radical is the chain-propagating radical. It first undergoes an interesting reaction in which it "unzips" to yield ethylene and a hydrogen atom in a process called β-**scission**:

First propagation step:

$$\overset{\frown}{\cdot CH_2} \overset{\frown}{\quad} CH_2 \overset{\frown}{\quad} H \quad \xrightarrow{\;\beta\text{-scission}\;} \quad CH_2 \!=\! CH_2 + H\cdot \qquad (5.63a)$$

ethyl radical   **ethylene**

β-Scission is another typical reaction of free radicals. The word "scission" means "cleavage" (it is derived from the same root as "scissors"). The Greek letter beta (β) refers to the fact that the cleavage occurs one carbon away from the radical site. (The Greek letters α, alpha; β, beta; γ, gamma; and so on are sometimes used to indicate the relative positions of groups on carbon chains.)

cleavage occurs here

$$\cdot \underset{\alpha}{CH_2} - \underset{\beta}{CH_2} \!\!\mid\!\! H$$

Although β-scission might look like a new reaction, actually it is simply the reverse of an addition; in this case it is the reverse of the addition of a hydrogen atom to ethylene.

    The hydrogen atom produced in the first propagation step then abstracts a hydrogen atom from another molecule of ethane to give a new ethyl radical:

Second propagation step:

$$H \cdot \overset{\frown}{\quad} H \overset{\frown}{\quad} CH_2 - CH_3 \quad \longrightarrow \quad H_2 \; + \; \cdot CH_2 - CH_3 \qquad (5.63b)$$

The ethyl radical then enters into the first propagation reaction, Eq. 5.63a, thus continuing the free-radical chain. The hydrogen that is a by-product of Eq. 5.63b is collected and used to produce ammonia by hydrogenation of nitrogen. The ammonia finds important use in agriculture as a fertilizer.

    You have now seen four types of reactions that free radicals can undergo. Three of these are reactions that produce other radicals:

1. addition to a double bond
2. atom abstraction
3. β-scission.

The other reaction destroys free radicals:

4. recombination of two free radicals to form a covalent bond.

These reactions constitute a "toolbox" of fundamental free-radical processes. Most free-radical processes are examples of these reactions or combinations of them.

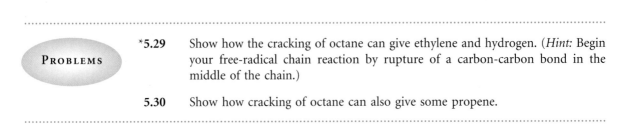

PROBLEMS

*5.29   Show how the cracking of octane can give ethylene and hydrogen. (*Hint:* Begin your free-radical chain reaction by rupture of a carbon-carbon bond in the middle of the chain.)

5.30   Show how cracking of octane can also give some propene.

C. Commercial Importance of Ethylene and Propene

More ethylene is produced than any other organic compound. In the United States, it has ranked fourth in industrial production of all chemicals (behind sulfuric acid, nitrogen, and oxygen) for a number of years. More than 40 billion pounds of ethylene is produced annually. Propene (known industrially by its older name *propylene*) is not far behind, ranking ninth in all industrially produced chemicals, with an annual U. S. output of more than 22 billion pounds.

The major uses of ethylene and propene are the production of polyethylene and polypropylene, respectively (Sec. 5.7A). Ethylene is also oxidized to give ethylene glycol, $HO-CH_2-CH_2-OH$, which is used in automotive antifreeze; it is hydrated to give industrial ethanol (Sec. 4.9B); and it is used, along with benzene, to produce polystyrene, an important polymer. Propene is a key compound in the production of phenol, which is used in adhesives, and acetone, a commercially important solvent. Some of these chemical interrelationships are shown in Fig. 5.3. We'll return to this figure periodically as other industrially important processes are discussed.

ETHYLENE AS A FRUIT RIPENER

An intriguing use of ethylene in nature has been put to commercial advantage. It was discovered not long ago that ethylene produced by plants causes fruit to ripen. Plants produce this *ripening hormone* by the degradation of a relatively rare amino acid:

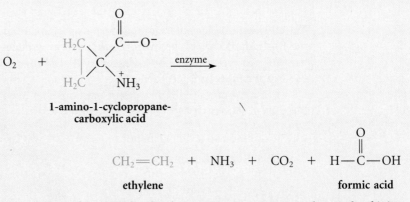

It is ethylene, for example, that brings green tomatoes to that peak of juicy redness in the home garden. Commercial growers have made use of this knowledge by picking and transporting fruit before it is ripe, and then ripening it "on location" with ethylene!

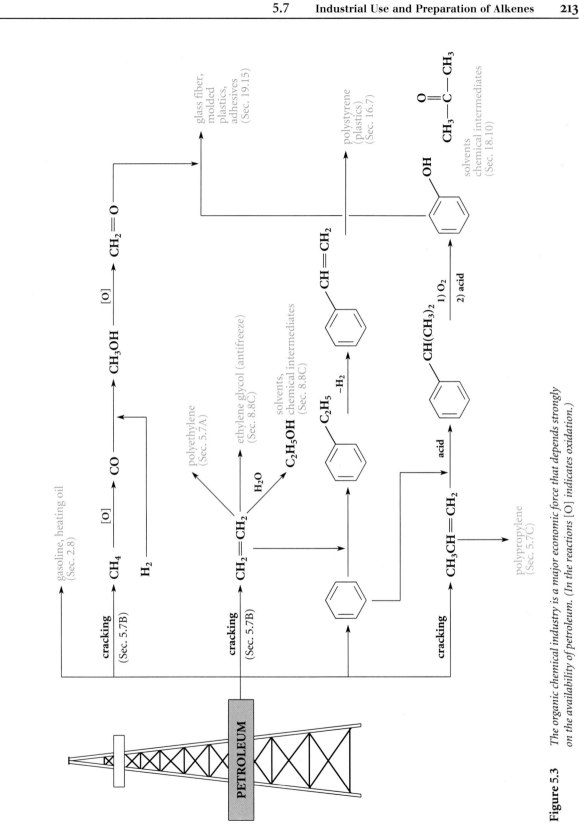

Figure 5.3   *The organic chemical industry is a major economic force that depends strongly on the availability of petroleum. (In the reactions [O] indicates oxidation.)*

KEY IDEAS IN CHAPTER 5

The characteristic reaction of alkenes is addition to the double bond.

Addition to the alkene double bond occurs by a variety of mechanisms:
1. by mechanisms involving carbocation intermediates (addition of hydrogen halides, hydration)
2. by mechanisms involving cyclic ion intermediates (oxymercuration, halogenation)
3. by concerted mechanisms (hydroboration, glycol formation, ozonolysis)
4. by mechanisms involving free-radical intermediates (free-radical addition of HBr, polymerization).

Addition reactions are typically regioselective; their regioselectivity is a consequence of their mechanisms.

Some useful transformations of alkenes involve additions followed by other transformations. These include oxymercuration-reduction and hydroboration-oxidation, which give alcohols; ozonolysis followed by treatment with $(CH_3)_2S$ or H_2O_2, which gives aldehydes, ketones, or carboxylic acids by cleavage of the double bond; and addition of OsO_4 or $KMnO_4$ followed by hydrolysis to give glycols.

Typical reactions of free radicals are
1. addition to a double bond
2. atom abstraction
3. β-scission (the reverse of addition to a double bond)
4. recombination (the reverse of bond rupture).

Reactions that occur by free-radical chain mechanisms are typically promoted by free-radical initiators (peroxides, AIBN), heat, or light.

The stability of free radicals is in the order tertiary > secondary > primary, but the effect of branching on free-radical stability is considerably smaller than the effect of branching on carbocation stability.

The reversal of the regioselectivity of HBr addition in the presence of peroxides (the "peroxide effect") is a consequence of the free-radical mechanism of the reaction. The key step in this mechanism is the reaction of a bromine atom at the less branched carbon of a double bond to give the more branched, and hence more stable, free radical intermediate.

The bond dissociation energy of a covalent bond measures the energy required to break the bond homolytically to form two free radicals. The $\Delta H°$ of a reaction can be calculated by subtracting the bond dissociation energies of the bonds formed from those of the bonds broken.

Ethylene is the organic compound produced industrially in the greatest amount. Alkenes such as ethylene and propene are produced by cracking alkanes at high temperature. A major use of alkenes is in the formation of addition polymers.

....

ADDITIONAL PROBLEMS

*5.31   Give the principal organic products expected when 1-ethylcyclopentene reacts with each of the following reagents.
(a) Br_2 in CCl_4 solvent   (b) O_3, $-78°$
(c) product of (b) with $(CH_3)_2S$ (d) product of (b) with H_2O_2
(e) O_2, flame   (f) HBr
(g) OsO_4, then H_2O, $NaHSO_3$ (h) I_2, H_2O
(i) H_2, Pt/C   (j) HBr, peroxides
(k) BH_3 in tetrahydrofuran (THF)   (l) product of (k) with NaOH, H_2O_2
(m) $Hg(OAc)_2$, H_2O (n) product of (m) with $NaBH_4$
(o) HI   (p) HI, AIBN

5.32   Repeat Problem 5.31 for 1-butene.

*5.33   Draw the structure of
(a) a six-carbon alkene that would give the same product from reaction with HBr whether peroxides are present or not
(b) a compound C_5H_{10} that would *not* react with alkaline $KMnO_4$
(c) four compounds of formula $C_{10}H_{16}$ that would undergo catalytic hydrogenation to give decalin

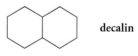

decalin

(d) two alkenes that would yield 1-methylcyclohexanol when treated with $Hg(OAc)_2$ in water, then $NaBH_4$:

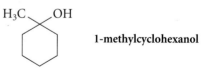

1-methylcyclohexanol

(e) an alkene with one double bond that would give the following compound as the *only* product after ozonolysis followed by H_2O_2:

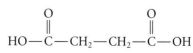

$$HO-\overset{\overset{\displaystyle O}{\|}}{C}-CH_2-CH_2-\overset{\overset{\displaystyle O}{\|}}{C}-OH$$

(*Problem 5.33 continues*)

(f) two stereoisomeric alkenes that would give the following alcohol as the major product of hydroboration followed by treatment with alkaline H_2O_2:

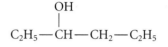

(g) an alkene of five carbons that would give the same product as a result of *either* oxymercuration-reduction *or* hydroboration-oxidation.

5.34 Draw the structure of

(a) a five-carbon alkene that would give the same product with HBr whether peroxides are present or not

(b) a compound C_6H_{12} that would not react with OsO_4

(c) four compounds of formula C_7H_{12} that would undergo catalytic hydrogenation to give methylcyclohexane

(d) two alkenes that would give the following alcohol when treated with $Hg(OAc)_2$ in water, then $NaBH_4$:

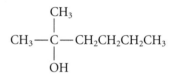

(e) an alkene that would give 2-pentanone as the only product after ozonolysis followed by treatment with H_2O_2:

(f) the alkene that would give the following alcohol after hydroboration-oxidation:

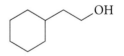

(g) an alkene C_6H_{12} that would give the same product from either hydroboration-oxidation or oxymercuration-reduction.

5.35 Give the missing reactant or product in each of the following equations.

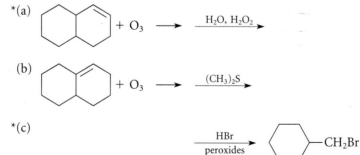

(d)

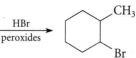 (only product)

5.36 Outline a laboratory preparation of each of the following compounds. Each should be prepared from an alkene *with the same number of carbon atoms* and any other reagents. The reactions and starting materials used should be chosen so that each compound is virtually uncontaminated by constitutional isomers.

*(a)

$$CH_3CH_2 - \underset{\underset{CH_2CH_3}{|}}{\overset{\overset{OH}{|}}{C}} - CH_2CH_3$$

(b)

$$\overset{OH}{\underset{|}{CH_3CHCH_2CH_2CH_2CH_3}}$$

*(c) $HO - CH_2CH_2CH_2CH_2CH_2CH_3$

(d)

$$\overset{OH}{\underset{|}{CH_3CH_2CHCH_2CH_2CH_3}}$$

*(e)

$$\overset{Br}{\underset{|}{CH_3CH}} - \overset{Br}{\underset{\underset{CH_3}{|}}{\overset{|}{C}}} - CH_2CH_3$$

*(f)

$$\overset{OH}{\underset{\underset{CH_2OH}{|}}{CH_3CH_2CCH_2CH_3}}$$

(g)

$$\overset{OH}{\underset{\underset{CH_2Br}{|}}{CH_3CH_2CCH_2CH_3}}$$

*(h)

$$\overset{O}{\overset{\|}{H - C}} (CH_2)_4 \overset{O}{\overset{\|}{C}} - CH_3$$

(i)

$$\overset{O}{\overset{\|}{HO - C}} (CH_2)_3 \overset{O}{\overset{\|}{C}} - OH$$

*(j) $Br - CH_2CH_2CH_2CH_2CH_3$

(k)

$$\overset{Br}{\underset{|}{CH_3CHCH_2CH_2CH_3}}$$

*(l) $\overset{}{\underset{\underset{CH_2CH_3}{|}}{CH_3CH_2CHCH_2CH_3}}$

(m) another alkene that could be used to prepare the compound in part (l) by the same method.

*5.37 Deuterium (D, or ^2H) is an isotope of hydrogen with atomic mass = 2. Deuterium can be introduced into organic compounds by using reagents in which hydrogen has been replaced by deuterium. Outline preparations of both isotopically labeled compounds from the same alkene.

(a) $HO - CH_2 - \underset{\underset{D}{|}}{CH} - CH_2 - \underset{\underset{CH_3}{|}}{CH} - CH_3$

(b) $D - CH_2 - \underset{\underset{OH}{|}}{CH} - CH_2 - \underset{\underset{CH_3}{|}}{CH} - CH_3$

5.38   When the isotopically labeled compound below is subjected to hydroboration–oxidation, two isotopically labeled alcohols are formed in essentially equal amounts. Give the structures of these products. (See the note about deuterium in Problem 5.37.)

*5.39   Using the mechanism of halogen addition to alkenes, predict the product(s) that would be obtained when 2-methyl-1-butene is subjected to each of the following conditions. Explain your answers.
(a) Br_2 in CH_2Cl_2 (an inert solvent)   (b) Br_2 in H_2O
(c) Br_2 in CH_3OH solvent
(d) Br_2 in CH_3OH solvent containing concentrated $Li^+\ Br^-$

5.40   Using the mechanism of the oxymercuration reaction, predict the product(s) that would be obtained when 1-hexene is treated with mercuric acetate in each of the following solvents, and the resulting products are treated with $NaBH_4$. Explain your answers and tell what functional groups are present in each of the products.
(a) CH_3OH (methanol)   (b) $(CH_3)_2CH$—OH (isopropyl alcohol)

(c)   :O:
                  ‖
      H—Ö—C—CH_3 (acetic acid)

*5.41   In the addition of HBr to 3,3-dimethyl-1-butene, the following results are observed:

$$CH_3\!-\!\underset{\underset{CH_3}{|}}{\overset{\overset{CH_3}{|}}{C}}\!-\!CH\!=\!CH_2\ +\ HBr\ \longrightarrow$$

$$CH_3\!-\!\underset{\underset{Br}{|}}{\overset{\overset{CH_3}{|}}{C}}\!-\!CH(CH_3)_2\ +\ CH_3\!-\!\underset{\underset{CH_3}{|}}{\overset{\overset{CH_3}{|}}{C}}\!-\!\underset{\underset{Br}{|}}{CH}\!-\!CH_3\ +\ CH_3\!-\!\underset{\underset{CH_3}{|}}{\overset{\overset{CH_3}{|}}{C}}\!-\!CH_2CH_2Br$$

	71%	29%	none
no peroxides:	71%	29%	none
with peroxides:	trace	trace	100%

(a)   Explain why the different conditions give different product distributions.
(b)   Write a detailed mechanism for each reaction that explains the origin of all products.
(c)   Which conditions give the faster reaction? Explain.

*5.42   (a)   Give the structures of the two products formed when (*E*)-4,4-dimethyl-2-pentene is treated with HBr in the presence of peroxides.

(b) One of these products is formed in greater amount than the other. After considering the structure of the transition state for peroxide-promoted HBr addition (see Fig. 5.1), suggest which product is the predominant one. Explain.

5.43   Give the structures of both the reactive intermediate and the product in each of the following reactions:

*(a) $CH_3CH_2CCH_3$ + Br_2 $\longrightarrow$ (b) $CH_3CH_2CCH_3$ + HBr $\longrightarrow$
          ‖   ‖
         CH_2 CH_2

*(c) $CH_3CH_2CCH_3$ + $Hg(OAc)_2$ + H_2O $\longrightarrow$
          ‖
         CH_2

5.44   Trifluoroiodomethane undergoes an addition to alkenes in the presence of light by a free-radical chain mechanism.

$$CH_3CH_2CH_2CH{=}CH_2 + CF_3I \xrightarrow{\text{light}} CH_3CH_2CH_2CH-CH_2-CF_3$$
$$|$$
$$I$$

The initiation step of this reaction is the light-induced homolysis of the C—I bond:

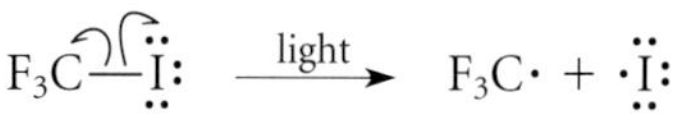

$$F_3C-I: \xrightarrow{\text{light}} F_3C\cdot + \cdot\ddot{I}:$$

☞
Study Guide Link:
✓**5.4**
Writing Free-Radical Chain Mechanisms

(a) Using the fishhook formalism, write the propagation steps of a free-radical chain mechanism for this reaction.

(b) Predict which alkene would react more rapidly with CF_3I in the presence of light: 2-methyl-1-pentene or (E)-4-methyl-2-pentene. Explain your choice.

5.45   (a) In the thermal cracking of 2,2,3,3-tetramethylbutane, which carbon-carbon bond would be most likely to break? Explain.

(b) Which compound, 2,2,3,3-tetramethylbutane or ethane, undergoes thermal cracking at the lower temperature? Explain.

(c) Check your answer to (b) by calculating the $\Delta H°$ for the initial carbon-carbon bond breaking. Use the $\Delta H_f°$ values in Table 5.2 as well as the $\Delta H_f°$ of 2,2,3,3-tetramethylbutane (−225.9 kJ/mol, −53.99 kcal/mol) and ethane (−84.7 kJ/mol, −20.24 kcal/mol).

5.46   Natural rubber is a polymer with the following structure:

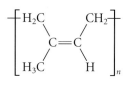

(Problem 5.46 continues)

(a) What product would be obtained from ozonolysis of natural rubber, followed by reaction with H_2O_2? (*Hint:* Write out two units of the polymer structure.)

(b) *Gutta-percha* is a natural polymer that gives the same ozonolysis product as natural rubber. Suggest a structure for gutta-percha.

*5.47   Although the addition of H—CN to an alkene could be envisioned to occur by a free-radical chain mechanism, this reaction is not observed. Justify each of the following reasons with appropriate calculations using bond dissociation energies.

(a) The reaction of H—CN with initiating $(CH_3)_3C$—O· radicals is not a good source of ·CN radicals.

(b) The second propagation step of the free-radical addition is energetically unfavorable:

$$R—\overset{\cdot}{C}H—CH_2—CN + H—CN \longrightarrow R—CH_2—CH_2—CN + ·CN$$

(This is the same reason that HCl does not undergo free-radical addition to alkenes.)

5.48   Free-radical addition of thiols (molecules of the general structure R—SH) to alkenes is a well-known reaction and is initiated by peroxides. Use bond dissociation energies to show that the following initiation reaction should be a good source of ·SR radicals:

$$(CH_3)_3C—O· + H—SR \longrightarrow (CH_3)_3C—O—H + ·SR$$
(from homolysis
of a peroxide)

*5.49   The halogenation of methane in the gas phase is an industrial method for the preparation of certain alkyl halides and takes place by the following equation (X = halogen):

$$X_2 + CH_4 \longrightarrow CH_3—X + H—X$$

(a) This reaction takes place readily when X = Br or X = Cl, but not when X = I. Show that these observations are expected from the $\Delta H°$ values of the reactions. Calculate the $\Delta H°$ values from appropriate bond dissociation energies.

(b) Explain why samples of methyl iodide (CH_3—I) that are contaminated with traces of acid darken with the color of iodine on standing for long periods of time.

*5.50   (a) Draw the structure of *polystyrene*, the polymer obtained from the free-radical polymerization of styrene.

(b) How would the structure of the polymer product differ from the one in (a) if a few percent of 1,4-divinylbenzene were included in the reaction mixture?

$$CH_2{=}CH{-}\hspace{-0.3em}\langle\!\langle\ \rangle\!\rangle\hspace{-0.3em}{-}CH{=}CH_2 \quad \textbf{1,4-divinylbenzene}$$

*5.51   In a laboratory a bottle was found containing a clear liquid *A*. The bottle was labeled, "Isolated from a pine tree." You have been offered a substantial reward by the Department of Agriculture to identify this substance. Elemental analysis reveals that *A* contains 88.16% C and 11.84% H. Compound *A* decolorizes Br_2 in CCl_4, and gives a brown precipitate with alkaline $KMnO_4$. When *A* is hydrogenated over a catalyst, two equivalents of H_2 are consumed and the product is found to be 4-isopropyl-1-methylcyclohexane. Ozonolysis of *A* followed by treatment of the reaction mixture with H_2O_2 gives the following compound as a major product:

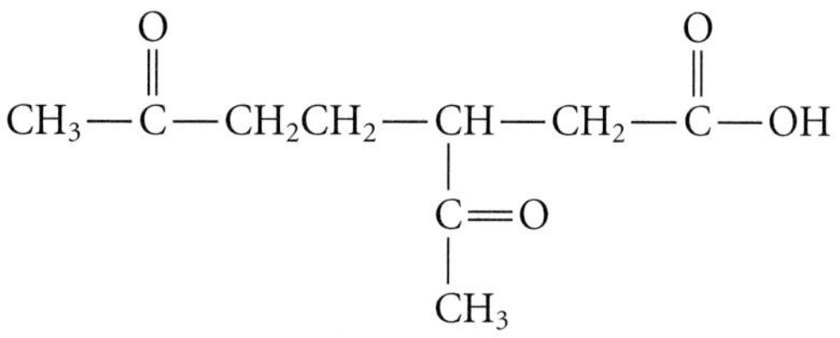

$$CH_3-\overset{\overset{O}{\|}}{C}-CH_2CH_2-\underset{\underset{CH_3}{\overset{|}{C=O}}}{CH}-CH_2-\overset{\overset{O}{\|}}{C}-OH$$

Suggest a structure for *A* and explain all observations.

Alkene

Brown Precipitate
KMnO4

STUDY GUIDE LINK:
✓5.5
Solving Structure Problems

5.52   A compound *A* was found by elemental analysis to contain 85.60% carbon and 14.40% hydrogen. The compound decolorized Br_2 in CCl_4 and gave a brown precipitate when treated with $KMnO_4$. Catalytic hydrogenation of *A* gave octane as the product. Treatment of the unknown with O_3, followed by H_2O_2, gave butanoic acid, *B*. Suggest a structure for *A*. What part of the structure is not determined by the data?

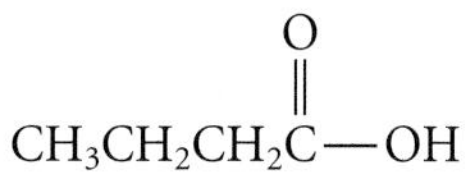

$$CH_3CH_2CH_2\overset{\overset{O}{\|}}{C}-OH \quad B$$

5.53   Using the curved-arrow formalism, suggest mechanisms for each of the following reactions. For help, see Study Guide Link 4.7.
*(a)

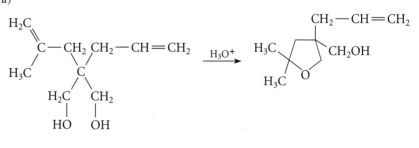

(b)

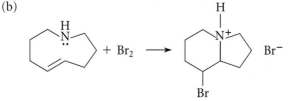

(Problem 5.53 continues)

*(c) For the following reaction, give the curved-arrow formalism only for the reaction of the alkene with $Hg(OAc)_2$ and H_2O. Then show that the compounds that you obtain from this mechanistic reasoning can be converted into the observed products by the $NaBH_4$ reduction.

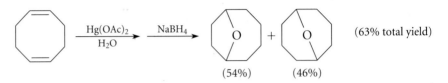

(54%)   (46%)   (63% total yield)

*(d) For this and the following part, consult Study Guide Link 5.4 if you need help.

(e)

$$CH_2{=}CH(CH_2)_5CH_3 + CBr_4 \xrightarrow[\text{light}]{\text{peroxides}} Br_3C{-}CH_2{-}\underset{\underset{Br}{|}}{CH}(CH_2)_5CH_3$$

(96% yield)

*5.54 (a) Using O—Cl homolysis as an initiation step, give a free-radical chain mechanism for the following reaction:

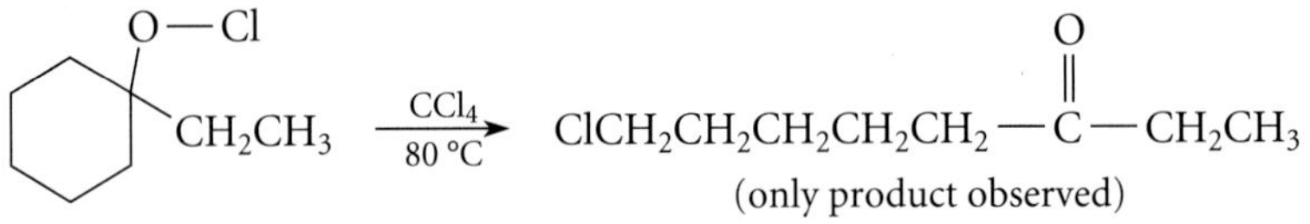

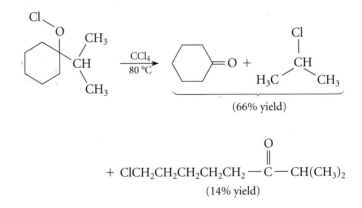

(b) Account for all three products in the following reaction. Suggest a reason why ring opening, observed in the reaction of part (a), is only a minor pathway in the following reaction:

$$+ ClCH_2CH_2CH_2CH_2CH_2{-}\overset{O}{\overset{\|}{C}}{-}CH(CH_3)_2$$

(14% yield)

(*Hint:* Compare the free radicals formed in the two competing reactions.)

*5.55   In the following sequence, the second reaction is unfamiliar. Nevertheless, iden-
tify compounds A and B from the information provided.

$$(CH_3)_3C—CH{=}CH_2 + HBr \longrightarrow A + \text{other compound(s)}$$
$$\text{(mostly)}$$

$$A + (CH_3)_3C—\overset{..}{\underset{..}{O}}\!{:}^- K^+ \longrightarrow (CH_3)_3C—\overset{..}{\underset{..}{O}}H + KBr + B$$
$$\text{(a strong base)}$$

$$B + O_3 \longrightarrow \xrightarrow{(CH_3)_2S} (CH_3)_2C{=}O \quad \text{(only compound formed)}$$

Compound B contains 85.60% C and 14.40% H, decolorizes Br_2 in CCl_4, and
takes up one equivalent of H_2 over a Pt/C catalyst. Once you have identified B,
try to give an arrow formalism for its formation from A in one step.

6

Introduction to Stereochemistry

Most of us are conscious of symmetry in objects that surround us. In the submicroscopic world of molecules symmetry can also be found, and the absence of molecular symmetry has some remarkable and interesting consequences.

This chapter deals with stereoisomers and their properties. **Stereoisomers** are compounds that have the same atomic connectivity but a different arrangement of atoms in space. The study of stereoisomers and the chemical effects of stereoisomerism is called **stereochemistry**. Through your study of stereochemistry you will see the consequences of symmetry—and its absence—at the molecular level.

The first discussion of stereochemistry dealt with *cis–trans* (or *E,Z*) isomerism of alkenes (Sec. 4.1B). This chapter delves into stereochemistry in more detail by concentrating on the basic definitions and principles of stereochemistry. You'll see how stereochemistry played a key role in the determination of the geometry of tetravalent carbon. Chapter 7 continues the discussion of stereochemistry by considering both the stereochemical aspects of cyclic compounds and how stereochemical principles apply to chemical reactions.

The use of molecular models during the study of this chapter is very important. Use of models will help you develop the ability to visualize three-dimensional structures, and if you use models, your reliance on them will eventually decrease.

6.1 Enantiomers, Chirality, and Symmetry

A. Enantiomers and Chirality

Any molecule—indeed, any object—has a mirror image. Some molecules are **congruent** to their mirror images. This means that all atoms and bonds in a molecule can be simultaneously aligned with identical atoms and bonds in its mirror image. An example of such a molecule is ethanol, or ethyl alcohol, CH_3—CH_2—OH (Fig. 6.1). Construct a model of ethanol and another model of its mirror image and use the following procedure to show that these two models are congruent. For simplicity, use a single colored ball to represent the methyl group and a single ball of another color to represent the hydroxy (—OH) group. Place the two central carbons side-by-side and align the methyl and

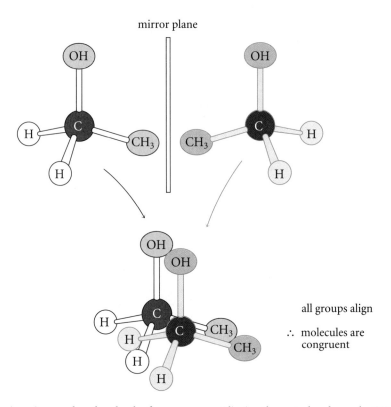

Figure 6.1   *Testing mirror-image ethanol molecules for congruence. Aligning the central carbons, the CH_3 groups, and the OH groups causes the two hydrogens to align as well.*

hydroxy groups, as shown in Fig. 6.1. *The hydrogens should then align as well. The congruence of an ethanol molecule and its mirror image shows that they are identical.*

Some molecules, however, are not congruent to their mirror images. An example is the 2-butanol molecule (Fig. 6.2).

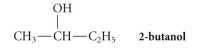

$$CH_3 - \overset{\displaystyle OH}{\underset{\displaystyle |}{CH}} - C_2H_5 \qquad \textbf{2-butanol}$$

Build a model of 2-butanol and a second model of its mirror image. If you align the central carbon and any two of its attached groups, the other two groups do not align. Hence, *a 2-butanol molecule and its mirror image are noncongruent and are therefore different molecules.* Because these two molecules have the identical connectivities, then by definition they are *stereoisomers.* Molecules that are noncongruent mirror images are called **enantiomers.** Thus, the two 2-butanol stereoisomers are enantiomers; they have an **enantiomeric** relationship.

Notice that enantiomers must not only be mirror images; they must also be *noncongruent* mirror images. Thus, ethanol (Fig. 6.1) has no enantiomer because an ethanol molecule and its mirror image are congruent.

Molecules (or other objects) that can exist as enantiomers are said to be **chiral** (pronounced kı′ rəl); they possess the property of **chirality,** or handedness (Fig. 6.3). (*Chiral* comes from the Greek word for hand.) Enantiomeric molecules have the same

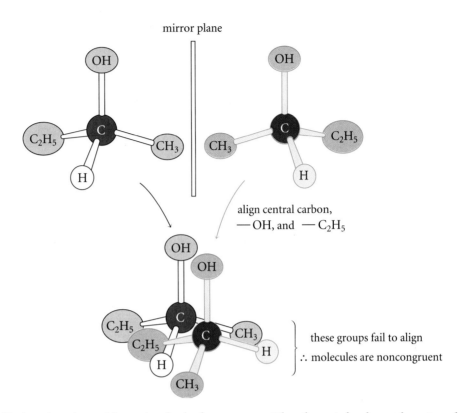

Figure 6.2   *Testing mirror-image 2-butanol molecules for congruence. When the central carbon and any two of the groups attached to it (OH and C_2H_5 in this figure) are aligned, the remaining groups do not align.*

relationship as the right and left hands—the relationship of object and noncongruent mirror image. Thus, 2-butanol is a chiral molecule. Molecules (or other objects) that are not chiral are said to be **achiral**, without chirality. Enthanol is an achiral molecule. Both chiral and achiral objects are matters of everyday acquaintance. A foot or a hand is chiral; the helical thread of a screw gives it chirality. Achiral objects include a ball and a soda straw.

IMPORTANCE OF CHIRALITY

Chiral molecules occur widely throughout all of nature. For example, glucose, an important sugar and energy source, is a chiral molecule; the enantiomer of naturally occurring glucose cannot be utilized as a food source. All sugars, proteins, and nucleic acids are chiral and occur naturally in only one enantiomeric form. Chirality is important in medicine as well. Over half of the organic compounds used as drugs are chiral, and in most cases only one enantiomer is active; in rare cases, the inactive enantiomer is toxic. The safety and effectiveness of synthetically prepared chiral drug molecules have become an issue of increasing concern for both pharmaceutical manufacturers and the Food and Drug Administration.

Figure 6.3   *Concept of chirality, as conceived by the Swiss artist Hans Erni.*

B. Asymmetric Carbon and Stereocenters

Many chiral molecules contain one or more asymmetric carbon atoms. An **asymmetric carbon atom** is a carbon to which *four different groups* are bound. Thus, 2-butanol (Fig. 6.2), a chiral molecule, contains an asymmetric carbon atom; this is the carbon that bears the four different groups $-CH_3$, $-C_2H_5$, $-H$, and $-OH$. Ethanol, an achiral molecule, has no asymmetric carbon. *A molecule that contains only one asymmetric carbon is chiral.* No generalization can be made, however, for molecules with more than one asymmetric carbon. Although most molecules with two or more asymmetric carbons are indeed chiral, not all of them are (Sec. 6.7). Moreover, an asymmetric carbon atom (or other asymmetric atom) is not a *necessary* condition for chirality; some chiral molecules have no asymmetric carbon at all (Sec. 6.9). Despite these caveats, it is important to recognize asymmetric carbon atoms because so many chiral organic compounds contain them.

STUDY PROBLEM 6.1

Identify the asymmetric carbon(s) in the following structure:

$$CH_3$$
$$|$$
$$CH_3CH_2CH_2CHCH_2CH_2CH_2CH_3$$

Solution   The asymmetric carbon is asterisked:

$$CH_3$$
$$|$$
$$CH_3CH_2CH_2\overset{*}{C}HCH_2CH_2CH_2CH_3$$

STUDY GUIDE LINK:
✓6.1
Finding Asymmetric Carbons in Rings

This is an asymmetric carbon because it bears four different groups: H, CH_3, $CH_3CH_2CH_2$, and $CH_2CH_2CH_2CH_3$. Notice that the last two groups (propyl and butyl) are not different at the point of attachment—both have CH_2 groups at that point, as well as at the next carbon removed. The difference is found at the ends of the groups. The point is that two groups are different even when the difference is remote from the carbon in question.

An asymmetric carbon atom is another type of *stereocenter,* or *stereogenic atom.* Recall (Sec. 4.1B) that a **stereocenter** is an atom at which the interchange of two groups gives a stereoisomer. For example, in Fig. 6.2, interchanging the methyl and ethyl groups in one enantiomer of 2-butanol gives the other enantiomer. If this point is not clear from Fig. 6.2, *use models to demonstrate this to yourself.* To do this, you need to build *two models.* First construct a model of either enantiomer, and then construct a model of its mirror image. Then show that the interchange of *any* two groups on one model gives the other model.

Asymmetric carbon atoms are not the only type of carbon stereocenters. Recall (Sec. 4.1B) that the carbons involved in the double bonds of *E-Z* isomers are also stereocenters; such carbons are not asymmetric carbons, because they are not connected to four different groups. In other words, the term *stereocenter* is not associated solely with chiral molecules. *All asymmetric atoms are stereocenters, but not all stereocenters are asymmetric atoms.*

STUDY GUIDE LINK: ✓6.2 *Stereocenters and Asymmetric Carbons*

C. Chirality and Symmetry

What causes chirality? *Chiral molecules lack certain types of symmetry.* The symmetry of any object (including a molecule) can be described by certain **symmetry elements,** which are lines, points, or planes that relate equivalent parts of an object. A very important symmetry element is a **plane of symmetry,** sometimes called an **internal mirror plane.** This is a plane that divides an object into halves that are exact mirror images. For example, the cup in Fig. 6.4a has a plane of symmetry. Similarly, the ethanol molecule shown in Fig. 6.4b also has a plane of symmetry. *A molecule or other object that has a plane of*

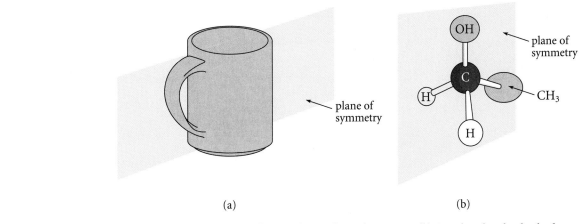

(a)   (b)

Figure 6.4   *Plane of symmetry. (a) The coffee mug has a plane of symmetry. (b) An ethanol molecule also has a plane of symmetry.*

symmetry is achiral. Thus, the ethanol molecule and the cup in Fig. 6.4 are achiral. Chiral molecules and other chiral objects *do not* have planes of symmetry. The chiral molecule 2-butanol, analyzed in Fig. 6.2, has no plane of symmetry. A human hand, also a chiral object, likewise has no plane of symmetry.

A plane of symmetry is the most common symmetry element found in achiral objects, but it is not the only one. Some achiral objects lack planes of symmetry but contain other types of symmetry not discussed here.

How do you know whether a molecule or other object is chiral? Let's summarize. If a molecule has a single asymmetric carbon, it is chiral. If a molecule has a plane of symmetry, it is not chiral. For other situations, the best test for chirality is to build two models, one of the molecule and the other of its mirror image, and then test the two for congruence. If the two mirror images are congruent, the molecule is achiral; if not, the molecule is chiral.

PROBLEMS

6.1   State whether each of the following molecules is achiral or chiral.

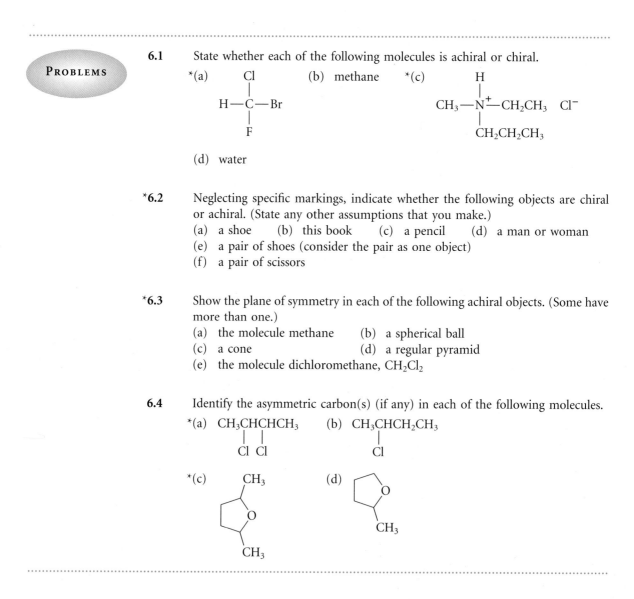

*(a)

$$H-\overset{\underset{\displaystyle F}{|}}{\underset{}{\overset{\displaystyle Cl}{|}}}{C}-Br$$

(b) methane

*(c)

$$CH_3-\overset{\underset{\displaystyle CH_2CH_2CH_3}{|}}{\overset{\displaystyle H}{|}}{\overset{+}{N}}-CH_2CH_3 \quad Cl^-$$

(d) water

*6.2   Neglecting specific markings, indicate whether the following objects are chiral or achiral. (State any other assumptions that you make.)
(a) a shoe   (b) this book   (c) a pencil   (d) a man or woman
(e) a pair of shoes (consider the pair as one object)
(f) a pair of scissors

*6.3   Show the plane of symmetry in each of the following achiral objects. (Some have more than one.)
(a) the molecule methane   (b) a spherical ball
(c) a cone   (d) a regular pyramid
(e) the molecule dichloromethane, CH_2Cl_2

6.4   Identify the asymmetric carbon(s) (if any) in each of the following molecules.

*(a) $CH_3CHCHCH_3$ (b) $CH_3CHCH_2CH_3$
            $|\ \ |$ $|$
            $Cl\ Cl$ Cl

*(c) CH_3 (d)

6.2 Nomenclature of Enantiomers: The *R,S* System

The existence of enantiomers poses a special problem of nomenclature. For example, suppose you are holding a model of 2-butanol. How do you indicate in the name of this compound which enantiomer it is? This can be done quite easily with the same Cahn-Ingold-Prelog priority rules used to assign *E* and *Z* conformations to alkene stereoisomers (Sec. 4.2B). A *stereochemical configuration*, or arrangement of atoms, at each asymmetric carbon in a molecule can be assigned using the following steps, which are illustrated in Fig. 6.5.

1. Identify an asymmetric carbon and the four different groups bound to it.
2. Assign priorities to the four different groups according to the rules given in Sec. 4.2B. The convention used in this text is that the highest priority = 1.
3. View the molecule along the bond *from the asymmetric carbon to the group of lowest priority*, that is, with the asymmetric carbon nearer and the low-priority group farther away.
4. Consider the clockwise or counterclockwise order of the remaining group priorities. If the priorities of these groups decrease in the *clockwise* direction the asymmetric carbon is said to have the *R* configuration. If the priorities of these groups decrease in the *counterclockwise* direction, the asymmetric carbon is said to have the *S* configuration.

STUDY PROBLEM 6.2

Determine the stereochemical configuration of the following enantiomer of 3-chloro-1-pentene:

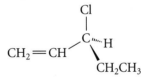

STUDY GUIDE LINK:
✓**6.3**
Using Perspective Structures

Solution   First assign relative priorities to the four groups attached to the asymmetric carbon. These are (1) —Cl; (2) CH_2=CH—; (3) —CH_2CH_3; (4) —H. Then, *using a model if necessary*, sight along the bond from asymmetric carbon to the lowest-priority group, in this case H. The resulting view is essentially a Newman projection along the C—H bond:

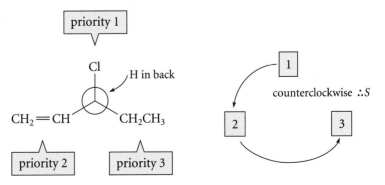

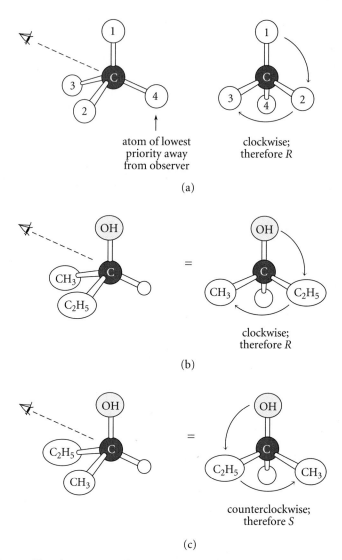

atom of lowest
priority away
from observer

clockwise;
therefore *R*

(a)

=

clockwise;
therefore *R*

(b)

=

counterclockwise;
therefore *S*

(c)

Figure 6.5   *Use of the Cahn-Ingold-Prelog system to designate stereochemistry (a) of a general asymmetric carbon atom; (b) of (R)-2-butanol; (c) of (S)-2-butanol. The direction of observation is shown on the left of each part, and what the observer sees is shown on the right. Priority 1 is highest, and priority 4 is lowest.*

Because the priorities of the first three groups descend in a counterclockwise direction, this is the *S* enantiomer of 3-chloro-1-pentene.

A stereoisomer is named by indicating the configuration of each asymmetric carbon before the systematic name of the compound, as in the following examples:

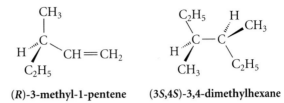

(**R**)-3-methyl-1-pentene   (**3S,4S**)-3,4-dimethylhexane

As illustrated by the second example, numbers are used with the *R,S* designations when there is more than one asymmetric carbon.

When you assign an *R* or *S* configuration to every asymmetric carbon in a molecule, you have specified its **absolute stereochemical configuration**, or **absolute stereochemistry**. Another, older system for specifying absolute stereochemistry, the D,L system, is still used in amino acid and carbohydrate chemistry (Chapters 26 and 27). With this exception, the *R,S* system has gained virtually complete acceptance.

IS *R* RIGHT, OR IS IT PROPER?

Choice of the letter *R* presented a problem for Cahn, Ingold, and Prelog, the scientists who devised the *R,S* system. The letter *S* stands for *sinister*, one of the Latin words for *left*. However, the Latin word for *right* (in the directional sense) is *dexter*, and unfortunately the letter *D* was already being used in another system of configuration (the D,L system). It was difficulties with this latter system that led to the need for a new system, and the last thing anyone needed was to confuse the two! Fortunately, Latin provided another word for *right*: the participle *rectus*. But this "right" does not indicate direction: it means *proper*, or *correct*. (The English word *rectify* comes from the same root.) Although the Latin wasn't quite proper, it solved the problem!

PROBLEMS

6.5   Draw perspective representations for each of the following chiral molecules. Use models if necessary.

*(a)   (*S*)-CH₃—CH—OH
                              |
                              D

(b)   (*S*)-CH₃—CH—NH₂
                              |
                              C=O
                              |
                              OH

*(c)   (*R*)-4-methyl-(*Z*)-2-hexene

(d)   (*S*)-4-bromo-(*E*)-2-pentene

6.6   Indicate whether the asymmetric atom in each of the following compounds has the *R* or *S* configuration.

(Problem 6.6 continues)

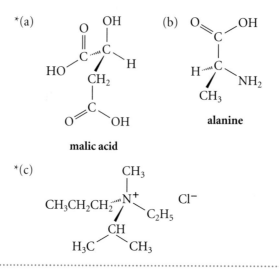

*(a)

malic acid

(b)

alanine

*(c)

6.3   Physical Properties of Enantiomers: Optical Activity

Most of the chemical and physical properties of a pair of enantiomers are identical. For example, both (R)- and (S)-2-butanol have the same boiling point, 99.5 °C. Likewise, (R)- and (S)-lactic acid have the same melting point, 53 °C.

$$CH_3-CH-C-OH$$ **lactic acid**

A pair of enantiomers also have identical densities, indices of refraction, heats of formation, standard free energies, and many other properties.

If a pair of enantiomers in fact have so many identical properties, how can we tell one enantiomer from the other? *A compound and its enantiomer can be distinguished by their effects on polarized light.* Understanding these phenomena requires an introduction to the nature of polarized light.

A. Polarized Light

Light is a wave motion that consists of oscillating electric and magnetic fields. The electric field of ordinary light oscillates in all planes. However, it is possible to obtain light with an electric field that oscillates in only one plane. This is called **plane-polarized light**, or simply, **polarized light** (Fig. 6.6).

Polarized light is obtained by passing ordinary light through a polarizer, such as a Nicol prism. The orientation of the polarizer's axis of polarization determines the plane of the resulting polarized light. If plane-polarized light is subjected to a second polarizer whose axis of polarization is perpendicular to that of the first, no light passes the second polarizer (Fig. 6.7a). This same effect can be observed with two pairs of Polaroid® sunglasses (Fig. 6.7b). When the lenses are oriented in the same direction, light will pass.

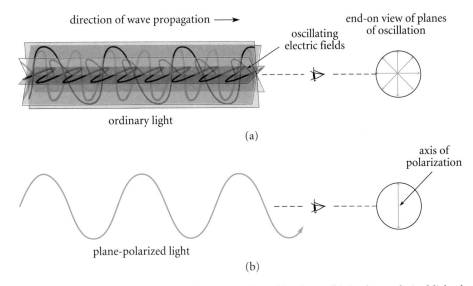

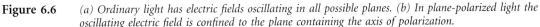

Figure 6.6   *(a) Ordinary light has electric fields oscillating in all possible planes. (b) In plane-polarized light the oscillating electric field is confined to the plane containing the axis of polarization.*

When the lenses are turned at right angles, their axes of polarization are crossed, no light is transmitted, and the lenses appear dark.

> Photography enthusiasts will also recognize the same phenomenon at work in a polarizing filter. A great deal of the glare in indirect skylight is polarized light. This can be filtered out by turning a polarizing filter so that the image has minimum intensity. The resulting photograph has much greater contrast than the same picture taken without the filter.

B. Optical Activity

If plane-polarized light is passed through one enantiomer of a chiral substance (either the pure enantiomer or a solution of it), *the plane of polarization of the emergent light is rotated.* A substance that rotates the plane of polarized light is said to be **optically active**. In general, *individual enantiomers of chiral substances are optically active.*

Optical activity is measured in a device called a **polarimeter** (Fig. 6.8), which is basically the system of two polarizers shown in Fig. 6.7. The sample to be studied is placed in the light beam between the two polarizers. Because optical activity changes with the wavelength (color) of the light, monochromatic light—light of a single color—is utilized to measure optical activity. The yellow light from a sodium arc (the sodium D-line with a wavelength of 5893 Å) is often used in this type of experiment. An optically inactive sample (such as air or solvent) is placed in the light beam. Light polarized by the first polarizer passes through the sample, and the analyzer is turned to establish a dark field. This setting of the analyzer defines the zero of optical rotation. Next, the sample whose optical activity is to be measured is placed in the light beam. The number of degrees α that the analyzer must be turned to reestablish the dark field is the **optical rotation** of the sample. If the sample rotates the plane of polarized light in the clockwise direction, the optical rotation is given a plus sign. Such a sample is said to be **dextrorotatory**.

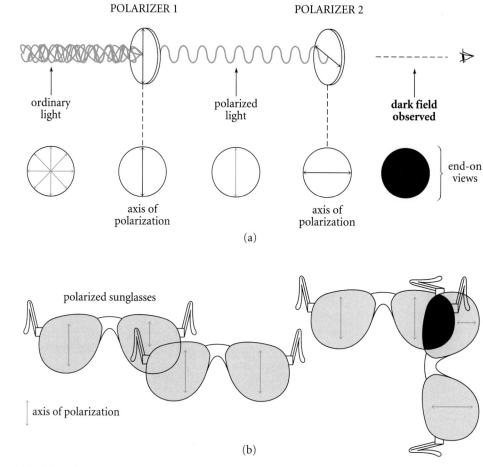

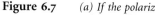

Figure 6.7   *(a) If the polarization axes of two polarizers are at right angles, no light passes through the second polarizer. (b) The same phenomenon can be observed using two pairs of polarized sunglasses.*

If the sample rotates the plane of polarized light in the counterclockwise direction, the optical rotation is given a minus sign, and the sample is said to be **levorotatory**.

The optical rotation of a sample is the quantitative measure of its optical activity. The observed optical rotation α is proportional to the number of optically active molecules present in the light beam. Thus, α is proportional to both the concentration c of the optically active compound in the sample as well as the length l of the sample container:

$$\alpha = [\alpha]cl \tag{6.1}$$

The constant of proportionality, $[\alpha]$, is called the **specific rotation**. The concentration of the sample is in g/mL, and the path length in dm (decimeters). (For a pure liquid, c is taken as the density). Thus, the specific rotation is equal to the observed rotation at a concentration of 1 g/mL and a path length of 1 dm. The specific rotation is used as the standard measure of optical activity. It is conventionally reported with a subscript that indicates the wavelength of light used and a superscript that indicates the temperature. Thus, a specific rotation reported as $[\alpha]_D^{20}$ has been determined at 20 °C using the sodium D-line.

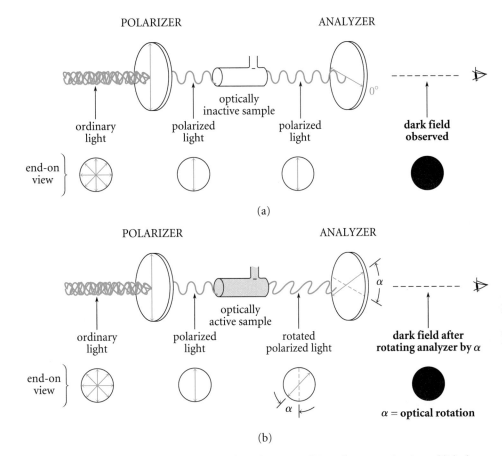

Figure 6.8   *Determination of optical rotation. (a) First, the reference condition of zero rotation is established as a dark field. (b) Next, the polarized light is passed through an optically active sample with observed rotation α. The analyzer is rotated to establish the dark-field condition again. The optical rotation α can be read from the calibrated scale on the analyzer.*

STUDY PROBLEM 6.3

A sample of (S)-2-butanol has a specific rotation $[\alpha]_D^{20}$ of $+13.9°$. What is the observed optical rotation of a 1.0 *M* solution of (S)-2-butanol taken in a sample container that is 10 cm long?

Solution   In order to use the specific rotation, the sample concentration in g/mL must be determined. Since the molecular mass of 2-butanol is 74.12, then the solution contains 74.1 g/L or 0.0741 g/mL of 2-butanol. This is the value of *c* used in Eq. 6.1. The value of *l* is 1 dm. Substituting in Eq. 6.1, $\alpha = (+13.9°)(0.0741)(1) = +1.03°$. This is the observed optical rotation of the sample.

Notice that, for dimensional consistency, the units of the specific rotation $[\alpha]_D^{20}$ must be deg $\cdot$ mL $\cdot$ g^{-1} $\cdot$ dm^{-1}. However, specific rotations are generally cited in degrees, as in Study Problem 6.3. The remaining units are omitted for convenience.

C. Optical Activities of Enantiomers

A pair of enantiomers are distinguished by their optical activities because *a pair of enantiomers rotate the plane of polarized light by equal amounts in opposite directions*. For example, the specific rotation $[\alpha]_D^{20}$ of (S)-2-butanol is $+13.9°$. The specific rotation of its enantiomer (R)-2-butanol is $-13.9°$. If a solution of (S)-2-butanol has an observed rotation of $+3.5°$, then a solution of (R)-2-butanol under the same conditions will have an observed rotation of $-3.5°$.

A sample of a pure chiral compound uncontaminated by its enantiomer is said to be **enantiomerically pure**. (The term *optically pure* was used in older literature.) In a mixture of the two enantiomers, each contributes to the optical rotation in proportion to its concentration. It follows that a sample containing equal amounts of two enantiomers must have an observed optical rotation of zero.

Sometimes a plus or minus sign is used with the name of a chiral compound to indicate the sign of its optical rotation. Thus, (S)-2-butanol is sometimes called (S)-(+)-2-butanol because it has a positive optical rotation. Similarly, (R)-2-butanol would be termed (R)-(−)-2-butanol.

There is no simple relationship between the sign of optical rotation and absolute configuration. Thus, some compounds with the *S* configuration have positive rotations, and others have negative rotations. Although the *S* enantiomer of 2-butanol is dextrorotatory ($[\alpha]_D^{20} = +13.9°$), the dextrorotatory enantiomer of glyceraldehyde has the *R* configuration.

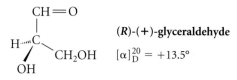

(R)-(+)-glyceraldehyde $[\alpha]_D^{20} = +13.5°$

Why don't achiral molecules show optical activity? When polarized light passes through a sample of a compound, each molecule makes its own tiny contribution to optical activity, because the electrons in the molecule interact with the electric field of the light wave. For example, when polarized light passes through a sample of ethanol, the light may "strike" one ethanol molecule in a certain orientation with respect to the light beam. This molecule may actually cause a tiny optical rotation to occur. However, it is equally likely that the light will "strike" another ethanol molecule in a mirror-image orientation. This orientation is equally likely because ethanol is achiral. Because the optical rotation of the second orientation will be equal in magnitude and opposite in sign to that of the first, the two will cancel. In contrast, two identical chiral molecules *cannot* assume mirror-image orientations with respect to the light beam precisely because they lack the appropriate symmetry—they are chiral. Thus, a sample of any chiral compound shows net optical activity, provided, of course, that an equal amount of its enantiomer is not present.

PROBLEM

*6.7   Suppose a sample of an optically active substance has an observed rotation of $+10°$. The scale on the analyzer of a polarimeter is circular; $+10°$ is the same as $-350°$, or $+370°$. How would you determine whether the observed rotation is $+10°$ or some other value?

6.4

STUDY GUIDE LINK:
6.4
Terminology of Racemates

Racemates

A mixture containing equal amounts of two enantiomers is encountered so commonly that it is given a special name. Such a mixture is called a **racemate** or **racemic mixture**. (In older literature the term *racemic modification* was used.) A racemate is referred to by name in two ways. The racemate of 2-butanol, for example, can be called either racemic 2-butanol or (±)-2-butanol.

Racemates typically have physical properties that are different from those of the pure enantiomers. For example, the melting point of either enantiomer of lactic acid is 53 °C, but the melting point of racemic lactic acid is 18 °C. The optical rotation of the racemate is zero because a racemate contains equal amounts of two enantiomers whose equal optical rotations of opposite sign exactly cancel each other.

One point of terminology is worth special notice. It is sometimes said, incorrectly, that racemates are achiral. This statement is wrong because chirality is a property of individual molecules (or other objects). A chiral molecule has this property whether it is mixed with its enantiomer or not. Thus, 2-butanol is a chiral molecule because it exists in two enantiomeric forms. The fact that it might also occur as a mixture of two enantiomers does not alter the fact that it is chiral. What *is* true about racemates is that they are *optically inactive*. Optical activity is a physical property; chirality is a structural attribute. Optical activity requires a chiral sample; however, a sample that contains chiral molecules (a racemate, for example) may not be optically active. Although optical activity and chirality are closely associated, they should not be confused.

The process of forming a racemate from a pure enantiomer is called **racemization**. The simplest method of racemization is to mix equal amounts of enantiomers. As you will learn, racemization can also occur as a result of conformational changes or chemical reactions.

Because the enantiomers of a pair have the same boiling points and the same melting points, perhaps you can appreciate that the separation of enantiomers poses a special problem. Enantiomers cannot ordinarily be separated by fractional distillation or crystallization, because these techniques depend on differences in boiling points, melting points, or solubilities. The separation of a pair of enantiomers, called an **enantiomeric resolution**, requires special methods that are considered in Sec. 6.8.

PROBLEMS

*6.8 A 0.1 *M* solution of an enantiomerically pure chiral compound *D* has an observed rotation of +0.20° in a 1-dm sample container. The molecular mass of the compound is 150.

(a) What is the specific rotation of *D*?

(b) What is the observed rotation if this solution is mixed with an equal volume of a solution that is 0.1 *M* in *L*, the enantiomer of *D*?

(c) What is the observed rotation if the solution of *D* is diluted with an equal volume of solvent?

(d) What is the specific rotation of *D* after the dilution described in (c)?

(e) What is the specific rotation of *L*, the enantiomer of *D*?

(f) What is the observed rotation of 100 mL of a solution that contains 0.01 mole of *D* and 0.005 mole of *L*? (Assume a 1 dm path length.)

6.9   What observed rotation is expected when a 1.5 *M* solution of (*R*)-2-butanol is mixed with an equal volume of a 0.75 *M* solution of racemic 2-butanol, and the resulting solution is analyzed in a sample container that is 1 dm long? The specific rotation of (*R*)-2-butanol is −13.9°.

6.5   Stereochemical Correlation

You have learned how to assign the *R* or *S* configuration to compounds with known three-dimensional structures (Sec. 6.2). But how is a three-dimensional structure determined in the first place? (Recall that the sign of optical rotation *cannot* be used to assign an *R* or *S* configuration; Sec. 6.3C). One way to determine experimentally the absolute stereochemistry of a compound is to use a variation of X-ray crystallography, called *anomalous dispersion*. This technique, however, requires special instruments, is not readily available in the average laboratory, and has been applied to relatively few compounds. The absolute configurations of most organic compounds are determined instead by using chemical reactions to correlate them with other compounds of known absolute configurations. This process is called **stereochemical correlation**.

To illustrate a stereochemical correlation, suppose you have in hand an enantiomerically pure sample of the following alkene. Also suppose that you find from earlier work published in the chemical literature that the *R* enantiomer of this alkene has a negative optical rotation.

$$Ph-CH-CH=CH_2 \quad R \text{ enantiomer has } [\alpha]_D^{22} = -6.39°$$
$$| $$
$$CH_3$$

Suppose further that you subject this alkene to hydroboration-oxidation and obtain an alcohol that has negative optical rotation.

$$(R)\text{-}(-)\text{-}Ph-CH-CH=CH_2 \xrightarrow{BH_3} \xrightarrow{H_2O_2/OH^-} (-)\text{-}Ph-CH-CH_2CH_2-OH \quad (6.2)$$
$$| \qquad\qquad\qquad\qquad\qquad\qquad\qquad\qquad\qquad | $$
$$CH_3 \qquad\qquad\qquad\qquad\qquad\qquad\qquad\qquad\qquad CH_3$$

known configuration and optical rotation · unknown configuration but known optical rotation

Notice that *this reaction does not break any of the bonds to the asymmetric carbon atom.* Thus, the way that corresponding groups are arranged about the asymmetric carbon must be the same in both reactant and product. Hence, the configuration of the (−)-alcohol can be determined from this experiment, because the —CH₂CH₂OH group of the product and the —CH=CH₂ group of the alkene are in the same stereochemical positions. Drawing the alcohol configuration so that it corresponds to the alkene configuration, and applying the *R,S* system to the resulting structure shows that the alcohol has the *R* configuration.

$$(6.3)$$

If the configuration of the (−)-alcohol were previously unknown, this reaction would establish it as *R*. It follows from the same experiment that the (+)-alcohol has the *S* configuration. Consequently, the hydroboration-oxidation reaction of the alkene serves to *correlate* the configuration of the alcohol with its sign of optical rotation. Without such an *experimental* correlation, chemists would have no way of assigning a configuration to the alcohol solely from the sign of its optical rotation.

Although both reactant and product in this example have the same *R,S* designations, this does not have to be true in general. It is possible for the *R,S* designations of reactant and product to differ if a reaction results in a change in the relative priority of groups at the asymmetric carbon, as in Problem 6.10.

> The most secure way of relating absolute configurations is to use reactions that do not break the bonds at the asymmetric carbon, as shown above. A reaction that breaks these bonds, however, may also be used provided that the stereochemical outcome of such a reaction has been established previously on a number of related compounds.

PROBLEMS

*6.10   From the outcome of the following transformation, indicate whether the levorotatory enantiomer of the product has the *R* or *S* configuration. Draw a structure of the product that shows its absolute configuration. (*Hint:* The phenyl group has a higher priority than the vinyl group in the *R,S* system.)

$$(S)\text{-}(+)\text{-Ph}-\underset{\underset{\text{CH}_2\text{CH}_2\text{CH}_3}{|}}{\text{CH}}-\text{CH}=\text{CH}_2 + \text{H}_2 \xrightarrow{\text{Pd/C}} (-)\text{-Ph}-\underset{\underset{\text{CH}_2\text{CH}_2\text{CH}_3}{|}}{\text{CH}}-\text{CH}_2\text{CH}_3$$

6.11   Explain how you would use the alkene starting material in Eq. 6.2 to determine the absolute configuration of the dextrorotatory enantiomer of the following hydrocarbon:

$$\text{Ph}-\underset{\underset{\text{CH}_3}{|}}{\text{CH}}-\text{CH}_2-\text{CH}_3$$

Outline the possible results of your experiment and how you would interpret them.

6.6   Diastereomers

Up to this point, our discussion has focused on molecules with only one asymmetric carbon. What happens when a molecule has two or more asymmetric carbons? This situation is illustrated by 2,3-pentanediol.

$$\underset{1 \quad\quad 2 \quad\quad 3 \quad\quad 4 \quad 5}{\underset{\text{carbon number}}{\text{CH}_3-\overset{\overset{\text{OH}}{|}}{\text{CH}}-\overset{\overset{\text{OH}}{|}}{\text{CH}}-\text{CH}_2\text{CH}_3}} \quad \text{2,3-pentanediol}$$

The 2,3-pentanediol molecule has two asymmetric carbons: carbons 2 and 3. Each might have the *R* or *S* configuration. With two possible configurations at each carbon, four stereoisomers are possible:

$$(2S,3S) \qquad (2R,3R)$$
$$(2S,3R) \qquad (2R,3S)$$

These possibilities are shown as ball-and-stick models in Fig. 6.9. What are the relationships among these stereoisomers?

The 2*S*,3*S* and 2*R*,3*R* isomers are a pair of enantiomers, because they are noncongruent mirror images; the 2*S*,3*R* and 2*R*,3*S* isomers are also an enantiomeric pair. (Demonstrate this point to yourself with models.) These structures illustrate the following generalization: *In order for a pair of chiral molecules with more than one asymmetric carbon to be enantiomers, they must have different configurations at every asymmetric carbon.*

Because neither the 2*S*,3*S* and 2*S*,3*R* pair nor the 2*R*,3*R* and 2*R*,3*S* pair are enantiomers, they must have a different stereochemical relationship. Stereoisomers that are not enantiomers are called **diastereoisomers**, or, more simply, **diastereomers**. They have a **diastereomeric** relationship. Notice that diastereomers are not mirror images. All of the relationships among the stereoisomeric 2,3-pentanediols are shown in Fig. 6.10.

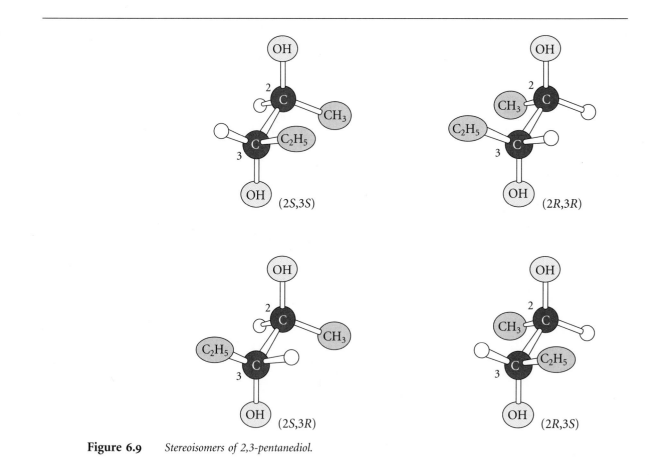

Figure 6.9   *Stereoisomers of 2,3-pentanediol.*

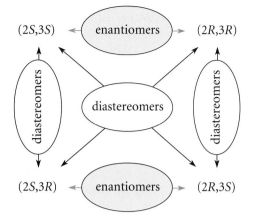

Figure 6.10   *Relationships among the stereoisomers of 2,3-pentanediol. Any pair of stereoisomers at opposite ends of a double-headed arrow have the relationship indicated within the arrow. Notice that enantiomers have different configurations at both asymmetric carbons.*

Diastereomers differ in all of their physical properties. Thus, diastereomers have different melting points, boiling points, heats of formation, and standard free energies. Because diastereomers differ in all of their physical properties, they can in principle be separated by conventional means, for example, by fractional distillation or crystallization. If diastereomers happen to be chiral, they can be expected to have different specific rotations, and their specific rotations will have no relationship. These points are illustrated in Table 6.1, which gives some physical properties for four stereoisomers and their racemates.

You have now seen an example of every common type of isomerism. To summarize:

1. *Isomers* have the same molecular formula.

2. *Constitutional isomers* have different atomic connectivities.

3. *Stereoisomers* have identical atomic connectivities. There are *only two* types of stereoisomers:
   a. *Enantiomers* are noncongruent mirror images.
   b. *Diastereomers* are stereoisomers that are not enantiomers.

The structural relationships among molecules are analyzed by working with *one pair of molecules at a time.* The flow chart in Fig. 6.11 provides a systematic way to determine the isomeric relationship, if one exists, between two nonidentical molecules. Given a pair of molecules, work down the chart from top to bottom asking each question in order and following the appropriate branch. When you get to a box labeled "END," the isomeric relationship is determined. The following Study Problem illustrates the use of Fig. 6.11.

STUDY PROBLEM 6.4

Determine the isomeric relationship between the following two molecules:

$$CH_3CH_2 \quad CH_2CH_3 \qquad H \quad CH_2CH_3$$
$$C{=}C \qquad\qquad C{=}C$$
$$H \qquad H \qquad CH_3CH_2 \quad H$$

Table 6.1   **Properties of Four Chiral Stereoisomers**

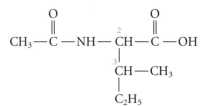

Configuration	Specific rotation (25°, ethanol)	Melting point, °C	Relationship	
(2S,3S)	+15°	150–151°	Enantiomers	Diastereomers
(2R,3R)	−15°	150–151°		
(2S,3R)	+21.5°	155–156°	Enantiomers	
(2R,3S)	−21.5°	155–156°		
racemate of (2S,3S) and (2R,3R)	0°	117–123°		
racemate of (2S,3R) and (2R,3S)	0°	165–166°		

Solution   Work from the top of Fig. 6.11 and answer each question in turn. These two molecules have the same molecular formula; hence, they are isomers. They have the same atomic connectivity; hence, they are stereoisomers. (In fact, they are the *E* and *Z* isomers of 3-hexene.) Because the molecules are not mirror images, they must be diastereomers. Thus, (*E*)- and (*Z*)-3-hexene are diastereomers.

Notice from this study problem that *cis-trans* (or *E,Z*) isomerism is one type of diastereomeric relationship. The fact that neither (*E*)- nor (*Z*)-3-hexene is chiral shows that some diastereomers are *not* chiral. On the other hand, some diastereomers are chiral, as in the case of the diastereomeric 2,3-pentanediols (Fig. 6.9).

6.7   *Meso* Compounds

Until now, each molecule containing one or more asymmetric carbon atoms has been chiral. However, certain compounds containing two or more asymmetric carbons are achiral. The compound 2,3-butanediol is an example:

$$\underset{1}{CH_3} - \underset{2}{\overset{\overset{\displaystyle OH}{|}}{CH}} - \underset{3}{\overset{\overset{\displaystyle OH}{|}}{CH}} - \underset{4}{CH_3} \qquad \textbf{2,3-butanediol}$$

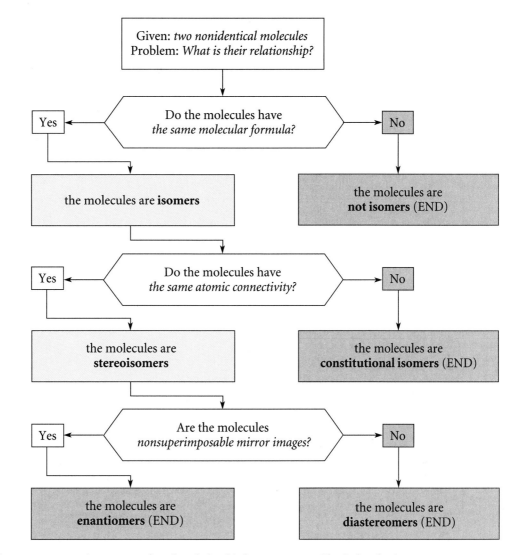

Figure 6.11   *A systematic way to analyze the relationship between two nonidentical molecules.*

As with 2,3-pentanediol in the previous section, there appear to be four stereochemical possibilities:

(2*S*,3*S*)   (2*R*,3*R*)

(2*S*,3*R*)   (2*R*,3*S*)

Ball-and-stick models of these molecules are shown in Fig. 6.12. Consider the relationships among these structures. As you can see from the top row of Fig. 6.12, the 2*S*,3*S* and 2*R*,3*R* structures are noncongruent mirror images, and are thus enantiomers.

What is the relationship of the other two molecules—the 2*S*,3*R* and 2*R*,3*S* pair? These structures, as they are drawn in Fig. 6.12, are mirror images. However, they are

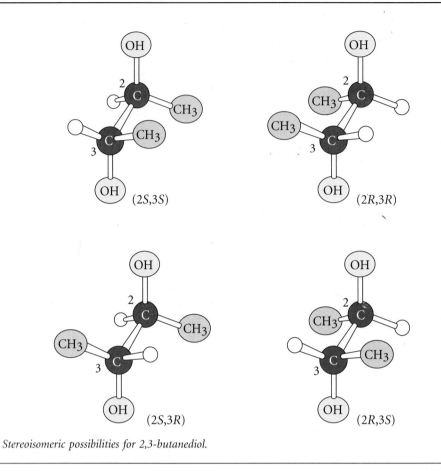

Figure 6.12   *Stereoisomeric possibilities for 2,3-butanediol.*

congruent, and thus *they are the same molecule!* You can show this by rotating the 2*R*,3*S* structure 180° about an axis perpendicular to the C2-C3 bond:

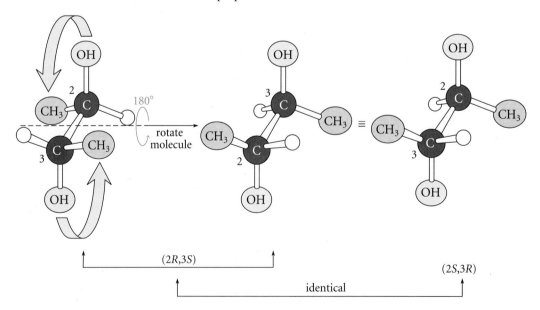

This stereoisomer of 2,3-butanediol is an example of a *meso* compound, *meso*-2,3-butanediol. A ***meso*** **compound** is an achiral compound that contains asymmetric atoms (in this case, asymmetric carbons). Because a *meso* compound is congruent to its mirror image, *it is not chiral*. Because it is not chiral, a *meso* compound cannot be optically active. *Meso*-2,3-butanediol is a diastereomer of the (2S,3S)- and (2R,3R)-butanediols. A summary of the relationships between the stereoisomers of 2,3-butanediol is given in Fig. 6.13.

STUDY GUIDE LINK:
6.5
Center of Symmetry

Notice carefully the difference between a *meso* compound and a racemate. Although both are optically inactive, a *meso* compound is a *single achiral compound*, but a racemate is a *mixture of chiral compounds*—specifically, an equimolar mixture of enantiomers.

The existence of *meso* compounds shows that *some achiral compounds have asymmetric carbons*. Thus, the presence of asymmetric carbons in a molecule is *not* a sufficient condition for it to be chiral, unless it has only *one* asymmetric carbon. If a molecule contains n asymmetric carbons, then it has 2^n stereoisomers unless there are *meso* compounds. If there are *meso* compounds, then there are fewer than 2^n stereoisomers.

Notice that the only real difference between a *meso* compound and any other type of achiral stereoisomer is that a *meso* compound by definition contains *asymmetric atoms*. Thus, neither *cis*-2-butene nor *trans*-2-butene is a *meso* compound. Although they are achiral stereoisomers, these compounds have no asymmetric atoms.

How do you know whether a molecule possesses a *meso* stereoisomer? A *meso* compound is possible only when a molecule can be divided into structurally identical halves. (The word *meso* means "in the middle.")

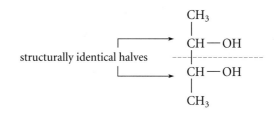

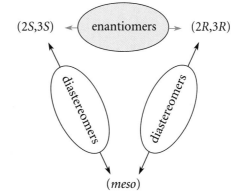

Figure 6.13   *Relationships among the stereoisomers of 2,3-butanediol. Any pair of compounds at opposite ends of the arrow have the relationship indicated within the arrow. Notice that the meso stereoisomer is achiral and thus has no enantiomer.*

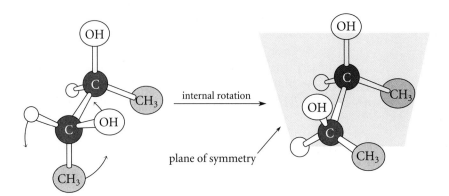

Figure 6.14   *Internal rotation to give an eclipsed form with a plane of symmetry (internal mirror plane) identifies a meso compound.*

Once you recognize the possibility of a *meso* compound, how do you know which stereoisomers are *meso* and which are chiral? First, in a *meso* compound, the corresponding asymmetric carbons in each half of the molecule must have *opposite stereochemical configurations*:

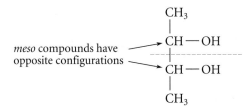

Thus, one asymmetric carbon in 2,3-butanediol (Fig. 6.12) is *R* and the other is *S*. Another simple way to identify a *meso* compound without assigning stereochemical configurations is shown in Fig. 6.14. If any conformation, even an *eclipsed* conformation, has a plane of symmetry, the molecule is *meso*. This method works even though the molecule does not exist in an eclipsed conformation.

PROBLEMS

6.12   Tell whether each of the following molecules has a *meso* stereoisomer.

*(a)
$$CH_3CHCH_2CHCH_3$$
with Cl, Cl substituents

(b) $CH_3CH_2CHCHCH_2CH_3$
with Cl Cl substituents

*(c) *trans*-2-hexene   (d) *cis*-3-hexene

*6.13   Explain why the following compound has two *meso* stereoisomers.

$$CH_3-CH-CH-CH-CH_3$$
with OH OH OH substituents

(*Hint:* The plane that divides the molecule into structurally identical halves can go *through* one or more atoms.)

6.8

Enantiomeric Resolution

As already noted, the isolation of the pure enantiomers from a racemate (an *enantiomeric resolution*) poses a special problem. Because a pair of enantiomers have identical melting points, boiling points, and solubilities, we cannot exploit these properties for resolution of enantiomers as we might for the separation of other compounds. How, then, are enantiomers separated?

The resolution of enantiomers takes advantage of the fact that *diastereomers, unlike enantiomers, have different physical properties*. The strategy used is to convert a racemate *temporarily* into a mixture of diastereomers by allowing the racemate to combine with an enantiomerically pure chiral compound called a **resolving agent**. The resulting diastereomers are separated, and each diastereomer is then converted back into the free resolving agent and a pure enantiomer of the compound of interest.

ANALOGY FOR A RESOLVING AGENT

Suppose you are blindfolded and asked to sort a pile of one hundred gloves into separate piles of right- and left-handed gloves. (Never mind how you got into this predicament!) The gloves are identical except that fifty are right-handed, and fifty are left-handed. The mixture of gloves is a "racemate." How would you separate them? You can't do it by weight, by smell, or by any other simple physical property, because right- and left-handed gloves have the same properties. The way you do it, of course, is by trying each glove on your right (or left) hand. Your hand thus acts as an "enantiomerically pure" chiral resolving agent. A right-handed glove on your right hand generates a certain feeling (which we describe by saying "it fits"), and a left-handed glove on the right hand generates a totally different feeling. You allow the hand (resolving agent) to interact with each glove, and you segregate the gloves on the basis of the resulting sensation. You then break the hand-glove interaction (remove the glove) and put the glove in the appropriate pile.

An example of an enantiomeric resolution is the separation of the racemate of α-phenethylamine into its enantiomers.

$$\overset{\displaystyle :NH_2}{\underset{\displaystyle Ph-CH-CH_3}{|}} \quad \textbf{α-phenethylamine}$$

This process takes advantage of the fact that amines, like ammonia, are bases, and they react rapidly and quantitatively with carboxylic acids to form salts:

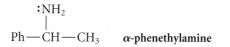

$$\text{(6.4)}$$

an amine   a carboxylic acid   a salt
(a base)   (an acid)

The resolving agent is an *enantiomerically pure* carboxylic acid. In many cases, enantiomerically pure compounds used for this purpose can be obtained from natural sources. One such compound is (2R,3R)-(+)-tartaric acid:

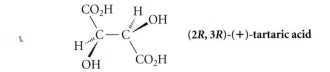

(**2R, 3R**)-(**+**)-**tartaric acid**

When (+)-tartaric acid reacts with the racemic amine, a mixture of two *diastereomeric* salts is formed:

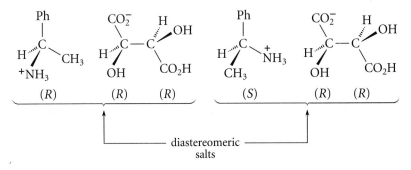

diastereomeric salts

These salts are diastereomers because they differ in configuration at *only one* of their three asymmetric carbons. (Enantiomers would differ at *every* asymmetric carbon.) Because these salts are diastereomers, they have different physical properties. In this case, they have significantly different solubilities in methanol, a commonly used alcohol solvent. The (*S,R,R*) diastereomer crystallizes from methanol, leaving the (*R,R,R*) diastereomer in solution, from which it may be recovered. Once either pure diastereomer is in hand, the salt can be decomposed with base to liberate the water-insoluble, optically active amine, leaving the tartaric acid in solution as its conjugate-base dianion.

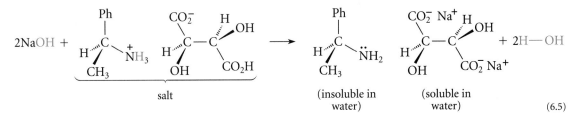

salt   (insoluble in water)   (soluble in water)   (6.5)

Salt formation is such a convenient reaction that it is often used for the enantiomeric resolution of amines and carboxylic acids. Although the reaction involved in the optical resolutions of other types of compounds may differ, the principle involved is always the same: an enantiomerically pure resolving agent is used to form *temporarily* a mixture of diastereomers from a racemate.

PROBLEMS

*6.14 Which of the following amines could in principle be used as a resolving agent for a racemic carboxylic acid?

$$(-)\text{-Ph}-\overset{\cdot\cdot}{\underset{|}{\text{CH}}}-\overset{\cdot\cdot}{\text{NH}}_2 \qquad (\pm)\text{-Ph}-\overset{\cdot\cdot}{\underset{|}{\text{CH}}}-\overset{\cdot\cdot}{\text{NH}}_2 \qquad \text{CH}_3-\overset{\cdot\cdot}{\text{NH}}_2$$
$$\overset{}{\underset{}{\text{CH}_3}} \qquad\qquad\qquad \overset{}{\underset{}{\text{CH}_3}} \qquad\qquad\qquad C$$
$$\qquad\qquad A \qquad\qquad\qquad\qquad\qquad B$$

6.15   Which of the following carboxylic acids could in principle be used as a resolving agent for a racemic amine?

$$(+)\text{-CH}_3\text{CH}_2\overset{\overset{\displaystyle O}{\|}}{\underset{\underset{\displaystyle \text{CH}_3}{|}}{\text{CHC}}}-\text{OH} \qquad \text{CH}_3\text{CH}_2\text{CH}_2\overset{\overset{\displaystyle O}{\|}}{\text{C}}-\text{OH} \qquad (\pm)\text{-CH}_3\text{CH}_2\overset{\overset{\displaystyle O}{\|}}{\underset{\underset{\displaystyle \text{CH}_3}{|}}{\text{CHC}}}-\text{OH}$$

$$A \qquad\qquad\qquad\qquad B \qquad\qquad\qquad\qquad C$$

6.9

Chiral Molecules without Asymmetric Atoms

The existence of *meso* compounds shows that the presence of asymmetric carbons is not a *sufficient* condition for the chirality of a molecule. This section will show that the presence of an asymmetric atom is also not *necessary* for chirality. In other words, *some chiral molecules contain no asymmetric atoms*. An example is the pair of molecules shown in Fig. 6.15. Notice that these two molecules are *noncongruent mirror images*, and are therefore enantiomers. (If necessary, build models of them and convince yourself that this is so.)

Although the molecules in Fig. 6.15 contain no asymmetric carbons, *each contains three carbon stereocenters*.

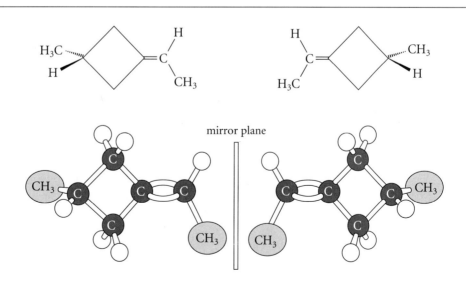

Figure 6.15   *Enantiomers of a chiral molecule that has no asymmetric carbon atoms. The enantiomers are drawn to show their mirror-image relationship. Although these molecules contain no asymmetric carbon atoms, they do contain stereocenters (stereogenic atoms).*

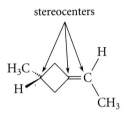

stereocenters

You can verify that any one of these carbons is a stereocenter by interchanging any two groups bound to it; this interchange generates the other enantiomer. To illustrate, let's interchange the two ring bonds at the stereocenter in the middle of the molecule.

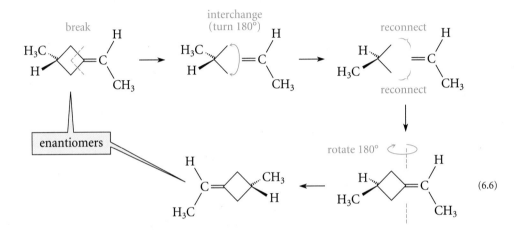

(6.6)

(Be sure to demonstrate that a group interchange at each of the other two stereocenters also gives enantiomers.)

Molecules such as this one are important because they demonstrate the phenomenon of chirality without asymmetric atoms. Nevertheless, such cases are relatively rare. Most of the chiral molecules you'll encounter will contain one or more asymmetric carbons.

PROBLEM

6.16 Indicate whether each compound is chiral. Identify the asymmetric carbons and stereocenters (if any) in each.

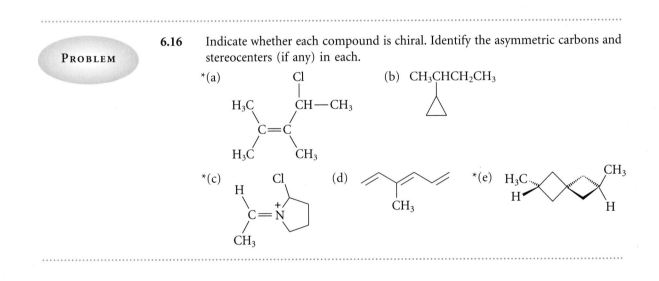

6.10   Conformational Stereoisomers

A.   Stereoisomers Interconverted by Internal Rotations

If you examine the structure of butane, $CH_3CH_2CH_2CH_3$, you might conclude that it has no stereocenters, and that butane cannot exist in stereoisomeric forms. However, an examination of the individual conformations of butane leads to a different conclusion. As shown in Fig. 6.16, the two *gauche* conformations of butane are noncongruent mirror images, or enantiomers; consequently, *gauche*-butane is chiral! (This is another situation in which chirality exists in the absence of an asymmetric carbon.) The two *gauche* conformations of butane are examples of **conformational enantiomers**: enantiomers that are interconverted by a conformational change. The "conformational change" in this case is an internal rotation. Similarly, the *anti* conformation of butane is a diastereomer of either one of the *gauche* conformations. *Anti*- and either *gauche*-butane are therefore **conformational diastereomers**: diastereomers that are interconverted by a conformational change.

Despite the chirality of any one *gauche* conformation of butane, the compound butane is not optically active because the two *gauche* conformations are present in equal amounts. The optical activity of one *gauche* enantiomer thus cancels the optical activity of the other. (The *anti* conformation, because it is achiral, would not be optically active even if it were present alone.) However, imagine an amusing experiment in which the two *gauche* conformations of butane are separated (by an as yet undisclosed method!) at such a low temperature that the interconversion between the *gauche* and *anti* conformations of butane is very slow. Each *gauche* butane isomer would then be optically active! The two *gauche* isomers would have equal specific rotations of opposite signs, but many of their other properties would be the same. Because *anti*-butane is achiral, it would have

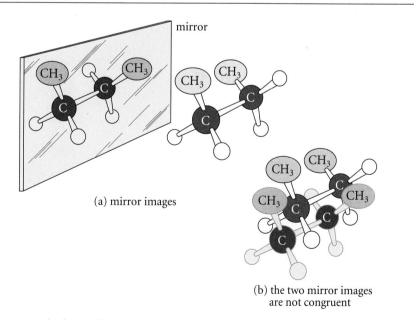

(a) mirror images

(b) the two mirror images
are not congruent

Figure 6.16   *(a) The two gauche forms of butane are mirror images. (b) Because these mirror images are not congruent, they are enantiomers.*

zero optical rotation, and all of its properties would differ from those of *gauche*-butane. Of course, at room temperature, the isolation of individual conformations would be impossible, because the butane isomers come to equilibrium within 10^{-9} second by rotation about the central carbon-carbon bond. (This is another example of *racemization*; Sec. 6.4.) It is conceivable, though, that on some planet with a temperature near absolute zero, *gauche*- and *anti*-butanes exist as separate compounds. (It would also be interesting to meet any inhabitants of such a planet capable of appreciating this fact!)

Is butane a chiral molecule? Although it contains chiral conformations, butane *behaves as if* it is achiral, because it has zero optical activity and it cannot be separated into enantiomers on any reasonable time scale. If a molecule is made up of enantiomeric pairs that are rapidly interconverting under ordinary conditions, the molecule is considered to be achiral.

STUDY GUIDE LINK:
6.6
Conformational Enantiomers

PROBLEMS

6.17   Taking *(a) the *anti* conformation, (b) a *gauche* conformation of butane as an isolated species, identify the stereocenters in each structure.

*6.18   (a)   What are the stereochemical relationships among the three conformations of *meso*-2,3-butanediol (the compound discussed in Sec. 6.7)?

(b)   Explain why *meso*-2,3-butanediol is achiral even though it can exist in chiral conformations.

B. Asymmetric Nitrogen: Amine Inversion

Some amines also undergo a rapid interconversion of stereoisomers. *Amines* are derivatives of ammonia in which one or more of the hydrogen atoms have been replaced by an organic group. An example is ethylmethylamine.

$$CH_3 - \overset{\displaystyle H}{\underset{\displaystyle C_2H_5}{N:}}$$ **ethylmethylamine**

Ethylmethylamine has four different groups around the nitrogen: a hydrogen, an ethyl group, a methyl group, and an electron pair. Since the geometry of this molecule is essentially tetrahedral, ethylmethylamine should be a chiral molecule—it should exist as two enantiomers. The asymmetric atom in this molecule is a nitrogen.

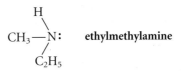

enantiomers of
ethylmethylamine

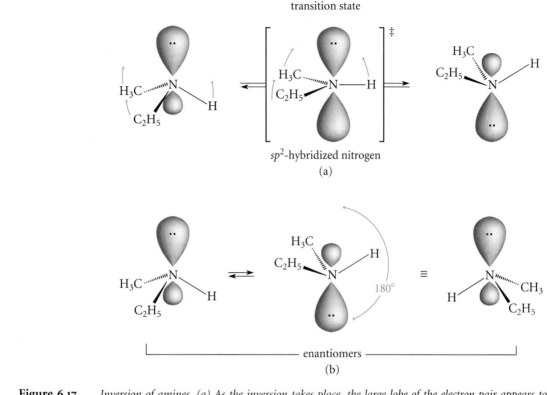

Figure 6.17   *Inversion of amines. (a) As the inversion takes place, the large lobe of the electron pair appears to push through the nitrogen to the other side. As this occurs, the three other groups move first into a common plane with the nitrogen, then to the other side (colored arrows). (b) The enantiomeric relationship of the inverted amines is shown by turning either molecule 180° in the plane of the page.*

In fact, the two enantiomers of amines such as ethylmethylamine cannot be separated, because they rapidly interconvert by a process called **amine inversion**, shown in Fig. 6.17. In this process, the larger lobe of the electron pair seems to push through the nucleus to emerge on the other side. Notice that the molecule is not simply turning over; it is actually turning itself inside out! This is something like what happens when an umbrella turns inside out in the wind. This process occurs through a transition state in which the amine nitrogen becomes sp^2-hybridized. Figure 6.17b shows that this inversion process interconverts the enantiomeric forms of the amine. Because this process is rapid at room temperature, it is impossible to separate the enantiomers. Therefore, ethylmethylamine is a mixture of rapidly interconverting enantiomers. Amine inversion is yet another example of *racemization* (Sec. 6.4).

PROBLEM

*6.19   Assume that the following compound has the *S* configuration at its asymmetric carbon.

(Problem 6.19 continues)

$$C_2H_5\!-\!\overset{\displaystyle CH_3}{\underset{\displaystyle CH_3}{CH}}\!-\!\overset{\displaystyle CH_3}{\underset{\displaystyle C_2H_5}{N}}\!:$$

(a) What is the isomeric relationship between the two forms of this compound that are interconverted by amine inversion?

(b) Could this compound be resolved into enantiomers?

· ·

6.11   Fischer Projections

As you've seen, the three-dimensional structures of molecules can be represented on a two-dimensional page using perspective drawings that employ lines and wedges. It is simple enough to draw a perspective structure for a compound with a single asymmetric carbon, but when a molecule contains several asymmetric carbons, the drawing of perspective structures can be very tedious. For this reason, chemists have adopted another way of writing three-dimensional structures on a two-dimensional surface (paper or blackboard). The resulting structures are called **Fischer projections**, after the German chemist Emil Fischer.

To represent a molecule in a Fischer projection, view each asymmetric carbon in such a way that two of the bonds to this carbon are vertical and pointing away from you, and two are horizontal and pointing toward you. The Fischer projection is the structure obtained when this view is projected on a plane (Fig. 6.18a). The asymmetric carbons themselves are not drawn, but are assumed to be located at the intersections of vertical and horizontal bonds. Such a projection is, in effect, a flattened-out picture of the molecule. (As one student pointed out, the Fischer projection is the way that the molecule would look if we were to put it on the floor and step on it!)

The most useful applications of Fischer projections involve molecules that contain two or more asymmetric carbons that are part of a continuous carbon chain. In such a case, a molecule is first placed (or imagined) in an eclipsed conformation such that the chain of asymmetric carbons will be vertical in the resulting projection. In effect, the carbon backbone can be imagined to be written on a curved, convex surface as shown in Fig. 6.18c. The projection is derived by viewing each carbon in the chain as in Fig. 6.18a and projecting all bonds onto this surface. Mentally cutting the surface and flattening it gives the Fischer projection.

> Although the Fischer projection is derived from an eclipsed conformation, this does *not* mean that the molecule actually has such a conformation. As you've learned, most molecules actually assume staggered conformations (Sec. 2.3). The use of an eclipsed conformation to draw the Fischer projection is a convention for showing the *stereochemical configuration* of each asymmetric carbon; it is not meant to convey the actual *conformation* of the molecule.

To derive a three-dimensional model of a molecule from its Fischer projection, reverse the process just described. Always remember that the vertical bonds in the Fischer projection are *away from* you, and the horizontal bonds are *toward* you.

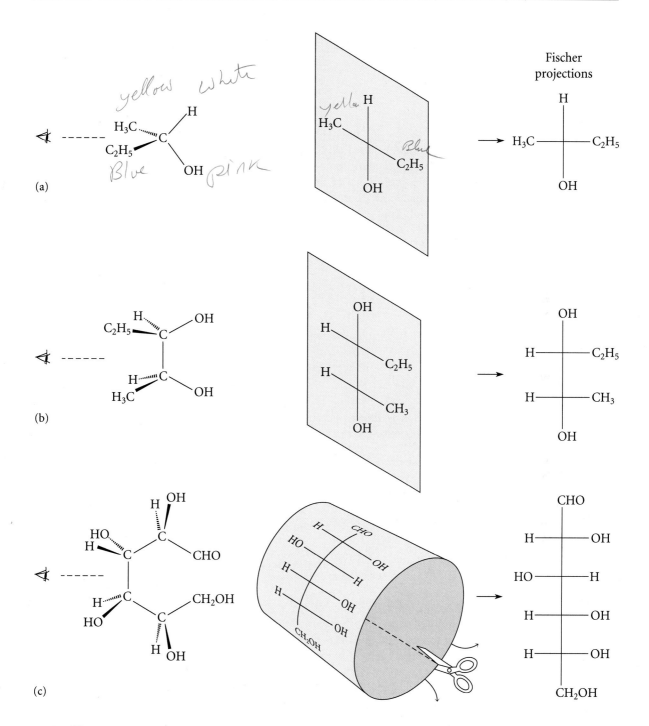

(a)

(b)

(c)

Fischer
projections

Figure 6.18   *Formation of Fischer projections from perspective structures for stereoisomers of (a) 2-butanol, a mol-*
ecule with one asymmetric carbon; (b) 2,3-pentanediol, a molecule with two asymmetric carbons;
and (c) glucose, a molecule with four asymmetric carbons. Notice that groups remote from the
observer are vertical in the Fischer projection, and groups near the observer are horizontal.

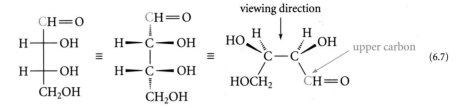

Because several eclipsed conformations can be drawn for any one molecule, *several different Fischer projections can be written.* Consequently, it is useful to be able to draw different Fischer projections of the same molecule without going back and forth to a three-dimensional model. For this purpose some rules for manipulation of Fischer projections are helpful. You should use models to convince yourself of the validity of these rules.

1. A Fischer projection may be turned 180° in the plane of the paper.

By this rule, the following two Fischer projections represent the same stereoisomer.

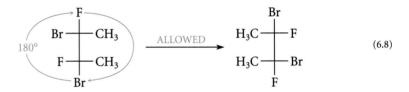

This manipulation is allowed because it leaves horizontal bonds horizontal and vertical bonds vertical; therefore, it does not alter the meaning of the Fischer projection.

2. A Fischer projection may *not* be turned 90°.

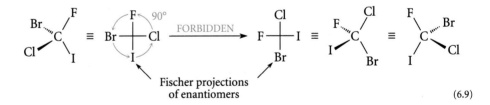

The reason this operation is forbidden is that it causes "out" bonds to be portrayed as "back" and vice versa. In other words, the original structure is converted into its enantiomer by this operation. This is disastrous, since the whole idea of Fischer projections is to convey stereochemical information. The following rule has a similar rationale.

3. A Fischer projection may *not* be lifted from the plane of the paper and turned over.

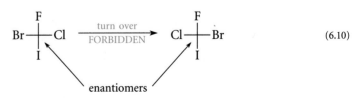

4. The three groups at either end of a Fischer projection may be interchanged in a *cyclic permutation*. That is, all three groups can be moved at the same time in a closed loop so that each occupies an adjacent position.

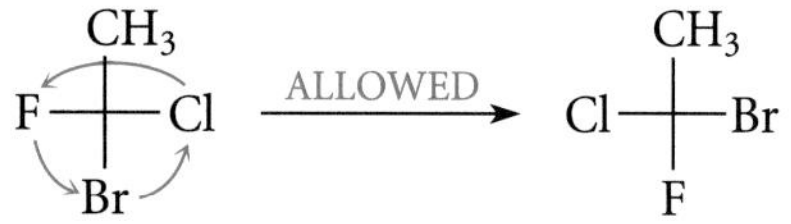

(6.11)

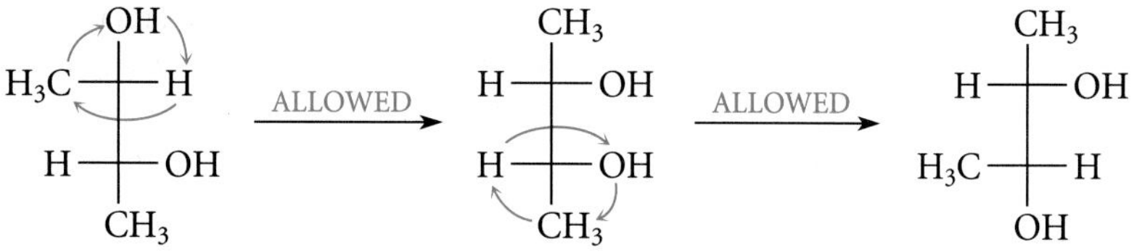

(6.12)

Fischer projections of the same molecule

This operation is equivalent to an internal rotation. This will become clear if you convert any one of the structures in Eq. 6.12 into a model. Leaving the model in an eclipsed conformation, carry out an internal rotation of 120° about the central carbon-carbon bond as shown by the colored arrows in Eq. 6.12 and form a new Fischer projection from the resulting structure. Each 120° internal rotation is equivalent to one cyclic permutation described by Rule 4. A different Fischer projection of the same molecule results from each different eclipsed conformation.

It is particularly easy to recognize enantiomers and *meso* compounds from the appropriate Fischer projections, because planes of symmetry in the actual molecules reduce to lines of symmetry in their projections.

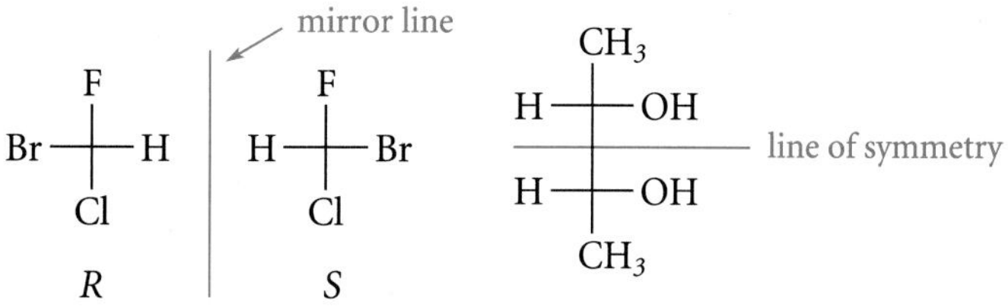

(6.13)

a *meso* compound

enantiomers

The *R*,*S* system can be applied to a Fischer projection without using a model. First draw an equivalent Fischer projection (if necessary) in which the group of lowest priority is in either of the two vertical positions. Then apply the priority rules to the remaining three groups.

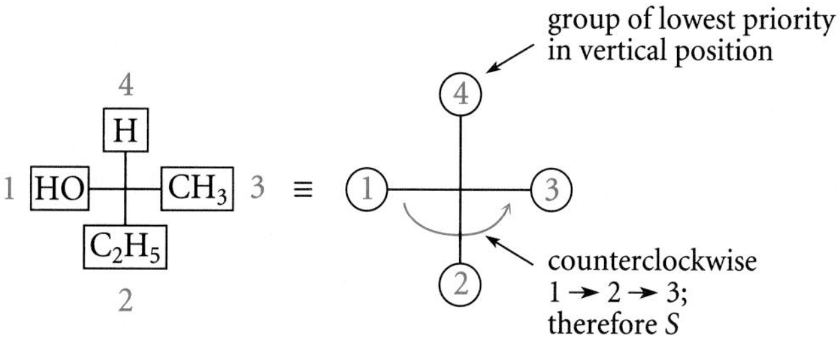

(6.14)

☞
STUDY GUIDE LINK:
✓6.7
*Additional
Manipulations with
Fischer Projections*

This method works because if the lowest-priority group is in a vertical position in the Fischer projection, it is oriented away from the observer as required for application of the priority rules.

PROBLEMS

6.20   Draw at least two Fischer projections for each of the following molecules.

*(a)   *R   S   S*   (b) (*S*)-2-butanol (see Fig. 6.2)

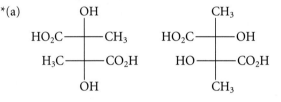

6.21   Indicate whether the structures in each of the following pairs are enantiomers, diastereomers, or identical molecules.

*(a)

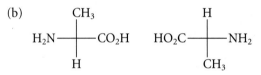

(b)

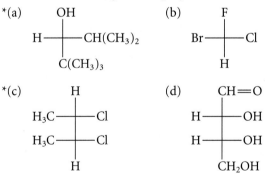

6.22   State whether the configuration of each asymmetric carbon atom is *R* or *S* in each of the following Fischer projections.

*(a)   OH   (b)   F

H———CH(CH₃)₂   Br———Cl

C(CH₃)₃   H

*(c)   H   (d)   CH═O

H₃C———Cl   H———OH

H₃C———Cl   H———OH

H   CH₂OH

6.12   The Postulation of Tetrahedral Carbon

Chemists recognized the tetrahedral configuration of tetracoordinate carbon almost one-half century before physical methods confirmed the idea with direct evidence. This section shows how the phenomena of optical activity and chirality played key roles in this development, one of the most important in the history of organic chemistry.

The first chemical substance in which optical activity was observed was quartz. It was discovered that when a quartz crystal is cut in a certain way and exposed to polarized light along a particular axis, the plane of polarization of the light is rotated. In 1815, the French chemist Jean-Baptiste Biot (1774–1862) showed that quartz exists as both levorotatory and dextrorotatory crystals. The Abbe Rene Just Hauy (1743–1822), a French crystallographer, had earlier shown that there are two kinds of quartz crystals related as object and nonsuperimposable mirror image. Sir John E. W. Herschel (1792–1871), a British astronomer, found a correlation between these crystal forms and their optical activities: one of these forms of quartz is dextrorotatory and the other levorotatory.

During the period 1815–1838, Biot examined several organic substances, both pure and in solution, for optical activity. He found that some (for example, oil of terpentine) show optical activity, and others do not. He recognized that since optical activity can be displayed by compounds in solution, *it must be a property of the molecules themselves.* (The dependence of optical activity on concentration, Eq. 6.1, is sometimes called *Biot's law.*) What Biot did *not* have a chance to observe is that some organic molecules exist in both dextrorotatory and levorotatory forms. The reason Biot never made this observation is undoubtedly that many optically active compounds are obtained from a given natural source as single enantiomers.

The next substance to figure prominently in the history of stereochemistry was tartaric acid:

$$
\underset{\text{tartaric acid}}{
\text{HO}-\overset{\overset{\text{O}}{\|}}{\text{C}}-\overset{\overset{\text{OH}}{|}}{\text{CH}}-\overset{\overset{\text{OH}}{|}}{\text{CH}}-\overset{\overset{\text{O}}{\|}}{\text{C}}-\text{OH}
}
$$

This substance had been known by the ancient Romans as its monopotassium salt, *tartar*, which deposits from fermenting grape juice. Tartaric acid derived from tartar was one of the compounds examined by Biot for optical activity; he found that it has a positive rotation. An isomer of tartaric acid discovered in crude tartar, called *racemic acid* (*racemus*, Latin, "a bunch of grapes"), was also studied by Biot and found to be optically inactive. The exact structural relationship of (+)-tartaric acid and its isomer racemic acid remained obscure.

All of these observations were known to Louis Pasteur (1822–1895), the French chemist and biologist. One day the young Pasteur was viewing crystals of the sodium ammonium double salts of (+)-tartaric acid and racemic acid under the microscope. Pasteur noted that the crystals of the salt derived from (+)-tartaric acid were hemihedral (chiral). But he noted that the racemic acid salt was not a single type of crystal, but was actually a mixture of hemihedral crystals: some crystals were "right-handed," like those in the corresponding salt of (+)-tartaric acid, and some were "left-handed" (Fig. 6.19a; thus the name "racemic mixture"). Pasteur meticulously separated the two types of crystals with a pair of tweezers; he found that the right-handed crystals were identical in every way to the crystals of the salt of (+)-tartaric acid. When equally concentrated solutions of the two types of crystals were prepared, he found that the optical rotations of the left- and right-handed crystals were equal in magnitude, but opposite in sign. Pasteur had thus performed the first enantiomeric resolution by human hands! Racemic acid, then, was the first organic compound shown to exist as enantiomers—object and noncongruent mirror image. One of these mirror-image molecules was identical to (+)-tartaric acid, but the other was previously unknown. Pasteur's own words tell us what then took place.

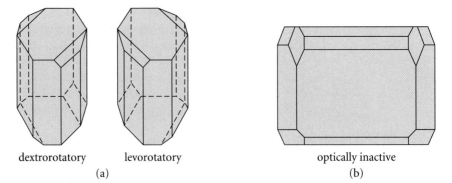

dextrorotatory   levorotatory

(a)

optically inactive

(b)

Figure 6.19   *Diagrams of the crystals of the tartaric acid isomers that figured prominently in the history of stereochemistry. (a) The hemihedral crystals of sodium ammonium tartrate separated by Pasteur. (b) The holohedral crystal of sodium ammonium racemate that crystallizes at a higher temperature.*

The announcement of the above facts naturally placed me in communication with Biot, who was not without doubts concerning their accuracy. Being charged with giving an account of them to the Academy, he made me come to him and repeat before his very eyes the decisive experiment. He handed over to me some racemic acid that he himself had studied with particular care, and that he found to be perfectly indifferent to polarized light. I prepared the double salt in his presence with soda and ammonia that he also desired to provide. The liquid was set aside for slow evaporation in one of his rooms. When it had furnished about thirty to forty grams of crystals, he asked me to call at the *College de France* in order to collect them and isolate, before his very eyes, by recognition of their crystallographic character, the right and left crystals, requesting me to state once more whether I really affirmed that the crystals that I should place at his right would really deviate [the plane of polarized light] to the right and the others to the left. This done, he told me that he would undertake the rest. He prepared the solution with carefully measured quantities, and when ready to examine them in the polarizing apparatus, he once more invited me to come into his room. He first placed in the apparatus the more interesting solution, that which should deviate to the left [previously unknown]. Without even making a measurement, he saw by the tints of the images . . . in the analyzer that there was a strong deviation to the left. Then, very visibly affected, the illustrious old man took me by the arm and said, "My dear child, I have loved science so much all my life that this makes my heart throb!"

Pasteur's discovery of the two types of crystals of racemic acid was serendipitous (accidental). It is now known that the sodium ammonium salt of racemic acid forms separate right- and left-handed crystals only at temperatures below 26 °C. Had Pasteur's laboratories been warmer, he would not have made the discovery. Above this temperature, this salt forms only one type of crystal: a holohedral (achiral) crystal of the racemate (Fig. 6.19b)! From his discovery, Pasteur recognized that some molecules could, like the quartz crystals, have an enantiomeric relationship, but he was never able to deduce a structural basis for this relationship.

PROBLEM

*6.23   As described in the previous account, Pasteur discovered two stereoisomers of tartaric acid. Draw their structures [you cannot, of course, tell which is (+) and which is (−)]. Which stereoisomer of tartaric acid was yet to be discovered? (It was discovered in 1906.) What can you say about its optical activity?

In 1874, Jacobus Hendricus van't Hoff (1852–1911), a professor at the veterinary college at Utrecht, The Netherlands, and Achille Le Bel (1847–1930), a French chemist, independently arrived at the idea that if a molecule contains a carbon atom bearing four different groups, these groups can be arranged in different ways to give enantiomers. Van't Hoff suggested a tetrahedral arrangement of groups about the central carbon, but Le Bel was less specific. Van't Hoff's conclusions, published in a treatise of eleven pages entitled "*La chemie dans l'espace*," were not immediately accepted. A caustic reply, reproduced in considerably censored form below, came from the famous German chemist Hermann Kolbe:

> A Dr. van't Hoff of the veterinary college, Utrecht, appears to have no taste for exact chemical research. Instead, he finds it a less arduous task to mount his Pegasus (evidently borrowed from the stables of the College) and soar to his chemical Parnassus, there to reveal in his "*La chemie dans l'espace*" how he finds atoms situated in universal space. This paper is fanciful nonsense! What times are these, that an unknown chemist should be given such attention!

Kolbe's reply notwithstanding, van't Hoff's ideas prevailed to become a cornerstone of organic chemistry.

How can the existence of enantiomers be used to deduce a tetrahedral arrangement of groups around carbon? Let's examine some other possible carbon geometries to see the sort of reasoning that was used by van't Hoff and Le Bel. Consider a general molecule in which the carbon and its four groups lie in a single plane:

$$Br—C—F \quad \text{all atoms in the same plane}$$

with Cl above C and H below C.

Because the mirror image of such a *planar* molecule is congruent (show this!), enantiomeric forms are not possible. The existence of enantiomers thus *rules out* this planar geometry.

However, other, nonplanar nontetrahedral structures could exist as enantiomers. One structure has a pyramidal geometry:

(Convince yourself that such a structure can have an enantiomer.) This geometry could not, however, account for other facts. Consider, for example, the compound methylene chloride (CH_2Cl_2). In the pyramidal geometry, two *diastereomers* would be known. In one, the chlorines are on opposite corners of the pyramid; in the other, the chlorines are adjacent. (Why are these diastereomers?)

These molecules should be separable because diastereomers have different properties. Yet in the entire history of chemistry, only one isomer of CH_2Cl_2, CH_2Br_2, or any similar molecule, has ever been found. Now, this is *negative* evidence. To take this evidence as conclusive would be like saying to the Wright brothers in 1902, "No one has ever seen an airplane fly; therefore airplanes can't fly." Yet this evidence is certainly suggestive, and other types of experiments (Problems 6.48 and 6.49) were later carried out that could only be interpreted in terms of tetrahedral carbon. Indeed, modern methods of structure determination have shown that van't Hoff's original proposal—tetrahedral geometry—is essentially correct.

KEY IDEAS IN CHAPTER 6

Stereoisomers are molecules that have the same atomic connectivity but differ in the arrangement of their atoms in space.

Two types of stereoisomers are:
1. enantiomers—molecules that are related as object and noncongruent mirror image;
2. diastereomers—stereoisomers that are not enantiomers.

A molecule that has an enantiomer is said to be chiral. Chiral molecules lack certain symmetry elements such as a plane of symmetry.

The absolute configurations of chiral compounds are designated by the *R,S* system, which is based on the relative arrangement of group priorities. The priorities are assigned as in the *E,Z* system. The absolute configurations of many compounds are determined experimentally by correlating them chemically with other chiral compounds of known absolute configuration.

A pair of enantiomers have the same physical properties except for their optical activities. The optical rotations of a pair of enantiomers have equal magnitudes but opposite signs.

An equimolar mixture of enantiomers is called a racemate, or racemic mixture.

Diastereomers in general differ in their physical properties.

A pair of enantiomers are typically separated by an enantiomeric resolution. In this process, they are allowed to react with a chiral resolving agent, a process that forms diastereomers. After separating the resulting diastereomers, the pure enantiomers are separated from the resolving agent.

An asymmetric carbon is a carbon bonded to four different groups. All asymmetric carbons are stereocenters, but not all stereocenters are asymmetric carbons.

A *meso* compound is an achiral compound with asymmetric atoms. Some chiral molecules contain no asymmetric atoms.

Some achiral molecules consist of chiral conformational stereoisomers that interconvert very rapidly.

The structures of chiral compounds can be drawn in planar representations called Fischer projections. In a Fischer projection, all vertical bonds are assumed to be oriented away from the observer, and all horizontal bonds toward the observer. Several valid Fischer projections can be drawn for most chiral molecules. These are derived by the rules in Sec. 6.11.

Optical activity and chirality formed the logical foundation for the postulate of tetrahedral bonding geometry at carbon.

ADDITIONAL PROBLEMS

6.24 Point out the carbon stereocenters and the asymmetric carbons (if any) in each of the following structures.
*(a) (E)-4-methyl-2-hexene   (b) 4-methyl-1-pentene
*(c) 3-methylcyclohexene   (d) (Z)-4-methyl-2-pentene

*6.25 (a) How many stereoisomers are there of 3,4,5,6-tetramethyl-4-octene?
(b) Show all of the carbon stereocenters in the structure of this compound.
(c) Show all of the asymmetric carbons in the structure of this compound.

6.26 (a) How many stereoisomers are there of 3,4-dimethyl-2-hexene?
(b) Show all of the carbon stereocenters in the structure of this compound.
(c) Show all of the asymmetric carbons in the structure of this compound.

6.27 Identify all of the asymmetric carbon atoms in each of the following structures.

*(a)

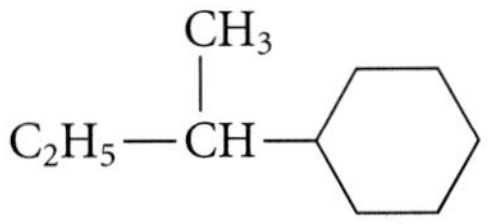

$$CH_3$$
$$|$$
$$C_2H_5-CH-\!\!\bigcirc$$

(b) $CH_3-CH-CH_2OH$
$$|$$
$$NH_2$$

(Problem 6.27 continues)

*(c)   (d)

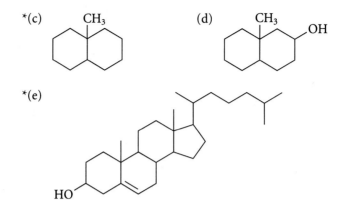

*(e)

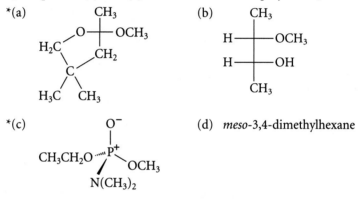

6.28   Give the configuration of each asymmetric atom in the following compounds. [Compounds (a) and (b) are drawn in Fischer projection.]

*(a)   (b)

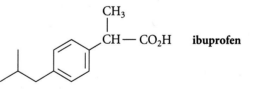

*(c)   (d) *meso*-3,4-dimethylhexane

$$CH_3CH_2O \overset{O^-}{\underset{N(CH_3)_2}{\overset{|}{P^+}}} OCH_3$$

*6.29   (a) Using lines, wedges, and dashed wedges as appropriate, draw perspective structures of the two stereoisomers of ibuprofen, a well known analgesic.

$$\overset{CH_3}{\underset{}{\overset{|}{CH}}} - CO_2H \quad \textbf{ibuprofen}$$

(b) Indicate which one of the structures you drew in (a) is the *S* enantiomer. (Only this stereoisomer is biologically active.)

*6.30   Draw all the allowed Fischer projections of (2*S*,3*R*)-2,3-pentanediol.

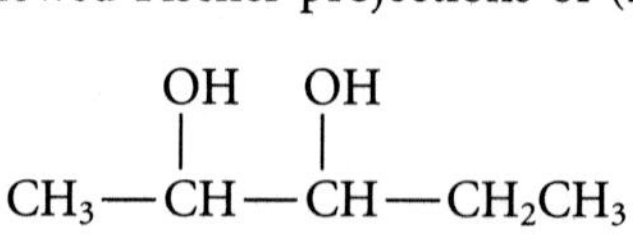

$$\overset{OH}{\underset{}{\overset{|}{}}} \quad \overset{OH}{\underset{}{\overset{|}{}}}$$
$$CH_3 - CH - CH - CH_2CH_3 \quad \textbf{2,3-pentanediol}$$

6.31   Draw a Fischer projection for each stereoisomer of 2,3-dibromobutane. Using Fig. 6.11, indicate the isomeric relationship between each pair of stereoisomers.

$$\begin{array}{cc} Br & Br \\ | & | \\ CH_3-CH-CH-CH_3 \end{array} \quad \textbf{2,3-dibromobutane}$$

*6.32   Draw the structure of the chiral alkane of lowest molecular mass not containing a ring. (No isotopes are allowed.)

6.33   Draw the structure of the chiral cyclic alkane of lowest molecular mass. (No isotopes are allowed.)

6.34   Indicate whether each of the following statements is true or false. If false, explain why.
    *(a) In some cases, constitutional isomers are chiral.
    (b) In every case, a pair of enantiomers have a mirror-image relationship.
    *(c) Mirror-image molecules are in all cases enantiomers.
    (d) If a compound has an enantiomer it must be chiral.
    *(e) Every chiral compound has a diastereomer.
    (f) If a compound has a diastereomer it must be chiral.
    *(g) Any molecule containing an asymmetric carbon must be chiral.
    (h) Any molecule containing a stereocenter must be chiral.
    *(i) Any molecule with a stereocenter must have a stereoisomer.
    (j) Some diastereomers have a mirror-image relationship.
    *(k) Some chiral compounds are optically inactive.
    (l) Any chiral compound with a single asymmetric carbon must have a positive optical rotation if the compound has the R configuration.
    *(m) If a structure has no plane of symmetry it is chiral.
    (n) All asymmetric carbons are stereocenters.
    *(o) All chiral molecules have no plane of symmetry.

*6.35   Imagine substituting each hydrogen atom of 3-methylpentane, in turn, with a chlorine atom to give a series of isomers with molecular formula $C_6H_{13}Cl$. Give the structure of each of these isomers. Which of these are chiral? Classify the relationship within every pair of these stereoisomers.

6.36   Draw the structures of all *meso* compounds with the formula $C_6H_{12}Cl_2$.

*6.37   Construct models or draw Newman projections (Sec. 2.3A) of the three staggered conformations of 2-methylbutane (isopentane) that result from rotation about the C2-C3 bond.
    (a) Identify the conformations that are chiral.
    (b) Explain why 2-methylbutane is not a chiral compound, even though it has chiral conformations.
    (c) Suppose each of the three conformations in part (a) could be isolated and their heats of formation determined. Rank these isomers in order of increasing heat of formation (that is, smallest first). Explain your choice. Indicate whether the ΔH_f° values for any of the isomers are equal and why.

containing the phosphorus atom (shaded in the structure below), and the bonds to the other two groups (called *axial* groups) are perpendicular to this plane:

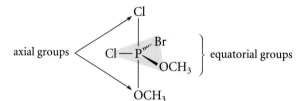

Is this compound chiral? Explain.

*6.47   Identify the stereocenters in each of the following structures, and tell whether each structure is chiral.

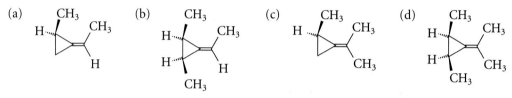

*6.48   In 1914, the chemist Emil Fischer carried out the following conversion in which optically active starting material was transformed into a product with an identical melting point and an optical rotation of equal magnitude and opposite sign. No bonds to the asymmetric carbon were broken in the process.

$$
\begin{array}{ccc}
\text{CH}_2\text{CH}_2\text{CH}_3 & & \text{CH}_2\text{CH}_2\text{CH}_3 \\
| & & | \\
\text{H}-\text{C}-\text{CONH}_2 & \xrightarrow[\text{reactions}]{\text{several}} & \text{H}-\text{C}-\text{CO}_2\text{H} \\
| & & | \\
\text{CO}_2\text{H} & & \text{CONH}_2
\end{array}
$$

Show that this result is consistent with *either* tetrahedral or pyramidal geometry at the asymmetric carbon.

*6.49   Fischer also carried out the following pair of conversions. Again, no bonds to the asymmetric carbon were broken. Explain why this *pair* of conversions (but not either one alone) and the associated optical activities rule out pyramidal geometry at the asymmetric carbon but are consistent with tetrahedral geometry.

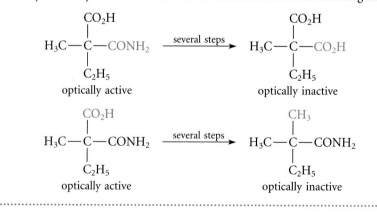

7

Cyclic Compounds and Stereochemistry of Reactions

Compounds with cyclic structures present some unique problems of stereochemistry and conformation. This chapter deals with the stereochemical aspects of cyclic compounds and their derivatives, followed by a discussion of how stereochemistry enters into chemical reactions. You have already learned about *regioselective reactions*, which yield one *constitutional isomer* in preference to another (for example, HBr addition to alkenes). Many reactions also yield certain *stereoisomers* to the exclusion of others. Several such reactions will be examined so that you can understand some of the principles that govern the formation of stereoisomers. You'll also see how the stereochemistry of a reaction can be used to understand its mechanism.

7.1 Relative Stabilities of the Monocyclic Alkanes

A compound that contains a single ring is called a **monocyclic** compound. Cyclohexane, cyclopentane, and methylcyclohexane are all examples of monocyclic alkanes.

The relative stabilities of the monocyclic alkanes can be determined from their heats of formation, given in Table 7.1. Although the monocyclic alkanes are not isomers, they have the same empirical formula, CH_2. Consequently, the stabilities of the monocyclic alkanes can be compared on a *per carbon basis* by dividing the heat of formation of each compound by its number of carbons. The data in Table 7.1 show that, of the cycloalkanes with ten or fewer carbons, cyclohexane has the smallest (that is, the largest negative) heat of formation per CH_2. Thus, *cyclohexane is the most stable of these cycloalkanes.*

Further insight into the stability of cyclohexane comes from a comparison of its stability with that of a typical *noncyclic* alkane. The heats of formation of pentane, hexane, and heptane are -146.4, -167.2, and -187.8 kJ/mol (-35.00, -39.96, and -44.88 kcal/mol), respectively. These data show that heats of formation, like other physical properties, change regularly within a series; each CH_2 group contributes about -20.7 kJ/mol (-4.95 kcal/mol) to the heat of formation. The data for cyclohexane in Table 7.1 show that a CH_2 group in cyclohexane makes exactly the same contribution to its heat of formation

271

Table 7.1   Heats of Formation per —CH$_2$— for Some Cycloalkanes (*n* = number of carbon atoms)

n	*Compound*	$\Delta H_f^\circ/n$		*n*	*Compound*	$\Delta H_f^\circ/n$	
		kJ/mol	*kcal/mol*			*kJ/mol*	*kcal/mol*
3	cyclopropane	+17.8	+4.25	9	cyclononane	−16.5	−3.95
4	cyclobutane	+6.90	+1.65	10	cyclodecane	−15.7	−3.75
5	cyclopentane	−15.3	−3.65	11	cycloundecane	−20.7	−4.95
6	cyclohexane	−20.7	−4.95	12	cyclododecane	−19.5	−4.65
7	cycloheptane	−17.0	−4.05	13	cyclotridecane	−19.0	−4.55
8	cyclooctane	−15.7	−3.75	14	cyclotetradecane	−20.7	−4.95

(−20.7 kJ/mol or −4.95 kcal/mol). This means that *cyclohexane has the same stability as a typical unbranched alkane.*

Cyclohexane is the most widely occurring ring in compounds of natural origin. Its prevalence, undoubtedly a consequence of its stability, makes it the most important of the cycloalkanes. The stability of cyclohexane is due to its conformation, which is the subject of the next section. The stabilities and conformations of other cycloalkanes are discussed in Sec. 7.5.

7.2   Conformations of Cyclohexane

A. The Chair Conformation

The most stable conformation of cyclohexane, shown in Fig. 7.1, is called the **chair conformation** because of its resemblance to a lawn chair. You should construct a model of cyclohexane *now* and use it to follow the subsequent discussion.

1. A cyclohexane ring is usually drawn in a slightly tilted and rotated perspective so that all of its bonds are visible. That is, in the usual skeletal structure, if we number the carbons as shown below and imagine that carbons 1 and 4 are in the plane of the page, carbons 2 and 3 are behind the page, and carbons 5 and 6 are in front of the page:

means   (7.1)

2. Three of the carbons (carbons 1, 3, and 5 in the structure above) define a plane that is above the plane defined by the other three (carbons 2, 4, and 6). Carbons such as 1, 3, and 5 will be referred to as *up carbons*, and carbons such as 2, 4, and 6 as *down carbons*.

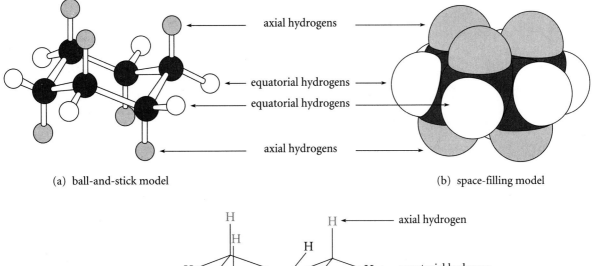

(a) ball-and-stick model

(b) space-filling model

(c) skeletal structure with hydrogens shown

Figure 7.1   *Chair conformation of cyclohexane shown as (a) a ball-and-stick model, (b) a space-filling model, and (c) a skeletal structure. In (a) and (b), the axial hydrogens are shaded in gray, and in (c), they are shown in shaded type.*

3. Bonds on opposite sides of the ring are parallel:

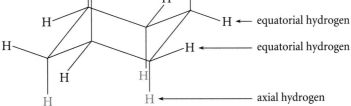

4. Two perspectives are commonly used for cyclohexane rings. In one, the leftmost carbon is below the rightmost carbon; and in the other, the leftmost carbon is above the rightmost carbon:

5. A rotation of either perspective by an odd multiple of 60° (that is, 60°, 180°, etc.) about the axis shown on page 274 gives the other perspective:

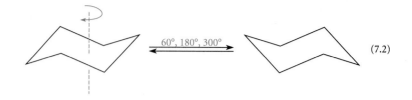

$$(7.2)$$

It is important for you to be able to draw a cyclohexane ring. Once you've examined the points above, practice drawing some cyclohexane rings in the two perspectives. Use the following three steps:

Step 1   Begin by drawing two parallel bonds slanted to the left for one perspective, and slanted to the right for the other.

Step 2   Connect the tops of the slanted bonds with two more bonds in a "V" arrangement.

Step 3   Connect the bottoms of the slanted bonds with the remaining two bonds in an "inverted V" arrangement.

To summarize:

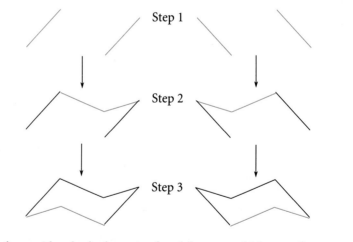

$$(7.3)$$

Now let's consider the hydrogens of cyclohexane, which are of two types. If you place your model of cyclohexane on a tabletop (you did build it, didn't you?), you'll find that six C—H bonds are perpendicular to the plane of the table. (Your model should be resting on three such hydrogens.) These hydrogens, shown in gray shading in Fig. 7.1, are called **axial** hydrogens. The remaining C—H bonds point outward along the periphery of the ring. These hydrogens, shown in white in Fig. 7.1, are called **equatorial** hydrogens. Of course, other groups can be substituted for the hydrogens, and these groups also can exist in either axial or equatorial arrangements.

In the chair conformation, all bonds are staggered. You should be able to see this from your model by looking down any C—C bond. As you learned when you studied the conformations of ethane and butane (Sec. 2.3), staggered bonds are preferred to eclipsed bonds. The stability of cyclohexane (Sec. 7.1) is due to the fact that all of its bonds can be staggered without compromising the tetrahedral carbon geometry.

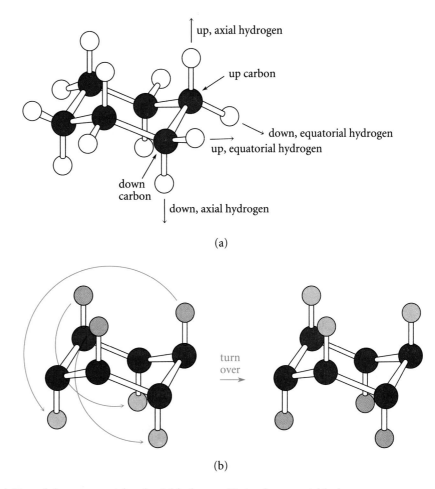

(a)

(b)

Figure 7.2   *(a) Up and down equatorial and axial hydrogens. Notice that up axial hydrogens are on up car-bons, and down axial hydrogens are on down carbons. The opposite is true for equatorial hydrogens. (b) The up and down axial hydrogens are equivalent. This can be demonstrated by turning the ring (colored arrows). The up axial hydrogens (color) become equivalent to the down axial hydrogens (gray). The same operation shows the equivalence of the up and down equatorial hydrogens.*

Once you have mastered drawing the cyclohexane ring, it's time to add the C—H bonds to the ring. Drawing the axial bonds is fairly easy: they are simply vertical lines. However, drawing the equatorial bonds can be a little tricky. Notice that pairs of equatorial bonds are parallel to pairs of ring bonds:

(7.4)

(Notice also how all the equatorial bonds in Fig. 7.1 adhere to this convention.)

You should notice a few other things about the cyclohexane ring and its bonds. First, the three axial hydrogens on up carbons point up, and the three axial hydrogens on down carbons point down. In contrast, the three equatorial hydrogens on up carbons point down, and the three equatorial hydrogens on down carbons point up (Fig. 7.2a).

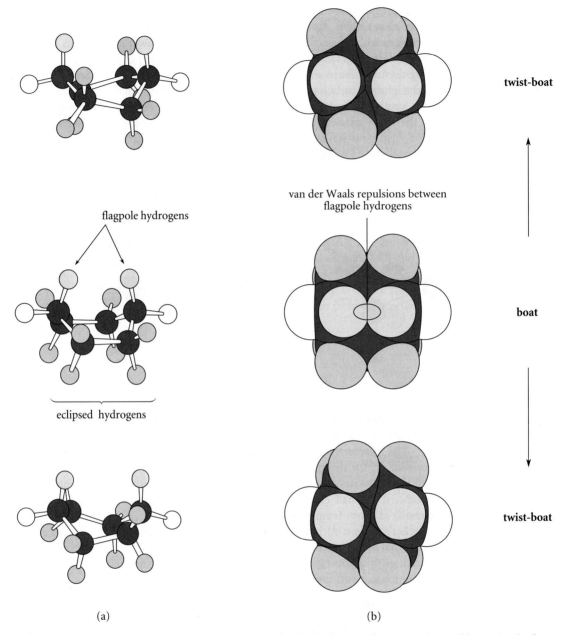

flagpole hydrogens

van der Waals repulsions between
flagpole hydrogens

twist-boat

boat

twist-boat

eclipsed hydrogens

(a)

(b)

Figure 7.4   *Boat cyclohexane (center) and its two related twist-boat conformations (top and bottom). The flag-pole hydrogens are shown in color, and the hydrogens that are eclipsed in the boat conformation are shaded gray. (a) Ball-and-stick models. Note the eclipsed relationship in the boat conformation among the pairs of hydrogens that are shaded gray. This eclipsing is reduced in the twist-boat confor-mation. (b) Space-filling models viewed from above the flagpole hydrogens. Note the van der Waals repulsion between the flagpole hydrogens in the boat conformation. This unfavorable interaction is reduced in the twist-boat conformations because the flagpole hydrogens (color) are further apart.*

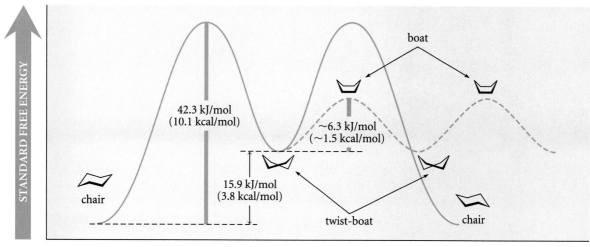

Figure 7.5   *Relative free energies of cyclohexane conformations.*

stable conformation of cyclohexane called a **twist-boat conformation**. To see the conversion of a boat into a twist-boat, view a model of the boat conformation from above the flagpole hydrogens, as shown in Fig. 7.4b. Grasping the model by its flagpole hydrogens, nudge one flagpole hydrogen up and the other down to obtain a twist-boat conformation. As shown in Fig. 7.4, this motion can occur in either of two ways, so that there are two twist-boats related to any one boat conformation.

The free-energy relationships among the conformations of cyclohexane are shown in Fig. 7.5. Although the twist-boat conformation is at an energy minimum, it is less stable than the chair conformation by about 15.9 kJ/mol (3.8 kcal/mol) in standard free energy, or by about 23 kJ/mol (5.5 kcal/mol) in standard enthalpy. Hence, a sample of cyclohexane has very little twist-boat conformation present at equilibrium. The boat conformation itself can be thought of as the transition state for interconversion of the twist-boat conformations.

PROBLEM

⋆7.1   Using the information in Fig. 7.5, determine the relative amounts of the chair and twist-boat forms of cyclohexane in equilibrium at 25 °C.

7.3   Monosubstituted Cyclohexanes and Conformational Analysis

A substituent group in a substituted cyclohexane, such as the methyl group in methylcyclohexane, can be in either an axial or an equatorial position.

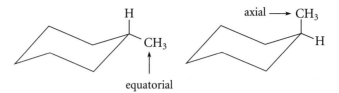

These two compounds are obviously stereoisomers, and since they are not enantiomers, they must be diastereomers. Substituted cyclohexanes such as methylcyclohexane also undergo the chair flip. As Fig. 7.6 shows, axial methylcyclohexane and equatorial methylcyclohexane are interconverted by this process. Note that in this interconversion, a down methyl remains down and an up methyl remains up. (Demonstrate this to yourself with models!) Because this process is rapid at room temperature, methylcyclohexane is a mixture of two *conformational diastereomers* (Sec. 6.10A). Because diastereomers have different energies, one form is more stable than the other, and the more stable form is present in greater amount. Which form is more stable?

Equatorial methylcyclohexane is more stable than axial methylcyclohexane. In fact, it is usually the case that *the equatorial conformation of a substituted cyclohexane is more stable than the axial conformation.* Why should this be so?

Examination of a space-filling model of axial methylcyclohexane (Fig. 7.7) shows that van der Waals repulsions occur between one of the methyl hydrogens and the two axial hydrogens on the same face of the ring. Such unfavorable interactions between axial groups are called **1,3-diaxial interactions**. These van der Waals repulsions destabilize the axial conformation relative to the equatorial conformation, in which such van der Waals repulsions are absent.

Equatorial methylcyclohexane is 7.4 kJ/mol (1.8 kcal/mol) more stable than axial methylcyclohexane (Fig. 7.8). This energy difference is *twice* the energy difference between the *gauche* and *anti* conformations of butane (see Fig. 2.5). This correspondence is not a coincidence, because the 1,3-diaxial interaction of a methyl group with a hydrogen in cyclohexane is almost exactly the same as the van der Waals repulsion between the methyl groups in *gauche*-butane (Fig. 7.9, p. 282). The energy "cost" of this interaction in both axial methylcyclohexane and *gauche*-butane is about 3.7 kJ/mol (0.9 kcal/mol). Since there are *two* such interactions in axial methylcyclohexane, the energy "cost" is twice as

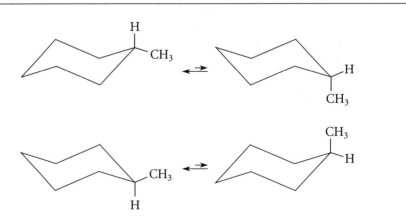

Figure 7.6   *The chair flip interconverts equatorial (left) and axial (right) conformations of methylcyclohexane. The conversion is shown with two different ring perspectives.*

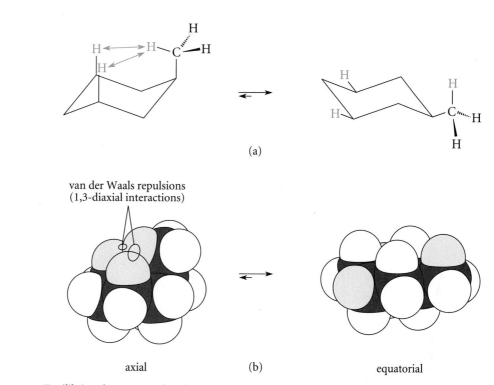

Figure 7.7   *Equilibrium between axial and equatorial conformations of methylcyclohexane shown with (a) skeletal structures and (b) space-filling models. The hydrogens involved in 1,3-diaxial interactions in the axial conformation are shown in color.*

great: $2 \times 3.7 = 7.4$ kJ/mol (or $2 \times 0.9 = 1.8$ kcal/mol). In other words, the 1,3-diaxial interactions in axial methylcyclohexane can be thought of as two "*gauche*-butane" interactions. This observation illustrates an important philosophical point: *We can learn much about complex molecules by studying simpler molecules with similar features.* In this situation, *gauche*-butane serves as the "model" for a calculation of the relative energy of axial methylcyclohexane. Scientists, in an effort to simplify problems as much as possible, usually treat complex cases in terms of the principles learned in simpler ones. You will

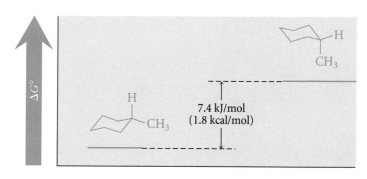

Figure 7.8   *Relative free energies of axial and equatorial methylcyclohexane. Notice that the energy difference between the two conformations is twice the difference between anti and gauche butane (7.4 kJ/mol or 1.8 kcal/mol).*

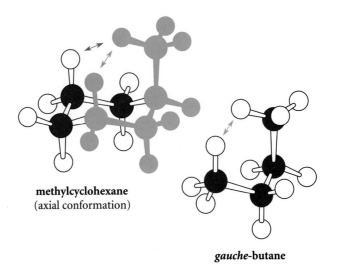

methylcyclohexane
(axial conformation)

gauche-**butane**

Figure 7.9 *The relationship between the axial conformation of methylcyclohexane and gauche-butane. The gauche-butane part of methylcyclohexane is shown in color, and the corresponding van der Waals repulsion is shown with a colored double-headed arrow. The second gauche-butane interaction in methylcyclohexane is shown with the gray arrow.*

find that the simple analysis used here can be applied to even more complex molecules with excellent results. All you have to remember is that a methyl-hydrogen 1,3-diaxial interaction "costs" the same as a *gauche*-butane—3.7 kJ/mol (0.9 kcal/mol).

The investigation of molecular conformations and their relative energies is called **conformational analysis**. We have just carried out a conformational analysis of methyl-cyclohexane. The conformational analyses of many different substituted cyclohexanes have been performed. As might be expected, the larger a group is, the more severe are the 1,3-diaxial interactions it suffers when it assumes an axial position. Thus, *the larger the substituent group on a cyclohexane ring, the more the equatorial conformation is favored.* For example, the equatorial conformation of *tert*-butylcyclohexane is favored over the axial conformation by 21 to 25 kJ/mol (5 to 6 kcal/mol). The relative energy cost of placing the methyl and *tert*-butyl groups in an axial position are thus in direct relation to their sizes.

SEPARATION OF CHAIR CONFORMATIONS

The chair flip of a monosubstituted cyclohexane interconverts diastereomers. Therefore, if the two chair forms could be separated, they would have different physical properties. In the late 1960s, C. Hackett Bushweller, then a graduate student at the University of California, Berkeley, cooled a solution of chloro-cyclohexane in an inert solvent to −150 °C. Crystals suddenly appeared in the solution. He filtered the crystals at low temperature; subsequent investigations showed that he had selectively crystallized the equatorial form of chlorocyclohexane!

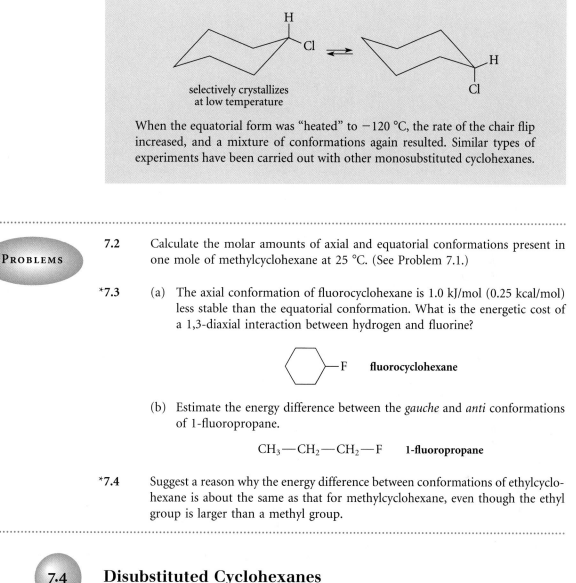

selectively crystallizes
at low temperature

When the equatorial form was "heated" to −120 °C, the rate of the chair flip increased, and a mixture of conformations again resulted. Similar types of experiments have been carried out with other monosubstituted cyclohexanes.

PROBLEMS

7.2   Calculate the molar amounts of axial and equatorial conformations present in one mole of methylcyclohexane at 25 °C. (See Problem 7.1.)

*7.3   (a) The axial conformation of fluorocyclohexane is 1.0 kJ/mol (0.25 kcal/mol) less stable than the equatorial conformation. What is the energetic cost of a 1,3-diaxial interaction between hydrogen and fluorine?

⬡—F   **fluorocyclohexane**

(b) Estimate the energy difference between the *gauche* and *anti* conformations of 1-fluoropropane.

$CH_3—CH_2—CH_2—F$ **1-fluoropropane**

*7.4   Suggest a reason why the energy difference between conformations of ethylcyclohexane is about the same as that for methylcyclohexane, even though the ethyl group is larger than a methyl group.

7.4   Disubstituted Cyclohexanes

A. *Cis-Trans* Isomerism

Consider a typical disubstituted cyclohexane, 1-chloro-2-methylcyclohexane.

1-chloro-2-methylcyclohexane

In one stereoisomer of this compound, both the chloro and methyl groups assume equatorial positions. This compound is in rapid equilibrium with a conformational diastereomer in which both the chloro and methyl groups assume axial positions.

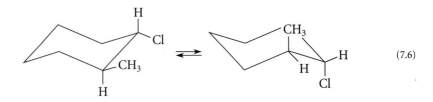

$$\textbf{\textit{trans}-1-chloro-2-methylcyclohexane}$$

Either conformation (or the mixture of them) is called *trans*-1-chloro-2-methylcyclo-hexane. The *trans* designation is used with cyclic compounds when two substituents have an up-down relationship.

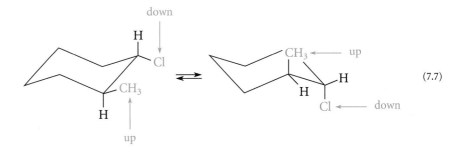

Notice that the up-down relationship is not affected by the conformational equilibrium.

1-Chloro-2-methylcyclohexane also exists as a stereoisomer in which the chloro and methyl groups occupy adjacent equatorial and axial positions.

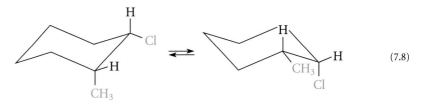

$$\textbf{\textit{cis}-1-chloro-2-methylcyclohexane}$$

This compound, called *cis*-1-chloro-2-methylcyclohexane, is also a rapidly equilibrating mixture of conformational diastereomers. In a *cis*-disubstituted cycloalkane, the substituents have an up-up or a down-down relationship.

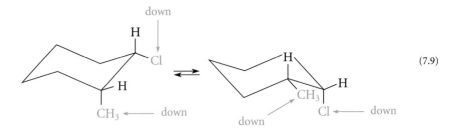

Again, the *cis* relationship is not altered by the conformational equilibrium.

The same definition of *cis* and *trans* substitution can be applied to substituent groups in other positions of a cyclohexane ring, as illustrated by the following study problem.

STUDY
PROBLEM
7.1

Draw structures of the two chair conformations of *trans*-1,3-dimethylcyclohexane.

Solution   In a *trans*-disubstituted cyclohexane, the two substituent groups have an up-down relationship. It doesn't matter which chair conformation is drawn first, because the chair flip does not affect the *trans* relationship of the two methyl groups:

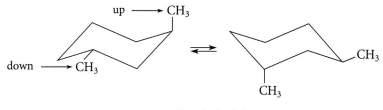

trans-1,3-dimethylcyclohexane

Notice that the *cis* and *trans* designations specify the *relative* stereochemical configurations of two asymmetric carbons in the ring, but they say nothing about the absolute configurations of these carbons. Thus, there are two enantiomers of *cis*-1-chloro-2-methylcyclohexane.

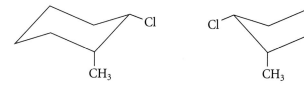

(1S,2R)-1-chloro-2-methylcyclohexane   (1R,2S)-1-chloro-2-methylcyclohexane

└──────────── enantiomers of *cis*-1-chloro-2-methylcyclohexane ────────────┘

The *cis-trans* nomenclature is used when the absolute configuration of the molecule is not important to the task at hand. This nomenclature is particularly convenient to express *relative* configurations because it is more intuitive than the *R* and *S* notation.

When a ring contains more than two substituents, *cis-trans* nomenclature is cumbersome. In such cases, the *R,S* system must be used to indicate configuration.

PROBLEMS

7.5   For each of the following compounds, draw the two chair conformations that are in equilibrium.
*(a) *cis*-1,3-dimethylcyclohexane   (b) *trans*-1-ethyl-4-isopropylcyclohexane

*7.6   For each of the compounds in Problem 7.5, draw a boat conformation.

B. Conformational Analysis

Disubstituted cyclohexanes, like monosubstituted cyclohexanes, can be subjected to conformational analysis. The relative stability of the two chair conformations is determined by analyzing the 1,3-diaxial interactions (or *gauche*-butane interactions) in each conformation and adding them up at 3.7 kJ/mol (or 0.9 kcal/mol) each. Such an analysis is illustrated in Study Problem 7.2.

STUDY PROBLEM 7.2

Determine the relative energies of the two chair conformations of *trans*-1,2-dimethylcyclohexane. Which is more stable?

Solution   The first step in solving any problem is to draw the structures of the species involved. The two chair conformations of *trans*-1,2-dimethylcyclohexane are as follows:

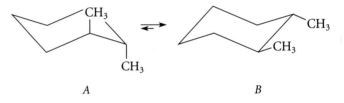

A   *B*

Notice that conformation *A* has the greater number of axial groups and should therefore be the less stable conformation—but by how much? Conformation *A* has four 1,3-diaxial methyl-hydrogen interactions (show these!), which contribute 4 × 3.7 = 14.8 kJ/mol (4 × 0.9 = 3.6 kcal/mol) to its energy. What about *B*? You might be tempted to say that *B* has no unfavorable interactions because it has no axial groups, but in fact *B* does have one *gauche*-butane interaction—the one between the two methyl groups themselves, which have a dihedral angle between them of 60°, just as in *gauche*-butane. This interaction is easy to see in a Newman projection looking down the bond between the carbons bearing the methyl groups:

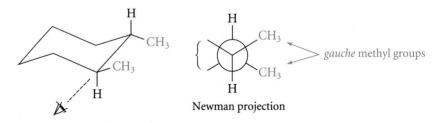

Newman projection

This *gauche*-butane interaction contributes 3.7 kJ/mol (0.9 kcal/mol) to the energy of conformation *B*. The relative energy of the two conformations is the difference between their methyl-hydrogen interactions: 14.8 − 3.7 = 11.1 kJ/mol (or 3.6 − 0.9 = 2.7 kcal/mol). The diaxial conformation *A* is less stable by this amount of energy.

When two groups on a substituted cyclohexane conflict in their preference for the equatorial position, the preferred conformation can usually be predicted from the relative conformational preferences of the two groups. An extreme example of this situation occurs in a cyclohexane substituted with a *tert*-butyl group, for example, *cis*-1-*tert*-butyl-4-methylcyclohexane.

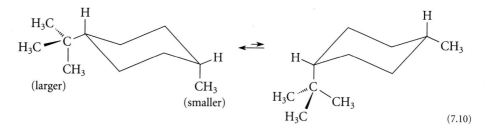

(7.10)

The *tert*-butyl group is so large that its van der Waals repulsions control the conformational equilibrium. Hence, the chair conformation in which the *tert*-butyl group assumes the equatorial position is strongly favored. The methyl group is thus forced to take up the axial position.

> There is so little of the conformation with an *axial tert*-butyl group that chemists say sometimes that the conformational equilibrium is "locked." This statement is somewhat misleading because it implies that the two conformations are not at equilibrium. The equilibrium indeed occurs rapidly, but simply contains very little of the conformation in which the *tert*-butyl group is axial.

PROBLEMS

*7.7   The principles of conformational analysis can also be applied to two compounds that are not conformational isomers. Calculate the energy difference between the following conformations of *cis*- and *trans*-1,3-dimethylcyclohexane.

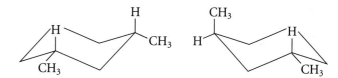

7.8   Calculate the energy difference between the two chair conformations of *trans*-1,4-dimethylcyclohexane.

C. Use of Planar Structures for Cyclic Compounds

Although many cyclic compounds (such as cyclohexane and its derivatives) have nonplanar conformations, planar structures of these compounds can be used for situations in which conformational details are not important. In this notation, the structures of cyclic compounds are represented by planar polygons with the stereochemistry of substituents indicated by dashed or solid wedges. In this type of structure, imagine viewing the ring from above one face. If a substituent is up, the bond to it is represented by a solid wedge; if it is down, the bond to it is represented as a dashed wedge. In this convention, one enantiomer of *trans*-1,2-dimethylcyclohexane can be drawn as follows:

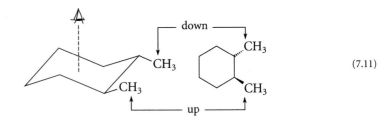

(7.11)

A planar structure for *cis*-1,2-dimethylcyclohexane can be drawn in either of two ways:

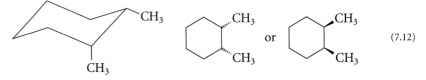

(7.12)

***cis*-1,2-dimethylcyclohexane**

The two planar structures are derived respectively by viewing the ring from each of its two faces. That the two structures are equivalent can be further demonstrated by turning either one over:

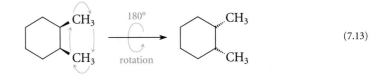

(7.13)

A planar structure, of course, does not convey any conformational information. That is, so long as all conformations are viewed from the same face of the ring, the chair flip does not interchange wedges and dashed lines because it does not change the up or down relationship of the ring substituents.

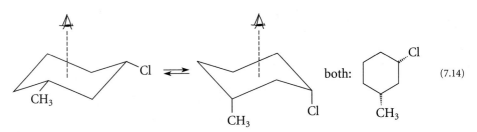

(7.14)

....

PROBLEMS

7.9 Draw a planar structure for each of the following compounds using dashed or solid wedges to show the stereochemistry of substituent groups.

*(a) *cis*-1,3-dimethylcyclohexane

(b) *trans*-1,3-dimethylcyclohexane

*(c) (1*R*,2*S*,3*R*)-2-chloro-1-ethyl-3-methylcyclohexane

(d) (1*S*,2*R*)-1-chloro-2-methylcyclopentane

7.10 Draw the more stable chair conformation for each of the following compounds:

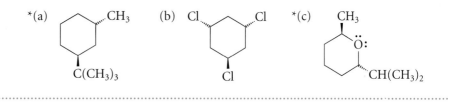

D. Stereochemical Consequences of the Chair Flip

The chair flip has some interesting stereochemical consequences that can be illustrated with some dimethylcyclohexanes. Consider first *cis*-1,2-dimethylcyclohexane. The chirality of either conformation can be demonstrated by showing, as usual, that its mirror image is noncongruent.

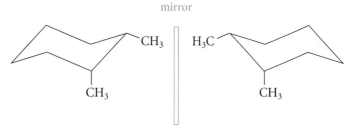

noncongruent mirror images of *cis*-1,2-dimethylcyclohexane

However, *the chair flip interconverts these enantiomers*:

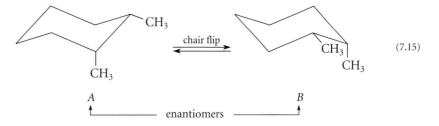

(7.15)

Structures *A* and *B* may not look like enantiomers, but they are! You can see this by turning structure *B* 120° about a vertical axis:

STUDY GUIDE LINK:
✓7.1
*Manipulating
Cyclohexane Rings*

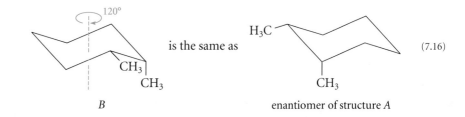

(7.16)

In other words, *cis*-1,2-dimethylcyclohexane is a mixture of *conformational enantiomers* (Sec. 6.10A). Because these enantiomers are interconverted very rapidly, they cannot be separated at ordinary temperatures. Consequently, *cis*-1,2-dimethylcyclohexane cannot be optically active. Recall that molecules which consist of rapidly interconverting enantiomers—molecules like *cis*-1,2-dimethylcyclohexane—are considered to be achiral (Sec. 6.10A). Because *cis*-1,2-dimethylcyclohexane is an achiral molecule that contains asymmetric carbons, it is a *meso* compound.

It is not necessary to draw and manipulate its conformations to tell whether a cyclic compound is *meso*. A cyclic compound is *meso* when its *planar structure* can be divided into mirror-image halves.

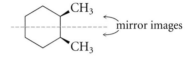

Equivalently, a cyclic compound is *meso* when a *planar structure* is congruent to its mirror image.

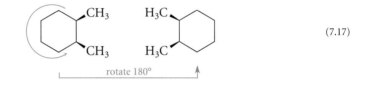

(7.17)

The enantiomers of *trans*-1,2-dimethylcyclohexane represent a different situation.

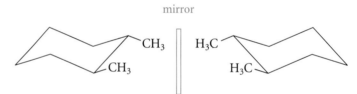

enantiomers of *trans*-1,2-dimethylcyclohexane

These two molecules cannot be interconverted by the chair flip, because the chair flip causes equatorial methyls to become axial methyls. Indeed, the chair flip converts each enantiomer into a conformational diastereomer:

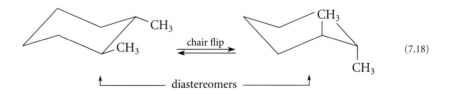

(7.18)

Hence, *trans*-1,2-dimethylcyclohexane is truly chiral, and each of its two enantiomers is a rapidly interconverting mixture of conformational diastereomers. Because each enantiomer is capable of independent existence, *trans*-1,2-dimethylcyclohexane can be isolated in optically active form.

The chirality of *trans*-1,2-dimethylcyclohexane can be seen from its planar structures. A cyclic compound is chiral when a planar structure and its mirror image are noncongruent.

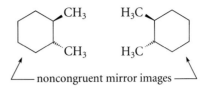

noncongruent mirror images

Cis-1,3-dimethylcyclohexane, another molecule with two asymmetric carbon atoms, presents yet a different situation. This compound too is a mixture of conformational diastereomers:

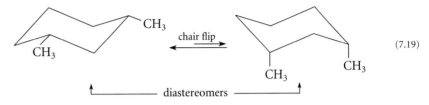

(7.19)

diastereomers

Yet even though each conformation has two asymmetric carbons, neither conformation is chiral because each has an internal plane of symmetry.

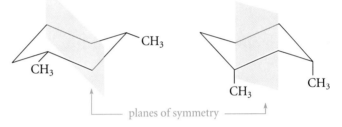

planes of symmetry

In other words, *both conformations are meso.* Cis-1,3-dimethylcyclohexane is thus a rapidly interconverting mixture of two different *meso* compounds, each a diastereomer of the other.

Notice that this case is subtly different from that of *cis*-1,2-dimethylcyclohexane: *cis*-1,3-dimethylcyclohexane cannot be separated into enantiomers *even at very low temperature* because *each conformation is achiral.* Recall that *cis*-1,2-dimethylcyclohexane, in contrast, consists of two chiral conformations in rapid equilibrium, and thus could in principle be separated into enantiomers at very low temperature. Although planar representations show the *meso* character of both compounds clearly, they conceal this subtle difference.

PROBLEMS

7.11   Determine whether each of the following compounds can in principle be isolated in optically active form under ordinary conditions.
*(a)   *trans*-1,3-dimethylcyclohexane   (b)   *cis*-1,3-dimethylcyclohexane
*(c)   *cis*-1,4-dimethylcyclohexane   (d)   1,1-dimethylcyclohexane
*(e)   *cis*-1-ethyl-3-methylcyclohexane

*7.12   (a)   Does *trans*-1,4-dimethylcyclohexane contain asymmetric carbons? If so, identify them.

(Problem 7.12 continues)

(b)   Does *trans*-1,4-dimethylcyclohexane have any stereocenters? If so, identify them.

(c)   What is the stereochemical relationship of the two chair conformations of *trans*-1,4-dimethylcyclohexane?

(d)   Is *trans*-1,4-dimethylcyclohexane chiral?

7.13   What is the relationship of the two structures in each set (identical molecules, enantiomers, or diastereomers)?

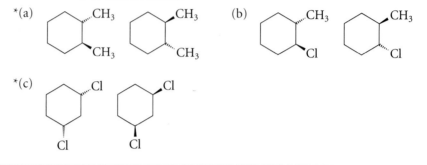

7.5   Cyclopentane, Cyclobutane, and Cyclopropane

A.   Cyclopentane

Cyclopentane, like cyclohexane, exists in a puckered conformation, called the **envelope conformation** (Fig. 7.10). This conformation undergoes rapid conformational changes in which each carbon alternates as the "point" of the envelope.

The data in Table 7.1 show that cyclopentane has somewhat higher energy than cyclohexane. The higher energy of cyclopentane is due to eclipsing between hydrogen atoms, which is also shown in Fig. 7.10.

Substituted cyclopentanes also exist in envelope conformations, but the substituents adopt positions that minimize van der Waals repulsions with neighboring groups. For

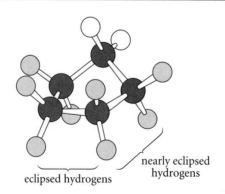

eclipsed hydrogens

nearly eclipsed hydrogens

Figure 7.10   *A ball-and-stick model of the envelope conformation of cyclopentane. The hydrogens shown in gray are either eclipsed or nearly eclipsed.*

example, in methylcyclopentane, the methyl group assumes an equatorial position at the point of the envelope.

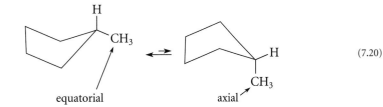

(7.20)

When a cyclopentane ring has two or more substituent groups, *cis* and *trans* relationships between the groups are possible, just as in cyclohexane.

cis-**1,2-dimethylcyclopentane**

trans-**1-ethyl-3-isopropylcyclopentane**

B. Cyclobutane and Cyclopropane

The data in Table 7.1 show that cyclobutane and cyclopropane are the least stable of the monocyclic alkanes. In each compound, the angles between carbon-carbon bonds are constrained by the size of the ring to be much smaller than the optimum tetrahedral angle of 109.5°. When the bond angles in a molecule deviate from their ideal values, the energy of the molecule is raised in the same sense that squeezing the handles of a hand exerciser increases the potential energy of the resisting spring. This excess energy, which is reflected in a greater heat of formation, is termed **angle strain**. (When angle strain occurs in a ring, it is sometimes called **ring strain**.) Hence, angle strain contributes significantly to the high energies of both cyclobutane and cyclopropane.

Puckering of the cyclobutane ring avoids complete eclipsing between hydrogens. Cyclobutane consists of two puckered conformations in rapid equilibrium (Fig. 7.11).

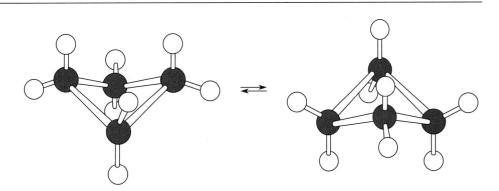

Figure 7.11   *Cyclobutane consists of two identical puckered conformations in rapid equilibrium.*

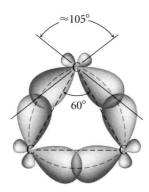

Figure 7.12   *The orbitals that overlap to form each C—C bond in cyclopropane do not lie along the straight line between the carbon atoms. These carbon-carbon bonds are sometimes called "bent" or "banana" bonds.*

Because three carbons define a plane, cyclopropane is planar; thus neither its angle strain nor the eclipsing interactions between its hydrogens can be relieved by puckering. As the data in Table 7.1 show, cyclopropane is the least stable of the cyclic alkanes. The carbon-carbon bonds of cyclopropane are bent in a "banana" shape around the periphery of the ring. Such "bent bonds" allow angles between the carbon orbitals that are on the order of 105°, closer to the ideal tetrahedral value of 109.5° (Fig. 7.12). Although bent bonds reduce angle strain, they do so at a cost of less effective overlap between the carbon orbitals.

☞
Study Guide Link:
7.2
Alkenelike Behavior of Cyclopropanes

Problems

7.14   ⋆(a)   The dipole moment of *trans*-1,3-dibromocyclobutane is 1.1 D. Explain why a nonzero dipole moment supports a puckered structure rather than a planar structure for this compound.

(b)   Draw a structure for the more stable conformation of *trans*-1,2-dimethylcyclobutane.

7.15   Tell whether each of the following compounds is chiral.
⋆(a)   *cis*-1,2-dimethylcyclopropane   (b)   *trans*-1,2-dimethylcyclopropane

7.6   # Bicyclic and Polycyclic Compounds

A. Classification and Nomenclature

Cyclic compounds may, of course, contain more than one ring. If two rings share two or more common atoms, the compound is called a **bicyclic** compound. If two rings have a single common atom, the compound is called a **spirocyclic** compound.

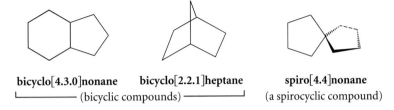

bicyclo[4.3.0]nonane   **bicyclo[2.2.1]heptane**   **spiro[4.4]nonane**
└─────────── (bicyclic compounds) ───────────┘   (a spirocyclic compound)

Bicyclic compounds are further classified according to the relationship of the **bridge-head carbons**: the carbons at which the two rings are joined. When the bridgehead carbons of a bicyclic compound are adjacent, the compound is classified as a **fused bicyclic compound**:

When the bridgehead carbons are not adjacent, the compound is classified as a **bridged bicyclic compound**:

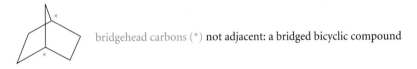

The nomenclature of bicyclic hydrocarbons is best illustrated by example:

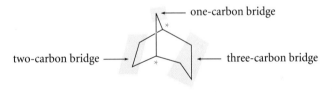

bicyclo[3.2.1]octane
(* = bridgehead carbons)

This compound is named as a bicyclooctane because it is a bicyclic compound containing a total of eight carbon atoms. The numbers in brackets represent the number of carbon atoms in the respective bridges, in order of decreasing size.

....

**STUDY
PROBLEM
7·3**

Name the following compound:

Solution   The compound has two fused rings that contain a total of ten carbons, and is therefore named as a bicyclodecane. There are three bridges connecting the bridgehead carbons: two contain four carbons, and *one contains zero carbons*. (The bond connecting the bridgehead carbons is considered to be a bridge with zero carbons.)

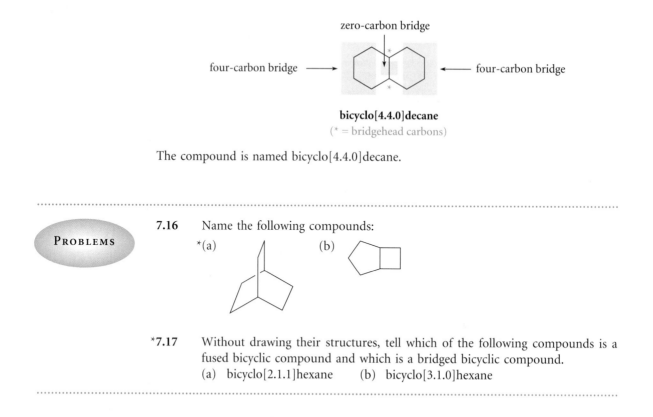

bicyclo[4.4.0]decane

(* = bridgehead carbons)

The compound is named bicyclo[4.4.0]decane.

PROBLEMS

7.16   Name the following compounds:

*(a)   (b)

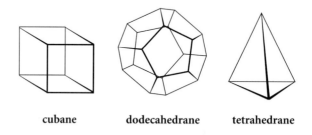

*7.17   Without drawing their structures, tell which of the following compounds is a fused bicyclic compound and which is a bridged bicyclic compound.
(a) bicyclo[2.1.1]hexane   (b) bicyclo[3.1.0]hexane

Some organic compounds contain many rings joined at common atoms; these compounds are called **polycyclic compounds**. Among the more intriguing polycyclic compounds are those that have the shapes of regular geometric solids. Three of the more spectacular examples are cubane, dodecahedrane, and tetrahedrane.

cubane   **dodecahedrane**   **tetrahedrane**

Cubane, which contains eight —CH— groups at the corners of a cube, was first synthesized in 1964 by Professor Philip Eaton and his associate Thomas W. Cole at the University of Chicago. Dodecahedrane, in which twenty —CH— groups occupy the corners of a dodecahedron, was synthesized in 1982 by a team of organic chemists led by Professor Leo Paquette of Ohio State University. Tetrahedrane itself has not yielded to synthesis, although a derivative containing *tert*-butyl substituent groups at each corner has been prepared. Chemists tackle the syntheses of these very pretty molecules not only because they represent interesting problems in chemical bonding, but also because of the sheer challenge of the endeavor.

B. *Cis* and *Trans* Ring Fusion

Two rings in a fused bicyclic compound can be joined in more than one way. Consider, for example, bicyclo[4.4.0]decane, which has the common name *decalin*.

decalin
(bicyclo[4.4.0]decane)

There are two stereoisomers of decalin. In *cis*-decalin, two —CH$_2$— groups of ring *B* (circles) are *cis* substituents on ring *A*; likewise, two —CH$_2$— groups of ring *A* (squares) are *cis* substituents on ring *B*.

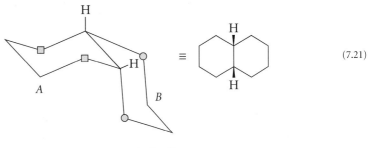

(7.21)

cis-decalin

Notice that the *cis* ring fusion is shown in a planar structure by showing the *cis* arrangements of the bridgehead hydrogens.

In *trans*-decalin, the —CH$_2$— groups adjacent to the ring fusion are in a *trans*-diequatorial arrangement. The bridgehead hydrogens are *trans*-diaxial.

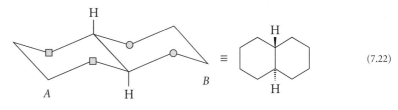

(7.22)

trans-decalin

Both *cis*- and *trans*-decalin have two equivalent planar structures:

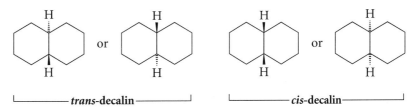

|— **trans-decalin** —| |— **cis-decalin** —|

Each cyclohexane ring in *cis*-decalin can undergo the chair flip, like cyclohexane itself. However, in *trans*-decalin, the six-membered rings can assume twist-boat conformations, but they cannot flip into their alternate chair conformations. You should try the chair flip with models of *cis*- and *trans*-decalin to see for yourself the validity of these statements. Focus on ring *B* of the trans-decalin structure in Eq. 7.22. Notice that the two circles represent carbons that are in effect *equatorial* substituents on ring *A*. If ring *A* were to flip into the other chair conformation, these two carbons in ring *B* would have to assume *axial* positions, because the chair flip converts equatorial groups into axial groups. When these two carbons are in axial positions they are much farther apart than they are in equatorial positions; the distance between them is simply too great to be spanned easily by the remaining two carbons of ring *B*. As a result, the chair flip introduces so much ring strain into ring *B* that the flip cannot occur. Of course, exactly the same problem occurs with ring *A* when ring *B* undergoes the chair flip.

PROBLEM

*7.18   How many 1,3-diaxial interactions are there in *cis*-decalin? In *trans*-decalin? Which compound has the lower energy and by how much? (*Hint:* Use your models, and don't count the same 1,3-diaxial interaction twice.)

Trans-decalin is more stable than *cis*-decalin because it has fewer 1,3-diaxial interactions (Problem 7.18). *Trans* ring fusion, however, is not the more stable way of joining rings in all fused bicyclic molecules. In fact, if both of the rings are small, *trans* ring fusion is virtually impossible. For example, only the *cis* isomers of the following two compounds are known:

bicyclo[1.1.0]butane   **bicyclo[3.1.0]hexane**

Attempting to join two small rings with a *trans* ring junction introduces too much ring strain. The best way to see this is with models, using the following exercise as your guide.

PROBLEM

7.19   *(a)   Compare the difficulty of making models of the *cis* and *trans* isomers of bicyclo[3.1.0]hexane. (Don't break your models!) Which is easier to make? Why?

(b)   Compare the difficulty of making models of *trans*-bicyclo[3.1.0]hexane and *trans*-bicyclo[5.3.0]decane. Which is easier to make? Explain.

In summary:

1. Two rings may be fused in a *cis* or *trans* arrangement.

2. When the rings are small, only *cis* fusion is observed because *trans* fusion introduces too much ring strain.

3. In larger rings, both *cis*- and *trans*-fused isomers are well known, but the *trans*-fused ones are more stable because 1,3-diaxial interactions are minimized (as in the decalins).

Effects (2) and (3) are about equally balanced in the *hydrindanes* (bicyclo[4.3.0]nonanes); the *trans* isomer is only 2.8 kJ/mol (0.66 kcal/mol) more stable than the *cis* isomer.

hydrindane
(bicyclo[4.3.0]nonane)

C. *Trans*-Cycloalkenes and Bredt's Rule

Cyclohexene and other cycloalkenes with small rings clearly have *cis* (or *Z*) stereochemistry at the double bond. Is there a *trans*-cyclohexene? The answer is that the *trans*-cycloalkenes with six or fewer carbons have never been observed. The reason is obvious if you try to build a model of *trans*-cyclohexene. In this molecule the carbons attached to the double bond are so far apart that it is difficult to connect them with only two other carbon atoms. To do so either introduces a great amount of strain, or requires twisting the molecule about the double bond, thus weakening the overlap of *p* orbitals involved in the π bond. *Trans*-cyclooctene is the smallest *trans*-cycloalkene that can be isolated under ordinary conditions; however, it is 38–46 kJ/mol (9–11 kcal/mol) less stable than its *cis* isomer.

Closely related to the instability of *trans*-cycloalkenes is the instability of any small bridged bicyclic compound that has a double bond at a bridgehead atom. The following compound, for example, is very unstable and has never been isolated:

bridgehead

bicyclo[2.2.1]hept-1(2)-ene
(unknown)

Compounds such as this are said to violate **Bredt's rule**: *In a bicyclic compound, a bridgehead atom contained solely within small rings cannot be part of a double bond.* ("Small rings," for purposes of Bredt's rule, contain seven or fewer atoms.)

The reason for Bredt's rule is that double bonds at bridgehead carbons within small rings are twisted; that is, the atoms directly attached to such double bonds *cannot* lie in the same plane. To see this, try to construct a model of the above compound. You will see that the bicyclic ring system cannot be completed without twisting the double bond. This is similar to the double-bond twisting that would occur in *trans*-cyclohexene. Like the corresponding *trans*-cycloalkenes, bicyclic compounds containing bridgehead double bonds solely within small rings are too unstable to isolate. Bicyclic compounds that have bridgehead double bonds within larger rings are more stable and can be isolated.

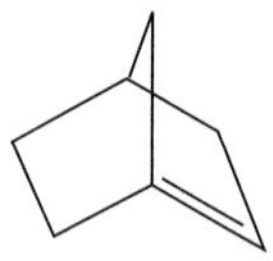

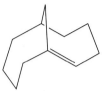

bicyclo[2.2.1]hept-1(2)-ene
too unstable to isolate

bicyclo[4.4.1]undec-1(2)-ene
stable enough to isolate

PROBLEM

7.20   Use models if necessary to help you decide which compound within each pair should have the greater heat of formation. Explain.

*(a)

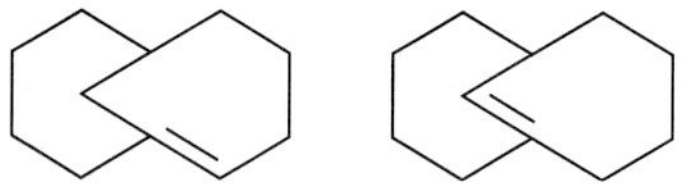

A   *B*

(b)

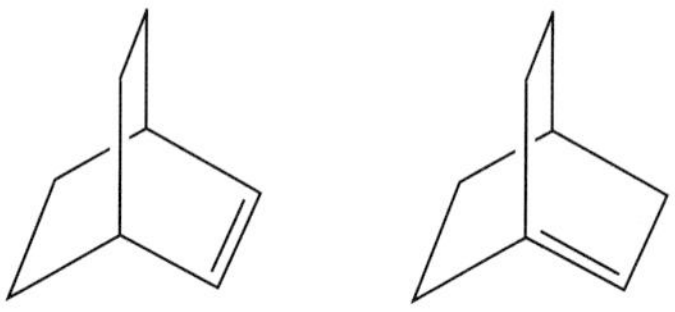

A   *B*

D. Steroids

Of the many naturally occurring compounds with fused rings, the class of compounds called the **steroids** is particularly important. The fundamental ring system of the steroids, with its special numbering, is as follows:

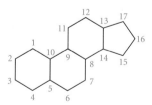

The various steroids differ in the types of functional groups that are present on this carbon skeleton.

Two structural features are particularly common in naturally occurring steroids. The first is that in many cases all ring fusions are *trans*. Because *trans*-fused cyclohexane rings cannot undergo the chair flip (see Eq. 7.22 and subsequent discussion), all-*trans* ring fusion causes a steroid to be conformationally rigid and relatively flat. Second, many steroids have methyl groups, called *angular methyls*, at carbons 10 and 13.

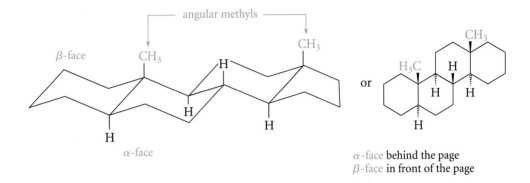

or

α-face behind the page
β-face in front of the page

These methyl groups occupy axial positions over one face, called the *β*-face, of the steroid molecule. The other face is called the *α*-face.

Many important hormones and other natural products are steroids. Cholesterol occurs widely and was the first steroid to be discovered (1775). The corticosteroids and the sex hormones represent two biologically important classes of steroid hormones.

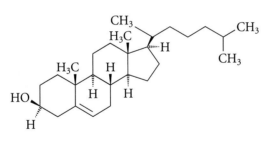

cholesterol
(important component of membranes:
principal component of gallstones)

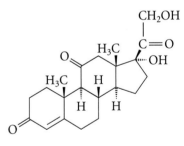

cortisone
(anti-inflammatory hormone)

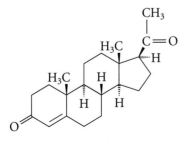

progesterone
(human female sex hormone)

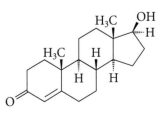

testosterone
(human male sex hormone)

SOURCE OF STEROIDS

Prior to 1940, steroids were isolated from such inconvenient sources as sows' ovaries or the urine of pregnant mares, and were scarce and expensive. In the 1940s, however, a Pennsylvania State University chemist, Russell Marker (b. 1902), developed a process that could bring about the conversion of a naturally occurring compound called *diosgenin* into progesterone.

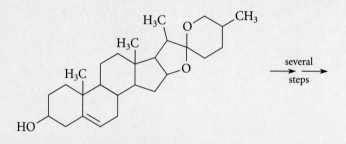

diosgenin

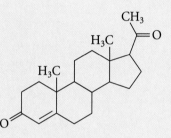

progesterone

The natural source of diosgenin is the root of a vine, *cabeza de negro*, genus *Dioscorea*, indigenous to Mexico. *Dioscorea*, which takes three years to mature, is now grown on plantations, and the Mexican government justifiably regards this plant as a valuable natural resource.

7.7 Relative Reactivities of Stereoisomers

The remainder of this chapter focuses on the importance of stereochemistry in organic reactions. To begin, this section develops some general principles concerning the relative reactivities of stereoisomers.

A. Relative Reactivities of Enantiomers

Imagine subjecting a pair of enantiomers to the same reaction conditions. Will the reactivities of the enantiomers differ? As an example, consider the following reaction, in which a chiral alkene reacts with borane:

$$3 \text{ Ph} - \underset{\underset{\text{H}}{|}}{\overset{\overset{\text{CH}_3}{|}}{\text{C}}} - \text{CH} = \text{CH}_2 + \text{BH}_3 \xrightarrow{\text{THF}} \left(\text{Ph} - \underset{\underset{\text{H}}{|}}{\overset{\overset{\text{CH}_3}{|}}{\text{C}}} - \text{CH}_2 - \text{CH}_2 \right)_{3}\! \text{B} \qquad (7.23)$$

(a chiral alkene)

Do the *R* and *S* enantiomers of this alkene have different reactivities? A general principle applies to situations like this. *A pair of enantiomers react at the same rates with an achiral reagent.* This means that the enantiomers of the alkene in Eq. 7.23 react with borane, an achiral reagent, at exactly the same rates to give their respective products in exactly the same yield.

An analogy from common experience can help you understand why this should be so. Consider your feet, an enantiomeric pair of objects. Imagine placing first your right foot, then your left, in a perfectly square box—an achiral object. Each foot will fit this box in exactly the same way. If the box pinches the big toe on your right foot, it will also pinch the big toe on your left foot in the same way. Just as your feet interact in the same way with the achiral box, so enantiomeric molecules react in exactly the same way with achiral reagents. Because borane is an achiral reagent, the two alkene enantiomers in Eq. 7.23 react with borane in exactly the same way.

Enantiomers have identical reactivities because *enantiomers have the same physical properties* (Sec. 6.3). Such properties include not only things like melting points and boiling points, but also free energies. Both the starting materials in Eq. 7.23 and their respective transition states are enantiomeric. The enantiomeric transition states have identical free energies, as do the enantiomeric starting materials. Because relative reactivity is determined by the difference in free energies of transition state and starting material, and since this difference is identical for both enantiomers, enantiomers react at identical rates.

Suppose, though, that each enantiomer of the alkene in Eq. 7.23 reacts in turn with an enantiomerically pure *chiral* reagent, such as the following borane:

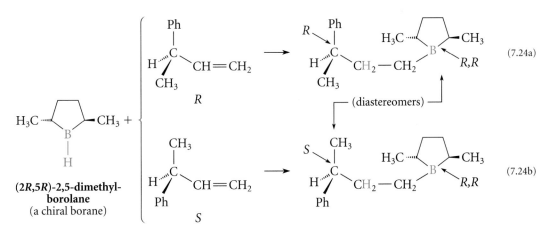

(2*R*,5*R*)-2,5-dimethyl-
borolane
(a chiral borane)

The following general principle applies to this situation: *A pair of enantiomers react at different rates with a chiral reagent.* This means that if one mole of the alkene racemate (one-half mole of each enantiomer) were to react with one-half mole of the chiral borane, one of the diastereomeric products in Eq. 7.24 would be formed in a greater amount than the other; the alkene remaining after the borane is consumed would be enriched in the less reactive enantiomer.

Another analogy might help you see why this result is reasonable. Imagine placing first your right foot, then your left, in your right shoe—a chiral object. Your right and left feet interact differently with your right shoe; the shoe fits one foot and not the other. The enantiomeric objects (feet) interact differently with the chiral shoe. Likewise, a pair of chiral molecules interact differently with a chiral reagent. The interaction of one alkene enantiomer in Eq. 7.24 with the chiral borane is more favorable than the interaction of the other.

Enantiomers have different reactivities with a chiral reagent because *diastereomers have different properties* (Sec. 6.6). These different properties encompass not only things like melting points and boiling points, but also free energies. In this case, the transition state for the reaction of one enantiomer (Eq. 7.24a) is the *diastereomer* of the transition state for the reaction of the other (Eq. 7.24b). Because diastereomeric transition states have different energies, the reaction of one enantiomer occurs more rapidly than the reaction of the other. (Note that we may not be able to predict which enantiomer will be more reactive.)

The reactivity of enantiomers can be generalized in two equivalent ways:

1. *A pair of enantiomers differ in their chemical or physical behavior only when they interact with other chiral objects or forces.*

2. *A pair of enantiomers behave differently only under conditions that cause them to be involved in diastereomeric interactions.*

Study Guide Link:
7.3
Optical Activity

Enantiomeric resolution (Sec. 6.8) is another important application of these principles. The pair of enantiomers to be resolved assume different properties when they interact with an enantiomerically pure resolving agent (the "chiral object") to form diastereomeric salts.

Problem

7.21 Apply the principles of this section to solve each of the following problems.
*(a) Assuming equal strength in both hands, would your right and left hands differ in their ability to drive a nail? To drive a screw with a screwdriver?
(b) Imagine that a certain Mr. D. has been visited by a certain Mr. L. from elsewhere in the universe. Mr. D. and Mr. L. are alike in every way, except that they are noncongruent mirror images! You have to introduce each of them at an international press conference, but neither of them will agree to give their names. How would you tell them apart? (There may be several ways.)

B. Relative Reactivities of Diastereomers

Diastereomers in general have different reactivities toward *any* reagent, whether the reagent is chiral or achiral. The reason is that diastereomers have different free energies.

Consequently, their standard free energies of activation, and hence their reaction rates, must differ in principle. Thus, the two diastereomeric alkenes *cis-* and *trans*-2-butene react at different rates with all reagents. We may not be able to predict which alkene is more reactive or by how much, but we can be sure that the two alkenes will not be equally reactive.

7.8   Reactions That Form Stereoisomers

The previous section discussed the reactivity of stereoisomers when they are subjected to the same chemical reaction. This section considers what happens when a reaction results in the formation of stereoisomers. Are the stereoisomers formed at the same rates? Are they formed in the same amounts?

A.   Reactions of Achiral Compounds That Give Enantiomeric Products

Suppose a chemical reaction of *achiral* starting materials yields a *chiral* product. An example of such a reaction is the addition of HBr to styrene.

$$\text{Ph}-\text{CH}=\text{CH}_2 + \text{HBr} \longrightarrow \text{Ph}-\underset{\underset{\text{Br}}{\mid}}{\text{CH}}-\text{CH}_3 \qquad (7.25)$$

styrene

(1-bromoethyl)benzene

Neither of the reactants—styrene nor HBr—is chiral. However, the product of the reaction, (1-bromoethyl)benzene, *is* chiral. This product could be either of two enantiomers or both. Thus, a question of stereochemistry arises: Which of these stereoisomers is formed? A general principle applies to this situation: *When chiral products are formed from achiral starting materials, both enantiomers of a pair are always formed at identical rates.* That is, the product is always the *racemate*. Thus, in the example of Eq. 7.25, equal amounts of (*R*)- and (*S*)-(1-bromoethyl)benzene are formed. Another way of expressing the same conclusion is that *optical activity never arises spontaneously in the reactions of achiral compounds.*

An understanding of this idea hinges on the fact that a pair of enantiomers have identical energies. That is, for each chiral transition state in the reaction pathway, there is an enantiomeric transition state of equal energy. Since the enantiomeric transition states are formed from the same starting material, the two enantiomers of the product are formed with identical free energies of activation. Hence, they are formed at identical rates, and therefore in identical amounts. As a result, the product is a 1:1 mixture of *R* and *S* enantiomers—the racemate.

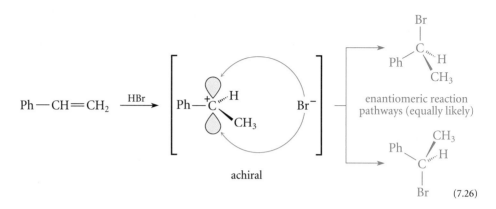

$$\text{(7.26)}$$

When any sort of chiral influence is present—for example, a chiral solvent, or a chiral catalyst—then the situation changes. In such a situation, two enantiomers of a pair are formed in different amounts. Enzyme catalysis provides many important examples of this phenomenon. Consider, for example, the conversion of fumarate and water, both achiral compounds, into malate, a chiral compound:

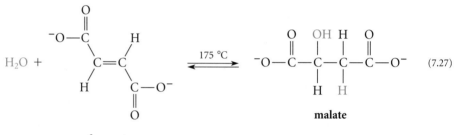

$$\text{(7.27)}$$

If this reaction is carried out in the laboratory at 175 °C, the malate produced in this reaction is the racemate, as expected from the discussion above. In biological systems, however, this reaction is catalyzed at a much lower temperature by the enzyme fumarase (Sec. 4.9C), *and the product malate is the enantiomerically pure S enantiomer.*

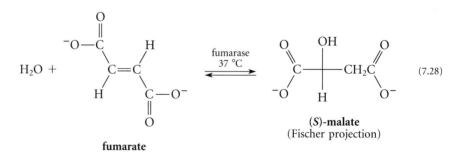

$$\text{(7.28)}$$

Why should the enzyme catalyze the formation of one enantiomer and not the other?

The reason is that fumarase and all other enzymes are *enantiomerically pure chiral molecules.* Because the transition state for the reaction of water and fumarate includes the chiral enzyme, then the *R* and the *S* transition states, which are enantiomers in the *absence* of the enzyme, are diastereomers in the *presence* of the enzyme.

$$[(R)\text{-transition state} + \text{chiral enzyme}]$$

enzyme + H_2O + fumarate   diastereomers   (7.29)

$$[(S)\text{-transition state} + \text{chiral enzyme}] \longrightarrow (S)\text{-malate} + \text{enzyme}$$

Remember that *diastereomeric transition states have different free energies.* Because the S enantiomer of malate is formed more rapidly (so much more rapidly that the R enantiomer is not formed at all), the transition state leading to malate of S configuration must have lower free energy.

If a catalyst accelerates a reaction in one direction, then it must also accelerate the same reaction in the reverse direction (Sec. 4.9C). The fumarase-catalyzed reaction is reversible, and indeed fumarase accelerates the conversion of malate into fumarate. However, because *only* (S)-malate is formed in the forward reaction, *only* the reaction of (S)-malate is catalyzed by the enzyme in the reverse reaction; (R)-malate is inert! This situation illustrates the principle in Sec. 7.7A: two enantiomers (R- and S-malate) differ when they interact with a chiral object (the enzyme). Indeed, *most enzymes accept only one enantiomer of a chiral substrate.*

Thus, enzymes catalyze the selective formation of specific enantiomers in biological reactions. Likewise, when confronted with enantiomeric reactants, enzymes in most cases will catalyze the reaction of only one enantiomer. Although enzymes are the best known examples of chiral catalysts, chemists have also produced a variety of synthetic chiral catalysts and reagents. Hence, the stereochemical selectivity of enzymes is not particularly unique, although the degree of selectivity of enzymes is greater than that of most synthetic catalysts.

To summarize: When chiral compounds are formed from achiral starting materials, the product is *always* racemic *unless* the reaction is carried out under the influence of a chiral environment such as a chiral solvent or a chiral catalyst. In that case, the predominance of one enantiomer can be expected.

STUDY GUIDE LINK:
✓7.4
Reactions of Chiral Molecules

B. Reactions That Give Diastereomeric Products

Some reactions can in principle give pairs of diastereomeric products, as in the following example:

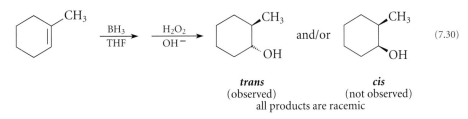

trans
(observed)

cis
(not observed)

all products are racemic

(7.30)

In this case either the *cis* or *trans* diastereomer of the product might have been formed. In general *when diastereomeric products can be formed in a reaction, they are always formed at different rates, and different amounts of each product are formed.*

Without knowing more about the reaction, however, we might not be able to predict *which* diastereomer is the major one, but we can always expect one product to be formed in greater amount than the other. (In Eq. 7.30, the *trans* isomer is formed exclusively; we'll see why in Sec. 7.9D.)

Diastereomers are formed in different amounts because they are formed through diastereomeric transition states. In general, one transition state has lower standard free energy than its diastereomer. The diastereomeric reaction pathways thus have different standard free energies of activation and therefore different rates, and their respective products are formed in different amounts.

Note that although diastereomers are formed in different amounts, the differences in their amounts might be beyond detection: for example, one might be formed as 49.99% and the other as 50.01% of the product mixture. The important point is that *in principle* the two are formed in different amounts.

Note also that when the starting materials are achiral, each diastereomer of the product will be formed as a pair of enantiomers (the racemate) by the principle of Sec. 7.8A. This is the situation, for example, in Eq. 7.30. For convenience we sometimes draw only one enantiomer of each product formed, as in this equation, but in situations like this it is understood that each of these diastereomers *must* be racemic.

Study Problem 7.4

What stereoisomeric products could be formed in the addition of bromine to cyclohexene? Which should be formed in the same amounts? Which should be formed in different amounts?

Solution   Before dealing with any issue involving the stereochemistry of any reaction, first be sure you understand the reaction itself. Bromine addition to cyclohexene gives 1,2-dibromocyclohexane:

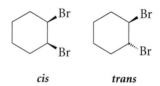

1,2-dibromocyclohexane

Next, enumerate the possible stereoisomers of the product that might be formed. The product, 1,2-dibromocyclohexane, can exist as a pair of diastereomers:

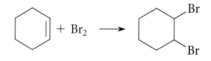

cis   *trans*

The *trans* diastereomer can exist as a pair of enantiomers. The *cis* diastereomer is *meso*, and thus is achiral (Sec. 7.4D). Hence, three stereoisomers could be formed: the *cis* isomer and the two enantiomers of the *trans* isomer. Because the *cis* and *trans* isomers are diastereomers *they are formed in different amounts*. (You can't predict at this point which one predominates, but we'll return to that issue in Sec. 7.9C.) The two enantiomers of the *trans* diastereomer *must be formed in identical amounts*. Thus, whatever the amount of the *trans* isomer we obtain from the reaction, it is a 50:50 mixture of the two enantiomers.

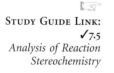

Study Guide Link:
✓7.5
Analysis of Reaction Stereochemistry

PROBLEMS

*7.22   What stereoisomeric products are possible when *cis*-2-butene undergoes bromine addition? Which are formed in different amounts? Which are formed in the same amounts?

7.23   What stereoisomeric products are possible when *trans*-2-butene undergoes hydroboration-oxidation? Which are formed in different amounts? Which are formed in the same amounts?

*7.24   Write all the possible products that might form when racemic 3-methylcyclohexene reacts with Br_2. What is the relationship of each pair? Which compounds should in principle be formed in the same amounts, and which in different amounts? Explain.

THE ORIGIN OF OPTICALLY ACTIVE COMPOUNDS

When a chiral compound occurs in nature, typically only one of its two enantiomers is found. That is, *nature is a source of optically active compounds.* For example, the sugar glucose occurs only as the dextrorotatory form shown below; the naturally occurring amino acid leucine is the levorotatory enantiomer.

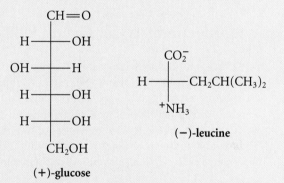

(+)-glucose

(−)-leucine

Many scientists hypothesize that eons ago the first chiral compounds were formed from simple, achiral starting materials such as methane, water, and HCN. This hypothesis presents a problem. As shown in Sec. 7.8A, reactions that give chiral products from achiral starting materials always give the racemate; net optical activity cannot be generated in the reactions of achiral molecules. If the biological starting materials are all achiral, why is the world full of optically active compounds? Instead it should be full of racemates! (This would mean that somewhere in the world your noncongruent mirror image is studying organic chemistry!) The only way out of this dilemma is to postulate that at some point in geologic time an *enantiomeric resolution* must have occurred. How could this have happened?

(continues)

This question has generated much speculation. However, many scientists believe that the first optical resolution occurred purely by chance. Although a spontaneous enantiomeric resolution is *highly improbable*, it is not strictly *impossible*. In fact, such chance enantiomeric resolutions have been known to occur in the laboratory. For example, one enantiomer of Pasteur's sodium ammonium tartrate (Sec. 6.12) sometimes crystallizes from a solution of the racemate, particularly if the solution is seeded by a crystal of the enantiomer. Perhaps the spontaneous crystallization of a pure enantiomer took place on the prebiotic earth, seeded by a speck of dust with just the right shape. The question is an intriguing one, and no one really knows the answer.

Given that one or more enantiomeric resolutions occurred by chance at some time during the course of natural history, it is not difficult to understand how nature continues to manufacture optically active compounds. Recall that naturally occurring chiral catalysts—enzymes—perpetuate the formation of optically active compounds, such as malate, from achiral starting materials and accept only one enantiomer of a chiral reactant (Sec. 7.8A). Such catalytic discrimination guarantees a high degree of enantiomeric purity in naturally occurring compounds.

7.9 Stereochemistry of Chemical Reactions

At this point it may seem that stereochemistry adds a complicated new dimension to the study and practice of organic chemistry. To some extent this is true. No chemical structure is complete without stereochemical detail, and no chemical synthesis can be planned without considering problems of stereochemistry that might arise. This section examines the possible stereochemical outcomes of two general types of reactions, addition reactions and substitution reactions. Then some addition reactions covered in Chapter 5 will be revisited with particular attention to their stereochemistry.

A. Stereochemistry of Addition Reactions

Recall that an *addition reaction* is a reaction in which a general species X—Y adds to each end of a double bond or triple bond:

$$
\begin{array}{c}
\diagup \\
C = C \\
\diagup
\end{array}
\ + \ X-Y \ \longrightarrow \
\begin{array}{c}
\diagup \\
C - C \\
| \ \ \ | \\
X \ \ \ Y
\end{array}
\tag{7.31}
$$

An addition reaction can occur in either of two stereochemically different ways. These will be illustrated with cyclohexene and a general reagent X—Y.

In a *syn addition*, two groups add to a double bond from the same side or face:

Syn addition:

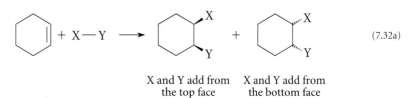

$$\text{X and Y add from} \quad\quad \text{X and Y add from}$$
$$\text{the top face} \quad\quad\quad \text{the bottom face}$$

(7.32a)

In an ***anti*** **addition**, two groups add to a double bond from opposite sides or faces:

Anti addition:

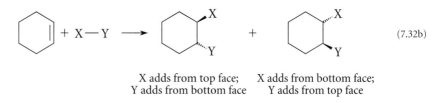

$$\text{X adds from top face;} \quad\quad \text{X adds from bottom face;}$$
$$\text{Y adds from bottom face} \quad\quad \text{Y adds from top face}$$

(7.32b)

Of course, an addition might occur as a mixture of *syn* and *anti* modes. That is, some molecules might undergo *syn* addition, and others *anti* addition. In such a reaction, the products would be a mixture of all of the products in both Eqs. 7.32a and b. Examples of both *syn* and *anti* additions, as well as mixed additions, will be examined later in this section.

As Eqs. 7.32a and b suggest, the *syn* and *anti* modes of addition can be distinguished *by analyzing the stereochemistry of the products.* In Eq. 7.32a, for example, the *cis* relationship of the groups X and Y in the product would tell us that a *syn* addition has occured. Thus, the stereochemistry of an addition can be determined *only when the stereochemically different modes of addition give rise to stereochemically different products.* Thus, when two groups X and Y are added to ethylene ($CH_2{=}CH_2$), the same product ($X{-}CH_2{-}CH_2{-}Y$) results whether the reaction is a *syn* or an *anti* addition. Because this product can't exist as stereoisomers, we can't tell whether the addition is *syn* or *anti.* A more general way of stating the same point is to say that *syn* and *anti* additions give different products only when *both* carbons of the double bond become carbon stereocenters in the product. If you stop and think about it, this should make sense, because the question of *syn* and *anti* addition is a question of the relative stereochemistry at *both* carbons, and the relative stereochemistry can't be determined if both carbons aren't stereocenters. Furthermore, in additions to alkenes, the two carbons of the double bond can't be stereocenters in the product unless they are stereocenters in the alkene starting material in the first place.

B. **Stereochemistry of Substitution Reactions**

In a **substitution reaction**, one group is replaced by another. In the following substitution reaction, for example, the Br is replaced by OH:

$$CH_3{-}\ddot{\overset{..}{Br}}: \ + \ {}^{-}\!\!:\!\ddot{\overset{..}{O}}H \ \longrightarrow \ CH_3{-}\ddot{\overset{..}{O}}H \ + \ :\!\ddot{\overset{..}{Br}}:^{-}$$

(7.33)

The oxidation step of hydroboration–oxidation is also a substitution reaction in which the boron is replaced by an OH group.

$$OH^- + 3\ HO{-}OH + (CH_3CH_2)_3B \longrightarrow 3\ CH_3CH_2{-}OH + {}^-B(OH)_4 \qquad (7.34)$$

A substitution reaction can occur in two stereochemically different ways. When a group X′ replaces another group X with **retention of configuration**, then X and X′ have the same relative stereochemical positions. Thus, in the following example, if X is *cis* to Y, then X′ is also *cis* to Y.

Substitution with retention of configuration:

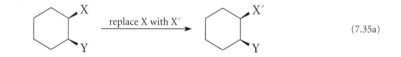

$$(7.35a)$$

Substitution with retention also implies that if X and X′ have the same relative priorities in the *R,S* system, then the carbon that undergoes substitution will have (for example) the same configuration in the reactant and the product. Thus, if this carbon has the *R* configuration in the starting material, it has the same, or *R*, configuration in the product.

When substitution occurs with **inversion of configuration**, then X and X′ have different relative stereochemical positions. Thus, if X is *cis* to Y in the starting material, X′ is *trans* to Y in the product:

Substitution with inversion of configuration:

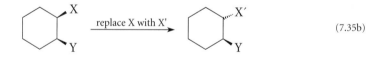

$$(7.35b)$$

Substitution with inversion also implies that if X and X′ have the same relative priorities in the *R,S* system, then the carbon that undergoes substitution must have opposite configurations in the reactant and the product. Thus, if this carbon has (for example) the *R* configuration in the starting material, it has the opposite, or *S*, configuration in the product.

As with addition, it is also possible that a reaction might occur so that both retention and inversion can occur at comparable rates in a substitution reaction. In such a case, stereoisomeric products corresponding to both pathways will be formed. Examples of substitution reactions with inversion, retention, and mixed stereochemistry are well known.

As Eqs. 7.35a and b suggest, analysis of the stereochemistry of substitution requires that the carbon which undergoes substitution must be a stereocenter in both the reactants and the products. For example, in the following situation, the stereochemistry of substitution cannot be determined.

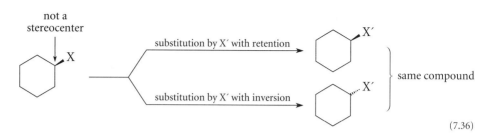

(7.36)

Because the carbon that undergoes substitution is not a stereocenter, the same product is obtained from both retention and inversion modes of substitution.

A reaction in which particular stereoisomer(s) of the product are formed to the exclusion of other(s) is said to be a **stereoselective reaction**. Thus, an addition that occurs only with *anti* stereochemistry as shown in Eq. 7.32a is a stereoselective reaction because only one pair of enantiomers is formed to the exclusion of a diastereomeric pair. A substitution that occurs only with inversion as shown in Eq. 7.35b is also a stereoselective reaction because one diastereomer of the product is formed to the exclusion of the other.

This section has established the stereochemical possibilities that might be expected in two types of reactions, additions and substitutions. The remaining sections apply these ideas in discussing the stereochemical aspects of several reactions that were first introduced in Chapter 5.

C. **Stereochemistry of Bromine Addition**

The addition of bromine to alkenes (Sec. 5.1A) is in many cases a stereoselective reaction. The addition of bromine to *cis*- and *trans*-2-butene can be used to apply the ideas of the previous section as well as to show how the stereochemistry of a reaction can be used to understand its mechanism.

When *cis*-2-butene reacts with Br_2, the product is 2,3-dibromobutane.

$$
\underset{\textit{\textbf{cis}-2-butene}}{\begin{array}{c} H_3C \qquad CH_3 \\ \diagdown \qquad \diagup \\ C{=}C \\ \diagup \qquad \diagdown \\ H \qquad\quad H \end{array}} + Br_2 \ \longrightarrow\ \underset{\textbf{2,3-dibromobutane}}{CH_3{-}CHBr{-}CHBr{-}CH_3} \qquad (7.37)
$$

You should now realize that several stereoisomers of this product are possible: a pair of enantiomers and the *meso* compound. The *meso* compound and the enantiomeric pair should be formed in different amounts (Sec. 7.8B). If the enantiomers are formed, they should be formed as the racemate because the starting materials are achiral (Sec. 7.8A).

When bromine addition to *cis*-2-butene is carried out in the laboratory, it is found that the product is virtually all (99%) racemate. Bromine addition to the *trans*-alkene, in contrast, gives exclusively the *meso* compound. A summary follows.

Experimental facts: $CH_3{-}CH{=}CH{-}CH_3 \ \xrightarrow{\ Br_2\ }\ CH_3{-}CHBr{-}CHBr{-}CH_3$ (7.38)

cis $\longrightarrow$ racemate

trans $\longrightarrow$ *meso*

This information indicates that addition reactions of bromine to both *cis*- and *trans*-2-butene are stereoselective. Are these additions *syn* or *anti*? Because the alkene is not cyclic (as it is in Eq. 7.32), the answer may not be obvious. The following Study Problem illustrates how to analyze the result systematically to get the answer.

STUDY PROBLEM 7.5

According to the experimental results in Eq. 7.38, is the addition of bromine to *cis*-2-butene a *syn* or an *anti* addition?

Solution   To answer this question, you should imagine *both syn* and *anti* additions to *cis*-2-butene and see what results would be obtained for each. Comparison of these results with experiment then shows us which alternative is correct.

If bromine addition were *syn*, the Br$_2$ could add to either face of the double bond. (In the following structures we are viewing the alkene edge-on.)

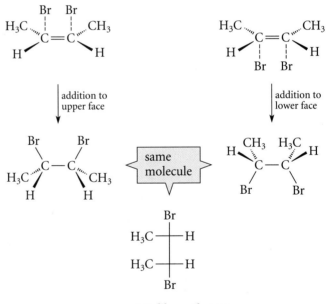

meso-2,3-dibromobutane   (7.39)

This analysis shows that *syn*-addition from either direction gives the *meso* diastereomer. Because the experimental facts (Eq. 7.38) show that *cis*-2-butene does *not* give the *meso* isomer, the two bromine atoms *cannot* be adding from the same face of the molecule. Therefore *syn* addition does not occur.

Consider next the *anti* addition of the two bromines to *cis*-2-butene. This addition, too, can occur in two equally probable ways.

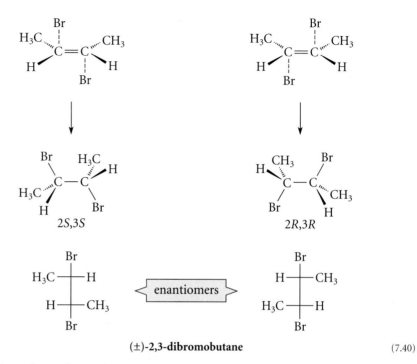

(±)-2,3-dibromobutane   (7.40)

This analysis shows that each mode of addition gives the enantiomer of the other; that is, the two modes of *anti* addition operating at the same time should give the racemate. Because the experimental facts of Eq. 7.38 show that bromine addition to *cis*-2-butene indeed gives the racemate, this reaction is an *anti* addition.

It is very important that you analyze in a similar manner the addition of bromine to *trans*-2-butene to show that this addition, too, is an *anti* addition.

- -

As suggested at the end of Study Problem 7.5, you should have demonstrated to yourself that the addition of bromine to *trans*-2-butene is also a stereoselective *anti* addition. In fact, the bromine addition to most simple alkenes occurs with *anti* stereochemistry. Bromine addition is therefore a *stereoselective reaction*.

The stereochemistry of bromine addition is one of the main reasons that the bromonium-ion mechanism, shown in Eqs. 5.3–5.4, was postulated. Let's see how this mechanism can account for the observed results. First, the bromonium ion can form at either face of the alkene. (Attack at one face is shown in the following equation; you should show attack at the other face and take your structures through the subsequent discussion.)

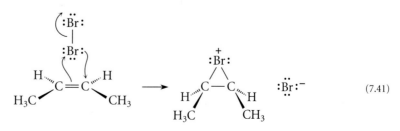

(7.41)

Bromonium ion formation as represented here is a *syn* addition, because even though only one group has added to the double bond, the methyl groups and hydrogens have the same (*cis*) relationship in both reactant and product.

If formation of the bromonium ion is a *syn* addition, then the *anti* addition observed in the overall reaction with bromine must be established by the stereochemistry of the attack of bromide ion. Suppose that the bromonium ion undergoes **backside attack**. This means that the bromide ion attacks a carbon of the bromonium ion at the face opposite the carbon-bromine bond. If backside attack occurs, the bromonium ion is opened to give the observed product. Notice that the attack of bromide ion is a *substitution reaction* that occurs with *inversion of configuration* (Sec. 7.9B). As this reaction takes place, the methyl and the hydrogen move up (colored arrows) to maintain the tetrahedral configuration of carbon. Attack of the bromide ion at one carbon yields one enantiomer; attack at the other carbon yields the other enantiomer.

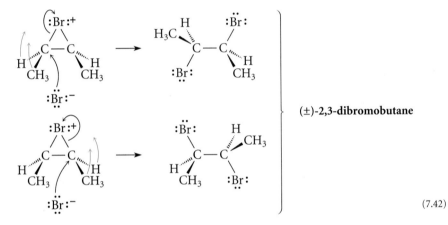

(±)-2,3-dibromobutane

(7.42)

Thus, formation of a bromonium ion followed by *backside attack* of bromide is a mechanism that accounts for the observed *anti* addition of Br_2 to alkenes. In general, when nucleophiles attack saturated carbon atoms, backside attack is always observed. (This idea is explored in more detail in Chapter 9.)

Might there be other mechanisms that would be consistent with the *anti* stereochemistry of bromine addition? Let's see what sort of prediction a carbocation mechanism makes about the stereochemistry of the reaction.

Imagine the addition of Br_2 to either face of *cis*-2-butene to give a carbocation intermediate.

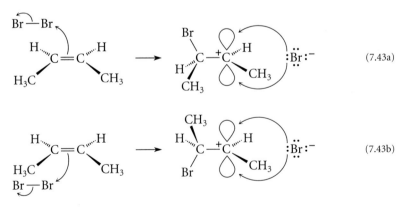

(7.43a)

(7.43b)

Attack of Br⁻ could occur at either face of this carbocation; this would give a mixture of *meso* and racemic diastereomers. Such a reaction would not be stereoselective. Because this result is not observed (Eq. 7.38), a carbocation mechanism is not in accord with the data. This mechanism also is not in accord with the observed absence of rearrangements in bromine addition. The bromonium-ion mechanism, however, accounts for the results in a direct and simple way. The credibility of this mechanism has been enhanced by the direct observation of bromonium ions under special conditions. In 1985, the structure of a bromonium ion was proved by X-ray crystallography.

Does the observation of *anti* stereochemistry *prove* the bromonium-ion mechanism? The answer is no. *No mechanism is ever proved.* Chemists deduce a mechanism by gathering as much information as possible about a reaction, such as its stereochemistry, presence and absence of rearrangements, etc., and ruling out all mechanisms that do not fit the experimental facts. If someone can think of another mechanism that explains the facts, then that mechanism is just as good until someone finds a way to decide between the two by a new experiment.

STUDY GUIDE LINK:
7.6
Stereoselective and
Stereospecific
Reactions

PROBLEM

7.25   *(a)   Assuming the operation of the bromonium-ion mechanism, give the structure of the product(s) (including all stereoisomers) expected from bromine addition to cyclohexene. (See Study Problem 7.4.)

   (b)   In view of the bromonium-ion mechanism, which of the products in your answer to Problem 7.24 are likely to be the major ones?

D. Stereochemistry of Hydroboration-Oxidation

Because hydroboration-oxidation involves two distinct reactions, its stereochemical outcome is a consequence of the stereochemistry of both reactions.

Hydroboration is a *syn* addition.

(7.44)

(both enantiomers formed)

Even though just one enantiomer of the product is shown, the product is racemic because the starting materials are achiral; Sec. 7.8A.

The *syn* addition of borane is a direct consequence of the concerted mechanism of the reaction.

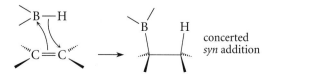

concerted
syn addition

(7.45a)

Occurrence of an *anti* addition by the same mechanism would be difficult, because it would require an abnormally long B—H bond to bridge opposite faces of the alkene π bond.

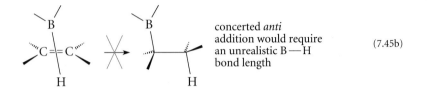

concerted *anti* addition would require an unrealistic B—H bond length   (7.45b)

The oxidation of organoboranes is a *substitution reaction* that occurs with *retention of stereochemical configuration.*

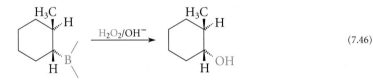

(7.46)

***trans*-2-methylcyclohexanol**

The results from Eqs. 7.44 and 7.46 taken together show that *hydroboration-oxidation of an alkene brings about the net syn addition of the elements of H—OH to the double bond.*

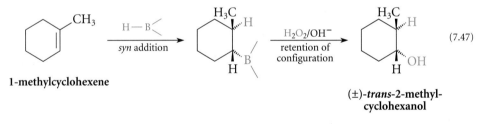

(7.47)

1-methylcyclohexene

(±)-*trans*-2-methyl-cyclohexanol

As far as is known, *all* hydroboration-oxidation reactions of alkenes are *syn* additions.

> Notice carefully that it is the —H and —OH that have been added in a *syn* manner. The *trans* designation in the name of the product of Eq. 7.47 has nothing to do with the groups that have added—it refers to the relationship of the methyl group, which was part of the alkene starting material, and the —OH group.

PROBLEMS

7.26   What products, including their stereochemistry, should be obtained when each of the following alkenes is subjected to hydroboration-oxidation? (D = deuterium = ^{2}H.)

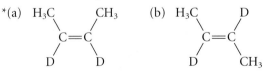

*7.27   Contrast the results in the previous problem with those to be expected when *cis-* and *trans*-2-butene (not isotopically labeled) are subjected to the same reaction conditions.

E. Stereochemistry of Other Addition Reactions

Catalytic Hydrogenation   Catalytic hydrogenation of most alkenes (Sec. 4.9A) is a *syn* addition. The following example is illustrative:

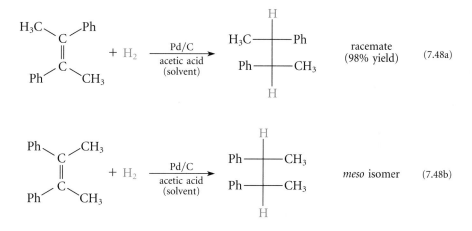

Be sure you understand why the results indicate a *syn* addition. Do this by using the procedure introduced in Study Problem 7.5. Construct a model of the *syn* addition product in each case and show that this model can be converted into the Fischer projection on the right. Results like these show that the two hydrogen atoms are delivered from the catalyst to the same face of the double bond. The stereoselectivity of catalytic hydrogenation is one reason that the reaction is so important in organic chemistry.

Oxymercuration–Reduction   Oxymercuration of alkenes (Sec. 5.3A) is typically a stereoselective *anti* addition.

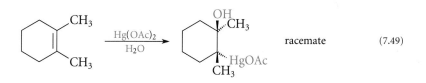

Since this reaction occurs by a cyclic-ion mechanism much like bromine addition, it should not be surprising that the stereochemical course of the reaction is the same. In the reaction of the mercury-containing product with $NaBH_4$, however, the mercury is replaced by hydrogen with *loss of stereochemical configuration*.

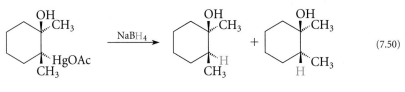

(7.50)

mixture of diastereomers

Hence, oxymercuration-reduction is *not* a stereoselective reaction. Despite its lack of stereoselectivity, the reaction is *regioselective* and is very useful in situations in which stereoselectivity is not an issue—that is, situations in which the carbons of the double bond in the alkene starting material are not converted into stereocenters as a result of the reaction.

STUDY GUIDE LINK:
✓7.7
When
Stereochemistry
Matters

Glycol Formation with OsO$_4$ or KMnO$_4$ The formation of glycols from alkenes with either alkaline KMnO$_4$ or OsO$_4$ is a stereoselective *syn* addition.

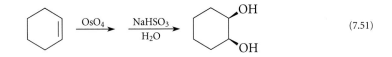

(7.51)

The mechanism of this reaction suggests a reason for this stereochemistry (Eq. 5.35a). The five-membered osmate ester ring is easily formed when the two oxygens are added from the same face of the double bond by a concerted mechanism. Hydrolysis of the osmate ester gives the glycol.

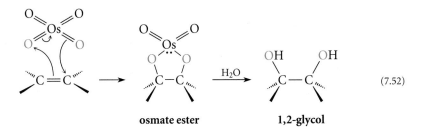

(7.52)

osmate ester   **1,2-glycol**

On the other hand, an *anti* addition by a concerted mechanism would be very difficult: the two oxygens of OsO$_4$ cannot simultaneously bridge the upper and lower faces of the π bond. *Syn* addition of the two —OH groups with KMnO$_4$ occurs for similar reasons.

PROBLEM

7.28   What products are formed (including their stereochemistry) when alkaline KMnO$_4$ reacts with each of the following alkenes?
*(a) 1-methylcyclopentene   (b) 1-ethyl-2-methylcyclohexene
*(c) *cis*-2-butene   (d) *trans*-2-butene

KEY IDEAS IN CHAPTER 7

- Except for cyclopropane, the cycloalkanes have puckered carbon skeletons.

- Of the cycloalkanes containing relatively small rings, cyclohexane has the greatest stability because it has no angle strain and it can adopt a conformation in which all bonds are eclipsed.

- The most stable conformation of cyclohexane is the chair conformation, in which hydrogens or substituent groups assume axial or equatorial positions. Cyclohexanes and substituted cyclohexanes undergo the chair flip, in which equatorial groups become axial, and vice-versa. The twist-boat conformation is a less stable conformation of cyclohexane derivatives. Twist-boat conformations are interconverted through boat transition states.

- A cyclohexane conformation with an axial substituent is typically less stable than a conformation with the same substituent in an equatorial position because of unfavorable van der Waals interactions (1,3-diaxial interactions) of the axial substituent and the two axial hydrogens on the same face of the ring. The quantitative analysis of the relative stabilities of various conformations is called conformational analysis.

- Cyclopentane exists in an envelope conformation. Cyclopentane has a greater heat of formation per CH_2 than cyclohexane because of eclipsing between hydrogen atoms.

- Cyclobutane and cyclopropane contain significant angle strain because their bonds are forced to deviate significantly from the ideal tetrahedral angle. Cyclopropane has bent carbon-carbon bonds. Cyclobutane and cyclopropane are the least stable cycloalkanes.

- Cycloalkanes can be represented by planar polygons in which the stereochemistry of substituents is indicated by dashed or solid wedges. Such structures are particularly useful for assessing the chirality of cyclic compounds.

- Bicyclic compounds contain two rings joined at two common atoms, called bridgehead atoms. If the bridgehead atoms are adjacent, the compound is a fused bicyclic compound; if the bridgehead atoms are not adjacent, the compound is a bridged bicyclic compound. Either *cis* or *trans* ring fusion is possible. *Trans* fusion, which avoids 1,3-diaxial interactions, is the most stable way to connect larger rings; *cis* fusion, which minimizes angle strain, is the most stable way to connect smaller rings. Polycyclic compounds contain many fused and/or bridged rings.

(continues)

🧪 *Trans* cycloalkenes containing rings with fewer than eight members are too unstable to exist under normal circumstances.

🧪 Bicyclic compounds consisting of small rings containing bridgehead double bonds are also unstable (Bredt's rule) because such compounds incorporate a highly twisted double bond.

🧪 Certain fundamental principles govern reactions involving stereoisomers.
1. A pair of enantiomers have identical reactivities unless the reaction conditions cause them to be involved in diastereomeric interactions (for example, chiral catalyst, chiral solvent, etc.).
2. Diastereomers in general have different reactivities.
3. Chiral products are always formed as racemates in a chemical reaction involving achiral starting materials unless the reaction conditions create diastereomeric interactions (for example, chiral catalyst, chiral solvent, etc.)
4. Diastereomeric products of chemical reactions are formed at different rates and in unequal amounts.

🧪 Addition reactions can occur with *syn* or *anti* stereochemistry. Substitution reactions can occur with retention or inversion of configuration. The stereochemistry of a reaction is determined from the stereochemistry of the product(s). Each carbon at which a chemical change occurs must be a stereocenter in the product in order for the stereochemistry of the reaction to be determined.

🧪 In a stereoselective reaction some stereoisomers of the product are formed to the exclusion of others.

🧪 Bromine addition to simple alkenes is a stereoselective *anti* addition in which the *syn* addition of one bromine to give a bromonium ion is followed by attack of bromide ion with inversion of configuration.

🧪 Hydroboration of alkenes is a stereoselective *syn* addition, and the subsequent oxidation of organoboranes is a substitution that occurs with retention of configuration. Thus, hydroboration-oxidation of alkenes is an overall *syn* addition of the elements of H—OH to alkenes.

🧪 Catalytic hydrogenation, as well as OsO_4 and $KMnO_4$ oxidations, are stereoselective *syn* additions. Oxymercuration-reduction is not stereoselective because the replacement of mercury with hydrogen occurs with mixed stereochemistry.

ADDITIONAL PROBLEMS

*7.29 Draw the structures of the following compounds.
(a) a bicyclic alkane with six carbon atoms
(b) (*S*)-4-cyclobutylcyclohexene
Name the compound whose structure you drew in (a).

7.30   Draw the structure of the following compounds.
(a) a bicyclic alkane with six carbon atoms other than the one you drew in Problem 7.29(a)
(b) (*R*)-3-ethylcyclobutene
Name the compound whose structure you drew in (a).

*7.31   Which of the following would distinguish (in principle) between methylcyclohexane and (*E*)-4-methyl-2-hexene?
(a) molecular mass determination
(b) uptake of H_2 in the presence of a catalyst
(c) reaction with alkaline $KMnO_4$
(d) determination of the empirical formula
(e) determination of the heat of formation
(f) enantiomeric resolution
(g) bromine addition with Br_2 in CCl_4

7.32   Draw the structure of each of the following molecules after it undergoes the chair flip.

*(a)   (b)

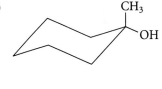

*(c)

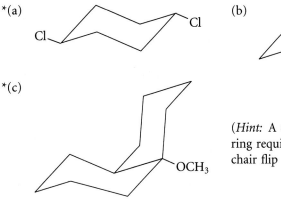

(*Hint*: A chair flip in one ring requires a simultaneous chair flip in the other.)

7.33   Draw a structure for each of the following compounds in its more stable chair conformation.

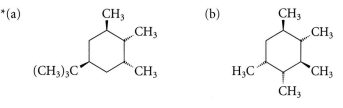

*(a)   (b)

7.34   *(a) Chlorocyclohexane contains 2.07 times as much of the equatorial form as the axial form at equilibrium at 25 °C. What is the standard free-energy difference between the two forms? Which is more stable?
(b) The standard free-energy difference between the two chair conformations of isopropylcyclohexane is 9.2 kJ/mol (2.2 kcal/mol). What is the ratio of concentrations of the two conformations at 25 °C?

7.35 Which of the following alcohols can be synthesized relatively free of constitutional isomers and diastereomers by

*(a) hydroboration-oxidation; (b) oxymercuration-reduction?

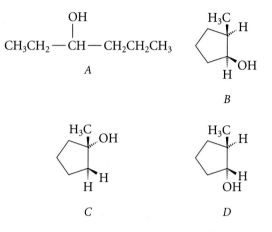

$$CH_3CH_2-\underset{\underset{A}{|}}{\overset{\overset{OH}{|}}{CH}}-CH_2CH_2CH_3$$

A

B

C *D*

7.36 For each of the following reactions, provide the following information.

(a) Give the structures of all products (including stereoisomers).

(b) If more than one product is formed, give the stereochemical relationship (if any) of each pair of products.

(c) If more than one product is formed, indicate which products are formed in identical amounts and which in different amounts.

(d) If more than one product is formed, indicate which products have different physical properties (melting point or boiling point).

*(1) $CH_3CH_2CH_2\underset{\underset{CH_2CH_3}{|}}{C}{=}CH_2$ + HBr $\longrightarrow$

(2) $(S)\text{-}CH_3CH_2\underset{\underset{CH_3}{|}}{CH}-\underset{\underset{CH_2CH_3}{|}}{C}{=}CH_2$ + HBr $\longrightarrow$

*(3) $(R)\text{-}CH_3CH_2\underset{\underset{CH_3}{|}}{CH}-\underset{\underset{CH_3}{|}}{C}{=}CH_2$ $\xrightarrow[\text{THF}]{BH_3}$ $\xrightarrow[\text{NaOH}]{H_2O_2}$

(4) $CH_3CH_2CH{=}CH_2$ + Br_2 $\xrightarrow{CH_2Cl_2}$

*(5) $(\pm)\text{-}CH_3\underset{\underset{Ph}{|}}{CH}CH{=}CH_2$ + Br_2 $\xrightarrow{CH_2Cl_2}$

(6) $(\pm)\text{-}CH_3CH_2\underset{\underset{CH_3}{|}}{CH}-\underset{\underset{CH_3}{|}}{C}{=}CH_2$ $\xrightarrow[\text{THF}]{BH_3}$ $\xrightarrow[\text{NaOH}]{H_2O_2}$

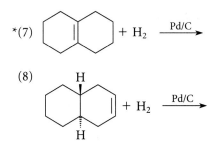

7.37 Draw the structures of the following compounds. (Some parts may have more than one correct answer.)

*(a) a dimethylcyclohexane with two identical chair conformations

(b) an achiral dimethylcyclohexane with two chair forms that are conformational diastereomers

*(c) a chiral dimethylcyclohexane with two chair forms that are conformational diastereomers

(d) a dimethylcyclohexane with chair forms that are conformational enantiomers

7.38 *(a) Draw a conformational representation of the following steroid. Show the α- and β-faces of the steroid, and label the angular methyl groups.

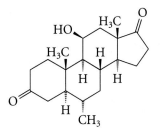

(b) Draw the two chair forms of the sugar α-(+)-glucopyranose, one form of the sugar glucose. Which of these two forms is the major one at equilibrium?

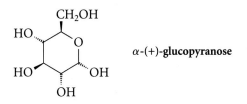

α-(+)-glucopyranose

7.39 From your knowledge of the mechanism of bromine addition to alkenes, give the structure and stereochemistry of the product(s) expected when

*(a) (3R,5R)-3,5-dimethylcyclopentene adds bromine;

(b) cyclopentene reacts with bromine in the presence of water. (*Hint:* See Sec. 5.1B.)

*7.40   *Anti* addition of bromine to the following bicyclic alkene gives two separable dibromides. Suggest structures for each. (Remember that *trans*-decalin derivatives cannot undergo the chair flip.)

7.41   When 1,4-cyclohexadiene reacts with two equivalents of Br_2, a solid that melts at 174–178° is formed. The wide melting range indicates a mixture of compounds. If this solid is dissolved in $CHCl_3$ and the solvent is slowly evaporated, two different types of crystals separate. Each type melts sharply, one at 188° and the other at 218°. If the crystals are mixed and melted together, then cooled, the solid that melts at 174–178° is formed again. Account for these observations.

*7.42   An optically active compound *A* with molecular formula C_8H_{14} undergoes catalytic hydrogenation to give an optically inactive product. Which of the following structures for *A* is(are) consistent with all the data?

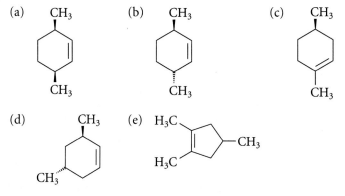

*7.43   State whether you would expect each of the following properties to be identical or different for the two enantiomers of 2-pentanol. Explain.

$$\underset{\textbf{2-pentanol}}{CH_3-\overset{\overset{\displaystyle OH}{|}}{CH}-CH_2CH_2CH_3}$$

(a) boiling point   (b) melting point
(c) rotation of the plane of polarized light   (d) solubility in hexane
(e) density
(f) solubility in (S)-3-methylhexane
(g) dipole moment
(h) taste (*Hint:* Your taste buds are chiral.)

*7.44   Draw a chair conformation for 3-methylpiperidine showing the sp^3 orbital that contains the nitrogen unshared electron pair. How many chair conformations of this compound are in rapid equilibrium? (*Hint:* See Sec. 6.10B.)

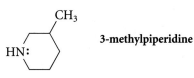

3-methylpiperidine

7.45 Which of the following compounds could be resolved into enantiomers at room temperature? Explain.

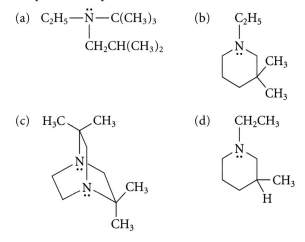

★7.46 Explain why 1-methylaziridine undergoes amine inversion much more slowly than 1-methylpyrrolidine. (*Hint:* What are the hybridization and bond angles at nitrogen in the transition state for inversion?)

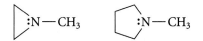

1-methylaziridine   1-methylpyrrolidine

★7.47 An alkene *A* of unknown stereochemistry was oxidized with alkaline $KMnO_4$ and the *meso* stereoisomer of the product was obtained:

$$CH_3(CH_2)_7CH{=}CH(CH_2)_7CH_3 \xrightarrow[OH^-]{KMnO_4} CH_3(CH_2)_7\overset{\underset{|}{OH}}{CH}{-}\overset{\underset{|}{OH}}{CH}(CH_2)_7CH_3$$

A *meso*

What stereoisomer of *A* was used in the reaction?

7.48 For each of the following alkenes, state whether a reaction with OsO_4 followed by aqueous $NaHSO_3$ will give a racemic mixture of products that can (in principle) be resolved into enantiomers under ordinary conditions.
(a) ethylene   (b) *cis*-2-butene   (c) *trans*-2-butene
(d) *cis*-2-pentene   (e) cyclopentene

*7.49   When fumarate reacts with D_2O in the presence of the enzyme *fumarase* (Sec. 4.9C and 7.8A), only one stereoisomer of deuterated malate is formed.

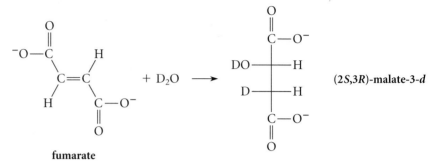

fumarate

Is this a *syn* or an *anti* addition? Explain.

*7.50   By answering the following questions, indicate the relationship between the two structures in each of the pairs below. Are they chair conformations of the same molecule? If so, are they conformational diastereomers, conformational enantiomers, or identical? If not, what is their stereochemical relationship? (*Hint:* Use planar structures to help you.)

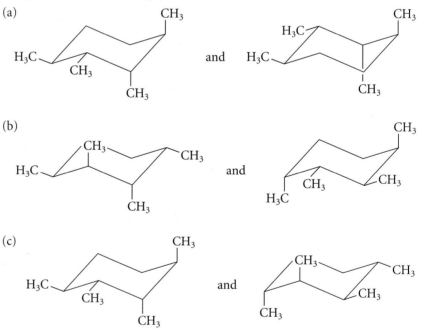

(a)

(b)

(c)

*7.51   When 1-methylcyclohexene is hydrated in D_2O, the product is a mixture of diastereomers; the hydration is thus *not* a stereoselective reaction.

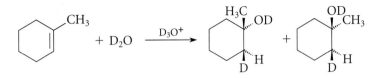

both compounds are racemates

(a) Show why the accepted mechanism for this reaction is consistent with these stereochemical results.

(b) Why must D_2O (rather than H_2O) be used to investigate the stereoselectivity of this addition?

(c) What isotopic substitution could be made in the starting material 1-methylcyclohexene that would allow investigation of the stereoselectivity of this addition with H_2O?

*7.52   Consider the following compound.

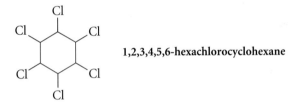

1,2,3,4,5,6-hexachlorocyclohexane

(a) Of the nine stereoisomers of this compound, only two can be isolated in optically active form under ordinary conditions. Give the structures of these enantiomers.

(b) Give the structures of the two stereoisomers that each have two identical chair conformations.

7.53   Give the structure of every stereoisomer of 1,2,3-trimethylcyclohexane. Label the enantiomeric pairs and show the plane of symmetry in each of the achiral stereoisomers.

*7.54   Which of the following statements about *cis*- and *trans*-decalin are true? Explain your answers.

(a) They are different conformations of the same molecule.

(b) They are constitutional isomers.

(c) They are diastereomers.

(d) At least one chemical bond would have to be broken in order to convert one into the other.

(e) They are enantiomers.

(f) They interconvert rapidly.

*7.55   Which *one* of the following compounds would be most likely to exist with one of its cyclohexane rings in a twist-boat conformation? Explain.

(*Problem 7.55 continues*)

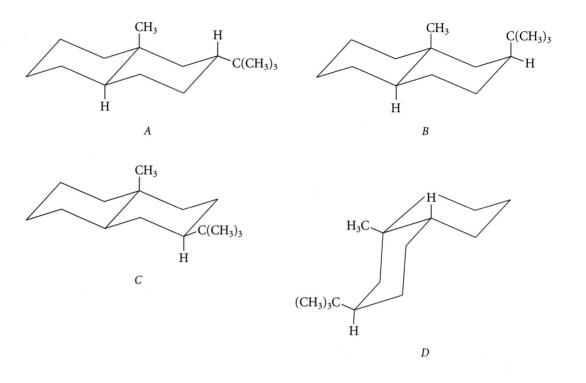

*7.56   The claim has been made that the more stable conformation of the following compound is *B*, in which one ring assumes a twist-boat conformation. Using models if necessary, explain why such a claim is reasonable.

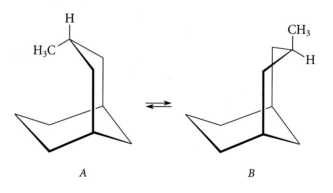

7.57   It has been argued that the energy difference between *cis*- and *trans*-1,3-di-*tert*-butylcyclohexane is a good approximation for the energy difference between the chair and twist-boat forms of cyclohexane. Using models to assist you, explain why this view is reasonable.

*7.58   Rank the following compounds according to their heats of formation, lowest first, and estimate the $\Delta H°$ difference between each pair.

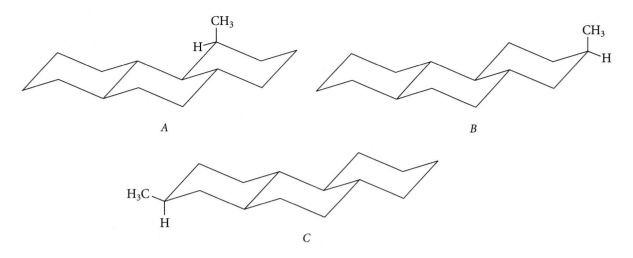

A

B

C

*7.59   When (2*R*,5*R*)-2,5-dimethylborolane (structure below) is used to hydroborate *cis*-2-butene, and the product borane is treated with alkaline H$_2$O$_2$, mostly a single enantiomer of the product alcohol is formed. What is the absolute configuration of this alcohol? Explain why the other enantiomer is not formed. (*Hint:* Build models of the borane and the alkenes. Let the borane model approach the alkene model from one face of the π bond, then the other. Decide which reaction is preferred by analyzing van der Waals repulsions in the transition state in each case.)

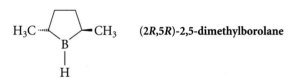

(2*R*,5*R*)-2,5-dimethylborolane

7.60   (a) What two diastereomeric products could be formed in the hydroboration-oxidation of the following alkene?

(b) Considering the effect of the methyl group on the approach of the borane-THF reagent to the double bond, suggest which of the two products in (a) should be the major product.

*7.61   Rank the compounds within each of the following sets according to their heats of formation, lowest first. Explain.

(a)

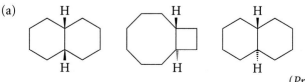

(*Problem 7.61 continues*)

(b)

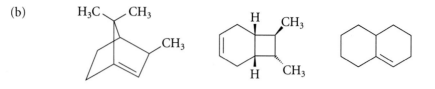

*7.62 (a) The $\Delta G°$ for the following equilibrium is 8.4 kJ/mol (2.0 kcal/mol). (Conformation A has lower energy.)

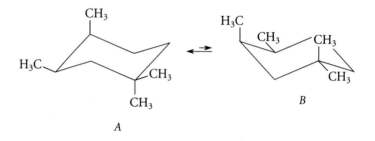

Use this information to estimate the energy cost of a 1,3-diaxial interaction between two methyl groups:

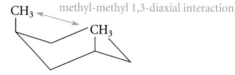

(b) Using the result in (a), estimate the $\Delta G°$ for the following equilibrium:

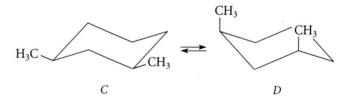

7.63 The $\Delta G°$ for the following equilibrium is 4.73 kJ/mol (1.13 kcal/mol). (The equilibrium favors conformation A.)

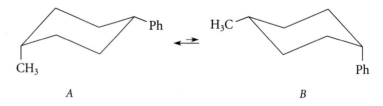

(a) Which behaves as if it is larger, methyl or phenyl (Ph)? Why is this reasonable?

(b) Use the $\Delta G°$ given above along with any other appropriate data to estimate $\Delta G°$ for the following two equilibria:

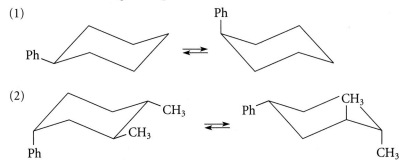

(1)

(2)

*7.64 (a) In how many stereochemically different ways can the two rings in a *bridged* bicyclic compound be joined?

(b) For which one of the following bridged bicyclic compounds are all such stereoisomers likely to be stable enough to isolate? Explain.

(1) bicyclo[2.2.2]octane

(2) bicyclo[25.25.25]heptaheptacontane

(A heptaheptacontane has 77 carbons.)

8

Introduction to Alkyl Halides, Alcohols, Ethers, Thiols, and Sulfides

This chapter covers the nomenclature and properties of several classes of compounds. They are considered together because their chemical reactions are closely related.

In an **alkyl halide**, a halogen atom is bonded to the carbon of an alkyl group. Alkyl halides are classified as methyl, primary, secondary, or tertiary, depending on the number of alkyl groups (color in the structures below) attached to the *carbon bearing the halogen.* A methyl halide has no alkyl groups, a primary halide has one, a secondary halide has two, and a tertiary halide has three.

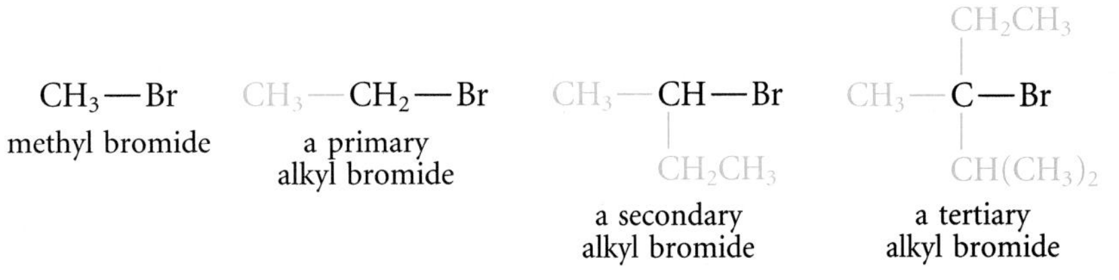

$$CH_3 \!-\! Br \qquad CH_3 \!-\! CH_2 \!-\! Br \qquad CH_3 \!-\! \underset{\underset{CH_2CH_3}{|}}{CH} \!-\! Br \qquad CH_3 \!-\! \underset{\underset{CH(CH_3)_2}{|}}{\overset{\overset{CH_2CH_3}{|}}{C}} \!-\! Br$$

methyl bromide a primary alkyl bromide a secondary alkyl bromide a tertiary alkyl bromide

In an **alcohol**, a **hydroxy group**, —OH, is bonded to the carbon of an alkyl group. Alcohols, too, are classified as methyl, primary, secondary, or tertiary.

$$CH_3 \!-\! OH \qquad CH_3 \!-\! CH_2 \!-\! OH \qquad CH_3 \!-\! \underset{\underset{CH_2CH_3}{|}}{CH} \!-\! OH \qquad CH_3 \!-\! \underset{\underset{CH(CH_3)_2}{|}}{\overset{\overset{CH_2CH_3}{|}}{C}} \!-\! OH$$

methyl alcohol a primary alcohol a secondary alcohol a tertiary alcohol

335

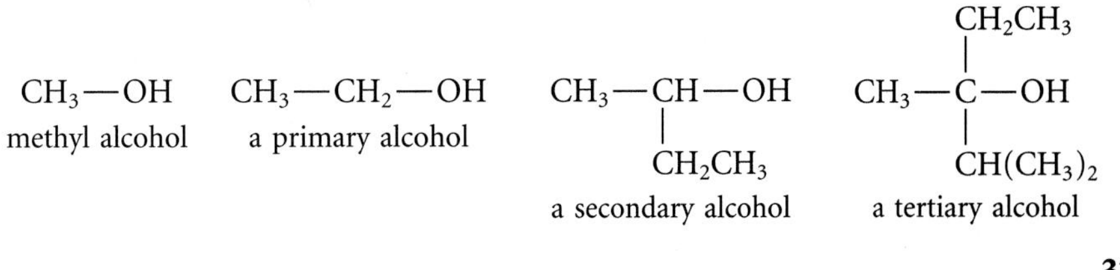

Compounds that contain two or more hydroxy groups are called **glycols**. If the hydroxy groups are on adjacent carbons, the compound is a **vicinal glycol**, the most important type of glycol.

$$HO-CH_2-CH_2-OH \qquad \text{ethylene glycol (a vicinal glycol)}$$

Thiols, sometimes called **mercaptans**, are the sulfur analogs of alcohols. In a thiol, a **sulfhydryl group**, —SH, also called a **mercapto group**, is bonded to an alkyl group. An example of a thiol is ethanethiol (ethyl mercaptan), CH_3CH_2-SH.

In an **ether**, an oxygen is bonded to two carbon groups, which may or may not be the same. A **thioether**, or **sulfide**, is the sulfur analog of an ether.

$$CH_3CH_2-O-CH_2CH_3 \qquad CH_3CH_2-O-CH(CH_3)_2 \qquad CH_3-S-CH_2CH_3$$

diethyl ether	**ethyl isopropyl ether**	**ethyl methyl sulfide**

The introduction to the functional groups in this chapter will be followed by subsequent chapters that describe in turn the chemistry of each group.

8.1   Nomenclature

Several systems are recognized by the IUPAC for the nomenclature of organic compounds. **Substitutive nomenclature**, the most broadly applicable system, was introduced in the nomenclature of both alkanes (Sec. 2.4C) and alkenes (Sec. 4.2A), and will be applied to the compound classes in this chapter as well. Another widely used system that will be introduced in this chapter is called **radicofunctional nomenclature** by the IUPAC; for simplicity, this system will be called **common nomenclature**. Common nomenclature is generally used only for the simplest and most common compounds. Although the adoption of a single nomenclature system might seem desirable, historical usage and other factors have dictated the use of both common and substitutive names.

A. Nomenclature of Alkyl Halides

Common Nomenclature   The common name of an alkyl halide is constructed from the name of the alkyl group (see Table 2.2) followed by the name of the halide as a separate word.

STUDY GUIDE LINK:
✓8.1
*Common
Nomenclature*

$$CH_3CH_2-Cl \qquad\qquad CH_2Cl_2$$

ethyl chloride	**methylene chloride**
	($-CH_2-$ group = methylene group)

$$CH_3CH_2CH_2CH_2-Br \qquad (CH_3)_2CH-I$$

butyl bromide	**isopropyl iodide**

The common names of the following compounds should be learned.

$$CH_2=CH-CH_2-Cl \qquad Ph-CH_2-Br \qquad CH_2=CH-Cl \qquad CCl_4$$

allyl chloride   **benzyl bromide**   **vinyl chloride**   **carbon tetrachloride**

(Compounds with halogens attached to alkene carbons, such as vinyl chloride, are not alkyl halides, but it is convenient to discuss their nomenclature here.)

The **allyl group**, as the name above implies, is the $CH_2=CH-CH_2-$ group. This should not be confused with the **vinyl group**, $CH_2=CH-$, which lacks the additional $-CH_2-$. Similarly, the **benzyl group**, $Ph-CH_2-$, should not be confused with the **phenyl group**.

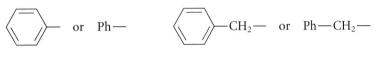

phenyl group   **benzyl group**

The **haloforms** are the methyl trihalides.

$$HCCl_3 \qquad HCBr_3 \qquad HCI_3$$

chloroform   **bromoform**   **iodoform**

Substitutive Nomenclature The IUPAC substitutive name of an alkyl halide is constructed by applying the rules of alkane and alkene nomenclature (Sec. 2.4C, 4.2A). Halogens are always treated as substituents; the halogen substituents are named fluoro, chloro, bromo, or iodo, respectively.

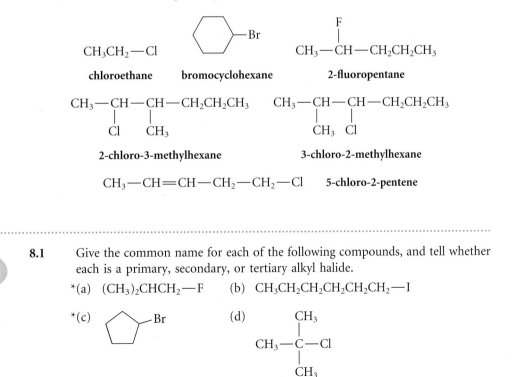

PROBLEMS

8.1 Give the common name for each of the following compounds, and tell whether each is a primary, secondary, or tertiary alkyl halide.

*(a) $(CH_3)_2CHCH_2-F$ (b) $CH_3CH_2CH_2CH_2CH_2CH_2-I$

*(c) (cyclopentyl)—Br   (d)
$$CH_3-\underset{\underset{CH_3}{|}}{\overset{\overset{CH_3}{|}}{C}}-Cl$$

8.2   Give the structure of each of the following compounds.

*(a) 2,2-dichloro-5-methylhexane   (b) chlorocyclopropane

*(c) 6-bromo-1-chloro-3-methylcyclohexene (d) methylene iodide

8.3   Give the substitutive name for each of the following compounds.

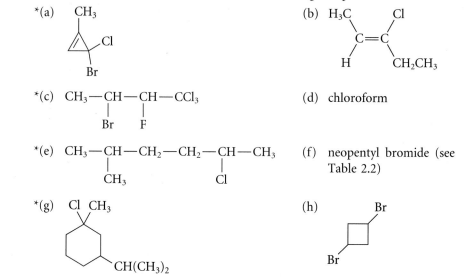

*(c) CH_3—CH—CH—CCl_3 (d) chloroform
             | |
             Br   F

*(e) CH_3—CH—CH_2—CH_2—CH—CH_3 (f) neopentyl bromide (see
             | | Table 2.2)
             CH_3 Cl

B. Nomenclature of Alcohols and Thiols

Common Nomenclature   The common name of an alcohol is derived by specifying the alkyl group to which the —OH group is attached, followed by the separate word *alcohol*.

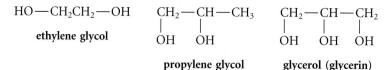

CH_3—OH $(CH_3)_2$CH—OH   ⬡—OH $CH_3CH_2CH_2$—OH

methyl alcohol   **isopropyl alcohol**   **propyl alcohol**

cyclohexyl alcohol

CH_2=CH—CH_2—OH   Ph—CH_2—OH

allyl alcohol   **benzyl alcohol**

A few glycols have important traditional names.

HO—CH_2CH_2—OH CH_2—CH—CH_3 CH_2—CH—CH_2
                              | | | | |
ethylene glycol   OH   OH   OH   OH   OH

                         propylene glycol   **glycerol (glycerin)**

Thiols are named in the common system as *mercaptans*; this name, which means "captures mercury," comes from the fact that thiols readily form heavy-metal derivatives (Sec. 8.5A).

CH_3CH_2—SH   **ethyl mercaptan**

Substitutive Nomenclature   The substitutive nomenclature of alcohols and thiols involves an important concept of nomenclature called the **principal group**. The principal group is the chemical group on which the name is based, *and is always cited as a suffix in the name.* For example, in a simple alcohol, the —OH group is the principal group, and its suffix is *ol.* The name of an alcohol is constructed by dropping the final *e* from the name of the parent alkane and adding this suffix.

$$CH_3—CH_2—OH \qquad \text{ethan}\cancel{e} + ol = \textbf{ethanol}$$

(The final *e* is generally dropped when the suffix begins with a vowel and retained otherwise.)

For simple thiols, the —SH group is the principal group, and its suffix is *thiol.* The name is constructed by adding this suffix to the name of the parent alkane. Note that because the suffix begins with a consonant, the final *e* of the alkane name is retained.

$$CH_3—CH_2—SH \qquad \textbf{ethanethiol}$$

Only certain groups are cited as principal groups. The —OH and —SH groups are the only ones in the compound classes considered so far, but others will be added in later chapters. If a compound does not contain a principal group, it is named as a substituted hydrocarbon in the manner illustrated for the alkyl halides in the previous section.

The *principal group* and the *principal chain* are the key concepts used in the construction of a substitutive name according to the *general rules for substitutive nomenclature of organic compounds*, which follow. The simplest way to learn these rules is to skip to the study problems and examples that follow, and let them guide you through the application of the rules to specific cases.

1. *Identify the principal group.*
   When a structure has several candidates for the principal group, the group chosen is the one given the highest priority by the IUPAC. The IUPAC specifies that the —OH group receives precedence over an —SH group:

   $$\textit{Priority as principal group:} \quad —OH > —SH \qquad \text{(8.1)}$$

   (The complete list of principal groups and their relative priorities are summarized in Appendix I.)

2. *Identify the principal carbon chain.*
   The principal chain is the carbon chain on which the name is based (Sec. 2.4C). The principal chain is identified by applying the following criteria *in order* until a decision can be made:
   a. *greatest number of principal groups;*
   b. *greatest number of double and triple bonds;*
   c. *greatest length;*
   d. *greatest number of other substituents.*
   These criteria cover most of the cases you'll encounter. A more extensive list is given in Appendix I.

3. *Number the principal chain consecutively from one end.*
   In numbering the principal chain, apply the following criteria *in order* until there is no ambiguity:
   a. *lowest numbers for the principal groups;*
   b. *lowest numbers for multiple bonds, with double bonds having priority over triple bonds;*
   c. *lowest numbers for other substituents;*
   d. *lowest number for the substituent cited first in the name.*

4. *Begin construction of the name with the name of the hydrocarbon corresponding to the principal chain.*
   a. *Cite the principal group by its suffix and number; its number is the last one cited in the name.* (See the examples below.)
   b. *If there is no principal group, name the compound as a substituted hydrocarbon* (Sec. 2.4C, 4.2A).
   c. *Cite the names and numbers of the other substituents in alphabetical order at the beginning of the name.*

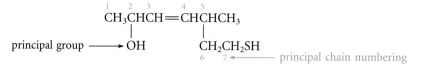

STUDY PROBLEM 8.1

Provide an IUPAC substitutive name for each of the following compounds.

(a) $CH_3CH_2CHCH_3$
         |
         OH

(b) $CH_3CHCH=CHCHCH_3$
         | |
         OH CH_2CH_2SH

(handwritten: 3 heptlanol, 5 methyl, 7 mercapto)

(handwritten left margin: butanol)

Solution   *(handwritten: 3 hepten 2 ol)*

(a) Applying Rule 1, the principal group is the —OH group. Because there is only one possibility for the principal chain, Rule 2 does not enter the picture. Applying Rule 3, the principal group is located at carbon-2. Applying Rule 4a, the name is based on the four-carbon hydrocarbon, butane. After dropping the final *e* and adding the suffix *-ol*, the name is obtained: 2-butanol.

$$\overset{4}{C}H_3\overset{3}{C}H_2\,\overset{2}{C}H\overset{1}{C}H_3$$
$$|$$
$$OH$$

2-butanol

(b) Applying Rule 1, the principal group is again the —OH group, because —OH has precedence over —SH. Applying Rule 2, the principal chain is the longest one containing the —OH group, and therefore has seven carbons. Numbering the principal chain in accord with Rule 3a gives the —OH group the lowest number at carbon-2 and a double bond at carbon-3:

$$\overset{1}{C}H_3\overset{2}{C}H\overset{3}{C}H=\overset{4}{C}H\overset{5}{C}HCH_3$$

principal group ⟶ OH CH_2CH_2SH
                                              6   7 ⟵ principal chain numbering

Applying Rule 4a, the parent hydrocarbon is 3-heptene, from which we drop the final *e* and add the suffix *-ol*, to give 3-hepten-2-ol as the final part of the name. (Notice that because we have to cite the number of the double bond, the number for the —OH principal group is located before the final suffix *-ol*.) Applying Rule 4c, the methyl group at carbon-5 and the —SH group at carbon-7 are cited as ordinary substituents. (The substituent name of the —SH group is the *mercapto* group.) The name is

substituent numbers; note alphabetical
citation of substituents

7-mercapto-5-methyl-3-hepten-2-ol

number of the principal group

number of the double bond

To name an alcohol containing more than one —OH group, the suffixes *-diol*, *-triol*, etc., are added to the name of the appropriate alkane *without* dropping the final *e*.

$$\overset{1}{CH_3}-\overset{2}{CH}-\overset{3}{CH}-\overset{4}{CH_2}-\overset{5}{CH_3}$$
$$\quad\;\; |\quad\;\; |$$
$$\quad\;\; OH\;\; OH$$

2,3-pentanediol

Name the following compound.

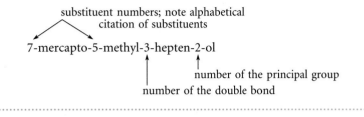

4 cyclo hexene

4 cyclohene 1,3 diol

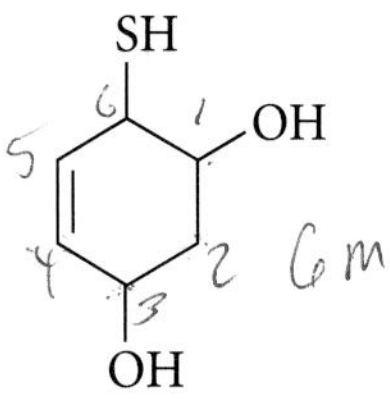

2 6mercapto

6mercapto 4cyclohexen 1,3diol

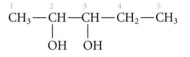

Solution Applying Rule 1, the principal groups are the —OH groups. By Rule 3a these are given numerical precedence; thus they receive the numbers 1 and 3. There are two numbering schemes that give these groups the numbers 1 and 3; choose the scheme that gives the double bond the lower number, by Rule 3b. Applying Rule 4a, the parent hydrocarbon is cyclohexene, and since the suffix is *-diol*, the final *e* is retained to give 4-cyclohexene-1,3-diol. Finally, notice that because the —SH group has been eliminated from consideration as the principal group, it is treated as an ordinary substituent group by Rule 4c. The completed name is thus

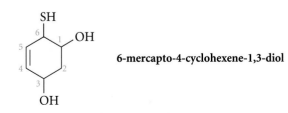

6-mercapto-4-cyclohexene-1,3-diol

Common and substitutive nomenclature should not be mixed. The following compounds are frequently named incorrectly:

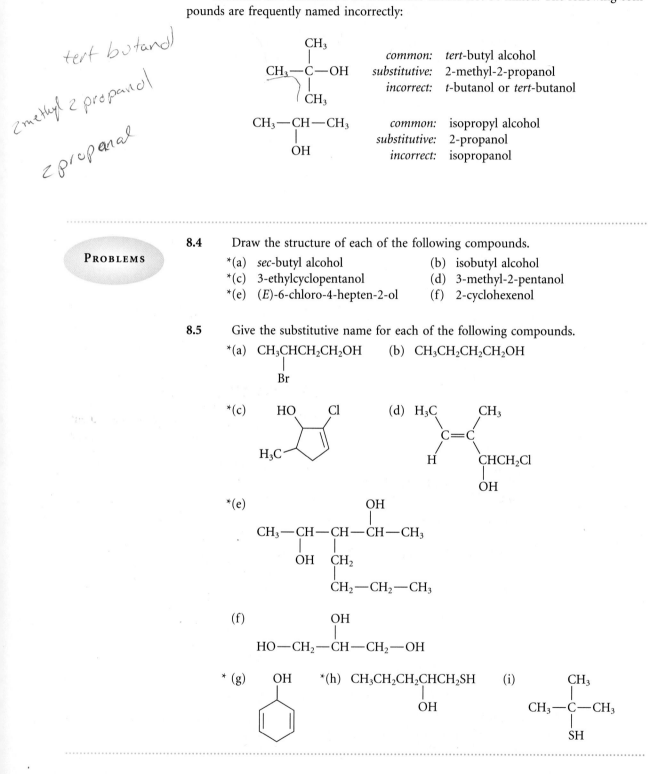

tert butanol

2methyl 2 propanol

2 propanol

common:	*tert*-butyl alcohol
substitutive:	2-methyl-2-propanol
incorrect:	*t*-butanol or *tert*-butanol

common:	isopropyl alcohol
substitutive:	2-propanol
incorrect:	isopropanol

PROBLEMS

8.4   Draw the structure of each of the following compounds.

*(a) *sec*-butyl alcohol   (b) isobutyl alcohol
*(c) 3-ethylcyclopentanol   (d) 3-methyl-2-pentanol
*(e) (*E*)-6-chloro-4-hepten-2-ol   (f) 2-cyclohexenol

8.5   Give the substitutive name for each of the following compounds.

*(a) CH₃CHCH₂CH₂OH   (b) CH₃CH₂CH₂CH₂OH
            |
            Br

*(c)   (d)

*(e)

(f)

* (g)   *(h) CH₃CH₂CH₂CHCH₂SH   (i)
                                         |
                                        OH

C. Nomenclature of Ethers and Sulfides

Common Nomenclature The common name of an ether is constructed by citing as separate words the two groups attached to the ether oxygen in alphabetical order, followed by the word *ether*.

$$CH_3CH_2-O-CH_2CH_3 \qquad CH_3-O-C_2H_5$$

diethyl ether   **ethyl methyl ether**
(also called ethyl ether or
simply ether)

A sulfide is named in a similar manner, using the word *sulfide*. (In older literature, the word *thioether* was also used.)

$$CH_3CH_2-S-CH_3 \qquad (CH_3)_2CH-S-CH(CH_3)_2$$

ethyl methyl sulfide   **diisopropyl sulfide**
(also ethyl methyl thioether)

Substitutive Nomenclature In substitutive nomenclature, ethers and sulfides are never cited as principal groups. *Alkoxy groups* (RO—) and *alkylthio groups* (RS—) are always cited as substituents.

ethoxy substituent ⟶ CH_3CH_2O CH_3 ⟵ methyl substituent

principal chain ⟶ $CH_3CHCH_2CH_2CHCH_3$ 2-ethoxy-5-methylhexane

In this example, the principal chain is a six-carbon chain. Hence, the compound is named as a hexane, and the C_2H_5O— group and the methyl group are treated as substituents. The C_2H_5O— group is named by dropping the final *yl* from the name of the alkyl group and adding the suffix *-oxy*. Thus, the C_2H_5O— group is the (ethy̶l̶ + oxy) = ethoxy group. The numbering follows from nomenclature Rule 3d, Sec. 8.1B.

The nomenclature of sulfides is similar. An RS— group is named by adding the suffix *-thio* to the name of the R group; the final *yl* is not dropped.

substituent ⟶ SCH_3

principal chain ⟶ $CH_3CHCH_2CH_2CH_2CH_3$ 2-(methylthio)hexane

                                  1   2   3   4   5   6

The parentheses in the name are used to indicate that "thio" is associated with "methyl" rather than with "hexane."

STUDY PROBLEM 8.3

Name the following compound.

$$CH_3CH_2CH_2CH_2-O-CH_2CH_2CH_2-OH$$

Solution The —OH group is cited as the principal group, and the principal chain is the chain containing this group. Consequently, the $CH_3CH_2CH_2CH_2O$— group is cited as a butoxy substituent (buty̶l̶ + oxy) at carbon-3 of the principal chain:

$$CH_3CH_2CH_2CH_2-O-\underbrace{\overset{3}{CH_2}\overset{2}{CH_2}\overset{1}{CH_2}}-OH \qquad \textbf{3-butoxy-1-propanol}$$

principal chain
(contains principal group —OH)

Heterocyclic Nomenclature

A number of important ethers and sulfides contain an oxygen or sulfur atom within a ring. Cyclic compounds with rings that contain more than one type of atom are called **heterocyclic compounds**. The names of some common heterocyclic ethers and sulfides should be learned.

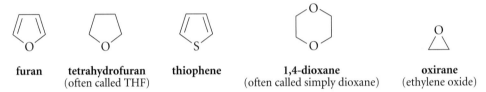

furan   **tetrahydrofuran**   **thiophene**   **1,4-dioxane**   **oxirane**
(often called THF)   (often called simply dioxane)   (ethylene oxide)

(The IUPAC name for tetrahydrofuran, *oxolane*, is not commonly used except in indexes such as *Chemical Abstracts*.)

Oxirane is the parent compound of a special class of heterocyclic ethers, called **epoxides**, that contain three-membered rings. A few epoxides are named traditionally as oxides of the corresponding alkenes:

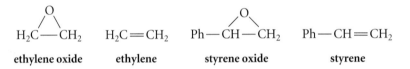

ethylene oxide   **ethylene**   **styrene oxide**   **styrene**

However, most epoxides are named substitutively as derivatives of oxirane. The atoms of the epoxide ring are numbered consecutively with the oxygen receiving the number 1 *regardless of the substituents present.*

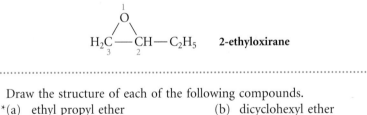

$$\underset{3}{H_2C}-\underset{2}{CH}-C_2H_5 \qquad \textbf{2-ethyloxirane}$$

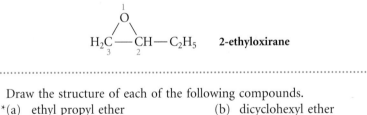

PROBLEMS

8.6   Draw the structure of each of the following compounds.
⋆(a) ethyl propyl ether   (b) dicyclohexyl ether
⋆(c) dicyclopentyl sulfide   (d) *tert*-butyl isopropyl sulfide
⋆(e) allyl benzyl ether   (f) phenyl vinyl ether
⋆(g) (2R,3R)-2,3-dimethyloxirane   ⋆(h) 5-(ethylthio)-2-methylheptane

8.7   Give a substitutive name for each of the following compounds.
⋆(a) $CH_3CH_2-O-CH_2CH_2-OH$ (b) $(CH_3)_3C-O-CH_3$

⋆(c) CH_3OCH_2 ... (structure)   ⋆(d) (structure)

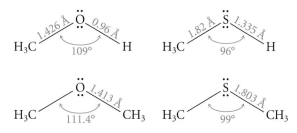

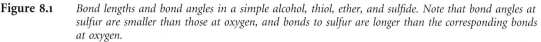

Figure 8.1 *Bond lengths and bond angles in a simple alcohol, thiol, ether, and sulfide. Note that bond angles at sulfur are smaller than those at oxygen, and bonds to sulfur are longer than the corresponding bonds at oxygen.*

 ## Structures

In all the compounds covered in this chapter, the bond angles at carbon are very nearly tetrahedral. For example, in the simple methyl derivatives—the methyl halides, methanol, methanethiol, dimethyl ether, and dimethyl sulfide—the H—C—H bond angle in the methyl group does not deviate more than a degree or so from 109.5°. In an alcohol, thiol, ether, or sulfide the bond angle at oxygen or sulfur further defines the shape of the molecule. You learned in Sec. 1.3B that the shapes of such molecules can be predicted by thinking of an unshared electron pair as a bond without an atom at the end. This means that the oxygen or sulfur has four "groups": two electron pairs and two alkyl groups or hydrogens. These molecules are therefore bent at oxygen and sulfur, as you can see from the structures in Fig. 8.1. The angle at sulfur is generally found to be closer to 90° than the angle at oxygen. One reason for this trend is that the unshared electron pairs on sulfur occupy orbitals derived from quantum level 3 that take up more space than those on oxygen, which are derived from level 2. The repulsion between these unshared pairs and the electrons in the chemical bonds forces the bonds closer together than they are on oxygen.

The lengths of bonds between carbon and other atoms follow the trends discussed in Sec. 1.3B. Within a column of the periodic table, bonds to atoms of higher atomic number are longer. Thus, the C—S bond of methanethiol is longer than the C—O bond of methanol (Fig. 8.1, Table 8.1). Within a row, bond lengths decrease toward higher atomic number (that is, to the right). Thus, the C—O bond in methanol is longer

Table 8.1   Bond Lengths (in Angstroms) in Some Methyl Derivatives

CH_3—CH_3	CH_3—NH_2	CH_3—OH	CH_3—F
1.536	1.474	1.426	1.391
		CH_3—SH	CH_3—Cl
		1.82	1.781
	— Increasing electronegativity →		CH_3—Br
	|		1.939
	Increasing atomic radius		CH_3—I
	↓		2.129

than the C—F bond in methyl fluoride (Table 8.1); the C—S bond in methanethiol is longer than the C—Cl bond in methyl chloride.

PROBLEM

*8.8   Using the data in Table 8.1, estimate the carbon-selenium bond length in CH_3—Se—CH_3.

8.3   Effect of Molecular Polarity and Hydrogen Bonding on Physical Properties

A. Boiling Points of Ethers and Alkyl Halides

Most alkyl halides, alcohols, and ethers are *polar molecules*; that is, they have permanent dipole moments (Sec. 1.2D). The following examples are typical.

	CH_3—F	CH_3—Cl	CH_3—O—CH_3	CH_3—OH
dipole moment	1.82 D	1.94 D	1.31 D	1.7 D

The polarity of a compound affects its boiling point. When the boiling points of two molecules with the same shape and molecular mass are compared, in many cases the more polar molecule has the higher boiling point.

dipole moment	1.31 D	≈0
boiling point	−23.7°	−42.1°

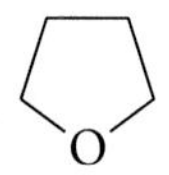

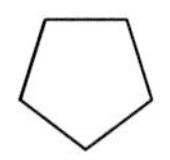

dipole moment	1.7 D	0 D
boiling point	66°	49.3°

What is the reason for this effect? A higher boiling point results from *greater attraction between molecules in the liquid state* (Sec. 2.6A). Polar molecules are attracted to each other because they can align in such a way that the negative end of one dipole is attracted to the positive end of the other.

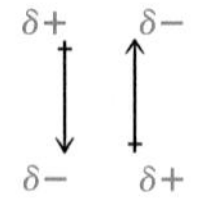

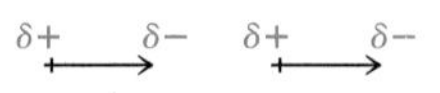

two ways in which dipoles can align attractively

Of course, since molecules in the liquid state are in constant motion, their relative positions are changing constantly; however, on the average, this attraction exists and serves to raise the boiling point of a polar compound.

When a polar molecule contains a hydrocarbon portion of even moderate size its polarity has little effect on its physical properties; it is sufficiently alkanelike that its properties resemble those of an alkane.

	$CH_3OCH_2CH_2CH_2CH_2CH_3$	$CH_3CH_2CH_2CH_2CH_2CH_2CH_3$
boiling point	99°	98°

From the discussion above, you might expect that an alkyl halide should have a higher boiling point than an alkane of the same molecular mass. However, this is not so. Alkyl chlorides have about the same boiling points as alkanes of the same molecular mass, and alkyl bromides and iodides have lower boiling points than the alkanes of about the same molecular mass.

	$CH_3CH_2CH_2CH_2Cl$	$CH_3CH_2CH_2CH_2CH_2CH_3$
molecular mass	92.6	86.2
boiling point	78.4°	68.7°
density	0.886 g/mL	0.660 g/mL

	CH_3CH_2Br	$CH_3CH_2CH_2CH_2CH_2CH_2CH_3$
molecular mass	109	100.2
boiling point	38.4°	98.4°
density	1.46 g/mL	0.684 g/mL

	CH_3I	$CH_3CH_2CH_2CH_2CH_2CH_2CH_2CH_2CH_2CH_3$
molecular mass	142	142
boiling point	42.5°	174°
density	2.28 g/mL	0.73 g/mL

The key to understanding these trends is to realize that although the molecules compared in each row have similar molecular masses, they have *very different molecular sizes and shapes*. From their relatively high densities, it is apparent that alkyl halide molecules have large masses within relatively small volumes. Thus, *for a given molecular mass*, alkyl halide molecules have smaller volumes than alkane molecules. Recall that the attractive forces between molecules—van der Waals forces, or dispersion forces—are greater for larger molecules (Sec. 2.6A). Larger intermolecular attractions translate into higher boiling points. The greater molecular volumes of alkanes, then, should cause them to have *higher* boiling points than alkyl halides. The polarity of alkyl halides, in contrast, has the opposite effect on boiling points: if polarity were the only effect, alkanes would have *lower* boiling points than alkyl halides. Thus, the effects of molecular volumes and polarity oppose each other. They nearly cancel in the case of alkyl chlorides, which have about the same boiling points as alkanes of about the same molecular mass. However, alkane molecules are so much larger than alkyl bromide and alkyl iodide molecules of the same molecular mass that the volume effect dominates, and alkanes have higher boiling points.

PROBLEMS

*8.9   (a) The dipole moment of one of the 2-butene stereoisomers is zero, and the dipole moment of the other is 0.25 D. Which is which? Explain.

(b) Which of the stereoisomeric 2-butenes has the higher boiling point? Explain your choice.

8.10   The boiling points of the 1,2-dichloroethylene stereoisomers are 47.4 °C and 60.3 °C. Give the structure of the stereoisomer with the higher boiling point. Explain.

B. Boiling Points of Alcohols

The boiling points of alcohols, especially alcohols of lower molecular mass, are unusually high in comparison to those of other organic compounds. For example, ethanol has a much higher boiling point than other organic compounds of about the same shape and molecular mass.

compound	CH_3CH_2—OH ethanol	$CH_3CH_2CH_3$ propane	CH_3—O—CH_3 dimethyl ether	CH_3CH_2—F ethyl fluoride
boiling point	78°	−42°	−24°	−38°
dipole moment	1.7 D	0 D	1.3 D	1.8 D

The contrast between ethanol and the last two compounds is particularly striking: all have similar dipole moments, and yet the boiling point of ethanol is much higher. The fact that something is unusual about the boiling points of alcohols is also apparent from a comparison of the boiling points of ethanol, methanol, and the simplest "alcohol," water.

	C_2H_5—OH	CH_3—OH	H—OH
boiling point	78°	65°	100°

Generally, each additional —CH_2— group results in a 20–30° difference in the boiling points of successive compounds in a series (Sec. 2.6A). Yet the difference in the boiling points of methanol and ethanol is only 13°; and water, although the "alcohol" of lowest molecular mass, has the highest boiling point of the three compounds.

C. Hydrogen Bonding

The unusual trends in the boiling points of alcohols are the result of a phenomenon called **hydrogen bonding**. In the case of alcohols, hydrogen bonding is a weak association of the O—H proton of one molecule with the oxygen of another.

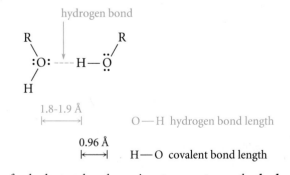

Formation of a hydrogen bond requires two partners, the **hydrogen-bond donor** (the atom to which the hydrogen is fully bonded) and the **hydrogen-bond acceptor** (the atom to which the hydrogen is partially bonded).

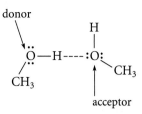

In a classical Lewis sense, a proton can only share two electrons. Thus, a hydrogen bond is difficult to describe with conventional Lewis structures. The hydrogen bond results from the combination of two factors: first, a weak covalent interaction between a hydrogen on the donor atom and unshared electron pairs on the acceptor atom; and second, an electrostatic attraction between oppositely charged ends of two dipoles. Opinions differ as to the relative importance of these two factors.

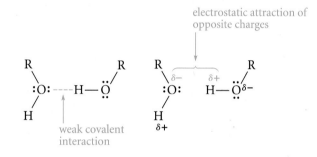

The hydrogen bond between two molecules resembles the same two molecules poised to undergo a Brønsted acid-base reaction:

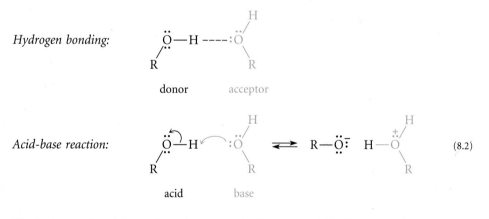

$$(8.2)$$

The hydrogen-bond donor is analogous to the Brønsted acid in Eq. 8.2, and the acceptor is analogous to the Brønsted base. In fact, it is not a bad analogy to think of the hydrogen bond as an acid-base reaction that has not quite started! In an acid-base reaction, the proton is fully transferred from the acid to the base; in a hydrogen bond, the proton interacts weakly with the acceptor but remains covalently bound to the donor.

The best hydrogen-bond donor atoms in neutral molecules are oxygens, nitrogens, and halogens. In addition, as might be expected from the similarity between hydrogen-bond interactions and Brønsted acid-base reactions, all strong proton acids are also good hydrogen-bond donors. The best hydrogen-bond acceptors in neutral molecules are the electronegative first-row atoms oxygen, nitrogen, and fluorine. All strong Brønsted bases are also good hydrogen-bond acceptors.

Sometimes an atom can act as both a donor and an acceptor of hydrogen bonds. For example, because the oxygen atoms in water or alcohols can act as both donors and acceptors, some of the molecules in liquid water and alcohols exist in hydrogen-bonded chains.

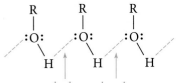

In contrast, the oxygen atom of an ether is a hydrogen-bond acceptor, but it is not a donor because it has no hydrogen to donate. Finally, some atoms are donors but not acceptors. The ammonium ion, $^{+}NH_4$, is a good hydrogen-bond donor; but because the nitrogen has no unshared electron pair, it is not a hydrogen-bond acceptor.

How can hydrogen bonding account for the unusually high boiling points of alcohols? In the liquid state, hydrogen bonding is a force of attraction between molecules. In the gas phase, hydrogen bonding is much less important and, at low pressures, does not exist. In order to vaporize a hydrogen-bonded liquid, energy is required to break the hydrogen bonds between molecules. This energy is manifested as an unusually high boiling point for hydrogen-bonded compounds such as alcohols.

Hydrogen bonding is also important in other ways. You'll see in Sec. 8.4B how it can affect the solubility of organic compounds. It is also a very important phenomenon in biology. Hydrogen bonds have critical roles in maintaining the structures of proteins and nucleic acids (for example, enzymes and genes). Without hydrogen bonds, life as we know it would not exist.

In summary, the tendency of molecules to associate noncovalently in the liquid state increases their boiling points. The most important forces involved in these intermolecular associations are

1. hydrogen bonding: hydrogen-bonded molecules have greater boiling points;
2. attractive van der Waals forces, which are influenced by
   a. molecular size: larger molecules have greater boiling points;
   b. molecular shape: more extended, less spherical molecules have greater boiling points;
3. attractive interactions between permanent dipoles: molecules with permanent dipole moments have higher boiling points.

An understanding of these factors will allow you to predict trends in boiling points within a group of compounds, as illustrated in the following study problem.

Study Problem 8.4

Arrange the following compounds in order of increasing boiling point: 1-hexanol, 1-butanol, *tert*-butyl alcohol, pentane

Solution   1-Butanol and pentane have the same molecular mass and about the same size and shape. However, because 1-butanol is a polar molecule that can both donate and accept hydrogen bonds, it has a considerably higher boiling point than pentane. Because 1-hexanol, also a primary alcohol, is a larger molecule than 1-butanol, its boiling point is the highest of the three. So far, the order of increasing boiling points is: pentane < 1-butanol < 1-hexanol. *Tert*-butyl alcohol has about the same molecular mass as pentane, but the alcohol has a higher boiling point because of its polarity and hydrogen bonding. However, a *tert*-butyl alcohol molecule is more branched and more nearly spherical than the isomeric 1-butanol molecule; thus, the boiling point of *tert*-butyl alcohol should be lower than that of 1-butanol. Therefore, the correct order of boiling points is: pentane < *tert*-butyl alcohol < 1-butanol < 1-hexanol. (The respective boiling points are [in °C] 36, 82, 118, 157.)

PROBLEMS

8.11   Within each set, arrange the compounds in order of increasing boiling point.
*(a)   4-ethylheptane, 2-bromopropane, 4-ethyloctane
(b)   1-butanol, 1-pentene, chloromethane

8.12   Label each of the following molecules as a hydrogen-bond acceptor, donor, or both. Indicate the hydrogen that is donated and/or the atom that serves as the hydrogen-bond acceptor.

*(a)   H—Br:

(b)   H—F:

*(c)

$$CH_3-\overset{\overset{\displaystyle :O:}{\|}}{C}-CH_3$$

(d)

$$CH_3-\overset{\overset{\displaystyle :O:}{\|}}{C}-\overset{..}{N}H-CH_3$$

*(e)

—ÖH

(f) $CH_3-CH_2-\overset{+}{N}H_3$

8.4   Solvents in Organic Chemistry

A **solvent** is a liquid used to dissolve one or more compounds. Solvents have tremendous practical importance. They affect the acidities and basicities of solutes. In some cases, the choice of a solvent can have dramatic effects on reaction rates. Understanding effects like these requires a classification of solvent types, to which the first part of this section is devoted.

The rational choice of a solvent requires an understanding of *solubility*, that is, the factors that determine whether a given compound will dissolve in a particular solvent. The second part of this section discusses the principles that will allow you make general predictions about the solubilities of organic compounds in different solvents. The effects of solvents on chemical reactions are closely tied to the principles of solubility. Solubility is also important in biology. For example, the solubilities of drugs determine their dosage forms and such important characteristics as whether they are absorbed from the gut, and whether they pass from the bloodstream into the brain.

Because certain alcohols, alkyl halides, and ethers are among the most important organic solvents, this is a good point in your study of organic chemistry to study solvent properties.

A. Classification of Solvents

Solvents can be classified in three ways, which are not mutually exclusive.

1. A solvent can be *protic* or *aprotic*.
2. A solvent can be *polar* or *apolar*.
3. A solvent can be a *donor* or a *nondonor*.

A **protic solvent** consists of molecules that can act as hydrogen-bond donors. Water, alcohols, and carboxylic acids are examples of protic solvents. Solvents that cannot act as hydrogen-bond donors are termed **aprotic solvents**. Ether, carbon tetrachloride, and hexane are examples of aprotic solvents.

A *polar solvent* separates, or shields, ions effectively from each other. That is, an ion dissolved in a polar solvent has a weaker interaction with other ions than it does in an *apolar solvent*. A solvent's polarity is defined by the magnitude of its *dielectric constant*: a **polar solvent** has a high dielectric constant; an **apolar solvent** has a low dielectric constant. The **dielectric constant** is a measure of how well a solvent reduces the interactions between ions in solution. The interaction energy E of two ions with respective charges q_1 and q_2 separated by a distance r is given by the *electrostatic law*:

$$E = k\frac{q_1 q_2}{\epsilon r} \tag{8.3}$$

In this equation, k is a proportionality constant and ϵ is the dielectric constant of the solvent in which the two ions are imbedded. This equation shows that when the dielectric constant ϵ is large, the magnitude of E, the energy of interaction between the ions, is small. If a solvent has a dielectric constant of about 15 or greater, it is polar. Water ($\epsilon = 78$), methanol ($\epsilon = 33$), and formic acid ($\epsilon = 59$) are polar solvents. Hexane ($\epsilon = 2$), ether ($\epsilon = 4$), and acetic acid ($\epsilon = 6$) are apolar solvents.

Unfortunately, the word *polar* has a double usage in organic chemistry. When we say that a *molecule* is polar, we mean that it has a significant dipole moment, μ (Sec. 1.2D). When we say that a *solvent* is polar, we mean that it has a high dielectric constant. In other words, solvent polarity, or dielectric constant, is a property of many molecules acting together, but molecular polarity, or dipole moment, is a property of individual molecules. While it is true that all polar *solvents* consist of polar *molecules*, the converse is not true. The contrast between acetic acid and formic acid is particularly striking:

$$
\begin{array}{cc}
\mathrm{CH_3-\overset{\displaystyle O}{\overset{\displaystyle \|}{C}}-OH} & \mathrm{H-\overset{\displaystyle O}{\overset{\displaystyle \|}{C}}-OH} \\[4pt]
\textbf{acetic acid} & \textbf{formic acid} \\
\mu = 1.5\text{--}1.7\text{ D} & \mu = 1.6\text{--}1.8\text{ D} \\
\epsilon = 6.1 & \epsilon = 59
\end{array}
$$

These two compounds contain identical functional groups and have very similar structures and dipole moments. Both are polar *molecules*. Yet they differ substantially in their dielectric constants *and in their solvent properties*! Formic acid is a polar solvent; acetic acid is not.

Donor solvents consist of molecules that can donate unshared electron pairs—that is, molecules that can act as Lewis bases. Ether, THF, and methanol are donor solvents. **Nondonor solvents** cannot act as Lewis bases; pentane and benzene are nondonor solvents.

Table 8.2 on pages 354–355 lists some common solvents used in organic chemistry and their classifications. As this table shows, a solvent can have a combination of properties. For example, some polar solvents are protic (such as water and methanol), but other polar solvents are aprotic (such as acetone).

PROBLEM

8.13   Classify each of the following substances according to their solvent properties (as in Table 8.2).

*(a)   2-methoxyethanol ($\epsilon = 17$)   (b)   2,2,2-trifluoroethanol ($\epsilon = 26$)

*(c)

$$CH_3-\overset{\overset{\displaystyle O}{\|}}{C}-CH_2CH_3 \ (\epsilon = 19)$$

(d)   2,2,4-trimethylpentane ($\epsilon = 2$)

B.   Solubility

One role of a solvent is simply to dissolve compounds of interest. Although finding a suitable solvent sometimes involves trial and error, certain principles can help us choose a solvent rationally. The discussion of solubility is divided into two parts: the solubility of covalent compounds, and the solubility of ionic compounds.

Solubility of Covalent Compounds   In determining a solvent for a covalent compound, the best rule of thumb is *like dissolves like*. That is, a good solvent usually has at least some of the molecular characteristics of the compound to be dissolved. To illustrate, consider the water solubility of the following compounds of comparable size and molecular mass:

$$\underbrace{CH_3CH_2CH_2CH_3 \qquad CH_3CH_2Cl}_{} \qquad CH_3CH_2-O-CH_3 \qquad CH_3CH_2CH_2-OH$$

water solubility: virtually insoluble   soluble   miscible

Of these compounds, the alcohol is most soluble; in fact, it is **miscible** with water. This means that a solution is obtained when the alcohol is mixed with water in any proportion.

You can understand these observations using what you know about water and alcohols. In order for a substance to dissolve in water, it must interact favorably with the solvent water molecules. The most favorable solute-water interactions occur when the solute can both donate hydrogen bonds to water and accept hydrogen bonds from water. An alcohol is soluble in water because its —OH group can both donate and accept hydrogen bonds. Since an ether can only accept hydrogen bonds, ethers are less soluble

Table 8.2   Properties of Some Common Organic Solvents
(Listed in order of increasing dielectric constant)

Solvent	Structure	Common abbreviation	Boiling point, °C	Dielectric constant ϵ^a	Class		
					Polar	Protic	Donor
hexane	$CH_3(CH_2)_4CH_3$	—	68.7	1.9			
1,4-dioxane[b]	(ring with two O)	—	101.3	2.2			x
carbon tetrachloride[c]	CCl_4	—	76.8	2.2			
benzene[b]	(benzene ring)	—	80.1	2.3			
diethyl ether	$(C_2H_5)_2O$	Et_2O	34.6	4.3			x
chloroform	$CHCl_3$	—	61.2	4.8			
ethyl acetate	$\overset{\overset{O}{\|\|}}{CH_3COC_2H_5}$	EtOAc	77.1	6.0			x
acetic acid	$\overset{\overset{O}{\|\|}}{CH_3COH}$	HOAc	117.9	6.1		x	x
tetrahydrofuran	(ring with O)	THF	66	7.6			x
methylene chloride	CH_2Cl_2	—	39.8	8.9			
acetone	$\overset{\overset{O}{\|\|}}{CH_3CCH_3}$	Me_2CO, DMK	56.3	21	x		x

[a] Most values are at or near 25°   [b] Known carcinogen   [c] Production to be banned in 1996.

in water than are alcohols. Finally, an alkane and an alkyl chloride can neither donate nor accept hydrogen bonds; consequently, these compounds are very insoluble. Thus, an alcohol molecule is most like water, and therefore dissolves in water to a greater extent; an alkane molecule is least like water, and is insoluble.

The same effect occurs in the following series:

CH_3OH C_2H_5OH $CH_3CH_2CH_2OH$	$CH_3CH_2CH_2CH_2OH$	$CH_3CH_2CH_2CH_2CH_2CH_2OH$
water solubility: miscible	7.7 mass %	0.58 mass %

Table 8.2   Properties of Some Common Organic Solvents (*continued*)
(**Listed in order of increasing dielectric constant**)

Solvent	Structure	Common abbreviation	Boiling point, °C	Dielectric constant ϵ^a	Class		
					Polar	Protic	Donor
ethanol	C_2H_5OH	EtOH	78.3	25	x	x	x
hexamethylphosphoric triamide[b]	$[(CH_3)_2N]_3P{=}O$	HMPA, HMPT	233	30	x		x
methanol	CH_3OH	MeOH	64.7	33	x	x	x
nitromethane	CH_3NO_2	$MeNO_2$	101.2	36	x		x
*N,N-*dimethylformamide	$\overset{\displaystyle O}{\overset{\|}{HCN(CH_3)_2}}$	DMF	153.0	37	x		x
acetonitrile	$CH_3C{\equiv}N$	MeCN	81.6	38	x		x
sulfolane	![sulfolane structure]	—	287 (dec)	43	x		x
dimethylsulfoxide	$\overset{\displaystyle O}{\overset{\|}{CH_3SCH_3}}$	DMSO	189	47	x		x
formic acid	$\overset{\displaystyle O}{\overset{\|}{HCOH}}$	—	100.6	59	x	x	x
water	H_2O	—	100.0	78	x	x	x
formamide	$\overset{\displaystyle O}{\overset{\|}{HCNH_2}}$	—	211 (dec)	111	x	x	x

[a] Most values are at or near 25°   [b] Known carcinogen

Alcohols with longer hydrocarbon chains are more like alkanes—and alkanes are insoluble in water. Hence, alcohols (as well as any other organic compounds) with long hydrocarbon chains are relatively insoluble.

In contrast, hydrocarbons dissolve well in other hydrocarbons. Because the attractive interactions between molecules of a hydrocarbon are mainly attractive van der Waals forces, the energy cost for substituting a hydrocarbon molecule with another hydrocarbon is small.

Solvents consisting of polar molecules lie in between the extremes of water on the one hand and hydrocarbons on the other. For example, consider the widely used solvent diethyl ether. Because ether can accept hydrogen bonds, it dissolves many alcohols. Because its dipole moment can interact favorably with other dipoles, it also dissolves polar compounds (for example, alkyl halides). On the other hand, because its hydrocarbon portion can take part in attractive van der Waals interactions, it also dissolves hydrocarbons.

The solvent tetrahydrofuran (THF) is even more versatile. It dissolves everything ether dissolves, but it also dissolves more water than does ether. As a solvent for the reaction of a water-insoluble compound with water, THF is typically an excellent choice, because it dissolves both compounds. For example, THF is the solvent of choice in oxymercuration of alkenes (Sec. 5.3A); it dissolves both water and alkenes.

What you should begin to see from discussions of this sort are the *trends* to be expected in the solubility behavior of various compounds. You cannot be expected to remember absolute solubilities, but you should be able to make an intelligent guess about the relative solubilities of a given compound in different solvents, or the relative solubilities of a series of compounds in a given solvent. This ability, for example, is required to solve the following problems.

PROBLEMS

*8.14   In which of the following solvents should hexane be *least* soluble: diethyl ether, methylene chloride (CH_2Cl_2), ethanol, or 1-octanol? Explain.

8.15   Which of the following substances should be least soluble in ethanol: diethyl ether, methylene chloride, hexane, or 1-octanol? Explain.

*8.16   A widely used undergraduate experiment is the recrystallization of acetanilide from water. Acetanilide is moderately soluble in hot water, but much less soluble in cold water. Identify one structural feature of the acetanilide molecule that would be expected to contribute positively to its solubility in water and one that would be expected to contribute negatively.

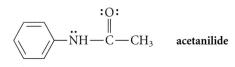

Solubility of Ionic Compounds   Because of the importance of both ionic reagents and ionic reactive intermediates in organic chemistry, the solubility of ionic compounds is worth special attention. Ionic compounds in solution can exist in several forms, two of which, *ion pairs* and *dissociated ions*, are shown in Fig. 8.2. In an **ion pair**, each ion is closely associated with an ion of opposite charge. In contrast, **dissociated ions** move more or less independently in solution, and are surrounded by several solvent molecules, called collectively the **solvation shell** or **solvent cage** of the ion. **Solvation** is a term used to describe the favorable interaction of a dissolved molecule with solvent. When solvent molecules interact favorably with an ion, they are said to **solvate** the ion.

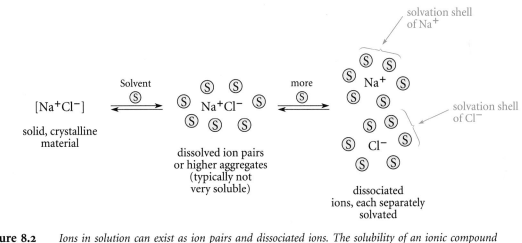

Figure 8.2   *Ions in solution can exist as ion pairs and dissociated ions. The solubility of an ionic compound depends on the ability of the solvent to break the electrostatic attractions between ions and form separate solvation shells around the dissociated ions. (The colored circles are solvent molecules.)*

Ion separation and ion solvation are mechanisms by which ions are stabilized in solution. If you think of the sequence in Fig. 8.2 as an ordinary chemical equilibrium, you can see that anything that favors the right side of this equilibrium tends to solubilize ions. The separation and solvation of ions reduce the tendency of the ions to associate into aggregates and ultimately, to precipitate as solids from solution. Hence, ionic compounds are relatively soluble in solvents in which ions are well separated and solvated.

The ability of a solvent to *separate ions* is measured by its dielectric constant ϵ in Eq. 8.3 on p. 352. Look carefully at this equation again. The energy of attraction of two ions of opposite charge is reduced in a solvent with a high dielectric constant. Hence, ions in solvents with high dielectric constants have a reduced tendency to associate, and thus a higher solubility.

Solvent molecules *solvate ions* in several ways, illustrated below for the solvation of sodium and chloride ions by water molecules:

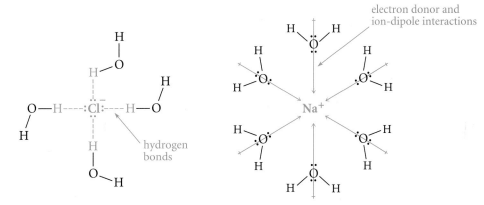

Anions, such as the chloride ion, are solvated by *hydrogen bonding*; that is, they accept hydrogen bonds from the solvent. Cations, such as the sodium ion, are solvated by what are called collectively *donor interactions*. In one type of donor interaction, an atom with an unshared electron pair on a solvent molecule (such as the oxygen of water) acts as a

Lewis base (electron donor) toward an electron-deficient cation. In the second type of donor interaction, a solvent molecule aligns itself so that the negative end of its dipole moment vector is pointed towards the cation. This type of interaction is called an *ion-dipole attraction*.

To summarize: three solvent properties contribute to the solubility of ionic compounds: *polarity* (high dielectric constant), by which solvent molecules separate ions of opposite charge; *proticity* (hydrogen-bond donor capability), by which solvent molecules solvate anions; and *electron-donor ability*, by which solvent molecules solvate cations through Lewis-base and ion-dipole interactions. It follows, then, that *the best solvents for dissolving ionic compounds are polar, protic, donor solvents.*

Thus, water is the ideal solvent for ionic compounds, something you probably know from experience. First, because it is polar—it has a very large dielectric constant—it is effective in separating ions of opposite charge. Second, because it is a donor solvent—a good Lewis base—it readily solvates cations. Finally, because it is protic—a good hydrogen-bond donor—it readily solvates anions. In contrast, hydrocarbons such as hexane do not dissolve ordinary ionic compounds because such solvents are apolar, aprotic, and nondonor solvents. Some ionic compounds, however, have appreciable solubilities in *polar aprotic* solvents such as acetone or DMSO. Although these solvents lack the protic character that solvates anions, their donor capacity solvates cations and their polarity separates ions of opposite charge. However, it is not surprising that because polar aprotic solvents lack the protic character that stabilizes anions, most salts are less soluble in these solvents than in water, and salts dissolved in polar aprotic solvents exist to a greater extent as ion pairs (Fig. 8.2).

C. Cation-Binding Molecules

Ionophores are molecules that form strong complexes with specific ions. (The word ionophore means "ion-bearing.") The *crown ethers*, a class of synthetic ionophores, and the *ionophore antibiotics*, ionophores found in nature, are intriguing because they interact with cations through the same mechanism used by donor solvents. Study of these ionophores provides additional insight into the mechanism of ionic solvation.

Crown Ethers   Some metal cations form stable complexes with a class of synthetic ionophores known as **crown ethers**, which were first prepared in 1967. These compounds are heterocyclic ethers containing a number of regularly spaced oxygen atoms. Some examples of crown ethers are the following:

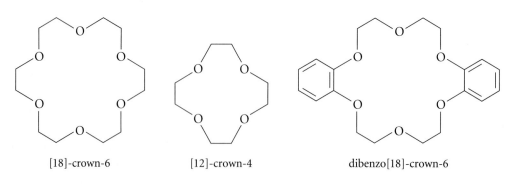

[18]-crown-6   [12]-crown-4   dibenzo[18]-crown-6

(The number in brackets indicates the number of atoms in the ring containing the oxygens, and the number following the hyphen indicates the number of oxygens.) The term "crown" was suggested by the three-dimensional shape of these molecules, shown in Fig. 8.3 for the complex of dibenzo[18]-crown-6 with the rubidium ion (Rb^+). The oxygens of the "host" crown ether wrap around the "guest" metal cation, complexing it within the cavity of the ether using the donor interactions discussed in the previous section: Lewis-base-type electron donation and ion-dipole interactions. In fact, you can think of a crown ether molecule as a "synthetic solvation shell" for a cation. Because the metal ion must fit within the cavity, the crown ethers have some selectivity for metal ions according to size. For example, dibenzo[18]-crown-6 forms the strongest complexes with potassium ion, somewhat weaker complexes with sodium, cesium, and rubidium ions, and does not complex lithium or ammonium ions appreciably. On the other hand, [12]-crown-4, with its smaller cavity, specifically complexes the lithium ion.

Because their structures contain hydrocarbon groups, crown ethers have significant solubilities in hydrocarbon solvents such as hexane or benzene. The remarkable thing about the crown ethers is that they can cause inorganic salts to dissolve in hydrocarbons—solvents in which these salts otherwise have no solubility whatsoever. For example, when potassium permanganate is added by itself to the hydrocarbon benzene, the $KMnO_4$ remains suspended, undissolved. Upon addition of a little dibenzo[18]-crown-6, which complexes potassium ion, the benzene takes on the purple color of a $KMnO_4$ solution and this solution (nicknamed "purple benzene") acquires the oxidizing power typical of $KMnO_4$. What happens is that the crown ether complexes the potassium cation and dissolves it in benzene; electrical neutrality demands that the permanganate ion accompany the complexed potassium ion into solution. The stabilization of the potassium ion by the crown ether compensates for the fact that the permanganate anion is essentially unsolvated, or "naked." Other potassium salts can be dissolved in hydrocarbon solvents in a similar manner. For example, KCl and KBr can be dissolved in hydrocarbons in the presence of crown ethers to give solutions of "naked chloride" and "naked bromide," respectively.

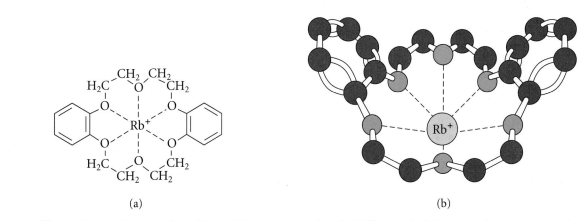

(a)   (b)

Figure 8.3   *Structure of the dibenzo[18]-crown-6 complex of rubidium ion in (a) a conventional structure and (b) a perspective drawing based on the three-dimensional structure determined by X-ray crystallography. In (b), the metal ion is shaded and the oxygens are shown in color. None of the hydrogen atoms are shown. Crown ethers can be regarded as "synthetic solvation shells" for cations.*

HOST-GUEST CHEMISTRY

As noted above, crown ethers can discriminate among various cations on the basis of a structural attribute: *ionic size*. As a result, crown ethers bind ions with a degree of *selectivity*. In recent years chemists have designed other classes of molecules that can "recognize" and bind more complicated compounds on the basis of their precise structures. This type of work has been spurred, at least in part, by a desire to understand and duplicate synthetically the highly specific binding that is characteristic of biological molecules such as enzymes and receptors. This general field, called **host-guest chemistry** or **molecular recognition**, was recognized with the 1987 Nobel Prize in Chemistry, which was awarded to three of its pioneers: the late C. J. Pedersen, then a chemist with du Pont, who invented the crown ethers; Donald J. Cram, Professor Emeritus of Chemistry at the University of California, Los Angeles; and Jean-Marie Lehn, Professor of Chemistry at Universite Louis Pasteur in Strasbourg, France, and the College de France in Paris.

Ionophore Antibiotics   Closely related to the crown ethers are the *ionophore antibiotics*. An *antibiotic* is a compound found in nature (or a synthetically prepared analog) that interferes with the growth or survival of one or more microorganisms. The ionophore antibiotics form strong complexes with metal ions in much the same way as crown ethers. Nonactin is an example of an ionophore antibiotic:

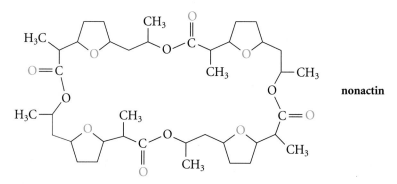

nonactin

Nonactin has a strong affinity for the potassium ion. As shown in Fig. 8.4, the molecule contains a cavity in which the colored oxygen atoms in the structure form a complex with the ion. The ion-binding properties of nonactin and related compounds are the basis of their action as antibiotics.

8.5   Acidity of Alcohols and Thiols

Alcohols and thiols are weak acids. In view of the similarity between the structures of water and alcohols, it may come as no surprise that their acidities are about the same.

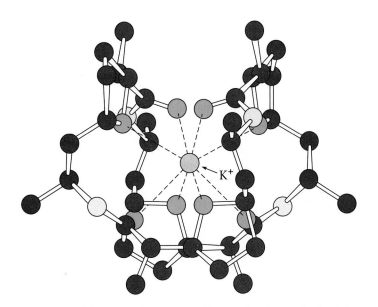

Figure 8.4   *Structure of the complex of the antibiotic nonactin with potassium ion. Only the carbons and oxygens are shown; hydrogen atoms are omitted. Notice that the nonactin molecule wraps around the potassium ion like a hand holding a ball. The oxygen atoms in direct contact with the potassium ion are shown in heavy color, and the other oxygens in lighter color.*

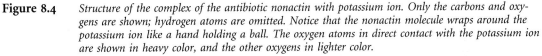

$$\text{p}K_a \qquad 15.9 \qquad\qquad 15.7$$

The conjugate bases of alcohols are generally called *alkoxides*. The common name of an alkoxide is constructed by deleting the final *yl* from the name of the alkyl group and adding the suffix *oxide*. In substitutive nomenclature, the suffix *-ate* is simply added to the name of the alcohol.

$$\text{CH}_3\text{CH}_2\ddot{\text{O}}\text{:}^- \ \text{Na}^+$$

common: sodium ethoxide
substitutive: sodium ethanolate

The relative acidities of alcohols and thiols are a reflection of the *element effect* described in Sec. 3.6A. Thiols, with pK_a values near 10, are substantially more acidic than alcohols. For example, the pK_a of ethanethiol, $\text{CH}_3\text{CH}_2\text{SH}$, is 10.5.

The conjugate bases of thiols are called *mercaptides* in common nomenclature and *thiolates* in substitutive nomenclature.

$$\text{CH}_3\ddot{\text{S}}\text{:}^- \ \text{Na}^+$$

common: sodium methyl mercaptide
substitutive: sodium methanethiolate

PROBLEMS

8.17 Give the structure of each of the following compounds.
*(a) potassium *tert*-butoxide   (b) sodium isopropoxide
*(c) magnesium 2,2-dimethyl-1-butanolate   (d) lithium methanolate

8.18 Name the following compounds.
*(a) $Cu—S—C_2H_5$ (b) $Ca(OCH_3)_2$

A. Formation of Alkoxides and Mercaptides

Because the acidity of a typical alcohol is about the same as that of water, an alcohol *cannot* be converted completely into its alkoxide conjugate base in an aqueous NaOH solution.

$$C_2H_5—\ddot{\underset{..}{O}}H + {}^-\!\!:\!\ddot{\underset{..}{O}}H \rightleftharpoons C_2H_5\ddot{\underset{..}{O}}{:}^- + H_2\ddot{\underset{..}{O}}{:}$$ (8.4)

$$pK_a = 15.9 \qquad\qquad\qquad pK_a = 15.7$$

You can see why this is true from the relative pK_a values: since these are nearly the same for ethanol and water, both sides of the equation contribute significantly at equilibrium. In other words, *hydroxide is not a strong enough base to convert an alcohol completely into its conjugate-base alkoxide.*

Alkoxides can be formed from alcohols with stronger bases. One convenient base used for this purpose is sodium hydride, NaH, which is a source of the *hydride ion, $H{:}^-$*. Because hydride ion is a very strong base (the pK_a of its conjugate acid, H_2, is estimated to be 42), its reaction with alcohols goes essentially to completion. In addition, when NaH reacts with an alcohol, the reaction cannot be reversed because the by-product, hydrogen gas, simply bubbles out of solution.

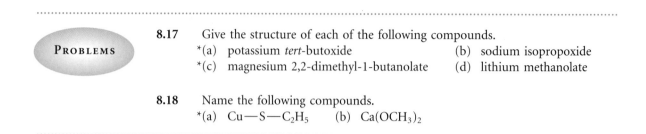

$$\text{Na}^+ \ H{:}^- + H—\ddot{\underset{..}{O}}—\underset{\underset{CH_3}{|}}{CHCH_2CH_3} \longrightarrow \text{Na}^+ \ {}^-\!\!:\!\ddot{\underset{..}{O}}—\underset{\underset{CH_3}{|}}{CHCH_2CH_3} + H_2\uparrow$$ (8.5)

quantitative yield

> Potassium hydride and sodium hydride are supplied as dispersions in mineral oil to protect them from reaction with moisture. When these compounds are used to convert an alcohol into an alkoxide, the mineral oil is rinsed away with pentane, a solvent such as ether or THF is added, and the alcohol is introduced cautiously with stirring. Hydrogen is evolved vigorously and a solution or suspension of the pure sodium or potassium alkoxide is formed.

Solutions of alkoxides in their conjugate-acid alcohols find wide use in organic chemistry. The reaction used to prepare such solutions is analogous to a reaction of water you may have observed. Sodium reacts with water to give an aqueous sodium hydroxide solution:

$$2H—\ddot{\underset{..}{O}}H + 2Na \longrightarrow 2Na^+ \ {}^-\!\!:\!\ddot{\underset{..}{O}}H + H_2\uparrow$$ (8.6)

The analogous reaction occurs with many alcohols. Thus, sodium metal reacts with an alcohol to afford a solution of the corresponding sodium alkoxide:

$$2R \overset{\cdot\cdot}{-\underset{\cdot\cdot}{O}}H + 2Na \longrightarrow 2Na^+ \ \overset{\cdot\cdot}{\underset{\cdot\cdot}{:O}}R + H_2\uparrow \qquad (8.7)$$

sodium alkoxide

The rate of this reaction depends strongly on the alcohol. The reactions of sodium with anhydrous (water-free) ethanol and methanol are vigorous, but not violent. However, the reactions of sodium with some alcohols, such as *tert*-butyl alcohol, are rather slow. The alkoxides of such alcohols can be formed more rapidly with the more reactive potassium metal.

Because thiols are much more acidic than water or alcohols, they, unlike alcohols, can be converted completely into their conjugate-base mercaptide anions by reaction with one equivalent of hydroxide or alkoxide. In fact, a common method of forming alkali-metal mercaptides is to dissolve them in ethanol containing one equivalent of sodium ethoxide:

$$C_2H_5\overset{\cdot\cdot}{\underset{\cdot\cdot}{S}}H \quad + \quad C_2H_5\overset{\cdot\cdot}{\underset{\cdot\cdot}{O}}{:}^- \ \rightleftharpoons \ C_2H_5\overset{\cdot\cdot}{\underset{\cdot\cdot}{S}}{:}^- \quad + \quad C_2H_5\overset{\cdot\cdot}{\underset{\cdot\cdot}{O}}H \qquad (8.8)$$

ethanethiol	ethoxide ion	ethanethiolate	ethanol
$pK_a = 10.5$		ion	$pK_a = 15.9$

Because the equilibrium constant for this reaction is $>10^5$ (how do you know this?), the reaction goes essentially to completion.

Although alkali-metal mercaptides are soluble in water and alcohols, thiols form insoluble mercaptides with many heavy-metal ions, such as Hg^{2+}, Cu^{2+}, and Pb^{2+}.

$$2CH_3(CH_2)_9 \overset{}{-}SH + PbCl_2 \ \xrightarrow{\ C_2H_5OH\ } \ [CH_3(CH_2)_9S]_2Pb \ + \ 2HCl \qquad (8.9)$$

decanethiol   lead(II) decanethiolate
(87% yield)

$$2PhSH + HgCl_2 \ \longrightarrow \ (PhS)_2Hg + 2HCl \qquad (8.10)$$

(98% yield)

The insolubility of heavy-metal mercaptides is analogous to the insolubility of heavy-metal sulfides (for example, PbS), which are among the most insoluble inorganic compounds known. One reason for the toxicity of heavy-metal salts is that they form tight mercaptide complexes with important biomolecules that normally contain free thiol groups.

CURING A DISEASE WITH MERCAPTIDES

A relatively rare inherited disease of copper metabolism, Wilson's disease, can be treated by utilizing the tendency of thiols to form complexes with copper ions. Accumulation of toxic levels of copper in the brain and liver causes the disease. Penicillamine is administered to complex the Cu^{+2} ions:

(*continues*)

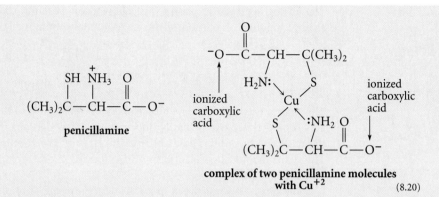

complex of two penicillamine molecules
with Cu^{+2}
                                                    (8.20)

The penicillamine-copper complex, unlike ordinary cupric thiolates, is relatively soluble in water because of the ionized carboxylic acid groups, and for this reason can be excreted by the kidneys.

B. Polar Effects on Alcohol Acidity

Substituted alcohols and thiols show the same type of polar effects on acidity as do substituted carboxylic acids (Sec. 3.6B). For example, alcohols containing electronegative substituent groups have enhanced acidity. Thus, 2,2,2-trifluoroethanol is more than three pK_a units more acidic than ethanol itself.

Relative acidity:

$$CH_3\text{—}CH_2\text{—}OH < CF_3\text{—}CH_2\text{—}OH \qquad (8.11)$$
$$pK_a \qquad\quad 15.9 \qquad\qquad\quad 12.4$$

The polar effects of electronegative groups are more important when the groups are closer to the —OH group:

Relative acidity:

$$CF_3\text{—}CH_2\text{—}CH_2\text{—}CH_2\text{—}OH < CF_3\text{—}CH_2\text{—}CH_2\text{—}OH < CF_3\text{—}CH_2\text{—}OH \qquad (8.12)$$
$$pK_a \qquad\quad 15.4 \qquad\qquad\qquad\qquad 14.6 \qquad\qquad\qquad 12.4$$

The fluorines have a negligible effect on acidity when they are separated from the —OH group by four or more carbons.

PROBLEM

8.19   In each of the following sets, arrange the compounds in order of increasing acidity (decreasing pK_a). Explain your choices.

*(a) $ClCH_2CH_2OH$, Cl_2CHCH_2OH, $Cl(CH_2)_3OH$

(b) HCF_2CH_2OH, $CF_3CF_2CH_2OH$, $HCF_2CH_2CH_2OH$

*(c) $ClCH_2CH_2SH$, $ClCH_2CH_2OH$, CH_3CH_2OH

(d) $CH_3CH_2CH_2CH_2OH$, $CH_3OCH_2CH_2OH$

C. Role of the Solvent in Alcohol Acidity

Primary, secondary, and tertiary alcohols differ significantly in their acidities; some relevant pK_a values are shown in Table 8.3. The data in this table show that the acidities of alcohols are in the order methyl > primary > secondary > tertiary. For many years chemists thought that this order was due to some sort of polar effect (Sec. 3.6B) of the alkyl groups around the alcohol oxygen. However, in relatively recent times, chemists were fascinated to learn that in the *gas phase*—in the absence of solvent—the order of acidity of alcohols is exactly reversed.

Relative gas-phase acidity:

$$(CH_3)_3COH > (CH_3)_2CHOH > CH_3CH_2OH > CH_3OH \qquad (8.13)$$

Notice carefully what is being stated here. The *relative order* of acidity of different types of alcohols is reversed in the gas phase compared to the *relative order* of acidity in solution. It is *not* true that alcohols are more acidic in the gas phase than they are in solution; rather, all alcohols are *much* more acidic in solution than they are in the gas phase.

Branched alcohols are more acidic than unbranched ones in the *gas phase* because α-alkyl substituents stabilize alkoxide ions. (Recall that stabilization of a conjugate-base anion increases acidity; Sec. 3.6B, Fig. 3.2). This stabilization occurs by a polarization mechanism. That is, the electron clouds of each alkyl group distort so that electron density moves away from the negative charge on the alkoxide oxygen, leaving a partial positive charge nearby. The anion is stabilized by its favorable electrostatic interaction with these partial positive charges.

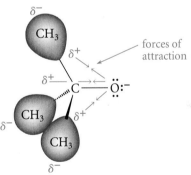

Because a tertiary alcohol has more α-alkyl substituents than a primary alcohol, a tertiary alkoxide is stabilized by this polarization effect more than a primary alkoxide. Consequently, tertiary alcohols are more acidic in the gas phase.

Table 8.3   Acidities of Alcohols in Aqueous Solution

Alcohol	pK_a	Alcohol	pK_a
CH_3OH	15.1	$(CH_3)_2CHOH$	17.1
CH_3CH_2OH	15.9	$(CH_3)_3COH$	19.2

The same polarization effect is present in solution, but the different acidity order in solution shows that another effect is operating as well. The acidity order in solution is due to the effectiveness with which alcohol molecules *solvate* their conjugate-base anions. Recall (Sec. 8.4B) that anions are solvated, or stabilized in solution, by hydrogen bonding with the solvent. Such hydrogen bonding is relatively unimportant in the gas phase. It is thought that the alkyl groups of a tertiary alkoxide somehow interfere with the solvation of the alkoxide oxygen, probably by disrupting the network of nearby hydrogen bonds. Reducing the solvation of the tertiary alkoxide increases its energy, and therefore increases its basicity. Because primary alkoxides do not have so many alkyl branches, solvation of primary alkoxides is more effective. Consequently, their basicities are lower. To summarize: *tertiary alkoxides are more basic than primary alkoxides.* An equivalent statement is that *primary alcohols are more acidic than tertiary alcohols.*

This discussion has shown that the solvent is not an idle bystander in the acid-base reaction; rather, it takes an active role in stabilizing the molecules involved, especially the charged species.

8.6 Basicity of Alcohols and Ethers

Just as water can accept a proton to form the hydronium ion, alcohols, ethers, thiols, and sulfides can also be protonated to form positively charged conjugate acids. Alcohols and ethers do not differ greatly from water in their basicities; thiols and sulfides, however, are much less basic.

| hydronium ion | conjugate acid of ethanol | conjugate acid of diethyl ether | conjugate acid of ethanethiol | conjugate acid of diethyl sulfide |

pK_a -1.74 -2 to -3 -5 to -7

The negative pK_a values mean that these protonated species are very strong acids, and that their neutral conjugate bases are rather weak. Nevertheless, the ability of alcohols, ethers, and their sulfur analogs to accept a proton plays a very important role in many of their reactions, particularly those that take place in acidic solutions.

Notice carefully that the pK_a values above refer in each case to the conjugate acid of the *neutral* base. Alcohols and thiols, like water, are *amphoteric* substances, that is, they can both gain and lose a proton. Thus, two acid-base equilibria are associated with an alcohol:

Loss of a proton:

$$C_2H_5-\overset{..}{\underset{..}{O}}-H + :\overset{..}{\underset{..}{O}}H \rightleftharpoons C_2H_5-\overset{..}{\underset{..}{O}}:^- + H-\overset{..}{\underset{..}{O}}H \qquad (8.14a)$$
$$pK_a = 15.9$$

Gain of a proton:

$$C_2H_5-\overset{..}{\underset{..}{O}}-H + H_3\overset{..}{O}^+ \rightleftharpoons C_2H_5-\overset{\overset{\displaystyle H}{|}}{\underset{+}{\underset{..}{O}}}-H + H_2\overset{..}{O}: \qquad (8.14b)$$
$$pK_a = -2 \text{ to } -3$$

The acidity of an alcohol—the loss of a proton—is exemplified by the reaction in Eq. 8.14a. Because alcohols are weak acids, this reaction is usually significant only in the presence of strong bases. The basicity of alcohols—the gain of a proton—is exemplified by the reaction in Eq. 8.14b. Because alcohols are weak bases, this reaction is usually significant only in the presence of strong acids.

Ethers are also important Lewis bases. For example, the Lewis acid-Lewis base complex of boron trifluoride and diethyl ether is stable enough that it can be distilled (bp 126°). This complex provides a convenient way to handle BF_3.

$$F_3\bar{B}\overset{+}{\underset{\diagdown}{\overset{\diagup}{-}}}\overset{C_2H_5}{\underset{C_2H_5}{\ddot{O}}}:$$ **boron trifluoride etherate**

Another example of the Lewis basicity of ethers is the complexation of BH_3 by tetrahydrofuran (THF), the solvent used in hydroboration (Sec. 5.3B).

Water and alcohols are also excellent Lewis bases, but in a practical sense they often cannot be used for forming complexes with Lewis acids. The reason is that the protons on water and alcohols can react further and, as a result, the complex is destroyed.

$$C_2H_5\overset{..}{\underset{..}{O}}-H + BF_3 \longrightarrow C_2H_5\overset{..}{\underset{\underset{H}{|}}{\overset{+}{O}}}-\bar{B}F_3 \longrightarrow C_2H_5\overset{..}{\underset{..}{O}}-BF_2 + H-F \qquad (8.15)$$
$$\text{(reacts further}$$
$$\text{with } C_2H_5OH)$$

Because ethers lack these protons, their complexes with Lewis acids do not react further.

8.7 Grignard and Organolithium Reagents

Compounds that contain carbon-metal bonds are called **organometallic compounds**. Two of the most useful types of organometallic compounds are Grignard reagents and organolithium reagents. A **Grignard reagent** is a compound of the form R—Mg—X, where X = Br, Cl, or I.

Examples of Grignard reagents:

carbon-metal bond
$$CH_3CH_2\overset{\downarrow}{-}Mg-Br$$
ethylmagnesium bromide

$$\langle\ \rangle-Mg-Cl$$
cyclohexylmagnesium chloride

DEVELOPMENT OF GRIGNARD REAGENTS

Well known for many years, Grignard reagents are among the most versatile and important reagents in organic chemistry. The utility of these reagents was originally investigated by Professor François Phillipe Antoine Barbier

(continues)

(1848–1922) of the University of Lyon in France. However, it was Barbier's successor at Lyon, Victor Grignard (1871–1935), who developed many applications of organomagnesium halides during the early part of the twentieth century. For this work, Grignard received the Nobel Prize in 1912.

Organolithium reagents are compounds of the form R—Li.

Examples of organolithium reagents:

carbon-metal bond

$$CH_3CH_2CH_2CH_2\text{—}Li$$

butyllithium

—Li

phenyllithium

Although the organolithium reagents are pictured for convenience as R—Li, many studies have shown that these reagents in solution are complexes of several molecules.

Other examples of organometallic compounds include the oxymercuration adducts of alkenes and mercuric acetate (Sec. 5.3A), as well as the organoboranes formed by addition of BH_3 to alkenes (Sec. 5.3B).

A. Formation of Grignard and Organolithium Reagents

Both Grignard and organolithium reagents are formed by adding the corresponding alkyl or aryl halides to rapidly stirred suspensions of the appropriate metal. Ether solvents must be used for the formation of Grignard reagents:

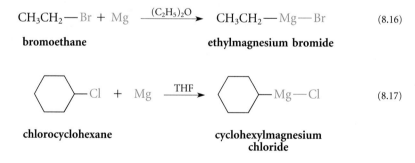

$$CH_3CH_2\text{—}Br + Mg \xrightarrow{(C_2H_5)_2O} CH_3CH_2\text{—}Mg\text{—}Br \qquad (8.16)$$

bromoethane   **ethylmagnesium bromide**

chlorocyclohexane + Mg $\xrightarrow{THF}$ cyclohexylmagnesium chloride   (8.17)

The solubility of Grignard reagents in ether solvents plays a crucial role in their formation. Grignard reagents are formed on the surface of the magnesium metal. As they form, these reagents are dissolved from the metal surface by the ether solvent. As a result, a fresh metal surface is continuously exposed to the alkyl halide. Grignard reagents are soluble in ether solvents because the ether associates with the metal in a *Lewis acid-base interaction*.

$$
\begin{array}{c}
\underset{\substack{C_2H_5 \diagdown \ddot{O} \diagup C_2H_5 \\ \downarrow \\ R-Mg-Br \\ \uparrow \\ C_2H_5 \diagup \ddot{O} \diagdown C_2H_5}}{} \equiv
\left[
\begin{array}{ccccc}
\underset{\substack{C_2H_5 \diagdown \ddot{O} \diagup C_2H_5 \\ R-Mg-Br \\ C_2H_5 \diagup \ddot{O} \diagdown C_2H_5}}{}
& \longleftrightarrow &
\underset{\substack{C_2H_5 \diagdown \overset{+}{\underset{..}{O}} \diagup C_2H_5 \\ R-Mg^--Br \\ C_2H_5 \diagup \ddot{O} \diagdown C_2H_5}}{}
& \longleftrightarrow &
\underset{\substack{C_2H_5 \diagdown \ddot{O} \diagup C_2H_5 \\ R-Mg^--Br \\ C_2H_5 \diagup \overset{+}{O} \diagdown C_2H_5}}{}
\end{array}
\right] \quad (8.18)
\end{array}
$$

The magnesium of the Grignard reagent is two electron pairs short of an octet, and the oxygen of an ether can donate an electron pair to the metal. (This interaction is very similar to the donor interactions that stabilize cations in solution; Sec. 8.4B.)

Lithium reagents are typically formed in hydrocarbon solvents such as hexane:

$$
\underset{\textbf{1-chlorobutane}}{CH_3CH_2CH_2CH_2-Cl} + 2Li \xrightarrow{\text{hexane}} \underset{\textbf{butyllithium}}{CH_3CH_2CH_2CH_2-Li} + LiCl \quad (8.19)
$$

Because organolithium reagents are soluble in hydrocarbons, ether solvents are not required for their formation.

Although the issue is still debated, many chemists believe that the formation of Grignard reagents involves radical intermediates. Magnesium atoms of the metal first abstract a bromine atom from the alkyl halide to form an alkyl radical $R\cdot$ at or near the magnesium surface:

$$
\underset{\substack{\text{(on the} \\ \text{metal surface)}}}{R-\ddot{\ddot{Br}}: \quad \cdot Mg} \longrightarrow R\cdot \quad \cdot Mg-\ddot{\ddot{Br}}: \quad (8.20a)
$$

The resulting radicals combine before the alkyl radical can diffuse away:

$$
R\cdot \quad \cdot Mg-\ddot{\ddot{Br}}: \longrightarrow \underset{\substack{\text{(dissolved by} \\ \text{the ether solvent)}}}{R-Mg-\ddot{\ddot{Br}}:} \quad (8.20b)
$$

This mechanism is different from most free-radical mechanisms because it is *not a chain reaction*. Formation of organolithium reagents may occur in a similar manner.

Grignard and organolithium reagents react violently with oxygen and (as shown in the next section) with water. For this reason these reagents must be prepared under rigorously oxygen-free and moisture-free conditions. In the case of Grignard reagents, exclusion of oxygen is easily assured by the low boiling points of the ether solvents that are normally used. As the Grignard reagent begins to form, heat is liberated and the ether boils. Because the reaction flask is filled with ether vapor, oxygen is virtually excluded.

PROBLEMS

8.20 Give an equation showing the preparation of each of the following organometallic compounds.

*(a) $(CH_3)_2CH-MgBr$ (b) CH_3-MgI

*(c) $Ph-Li$ (d) $(CH_3)_2CH-Li$

*(e) $[(CH_3)_2CH-CH_2]_3B$

8.21   Complete each of the following equations.

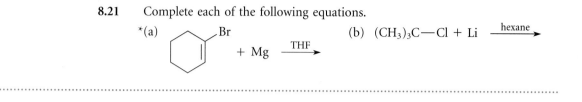

B. Protonolysis of Grignard and Organolithium Reagents

All reactions of Grignard and organolithium reagents can be understood in terms of the polarity of the carbon-metal bond. Because carbon is more electronegative than either magnesium or lithium, *the negative end of the carbon-metal bond is the carbon atom.*

Imagine breaking the carbon-metal bond of a Grignard or organolithium reagent so that the metal becomes positively charged and electron-deficient, and the pair of electrons in the bond ends up on carbon. Such a carbon, bearing three bonds, an unshared electron pair, and a negative formal charge, is termed a carbon anion, or **carbanion**. *Grignard and organolithium reagents react as if they were carbanions:*

$$-\overset{|}{\underset{|}{C}}-MgX \qquad \text{reacts as if it were} \qquad -\overset{|}{\underset{|}{C}}\!:^- \; \overset{+}{MgX} \qquad\qquad (8.21)$$

a carbon anion,
or carbanion

Grignard and organolithium reagents are not *true* carbanions because they have covalent carbon-metal bonds. However, we can predict their reactivity by thinking of them *conceptually* as carbanions.

For example, the view of Grignard and organolithium reagents as carbanions predicts the outcome of simple Brønsted acid-base reactions. Carbanions are powerful Brønsted bases because their conjugate acids, the corresponding alkanes, are extremely weak acids, with pK_a values estimated to be in the 55–60 range. The logic, then, is

1. R—H is a very weak acid ($pK_a = 55$–60); therefore,

2. R:$^-$ is a very strong base; therefore,

3. R—MgX and R—Li are also strong bases.

In fact, any Grignard or organolithium reagent reacts vigorously with even relatively weak acids such as water and alcohols to give the conjugate-base hydroxide or alkoxides and the conjugate-acid hydrocarbon of the carbanion.

$$CH_3CH_2-MgBr + H-OH \longrightarrow CH_3CH_2-H + HO-MgBr \qquad (8.22)$$

$$(CH_3)_3C-Li + H_2O \longrightarrow (CH_3)_3C-H + LiOH \qquad (8.23)$$

$$CH_3CH_2CH_2-MgBr + CH_3OH \longrightarrow CH_3CH_2CH_3 + CH_3O-MgBr \qquad (8.24)$$

Each of these reactions can be viewed as the reaction of a carbanion base with the proton of water or alcohol:

$$CH_3\overset{..}{C}H_2 \quad \overset{+}{M}gX \quad H—\overset{..}{\underset{..}{O}}R \longrightarrow CH_3CH_2—H + R\overset{..}{\underset{..}{O}}\!:^- \;\; ^+MgX \qquad (8.25)$$

$$\underset{\substack{\text{conjugate acid} \\ \text{of } CH_3CH_2:^-}}{} \qquad \underset{\substack{\text{conjugate base} \\ \text{of } RO—H}}{}$$

This is an example of a **protonolysis**, which is a reaction with the proton of an acid that breaks chemical bonds. In this case, the carbon-metal bond of the Grignard reagent is broken. The protonolysis reaction can be an annoyance, since, because of it, Grignard and organolithium reagents must be prepared in the absence of moisture. However, the protonolysis reaction is also useful, because it provides a *method for the preparation of hydrocarbons from alkyl halides*. Notice, for example, in Eq. 8.22 that ethane (a hydrocarbon) is produced from ethylmagnesium bromide, which, in turn, comes from ethyl bromide (an alkyl halide). Although one would not normally prepare an ordinary hydrocarbon by protonolysis, a particularly useful variation of this reaction is the preparation of hydrocarbons labeled with the hydrogen isotopes deuterium (D, or ^{2}H) or tritium (T, or ^{3}H) by reaction of a Grignard reagent with the corresponding isotopically labeled water.

$$(CH_3)_3CCH_2—Br \xrightarrow[\text{ether}]{Mg} (CH_3)_3CCH_2—MgBr \xrightarrow{D_2O} (CH_3)_3CCH_2—D \qquad (8.26)$$

PROBLEMS

8.22   Give the products of the following reactions. Show the curved-arrow formalism for each.

   *(a) $CH_3—Li + CH_3OH \longrightarrow$

   (b) $(CH_3)_2CHCH_2—MgCl + H_2O \longrightarrow$

8.23   *(a) Give the structures of two isomeric alkylmagnesium bromides that would react with water to give propane.

   (b) What compounds would be formed from the reactions of the reagents in (a) with D_2O?

8.8   Industrial Preparation and Use of Alkyl Halides, Alcohols, and Ethers

A. Free-Radical Halogenation of Alkanes

Among the methods used in industry to produce simple alkyl halides is direct halogenation of alkanes. When an alkane such as methane is treated with Cl_2 or Br_2 in the presence of heat or light, a mixture of alkyl halides is formed by successive chlorination reactions.

$$CH_4 + Cl_2 \xrightarrow{\text{heat or light}} CH_3Cl + HCl \qquad (8.27a)$$

$$CH_3Cl + Cl_2 \xrightarrow{\text{heat or light}} CH_2Cl_2 + HCl \qquad (8.27b)$$

$$CH_2Cl_2 + Cl_2 \xrightarrow{\text{heat or light}} CHCl_3 + HCl \qquad (8.27c)$$

$$CHCl_3 + Cl_2 \xrightarrow{\text{heat or light}} CCl_4 + HCl \qquad (8.27d)$$

The relative amounts of the various products can be controlled by varying the reaction conditions. This reaction is particularly useful for preparing extensively halogenated compounds such as carbon tetrachloride.

The products in Eq. 8.27a–d are formed in a series of *substitution* reactions (Sec. 7.9B). For example, CH_3Cl is formed by the substitution of a hydrogen atom in methane by a chlorine atom:

$$
\underset{\substack{| \\ H}}{\overset{\substack{H \\ |}}{H-C-H}} + Cl_2 \xrightarrow{\text{heat or light}} \underset{\substack{| \\ H}}{\overset{\substack{H \\ |}}{H-C-Cl}} + HCl \tag{8.28}
$$

The conditions of this reaction (initiation by heat or light) suggest the involvement of free-radical intermediates (Sec. 5.6C). The mechanism of this reaction in fact follows the typical pattern of other *free-radical chain reactions*; it has initiation, propagation, and termination steps. The reaction is initiated when a halogen molecule absorbs energy from the heat or light and dissociates homolytically into halogen atoms:

$$
:\ddot{\overset{..}{Cl}} - \ddot{\overset{..}{Cl}}: \xrightarrow{\text{light}} :\ddot{\overset{..}{Cl}}\cdot + \cdot\ddot{\overset{..}{Cl}}: \tag{8.29}
$$

The ensuing chain reaction has the following propagation steps:

$$
:\ddot{\overset{..}{Cl}}\cdot \quad H \! - \! CH_3 \longrightarrow :\ddot{\overset{..}{Cl}} - H + \cdot CH_3 \tag{8.30}
$$

methyl radical

$$
:\ddot{\overset{..}{Cl}} - \ddot{\overset{..}{Cl}}: \quad \cdot CH_3 \longrightarrow :\ddot{\overset{..}{Cl}}\cdot + :\ddot{\overset{..}{Cl}} - CH_3 \tag{8.31}
$$

Termination steps result from the recombination of radical species (Problem 8.27).

The halogenation of alkanes by a free-radical mechanism is an example of a **free-radical substitution** reaction: a substitution reaction that occurs by a free-radical chain mechanism. (Contrast this with the *free-radical addition* mechanism for peroxide-mediated addition to alkenes in Sec. 5.6C.)

Free-radical halogenations with chlorine and bromine proceed smoothly, halogenation with fluorine is violent, and halogenation with iodine does not occur. These observations correlate with the $\Delta H°$ values for halogenation of methane by each halogen (see Problem 5.49). Halogenation by fluorine is so strongly exothermic ($\Delta H° = -424$ kJ/mol, -101 kcal/mol) that the reaction is difficult to control; that is, the temperature of the reaction mixture rises more rapidly than the heat can be dissipated. Iodination is endothermic ($\Delta H° = +54$ kJ/mol, $+13$ kcal/mol); the reaction is so unfavorable energetically that it does not proceed to a useful extent. Chlorination ($\Delta H° = -106$ kJ/mol, -25 kcal/mol) and bromination ($\Delta H° = -30$ kJ/mol, -7 kcal/mol) are mildly exothermic, and proceed to completion without becoming violent.

PROBLEMS

*8.24   Give the free-radical chain mechanism for the formation of methylene chloride from methane and chlorine in the presence of heat or light.

8.25   Give the free-radical chain mechanism for the formation of ethyl bromide from ethane and bromine in the presence of light.

*8.26   Explain why ethane is formed as a minor by-product in the free-radical chlorination of methane.

8.27   Explain why butane is formed as a minor by-product in the free-radical bromination of ethane.

B.   Uses of Halogen-Containing Compounds

Very few halogen-containing organic compounds occur naturally. Those that do occur are produced by marine organisms that inhabit salt water, in which the concentration of halide ions is relatively high.

Alkyl halides and other halogen-containing organic compounds have many practical uses. Methylene chloride, chloroform, and carbon tetrachloride are important solvents (Table 8.2) that do not pose the flammability hazard of ethers. Tetrachloroethylene, trichlorofluoroethane, and trichloroethylene are used industrially as dry-cleaning solvents. A number of halogen-containing alkenes serve as monomers for the synthesis of useful polymers, for example, PVC, Teflon, and Kel-F (Table 5.4). Bromotrifluoromethane and a number of other brominated organic compounds are used as commercial flame retardants. The compound 2,4-dichlorophenoxyacetic acid (sold as 2,4-D) mimics a plant growth hormone, and causes broadleaved weeds to overgrow and eventually die. This is the dandelion killer used in commercial lawn fertilizers.

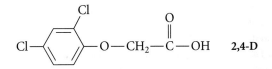

Some alkyl halides have medical uses. Halothane, $ClBrCH—CF_3$, and methoxyflurane, $Cl_2CH—CF_2—OCH_3$, are safe and inert general anesthetics that have largely supplanted the highly flammable compounds ether and cyclopropane. Certain fluorocarbons dissolve substantial amounts of oxygen, and there have been reports that such compounds have been used experimentally as artificial blood in surgical applications.

Because alkyl halides are rarely found in nature, and because many are not biologically degraded, it is perhaps not surprising that some alkyl halides released into the environment have become the focus of concern. The chlorofluorocarbons (Freons, or CFCs) such as F_2CCl_2, $HCClF_2$, and $HCCl_2F$, are among the most noteworthy examples. Until relatively recent times these compounds were the only ones used as refrigerants in commercial cooling systems, and they were also widely used as propellants in aerosol products. Nontoxic and nonflammable, and with properties ideally suited to their applications, they seemed to be ideal industrial chemicals. During the 1970s a number of studies implicated them in the destruction of stratospheric ozone, a process that occurs by a series of free-radical chain reactions. In October 1978 the United States government banned their use in virtually all but certain medically essential aerosol products, and a production phaseout of CFCs and certain other halogenated compounds is planned for January 1996. The search for satisfactory CFC replacements has not been wholly successful, but many of the most promising substitutes are other alkyl halides.

Some potent and effective insecticides are organohalogen compounds.

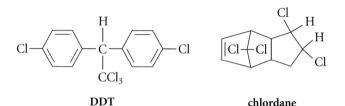

DDT   **chlordane**

DDT, which once played a key role in the elimination of malaria in the southern United States, was found to accumulate in the fatty tissues of birds and fish, and its use has been curtailed. Chlordane, too, has been banned except for certain specialized applications.

The conflict between the use of chemistry to improve humanity's living conditions and the generation of new problems caused by the release of chemicals into the environment finds real focus in the controversies surrounding the use of many organohalogen compounds. Ultimately, the potential benefit of any chemical product must be weighed against its hazards. As one official has said, those who shoulder the responsibility for risk-benefit analysis must ultimately make a subjective judgment over which reasonable persons may differ.

A NINETEENTH-CENTURY BALL ENDED BY AN ALKYL HALIDE REACTION

Perhaps the first recorded instance of an adverse environmental effect caused by alkyl halides occurred during the reign of Charles X of France. The French chemist Jean-Baptiste Andre Dumas (1800–1884) was asked to investigate something unusual that occurred during a ball given at the Tuileries. The candles used at the ball had sputtered and had given off noxious fumes, driving the guests from the ballroom. Dumas found that the beeswax used to make the candles had been bleached with chlorine gas. (Beeswax contains large numbers of double bonds. What reaction with chlorine took place?) The heat from the candle flame caused the chlorinated beeswax to decompose, liberating HCl gas—the noxious fumes.

C. Production and Use of Alcohols and Ethers

Ethanol   A number of alcohols are important articles of commerce. Most industrial ethanol is made by the hydration of ethylene (Sec. 4.9B).

$$CH_2{=}CH_2 + H_2O \xrightarrow[300°]{H_3PO_4} CH_3CH_2OH \tag{8.32}$$

ethylene   **ethanol**

Ethanol obtained from this reaction, called 95% ethanol, is 95.6 mass percent pure; the remainder is water. Anhydrous ethanol, or *absolute ethanol*, is obtained by further drying.

Industrial alcohol is chemically pure ethanol used as a starting material for the preparation of other compounds. It is subject to extensive federal controls to ensure that it is not diverted for illicit use in beverages. *Denatured alcohol*, used as a solvent for inks, fragrances, and the like, is ethanol that has been made unfit for human consumption by the addition of certain toxic additives, such as methanol. About 55% of the ethanol produced synthetically finds use in the formulation of solvents; about 35% is used for other industrial processes.

Beverage alcohol is produced by the fermentation of malt, barley, grape juice, corn mash, or other sources of natural sugar. Beverage alcohol is not isolated; rather, alcoholic beverages are the mixtures of ethanol, water, and the natural colors and flavorings produced in the fermentation process and purified by sedimentation (as in wine) or distillation (as in brandy or whiskey). Industrial alcohol is not used to alter the alcoholic composition of beverages.

Ethanol is a drug, and like many useful drugs, when consumed in excess it is toxic. Ethanol is the most abused drug in the world.

Hydrocarbons are the principal raw materials of the chemical industry, and ethylene is the most important industrial hydrocarbon. When the price of crude oil rises, however, as it did in the late 1970s, production of ethanol by fermentation becomes increasingly attractive as a source of carbon. The production of gasohol, a mixture of ethanol and gasoline, has been taking place for nearly two decades. The ethanol used in gasohol is produced by the fermentation of the sugars in corn. This process provides farmers with an additional market for their surplus corn crop, and offers an alternative to imported oil. Although gasohol production is relatively small at present, it seems likely to increase. Brazil operates essentially on an ethanol-based economy rather than a hydrocarbon-based economy, producing about 4 billion gallons of ethanol annually by fermentation of sugars from plants.

ENVIRONMENTAL BENEFITS OF BIOFUELS

A very real problem in the combustion of hydrocarbons for fuel is the formation of CO_2. The CO_2 content in the atmosphere has risen in the last 100 years from 290 to 360 parts per million, with 20% of this increase occurring in the last ten years (Fig. 8.5, page 376). Although the exact consequences of this increase cannot be predicted with certainty, some scientists have argued that it might cause a gradual warming of the earth followed by return to another ice age. At the very least, a gigantic chemical experiment of uncertain outcome is being conducted on the environment. From this point of view, the production of fuels from fermentation-derived ethanol is particularly attractive. Ethanol is fermented from sugars such as glucose. When plants synthesize glucose, they remove CO_2 from the atmosphere. The energy for the plant synthesis of glucose is derived from the sun (photosynthesis). If the reactions are added together for the plant synthesis of glucose, its conversion

(*continues*)

into ethanol, and the combustion of ethanol (or anything derived from it) as fuel, no net change is effected in the CO_2 content in the environment.

$$6CO_2 + 6H_2O \xrightarrow{\text{light}} C_6H_{12}O_6 + 6O_2 \quad \text{(biosynthesis of glucose, } C_6H_{12}O_6\text{)}$$
$$C_6H_{12}O_6 \longrightarrow 2C_2H_5OH + 2CO_2 \quad \text{(anaerobic fermentation)}$$
$$2C_2H_5OH + 6O_2 \longrightarrow 4CO_2 + 6H_2O \quad \text{(combustion of ethanol)}$$

Sum: No Net Change $\hspace{8cm}$ (8.33)

The idea that producing fuels from biomass is environmentally beneficial is not without its critics. It has been argued that the CO_2 produced and the energy consumed from burning the fossil fuels used to cultivate sugar-producing plants make the use of biomass to produce fuels much less attractive than it might seem.

Methanol and *tert*-Butyl Methyl Ether Methanol is formed from a mixture of carbon monoxide and hydrogen, called *synthesis gas*, at high temperature over special catalysts.

$$\underbrace{CO + H_2}_{\text{synthesis gas}} \xrightarrow[\substack{\text{Cr-Zn catalyst} \\ \text{100–600 atm}}]{250-400°} CH_3OH \hspace{4cm} (8.34)$$

Synthesis gas comes from the partial oxidation of methane, which is, in turn, derived from the cracking of hydrocarbons (Sec. 5.7B), or from the gasification of coal. If petroleum prices rise sharply, synthesis gas from coal may become another important source of carbon.

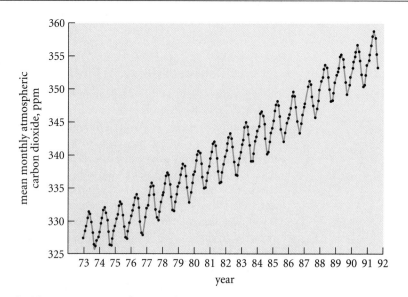

Figure 8.5   *Carbon dioxide concentration in the atmosphere as monitored by a National Oceanic and Atmospheric Administration station at Mauna Loa, Hawaii. Although seasonal fluctuations occur (small peaks), a general trend toward higher CO_2 concentration is clearly apparent.*

In 1992, 1.4 billion gallons of methanol valued at about \$550 million were produced in the United States. Important uses of methanol include oxidation to formaldehyde (CH_2=O) and reaction with carbon monoxide over special catalysts to give acetic acid (CH_3CO_2H). An important newer use of methanol is its reaction with 2-methylpropene to give *tert*-butyl methyl ether (MTBE).

$$
\begin{array}{c}
CH_3 \\
\backslash \\
C = CH_2 \quad + \quad CH_3OH \quad \xrightarrow{\;H_2SO_4\;} \quad CH_3 - \underset{\underset{CH_3}{|}}{\overset{\overset{OCH_3}{|}}{C}} - CH_3 \\
/ \\
CH_3
\end{array}
\qquad (8.35)
$$

2-methylpropene   **methanol**   **MTBE**

MTBE has replaced tetraethyllead as an antiknock additive in gasoline (Sec. 2.8), and is also used, along with *tert*-butyl ethyl ether, as an additive to decrease the volatility of gasoline (another environmental concern). Consumption of this ether grows significantly each year; about 1.8 billion gallons valued at \$1.6 billion are produced annually. Methanol, which has an octane rating of 116, also has a largely unrealized potential for use as a motor fuel. (It has been used as a fuel in racing engines for years.) In fact, California has pioneered the introduction of automobile engines that burn only methanol.

Ethylene Oxide and Ethylene Glycol   Ethylene oxide, produced by oxidation of ethylene over a silver catalyst, is one of the most important industrial derivatives of ethylene:

$$
2CH_2 = CH_2 + O_2 \quad \xrightarrow{\;Ag\ (catalyst)\;} \quad H_2C \overset{O}{\overset{\diagup\diagdown}{\underline{\quad\quad}}} CH_2 \qquad (8.36)
$$

About 5.6 billion pounds of ethylene oxide are produced annually in the United States. The most important single use of ethylene oxide is its reaction with water to give ethylene glycol:

$$
H_2C \overset{O}{\overset{\diagup\diagdown}{\underline{\quad\quad}}} CH_2 + H_2O \quad \longrightarrow \quad \underset{CH_2 - CH_2}{\overset{OH \quad\ OH}{\overset{|\qquad |}{}}} \qquad (8.37)
$$

Ethylene glycol is used both for automotive antifreeze and as a starting material for polyester fibers and films. About 5.1 billion pounds of ethylene glycol are produced annually in the United States. When automotive sales in the United States are depressed, the ethylene glycol and ethylene oxide markets are also depressed by the reduced demand for antifreeze.

D. Safety Hazards of Ethers

Two safety hazards are generally associated with the use of ethers. The first is peroxide formation. On standing in air, ethers undergo **autoxidation**, or spontaneous oxidation by oxygen in air. Samples of ethers can accumulate dangerous quantities of explosive peroxides and hydroperoxides by autoxidation.

$$\text{CH}_3\text{CH}_2\text{—O—CH}_2\text{CH}_3 + \text{O}_2 \longrightarrow \underset{\underset{\text{O—O—H}}{|}}{\text{CH}_3\text{CH}_2\text{O—CH—CH}_3} \longrightarrow \text{other polymeric peroxides} \qquad (8.38)$$

a hydroperoxide

$$\downarrow$$

$$\text{CH}_3\text{CH}_2\text{—O—O—CH}_2\text{CH}_3}$$

diethyl peroxide

These peroxides can form by free-radical processes in samples of anhydrous diethyl ether or tetrahydrofuran (THF) within less than two weeks. For this reason, some ethers are sold with small amounts of free-radical inhibitors such as hydroquinone, which can be removed by distilling the ether. Because peroxides are particularly explosive when heated, it is a good practice not to distill ethers to dryness. Peroxides in an ether can be detected by shaking a portion of the ether with 10% aqueous potassium iodide solution. If peroxides are present, they oxidize the iodide to iodine, which imparts a yellow tinge to the solution. Small amounts of peroxides can be removed by distillation of ethers from lithium aluminum hydride (LiAlH_4), which both reduces the peroxides and removes contaminating water and alcohols.

The second ether hazard is the high flammability of diethyl ether, the ether most commonly used in the laboratory. Its flammability is indicated by its very low flash point of -45 °C. The **flash point** of a material is the minimum temperature at which it is ignited by a small flame under certain standard conditions. In contrast, the flash point of THF is -14 °C. Compounding the flammability hazard of ether is the fact that its vapor is 2.6 times as dense as air. This means that ether vapors from an open vessel will accumulate in a heavy layer along a laboratory floor or benchtop. For this reason flames can ignite ether vapors that have spread from a remote source. Good safety practice demands that open flames or sparks not be permitted anywhere in a laboratory in which ether is in active use. Even the spark from an electric switch (such as that on a hot plate) can ignite ether vapors. A steam bath is therefore one of the safest ways to heat ether.

The flammability of ether, however, has its uses. Ether is one of the active ingredients in the starter fluid used to start automotive engines on very cold mornings.

KEY IDEAS IN CHAPTER 8

Organic compounds are named by both common and substitutive nomenclature. In substitutive nomenclature, the name is based on the principal group and principal chain. The group designated as the principal group according to a list of priorities is cited as a suffix in the name. Other groups are cited as substituents. Hydroxy (—OH) and thiol (—SH) groups can be cited as principal groups. Halogens, alkoxy (—OR) groups, and alkylthio (—SR) groups are always cited as substituents.

🝧 The noncovalent association of molecules in the liquid state raises their boiling points. Such molecular association can result from hydrogen bonding; attractive van der Waals forces, which are greatest for larger, more extended molecules; and the interaction of dipoles associated with polar molecules.

🝧 Alcohols and thiols are weakly acidic. The conjugate bases of alcohols are called alkoxides, or alcoholates; and the conjugate bases of thiols are called mercaptides, or thiolates.

🝧 Typical primary alcohols have pK_a values near 15–16 in aqueous solution. The acidity of alcohols is reduced by branching near the —OH group and increased by electron-withdrawing substituents. Alkoxides are formed by reaction of alcohols with strong bases such as sodium hydride (NaH), or by reaction with alkali metals.

🝧 Typical thiols have pK_a values near 10–11. Solutions of thiolates can be formed by reaction of thiols with NaOH in alcohol solvents.

🝧 Alcohols, thiols, and ethers are weak Brønsted bases, and react with strong acids to form positively charged conjugate-acid cations that have negative pK_a values. The Lewis basicities of ethers account for their formation of stable complexes with Lewis acids such as boron compounds and Grignard reagents.

🝧 A solvent is classified as protic or aprotic, depending on its ability to donate hydrogen bonds; polar or apolar, depending on the size of its dielectric constant; donor or nondonor, depending on its ability to act as a Lewis base.

🝧 The solubility of covalent compounds follows the "like-dissolves-like" rule. The solubility of ionic compounds tends to be greatest in solvents that have high dielectric constants, and in solvents that can solvate anions by hydrogen bonding and cations by donor interactions. Crown ethers and other ionophores form complexes with cations by creating artificial solvation shells for them.

🝧 Reaction of alkyl halides with magnesium metal yields Grignard reagents; reaction with lithium yields organolithium reagents. Both types of reagents behave as strong Brønsted bases and react readily with acids, including water and alcohols, to give alkanes.

🝧 Alkanes react with bromine and chlorine in the presence of heat or light in free-radical substitution reactions to give alkyl halides.

ADDITIONAL PROBLEMS

*8.28   (a)   Give the structures of all alcohols with the molecular formula $C_5H_{11}OH$.

           (b)   Which ones are chiral?

(Problem 8.28 continues)

(c) Name each compound using IUPAC substitutive nomenclature.

(d) Classify each as a primary, secondary, or tertiary alcohol.

8.29 (a) Give the structures of all alcohols with the molecular formula C_4H_9OH.

(b) Which ones are chiral?

(c) Name each compound using IUPAC substitutive nomenclature.

(d) Classify each as a primary, secondary, or tertiary alcohol.

8.30 Give the IUPAC substitutive name for each of the following compounds, which are used as general anesthetics.

*(a)
$$\underset{\underset{\displaystyle Cl}{|}}{\overset{\overset{\displaystyle Br}{|}}{H-C-CF_3}}$$

halothane

(b) $Cl_2CH-CF_2-OCH_3$

methoxyflurane

·**8.31** Thiols of low molecular mass are known for their extremely foul odors. In fact, the following two thiols are the active components in the scent of the skunk. Give the IUPAC substitutive names for these compounds.

*(a) $CH_3CH=CHCH_2-SH$

(b) $\underset{\underset{\displaystyle CH_3}{|}}{CH_3CHCH_2CH_2-SH}$

8.32 Without consulting tables, arrange the compounds within each of the following sets in order of increasing boiling point, and give your reasoning.

*(a) 1-hexanol, 2-pentanol, *tert*-butyl alcohol

(b) butyl alcohol, *tert*-butyl alcohol

*(c) 1-hexanol, 1-hexene, 1-chloropentane

(d) 1-chlorobutane, 1-chloropropane, 1-pentanol

*(e) diethyl ether, propane, 1,2-propanediol

(f) cyclooctane, chlorocyclobutane, cyclobutane

8.33 In each part below, explain why the first compound has a higher boiling point than the second, despite a lower molecular mass.

*(a)
$$CH_3-\overset{\overset{\displaystyle O}{\|}}{C}-OH \quad (bp\ 118°) \qquad CH_3-\overset{\overset{\displaystyle O}{\|}}{C}-OCH_2CH_3 \quad (bp\ 77°)$$

(b)
$$CH_3-\overset{\overset{\displaystyle O}{\|}}{C}-NH_2 \quad (bp\ 221°) \qquad CH_3-\overset{\overset{\displaystyle O}{\|}}{C}-N(CH_3)_2 \quad (bp\ 166°)$$

8.34 Give the structures of the following compounds.

*(a) a chiral ether $C_5H_{10}O$ that has no double bonds

(b) another compound that meets the criteria in (a)

*(c) a chiral alcohol C_4H_6O

(d) another compound that meets the criteria in (c)

*(e) a vicinal glycol $C_6H_{10}O_2$ that cannot be optically active at room temperature

*(f) a diol $C_4H_{10}O_2$ that exists in only three stereoisomeric forms

(g) a diol $C_4H_{10}O_2$ that exists in only two stereoisomeric forms

*(h) the *six* epoxides (counting stereoisomers) with the molecular formula C_4H_8O

(i) four compounds (counting stereoisomers) not containing double bonds with the formula C_3H_6O

8.35 Identify the gas evolved in each of the following reactions.

*(a) $K + H_2O \longrightarrow$

(b) $Na + D_2O \longrightarrow$

*(c) $CH_3CH_2MgBr + H_2O \longrightarrow$

(d) $CH_3MgI + H_2O \longrightarrow$

*(e)

$$NaH + CH_3\!-\!\overset{\displaystyle OH}{\underset{\displaystyle |}{CH}}\!-\!CH_3 \longrightarrow$$

(f)

$$CH_3\!-\!\overset{\displaystyle MgBr}{\underset{\displaystyle |}{CH}}\!-\!CH_3 + CH_3\!-\!\overset{\displaystyle OH}{\underset{\displaystyle |}{CH}}\!-\!CH_3 \longrightarrow$$

*8.36 Identify the correct compound(s) in each case.

(a) A compound believed to be either diethyl ether or propyl alcohol is miscible in water.

(b) A compound believed to be either allyl methyl ether or propyl alcohol decolorizes a solution of Br_2 in CCl_4.

(c) Four stereoisomeric compounds C_4H_8O, all optically active, contain no double bonds and evolve a gas when treated with CH_3MgI.

8.37 Identify the correct compound in each case.

(a) A compound believed to be either cyclohexyl methyl ether or 2-methylcyclo-hexanol evolves a gas when treated with NaH.

(b) A compound believed to be one of the two isomers $CH_3OCH_2CH_2SH$ or $CH_3SCH_2CH_2OH$ is soluble in water.

*8.38 (a) One of the following compounds is an unusual example of a salt that is soluble in hydrocarbon solvents. Which one is it? Explain your choice.

$$CH_3(CH_2)_{15}\!-\!\overset{\displaystyle CH_2CH_2CH_2CH_3}{\underset{\displaystyle CH_2CH_2CH_2CH_3}{\overset{+}{N}}}\!-\!CH_2CH_2CH_2CH_3 \quad Br^- \qquad \overset{+}{N}H_4 \quad Cl^-$$

$$A \qquad\qquad\qquad\qquad\qquad\qquad B$$

(b) Which of the following would be present in greater amount in a hexane solution of the compound in part (a): separately solvated ions, or ion pairs and higher aggregates? Explain your reasoning.

8.39 In each case, give the structure of the hydrocarbon that would react with Cl_2 and light to give the indicated products.

(Problem 8.39 continues)

*(a) C_5H_{12}, which gives only one monochlorination product

(b) C_4H_{10}, which gives two and only two monochlorination products, both achiral

8.40 Arrange the compounds within each set in order of increasing acidity (decreasing pK_a) in solution. Explain your reasoning.

*(a) propyl alcohol, isopropyl alcohol, *tert*-butyl alcohol, 1-propanethiol

(b) 2-chloro-1-propanethiol, 2-chloroethanol, 3-chloro-1-propanethiol

*(c) $CH_3NH—CH_2CH_2—OH$, $CH_3NH—CH_2CH_2CH_2—\overset{+}{}OH$,

$(CH_3)_3\overset{+}{N}—CH_2CH_2—OH$

(d) $CH_3O—CH_2CH_2—OH$, $^-O—CH_2CH_2—OH$, $CH_3CH_2CH_2OH$

*(e)

$$CH_3CH_2OH, \quad CH_3\overset{\overset{H}{|}}{\underset{+}{S}}CH_3, \quad CH_3CH_2\overset{\overset{H}{|}}{\underset{+}{O}}CH_2CH_3$$

*8.41** The pK_a of water is 15.7. Titration of an aqueous solution containing Cu^{+2} ion suggests the presence of a species that acts as a Brønsted acid with $pK_a = 8.3$. Suggest a structure for this species. (*Hint:* Cu^{+2} is a Lewis acid.)

*8.42** Normally, dibutyl ether is much more soluble in benzene than it is in water. Explain why this ether can be extracted from benzene into water if the aqueous solution contains moderately concentrated nitric acid. (*Hint:* Many ionic compounds are soluble in water.)

*8.43** Explain why the complex of the crown ether [18]-crown-6 with potassium ion has a much larger dissociation constant in water than it does in ether.

*8.44** Although Grignard reagents are normally insoluble in hydrocarbon solvents, they can be dissolved in such solvents if a tertiary amine (a compound with the general structure $R_3N:$) is added. Explain.

*8.45** Ethyl alcohol in the solvent CCl_4 forms a hydrogen-bonded complex with an equilibrium constant $K_{eq} = 11$.

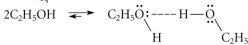

(a) What happens to the concentration of the complex as the concentration of ethanol is increased?

(b) What is the standard free energy change for this reaction at 25 °C?

(c) If one mole of ethanol is dissolved in one liter of CCl_4, what are the concentrations of free ethanol and of complex?

8.46 The equilibrium constant for the reaction of ethanethiol analogous to the one for ethanol in Problem 8.45 is 0.004.

(a) Which has the stronger hydrogen bond: ethanethiol or ethanol?

(b) What is the standard free energy change for this reaction at 25°?

(c) If one mole of ethanethiol is dissolved in one liter of CCl_4, what are the concentrations of free ethanethiol and of the complex?

*8.47   A student, Flick Flaskflinger, in his twelfth year of graduate work, needed to prepare ethylmagnesium bromide from ethyl bromide and magnesium, but found that his laboratory was out of diethyl ether. From his years of accumulated knowledge he recalled that Grignard reagents will form in other ether solvents. He therefore attempted to form ethylmagnesium bromide in the ether C_2H_5—O—CH_2CH_2OH and was shocked to find that no Grignard reagent was present after several hours' stirring. Explain why Flick's reaction failed.

*8.48   (a)   The bromination of 2-methylpropane in the presence of light can give two monobromination products; give their structures.
       (b)   In fact, only *one* of these compounds is formed in appreciable amount. Write the mechanism for formation of each compound, and use what you know about the stabilities of free radicals to predict which compound should be the major one formed.

*8.49   In a laboratory three alkyl halides, each with the formula $C_7H_{15}Br$, are discovered and found to have different boiling points. One of the compounds is optically active. Following reaction with Mg in ether, then with water, each compound gives 2,4-dimethylpentane. After the same reaction with D_2O, a different product is obtained from each compound. Suggest a structure for each of the three alkyl halides.

*8.50   When *sec*-butylbenzene undergoes free-radical bromination, one major product is formed, as follows:

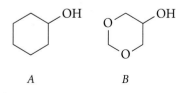

sec-butylbenzene	(1-bromo-1-methylpropyl)benzene

       If the starting material is optically active, predict whether the substitution product should also be optically active. (*Hint:* Consider the geometry of the free radical intermediate; see Fig. 5.2.)

8.51   Offer an explanation for each of the following observations.
     *(a)   Compound *A* exists preferentially in a chair conformation with an equatorial —OH group, but compound *B* prefers a chair conformation with an axial —OH group.

A   B

       (b)   Ethylene glycol contains a greater fraction of *gauche* conformation than butane in carbon tetrachloride solution.

(*Problem 8.51 continues*)

*(c) The racemate of 2,2,5,5-tetramethyl-3,4-hexanediol exists with a strong intramolecular hydrogen bond, but the *meso* compound has no intramolecular hydrogen bond.

*8.52   It has been argued that a covalently bound chlorine has about the same effective size as a methyl group. Assuming this is true, explain why the energy barrier for interconversion of the two *gauche* conformations of 1,2-dichloroethane (39 kJ/mol, 9.3 kcal/mol) is considerably larger than the analogous barrier for butane (19 kJ/mol, 4.5 kcal/mol). (*Hint:* Consider the relationship of the C—Cl bond dipoles in the eclipsed conformation. What is the effect of this relationship on the energy of the molecule?)

*8.53   (a) Show that the dipole moment of 1,4-dioxane (Sec. 8.1C) should be zero if the molecule exists solely in a chair conformation.

   (b) Account for the fact that the dipole moment of 1,4-dioxane, although small, is definitely not zero. (It is 0.38 D.)

8.54   *(a) By considering the relative bond lengths of the C—C and C—O bonds, predict which of the following two equilibria lies farther to the right. (That is, predict which of the two compounds contains more of the conformation with the axial methyl group.)

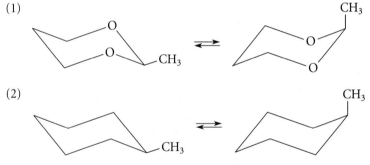

   (b) Which one of the following compounds contains the greater amount of *gauche* conformation for internal rotation about the bond shown? Explain.

$$CH_3CH_2 \overset{}{\frown} OCH_3 \qquad CH_3CH_2 \overset{}{\frown} CH_2CH_3$$

$$A \qquad\qquad\qquad B$$

*8.55   Give a mechanism for the following reaction, which takes place in several steps. Use the curved-arrow formalism. Use only Lewis acid-base associations, Lewis acid-base dissociations, and Brønsted acid-base reactions in your mechanism, and write *one reaction per step*. (*Hint:* See Eq. 8.15.)

$$3C_2H_5\overset{..}{\underset{..}{O}}H + BF_3 \longrightarrow C_2H_5\overset{..}{\underset{..}{O}}-\overset{\displaystyle :\overset{..}{O}C_2H_5}{\underset{\displaystyle :\underset{..}{O}C_2H_5}{B}} + 3H-F$$

9

Chemistry of Alkyl Halides

This chapter covers two very important types of alkyl halide reactions, *nucleophilic substitution reactions* and *β-elimination reactions*. These are among the most common and important reactions in organic chemistry. This chapter also introduces another class of reactive intermediates: *carbenes*.

9.1 An Overview of Nucleophilic Substitution and β-Elimination Reactions

A. Nucleophilic Substitution Reactions

When a *methyl halide* or a *primary alkyl halide* reacts with a Lewis base such as sodium ethoxide, a reaction occurs in which the Lewis base replaces the halogen, which is expelled as halide ion.

$$Na^+ \ CH_3CH_2\ddot{O}:^- \ + \ :\ddot{B}r\!-\!CH_2CH_3 \ \longrightarrow \ CH_3CH_2\ddot{O}\!-\!CH_2CH_3 \ + \ Na^+ \ :\ddot{B}r:^- \qquad (9.1)$$

sodium ethoxide   **ethyl bromide**   **diethyl ether**   **sodium bromide**

This is an example of a very general type of reaction, called a **nucleophilic substitution reaction**, or **nucleophilic displacement reaction**. It is a *substitution* because one group, ethoxy, is substituted for (or displaces) the bromo group, which departs as bromide ion. It is a *nucleophilic* substitution because ethoxide ion acts as a *nucleophile*, or Lewis base. The group that is displaced in a nucleophilic substitution reaction—the halide in Eq. 9.1—is termed a **leaving group**. Notice that the roles of nucleophile and leaving group would be reversed if the reaction could be run in the reverse direction.

nucleophile
↓

$$Na^+ \ CH_3CH_2\ddot{O}:^- \ + \ :\ddot{B}r\!-\!CH_2CH_3 \ \longrightarrow \ CH_3CH_2\ddot{O}\!-\!CH_2CH_3 \ + \ Na^+ \ :\ddot{B}r:^- \qquad (9.2)$$

↑
leaving group

Table 9.1   Some Nucleophilic Substitution Reactions

(X = halogen or other leaving group; R,R′ = alkyl groups)

R—$\ddot{\underset{..}{X}}$: + *Nucleophile (name)*	⟶	:$\ddot{\underset{..}{X}}$:⁻ + *Product (name)*
R—$\ddot{\underset{..}{X}}$: + :$\ddot{\underset{..}{Y}}$:⁻ (another halide)	⟶	:$\ddot{\underset{..}{X}}$:⁻ + R—$\ddot{\underset{..}{Y}}$: (another alkyl halide)
+ :⁻C≡N: (cyanide)	⟶	+ R—C≡N: (nitrile)
+ :⁻$\ddot{\underset{..}{O}}$H (hydroxide)	⟶	+ R—$\ddot{\underset{..}{O}}$H (alcohol)
+ :⁻$\ddot{\underset{..}{O}}$R′ (alkoxide)	⟶	+ R—$\ddot{\underset{..}{O}}$—R′ (ether)
+ ⁻N₃ (azide = :$\ddot{N}$=$\overset{+}{N}$=$\ddot{\underset{..}{N}}$:)	⟶	+ R—N₃ (alkyl azide)
+ :⁻$\ddot{\underset{..}{S}}$R′ (alkanethiolate)	⟶	+ R—$\ddot{\underset{..}{S}}$—R′ (thioether or sulfide)
+ :NR′₃ (amine)	⟶	R—$\overset{+}{N}$R′₃ :$\ddot{\underset{..}{X}}$:⁻ (alkylammonium salt)
+ :$\ddot{O}$H₂ (water)	⟶	R—$\overset{\overset{+}{\cdot\cdot}}{\underset{\mid}{O}}$—H :$\ddot{\underset{..}{X}}$:⁻ ⇌ R—$\ddot{\underset{..}{O}}$—H + H$\ddot{\underset{..}{X}}$: (alcohol) 	H
+ :$\ddot{O}$—R′ (alcohol) 	H	⟶	R—$\overset{\overset{+}{\cdot\cdot}}{\underset{\mid}{O}}$—R′ :$\ddot{\underset{..}{X}}$:⁻ ⇌ R—$\ddot{\underset{..}{O}}$—R′ + H$\ddot{\underset{..}{X}}$: (ether) 	H

Although many nucleophiles are anions, others are uncharged, or in a few cases, even positively charged. The following equation contains an example of an uncharged nucleophile. In addition, it illustrates an *intramolecular substitution reaction*—a reaction in which the nucleophile and the leaving group are part of the same molecule. In this case, the nucleophilic substitution reaction causes a ring to form.

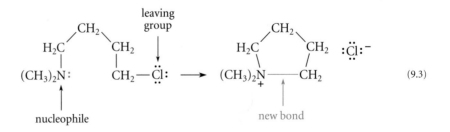

(9.3)

Nucleophilic substitution reactions can involve many different nucleophiles, a few of which are listed in Table 9.1. Notice from this table that nucleophilic substitution reactions can be used to transform alkyl halides into a wide variety of other functional groups. Moreover, we'll show in subsequent chapters that groups other than halides can act as leaving groups.

9.1   What is the expected nucleophilic substitution product when
*(a)   ethyl iodide reacts with ammonia?
(b)   methyl iodide reacts with Na^+ $CH_3CH_2CH_2S^-$?

B. β-Elimination Reactions

When a *tertiary alkyl halide* reacts with a base such as sodium ethoxide, a very different type of reaction is observed.

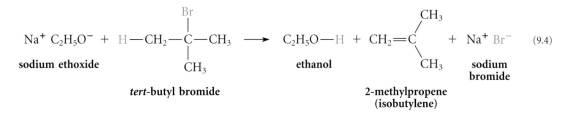

$$Na^+ \, C_2H_5O^- \; + \; \underset{\text{tert-butyl bromide}}{H-CH_2-\overset{\overset{\displaystyle Br}{|}}{\underset{\underset{\displaystyle CH_3}{|}}{C}}-CH_3} \; \longrightarrow \; \underset{\text{ethanol}}{C_2H_5O-H} \; + \; \underset{\substack{\text{2-methylpropene} \\ \text{(isobutylene)}}}{CH_2=\overset{\overset{\displaystyle CH_3}{\diagup}}{\underset{\diagdown CH_3}{C}}} \; + \; \underset{\substack{\text{sodium} \\ \text{bromide}}}{Na^+ \, Br^-} \qquad (9.4)$$

sodium ethoxide

This is an example of an **elimination reaction**: a reaction in which two or more groups (in this case H and Br) are lost from within the same molecule.

In an alkyl halide, the carbon bearing the halogen is often referred to as the **α-carbon**, and the adjacent carbons are referred to as the **β-carbons**. Notice in Eq. 9.4 that the halide is lost from the α-carbon and the hydrogen from a β-carbon.

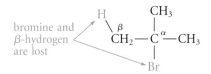

bromine and
β-hydrogen
are lost

An elimination that involves loss of two groups from adjacent carbons is termed a **β-elimination**. This is by far the most common type of elimination reaction in organic chemistry. Notice that a β-elimination reaction is conceptually the reverse of an addition to an alkene.

Strong bases promote the β-elimination reactions of alkyl halides. Among the most frequently used bases are alkoxides, such as sodium ethoxide (Na^+ $C_2H_5O^-$) and potassium *tert*-butoxide (K^+ $(CH_3)_3C-O^-$). Often the conjugate-acid alcohols of these bases are used as solvents, for example, sodium ethoxide in ethanol, or potassium *tert*-butoxide in *tert*-butyl alcohol.

The role of the base in a β-elimination reaction is quite different from that in a nucleophilic substitution reaction, but the role of the halogen is similar. In a base-promoted β-elimination reaction, a Brønsted base attacks a β-hydrogen of the alkyl halide, not a carbon atom as in a nucleophilic substitution reaction. However, the halogen is expelled as a halide ion *leaving group* in both types of reactions.

If the reacting alkyl halide has more than one type of β-hydrogen atom, then more than one β-elimination reaction is possible. It often happens that these different reactions occur at comparable rates so that more than one alkene product is formed, as in the following example.

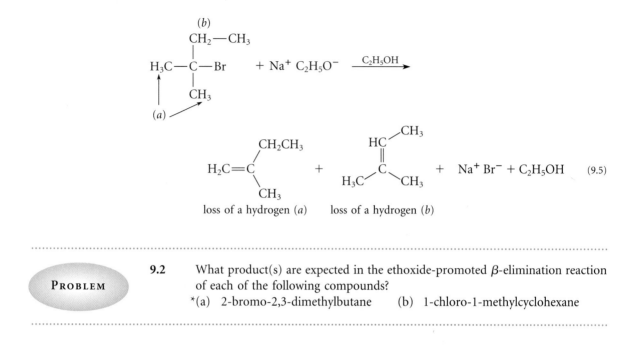

loss of a hydrogen (*a*)   loss of a hydrogen (*b*)

9.2 What product(s) are expected in the ethoxide-promoted β-elimination reaction of each of the following compounds?
*(a) 2-bromo-2,3-dimethylbutane   (b) 1-chloro-1-methylcyclohexane

C. Competition between Nucleophilic Substitution and β-Elimination Reactions

In the presence of a strong base such as ethoxide, the nucleophilic substitution reaction is a typical one for primary alkyl halides, and a β-elimination reaction is observed for tertiary alkyl halides. What about secondary alkyl halides? A typical secondary alkyl halide under the same conditions undergoes both reactions.

$$\underset{\substack{\textbf{isopropyl bromide}}}{\underset{\substack{|\\ \textbf{Br}}}{CH_3-CH-CH_3}} + \underset{\substack{\textbf{ethoxide}}}{C_2H_5O^-} \xrightarrow{C_2H_5OH} \underset{\substack{\textbf{ethyl isopropyl ether}\\ \textbf{substitution product}\\ (\text{about 50\%})}}{\underset{\substack{|\\ OC_2H_5}}{CH_3-CH-CH_3}} + \underset{\substack{\textbf{propene}\\ \textbf{elimination product}\\ (\text{about 50\%})}}{CH_2=CH-CH_3} \quad (9.6)$$

In other words, some molecules of the alkyl halide undergo substitution, while others undergo elimination. This means that the two reactions occur at comparable *rates*; in other words, the reactions are *in competition*. In fact, nucleophilic substitution and base-promoted β-elimination reactions are in competition for *all* alkyl halides with β-hydrogens, even primary and tertiary halides. It happens that in the presence of a strong Brønsted base, nucleophilic substitution is a faster reaction (it "wins the competition") for many primary alkyl halides, and in most cases β-elimination is a faster reaction for tertiary halides; that is why substitution predominates in the former case and elimination in the latter. However, under some conditions, the results of the competition can be changed. For example, under certain conditions even primary alkyl halides give mostly elimination products.

In the following sections we'll focus first on nucleophilic substitution reactions, then on β-elimination reactions. We'll discuss the factors that govern the reactivities of alkyl halides in each of these reaction types. Although each type of reaction is considered in isolation, keep in mind that substitutions and eliminations are always in competition.

PROBLEM

9.3   What are the substitution and elimination products that might be obtained when each of the following alkyl halides is treated with sodium methoxide in methanol?
*(a) 2-bromobutane   (b) bromocyclopentane
*(c) methyl iodide   (d) *trans*-1-bromo-3-methylcyclohexane

9.2   Equilibrium in Nucleophilic Substitution Reactions

Table 9.1 (p. 386) shows some of the many possible nucleophilic substitution reactions. How can you know whether the equilibrium for a given substitution is favorable? This problem is illustrated by the reaction of a cyanide ion with methyl iodide, which has an equilibrium constant that favors the product acetonitrile by many powers of ten.

$$^-:C\equiv N: + CH_3\ddot{I}: \longrightarrow CH_3C\equiv N: + :\ddot{I}:^- \tag{9.7}$$

acetonitrile

Other substitution reactions, however, are reversible or even unfavorable.

$$:\ddot{I}:^- + CH_3-\ddot{B}r: \rightleftarrows CH_3-\ddot{I}: + :\ddot{B}r:^- \tag{9.8}$$

$$:\ddot{I}:^- + CH_3-\ddot{O}H \longleftarrow CH_3-\ddot{I}: + ^-:\ddot{O}H \quad \text{(does not proceed to the right)} \tag{9.9}$$

Results such as these can be predicted by recognizing that each nucleophilic substitution reaction is conceptually similar to a Brønsted acid-base reaction. That is, if the alkyl group of the alkyl halide is replaced with a hydrogen, the substitution looks like an acid-base reaction.

$$^-OH + HI \longrightarrow H-OH + I^- \quad \text{(acid-base reaction)} \tag{9.10}$$

$$^-OH + CH_3I \longrightarrow CH_3OH + I^- \quad \text{(substitution reaction)} \tag{9.11}$$

In the acid-base reaction, ^-OH displaces I^- from the *proton*; in the substitution reaction, ^-OH displaces I^- from *carbon*. What makes this analogy particularly useful is that *whether the equilibrium in a nucleophilic substitution reaction is favorable can be predicted from an analysis of the corresponding Brønsted acid-base reaction.* Thus, if the Brønsted acid-base reaction

$$H-X + Y:^- \rightleftarrows Y-H + X:^- \tag{9.12a}$$

strongly favors the right side of the equation, then the analogous substitution reaction

$$R-X + Y:^- \rightleftarrows Y-R + X:^- \tag{9.12b}$$

likewise favors the right side of the equation. This means that *the equilibrium in any nucleophilic substitution reaction, as in an acid-base reaction, favors release of the weaker base.* This principle, for example, shows why I^- will *not* displace ^-OH from CH_3OH in Eq. 9.9: I^- is a much weaker base than ^-OH (Table 3.1). In fact, just the opposite reaction occurs: ^-OH readily displaces I^- from CH_3I. This example illustrates how mastery of the acid-base principles discussed in Chapter 3 can prove very useful in understanding nucleophilic substitution reactions.

> The analogy between nucleophilic substitution reactions and Brønsted acid-base reactions, as useful as it is, is not perfect, because in the former case we are dealing with bonds to *carbon*, whereas in the latter, we are dealing with bonds to *hydrogen*. Another deficiency in this analogy is that nucleophilic substitution reactions are sometimes run in solvents in which basicities are quite different from those in water (the solvent in which the pK_a values in Table 3.1 apply). Nevertheless, in many cases the difference in the basicities of nucleophiles and leaving groups is so great that this analogy is very useful.

Some equilibria that are not too unfavorable can be driven to completion by applying **LeChatelier's principle**, a central principle of chemical equilibrium. LeChatelier's principle states that if an equilibrium is disturbed, the components of the equilibrium will react so as to offset the effect of the disturbance. For example, alkyl chlorides normally do not react to completion with iodide ion because iodide is a weaker base than chloride; the equilibrium favors the formation of the weaker base, iodide. However, in the solvent acetone, it happens that potassium iodide is relatively soluble and potassium chloride is relatively insoluble. Thus, when an alkyl chloride reacts with KI in acetone, KCl precipitates, and the equilibrium compensates for this disturbance—the loss of KCl—by forming more of it, along with, of course, more alkyl iodide.

$$R\text{—}Cl + KI \longrightarrow R\text{—}I + KCl \quad \text{(precipitates in acetone)} \qquad (9.13)$$

PROBLEM

9.4   Tell whether each of the following reactions favors reactants or products at equilibrium. (Assume that all reactants and products are soluble.)

*(a) $CH_3Cl + I^- \longrightarrow CH_3I + Cl^-$

(b) $CH_3Cl + F^- \longrightarrow CH_3F + Cl^-$

*(c) $CH_3Cl + N_3^- \longrightarrow CH_3N_3 + Cl^-$ (*Hint:* the pK_a of HN$_3$ is 4.72.)

(d) $CH_3Cl + {}^-OCH_3 \longrightarrow CH_3OCH_3 + Cl^-$

9.3   Reaction Rates

The last section showed how to determine whether the equilibrium for a nucleophilic substitution reaction is favorable. Knowledge of the equilibrium constant for a reaction provides no information about the rate at which the reaction takes place (Sec. 4.8A). Although some substitution reactions with favorable equilibria proceed rapidly, others proceed slowly. For example, the reaction of methyl iodide with cyanide is a relatively fast reaction, whereas the reaction of cyanide with neopentyl iodide is so slow that it is virtually useless:

$$^-{:}C{\equiv}N{:} + CH_3\text{—}\ddot{\underset{\cdot\cdot}{I}}{:} \xrightarrow[\text{(rapid)}]{} CH_3\text{—}C{\equiv}N{:} + {:}\ddot{\underset{\cdot\cdot}{I}}{:}^- \qquad (9.14)$$

$$^-{:}C{\equiv}N{:} + CH_3\text{—}\underset{\underset{CH_3}{|}}{\overset{\overset{CH_3}{|}}{C}}\text{—}CH_2\text{—}\ddot{\underset{\cdot\cdot}{I}}{:} \xrightarrow[\text{(very slow)}]{} CH_3\text{—}\underset{\underset{CH_3}{|}}{\overset{\overset{CH_3}{|}}{C}}\text{—}CH_2\text{—}C{\equiv}N{:} + {:}\ddot{\underset{\cdot\cdot}{I}}{:}^- \qquad (9.15)$$

Why do reactions that are so similar conceptually differ so drastically in their rates? In other words, what determines the *reactivity* of a given alkyl halide in a nucleophilic substitution reaction? Because this question deals with reaction rates and the concept of the transition state, you should review the introduction to these subjects in Sec. 4.8.

A. Definition of Reaction Rate

The term *rate* implies that something is changing with time. For example, in the rate of travel, or the *velocity*, of a car, the "something" that is changing is the car's position:

$$\text{velocity} = v = \frac{\text{change in position}}{\text{corresponding change in time}} \tag{9.16}$$

The quantities that change with time in a chemical reaction are the *concentrations of the reactants and products.*

$$\text{reaction rate} = \frac{\text{change in product concentration}}{\text{corresponding change in time}} \tag{9.17a}$$

$$= -\frac{\text{change in reactant concentration}}{\text{corresponding change in time}} \tag{9.17b}$$

(The reason for the different signs is that the concentrations of the reactants decrease with time, and the concentrations of the products increase.)

In mechanics, a rate has the dimensions of length per unit time, for example, meters per second. A reaction rate, by analogy, has the dimensions of concentration per unit time. When the concentration unit is the mole per liter (M), and if time is measured in seconds, then the unit of reaction rate is

$$\frac{\text{concentration}}{\text{time}} = \frac{\text{mol/L}}{\text{sec}} = M/\text{sec} = M \text{ sec}^{-1} \tag{9.18}$$

B. The Rate Law

In order for molecules to react with one another, they must "get together," or collide. Because molecules at high concentration are more likely to collide than molecules at low concentration, the rate of a reaction should be a function of the concentrations of the reactants. The mathematical statement of how a reaction rate depends on concentration is called the **rate law**. A rate law is determined experimentally by varying the concentration of each reactant (including any catalysts) independently and measuring the resulting effect on the rate. Each reaction has its own characteristic rate law. For example, suppose that for the reaction $A + B \rightarrow C$ the reaction rate doubles if *either* $[A]$ or $[B]$ is doubled, and increases by a factor of four if *both* $[A]$ and $[B]$ are doubled. The rate law for this reaction is then

$$\text{rate} = k[A][B] \tag{9.19a}$$

If in another reaction $D + E \rightarrow F$, the rate doubles only if the concentration of D is doubled, and changing the concentration of E has no effect, the rate law is then

$$\text{rate} = k[D] \tag{9.19b}$$

The concentrations in the rate law are the concentrations of reactants at any time during the reaction, and the rate is the velocity of the reaction at that same time. The constant of proportionality, k, is called the **rate constant**. In general, the rate constant is different for every reaction, and it is a *fundamental physical constant* for a given reaction under particular conditions of temperature, pressure, solvent, and so on. As Eqs. 9.19a and b show, the rate constant is numerically equal to the rate of the reaction when all reactants are present at 1 M concentration; that is, the rate constant is the rate of the reaction under standard conditions of unit concentration. *The rates of two reactions are compared by comparing their rate constants.*

An important aspect of a reaction is its **kinetic order**. The **overall kinetic order** for a reaction is the sum of the powers of all the concentrations in the rate law. For a reaction described by the rate law in Eq. 9.19a, the overall kinetic order is two; the reaction described by this rate law is said to be a *second-order reaction*. The overall kinetic order of a reaction having the rate law in Eq. 9.19b is one; such a reaction is thus a *first-order reaction*. The **kinetic order in each reactant** is the power to which its concentration is raised in the rate law. Thus, the reaction described by the rate law in Eq. 9.19a is said to be *first order in each reactant*. A reaction with the rate law in Eq. 9.19b is first order in D and zero order in E.

The *dimensions* of the rate constant depend on the kinetic order of the reaction. With concentrations in moles/liter, and time in seconds, the rate of any reaction has the dimensions of M/sec (Eq. 9.18). For a second-order reaction, then, dimensional consistency requires that the rate constant have the dimensions of $M^{-1}\text{sec}^{-1}$.

$$\text{rate} = k[A][B] \tag{9.20}$$
$$= M^{-1}\text{sec}^{-1} \cdot M \cdot M = M/\text{sec}$$

Similarly, the rate constant for a first-order reaction has dimensions of sec^{-1}.

C.   Reaction Rate and the Standard Free Energy of Activation

According to transition-state theory in Sec. 4.8, the standard free energy of activation, or energy barrier, determines the rate of a reaction under standard conditions. Because the rate constant *is* the reaction rate under standard conditions, it follows that the rate constant is related to the standard free energy of activation $\Delta G^{\circ\ddagger}$. If $\Delta G^{\circ\ddagger}$ is large for a reaction, the reaction is relatively slow, and the rate constant is small. If $\Delta G^{\circ\ddagger}$ is small, the reaction is relatively fast, and the rate constant is large. This relationship is shown in Fig. 9.1. The mathematical relationship between rate and $\Delta G^{\circ\ddagger}$ is a logarithmic one, much like the relationship between ΔG° and equilibrium constant. That is, if the rates of two reactions A and B are compared, the relationship between their rate constants and their standard free energies of activation is

$$\log (k_A/k_B) = \frac{\Delta G_A^{\circ\ddagger} - \Delta G_B^{\circ\ddagger}}{-2.3RT} \tag{9.21}$$

This equation says that the reaction rates of two reactions differ by a factor of 10 (that is, one log unit) for every increment of $2.3RT$ (5.7 kJ/mol or 1.4 kcal/mol at 298 K) difference in their standard free energies of activation.

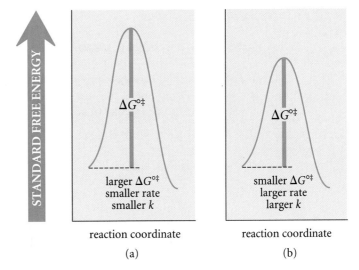

Figure 9.1   *Relationship between standard free energy of activation ($\Delta G^{\circ\ddagger}$), reaction rate, and rate constant, k. (a) A reaction with a larger $\Delta G^{\circ\ddagger}$ has a smaller rate and a smaller rate constant. (b) A reaction with a smaller $\Delta G^{\circ\ddagger}$ has a larger rate and a larger rate constant.*

PROBLEMS

9.5   For each of the following reactions, (1) what is the overall kinetic order of the reaction, (2) what is the order in each reactant, and (3) what are the dimensions of the rate constant?

*(a)   an addition reaction of bromine to an alkene with the rate law

$$\text{rate} = k[\text{alkene}][\text{Br}_2]^2$$

(b)   a substitution reaction of an alkyl halide with the rate law

$$\text{rate} = k[\text{alkyl halide}]$$

9.6   *(a)   What is the ratio of rate constants k_A/k_B at 25 °C for two reactions A and B if the standard free energy of activation of reaction A is 14 kJ/mol (3.4 kcal/mol) less than that of reaction B?

(b)   What is the difference in the standard free energies of activation at 25 °C of two reactions A and B if reaction B is 450 times as fast as reaction A? Which reaction has the greater $\Delta G^{\circ\ddagger}$?

*9.7   What prediction does the rate law in Eq. 9.19a make about how the rate of the reaction changes as the reactants *A* and *B* are converted into *C* over time? Does the rate increase, decrease, or stay the same? Explain. Use your answer to sketch a plot of the concentrations of starting materials and products against time.

The S_N2 Reaction

A. Rate Law and Mechanism of the S_N2 Reaction

The rate law is important because it provides fundamental information about the mechanism of a reaction. For example, consider the nucleophilic substitution reaction of ethoxide ion with methyl iodide in ethanol at 25°.

$$C_2H_5O^- + CH_3—I \longrightarrow C_2H_5O—CH_3 + I^- \tag{9.22}$$

The following rate law for this reaction was experimentally determined:

$$\text{rate} = k[CH_3I][C_2H_5O^-] \tag{9.23}$$

with $k = 6.0 \times 10^{-4} \ M^{-1}\text{sec}^{-1}$. That is, this is a *second-order reaction, first order in each reactant.*

The concentration terms of the rate law indicate *what atoms are present in the transition state of the rate-limiting step.* Hence, the transition state of reaction 9.22 consists of the elements of one methyl iodide molecule and one ethoxide ion. The rate law excludes some mechanisms from consideration. For example, any mechanism in which the transition state contains two molecules of ethoxide is *ruled out* by the rate law, because the rate law for such a mechanism would have to be second order in ethoxide.

The simplest possible mechanism consistent with the rate law is one in which the ethoxide ion *directly displaces* the iodide ion from the methyl carbon:

$$C_2H_5\ddot{\underset{..}{O}}: \overset{\curvearrowright}{} CH_3 \overset{\curvearrowleft}{—} \ddot{\underset{..}{I}}: \longrightarrow \left[C_2H_5\ddot{\underset{..}{O}}: \overset{\delta-}{\cdots} \underset{\underset{H \quad H}{\diagup \ \diagdown}}{\overset{\overset{H}{|}}{C}} \cdots \overset{\delta-}{:}\ddot{\underset{..}{I}}: \right]^{\ddagger} \longrightarrow C_2H_5\ddot{\underset{..}{O}}—CH_3 + :\ddot{\underset{..}{I}}:^- \tag{9.24}$$

transition state

Mechanisms like this account for many nucleophilic substitution reactions. A mechanism in which attack of a nucleophile on an atom (usually carbon) displaces a leaving group from the same atom in a concerted manner is called an **S_N2 mechanism**. Reactions that occur by S_N2 mechanisms are called **S_N2 reactions**. The meaning of the "nickname" S_N2 is as follows:

(The word *bimolecular* means that the transition state of the reaction involves two species, in this case, one methyl iodide molecule and one ethoxide ion.) Notice that an S_N2 reaction, because it is concerted, involves no reactive intermediates.

The rate law does not reveal the exact mechanism of a reaction. *Although the rate law indicates what atoms are present in the transition state, it provides no information about how they are arranged.* Thus, the following two mechanisms for the S_N2 reaction of ethoxide ion with methyl iodide are equally consistent with the rate law.

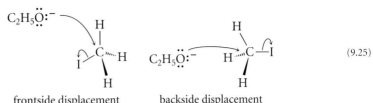

(9.25)

frontside displacement   backside displacement

STUDY GUIDE LINK:
✓9.1
Deducing
Mechanisms from
Rate Laws

As far as the rate law is concerned, either mechanism is acceptable. To decide between these two possibilities, other types of experiments are needed (Sec. 9.4B).

Let's summarize the relationship between the rate law and the mechanism of a reaction.

1. The concentration terms of the rate law indicate what atoms are involved in the rate-limiting transition state.

2. Mechanisms not consistent with the rate law are ruled out.

3. Of the chemically reasonable mechanisms consistent with the rate law, the simplest one is provisionally adopted.

4. The mechanism of a reaction is modified or refined if required by subsequent experiments.

Point (4) may seem disturbing; after all, it means that a mechanism can be changed at a later time. Perhaps it seems that an "absolutely true" mechanism should exist for every reaction. However, the value of a mechanism lies not in its absolute truth but rather in its validity as a conceptual framework, or theory, that generalizes the results of many experiments and predicts the outcome of others. Mechanisms allow us to place reactions into categories and thus impose a conceptual order on chemical observations. Thus, when someone observes an experimental result different from that predicted by a mechanism, then the mechanism must be modified to accommodate both the previously known facts and the new facts. The evolution of mechanisms is no different than the evolution of science in general. Knowledge is dynamic: theories (mechanisms) predict the results of experiments, a test of these theories may lead to new theories, and so on.

PROBLEMS

*9.8   The reaction of acetic acid with ammonia is very rapid and follows the simple rate law below. Propose a mechanism that is consistent with this rate law.

$$CH_3-\overset{\displaystyle O}{\overset{\|}{C}}-\ddot{\underset{\cdot\cdot}{O}}-H + :NH_3 \;\rightleftharpoons\; CH_3-\overset{\displaystyle O}{\overset{\|}{C}}-\ddot{\underset{\cdot\cdot}{O}}:^- + \overset{+}{N}H_4$$

acetic acid

$$\text{rate} = k[CH_3-\overset{\displaystyle O}{\overset{\|}{C}}-OH][NH_3]$$

9.9   What rate law would be expected for the reaction of cyanide ion ($^-$:CN) with ethyl bromide by the S$_N$2 mechanism?

Table 9.2 **Effect of Branching in the Alkyl Halide on the Rate of a Typical S_N2 Reaction**

$$R-Br + I^- \xrightarrow{\text{acetone, 25}^\circ} R-I + Br^-$$

R—	Name of R	Relative rate[a]
CH_3-	methyl	145
Increased branching at the β-carbon:		
$CH_3CH_2CH_2-$	propyl	0.82
$(CH_3)_2CHCH_2-$	isobutyl	0.036
$(CH_3)_3CCH_2-$	neopentyl	0.000012
Increased branching at the α-carbon:		
CH_3CH_2-	ethyl	1.0
$(CH_3)_2CH-$	isopropyl	0.0078
$(CH_3)_3C-$	*tert*-butyl	~0.0005[b]

[a] All rates are relative to that of ethyl bromide.
[b] Estimated from the rates of closely related reactions.

structure. If an alkyl halide is very reactive, its S_N2 reactions occur rapidly under mild conditions. If an alkyl halide is relatively unreactive, then the severity of the reaction conditions (for example, the temperature) must be increased in order for the reaction to proceed at a reasonable rate.

Alkyl halides differ, in some cases by many orders of magnitude, in the rates with which they undergo a given S_N2 reaction. Typical reactivity data are given in Table 9.2. To put these data in some perspective: if the reaction of a methyl halide takes about one *minute*, then the reaction of a neopentyl halide under the same conditions takes about *23 years*!

The data in Table 9.2 show, first, that *increased branching at the β-carbon retards an S_N2 reaction.* As Fig. 9.3 shows, these data are consistent with a backside displacement mechanism. When a methyl halide undergoes substitution, approach of the nucleophile and departure of the leaving group are relatively unrestricted. However, when a neopentyl halide is attacked by a nucleophile, both the nucleophile and the leaving group experience severe van der Waals repulsions with hydrogens of the methyl branches. These van der Waals repulsions raise the energy of the transition state and therefore reduce the reaction rate. This is another example of a *steric effect* (Sec. 5.6D). Thus, S_N2 reactions of branched alkyl halides are retarded by a steric effect. Indeed, S_N2 reactions of neopentyl halides are so slow that they are not practically useful.

Table 9.2 shows that branching at the α-carbon also decreases the rate of the S_N2 reaction. Thus, the secondary alkyl halide, isopropyl bromide, reacts very slowly, and the tertiary alkyl halide, *tert*-butyl bromide, more slowly still. In general, secondary alkyl halides undergo S_N2 reactions much more slowly than typical primary alkyl halides, and tertiary alkyl halides are even less reactive.

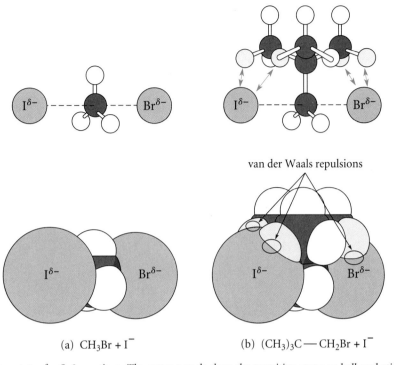

van der Waals repulsions

Figure 9.3 *Transition states for S_N2 reactions. The upper panels show the transition states as ball-and-stick models, and the lower panels show them as space-filling models. (a) The reaction of methyl bromide with iodide ion. (b) The reaction of neopentyl bromide with iodide ion. The S_N2 reactions of neopentyl bromide are very slow because of the severe van der Waals repulsions of the nucleophile and the leaving group with the colored hydrogens of the methyl branches. These repulsions are indicated with double-headed colored arrows in the ball-and-stick models.*

The data in Table 9.2 help explain why elimination reactions compete with the S_N2 reactions of secondary and tertiary alkyl halides (Sec. 9.1C): these halides react so slowly in S_N2 reactions that the rates of elimination reactions are competitive with the rates of substitution. The rates of the S_N2 reactions of tertiary alkyl halides are so slow that, except in certain very special cases, elimination is the only reaction observed. The competition between E2 and S_N2 reactions will be considered in more detail in Sec. 9.5F.

D. Solvent Effects on Nucleophilicity in the S_N2 Reaction

As Table 9.1 illustrates, the S_N2 reaction is especially useful because of the variety of nucleophiles that can be employed. However, nucleophiles differ significantly in their reactivities, and the reactivity of a given nucleophile in an S_N2 reaction can be significantly affected by the solvent used for the reaction. This section deals with nucleophilic reactivity in S_N2 reactions and the effect of solvents on nucleophilic reactivity.

Nucleophilic reactivity is termed **nucleophilicity**. Thus, a more reactive nucleophile has greater nucleophilicity than a less reactive one. Notice that even though all nucleophiles are Lewis bases, nucleophilicity and basicity are somewhat different concepts. Basicity, or strength, of a base is determined by the magnitude of the *equilibrium constant* for its

Table 9.3   **Solvent Effects on the Rates of S_N2 Reactions of Different Nucleophiles with Methyl Iodide**

$$\left.\begin{array}{c} X^- \\ \text{or } XH \end{array}\right\} + CH_3—I \longrightarrow X—CH_3 + \left\{\begin{array}{l} I^- \\ \text{or } HI \end{array}\right.$$

Nucleophile X^- or XH	pK_a of conjugate acid[a]	log k in methanol ($\epsilon = 33$)[c]	DMF[b] ($\epsilon = 37$)[c]	log ($k_{DMF}/k_{methanol}$)
CH_3O^-	15.1	-3.6	1.8^d	5.4^d
^-CN	9.4	-3.2	2.5	5.7
^-OAc	4.76	-5.6	1.3	6.9
^-SCN	4.00	-3.3	-1.1	2.2
F^-	3.2	-7.3^e	$>0.4^e$	$>7.7^e$
CH_3OH	-2.5	negligible	negligible	—
Cl^-	-6 to -7	-5.5	0.4	5.9
Br^-	-8 to -9.5	-4.1	0.1	4.2
I^-	-9.5 to -10	-2.5	-0.4	2.1

[a] pK_a values in aqueous solution.
[b] DMF = N,N-dimethylformamide
[c] ϵ = solvent dielectric constant. Notice that the two solvents do not differ appreciably in polarity.
[d] In dimethyl sulfoxide (DMSO) ($\epsilon = 47$); methoxide ion reacts with DMF.
[e] estimated

reaction with a standard acid. In contrast, the nucleophilicity of a base is determined by the *rate* of its reaction with a standard Lewis acid. The standard Lewis acid used in this discussion will be methyl iodide.

Some nucleophilicity data for the S_N2 reaction of methyl iodide with different nucleophiles in two different solvents, methanol and N,N-dimethylformamide (DMF), are given in Table 9.3. Notice that because the rates of the S_N2 reactions in this table vary over many orders of magnitude, they are given in logarithmic form; the numbers are the logarithms of the S_N2 rate constants. For example, looking at the second entry (for ^-CN), the log of the reaction rate in methanol is -3.2; this means that the rate constant is $10^{-3.2}$ or 6.3×10^{-4} $M^{-1}sec^{-1}$. The last column compares the rate of each reaction in DMF relative to its rate in methanol. For ^-CN, the logarithm of this ratio is 5.7; in other words, the reaction of methyl iodide with cyanide ion is $10^{5.7}$, or 5×10^5, or *500,000 times as fast* in DMF as it is in methanol.

What are the characteristics of a good nucleophile? First, when the attacking atom is the same, nucleophilicity correlates roughly with Brønsted basicity. Consider, for example, the relative rates of reaction of methyl iodide with the following three nucleophiles (in each case an oxygen is the nucleophilic atom):

Relative rate of reaction with methyl iodide:

$$CH_3OH \ll {^-OAc} < {^-OCH_3} \tag{9.28}$$

The rates increase in the same order as the Brønsted basicity of the nucleophiles. In fact, alcohols such as methanol are such poor nucleophiles that a primary alkyl halide such as methyl iodide or ethyl iodide dissolved in methanol solvent does not react at all unless it is left for several days, or the solution is heated strongly. Yet, if a strong base such as methoxide is added to the solution, the methyl iodide reacts at a convenient rate. Since Brønsted basicity and nucleophilicity both involve donation of an electron pair, the relationship between these two properties should not be surprising.

When the data of Table 9.3 are examined for a more general relationship of nucleophilicity and basicity, the result is that *the relationship between nucleophilicity and basicity depends strongly on the solvent.* Consider, for example, the reactivities of nucleophiles in methanol within a column of the periodic table.

Nucleophilic reactivities of halide ions in methanol:

$$F^- \ll Cl^- < Br^- < I^- \tag{9.29a}$$

Notice that the reactivities of these nucleophiles are *exactly the opposite* of their basicities. Fluoride, the most basic halide, is so unreactive that it is virtually useless as a nucleophile in methanol. In contrast, iodide, the least basic halide, is a good nucleophile. To generalize: *In protic solvents nucleophilicity increases toward greater atomic number within a group of the periodic table.*

In DMF, however, the relative nucleophilic reactivities of the halides follow their relative basicities.

Nucleophilic reactivities of halide ions in DMF:

$$F^- > Cl^- > Br^- > I^- \tag{9.29b}$$

However, as Table 9.3 shows, the differences in their reactivities, in contrast to the differences in methanol, are not large.

The most significant conclusion from Table 9.3 is that the reactivities of virtually all nucleophilic anions are significantly greater in DMF than in methanol. To generalize: *The nucleophilicity of an anion is much greater in a polar aprotic solvent* (such as acetone, DMF, or dimethyl sulfoxide) *than in a protic solvent* (such as water, methanol, or ethanol). To see just how dramatic this solvent effect is, consider that the reaction of cyanide ion ($^-$CN) with methyl iodide in methanol solvent is 90% complete in about an hour. In DMF, with the same concentrations of reactants, the reaction is 90% complete in about 0.01 second!

Both of these trends have a common explanation. In a protic solvent, *hydrogen bonding* occurs between the protic solvent molecules (as hydrogen bond donors) and the nucleophilic anions (as hydrogen bond acceptors). The strongest *Brønsted* bases are the best hydrogen bond acceptors. For example, fluoride ion forms much stronger hydrogen bonds than iodide ion. When the electron pairs of a nucleophile are involved in hydrogen bonding, they are not available for donation to carbon in an S$_N$2 reaction. In order for the S$_N$2 reaction to take place, *the hydrogen bonds between the solvent and the nucleophile must be broken* (Fig. 9.4). More energy is required to break a strong hydrogen bond to fluoride ion than is required to break a relatively weak hydrogen bond to iodide ion. This extra energy is reflected in a greater free energy of activation—the energy barrier—and, as a result, the reaction of fluoride ion is slower. To use a football analogy, the attack of a strongly hydrogen-bonded anion on an alkyl halide is about as likely as a tackler bringing down a ball carrier when both of the tackler's arms are being held by opposing linemen.

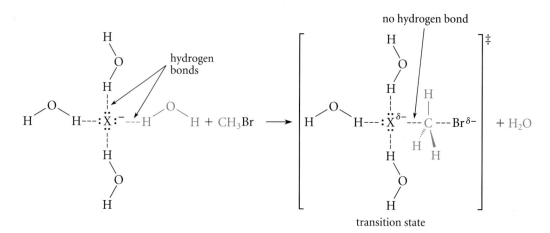

Figure 9.4   *An S_N2 reaction involving a halide nucleophile ($\ddot{\text{:X:}}^{-}$) in a protic solvent requires breaking a hydrogen bond to the nucleophile. When $X^{-} = F^{-}$, the hydrogen bond is strong, but when $X^{-} = I^{-}$, it is weak. For this reason more energy is required to reach the transition state when $X^{-} = F^{-}$, and, as a result, the reaction is slower.*

In aprotic solvents, there is no hydrogen bonding to reduce the nucleophilicity of dissolved anions; consequently, anions dissolved in such solvents are considerably more nucleophilic than they are in protic solvents. Realistically, the only aprotic solvents that can dissolve useful quantities of ionic compounds are the polar aprotics (Sec. 8.4B). This accounts for their utility as solvents in the S_N2 reaction.

Because uncharged nucleophiles are much less strongly hydrogen-bonded than anionic ones, uncharged nucleophiles show a much weaker dependence on solvent. For example, the uncharged nucleophile dimethyl sulfide, $(CH_3)_2S$, reacts only about three times faster with methyl bromide in a polar aprotic solvent than it does in methanol.

In a practical sense, what are the best conditions for an S_N2 reaction? First, the solvent must be able to dissolve the nucleophile, which in many cases is the anion of an ionic compound. For this reason, water, alcohols, and mixtures of water and other solvents are frequently used, even though S_N2 reactions are relatively slow in these protic solvents. Part of the popularity of these solvents also derives from the fact that they are cheap, easy to handle, and easy to remove from products when the reaction is done. However, polar aprotic solvents are particularly useful provided that they dissolve the desired nucleophile. An alkyl halide-nucleophile combination that reacts sluggishly in a protic solvent can react rapidly in a polar aprotic solvent. The reactions of KBr and KI with alkyl chlorides in acetone (Eq. 9.13), for example, take advantage not only of the insolubility of the KCl product, but also of the enhanced nucleophilicity of bromide and iodide ions in acetone, a polar aprotic solvent.

PROBLEMS

9.11   In each pair, choose the species that reacts more rapidly with methyl bromide in ethanol solvent, and explain your choices.
*(a) $^{-}OC_2H_5$ or C_2H_5OH (b) F^{-} or I^{-}

9.12   In each pair, choose the combination of nucleophile and solvent that would give the faster S_N2 reaction with ethyl iodide. (Consult Table 8.2, if necessary.)

*(a)   Na$^+$ $^-$OAc in formamide, or Na$^+$ $^-$OAc in DMF
(b)   Na$^+$ $^-$OAc in acetic acid, or Na$^+$ $^-$OAc in DMF

E. Leaving-Group Effects in the S$_N$2 Reaction

In many cases, when an alkyl halide is to be used as a starting material in an S$_N$2 reaction, a choice of leaving group is possible. That is, an alkyl halide might be readily available as either an alkyl chloride, alkyl bromide, or alkyl iodide. In such a case, the halide that reacts most rapidly is usually preferred. The reactivities of alkyl halides can be predicted from the close analogy between S$_N$2 reactions and Brønsted acid-base reactions. Recall that the ease of dissociating an H—X bond within the series of hydrogen halides depends mostly on the H—X bond energy (Sec. 3.6A), and for this reason, H—I is the strongest acid among the hydrogen halides. Likewise, S$_N$2 reactivity depends primarily on the *carbon*-halogen bond energy, which follows the same trend: alkyl iodides are the most reactive alkyl halides, and alkyl fluorides are the least reactive.

Relative reactivities in S$_N$2 reactions:

$$R—F \ll R—Cl < R—Br < R—I \qquad (9.30)$$

In other words, *the best leaving group reacts to give the weakest base.* Fluorides are useless as leaving groups in most S$_N$2 reactions, but chlorides, bromides, and iodides all have acceptable reactivities. Typically, an alkyl bromide is 50 times as reactive as an alkyl chloride, which is about 200 times as reactive as an alkyl fluoride. However, alkyl iodides are only 3 to 5 times as reactive as alkyl bromides. On a laboratory scale, alkyl bromides, which are in most cases less expensive than alkyl iodides, usually represent the best compromise between expense and reactivity. On a large scale, the lower cost of alkyl chlorides offsets the disadvantage of their lower reactivity.

Halides are not the only groups that can be used as leaving groups in S$_N$2 reactions. The next chapter will introduce a variety of alcohol derivatives that can also be used as starting materials for S$_N$2 reactions.

F. Summary of the S$_N$2 Reaction

Primary and some secondary alkyl halides undergo nucleophilic substitution by the S$_N$2 mechanism. Let's summarize six of the characteristic features of this mechanism.

1. The reaction rate is second order overall: first order in the nucleophile and first order in the alkyl halide.
2. The mechanism involves backside attack of the nucleophile on the alkyl halide and inversion of stereochemical configuration.
3. The reaction rate is retarded by branching at both the α- and β-carbon atoms; neopentyl halides are virtually unreactive.
4. The strongest bases are generally the most reactive nucleophiles; however, the reverse is true in protic solvents for attacking atoms within a group (column) of the periodic table.

5. The nucleophilicities of anionic nucleophiles are much greater in a polar aprotic solvent than in a protic solvent.

6. The fastest S_N2 reactions involve leaving groups that have the weakest bonds to carbon—those that give the weakest bases as products.

9.5   The E2 Reaction

This section discusses base-promoted β-elimination, which is a second important reaction of alkyl halides. An example of such a reaction is the elimination of the elements of HBr from *tert*-butyl bromide:

$$CH_3-\underset{\underset{CH_3}{|}}{\overset{\overset{CH_3}{|}}{C}}-Br + Na^+\ C_2H_5O^- \xrightarrow[\text{ethanol}]{25°} CH_2=\underset{CH_3}{\overset{CH_3}{C}} + C_2H_5OH + Na^+\ Br^- \qquad (9.31)$$

Recall (Sec. 9.1B) that this type of elimination is a dominant reaction of tertiary alkyl halides in the presence of a strong base, and it competes with the S_N2 reaction in the case of secondary and primary alkyl halides.

A.   Rate Law and Mechanism of the E2 Reaction

Base-promoted β-elimination reactions typically follow a rate law that is second order overall and first order in each reactant:

$$\text{rate} = k[(CH_3)_3C-Br][C_2H_5O^-] \qquad (9.32)$$

A mechanism consistent with this rate law is the following:

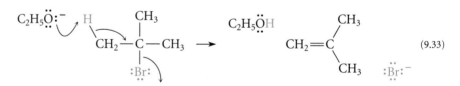

$$(9.33)$$

This type of mechanism, involving concerted removal of a β-proton by a base and loss of a halide ion, is called an **E2 mechanism**. Reactions that occur by the E2 mechanism are called **E2 reactions**. The meaning of the "nickname" E2 is as follows:

B.   Leaving-Group Effects on the E2 Reaction

In the mechanism of the E2 reaction, the role of the leaving halide is much the same as it is in the S_N2 reaction. Consequently, it should not be surprising to find that the rates of S_N2 and E2 reactions are affected in similar ways by changing the halide leaving group:

Relative reactivities in E2 reactions:

$$R—Cl < R—Br < R—I \qquad (9.34)$$

As in the S_N2 reaction, the reactivity difference between alkyl bromides and iodides is not great. Alkyl bromides are usually used in the laboratory for E2 reactions as the best compromise of reactivity and expense, and the less expensive alkyl chlorides are used in large-scale reactions.

C. Deuterium Isotope Effects in the E2 Reaction

The mechanism in Eq. 9.33 implies that a proton is removed in the transition state of the E2 reaction. This aspect of the mechanism can be tested in an interesting way. When a hydrogen is transferred in the rate-limiting step of a reaction, a compound in which that hydrogen is replaced by its isotope deuterium will react more slowly in the same reaction. This effect of isotopic substitution on reaction rates is called a **primary deuterium isotope effect**. For example, suppose the rate constant for the following E2 reaction of 2-phenyl-1-bromoethane is k_H, and the rate constant for the reaction of its β-deuterium analog is k_D:

$$Ph—CH_2—CH_2—Br + C_2H_5O^- \xrightarrow[\text{rate constant } k_H]{C_2H_5OH} Ph—CH{=}CH_2 + Br^- + C_2H_5OH \qquad (9.35a)$$

$$Ph—CD_2—CH_2—Br + C_2H_5O^- \xrightarrow[\text{rate constant } k_D]{C_2H_5OH} Ph—CH{=}CH_2 + Br^- + C_2H_5OD \qquad (9.35b)$$

The primary deuterium isotope effect is the ratio k_H/k_D; typically such isotope effects are in the range 2.5–8. In fact, k_H/k_D for the reactions in Eq. 9.35 is 7.1. The observation of a primary isotope effect of this magnitude shows that the bond to a β-hydrogen is broken in the rate-limiting step of this reaction.

The theoretical basis for the primary isotope effect lies in the comparative strengths of C—H and C—D bonds. In the starting material, the bond to the heavier isotope D is stronger (and thus requires more energy to break; Sec. 5.6E) than the bond to the lighter isotope H. However, in the transition states for both reactions, the bond from H or D to carbon is partly broken, and the bond from H or D to the attacking group is partly formed. To a crude approximation, the isotope undergoing transfer is not bonded to anything—it is "in flight." Because there is no bond, there is no bond-energy difference between the two isotopes. Therefore, the compound with the C—D bond starts out at a lower energy than the compound with the C—H bond, and requires more energy to achieve the transition state (Fig. 9.5). In other words, the energy barrier, or free energy of activation, for the compound with the C—D bond is greater; as a result, its rate of reaction is smaller.

Be sure you understand that a primary deuterium isotope effect is observed only when the *hydrogen that is transferred* in the rate-determining step is substituted by deuterium. Substitution of other hydrogens with deuterium usually has little or no effect on the rate of the reaction.

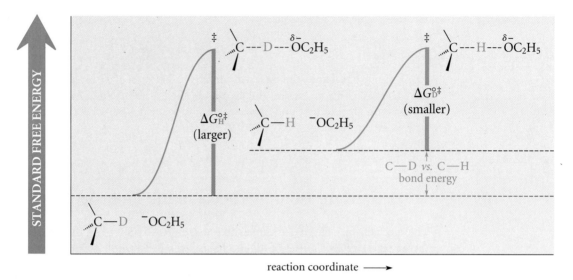

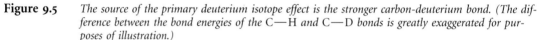

Figure 9.5 *The source of the primary deuterium isotope effect is the stronger carbon-deuterium bond. (The difference between the bond energies of the C—H and C—D bonds is greatly exaggerated for purposes of illustration.)*

USE OF THE PRIMARY DEUTERIUM ISOTOPE EFFECT IN DRUG DESIGN

An ingenious practical application of the primary deuterium isotope effect was used in the design of an experimental antibiotic, 3-fluoroalanine. It was suspected that the drug is destroyed in the body by a β-elimination reaction:

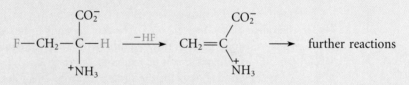

3-fluoroalanine (9.36)

On the possibility that such an elimination reaction might be retarded by a deuterium isotope effect (in the same sense that the E2 reaction of alkyl halides is retarded), a new drug was synthesized in which the hydrogen shown in color was replaced by deuterium. It was found that the isotopically substituted drug is much more effective as an antibiotic, presumably because its metabolic destruction is slower and it lasts longer in the body.

PROBLEMS

9.13 In each of the following series, arrange the compounds in order of increasing reactivity in the E2 reaction with Na^+ $C_2H_5O^-$.

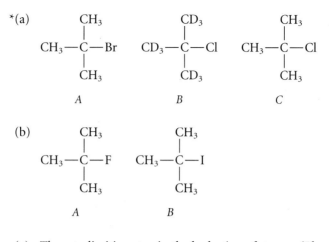

*(a)

A   B   C

(b)

A   B

*9.14   (a)   The rate-limiting step in the hydration of styrene (Ph—CH=CH$_2$) is the initial transfer of the proton from H$_3$O$^+$ to the alkene (Sec. 4.9B). How would you expect the rate of the reaction to change if the reaction were run in D$_2$O/D$_3$O$^+$ instead of H$_2$O/H$_3$O$^+$? Would the product be the same?

   (b)   How would the rate of styrene hydration in H$_2$O/H$_3$O$^+$ differ from that of an isotopically substituted styrene Ph—CH=CD$_2$? Explain.

D. Stereochemistry of the E2 Reaction

An elimination reaction might occur in two stereochemically different ways, illustrated as follows for the elimination of H—X from a general alkyl halide:

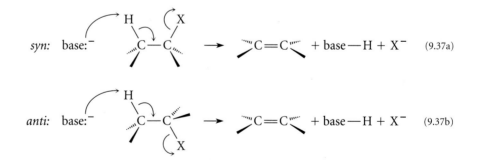

$$syn: \quad base:^- \qquad \qquad \longrightarrow \qquad C=C + base—H + X^- \qquad (9.37a)$$

$$anti: \quad base:^- \qquad \qquad \longrightarrow \qquad C=C + base—H + X^- \qquad (9.37b)$$

In a *syn* elimination, H and X leave the alkyl halide molecule from the same side; in an *anti* elimination, H and X leave from opposite sides.

Recall that the terms *syn* and *anti* were used in discussing the stereochemistry of additions to double bonds (Sec. 7.9A). Notice that *syn* elimination is conceptually the reverse of a *syn* addition, and *anti* elimination is conceptually the reverse of an *anti* addition.

Investigation of the stereochemistry of an elimination reaction requires that the α- and β-carbons be stereocenters in both the starting alkyl halide and the product alkene.

In such cases, it is found experimentally that most E2 reactions are stereoselective *anti* eliminations, as in the following example.

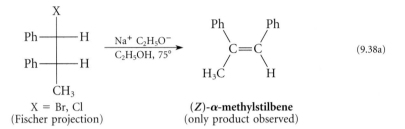

$$(9.38a)$$

$X = Br, Cl$
(Fischer projection)

(Z)-α-methylstilbene
(only product observed)

To see that this is an *anti* elimination, draw the alkyl halide molecule in a conformation in which the hydrogen and the halogen to be eliminated are *anti*, that is, situated at a dihedral angle of 180°.

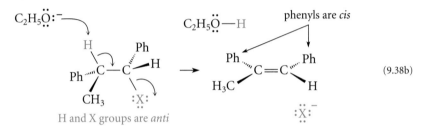

$$(9.38b)$$

Notice that when the hydrogen and halogen are eliminated from these positions, the phenyl groups (Ph) are on the same side of the molecule, and therefore must end up in a *cis* relationship in the product alkene. Notice also that a *syn* elimination would give the other alkene stereoisomer:

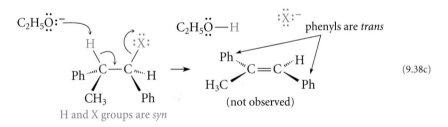

$$(9.38c)$$

Anti elimination is preferred for three reasons. First, *syn* elimination occurs through a transition state that has an eclipsed conformation (Fig. 9.6a), whereas *anti* elimination occurs through a transition state that has a staggered conformation (Fig. 9.6b). Because eclipsed conformations are unstable, the transition state for *syn* elimination is less stable than that for *anti* elimination. As a consequence, *anti* elimination is faster. The second reason that *anti* elimination is preferred is that the base and leaving group are on opposite sides of the molecule, out of each other's way. In *syn* elimination, they are on the same side of the molecule, and can interfere sterically with each other. Finally, quantum-mechanical calculations suggest that *anti* elimination is more favorable because *anti* elimination involves all-backside electronic displacements. In contrast, *syn* elimination requires a frontside electronic displacement on the carbon-halogen bond. (Recall that direct displacements occur by backside attack; Sec. 9.4B.)

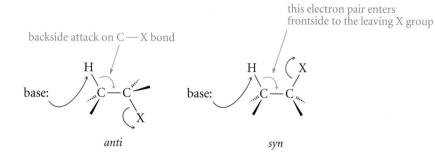

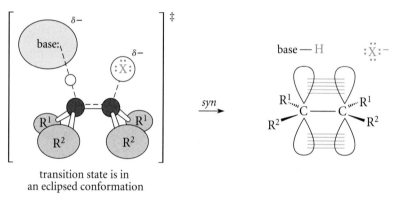

(a) *Syn* elimination

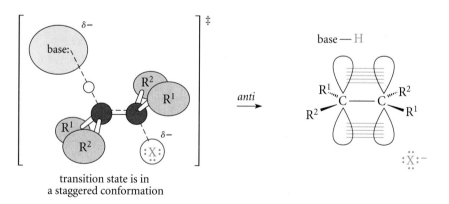

(b) *Anti* elimination

Figure 9.6   *Stereochemistry of β-elimination reactions. (a) In a syn elimination, the leaving group and base are situated on the same face of the molecule, and the transition state is in an eclipsed conformation. (b) In an anti elimination, the leaving group and base are situated on opposite faces, and the transition state is in a staggered conformation.*

9.15   Predict the products, including their stereochemistry, from the E2 reactions of the following diastereomers of stilbene dibromide with sodium ethoxide in ethanol. Assume that one equivalent of HBr is eliminated in each case.

*(a) $(\pm)$-Ph—CHBr—CHBr—Ph   (b)   *meso*-Ph—CHBr—CHBr—Ph

*9.16   Draw the structure of the starting material that would give the *E* isomer of the alkene product in the E2 reaction of Eq. 9.38a.

E. Regioselectivity of the E2 Reaction

When an alkyl halide has more than one type of β-hydrogen, more than one alkene product can be formed (Sec. 9.1B).

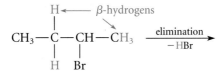

2-bromobutane

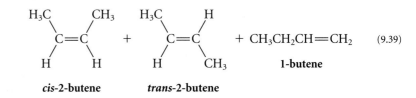

cis-2-butene   *trans*-2-butene   **1-butene**   (9.39)

This section focuses on which of the possible products is preferred and why.

When simple alkoxide bases such as methoxide and ethoxide are used, *the predominant product of an E2 reaction is usually the most stable alkene isomer.* Recall that the most stable alkene isomers are generally those with the most alkyl branches at the carbons of the double bond (Sec. 4.5B). These isomers, then, are the ones formed in greatest amount.

$$CH_3CH_2C(CH_3)_2 \xrightarrow[C_2H_5O^- \ K^+]{C_2H_5OH} CH_3CH=C(CH_3)_2 + CH_3CH_2C\genfrac{}{}{0pt}{}{\overset{CH_2}{\diagup}}{\diagdown CH_3} \qquad (9.40)$$
$$\underset{Br}{|} \qquad\qquad (70\%) \qquad\qquad (30\%)$$

Notice that in this reaction, the alkene isomer formed in smaller amount would actually be favored on statistical grounds: six equivalent hydrogens can be lost to give this product, but only two can be lost to give the other product. In the absence of a structural effect on the product distribution, three times as much of the 1-alkene would have been formed. The fact that the other alkene is the major one shows that some other factor is at work.

Transition-state theory explains why the more stable alkene is formed. The transition state for the E2 reaction can be visualized as a structure that lies somewhere between alkyl halide and alkene (plus the other species present). To the extent that the transition state resembles alkene, it is stabilized by the same factors that stabilize alkenes—and one such factor is branching at the double bond. A reaction that can give two alkene products is really two reactions in competition, each with its own transition state. The reaction with the transition state of lower energy—the one with more branching at the developing double bond—is the faster reaction. Hence, more product is formed through this transition state.

Could the predominance of the more stable alkene isomer be due to equilibration of the alkenes themselves, and have little or nothing to do with the structures of the transition states? The answer is no, and the reason is that *the alkene products are stable under the conditions of the reaction*. Because the product mixture, once formed, does not change, the distribution of products must reflect the relative energies of the competing transition states.

SAYTZEFF'S RULE

An elimination reaction that forms predominantly the most highly branched alkene isomers is sometimes called a **Saytzeff elimination**, after Alexander Saytzeff, a Russian chemist who observed this phenomenon in 1875. Just as the "Markownikoff rule" describes the regioselectivity of hydrogen halide addition to alkenes, the "Saytzeff rule" describes the regioselectivity of elimination reactions.

When an alkyl halide has more than one type of β-hydrogen, a mixture of alkenes is generally formed in its E2 reaction. The formation of a mixture means that the yield of the desired alkene isomer is reduced. Furthermore, since the alkenes in such mixtures are isomers of closely related structure, they generally have similar boiling points and are therefore difficult to separate. Consequently, the greatest use of the E2 elimination for the preparation of alkenes is when the alkyl halide has only one type of β-hydrogen, and only one alkene product is possible.

F.   Competition between the E2 and S_N2 Reactions: a Closer Look

Nucleophilic substitution reactions and base-promoted elimination reactions are *competing* processes (Sec 9.1C). In other words, whenever an S_N2 reaction is carried out, there is the possibility that an E2 reaction can also occur, and vice-versa.

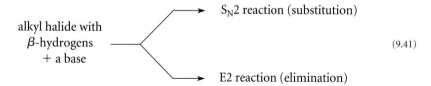

$$(9.41)$$

This competition is a matter of *relative rates*: the reaction pathway that occurs most rapidly is the one that predominates.

What determines which reaction—the S_N2 reaction or the E2 reaction—will be the major process observed in a given case? Two variables affect this competition:

1. the structure of the alkyl halide; and

2. the structure of the base.

A structural effect in the alkyl halide that determines the amount of elimination *vs.* substitution is the *degree of branching* at both the α- and β-carbons. In order for the S_N2 reaction to occur, the base (nucleophile) must attack a *carbon* atom. When attack at the α-carbon is retarded by unfavorable van der Waals repulsions, as in a tertiary alkyl halide, or in any alkyl halide with significant branching at the β-carbon, attack at a less hindered β-hydrogen—elimination—occurs instead.

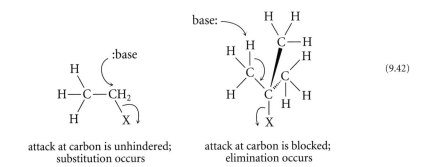

(9.42)

attack at carbon is unhindered;
substitution occurs

attack at carbon is blocked;
elimination occurs

Another reason that branching promotes the E2 reaction is that the standard free energy of the E2 transition state, like that of an alkene, is lowered by branching (previous section). Consequently, the rate of the E2 reaction is *increased by branching*. Two effects of branching, then, favor the E2 reaction: the rate of the S_N2 reaction is *decreased*, and the rate of the E2 reaction is *increased*.

These same effects can be seen not only in tertiary alkyl halides, but in secondary and even primary alkyl halides as well. Notice in the following examples that the alkyl halides with more β-branches show a greater proportion of elimination.

Secondary alkyl halides:

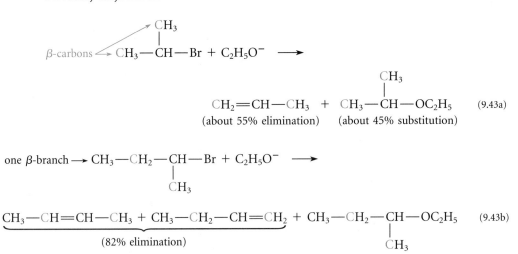

CH₃—CH=CH—CH₃ + CH₃—CH₂—CH=CH₂ + CH₃—CH₂—CH—OC₂H₅   (9.43b)

(82% elimination)

Primary alkyl halides:

β-carbon $\longrightarrow$ CH_3—CH_2—Br + $C_2H_5O^-$ $\longrightarrow$

$$CH_2{=}CH_2 \;+\; CH_3{-}CH_2{-}OC_2H_5 \qquad (9.44a)$$
(1% elimination)   (99% substitution)

one β-branch $\longrightarrow$ CH_3—CH_2—CH_2—Br + $C_2H_5O^-$ $\longrightarrow$

$$CH_3{-}CH{=}CH_2 \;+\; CH_3{-}CH_2{-}CH_2{-}OC_2H_5 \qquad (9.44b)$$
(10% elimination)   (90% substitution)

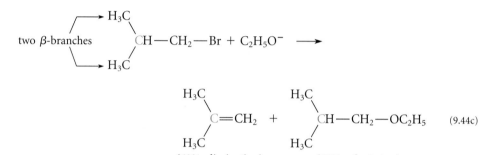

(62% elimination)   (38% substituion)

The structure of the base is the second variable that determines whether the E2 reaction or the S_N2 reaction is faster in a given case. First of all, a *highly branched base* such as *tert*-butoxide increases the proportion of elimination relative to substitution.

$(CH_3)_2CHCH_2$—Br + $^-OCH_2CH_3$ $\xrightarrow{\;C_2H_5OH\;}$ $(CH_3)_2C{=}CH_2$ + $(CH_3)_2CHCH_2$—OCH_2CH_3 $\qquad (9.45a)$

ethoxide
(a primary,
unbranched
alkoxide base)

(62% elimination)   (38% substitution)

$(CH_3)_2CHCH_2$—Br + $^-O{-}\overset{\displaystyle CH_3}{\underset{\displaystyle CH_3}{C}}{-}CH_3$ $\xrightarrow{\;(CH_3)_3COH\;}$ $(CH_3)_2C{=}CH_2$ + $(CH_3)_2CHCH_2$—$O{-}\overset{\displaystyle CH_3}{\underset{\displaystyle CH_3}{C}}{-}CH_3$

(92% elimination)

tert-butoxide
(a tertiary, branched
alkoxide base)

(8% substitution)

$(9.45b)$

When a highly branched base attacks the *α-carbon* to give a substitution product, the alkyl branches of the base suffer van der Waals repulsions with the hydrogens in the alkyl halide molecule; these repulsions raise the energy of the transition state for substitution. When such a base attacks a *β-proton* to give the elimination product, the base is further removed from the offending hydrogens in the alkyl halide, and van der Waals repulsions are less severe, as shown in Eq. 9.42. Consequently, the S_N2 reaction is retarded more than the E2 reaction by branching in the base, and elimination becomes the predominant reaction.

This effect of base structure is significant enough that the E2 reaction is useful for the synthesis of alkenes from even primary alkyl halides if a highly branched strong base such as potassium *tert*-butoxide is used. This fact is particularly useful in the synthesis of alkenes, because only one alkene product is possible when a 1-haloalkane is used in the E2 reaction.

A further effect of base structure on the E2-S_N2 competition has to do with its Brønsted basicity *vs.* its nucleophilicity. Recall that nucleophilicity, not Brønsted basicity, governs the rate of an S_N2 reaction (Sec. 9.4D). Recall also that within a group of the periodic table, species with attacking atoms of greater atomic number are excellent nucleophiles, even though they are weaker Brønsted bases than species with attacking atoms of smaller atomic number. Consequently, a greater fraction of S_N2 reaction is observed when the attacking atom is a good nucleophile but is a relatively weak base. For example, the reaction of potassium iodide with isobutyl bromide in acetone gives mostly substitution product and little elimination, because iodide is an excellent nucleophile and a weak base:

$$\underset{CH_3}{\overset{CH_3}{\diagdown}}\!\!\!\!\!\!\underset{\diagup}{\overset{\diagup}{CH}}\!\!-\!CH_2\!-\!Br\ +\ Na^+\ I^-\ \xrightarrow{\ acetone\ }\ \underset{CH_3}{\overset{CH_3}{\diagdown}}\!\!\!\!\!\!\underset{\diagup}{\overset{\diagup}{CH}}\!\!-\!CH_2\!-\!I\ +\ Na^+\ Br^-\!\!\downarrow \qquad (9.46)$$

Contrast this reaction with that in Eq. 9.44c, in which sodium ethoxide reacts with the same alkyl halide. Ethoxide, a strong Brønsted base, gives a significant percentage of alkene and a smaller percentage of substitution product.

Let's summarize the effects that govern the competition between the S_N2 and E2 reactions.

1. *Structure of the alkyl halide:*
   a. Alkyl halides with greater amounts of branching at the α-carbon give greater amounts of elimination. Consequently, tertiary alkyl halides give more elimination than secondary, which give more than primary.
   b. Alkyl halides with greater amounts of branching at the β-carbon give greater amounts of elimination.

2. *Structure of the base:*
   a. More highly branched bases give a greater fraction of elimination than unbranched ones.
   b. Weaker bases that are good nucleophiles give a greater fraction of substitution.

The application of these ideas is illustrated in the following study problem.

Which alkyl halide and what conditions should be used to prepare the following alkene in good yield by an E2 elimination?

methylenecyclohexane

Solution If this alkene is to be produced in an E2 reaction from an alkyl halide, the halide must be located at one of the two carbons that eventually become carbons of the double bond. This means that there are two choices for the starting alkyl halide:

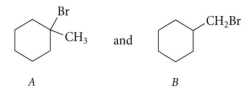

The advantage of alkyl halide *A* is that, because it is tertiary, it poses no significant competition from the S_N2 reaction. The disadvantage of alkyl halide *A* is that it contains more than one type of β-hydrogen, and consequently, more than one alkene product could be formed:

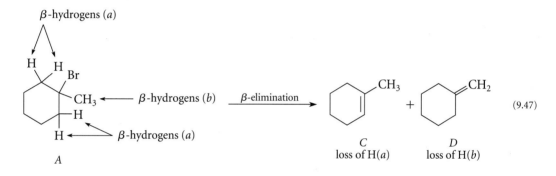

Product *C* is the more stable alkene because its double bond has three alkyl branches; hence, if *A* is used as the starting material, a major amount of this undesired alkene will be formed. If alkyl halide *B* is the starting material, then the desired product *D* is the *only* possible product of β-elimination. Because this alkyl halide is primary, however, it is possible that some by-product derived from the S_N2 reaction will be formed. The way to minimize the S_N2 reaction is to use a strong, highly branched base, such as *tert*-butoxide. In addition, the β-branching in alkyl halide *B* should also minimize the substitution reaction. Hence, a reasonable preparation of the desired alkene is the following:

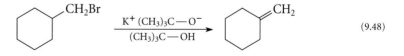

STUDY GUIDE LINK:
✓9.3
Branching in Cyclic Compounds

PROBLEMS

9.17 Suggest conditions (base or nucleophile) for conversion of isobutyl bromide in good yield into each of the following compounds.

*(a) $(CH_3)_2CHCH_2\overset{+}{S}(CH_3)_2\ Br^-$ (b) $(CH_3)_2CHCH_2SCH_2CH_3$

*(c) $(CH_3)_2C{=}CH_2$

*9.18   Arrange the following four alkyl halides in descending order with respect to the ratio of E2 elimination to S_N2 substitution products expected in their reactions with sodium ethoxide in ethyl alcohol. Explain your answers.

$$CH_3I \qquad (CH_3)_2CHCH_2-Br \qquad (CH_3)_3CCH_2CH_2CH_2-Br \qquad (CH_3)_2CHCH-Br$$

A   B   C $\overset{|}{CH_3}$

                                                                                        D

9.19   Arrange the following three alkoxide bases in descending order with respect to the ratio of E2 elimination to S_N2 substitution products expected when they react with isobutyl bromide. Explain your answers.

$$(CH_3)_2CH-O^- \qquad CH_3O^- \qquad (C_2H_5)_3C-O^-$$

A   B   C

G. Summary of the E2 Reaction

The E2 reaction is a β-elimination reaction of alkyl halides that is promoted by strong bases. To summarize the key points about this reaction:

1. The rates of E2 reactions are second order overall: first order in base and first order in the alkyl halide.

2. E2 reactions are normally *anti* eliminations.

3. The best leaving groups give the weakest bases as products.

4. The rates of E2 reactions show substantial primary deuterium isotope effects at the β-hydrogen atoms.

5. When an alkyl halide has more than one type of β-hydrogen, more than one alkene product can be formed; the most stable alkenes (the alkenes with the greatest numbers of alkyl substituents at their double bonds) are formed in greatest amount.

6. E2 reactions compete with S_N2 reactions. Elimination is favored by alkyl branches in the alkyl halide at the α- or β-carbon atoms, by alkyl branches in the base, and by stronger bases.

9.6   ## The S_N1 and E1 Reactions

The discussion up to this point has stressed the reactions of alkyl halides with species that are either strong bases or good nucleophiles. When a primary alkyl halide is dissolved in an alcohol solvent with no added base, no reaction takes place. When a tertiary alkyl halide such as *tert*-butyl bromide is subjected to the same conditions, however, both substitution and elimination reactions are observed.

$$CH_3-\underset{\underset{\displaystyle CH_3}{|}}{\overset{\overset{\displaystyle CH_3}{|}}{C}}-Br \;+\; HO-C_2H_5 \;\xrightarrow{55^\circ}$$

tert-butyl bromide

$$CH_3-\underset{\underset{\displaystyle CH_3}{|}}{\overset{\overset{\displaystyle CH_3}{|}}{C}}-O-C_2H_5 \;+\; \underset{\underset{\displaystyle CH_3}{\diagup}}{\overset{\overset{\displaystyle CH_3}{\diagdown}}{C}}=CH_2 \;+\; C_2H_5\overset{+}{O}H_2 \; Br^- \qquad (9.49)$$

tert-butyl ethyl ether   2-methylpropene   (ionized form
(72%)   (28%)   of HBr in ethanol)

The reaction of an alkyl halide with a solvent in which no base or nucleophile has been added is termed a **solvolysis** (literally, bond breaking by solvent). The substitution that occurs in the solvolysis of *tert*-butyl bromide cannot involve an S$_N$2 mechanism, because chain branching at the α-carbon retards the S$_N$2 reaction. The elimination that occurs in this solvolysis cannot occur by an E2 mechanism because a strong base is not present. Because both substitution and elimination reactions occur readily, they must then involve mechanisms that are different from the S$_N$2 and E2 mechanisms.

A. Rate Law and Mechanism of S$_N$1 and E1 Reactions

The solvolysis of *tert*-butyl bromide follows a *first-order rate law*:

$$\text{rate} = k[(CH_3)_3C-Br] \qquad (9.50)$$

Any involvement of solvent in the reaction cannot be detected in the rate law because the concentration of the solvent cannot be changed. However, the nature of the solvent does play a critical role in this reaction. The solvolysis reactions of tertiary alkyl halides are fastest in *polar, protic, donor solvents* such as alcohols, formic acid, and mixtures of water with solvents in which the alkyl halide is soluble, for example, aqueous acetone.

The occurrence of both substitution and elimination products shows that two *competing reactions* are involved. The first step in *both* reactions involves the ionization of the alkyl halide to a carbocation and a halide ion:

$$(CH_3)_3C\overset{\frown}{-}\ddot{B}r\colon \;\; \xrightleftharpoons{} \;\; (CH_3)_3C^+ \;\; \colon\!\ddot{B}r\colon^- \qquad \text{(rate-limiting step)} \qquad (9.51a)$$

carbocation
intermediate

This step, which is a *Lewis acid-base dissociation* (Sec. 3.1C), is the rate-limiting step of both the substitution and elimination reactions. In other words, when a tertiary alkyl halide is dissolved in a polar, protic solvent such as ethanol, it reacts by dissociating slowly into a carbocation and a halide ion; the carbocation then rapidly reacts to give both substitution and elimination products. Thus, substitution and elimination products arise from *competing reactions of the carbocation.*

Consider first the formation of the substitution product. This product is formed by attack of a solvent molecule on the carbocation. Even though the solvent is a poor nucleophile, the reaction occurs rapidly because the solvent is present in very high concentration, and because the carbocation is a very powerful Lewis acid.

$$(CH_3)_3C^+ \quad H\ddot{O}C_2H_5 \quad \rightleftharpoons \quad (CH_3)_3C—\overset{\pm}{\underset{H}{\ddot{O}C_2H_5}} \quad :\ddot{Br}:^- \qquad (9.51b)$$

$$:\ddot{Br}:^-$$

Notice that the nucleophile which attacks the carbocation is *not* ethoxide ion; such a strong base is not present in a solvolysis reaction, and if significant amounts of such a base were added, the E2 reaction would be observed exclusively.

The final step of the substitution reaction is loss of a proton to solvent from the product of Eq. 9.51b, the conjugate acid of an ether and a strong acid:

$$(CH_3)_3C—\overset{+}{\underset{\underset{H}{\overset{\curvearrowleft}{|}}}{\ddot{O}C_2H_5}} \quad Br^- \quad \rightleftharpoons \quad (CH_3)_3C—\ddot{O}C_2H_5 + H\overset{+}{\underset{H}{\ddot{O}C_2H_5}} \quad Br^- \qquad (9.51c)$$

$$H\overset{\curvearrowright}{\underset{}{\ddot{O}C_2H_5}}$$

(ionized form
of HBr in ethanol)

Again, the base involved in this step is ethanol, not ethoxide ion.

A substitution mechanism that involves a carbocation intermediate is called an **S_N1 mechanism.** Substitution reactions that take place by the S_N1 mechanism are called **S_N1 reactions.** The meaning of the S_N1 "nickname" is as follows:

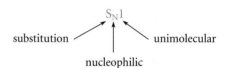

The word *unimolecular* means that a single molecule—the alkyl halide—is involved in the rate-limiting step.

Now consider the formation of the elimination product of Eq. 9.49, which involves a different reaction of the carbocation intermediate. Loss of a β-proton (a proton from the carbon adjacent to the electron-deficient carbon) gives the alkene.

$$H—\overset{\curvearrowleft}{\underset{\beta}{CH_2}}—\overset{}{\underset{\underset{CH_3}{\overset{}{\underset{\alpha}{|}}}}{C^+}} \quad :\ddot{Br}:^- \quad \longrightarrow \quad CH_2{=}\overset{\overset{CH_3}{}}{\underset{CH_3}{C}} \quad + \quad \overset{H}{\underset{H}{\overset{+}{\ddot{O}}{-}C_2H_5}} \quad :\ddot{Br}:^- \qquad (9.52)$$

$$H—\ddot{O}—C_2H_5$$

(ionized form of
HBr in ethanol)

The base that removes a β-proton from the carbocation is typically a solvent molecule. Although ethanol is a very weak base, the reaction occurs readily because ethanol, as the solvent, is present in very high concentration, and because the carbocation is a very strong Brønsted acid (its pK_a has been estimated to be about -8). Notice that the base is *not* ethoxide ion.

An elimination mechanism that involves carbocation intermediates is termed an **E1 mechanism**; reactions that occur by E1 mechanisms are called **E1 reactions**. The meaning of the E1 "nickname" is as follows:

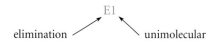

elimination   unimolecular

B.   Rate-Limiting and Product-Determining Steps

The S$_N$1 and E1 reactions have *a common rate-limiting step*. That is, the rate at which the alkyl halide disappears as it undergoes both competing reactions is determined by its *rate of ionization*—the rate at which it forms the carbocation. The relative amounts of substitution and elimination products are determined by the relative rates of the steps that follow the rate-limiting step: attack of the solvent as a nucleophile on the carbocation to give a substitution product, and loss of a β-proton to solvent from the carbocation to give the elimination product. For example, more substitution than elimination product is formed in Eq. 9.49. This means that the rate of formation of the substitution product from the carbocation is greater than the rate of formation of the elimination product. Because the relative rates of these steps determine the ratio of products, they are said to be the **product-determining steps**. Notice that *the rates of the product-determining steps have nothing to do with the rate at which the alkyl halide reacts.*

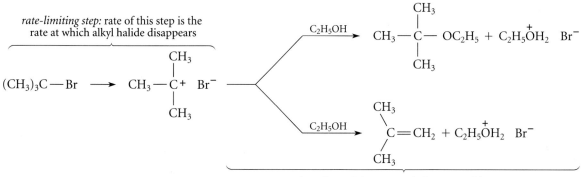

product-determining steps: relative rates of these steps
determine the relative amounts of different products   (9.53)

The reaction-free energy diagram in Fig. 9.7 summarizes these ideas. The first step, ionization of the alkyl halide to a carbocation, is the rate-limiting step and thus has the transition state of highest free energy. The rate of this step is the rate at which the alkyl halide reacts. The relative free-energy barriers for the product-determining steps determine the relative amounts of products formed.

ANALOGY FOR PRODUCT-DETERMINING STEPS

Imagine a very slow toll collector on a very busy freeway near Chicago. Suppose that the rate at which cars pass through the toll station determines how rapidly the collector works. Toll-taking is thus the rate-limiting step in

(continues)

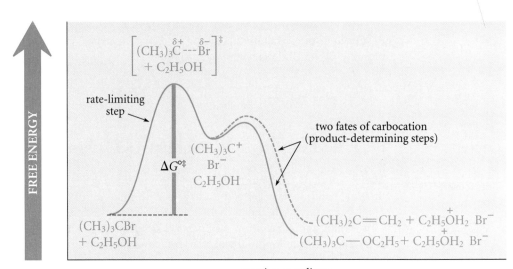

Figure 9.7   *Energetics of the reaction of $(CH_3)_3C$—Br with ethanol. Notice that the rate-limiting step—the step with the transition state of highest free energy—is ionization of the alkyl halide. Because more substitution than elimination product is observed, the energy barrier leading to the substitution product is smaller, and the rate of substitution is greater than the rate of elimination.*

the progress of cars past the station. The relative numbers of drivers that take the turnoffs to Wisconsin and Indiana determine the relative numbers of cars that arrive at the two destinations. Entering a turnoff is analogous to a product-determining step. If more drivers turn off for Indiana, the rate at which cars arrive in Indiana is greater than the rate at which cars arrive in Wisconsin. However, the total rate at which cars reach both destinations is determined only by toll-taking—the rate-limiting step.

The competition between the S_N1 and E1 reactions is somewhat different from the competition between the S_N2 and E2 reactions. The latter two reactions share nothing in common but starting materials; they follow completely separate reaction pathways with no common intermediates.

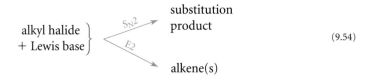

(9.54)

In contrast, the S_N1 and E1 reactions of an alkyl halide share not only common starting materials, but also a common rate-limiting step, and hence *a common intermediate*—the carbocation.

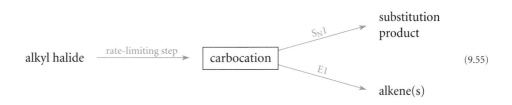

(9.55)

In the E1 reaction, the proton is not removed from the alkyl halide, as it is in the E2 reaction, but from the carbocation. Because the carbocation is a strong acid, a strong base is not required for the E1 reaction as it is for the E2 reaction.

C. Reactivity and Product Distributions in S$_N$1-E1 Reactions

S$_N$1-E1 reactions are most rapid with tertiary alkyl halides, they occur more slowly with secondary alkyl halides, and they are almost never observed with primary alkyl halides.

Reactivity of alkyl halides in S$_N$1 or E1 reactions:

$$\text{tertiary} > \text{secondary} \gg \text{primary} \tag{9.56}$$

Notice that this reactivity order is expected from the relative stability of the corresponding carbocation intermediates. Hammond's postulate (Sec. 4.8C) suggests that the rate-limiting transition state of an S$_N$1 or E1 reaction should closely resemble a carbocation.

The reactivity order of the alkyl halides in S$_N$1-E1 reactions is fluorides $\ll$ chlorides $<$ bromides $<$ iodides. This is the same reactivity order observed in S$_N$2 and E2 reactions, and is expected because the leaving group in the S$_N$1-E1 reaction has much the same role as it does in the E2 and S$_N$2 reactions.

S$_N$1-E1 reactions are fastest in polar, protic, donor solvents. This is the result expected in a reaction for which the rate-limiting step is a dissociation of a neutral molecule into ions of opposite charge. Ionic dissociation is favored by solvents that *separate* ions (that is, polar solvents—solvents with a high dielectric constant), and by solvents that *solvate* ions (that is, protic, donor solvents). The rate-limiting step of an S$_N$1-E1 reaction is not very different conceptually from the dissolution of an ionic compound (Sec. 8.4B); both processes hinge on the stabilization of ionic species by the solvent. The critical role of solvent shows that S$_N$1 and E1 reactions cannot truly be unimolecular processes as suggested by their nicknames. In the transition states of these reactions, solvent molecules must be actively involved in solvating the developing ions.

When an alkyl halide contains more than one type of β-hydrogen, more than one type of elimination product can be formed. As in the E2 reaction, the alkene with the greatest number of alkyl substituents at the double bond is usually formed in greatest amount; and the ratio of alkene (E1) to substitution product (S$_N$1) is greater when the alkene formed contains more than two alkyl substituents at the double bond. The following examples illustrate both of these points.

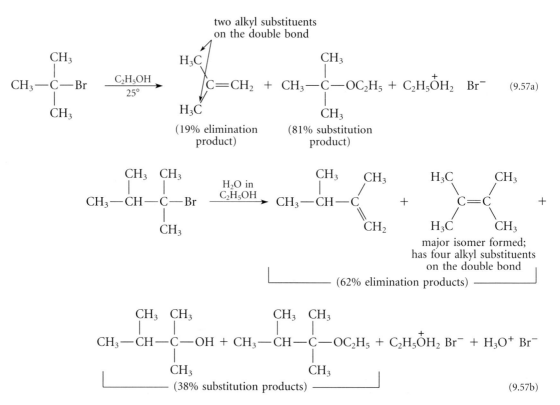

In Eq. 9.57a, relatively little alkene is formed. In Eq. 9.57b, a greater proportion of alkene is formed. In addition, in Eq. 9.57b, two alkenes are formed corresponding to loss of the two types of β-hydrogens in the alkyl halide starting material; and the alkene formed in major amount is the one with the greatest number of alkyl substituents on the double bond.

Finally, rearrangements are observed in certain solvolysis reactions.

Recall that rearrangements are a telltale sign of carbocation intermediates (Sec. 4.7D). For example, the secondary carbocation intermediate initially formed in Eq. 9.58 rearranges to a more stable tertiary carbocation; attack of solvent on this carbocation accounts for the product shown (Problem 9.21).

The different products that can be formed in S_N1-E1 reactions reflect three reactions of carbocation intermediates that you have now studied:

1. reaction with a nucleophile;

2. loss of a β-proton;

3. rearrangement to a new carbocation followed by (1) or (2).

Although solvolysis reactions of alkyl halides and related compounds have been extensively studied because of their central role in the development of carbocation theory, as a practical matter S_N1-E1 reactions of alkyl halides are not very useful for preparative purposes because mixtures of products are invariably formed (unless the alkyl halide has no β-hydrogens). However, an understanding of the S_N1 and E1 mechanisms is important because these mechanisms occur in many reactions of alcohols, ethers, and amines that *are* very useful.

PROBLEMS

9.20   Give all the products that might be formed when each of the following alkyl halides undergoes solvolysis in aqueous ethanol. Of the alkenes formed, which should be the major one(s)?
*(a)   3-chloro-2,2-dimethylbutane (the alkyl halide in Eq. 9.58)
(b)   2-bromo-2-methylbutane

*9.21   Write the mechanism for formation of the rearrangement product shown in Eq. 9.58 using the curved-arrow formalism.

D. Summary of the S_N1 and E1 Reactions

Let's summarize the important characteristics of the S_N1 and E1 reactions.

1. Tertiary and secondary alkyl halides undergo solvolysis reactions by the S_N1 and E1 mechanisms; tertiary alkyl halides are more reactive.

2. If an alkyl halide has β-hydrogens, elimination products formed by the E1 reaction accompany substitution products formed by the S_N1 mechanism.

3. Both S_N1 and E1 reactions of a given alkyl halide share the same rate-limiting step: ionization of the alkyl halide to form a carbocation.

4. The S_N1 and E1 reactions are first order in the alkyl halide.

5. S_N1 and E1 reactions differ in their product-determining steps. The product-determining step in the S_N1 reaction is attack of a nucleophile on the carbocation intermediate, and in the E1 reaction, loss of a β-proton from the carbocation intermediate.

6. Carbocation rearrangements occur when the initially formed carbocation intermediate can rearrange to a more stable carbocation.

7. The best leaving groups give the weakest bases as products.

8. The reactions are accelerated by polar, protic, donor solvents.

9.7   Summary of Substitution and Elimination Reactions of Alkyl Halides

This chapter has shown that substitution and elimination reactions of alkyl halides can occur by a variety of mechanisms. Although each type of reaction has been considered separately, a practical question to ask is what type of reaction is likely to occur when a given alkyl halide is subjected to a particular set of conditions.

Table 9.4   Predicting Substitution and Elimination Reactions of Alkyl Halides

Entry no.	Alkyl halide structure	Good nucleophile?	Strong Brønsted base?	Type of solvent?[a]	Major reaction(s) expected
1	Methyl	Yes	Yes or No	PP or PA	S_N2
2	Primary, unbranched	Yes	No	PP or PA	S_N2
3		Yes	Yes, unbranched	PP or PA	S_N2
4	Primary with β-branching	Yes	Yes, unbranched	PP or PA	$E2 + S_N2$
5	Any primary	Yes	Yes, branched	PP or PA	E2
6		No	No	PP or PA	no reaction
7	Secondary	Yes	Yes	PP or PA	E2; some S_N2 with isopropyl halides; only E2 with a branched base
8		Yes	No	PA	S_N2
9		No	No	PP	S_N1/E1
10		No	No	PA	no reaction
11	Tertiary	Yes	Yes	PP or PA	E2
12		Yes	No	PP	S_N1/E1
13		Yes	No	PA	no reaction, or very slow S_N2
14		No	No	PP	S_N1/E1
15		No	No	PA	no reaction

[a] Solvent types are PP = polar protic; PA = polar aprotic. The S_N2, E2, S_N1, and E1 reactions are rarely if ever run in apolar aprotic solvents except with the most reactive alkyl halides. In these cases, the results to be expected are similar to those above with polar aprotic (PA) solvents.

When asked to predict how a given alkyl halide will react, you must first answer three major questions.

1. *Is the alkyl halide primary, secondary, or tertiary? If primary or secondary, is there a significant amount of β-branching?*

2. *Is a Lewis base present? If so, is it a good nucleophile, a strong Brønsted base, or both?* Remember that most strong Brønsted bases such as ethoxide are good nucleophiles; but some excellent nucleophiles, such as iodide ion, are relatively weak Brønsted bases.

3. *What is the solvent?* Remember that the practical choices here are limited for the most part to polar protic solvents, polar aprotic solvents, or mixtures of them.

Once these questions have been answered, a satisfactory prediction in most cases can be obtained from Table 9.4, which is in essence a summary of this chapter. *Before using this table, you should consider each case and why the conclusions are reasonable, returning to review the material in this chapter when necessary.* The following study problem illustrates the practical application of the table.

STUDY PROBLEM 9.2

What products are formed, and by what mechanisms, in each of the following cases?
(a) methyl iodide and sodium cyanide (NaCN) in ethanol
(b) 2-bromo-3-methylbutane in ethanol
(c) 2-bromo-3-methylbutane in anhydrous acetone
(d) 2-bromo-3-methylbutane in ethanol containing an excess of sodium ethoxide
(e) 2-bromo-2-methylbutane in ethanol containing an excess of sodium iodide
(f) neopentyl bromide in ethanol containing an excess of sodium ethoxide

Solution
(a) Methyl iodide and sodium cyanide (NaCN) in ethanol. This case corresponds to entry 1 in Table 9.4. Because a methyl halide has no β-hydrogens, it cannot undergo a β-elimination reaction. Consequently, the only possible reaction is an S_N2 reaction. Because a good nucleophile cyanide is present (see Table 9.3), the product is CH_3—CN (acetonitrile), which is formed by the S_N2 mechanism. Although protic solvents are not as effective as polar aprotic ones for the S_N2 reaction, they are useful for reactive alkyl halides such as methyl iodide. Of course, the reaction would be faster if it were carried out in a polar aprotic solvent.

(b) 2-Bromo-3-methylbutane in ethanol. This is a secondary alkyl halide. (Draw its structure if you have not done so!) The conditions involve no nucleophile or base other than the solvent, and a polar protic solvent. This situation is covered by entry 9 in Table 9.4. Because the solvent ethanol is a poor nucleophile and a weak base, neither S_N2 nor E2 reactions can occur. Because polar protic solvents promote the S_N1 and E1 reactions, these will be the only reactions observed:

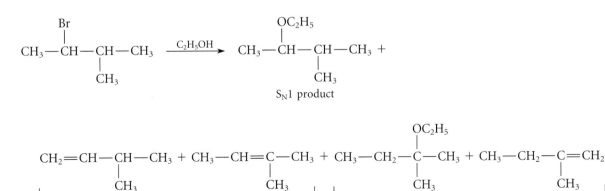

(Study Problem 9.2 continues)

Notice the rearrangement products. (You should show how these arise from the initially formed carbocation intermediate.) Any time the S_N1 or E1 reaction is expected, the possibility of rearrangements should be considered, especially when the initially formed carbocation is secondary.

(c) 2-Bromo-3-methylbutane in anhydrous acetone. The alkyl halide from part (b) is subjected to conditions in which a good nucleophile is not present (no S_N2 possible), no strong base has been added (no E2 possible), and a polar aprotic solvent is used. In this type of solvent carbocations do not form; hence, the S_N1 and E1 reactions cannot take place. Entry 10 in Table 9.4 predicts that no reaction will occur.

(d) 2-Bromo-3-methylbutane in ethanol containing an excess of sodium ethoxide. The alkyl halide from parts (b) and (c) is subjected to a strong base in a protic solvent. This situation is covered by entry 7 in Table 9.4. The S_N2 reaction is retarded by both α- and β-branching, but the E2 reaction can take place. Although an S_N1-E1 reaction is promoted by the protic solvent, the rate of the E2 reaction is greater because of the high base concentration (Eq. 9.32); the rates of the S_N1 and E1 reactions are unaffected by the base concentration (Eq. 9.50). The products are the following two alkenes:

$$CH_2\!=\!CH\!-\!CH\!-\!CH_3 \;+\; CH_3\!-\!CH\!=\!C\!-\!CH_3$$
$$\qquad\qquad\;\; | \qquad\qquad\qquad\qquad\quad |$$
$$\qquad\qquad CH_3 \qquad\qquad\qquad\qquad CH_3$$

The second of these predominates because of the greater number of alkyl substituents at the double bond.

(e) 2-Bromo-2-methylbutane in ethanol containing an excess of sodium iodide. This is a tertiary alkyl halide in a polar protic solvent containing a good nucleophile but a weak base (iodide ion). Entry 12 of Table 9.4 covers this situation. The polar protic solvent promotes carbocation formation, and hence the S_N1 and E1 reactions are observed. The S_N1 products are the following:

$$
\begin{array}{ccccccc}
CH_3 & & & CH_3 & & CH_3 & \\
| & & & | & & | & \\
CH_3CH_2C\!-\!Br + NaI & \xrightarrow{\;C_2H_5OH\;} & & CH_3CH_2C\!-\!I & + & CH_3CH_2C\!-\!OC_2H_5 & + \; C_2H_5\overset{+}{O}H_2 \;\; Br^- \\
| & & & | & & | & \\
CH_3 & & & CH_3 & & CH_3 & \text{(ionized form of} \\
 & & & A & & B & \text{HBr in ethanol)}
\end{array}
$$

Product *A* arises from attack of the nucleophile I^- on the carbocation intermediate, and product *B* is the solvolysis product that results from attack of the solvent on the same carbocation, with ionized HBr being formed as a byproduct. Which product (*A* or *B*) is formed in greater amount? It depends on how much iodide ion is present. The more iodide there is, the more effective it will be in competing with the solvent ethanol for the carbocation. Because the E1 reaction always accompanies the S_N1 reaction of an alkyl halide with β-hydrogens, some alkenes are also formed; you should draw their structures. No rearrangement products are predicted, because the carbocation intermediate is tertiary.

(f) Neopentyl bromide in ethanol containing an excess of sodium ethoxide. This is a primary alkyl halide with *three* β-branches [$(CH_3)_3C\!-\!CH_2\!-\!Br$]. Without thinking further about the structure of this alkyl halide, you might conclude that entry 4 of Table 9.4 would cover this case. However, because there are no β-hydrogens, no elimination is possible. Neopentyl halides are essentially unreactive in S_N2 reactions (Table 9.1), and, because primary alkyl halides do not form carbocations, neither an E1 nor an S_N1 reaction is possible. Thus, this alkyl halide is essentially inert. If the reaction mixture were heated strongly, some S_N2 reaction might occur after a few days, but the correct prediction is "no reaction."

PROBLEM

9.22   Predict the products expected from each situation below, and show the mechanism of any reaction that takes place using the curved-arrow formalism.

  *(a)   1-bromobutane in ethanol containing a large excess of sodium methoxide

  (b)   2-bromobutane in *tert*-butyl alcohol containing a large excess of potassium *tert*-butoxide

  *(c)   2-bromo-1,1-dimethylcyclopentane in ethanol

  (d)   bromocyclohexane in methanol

9.8   Carbenes and Carbenoids

A. α-Elimination Reactions

β-Elimination is one of the reactions that can occur when certain alkyl halides containing β-hydrogens are treated with base.

$$
\underset{\ddot{X}:}{\overset{H}{\underset{|}{\overset{|}{C}}}\!-\!\overset{|}{C}} \quad \longrightarrow \quad \overset{\diagdown}{\underset{\diagup}{C}}\!=\!\overset{\diagup}{\underset{\diagdown}{C}} \ + \ H\!-\!\ddot{X}\!: \quad (\beta\text{-elimination}) \tag{9.59}
$$

When an alkyl halide contains no β-hydrogens but has an α-hydrogen, a different sort of base-promoted elimination is sometimes observed. Chloroform is an alkyl halide that undergoes such a reaction. When chloroform, a weak acid with $pK_a = 25$, is treated with an alkoxide base such as potassium *tert*-butoxide, a small amount of its conjugate-base anion is formed.

$$
(CH_3)_3C\!-\!\ddot{\underset{\cdot\cdot}{O}}\!:^{-} \ \ H\!-\!CCl_3 \quad \rightleftharpoons \quad (CH_3)_3C\!-\!OH \ + \ ^{-}\!:CCl_3 \tag{9.60a}
$$

   tert-butoxide   chloroform   *tert*-butyl alcohol   trichloromethyl
                                                    anion

This anion can lose a chloride ion to give a neutral species called *dichloromethylene*.

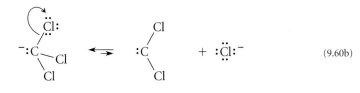

$$
^{-}\!:\!C\!\underset{Cl}{\overset{Cl}{\diagup}} \quad \longleftrightarrow \quad :C\!\underset{Cl}{\overset{Cl}{\diagup}} \ + \ :\ddot{\underset{\cdot\cdot}{Cl}}\!:^{-} \tag{9.60b}
$$

      trichloromethyl   dichloromethylene
         anion

Dichloromethylene is an example of a **carbene**—a species with a divalent carbon atom. Carbenes are unstable and highly reactive species.

The formation of dichloromethylene shown in Eqs. 9.60a and 9.60b involves an elimination of the elements of HCl from the *same* carbon atom. An elimination of two groups from the same atom is termed an **α-elimination**.

$$\underset{R^2}{\overset{R^1}{\diagdown}}\underset{\ddot{X}:}{\overset{H}{\diagup}}C \longrightarrow \underset{R^2}{\overset{R^1}{\diagdown}}C: \; + \; H\text{—}\ddot{X}: \qquad (\alpha\text{-elimination}) \qquad (9.61)$$

Chloroform cannot undergo a β-elimination because it has no β-hydrogens. When an alkyl halide has β-hydrogens, β-elimination occurs in preference to α-elimination because alkenes, the products of β-elimination, are much more stable than carbenes, the products of α-elimination. For example, CH_3CHCl_2 reacts with base to form the alkene $CH_2\text{=}CHCl$ rather than the carbene $CH_3\text{—}\overset{..}{C}\text{—}Cl$.

The reactivity of dichloromethylene follows from its electronic structure. Including the unshared pair of electrons, the carbon atom of dichloromethylene bears three groups (two chlorines and the lone pair), and therefore has approximately trigonal-planar geometry. Because trigonal-planar carbon atoms are sp^2-hybridized, the Cl—C—Cl bond angle is bent rather than linear, the unshared pair of electrons occupies an sp^2 orbital, and the p orbital is vacant:

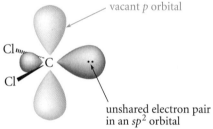

vacant p orbital

unshared electron pair in an sp^2 orbital

Because dichloromethylene lacks an electronic octet, it is an *electron-deficient compound* and can accept an electron pair; in other words, dichloromethylene is a powerful electrophile. On the other hand, an atom with an unshared electron pair reacts as a Lewis base, or nucleophile. The divalent carbon of dichloromethylene, with its unshared electron pair, appears to fit into this category as well. Indeed, it seems that the divalent carbon of a carbene could act as a nucleophile and an electrophile at the same time!

An important reaction of carbenes that fits this analysis is cyclopropane formation. When dichloromethylene is generated in the presence of an alkene, a cyclopropane is formed.

$$HCCl_3 \; + \; (CH_3)_3C\text{—}\overset{..}{\underset{..}{O}}:^- \; K^+ \; + \; (CH_3)_2C\text{=}CH_2 \longrightarrow$$

chloroform   **potassium**   **2-methylpropene**
                tert-butoxide

$$\underset{(CH_3)_2C\text{——}CH_2}{\overset{Cl \diagdown \; \diagup Cl}{\underset{\diagup \quad \diagdown}{C}}} \quad + \quad KCl \; + \; (CH_3)_3C\text{—}OH \qquad (9.62)$$

1,1-dichloro-2,2-dimethyl-cyclopropane

In general, reaction of a haloform with base in the presence of an alkene yields a 1,1-dihalocyclopropane. In the arrow formalism for cyclopropane formation, the π electrons of the alkene attack the empty orbital of the carbene (the carbene acts as a Lewis acid) while the unshared electron pair of the carbene attacks an alkene carbon (the carbene acts as a Lewis base.)

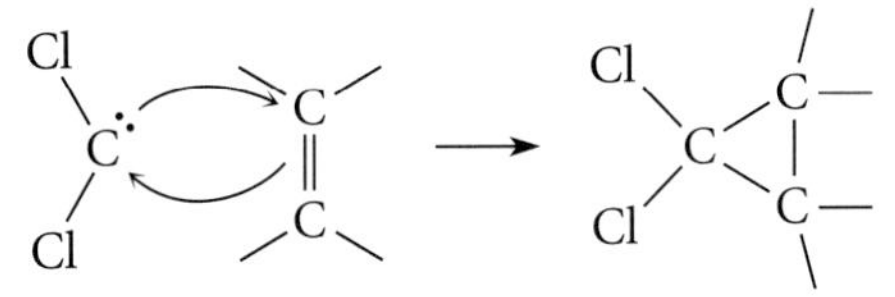

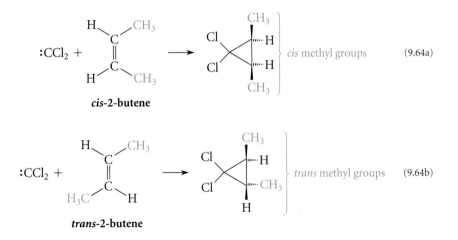

$$(9.63)$$

As the formalism in Eq. 9.63 suggests, the addition of dichloromethylene to an alkene is a *concerted reaction* (Sec. 5.3B). The stereochemistry of the reaction is that expected for this type of mechanism: dichloromethylene addition to an alkene is a *syn* addition. That is, groups that are *cis* in the reacting alkene are also *cis* in the cyclopropane product.

$$(9.64a)$$

$$(9.64b)$$

A concerted *anti* addition would require that the empty orbital of the divalent carbon and its unshared electron pair react simultaneously at opposite faces of the alkene, a stereochemically impossible situation.

PROBLEMS

9.23   What alkyl halide and alkene would yield each of the following cyclopropane derivatives in the presence of a strong base?

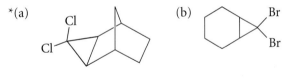

*(a)   (b)

(*Problem 9.23 continues*)

*(c)

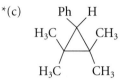

(*Hint:* The hydrogens on a carbon next to a benzene ring, or Ph group, are particularly acidic.)

9.24   Predict the products when each of the following alkenes reacts with chloroform and potassium *tert*-butoxide. Give the structures of all product stereoisomers, and, if more than one stereoisomer is formed, indicate whether they are formed in the same or different amounts.

*(a) cyclopentene   (b) *cis*-2-pentene   *(c) (*R*)-3-methylcyclohexene

B. The Simmons-Smith Reaction

Cyclopropanes without halogen atoms can be prepared by allowing alkenes to react with methylene iodide (CH_2I_2) in the presence of a copper-activated zinc preparation called a *zinc-copper couple.*

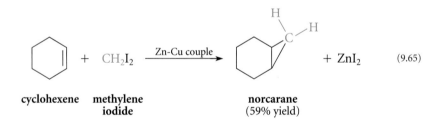

cyclohexene	**methylene iodide**	**norcarane** (59% yield)	(9.65)

This reaction is called the **Simmons-Smith reaction** in honor of the two Du Pont chemists who developed it in 1959, Howard E. Simmons and Ronald D. Smith.

The active reagent in the Simmons-Smith reaction is believed to be an α-halo organometallic compound, a compound with halogen and a metal on the same carbon. This species can form by a reaction analogous to the formation of a Grignard reagent (Sec. 8.7A):

$$CH_2I_2 + Zn \longrightarrow I-CH_2-ZnI$$
$$\text{Simmons-Smith reagent} \qquad (9.66)$$

From the discussion of the reactivity of carbenes with alkenes in the last section, the cyclopropane product of the Simmons-Smith reaction is what would be expected if the parent carbene **methylene** ($:CH_2$) were a reactive intermediate. *Free* methylene is not involved in the reaction, because free methylene generated in other ways gives not only cyclopropanes, but other products as well. However, the Simmons-Smith reagent can be conceptualized as methylene coordinated to the Zn atom. This view is reasonable, first, because the carbon-zinc bond polarity is the same as the carbon-magnesium bond polarity in a Grignard reagent (Sec. 8.7B):

$$I-\overset{\curvearrowleft}{CH_2}-ZnI \quad \text{reacts as if it were} \quad I-\overset{..}{\underset{\uparrow}{C}}H_2 \quad \overset{+}{ZnI} \tag{9.67}$$

an α-halo carbanion

Second, an α-halo carbanion loses halide ion to give a carbene (see Eq. 9.60b):

methylene

$$:\overset{..}{\underset{..}{I}}\!\!-\!\overset{..}{C}H_2 \quad \overset{+}{ZnI} \quad \longrightarrow \quad :\overset{..}{\underset{..}{I}}: \quad \overset{..}{C}H_2 \quad \overset{+}{ZnI} \tag{9.68}$$

coordinated
to the Zn

Reaction of this "coordinated methylene" with the alkene double bond gives a cyclopropane. Because they show carbenelike reactivity, α-halo organometallic compounds are sometimes termed **carbenoids**. A carbenoid is a reagent that is not a free carbene but has carbenelike reactivity.

Addition reactions of methylene from Simmons-Smith reagents to alkenes, like the reactions of dichloromethylene, are *syn* additions.

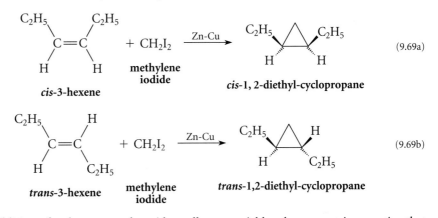

(9.69a)

(9.69b)

Addition of carbenes or carbenoids to alkenes to yield cyclopropanes is *a reaction that forms new carbon-carbon bonds*. Reactions that form carbon-carbon bonds are especially important in organic chemistry because they can be used to build up larger carbon skeletons from smaller ones.

PROBLEMS

9.25   Give the structure of the organic product expected when CH_2I_2 reacts with each of the following alkenes in the presence of a Zn-Cu couple:

*(a)   (Z)-3-methyl-2-pentene   (b)   *cis*-2-butene

*(c)   ◇=CH—CH₃   (d)   ⬡=CH₂

9.26   From which alkene could each of the following cyclopropane derivatives be prepared using the Simmons-Smith reaction?

(*Problem 9.26 continues*)

*(a)   (b)

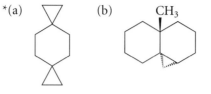

KEY IDEAS IN CHAPTER 9

🧪 Two of the most important types of alkyl halide reactions are nucleophilic substitution and β-elimination.

🧪 Nucleophilic substitution reactions occur by two mechanisms.
1. The S_N2 reaction occurs in a single step with inversion of stereochemical configuration, and is characterized by a second-order rate law. It occurs when an alkyl halide reacts with good nucleophiles, and is especially rapid in polar aprotic solvents. Branching in the alkyl halide at the α- or β-carbons retards the S_N2 reaction and, if a good Brønsted base is present, favors competing elimination by the E2 mechanism. The S_N2 reaction is most commonly observed with methyl, primary, and unbranched secondary alkyl halides.
2. The S_N1 reaction is characterized by a first-order rate law that contains only a term in alkyl halide concentration. This type of reaction occurs mostly in polar protic solvents. Because the reaction involves a carbocation intermediate, it is promoted by branching at the α-carbon. Consequently, the S_N1 reaction is observed mostly with tertiary and secondary alkyl halides.

🧪 β-Elimination reactions also occur by two mechanisms.
1. The E2 mechanism competes with the S_N2 mechanism, has a second-order rate law, and occurs with *anti* stereochemistry. It is favored both by use of a strong Brønsted base and by α- and β-branching in both the alkyl halide and the base. It is the major reaction of tertiary and secondary alkyl halides in the presence of a strong Brønsted base. When an alkyl halide has more than one type of β-hydrogen, more than one alkene product are generally obtained. The alkene with the most branching at the double bond is generally the predominant product.
2. The E1 mechanism is an alternative product-determining step of the S_N1 mechanism in which a carbocation intermediate loses a β-proton to form an alkene. The alkene with the greatest number of alkyl substituents on the double bond predominates.

🧪 The rate law indicates the species involved in the rate-limiting transition state of a reaction, but not how they are arranged.

⚗ A primary deuterium isotope effect indicates that a proton transfer takes place in the rate-limiting step of a reaction.

⚗ A haloform reacts with base in an α-elimination reaction to give dihalomethylene, a carbene. Carbenes are unstable species containing divalent carbon. Dihalomethylene undergoes *syn* additions with alkenes to give dihalocyclopropanes.

⚗ Methylene iodide reacts with a zinc-copper couple to give a carbenoid organometallic reagent. This reagent undergoes *syn* additions with alkenes to give cyclopropanes.

ADDITIONAL PROBLEMS

9.27 Choose the alkyl halide(s) from the following list of $C_6H_{13}Br$ isomers that meet each criterion below.

(1) 1-bromohexane   (2) 3-bromo-3-methylpentane
(3) 1-bromo-2,2-dimethylbutane   (4) 3-bromo-2-methylpentane
(5) 2-bromo-3-methylpentane

*(a) the compound(s) that can exist as enantiomers
(b) the compound(s) that can exist as diastereomers
*(c) the compound that gives the fastest S_N2 reaction with sodium methoxide
(d) the compound that is least reactive to sodium methoxide in methanol
*(e) the compound(s) that give only one alkene in the E2 reaction
(f) the compound(s) that give an E2 but no S_N2 reaction with sodium methoxide in methanol
*(g) the compound(s) that undergo an S_N1 reaction to give rearranged products
(h) the compound that gives the fastest S_N1 reaction

*9.28 Give the products expected when isopentyl bromide (1-bromo-3-methylbutane) or the other substances indicated react with the following reagents.

(a) KI in aqueous acetone   (b) KOH in aqueous ethanol
(c) K^+ $(CH_3)_3C—O^-$ in $(CH_3)_3C—OH$ (d) product of (c) + HBr
(e) product of (c) + chloroform + potassium *tert*-butoxide
(f) product of (c) + CH_2I_2 in the presence of a Zn-Cu couple
(g) Li in hexane, then ethanol
(h) sodium methoxide in methanol

9.29 Give the products expected when 1-iodohexane or the other substances indicated react with each of the following reagents.

(a) fluoride ion in DMF   (b) KOH in aqueous ethanol
(c) K^+ $(CH_3)_3C—O^-$ in $(CH_3)_3C—OH$
(d) product of (c) + bromoform + potassium *tert*-butoxide
(e) product of (c) + CH_2I_2 in the presence of a Zn-Cu couple
(f) Mg and ether, then water

*9.30 Give the products expected when 2-bromo-2-methylhexane or the other substances indicated react with the following reagents.
(a) 1:1 ethanol-water   (b) sodium ethoxide in ethanol
(c) KI in aqueous acetone
(d) product(s) of (b) + HBr in the presence of peroxides

9.31 Give the products expected when 3-chloro-3-ethylpentane or the other substances indicated react with the following reagents.
(a) 1:1 methanol-water   (b) sodium methoxide in methanol
(c) aqueous acetone
(d) product(s) of (b) + $Hg(OAc)_2$ in THF-water, followed by $NaBH_4$

*9.32 Rank the following compounds in order of increasing S_N2 reaction rate with KI in acetone.

$$(CH_3)_3CCl \qquad (CH_3)_2CHCl \qquad (CH_3)_2CHCH_2Cl$$
$$A \qquad\qquad B \qquad\qquad C$$

$$CH_3CH_2CH_2CH_2Br \qquad CH_3CH_2CH_2CH_2Cl$$
$$D \qquad\qquad E$$

9.33 Rank the following compounds in order of increasing S_N2 reaction rate with KI in acetone.

methyl bromide   *sec*-butyl bromide   3-(bromomethyl)-3-methylpentane
$$A \qquad\qquad B \qquad\qquad C$$

1-bromopentane   1-bromo-2-methylbutane
$$D \qquad\qquad E$$

9.34 Give the structure of the nucleophile that could be used to convert iodoethane into each of the following compounds in an S_N2 reaction.
*(a) $CH_3CH_2OCH_2CH_2CH_2OCH_3$ (b) $CH_3CH_2OCH_2CH_2CH_3$

*(c) $(CH_3)_3\overset{+}{N}CH_2CH_3\ I^-$ (d) $CH_3CH_2\overset{+}{P}Ph_3\ I^-$

*(e)   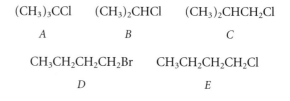   (f) CH_3CH_2CN

9.35 Give all the product(s) expected, including pertinent stereochemistry, when each of the following compounds reacts with sodium ethoxide in ethanol.
*(a)   (b) (R)-2-bromopentane

$$CH_3CH_2CH_2 \overset{\overset{\textstyle H}{\big|}}{\underset{\underset{\textstyle D}{\big|}}{-\!\!\!-\!\!\!-}} Br$$

*9.36   From the geometry of the reactive intermediate involved, predict the stereochemistry of the alcohol product in the following solvolysis.

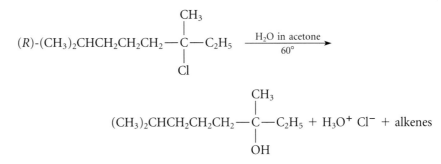

9.37   Tell which of the following alkyl halides can give only one alkene, and which can give a mixture of alkenes in the E2 reaction.

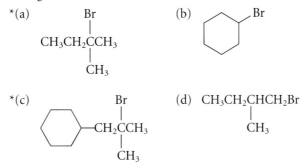

*9.38   Suppose that CH_3I is added to a methanol solution containing an excess of *both* K^+ $CH_3CH_2CH_2O^-$ and K^+ $CH_3CH_2CH_2S^-$ in equimolar amounts.
(a)   What is the major product that will be isolated from the reaction? Explain.
(b)   How would your answer change (if at all) if the experiment were conducted in anhydrous DMSO, a polar aprotic solvent?

*9.39   In the *Williamson ether synthesis*, an alkoxide reacts with an alkyl halide to give an ether.

$$R—\ddot{\underset{\cdot\cdot}{O}}:^- + R'—\ddot{\underset{\cdot\cdot}{X}}: \longrightarrow R—\ddot{\underset{\cdot\cdot}{O}}—R' + :\ddot{\underset{\cdot\cdot}{X}}:^-$$

You are in charge of a research group for a large company, Ethers Unlimited, and you have been assigned the task of synthesizing *tert*-butyl methyl ether, $(CH_3)_3C—O—CH_3$. You have decided to delegate this task to two of your staff chemists. One chemist, Ima Smart, allows $(CH_3)_3C—\ddot{O}:^-$ to react with $CH_3—I$ and indeed obtains a good yield of the desired ether. The other chemist, Notso Bright, allows $CH_3\ddot{O}:^-$ to react with $(CH_3)_3C—Br$. To his surprise, no ether was obtained. Explain why Notso Bright's reaction failed.

9.40   Propose a synthesis of ethyl isopropyl ether using an S_N2 reaction of an alkyl halide and any other reagents.

*9.41   When benzyl bromide (Ph—CH$_2$—Br) is added to a suspension of potassium fluoride in benzene, no reaction occurs. However, when a *catalytic* amount of the crown ether [18]-crown-6 (Sec. 8.4C) is added to the solution, benzyl fluoride can be isolated in high yield. If lithium fluoride is substituted for potassium fluoride, there is no reaction even in the presence of the crown ether. Explain these observations.

*9.42   Which of the following secondary alkyl halides reacts faster with ⁻CN in the S$_N$2 reaction? (*Hint:* Consider the hybridization and geometry of the S$_N$2 transition state.)

(a) 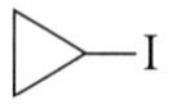 ▷—I   (b) (CH$_3$)$_2$CH—I

*9.43   The insecticide chlordane is reported to lose some of its chlorine and to be converted into other compounds when exposed to alkaline conditions. Explain.

principal component of chlordane

9.44   *(a) Explain why the following compound reacts to give a mixture of alkene stereoisomers in which only the *Z* isomer contains deuterium.

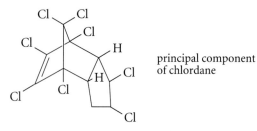

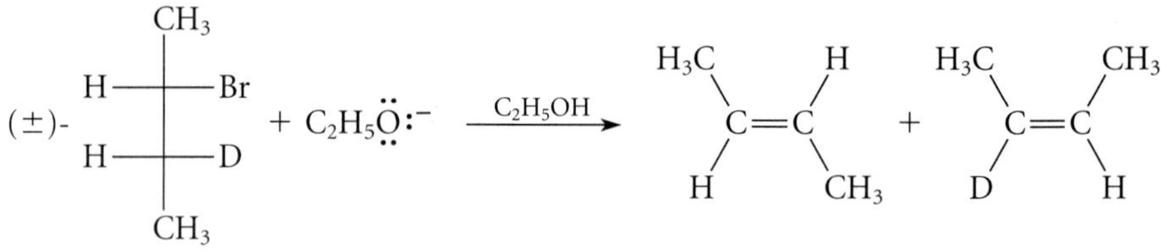

(b) Explain why the following compound reacts to give a mixture of alkene stereoisomers in which only the *E* isomer contains deuterium.

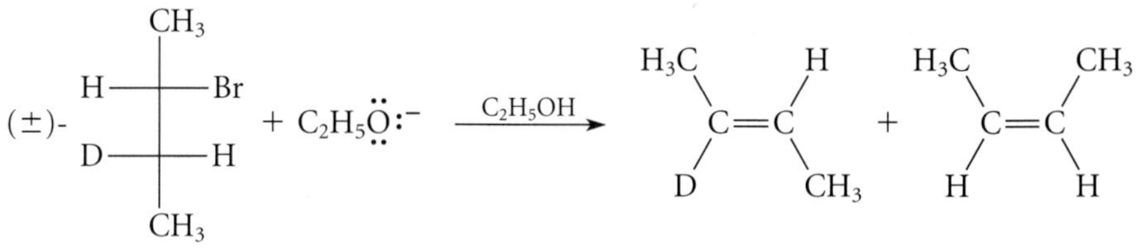

*9.45   *Tert*-butyl chloride undergoes solvolysis in either acetic acid or formic acid.

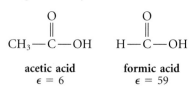

acetic acid
ϵ = 6

formic acid
ϵ = 59

Notice that both solvents are protic, donor solvents, but they differ substantially in their dielectric constants ϵ.

(a) What is the S_N1 solvolysis product in each solvent?

(b) In one solvent the S_N1 reaction is 5000 times as fast as it is in the other. In which solvent is the reaction more rapid, and why?

*9.46   Consider the following equilibrium:

$$(CH_3)_2\ddot{S}\colon + \ CH_3\!-\!\ddot{\underset{\cdot\cdot}{Br}}\colon \ \rightleftharpoons \ (CH_3)_2\overset{+}{\ddot{S}}\!-\!CH_3 + \colon\!\ddot{\underset{\cdot\cdot}{Br}}\colon^-$$

In each case (a) and (b), choose the solvent in which the equilibrium would lie farther *to the right*. Explain. (Assume that the products are soluble in all solvents considered.)

(a) ethanol or diethyl ether

(b) dimethylacetamide (a polar, aprotic solvent, $\epsilon = 38$) or a mixture of water and methanol that has the same dielectric constant

*9.47   Consider the following experiments with trityl chloride, $Ph_3C\!-\!Cl$, a very reactive tertiary alkyl halide:

(1) In aqueous acetone the reaction of trityl chloride follows a rate law that is first order in the alkyl halide, and the product is trityl alcohol, $Ph_3C\!-\!OH$.

(2) In another reaction, when one equivalent of sodium azide ($Na^+\ N_3^-$) is added to a solution that is otherwise identical to that used in experiment (1), the reaction rate is virtually the same as in (1); however, the product isolated in good yield is trityl azide, $Ph_3C\!-\!N_3$.

(3) In a reaction mixture in which both sodium azide and sodium hydroxide are present in equal concentrations, both trityl alcohol and trityl azide are formed, but the reaction rate is again unchanged.

Explain why the reaction rate is the same but the products are different in these three experiments. (The pK_a of HN_3 is 4.72, and the pK_a of H_2O is 15.7.)

*9.48   In a laboratory has been found an optically active compound *A* that has the following elemental analysis: C, 50.81%; H, 6.93%; Br, 42.26%. Compound *A* gives no reaction with Br_2 in CCl_4, but it reacts with $K^+\ (CH_3)_3C\!-\!O^-$ to give a single new compound *B* in good yield. Compound *B* decolorizes Br_2 in CCl_4 and takes up hydrogen over a catalyst. When compound *B* is treated with ozone followed by aqueous H_2O_2, dicarboxylic acid *C* is isolated in excellent yield; notice its *cis* stereochemistry.

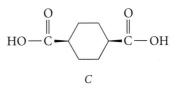

C

Identify compounds *A* and *B* and account for all observations. (If you need a refresher on how to solve this type of problem, see Study Guide Links 4.6 and 5.5.)

*9.49   In a laboratory two liquids, *A* and *B*, were found in a box labeled only "isomeric alkyl halides $C_5H_{11}Br$." You have been employed to deduce the structures of these compounds from the following data left in an accompanying laboratory notebook. Reaction of each compound with Mg in ether, followed by water, gives the same hydrocarbon. Compound *A*, when dissolved in ethanol, reacts to give an ethyl ether *C* and an acidic solution in a few minutes. Compound *B* reacts more slowly, but eventually gives the *same* ether *C* and an acidic solution under the same conditions. Both acidic solutions, when tested with $AgNO_3$ solution, give a light yellow precipitate of AgBr. Reaction of compound *B* with sodium ethoxide in ethanol gives an alkene that reacts with O_3, then aqueous H_2O_2, to give acetone $(CH_3)_2C{=}O$ as one product. Give the structures of *A* and *B*, and explain your reasoning.

*9.50   In the laboratories of the firm "Halides 'R' Us" has been found a compound *A* in a vial labeled only "achiral alkyl halide $C_{10}H_{17}Br$." The management feels that the compound might be useful as a pesticide, but they need to know its structure. You have been called in as a consultant at a handsome fee. Compound *A*, when treated with KOH in ethanol, yields two compounds *B* and *C*, each with the molecular formula $C_{10}H_{16}$. Compound *A* rapidly reacts in aqueous ethanol to give an acidic solution which, in turn, gives a precipitate of AgBr when tested with $AgNO_3$ solution. Ozonolysis of *A* followed by treatment with $(CH_3)_2S$ affords $(CH_3)_2C{=}O$ (acetone) as one of the products plus unidentified halogen-containing material. Hydrogenation of either *B* or *C* gives a mixture of both *trans-* and *cis*-1-isopropyl-4-methylcyclohexane. Compound *A* reacts with one equivalent of Br_2 to give a mixture of two separable compounds, *D* and *E*, both of which can be shown to be achiral compounds. Finally, ozonolysis of compound *B* followed by treatment with aqueous H_2O_2 gives acetone and the diketone *F*.

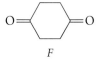

F

Propose structures for compounds *A* through *E* that best fit the data (and collect your fee).

*9.51   When menthyl chloride (see following structure) is treated with sodium ethoxide in ethanol, 2-menthene is the only alkene product observed. When neomenthyl chloride is subjected to the same conditions, the alkene products are mostly 3-menthene (78%) along with some 2-menthene (22%). Explain why different alkene products are formed from the different alkyl halides, and why 3-menthene is the major product in the second reaction. (*Hint:* Remember the stereochemistry of the E2 reaction, and don't forget about the chair flip of cyclohexanes.)

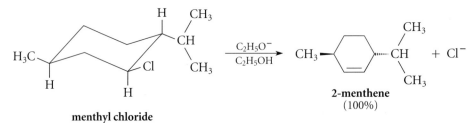

menthyl chloride   2-menthene
                          (100%)

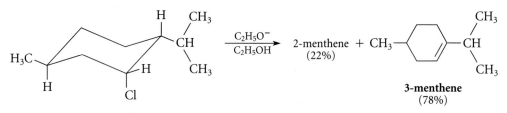

neomenthyl chloride

9.52 Tell whether each of the following eliminations is *syn* or *anti*.

*(a)

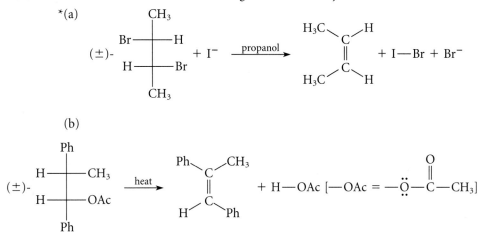

(b)

9.53 Explain why each alkyl halide stereoisomer gives a different alkene in the E2 reactions shown. It will probably help to build models or draw conformational structures of the two starting materials.

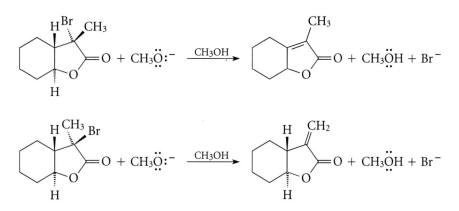

*9.54 (a) The reagent tributyltin hydride, $(C_4H_9)_3Sn$—H, brings about the rapid conversion of 1-bromo-1-methylcyclohexane into methylcyclohexane. The reaction is particularly fast in the presence of AIBN (Sec. 5.6C). Suggest a mechanism for this reaction. (*Hint:* The Sn—H bond is relatively weak.)

(b) Suggest two other reaction sequences using other reagents that would bring about the same overall transformation.

*9.55   The reaction of butylamine, $CH_3(CH_2)_3NH_2$, with 1-bromobutane in 60% aqueous ethanol follows the rate law

$$rate = k[\text{butylamine}][\text{1-bromobutane}]$$

The product of the reaction is $(CH_3CH_2CH_2CH_2)_2\overset{+}{N}H_2\ Br^-$. The following very similar reaction, however, has a first-order rate law:

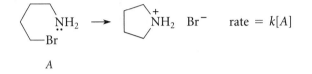

Give a mechanism for each reaction that is consistent with its rate law and with the other facts about nucleophilic substitution reactions. Use the curved-arrow formalism.

*9.56   The *cis* and *trans* stereoisomers of 4-chlorocyclohexanol give different products (shown below) when they react with OH^-.
(a) Give a curved-arrow mechanism for the formation of each product.
(b) Explain why the bicyclic material *B* is observed in the reaction of the *trans* isomer, but not in that of the *cis* isomer.

trans-4-chlorocyclohexanol + OH^- $\longrightarrow$ [product *A*] + [product *B*]

cis-4-chlorocyclohexanol + OH^- $\longrightarrow$ [product *A*] + [product *C*]

9.57   Consider the following reaction sequence.
(Bu— = butyl group = $CH_3CH_2CH_2CH_2$—)

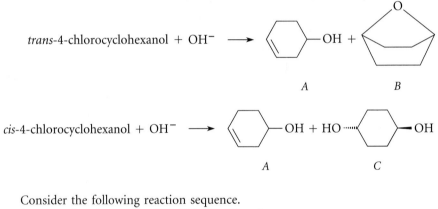

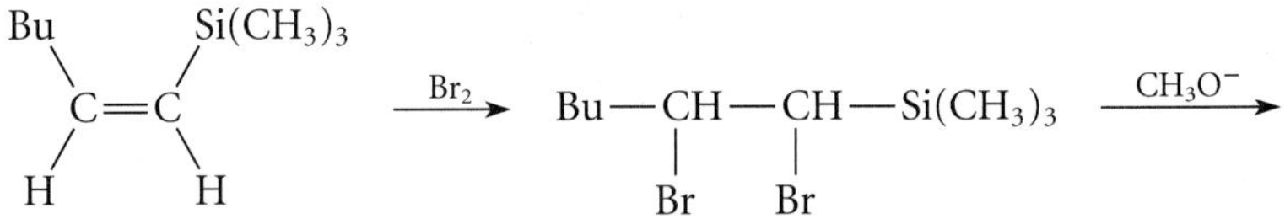

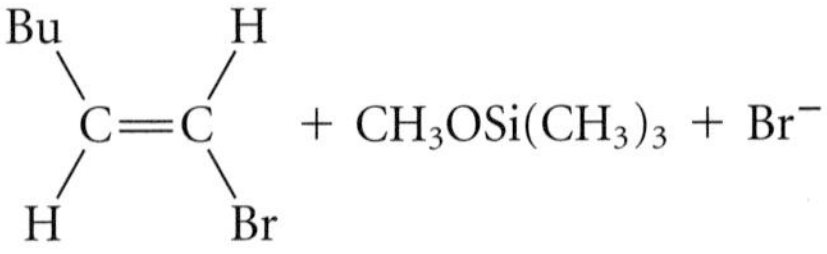

*(a) Use what you know about the stereochemistry of bromine addition to propose the stereochemistry of compound *B*.

*(b) Is the *B* → *C* reaction a *syn* or an *anti* elimination?

(c) How would the stereochemistry of products change if the *E* isomer of compound *A* were carried through the same sequence of reactions? Explain.

*9.58 Account for each of the following results with a mechanism: (*Hint for (b)*: Organolithium reagents are strong bases, and the hydrogens on a carbon adjacent to a benzene ring are relatively acidic.)

(a) $HCCl_3 + Na^+ I^- \xrightarrow[35°]{NaOH/H_2O} HCCl_2I + Na^+ Cl^-$

(This reaction is not observed in the absence of NaOH.)

(b)

$$Ph—CH_2—Cl + CH_3CH_2CH_2CH_2—Li + \bigcirc\!\!\parallel \longrightarrow$$

$$\triangleright\!\!—Ph + LiCl + CH_3CH_2CH_2CH_3$$

*9.59 On standing, 2-bromo-3-methylbutane is converted into 2-bromo-2-methylbutane. Propose a curved-arrow mechanism for this transformation.

Chemistry of Alcohols, Glycols, and Thiols

This chapter focuses on the reactions of alcohols, thiols, and glycols. Like alkyl halides, alcohols undergo substitution and elimination reactions. However, unlike alkyl halides, alcohols and thiols undergo *oxidation reactions*. This chapter explains how to recognize oxidations, and it presents some of the ways that oxidations of alcohols and thiols are carried out in the laboratory. A consideration of alcohol oxidation in nature leads to a discussion of stereochemical relationships of groups within molecules. Finally, the strategy used in planning organic syntheses is introduced.

10.1 Dehydration of Alcohols

Strong acids such as H_2SO_4 and H_3PO_4 catalyze a β-elimination reaction in which water is lost from a secondary or tertiary alcohol to give an alkene:

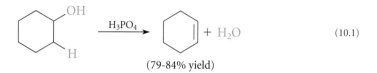

$$(10.1)$$

(79-84% yield)

A reaction such as this, in which the elements of water are lost from the starting material, is called a **dehydration**. Lewis acids such as alumina (aluminum oxide, Al_2O_3) and/or heat can also be used to catalyze or promote dehydration reactions.

Most acid-catalyzed dehydrations of alcohols are reversible reactions. However, these reactions can easily be driven toward the alkene products by applying LeChatelier's principle (Sec. 9.2). For example, in Eq. 10.1, the equilibrium is driven toward the alkene product because the water that is formed complexes tightly with the strong acid H_3PO_4, and the cyclohexene product is distilled from the reaction mixture. The dehydration of alcohols to alkenes is easily carried out in the laboratory and is an important procedure for the preparation of some alkenes.

This reaction occurs by a three-step mechanism involving a carbocation intermediate. In the first step, the —OH group of the alcohol accepts a proton from the catalyzing acid in a Brønsted acid-base reaction:

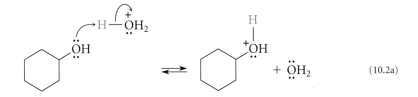

$$(10.2a)$$

Thus, the basicity of alcohols (Sec. 8.6) is important to the success of the dehydration reaction. Next, the carbon-oxygen bond of the alcohol breaks in a Lewis acid-base dissociation to give water and a carbocation:

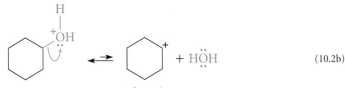

$$(10.2b)$$

a carbocation

Finally, water, the conjugate base of the catalyzing acid H_3O^+, removes a β-proton from the carbocation in another Brønsted acid-base reaction:

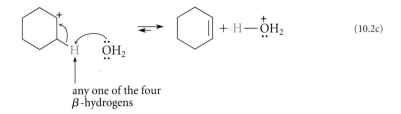

$$(10.2c)$$

any one of the four
β-hydrogens

This step generates the alkene product and regenerates the catalyzing acid H_3O^+.

We've discussed a mechanism like this twice before. First, alcohol dehydration is essentially an E1 reaction. Once the —OH group of the alcohol is protonated, it becomes a very good leaving group (water), and, like a halide leaving group in the E1 reaction, the protonated —OH departs to give a carbocation, which then loses a β-proton to give an alkene.

Alcohol dehydration: *E1 reaction of an alkyl halide:*

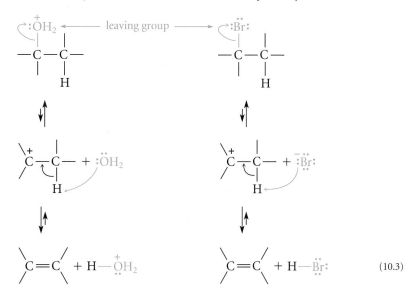

$$(10.3)$$

STUDY GUIDE LINK:
✓**10.1**
*Dehydration of
Alcohols*

Second, the dehydration of alcohols is the reverse of the hydration of alkenes (Sec. 4.9B). *Hydration of alkenes and dehydration of alcohols are the forward and reverse of the same reaction.*

It is important to recognize that *any reaction and its reverse proceed by the forward and reverse of the same mechanism.* This statement is known as the **principle of microscopic reversibility**. It follows from this principle that forward and reverse reactions must have the same intermediates and the same rate-limiting transition states. Thus, because protonation of the alkene is the rate-limiting step in alkene hydration, the reverse of this step—loss of the proton from the carbocation intermediate (Eq. 10.2c)—is rate-limiting in alcohol dehydration. This principle also requires that any reaction catalyzed in one direction is also catalyzed in the other. Thus, both hydration of alkenes to alcohols and dehydration of alcohols to alkenes are catalyzed by acids.

The involvement of carbocation intermediates explains several experimental facts about alcohol dehydration. First, the relative rates of alcohol dehydration are in the order tertiary > secondary > primary. Application of Hammond's postulate (Sec. 4.8C) suggests that the transition state of a dehydration reaction closely resembles the corresponding carbocation intermediate. Because tertiary carbocations are the most stable carbocations, dehydration reactions involving tertiary carbocations should be faster than those involving either secondary or primary carbocations, as observed. In fact, dehydration of primary alcohols is generally not a useful laboratory procedure for the preparation of alkenes. (Primary alcohols react in other ways with H_2SO_4; see Problem 10.57.)

Second, if the alcohol has more than one type of β-hydrogen, then a mixture of alkene products can be expected. As in the E1 reaction of alkyl halides, the most stable alkene—the one with the greatest number of branches at the double bond—is the alkene formed in greatest amount:

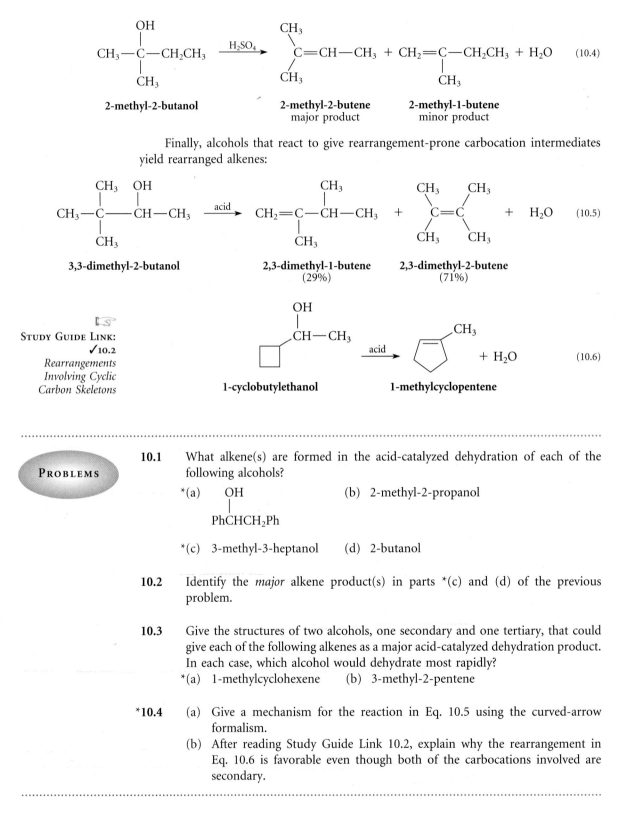

Finally, alcohols that react to give rearrangement-prone carbocation intermediates yield rearranged alkenes:

STUDY GUIDE LINK:
✔10.2
Rearrangements Involving Cyclic Carbon Skeletons

PROBLEMS

10.1 What alkene(s) are formed in the acid-catalyzed dehydration of each of the following alcohols?

*(a)   OH
       |
       PhCHCH₂Ph

(b) 2-methyl-2-propanol

*(c) 3-methyl-3-heptanol   (d) 2-butanol

10.2 Identify the *major* alkene product(s) in parts *(c) and (d) of the previous problem.

10.3 Give the structures of two alcohols, one secondary and one tertiary, that could give each of the following alkenes as a major acid-catalyzed dehydration product. In each case, which alcohol would dehydrate most rapidly?
*(a) 1-methylcyclohexene   (b) 3-methyl-2-pentene

*10.4 (a) Give a mechanism for the reaction in Eq. 10.5 using the curved-arrow formalism.
(b) After reading Study Guide Link 10.2, explain why the rearrangement in Eq. 10.6 is favorable even though both of the carbocations involved are secondary.

10.2   Reactions of Alcohols with Hydrogen Halides

Alcohols react with hydrogen halides to give alkyl halides:

$$(CH_3)_2CHCH_2CH_2—OH + HBr \xrightarrow[\text{heat, 5–6 hr}]{H_2SO_4} (CH_3)_2CHCH_2CH_2—Br + H_2O \qquad (10.7)$$

3-methyl-1-butanol   **1-bromo-3-methylbutane**
(93% yield)

$$(CH_3)_3C—OH + HCl \xrightarrow[\text{25°, 20 min}]{H_2O} (CH_3)_3C—Cl + H_2O \qquad (10.8)$$

tert-**butyl alcohol**   *tert*-**butyl chloride**
(almost
quantitative)

Because the equilibrium constant for formation of alkyl halides from alcohols is not very favorable, the successful preparation of alkyl halids from alcohols, like the dehydration of alcohols to alkenes, usually depends on the application of LeChatelier's principle (Sec. 9.2). For example, in both Eq. 10.7 and 10.8, the reactant alcohols are soluble in the reaction solvent, which is an aqueous acid, but the product alkyl halides are not. Separation of the alkyl halide products from the reaction mixture drives both reactions to completion.

The mechanism of alkyl halide formation depends on the type of alcohol used as the starting material. In the reactions of tertiary alcohols, protonation of the alcohol oxygen is followed by carbocation formation. The carbocation reacts with the halide ion, which is present in great excess:

$$(CH_3)_3C—\ddot{O}H + H—\ddot{\underset{..}{Cl}}: \rightleftharpoons (CH_3)_3C—\overset{H}{\underset{|}{\overset{+}{O}}}—H + :\ddot{\underset{..}{Cl}}:^- \qquad (10.9a)$$

$$(CH_3)_3C—\overset{+}{\ddot{O}}H_2 \rightleftharpoons (CH_3)_3C^+ + H_2\ddot{O}: \left.\begin{array}{c} \\ \\ \end{array}\right\} S_N1 \text{ reaction} \qquad (10.9b)$$

$$(CH_3)_3C^+ + :\ddot{\underset{..}{Cl}}:^- \rightleftharpoons (CH_3)_3C—\ddot{\underset{..}{Cl}}: \qquad (10.9c)$$

Notice that once the alcohol is protonated, *the reaction is essentially an* S_N1 *reaction with* H_2O *as the leaving group.*

When a primary alcohol is the starting material, the reaction occurs as a concerted displacement of water from the protonated alcohol by halide ion.

$$(CH_3)_2CHCH_2CH_2—\ddot{\underset{..}{O}}H + H—\ddot{\underset{..}{Br}}: \rightleftharpoons (CH_3)_2CHCH_2CH_2—\overset{+}{\ddot{O}}H_2 + :\ddot{\underset{..}{Br}}:^- \qquad (10.10a)$$

$$(CH_3)_2CHCH_2CH_2—\overset{+}{\ddot{O}}H_2 \rightleftharpoons (CH_3)_2CHCH_2CH_2—\ddot{\underset{..}{Br}}: + H_2\ddot{O}: \quad S_N2 \text{ reaction} \qquad (10.10b)$$
$$:\ddot{\underset{..}{Br}}:^-$$

In other words, *the reaction is an* S_N2 *reaction in which water is the leaving group.* Notice that the initial step of both S_N1 and S_N2 mechanisms is protonation of the —OH group. Secondary alcohols can react by either the S_N1 or S_N2 mechanism, or both.

The reactions of tertiary alcohols with hydrogen halides are much faster than the reactions of primary alcohols. Typically, tertiary alcohols react with hydrogen halides rapidly at room temperature, whereas the reactions of primary alcohols require heating for several hours. The reactions of primary alcohols with HBr and HI are satisfactory, but their reactions with HCl are very slow. Although reactions of alcohols with HCl can be accelerated with certain catalysts, other methods for preparing primary alkyl chlorides (discussed in the following section) are better.

When carbocation intermediates are involved in the reactions of alcohols with hydrogen halides, rearrangements occur in appropriate cases:

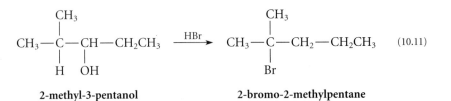

$$\text{2-methyl-3-pentanol} \qquad\qquad \text{2-bromo-2-methylpentane}$$

PROBLEMS

10.5   Suggest an alcohol starting material and the conditions for the preparation of each of the following alkyl halides.

*(a) I—CH₂CH₂CH₂CH₂CH₂—I   (b)

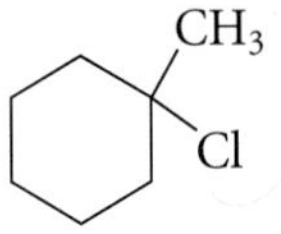

*10.6**   Using the curved-arrow formalism, give a mechanism for the reaction in Eq. 10.11.

10.7   Give the structure of the alkyl halide product expected (if any) in each of the following reactions.

*(a) HOCH₂   CH₂OH

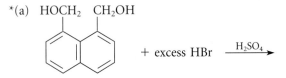

+ excess HBr $\xrightarrow{\text{H}_2\text{SO}_4}$

(b) HOCH₂CH₂CH₂OH + excess HI $\xrightarrow{\text{heat}}$

*(c)

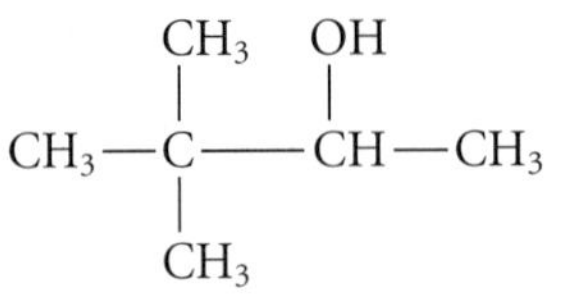

$$\underset{\underset{\displaystyle CH_3}{|}}{\overset{\overset{\displaystyle CH_3}{|}}{CH_3-C}}\!-\!\!\overset{\overset{\displaystyle OH}{|}}{CH}-CH_3 + \text{excess HBr} \xrightarrow{\text{heat}}$$

*(d) (CH₃)₃CCH₂OH + HCl $\xrightarrow{25°}$

(*Hint:* See Fig. 9.3, Sec. 9.4C.)

The dehydration of alcohols to alkenes and the reactions of alcohols with hydrogen halides have some important things in common. Both take place in very acidic solution; in both reactions the acid converts the —OH group into a good leaving group. If acid were not present, the halide ion would have to displace ⁻OH in order to form the alkyl halide. This reaction does not take place because ⁻OH is a much stronger base than any halide ion (Table 3.1 and Sec. 9.5B).

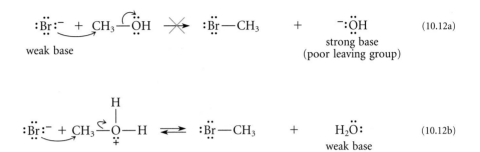

$$\text{:Br:}^- + \text{CH}_3\text{—OH} \;\;\not\!\!\rightarrow\;\; \text{:Br—CH}_3 \;+\; {}^-\text{:OH} \qquad (10.12\text{a})$$

weak base   strong base
                                                    (poor leaving group)

$$\text{:Br:}^- + \text{CH}_3\text{—O}^+\text{—H} \;\rightleftharpoons\; \text{:Br—CH}_3 \;+\; \text{H}_2\text{O:} \qquad (10.12\text{b})$$

weak base
(good leaving group)

This analysis shows that *substitution and elimination reactions of alcohols are possible if the* —OH *group is first converted into a better leaving group.* Other applications of this important point are presented in the following sections.

Notice that the formation of secondary and tertiary alkyl halides and dehydration of secondary and tertiary alcohols have the same initial steps, protonation of the alcohol oxygen and formation of a carbocation:

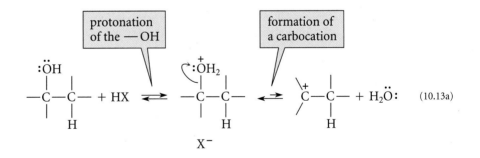

protonation
of the — OH

formation of
a carbocation

$$(10.13\text{a})$$

The two reactions differ in the fate of this carbocation, which in turn is governed by the conditions of the reaction. In the presence of a hydrogen halide, the halide ion attacks the carbocation, and the conditions are designed to cause the alkyl halide to separate from the solution. A carbocation can also lose a β-proton to form an alkene (Sec. 9.6); however, if an alkene does form, it can react with the large excess of hydrogen halide present to give alkyl halide (Sec. 4.7). In dehydration, no halide ion is present, and when the alkene forms by loss of a β-proton from the carbocation, the conditions of the dehydration reaction force the removal of the alkene product and the water by-product from the reaction mixture. It follows, then, that *alkyl halide formation and dehydration to alkenes are alternative branches of a common mechanism*:

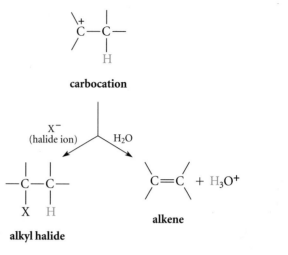

carbocation

alkyl halide

alkene

(10.13b)

Notice that the principles developed for substitutions and eliminations of alkyl halides are valid for other functional groups, in this case, alcohols.

10.3 Sulfonate and Inorganic Ester Derivatives of Alcohols

When an alkyl halide is prepared from an alcohol and a hydrogen halide, protonation converts the —OH group into a good leaving group. However, if the alcohol molecule contains a group that might be sensitive to strongly acidic conditions, or if milder or even nonacidic conditions must be used for other reasons, different ways of converting the —OH group into a good leaving group are required. Methods for accomplishing this objective are the subject of this section.

A. Sulfonate Ester Derivatives of Alcohols

Structures of Sulfonate Esters   An important method of activating alcohols toward nucleophilic substitution and β-elimination reactions is to convert them into *sulfonate esters*. Sulfonate esters are derivatives of **sulfonic acids**, which are compounds of the form R—SO₃H. Some typical sulfonic acids are the following:

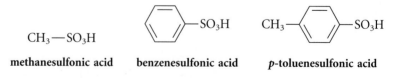

methanesulfonic acid   **benzenesulfonic acid**   ***p*-toluenesulfonic acid**

(The *p* in the name of the last compound stands for *para*, which indicates the relative positions of the two groups on the ring. This type of nomenclature is discussed in

Chapter 16.) A *sulfonate ester* is a compound in which the acidic hydrogen of a sulfonic acid is replaced by an alkyl or aryl group. Thus, in ethyl benzenesulfonate, the acidic hydrogen of benzenesulfonic acid is replaced by an ethyl group.

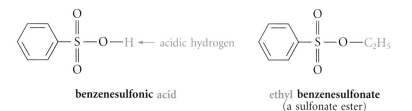

benzenesulfonic acid

ethyl **benzenesulfonate**
(a sulfonate ester)

Organic chemists use abbreviated structures and names for certain sulfonate esters. Esters of methanesulfonic acid are called *mesylates* (abbreviated R—OMs), and esters of *p*-toluenesulfonic acid are called *tosylates* (abbreviated R—OTs).

is the same as C_2H_5—OMs

ethyl mesylate

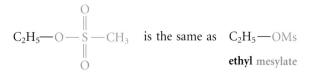

ethyl methanesulfonate

is the same as

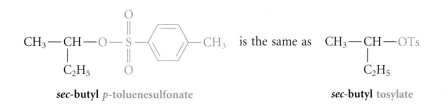

***sec*-butyl** *p*-toluenesulfonate

***sec*-butyl** tosylate

PROBLEM

10.8 Draw the complete structure and give another name for each of the following compounds.

*(a) isopropyl methanesulfonate
(b) methyl *p*-toluenesulfonate
*(c) phenyl tosylate
(d) cyclohexyl mesylate

Formation of Sulfonate Esters Sulfonate esters are prepared from alcohols and other sulfonic acid derivatives called sulfonyl chlorides. For example, *p*-toluenesulfonyl chloride, often known as *tosyl chloride*, and abbreviated TsCl, is the sulfonyl chloride used to prepare tosylate esters.

The E2 reactions of sulfonate esters, like the analogous reactions of alkyl halides, can be used to prepare alkenes:

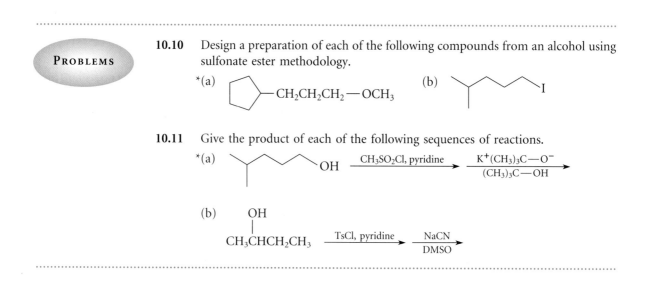

(83% yield)

This reaction is especially useful for cases in which the acidic conditions of alcohol dehydration lead to rearrangements or other side reactions, or for primary alcohols in which dehydration is not an option.

To summarize: An alcohol can be made to undergo substitution and elimination reactions typical of the corresponding alkyl halides by converting it into a sulfonate ester.

PROBLEMS

10.10   Design a preparation of each of the following compounds from an alcohol using sulfonate ester methodology.

*(a)

(b)

10.11   Give the product of each of the following sequences of reactions.

*(a)

(b)

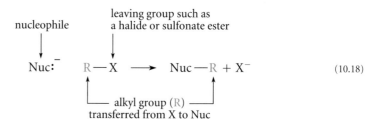

B. Alkylating Agents

As you've learned, alkyl halides, alkyl tosylates, and other sulfonate esters are reactive in nucleophilic substitution reactions. In a nucleophilic substitution, an alkyl group is transferred from the leaving group to the nucleophile.

$$ \text{Nuc:}^- \quad \text{R—X} \longrightarrow \text{Nuc—R} + \text{X}^- \tag{10.18} $$

The nucleophile is said to be **alkylated** by the alkyl halide or the sulfonate ester in the same sense that a Brønsted base is *protonated* by a strong acid. For this reason, alkyl halides, sulfonate esters, and related compounds are sometimes referred to as **alkylating agents**. To say that a compound is a *good alkylating agent* usually means that it reacts rapidly with nucleophiles in substitution reactions—that is, in S_N2 and S_N1 reactions.

C.   Ester Derivatives of Strong Inorganic Acids

Esters of strong inorganic acids are well known compounds. The structure of such an ester is derived conceptually by replacing the acidic hydrogen(s) of a strong acid with alkyl or aryl group(s). For example, dimethyl sulfate is an ester in which the acidic hydrogens of sulfuric acid are replaced by methyl groups.

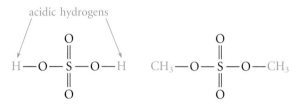

sulfuric acid   dimethyl **sulfate**

Because dimethyl sulfate can be prepared from methanol, it can also be viewed as a methanol derivative.

$$2CH_3\!-\!OH + HO\!-\!\overset{\displaystyle O}{\underset{\displaystyle O}{\overset{\|}{\underset{\|}{S}}}}\!-\!OH \longrightarrow CH_3\!-\!O\!-\!\overset{\displaystyle O}{\underset{\displaystyle O}{\overset{\|}{\underset{\|}{S}}}}\!-\!O\!-\!CH_3 + 2H_2O \qquad (10.19)$$

Alkyl esters of strong inorganic acids are typically very potent alkylating agents, because they contain leaving groups that are very weak bases. For example, dimethyl sulfate is a very effective methylating agent, as shown in the following example.

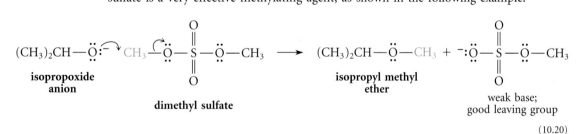

isopropoxide anion

dimethyl sulfate

isopropyl methyl ether

weak base; good leaving group

(10.20)

Dimethyl sulfate and diethyl sulfate are available commercially. These reagents, like other reactive alkylating agents, are toxic because they react with nucleophilic functional groups on proteins and nucleic acids.

Certain monoalkyl esters of phosphoric acid are utilized in nature as alkylating agents (Sec. 17.6B). DNA and RNA themselves are polymerized dialkyl esters of phosphoric acid (Sec. 27.11B).

Along the same line, alkyl halides can be thought of as alkyl esters of the halogen acids. Methyl bromide, for example, is conceptually derived by replacing the acidic hydrogen of HBr with a methyl group. As you have learned, this "ester" is an effective alkylating agent.

PROBLEMS

10.12   Phosphoric acid, H_3PO_4, has the following Lewis structure.

$$\underset{\underset{OH}{|}}{\overset{\overset{O}{\|}}{\underset{HO}{\diagup}\overset{P}{\underset{}{}}\overset{}{\diagdown}OH}}$$

*(a)   Draw the structure of the monoethyl ester of phosphoric acid.
(b)   Draw the structure of trimethyl phosphate.

10.13   Predict the products in the reaction of dimethyl sulfate with each of the following nucleophiles.
*(a)   water   (b) $CH_3\overset{..}{N}H_2$ (methylamine)
*(c)   sodium 1-propanethiolate   (d)   sodium ethoxide

D. Reactions of Alcohols with Thionyl Chloride

In most cases the preparation of primary alkyl chlorides from alcohols with HCl is not as satisfactory as the preparation of the analogous alkyl bromides with HBr (Sec. 10.2). A better method for the preparation of primary alkyl chlorides is the reaction of alcohols with thionyl chloride:

$$CH_3(CH_2)_6CH_2OH + SOCl_2 \xrightarrow{\text{pyridine}} CH_3(CH_2)_6CH_2Cl + SO_2\uparrow + HCl \qquad (10.21)$$

| | | | (reacts with |
| **1-octanol** | **thionyl chloride** | **1-chlorooctane** (80% yield) | pyridine) |

Thionyl chloride is a dense, fuming liquid (bp 75–76°). One advantage of using thionyl chloride for the preparation of alkyl chlorides is that the by-products of the reaction are HCl, which reacts with the base pyridine, and SO_2, a gas. Consequently, there are no separation problems in the purification of the product alkyl chlorides.

The preparation of an alkyl chloride from an alcohol with thionyl chloride, like the use of a sulfonate ester, involves the conversion of the alcohol —OH group into a good leaving group. When an alcohol reacts with thionyl chloride, a *chlorosulfite ester* intermediate is formed. (This reaction is analogous to that in Eq. 10.14.)

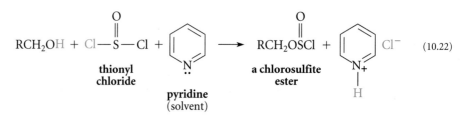

(10.22)

The chlorosulfite ester reacts readily with nucleophiles because the chlorosulfite group, —O—SO—Cl, is a very good leaving group. The chlorosulfite ester is usually not isolated, but reacts with the chloride ion formed in Eq. 10.22 to give the alkyl chloride. The displaced $^-$O—SO—Cl ion is unstable and decomposes to SO_2 and Cl^-.

$$R—CH_2\!\!-\!\!\overset{..}{\underset{..}{O}}\!\!-\!\!\overset{\overset{\displaystyle :O:}{\|}}{S}\!\!-\!\!Cl \longrightarrow R—CH_2 + \ ^-\!\!:\!\!\overset{..}{\underset{..}{O}}\!\!-\!\!\overset{\overset{\displaystyle :O:}{\|}}{S}\!\!-\!\!\overset{..}{\underset{..}{Cl}}: \longrightarrow \quad \overset{..}{\underset{..}{O}}\!\!\diagdown\!\!\overset{\displaystyle S}{}\!\!\diagup\!\!\overset{..}{\underset{..}{O}}: + \ :\!\overset{..}{\underset{..}{Cl}}:^- \qquad (10.23)$$

Although the thionyl chloride method is most useful with primary alcohols, it can also be used with secondary alcohols, although rearrangements in such cases have been known to occur. Rearrangements are best avoided in the preparation of secondary alkyl halides by using the reaction of a halide ion with a sulfonate ester in a polar aprotic solvent (as in Study Problem 10.1).

PROBLEMS

*10.14   Give two methods that can be used to convert 1-butanol into 1-chlorobutane.

10.15   According to the mechanism of the reaction of thionyl chloride with alcohols, what product, including its stereochemistry, should be obtained in the reaction of thionyl chloride with (S)-$CH_3CH_2CH_2CHD$—OH?

*10.16   Phosphorus tribromide, PBr_3, can be used to convert primary and secondary alcohols into alkyl bromides under mild conditions.

$$R—OH + PBr_3 \longrightarrow R—Br + HO—PBr_2$$

Give a curved-arrow mechanism for this transformation. (*Hint:* The mechanism follows the same general pattern as the mechanism of the reaction of alcohols with thionyl chloride.)

10.4   Conversion of Alcohols into Alkyl Halides: Summary

You have now studied a variety of reactions that can be used to convert alcohols into alkyl halides. These are

1. reaction with hydrogen halides
2. formation of sulfonate esters and displacement with halide ions
3. reaction with $SOCl_2$.

The method of choice depends on the structure of the alcohol and on the type of alkyl halide (chloride, bromide, iodide) to be prepared.

Primary Alcohols: Alkyl bromides are prepared from primary alcohols (for example, 1-hexanol) by the reaction of the alcohol with concentrated HBr. Analogously, HI can be used to form primary alkyl iodides; the HI is usually supplied by mixing an iodide salt such as KI with a strong acid such as phosphoric acid. Thionyl chloride is the method of choice for the preparation of alkyl chlorides, unless the alcohol is unusually reactive, in which case concentrated HCl can be used. The sulfonate ester method works well with primary alcohols, but requires two separate reactions (formation of the sulfonate ester, then reaction of the ester with halide ion); hence, this method is less convenient than the other methods. Because all these methods have an S_N2 mechanism as their basis, alcohols with a large amount of β-branching, such as neopentyl alcohol, do not react under the usual conditions.

Tertiary Alcohols: Tertiary alcohols (for example, *tert*-butyl alcohol) react rapidly with HCl or HBr under mild conditions to give the corresponding alkyl halide. The sulfonate ester method shown in Study Problem 10.1 is not used with tertiary alcohols because tertiary sulfonates do not undergo S_N2 reactions. Of course, if elimination is desired, such sulfonates readily undergo the E2 reaction with alkoxide bases.

Secondary Alcohols: If the secondary alcohol has relatively little β-branching, the thionyl chloride method can be used to prepare alkyl chlorides. To avoid rearrangements, the alcohol should be converted into a sulfonate ester which, in turn, should be treated with the appropriate halide ion (Cl^-, Br^-, or I^-) in a polar aprotic solvent. The HBr method can be used if the desired alkyl halide product is derived from a carbocation rearrangement; otherwise this method is not very satisfactory. Specialized methods that have not been discussed are required for primary and secondary alcohols that have significant β-branching.

The discussion in this chapter has shown that alcohols and their derivatives can undergo substitution and elimination reactions. However, the —OH group itself cannot act as a leaving group; it is far too basic. In order to break the carbon-oxygen bond, the —OH group must first be converted into a good leaving group. Several strategies can be used for this purpose:

1. Protonation: protonated alcohols are intermediates in both dehydration to alkenes and substitution to give alkyl halides.

2. Conversion into sulfonate esters or inorganic esters: these esters, to a useful approximation, react like alkyl halides.

3. Reaction with thionyl chloride: this reagent effects the conversion of alcohols into chlorosulfite esters, which are converted within the reaction mixture into alkyl chlorides.

PROBLEM

10.17   Suggest conditions for carrying out each of the following conversions to yield a product that is as free of isomers as possible.

*(a)

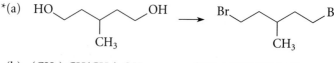

(b) $(CH_3)_2CH(CH_2)_4OH \longrightarrow (CH_3)_2CH(CH_2)_4Cl$

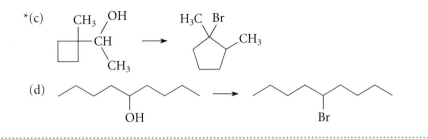

10.5   Oxidation and Reduction in Organic Chemistry

The previous sections have discussed substitution and elimination reactions of alcohols and their derivatives. These reactions have much in common with the analogous reactions of alkyl halides. Now we turn to a different class of reactions: oxidations. Oxidation is a reaction of alcohols that has no simple analogy in alkyl halide chemistry.

A. Oxidation Numbers

In organic chemistry, whether a transformation is an oxidation or a reduction is determined by the *oxidation numbers* of the reactants and products. The calculation and use of oxidation numbers is a "bookkeeping" process that involves three steps. After glancing over these steps, read carefully through Study Problem 10.2, which gives an example of the entire process.

1. Assign an **oxidation level** to each carbon that undergoes a change between reactant and product by the following method:
    a. For every bond from the carbon to a less electronegative element (including hydrogen), and for every negative charge on the carbon, assign a −1.
    b. For every bond from the carbon to another carbon atom, and for every unpaired electron on the carbon, assign a zero.
    c. For every bond from the carbon to a more electronegative element, and for every positive charge on the carbon, assign a +1.
    d. Add the numbers assigned under (a), (b), and (c) to obtain the oxidation level of the carbon under consideration.

2. Determine the **oxidation number** N_{ox} of both the reactant and product by adding, within each compound, the oxidation levels of all the carbons computed in Step 1. Remember: consider only the carbons that undergo a change in the reaction.

3. Compute the difference $N_{ox}(\text{product}) - N_{ox}(\text{reactant})$ to determine whether the transformation is an oxidation, reduction, or neither.
    a. If the difference is a positive number, the transformation is an **oxidation**.
    b. If the difference is a negative number, the transformation is a **reduction**.
    c. If the difference is zero, the transformation is neither an oxidation nor a reduction.

Decide whether the following transformation is an oxidation, a reduction, or neither.

$$\underset{\text{isopropyl alcohol}}{CH_3-\underset{\underset{\displaystyle \;}{|}}{\overset{\overset{\displaystyle OH}{|}}{CH}}-CH_3} \xrightarrow{H_2CrO_4} \underset{\text{acetone}}{CH_3-\overset{\overset{\displaystyle O}{\parallel}}{C}-CH_3} \qquad (10.24)$$

Solution

Step 1   For both the reactant and the product, compute the oxidation level of each carbon that undergoes a change. Since the two methyl groups are unchanged, do not assign oxidation levels to these carbons. Only one carbon is changed. For this carbon, −1 is assigned for each bond to hydrogen (Rule 1a); 0 is assigned for each bond to another carbon (Rule 1b); and +1 is assigned for each bond to oxygen (Rule 1c). Add the resulting numbers (color).

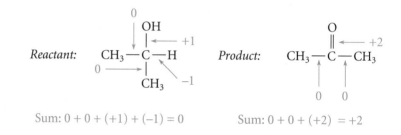

$$\text{Sum: } 0 + 0 + (+1) + (-1) = 0 \qquad\qquad \text{Sum: } 0 + 0 + (+2) = +2$$

Notice that the C=O double bond in the product acetone is treated as *two bonds*, each receiving a +1 for a total of +2 for the double bond.

Step 2   Add the oxidation levels for each carbon that changes to determine the oxidation number. Because only one carbon changes in Eq. 10.24, the oxidation level of this carbon, computed in Step 1, is the only one to be considered. Hence, N_{ox}(reactant), the oxidation number of the reactant, isopropyl alcohol, is 0. Similarly, N_{ox}(product), the oxidation number of the product, acetone, is +2.

Step 3   Compute the difference N_{ox}(product) − N_{ox}(reactant), which is +2 − 0 = +2. Because this difference is positive, the transformation of isopropyl alcohol to acetone is an oxidation.

Notice that oxidation numbers are calculated for only the organic starting material and the corresponding product. The other reactant(s) (H_2CrO_4 in Study Problem 10.2) are not involved in the calculation.

Verify that the acid-catalyzed hydration of 2-methylpropene is neither an oxidation nor a reduction.

Solution   First, write the structures involved in the transformation:

2-methylpropene   *tert*-**butyl alcohol**

The oxidation number of the organic reactant, 2-methylpropene, is -2.

$$\underset{H_3C}{\overset{0}{\underset{\diagup}{\overset{\diagdown}{C}}}}=\overset{-2}{CH_2} \qquad N_{ox} = 0 + (-2) = -2$$

The oxidation number of the organic product, *tert*-butyl alcohol, is also -2:

$$CH_3 \overset{+1}{\underset{\overset{|}{OH}}{-\,C\,-}} \overset{-3}{CH_3} \qquad N_{ox} = +1 + (-3) = -2$$

Notice that an oxidation level is computed for only the one methyl group that was formed as a result of the transformation. Because the oxidation numbers of the reactant and product are equal, the hydration reaction is neither an oxidation nor a reduction. The same conclusion, of course, applies to the reverse reaction, dehydration of the alcohol to the alkene.

Notice in Study Problem 10.3 that one of the carbons of 2-methylpropene is reduced and one is oxidized; however, the net change in oxidation number for the overall transformation is zero.

The methods described here show that the addition of Br_2 to an alkene is an oxidation (the change in oxidation number is $+2$):

$$\underset{N_{ox} = -2}{R-CH=CH-R} + Br_2 \longrightarrow \underset{N_{ox} = 0}{R-\overset{\overset{\displaystyle Br}{|}}{CH}-\overset{\overset{\displaystyle Br}{|}}{CH}-R} \qquad (10.25)$$

Study Guide Link:
✓**10.4**
*Oxidations and
Reductions*

Thus, whether a reaction is an oxidation or reduction does not necessarily depend on the introduction or loss of oxygen. However, in most oxidations of organic compounds, either a hydrogen in a C—H bond or a carbon in a C—C bond is replaced by a more electronegative element, which *may* be oxygen, but which may also be another element such as a halogen.

Study Problem 10.3 shows that a process which involves introduction of an oxygen (or other electronegative element) at one carbon atom is not an oxidation if another carbon atom is reduced at the same time. That is, the oxidation state of a molecule is determined by the sum of the oxidation states of its individual carbon atoms.

The oxidation-number formalism can also be used to relate organic oxidations and reductions to the definition of oxidation and reduction that you may have learned in general chemistry: *An oxidation is a transformation in which electrons are lost, and a reduction is a transformation in which electrons are gained.* To see how oxidation numbers are related to the loss or gain of electrons, consider the transformation of ethanol to acetic acid.

$$
\mathrm{CH_3-CH_2-OH} \longrightarrow \underset{\text{acetic acid}}{\mathrm{CH_3-\overset{\displaystyle O}{\overset{\displaystyle \|}{C}}-OH}} \tag{10.26}
$$

$$\underset{\text{ethanol}}{}$$

The change in oxidation number for this transformation is +4 (confirm this!); hence, this is an oxidation. Let's now view this transformation in another way by writing it as a *balanced half-reaction* that shows the loss of electrons. This process involves three steps:

1. Use H_2O to balance missing oxygens.
2. Use protons (that is, H^+) to balance missing hydrogens.
3. Use "dummy electrons" to balance charges.

This process is illustrated in Study Problem 10.4.

STUDY PROBLEM 10.4

Write the transformation of Eq. 10.26 as a balanced half-reaction.

Solution First, balance the extra oxygen on the right with a water on the left:

$$
\mathrm{CH_3CH_2OH + H_2O} \longrightarrow \mathrm{CH_3\overset{\displaystyle O}{\overset{\displaystyle \|}{C}}-OH} \quad \text{(oxygens are balanced)} \tag{10.27a}
$$

Next, balance the extra hydrogens on the left with four protons on the right:

$$
\mathrm{CH_3CH_2OH + H_2O} \longrightarrow \mathrm{CH_3\overset{\displaystyle O}{\overset{\displaystyle \|}{C}}-OH + 4H^+} \quad \text{(hydrogens and oxygens are balanced)} \tag{10.27b}
$$

Finally, balance the extra positive charges on the right with "dummy electrons" so that the charges on both sides of the equation are equal:

$$
\mathrm{CH_3CH_2OH + H_2O} \longrightarrow \mathrm{CH_3\overset{\displaystyle O}{\overset{\displaystyle \|}{C}}-OH + 4H^+ + 4\mathit{e}^-} \quad \text{(everything is balanced)} \tag{10.27c}
$$

The result is the balanced half-reaction.

According to this half-reaction, *four electrons are "lost" from the ethanol molecule when acetic acid is formed.* Of course, electrons really aren't lost, because a corresponding number of electrons are "gained" by the species that brings about the oxidation. Nevertheless, on the basis of this half-reaction, it can be said that *the oxidation of ethanol to acetic acid is a four-electron oxidation.* This type of terminology is frequently used in biochemistry.

Recall now that the change in oxidation number for the ethanol-to-acetic acid transformation is $+4$. Notice the correspondence between the change of oxidation number ($+4$) and the "electrons lost" (4) in the corresponding balanced half-reaction. *This correspondence is general.* That is, *the change in oxidation number is equal to the number of electrons lost or gained in the corresponding half-reaction.*

PROBLEMS

10.18 Classify each of the following transformations, some of which may be unfamiliar, as an oxidation, reduction, or neither. For those that are oxidations or reductions, tell how many electrons are gained or lost.

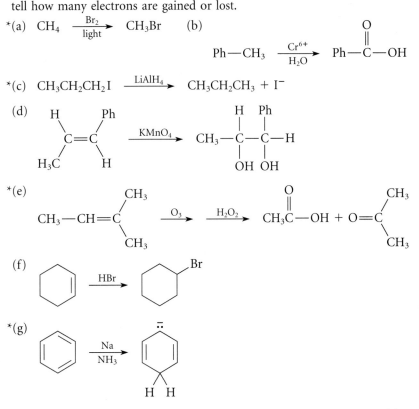

*(a) $CH_4 \xrightarrow[\text{light}]{Br_2} CH_3Br$ (b)

$Ph—CH_3 \xrightarrow[H_2O]{Cr^{6+}} Ph—\overset{\overset{\displaystyle O}{\|}}{C}—OH$

*(c) $CH_3CH_2CH_2I \xrightarrow{LiAlH_4} CH_3CH_2CH_3 + I^-$

(d)

$\underset{H_3C}{\overset{H}{\diagdown}}C=C\underset{H}{\overset{Ph}{\diagup}} \xrightarrow{KMnO_4} CH_3—\overset{\overset{\displaystyle H}{|}}{\underset{\underset{\displaystyle OH}{|}}{C}}—\overset{\overset{\displaystyle Ph}{|}}{\underset{\underset{\displaystyle OH}{|}}{C}}—H$

*(e)

$CH_3—CH=C\underset{CH_3}{\overset{CH_3}{\diagup\diagdown}} \xrightarrow{O_3} \xrightarrow{H_2O_2} CH_3\overset{\overset{\displaystyle O}{\|}}{C}—OH + O=C\underset{CH_3}{\overset{CH_3}{\diagup\diagdown}}$

(f)

HBr

*(g)

$\xrightarrow[NH_3]{Na}$

10.19 Write the transformations in *part (d) and part (b) of the previous problem as balanced half-reactions.

B. Oxidizing and Reducing Agents

Like acid-base reactions, oxidations and reductions always occur in pairs. Therefore, *whenever something is oxidized, something else is reduced.* When an organic compound is oxidized, the reagent that brings about the transformation is called an **oxidizing agent**. Similarly, when an organic compound is reduced, the reagent that effects the transformation is called a **reducing agent**. For example, suppose that chromate ion (CrO_4^{2-}) is used to bring about the oxidation of ethanol to acetic acid in Eq. 10.27c; in this reaction, chromate ion is reduced to Cr^{3+}.

The corresponding half-reaction for the reduction of chromate is

$$8H^+ + 3e^- + CrO_4^{2-} \longrightarrow Cr^{3+} + 4H_2O \qquad (10.28)$$

Three electrons are gained in the reduction of chromate to Cr^{3+}. A complete, balanced reaction for the oxidation of ethanol to acetic acid by chromate is obtained by reconciling the "dummy electrons" in the half-reactions given by Eq. 10.27c and Eq. 10.28 and adding the two equations. This process is illustrated in the following study problem.

STUDY PROBLEM 10.5

Give a complete balanced equation for the oxidation of ethanol to acetic acid by chromate ion.

Solution   The two half-reactions are

$$CH_3CH_2OH + H_2O \longrightarrow CH_3CO_2H + 4H^+ + 4e^- \qquad (10.29a)$$

$$8H^+ + 3e^- + CrO_4^{2-} \longrightarrow Cr^{3+} + 4H_2O \qquad (10.29b)$$

Multiply each equation by a factor that gives the same number of "dummy electrons" in both half-reactions. Thus, multiplying Eq. 10.29a by 3 and Eq. 10.29b by 4 gives twelve "dummy electrons" in both reactions:

$$3CH_3CH_2OH + 3H_2O \longrightarrow 3CH_3CO_2H + 12H^+ + 12e^- \qquad (10.30a)$$

$$32H^+ + 12e^- + 4CrO_4^{2-} \longrightarrow 4Cr^{3+} + 16H_2O \qquad (10.30b)$$

Add these equations, cancelling like terms on both sides. Thus, all "dummy electrons" cancel; the three water molecules on the left are cancelled by three of those on the right to leave thirteen water molecules on the right; and twelve protons on the right are cancelled by twelve of those on the left, leaving twenty protons on the left. Hence, the fully balanced equation is

STUDY GUIDE LINK:
✓10.5
More on Half-Reactions

$$20H^+ + 4CrO_4^{2-} + 3CH_3CH_2OH \longrightarrow 4Cr^{3+} + 3CH_3CO_2H + 13H_2O \qquad (10.31)$$

This equation shows that three ethanol molecules are oxidized for every four chromate ions reduced.

By considering the change in oxidation number for a transformation, you can tell whether an oxidizing or reducing agent is required to bring about the reaction. For example, the following unfamiliar transformation is neither an oxidation nor a reduction (verify this statement):

$$
\begin{array}{ccc}
& CH_3 \quad CH_3 & \qquad\qquad CH_3 \quad O \\
& \;|\qquad\;| & \qquad\qquad\;\;|\qquad\;\| \\
CH_3\!-\!C\!-\!\!-\!\!-\!C\!-\!CH_3 & \longrightarrow & CH_3\!-\!C\!-\!\!-\!\!-\!C\!-\!CH_3 \\
\;|\qquad\;| & & \;\;| \\
OH \quad OH & & CH_3
\end{array}
\qquad (10.32)
$$

Although one carbon is oxidized, another is reduced. Even though you might know nothing else about the reaction, it is clear that an oxidizing or reducing agent alone would not effect this transformation. (In fact, the reaction is brought about by strong acid.)

The oxidation-number concept can be used to organize organic compounds into functional groups with the same oxidation level, as shown in Table 10.1 on page 466. Compounds within a given box are generally interconverted by reagents that are neither oxidizing nor reducing agents. For example, alcohols can be converted into alkyl halides with HBr, which is neither an oxidizing nor a reducing agent. On the other hand, conversion of an alcohol into a carboxylic acid involves a change in oxidation level, and indeed this transformation requires an oxidizing agent. Notice also in Table 10.1 that carbons with larger numbers of hydrogens have a greater number of possible oxidation states. Thus, a tertiary alcohol cannot be oxidized at the α-carbon (without breaking carbon-carbon bonds) because this carbon bears no hydrogens. Methane, on the other hand, can be oxidized to CO_2. (Of course, any hydrocarbon can be oxidized to CO_2 if carbon-carbon bonds are broken; Sec. 2.7.)

PROBLEMS

10.20   Indicate which of the following balanced reactions are oxidation-reduction reactions and which are not. For those involving oxidation-reduction, indicate which compound(s) are oxidized and which are reduced. (*Hint:* Consider the organic compounds in each reaction first.)

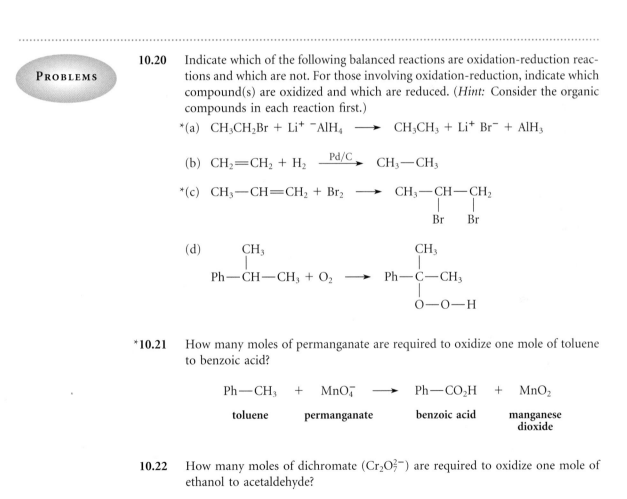

*10.21   How many moles of permanganate are required to oxidize one mole of toluene to benzoic acid?

$$Ph-CH_3 \quad + \quad MnO_4^- \quad \longrightarrow \quad Ph-CO_2H \quad + \quad MnO_2$$

    toluene   permanganate   benzoic acid   manganese
                                                            dioxide

10.22   How many moles of dichromate ($Cr_2O_7^{2-}$) are required to oxidize one mole of ethanol to acetaldehyde?

$$CH_3CH_2OH \quad + \quad Cr_2O_7^{2-} \quad \longrightarrow \quad CH_3CH=O \quad + \quad Cr^{3+}$$

      ethanol   dichromate   acetaldehyde

Table 10.1   Comparison of Oxidation States of Various Functional Groups

All molecules in the same box have the same oxidation number.

X = an electronegative group such as halogen, etc.

Methane ——————— increasing oxidation number ——————→

CH_4	$CH_3—OH$ $CH_3—X$	$H_2C{=}O$ H_2CX_2	$\overset{\displaystyle O}{\underset{\displaystyle OH}{H—C}}$ $H—CX_3$	$O{=}C{=}O$ CX_4

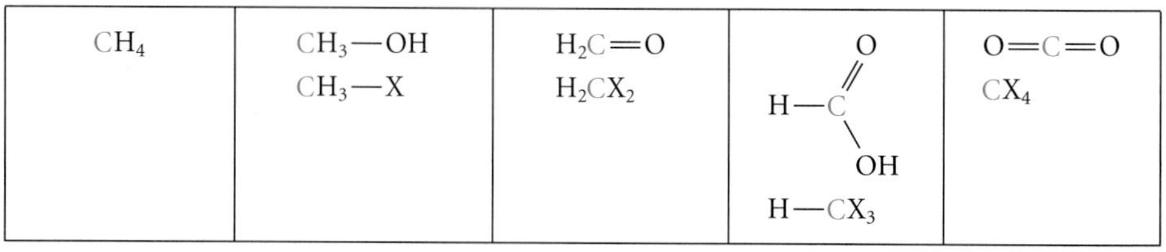

Primary Carbon ——————— increasing oxidation number ——————→

$R—CH_3$	$R—\underset{\displaystyle OH}{CH_2}$ $R—\underset{\displaystyle X}{CH_2}$	$R—CH{=}O$ $R—CHX_2$	$R—\underset{\displaystyle OH}{C{=}O}$ $R—CX_3$

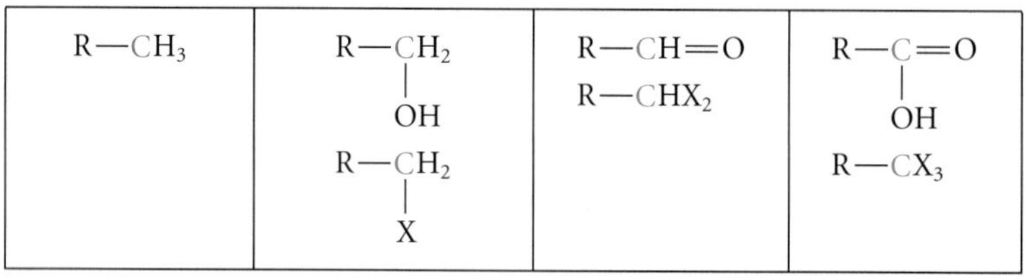

Secondary Carbon ——————— increasing oxidation number ——————→

$R—\underset{\displaystyle R}{CH_2}$	$R—\underset{\displaystyle R}{CH}—OH$ $R—\underset{\displaystyle R}{CH}—X$	$R—\underset{\displaystyle R}{C}{=}O$ $R—\underset{\displaystyle R}{CX_2}$

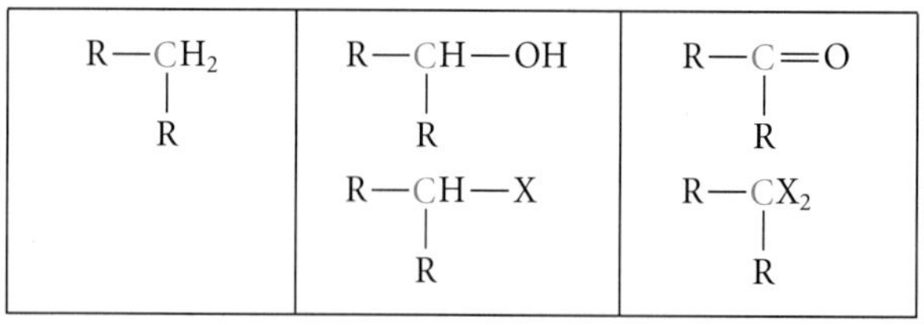

Tertiary Carbon ——————— increasing oxidation number ——————→

$R—\underset{\displaystyle CH_2R}{CH}—R$	$R—\underset{\displaystyle R}{\overset{\displaystyle R}{C}}—OH$ $R—\underset{\displaystyle R}{\overset{\displaystyle R}{C}}—X$

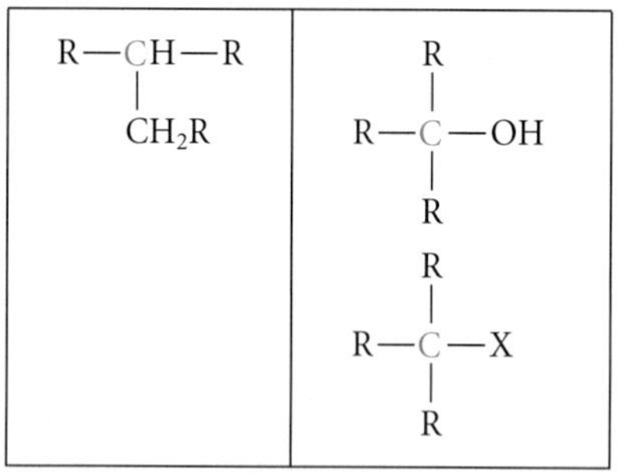

10.6   Oxidation of Alcohols

A. Oxidation to Aldehydes and Ketones

Primary and secondary alcohols are oxidized by reagents containing Cr(VI), that is, chromium in the +6 oxidation state, to give *carbonyl compounds* (compounds containing the carbonyl group, $\diagdown C{=}O$). For example, secondary alcohols are oxidized to ketones:

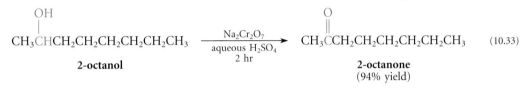

$$
\begin{array}{c} \text{OH} \\ | \\ \text{CH}_3\text{CHCH}_2\text{CH}_2\text{CH}_2\text{CH}_2\text{CH}_2\text{CH}_3 \end{array} \xrightarrow[\substack{\text{aqueous H}_2\text{SO}_4 \\ \text{2 hr}}]{\text{Na}_2\text{Cr}_2\text{O}_7} \begin{array}{c} \text{O} \\ \| \\ \text{CH}_3\text{CCH}_2\text{CH}_2\text{CH}_2\text{CH}_2\text{CH}_2\text{CH}_3 \end{array} \qquad (10.33)
$$

2-octanol   2-octanone
(94% yield)

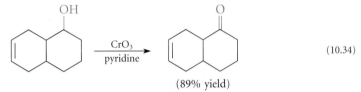

(10.34)

(89% yield)

Several forms of Cr(VI) can be used to convert secondary alcohols into ketones. Three of these are chromate (CrO_4^{2-}), dichromate ($Cr_2O_7^{2-}$), and chromic anhydride or chromium trioxide (CrO_3). The first two reagents are customarily used under strongly acidic conditions; the last is often used in pyridine.

Primary alcohols react with Cr(VI) reagents to give aldehydes, but if water is present, aldehydes are further oxidized to carboxylic acids:

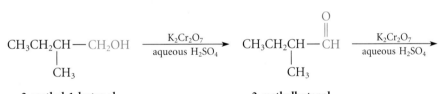

$$
\begin{array}{c} \text{CH}_3\text{CH}_2\text{CH}{-}\text{CH}_2\text{OH} \\ | \\ \text{CH}_3 \end{array} \xrightarrow[\text{aqueous H}_2\text{SO}_4]{\text{K}_2\text{Cr}_2\text{O}_7} \begin{array}{c} \text{O} \\ \| \\ \text{CH}_3\text{CH}_2\text{CH}{-}\text{CH} \\ | \\ \text{CH}_3 \end{array} \xrightarrow[\text{aqueous H}_2\text{SO}_4]{\text{K}_2\text{Cr}_2\text{O}_7}
$$

2-methyl-1-butanol   2-methylbutanal

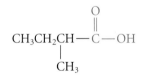

$$
\begin{array}{c} \text{O} \\ \| \\ \text{CH}_3\text{CH}_2\text{CH}{-}\text{C}{-}\text{OH} \\ | \\ \text{CH}_3 \end{array}
$$

2-methylbutanoic acid   (10.35)

For this reason, anhydrous preparations of Cr(VI) are generally used for the laboratory preparation of aldehydes from primary alcohols. One frequently used reagent of this type is a complex between chromium trioxide and two molecules of pyridine in methylene chloride solvent, commonly known as *Collins' reagent*:

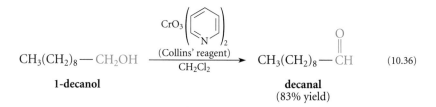

$$\begin{array}{c} \text{(10.36)} \end{array}$$

1-decanol

decanal
(83% yield)

Water promotes the transformation of aldehydes into carboxylic acids because, in water, aldehydes are in equilibrium with hydrates formed by addition of water across the C=O double bond.

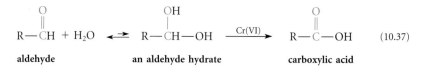

(10.37)

aldehyde   **an aldehyde hydrate**   **carboxylic acid**

Aldehyde hydrates are really alcohols and therefore can be oxidized just like secondary alcohols. Because of the absence of water in anhydrous reagents such as Collins' reagent, no 1,1-diol forms, and the reaction stops at the aldehyde.

Tertiary alcohols are not readily oxidized under the usual conditions. *Notice that a carbon atom must bear a hydrogen atom in order for oxidation of an alcohol to an aldehyde, a ketone, or a carboxylic acid to occur.*

The mechanism of alcohol oxidation by Cr(VI) involves several steps that have close analogies to other reactions. Consider, for example, the oxidation of isopropyl alcohol to the ketone acetone by chromic acid (H_2CrO_4).

The first steps of the reaction involve an acid-catalyzed displacement of water from chromic acid by the alcohol to form a *chromate ester*. (This ester is analogous to ester derivatives of other strong acids; Sec. 10.3C.)

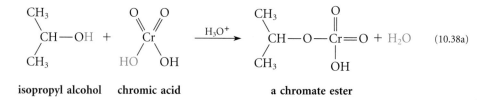

(10.38a)

isopropyl alcohol   chromic acid   **a chromate ester**

After protonation of the chromate ester (Eq. 10.38b), it decomposes in a β-elimination reaction (Eq. 10.38c).

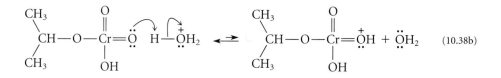

(10.38b)

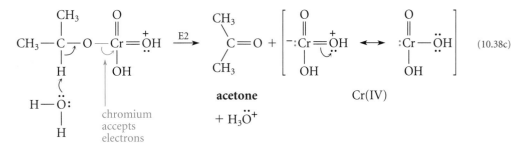

(10.38c)

This last step is much like an E2 reaction, except that it does not involve a strong base. In this step, chromium accepts an electron pair and is thus reduced. In the resulting H_2CrO_3 by-product, chromium is in a +4 oxidation state. The ultimate by-product is Cr^{3+} because, in subsequent reactions, Cr(IV) and Cr(VI) react to give two equivalents of a Cr(V) species, which then oxidizes an additional molecule of alcohol.

$$Cr(IV) + Cr(VI) \longrightarrow 2Cr(V) \tag{10.39a}$$

$$Cr(V) + (CH_3)_2CH-OH \longrightarrow Cr(III) + (CH_3)_2C=O + 2H^+ \tag{10.39b}$$

THE BREATHALYZER TEST

The oxidation of alcohols by Cr(VI) is the chemical basis of the breathalyzer test used by law-enforcement personnel to determine whether a person is under the influence of alcohol. In the lungs, a known (small) fraction of ethanol in the blood escapes into the air that is subsequently expired. A sample of this air is collected and is allowed to react with acidic potassium dichromate ($K_2Cr_2O_7$), reducing the chromium to Cr^{3+}. The resulting change in color of the chromium from the yellow-orange of the Cr(VI) oxidation state to the blue-green of the Cr(III) oxidation state is detected by a simple spectrometer. The amount of Cr(VI) reduced is calibrated in terms of percent blood alcohol. (See Problem 10.45 at the end of this chapter.)

B. Oxidation to Carboxylic Acids

As noted in the previous section (Eq. 10.35), primary alcohols can be oxidized to carboxylic acids using *aqueous* solutions of Cr(VI) such as aqueous potassium dichromate ($K_2Cr_2O_7$) in acid. Another useful reagent for oxidizing primary alcohols to carboxylic acids is potassium permanganate ($KMnO_4$) in basic solution:

$$CH_3CH_2CH_2CH_2\overset{\overset{\displaystyle C_2H_5}{|}}{C}HCH_2OH \xrightarrow[OH^-]{KMnO_4} \xrightarrow{H_3O^+} CH_3CH_2CH_2CH_2\overset{\overset{\displaystyle C_2H_5}{|}}{C}HCO_2H \tag{10.40}$$

2-ethylhexanol   **2-ethylhexanoic acid**
                                        (74% yield)

Manganese in $KMnO_4$ is in the Mn(VII) oxidation state; in this reaction, it is reduced to MnO_2, a common form of Mn(IV). Because $KMnO_4$ reacts with alkene double bonds (Sec. 5.5), Cr(VI) is preferred for the oxidation of alcohols that contain double or triple bonds (see Eq. 10.34).

Potassium permanganate is not used for the oxidation of secondary alcohols to ketones because many ketones react further with the alkaline permanganate reagent.

PROBLEMS

10.23 Complete each of the following reactions by giving the major organic product.

*(a)

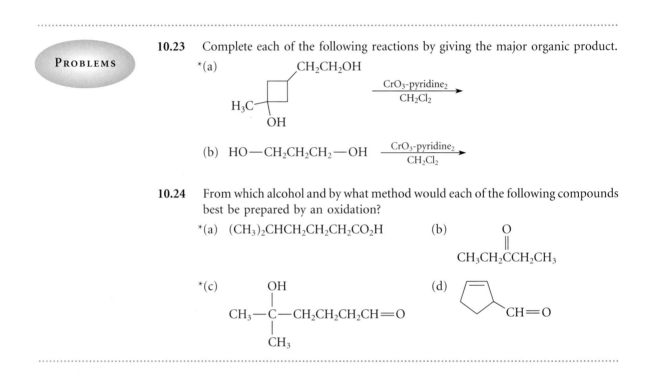

(b) $HO-CH_2CH_2CH_2-OH$ $\xrightarrow[CH_2Cl_2]{CrO_3\text{-pyridine}_2}$

10.24 From which alcohol and by what method would each of the following compounds best be prepared by an oxidation?

*(a) $(CH_3)_2CHCH_2CH_2CH_2CO_2H$

(b) $$CH_3CH_2\overset{\overset{\displaystyle O}{\|}}{C}CH_2CH_3$$

*(c)

$$CH_3-\overset{\overset{\displaystyle OH}{|}}{\underset{\underset{\displaystyle CH_3}{|}}{C}}-CH_2CH_2CH_2CH=O$$

(d)

C. Oxidative Cleavage of Glycols

The carbon-carbon bond between the —OH groups of a vicinal glycol can be cleaved with periodic acid to give two carbonyl compounds:

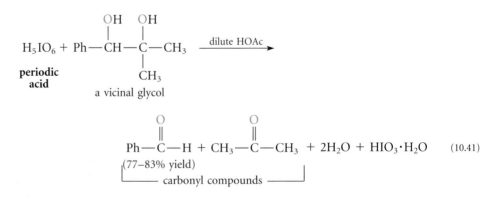

$$Ph-\overset{\overset{\displaystyle O}{\|}}{C}-H + CH_3-\overset{\overset{\displaystyle O}{\|}}{C}-CH_3 + 2H_2O + HIO_3 \cdot H_2O \qquad (10.41)$$

(77–83% yield)
carbonyl compounds

Periodic acid is the iodine analog of perchloric acid.

$$HClO_4 \qquad\qquad HIO_4$$

perchloric acid   **periodic acid**

Periodic acid is commercially available as the dihydrate, $HIO_4 \cdot 2H_2O$, often abbreviated, as in Eq. 10.41, as H_5IO_6 (sometimes called *para-periodic acid*). Its sodium salt, $NaIO_4$ (sodium metaperiodate), is sometimes also used. Periodic acid is a fairly strong acid ($pK_a = -1.6$). Because periodate can be determined by titration, the periodate cleavage reaction has been used as a test for glycols as well as for synthesis. The formulas HIO_4 or H_5IO_6 are used interchangeably for periodic acid.

The cleavage of glycols with periodic acid takes place through a cyclic periodate ester intermediate (Sec. 10.3C) that forms when the glycol displaces water from H_5IO_6.

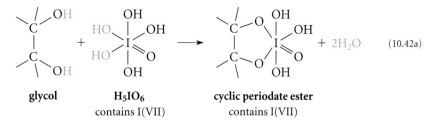

        glycol   **H_5IO_6**   **cyclic periodate ester**
                     contains I(VII)   contains I(VII)

The cyclic ester spontaneously breaks down by a cyclic flow of electrons in which the iodine accepts an electron pair.

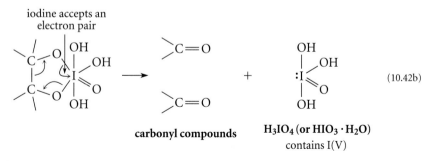

                 carbonyl compounds   **H_3IO_4 (or $HIO_3 \cdot H_2O$)**
                                           contains I(V)

A glycol that cannot form a cyclic ester intermediate is not cleaved by periodic acid. For example, the following compound is not cleaved because it is impossible for both oxygens to be part of the same cyclic periodate ester. (If you can't see why, build a model and try connecting the two oxygens with one other atom.)

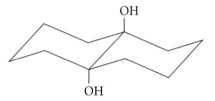

Do not confuse permanganate, osmium tetroxide, and periodate oxidations, all of which occur through cyclic ester intermediates (see Sec. 5.5). Periodate oxidizes *glycols*, but the other reagents oxidize *alkenes* to give glycols. In all of these reactions, oxidation occurs because an atom in a highly positive oxidation state can accept an additional pair of electrons. In the periodate oxidation, the reduction of the iodine occurs during the *breakdown* of the cyclic ester; in the permanganate and osmium textroxide oxidations, the metals are reduced during the *formation* of the cyclic ester.

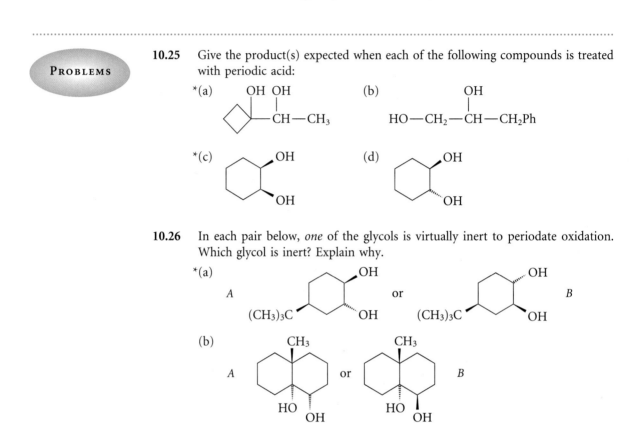

PROBLEMS

10.25 Give the product(s) expected when each of the following compounds is treated with periodic acid:

*(a)

(b)

*(c)

(d)

10.26 In each pair below, *one* of the glycols is virtually inert to periodate oxidation. Which glycol is inert? Explain why.

*(a)   A   or   B

(b)   A   or   B

10.7   Biological Oxidation of Ethanol

Oxidation and reduction reactions are extremely important in living systems. A typical biological oxidation is the conversion of ethanol into acetaldehyde, the principal reaction by which ethanol is removed from the bloodstream.

$$CH_3CH_2OH \xrightarrow{\text{biological oxidation}} CH_3CH{=}O \qquad (10.43)$$

ethanol   **acetaldehyde**

The reaction is carried out in the liver and is catalyzed by an enzyme called *alcohol dehydrogenase*. (Recall from Sec. 4.9C that enzymes are biological catalysts.) The oxidizing agent is not the enzyme, but a complex-looking molecule called *nicotinamide adenine dinucleotide*, abbreviated NAD^+; the structure of NAD^+ and a convenient abbreviated structure for it are shown in Figure 10.1. When ethanol is oxidized, NAD^+ is reduced to a product called NADH. The hydrogen removed from carbon-1 of the ethanol ends up in the NADH; the —OH hydrogen is lost as a proton.

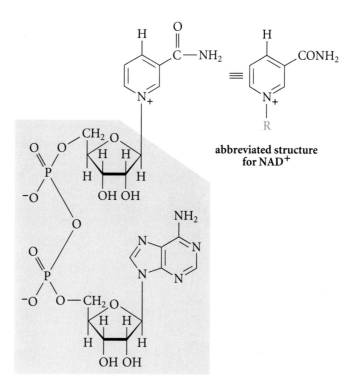

NAD⁺

Figure 10.1   *Structure of NAD⁺. The portion of the structure in the colored area is abbreviated R.*

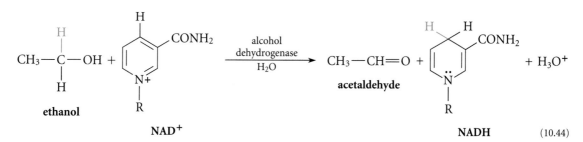

(10.44)

The compound NAD⁺ is one of nature's most important oxidizing agents. (It might be called "nature's substitute for Cr(VI).") NAD⁺ is an example of a **coenzyme**. Coenzymes are molecules required, along with enzymes, for certain biological reactions to occur. For example, ethanol cannot be oxidized by an enzyme unless the coenzyme NAD⁺ is also present, because NAD⁺ is one of the reactants. Thus, an ethanol molecule and an NAD⁺ molecule are juxtaposed when they bind noncovalently to alcohol dehydrogenase, the enzyme that catalyzes ethanol oxidation. It is within the complex of these three molecules that ethanol is oxidized to acetaldehyde and NAD⁺ is reduced to NADH.

The coenzymes NAD⁺ and NADH are derived from the vitamin *niacin*, a deficiency of which is associated with the disease pellagra (black tongue). Many biochemical processes employ the NAD⁺ $\rightleftarrows$ NADH interconversion, some of which reoxidize the NADH formed in ethanol oxidation back to NAD⁺.

FERMENTATION

The human body uses the ethanol-to-acetaldehyde reaction of Eq. 10.44 to remove ethanol, but yeast cells use the reaction in reverse as the last step in the production of ethanol. Thus, yeast added to dough produces ethanol by the reduction of acetaldehyde, which, in turn, is produced in other reactions from sugars in the dough. Ethanol vapors, wafted away from the bread by CO_2 produced in other reactions, give rising bread its pleasant odor. Special strains of yeast ferment the sugars in corn syrup, grape juice, or malt and barley to whiskey, wine, or beer. In order for the fermentation reaction to take place, the reaction must occur in the absence of oxygen. Otherwise, acetic acid, CH_3CO_2H (vinegar, or "spoiled wine"), is formed instead by other reactions. In winemaking, air is excluded by trapping the CO_2 formed during fermentation as a blanket in the fermentation vessel. Because the production of alcohol by yeast occurs in the absence of air, it is called *anaerobic fermentation*. This is one of the oldest chemical reactions known to civilization.

How does NAD^+ work as an oxidizing agent? The resonance structure of NAD^+ shows that it has the character of a carbocation:

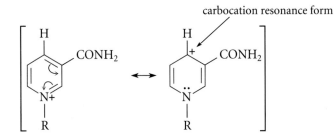

$$(10.45)$$

The electron-deficient carbon of NAD^+ and an α-hydrogen of ethanol (color) are held in proximity by the enzyme. The carbocation removes a hydride (a hydrogen with two electrons) from the α-carbon of ethanol:

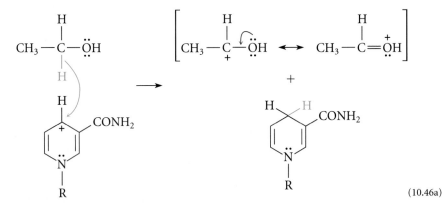

$$(10.46a)$$

Although this reaction may initially look strange, it is really just like a carbocation rearrangement involving the migration of a hydride (Sec. 4.7D), except that in this case the hydride moves to a different molecule. As a result, NADH and a new carbocation are formed. By loss of the proton bound to oxygen, acetaldehyde is formed.

$$CH_3-\overset{\overset{\displaystyle H}{|}}{C}=\overset{+}{\underset{\cdot\cdot}{O}}-H \quad :\ddot{O}H_2 \quad \longrightarrow \quad CH_3-\overset{\overset{\displaystyle H}{|}}{C}=\underset{\cdot\cdot}{\ddot{O}} + H_3\overset{+}{O}: \qquad (10.46b)$$

The acetaldehyde and NADH dissociate from the enzyme, which is then ready for another round of catalysis.

Despite the fact that the NAD^+ molecule looks large and complicated, the chemical changes that occur when it serves as an oxidizing agent take place in a relatively small part of the molecule. A number of other coenzymes also have complex structures but undergo simple reactions. The part of the molecule abbreviated by "R" in Fig. 10.1 provides the groups that cause it to bind tightly to the enzyme catalyst, but this part of the molecule remains unchanged in the oxidation reaction.

This section has shown that the chemical changes which occur in NAD^+-promoted oxidations have analogies in common laboratory reactions. Most other biochemical reactions have common laboratory analogies as well. Even though the molecules involved may be complex, their chemical transformations are in most cases relatively simple. Thus, *an understanding of the fundamental types of organic reactions and their mechanisms is useful in the study of biochemical processes.*

PROBLEM

*10.27   Give a curved-arrow mechanism for the following oxidation of 2-heptanol, which proceeds in 82% yield.

$$\overset{\overset{\displaystyle OH}{|}}{CH_3CHCH_2CH_2CH_2CH_2CH_3} \quad + \quad Ph_3C^+ \ BF_4^- \quad \longrightarrow$$

2-heptanol   a relatively stable carbocation

$$\overset{\overset{\displaystyle O}{\parallel}}{CH_3CCH_2CH_2CH_2CH_2CH_3} + HBF_4 + Ph_3CH$$

2-heptanone

10.8   Chemical and Stereochemical Group Equivalence

Different molecules with the same molecular formula—isomers—can have various relationships: they can be constitutional isomers, or they can be stereoisomers, which in turn can be diastereomers or enantiomers (Chapter 6). The subject of this section is the relationships that groups *within* molecules can have. This subject is particularly important in two areas. First, it is important in understanding the stereochemical aspects of enzyme catalysis. Second, the subject of group relationships is important in spectroscopy, particularly nuclear magnetic resonance spectroscopy, which is discussed in Chapter 13.

A. Chemical Equivalence and Nonequivalence

It is sometimes important to know when two groups in a molecule are **chemically equivalent**. That is, when do two groups behave in exactly the same way towards a chemical reagent?

An understanding of chemical equivalence hinges, first, on the concept of *constitutional equivalence*. Groups within a molecule are **constitutionally equivalent** when they have the same connectivity relationship to all other atoms in the molecule. Thus, within each of the following molecules, the hydrogens shown in color are constitutionally equivalent:

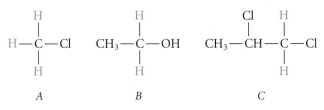

For example, in compound *C*, each of the colored hydrogens is connected to a carbon that is connected to a chlorine and a CH_3CHCl- group. Each of these hydrogens has the same connectivity relationship to the other atoms in the molecule. On the other hand, the CH_3- and $-CH_2-$ hydrogens in the ethanol molecule *B* are constitutionally nonequivalent. The CH_3- hydrogens are connected to a carbon that is connected to a $-CH_2OH$, but the $-CH_2-$ hydrogens are connected to a carbon that is connected to an $-OH$ and a $-CH_3$. However, the two $-CH_2-$ hydrogens of ethanol (color) are constitutionally equivalent, as are the three CH_3- hydrogens.

In general, *constitutionally nonequivalent groups are chemically nonequivalent*. This means that constitutionally nonequivalent groups have different chemical behavior. Thus, the CH_3- and $-CH_2-$ hydrogens of ethyl alcohol, which are constitutionally non-equivalent, have different reactivities toward chemical reagents. A reagent that reacts with one type of hydrogens will in general have a different reactivity (or perhaps none at all) with the other type. For example, oxidation of ethanol with Cr(VI) reagents causes the loss of a $-CH_2-$ hydrogen, but the CH_3- hydrogens are unaffected. Consequently, the two types of hydrogens have very different reactivities with the oxidizing agent.

Constitutional nonequivalence is a sufficient but not a necessary condition for chemical nonequivalence. That is, some constitutionally equivalent groups are chemically non-equivalent. *Whether two constitutionally equivalent groups are chemically equivalent depends on their stereochemical relationship*. The stereochemical relationship between constitutionally equivalent groups is revealed by a *substitution test*. Substitute each constitutionally equivalent group in turn with a fictitious circled group and compare the resulting molecules. Their stereochemical relationship determines the relationship of the circled groups. This process is best illustrated by example, starting with the molecules *A*, *B*, and *C* above. Substitute each hydrogen of molecule *A* with a circled hydrogen:

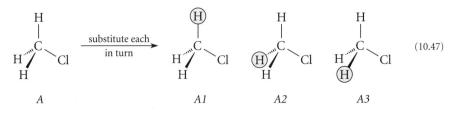

(10.47)

Each of these "new" molecules is congruent to the other. For example, the identity of *A1* and *A2* is shown in the following way:

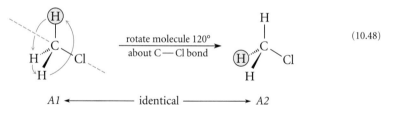

(10.48)

$$A1 \longleftarrow \text{identical} \longrightarrow A2$$

When the substitution test gives identical molecules, as in this example, the constitution-ally equivalent groups are said to be **homotopic**. *Homotopic groups are chemically equivalent and indistinguishable under all circumstances.* Thus, the homotopic hydrogens of methyl chloride all have the the same reactivity toward any chemical reagent; there is no way to distinguish among these protons, or to tell them apart.

Substitution of each of the constitutionally equivalent hydrogens in molecule *B* (ethanol) gives *enantiomers*:

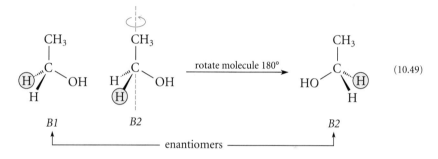

(10.49)

$$B1 \qquad\qquad B2 \qquad\qquad\qquad\qquad B2$$
$$\underset{\text{enantiomers}}{\underbrace{\qquad\qquad\qquad\qquad\qquad\qquad\qquad\qquad\qquad}}$$

When the substitution test gives enantiomers, the constitutionally equivalent groups are said to be **enantiotopic**. *Enantiotopic groups are chemically nonequivalent toward chiral reagents, but are chemically equivalent toward achiral reagents.*

Because enzymes are chiral, they can generally distinguish between enantiotopic groups within a molecule. For example, in the enzyme-catalyzed oxidation of ethanol, one of the two enantiotopic α-hydrogens is selectively removed. (This is further described in the next section.) Achiral reagents, however, cannot distinguish between enantiotopic groups. Thus, in the oxidation of ethanol to acetaldehyde by chromic acid, an achiral reagent, the α-hydrogens of ethanol are removed indiscriminately.

Finally, substitution of each of the constitutionally equivalent —CH_2— hydrogens in molecule *C* gives *diastereomers*.

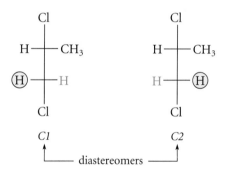

$$C1 \qquad\qquad\qquad C2$$
$$\underset{\text{diastereomers}}{\underbrace{\qquad\qquad\qquad\qquad\qquad}}$$

When the substitution test gives diastereomers, the constitutionally equivalent groups are said to be **diastereotopic**. *Diastereotopic groups are chemically nonequivalent under all conditions.* For example, the hydrogens labeled H^a and H^b in 2-bromobutane (Eq. 10.50) are diastereotopic. In the E2 reaction of this compound, *anti* elimination of H^b and Br gives *cis*-2-butene, and *anti* elimination of H^a and Br gives *trans*-2-butene. (Verify these statements using models, if necessary.)

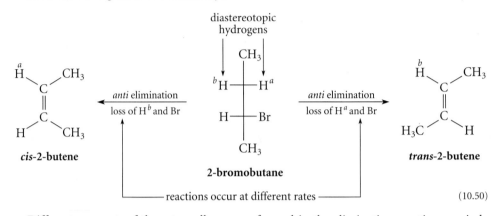

(10.50)

Different amounts of these two alkenes are formed in the elimination reaction precisely because the two diastereotopic hydrogens are removed at different rates—that is, these hydrogens are distinguished by the base that promotes the elimination.

Diastereotopic groups are easily recognized at a glance in two situations. The first occurs when two constitutionally equivalent groups are present in a molecule that contains an asymmetric carbon:

The second situation occurs when two groups on one carbon of a double bond are the same and the two groups on the other carbon are different:

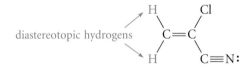

As you should readily verify, the substitution test on the colored hydrogens gives E,Z isomers, which are diastereomers.

Just as Fig. 6.11 can be used to summarize the relationships between isomeric molecules, Fig. 10.2 can be used to summarize the relationships of groups within a molecule. Notice the close analogy between the relationships between *different molecules* and the relationships between *groups within a molecule*. Just as two broad classes of isomers are based on connectivity—constitutional isomers (isomers with different connectivities) and stereoisomers (isomers with the same connectivities)—the two broad classes of groups *within* a molecule are also based on connectivity: constitutionally equivalent groups and constitutionally nonequivalent groups. Just as two different *structures* can be identical, two *constitutionally equivalent groups* within the same molecule can be

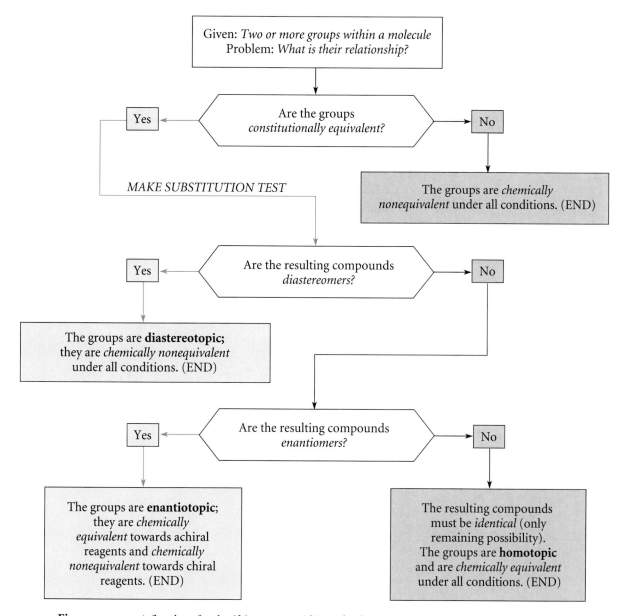

Figure 10.2   *A flowchart for classifying groups within molecules.*

homotopic. Just as there are classes of stereoisomeric relationships between *molecules*—enantiomers and diastereomers—there are corresponding relationships between *constitutionally equivalent groups within molecules*—enantiotopic and diastereotopic relationships. Just as *enantiomers* have different reactivities only with chiral reagents, *enantiotopic groups* also have different reactivities only with chiral reagents. Just as *diastereomers* have different reativities with any reagent, *diastereotopic groups* have different reactivities with any reagent.

Let's summarize the answer to the question posed at the beginning of this section: When are two groups in a molecule chemically equivalent?

☞

STUDY GUIDE LINK:
10.6
Symmetry Relationships Among Constitutionally Equivalent Groups

1. Constitutionally nonequivalent groups are chemically nonequivalent.

2. Homotopic groups are chemically equivalent in all situations.

3. Enantiotopic groups are chemically equivalent towards achiral reagents, but are chemically nonequivalent towards chiral reagents (such as enzymes).

4. Diastereotopic groups are chemically nonequivalent.

PROBLEM

10.28 For each of the following molecules state whether the groups indicated by italic letters are constitutionally equivalent or nonequivalent. If they are constitutionally equivalent, classify them as homotopic, enantiotopic, or diastereotopic.

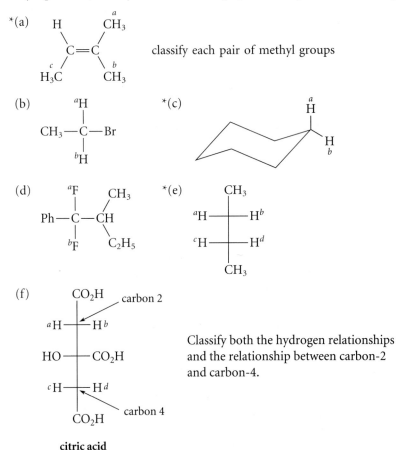

*(a) classify each pair of methyl groups

(b)

*(c)

(d)

*(e)

(f) Classify both the hydrogen relationships and the relationship between carbon-2 and carbon-4.

citric acid

B. Stereochemistry of the Alcohol Dehydrogenase Reaction

The alcohol dehydrogenase reaction provides an example which demonstrates not only that a chiral reagent can distinguish between enantiotopic groups, but also *how this discrimination can be detected*. Suppose each of the enantiotopic α-hydrogens of ethanol is replaced in turn with the isotope deuterium (^{2}H, or D). Replacing one hydrogen, called the pro-(R)-hydrogen, with deuterium, gives (R)-1-deuterioethanol; replacing the other, called the pro-(S)-hydrogen, gives (S)-1-deuterioethanol. Although ethanol itself is *not* chiral, the deuterium-substituted analogs *are* chiral; they are a pair of enantiomers:

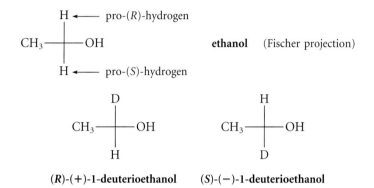

The deuterium isotope, then, provides a subtle way for the experimentalist to "label" the enantiotopic α-hydrogens of ethanol, and thus distinguish between them.

If the alcohol dehydrogenase reaction is carried out with (R)-1-deuterioethanol, *only* the deuterium is transferred from the alcohol to the NAD$^+$:

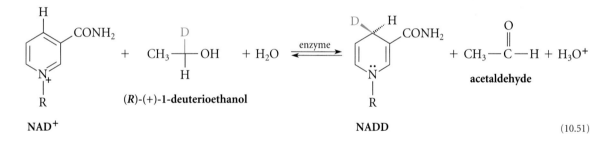

$$(10.51)$$

However, if the alcohol dehydrogenase reaction is carried out on (S)-1-deuterioethanol, *only* the hydrogen is transferred.

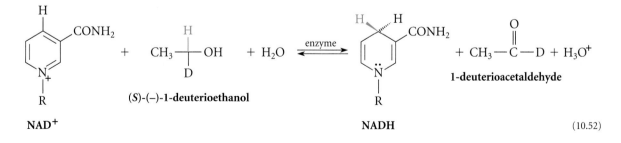

$$(10.52)$$

These two experiments show that *the enzyme distinguishes between the two α-hydrogens of ethanol*. These results cannot be attributed to a primary deuterium isotope effect (Sec. 9.5C), because an isotope effect would cause the enzyme to transfer the hydrogen in preference to the deuterium in both cases. Although the isotope is used to see the preference for transfer of one hydrogen and not the other, this experiment requires that even in the absence of the isotope the enzyme transfers the pro-(R)-hydrogen of ethanol.

The enzyme can distinguish between the two α-hydrogens of ethanol because the enzyme is chiral, and the two α-hydrogens are *enantiotopic*. Enantiotopic groups react differently with chiral reagents (Sec. 10.8A).

Another interesting stereochemical point about the alcohol dehydrogenase reaction is shown in Eqs. 10.51 and 10.52: the deuterium (or hydrogen) that is removed from the isotopically substituted ethanol molecule is transferred specifically to one particular face, or side, of the NAD$^+$ molecule. That is, the deuterium in the product NADD (color) occupies the position above the plane of the page. This result and the principle of microscopic reversibility suggest that if acetaldehyde and the NADD stereoisomer shown on the right of Eq. 10.51 were used as starting materials and the reaction run in reverse, only the deuterium should be transferred to the acetaldehyde, and (R)-1-deuterioethanol should be formed. Indeed, Eq. 10.51 *can* be run in reverse, and the experimental result is as predicted. No matter how many times the reaction runs back and forth, the H and the D on both the ethanol and the NADD molecules are never "scrambled"; they maintain their respective stereochemical positions. The two CH$_2$ hydrogens in NADH are in fact *diastereotopic*; they are distinguished not only by the enzyme, but would in principle be distinguished even without the chiral enzyme present (although without the enzyme they might not be distinguished as effectively).

PROBLEM

10.29   In each of the following cases, imagine that the two reactants shown are allowed to react in the presence of alcohol dehydrogenase. Tell whether the ethanol formed is chiral. If the ethanol is chiral, draw a Fischer projection of the enantiomer that is formed.

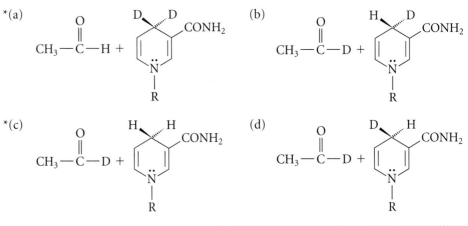

10.9   ## Oxidation of Thiols

Much of the chemistry of thiols is closely analogous to the chemistry of alcohols, because sulfur and oxygen are in the same group of the periodic table. For oxidation reactions,

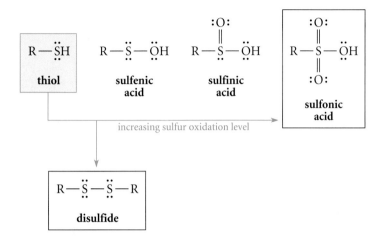

Figure 10.3   *Oxidation of thiols can give several possible products. Of these, disulfides and sulfonic acids (boxed) are the most common.*

however, this similarity disappears. Oxidation of an alcohol (Sec. 10.6) occurs at the *carbon* atom bearing the —OH group. However, oxidation of a thiol takes place not at the carbon, but at the *sulfur*. Although sulfur analogs of aldehydes, ketones, and acids are known, they are *not* obtained by simple oxidation of thiols:

$$RCH_2OH \xrightarrow{\text{oxidation}} RCH=O \xrightarrow{\text{oxidation}} RCO_2H \qquad (10.53a)$$

$$RCH_2SH \xcancel{\xrightarrow{\text{oxidation}}} RCH=S \xcancel{\xrightarrow{\text{oxidation}}} R-\overset{O}{\underset{\|}{C}}-SH, \ R-\overset{S}{\underset{\|}{C}}-OH \qquad (10.53b)$$

Some oxidation products of thiols are given in Fig. 10.3. The most commonly occurring oxidation products of thiols are disulfides and sulfonic acids (boxed in the figure). The same oxidation-number formalism used for carbon (Sec. 10.5A) can be applied to oxidation at sulfur.

PROBLEM

10.30   How many electrons are involved in the oxidation of 1-propanethiol to each of the following compounds? (See Fig. 10.3 for a more detailed description of the bonding at sulfur in the oxidation products.)
  *(a)   1-propanesulfonic acid, $CH_3CH_2CH_2SO_3H$
  (b)   1-propanesulfinic acid, $CH_3CH_2CH_2SO_2H$

Multiple oxidation states are common for elements in periods of the periodic table beyond the second. The various oxidation products of thiols exemplify the multiple oxidation states available to sulfur. The Lewis structures of these derivatives require either violation of the octet rule or separation of formal charge:

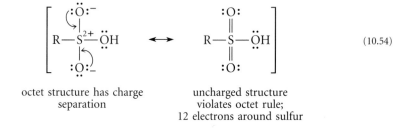

$$\tag{10.54}$$

octet structure has charge
separation

uncharged structure
violates octet rule;
12 electrons around sulfur

Sulfur can accommodate more than eight valence electrons because, in addition to its $3s$ and $3p$ orbitals, it has unfilled $3d$ orbitals of relatively low energy (Fig. 1.12). The overlap of an oxygen electron pair in a sulfonic acid with a sulfur d orbital is shown in Fig. 10.4; this is essentially an orbital picture of the $S{=}O$ double bond.

Sulfonic acids are formed by vigorous oxidation of thiols or disulfides with $KMnO_4$ or nitric acid (HNO_3).

$$CH_3CH_2CH_2CH_2{-}SH \xrightarrow{\text{conc. } HNO_3} CH_3CH_2CH_2CH_2{-}SO_3H \tag{10.55}$$

1-butanethiol
1-butanesulfonic acid
(72–96% yield)

Recall that sulfonate esters (Sec. 10.3) are derivatives of sulfonic acids; other sulfonic acid chemistry is considered in Chapters 16 and 20.

Many thiols spontaneously oxidize to disulfides merely on standing in air (O_2). Thiols can also be converted into disulfides by mild oxidants such as I_2 in base or Br_2 in CCl_4:

$$CH_3(CH_2)_4SH + I_2 + 2NaOH \longrightarrow CH_3(CH_2)_4S{-}S(CH_2)_4CH_3 + 2NaI + 2H_2O \tag{10.56}$$
(70% yield)

$$2C_2H_5SH + Br_2 \longrightarrow C_2H_5S{-}SC_2H_5 + 2HBr \tag{10.57}$$

ethanethiol
diethyl disulfide
(nearly quantitative yield)

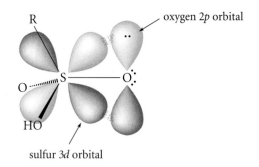

Figure 10.4   *Bonding in the higher oxidation states of sulfur involves sulfur d orbitals. In a sulfonic acid (RSO_3H), an electron pair on oxygen overlaps with one of several sulfur 3d orbitals.*

These reactions can be viewed as a series of S_N2 reactions in which halogen and sulfur are attacked by thiolate-anion nucleophiles.

$$R—\ddot{S}—H + ^-\!:\ddot{O}H \longrightarrow R—\ddot{S}:^- + H_2\ddot{O}: \qquad (10.58a)$$

$$R—\ddot{S}:^- \quad :\ddot{I}—\ddot{I}: \longrightarrow R—\ddot{S}—\ddot{I}: + :\ddot{I}:^- \qquad (10.58b)$$

$$R—\ddot{S}:^- \quad R—\ddot{S}—\ddot{I}: \longrightarrow R—\ddot{S}—\ddot{S}—R + :\ddot{I}:^- \qquad (10.58c)$$

When thiols and disulfides are present together in the same solution, an equilibrum among them is rapidly established. For example, if ethanethiol and dipropyl disulfide are combined, they react to give a mixture of all possible thiols and disulfides:

CH$_3$CH$_2$SH + CH$_3$CH$_2$CH$_2$S—SCH$_2$CH$_2$CH$_3$ ⇌

ethanethiol   **dipropyl disulfide**

CH$_3$CH$_2$S—SCH$_2$CH$_2$CH$_3$ + CH$_3$CH$_2$CH$_2$SH   (10.59a)

ethyl propyl disulfide   **propanethiol**

CH$_3$CH$_2$SH + CH$_3$CH$_2$S—SCH$_2$CH$_2$CH$_3$ ⇌ CH$_3$CH$_2$S—SCH$_2$CH$_3$ + CH$_3$CH$_2$CH$_2$SH   (10.59b)

ethanethiol   **ethyl propyl disulfide**   **diethyl disulfide**   **propanethiol**

Because the compounds on both sides of these equilibria have very similar structures, the equilibrium constants for these reactions are close to 1.0.

Thiols and sulfides are very important in biology. Thiol groups in many enzymes have catalytically essential functions, and disulfide bonds help to stabilize protein three-dimensional structures (Sec. 26.8A).

PROBLEM

*10.31   The rates of the reactions in Eqs. 10.59 are increased when the thiol is ionized, that is, in the presence of a base such as sodium ethoxide. Suggest a mechanism for Eq. 10.59a that is consistent with this observation, and explain why the presence of base makes the reaction faster.

10.10   Synthesis of Alcohols and Glycols

The preparation of organic compounds from other organic compounds by the use of one or more reactions is called **organic synthesis**. This section reviews the methods presented in earlier chapters for the synthesis of alcohols and glycols. All of these methods begin with alkene starting materials:

1. Hydroboration-oxidation of alkenes (Sec. 5.3B).

$$3RCH{=}CH_2 + BH_3 \xrightarrow{\text{THF}} (RCH_2CH_2)_3B \xrightarrow[\text{OH}^-]{\text{H}_2\text{O}_2} 3RCH_2CH_2OH \qquad (10.60)$$

2. Oxymercuration-reduction of alkenes (Sec. 5.3A).

$$R-CH=CH_2 + Hg(OAc)_2 \xrightarrow{\;H_2O\;} R-\overset{\displaystyle OH}{\underset{\displaystyle |}{C}}H-CH_2-HgOAc \xrightarrow{\;NaBH_4\;} R-\overset{\displaystyle OH}{\underset{\displaystyle |}{C}}H-CH_3 \qquad (10.61)$$

Acid-catalyzed hydration of alkenes (Sec. 4.9B) is used industrially to prepare certain alcohols, but this is not an important laboratory method. In principle, the S_N2 reaction of ^-OH with alkyl halides can also be used to prepare alcohols from alkyl halides. This method is of little practical importance, however, because alkyl halides are generally prepared from alcohols themselves.

Some of the most important methods for the synthesis of alcohols involve the reduction of carbonyl compounds (aldehydes, ketones, or carboxylic acids and their derivatives), as well as the reactions of carbonyl compounds with Grignard or organolithium reagents. These methods are presented in Chapters 19, 20, and 21. A summary of methods used to prepare alcohols is found in Appendix IV.

Only one method for preparing glycols has been presented, namely, oxidation of alkenes with either OsO_4 or $KMnO_4$ (Sec. 5.5):

$$3 \; \overset{\diagdown}{\underset{\diagup}{C}}=\overset{\diagup}{\underset{\diagdown}{C}} + 2KMnO_4 + 4H_2O \longrightarrow 3 -\overset{|}{\underset{\underset{\displaystyle OH}{|}}{C}}-\overset{|}{\underset{\underset{\displaystyle OH}{|}}{C}}- + 2MnO_2 + 2KOH \qquad (10.62)$$

$$\overset{\diagdown}{\underset{\diagup}{C}}=\overset{\diagup}{\underset{\diagdown}{C}} \xrightarrow{\;OsO_4\;} \xrightarrow[NaHSO_3]{\;H_2O\;} -\overset{|}{\underset{\underset{\displaystyle OH}{|}}{C}}-\overset{|}{\underset{\underset{\displaystyle OH}{|}}{C}}- \qquad (10.63)$$

Glycols can also be prepared from epoxides; this reaction is discussed in Sec. 11.4B.

10.11   Design of Organic Synthesis

Although many useful organic syntheses consist of only one reaction, more typically several reactions are required for the transformation of one organic compound into another. Let's examine the logic used in planning such a multistep synthesis.

The molecule to be synthesized is called the **target molecule**. In order to assess the best route to the target molecule from the starting material, you should take the same approach that a military officer might take in planning the assault on an objective, namely, *work backward from the target towards the starting material*. Just as the officer considers secondary objectives—a hill here, a tree there—from which the final assault on the target can be launched, in planning a synthesis you should first assess what compound can be used as the immediate precursor of the target. You should then continue to work backward from this precursor step-by-step until the route from the starting material becomes clear. Sometimes more than one synthetic route will be possible. In such a case, each synthesis is evaluated in terms of yield, limitations, expense, and so on. It is not unusual to find (both in practice and on examinations) that one synthesis is as good as another.

The following study problem illustrates this strategy.

Outline a synthesis of hexanal from 1-hexene.

$$CH_3CH_2CH_2CH_2CH{=}CH_2 \longrightarrow CH_3CH_2CH_2CH_2CH_2CH{=}O$$

1-hexene   **hexanal**

Solution   To "outline a synthesis" means to suggest the reagents and conditions required for each step of the synthesis, along with the structures of each intermediate compound. Begin by working backward from hexanal, the target molecule. First, ask whether aldehydes can be directly prepared from alkenes. The answer is yes. Ozonolysis (Sec. 5.4) can be used to transform alkenes into aldehydes and ketones. However, ozonolysis breaks a carbon-carbon double bond, and certainly would not work for preparing an aldehyde from an alkene *with the same number of carbon atoms*, because at least one carbon is lost when the double bond is broken. No other ways of preparing aldehydes directly from alkenes have been covered. The next step is to ask how aldehydes can be prepared from other starting materials. Two ways have been presented: cleavage of glycols (Sec. 10.6C), and oxidation of primary alcohols (Sec. 10.6A). The cleavage of a glycol, like ozonolysis of an alkene, breaks a carbon-carbon bond, and is unsatisfactory for the present objective because it requires the loss of at least one carbon atom. However, the oxidation of a primary alcohol could be a final step in a satisfactory synthesis:

$$CH_3CH_2CH_2CH_2CH_2CH_2{-}OH \xrightarrow[CH_2Cl_2]{CrO_3\text{-pyridine}_2} CH_3CH_2CH_2CH_2CH_2CH{=}O \qquad (10.64)$$

1-hexanol   **hexanal**

Now ask whether it is possible to prepare 1-hexanol from 1-hexene; the answer is yes. Hydroboration-oxidation will convert 1-hexene into 1-hexanol. The synthesis is now complete:

$$CH_3CH_2CH_2CH_2CH{=}CH_2 \xrightarrow[THF]{BH_3} \xrightarrow[NaOH]{H_2O_2} CH_3CH_2CH_2CH_2CH_2CH_2{-}OH$$

1-hexene   **1-hexanol**

$$\Bigg\downarrow \begin{array}{l} CrO_3\text{-pyridine}_2 \\ CH_2Cl_2 \end{array}$$

$$CH_3CH_2CH_2CH_2CH_2CH{=}O$$

hexanal   (10.65)

Notice carefully the process of working backward from the target molecule one step at a time.

   Working problems in organic synthesis is one of the best ways to master organic chemistry. It is akin to mastering a language: it is relatively easy to learn to read a language (be it a foreign language, English, or even a computer language), but writing it requires a more thorough understanding. Similarly, it is relatively easy to follow individual organic reactions, but to integrate them and use them out of context requires more understanding. One way to bring together organic reactions and study them systematically is to go back through the text and write a representative reaction for each of the methods used for preparing each functional group. For example, which reactions can be used to prepare alkanes? Carboxylic acids? Jot down some notes describing the stereochemistry of each reaction (if known) as well as its limitations—that is, the situations in which the reaction would not be expected to work. For example, dehydration of tertiary and secondary alcohols is a good laboratory method for preparking alkenes, but dehydration of primary alochols is not. (Do you understand the reason for this limitation?) This process should be continued throughout future chapters.

The summary in Appendix IV can be a starting point for this type of study. In Appendix IV are presented lists of reactions, in the order that they occur in the text, which can be used to prepare compounds containing each functional group. It is very important to integrate what you have learned about organic chemistry, and problems in organic synthesis will help you achieve that goal.

PROBLEMS

10.32 Outline a synthesis of each of the following compounds from the indicated starting material.

*(a) 2-methyl-3-pentanol from 2-methyl-2-pentanol

(b) hexane from 1-hexanol

*(c)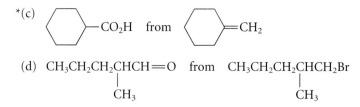

(d) $CH_3CH_2CH_2CHCH=O$ from $CH_3CH_2CH_2CHCH_2Br$
           | |
          CH_3 CH_3

***10.33** The reaction of alkyl chlorides with KI in acetone is one way to prepare alkyl iodides. What is the stereochemistry of this reaction? What is one limitation on the structure of the alkyl chloride that can be used in this reaction? Explain the reasons for your answers to both of these questions.

KEY IDEAS IN CHAPTER 10

Several reactions of alcohols involve breaking the C—O bond.
1. In dehydration and the reaction with hydrogen halides, the —OH group of an alcohol is converted by protonation into a good leaving group. The protonated —OH is eliminated as water (in dehydration) to give an alkene, or displaced by halide (in reaction with hydrogen halides) to give an alkyl halide.
2. In the reaction of an alcohol with $SOCl_2$, the reagent itself converts the —OH into a good leaving group, which is displaced in a subsequent substitution reaction to form an alkyl halide.
3. In the reaction with a sulfonyl chloride, the alcohol is converted into a sulfonate ester, such as a tosylate or a mesylate. The sulfonate group serves as an excellent leaving group in substitution or elimination reactions.

An oxidation in organic chemistry is a reaction in which the product has a greater (more positive, less negative) oxidation number than the reactant. A reduction is a reaction in which the product has a smaller (less positive, more negative) oxidation number than the reactant. The change in oxidation number is the same as the number of electrons lost or gained in the corresponding half-reaction.

Alcohols can be oxidized to carbonyl compounds with Cr(VI). Primary alcohols are oxidized to aldehydes (in the absence of water) or carboxylic acids (in the presence of water), secondary alcohols are oxidized to ketones, and tertiary alcohols are not oxidized. Primary alcohols are oxidized to carboxylic acids with $KMnO_4$. Vicinal glycols react with periodic acid to give aldehydes or ketones derived from cleavage of the carbon-carbon bond between the —OH groups.

Thiols are oxidized at sulfur, rather than at the α-carbon. Disulfides and sulfonic acids are two common oxidation products of thiols.

An example of a naturally occurring oxidation is the conversion of ethanol into acetaldehyde by NAD^+; this reaction is catalyzed by the enzyme alcohol dehydrogenase.

Just as two isomeric molecules can be classified as constitutional isomers or stereoisomers depending on their connectivities, two groups within a molecule can be classified as constitutionally equivalent or constitutionally nonequivalent. In general, constitutionally nonequivalent groups are chemically distinguishable. Constitutionally equivalent groups are of three types.
1. Homotopic groups, which are chemically equivalent under all conditions.
2. Enantiotopic groups, which are chemically nonequivalent in reactions with chiral reagents such as enzymes, but chemically equivalent in reactions with achiral reagents.
3. Diastereotopic groups, which are chemically nonequivalent under all conditions.

A useful strategy for organic synthesis is to work backward from the target compound systematically one step at a time.

ADDITIONAL PROBLEMS

*10.34   Give the product expected, if any, when 1-butanol (or other compound indicated) reacts with each of the following reagents.
(a) concentrated aqueous HBr, heat   (b) aqueous H_2SO_4, cold
(c) CrO_3-pyridine$_2$ complex in CH_2Cl_2 (d) NaH

(Problem 10.34 continues)

(e)   product of (d) + CH_3I in DMSO
(f)   *p*-toluenesulfonyl chloride in pyridine
(g) $CH_3CH_2CH_2MgBr$ in anhydrous ether
(h) $SOCl_2$ in pyridine
(i)   product of (a) + Mg in dry ether
(j)   product of (f) + K^+ $(CH_3)_3C-O^-$ in $(CH_3)_3COH$
(k)   product of (j) + OsO_4, then aqueous $NaHSO_3$

10.35   Give the product expected, if any, when 2-methyl-2-propanol (or other compound indicated) reacts with each of the following reagents.
(a)   conc. aqueous HCl   (b) CrO_3 in pyridine
(c) H_2SO_4, heat   (d) Br_2 in CH_2Cl_2 (dark)
(e)   potassium metal   (f)   methanesulfonyl chloride in pyridine
(g)   product of (f) + NaOH in DMSO
(h)   product of (e) + product of (a)
(i)   product of (g) + alkaline $KMnO_4$
(j)   product of (i) + periodic acid

10.36   Give the structure of a compound that satisfies the criterion given in each case. (There may be more than one correct answer.)
*(a)   a seven-carbon tertiary alcohol that yields a *single* alkene after acid-catalyzed dehydration
(b)   an eight-carbon secondary alcohol that yields a *single* alkene after acid-catalyzed dehydration
*(c)   an alcohol, which, after acid-catalyzed dehydration, yields an alkene that in turn, upon ozonolysis and treatment with $(CH_3)_2S$, gives only benzaldehyde, $Ph-CH=O$
(d)   an optically active glycol that yields two equivalents of acetaldehyde on treatment with periodic acid
*(e)   an alcohol that gives the same product when it reacts with $KMnO_4$ as is obtained from the ozonolysis of *trans*-3,6-dimethyl-4-octene followed by treatment with H_2O_2

*10.37   The following ester is a powerful explosive, but is also a medication for angina pectoris (chest pain). From what inorganic acid and what alcohol is it derived?

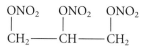

10.38   In each compound, identify (1) the diastereotopic fluorines, (2) the enantiotopic fluorines, (3) the homotopic fluorines, and (4) the constitutionally nonequivalent fluorines.

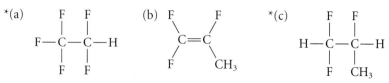

10.39 How many chemically nonequivalent sets of chemically equivalent hydrogens are in each of the following structures?

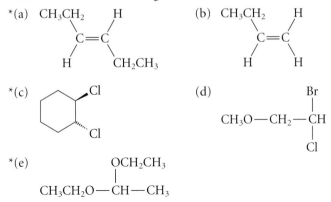

*(a)   CH₃CH₂ ... H / C=C / H   CH₂CH₃

(b)   CH₃CH₂ ... H / C=C / H   H

*(c)

(d)   CH₃O—CH₂—CH with Br and Cl

*(e)
OCH₂CH₃
|
CH₃CH₂O—CH—CH₃

*10.40 When *tert*-butyl alcohol is treated with H₂¹⁸O (water containing the heavy oxygen isotope ¹⁸O) in the presence of a small amount of acid, and the *tert*-butyl alcohol is reisolated, it is found to contain ¹⁸O. Write a curved-arrow mechanism that explains how the isotope is incorporated into the alcohol.

10.41 Indicate whether each of the following transformations is an oxidation, a reduction, or neither, and how many electrons are involved in each oxidation or reduction process.

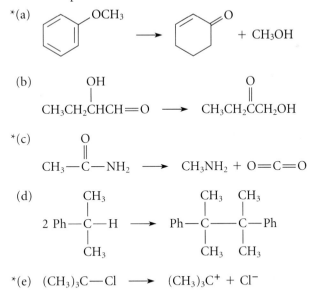

*(a)

→   + CH₃OH

(b)
OH   O
| ‖
CH₃CH₂CHCH=O → CH₃CH₂CCH₂OH

*(c)
O
‖
CH₃—C—NH₂ → CH₃NH₂ + O=C=O

(d)
CH₃   CH₃ CH₃
| | |
2 Ph—C—H → Ph—C——C—Ph
| | |
CH₃   CH₃ CH₃

*(e) (CH₃)₃C—Cl → (CH₃)₃C⁺ + Cl⁻

10.42 Outline a synthesis for the conversion of enantiomerically pure (*R*)-CH₃CH₂CHD—OH into each of the following isotopically labeled compounds. Assume that Na¹⁸OH or H₂¹⁸O is available as needed.

*(a) (*R*)-CH₃CH₂CHD—¹⁸OH   (*Hint:* two inversions of configuration correspond to a retention of configuration.)

(b) (*S*)-CH₃CH₂CHD—¹⁸OH

10.43 Outline a synthesis for each of the following compounds from the indicated starting material and any other reagents.

*(a) CH₃—CH—CH₂CH₂CH₃ from 1-pentanol
        |
        D

(b) D—CH₂CH₂CH₂CH₂CH₃ from 1-pentanol

*(c)  —CH₂CO₂H from ⬠—CH=CH₂ (cyclopentylethylene)

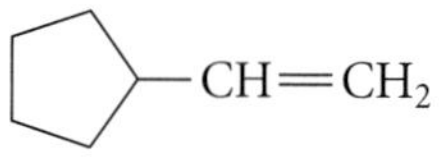

(d) 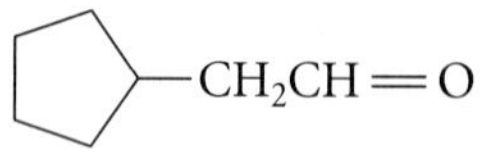 ⬠—CH₂CH=O from cyclopentylethylene

*(e) 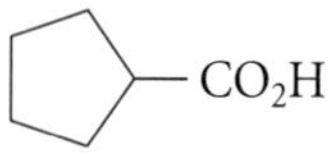 ⬠—CO₂H from cyclopentylethylene

(f) ethylcyclopentane from cyclopentylethylene
*(g) CH₃CH₂CH₂CH₂—CN from 1-butene (*Hint:* See Table 9.1)

(h) 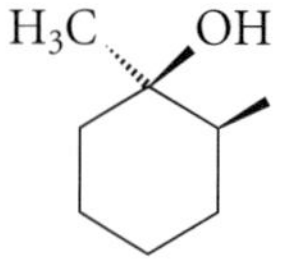 H₃C   OH
          ⬡   OH from 1-methylcyclohexene

10.44 Tell which of the two sulfonate esters in each of the following pairs reacts more rapidly in an S_N2 reaction with sodium methoxide in methanol and explain your reasoning.

*(a)

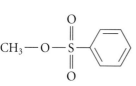

methyl 3, 5-dibromobenzenesulfonate   **methyl benzenesulfonate**

(b)

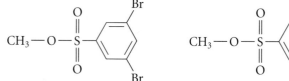

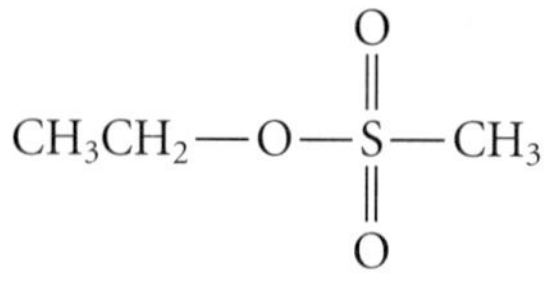

ethyl mesylate 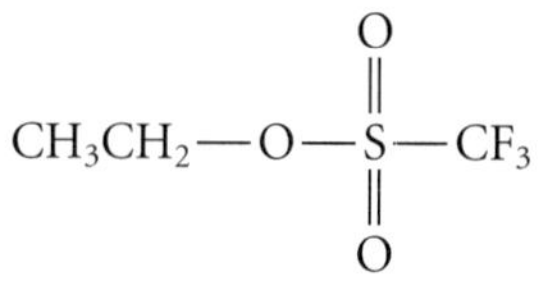   **ethyl triflate**

***10.45** A police officer, Lawin Order, has detained a driver, Bobbin Weaver, after observing erratic driving behavior. Administering a breathalyzer test, Officer Order collects 52.5 mL of expired air from Weaver and finds that the air reduces 0.507×10^{-6} mole of $K_2Cr_2O_7$ to Cr^{3+}. Assuming that 2100 mL of air contain the same amount of ethanol as 1 mL of blood, calculate the "percent blood

alcohol content" (BAC), expressed as (grams of ethanol per mL of blood) ×
100. If 0.10% BAC is the lower limit of legal intoxication, should Officer Order
make an arrest?

10.46 How many grams of CrO_3 are required to oxidize 10 g of 2-heptanol to the
corresponding ketone?

*10.47 The primary alcohol 2-methoxyethanol, CH_3O—CH_2CH_2—OH, can be oxidized to the corresponding carboxylic acid with aqueous nitric acid (HNO_3).
The by-product of the oxidation is nitric oxide, NO. How many moles of HNO_3
are required to oxidize 0.1 mole of the alcohol?

*10.48 Chemist Stench Thiall, intending to prepare the disulfide A, has mixed one mole
each of 1-butanethiol and 2-octanethiol with I_2 and base. Stench is surprised at
the low yield of the desired compound and has come to you for an explanation.
Explain why Stench should not have expected a good yield in this reaction.

$$A \quad CH_3CH_2CH_2CH_2-S-S-CH(CH_2)_5CH_3$$
$$\underset{\displaystyle CH_3}{|}$$

*10.49 Compound A, C_7H_{14}, decolorizes Br_2 in CH_2Cl_2 and reacts with BH_3 in THF
followed by H_2O_2/OH^- to yield a compound B. When treated with $KMnO_4$, B
is oxidized to a carboxylic acid C that can be resolved into enantiomers. Compound A, after ozonolysis and workup with H_2O_2, yields the same compound
D formed by oxidation of 3-hexanol with chromic acid. Identify compounds A,
B, C, and D.

10.50 A certain hydrocarbon A (C_8H_{14}) undergoes a reaction with OsO_4, followed by
aqueous $NaHSO_3$, to give a glycol B, which, when treated with periodic acid,
gives cyclopentanone and acetone. Propose a structure for compounds A and B.

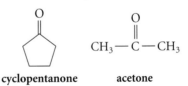

cyclopentanone   acetone

*10.51 Match each of the structures below with one of the compounds A–D on the
basis of the following experimental facts. Compounds A, B, and C are optically
active, but compound D is not. Compound C gives the same products as compound D on treatment with periodic acid, but compound B gives a different
product. Compound A does not react with periodic acid.

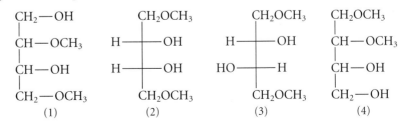

*10.52   In a laboratory are found two different compounds, A (melting point −4.7°) and B (melting point −1°). Both compounds have the same molecular formula ($C_7H_{14}O$) and both can be resolved into enantiomers. Both compounds give off a gas when treated with NaH. Treatment of either A or B with tosyl chloride in pyridine yields a tosylate ester, and treatment of either tosylate with potassium *tert*-butoxide gives a mixture of the same two alkenes, C and D. However, reaction of the tosylate of A with potassium *tert*-butoxide to give these alkenes is noticeably slower than the corresponding reaction of the tosylate of B. When either *optically active A* or *optically active B* is subjected to the same treatment, *both* alkene products C and D are optically active. Treatment of either C or D with H_2 over a catalyst yields methylcyclohexane. Identify all unknown compounds and explain your reasoning.

10.53   Complete each of the following reactions by giving the principal organic product(s) formed in each case.

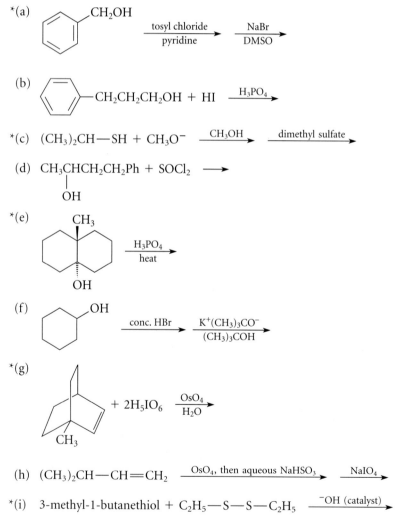

*(a)   [benzene ring]—CH_2OH $\xrightarrow[\text{pyridine}]{\text{tosyl chloride}}$ $\xrightarrow[\text{DMSO}]{\text{NaBr}}$

(b)   [cyclohexadiene]—$CH_2CH_2CH_2OH$ + HI $\xrightarrow{H_3PO_4}$

*(c) $(CH_3)_2CH$—SH + CH_3O^- $\xrightarrow{CH_3OH}$ $\xrightarrow{\text{dimethyl sulfate}}$

(d) $CH_3CHCH_2CH_2Ph$ + $SOCl_2$ $\longrightarrow$
          |
          OH

*(e)   [decalin structure with CH₃ and OH] $\xrightarrow[\text{heat}]{H_3PO_4}$

(f)   [cyclohexane with OH] $\xrightarrow{\text{conc. HBr}}$ $\xrightarrow[\text{(CH}_3)_3\text{COH}]{K^+(CH_3)_3CO^-}$

*(g)   [norbornene with CH₃]   + $2H_5IO_6$ $\xrightarrow[H_2O]{OsO_4}$

(h) $(CH_3)_2CH$—CH=CH_2 $\xrightarrow{\text{OsO}_4, \text{ then aqueous NaHSO}_3}$ $\xrightarrow{\text{NaIO}_4}$

*(i)   3-methyl-1-butanethiol + C_2H_5—S—S—C_2H_5 $\xrightarrow{^-\text{OH (catalyst)}}$

*10.54 (a) When the rate of oxidation of isopropyl alcohol to acetone is compared with the rate of oxidation of a deuterated derivative, an isotope effect is observed.

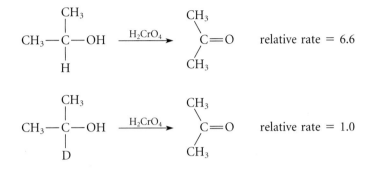

Which step in the mechanism of Eq. 10.38 is rate-limiting?

(b) When either (S)- or (R)-1-deuterioethanol is oxidized with CrO₃-pyridine₂ in CH₂Cl₂, the same product mixture results and it contains significantly more of the deuterated aldehyde A than the undeuterated aldehyde B.

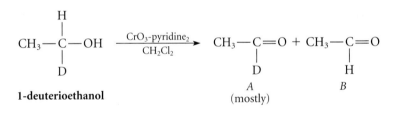

Explain why the deuterated aldehyde is the major product.

(c) Explain why, in the oxidation of (R)-1-deuterioethanol with alcohol dehydrogenase and NAD⁺, *none* of the deuterated aldehyde is obtained.

10.55 When the hydration of fumarate is catalyzed by the enzyme *fumarase* in D_2O, only (2S,3R)-3-deuteriomalate is formed as the product.

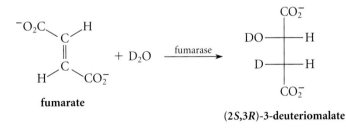

This reaction can also be run in reverse. By applying the principle of microscopic reversibility, predict the product (if any) when each of the following compounds is treated with fumarase in H_2O:

(*Problem 10.55 continues*)

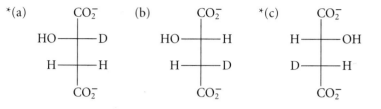

*10.56   Buster Bluelip, a student repeating organic chemistry for the fifth time, has observed that alcohols can be converted into alkyl bromides by treatment with concentrated HBr, and has proposed that, by analogy, alcohols should be converted into nitriles (organic cyanides, R—CN) by treatment with concentrated HCN. Upon running the reaction, Bluelip finds that the alcohol does not react. Another student has suggested that the reason the reaction failed is the absence of an acid catalyst. Following this suggestion, Bluelip runs the reaction in the presence of H_2SO_4, and again observes no reaction of the alcohol. Explain the difference in the reaction of alcohols with HBr and HCN, that is, why the latter fails but the former succeeds. (*Hint:* see Table 3.1.)

*10.57   Primary alcohols, when treated with H_2SO_4, do not dehydrate to alkenes under the usual conditions. However, they do undergo another type of "dehydration" to form ethers if heated strongly in the presence of H_2SO_4. The reaction of ethyl alcohol to give diethyl ether is typical:

$$2\ CH_3CH_2OH \xrightarrow[140°]{H_2SO_4} CH_3CH_2{-}O{-}CH_2CH_3 + H_2O$$

Using the curved-arrow formalism, show mechanistically how this reaction takes place.

10.58   Using the curved-arrow formalism, give a mechanism for each of the following known conversions. (If necessary, re-read Study Guide Link 10.2.)

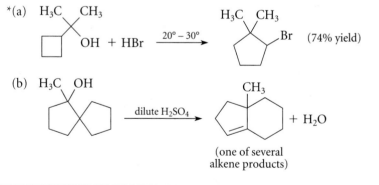

11

Chemistry of Ethers, Epoxides, and Sulfides

The chemistry of ethers is closely related to the chemistry of alkyl halides, alcohols, and alkenes. Ethers, however, are considerably less reactive than these other types of compounds. This chapter covers the synthesis of ethers and shows why the ether linkage is relatively unreactive.

Epoxides are heterocyclic compounds in which the ether linkage is part of a three-membered ring. Unlike ordinary ethers, epoxides are very reactive because of the great degree of strain in the epoxide ring. This chapter also presents the synthesis and reactions of epoxides.

Sulfides (thioethers), sulfur analogs of ethers, are also discussed briefly in this chapter. Although sulfides share some chemistry with their ether counterparts, they differ from ethers in the way they react in oxidation reactions, just as thiols differ from alcohols.

Finally, the strategy of organic synthesis will be revisited with a classification of reactions according to the way they are used in synthesis, and a further discussion of how to plan multistep syntheses.

11.1 Synthesis of Ethers and Sulfides

A. Williamson Ether Synthesis

Some ethers can be prepared by the reaction of alkoxides with methyl halides, primary alkyl halides, or the corresponding sulfonate esters.

$$
\underset{\text{an alkoxide}}{\overset{\overset{\textstyle O^- \, Na^+}{|}}{Ph-CH-CH_3}} + CH_3-I \longrightarrow \underset{\text{(90\% yield)}}{\overset{\overset{\textstyle O-CH_3}{|}}{Ph-CH-CH_3}} + Na^+ \, I^- \qquad (11.1)
$$

Some sulfides can be prepared in a similar manner from thiolates, the conjugate bases of thiols.

$$CH_3(CH_2)_3—SH \xrightarrow[\text{CH}_3\text{OH}]{^{-}\text{OH}} CH_3(CH_2)_3—S^{-} \xrightarrow{\text{CH}_3\text{CH}_2—OTs} CH_3(CH_2)_3—S—CH_2CH_3 + {}^{-}OTs$$

1-butanethiol   **1-butanethiolate**   **butyl ethyl sulfide, or**
                                                                            (1-ethylthio)butane
                                                                            (78% yield)   (11.2)

Both of these reactions are examples of the **Williamson ether synthesis**, named for Alexander William Williamson (1824–1904), Professor of Chemistry at the University of London. (Williamson's synthesis of diethyl ether and ethyl methyl ether in 1850 settled a controversy over the structures and relationship of alcohols and ethers.) The Williamson ether synthesis is an important practical example of the S_N2 reaction (Table 9.1). In this reaction the conjugate base of an alcohol or thiol acts as a nucleophile; an ether is formed by displacement of a halide or other leaving group.

$$R^1—\ddot{\underset{..}{O}}\!:^{-} + R^2—\ddot{\underset{..}{I}}\!: \longrightarrow R^1—\ddot{\underset{..}{O}}—R^2 + :\ddot{\underset{..}{I}}\!:^{-} \qquad (11.3)$$

Tertiary and many secondary alkyl halides cannot be used in this reaction. (Why?)

In principle, two different Williamson syntheses are possible for any ether with two different alkyl groups.

$$\left.\begin{array}{l} R^1—\ddot{\underset{..}{O}}\!:^{-} + R^2—X \\ R^2—\ddot{\underset{..}{O}}\!:^{-} + R^1—X \end{array}\right\} \longrightarrow R^1—\ddot{\underset{..}{O}}—R^2 + X^{-} \qquad (11.4)$$

The preferred synthesis is usually the one that involves the alkyl halide with the greater S_N2 reactivity. This point is illustrated by the following study problem.

STUDY PROBLEM 11.1

Outline a Williamson ether synthesis for *tert*-butyl methyl ether.

$$\begin{array}{c} CH_3 \\ | \\ CH_3—C—O—CH_3 \\ | \\ CH_3 \end{array}$$

***tert*-butyl methyl ether**

Solution   From Eq. 11.4, two possibilities for preparing this compound are the reaction of methyl bromide with potassium *tert*-butoxide and the reaction of *tert*-butyl bromide with sodium methoxide. Only the former combination will work.

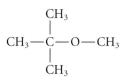

$$(CH_3)_3C—O^{-} K^{+} + CH_3—Br \qquad\qquad CH_3O^{-} Na^{+} + (CH_3)_3C—Br$$

satisfactory reaction   does not occur, why?

$$(CH_3)_3C—O—CH_3 \qquad (11.5)$$

Do you know why sodium methoxide and *tert*-butyl bromide would not work? (See Sec. 9.5F.)

PROBLEMS

11.1   Complete the following reactions.

*(a) $(CH_3)_2CHOH + Na$, then CH_3—I $\xrightarrow[\text{alcohol}]{\text{isopropyl}}$

(b) $CH_3SH + CH_2{=}CH$—CH_2—Cl $\xrightarrow[CH_3OH]{NaOH}$

*(c) $CH_3O^- Na^+ + (CH_3)_3C$—Br $\xrightarrow{CH_3OH}$

(d) $C_2H_5O^- K^+ + (CH_3)_3CCH_2$—OTs $\xrightarrow[25°]{C_2H_5OH}$

11.2   Suggest a Williamson ether synthesis, if one is possible, for each of the following compounds. If no Williamson ether synthesis is possible, explain why.

*(a)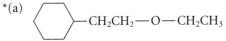

(b) $(CH_3)_2CH$—S—CH_3

*(c) $(CH_3)_3C$—O—$C(CH_3)_3$

B. Alkoxymercuration-Reduction of Alkenes

Another method for the preparation of ethers is a variation on a synthesis of alcohols, which can be prepared from alkenes by oxymercuration-reduction in *aqueous* solution (Sec. 5.3A). If the reaction is carried out with an *alcohol* solvent, an ether is formed instead. This process is called **alkoxymercuration-reduction**:

$$CH_2{=}CHCH_2CH_2CH_2CH_3 + Hg(OAc)_2 + (CH_3)_2CHOH \longrightarrow$$

1-hexene   (solvent)

$$\underset{\underset{O-CH(CH_3)_2}{|}}{AcOHg-CH_2CH}-CH_2CH_2CH_2CH_3 + H-OAc$$

1-acetoxymercuri-2-isopropoxyhexane   (11.6a)

$$\underset{\underset{O-CH(CH_3)_2}{|}}{AcOHg-CH_2CH}-CH_2CH_2CH_2CH_3 + NaBH_4 \longrightarrow \underset{\underset{O-CH(CH_3)_2}{|}}{CH_3CH}-CH_2CH_2CH_2CH_3 + Hg + borates$$

2-isopropoxyhexane
(91% yield)   (11.6b)

☞
STUDY GUIDE LINK:
  ✓11.1
  Learning New Reactions from Earlier Reactions

The mechanism of the reaction in Eq. 11.6a is completely analogous to the mechanism of oxymercuration, except that an alcohol instead of water is the nucleophile that attacks the mercurinium ion intermediate. Notice that the ether product of Eq. 11.6b could not have been prepared by a Williamson ether synthesis (Problem 11.3).

PROBLEMS

*11.3   (a) Give the mechanism of Eq. 11.6a and account for the regioselectivity of the reaction.

(b) Explain what would happen in an attempt to synthesize the ether product of Eq. 11.6b by a Williamson ether synthesis.

11.4   Complete the following reactions.

*(a) $CH_3CH{=}CHC_2H_5 + CH_3OH + Hg(OAc)_2 \longrightarrow \xrightarrow{NaBH_4}$

(b) $(CH_3)_2CH{-}CH{=}CH_2 + C_2H_5OH + Hg(OAc)_2 \longrightarrow \xrightarrow{NaBH_4}$

11.5   Outline a synthesis of each of the following ethers using alkoxymercuration-reduction:

*(a) dicyclohexyl ether   (b) *tert*-butyl isobutyl ether

C. Ethers from Alcohol Dehydration and Alkene Addition

In some cases, two molecules of an alcohol can undergo dehydration to give an ether.

$$2\ C_2H_5OH \xrightarrow[140°]{H_2SO_4} C_2H_5{-}O{-}C_2H_5 \tag{11.7}$$

ethanol   **diethyl ether**

This method is used industrially for the preparation of diethyl ether, and it can be used in the laboratory. However, it is generally restricted to the preparation of *symmetrical* ethers derived from *primary* alcohols. (A symmetrical ether is one in which both alkyl groups are the same.) Secondary and tertiary alcohols cannot be used because they undergo dehydration to alkenes (Sec. 10.1).

The formation of ethers from primary alcohols is an S_N2 reaction in which one alcohol displaces water from another molecule of *protonated* alcohol:

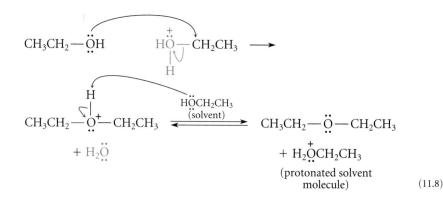

(11.8)

(See Problem 10.57, Chapter 10.)

Tertiary alcohols can be converted into *unsymmetrical* ethers by treating them with dilute mineral acids in an alcohol solvent. For example, *tert*-butyl ethyl ether can be prepared when *tert*-butyl alcohol is treated with dilute sulfuric acid in ethanol solvent:

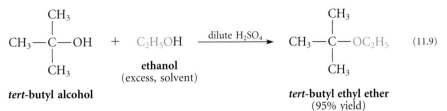

(11.9)

tert-butyl alcohol

ethanol
(excess, solvent)

tert-butyl ethyl ether
(95% yield)

The key to this reaction is that only one of the alcohol starting materials, in this case, *tert*-butyl alcohol, can readily lose water after protonation to form a relatively stable carbocation. The alcohol that is used in excess, in this case, ethanol, must be one that either *cannot* lose water after protonation to give a carbocation or should form a carbocation much less readily.

$$(CH_3)_3C\!-\!\ddot{O}H \;\rightleftharpoons\; (CH_3)_3C\!-\!\overset{+}{\underset{H}{\ddot{O}H}} \qquad CH_3CH_2\!-\!\overset{+}{\underset{H}{\ddot{O}H}} \;\rightleftharpoons\; CH_3CH_2\!-\!\ddot{O}H$$

$$H_2\ddot{O} + (CH_3)_3C^+ \qquad\qquad CH_3\overset{+}{C}H_2 + H_2\ddot{O}$$

tertiary   primary
carbocation   carbocation
(does not form)

(11.10)

When the carbocation derived from the tertiary alcohol is formed, it reacts rapidly with ethanol, which is present in large excess because it is the solvent.

$$(CH_3)_3\overset{+}{C} \quad H\ddot{O}\!-\!CH_2CH_3 \;\longrightarrow\; (CH_3)_3C\!-\!\overset{+}{\underset{H}{\ddot{O}}}\!-\!CH_2CH_3$$

(11.11)

(loses a proton to solvent to give the product)

There is an important relationship between this reaction and alcohol dehydration. Alcohols, especially tertiary alcohols, undergo dehydration to alkenes in the presence of mineral acids (Sec. 10.1). Ether formation from tertiary alcohols and the dehydration of tertiary alcohols are *alternative branches of a common mechanism.* Both ether formation and alkene formation involve carbocation intermediates; the conditions dictate which product is obtained. The dehydration of alcohols to alkenes involves relatively high temperatures and removal of the alkene and water products as they are formed. Ether formation from tertiary alcohols involves milder conditions under which alkenes are not removed from the reaction mixture. In addition, *a large excess of the second alcohol* (ethanol in Eq. 11.9) *is used as the solvent,* so that the major reaction of the carbocation intermediate is with this alcohol. Any alkene that does form is not removed but is reprotonated to give back the same carbocation, which eventually reacts with the alcohol solvent:

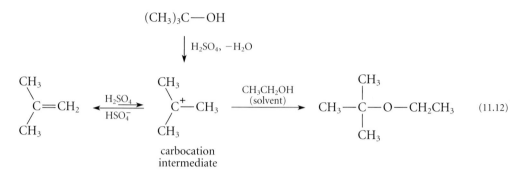

(11.12)

STUDY GUIDE LINK:
✓11.2
*Common
Intermediates from
Different Starting
Materials*

This analysis suggests that treatment of an alkene with a large excess of alcohol in the presence of an acid catalyst should also give an ether, provided that a relatively stable carbocation intermediate is involved. Indeed, such is the case; for example, the acid-catalyzed additions of methyl and ethyl alcohols to 2-methylpropene to give, respectively, *tert*-butyl methyl ether and *tert*-butyl ethyl ether are important industrial processes for the synthesis of these important gasoline additives (Sec. 2.8).

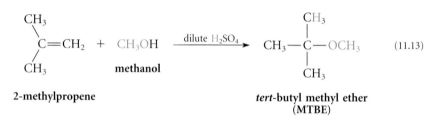

(11.13)

2-methylpropene   **methanol**   ***tert*-butyl methyl ether**
                                                              (MTBE)

Eqs. 11.9, 11.12, and 11.13 show that for the preparation of tertiary ethers, it makes no difference in principle whether the starting material from which the tertiary group is derived is an alkene or a tertiary alcohol.

. . . .

PROBLEMS

11.6   Complete each of the following reactions by giving the major organic product.

*(a)

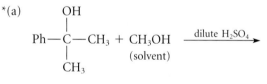

(b)

11.7   *(a)   Give the structure of an alkene which, when treated with dilute H_2SO_4 and methanol, will give the same ether product as the reaction in Problem 11.6a.

       (b)   Give the structure of two alkenes, either of which, when treated with dilute H_2SO_4 and ethanol, will give the same ether product as the reaction in Problem 11.6b.

11.8   Outline a synthesis of each ether using either alcohol dehydration or alkene addition, as appropriate.

*(a)   ClCH$_2$CH$_2$OCH$_2$CH$_2$Cl   (b)   dibutyl ether

*(c)   *tert*-butyl isopropyl ether   (d)   2-methoxy-2-methylbutane

11.2   Synthesis of Epoxides

A.   Oxidation of Alkenes with Peroxycarboxylic Acids

One of the best laboratory preparations of epoxides involves the direct oxidation of alkenes with peroxycarboxylic acids.

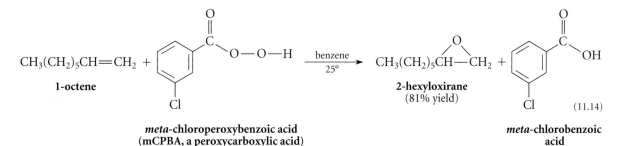

CH$_3$(CH$_2$)$_5$CH=CH$_2$ +

1-octene

meta-chloroperoxybenzoic acid
(mCPBA, a peroxycarboxylic acid)

→ (benzene, 25°)

CH$_3$(CH$_2$)$_5$CH—CH$_2$ +

2-hexyloxirane
(81% yield)

meta-chlorobenzoic acid

(11.14)

The use of alkenes as starting materials for epoxide synthesis is one reason that certain epoxides are named traditionally as oxidation products of the corresponding alkenes (Sec. 8.1C).

The oxidizing agent in Eq. 11.14, *meta*-chloroperoxybenzoic acid (abbreviated mCPBA), is an example of a peroxycarboxylic acid. A **peroxycarboxylic acid** is a carboxylic acid that contains an —O—O—H (hydroperoxy) group rather than an —OH (hydroxy) group.

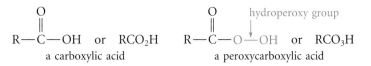

R—C—OH   or   RCO$_2$H   R—C—O—OH   or   RCO$_3$H
a carboxylic acid   a peroxycarboxylic acid

(with *hydroperoxy group* labeled)

(Note that the terms **peroxyacid** or **peracid** are sometimes used instead of *peroxycarboxylic acid*. These are actually more general terms that refer not only to peroxycarboxylic acids, but also to *any* acid containing an —O—O—H group instead of an —OH group.) Many peroxycarboxylic acids are unstable, but they can be formed just prior to use by mixing a carboxylic acid and hydrogen peroxide. In principle, any one of several peroxycarboxylic acids can be used for epoxidation of alkenes. The peroxyacid used in Eq. 11.14, mCPBA, has been popular because it is a crystalline solid that can be shipped commercially and stored in the laboratory. However, mCPBA, like most other peroxides, can be detonated if not handled carefully. A less hazardous peroxycarboxylic acid that has essentially the same reactivity is the magnesium salt of monoperoxyphthalic acid, abbreviated MMPP.

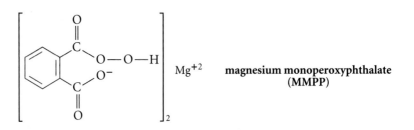

**magnesium monoperoxyphthalate
(MMPP)**

The formation of an epoxide from an alkene and a peroxycarboxylic acid is a *concerted addition reaction.*

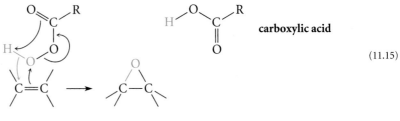

carboxylic acid

(11.15)

epoxide

STUDY GUIDE LINK:
✓ **11.3**
*Mechanism of
Epoxide Formation*

This mechanism is very similar to that for the formation of a bromonium ion in bromine addition to alkenes (Sec. 5.1A).

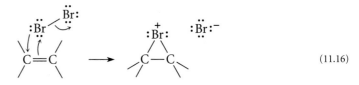

(11.16)

bromonium ion

The formation of epoxides with peroxycarboxylic acids is a *stereoselective* reaction; it takes place with complete retention of the alkene stereochemistry. That is, a *cis*-alkene gives a *cis*-substituted epoxide, and a *trans*-alkene gives a *trans*-substituted epoxide.

STUDY GUIDE LINK:
11.4
*Stereospecific
Reactions*

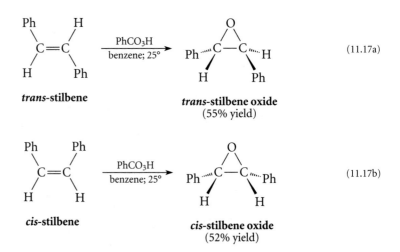

trans-stilbene   (11.17a)

trans-stilbene oxide
(55% yield)

cis-stilbene   (11.17b)

cis-stilbene oxide
(52% yield)

This is the result expected from a concerted reaction. The oxygen from the peroxycarboxylic acid must bond to both alkene carbons at the same face, since it cannot bridge both upper and lower faces simultaneously.

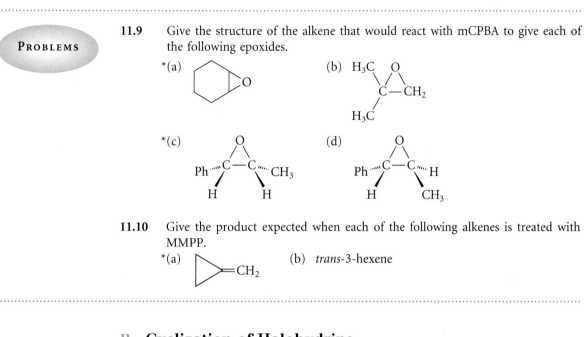

PROBLEMS

11.9 Give the structure of the alkene that would react with mCPBA to give each of the following epoxides.

*(a)

(b) H₃C   O
        \ /
         C—CH₂
        /
     H₃C

*(c)   O
            /   \
    Ph ⸝⸝⸝ C—C ⸝⸝⸝ CH₃
         | |
         H   H

(d)   O
           /   \
    Ph ⸝⸝⸝ C—C ⸝⸝⸝ H
         | |
         H   CH₃

11.10 Give the product expected when each of the following alkenes is treated with MMPP.

*(a) [triangle]=CH₂

(b) *trans*-3-hexene

B. Cyclization of Halohydrins

Epoxides can also be synthesized by the treatment of halohydrins (Sec. 5.1B) with base:

$$\begin{array}{c} OH \\ | \\ CH_3-C-CH_2-Br \\ | \\ CH_3 \end{array} + Na^+\,OH^- \xrightarrow{60\,°C} \begin{array}{c} H_3C\quad O \\ \diagdown\;\diagup \\ C-CH_2 \\ \diagup \\ H_3C \end{array} + Na^+\,Br^- + H-OH \qquad (11.18)$$

1-bromo-2-methyl-2-propanol
(a halohydrin)

2,2-dimethyloxirane
(81% yield)

This reaction is an intramolecular variation of the Williamson ether synthesis (Sec. 11.1A); in this case, the alcohol and the alkyl halide are part of the same molecule. The alkoxide anion, formed reversibly by reaction of the alcohol with NaOH, displaces halide ion from the neighboring carbon:

$$\begin{array}{c} :\ddot{O}-H \\ | \\ CH_3-C-CH_2-\ddot{Br}: \\ | \\ CH_3 \end{array} \rightleftharpoons \begin{array}{c} H-\ddot{O}H \\ :\ddot{O}:^- \\ | \\ CH_3-C-CH_2-\ddot{Br}: \\ | \\ CH_3 \end{array} \longrightarrow \begin{array}{c} H_3C\quad :O: \\ \diagdown\;\diagup \\ C-CH_2 \\ \diagup \\ H_3C \end{array} + :\ddot{Br}:^- \qquad (11.19)$$

Like other S_N2 reactions, this reaction takes place by *backside attack* of the nucleophilic oxygen anion at the halide-bearing carbon (Sec. 9.4B). Such a backside attack requires that the attacking oxygen and the leaving halide assume an *anti* relationship in the transition state of the reaction. For noncyclic halohydrins, this relationship can generally be achieved through a simple bond rotation.

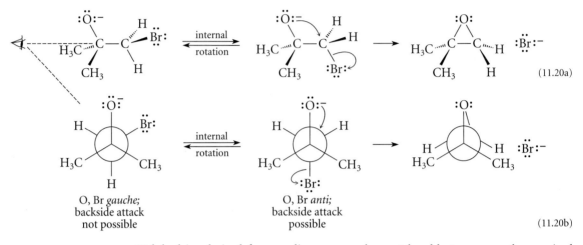

(11.20a)

O, Br *gauche*;
backside attack
not possible

O, Br *anti*;
backside attack
possible

(11.20b)

Halohydrins derived from cyclic compounds must be able to assume the required *anti* relationship through a conformational change if epoxide formation is to succeed. The cyclohexane derivative below, for example, must undergo the chair flip before epoxide formation can occur.

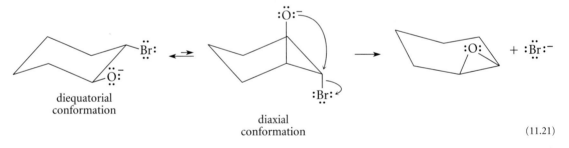

diequatorial
conformation

diaxial
conformation

(11.21)

Even though the diaxial conformation of the halohydrin is less stable than the diequatorial conformation, the two conformations are in rapid equilibrium. As the diaxial conformation reacts to give epoxide, it is replenished by the rapidly established conformational equilibrium.

PROBLEMS

*11.11 From models of the transition states for their reactions, predict which of the two following stereoisomers (shown in Fischer projection) should form an epoxide at the greater rate when treated with base. Explain.

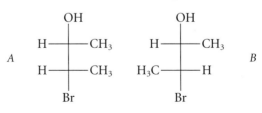

11.12 The chlorohydrin *trans*-2-chlorocyclohexanol reacts rapidly in base to form an epoxide. The *cis* stereoisomer, however, is relatively unreactive and does not give an epoxide. Explain why the two stereoisomers behave so differently.

11.3 Cleavage of Ethers

The ether linkage is relatively unreactive to many reagents. This is one reason ethers are widely used as solvents; that is, a great many reactions can be carried out in ether solvents without affecting the ether linkage. One reaction of ethers that does occur is the *cleavage* of the ether linkage by HI and HBr to give alcohols and alkyl halides.

When the ether contains only primary alkyl groups, strong acid and relatively harsh conditions are required to bring about ether cleavage.

$$\text{CH}_3\text{CH}_2\text{—O—CH}_2\text{CH}_3 + \text{H—I} \xrightarrow{\text{heat}} \text{CH}_3\text{CH}_2\text{—OH} + \text{I—CH}_2\text{CH}_3 \qquad (11.22)$$

$$\text{\textbf{diethyl ether}} \qquad\qquad\qquad \text{\textbf{ethanol}} \qquad \text{\textbf{ethyl iodide}}$$

The alcohol formed in the cleavage of an ether (ethanol in Eq. 11.22) can go on to react with HI to give a second molecule of alkyl halide (Sec. 10.2).

The mechanism of ether cleavage involves, first, protonation of the ether oxygen:

$$\text{CH}_3\text{CH}_2\text{—}\overset{..}{\underset{..}{\text{O}}}\text{—CH}_2\text{CH}_3 \quad \rightleftharpoons \quad \text{CH}_3\text{CH}_2\overset{H}{\underset{..}{\overset{|}{\text{—O}^+}}}\text{—CH}_2\text{CH}_3 \qquad (11.23a)$$

Then the iodide ion, which is a good nucleophile (Sec. 9.4D), attacks the protonated ether in an S_N2 reaction to form an alkyl halide and liberate an alcohol as a leaving group.

$$\text{CH}_3\text{CH}_2\overset{H}{\underset{..}{\overset{|}{\text{—O}^+}}}\text{—CH}_2\text{CH}_3 \quad \longrightarrow \quad \text{CH}_3\text{CH}_2\text{—}\overset{..}{\text{O}}\text{H} + \text{CH}_2\text{CH}_3 \qquad (11.23b)$$

If the ether contains secondary or tertiary alkyl groups, the cleavage occurs under milder conditions (lower temperatures, more dilute acid). The first step of the mechanism is the same—protonation of the ether linkage:

$$\text{CH}_3\overset{\text{CH}_3}{\underset{\text{CH}_3}{\overset{|}{\text{—C}}\underset{|}{}}}\text{—}\overset{..}{\underset{..}{\text{O}}}\text{—CH}_2\text{CH}_3 \quad \rightleftharpoons \quad \text{CH}_3\overset{\text{CH}_3}{\underset{\text{CH}_3}{\overset{|}{\text{—C}}\underset{|}{}}}\text{—}\overset{H}{\underset{..}{\overset{|}{\text{O}^+}}}\text{—CH}_2\text{CH}_3 \qquad (11.24a)$$

The formation of the alkyl iodide occurs by an S_N1 mechanism. A carbocation is formed by loss of the alcohol leaving group, and the carbocation is attacked by iodide ion.

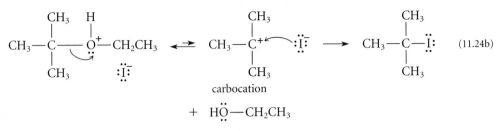

(11.24b)

carbocation

$$+ \quad H\ddot{O}\!-\!CH_2CH_3$$

Notice that the *tertiary* alkyl halide is formed along with the *primary* alcohol. Because the S_N1 reaction is faster than competing S_N2 processes, none of the primary alkyl iodide is formed.

Notice also the great similarity in the reactions of ethers and alcohols with halogen acids (Sec. 10.2). In the reaction of an alcohol, *water* acts as the leaving group. In the reaction of an ether, an *alcohol* acts as the leaving group. Otherwise the reactions are essentially the same.

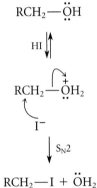

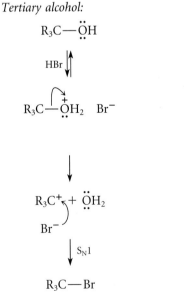

Methyl or primary alcohol:

$$RCH_2\!-\!\ddot{O}H$$

HI ⇅

$$RCH_2\!-\!\overset{+}{\ddot{O}}H_2$$

I^-

↓ S_N2

$$RCH_2\!-\!I \;+\; \ddot{O}H_2$$

Methyl or primary ether:

$$RCH_2\!-\!\ddot{O}\!-\!CH_2R$$

HI ⇅

$$RCH_2\!-\!\overset{+}{\underset{H}{\ddot{O}}}\!-\!CH_2R$$

I^-

↓ S_N2

$$RCH_2\!-\!I \;+\; H\ddot{O}\!-\!CH_2R$$

(11.25)

Tertiary alcohol:

$$R_3C\!-\!\ddot{O}H$$

HBr ⇅

$$R_3C\!-\!\overset{+}{\ddot{O}}H_2 \quad Br^-$$

↓

$$R_3C^+ + \ddot{O}H_2$$
$$Br^-$$

↓ S_N1

$$R_3C\!-\!Br$$

Tertiary ether:

$$R_3C\!-\!\ddot{O}\!-\!R$$

HBr ⇅

$$R_3C\!-\!\overset{+}{\underset{H}{\ddot{O}}}\!-\!R \quad Br^-$$

↓

$$R_3C^+ + H\ddot{O}\!-\!R$$
$$Br^-$$

↓ S_N1

$$R_3C\!-\!Br$$

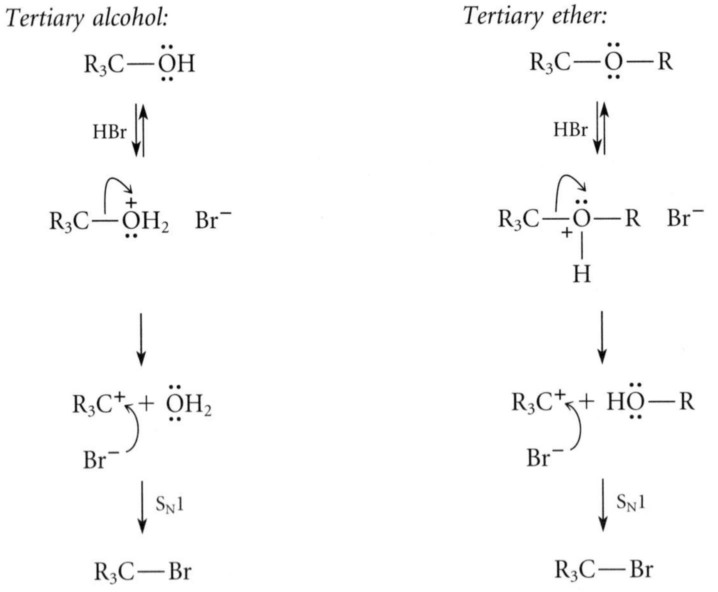

(11.26)

Although the cleavage of alkyl ethers gives alkyl halides and alcohols as products, this reaction is rarely used to prepare these compounds because ethers themselves are most often prepared from alkyl halides or alcohols.

PROBLEMS

*11.13   Explain the following facts with a mechanistic argument.

(a) When butyl methyl ether (1-methoxybutane) is treated with HI and heat, the initially formed products are mainly methyl iodide and 1-butanol; little or no methanol and butyl iodide are formed.

(b) When *tert*-butyl methyl ether is treated with HI, the products formed are *tert*-butyl iodide and methanol.

(c) When *tert*-butyl methyl ether is heated with sulfuric acid, methanol and 2-methylpropene distill from solution.

(d) *Tert*-butyl methyl ether cleaves much faster in HBr than its sulfur analog, *tert*-butyl methyl sulfide. (*Hint:* See Sec. 8.6.)

11.14   What products are formed when each of the following ethers reacts with concentrated aqueous HI?

*(a) 2-ethoxypropane   (b) diisopropyl ether

*(c) 2-ethoxy-2,3-dimethylbutane   (d) 1-methoxy-1-methylcyclohexane

11.4   Nucleophilic Substitution Reactions of Epoxides

A. Ring-Opening Reactions under Basic Conditions

Epoxides readily undergo reactions in which the epoxide ring is opened by nucleophiles.

$$
\begin{array}{c}
\text{H}_3\text{C}\quad\text{O} \\
\diagdown\diagup \\
\text{C}-\text{CH}_2 \\
\diagup \\
\text{H}_3\text{C}
\end{array}
\;+\; \text{C}_2\text{H}_5\text{OH} \;\xrightarrow[\text{5 hr, 80°}]{\text{Na}^+\,\text{C}_2\text{H}_5\text{O}^-}\;
\begin{array}{c}
\text{OH} \\
| \\
\text{CH}_3-\text{C}-\text{CH}_2-\text{OC}_2\text{H}_5 \\
| \\
\text{CH}_3
\end{array}
\quad (11.27)
$$

2,2-dimethyloxirane
(isobutylene oxide)

ethanol
(solvent)

1-ethoxy-2-methyl-2-propanol
(83% yield)

A reaction of this type is essentially *an* S_N2 *reaction in which the epoxide oxygen serves as the leaving group.* Of course, in this reaction, the leaving group does not depart as a separate entity, but rather remains within the same molecule.

leaving group

$$
\begin{array}{c}
\text{H}_3\text{C}\quad :\overset{..}{\text{O}}: \\
\diagdown\diagup \\
\text{C}-\text{CH}_2 \\
\diagup \\
\text{H}_3\text{C}\quad :\overset{..}{\text{O}}\text{C}_2\text{H}_5
\end{array}
\xrightarrow{S_N2}
\begin{array}{c}
:\overset{..}{\text{O}}:^{-} \\
| \\
\text{CH}_3-\text{C}-\text{CH}_2-\overset{..}{\underset{..}{\text{O}}}\text{C}_2\text{H}_5 \\
| \\
\text{CH}_3
\end{array}
\;\xrightleftharpoons[]{\text{H}-\overset{..}{\text{O}}\text{C}_2\text{H}_5}\;
\begin{array}{c}
:\overset{..}{\text{O}}\text{H} \\
| \\
\text{CH}_3-\text{C}-\text{CH}_2-\overset{..}{\underset{..}{\text{O}}}\text{C}_2\text{H}_5 \;+\; ^{-}:\overset{..}{\underset{..}{\text{O}}}\text{C}_2\text{H}_5 \\
| \\
\text{CH}_3
\end{array}
\quad (11.28)
$$

nucleophile

In an unsymmetrical epoxide, two ring-opening products could be formed corresponding to attack of the nucleophile at the two different carbons of the ring. As Eq. 11.28 illustrates, *nucleophiles typically attack unsymmetrical epoxides at the carbon with fewer branches.* This regioselectivity is expected from the effect of branching on the rates of S_N2 reactions (Sec. 9.4C). Branching retards the rate of attack; hence, attack at the unbranched carbon is faster and leads to the observed product.

Like other S_N2 reactions, the ring opening of epoxides by bases involves *backside attack* of the nucleophile on the epoxide carbon. When this carbon is a stereocenter, inversion of configuration occurs, as illustrated by the following study problem.

STUDY PROBLEM 11.2

What is the stereochemistry of the 2,3-butanediol formed when *meso*-2,3-dimethyloxirane reacts with aqueous sodium hydroxide?

Solution   First draw the structure of the epoxide. The *meso* stereoisomer of 2,3-dimethyloxirane has an internal plane of symmetry, and its two asymmetric carbons have opposite configurations.

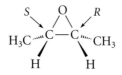

meso-2,3-dimethyloxirane

Because the two different carbons of the epoxide ring are substituted with the same groups, the hydroxide ion can attack either one. Backside attack on each carbon should occur with inversion of configuration.

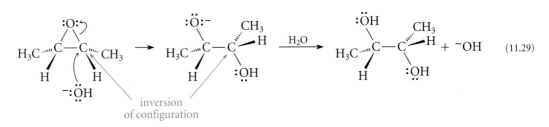

The product shown is the 2*S*,3*S* stereoisomer. Attack at the other carbon gives the 2*R*,3*R* stereoisomer. (Verify this point!) Because the starting materials are achiral, the two enantiomers of the product must be formed in equal amounts (Sec. 7.8A). Hence, the product of the reaction is racemic 2,3-butanediol. (This predicted result is observed in the laboratory.)

Because an epoxide is a type of ether, the ring opening of epoxides is an ether cleavage. However, ordinary ethers do *not* undergo cleavage in base.

$$CH_3CH_2-O-CH_2CH_3 + CH_3OH \xrightarrow{\text{CH}_3\text{O}^-} \text{no reaction} \qquad (11.30)$$

(Contrast this with the reaction in Eq. 11.27).

Why are epoxides so reactive? Epoxides, like their carbon analogs, the cyclopropanes, possess significant *ring strain* (Sec. 7.5B). Because of this strain, the bonds of an epoxide are weaker than those of an ordinary ether, and thus more easily broken. The opening of an epoxide relieves the strain of the three-membered ring just as the snapping of a twig relieves the strain of its bending.

11.15   Predict the products of the following reactions.

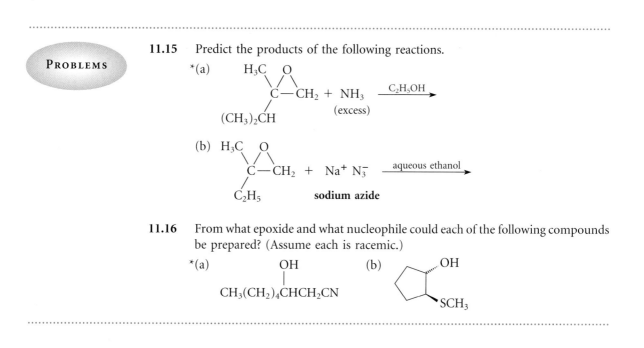

*(a)

H₃C O
\ /\
C—CH₂ + NH₃   ——C₂H₅OH——→
/
(CH₃)₂CH   (excess)

(b) H₃C   O
\ /\
C—CH₂ + Na⁺ N₃⁻   ——aqueous ethanol——→
/
C₂H₅   **sodium azide**

11.16   From what epoxide and what nucleophile could each of the following compounds be prepared? (Assume each is racemic.)

*(a)   OH   (b)

|
CH₃(CH₂)₄CHCH₂CN

B. Ring-Opening Reactions under Acidic Conditions

Ring-opening reactions of epoxides, like those of ordinary ethers, are catalyzed by acids. However, as might be expected from the previous section, epoxides are much more reactive than ethers under acidic conditions because of their ring strain.

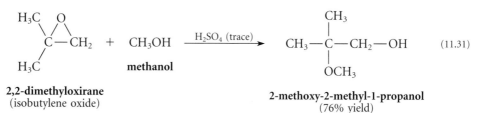

2,2-dimethyloxirane
(isobutylene oxide)

2-methoxy-2-methyl-1-propanol
(76% yield)

This reaction requires only a trace of the acid catalyst to proceed at a convenient rate.

   The regioselectivity of the ring-opening reaction is different under acidic and basic conditions. The structure of the product in Eq. 11.31 shows that the nucleophile methanol reacts at the *more branched carbon* of the epoxide. Contrast this with the result in Eq. 11.27, in which the nucleophile reacts at the *less branched carbon* under basic conditions. In general, if one of the carbons of an unsymmetrical epoxide is tertiary, nucleophiles react at this carbon under acidic conditions.

Some insight into why different regioselectivities are observed under different conditions comes from mechanistic considerations. The first step in the mechanism of Eq. 11.31, like the first step of ether cleavage, is protonation of the oxygen.

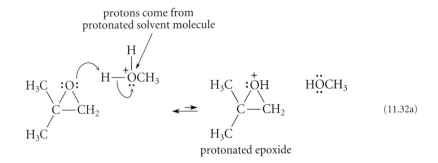

(11.32a)

Bonds to tertiary carbon atoms are weaker than bonds to primary carbon atoms (Table 5.3), and the protonated oxygen is a good leaving group. Consequently, the bond to the tertiary carbon in the protonated epoxide is very weak. In other words, the tertiary carbon has a great deal of carbocation character. A solvent molecule attacks this carbon as it would a carbocation to give the product after loss of a proton.

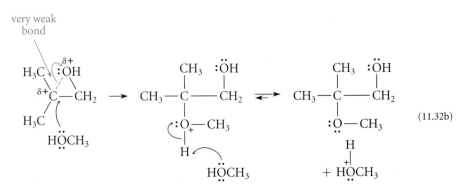

(11.32b)

If the tertiary carbon in the protonated epoxide has substantial carbocation character, then this carbon is essentially sp^2-hybridized and has nearly planar geometry. If so, the branches at this carbon are flattened and impede attack of the nucleophile much less than they would in an ordinary S_N2 reaction.

When the carbons of an unsymmetrical epoxide are secondary or primary, there is much less carbocation character at either carbon in the protonated epoxide, and acid-catalyzed ring opening reactions tend to give mixtures of products; the exact compositions of the mixtures vary from case to case.

$$CH_3 - \overset{O}{\overset{/ \backslash}{CH - CH_2}} + C_2H_5OH \xrightarrow[\text{H}_2\text{SO}_4]{0.8\%} CH_3 - \overset{OC_2H_5}{\underset{|}{CH}} - CH_2 - OH + CH_3 - \overset{OH}{\underset{|}{CH}} - CH_2 - OC_2H_5$$

secondary   primary   37% of product   63% of product

(11.33)

The mixture reflects the balance between opening of the weaker bond, which favors attack at the more branched carbon, and steric hindrance to nucleophilic attack, which favors attack at the less branched carbon.

Notice that the regioselectivities of acid-catalyzed epoxide ring opening and attack on bromonium ions are very similar (Sec. 5.1B). This is not surprising, since both types of reactions involve the opening of strained rings containing very electronegative leaving groups.

Acid-catalyzed ring-opening reactions of epoxides, like base-catalyzed ring-opening reactions, occur with *inversion of stereochemical configuration.*

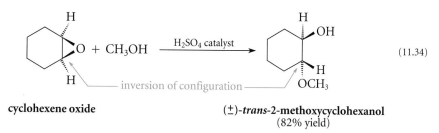

(11.34)

cyclohexene oxide   (±)-*trans*-2-methoxycyclohexanol
                                                          (82% yield)

When water is used as a nucleophile in acid-catalyzed epoxide ring opening, the product is a 1,2-diol, or *glycol.* Acid-catalyzed epoxide hydrolysis is generally a useful way to prepare glycols.

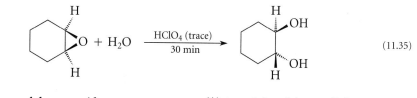

(11.35)

cyclohexene oxide   (±)-*trans*-1,2-cyclohexanediol
                                                          (80% yield)

Although base-catalyzed hydrolysis of epoxides also gives glycols (see Study Problem 11.2), polymerization sometimes occurs as a side reaction under the basic conditions (see Problem 11.48). Consequently, acid-catalyzed hydrolysis of epoxides is generally preferred for the preparation of glycols.

The acid-catalyzed hydrolysis of epoxides can be used as part of a second and complementary method for the preparation of glycols from alkenes, as shown by the following study problem.

STUDY PROBLEM 11.3

Outline preparations of *cis*-1,2-cyclohexanediol and (±)-*trans*-1,2-cyclohexanediol from cyclohexene.

Solution   The direct oxidation of cyclohexene by $KMnO_4$ in NaOH solution, or by OsO_4 followed by hydrolysis, yields the *cis*-diol through a *syn*-addition (Sec. 5.5).

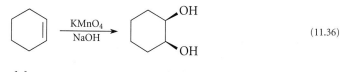

(11.36)

cyclohexene   *cis*-1,2-cyclohexanediol

In contrast, conversion of cyclohexene into the epoxide with a peroxycarboxylic acid (see Problem 11.9a), followed by acid-catalyzed hydrolysis (Eq. 11.35), gives the *trans*-diol.

Although epoxide formation is a *syn*-addition, epoxide hydrolysis gives the *trans*-diol because it occurs with *inversion of configuration*.

11.17 Predict the major product(s) of each of the following transformations.

*(a)

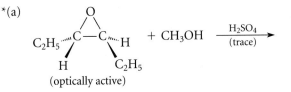

(optically active)

(b) The enantiomer of the epoxide in (a) + CH$_3$OH $\xrightarrow[\text{(trace)}]{\text{H}_2\text{SO}_4}$

Let's summarize the facts about the regioselectivity and stereoselectivity of epoxide ring-opening reactions:

1. Nucleophiles attack unsymmetrical epoxides under basic conditions at the less branched carbon, and inversion of configuration is observed if attack occurs at a stereocenter.

2. Nucleophiles attack unsymmetrical epoxides under acidic conditions at the tertiary carbon. If neither carbon is tertiary, a mixture of products is formed in most cases. Inversion of configuration is observed if attack occurs at a stereocenter.

These facts are used in the following study problem.

Predict the major product in each case that would be obtained when the following epoxide is hydrolyzed under (a) basic conditions; (b) acidic conditions. (The epoxide carbons are numbered for reference in the solution.)

Solution   As the summary above suggests, when attempting to predict the products of an epoxide ring-opening reaction, first decide whether the conditions of the reaction are basic or acidic. If basic, the nucleophile attacks the *less branched* carbon of the epoxide; if acidic, the nucleophile attacks the *tertiary carbon* of the epoxide. Then determine whether the carbon at which attack occurs is a stereocenter. If so, make sure to predict the product that results from inversion of configuration.

(a)   Under basic conditions, the hydroxide ion nucleophile will attack the *less branched* carbon of the epoxide—carbon-1. (If you have difficulty seeing why this is the less branched carbon, please read Study Guide Link 9.3.) Because this carbon is not a stereocenter, the stereochemistry of attack does not matter. Consequently, the reaction is

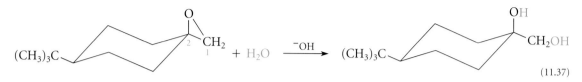

(11.37)

(b) Under acidic conditions, the nucleophile is water, which attacks the *protonated* epoxide at the more branched carbon (carbon-2). Notice that carbon-2 is a stereocenter (even though it is *not* an asymmetric carbon); nucleophilic attack at carbon-2 occurs with inversion of configuration. Consequently, the product of the reaction under acidic conditions is a diastereomer of the product obtained under basic conditions.

(11.38)

PROBLEM

*11.18 (a) Suppose 2,2-dimethyloxirane is hydrolyzed in water that has been enriched with the oxygen isotope ^{18}O. Indicate how the hydrolysis product would differ under acidic and basic conditions.

(b) Suppose you have carried out the reaction in (a) and want to test your prediction. Given that you can isolate any compound and have a way to determine its ^{18}O content, propose a scheme for analyzing the position of the isotope in your products. (*Hint:* What reaction of your products could be used to partition the two oxygens into separate compounds?)

C. Reaction of Ethylene Oxide with Grignard Reagents

Grignard reagents react with ethylene oxide to give, after a protonation step, primary alcohols:

$$CH_3CH_2CH_2CH_2CH_2CH_2MgBr \;+\; H_2C\!\!-\!\!CH_2 \xrightarrow[\text{2) } H_3O^+]{\text{1) ether, heat}} CH_3CH_2CH_2CH_2CH_2CH_2CH_2CH_2OH$$

hexylmagnesium bromide   **ethylene oxide**   **1-octanol**
(a Grignard reagent)   (71% yield)   (11.39)

This reaction is another epoxide ring-opening reaction. To understand this reaction, recall that the carbon in the C—Mg bond of the Grignard reagent has *carbanion* character and is therefore a very *basic* carbon (Sec. 8.7B). This carbon attacks the epoxide as a nucleophile. At the same time, the magnesium of the Grignard reagent, which is a Lewis acid, coordinates to the epoxide oxygen. (Recall that Grignard reagents coordinate strongly to ether oxygens; see Eq. 8.18.) Just as protonation of an oxygen makes it a better leaving group, coordination of an oxygen to a Lewis acid also makes it a better leaving group. Consequently, this coordination assists the ring opening of the epoxide in much the same way that Brønsted acids catalyze ring opening (Sec. 11.4B).

*11.26   Which of the two compounds in the previous problem would give the highest percentage of *intermolecular* reaction product, that is, would show the least amount of neighboring-group participation? Explain.

11.7   Oxidation of Ethers and Sulfides

Ethers are relatively inert toward many of the common oxidants used in organic chemistry if the reaction conditions are not too vigorous. For example, diethyl ether can be used as a solvent for oxidations with Cr(VI). On standing in air, however, ethers undergo the slow *autoxidation* discussed in Section 8.8D that leads to the formation of dangerously explosive peroxide contaminants.

The sulfur analogs of peroxides are disulfides, which are R—S—S—R oxidation products of thiols (Sec. 10.9). Disulfides are not explosive.

Like thiols, sulfides oxidize at *sulfur* rather than carbon when they react with common oxidizing agents. Sulfides can be oxidized to **sulfoxides** and **sulfones**:

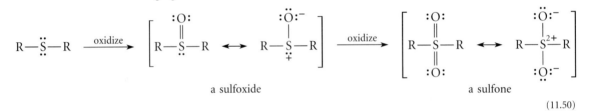

$$(11.50)$$

Dimethyl sulfoxide (DMSO) and sulfolane are well-known examples of a sulfoxide and a sulfone, respectively. (Both compounds are excellent dipolar aprotic solvents; see Table 8.2.)

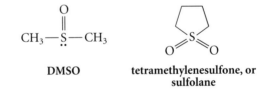

DMSO	**tetramethylenesulfone, or sulfolane**

Notice that nonionic Lewis structures for sulfoxides and sulfones cannot be written without violating the octet rule. The bonding at sulfur in such situations was discussed in Sec. 10.9.

Sulfoxides and sulfones can be prepared by the direct oxidation of sulfides with one and two equivalents, respectively, of hydrogen peroxide, H_2O_2:

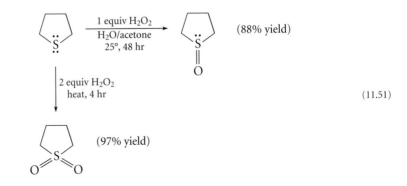

$$(11.51)$$

Other common oxidizing agents such as $KMnO_4$, HNO_3, and peroxyacids (Sec. 11.2A) readily oxidize sulfides.

11.8   The Three Fundamental Operations of Organic Synthesis

Section 10.11 introduced organic synthesis with a systematic approach to solving synthesis problems. This section continues this approach by classifying the types of operations involved in a typical synthesis. Most reactions used in organic synthesis involve one or more of *three fundamental operations.*

1. Functional-group transformation
2. Control of stereochemistry
3. Formation of carbon-carbon bonds

Functional-group transformation—the conversion of one functional group into another—is the most common type of synthetic operation. Most of the reactions you've studied so far involve functional-group transformation. For example, hydrolysis of epoxides transforms epoxides into glycols; hydroboration-oxidation converts alkenes into alcohols.

Control of stereochemistry is accomplished with stereoselective reactions. Whenever you have to prepare a compound that can exist as several stereoisomers you should think in terms of these reactions. Examples of stereoselective reactions include hydroboration-oxidation, which is a *syn* addition, and S_N2 reactions, which occur with inversion of configuration.

Reactions that bring about the *formation of carbon-carbon bonds* are particularly important, because these reactions must be used to add carbon atoms, and thus "grow" larger carbon chains from smaller ones. Only two reactions of this type have been presented:

1. Cyclopropane formation from carbenes or carbenoids and alkenes (Sec. 9.8)
2. Reaction of Grignard reagents with ethylene oxide (Sec. 11.4C)

Most reactions involve combinations of at least two of the three fundamental operations. For example, hydroboration-oxidation is a functional-group transformation that also allows control of stereochemistry; two operations can be accomplished at once with this reagent. The reaction of ethylene oxide with Grignard reagents effects both carbon-carbon bond formation and the transformation of an epoxide into an alcohol.

The following two study problems demonstrate how to use the three fundamental operations in planning an organic synthesis.

STUDY PROBLEM 11.5

Outline a synthesis of 1-hexanol from 1-butanol and any other reagents.

Solution   As usual, first write the problem in terms of structures:

$$CH_3CH_2CH_2CH_2\!-\!OH \xrightarrow{??} CH_3CH_2CH_2CH_2CH_2CH_2\!-\!OH$$

Next, analyze the types of operations needed. Two carbons must be added, but no issues of stereochemistry are involved. You should *not* assume that because both the starting

(g)

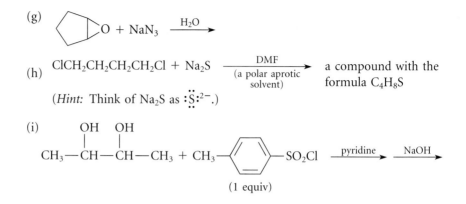

(h) $ClCH_2CH_2CH_2CH_2Cl + Na_2S$ $\xrightarrow[\text{(a polar aprotic solvent)}]{\text{DMF}}$ a compound with the formula C_4H_8S

(*Hint:* Think of Na_2S as $:\!\ddot{S}\!:^{2-}$.)

(i)

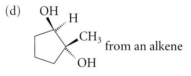

$\xrightarrow{\text{pyridine}}$ $\xrightarrow{\text{NaOH}}$

(1 equiv)

*11.44   Outline a synthesis for each of the following compounds from the indicated starting material and any other reagents. (All chiral compounds should be prepared as racemates.)

(a)   2-ethoxy-3-methylbutane from 3-methyl-1-butene

(b)   2-ethoxy-2-methylbutane from 2-methyl-2-butanol

(c)

$$\underset{\displaystyle C_2H_5—\overset{\displaystyle O}{\overset{\|}{S}}—CH_2CH_2CH_2CH_3}{}\text{ from compounds containing} \leq 2 \text{ carbons}$$

(d)   from an alkene

(e)   cyclohexyl isopropyl ether from cyclohexene

(f) $(CH_3)_2CHCH_2CH_2CH_2CH{=}O$ from 3-methyl-1-butene

(g)

$$\underset{H_3C}{\overset{CH_3CH_2}{>}}CH\overset{\displaystyle O}{\overset{\|}{C}}CH_2OCH_3 \text{ from 3-methyl-1-pentene}$$

(h)

$$CH_3—\overset{\displaystyle OCH_2CH_3}{\underset{\displaystyle CH_3}{\overset{\displaystyle |}{\underset{\displaystyle |}{C}}}}—CH{=}O \text{ from 2-methylpropene}$$

11.45   Outline a synthesis for each of the following compounds from (2*R*,3*R*)-2,3-dimethyloxirane:

*(a)   (2*R*,3*S*)-3-methoxy-2-butanol   (b)

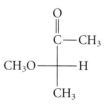

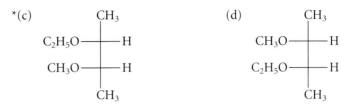

*(c) and (d) Fischer projection structures

*11.46 Compound *A*, C_8H_{16}, undergoes catalytic hydrogenation to give octane. When treated with *meta*-chloroperoxybenzoic acid, *A* gives an epoxide *B*, which, when treated with aqueous acid, gives a compound *C*, $C_8H_{18}O_2$, that can be resolved into enantiomers. When *A* is treated with OsO_4, an achiral compound *D*, a stereoisomer of *C*, forms. Identify all compounds, including stereochemistry where appropriate.

11.47 When $CH_3CH_2—\overset{..}{\underset{..}{S}}—CH_2CH_2—\overset{..}{\underset{..}{S}}—CH_2CH_3$ reacts with two equivalents of CH_3I, the following double sulfonium salt precipitates:

$$CH_3CH_2—\overset{\overset{CH_3}{|}}{\underset{+}{S}}—CH_2CH_2—\overset{\overset{CH_3}{|}}{\underset{+}{S}}—CH_2CH_3 \quad 2I^-$$

(a) Give a curved-arrow mechanism for the formation of this salt.
(b) Upon closer examination, this compound is found to be a mixture of two isomers with melting points of 123–124° and 154°, respectively. Explain why *two* compounds of this structure are formed. What is the relationship between these isomers? (*Hint:* Unlike amines, sulfonium salts do *not* undergo rapid inversion at sulfur.)

*11.48 One of the side reactions that take place when epoxides react with ^-OH is the formation of polymers. Propose a mechanism for the following polymerization reaction, using the curved-arrow formalism.

$$n\ CH_3—\overset{O}{\overset{/\backslash}{CH—CH_2}} \xrightarrow{\ ^-OH\ } \left[\overset{}{\underset{\overset{|}{CH_3}}{CH}}—CH_2—O\right]_n$$

*11.49 Account for the following observations with a mechanism:
(1) In 80% aqueous ethanol compound *A* reacts to give compound *B*. Notice that *trans-B* is the only stereoisomer of this compound that is formed.
(2) Optically active *A* gives completely racemic *B*.
(3) The reaction of *A* is about 10^5 times as fast as the analogous substitution reactions of both its stereoisomer *C* and chlorocyclohexane.

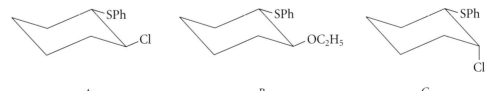

A $\qquad\qquad\qquad\qquad$ B $\qquad\qquad\qquad\qquad$ C

12.3   Use Figure 12.2 to answer the following questions.
*(a)   How does the energy of X-rays compare with that of blue light (greater or smaller)?
(b)   How does the energy of radar waves compare with that of red light (greater or smaller)?

B. Absorption Spectroscopy

The most common type of spectroscopy used for structure determination is **absorption spectroscopy**. The basis of absorption spectroscopy is that *matter can absorb energy from certain wavelengths of electromagnetic radiation.* In an absorption spectroscopy experiment, this absorption is determined as a function of wavelength, frequency, or energy in an instrument called a **spectrophotometer** or **spectrometer**. The rudiments of an absorption spectroscopy experiment are shown schematically in Fig. 12.3. The experiment requires, first, a *source* of electromagnetic radiation. (If the experiment measures the absorption of visible light, the source could be a common light bulb.) The material to be examined, the *sample*, is placed in the radiation beam. A *detector* measures the intensity of the radiation that passes through the sample unabsorbed; when this intensity is subtracted from the intensity of the source, the amount of radiation absorbed by the sample is known. The wavelength of the radiation falling on the sample is then varied, and the radiation absorbed at each wavelength is recorded as a graph of either radiation transmitted or radiation absorbed *vs.* wavelength or frequency. This graph is commonly called a **spectrum** of the sample.

The infrared spectrum of nonane shown in Figure 12.4 is an example of such a graph. It shows the radiation transmitted by a sample of nonane over a range of wavelengths in the infrared. (You'll learn how to read such a spectrum in more detail in the next section.)

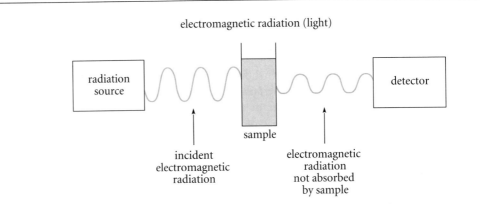

electromagnetic radiation (light)

radiation source

detector

sample

incident electromagnetic radiation

electromagnetic radiation not absorbed by sample

Figure 12.3   *The spectroscopy experiment. Electromagnetic radiation of a certain frequency from a source is passed through the sample and onto a detector. If radiation is absorbed by the sample, the radiation emerging from the sample has a lower intensity than the incident radiation. The intensity of the incident radiation is measured independently. Comparison of the intensity of the incident radiation and the radiation emerging from the sample tells how much radiation is absorbed by the sample. The spectrum is a plot of radiation absorbed or transmitted vs. wavelength or frequency of the radiation.*

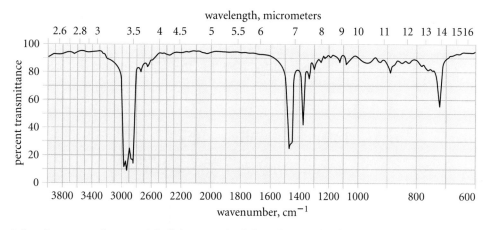

Figure 12.4 *Infrared spectrum of nonane. The light transmitted through a sample of nonane is plotted as a function of wavelength (upper horizontal axis) or wavenumber (lower horizontal axis). Absorptions are indicated by the inverted peaks.*

An Everyday Analogy to a Spectroscopy Experiment

You do not need experience with a spectrophotometer to appreciate the basic idea of a spectroscopy experiment. Imagine holding a piece of green glass up to the white light of the sun. The sun is the source, the glass is the sample, your eyes are the detector, and your brain provides the spectrum. White light is a mixture of all wavelengths. The glass appears green because only green light is transmitted through the glass; the other colors (wavelengths) in white light are absorbed. If you hold the same green glass up to red light, no light is transmitted to your eyes—the glass looks black—because the glass absorbs the red light.

The spectrum of a compound is determined by its structure. For this reason, spectroscopy can be used for structure determination. Chemists use many types of spectroscopy for this purpose. The three types of greatest use, and the general type of information each provides, are as follows:

1. Infrared (IR) spectroscopy provides information about what functional groups are present.

2. Nuclear magnetic resonance (NMR) spectroscopy provides information on the connectivity of carbons and hydrogens.

3. Ultraviolet-visible (UV-VIS) spectroscopy (often called simply UV spectroscopy) provides information about the types of π-electron systems that are present.

These types of spectroscopy differ conceptually only in the frequency of radiation used, although the instruments required for each are quite different. A fourth physical technique, mass spectrometry, which allows us to determine molecular masses, is also widely used for structure determination. It is not a type of absorption spectroscopy, and is thus fundamentally different from NMR, IR, and UV spectroscopy.

The remainder of this chapter is devoted to a description of IR spectroscopy and mass spectrometry. NMR spectroscopy is covered in Chapter 13 and UV spectroscopy in Chapter 15.

12.2 Infrared Spectroscopy

A. The Infrared Spectrum

An infrared spectrum, like any absorption spectrum, is a record of the light absorbed by a substance as a function of wavelength. The IR spectrum is measured in an instrument called an **infrared spectrophotometer**, described briefly in Sec. 12.5. In practice, the absorption of infrared radiation with wavelengths between 2.5×10^{-6} to 20×10^{-6} meters is of greatest interest to organic chemists. Let's consider the details of an IR spectrum by returning to the spectrum of nonane in Fig. 12.4.

The quantity plotted on the lower horizontal axis is the **wavenumber** $\tilde{\nu}$ of the light. The wavenumber, in units of *reciprocal centimeters* or *inverse centimeters* (cm^{-1}), is simply another way to express the wavelength or frequency of the radiation. Physically, the wavenumber is the number of wavelengths contained in one centimeter. Wavenumber is inversely proportional to the wavelength λ.

$$\left. \begin{aligned} \tilde{\nu} &= \frac{10^4}{\lambda} \\[2ex] \lambda &= \frac{10^4}{\tilde{\nu}} \end{aligned} \right\} \quad (\lambda \text{ in micrometers, } \tilde{\nu} \text{ in } cm^{-1})$$

(12.6a)

(12.6b)

In these equations the wavelength is in **micrometers**. A micrometer is 10^{-6} meter, and is abbreviated μm. Thus, according to Eq. 12.6a, a wavelength of 10 μm corresponds to a wavenumber of 1000 cm^{-1}. In Fig. 12.4, wavenumber, across the bottom of the spectrum, increases to the left; wavelength, across the top of the spectrum, increases to the right. Notice that the wavenumber scale is divided into three distinct regions in which the linear scale is different.

You may see older spectra that are calibrated in wavelength. In most cases, such spectra have a wavenumber scale as well, but if not, conversion between wavelengths and wavenumbers is a simple matter with Eq. 12.6b.

The relationship between wavenumber, energy, and frequency can be derived by combining Eqs. 12.1 and 12.3 with the definition of wavenumber.

$$E = \frac{hc}{\lambda}$$

(12.7a)

$$\nu = c\tilde{\nu}$$

(12.7b)

(When applying these equations, be sure to use consistent units, for example, λ in m, $\tilde{\nu}$ in m^{-1}, h in kJ sec mol^{-1}, c in m/sec, and E in kJ/mol.) Notice that energy, wavenumber, and frequency are proportional. Because of this proportionality, the wavenumber is sometimes loosely referred to as a frequency.

Referring again to the IR spectrum in Fig. 12.4, note that the quantity plotted on its vertical axis is *percent transmittance*. This is the percent of the radiation falling on the sample that is transmitted to the detector. If the sample absorbs all the radiation, then none is transmitted, and the sample has 0% transmittance. If the sample absorbs no radiation, then all of the radiation is transmitted, and the sample has 100% transmittance. Thus, absorptions in the IR spectrum are registered as downward deflections, that is, "upside-down peaks." Thus, absorptions in the spectrum of nonane occur at about 2920, 1470, 1380, and 720 cm^{-1}.

Sometimes an IR spectrum is not presented in graphical form, but is summarized completely or in part using descriptions of peak *positions*. Intensities are often expressed qualitatively using the designations vs (very strong), s (strong), m (moderate), or w (weak). Some peaks are sharp (narrow), whereas others are broad (wide). The spectrum of nonane can be summarized as follows:

$$\tilde{\nu} \text{ (cm}^{-1}\text{): } 2920 \text{ (vs); } 1470 \text{ (s); } 1380 \text{ (m); } 720 \text{ (w)}$$

PROBLEMS

12.4   *(a)   What is the wavenumber of light with a wavelength of 6.0 μm?
        (b)   What is the wavelength of light with a wavenumber of 1720 cm^{-1}?

*12.5   Show that Eq. 12.7b follows from Eqs. 12.7a and 12.1.

B. Physical Basis of IR Spectroscopy

Interpretation of an IR spectrum in terms of structure requires some understanding of why molecules absorb infrared radiation. The absorptions observed in an IR spectrum are the result of *vibrations* within a molecule. Atoms within a molecule are not stationary, but are constantly in motion. Consider, for example, the C—H bonds in a typical organic compound. The bonds undergo various stretching and bending motions. A useful analogy to the C—H stretching motion is the stretching and compression of a spring (Fig. 12.5). This vibration takes place with a certain frequency ν; that is, it occurs a certain number of times per second. Suppose that a C—H bond has a stretching frequency of 9×10^{13} times per second; this means that it undergoes a vibration every $1/(9 \times 10^{13})$ second or 1.1×10^{-14} second.

The colored line in Fig. 12.5 shows that the stretching of the C—H bond over time describes a wave motion. A wave of electromagnetic radiation can transfer its energy to the vibrational wave motion of the C—H bond only if *there is an exact match between the frequency of the radiation and the frequency of the vibration*. Thus, if a C—H vibration has a frequency of 9×10^{13} sec^{-1}, then it will absorb energy from radiation with the same frequency. From the relationship $\lambda = c/\nu$ (Eq. 12.1), the radiation must have a wavelength of

$$\lambda = (3 \times 10^8 \text{ m sec}^{-1})/(9 \times 10^{13} \text{ sec}^{-1})$$
$$= 3.33 \times 10^{-6} \text{ m} = 3.33 \ \mu\text{m}$$

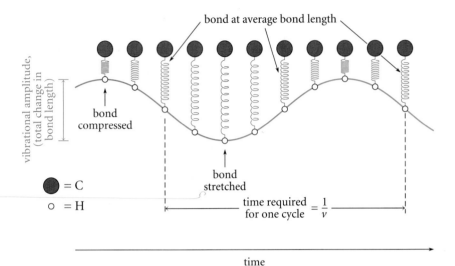

Figure 12.5   *Chemical bonds undergo a variety of vibrations. The one illustrated here is a stretching vibration. The bond, represented as a spring, is shown at various times. The bond stretches and compresses over time. The time required for one complete cycle of vibration is the reciprocal of the vibrational frequency.*

The corresponding wavenumber of this radiation (Eq. 12.6a) is 3000 cm^{-1}. When radiation of this wavelength interacts with a vibrating C—H bond, energy is absorbed and the intensity of the bond vibration increases. That is, after absorbing energy, the bond vibrates with the same wavelength but with a larger *amplitude* (a larger stretch and tighter compression; Fig. 12.6). This absorption gives rise to the peak in the IR spectrum.

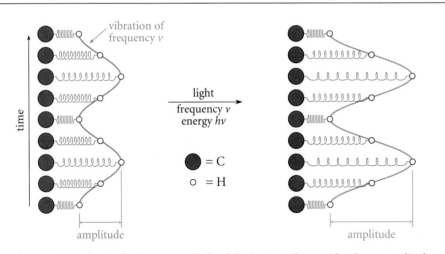

Figure 12.6   *Light absorption by a bond vibration causes the bond (spring) to vibrate with a larger amplitude at the same frequency. The frequency of the light must exactly match the frequency of the bond vibration for absorption to occur.*

AN ANALOGY TO ENERGY ABSORPTION
BY A VIBRATING BOND

Imagine a weight suspended from your finger by a rubber band. (An unopened twelve-ounce can of soda suspended by its pull-tab works nicely in this demonstration.) If you move your finger up and down very slowly the soda can scarcely moves. However, if you increase the rate of oscillation of your finger gradually, at some point the soda can will start to move up and down vigorously. The motion of your finger is analogous to electromagnetic radiation, and the rubber band-soda can combination is analogous to the vibrating bond. Energy is absorbed from the motion of your finger by the rubber band-soda can oscillator only when its natural oscillation frequency matches the oscillation frequency of your finger. Likewise, a vibrating bond absorbs energy from electromagnetic radiation only when there is a frequency match between the two oscillating systems.

In summary:

1. Bonds vibrate with characteristic frequencies.

2. Absorption of energy from infrared radiation can occur only when there is a match between the wavelength of the radiation and the wavelength of the bond vibration.

PROBLEM

*12.6   Given that the stretching vibration of a typical C—H bond has a frequency of about 9×10^{13} sec^{-1}, which peak(s) in the IR spectrum of nonane (Fig. 12.4) would you assign to the C—H stretching vibrations?

12.3   Infrared Absorption and Chemical Structure

Each peak in the IR spectrum of a molecule corresponds to absorption of energy by the vibration of a particular bond or group of bonds. The utility of IR spectroscopy for the chemist is that *in all compounds, a given type of functional group absorbs in the same general region of the IR spectrum.* The major regions of the IR spectrum are shown in Table 12.1.

The IR spectra of most compounds contain many more absorptions than can be readily interpreted. A major part of mastering IR spectroscopy is to learn which absorptions are important. Certain absorptions are *diagnostic*; that is, they indicate with reasonable certainty that a particular functional group is present. For example, an intense peak in the 1700–1750 cm^{-1} region is almost like a voice saying, "This molecule contains a carbonyl (C=O) group." Other peaks are *confirmatory*; that is, similar peaks can be found in other types of molecules, but their presence confirms a structural diagnosis made in other ways. For example, absorptions in the 1050–1200 cm^{-1} region of the IR spectrum due to a C—O bond could indicate the presence of an alcohol, an ether, an ester, or a carboxylic acid, among other things. However, if other evidence (perhaps

Table 12.1   Regions of the Infrared Spectrum

Wavenumber range, cm^{-1}	Type of Absorptions	Name of region
3400–2800	O—H, N—H, C—H stretching	Functional group
2250–2100	C≡N, C≡C stretching	
1850–1600	C=O, C=N, C=C stretching	
1600–1000	C—C, C—O, C—N stretching; various bending absorptions	Fingerprint
1000–600	C—H bending	C—H bending

obtained from other types of spectroscopy) suggests that the unknown molecule is, say, an ether, a peak in this region can serve to support this diagnosis. In the sections that follow, you'll learn about the relatively few absorptions that are important in IR spectroscopy.

Rarely if ever does an IR spectrum completely define a structure; rather, it provides information that restricts the possible structures under consideration. Once the structure for an unknown compound has been deduced, a comparison of its IR spectrum with that of an authentic sample can be used as a criterion of identity. Even subtle differences in structure generally give discernible differences in the IR spectrum, particularly in the region between 1000 cm^{-1} and 1600 cm^{-1}. Even though most of the absorptions in this region of the spectrum are generally not interpreted in detail, they serve as a valuable "molecular fingerprint." That is why this region of the spectrum is called the "fingerprint region" in Table 12.1.

A. Factors That Determine IR Absorption Position

One approach to the use of IR spectroscopy is simply to memorize the positions at which characteristic functional group absorbances appear, and look for peaks at these positions in the determination of unknown structures. However, you can use IR spectroscopy much more intelligently and learn the important peak positions much more easily if you understand a little more about the physical basis of IR spectroscopy. Two aspects of IR absorption peaks are particularly important. First is the *position* of the peak, that is, the wavenumber or wavelength at which it occurs. Second is the *intensity* of the peak, that is, how strong it is. Let's consider each of these aspects in turn.

What factors govern the position of IR absorption? Three considerations are most important.

1. Strength of the bond
2. Masses of the atoms involved in the bond
3. The type of vibration being observed

Focus for the moment on the first two of these factors, and think of the vibrating bond as a mechanical spring. How should bond strength affect the vibrational frequency? Intuitively, a stronger bond corresponds to a tighter spring. Objects connected by a tighter spring vibrate more rapidly, that is, with a higher frequency or wavenumber. Likewise, atoms connected by a stronger bond also vibrate at higher frequency. A simple measure of bond strength is the energy required to break the bond, that is, the *bond dissociation energy* (Table 5.3). *The higher the bond dissociation energy, the stronger the bond.* Thus: *the IR absorptions of stronger bonds—bonds with greater bond dissociation energies—occur at higher wavenumber.* The effect of bond strength is particularly clear from a comparison of the vibration frequencies of single, double, and triple bonds between the same atoms. As intuition and the bond dissociation energies in Table 5.3 dictate, C≡C bonds are stronger than C=C bonds, which, in turn, are stronger than C—C bonds. Thus, the wavenumbers for the IR absorptions of these stretching motions are in the order C≡C > C=C > C—C.

What about the mass effect? You may have observed that a heavier weight connected to a spring of a given tightness vibrates more slowly (with a lower frequency) than lighter weights connected to an identical spring. This effect is also evident in IR spectroscopy: *vibrations of heavier atoms occur at lower wavenumbers than vibrations of lighter atoms.* This effect is best illustrated by isotopic replacement. For example, the C—H and C—D bonds have nearly identical bond dissociation energies. However, the C—D stretching absorption occurs at considerably lower wavenumber (about 2250 cm^{-1}) than the C—H stretching absorption (about 3000 cm^{-1}).

Another important effect of atomic mass is that when the two atoms in a bond differ significantly in mass, the stretching frequency is governed primarily by the mass of the *lighter* atom. The importance of this idea is apparent from Table 12.1. All of the hydrogen stretching absorptions occur in the same general region of the IR spectrum; this is exactly what would be expected if the vibrational frequency is determined predominantly by the smaller atom. Similarly, all of the single-bond stretching absorptions between first-row atoms (C—C, C—O, C—N) also occur in the same general region of the IR spectrum.

AN ANALOGY FOR THE MASS EFFECT

An analogy is useful to show that the lighter atom has the major influence on the vibrational frequency. Imagine three situations: two identical light rubber balls connected by a spring; an identical rubber ball connected to a heavy cannonball with an identical spring; and an identical rubber ball connected to the Empire State Building by an identical spring. When the spring connecting the two rubber balls is stretched and released, both balls oscillate. That is, both masses are involved in the vibration. When the spring connecting the rubber ball and the cannonball is stretched and released, the rubber ball oscillates, and the cannonball remains almost stationary. When the spring connecting the rubber ball and the Empire State Building is stretched and released, only the rubber ball appears to oscillate; the motion of the building is imperceptible. In the last two cases, the vibrational frequencies are virtually identical, even though the masses attached to them differ by orders of magnitude. Consequently, for a given spring, the oscillation frequency of two connected objects of different mass depends more on the mass of the lighter object than on the mass of the heavier object.

STUDY GUIDE LINK:
12.1
IR Absorptions and the Vibrating Spring

*12.7   Which effect is more important in determining the absorption position in the following series, bond strength or atomic mass? How do you know?

bond:	H—C	H—O	H—F
wavenumber:	3000 cm^{-1}	3600 cm^{-1}	4000 cm^{-1}

The third factor that affects the absorption frequency is the *type of vibration*. A **stretching** vibration occurs along the line of the chemical bond. A **bending** vibration is any vibration that does not occur along the line of the chemical bond. A bending vibration can be envisioned as a ball hanging on a spring and swinging side-to-side. In general, *bending vibrations occur at lower frequencies (higher wavelengths) than stretching vibrations of the same groups.*

AN ANALOGY FOR BENDING VIBRATIONS

Imagine a ball attached to a stiff spring hanging from the ceiling. A gentle tap makes it swing back and forth. It takes considerably more energy to stretch the spring. Because the energy required to set the spring in motion is proportional to its frequency, then the swinging (bending) motion has a lower frequency than the stretching motion.

The only possible type of vibration in a diatomic molecule (for example, H—F) is a stretching vibration. However, when a molecule contains more than two atoms, then both stretching and bending vibrations are possible. The allowed vibrations of a molecule are termed its **normal vibrational modes**. The normal vibrational modes for a —CH_2— group are shown in Fig. 12.7. They serve as models for the kinds of vibrations that can be expected for other groups in organic molecules. The bending vibrations can be such that the hydrogens move *in the plane* of the —CH_2— group, or *out of the plane* of the —CH_2— group. Furthermore, stretching and bending vibrations can be *symmetrical* or *unsymmetrical* with respect to a plane between the two vibrating hydrogens. The bending motions have been given very graphic names (scissoring, wagging, and so on) that describe the type of motion involved. Each of these motions occurs with a particular frequency and can have an associated peak in the IR spectrum (although some peaks are weak or absent for reasons to be considered later). A —CH_2— group in a typical organic molecule undergoes all of these motions simultaneously. That is, while the C—H bonds are stretching, they are also bending. The IR spectrum of nonane (Fig. 12.4) shows absorptions for both C—H stretching and C—H bending vibrations. The peak at 2920 cm^{-1} is due to the C—H stretching vibrations; the peaks at 1470 and 1380 cm^{-1} are due to various bending modes of both —CH_2— and CH_3— groups; and the peak at 720 cm^{-1} is due to a different bending mode, the —CH_2— rocking vibration. Notice that all of the bending vibrations absorb at lower wavenumber (and therefore lower energy) than the stretching vibrations.

average
position

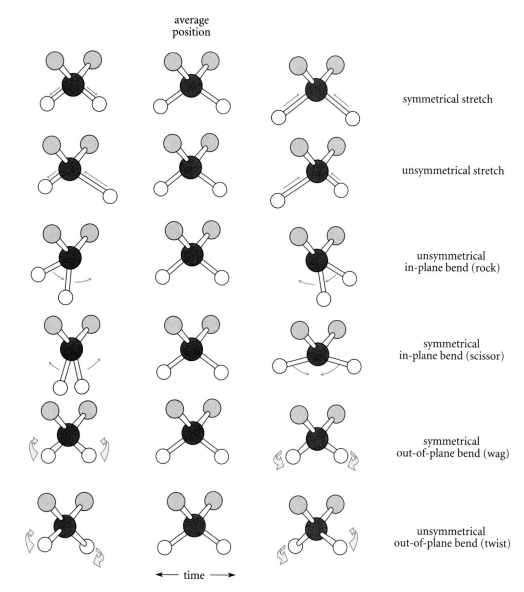

symmetrical stretch

unsymmetrical stretch

unsymmetrical
in-plane bend (rock)

symmetrical
in-plane bend (scissor)

symmetrical
out-of-plane bend (wag)

unsymmetrical
out-of-plane bend (twist)

⟵ time ⟶

Figure 12.7   *Normal vibrational modes of a typical —CH$_2$— group in an organic compound. Start at the center figure in each case and move left and right to see how the bonds change with time. The white atoms are hydrogens, the black atoms are carbons, and the gray groups are the other groups attached to carbon.*

B. Factors That Determine IR Absorption Intensity

The different peaks in an IR spectrum typically have very different intensities. Several factors affect absorption intensity. First, a greater number of molecules in the sample and more absorbing groups within a molecule give a more intense spectrum. Thus, a more concentrated sample gives a stronger spectrum than a less concentrated one, other things being equal. Similarly, at a given concentration, a compound such as nonane, which is rich in C—H bonds, has a stronger absorption for its C—H stretching vibrations than a compound of similar molecular mass with relatively few C—H bonds.

The dipole moment of a molecule also affects the intensity of an IR absorption. Spectroscopy theory shows that absorptions can be expected only for vibrations that cause a *change* in the molecular dipole moment. This explains why certain symmetrical (or nearly symmetrical) molecules lack IR absorptions that we otherwise might expect to observe. For example, the alkene 2,3-dimethyl-2-butene has a dipole moment of zero, and stretching the C=C bond does not impart a dipole moment to the molecule:

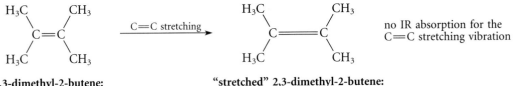

2,3-dimethyl-2-butene:
zero dipole moment

"stretched" 2,3-dimethyl-2-butene:
zero dipole moment

no IR absorption for the
C=C stretching vibration

Consequently, this alkene has no absorption in the 1600–1700 cm^{-1} region of the IR spectrum, the region in which C=C stretching absorptions occur for many other alkenes. (This compound does have other IR absorptions.) Note carefully that *the C=C stretching vibration occurs; it is simply not observed in the IR spectrum*. Molecular vibrations that occur but do not give rise to IR absorptions are said to be **infrared inactive**. (IR-inactive vibrations can be observed by a less common type of spectroscopy called *Raman spectroscopy*.) In contrast, any vibration that gives rise to an IR absorption is said to be **infrared active**.

Because the intensity of an IR absorption depends on the size of the dipole moment change that accompanies the corresponding vibration, IR absorptions differ widely in intensity. Chemists do not try to predict intensities; rather, they rely on collective experience to know which absorptions are weaker and which are stronger. Nevertheless, for symmetrical molecules with a zero dipole moment, we must be particularly aware of the possibility of IR-inactive vibrations that would be observed in less symmetrical molecules containing the same functional groups.

STUDY PROBLEM 12.1

Which one of the following molecular vibrations is infrared inactive? (a) the C=O symmetrical stretch of CO_2 (b) the C=O unsymmetrical stretch of CO_2. (See Fig. 12.7.)

Solution   First be sure you understand what is meant by the terms "symmetrical stretch" and "unsymmetrical stretch." These are defined by analogy to the C—H stretching vibrations in Fig. 12.7. In the symmetrical stretch, the two C=O bonds lengthen (or shorten) at the same time so that the molecule maintains its symmetry (the plane of symmetry is indicated by the dashed line):

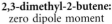

symmetrical stretch: maintains
the molecular symmetry

In the unsymmetrical stretch, one C=O bond shortens when the other lengthens:

$$O=C=O \rightleftharpoons O\!=\!\!=\!\!=\!\!=C=O$$
unsymmetrical stretch

Which of these vibrational modes results in a change of dipole moment? Since the CO_2 molecule is linear, the two C=O bond dipoles exactly oppose each other. Stretching a bond increases its bond dipole because the size of a bond dipole is proportional not only to the magnitudes of the partial charges at each end of the bond but also to *the distance by which the charges are separated*. Consequently, after the symmetrical stretch, both bond dipoles are increased; but since they exactly oppose each other, the dipole moment remains zero. Hence, the symmetrical stretching vibration is IR inactive.

In an unsymmetrical stretch, one C=O bond is reduced in length while the other is increased. Since the "long" C=O bond has a greater bond dipole than the "short" C=O bond, the two bond dipoles no longer cancel. Thus, the unsymmetrical stretch imparts a dipole moment to the CO_2 molecule. Consequently, this vibration is infrared active—it gives rise to an IR absorption.

PROBLEM

12.8 Which of the following vibrations should be infrared active and which should be infrared inactive (or nearly so)?

*(a) $CH_3CH_2CH_2CH_2$—C≡C—H   C≡C stretch

(b) $(CH_3)_2C=O$ C=O stretch

*(c) 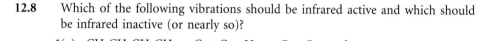 cyclohexane ring "breathing"
(simultaneous stretch of all C—C bonds)

(d) C_2H_5—C≡C—C_2H_5 C≡C stretch

*(e)

$$CH_3\!-\!\overset{+}{N}\!\overset{\ddot{O}:}{\underset{\ddot{O}:^-}{\big\backslash\!\!\big\backslash}}$$

symmetrical N—O stretch
(note resonance structures; p. 20)

(f) $(CH_3)_3C$—Cl $\overset{\backslash}{\underset{/}{-}}C$—Cl stretch

*(g) *trans*-3-hexene   (C=C stretch)

Functional-Group Infrared Absorptions

A. IR Spectra of Alkanes

The obvious structural features of alkanes are the carbon-carbon and carbon-hydrogen single bonds. The stretching of the carbon-carbon single bond is infrared inactive (or nearly so) because this vibration is associated with little or no change of the dipole moment. The stretching absorptions of alkyl C—H bonds are typically observed in the 2850–2960 cm^{-1} region. The peaks near 2920 cm^{-1} in the IR spectrum of nonane (Fig. 12.4) are examples of such absorptions. Various bending vibrations are also observed in the fingerprint region (1380 and 1470 cm^{-1} in nonane) and in the C—H bending region (720 cm^{-1} in nonane). Absorptions in these general regions can be expected for not only alkanes, but also any compounds that contain CH$_3$— and —CH$_2$— groups. Consequently, these absorptions are not often useful, but it is important to be aware of them so that they are not mistakenly attributed to other functional groups.

B. IR Spectra of Alkyl Halides

The carbon-halogen stretching absorption of alkyl chlorides, bromides, and iodides appear in the low-wavenumber end of the spectrum, but there are many interfering absorptions in this region. NMR spectroscopy and mass spectrometry are more useful than IR spectroscopy for determining the structures of alkyl halides.

C. IR Spectra of Alkenes

Unlike the spectra of alkanes and alkyl halides, the infrared spectra of alkenes are very useful and can help determine not only whether a carbon-carbon double bond is present, but also the branching pattern at the double bond. Typical alkene absorptions are given in Table 12.2. These fall into three categories: C=C stretching absorptions, =C—H stretching absorptions, and =C—H bending absorptions. The stretching vibration of the carbon-carbon double bond occurs in the 1640–1675 cm^{-1} range; the frequency of this absorption tends to be greater, and its intensity smaller, with increased branching at the double bond. The reason is the dipole moment effect discussed in the previous section. Thus, the C=C stretching absorption is clearly evident in the IR spectrum of 1-octene at 1642 cm^{-1} (Fig. 12.8a), but is virtually absent in the spectrum of the symmetrical alkene *trans*-3-hexene. The C=C stretching vibration is weak or absent even in unsymmetrical alkenes that have the same number of alkyl groups on each carbon of the double bond.

NMR spectroscopy is particularly useful for observing alkene hydrogens (Chapter 13). Nevertheless, a =C—H stretching absorption can often be used for confirmation of an alkene functional group. In general, the stretching absorptions of C—H bonds involving *sp*^{2}-hybridized carbons occur at wavenumbers *greater than* 3000 cm^{-1}, and the stretching absorptions of C—H bonds involving *sp*^{3}-hybridized carbons occur at wavenumbers *less than* 3000 cm^{-1}. Thus, 1-octene has a =C—H stretching absorption at 3080 cm^{-1} (Fig. 12.8a), and *trans*-3-hexene has a similar absorption which is barely discernible at 3030 cm^{-1} (Fig. 12.8b). The higher frequency of =C—H stretching absorptions is a manifestation of the bond-strength effect: bonds to *sp*^{2}-hybridized carbons are stronger (Table 5.3), and stronger bonds vibrate at higher frequencies.

The alkene =C—H bending absorptions that appear in the low-wavenumber region of the IR spectrum are in many cases very strong and can be used to determine the branching pattern at the double bond. The first three of these absorptions in Table 12.2—the ones for terminal vinyl, terminal methylene, and *trans*-alkene—are the most

Table 12.2   Important Infrared Absorptions of Alkenes

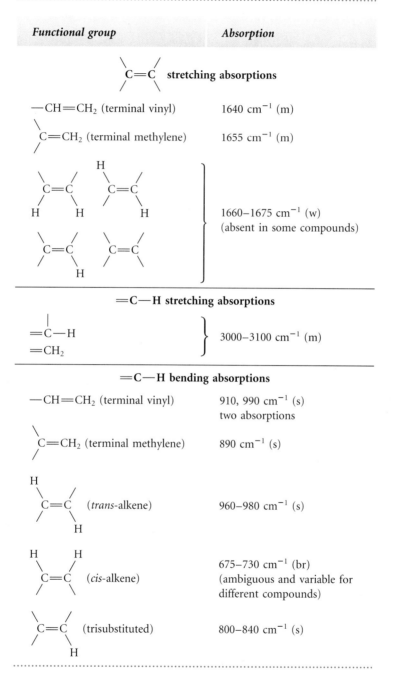

Functional group	Absorption
$C{=}C$ stretching absorptions	
$-CH{=}CH_2$ (terminal vinyl)	1640 cm^{-1} (m)
$C{=}CH_2$ (terminal methylene)	1655 cm^{-1} (m)
	1660–1675 cm^{-1} (w) (absent in some compounds)
$={=}C{-}H$ stretching absorptions	
$={=}C{-}H$ $={=}CH_2$	3000–3100 cm^{-1} (m)
$={=}C{-}H$ bending absorptions	
$-CH{=}CH_2$ (terminal vinyl)	910, 990 cm^{-1} (s) two absorptions
$C{=}CH_2$ (terminal methylene)	890 cm^{-1} (s)
(trans-alkene)	960–980 cm^{-1} (s)
(cis-alkene)	675–730 cm^{-1} (br) (ambiguous and variable for different compounds)
(trisubstituted)	800–840 cm^{-1} (s)

reliable. The 910 and 990 cm^{-1} terminal vinyl absorptions are illustrated in the IR spectrum of 1-octene (Fig. 12.8a), and the *trans*-alkene absorption is illustrated by the 965 cm^{-1} peak in the IR spectrum of *trans*-3-hexene (Fig. 12.8b).

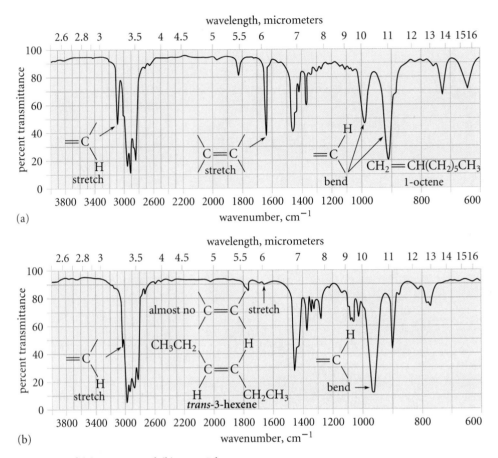

Figure 12.8   *IR spectra of (a) 1-octene and (b) trans-3-hexene.*

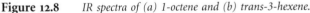

STUDY
PROBLEM
12.2

Each of three alkenes, *A*, *B*, and *C*, has the molecular formula C$_5$H$_{10}$, and each undergoes catalytic hydrogenation to yield pentane. Alkene *A* has IR absorptions at 1642, 990, and 911 cm^{-1}; alkene *B* has an IR absorption at 964 cm^{-1}, and no absorption in the 1600–1700 cm^{-1} region; and alkene *C* has absorptions at 1658 and 695 cm^{-1}. Identify the three alkenes.

Solution   In this problem, you can *write out all the possibilities* and then use the IR spectra to decide between them. The molecular formulas and the hydrogenation data show that the carbon chains of all the alkenes are unbranched and that all are isomeric pentenes. Hence, the *only* possibilities for compounds *A*, *B*, and *C* are the following:

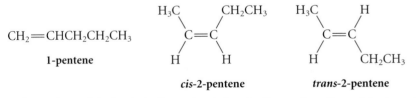

The IR absorptions of *A* clearly indicate that it is a 1-alkene; thus, it must be 1-pentene. The 964 cm^{-1} C—H bending absorption of *B* shows that it is *trans*-2-pentene. (Why is the C=C stretching vibration absent?) The remaining alkene *C* must be *cis*-2-pentene; the 1658 cm^{-1} C=C stretching absorption and the 695 cm^{-1} C—H bending absorption are consistent with this assignment.

Notice that you do not need the complete IR spectrum of each compound, but only the key absorptions, to solve this problem.

PROBLEM

*12.9   One of the spectra in Fig. 12.9 is that of *trans*-2-heptene and the other is that of 2-methyl-1-hexene. Which is which? Explain.

D. IR Spectra of Alcohols and Ethers

When O—H groups are not hydrogen-bonded to other groups, the O—H stretching absorption occurs near 3600 cm^{-1}. However, in most typical samples, O—H groups are strongly hydrogen-bonded and give a broad peak of moderate to strong intensity in

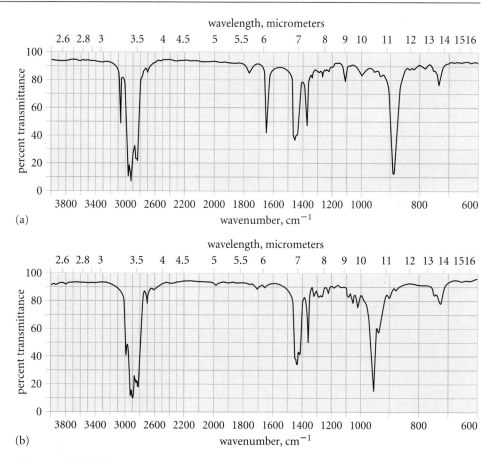

Figure 12.9   *IR spectra for Problem 12.9.*

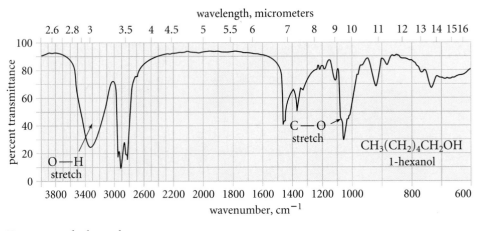

Figure 12.10 *IR spectrum of 1-hexanol.*

the 3200–3400 cm^{-1} region of the IR spectrum. Such an absorption, an important spectroscopic identifier for alcohols, is clearly evident in the IR spectrum of 1-hexanol (Fig. 12.10).

The other characteristic absorption of alcohols is a strong C—O stretching peak that occurs in the 1050–1200 cm^{-1} region of the spectrum; primary alcohols absorb near the low end of this range and tertiary alcohols near the high end. For example, this absorption occurs at about 1060 cm^{-1} in the spectrum of 1-hexanol. Because other functional groups (ethers, esters, carboxylic acids) also show C—O stretching absorptions in the same general region of the spectrum, the C—O stretching absorption is mainly used to support or confirm the presence of an alcohol diagnosed from the O—H absorption or from other spectroscopic evidence.

The most characteristic infrared absorption of ethers is the C—O stretching absorption, which, for the reasons just stated, is not very useful except for confirmation when an ether is already suspected from other data. For example, both dipropyl ether and an isomer 1-hexanol have strong C—O stretching absorptions near 1100 cm^{-1}.

The important IR absorptions of other functional groups are discussed in the respective chapters that cover these groups. In addition, a summary of key IR absorptions is given in Appendix II.

PROBLEMS

*12.10 Match the IR spectrum in Fig. 12.11 to one of the following three compounds: 2-methyl-1-octene, butyl methyl ether, or 1-pentanol.

*12.11 Explain why the IR spectra of some ethers have *two* C—O stretching absorptions. (*Hint:* See Fig. 12.7.)

*12.12 Explain why the O—H stretching absorption of the hydrogen-bonded —OH group is observed in the IR spectrum of an alcohol in concentrated solution, and why the O—H stretching absorption of the non-hydrogen-bonded —OH group is observed for an alcohol in dilute solution.

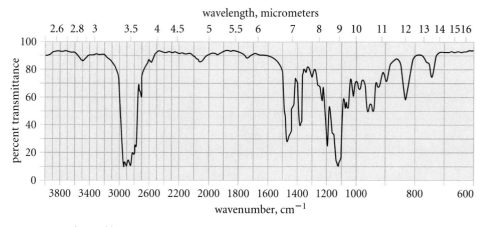

Figure 12.11   *IR spectrum for Problem 12.10.*

The Infrared Spectrometer

Study Guide Link:
12.2
FTIR Spectroscopy

Infrared spectrometers, the instruments with which IR spectra are obtained, are available in most chemical laboratories. A schematic diagram and description of an IR spectrometer are given in Fig. 12.12. In a conventional IR spectrometer, the slow scanning of wavelengths through the range of interest takes several minutes. In a newer type of IR spectrometer, called a *Fourier-transform infrared spectrometer* (FTIR spectrometer), the IR spectrum can be obtained in just a few seconds. The IR spectra in this text were obtained by FTIR.

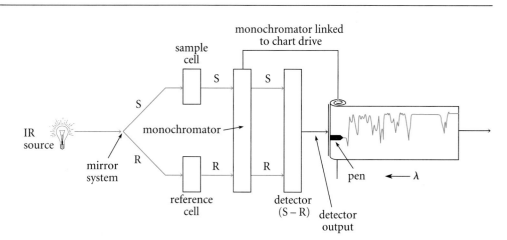

Figure 12.12   *Schematic diagram of an IR spectrometer. Infrared radiation, provided by the source, is split by a mirror into sample (S) and reference (R) beams. Beam S is passed through the sample, and beam R through a reference cell containing the same solvent as the sample cell. Each beam is transmitted to the monochromator, which can be a prism or grating. The monochromator sorts the transmitted radiation into separate wavelengths, and is slowly turned so that the detector receives light of varying wavelength over time. The detector measures the intensity of the radiation at each wavelength, and the signal from beam R is subtracted from that of beam S. This subtraction in principle removes interfering solvent or air absorptions. The chart recorder and the monochromator movement are synchronized so that the chart or pen moves along the horizontal axis as the wavelength changes. The detector output is sent to the pen and registered as a vertical deflection. The resulting trace is the IR spectrum.*

12.6 Introduction to Mass Spectrometry

In contrast to other spectroscopic techniques, mass spectrometry does not involve the absorption of electromagnetic radiation, but operates on a completely different principle. As the name implies, mass spectrometry is used to determine molecular masses, and it is the most important technique used for this purpose. It also has some use in determining molecular structure.

A. Production of a Mass Spectrum

The instrument used to obtain a mass spectrum is called a **mass spectrometer**. In this instrument a compound is vaporized in a vacuum and bombarded with an electron beam of high energy—typically, 70 electron-volts (more than 6,700 kJ/mol or 1,600 kcal/mol). Since this energy is much greater than the bond energies of chemical bonds, some fairly drastic things happen when a molecule is subjected to such conditions. One thing that happens is that an electron is ejected from the molecule. For example, if methane is treated in this manner, it loses an electron from one of the C—H bonds.

$$\overset{\overset{\displaystyle H}{..}}{\underset{\underset{\displaystyle H}{..}}{H:\overset{..}{C}:H}} + e^- \longrightarrow \overset{\overset{\displaystyle H}{..}}{\underset{\underset{\displaystyle H}{..}}{H:\overset{..}{C}{}^{+}H}} + 2e^- \tag{12.8}$$

The product of this reaction is sometimes abbreviated as follows:

$$\overset{\overset{\displaystyle H}{..}}{\underset{\underset{\displaystyle H}{..}}{H:\overset{..}{C}{}^{+}H}} \quad \text{abbreviated as} \quad CH_4{}^{+\cdot}$$

The symbol $^{+\cdot}$ means that the molecule is both a radical (a species with an unpaired electron) and a cation—a **radical cation**. The species $CH_4{}^{+\cdot}$ is called the *methane radical cation.*

Following its formation, the methane radical cation decomposes in a series of reactions called **fragmentation reactions**. In one such reaction, it loses a hydrogen *atom* to generate the methyl cation, a carbocation.

$$CH_4{}^{+\cdot} \longrightarrow {}^{+}CH_3 + H\cdot \tag{12.9}$$

$$\underset{\text{mass} = 16}{} \qquad \underset{\substack{\textbf{methyl cation} \\ \text{mass} = 15}}{}$$

The hydrogen atom carries the unpaired electron, and the methyl cation carries the charge. The process can be represented with the free-radical (fishhook) arrow formalism as follows:

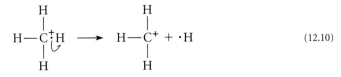

$$\tag{12.10}$$

Alternatively, the unpaired electron may remain associated with the carbon atom; in this case, the products of the fragmentation are a methyl radical and a proton.

$$CH_4^{+\bullet} \longrightarrow \ \bullet CH_3 \ + \ H^+ \tag{12.11}$$
$$\text{mass} = 16 \qquad \textbf{methyl radical} \quad \text{mass} = 1$$

Further decomposition reactions give fragments of progressively smaller mass. (Show how these occur by using the fishhook formalism.)

$$CH_3^+ \longrightarrow \ CH_2^{+\bullet} \ + \ H\bullet \tag{12.12a}$$
$$\text{mass} = 14$$

$$CH_2^{+\bullet} \longrightarrow \ CH^+ \ + \ H\bullet \tag{12.12b}$$
$$\text{mass} = 13$$

$$CH^+ \longrightarrow \ C^{+\bullet} \ + \ H\bullet \tag{12.12c}$$
$$\text{mass} = 12$$

Thus methane undergoes fragmentation in the mass spectrometer to give several positively charged **fragment ions** of differing mass: $CH_4^{+\bullet}$, CH_3^+, $CH_2^{+\bullet}$, CH^+, $C^{+\bullet}$, and H^+. In the mass spectrometer, these fragment ions are separated according to their **mass-to-charge ratio**, m/z (m = mass, z = the charge of the fragment). Since most ions formed in the mass spectrometer have unit charge, the m/z value can generally be taken as the mass of the ion. A **mass spectrum** is a graph of the relative amount of each ion (called the **relative abundance**) as a function of the ionic mass (or m/z). The mass spectrum of methane is shown in Figure 12.13. Note that only *ions* are detected by the mass spectrometer—neutral molecules and radicals do not appear as peaks in the mass spectrum. The mass spectrum of methane shows peaks at $m/z = 16, 15, 14, 13, 12,$ and 1, corresponding to the various ionic species that are produced from methane by electron ejection and fragmentation, as shown in Eqs. 12.8–12.12.

The mass spectrum can be determined for any molecule that can be vaporized in a high vacuum, and this includes most organic compounds. The utility of mass spectrometry is that (a) it can be used to determine the molecular mass of an unknown compound, and (b) it can be used to determine the structure (or a partial structure) of an unknown compound by an analysis of the fragment ions in the spectrum.

The ion derived from electron ejection before any fragmentation takes place is known as the **molecular ion**, and abbreviated M. *The molecular ion occurs at an m/z value equal to the molecular mass of the sample molecule.* Thus, in the mass spectrum of methane, the

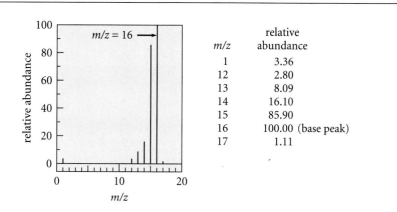

m/z	relative abundance
1	3.36
12	2.80
13	8.09
14	16.10
15	85.90
16	100.00 (base peak)
17	1.11

Figure 12.13 *Mass spectrum of methane.*

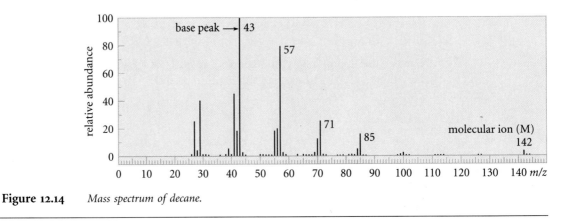

Figure 12.14   *Mass spectrum of decane.*

molecular ion occurs at $m/z = 16$. In the mass spectrum of decane (Fig. 12.14), the molecular ion occurs at $m/z = 142$. Except for peaks due to isotopes, discussed in Sec. 12.6B, the molecular ion peak is the peak of highest m/z in any ordinary mass spectrum.

The **base peak** is the ion of greatest relative abundance. The base peak is arbitrarily assigned a relative abundance of 100%, and the other peaks in the mass spectrum are scaled relative to it. In the mass spectrum of methane, the base peak is the same as the molecular ion, but in the mass spectrum of decane the base peak occurs at $m/z = 43$. In the latter spectrum and most others, the molecular ion and the base peak are different.

B. Isotopic Peaks

The mass spectrum of methane (Fig. 12.13) shows a small but real peak at $m/z = 17$, a mass one unit higher than the molecular mass. This peak is termed an M + 1 peak, because it occurs one mass unit higher than the molecular ion (M). This ion occurs because chemically pure methane is really a mixture of compounds containing the various isotopes of carbon and hydrogen.

$$\text{methane} = {}^{12}CH_4, \ {}^{13}CH_4, \ {}^{12}CDH_3, \text{ etc.}$$
$$m/z = \quad 16 \qquad 17 \qquad 17$$

The isotopes of several elements and their natural abundances are given in Table 12.3.

Possible sources of the $m/z = 17$ peak for methane are ${}^{13}CH_4$ and ${}^{12}CDH_3$. *Each isotopic compound contributes a peak with a relative abundance in proportion to its amount.* In turn, the amount of each isotopic compound is directly related to the natural abundance of the isotope involved. The relative abundance of a peak due to ${}^{12}C$ methane and the peak due to the presence of a ${}^{13}C$ isotope is then given by the following equation:

$$\text{relative abundance} = \left(\frac{\text{abundance of } {}^{13}C \text{ peak}}{\text{abundance of } {}^{12}C \text{ peak}}\right) \tag{12.13a}$$

$$= (\text{number of carbons}) \times \left(\frac{\text{natural abundance of } {}^{13}C}{\text{natural abundance of } {}^{12}C}\right)$$

$$= (\text{number of carbons}) \times \left(\frac{0.0110}{0.9890}\right)$$

$$= (\text{number of carbons}) \times 0.0111 \tag{12.13b}$$

Table 12.3   Exact Masses and Isotopic Abundances of Several Isotopes Important in Mass Spectrometry

Element	Isotope	Exact mass	Abundance, %
hydrogen	^{1}H	1.007825	99.985
deuterium	^{2}H	2.0140	0.015
carbon	^{12}C	12.0000	98.90
	^{13}C	13.00335	1.10
nitrogen	^{14}N	14.00307	99.63
	^{15}N	15.00011	0.37
oxygen	^{16}O	15.99491	99.759
	^{17}O	16.99913	0.037
	^{18}O	17.99916	0.204
fluorine	^{19}F	18.99840	100.
silicon	^{28}Si	27.97693	92.21
	^{29}Si	28.97649	4.67
	^{30}Si	29.97377	3.10
phosphorus	^{31}P	30.97376	100.
sulfur	^{32}S	31.97207	95.0
	^{33}S	32.97146	0.75
	^{34}S	33.96787	4.22
chlorine	^{35}Cl	34.96885	75.77
	^{37}Cl	36.96590	24.23
bromine	^{79}Br	78.91834	50.69
	^{81}Br	80.91629	49.31
iodine	^{127}I	126.90447	100.

Since methane has only one carbon, the $m/z = 17$ (M + 1) peak due to $^{13}CH_4$ is about 1.1% of the $m/z = 16$, or M, peak. A similar calculation can be made for deuterium.

$$\text{relative abundance} = (\text{number of hydrogens}) \times \left(\frac{\text{natural abundance of } ^2H}{\text{natural abundance of } ^1H} \right) \tag{12.14}$$

$$= (4) \times \left(\frac{0.00015}{0.99985} \right) = 0.0006$$

Thus, the CDH_3 naturally present in methane contributes 0.06% to the isotopic peak. Because the contribution of deuterium is negligible, ^{13}C is the major isotopic contributor to the M + 1 peak.

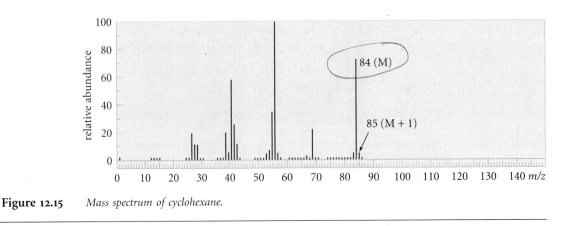

Figure 12.15   *Mass spectrum of cyclohexane.*

In a compound containing more than one carbon, the M + 1 peak is larger relative to the M peak because there is a 1.1% probability that *each carbon* in the molecule will be present as ^{13}C. For example, cyclohexane has six carbons, and the abundance of its M + 1 ion relative to that of its molecular ion should be 6(1.1) = 6.6%. In the mass spectrum of cyclohexane (Fig. 12.15), the molecular ion has a relative abundance of about 70%; that of the M + 1 ion is calculated to be (0.066)(70%) = 4.6%, which corresponds closely to the value observed. (With careful measurement, it is possible to use these isotopic peaks to estimate the number of carbons in an unknown compound; see Problem 12.37.) Not only the molecular ion peak, but also every other peak in the mass spectrum has isotopic peaks.

Several elements of importance in organic chemistry have isotopes with significant natural abundances. Table 12.3 shows that silicon has significant M + 1 and M + 2 contributions; sulfur has an M + 2 contribution; and the halogens chlorine and bromine have very important M + 2 contributions. In fact, the naturally occurring form of the element bromine consists of about equal amounts of ^{79}Br and ^{81}Br. The mixture of isotopes leaves a characteristic trail in the mass spectrum that can be used to diagnose the presence of the element.

Consider, for example, the mass spectrum of bromomethane, shown in Fig. 12.16. The peaks at m/z = 94 and 96 result from the contributions of the two bromine isotopes to the molecular ion. They are in the relative abundance ratio 100 : 98 = 1.02, which is

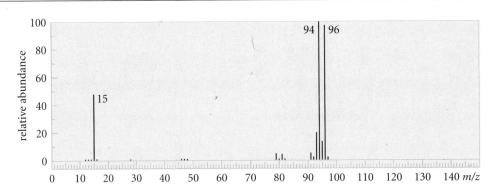

Figure 12.16   *Mass spectrum of bromomethane. Notice the two molecular ions of nearly equal abundance that result from the presence of the two isotopes ^{79}Br and ^{81}Br.*

in good agreement with the ratio of the relative natural abundances of the bromine isotopes (Table 12.2). This double molecular ion is a dead giveaway for a compound containing a single bromine. Notice that along with each major isotopic peak is a smaller isotopic peak one mass unit higher. These peaks are due to the isotope ^{13}C present naturally in bromomethane. For example, the $m/z = 95$ peak corresponds to bromomethane containing only ^{79}Br and one ^{13}C. The $m/z = 97$ peak arises from methyl bromide that contains only ^{81}Br and one ^{13}C.

Although isotopes such as ^{13}C and ^{18}O are normally present in small amounts in organic compounds, it is possible to synthesize compounds that are selectively enriched with these and other isotopes. Isotopes are especially useful because they provide specific labels at particular atoms without changing their chemical properties. One use of such compounds is to determine the fate of specific atoms in deciding between two mechanisms. (See, for example, Problem 11.18, Chapter 11.) Another use is to provide nonradioactive isotopes for biological metabolic studies (studies that deal with the fates of chemical compounds when they react in biological systems). When a compound has been isotopically enriched, isotopic peaks are much larger than they are normally. Mass spectrometry is used to measure quantitatively the amount of such isotopes present in labeled compounds.

PROBLEMS

*12.13   The mass spectrum of tetramethylsilane, $(CH_3)_4Si$, has a base peak at $m/z = 73$. Calculate the relative abundances of the isotopic peaks at $m/z = 74$ and 75.

12.14   From the information in Table 12.3, predict the appearance of the molecular ion peak(s) in the mass spectrum of chloromethane. (Assume that the molecular ion is the base peak.)

C. Fragmentation Mechanisms

The molecular ion is formed by loss of an electron. If this ion is stable, it decomposes slowly and is detected by the mass spectrometer as a peak of large relative abundance. If this ion is less stable, it decomposes, in some cases completely, into smaller pieces, which are then detected as smaller ions, called **fragment ions**. The relative abundances of the various fragments in a mass spectrum depend on their relative lifetimes, that is, the relative rates at which they break apart into smaller fragments. Typically the most stable ions have the longest lifetimes, and are detected as the largest peaks in a mass spectrum.

The fragment ions produced in the mass spectrometer are literally pieces of the whole molecule. Just as a picture can be reconstructed from a jigsaw puzzle, the structure of a molecule can in principle be reconstructed from its mass spectrum.

The simplest way to analyze a mass spectrum is to think of fragment ions as coherent pieces that result from the breaking of chemical bonds in the molecular ion. This approach successfully accounts for the mass spectrum of methane, analyzed in Sec. 12.6A, as well as the mass spectrum of decane (Fig. 12.14). Decane undergoes fragmentation at several of its carbon-carbon bonds.

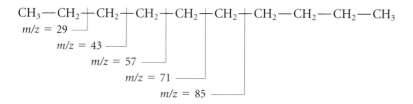

For example, the $m/z = 71$ fragment is formed in the following way:

Electron ejection to yield molecular ion:

$$CH_3(CH_2)_3—CH_2\!:\!CH_2—(CH_2)_3CH_3 \xrightarrow{-e^-} CH_3(CH_2)_3—CH_2\!\overset{+}{\cdot}CH_2—(CH_2)_3CH_3 \qquad (12.15a)$$

Fragmentation:

$$CH_3(CH_2)_3—CH_2\!\overset{+}{\underset{\curvearrowleft}{\cdot}}CH_2—(CH_2)_3CH_3 \longrightarrow CH_3(CH_2)_3—CH_2^+ + \cdot CH_2—(CH_2)_3CH_3 \qquad (12.15b)$$
$$m/z = 71$$

(Write a mechanism for formation of one or two other fragments.) Notice that the different fragments are not formed with the same relative abundances. Furthermore, some possible fragment peaks are weak or missing. Thus, there is no $m/z = 15$ fragment for $^+CH_3$, and the peaks corresponding to fragments at m/z greater than 85 are very weak. As illustrated by the following study problem, the reason for the predominance of certain fragments follows from their relative stabilities.

STUDY PROBLEM 12.3

The base peak in the mass spectrum of 2,2,5,5-tetramethylhexane is at $m/z = 57$, which corresponds to a composition C_4H_9. (a) Suggest a structure for the fragment that accounts for this peak. (b) Offer a reason why this fragment is so abundant. (c) Give a fragmentation mechanism that shows the formation of this fragment.

Solution   The first step is to draw the structure of 2,2,5,5-tetramethylhexane: $(CH_3)_3C—CH_2CH_2—C(CH_3)_3$.

(a)   A fragment with the composition C_4H_9 could be a *tert*-butyl cation formed by splitting the compound at the bond to either of the *tert*-butyl groups:

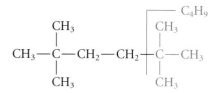

(b)   The most abundant peaks in the mass spectrum result from the most stable cationic fragments. Because a *tert*-butyl cation is a relatively stable carbocation (it is tertiary), it is formed in relatively high abundance.

(c)   To form this cation, one electron is ejected from the C—C bond, and the compound fragments so that the remaining electron remains on the methylene carbon (see Eq. 12.15a,b):

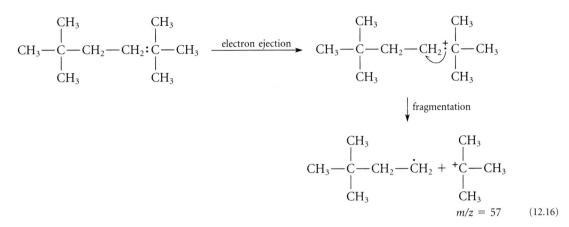

$$CH_3 - C - CH_2 - CH_2 - \overset{.}{C}H_2 \ + \ ^+C - CH_3$$

$$m/z = 57 \quad (12.16)$$

Notice that fragmentation might have occurred at the same bond so that the unpaired electron remains associated with the *tert*-butyl group and a primary carbocation is formed. (In other words, a *more stable* free radical and a *less stable* carbocation would be formed.) That this is not the major mode of fragmentation demonstrates that carbocation stability is more important than free-radical stability in determining fragmentation patterns.

The mass spectrum of *sec*-butyl isopropyl ether, shown in Fig. 12.17, illustrates two other important modes of fragmentation. The peaks at $m/z = 57$ and $m/z = 43$ come from obvious pieces of the molecule:

$$m/z = 57 \qquad m/z = 43$$
$$CH_3 - CH + O + CH(CH_3)_2$$
$$|$$
$$C_2H_5$$

Notice that the fragmentations that give these peaks occur at the ether oxygen. The first step of the fragmentation mechanism is loss of an electron from one of the unshared pairs on oxygen:

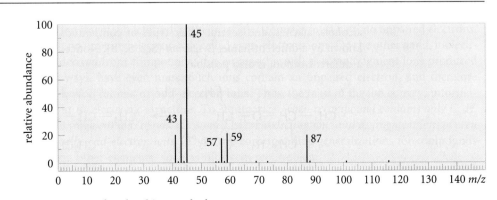

Figure 12.17   *Mass spectrum of sec-butyl isopropyl ether.*

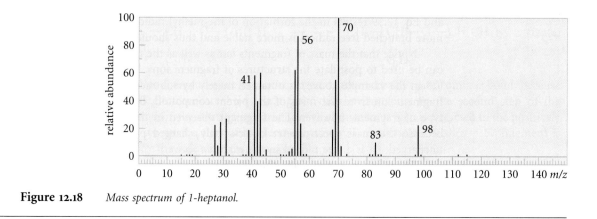

Figure 12.18   *Mass spectrum of 1-heptanol.*

and can be accounted for by the loss of water. A mechanism for this fragmentation begins with loss of an electron from one of the unshared pairs of oxygen. A hydrogen (color) then migrates from the β-carbon to this oxygen, and water leaves the molecule.

$$CH_3(CH_2)_4\!-\!\overset{\cdot}{C}H\!-\!\overset{+}{C}H_2 \quad\quad (12.21)$$

odd-electron ion, $m/z = 98$
(undergoes further fragmentation)

Hydrogen-atom transfer followed by loss of a stable neutral molecule is a very common mechanism for the formation of odd-electron ions. This mode of fragmentation, for example, is common in primary alcohols, alkyl halides, and other compounds.

PROBLEM

*12.17   The mass spectrum of 2-chlorobutane shows large and almost equally intense peaks at $m/z = 57$ and $m/z = 56$.
(a)   Which of these fragments is an even-electron ion? (The presence of chlorine in the molecular ion does not affect the generalizations above.)
(b)   What stable neutral molecule can be lost to give the odd-electron ion?
(c)   Suggest a mechanism for the origin of each of these fragments.

E. Identifying the Molecular Ion

The molecular ion is the most important peak in the mass spectrum because it provides the molecular mass of the molecule under study and thus is the basis for calculating the losses involved in fragmentation. However, in some mass spectra, the abundance of the molecular ion is so small that it is undetectable. Thus there is a problem: suppose you are dealing with a compound of unknown molecular mass. How do you know whether the peak of highest mass is due to the molecular ion or a fragment ion? Although there is no general answer to this question, some simple considerations can provide useful clues.

All compounds containing only the elements C, H, and O have even molecular masses, since carbon and oxygen have even atomic masses and there must be an even number of hydrogen atoms. Such compounds must therefore have a molecular ion of even mass. (Corresponding generalizations are possible for compounds containing other elements.) For example, suppose an alcohol *A* of unknown structure contains only C, H, and O, and has a peak of highest mass in its mass spectrum at $m/z = 87$. This peak cannot qualify as the molecular ion because of its odd mass.

Another test for the molecular ion involves the observed losses that are calculated from the candidate peak under the *assumption* that it is the molecular ion. Mass losses of 4–14, 21–25, 33, 37, and 38 from the molecular ion are relatively rare because combinations of atoms that give these losses simply do not exist. For example, in the mass spectrum of alcohol *A* in the previous paragraph, the base peak occurs at $m/z = 73$, fourteen mass units below the peak of highest mass at $m/z = 87$. Even if you did not know that this compound contains only C, H, and O, the loss of 14 mass units would cast serious doubt that the peak at $m/z = 87$ is the molecular ion.

Another way to determine the mass of the molecular ion is by chemical derivatization. For example, if the mass spectrum of an alcohol shows no molecular ion, a methyl ether that could be readily prepared from the alcohol might give one.

Yet another method of determining the mass of the molecular ion is to use a different ionization technique. The mass spectra discussed in this chapter were all obtained using the electron-bombardment technique discussed in Sec. 12.6A. Such spectra are termed **EI spectra**. (EI means "electron-impact.") Another ionization technique involves treating a sample in the gas phase with a proton source. In this technique, the sample molecule is protonated at its most basic site to give a conjugate-acid cation. Because a proton is added, the molecular ion appears at one mass unit higher than the molecular mass. This technique is called *chemical ionization*, and the resulting spectra are termed **CI spectra**. Much higher percentages of molecular ions, as well as different fragmentation patterns containing fewer peaks, are observed in CI spectra.

PROBLEM

*12.18   The alcohol *A* used as an example in this section gives a methyl ether with a strong M + 1 peak in its CI spectrum at $m/z = 117$. Compound *A*, known from other evidence to be a tertiary alcohol, has prominent fragments in its EI mass spectrum at $m/z = 87$ and $m/z = 73$ (base peak). Propose a structure for alcohol *A*.

F. The Mass Spectrometer

A diagram of a conventional mass spectrometer is shown in Fig. 12.19. The sample is ionized and the fragment ions are formed, detected, and analyzed within the instrument. A modern mass spectrometer is an extremely sensitive instrument and can readily produce a mass spectrum from amounts of material in the range of micrograms (10^{-6} g) to picograms (10^{-12} g). For this reason the instrument is very useful for the analysis of materials available in only trace quantities. It has played a key role in such projects as the analysis of drug levels in blood serum and the elucidation of the structures of insect pheromones (Sec. 14.9) that are available only in minuscule amounts.

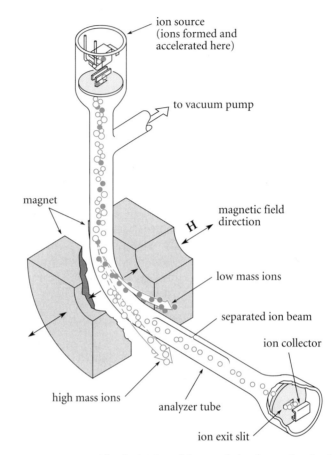

Figure 12.19   *Diagram of a mass spectrometer. After ionization of the sample by electron bombardment, the ions are accelerated by a high voltage and are passed into a magnetic field **H** along a path perpendicular to the field. The field bends the paths of the ions; the paths of lower-mass ions are bent more than those of higher-mass ions. As the field is progressively increased, ions of increasingly higher mass attain exactly the correct path to enter the detection slit.*

One of the operating characteristics of a mass spectrometer is its *resolution*, that is, how well it separates ions of different mass. A relatively simple mass spectrometer readily distinguishes, over a total m/z range of several hundred, ions that differ in mass by one unit. More complex mass spectrometers, called *high-resolution mass spectrometers*, can resolve ions that are separated in mass by only a few thousandths of a mass unit. Why is such high resolution useful? Suppose an unknown compound has a molecular ion at $m/z = 124$. Two possible formulas for this ion are $C_8H_{12}O$ and C_9H_{16}. Both formulas have the same **nominal mass** (mass to the nearest whole number). However, if the **exact mass** (the mass to four or more decimal places) is calculated for each formula (using the values of the most abundant isotopes in Table 12.2), then different results are obtained:

$C_8H_{12}O$, exact mass: 124.0888
C_9H_{16}, exact mass: 124.1252

The difference of 0.0364 mass units is easily resolved by a high-resolution mass spectrometer. Computers used with such instruments can be programmed to work backwards from the exact mass and provide *an elemental analysis of the molecular ion* (and therefore the compound of interest) *as well as the elemental analysis of each fragment in the mass spectrum*! Because a high-resolution mass spectrometer and its associated computer and other accessories can cost more than $500,000, it is generally shared by a large number of researchers.

Before a compound can be analyzed by mass spectrometry, it must be vaporized. This presents a difficult problem for large molecules that have negligible vapor pressure. Research in mass spectrometry has focused on novel ways to produce ions in the gas phase from large nonvolatile molecules, many of which are of biological interest. In one technique, called *fast-atom bombardment* (FAB), compounds in solution are converted directly into gas-phase ions by subjecting them to a beam of heavy atoms (such as xenon, argon, or cesium) that have been accelerated to high velocities. Ions with molecular masses up to about 5,000 can be produced by this technique. A method developed even more recently, called *electrospray mass spectrometry*, involves atomizing a solution of the sample within highly charged droplets. Samples treated in this way take on more than one charge. Since the resolution of a mass spectrometer is based not on mass alone, but on *m/z*, the mass-to-charge ratio, increasing the sample charge (*z*) increases the mass (*m*) that can be resolved. Electrospray mass spectrometry has been used to analyze molecules with molecular masses greater than 100,000.

STUDY GUIDE LINK:
12.3
*The Mass
Spectrometer*

KEY IDEAS IN CHAPTER 12

Spectroscopy deals with the interaction of matter and electromagnetic radiation. Electromagnetic radiation is characterized by its energy, wavelength, and frequency, which are interrelated by Eqs. 12.1, 12.3, and 12.7.

Infrared spectroscopy deals with the absorption of infrared radiation by molecular vibrations. An infrared spectrum is a plot of the infrared radiation transmitted through a sample as a function of the wavenumber or wavelength of the radiation.

The wavenumber of an absorption is greater for vibrations involving stronger bonds and smaller atomic masses.

The intensity of an absorption increases with the number of absorbing groups in the sample and the size of the dipole moment change that occurs in the molecule when the vibration occurs. Absorptions that result in no dipole moment change are infrared inactive.

The infrared spectrum provides information about the functional groups present in a molecule. The $=$C—H stretching and bending absorptions and the C$=$C stretching absorption are very useful for the identification of alkenes. The O—H stretching absorption is diagnostic for alcohols.

(continues)

In electron-impact mass spectrometry, a molecule loses an electron to form a radical cation called the molecular ion, which then decomposes to fragment ions. The relative abundances of the fragment ions are recorded as a function of their mass-to-charge ratios *m/z*, which, for most ions, equal their masses. Both molecular masses and partial structures can be derived from the masses of these ionic fragments.

Associated with each peak in a mass spectrum are other peaks at higher mass that arise from the presence of isotopes at their natural abundance. Such isotopic peaks are particularly useful for diagnosing the presence of elements that consist of more than one isotope with high natural abundance, such as chlorine and bromine.

Ionic fragments are of two types: even-electron ions, which contain no unpaired electrons; and odd-electron ions, which contain an unpaired electron. Even-electron ions are formed by such processes as α cleavage, inductive cleavage, and direct fragmentation at a σ bond. An important pathway for the formation of odd-electron ions is hydrogen transfer followed by loss of a small molecule such as water or hydrogen halide.

ADDITIONAL PROBLEMS

*12.19   List the factors that determine the wavenumber of an infrared absorption.

12.20   List two factors that determine the intensity of an infrared absorption.

12.21   Indicate how you would carry out each of the following chemical transformations. What are some of the changes in the infrared spectrum that could be used to indicate whether the reaction has proceeded as indicated? (Your answer can include disappearance as well as appearance of IR absorptions.)

*(a)

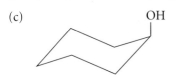

(b)   1-methylcyclohexene → methylcyclohexane
*(c)   1-hexanol → hexyl methyl ether   (d)   ethylene oxide → 1-butanol

*12.22   Which of the molecules in each of the following pairs should have identical IR spectra, and which should have different IR spectra (if only slightly different)? Explain your reasoning carefully.
(a)   3-pentanol and (±)-2-pentanol
(b)   (*R*)-2-pentanol and (*S*)-2-pentanol

(c)

and

*12.23   Match each of the IR spectra in Fig. 12.20 to one of the following compounds. (Notice that there is no spectrum for two of the compounds.)
(a) 1,5-hexadiene   (b) 1-methylcyclopentene   (c) 1-hexen-3-ol
(d) dipropyl ether   (e) *trans*-4-octene   (f) cyclohexane
(g) 3-hexanol

12.24   A former theological student, Heavn Hardley, has turned to chemistry and, during his eighth year of graduate study, has carried out the following reaction:

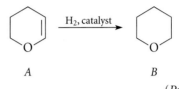

A   *B*

(*Problem 12.24 continues*)

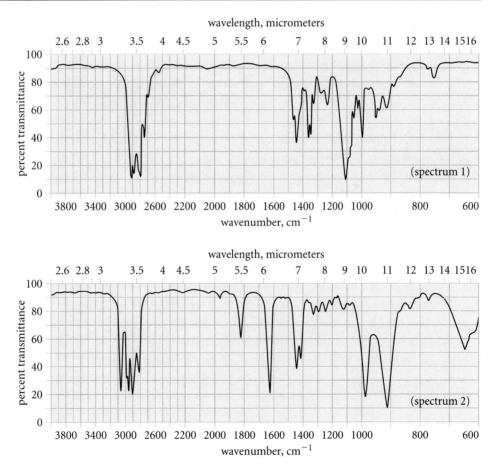

Figure 12.20   *IR spectra for Problem 12.23. (Figure continues on next page.)*

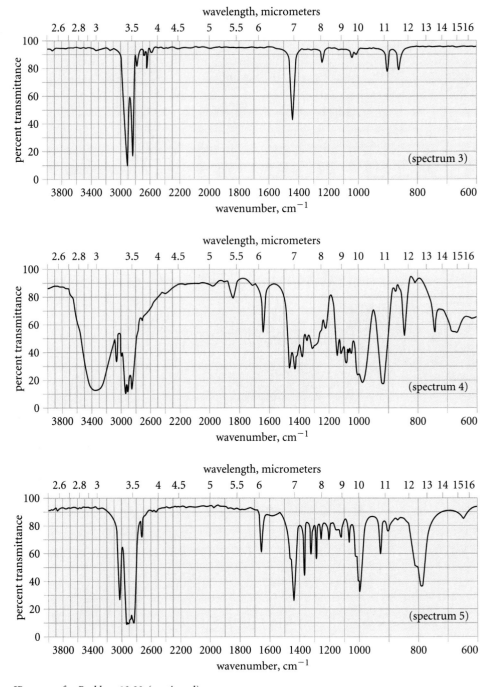

Figure 12.20 *IR spectra for Problem 12.23 (continued).*

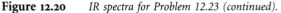

Unfortunately, Hardley thinks he may have mislabeled his samples of *A* and *B*, but has wisely decided to take an IR spectrum of each sample. The spectra are reproduced in Fig. 12.21. Which sample goes with which spectrum? How do you know?

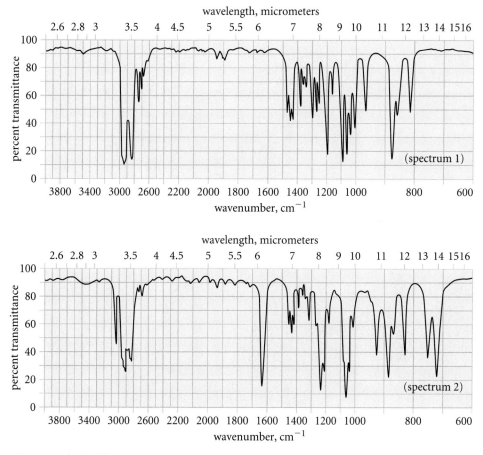

Figure 12.21   *IR spectra for Problem 12.24.*

*12.25   Given the stretching frequencies for the C—H bonds shown in color below, arrange the corresponding bonds in order of increasing strength. Explain your reasoning.

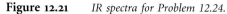

$$RCH=\overset{\overset{\displaystyle H}{|}}{C}H \qquad R\overset{\overset{\displaystyle H}{|}}{C}H_2 \qquad RC\equiv C-H$$
$$3080 \text{ cm}^{-1} \qquad 2850 \text{ cm}^{-1} \qquad 3300 \text{ cm}^{-1}$$

12.26   Arrange the following bonds in order of increasing stretching frequencies, and explain your reasoning.

$$C=C \qquad C\equiv C \qquad C=O \qquad C-C$$

*12.27   Give an upper and lower limit on the wavenumber range in which you would expect to find the C—F stretching absorption. Explain your reasoning carefully.

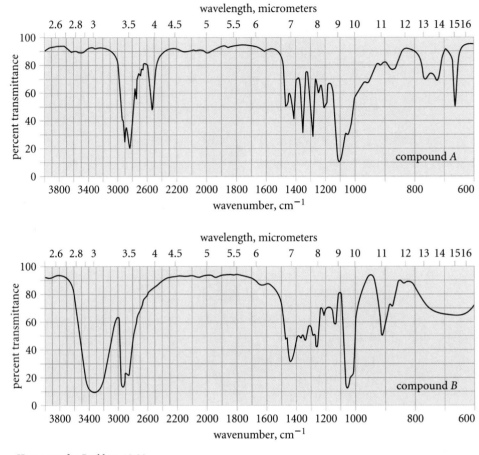

Figure 12.22   *IR spectra for Problem 12.28.*

*12.28   (a) Explain why the S—H stretching absorption in the IR spectrum of a thiol is less intense and occurs at lower frequency (2550 cm^{-1}) than the O—H stretching absorption of an alcohol.

(b) Two unlabeled bottles *A* and *B* contain liquids. Laboratory notes suggest that one compound is $(HSCH_2CH_2)_2O$ and the other is $(HOCH_2CH_2)_2S$. The IR spectra of the two compounds are given in Fig. 12.22. Identify *A* and *B*.

*12.29   (a) You have found in the laboratory two liquids *C* and *D* in unlabeled bottles that both smell like chloroform, but you suspect that one is deuterated chloroform ($CDCl_3$) and the other is ordinary chloroform ($CHCl_3$). Unfortunately, the mass spectrometer is not operating because the same person who failed to label the bottles has been recently using the mass spectrometer! From the IR spectra of the two compounds, shown in Fig. 12.23, indicate which compound is which. Explain.

(b) How would these compounds be distinguished by mass spectrometry?

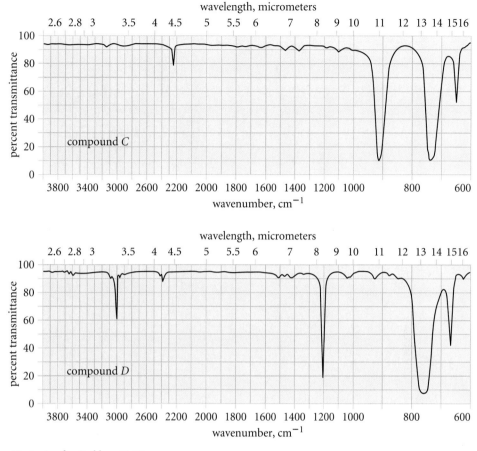

Figure 12.23   *IR spectra for Problem 12.29.*

12.30   Suppose a time machine were to take you back to the earliest days of IR spectroscopy. Imagine that you are the very first person to measure the IR spectrum of a 1-alkene, in this case, 2-methyl-1-pentene. From theory, you suspect that the sharp absorption peak at 3090 cm^{-1} is the $=$C$-$H stretching absorption, but you wish to gather experimental evidence to support your idea. Assuming you could prepare the necessary compounds, what isotopically substituted analog(s) of 2-methyl-1-pentene could be used to prove your point? Explain your answer.

12.31   Rationalize the indicated fragments in the mass spectrum of each of the following molecules by proposing a structure of the fragment and a mechanism by which it is produced.

    *(a)   3-methyl-3-hexanol, *m/z* = 73

(*Problem 12.31 continues*)

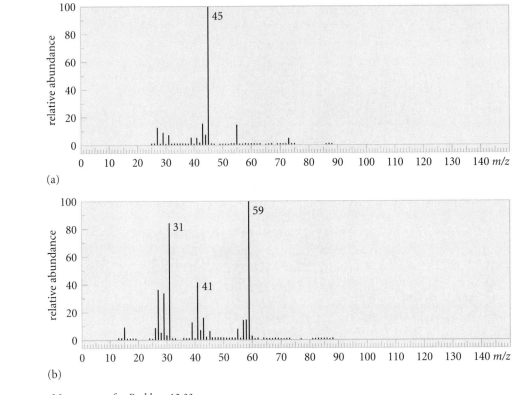

Figure 12.24   *Mass spectra for Problem 12.33.*

(b)

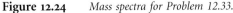

$$CH_3CH_2CH_2-\underset{\underset{CH_2CH_3}{|}}{\overset{\overset{CH_3}{|}}{C}}-\ddot{N}H_2, \; m/z = 72$$

*(c)   neopentane, $m/z = 57$ *(d) 1-pentanol, $m/z = 70$

12.32   Suggest structures for the following *neutral* molecules commonly lost in mass spectral fragmentation.
*(a)   mass $= 28$ from a compound containing only C and H
(b)   mass $= 18$ from a compound containing oxygen
*(c)   mass $= 36$ from a compound with an M $+$ 2 peak about one-third the size of the molecular ion.

*12.33   A chemist, Ilov Boronin, carried out a reaction of *trans*-2-pentene with BH$_3$ in THF followed by treatment with H$_2$O$_2$/OH$^-$. Two products were separated and isolated. Desperate to know their structures, Ilov took his compounds to the spectroscopy laboratory and found that only the mass spectrometer was operating. The mass spectra of the two products are given in Fig. 12.24. Suggest structures for the compounds, and indicate which mass spectrum goes with which compound.

12.34 Figure 12.25 shows the mass spectra of two isomeric ethers, 2-methoxybutane and 1-methoxybutane. Match each compound with its spectrum.

*12.35 Predict the relative intensities of the three peaks in the mass spectrum of dichloromethane at $m/z = 84$, 86, and 88.

12.36 Explain why the mass spectrum of dibromomethane has three peaks at $m/z = 172$, 174, and 176 in the approximate relative abundances $1 : 2 : 1$.

12.37 From the molecular masses and the relative intensities of their M and M + 1 peaks, suggest molecular formulas for the following compounds. (*Hint:* Assume the major contributor to the M + 1 peak is ^{13}C, and use the relative abundance of the M + 1 peak to calculate the number of carbons.)
   *(a) M ($m/z = 86$; 19%), M + 1 (1.06%); contains C, H, and O.
   (b) M ($m/z = 82$; 37%), M + 1 (2.5%); contains C and H.

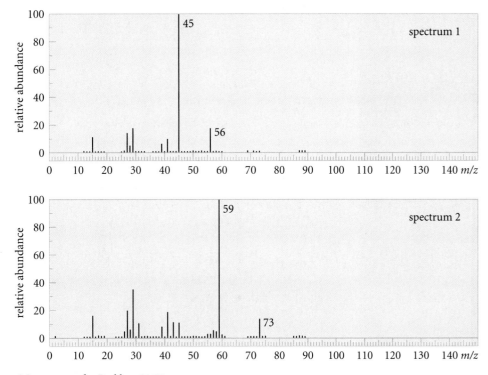

Figure 12.25   *Mass spectra for Problem 12.34.*

12.38   Suggest a structure for each of the ions corresponding to the following peaks in the mass spectrum of ethyl bromide, and give a mechanism for the formation of each ion. (The numbers in parentheses are the relative abundances.)
*(a) $m/z = 110$ (98%)   (b) $m/z = 108$ (100%)   *(c) $m/z = 81$ (5%)
(d) $m/z = 79$ (5%)   *(e) $m/z = 29$ (61%)   (f) $m/z = 28$ (25%)
*(g) $m/z = 27$ (53%)

12.39   A compound contains carbon, hydrogen, oxygen, and *one* nitrogen. Classify each of the following fragment ions derived from this compound as an odd-electron or an even-electron ion. Explain.
*(a)   the molecular ion
*(b)   a fragment ion of even mass containing nitrogen
(c)   a fragment ion of odd mass containing nitrogen

*12.40   (a)   Explain why ionization of a π electron requires less energy than ionization of a σ electron.
(b)   Using the ionization of a π electron to begin your fragmentation mechanism, account for the presence of a base peak at $m/z = 41$ in the mass spectrum of 1-heptene.
(c)   Compare the molecular ion of 1-heptene with the ion formed by the loss of water from 1-heptanol (Fig. 12.18, Eq. 12.21). Explain why the mass spectra of 1-heptanol and 1-heptene are nearly identical.

13

Nuclear Magnetic Resonance Spectroscopy

Infrared spectroscopy can be used to determine the functional groups present in a compound, and mass spectrometry provides the masses of a molecule and its coherent fragments. With rare exceptions, however, neither of these techniques gives enough information to define a complete structure. Another form of spectroscopy, *nuclear magnetic resonance* (NMR) spectroscopy, enables the chemist to probe molecular structure in much greater detail. Using NMR, sometimes in conjunction with other forms of spectroscopy, but often by itself, a chemist can usually determine a complete molecular structure in a very short time. In the period since its introduction in the late 1950s, NMR spectroscopy has revolutionized organic chemistry. This chapter presents the basic principles of NMR spectroscopy and shows how it is used in structure determination.

13.1 Introduction to NMR Spectroscopy

One of the most important uses of NMR spectroscopy is to examine the *hydrogens* in organic compounds. This type of NMR spectroscopy is based on the *magnetic properties of the hydrogen nucleus* that result from its *nuclear spin*.

Just as electrons have two allowed spin states, designated by the quantum numbers $+\frac{1}{2}$ and $-\frac{1}{2}$, some *nuclei* also have spin. The hydrogen nucleus ^1H, that is, the proton, has a nuclear spin that also can assume either of two values, designated by quantum numbers $+\frac{1}{2}$ and $-\frac{1}{2}$.

The physical significance of nuclear spin is that the nucleus acts like a tiny magnet. This means that hydrogen nuclei, whether alone or in a compound, respond to a magnetic field. When a compound containing hydrogens is placed in a magnetic field, its hydrogen nuclei become magnetized.

To see what causes this magnetization, let's represent the hydrogen nuclei in a chemical sample with arrows indicating their magnetic ("north-south") polarity. In the absence of a magnetic field, the nuclear magnetic poles are oriented randomly. After a magnetic field is applied, the magnetic poles of nuclei with spin of $+\frac{1}{2}$ are oriented parallel to the magnetic field, and those of nuclei with spin of $-\frac{1}{2}$ are oriented antiparallel to the field.

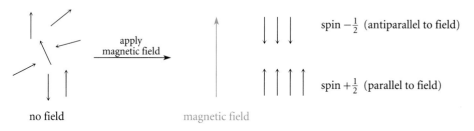

The most important effect of the magnetic field for NMR is how it affects the energies of the two spin states. In the absence of a field, the two spin states have the same energy. But when a magnetic field is applied, the two spin states have *different* energies: the $+\frac{1}{2}$ spin state has lower energy than the $-\frac{1}{2}$ spin state. In addition, the populations of the two spin states are in rapid equilibrium. The spin equilibrium, like a chemical equilibrium, favors the state with lower energy (Eq. 3.30). Thus, after the field is applied, more protons have spin $+\frac{1}{2}$ than spin $-\frac{1}{2}$. For every *million* protons in a sample, the state with spin $+\frac{1}{2}$ typically contains an excess of only *ten to twenty* protons—certainly not a large difference, but enough to be detected. Because the magnetic poles of the nuclei with spin $+\frac{1}{2}$ are aligned with the magnetic field, and because there are more of these nuclei, the sample has a small net magnetization in the direction of the field.

The energy difference ΔE between the two spin states for a given proton depends on the intensity of the magnetic field $\mathbf{H_p}$ at the proton. The larger the field, the greater is ΔE. This idea, an important one for NMR spectroscopy, is illustrated in Fig. 13.1.

Here is where things stand: Molecules of a sample are situated in a magnetic field; each nucleus is in one of two spin states that differ in energy by an amount ΔE; and

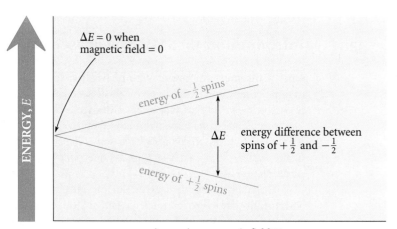

Figure 13.1   *Effect of increasing magnetic field at a proton on the energy difference between its $+\frac{1}{2}$ and $-\frac{1}{2}$ spin states. Notice that the two spin states have identical energies when the field is absent, and that the energy difference between the two spin states grows with increasing field.*

more nuclei have spin $+\frac{1}{2}$. If the sample is now subjected to electromagnetic radiation of energy E_0 exactly equal to ΔE, this energy is *absorbed* by some of the nuclei in the $+\frac{1}{2}$ spin state. The absorbed energy causes these nuclei to invert or "flip" their spins and assume a more energetic state with spin $-\frac{1}{2}$.

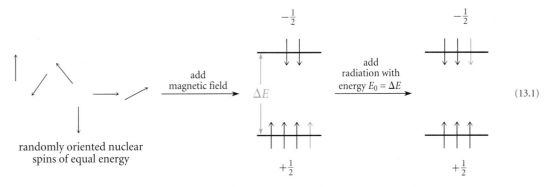

randomly oriented nuclear
spins of equal energy

(13.1)

This energy absorption by nuclei in a magnetic field is termed **nuclear magnetic resonance** and can be detected in a type of absorption spectrophotometer called a *nuclear magnetic resonance spectrometer*, or **NMR spectrometer** (Sec. 13.10). The study of this absorption is called **NMR spectroscopy**. (Note that the resonance phenomenon in NMR has *nothing whatsoever* to do with resonance structures discussed in Sec. 1.4.)

To summarize: for nuclei to absorb energy, they must have a nuclear spin and must be situated in a magnetic field. Once these two conditions are met, then the nuclei can be examined by an absorption spectroscopy experiment that is conceptually the same as the simple experiment shown in Fig. 12.3. The absorption of energy corresponds physically to the "flipping" of nuclear spins from the $+\frac{1}{2}$ to the $-\frac{1}{2}$ spin state.

Even with the large magnetic fields conventionally used in NMR spectroscopy, the energy of the radiation required for the absorption experiment is very small—on the order of 0.02 J/mol or 0.005 cal/mol (that's joules and calories, *not* kilojoules or kilocalories!). It follows from Eq. 12.3 and Fig. 12.2 that electromagnetic radiation with this energy lies in the FM area of the radiofrequency, or "rf," region of the electromagnetic spectrum. Consequently, FM radio waves are the radiation used in NMR spectroscopy. Typical values of the frequency used in NMR experiments are between 60×10^6 and 500×10^6 sec^{-1}, or between 60 and 500 megahertz (MHz). Indeed, an ordinary FM radio receiver located near an NMR spectrometer and tuned to the appropriate frequency can produce audible sounds associated with an NMR experiment.

Any nucleus with a spin can be detected by NMR. Since protons have nuclear spin, NMR can be used to detect and study the hydrogen nuclei in any compound. ^{12}C, however, has no nuclear spin and therefore does not give an NMR signal. Other nuclei with spin are ^{19}F, ^{31}P, and the carbon isotope ^{13}C. These nuclei all have spins of $\pm\frac{1}{2}$ and can be observed with NMR. Some nuclei also have spin values other than $\pm\frac{1}{2}$. ^{14}N and deuterium (^{2}H), for example, also can be detected by NMR, but these nuclei have three allowed spin states of ± 1 and 0. This chapter considers only the NMR spectroscopy of nuclei with spin of $\pm\frac{1}{2}$, and mostly the NMR spectroscopy of the hydrogen nucleus—the proton. Because almost all organic compounds contain hydrogens, *proton NMR spectroscopy* is especially useful to the organic chemist. We'll return in Sec. 13.8 to a discussion of the NMR of other nuclei, particularly ^{13}C. Despite the low abundance of this isotope, advances in instrumental techniques have made ^{13}C NMR a tool of major importance for the organic chemist.

STUDY GUIDE LINK:
13.1
*Quantitative Aspects
of NMR*

13.2　The NMR Spectrum: Chemical Shift and Integral

A.　Obtaining the NMR Spectrum

In a typical NMR experiment, radiation with a frequency of 60 MHz is applied to the sample. This is called the **operating frequency**, or **applied frequency**, and is symbolized by ν_0. Then the magnetic field of the instrument is applied and slowly increased over time. This increases the energy separation between the spin energy levels (Fig. 13.1). Now, remember that energy and frequency are related by $E = h\nu$ (Eq. 12.3). When the energy separation between the spin energy levels exactly corresponds to a frequency of 60 MHz, energy is absorbed from the applied frequency. This absorption is registered as a peak on the spectrum. Consequently, *an NMR spectrum is a plot of energy absorption vs. magnetic field strength.* (Alternatively, the magnetic field could be fixed and the applied frequency varied to get the proper match.)

When a chemical compound is subjected to a high magnetic field in an NMR spectrometer, the field $\mathbf{H_p}$ at a given proton within the compound is not the same as the field provided by the spectrometer, $\mathbf{H_0}$. The reason for this difference is that electrons circulating in the vicinity of a proton provide their own magnetic fields that oppose the external field. The external field $\mathbf{H_0}$ has to be strong enough to overcome the fields of the circulating electrons in order to make $\mathbf{H_p}$ large enough for absorption to occur. (You might say that the external magnetic field has to "fight through" the field of the circulating electrons.) In addition, the field provided by the circulating electrons is different for protons in different chemical environments. This means, in turn, that *the magnetic field strength required for NMR absorption depends on the chemical environment of the proton.* Turning this idea around, the strength of the field at which a proton absorbs energy gives information about its chemical environment.

B.　Chemical Shift Scales

Keeping the ideas of the previous section in mind, let's examine an actual NMR spectrum. The NMR spectrum of $CH_3O\!-\!CH_2\!-\!OCH_3$ (dimethoxymethane) is shown in Fig. 13.2. Absorption is plotted on the vertical axis in arbitrary units, increasing from bottom to top; that is, absorption peaks are registered as upward deflections in the spectrum. Absorption peaks are termed *absorptions*, *lines*, or *resonances*. The two lines near the middle of the spectrum are the NMR absorptions of dimethoxymethane. Notice that dimethoxymethane contains two distinguishable types of protons: the OCH_3 protons (six of them) and the CH_2 protons; consequently, the NMR spectrum of dimethoxymethane consists of two absorptions. Because more protons contribute to the CH_3 absorption, it is the larger of the two.

The small peak on the far right of the spectrum is the NMR absorption of another compound, tetramethylsilane (TMS; $[CH_3]_4Si$), which is added to the sample to provide an internal reference absorption from which the positions of all other absorptions are measured.

The strength of the applied field $\mathbf{H_0}$ is plotted on the horizontal axis of an NMR spectrum, and increases to the right, or to the *upfield direction*. Notice in Fig. 13.2 that both absorptions of dimethoxymethane occur at lower field than the absorption for TMS. These absorptions are thus said to be *downfield* from TMS.

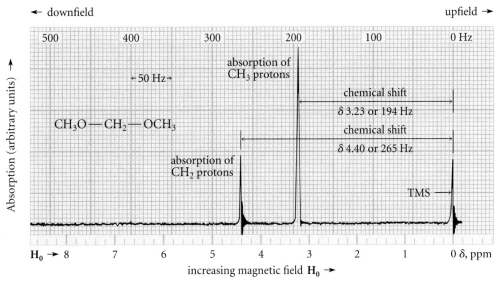

Figure 13.2   *NMR spectrum of dimethoxymethane.*

Absorption positions are always cited by their *chemical shifts.* If the absorption frequency for a proton of interest is ν, and that of the standard TMS is ν_{TMS}, then the **chemical shift** of the proton is defined as the difference between its absorption position and that of TMS:

$$\text{chemical shift (in Hz)} = \nu - \nu_{TMS} \qquad (13.2a)$$

Notice that when $\nu = \nu_{TMS}$ the chemical shift is zero. That is, the chemical shift of TMS is set to zero by definition. The chemical shift scale is shown on the upper horizontal axis of the spectrum in units of hertz (Hz). This scale defines the horizontal spacing of the grid within the spectrum; each small space in this grid equals 5 Hz. Thus, in Fig. 13.2, the chemical shift of the CH_3 protons is 194 Hz downfield from TMS, and that of the CH_2 protons is 265 Hz downfield from TMS.

Although Hz is a unit of frequency, it can also be used as a unit of field strength because the energy difference between spin energy levels is proportional to the field strength (Fig. 13.1). Remember that the energy of the radiation required to "flip" nuclear spins is equal to this energy difference (Eq. 13.1), and that the energy and frequency of this radiation are proportional according to $E = h\nu$.

The spectrum in Fig. 13.2 as well as most of the spectra in this text were taken with an operating frequency (ν_0) of sixty million Hz (60 megahertz, or 60 MHz). The use of larger operating frequencies (and correspondingly larger magnetic fields) is common. *The chemical shift in Hz is proportional to the operating frequency.*

$$\text{chemical shift (in Hz)} = \nu - \nu_{TMS} = \delta\nu_0 \qquad (13.2b)$$

In this equation, the frequencies ν and ν_{TMS} are in Hz, and ν_0 is in MHz. Each proton has a characteristic δ value that is independent of the operating frequency. This equation says, for example, that if ν_0 is tripled, then the chemical shifts of all protons are also tripled. Thus, at an operating frequency of 180 MHz, the chemical shifts of both absorptions in Fig. 13.2 would be three times as great as they are at an operating frequency of 60 MHz.

For example, the chemical shift of the larger resonance in Fig. 13.2 at $\nu_0 = 180$ MHz would be (3×194), or 582, Hz downfield from TMS.

The interpretation of NMR spectra often requires the use of tables of chemical shift data. Such tables are most useful when chemical shifts are presented in a form that is independent of the operating frequency. The quantity δ in Eq. 13.2b is used for this purpose. Dividing both sides of Eq. 13.2b by ν_0 gives

$$\delta = \frac{\nu - \nu_{TMS}}{\nu_0} = \frac{\text{chemical shift in Hz}}{\text{operating frequency in MHz}} \tag{13.2c}$$

Because the units of the numerator are Hz and the units of the denominator are MHz, the units of δ are *parts per million*, abbreviated ppm. The ppm scale is shown on the lower horizontal axis of the spectrum in Fig. 13.2. Notice that the ppm scale increases in the downfield direction, that is, toward lower magnetic field. Thus, the absorption of the CH_3 protons occurs at δ 3.23, or 3.23 ppm downfield from TMS. The absorption of the CH_2 protons occurs at δ 4.42, or 4.42 ppm downfield from TMS. NMR spectra taken at different operating frequencies have the same chemical shift scale in ppm but a different chemical shift scale in Hz. Thus, the chemical shift of the CH_3 protons of dimethoxymethane is 194 Hz at $\nu_0 = 60$ MHz, 970 Hz at $\nu_0 = 300$ MHz, but is δ 3.23 at any value of ν_0.

In most organic molecules, protons give NMR absorptions over a chemical shift range of about 0–10 ppm downfield from TMS. Indeed, TMS was chosen as a standard not only because it is volatile and inert, but also because its NMR absorption occurs at a higher field than the absorptions of most common organic compounds. (The rare absorptions that occur at higher field than TMS are assigned negative δ values.)

PROBLEMS

13.1 What is the chemical shift, in Hz downfield from TMS, of the CH_2 protons of dimethoxymethane (Fig. 13.2) in a spectrum taken at the following operating frequencies?
*(a) 90 MHz   (b) 360 MHz

13.2 What is the chemical-shift difference in ppm of two signals separated by 45 Hz at each of the following operating frequencies?
*(a) 60 MHz   (b) 300 MHz

C. The Number of Absorptions in an NMR Spectrum

Chemically nonequivalent protons in principle have different chemical shifts. (The qualifier "in principle" is used because it is possible for chemical-shift differences to be so small that they are not detectable.) For example, in dimethoxymethane (Fig. 13.2), the set of CH_2 protons has a different chemical shift from the set of CH_3 protons because the two sets of protons are chemically nonequivalent.

Chemically equivalent protons have identical chemical shifts. The two CH_2 protons of dimethoxymethane are chemically equivalent; one proton is indistinguishable from the other. Thus, these protons have the same chemical shift, δ 4.40. Likewise, the six protons comprising the two CH_3 groups are chemically equivalent, and all have the same chemical shift: δ 3.23.

The foregoing discussion shows that predicting chemical shift nonequivalence is essentially the same as recognizing chemical nonequivalence. Because chemically non-equivalent protons have different chemical shifts, *the minimum number of chemically nonequivalent sets of protons in a compound of unknown structure can be determined by counting the number of different resonances in its NMR spectrum.* (The "minimum" qualifier is used because of the possibility that the resonances of chemically different groups might accidentally overlap.)

In most cases, it is not difficult to determine whether protons are chemically equivalent or nonequivalent. However, in some cases, determining chemical equivalence requires application of the concepts introduced in Sec. 10.8A. Recall that for two protons to be *chemically equivalent*, they first must have the same connectivity relationship to the rest of the molecule; that is, they must be *constitutionally equivalent*. You can tell whether two protons are constitutionally equivalent by tracing their connectivity relationships to the rest of the molecule. For example, the CH_3 protons and the CH_2 protons of dimethoxymethane (Fig. 13.2) have a different connectivity relationship; consequently, they are chemically nonequivalent, and have different chemical shifts.

Recall also (Sec. 10.8A) that constitutional equivalence is necessary but not sufficient for chemical equivalence. Thus, *diastereotopic groups* are constitutionally equivalent but are *chemically nonequivalent*. It follows, then, that *diastereotopic protons in principle have different chemical shifts.*

In contrast, *enantiotopic protons are chemically equivalent* as long as they are in an achiral environment. Thus, in an achiral solvent such as CCl_4 or chloroform, enantiotopic protons have identical chemical shifts; but in an enantiomerically pure chiral solvent, enantiotopic protons in principle have different chemical shifts.

Finally, *homotopic protons are chemically equivalent.* Thus, homotopic protons have identical chemical shifts under all circumstances.

The following study problem illustrates how to use chemical equivalence to determine whether two protons should have different chemical shifts.

STUDY PROBLEM 13.1

Determine whether the labeled protons in each of the following cases are expected to have identical or different chemical shifts.

(a)

$$\begin{array}{c} H^a \quad\quad Cl \\ \diagdown \quad\quad\diagup \\ C = C \\ \diagup \quad\quad\diagdown \\ H^b \quad\quad CH_3 \end{array}$$

(b)

$$CH_3 - \overset{\displaystyle H^a}{\underset{\displaystyle H^b}{C}} - OH$$

(c)

$$CH_3 - \overset{\displaystyle H^a}{\underset{\displaystyle H^b}{C}} - \overset{\displaystyle Cl}{CH} - CH_3$$

(d)

$$CH_3O - \overset{\displaystyle H^a}{\underset{\displaystyle H^b}{C}} - OCH_3$$

Solution   Apply the principles of Sec. 10.8A to determine whether the protons in question are chemically equivalent. If the two protons are chemically equivalent, they have the same chemical shift. If not, their chemical shifts are different in principle.

The first step in each case is to ask whether the protons of interest are constitutionally equivalent. If not, they cannot be chemically equivalent. If so, then further analysis is required.

(a) Protons H^a and H^b are constitutionally equivalent, because they are both attached to the same carbon. Perform the *substitution test* (Sec. 10.8A) on protons H^a and H^b; that is, replace H^a and H^b in turn with a circled proton, and examine the relationship between the resulting compounds:

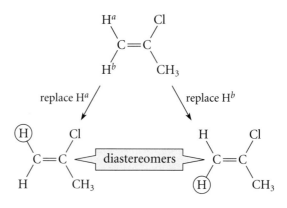

Remember to think of an "H" and a "circled H" as different atoms. The two compounds that result are *E,Z* isomers, and are therefore diastereomers. Consequently, H^a and H^b are diastereotopic, and are therefore *chemically nonequivalent*. They are expected to have different chemical shifts.

(b) Let's use Fischer projections to do the substitution test for the two α-protons of ethanol.

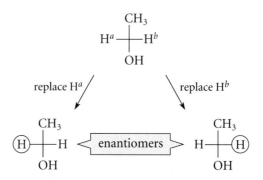

Thus, these two hydrogens are enantiotopic, and have the same chemical shift (in an ordinary achiral solvent).

(c) Using Fischer projections, you should be able to show that, in this example, protons H^a and H^b are diastereotopic, and hence chemically nonequivalent. These protons should have different chemical shifts.

(d) This is dimethoxymethane (Fig. 13.2). The two protons H^a and H^b are constitutionally equivalent, because they are connected to the same carbon. A substitution test shows that these two protons are homotopic. (Do this test!) Thus, they have identical chemical shifts. That is why a single resonance is observed for these protons in the NMR spectrum of dimethoxymethane (Fig. 13.2). Likewise, the two CH_3 groups are homotopic, and, within each of these groups, the three protons are also homotopic. Thus, a single resonance is observed for all six CH_3 protons.

Two of the most common situations in which diastereotopic protons are encountered are diastereotopic protons on a double bond, as in part (a) of Study Problem 13.1; and diastereotopic protons of methylene groups within a molecule containing an asymmetric carbon, as in part (c) of Study Problem 13.1. Unless you are particularly alert to these situations, it is easy to regard such groups mistakenly as chemically equivalent.

PROBLEM

13.3   Specify whether the labeled protons in each of the structures below would be expected to have the same or different chemical shifts.

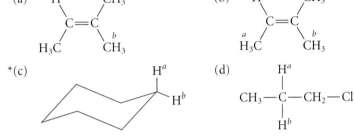

D. Relationship of Chemical Shift to Structure

Let's now return to the point with which Sec. 13.1A concluded: the chemical shift of a proton provides information about nearby groups. One of the most important factors that affects a proton chemical shift is the *electronegativities of nearby groups*. Some data that illustrate this idea are presented in Table 13.1. Examine these data using Problem 13.4 as your guide.

Table 13.1   Effect of Electronegativity on Proton Chemical Shift

Entry number	Compound	Chemical shift, δ	Electronegativity (Table 1.1)
1	CH_3F	4.26	F: 3.98
2	CH_3Cl	3.05	Cl: 3.16
3	CH_3Br	2.68	Br: 2.96
4	CH_3I	2.16	I: 2.66
5	CH_2Cl_2	5.30	
6	$CHCl_3$	7.27	
7	CH_3CCl_3	2.70	
8	$(CH_3)_4C$	0.86	C: 2.55
9	$(CH_3)_4Si$	0.00^a	Si: 1.90

a By definition.

*13.4   (a) Consider entries 1 through 4 of Table 13.1. How does the chemical shift of a proton vary with the electronegativity of the neighboring halogen?
       (b) Compare entries 2, 5, and 6 of Table 13.1. How does chemical shift vary with number of neighboring halogens?
       (c) Compare entries 6 and 7. How is the chemical shift of a proton affected by its distance from an electronegative group?
       (d) Explain why $(CH_3)_4Si$ absorbs at higher field (lower chemical shift) than the other molecules in the table. Can you think of a molecule with protons that would have a smaller chemical shift than TMS (that is, a negative δ value)?

You should have concluded from your examination of Table 13.1 that the following factors *increase* the proton chemical shift:

1. Increasing *electronegativity* of nearby groups
2. Increasing *number* of nearby electronegative groups
3. Decreasing *distance* between the proton and nearby electronegative groups

The effect of electropositive groups (such as Si) is, as expected, opposite that of electronegative groups.

The spectrum of dimethoxymethane (Fig. 13.2) also illustrates these points. The CH_2 protons are adjacent to two oxygens, and thus have a larger chemical shift than the CH_3 protons, which are adjacent to only one oxygen.

Electronegative groups affect the chemical shift of a proton through their effects on the magnetic fields caused by the nearby electrons. Circulation of electrons in the vicinity of a proton causes local fields that shield the proton from the effects of an external field. You can think of these local fields as a "magnetic umbrella" that makes it more difficult for the external magnetic field to exert its effects. When the electron density in the environment of a proton is reduced, the magnitude of these local magnetic fields, and thus the effectiveness of the "magnetic umbrella," is also reduced. For this reason, a proton is said to be **deshielded** by electronegative groups. Hence, the external magnetic field required to bring a deshielded proton into resonance is smaller than the field required for a proton that is not deshielded. A deshielded proton thus appears in the NMR spectrum at a lower field, that is, at greater chemical shift.

Chemical shift information for *methyl* (CH_3) and *methylene* (CH_2) protons in different chemical environments is summarized in Table 13.2. (This table contains some data for unfamiliar functional groups that will be useful in later parts of the text.) The first column lists various groups G that might be found in an organic compound. The second column lists the *approximate* chemical shift of the protons of a methyl group bound to group G, that is, the shift of CH_3—G. For example, when G = F, the chemical shift of CH_3—F is δ 4.3. The third column contains effective shift contributions, σ_G, that can be used to calculate the *approximate* chemical shift for the protons of a methylene group bound to two groups G_1 and G_2. To carry out such a calculation, the following simple equation is used:

$$\delta\,(G_1—CH_2—G_2) = 0.2 + \sigma_{G_1} + \sigma_{G_2} \tag{13.3}$$

For example, suppose you want to calculate the chemical shift of the protons in CH_2Cl_2. This compound is of the form G_1—CH_2—G_2, in which $G_1 = G_2 = Cl$. From Table 13.2, $\sigma_{Cl} = 2.5$; application of Eq. 13.3 gives a predicted chemical shift of δ 5.2. The actual chemical shift is δ 5.3. It is important to understand that chemical shifts calculated in this manner do not in most cases agree perfectly with experiment, but the estimates are usually good to within several tenths of a ppm.

Table 13.2   Contributions of Common Functional Groups to Chemical Shifts of CH_2 and CH_3 Groups

Group, G	δ for CH_3—G	Effective shift contribution σ_G for —CH_2—G[a]
—H	0.2	—
—CR_3 or —CR_2— (R = H, alkyl)	0.9	0.6
—F	4.3	3.6
—Cl	3.0	2.5
—Br	2.7	2.3
—I	2.2	1.8
—CR=CR_2 (R = H, alkyl)	1.8	1.3
—C≡C—R (R = H, alkyl)	2.0	1.4
—OH	3.5	2.6
—OR (R = alkyl)	3.3	2.4
—OR (R = aryl)	3.7	2.9
—SH, —SR	2.4	1.6
⬡ or —Ph (phenyl)	2.3	1.8
—C(=O)—R	2.1 (R = alkyl) / 2.6 (R = aryl)	1.5
—C(=O)—OR (R = alkyl, H)	2.1	1.5
—O—C(=O)—R	3.6 (R = alkyl) / 3.8 (R = aryl)	3.0

[a] Used with Eq. 13.3 in text.

(Table 13.2 continues)

Table 13.2   Contributions of Common Functional Groups to Chemical Shifts of CH$_2$ and CH$_3$ Groups

(continued)

Group, G	δ for CH$_3$—G	Effective shift contribution σ$_G$ for —CH$_2$—G^a
$\overset{\displaystyle O}{\overset{\|}{-C}}$—NR$_2$ (R = alkyl, H)	2.0	1.5
$-NR-\overset{\displaystyle O}{\overset{\|}{C}}-R$	2.8	—
—NR$_2$ (R = alkyl, H)	2.2	1.6
$-\overset{\displaystyle }{\underset{\|}{N}}-R$ (R = aryl)	2.9	—
—C≡N	2.0	1.7

^{a} Used with Eq. 13.3 in text.

The fact that a simple addition formula like Eq. 13.3 works fairly well demonstrates that, to a useful approximation, *the chemical shift contributions of different groups are additive.*

Chemical shift predictions can be refined with two other observations. First, a β-halogen or β-oxygen adds about 0.5 ppm to the chemical shift. For example, let's estimate the chemical shift of the colored protons in the following compound:

$$\text{Br—}C\!H_2\text{—CH}_2\text{—Cl} \longleftarrow \text{ chlorine is } \beta \text{ to colored protons}$$

For the effects of the adjacent CH$_2$ and Br groups, Eq. 13.3 and Table 13.2 give a contribution of $0.2 + 0.6 + 2.3 = 3.1$, to which another 0.5 ppm is added for the β-Cl. The predicted chemical shift of the indicated hydrogens is δ 3.6; the observed value is δ 3.3.

The other useful correlation is that the chemical shifts of methine (CH) protons are greater than those of methylene (CH$_2$) protons, which are greater than those of methyl (CH$_3$) protons. Protons of a —CH$_2$—G group have chemical shifts that are typically 0.3–0.9 ppm greater than the protons of a methyl group bearing the same group G. The chemical shift of an analogous methine proton, although less predictable, is in many cases 0.5–1.0 ppm larger still.

Study Guide Link:
✓13.2
Other Chemical Shift Tables

Problems

13.5   In the following sets, indicate the proton that has the greatest chemical shift.
*(a) (CH$_3$)$_3$C—C(CH$_3$)$_3$, Cl$_2$CH—CHCl$_2$, Cl$_3$C—CH$_3$
(b) CH$_2$Cl$_2$, CH$_2$I$_2$, CH$_3$I
*(c) (CH$_3$)$_4$Si, (CH$_3$)$_4$Sn (*Hint:* See Table 1.1.)

13.6   Using the data in Table 13.2 along with an estimate for the effect of a β-halogen, estimate the chemical shifts of the italicized protons in each of the following molecules.

*(a) $CH_3—CH_2—CH_3$ (b) $Br—CH_2—I$

*(c) $CH_3—CH_2—O—CH_2—CH_3$ (d) $CH_3—Ph$

*(e) $Br—CH_2—CH_2—CH=CH_2$ (f) $CH_3—CH_2—CH_2—Br$

*(g) $CH_3—CH_2—CH_2—Br$ (h) $CH_3—CH_2—CH_2—Br$

E. Counting Protons with the Integral

Re-examine Fig. 13.2 and notice that the two resonances of dimethoxymethane are not the same size. The reason for this is quite simple: the size of an NMR absorption is governed essentially by the number of protons contributing to it. None of the complicating factors that govern, for example, IR absorption intensities are present in NMR spectroscopy. The exact intensity of an NMR absorption is given not by its peak height, but rather by the total area under the peak. This quantity can be determined by mathematical integration of the peak using more or less the same integration procedures used in calculus to determine the area under a curve. NMR instruments are equipped with an integrating device that can be used to display the integral on the spectrum. Such a spectrum integral is illustrated in Fig. 13.3 for the dimethoxymethane spectrum as the curve superimposed on each peak. *The relative height of the integral (in any convenient units, such as chart spaces) is proportional to the number of protons contributing to the peak.* Thus the heights of the integrals in Fig. 13.3 are 21 and 7 vertical chart spaces, respectively, in a ratio of 3:1, and the relative numbers of protons are 6 and 2, respectively, also in the ratio 3:1.

It is important to remember that these values give the *relative* numbers of protons under the NMR peaks, not the absolute numbers. Thus a sample giving a spectrum with three peaks having relative integrals 1:2:3 might contain any multiple of six protons.

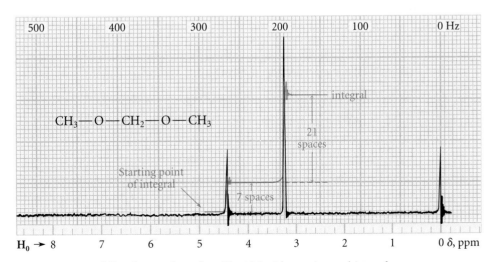

Figure 13.3   *NMR spectrum of dimethoxymethane from Fig. 13.2 with superimposed integral.*

F.   Using the Chemical Shift and Integral to Determine Unknown Structures

Let's summarize the main ideas of the previous sections that will be useful in applying NMR spectroscopy to the solution of unknown structures.

First, each chemically nonequivalent set of protons in a molecule gives (in principle) a different resonance in the NMR spectrum. Thus, the number of absorptions in principle indicates the number of chemically nonequivalent sets of protons. Next, the chemical shift of a set of protons provides information about what groups are nearby. Finally, the integral of each absorption is proportional to the number of contributing protons. Thus, the integral indicates the relative numbers of protons in each nonequivalent set. If the total number of protons is known (from an elemental analysis and molecular mass determination), then the absolute number of protons in each set can be calculated.

These ideas are incorporated into the following systematic approach for the determination of unknown structures by NMR spectroscopy:

1. Write down everything about the molecular structure that is known from the molecular formula, including the unsaturation number and the functional groups that might be present.

2. From the number of absorptions, determine how many chemically nonequivalent sets of hydrogens are in the unknown.

3. Use the total integral of the entire spectrum and the molecular formula to determine how many chart spaces per proton are in the integral.

4. Count the chart spaces in the integral of each absorption and use the result from (3) to determine the number of protons in each set.

5. From the chemical shift of each set, determine which set must be closest to each of the functional groups that are present.

6. Write down partial structures that are consistent with each piece of evidence, and then write down *all possible structures* that are consistent with *all* the evidence.

7. Use Eq. 13.3 and Table 13.2 to estimate the chemical shifts of the protons in each structure, and, if possible, choose the structure that best reconciles the predicted and observed chemical shifts.

This approach is illustrated in the following sample problem.

☞
Study Guide Link:
✓**13.3**
Approaches to Problem Solving

Study Problem 13.2

An unknown compound with the molecular formula $C_5H_{11}Br$ has an NMR spectrum consisting of two resonances, one at δ 1.01 (relative integral 4.5), and the other at δ 3.17 (relative integral 1.0). Propose a structure for this compound.

Solution   Follow the steps given above in the text.

Step 1   Because the unsaturation number is zero (Eq. 4.7), the compound has no rings or double bonds; it is a simple alkyl halide.

Step 2 Because the spectrum consists of two lines (absorptions), the compound contains (barring accidental overlaps) only two chemically nonequivalent sets of hydrogens.

Step 3 The total integral is $(4.5 + 1.0) = 5.5$ units; because the molecular formula indicates 11 hydrogens, the integral corresponds to $5.5/11 = 0.5$ units per hydrogen.

Step 4 The larger peak accounts for $(4.5/0.5) = 9$ protons; the smaller peak accounts for $(1.0/0.5) = 2$ protons. Notice that all eleven protons have been accounted for.

Step 5 Because of its greater chemical shift, the two-proton set must be closer to the bromine than the nine-proton set.

Step 6 A partial structure consistent with the conclusion of Step 5 is

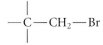

Step 7 Application of Table 13.2 and Eq. 13.3 predicts a chemical shift of δ 3.0 for the CH_2 protons in this partial structure (verify this); this prediction is acceptably close to the observed shift of δ 3.15.

Comparison of this partial structure with the molecular formula shows that three carbons and nine protons are missing from this partial structure. The nine protons are chemically equivalent. Adding three methyl groups to the above partial structure gives the correct structure:

$$(CH_3)_3C—CH_2—Br \quad \textbf{neopentyl bromide}$$

δ 1.02 (9 hydrogens) δ 3.15 (2 hydrogens)

PROBLEMS

13.7 How many different NMR absorptions would you expect to observe in the spectrum of each of the following compounds?

*(a) $(CH_3)_3C—\underset{\displaystyle Cl}{C}(CH_3)_2$ (b) *tert*-butyl bromide   *(c) *cis*-2-butene

13.8 In each case give a single structure that fits the data provided.

*(a) A compound $C_5H_9Cl_3$ has three NMR absorptions at δ 1.99, δ 4.31, and δ 6.55 with a relative integral of 6:2:1, respectively.

(b) A compound $C_7H_{15}Cl$ has two NMR absorptions at δ 1.08 and δ 1.59, with a relative integral of 3:2, respectively.

*13.9 (a) Strange results in the undergraduate organic laboratory have led to the admission by a teaching assistant, Thumbs Throckmorton, that he has accidentally mixed some *tert*-butyl bromide with the methyl iodide. The NMR spectrum of this mixture indeed contains two single resonances at δ 2.20 and δ 1.8 with relative integrals of 5:1. What is the mole percent of each compound in the mixture? (*Hint:* Be sure to assign each absorption to a compound before doing the analysis.)

(b) Which would be more easily detected by NMR: 1 mole percent CH_3I impurity in $(CH_3)_3C$—Br, or 1 mole percent $(CH_3)_3C$—Br impurity in CH_3I? Explain.

13.3   The NMR Spectrum: Spin-Spin Splitting

Although substantial information can be gleaned from the chemical shift and integral, another aspect of NMR spectra provides the most detailed information about chemical structures. Consider the compound ethyl bromide:

$$\overset{a}{CH_3}—\overset{b}{CH_2}—Br$$

This molecule has two chemically different sets of hydrogens, labeled *a* and *b*. We expect to find NMR absorptions for these two sets of protons in the integral ratio 3:2, respectively, with the absorption of protons *a* at higher field (smaller chemical shift). The NMR spectrum of ethyl bromide is shown in Fig. 13.4. This spectrum contains more lines than you might have expected—seven lines in all. Moreover, the lines fall into two distinct groups: a packet of three lines, or *triplet*, at high field; and a packet of four lines, or *quartet*, at low field. It turns out that all three lines of the triplet are the absorption for the CH_3 protons, and all four lines of the quartet are the absorption for the CH_2 protons. The chemical shift of each packet of lines, taken at its center, is in agreement with the predictions of Eq. 13.3 and Table 13.2. The low-field quartet and the high-field triplet have total integrals, respectively, in the ratio 2:3.

The NMR signal for each group of protons is said to be *split*. **Splitting** *arises from the effect that one set of protons has on the NMR absorption of neighboring protons.* The

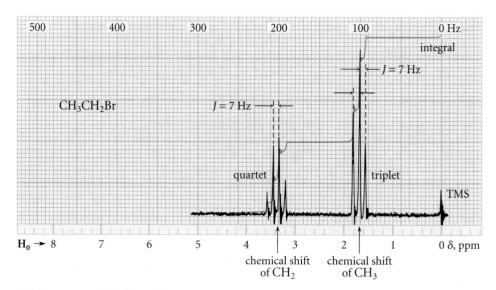

Figure 13.4   *NMR spectrum of ethyl bromide.*

physical reasons for splitting are considered in Sec. 13.3B. First, let's focus on the appearance of the splitting pattern and the information it provides about structure.

A. The *n* + 1 Splitting Rule

Recall that the integral gives a proton count for each resonance. The splitting pattern gives a different proton count: the number of protons *adjacent* to the protons being observed. The relationship between the number of lines in the splitting pattern for an observed proton and the number of adjacent protons is given by the following rule, called the **n + 1 rule**: *If there are n adjacent protons, the resonance of an observed proton is split into n + 1 lines.*

Let's see how the *n* + 1 rule accounts for the splitting patterns in the spectrum of ethyl bromide. Because the carbon *adjacent* to the CH_3 group has two protons, the signal for the CH_3 protons of ethyl bromide is split into a pattern of 2 + 1 = 3 lines, called a *triplet*. (The fact that there are also three methyl protons is a coincidence; the number of protons has *nothing* to do with their own splitting.) Because the carbon *adjacent* to the CH_2 group has three equivalent protons, the resonance for the CH_2 protons is split into a pattern of 3 + 1 = 4 lines, called a *quartet*.

$$3 \text{ Hs on } adjacent \text{ carbon; } 3 + 1 = 4 \text{ lines (a quartet)}$$
$$\downarrow$$
$$CH_3\!-\!CH_2\!-\!Br$$
$$\uparrow$$
$$2 \text{ Hs on } adjacent \text{ carbon; } 2 + 1 = 3 \text{ lines (a triplet)}$$

The CH_3 resonance is split by the CH_2 protons, and the CH_2 resonance is split by the CH_3 protons. Because these two sets of protons split each other, they are said to be **coupled protons.**

Why is no splitting observed in our previous examples of NMR spectra? The reason is that *with saturated carbon atoms, splitting is observed only between protons on adjacent carbon atoms.* Thus, since the protons in dimethoxymethane (Fig. 13.2) are on *nonadjacent* carbons, their splitting is negligible; the two absorptions in the NMR spectrum of this compound are unsplit singlets (single lines).

The spacing between adjacent peaks of a splitting pattern, measured in Hz, is called the **coupling constant** (abbreviated *J*). For ethyl bromide, this distance is about 7 Hz. *Two coupled protons must have the same value of J.* Thus, the coupling constants for both the CH_2 protons and the CH_3 protons of ethyl bromide are the same because these protons split each other. Letting the CH_3 protons be *a* and the CH_2 protons be *b*, then $J_{ab} = J_{ba}$. The coupling constant, unlike the chemical shift in Hz, does not vary with the operating frequency. Thus, the value of *J* for ethyl bromide is 7 Hz whether the spectrum is taken at 60 MHz or 300 MHz.

The chemical shift of a split resonance in most cases occurs at or near the midpoint of the splitting pattern. Thus, in the ethyl bromide spectrum, the chemical shift of the CH_2 protons can be taken to be the midpoint of the quartet, and that of the CH_3 protons is at the middle line of the triplet.

How do you know whether a group of lines is a single split resonance rather than several individual resonances with different intensities? Sometimes there is ambiguity, but in most cases a splitting pattern can be discerned by the relative intensities of its component lines. These intensities have well-defined ratios, shown in Table 13.3. For

Table 13.3   Relative Intensities of Lines within Common NMR Splitting Patterns

Number of equivalent adjacent protons	Number of lines in splitting pattern (name)	Relative line intensity within splitting pattern						
0	1 (singlet)				1			
1	2 (doublet)				1 1			
2	3 (triplet)			1	2	1		
3	4 (quartet)			1 3		3 1		
4	5 (quintet)		1	4	6	4	1	
5	6 (sextet)		1 5	10		10	5 1	
6	7 (septet)	1	6	15	20	15	6	1

example, the relative intensities of a triplet, such as the one in ethyl bromide, are in the ratio 1:2:1; the relative intensities of a quartet are 1:3:3:1.

Notice that the lines of most splitting patterns do not conform *exactly* to the intensity ratios in Table 13.3. This point is examined further in Fig. 13.5. In both the triplet and the quartet, the inner lines are a little taller than the outer lines, although, according to Table 13.3, they should be the same. This departure from the ratios in Table 13.3 is called *leaning*. Leaning is most severe when the chemical-shift difference between two absorptions that split each other is small. Leaning is less pronounced when the chemical-shift difference between the two absorptions is large. For example, Fig. 13.6 shows the NMR spectrum of a hypothetical compound XYCH—CHAB. According to the splitting rules, the spectrum of such a compound should consist of two doublets. In successive parts of the figure the coupling constant between the two protons is the same, but the chemical-shift difference between them is made progressively smaller. According to Table

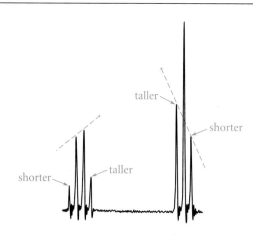

Figure 13.5 *The phenomenon of leaning in the spectrum of ethyl bromide. Notice that the inner lines of each pattern are taller than the outer lines, and thus depart from the equal intensities predicted by Table 13.3.*

13.3, each line of each doublet should have the same intensity. As the chemical-shift difference between the two protons is made smaller, the outer lines of each doublet become smaller relative to the inner lines. In fact, when the chemical shifts of the two protons are identical, as in Fig. 13.6d, the outer lines of the doublets disappear altogether and the large inner lines merge to become a single line. A single line, of course, is exactly what is observed for two protons that are chemically equivalent. This is why *no splitting is observed between two protons with identical chemical shifts*. Thus, the four chemically equivalent protons of 1,2-dichloroethane, $Cl—CH_2CH_2—Cl$, appear as a single line because they all have the same chemical shift. These protons *are coupled*, but the coupling is invisible in the NMR. The same behavior can be observed even when two sets of protons are not chemically equivalent. For example, in the following molecule, the two sets of protons H^a and H^b are clearly in different chemical environments:

$$N≡C—\overset{a}{C}H_2—\overset{b}{C}H_2—\overset{\overset{\displaystyle O}{\|}}{C}—OCH_3$$

Yet *by chance* they happen to have the same chemical shift (δ 2.68). Therefore both sets of protons appear as a single line at δ 2.68 and no splitting is observed. (This equivalence could not have been predicted in advance.)

A set of protons can be split by protons on more than one adjacent carbon. An example of this situation occurs in 1,3-dichloropropane, the NMR spectrum of which is shown in Fig. 13.7.

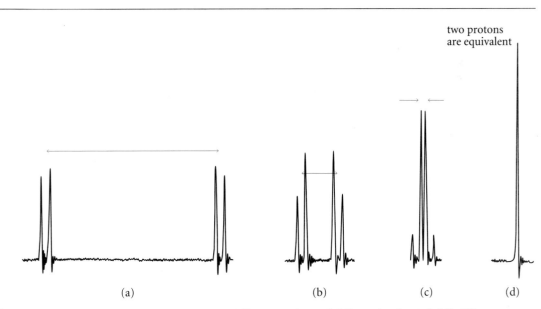

(a)   (b)   (c)   (d)

Figure 13.6   *Leaning increases with decreasing difference in chemical shift. As the chemical-shift difference between the doublets (colored horizontal arrow) decreases, the leaning becomes more severe. In (a), the chemical shifts of the coupled protons are well separated and the intensities within each doublet are close to 1:1. In (b) and (c), the chemical shifts are progressively closer, and the leaning is correspondingly more severe; in (c), the outer lines of each doublet are very small relative to the inner lines. In (d), the outer lines have disappeared altogether because the two protons are equivalent; that is, they appear as one line.*

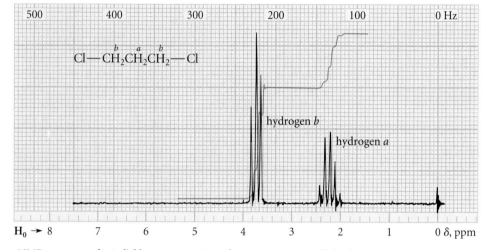

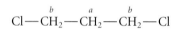

Figure 13.7   *NMR spectrum of 1,3-dichloropropane. Note that protons a are split by four protons b.*

$$\overset{b}{Cl-CH_2}-\overset{a}{CH_2}-\overset{b}{CH_2}-Cl$$

This molecule has two chemically nonequivalent sets of protons, labeled *a* and *b*. The key to understanding this spectrum is to recognize that all four protons H^b are chemically equivalent. The absorption for H^a is therefore split into a quintet because there are four protons on adjacent carbons; the fact that two of the protons are on one carbon and two are on the other makes no difference. The absorption for H^b is a triplet that appears at higher field.

PROBLEMS

13.10   Predict the NMR spectrum of each of the following compounds, including chemical shifts and splitting. (Assume that the coupling constants are about the same as those for ethyl bromide.) Indicate what effect leaning would have on the appearance of each spectrum.

*(a) CH_3-CHCl_2 (b) $ClCH_2-CHCl_2$

13.11   Predict the NMR spectra of the following compounds. Consult Table 13.2 for chemical shift information when necessary.

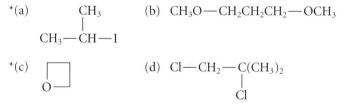

*(a)
$$CH_3$$
$$|$$
$$CH_3-CH-I$$

(b) $CH_3O-CH_2CH_2CH_2-OCH_3$

*(c) [structure]

(d) $Cl-CH_2-C(CH_3)_2$
$$|$$
$$Cl$$

B. Why Splitting Occurs

Splitting occurs because the magnetic field caused by the spin of a neighboring proton adds to or subtracts from the applied magnetic field and affects the total field experienced by an observed proton. To see this, let's analyze the absorption of a set of equivalent protons a (such as the methyl protons of ethyl bromide) adjacent to *two* equivalent neighboring protons b. What are the spin possibilities for the neighboring protons b? In any one molecule, these protons could *both* have spin $+\frac{1}{2}$; they could *both* have spin $-\frac{1}{2}$; or they could have differing spins.

$$
\begin{array}{ccc}
b \text{ protons} & b \text{ protons} & b \text{ protons} \\
1 \quad 2 & 1 \quad 2 & 1 \quad 2 \\
 & +\frac{1}{2} \quad -\frac{1}{2} & \\
+\frac{1}{2} \quad +\frac{1}{2} & -\frac{1}{2} \quad +\frac{1}{2} & -\frac{1}{2} \quad -\frac{1}{2}
\end{array}
\tag{13.4}
$$

(Notice that the spins of the b protons can differ in two ways: proton 1 can have spin $+\frac{1}{2}$ and proton 2 spin $-\frac{1}{2}$, or vice-versa.)

Suppose that the a protons are next to two b protons with a spin of $+\frac{1}{2}$. Because the spins of these b protons are parallel to the applied field, they augment the applied field, and thus the a protons are subjected to a slightly greater field than they would be in the absence of the b protons. Hence, a smaller applied field is required to bring the a protons into resonance. The result is a line in the splitting pattern at *lower* field (Fig. 13.8). Now suppose that the a protons are next to two b protons with a spin of $-\frac{1}{2}$. Since the spins of these b protons are antiparallel to the applied field, they subtract from the applied field, and as a result the a protons are subjected to a slightly smaller field than they would be in the absence of the b protons. Hence, a larger applied field is required to bring the a protons into resonance. The result is a line in the splitting pattern at *higher* field. Finally, suppose that the two b protons have opposite spins; in this case, the effects of the two b proton spins cancel, and protons a are subjected to the same field that they would be in the absence of protons b. The result is the line in the center of the splitting pattern. In a sample containing many millions of molecules, all three spin combinations occur in accord with the laws of probability. Consequently, there are three lines in the splitting pattern. The probability that both b protons have spin $+\frac{1}{2}$ is equal to the probability that they both have spin $-\frac{1}{2}$, but the probability that these protons have opposite spins is twice as high because this situation can occur in two ways. Hence, the center line of the splitting pattern is twice as large as the outer lines, and a 1:2:1 triplet is observed for the absorption of protons a.

> An analogy to the spin combinations is the combinations that can be rolled with a pair of dice. A 3 is twice as probable as a 2 because it can be rolled in two ways ($2 + 1$, $1 + 2$), but a 2 can only be rolled in one way ($1 + 1$).

Protons a are affected by protons b in the same way because splitting is mutual: if protons b split protons a, then protons a split protons b.

How does one proton "know" about the spin of adjacent protons? One of the most important ways that proton spins interact is through the electrons in the intervening chemical bonds. This interaction is weaker when the protons are separated by more chemical bonds. Thus, the coupling constant between hydrogens on adjacent saturated

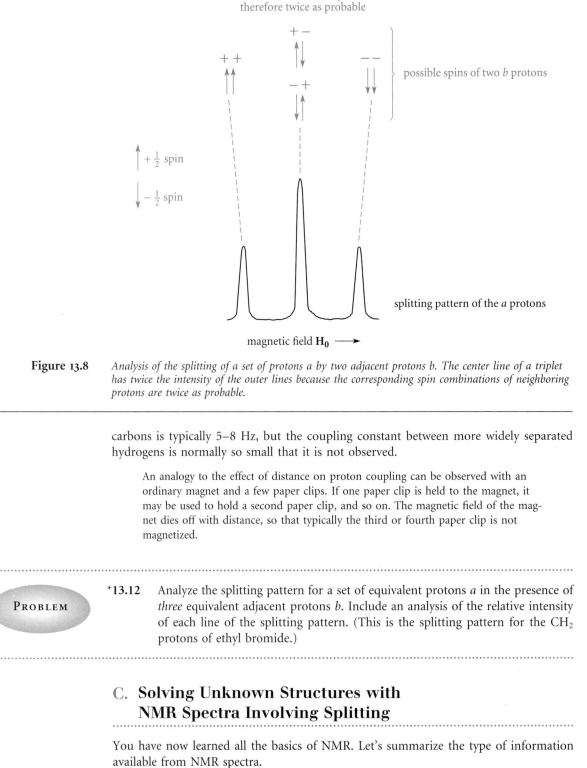

Figure 13.8 *Analysis of the splitting of a set of protons a by two adjacent protons b. The center line of a triplet has twice the intensity of the outer lines because the corresponding spin combinations of neighboring protons are twice as probable.*

carbons is typically 5–8 Hz, but the coupling constant between more widely separated hydrogens is normally so small that it is not observed.

> An analogy to the effect of distance on proton coupling can be observed with an ordinary magnet and a few paper clips. If one paper clip is held to the magnet, it may be used to hold a second paper clip, and so on. The magnetic field of the magnet dies off with distance, so that typically the third or fourth paper clip is not magnetized.

PROBLEM

*13.12 Analyze the splitting pattern for a set of equivalent protons *a* in the presence of *three* equivalent adjacent protons *b*. Include an analysis of the relative intensity of each line of the splitting pattern. (This is the splitting pattern for the CH_2 protons of ethyl bromide.)

C. Solving Unknown Structures with NMR Spectra Involving Splitting

You have now learned all the basics of NMR. Let's summarize the type of information available from NMR spectra.

1. The *chemical shift* provides information about functional groups that are near an observed proton.

2. The *integral* indicates the relative number of protons contributing to a given resonance.

3. The *splitting pattern* indicates the number of protons adjacent to an observed proton.

These three elements of an NMR spectrum can be put together like pieces of a puzzle to deduce a great deal about chemical structure; it is not unusual for a complete structure to be determined from the NMR spectrum alone.

Because NMR spectra consume a large amount of space, it is common to see NMR spectra recorded in books and journals in an abbreviated form. In the form used in this text, the chemical shift of each resonance is followed by its integral, its splitting, and (if split) its coupling constants, if known. Abbreviations used to indicate splitting patterns are s (singlet), d (doublet), t (triplet), and q (quartet); complex patterns in which the nature of the splitting is not clear are designated m, for multiplet. It is assumed that the relative intensities of each splitting pattern, except for leaning, approximately match those in Table 13.3. For example, the spectrum of ethyl bromide (Fig. 13.4) would be summarized as follows:

$$\delta\ 1.67\ (3H,\ \text{t},\ J = 7\ \text{Hz});\ \delta\ 3.43\ (2H,\ \text{q},\ J = 7\ \text{Hz})$$

coupling constant

type of splitting pattern (triplet; intensities in Table 13.3)

integral

chemical shift (ppm downfield from TMS)

You should now have the tools needed to determine some structures using NMR spectra that contain some splitting information. The general method of problem solving given in Sec. 13.2F remains valid when splitting information is involved. Just remember to take into account splitting information when writing out partial structures (Step 6). The following study problem illustrates this approach.

STUDY PROBLEM 13.3

Give the structure of a compound $C_4H_8Cl_2$ with the following NMR spectrum: $C_4H_8Cl_2$: δ 1.60 (3H, d, $J = 7.5$ Hz); δ 2.15 (2H, q, $J = 7.5$ Hz); δ 3.72 (2H, t, $J = 7.5$ Hz); δ 4.27 (1H, sextet, $J = 7.5$ Hz).

Solution   First, you may have noticed that all the coupling constants are 7.5 Hz. This does not mean that each set of protons is coupled to every other set. Couplings in the 7–8 Hz range are very common in saturated organic compounds; thus, the equality of the coupling constants is a coincidence.

To solve the problem, apply the procedure in Sec. 13.2F.

Step 1   The unsaturation number is zero; hence, the unknown contains no rings and no double or triple bonds.

Step 2   The unknown contains four sets of chemically nonequivalent protons.

Step 3   The numbers of protons in each resonance is given in the problem.

Step 4   The set of protons at δ 1.60 must be a methyl (CH_3) group because it contains three equivalent protons. The integrals also indicate that the sets at δ 2.15 and δ 3.72 are methylene (CH_2) protons, and the "set" at δ 4.27 is a methine proton (CH).

Step 5   Notice that the unknown contains two chlorines. The protons at δ 3.72 and δ 4.27 are closest to the chlorines.

Step 6   Steps 4 and 5 require the following two partial structures:

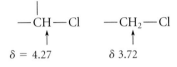

The splitting information shows that the proton at δ 4.27 is adjacent to five protons (its resonance is a sextet), and the protons at δ 3.72 are adjacent to a CH_2 group (their resonance is a triplet). The partial structures above can thus be modified as follows:

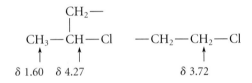

Notice that the methyl group at δ 1.60 has thus been accounted for in the first partial structure; the partial structure requires that the methyl resonance be a doublet, as observed. (It is common for spectra to contain redundant information that provides cross-checks.) The two partial structures contain a total of five carbons, but the unknown contains four. Hence, the unassigned CH_2 group must be common to both:

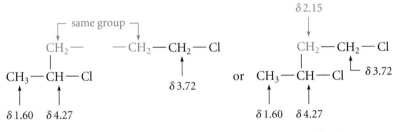

1, 3-dichlorobutane

The protons of this CH_2 group are adjacent to three other protons (two on one carbon, one on the other), and hence, their resonance should be a quartet, as observed. Thus, the unknown is 1,3-dichlorobutane.

Step 7   The —CH_2Cl protons, according to Eq. 13.3 and Table 13.2, should have a chemical shift of 0.2 + 2.5 (chlorine) + 0.6 (carbon) = δ 3.3; this is acceptably close to the δ 3.72 observed. The chemical shift of the other CH_2 group is calculated to be 0.2 + 0.5 + 0.5 (two β-chlorines) + 0.6 + 0.6 (two attached carbons) = δ 2.3, close to the δ 2.15 observed. The chemical shift of the methine proton cannot be calculated with Eq. 13.3, which only applies to methylene protons. However, this proton has the same proximity to the two chlorines as the $-CH_2-Cl$ protons (that is, one chlorine on the same carbon and one two carbons removed), and methine protons generally have chemical shifts that are 0.5–1.0 ppm greater than methylene protons in similar environments (Sec. 13.2D). Hence, the calculated chemical shifts for 1,3-dichlorobutane are consistent with those observed.

STUDY GUIDE LINK:
✓13.4
*More NMR Problem-
Solving Hints*

PROBLEM

13.13   Give structures for each of the following compounds.
*(a) C_3H_7Br: δ 1.03 (3*H*, t, *J* = 7 Hz); δ 1.88 (2*H*, sextet, *J* = 7 Hz); δ 3.40 (2*H*, t, *J* = 7 Hz)
(b) $C_2H_3Cl_3$: δ 3.98 (2*H*, d, *J* = 7 Hz); δ 5.87 (1*H*, t, *J* = 7 Hz)
*(c) $C_5H_8Br_4$: δ 3.6 (s; only line in the spectrum)

An important use of spectroscopy is to confirm structures that are suspected from other information. For example, if a well-known reaction is run on a known starting material, the structure of the major product can often be predicted from a knowledge of the reaction. NMR spectroscopy can be used to confirm (or refute) the predicted structure, as shown in the following problem.

PROBLEM

*13.14   When 3-bromopropene is allowed to react with HBr in the presence of peroxides, a compound *A* is formed that has the following NMR spectrum: δ 3.60 (4*H*, t, *J* = 6 Hz); δ 1.38 (2*H*, quintet, *J* = 6 Hz).
(a)   From the reaction, what do you think *A* is?
(b)   Use the NMR spectrum to confirm or refute your hypothesis. Identify *A*.

13.4   Complex NMR Spectra

The NMR spectra of some compounds contain splitting patterns that do not appear to be the simple ones predicted by the *n* + 1 rule. Two common sources of such complex spectra are, first, the splitting of one set of protons by more than one other set, called *multiplicative splitting*, and, second, the breakdown of the *n* + 1 rule in certain cases. This section discusses each of these situations, and shows how to deal with them.

A. Multiplicative Splitting

The NMR spectrum of the ester vinyl 2,2-dimethylpropanoate (vinyl pivalate), shown in Fig. 13.9, illustrates how multiplicative splitting can give complex spectra.

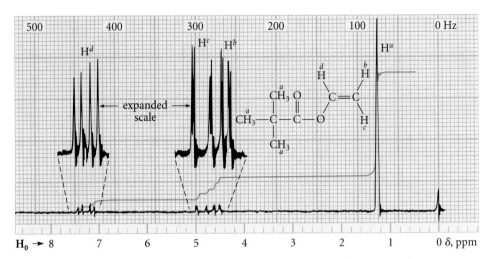

Figure 13.9   *NMR spectrum of vinyl pivalate. The insets shown the resonances for H^b, H^c, and H^d on an expanded vertical scale. These patterns result from multiplicative splitting, which is analyzed in Fig. 13.10.*

vinyl 2,2-dimethylpropanoate
(vinyl pivalate)

(Do not be concerned that there is an unfamiliar functional group in this molecule; the principles are the same.) The nine equivalent *tert*-butyl protons of vinyl pivalate (H^a) give the large singlet at δ 1.26. The interesting part of this spectrum is the region containing the resonances of the alkene protons (which generally have chemical shifts greater than 4.5 ppm; Sec. 13.6A). The protons H^b and H^c are farthest from the electronegative oxygen and therefore have the smallest chemical shifts; the complex resonances in the δ 4.5–5.0 region are from these protons. The four lines in the δ 7.0–7.5 region are all resonances of the one proton H^d.

The complexity of the alkene absorptions is the result of two factors. First, the three alkene protons are *chemically nonequivalent*, because protons H^b and H^c are diastereotopic and proton H^d is constitutionally nonequivalent to the others. Hence, the absorptions of all three protons occur at different chemical shifts. Second, *each proton is split by the other two with a different coupling constant*. When one set of protons is split by more than one chemically nonequivalent set, and *when all the coupling constants are different*, the splitting is in general *multiplicative*.

Multiplicative splitting can be visualized with the aid of a *splitting diagram*, which is given in Fig. 13.10 for the alkene protons of vinyl pivalate. In a **splitting diagram**, the resonance for each nonequivalent set of protons is drawn as a single, vertical unsplit line at the appropriate chemical shift with a height proportional to its integral. (In Fig. 13.10, the lines for the three alkene protons are the same size because each has the same integral.) The different splittings are then applied *successively* to each line using known values of the coupling constants to obtain the actual spectrum.

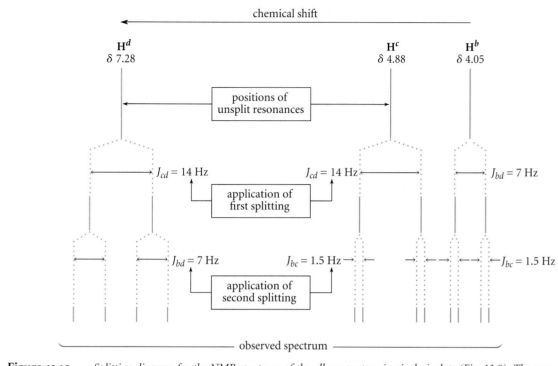

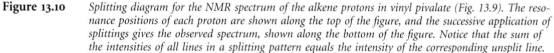

Figure 13.10   *Splitting diagram for the NMR spectrum of the alkene protons in vinyl pivalate (Fig. 13.9). The resonance positions of each proton are shown along the top of the figure, and the successive application of splittings gives the observed spectrum, shown along the bottom of the figure. Notice that the sum of the intensities of all lines in a splitting pattern equals the intensity of the corresponding unsplit line.*

Consider, for example, the splitting of the resonance for H^d. The single line for this proton is in the upper left-hand corner of Fig. 13.10. This proton is split by H^c into a doublet, and the coupling constant J_{cd} is 14 Hz. (Notice that the coupling constant is *not* being determined by this procedure; it is already known from other measurements.) Each line of the resulting splitting pattern has half the intensity of the original, and the distance of both lines from the original on the horizontal axis is the same. Then the splitting by H^b is applied to *each* of these new lines using the known coupling constant $J_{bd} = 7$ Hz. This gives two doublets, each with $J_{bd} = 7$ Hz, in which the intensity of each line is halved again. The resulting spectrum for H^d, then, is four lines of equal intensity, each with one-fourth the intensity of the original. The resonance for H^d is said to be a *doublet of doublets*, because the pattern results from successive application of two one-proton splittings. Notice that this is different from a quartet, in which the four lines have an intensity ratio of about 1:3:3:1 (Table 13.3).

The same process is illustrated in Fig. 13.10 for the other two alkene protons using the two coupling constants above along with $J_{bc} = 1.5$ Hz. Thus, the resonance for *each* proton appears as a doublet of doublets—that is, four lines each and a total of twelve for all alkene protons. (Compare this pattern with that in the actual spectrum; Fig. 13.9.)

The same end result is obtained if the splittings are applied in reverse order. Thus, for H^d, the 7 Hz splitting could have been applied first followed by the 14 Hz splitting. (Show that this is true by trying it yourself. Measure the splittings by using a ruler or graph paper.)

The analysis in Fig. 13.10 shows how a complex splitting pattern can arise from multiplicative splitting and demonstrates that such a pattern can be analyzed once the correct structure is known. What happens if you encounter a complex splitting pattern in the spectrum of an unknown? In such cases, a student who is just starting to learn how to interpret NMR spectra should bypass an analysis of the splitting and extract as much information as possible from the chemical shift and integral. Thus, if vinyl pivalate were an unknown, its NMR spectrum would indicate, from the integral of the δ 4–7.5 region, that three hydrogens are attached to a double bond, and from the nine-proton singlet at δ 1.26, that the molecule contains a *tert*-butyl group. This information, along with the molecular formula and the IR spectrum, would be sufficient to determine the structure. (An unknown involving complex splitting is solved in Study Problem 13.4 in the next section.)

Notice that complex multiplicative splitting arises because the various coupling constants involved are different. What if multiple splitting occurs, but the coupling constants involved happen to be identical? The NMR spectrum of 1-bromo-3-chloropropane is a situation of this type.

$$\overset{a}{Br-CH_2}-\overset{b}{CH_2}-\overset{c}{CH_2}-Cl$$

1-bromo-3-chloropropane

The set of protons H^b is split by the two nonequivalent sets H^a and H^c. If splitting of H^b were multiplicative, and if the two coupling constants were different, the triplet from splitting by H^a would be split again by H^c (or vice versa)—a triplet of triplets, or nine lines. However, as shown in Fig. 13.11a, the resonance for H^b consists of only five lines, exactly like the resonance for H^b in 1,3-dichloropropane (Fig. 13.7). The simpler spectrum occurs because the two coupling constants are equal; that is, $J_{ab} = J_{bc}$.

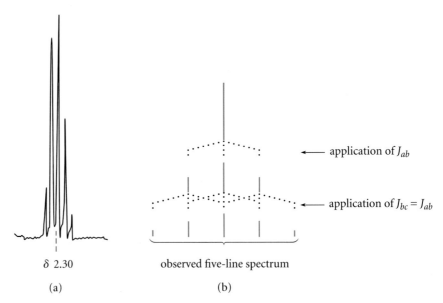

δ 2.30   observed five-line spectrum

← application of J_{ab}

← application of $J_{bc} = J_{ab}$

(a)   (b)

Figure 13.11   *(a) Observed NMR spectrum of the H^b protons in 1-bromo-3-chloropropane. (b) A splitting diagram for the H^b protons assuming equal coupling constants for their splitting with H^a and H^c.*

The reason that equal coupling constants result in a simplified spectrum is shown by the splitting diagram in Fig. 13.11b. Applying two successive splittings with equal J values indeed gives nine lines, but *many of the lines overlap*. The result is five lines with the characteristic intensities predicted by the $n + 1$ rule (Table 13.3).

To summarize: *When multiplicative splitting occurs and the coupling constants are equal, the splitting is as predicted by the $n + 1$ rule.* Although exceptions exist, it is common to find this simpler situation when the protons involved are on *adjacent saturated carbon atoms*. Notice, for example, that the unknown of Study Problem 13.3 is in this category; because all coupling constants are identical, the splittings are those predicted by the $n + 1$ rule.

PROBLEMS

*13.15   The three absorptions in the NMR spectrum of 1,1,2-trichloropropane have the following characteristics:

$$\overset{a}{Cl_2CH}-\overset{b}{CH}-\overset{c}{CH_3}$$
$$|$$
$$Cl$$

H^a: δ 5.82, J_{ab} = 3.5 Hz
H^b: δ 4.40, J_{ab} = 3.5 Hz, J_{bc} = 6.0 Hz
H^c: δ 1.78, J_{bc} = 6.0 Hz.

Using bars to represent lines in the spectrum and a splitting diagram to determine the appearance of the H^b absorption, sketch the appearance of the spectrum. (Graph paper is especially useful in constructing splitting diagrams.)

13.16   *(a)   Account for the fact that the resonance for H^a in 1,2-dichloropropane consists of sixteen lines.

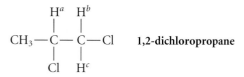

(b)   Assuming all lines could be observed, how many lines altogether would be present in the resonances of H^b and H^c? Explain.

B. Breakdown of the $n + 1$ Rule

Spectra in which all resonances conform to the $n + 1$ rule are called **first-order spectra**. In all the spectra discussed to this point, splitting patterns have been first-order, even the complex multiplicative patterns discussed in the previous section. The spectra of some compounds, however, contain splitting patterns that are more complex than predicted by the $n + 1$ rule. Although such spectra can be analyzed rigorously (in many cases) by special mathematical techniques, a great deal of information can be obtained from them without such sophisticated methods. The following study problem illustrates a situation of this sort.

Determine the structure of the compound with the formula C_4H_9Cl that has the NMR spectrum shown in Fig. 13.12.

Solution   The unknown compound has an unsaturation number of zero, and is therefore an alkyl halide. The spectrum contains a very complex splitting pattern in the δ 0.8–2.0 region that cannot be readily interpreted. Because the formula contains nine hydrogens, the integral of the entire compound (43.5 spaces) corresponds to 4.8 spaces/proton. One clear, first-order feature appears in the spectrum: the triplet at δ 3.55. Its chemical shift indicates that this triplet *must* arise from protons on the carbon that bears the chlorine. From its integral (9.5 spaces), this is a resonance for two protons. Its splitting shows that two protons are on the *adjacent* carbon. All this information together requires the partial structure —CH_2—CH_2—Cl. A compound with the formula C_4H_9Cl and this partial structure can *only* be 1-chlorobutane, $CH_3CH_2CH_2CH_2Cl$: structure solved!

Study Problem 13.4 shows that you don't have to interpret *every* splitting pattern in a spectrum to solve a structure because *most spectra contain redundant information.* (The same point was made in the previous section.)

Something very important about NMR, however, can be learned by asking why the NMR spectrum of 1-chlorobutane is so complex—why it is not first-order. It turns out that first-order NMR spectra are generally observed when *the chemical shift difference, in Hz, between coupled protons is much greater than their coupling constant.* If the difference in chemical shift of two resonances *a* and *b* is $\Delta\nu_{ab}$ (in Hz) and their coupling constant is J_{ab}, then this condition is simply expressed as follows:

Condition for first-order behavior: $\qquad\qquad\qquad \Delta\nu_{ab} \underset{\underset{\text{(in Hz)}}{\uparrow}}{\gg} J_{ab}$ $\qquad$ (13.5)

As a practical matter, if $\Delta\nu_{ab}$ is greater than J_{ab} by a factor of about ten or more, a first-order spectrum can be expected. Because the coupling constants in many aliphatic

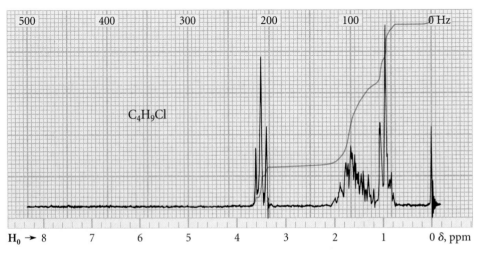

Figure 13.12   *NMR spectrum for Study Problem 13.4.*

compounds are on the order of 6–7 Hz, first-order behavior in their NMR spectra can be expected when the chemical shift difference between coupled protons is more than about 60 Hz, or 1 ppm at 60 MHz operating frequency. When the chemical shift difference of coupled protons is substantially less than this, non-first-order splitting patterns usually can be anticipated. In the spectrum of 1-chlorobutane (Fig. 13.12), the differences in the chemical shifts of H^b, H^c, and H^d are small (all occur within 1 ppm of each other).

$$\overset{d}{C}H_3\overset{c}{C}H_2\overset{b}{C}H_2\overset{a}{C}H_2 \!-\! Cl$$

very similar chemical shifts

Therefore, the $n + 1$ rule breaks down and a complex spectrum is observed. The resonance for H^a, on the other hand, is well separated from the remainder of the spectrum, and this resonance shows first-order splitting: It is a triplet, as predicted by the $n + 1$ rule.

A very useful fact about NMR can be utilized to simplify splitting patterns that are not first-order: *Coupling constants do not vary with the operating frequency.* Recall (Sec. 13.2B) that NMR experiments can be run at a variety of operating frequencies (ν_0 in Eq. 13.2b) and corresponding magnetic field strengths. Recall also that chemical shifts (*in Hz*) vary in proportion to the operating frequency used (Eq. 13.2b). Consequently, if a very large operating frequency and a correspondingly large magnetic field are used, the chemical shifts in Hz are much greater, but the coupling constants are unchanged. Consequently, the condition for first-order behavior in Eq. 13.5 is more likely to be met. This point is illustrated by the 360 MHz NMR spectrum of 1-chlorobutane (Fig. 13.13, p. 610); contrast this with the spectrum of the same compound taken at 60 MHz in Fig. 13.12. Notice that the spectrum in Fig. 13.13 is completely first-order; it conforms to the $n + 1$ splitting rule. The resonances for H^b and H^c are separated by about 0.3 ppm. At 60 MHz, this separation corresponds to only 18 Hz, a number not much larger than the coupling constant J_{bc} (about 7 Hz). At 360 MHz a 0.3 ppm chemical shift difference equals 108 Hz, more than ten times greater than J_{bc}, which remains unchanged at the higher operating frequency. Since at the higher frequency the condition of Eq. 13.5 for first-order spectra is met, a simplified spectrum is observed.

An additional (and related) benefit of using higher fields is that intensity ratios within splitting patterns more closely approach the idealized values in Table 13.3. Notice in Fig. 13.13 that leaning is much less pronounced. (Compare the triplet at δ 3.55 with the same absorption in Fig. 13.12.)

Instruments that operate at 300 MHz and higher are becoming increasingly common. Such instruments employ very high magnetic fields that are achieved with powerful superconducting magnets. The major use of such instruments is found not in obtaining the spectra of simple molecules, but rather in unraveling the structures of complex molecules whose NMR spectra would be hopelessly complicated when taken at lower field.

PROBLEM

13.17   *(a)   In a box labeled "C_4H_9Cl isomers" are two bottles, each containing a colorless liquid. The NMR of compound *A* is a singlet at δ 1.63, and the NMR of compound *B* is shown in Fig. 13.14 on page 610. Identify each of the compounds.

(b)   Predict how the spectrum of compound *B* would look if it were obtained on a high-field NMR spectrometer and the $n + 1$ splitting rule applied to all absorptions.

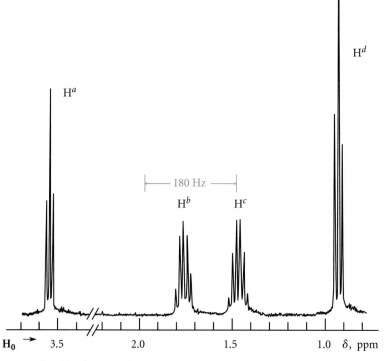

Figure 13.13 *NMR spectrum of 1-chlorobutane taken at an operating frequency of 360 MHz. Compare this spectrum with the 60 MHz spectrum of the same compound in Fig. 13.12.*

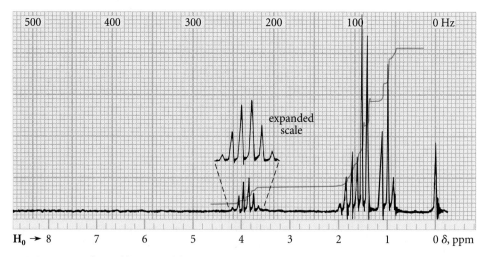

Figure 13.14 *NMR spectrum for Problem 13.17(a).*

13.5   Use of Deuterium in Proton NMR

Deuterium (^{2}H, or D) finds special use in proton (^{1}H) NMR. Although deuterium has a nuclear spin, deuterium NMR and proton NMR require different operating frequencies. Consequently, deuterium NMR absorptions are not detected under the conditions used for proton NMR. Therefore deuterium is effectively "silent" in proton NMR.

An important practical application of this fact is the use of deuterated solvents in NMR experiments. (Solvents are needed for solid and viscous liquid samples because, in the usual type of proton NMR experiment, the sample must be in a free-flowing liquid state.) In order that the solvent not interfere with the NMR spectrum of the sample, it must either be devoid of protons, or its protons must not have NMR absorptions that obscure the sample absorptions. Carbon tetrachloride (CCl_4) is a useful solvent because it has no protons, and therefore has no NMR absorption. However, many important organic compounds are not dissolved by carbon tetrachloride. Chloroform, $CHCl_3$, is an excellent solvent for many organic compounds, but it has an interfering NMR absorption. However, its deuterium analog, $CDCl_3$ (chloroform-*d*, or "deuterochloroform"), has no proton-NMR absorption, but has all the desirable solvent properties of chloroform. This solvent is so widely used for NMR spectra that it is a relatively inexpensive article of commerce. Other deuterated solvents are also available.

Another useful observation about deuterium is that the coupling constants for proton-deuterium splitting are very small. Even when H and D are on adjacent carbons, the H-D coupling is negligible. For this reason, deuterium substitution can be used to simplify NMR spectra and assign resonances. Although deuterium substitution is normally most useful in more complex molecules, let's see how it might be used to assign the resonances of ethyl bromide (Fig. 13.4). If you were to synthesize CH_3—CD_2—Br and record its NMR, you would find that the low-field quartet of ethyl bromide has disappeared from the NMR spectrum, and the high-field resonance is a singlet. This experiment would establish that the low-field quartet is the signal of the CH_2 group and that the high-field triplet is that of the CH_3 group.

PROBLEM

*13.18   The 60 MHz NMR spectrum of 1-chlorobutane, given in Fig. 13.12, is complex and not first-order. Assuming you could synthesize the needed compounds, explain how you would use deuterium substitution to determine the chemical shift of each chemically nonequivalent set of protons in the molecule. Explain what you would see and how you would interpret the results.

13.6   Characteristic Functional-Group NMR Absorptions

This section surveys the important NMR absorptions of the major functional groups that you've already studied. The NMR spectra of other functional groups will be considered in the chapters devoted to those groups. A summary table of chemical shift information is given in Appendix III.

A. NMR Spectra of Alkenes

Two characteristic proton NMR absorptions for alkenes are the absorptions for the protons on the double bond, called **vinylic protons** (color in the structures below), and the protons on carbons *adjacent* to the double bond, called **allylic protons**. Don't confuse these two types of protons. The shifts cited in line 7 of Table 13.2 are for protons *adjacent* to the alkene unit—the allylic protons. Typical alkene chemical shifts are illustrated in the structures below and are summarized in Table 13.4.

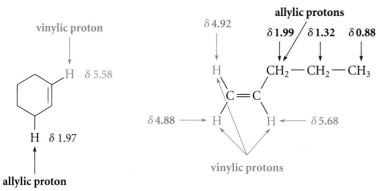

Notice that allyic protons have greater chemical shifts than ordinary alkyl protons, but considerably smaller chemical shifts than vinylic protons. Notice also that internal vinylic protons occur at somewhat lower field than terminal vinylic protons.

The chemical shifts of vinylic protons are much greater than would be predicted from the electronegativity of the alkene functional group and can be understood in the following way. Imagine that an alkene molecule in an NMR spectrometer is oriented with respect to the applied field $\mathbf{H_0}$ as shown in Figure 13.15. The applied field induces

Table 13.4   Chemical Shifts of Alkene Protons

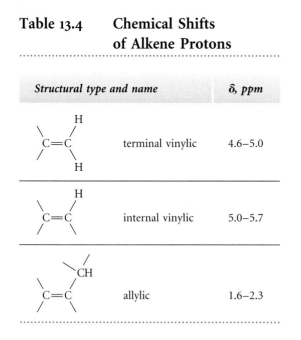

Structural type and name		δ, ppm
	terminal vinylic	4.6–5.0
	internal vinylic	5.0–5.7
	allylic	1.6–2.3

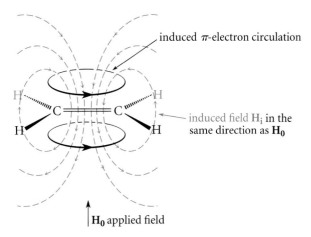

induced π-electron circulation

induced field H_i in the
same direction as H_0

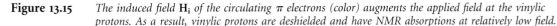

H_0 applied field

Figure 13.15   *The induced field H_i of the circulating π electrons (color) augments the applied field at the vinylic protons. As a result, vinylic protons are deshielded and have NMR absorptions at relatively low field.*

a circulation of the π electrons in closed loops above and below the plane of the alkene. This electron circulation gives rise to an induced magnetic field H_i that *opposes* the applied field H_0 at the center of the loop. This induced field can be described as contours of closed circles. Although the induced field opposes the applied field H_0 in the region of the π bond, the curvature of the induced field causes it to lie in the same direction as H_0 at the vinylic protons. The induced field therefore *augments* the applied field at the vinylic protons. As a result, the vinylic protons are *deshielded* and require a smaller external field to bring them into resonance. Consequently, their NMR absorptions occur at relatively low field.

> Of course, molecules in solution are constantly in motion, tumbling wildly. At any given time, only a fraction of the alkene molecules are oriented with respect to the external field as shown in Fig. 13.15. The chemical shift of a vinylic proton is an average over all orientations of the molecule. However, this particular orientation makes such a large contribution that it dominates the chemical shift.

Splitting between vinylic protons in alkenes depends strongly on the geometrical relationship of the coupled protons. Typical coupling constants are given in Table 13.5 on page 614. The spectra shown in Fig. 13.16 illustrate the very important observation that vinylic protons of a *cis*-alkene have smaller coupling constants than those of their *trans* isomers. (The same point is evident in the coupling constants of *cis* and *trans* protons shown in Figs. 13.9 and 13.10.) These coupling constants, along with the characteristic $=$C—H bending bands from IR spectroscopy (Sec. 12.4C), provide important ways to determine alkene stereochemistry. The very weak splitting (called *geminal splitting*) between vinylic protons on the *same* carbon stands in contrast to the much larger *cis* and *trans* splittings. Geminal splitting is also illustrated in Figs. 13.9 and 13.10.

The last two entries in Table 13.5 show that splitting in alkenes is sometimes observed between protons separated by more than three bonds. Recall that splitting over these distances is usually *not* observed in saturated compounds. These long-distance interactions between protons are transmitted by the π electrons.

In many spectra geminal, four-bond, and five-bond splittings are not readily discernible as clearly separated lines, but instead are manifested as perceptibly broadened peaks. Such is the case, for example, in the NMR spectrum in Fig. 13.17 (Problem 13.19).

Table 13.5   Coupling Constants for Proton Splitting in Alkenes

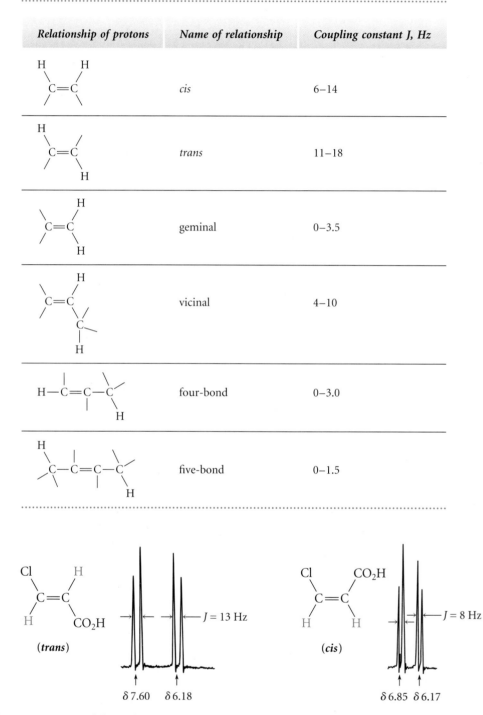

Relationship of protons	Name of relationship	Coupling constant J, Hz
	cis	6–14
	trans	11–18
	geminal	0–3.5
	vicinal	4–10
	four-bond	0–3.0
	five-bond	0–1.5

Figure 13.16   *NMR spectra of the vinylic protons (color) of cis-trans isomers. Notice the larger coupling constants for the trans protons.*

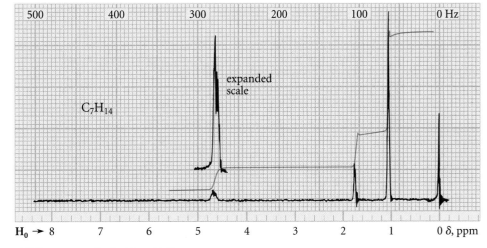

Figure 13.17   *NMR spectrum for Problem 13.19.*

*13.19   Propose a structure for a compound with the formula C_7H_{14} with the NMR spectrum shown in Fig. 13.17. Explain in detail how you arrived at your structure.

B. NMR Spectra of Alkanes and Cycloalkanes

Because all of the protons in a typical alkane are in very similar chemical environments, the NMR spectra of alkanes and cycloalkanes cover a very narrow range of chemical shifts, typically δ 0.7 to δ 1.5. Because of this narrow range, the splitting in such spectra shows extensive non-first-order behavior.

An exception to these generalizations is the chemical shifts of protons on a cyclopropane ring, which are unusual for alkanes; they absorb at rather high field, typically δ 0 to δ 0.5. Some even have resonances at *higher* field than TMS (that is, negative δ values). For example, the chemical shifts of the ring protons of *cis*-1,2-dimethylcyclopropane shown in color below are δ (−0.11).

The reason for this unusual chemical shift is an induced electron current in the cyclopropane ring and the resulting magnetic field, similar in principle to the one responsible for the large chemical shifts in alkenes (Fig. 13.15). In cyclopropanes, however, the induced field is oriented so that the chemical shifts of the cyclopropane protons are *decreased*.

C. NMR Spectra of Alkyl Halides and Ethers

Several NMR spectra of alkyl halides and ethers have been presented earlier in this chapter. The chemical shifts caused by the halogens are usually in proportion to their electronegativities. For the most part, chloro groups and ether oxygens have about the same chemical-shift effect on neighboring protons (Table 13.2). However, epoxides, like cyclopropanes, have considerably smaller chemical shifts than their open-chain analogs.

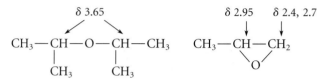

An interesting type of splitting is observed in the NMR spectra of compounds containing fluorine. The common isotope of fluorine (^{19}F) has a spin of $\pm\frac{1}{2}$. Proton resonances are split by neighboring fluorine in the same general way that they are split by neighboring protons; the same $n + 1$ splitting rule applies. For example, the proton in $HCCl_2F$ appears as a doublet centered at δ 7.43 with a large coupling constant J_{HF} of 54 Hz. Notice that this is *not* the NMR spectrum of the fluorine; it is the *splitting of the proton spectrum caused by the fluorine*. (It is also possible to do fluorine NMR, but this requires, for the same magnetic field, a different operating frequency; the spectra of ^{1}H and ^{19}F do not overlap.) Values of H—F coupling constants are larger than H—H coupling constants. The J_{HF} value in $(CH_3)_3C$—F is 20 Hz; a typical J_{HH} value over the same number of bonds is 6–8 Hz. Because J_{HF} values are so large, coupling between protons and fluorines can sometimes be observed over as many as four single bonds. The common isotopes of chlorine, bromine, and iodine also have nuclear spins, but they do not cause detectable proton splittings.

PROBLEMS

13.20 Suggest structures for compounds with the following proton NMR spectra.
*(a) $C_3H_5F_2Cl$: δ 1.75 (3*H*, t, *J* = 17.5 Hz); δ 3.63 (2*H*, t, *J* = 13 Hz)
(b) $C_4H_{10}O$: δ 1.13 (3*H*, t, *J* = 7 Hz); δ 3.38 (2*H*, q, *J* = 7 Hz)

*13.21 How would the NMR spectrum of ethyl fluoride differ from that of ethyl chloride?

D. NMR Spectra of Alcohols

Protons on the α-carbons of primary and secondary alcohols generally have chemical shifts in the same range as ethers, from δ 3.0 to δ 4.1 (see Table 13.2). Since tertiary alcohols have no α-protons, the observation of an OH stretching absorption in the IR spectrum accompanied by the *absence* of the CH—O absorption in the NMR is good evidence for a tertiary alcohol (or a phenol; see Sec. 16.3B).

$$CH_3\text{—}OH \qquad CH_3\text{—}CH_2\text{—}OH \qquad (CH_3)_2CH\text{—}OH \qquad (CH_3)_3C\text{—}OH$$

↑	↑	↑	no proton absorption in δ 3–4 region
δ 3.5	δ 3.6	δ 4.0	

The chemical shift of the OH proton in an alcohol depends on the degree to which the alcohol is involved in hydrogen bonding under the conditions that the spectrum is determined. For example, in pure ethanol, in which the alcohol molecules are extensively hydrogen bonded, the chemical shift of the OH proton is δ 5.3. When ethanol is dissolved in CCl_4, the ethanol molecules are more dilute and less extensively hydrogen bonded, and the OH absorption occurs at δ 2 to δ 3. In the gas phase, there is almost no hydrogen bonding, and the OH resonance of ethanol occurs at δ 0.8.

> The chemical shift of the O—H proton in the gas phase is not as large as might be expected for a proton bound to an electronegative atom such as oxygen. The unusually high-field absorption of unassociated OH protons is probably due to the induced field caused by circulation of the lone-pair electrons on oxygen. This field shields the OH proton from the applied field (Sec. 13.2D). Hydrogen-bonded protons, on the other hand, absorb at lower field because they bear less electron density and more positive charge (Sec. 8.3C).

The splitting between the OH proton of an alcohol and neighboring protons is interesting. The $n + 1$ splitting rule predicts that the OH resonance of ethanol should be a triplet, and the CH_2 resonance should be split by both the adjacent CH_3 and OH protons, and should therefore appear as (4×2) or eight lines (multiplicative splitting; Sec. 13.4A).

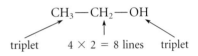

The NMR spectrum of very dry ethanol, shown in Fig. 13.18a, is as expected. However, when a trace of water, acid, or base is added to the ethanol, the spectrum changes, as shown in Fig. 13.18b. *The presence of water, acid, or base causes collapse of the* OH *resonance to a single line*, and obliterates all splitting associated with this proton. Thus, the CH_2 proton resonance becomes a quartet, apparently split only by the CH_3 protons. *This type of behavior is quite general for alcohols, amines, and other compounds with a proton bonded to an electronegative atom.*

This effect is caused by a phenomenon called **chemical exchange**: an equilibrium involving chemical reactions that take place very rapidly as the NMR spectrum is being determined. In this case, the chemical reaction is proton exchange between the protons of the alcohol and those of water (or other alcohol molecules):

$$R-\overset{\cdot\cdot}{\underset{H}{O}}{:} + H-\overset{+}{\overset{\cdot\cdot}{O}}H_2 \;\rightleftarrows\; R-\overset{\cdot\cdot}{\underset{H}{O}}-H + {:}\overset{\cdot\cdot}{O}H_2 \tag{13.6a}$$

$$R-\overset{+}{\underset{\underset{:\overset{\cdot\cdot}{O}H_2}{H}}{\overset{\cdot\cdot}{O}}}-H \;\rightleftarrows\; R-\overset{\cdot\cdot}{\overset{\cdot\cdot}{O}}-H + H-\overset{+}{\overset{\cdot\cdot}{O}}H_2 \tag{13.6b}$$

This exchange is, of course, nothing more than two successive acid-base reactions. (Write the mechanism for ⁻OH-catalyzed exchange.) For reasons that are discussed in Sec. 13.7, *rapidly exchanging protons do not show spin-spin splitting with neighboring protons.* Acid and base catalyze this exchange reaction, accelerating it enough that splitting is no longer

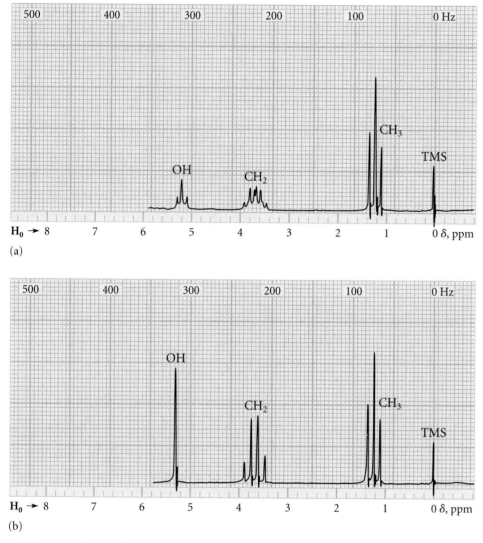

Figure 13.18   *NMR spectra of ethanol. (a) Absolute, or dry, ethanol. Notice that the CH$_2$ resonance is split by both the OH and the CH$_3$. (b) Wet, acidified ethanol. Notice that the CH$_2$ resonance in this spectrum is split only by the CH$_3$.*

observed. In the absence of acid or base, this exchange is much slower, and splitting of the OH proton and neighboring protons is observed.

As a practical matter, many NMR samples of alcohols contain just enough moisture to obliterate the α-proton splitting, but the exchange of the water and the OH proton of the alcohol is slow enough that the OH peak is broadened (that is, it is "fat"). The assignment of the OH proton can be confirmed by what is called "the D$_2$O shake." If a drop of D$_2$O is added to the NMR sample tube, and the tube is shaken, the OH protons rapidly exchange with the protons of D$_2$O to form OD groups on the alcohol and, as a result, become invisible in NMR. The HOD produced by the exchange floats to the top of the CDCl$_3$ or CCl$_4$ solvent, out of the area covered by the detector.

PROBLEMS

13.22   Suggest structures for each of the following compounds.

*(a) $C_5H_{10}O$: δ 1.71 (s); δ 1.78 (s), total integral of both resonances 6*H*; δ 2.31 (1*H*, broad s; disappears after D_2O shake); δ 4.14 (2*H*, d, *J* = 7 Hz); δ 5.41 (1*H*, broad t, *J* = 7 Hz)

(b) $C_4H_{10}O$; δ 1.20 (9*H*, s); δ 2.40 (1*H*, broad s; disappears after D_2O shake)

*13.23   In which solution does the OH proton of ethanol have the greater chemical shift: 2 *M* ethanol in CCl_4 or 0.2 *M* ethanol in CCl_4? Explain.

13.7   NMR Spectroscopy of Dynamic Systems

Consider the NMR spectrum of cyclohexane, a molecule that undergoes rapid conformational isomerization (the chair flip; Sec. 7.2B). The NMR spectrum of cyclohexane consists of a singlet at δ 1.4. Yet cyclohexane has two diastereotopic, and therefore chemically nonequivalent, sets of hydrogens: the *axial* hydrogens and the *equatorial* hydrogens. Why shouldn't cyclohexane have two resonances, one for each type of hydrogen? The reason has to do with the *rate* of the chair flip, which is so rapid that the NMR instrument detects only the average of the two conformations. Because the chair flip interchanges the positions of axial and equatorial protons (Eq. 7.5), only one proton resonance is observed. This is the resonance of the "average" proton in cyclohexane—one that is axial half the time and equatorial half the time. This example illustrates an important aspect of NMR spectroscopy: *the spectrum of a compound involved in a rapid equilibrium is a single spectrum that is the time-average of all species involved in the equilibrium.*

> Although there are equations that describe this phenomenon exactly, it can be understood by the use of an analogy from common experience. Imagine looking at a three-blade fan or propeller that is rotating at a speed of about 100 times per second (Fig. 13.19). Our eyes do not see the individual blades, but only a blur. The appearance of the blur is a time-average of the blades and the empty space between them. If we photograph the fan using a shutter speed of about 0.1 second, the fan appears as a blur in the resulting picture for the same reason: during the time the

(a) 0.1 sec; image totally blurred

(b) 0.001 sec; individual blades
      visible but blurred

(c) 0.00001 sec; individual blades
      visible and in sharp focus

Figure 13.19   *Imagine a three-blade propeller rotating at about 100 times per second (100 Hz). The diagrams show what would be seen in snapshots taken at a shutter speed of (a) 0.1 sec, (b) 0.001 sec, and (c) 0.00001 sec.*

camera shutter is open (0.1 sec) the blades make ten full revolutions (Fig. 13.19a). Now imagine taking a picture with a shutter speed of 0.001 second. While the shutter is open, the fan blades make only 0.1 revolution—about 36°. The fan blades are more distinct, but still somewhat blurred (Fig. 13.19b). Finally, imagine taking a picture with a shutter speed of 0.00001 second (very fast film). While the shutter is open, the fan blades traverse only $\frac{1}{1000}$ of a circle—about 0.36°. In the resulting picture the individual blades are visible and in relatively sharp focus (Fig. 13.19c). The rapid conformational flipping of cyclohexane is to the NMR spectrometer roughly what the rapidly rotating propeller is to the slow camera shutter. *The NMR spectrometer is intrinsically limited in its capacity to resolve events in time.*

Both types of cyclohexane protons can be observed if the rate of the chair flip is reduced by lowering the temperature. Imagine cooling a sample of cyclohexane in which all protons but one have been replaced by deuterium. (The use of deuterium significantly reduces splitting with neighboring protons; Sec. 13.5.)

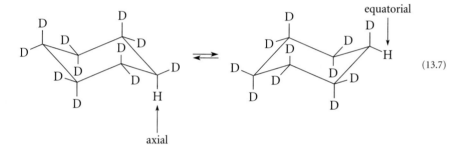

$$(13.7)$$

cyclohexane-d_{11}

In the chair flip, the single remaining proton alternates between axial and equatorial positions.

The NMR spectrum of cyclohexane-d_{11} at various temperatures is shown in Fig. 13.20. At room temperature, the spectrum consists of a single line, as in cyclohexane itself. As the temperature is lowered progressively, the resonance becomes broader until, just below $-60°$, it divides into two resonances equally spaced about the original one. When the temperature is lowered still further, the spectrum becomes two sharp single lines. Thus, lowering the temperature progressively retards the chair flip until, at low temperature, NMR spectrometry can detect both chair forms independently. This is analogous to taking pictures of the propeller in Fig. 13.19 with a constant shutter speed and slowing down the propeller until the blur disappears and the individual blades become clearly separated.

It is possible to use the information from these spectra at different temperatures to calculate the rate of the chair flip. The energy barrier for the chair flip shown in Fig. 7.5 was obtained from this type of calculation.

The time-averaging effect of NMR is not limited simply to conformational equilibria. The spectra of molecules undergoing any rapid process, such as a chemical reaction, are also averaged by NMR spectroscopy. This is the reason, for example, that the splitting associated with the OH protons of an alcohol is obliterated by chemical exchange (Sec. 13.6D). For example, consider the effects of chemical exchange on the spectrum of the CH_3 protons of methanol. In absolutely dry methanol, the resonance of these protons is split by the OH proton into a doublet. Recall (Fig. 13.8) that this splitting occurs because the adjacent OH proton can have either of two spins. If acid or base is added to the

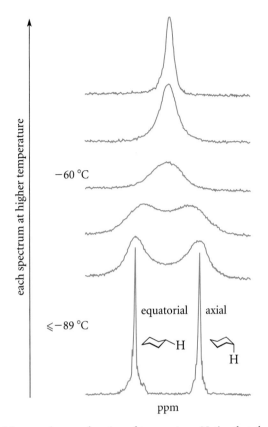

Figure 13.20   *NMR spectrum of cyclohexane-d$_{11}$ as a function of temperature. Notice that the axial and equatorial protons are separately observable at low temperature because the chair flip is slow.*

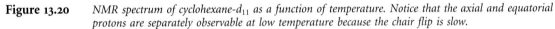

methanol, causing the OH protons to exchange rapidly, protons of different spins jump quickly on and off the OH. Thus, the CH$_3$ protons *on any one molecule* are next to an OH proton with spin $+\frac{1}{2}$ about half of the time, and an OH proton with spin $-\frac{1}{2}$ about half of the time. (The very small difference between the numbers of protons in the two spin states can be ignored.) In other words, the CH$_3$ protons "see" an adjacent OH proton with a spin that averages to zero over time. Because a proton is not split by an adjacent nucleus with zero spin, rapid exchange eliminates splitting of the CH$_3$ protons. Similar reasoning can be applied to the spectrum of the methanol OH proton, which, in a dry sample, is a quartet, but is a singlet in a sample containing traces of moisture.

PROBLEMS

*13.24   Suppose you were able to cool a sample of 1-bromo-1,1,2-trichloroethane enough that rotation about the carbon-carbon bond becomes slow on the NMR time scale. What changes in the NMR spectrum would you anticipate? Be explicit.

13.25   Describe in detail what changes you would expect to see in the NMR spectrum of the methyl group as 1-chloro-1-methylcyclohexane is cooled from room temperature to very low temperature.

13.8   Carbon NMR

STUDY GUIDE LINK:
13.5
*NMR of Other
Nuclei*

Although we've concentrated our attention on proton NMR, any nucleus with a nuclear spin can be studied by NMR spectroscopy. Table 13.6 lists a few other nuclei with spin $= \pm\frac{1}{2}$. For a given magnetic field strength, different nuclei absorb energy in different frequency ranges. For example, in the proton NMR spectrum of an alkyl fluoride, the NMR signals of protons are observed, but not those of the fluorines. (The proton splitting *caused* by the fluorines *is* observed, however; Sec. 13.6C.) To observe the NMR of the fluorines, a different frequency range is used, in which case the fluorine resonances but not the proton resonances are observed. (In this situation, the splitting of fluorine signals caused by nearby protons would be observed.)

Because organic compounds by definition contain carbon, the NMR spectroscopy of carbon, called **carbon NMR**, or **CMR**, has undergone rapid development. Unfortunately, as Table 13.6 shows, the only isotope of carbon that has a nuclear spin is ^{13}C. Organic compounds contain only about 1.1% of ^{13}C at each carbon position. The relative abundance of ^{13}C suggests that carbon NMR spectra should be about 1.1% as intense as proton NMR spectra. It happens that the resonance of a ^{13}C nucleus is also *intrinsically* weaker than that of a proton because of the magnetic properties of the carbon nucleus. Taking both natural abundance and intrinsic sensitivity into account, carbon spectra have only about 0.002 times the intensity of proton spectra. The weak ^{13}C NMR resonance at one time presented a serious detection problem, but advances in instrumentation have made it possible to obtain CMR spectra on compounds containing the *natural abundance* of ^{13}C on a routine basis. As a result, CMR has become a very important spectroscopic technique in organic chemistry.

Although the principles of CMR and proton NMR are essentially the same, some aspects of CMR are unique. First, coupling (splitting) between carbons is *not* generally observed. The reason is the low natural abundance of ^{13}C. Recall that CMR measures the resonance of ^{13}C, not the common isotope ^{12}C. If the probability of finding a ^{13}C at a given carbon is 0.0110, then the probability of finding ^{13}C at any two carbons in the same molecule is $(0.0110)^2$, or 0.00012. This means that *two ^{13}C atoms almost never occur together within the same molecule.* (The two ^{13}C atoms would have to occur in the

Table 13.6   Properties of Some Nuclei with Spin $\pm\frac{1}{2}$

Isotope	Relative sensitivity	Natural abundance, %	Magnetogyric ratio[a]
^1H	(1.00)	99.98	26,753
^{13}C	0.0159	1.10	6,728
^{19}F	0.834	100	25,179
^{31}P	0.0665	100	10,840

[a] In radians gauss^{-1} sec^{-1}; defined in Study Guide Link 13.5.

Table 13.7   Typical CMR Chemical Shift Ranges for Common Functional Groups

Functional group	δ, ppm[a]	Functional group	δ, ppm[a]
$-CH_3, -CH_2-,$ $-CH, -C-$	0–55	$-C-O-$	40–80
		$-C-N-$	25–70
$C=C$	100–145	$-\overset{O}{\overset{\|}{C}}-$	190–220
$-C\equiv C-$	65–85		
$C-$ (aromatic)	110–160		
$-\overset{\|}{\underset{\|}{C}}-Cl$	20–60	$-\overset{O}{\overset{\|}{C}}-O-$	160–190
$-\overset{\|}{\underset{\|}{C}}-Br$	10–50	$-\overset{O}{\overset{\|}{C}}-N$	160–180
$-\overset{\|}{\underset{\|}{C}}-I$	−15–25	$-C\equiv N$	110–130

[a] Downfield from TMS.

same molecule for coupling to be observed.) Of course, compounds that are isotopically enriched in ^{13}C can be prepared, in which case the regular splitting rules apply (see Problem 13.53).

A second important aspect of CMR is that the range of chemical shifts is very large compared to that in proton NMR. Typical carbon chemical shifts, shown in Table 13.7, cover a range of about 200 ppm. With a few exceptions, trends in carbon chemical shifts parallel those for proton chemical shifts; but chemical shifts in CMR are more sensitive to small changes in chemical environment. As a result, it is often possible to observe distinct resonances for two carbons in very similar chemical environments. This point is illustrated in the CMR spectrum of 3-methylpentane, in which each chemically nonequivalent set of carbons gives a separate, clearly discernible resonance:

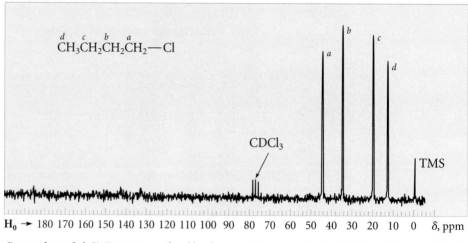

Figure 13.21   *Proton-decoupled CMR spectrum of 1-chlorobutane. (The reason for the CDCl$_3$ triplet is considered in Problem 13.43.)*

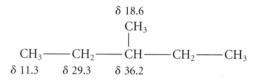

As this example shows, highly branched carbons typically have greater chemical shifts than unbranched ones.

A third aspect of CMR is that the splitting of ^{13}C resonances by protons (^{13}C—^{1}H splitting) is large; typical coupling constants are 120–200 Hz for directly attached protons. Furthermore, carbon NMR signals are also split by more remote protons. Although such splitting can sometimes be useful, more typically it presents a serious complication in the interpretation of CMR spectra. In most CMR work, splitting is eliminated by a special technique called *proton spin decoupling*. Spectra in which proton coupling has been eliminated are called **proton-decoupled CMR spectra**. In such spectra a *single unsplit line* is observed for each chemically nonequivalent set of carbon atoms.

These points are illustrated by the proton-decoupled CMR spectrum of 1-chlorobutane, shown in Fig. 13.21. The carbon spectrum consists of four single lines, one for each carbon of the molecule. The assignment of the lines in Fig. 13.21 shows that carbon chemical shifts, like proton chemical shifts, decrease with distance from the electronegative chlorine. (Contrast this very simple spectrum with the complex proton NMR spectrum of the same compound in Fig. 13.12.)

CMR is particularly useful in differentiating closely related compounds on the basis of their molecular symmetry. The basis of this idea is that symmetrical compounds have fewer chemically nonequivalent sets of carbons than less symmetrical isomers. This point is illustrated in the following study problem.

STUDY PROBLEM 13.5

How would you use CMR spectroscopy to differentiate the two isomers 1-chloropentane and 3-chloropentane?

Solution   The *first* thing to do is draw the structures:

$$CH_3CH_2CH_2CH_2CH_2Cl$$

1-chloropentane

$$\overset{\displaystyle Cl}{\overset{\displaystyle |}{CH_3CH_2CHCH_2CH_3}}$$

3-chloropentane

Assuming that a separate resonance is observed for each chemically nonequivalent set of carbons, then the proton-decoupled CMR spectrum of 1-chloropentane should consist of five lines, but that of 3-chloropentane should consist of only three lines; the two CH_3 carbons are chemically equivalent, and the two CH_2 carbons are equivalent. As this example shows, if a molecule has symmetry, it will have fewer absorptions than there are carbons.

PROBLEM

*13.26   The isomers 3-heptanol (*A*) and 4-heptanol (*B*) are difficult to distinguish by either their IR or proton NMR spectra. The proton-decoupled CMR spectra of these compounds are given in Fig. 13.22 on p. 626. Indicate which compound goes with each spectrum, and explain your reasoning.

CMR spectra are generally not integrated because the instrumental technique used for taking the spectra (Sec. 13.10) gives relative peak integrals that are governed by factors other than the number of carbons. For example, the decoupling technique enhances the peaks of carbons that bear hydrogens; hence, peaks for carbons that bear *no* hydrogens are usually smaller than those for other carbons.

The utility of CMR is enhanced by a number of techniques that provide a count of the protons directly attached to each carbon. In other words, it is possible to determine which of the carbon signals in a CMR spectrum come from methyl, methylene, methine, and quaternary carbons. The best current technique for making such a determination is known by the acronym **DEPT** (for "Distortionless Enhancement with Polarization Transfer"). The DEPT technique yields separate spectra for methyl, methylene, and methine carbons, and each line in these spectra corresponds to a line in the complete CMR spectrum. Lines in the complete CMR spectrum that do not appear in the DEPT spectra are assumed to arise from quaternary carbons. This technique is illustrated with the DEPT spectra of camphor (Fig. 13.23, p. 627).

STUDY PROBLEM 13.6

A compound $C_7H_{16}O_3$ has the following CMR-DEPT spectrum (the numbers in parentheses indicate the number of attached hydrogens):

$$\delta\ 15.2\ (3),\ \delta\ 59.5\ (2),\ \delta\ 112.9\ (1)$$

Propose a structure for this compound.

Solution   The compound has no rings or double bonds, because its unsaturation number is zero. The simplest assumption from the CMR spectrum is that the compound has three chemically nonequivalent sets of carbons, because there are three lines. One set

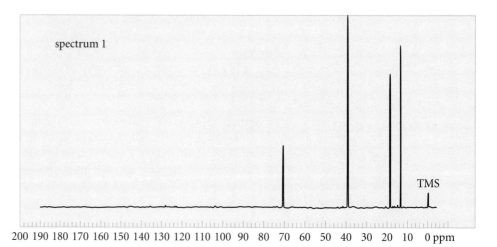

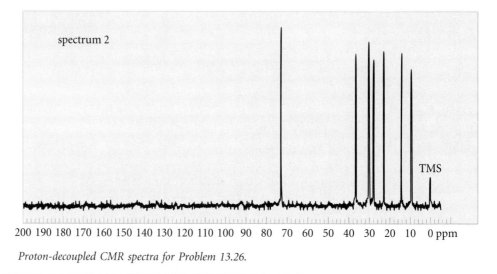

Figure 13.22   *Proton-decoupled CMR spectra for Problem 13.26.*

(δ 15.2) consists of methyl groups (three attached hydrogens) which, from their chemical shift, are not very close to the oxygens.

Another set consists of methylene (CH_2) groups, which are within the chemical shift range for the α-carbons of ethers (Table 13.7).

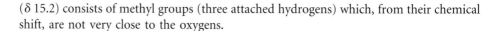

(*Study Problem 13.6 continues on p. 628*)

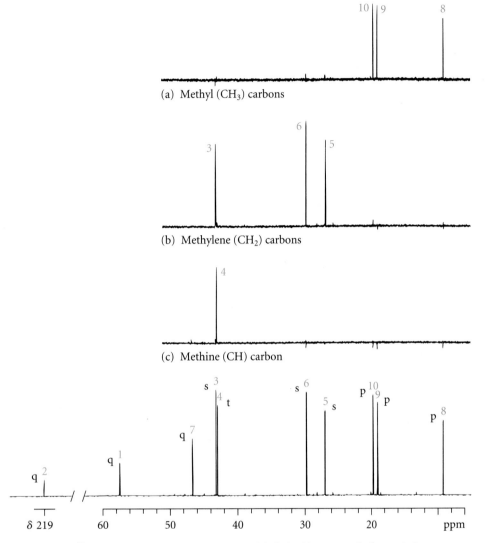

(a) Methyl (CH₃) carbons

(b) Methylene (CH₂) carbons

(c) Methine (CH) carbon

(d) Complete ^{13}C spectrum. The methyl carbons are labeled p (for primary), the methylene carbons s (for secondary), the methine carbon t (for tertiary), and the quarternary carbons q.

Figure 13.23   *The CMR spectrum of camphor (structure at right) edited by the DEPT technique. The absorptions for the methyl carbons are given in (a), the methylene carbons in (b), and the methine carbon in (c). Each peak in these three spectra corresponds to a peak in the full spectrum, shown in (d). Absorptions in the full spectrum that do not appear in (a), (b), or (c) are quaternary carbons. The number over each peak is the assignment using the carbon number in the camphor structure at the right. (Courtesy John Kozlowski, Purdue University.)*

The last set consists of one or more methine (CH) groups, which, from the chemical shift, must be bound to more than one oxygen.

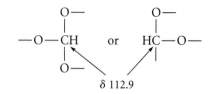

$$\delta \ 112.9$$

Only the structure of triethoxymethane gives only three absorptions and at the same time accommodates these partial structures:

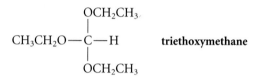

13.27   Explain why each of the following structures is *not* consistent with the CMR data in Study Problem 13.6.

*(a)

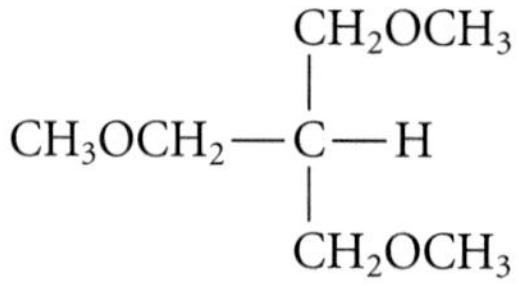

$$CH_3OCH_2 - \overset{\displaystyle CH_2OCH_3}{\underset{\displaystyle CH_2OCH_3}{\overset{|}{\underset{|}{C}}} - H}$$

(b)

$$HO - \overset{\displaystyle OCH_2CH_2CH_3}{\underset{\displaystyle OCH_2CH_2CH_3}{\overset{|}{\underset{|}{C}}} - H}$$

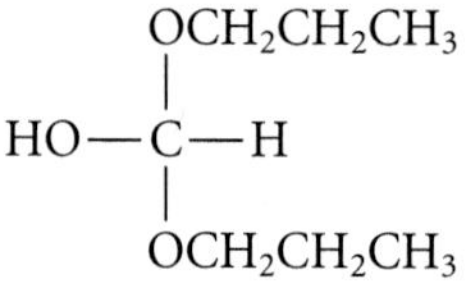

*(c)

$$CH_3CH_2O - \overset{\displaystyle OCH_2CH_3}{\underset{\displaystyle OCH_3}{\overset{|}{\underset{|}{C}}} - CH_3}$$

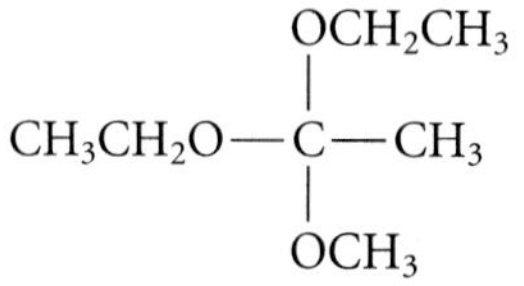

13.9   Solving Structure Problems with Spectroscopy

You are now ready to use what you know about IR, NMR, and mass spectrometry to solve some problems that use more than one of these techniques simultaneously. The following study problems illustrate the techniques involved. Although no simple method works in every case, the following suggestions should prove useful.

1. From the mass spectrum determine, if possible, the molecular mass.

2. If an elemental analysis is given, calculate the molecular formula and determine the unsaturation number.

3. Look for evidence in both the IR and NMR spectra for any functional groups that are consistent with the molecular formula: OH groups, alkenes, etc. Write down any structural fragments indicated by the spectra.

4. Use the CMR spectrum and, if possible, the proton NMR spectrum, to determine the number of nonequivalent sets of carbons and/or protons. If the proton NMR spectrum is complex, this may not be possible, but you should be able to set some limits.

5. Apply the suggestions in both Sec. 13.2F and Sec. 13.3C to complete your analysis by NMR. *Be sure to write out partial structures and all possible complete structures that are consistent with your spectra.* As you write out partial structures, notice how many carbons are unaccounted for; different partial structures may have carbons in common. Decide between possible structures by asking what features of the different spectra would be expected for each, and look for those features; it is sometimes easy to overlook some feature of a spectrum that will decide between structures.

6. Finally, rationalize all spectra for consistency with the proposed structure.

STUDY PROBLEM 13.7

Propose a structure for the compound with the IR, NMR, and mass spectra shown in Fig. 13.24 on p. 630.

Solution   The mass spectrum of this compound shows a *doublet* at $m/z = 90$ and 92, with the latter peak about one-third the size of the former. This pattern indicates the presence of chlorine. Furthermore, the base peak at $m/z = 55$ corresponds to a loss of Cl (35 and 37 mass units, respectively). Let's adopt the hypothesis that this is a chlorine-containing compound with molecular mass of 90 (for the ^{35}Cl isotope). In the IR spectrum, the peak at 1642 cm^{-1} suggests a C=C stretch, and, in the NMR spectrum, there is a complex signal in the vinylic proton region. Evidently this compound is a chlorine-containing alkene. If the molecular mass is indeed 90, then chlorine and two alkene carbons account for 59 mass units; ony 31 mass units remain to be accounted for.

In the NMR spectrum, the total integral is 43 spaces; the vinylic protons at δ 4.9–6.3 account for 17.5 spaces, or 40.7% of the integral. The quintet at δ 4.5 accounts for 6 integral spaces, or 14% of the integral. The doublet at δ 1.6 accounts for the remaining 19.5 integral spaces, or 45.3% of the total integral. The integrals of the three sets of resonances are (in order from lowest field) in the ratio 17.5:6:19.5, or about 3:1:3, to nearest whole numbers. The integral suggests some multiple of seven protons. If the compound has seven protons (7 mass units) and one chlorine (35 mass units), then the remaining 48 mass units can be accounted for by four carbons, two of which are part of an alkene double bond. A possible molecular formula is then C_4H_7Cl. (Would fourteen protons be a likely possibility? Why or why not?)

The unsaturation number for this formula is one. Hence, there can be only one double bond in the molecule. Since the NMR integral indicates three vinyl protons, then the molecule must contain a $-CH=CH_2$ group. In the IR spectrum the peaks at 930 and 990 cm^{-1} are consistent with such a group, although the former peak is at somewhat higher wavenumber than is usual for this type of alkene. The three-proton doublet at δ 1.6 suggests a methyl group adjacent to a CH group.

$$CH_3-CH-$$
$$\vert$$

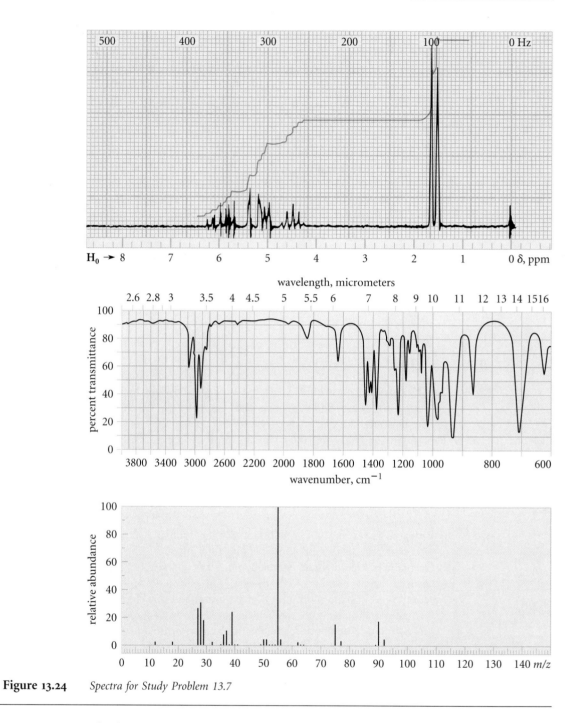

Figure 13.24 *Spectra for Study Problem 13.7*

The δ 4.5 absorption accounts for one proton, and its coupling constant matches that of the absorption at δ 1.6. The splitting and chemical shift of the δ 4.5 absorption fit the partial structure

$$CH_3\!-\!\underset{\underset{Cl}{|}}{CH}\!-\!CH\!=\!$$

With a molecular mass of 90 and three vinyl protons, the only possible complete structure is

**STUDY
PROBLEM
13.8**

A compound $C_8H_{18}O_2$ with a strong, broad infrared absorption at 3293 cm^{-1} has the following proton NMR spectrum:

$$\delta\ 1.22\ (12H,\ s);\ \delta\ 1.57\ (4H,\ s);\ \delta\ 1.96\ (2H,\ s)$$

(The resonance at δ 1.96 disappears when the sample is shaken with D_2O.) The proton-decoupled CMR of this compound consists of three lines, with the following chemical shifts and DEPT data (in parentheses) for attached protons:

$$\delta\ 29.4\ (3),\ \delta\ 37.8\ (2),\ \delta\ 70.5\ (0)$$

Identify the compound.

Solution   The IR spectrum indicates the presence of an alcohol, and the disappearance of the δ 1.96 NMR absorption after the "D_2O shake" (Sec. 13.6D) provides confirmation. Furthermore, because this absorption integrates for two protons, and because the formula contains two oxygens, the compound is a diol. Because the proton NMR spectrum contains no absorptions in the δ 3–4 region, both alcohols must be tertiary. The proton NMR indicates only three chemically nonequivalent sets of hydrogens, and the CMR indicates only three chemically nonequivalent sets of carbons, one of which must be the two α-carbons of the tertiary alcohol groups. The DEPT data confirm that one set of carbons indeed has no attached protons, as expected for a tertiary alcohol, and the chemical shift is consistent with that expected for the α-carbon of an alcohol. The presence of only *three* nonequivalent sets of protons and *three* nonequivalent sets of carbons requires a structure of considerable symmetry. The *only* structure that fits these data is

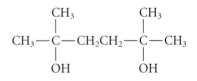

2,5-dimethyl-2,5-hexanediol

PROBLEMS

13.28   Tell why each of the following structures is not consistent with the data in Study Problem 13.7.
*(a) *trans*-1-chloro-2-butene   (b) 2-chloro-1-butene

13.29   Tell why each of the following structures is not consistent with the spectroscopic data in Study Problem 13.8.

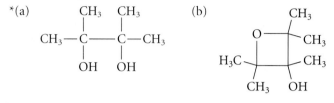

13.10   The NMR Spectrometer

The basic components of an NMR instrument are shown in Fig. 13.25. This diagram can be related to the description of the NMR experiment in Sec. 13.2A.

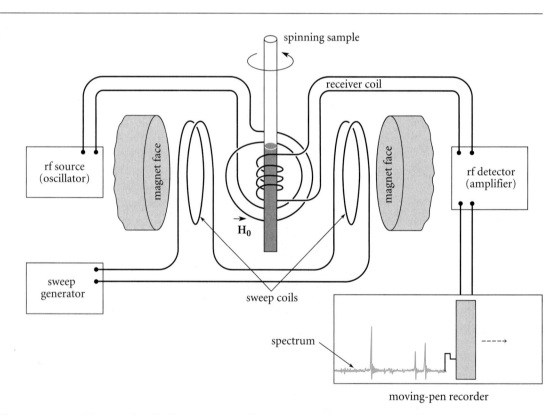

Figure 13.25   *Diagram of an NMR spectrometer. The magnet provides the magnetic field required for energy absorption by the sample. Radiofrequency (rf) energy at the operating frequency is absorbed from a coil connected to the rf source. The magnetic field is varied over the range of chemical shift by altering it slightly with current provided by a sweep generator and sent through the sweep coils. Energy absorption is detected by the receiver coil, amplified, and presented as a spectrum on a chart recorder or digitized and stored in a computer.*

Study Guide Link:
13.6
Fourier-Transform NMR

NMR spectrometers of recent vintage contain essentially the same components, but employ a technique for taking spectra called **pulse-Fourier transform NMR (FT-NMR)**. With FT-NMR, an entire spectrum can be obtained in less than a second. Consequently, a large number of spectra of a given sample (anywhere from 50 to 20,000, depending on the sample concentration and the isotope) can be recorded in a relatively short time. A computer stores and mathematically adds the spectra. Since electronic noise is random, it sums to zero when averaged over many spectra, while the resonances of the sample reinforce to give a much stronger spectrum than could be obtained in a single experiment. The FT-NMR technique made possible the routine use of CMR for structure determination. The cost of FT-NMR instruments has been reduced by the availability of relatively inexpensive dedicated small computers that are required for application of the FT-NMR technique.

The FT-NMR method was conceived by Richard Ernst of the ETH (Federal Technical Institute) in Zürich, Switzerland; for this contribution he was honored with the 1991 Nobel Prize in Chemistry.

13.11   Other Uses of NMR

The utility of NMR for structure determination should by now be fairly obvious. However, NMR also has many other uses. *Solid-state NMR* is being used to study the properties of important solid substances as diverse as drugs, coal, and industrial polymers. *Phosphorus NMR* (^{31}P NMR) is being used to study biological processes, in some cases using intact cells or even whole organisms. One of the most exciting clinical applications of NMR is *NMR tomography*, or *magnetic resonance imaging* (MRI). By monitoring the proton magnetic resonance signals from water in various parts of the body, clinical scientists can achieve organ imaging without using X-rays or other potentially harmful types of radiation. The brain image in Fig. 13.26 was obtained by this method.

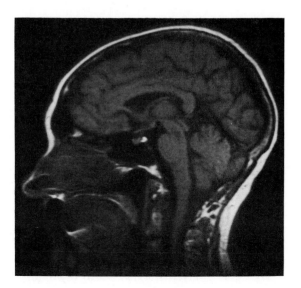

Figure 13.26   *Magnetic resonance imaging can show the details of soft tissue not visible in X-ray images, as in this brain scan of a 69-year-old female.*

Key Ideas in Chapter 13

◊ The NMR spectrum records the absorption of energy by nuclei from a radio-frequency (rf) source in the presence of a magnetic field.

◊ Only nuclei with spin give NMR spectra; protons have spin $\pm\frac{1}{2}$.

◊ The three elements of an NMR spectrum are the chemical shift, which provides information on the chemical environment of the observed nucleus; the integral, which indicates the relative number of nuclei being observed; and the splitting, which gives information on the number of nuclei on adjacent atoms.

◊ The chemical shifts of protons in methylene groups can be calculated with reasonable accuracy by Eq. 13.3 using the data in Table 13.2.

◊ Chemically nonequivalent nuclei in principle have different chemical shifts. Constitutionally nonequivalent nuclei and diastereotopic nuclei are chemically nonequivalent.

◊ The $n + 1$ rule determines the splitting observed in most spectra. When a nucleus is split by more than one chemically nonequivalent set of nuclei, the observed splitting is the result of successive applications of the splittings caused by each set.

◊ Nuclei of spin $\pm\frac{1}{2}$ such as the proton have splitting patterns in which the individual lines ideally have the relative intensities shown in Table 13.3. In practice, splitting patterns show leaning: they deviate from these intensities, and the deviation is greater as the difference in the chemical shifts of the mutually split protons is smaller.

◊ Splitting more complex than that predicted by the $n + 1$ rule is observed when the chemical shifts of two resonances (in Hz) differ by an amount that is not much greater than their mutual coupling constant. Because the chemical shift in Hz increases in proportion to the operating frequency, but coupling constants do not, many compounds that give non-first-order spectra at a lower operating frequency give first-order spectra at a higher operating frequency.

◊ Deuterium resonances are not observed in a proton NMR spectrum. Thus, shaking the solution of an alcohol with D_2O removes the resonance of the OH proton because of its rapid exchange for deuterium.

◊ A time-averaged NMR spectrum is observed for species involved in rapid equilibria. It is possible to observe absorptions for the individual species by retarding the reactions involved in the equilibria (for example, by lowering the temperature).

> The NMR of carbon atoms in a compound can be observed as a ^{13}C NMR (CMR) spectrum despite the low natural abundance of this isotope.

> In a proton-decoupled CMR spectrum, the carbon-proton couplings are removed; each chemically nonequivalent set of carbons appears as a single line. The number of protons attached to each carbon can be determined using the DEPT technique.

ADDITIONAL PROBLEMS

Note: In these problems, the term NMR refers to *proton* NMR unless otherwise indicated.

*13.30   What three pieces of information are available from an NMR spectrum? How is each used?

13.31   How would you distinguish among the compounds within each set below using their NMR spectra? Explain carefully and explicitly what features of the NMR spectrum you would use.
  *(a) cyclohexane and *trans*-2-hexene   (b) *trans*-3-hexene and 1-hexene
  *(c) 1,1-dichlorohexane, 1,6-dichlorohexane, and 1,2-dichlorohexane
  (d) 1,1,2,2-tetrabromoethane and 1,1,1,2-tetrabromoethane
  *(e) $Cl_3C-CH_2-CH_2-CHF_2$ and $CH_3-CH_2-CCl_2-CClF_2$
  (f) *tert*-butyl methyl ether and isopropyl methyl ether
  *(g) *cis*-$Cl-CH=CH-Br$, *trans*-$Cl-CH=CH-Br$, and $CH_2=CClBr$

*13.32   Answer each of the following questions as briefly as possible.
  (a) What is the relationship between chemical shift in Hz and operating frequency ν_0?
  (b) What is the relationship between coupling constant J and operating frequency?
  (c) Why does the chemical shift in ppm not change with operating frequency?
  (d) How does NMR spectroscopy differ conceptually from other forms of absorption spectroscopy?
  (e) What condition must be met for an NMR spectrum to be first order?
  (f) What happens physically when energy is absorbed by nuclei in an NMR experiment?

13.33   Give the structure of each of the following compounds.
  *(a) a six-carbon hydrocarbon, not an alkene, whose proton NMR spectrum consists entirely of one singlet
  (b) a six-carbon alkene whose proton NMR spectrum consists of one singlet
  *(c) an eight-carbon ether whose proton NMR spectrum consists of one singlet and whose CMR spectrum consists of two singlets.

(*Problem 13.33 continues*)

(d) a nine-carbon hydrocarbon whose proton NMR spectrum consists of two singlets

*(e) a seven-carbon hydrocarbon whose proton NMR spectrum consists of two singlets at δ 0.23 and δ 1.21 (relative integral 1:6) and whose proton-decoupled CMR spectrum consists of three absorptions

13.34   Give the structure that corresponds to each of the following molecular formulas and NMR spectra:

*(a) C_5H_{12}; δ 0.93, s   (b) C_5H_{10}; δ 1.5, s

The compounds in (c) and (d) can both be hydrogenated to give 2,2,4-trimethylpentane.

*(c) C_8H_{16}: NMR spectrum in Fig. 13.27a

(d) C_8H_{16}: NMR spectrum in Fig. 13.27b

*(e) $C_7H_{12}Cl_2$: δ 1.07 (s, 31); δ 2.28 (d, J = 6 Hz, 7.1); δ 5.77 (t, J = 6 Hz, 3.6). *Note:* The numbers in this spectrum are the actual values of the integral in chart spaces.

(f) $C_2H_2Br_2Cl_2$: δ 4.40, s

*(g) $C_2H_2Br_2F_2$: δ 4.02 (t, J = 16 Hz)

(h) $C_2H_2F_3I$: δ 3.56 (q, J = 10 Hz)

*(i) $C_7H_{16}O_4$: δ 1.93 (t, J = 6 Hz); δ 3.35 (s); δ 4.49 (t, J = 6 Hz); relative integral 1:6:1

(j) $C_2H_4OCl_2$: δ 3.7 (3H, s); δ 7.33 (1H, s)

*(k) C_3H_5Br: NMR spectrum in Fig. 13.27c

*13.35   Each of four bottles, *A*, *B*, *C*, and *D*, is labeled only "C_6H_{12}" and contains a colorless liquid. You have been called in as an expert to identify these compounds from their spectra:

Compound *A*:
NMR: δ 1.66 (s), decolorizes Br_2 in CCl_4; IR: no absorption in the range 1620–1700 cm^{-1}

Compound *B*:
NMR: δ 1.07 (6H, d, J = 7 Hz); δ 1.70 (3H, d, J = 1.5 Hz); δ 2.20 (1H, septet, J = 7 Hz); δ 4.60 (2H, d, J = 1.5 Hz)
IR: 1642, 891, 3080 cm^{-1}

Compound *C*:
NMR and IR spectra in Fig. 13.28 on p. 638

Compound *D*:
NMR: δ 1.40 (s); does not decolorize Br_2 in CCl_4.

*13.36   Suppose you wish to carry out the following reactions and you have the NMR spectrum of each starting material. In each case explain how the NMR spectra of the product and starting material would be expected to differ.

(a) $(CH_3)_2C{=}CH_2$ + HBr $\longrightarrow$ $(CH_3)_3CBr$

(b) $(CH_3)_2C{=}C(CH_3)_2$ + HCl $\longrightarrow$ $(CH_3)_2CHC(CH_3)_2$
                                                                      |
                                                                      Cl

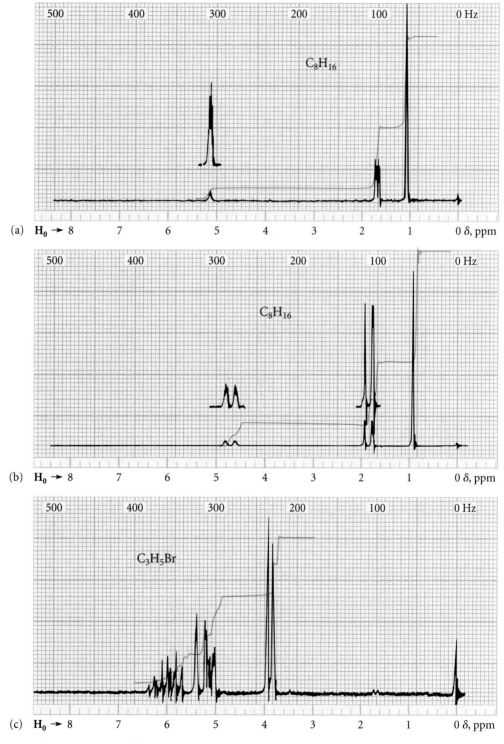

(a) $H_0 \rightarrow$

(b) $H_0 \rightarrow$

(c) $H_0 \rightarrow$

Figure 13.27   *NMR spectra for Problem 13.34.*

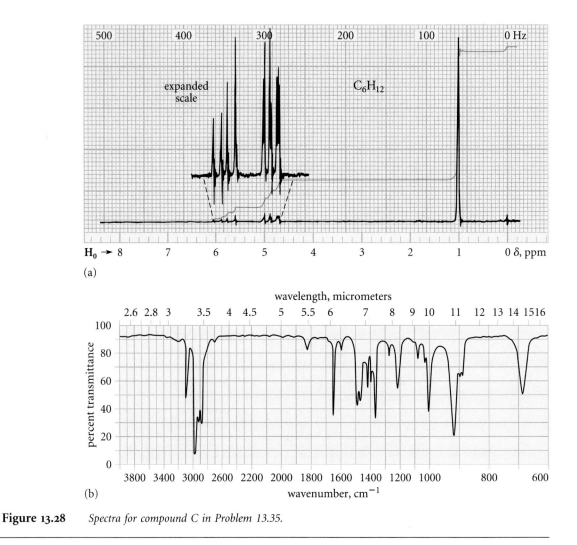

Figure 13.28   *Spectra for compound C in Problem 13.35.*

13.37   When 3-methyl-1-butene reacts with HCl, two products are observed: a normal addition product and a rearrangement product. Assume that these two products have been separated. Using what you know about HCl addition, postulate structures for these products and explain what you would look for in the NMR spectra of these products to identify them.

*13.38   In the back of a reagent cabinet in an industrial laboratory has been found a compound *A*. The NMR spectrum of *A* is shown in Fig. 13.29. Compound A reacts with H_2 over Pd/C to give methylcyclohexane. Al Keen, a chemist for the company, has deduced that the compound must be either 1-methylcyclohexene or 3-methylcyclohexene. You have been called in as a consultant to help Keen decide between these two structures. Give reasons for your answer.

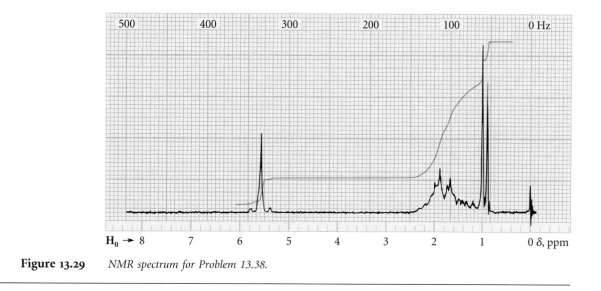

Figure 13.29   *NMR spectrum for Problem 13.38.*

13.39   To which of the compounds below does the NMR spectrum shown in Fig. 13.30 belong? Explain your choice carefully.

(1) *cis*-3-hexene   (2) (*Z*)-1-ethoxy-1-butene   (3) 2-ethyl-1-butene

(4)   Cl—CH—CH—Cl   (5) Cl_2CH—$CH(OCH_2CH_3)_2$
                       | |
         CH_3CH_2—O   O—CH_2CH_3

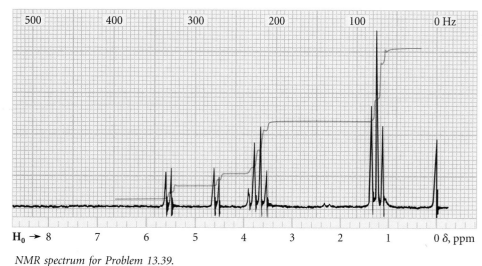

Figure 13.30   *NMR spectrum for Problem 13.39.*

*13.40 To which of the compounds below does the following CMR-DEPT spectrum belong (attached protons in parentheses):

δ 15.5 (3), δ 20.1 (3), δ 60.7 (2), δ 99.6 (1)

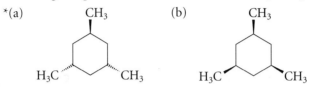

CH₃CH₂O—CH—OCH₂CH₃   CH₃CH₂OCH₂CH₂OCH₂CH₃   CH₃OCH₂—CH—CH₂OCH₃

      | *B* |

      CH₃   CH₃

       A   *C*

13.41 How many absorptions should be observed in the CMR spectrum of each of the following compounds? (Assume that the chair flip is rapid.)

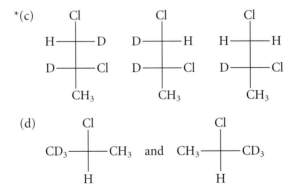

*(a)   CH₃   (b)   CH₃

H₃C⁂⁂⁂⁀⁀⁀CH₃   H₃C⁂⁂⁂⁀⁀⁀CH₃

13.42 Explain how the proton NMR spectra of the compounds within each of the following sets would differ, if at all.

*(a)   Cl—CD₂—CH₂—CH₂—Cl   and   Cl—CH₂—CH₂—CH₂—Cl

(b)   (CH₃)₂CH—Cl   and   (CH₃)₂CD—Cl

*(c)
$$
\begin{array}{ccc}
\text{Cl} & \text{Cl} & \text{Cl} \\
| & | & | \\
\text{H—}\!|\!\text{—D} & \text{D—}\!|\!\text{—H} & \text{H—}\!|\!\text{—H} \\
\text{D—}\!|\!\text{—Cl} & \text{D—}\!|\!\text{—Cl} & \text{D—}\!|\!\text{—Cl} \\
| & | & | \\
\text{CH}_3 & \text{CH}_3 & \text{CH}_3
\end{array}
$$

(d)
$$
\begin{array}{cc}
\text{Cl} & \text{Cl} \\
| & | \\
\text{CD}_3\text{—}\!|\!\text{—CH}_3 \;\;\text{and}\;\; \text{CH}_3\text{—}\!|\!\text{—CD}_3 \\
| & | \\
\text{H} & \text{H}
\end{array}
$$

13.43 Although this chapter has discussed only nuclei that have spin $\pm\frac{1}{2}$, several common nuclei such as ^{14}N and deuterium (^2H, or D) have a spin of 1. This means that the nuclear spin can be any of three values: $+1$, 0, and -1.

*(a)   How many lines would you expect to observe in the proton NMR of $^+NH_4$? What is the theoretical relative intensity of each line?

(b)   Explain why the CMR spectrum of $CDCl_3$ is a 1:1:1 triplet. (This triplet is sometimes observed in the spectra of compounds taken in $CDCl_3$ solvent; see, for example, Fig. 13.21.)

*(c)   The proton NMR spectrum of CHD_2—I is a quintet (five-line pattern) with relative intensities 1:2:3:2:1. Explain why this pattern should be expected.

(d)   How could you tell a sample of CH_2D—I apart from one of CHD_2—I by proton NMR? What other spectroscopic technique could be used for this determination?

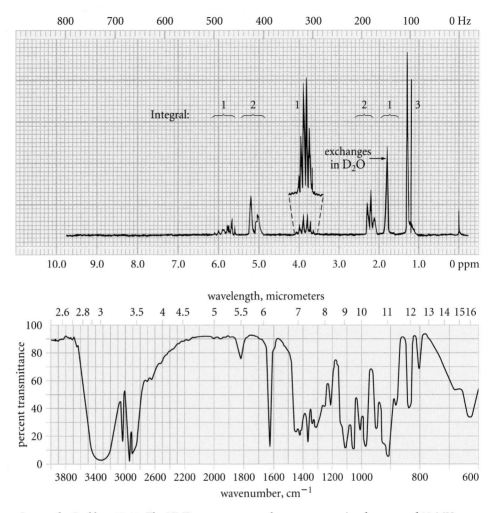

Figure 13.31   *Spectra for Problem 13.44. The NMR spectrum was taken at an operating frequency of 80 MHz.*

---

*13.44   (a) Propose a structure for a compound $C_5H_{10}O$ that can be resolved into enantiomers and has the NMR and IR spectra shown in Fig. 13.31.

(b) Rationalize the base peak at $m/z = 45$ in the mass spectrum of this compound.

13.45   Propose a structure for the alcohol $C_6H_{14}O$ that has the following NMR spectra:

Proton NMR: δ 0.90 (6*H*, d, *J* = 7 Hz); δ 1.10 (6*H*, s); δ 1.25 (1*H*, broad s, disappears after D_2O shake); δ 1.3–1.8 (1*H*, complex)

CMR (DEPT): δ 17.7 (3), δ 26.4 (3), δ 38.9 (1), δ 73.0 (0)

*13.46   A bottle has been found that contains liquid *A*, which has a penetrating, camphorlike odor. The IR, NMR, and mass spectra of *A* are given in Fig. 13.32 on pp. 642 and 643; the proton-decoupled CMR spectrum of *A* consists of three

(*Problem 13.46 continues*)

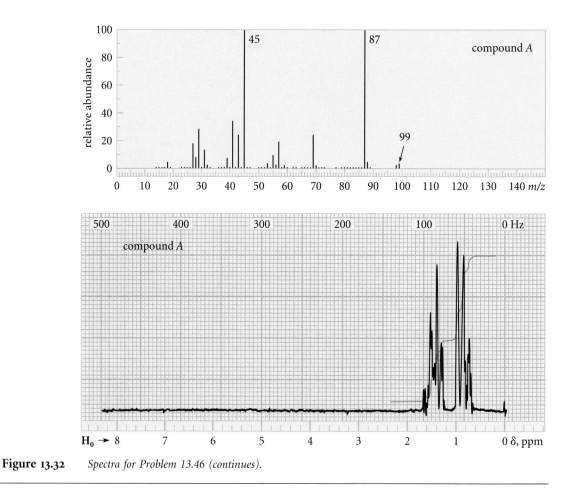

Figure 13.32   *Spectra for Problem 13.46 (continues).*

lines at δ 8, δ 31, and δ 75. On treatment with concentrated H_3PO_4, compound *A* is converted into compound *B* that has the mass spectrum also shown in Fig. 13.32. Because of your widely known expertise in spectroscopic methods you have been asked to determine the structure of *A*.

13.47   Propose a structure for the compound that has the following spectra:

NMR: δ 1.28 (3*H*, t, *J* = 7 Hz); δ 3.91 (2*H*, q, *J* = 7 Hz); δ 5.0 (1*H*, d, *J* = 4 Hz); δ 6.49 (1*H*, d, *J* = 4 Hz)

IR: 3100, 1644 (strong), 1104, 1166, 694 cm^{-1} (strong); no IR absorptions in the range 700–1100 or above 3100 cm^{-1}

Mass spectrum: *m/z* = 152, 150 (equal intensity; double molecular ion)

*13.48   You work for a well-known, reputable chemical supply house. An irate customer, Fly Ofterhandle, has called, alleging that a sample of 2,5-hexanediol obtained from your firm is impure. As evidence he cites the following CMR spectrum:

δ 23.2, δ 23.5, δ 35.1, δ 35.8, δ 67.4, δ 67.8

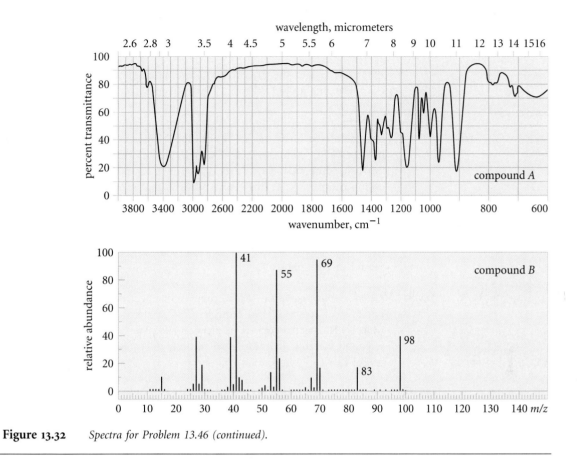

Figure 13.32 *Spectra for Problem 13.46 (continued).*

(Notice that the sample contains three pairs of closely spaced absorptions.) After verifying the CMR spectrum, you can confidently assure him that the sample he purchased contains *only* 2,5-hexanediol. Explain why the CMR spectrum is consistent with this claim.

*13.49 (a) How many different sets of proton absorptions (neglecting splitting) should be observed in the proton NMR spectrum of 4-methyl-1-penten-3-ol?

  (b) How many absorptions should there be in the CMR spectrum?

13.50 The CMR spectrum of 4-methyl-2-pentanol contains six absorptions; that of its isomer 4-methyl-1-pentanol shows five absorptions. Explain, assuming that no absorptions are concealed by overlapping peaks.

*13.51 ^{17}O is a rare isotope that has a nuclear spin. The ^{17}O NMR of a small amount of water dissolved in CCl_4 is a triplet (intensity ratio 1:2:1). When water is dissolved in the strongly acidic $HF-SbF_5$ solvent, its ^{17}O NMR becomes a 1:3:3:1 quartet. Suggest a reason for these observations. (*Hint:* Think of this solvent as $H^+ SbF_6^-$.)

13.52 *(a) How would the proton NMR spectrum of very dry 2-chloroethanol differ from that of the same compound containing a trace of aqueous acid? Explain your answer.

(b) How would the proton NMR spectrum of the compound in (a) change following a D_2O shake?

*13.53 Carbon-carbon splitting is not apparent in natural-abundance CMR spectra because of the rarity of the ^{13}C isotope. However, carbon-carbon splitting can be observed in compounds that are enriched in ^{13}C. A colleague, Buster Magnet, has just completed a synthesis of CH_3—CH_2—Br that contains 50% ^{13}C at each position. (What this really means is that some molecules contain no ^{13}C, some contain ^{13}C at one position, and some contain ^{13}C at both positions.) Buster does not know what to expect for the spectrum of this compound and has come to you for assistance. Describe the proton-decoupled ^{13}C NMR spectrum of this compound. (*Hint:* The spectrum of a mixture shows peaks for each compound in the mixture.)

*13.54 *Electron spin resonance spectroscopy* (ESR spectroscopy) is used to study the unpaired electrons in free radicals. (ESR spectroscopy is to unpaired electrons what NMR spectroscopy is to protons.) Explain why the ESR spectrum of the unpaired electron in the methyl radical, $\cdot CH_3$, is a quartet of four lines in a 1:3:3:1 ratio.

*13.55 The 60-MHz proton NMR spectrum of 2,2,3,3-tetrachlorobutane is a sharp singlet at 25 °C, but at −45 °C is two singlets of different intensities separated by about 10 Hz.

(a) Explain the changes in the spectrum as a function of temperature.

(b) Explain why the two lines observed at low temperature have different intensities.

14

Chemistry of Alkynes

Alkynes, or **acetylenes**, are hydrocarbons with carbon-carbon triple bonds; the simplest member of this family is **acetylene**, $H—C\equiv C—H$. The chemistry of the carbon-carbon triple bond is similar in many respects to that of the carbon-carbon double bond; indeed, alkynes and alkenes undergo many of the same addition reactions. Alkynes also have some unique chemistry, most of it associated with the bond between hydrogen and the triply bonded carbon, the $\equiv C—H$ bond.

14.1 Nomenclature of Alkynes

In common nomenclature, simple alkynes are named as derivatives of the parent compound acetylene.

$$CH_3—C\equiv C—H \qquad \textbf{methylacetylene}$$

$$CH_3—C\equiv C—CH_3 \qquad \textbf{dimethylacetylene}$$

$$CH_3CH_2—C\equiv C—CH_3 \qquad \textbf{ethylmethylacetylene}$$

Certain compounds are named as derivatives of the **propargyl group**, $HC\equiv C—CH_2—$, in the common system. Notice that the propargyl group is the triple-bond analog of the allyl group.

$$HC\equiv C—CH_2—Cl \qquad CH_2=CH—CH_2—Cl$$

$$\textbf{propargyl chloride} \qquad \textbf{allyl chloride}$$

The substitutive nomenclature of alkynes is much like that of alkenes. The suffix *-ane* in the name of the corresponding alkane is replaced by the suffix *-yne*, and the triple bond is given the lowest possible number.

$$CH_3—C\equiv C—H \qquad \textbf{propyne}$$

$$CH_3—CH_2—C\equiv C—H \qquad \textbf{1-butyne}$$

$$CH_3CH_2CH_2CH_2—C\equiv C—CH_3 \qquad \textbf{2-heptyne}$$

$$HC\equiv C—CH_2—CH_2—C\equiv C—CH_3 \qquad \textbf{1,5-heptadiyne}$$

$$\underset{\overset{|}{CH_3}}{CH_3—CH}—C\equiv C—CH_3 \qquad \textbf{4-methyl-2-pentyne}$$

When double bonds and triple bonds are present in the same molecule, the principal chain is the carbon chain containing the greatest number of double and triple bonds. The numerical precedence of a double or triple bond within the principal chain is decided by the *first point of difference* rule (Sec. 2.4C, Rule 8): precedence is given to the bond that gives the lowest number in the name of the compound, whether it is a double or triple bond. The double bond, however, is always *cited* first in the name by dropping the terminal *e* from the *-ene* suffix, as in the following examples:

double bond cited first

$$\overset{1}{HC}\equiv\overset{2}{C}—\overset{3}{CH_2}—\overset{4}{CH}=\overset{5}{CH}—\overset{6}{CH_3}$$

possible names: **4-hexen-1-yne** (correct)
2-hexen-5-yne (incorrect)

$$\overset{1}{CH_2}=\overset{2}{CH}—\overset{3}{CH_2}—\overset{4}{C}\equiv\overset{5}{C}—\overset{6}{CH_3}$$

1-hexen-4-yne

$$\underset{\overset{|}{\underset{CH=CH_2}{}}}{\overset{1}{CH_2}=\overset{2}{CH}—\overset{3}{CH}—\overset{4}{CH}=\overset{5}{CH}—\overset{6}{C}\equiv\overset{7}{C}—\overset{8}{CH_3}}$$

3-vinyl-1,4-octadien-6-yne
(principal chain contains most
double and triple bonds)

vinyl group

When the foregoing rules do not completely specify the name, the double bond takes precedence in numbering.

$$CH_2=CH—CH_2—C\equiv CH \qquad \textbf{1-penten-4-yne}$$

Substituent groups that contain a triple bond (called *alkynyl groups*) are named by replacing the final *e* in the name of the corresponding alkyne with the suffix *-yl*. (This is exactly analogous to the nomenclature of substituent groups containing double bonds; see Sec. 4.2A.) The alkynyl group is numbered from its point of attachment to the main chain:

$$HC\equiv C— \qquad \textbf{ethynyl group} \quad \text{(ethyne + yl)}$$

$$HC\equiv C—CH_2— \qquad \textbf{2-propynyl group}$$

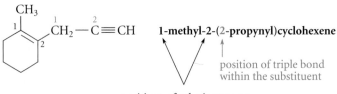

1-methyl-2-(2-propynyl)cyclohexene

position of triple bond
within the substituent

positions of substituents on
the cyclohexene ring

Finally, groups that can be cited as principal groups, such as the —OH group in the example below, are given numerical precedence over the triple bond. (See Appendix I.)

$$HC{\equiv}C{-}CH_2{-}\overset{\overset{\displaystyle OH}{|}}{CH}{-}CH_3 \qquad \textbf{4-pentyn-2-ol}$$

14.1   Draw a Lewis structure for each of the following alkynes.
*(a) isopropylacetylene   (b) *tert*-butylacetylene
*(c) cyclononyne   (d) 4-methyl-1-pentyne
*(e) 1-ethynylcyclohexanol   (f) 2-butoxy-3-heptyne
*(g) 1,3-hexadiyne

14.2   Provide the substitutive name for each of the following compounds. Also provide common names for (a) and (b).
*(a) $CH_3CH_2CH_2CH_2C{\equiv}CCH_2CH_2CH_2CH_3$

(b) $CH_3CH_2CH_2CH_2C{\equiv}CH$

*(c) $HC{\equiv}CCHCH_2CH_2CH_2CH_3$
$\phantom{*(c) HC{\equiv}CCH}|$
$CH_2CH_2{-}CH{=}CH{-}CH_2OCH_3$

(d) $\quad\quad OH$
$|$
$CH_3{-}C{-}C{\equiv}C{-}CH_3$
$|$
CH_3

14.2   Structure and Bonding in Alkynes

Because each carbon of acetylene is connected to two groups—a hydrogen and another carbon—the H—C≡C bond angle in acetylene is 180° (Sec. 1.3B); thus, the acetylene molecule is *linear*. The C≡C bond, with a bond length of 1.20 Å, is shorter than the C=C and C—C bonds, which have bond lengths of 1.33 Å and 1.54 Å, respectively.

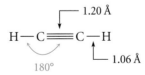

Because of the 180° bond angles at the carbon-carbon triple bond, *cis-trans* isomerism cannot occur in alkynes. Thus, although 2-butene can exist as *cis* and *trans* stereoisomers, 2-butyne cannot. Another consequence of this linear geometry is that cycloalkynes smaller than cyclooctyne cannot be isolated under ordinary conditions (see Problem 14.3).

You've learned that carbon hybridization and geometry are closely connected: tetrahedral carbon is typically sp^3-hybridized, and trigonal-planar carbon is sp^2-hybridized.

The linear geometry found in alkynes is characterized by a third type of carbon hybridization. Imagine that the 2s orbital and *one* 2p orbital on carbon mix to form two new hybrid orbitals. Because these two new orbitals are each one part s and one part p, they are called *sp* **hybrid orbitals**.

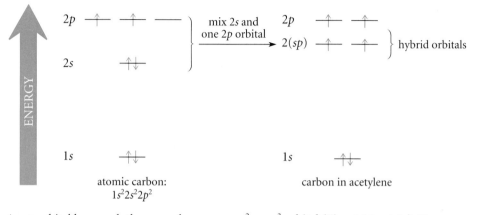

An *sp* orbital has much the same shape as an sp^2 or sp^3 orbital (Figs. 1.16a, 4.4a). However, electrons in an *sp* hybrid orbital are, on the average, somewhat closer to the carbon nucleus than they are in sp^2 or sp^3 hybrid orbitals. In other words, *sp* orbitals are more compact than sp^2 or sp^3 hybrid orbitals. An *sp*-hybridized carbon atom, shown in Fig. 14.1a, has two *sp* orbitals at a relative orientation of 180°. The two remaining unhybridized *p* orbitals lie along axes that are at right angles both to each other and to the *sp* orbitals.

The bonding in acetylene results from the combination of two *sp*-hybridized carbon atoms and two hydrogen atoms (Fig. 14.1b). One bond between the carbon atoms is a σ bond resulting from the overlap of two *sp* hybrid orbitals. This bonding molecular orbital contains two electrons. Two π bonds are formed by the side-to-side overlap of the *p* orbitals. These bonding π molecular orbitals, like the *p* orbitals of which they are composed, are mutually perpendicular, and each contains two electrons. The total electron density from all the π electrons taken together forms a cylinder, or barrel, about the axis

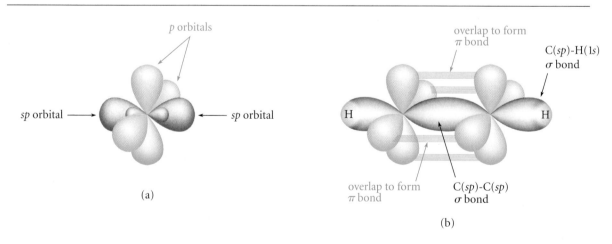

Figure 14.1   *(a) An sp-hybridized carbon atom. (b) Bonding molecular orbitals in acetylene resulting from the combination of two sp-hybridized carbon atoms and two hydrogen atoms.*

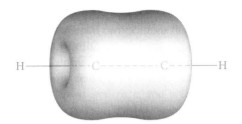

H————C - - - - - - - C - - -————H

Figure 14.2   *The π-electron density in acetylene lies in a cylinder about the axis of the molecule.*

of the molecule (Fig. 14.2). The remaining *sp* orbital on each carbon overlaps with a hydrogen 1*s* orbital to form a carbon-hydrogen σ-bond.

The following heats of formation show that alkynes are less stable than isomeric dienes:

$$\Delta H_f^\circ$$

| H—C≡C—CH₂CH₂CH₃ | CH₃—C≡C—CH₂CH₃ | CH₂=CH—CH₂—CH=CH₂ |

ΔH_f° H—C≡C—CH₂CH₂CH₃   CH₃—C≡C—CH₂CH₃   CH₂=CH—CH₂—CH=CH₂
        +144.3 kJ/mol   +128.9 kJ/mol   +106.4 kJ/mol
        (34.50 kcal/mol)   (30.80 kcal/mol)   (25.42 kcal/mol)

In other words, the *sp* hybridization state is inherently less stable than the *sp²* hybridization state, other things being equal. These heats of formation also show that a triple bond, like a double bond, is more stable in the interior of a carbon chain than at the end.

PROBLEMS

*14.3   Attempt to build a model of cyclohexyne. Explain why this compound is not stable.

14.4   Build a model of cyclodecyne. Compare its stability qualitatively to that of cyclohexyne; explain your answer.

14.3   Physical Properties of Alkynes

A.   Boiling Points and Solubilities

The boiling points of most alkynes are not very different from those of analogous alkenes and alkanes:

HC≡C(CH₂)₃CH₃   CH₂=CH(CH₂)₃CH₃   CH₃—CH₂(CH₂)₃CH₃

	1-hexyne	**1-hexene**	**hexane**
boiling point:	71.3°	63.4°	68.7°
density:	0.7155 g/mL	0.6731 g/mL	0.6603 g/mL

Like alkanes and alkenes, alkynes have much lower densities than water and are also insoluble in water.

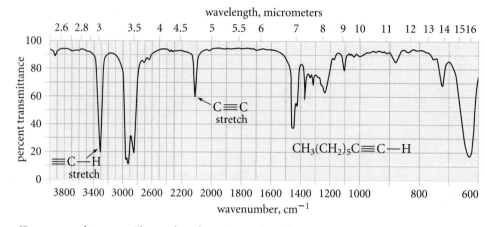

Figure 14.3   *IR spectrum of 1-octyne. The two key absorptions indicated are absent in the spectrum of 4-octyne.*

B. IR Spectroscopy of Alkynes

Many alkynes have a C≡C stretching absorption in the 2100–2200 cm^{-1} region of the infrared spectrum. This absorption is clearly evident, for example, at 2120 cm^{-1} in the IR spectrum of 1-octyne (Fig. 14.3). However, this absorption is very weak or absent in the IR spectra of many symmetrical, or nearly symmetrical, alkynes because of the dipole moment effect (Sec. 12.3B). For example, 4-octyne has no C≡C stretching absorption at all.

A very useful absorption of 1-alkynes is the ≡C—H stretching absorption, which occurs at about 3300 cm^{-1}. This absorption, very prominent in the spectrum of 1-octyne (Fig. 14.3), is well separated from other C—H absorptions. Of course, alkynes other than 1-alkynes lack the unique ≡C—H bond, and therefore do not show this absorption.

C. NMR Spectroscopy of Alkynes

Typical chemical shifts observed in the proton NMR spectra of alkynes are contrasted below with the analogous shifts for alkenes:

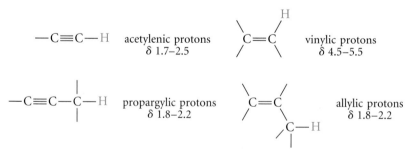

Although the chemical shifts of allylic and propargylic protons are very similar (as might be expected from the fact that both double and triple bonds involve π electrons), the chemical shifts of acetylenic protons are much smaller than those of vinylic protons.

Similar contrasts of chemical shifts are observed in CMR. Although carbons involved in double bonds have chemical shifts in the δ 100–145 range, carbons involved in triple bonds absorb at considerably higher field, in the δ 65–85 range. Propargylic carbons, like acetylenic hydrogens, are also shifted to higher field, typically by 5–15 ppm. The chemical shifts in 2-heptyne are typical:

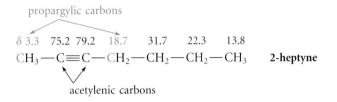

Compare, for example, the δ 3.3 chemical shift of the propargylic methyl carbon with the δ 13.8 chemical shift of the other methyl carbon, which is much like that of a methyl group in alkanes.

The explanation for the unusual carbon and proton chemical shifts observed in alkynes is closely related to the explanation for the chemical shifts of vinylic protons (Fig. 13.15), although the effect is in the opposite direction. If an alkyne is oriented in the applied field H_0 as shown in Figure 14.4, an induced electron circulation is set up in the cylinder of π electrons that encircles the alkyne molecule. The resulting induced field H_i *opposes* the applied field along the axis of this cylinder. Since the acetylenic proton lies along this axis, the net field at this proton is reduced. Thus, the acetylenic proton is

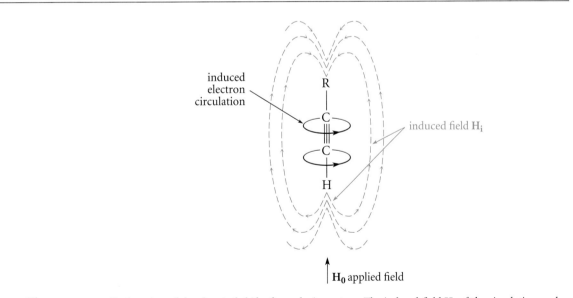

Figure 14.4   *Explanation of the chemical shift of acetylenic protons. The induced field H_i of the circulating π electrons opposes the applied field H_0 from the spectrometer in the region of space occupied by acetylenic protons. As a result, acetylenic protons are shielded and thus have NMR absorptions at relatively high field. The same effect accounts for the chemical shifts of acetylenic and propargylic carbons in the CMR spectra of alkynes.*

Conversion of Alkynes into Aldehydes and Ketones

A. Hydration of Alkynes

Water can be added to the triple bond. Although the reaction can be catalyzed by a strong acid, it is faster, and yields are higher, when a combination of dilute acid and mercuric ion (Hg^{2+}) catalysts is used.

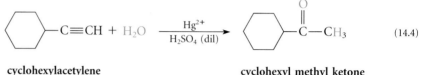

$$\text{cyclohexylacetylene} \qquad\qquad\qquad \text{cyclohexyl methyl ketone}$$
$$\text{(91\% yield)}$$

(14.4)

The addition of water to a triple bond, like the corresponding addition to a double bond, is called **hydration**. Hydration of alkynes gives ketones (except in the case of acetylene itself, which gives an aldehyde; see Study Problem 14.1, p. 656).

Let's contrast the hydration reactions of alkenes (Sec. 4.9B) and alkynes. The hydration of an alkene gives an *alcohol*.

$$R\text{—}CH\text{=}CH_2 + H_2O \xrightarrow{\ H_2SO_4\ } R\text{—}\underset{\underset{\textstyle OH}{|}}{CH}\text{—}CH_3$$

$$\text{an alkene} \qquad\qquad\qquad\qquad \text{an alcohol}$$

(14.5a)

Because addition reactions of alkenes and alkynes are closely analogous, it might seem that an alcohol should also be obtained from hydration of an alkyne:

$$R\text{—}C\text{≡}C\text{—}H + H_2O \xrightarrow{\ H_2SO_4,\ Hg^{2+}\ } R\text{—}\underset{\underset{\textstyle OH}{|}}{C}\text{=}CH_2$$

$$\text{an alkyne} \qquad\qquad\qquad\qquad\qquad \text{an enol}$$

(14.5b)

An alcohol containing an OH group on a carbon of a double bond is called an **enol**. In fact, enols *are* formed in the hydration of alkynes. However, they cannot be isolated because *enols are unstable and are rapidly converted into the corresponding aldehydes or ketones.*

$$R\text{—}\underset{\underset{\textstyle OH}{|}}{C}\text{=}CH_2 \rightleftharpoons R\text{—}\underset{\overset{\textstyle O}{\|}}{C}\text{—}CH_3$$

$$\text{enol} \qquad\qquad \text{ketone}$$

(14.5c)

Most aldehydes and ketones are in equilibrium with the corresponding enols, but the equilibrium concentrations of enols are in most cases minuscule—typically, one part in 10^8 or less. The relationships between aldehydes, ketones, and enols is explored in Chapter 22. The important point here is that, because most enols are unstable, *any synthesis designed to give an enol gives instead the corresponding aldehyde or ketone.*

The mechanism of alkyne hydration is very similar to that of oxymercuration of alkenes (Sec. 5.3A). In the first part of the mechanism, a mercuric ion and an OH group from the solvent water add to the triple bond.

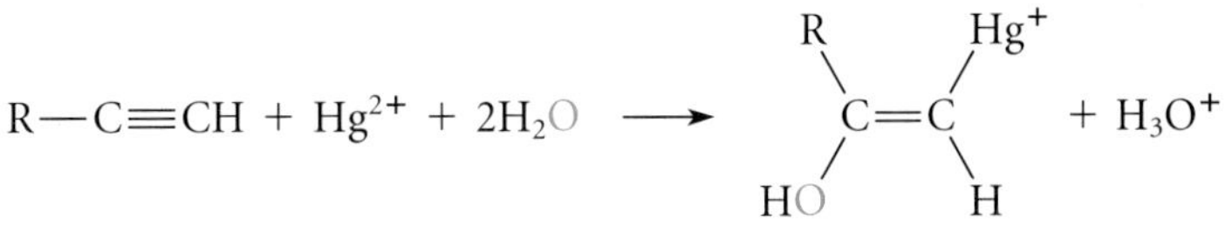

$$R-C{\equiv}CH + Hg^{2+} + 2H_2O \longrightarrow \underset{HO}{\overset{R}{\underset{}{}}}C{=}C\overset{Hg^+}{\underset{H}{}} + H_3O^+ \qquad (14.6a)$$

Notice that the OH group of water adds to the *carbon of the triple bond that bears the alkyl substituent* for the same reason that it adds to the more branched carbon of an alkene in oxymercuration: the bond between mercury and this carbon in the cyclic ion intermediate is weaker and thus more readily broken.

In oxymercuration of alkenes, the reducing agent $NaBH_4$ is the source of hydrogen that replaces the mercury. However, the use of $NaBH_4$ is unnecessary in hydration. The reason is that the presence of a double bond makes possible removal of the mercury by a protonolysis reaction. This protonolysis occurs under the conditions of hydration; a separate procedure is not required. The first step in the mechanism of this protonolysis reaction is protonation of the double bond. This protonation occurs at the carbon bearing the mercury because the resulting carbocation is resonance-stabilized.

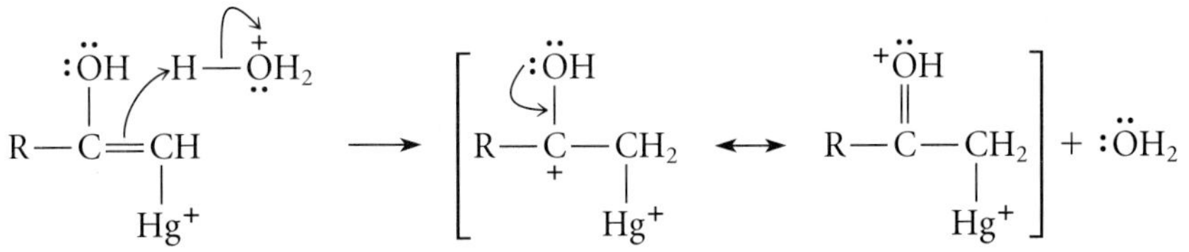

$$(14.6b)$$

resonance-stabilized carbocation

Dissociation of mercury from this carbocation liberates the catalyst Hg^{2+} along with the enol.

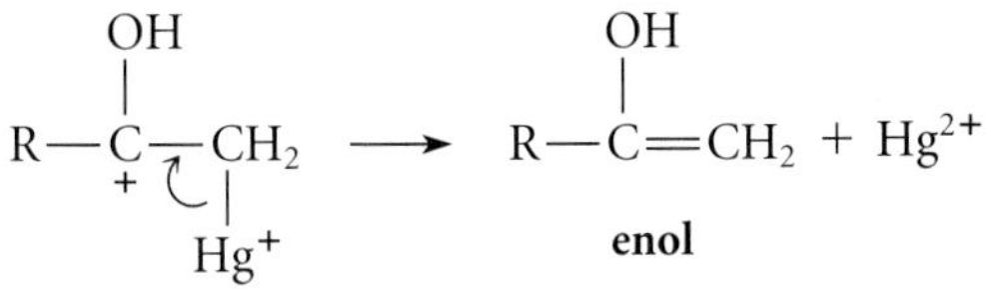

$$(14.6c)$$

enol

Conversion of the enol into the ketone is a rapid, acid-catalyzed process. Protonation of the double bond gives another resonance-stabilized carbocation:

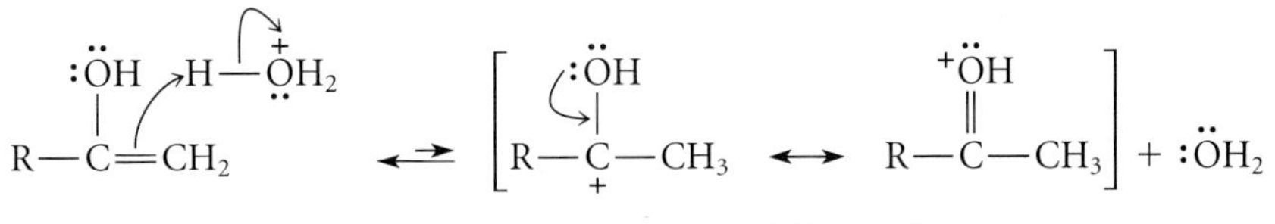

$$(14.6d)$$

resonance-stabilized carbocation

This carbocation is also the conjugate acid of a ketone. Loss of a proton gives the ketone product.

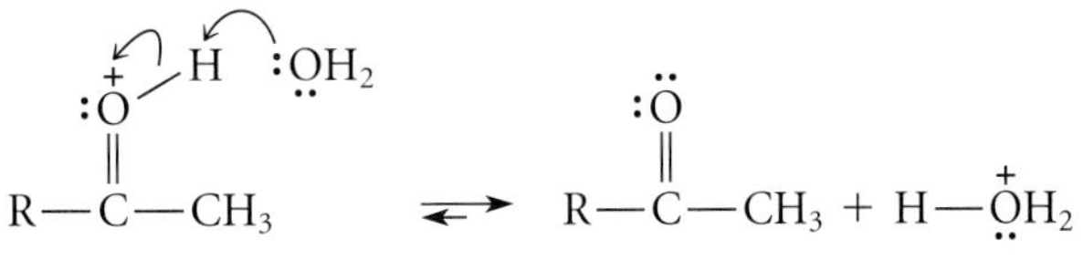

$$(14.6e)$$

The hydration of alkynes is a useful way to prepare ketones provided that the starting material is a 1-alkyne or a symmetrical alkyne (an alkyne with identical groups on each end of the triple bond). This point is explored in the following study problem.

STUDY PROBLEM 14.1

Which one of the following compounds could be prepared by hydration of alkynes so that it is uncontaminated by constitutional isomers? Explain your answer.

(a) $CH_3CH{=}O$

     acetaldehyde

(b)   O
            ‖
     $CH_3CH_2CCH_2CH_3$

         3-pentanone

Solution   First, if these compounds could be prepared by alkyne hydration, what alkyne starting materials would be required? The equations in the text show that the two carbons of the triple bond in the starting material correspond within the product to the carbon of the C=O group and an adjacent carbon. Thus, for (a), the only possible alkyne starting material is acetylene itself, $HC{\equiv}CH$. For (b), the only possible alkyne starting material is 2-pentyne, $CH_3C{\equiv}CCH_2CH_3$.

Next, it remains to be shown whether hydration of these alkynes gives the products in the problem. Hydration of acetylene indeed gives acetaldehyde. (Notice that acetaldehyde is the only aldehyde that can be prepared by hydration of an alkyne.) However, hydration of 2-pentyne gives a mixture of 2-pentanone and 3-pentanone, *because the carbons of 2-pentyne both have one alkyl substituent.* Thus, there is no basis on which attack of water on either carbon should be favored.

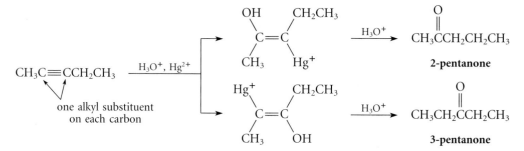

Hence, hydration would give a mixture of constitutional isomers which would have to be separated, and the yield of the desired product would be low. Consequently, hydration would *not* be a good way to prepare 3-pentanone. (However, 2-pentanone could be prepared by hydration of a different alkyne; see Problem 14.10a.)

PROBLEMS

14.10   From which alkyne could each of the following compounds be prepared by acid-catalyzed hydration?

*(a)   2-pentanone (See Study Problem 14.1.)

(b)   O
                 ‖
         $(CH_3)_3C{-}C{-}CH_3$

*(c)   O
                      ‖
     $CH_3CH_2CH_2CH_2{-}C{-}CH_2CH_2CH_2CH_2CH_3$

14.11   Hydration of an alkyne is *not* a reasonable preparative method for each of the following compounds. Explain why.

*(a) $CH_3CH_2CH{=}O$ (b) $(CH_3)_3C{-}\overset{\overset{\displaystyle O}{\|}}{C}{-}C(CH_3)_3$ *(c) cyclohexanone

*14.12   (a) Draw the structures of *all* enol forms of the following ketone, including stereoisomers.

$$CH_3CH_2{-}\overset{\overset{\displaystyle O}{\|}}{C}{-}CH(CH_3)_2$$

(b) Would alkyne hydration be a good preparative method for this compound? Explain.

B. Hydroboration-Oxidation of Alkynes

The hydroboration of alkynes is analogous to the same reaction of alkenes (Sec. 5.3B).

$$3C_2H_5{-}C{\equiv}C{-}C_2H_5 + BH_3 \xrightarrow{THF} \left(\begin{array}{c} C_2H_5 \\ C{=}C \\ H \end{array}\right)_3 B \qquad (14.7a)$$

As in the similar reaction of alkenes, oxidation of the organoborane with alkaline hydrogen peroxide yields the corresponding "alcohol," which in this case is an *enol*. As shown in Sec. 14.5A, enols react further to give the corresponding aldehydes or ketones.

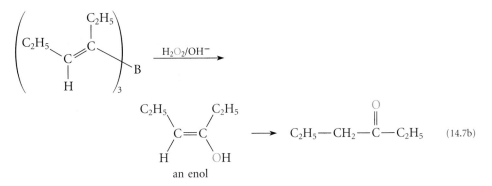

an enol $\xrightarrow{H_2O_2/OH^-}$ ... $\longrightarrow C_2H_5{-}CH_2{-}\overset{\overset{\displaystyle O}{\|}}{C}{-}C_2H_5$ (14.7b)

Because the organoborane product of Eq. 14.7a has a double bond, a second addition of BH_3 is in principle possible. However, the reaction conditions can be controlled so that only one addition takes place, as shown, provided that the alkyne is not a 1-alkyne.

A second addition of BH_3 cannot be prevented in the hydroboration of 1-alkynes. However, the hydroboration of 1-alkynes can be stopped after a single addition provided that an organoborane containing highly branched groups is used instead of BH_3. One reagent developed for this purpose is *disiamylborane*, abbreviated as shown in Eq. 14.8. (How would you synthesize disiamylborane? See Sec. 5.3B.)

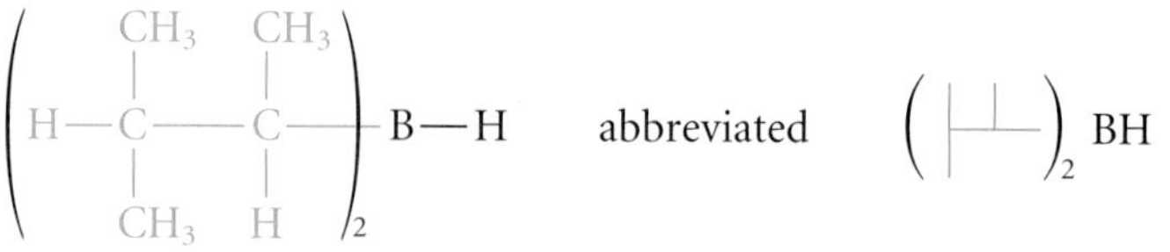

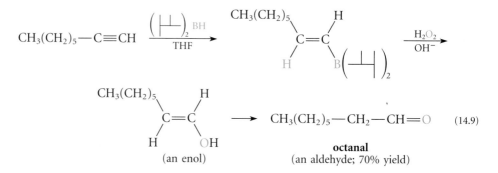

disiamylborane

The disiamylborane molecule is so large and highly branched that only one equivalent can react with a 1-alkyne; addition of a second molecule results in severe van der Waals repulsions. In many cases, van der Waals repulsions, or *steric effects*, interfere with a *desired* reaction; in this case, however, van der Waals repulsions are used to advantage, to prevent an *undesired* second addition from occurring:

Notice from this example that the regioselectivity of alkyne hydroboration is similar to that observed in alkene hydroboration (Sec. 5.3B): boron adds to the unbranched carbon atom of the triple bond, and hydrogen adds to the branched carbon.

Because hydroboration-oxidation and mercury-catalyzed hydration give different products when a 1-alkyne is used as the starting material (why?), these are *complementary* methods for the preparation of aldehydes and ketones in the same sense that hydroboration-oxidation and oxymercuration-reduction are complementary methods for the preparation of alcohols from alkenes.

STUDY GUIDE LINK:
✓14.1
Functional Group
Preparations

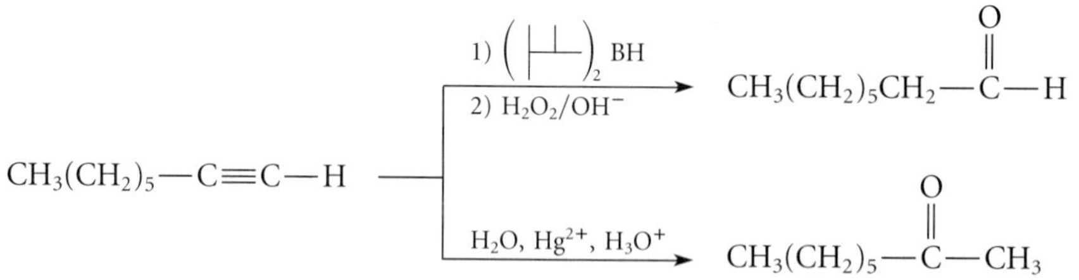

Notice that hydroboration-oxidation of a 1-alkyne gives an *aldehyde*; hydration of any 1-alkyne (other than acetylene itself) gives a *ketone*.

PROBLEM

14.13 Compare the results of hydroboration-oxidation and mercuric ion-catalyzed hydration for *(a) cyclohexylacetylene and (b) 2-butyne.

14.6 Reduction of Alkynes

A. Catalytic Hydrogenation of Alkynes

Alkynes, like alkenes (Sec. 4.9A), undergo catalytic hydrogenation. The first addition of hydrogen yields an alkene; a second addition of hydrogen gives an alkane.

$$R—C{\equiv}C—R \xrightarrow[\text{catalyst}]{H_2} R—CH{=}CH—R \xrightarrow[\text{catalyst}]{H_2} R—CH_2—CH_2—R \qquad (14.11)$$

The utility of catalytic hydrogenation is enhanced considerably by the fact that hydrogenation of an alkyne may be stopped at the alkene stage if the reaction mixture contains a **catalyst poison**: a compound that disrupts the action of a catalyst. Among the useful catalyst poisons are salts of Pb^{2+}, and certain nitrogen compounds, such as pyridine, quinoline, or other amines. These compounds *selectively* block the hydrogenation of alkenes without preventing the hydrogenation of alkynes to alkenes.

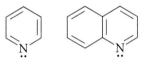

pyridine   **quinoline**

For example, a $Pd/CaCO_3$ catalyst can be washed with $Pb(OAc)_2$ to give a poisoned catalyst known as **Lindlar catalyst**. In the presence of Lindlar catalyst, an alkyne is hydrogenated to the corresponding alkene:

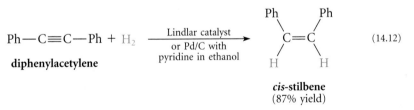

As Eq. 14.12 shows, hydrogenation of alkynes, like hydrogenation of alkenes (Sec. 7.9E), is a stereoselective *syn* addition. Thus, in the presence of a poisoned catalyst, hydrogenation of appropriate alkynes gives *cis*-alkenes. In fact, *catalytic hydrogenation of alkynes is one of the best ways to prepare cis-alkenes*.

In the absence of a catalyst poison, two equivalents of H_2 are added to the triple bond.

$$Ph—C{\equiv}C—Ph + 2H_2 \xrightarrow[\text{no poison}]{Pd/C} Ph—CH_2CH_2—Ph \qquad (14.13)$$

(95% yield)

Catalytic hydrogenation of alkynes can therefore be used to prepare alkenes or alkanes, respectively, by either including or omitting the catalyst poison. How catalyst poisons exert their inhibitory effect on the hydrogenation of alkenes is not well understood.

PROBLEM

14.14   Give the principal organic product formed in each of the following reactions.

*(a) $CH_3CH_2CH_2C{\equiv}C(CH_2)_4CH{=}CH_2 + H_2$  $\xrightarrow{\text{Lindlar catalyst}}$

(b)   Same as (a) with no poison   *(c)

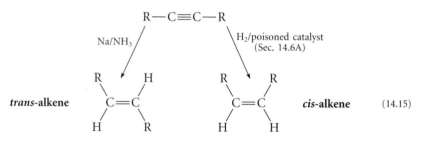

B. Reduction of Alkynes with Sodium in Liquid Ammonia

Reaction of an alkyne with a solution of an alkali metal (usually sodium) in liquid ammonia gives a *trans*-alkene.

$$CH_3CH_2CH_2{-}C{\equiv}C{-}CH_2CH_2CH_3 + 2Na + 2\ddot{N}H_3 \longrightarrow$$

$$\underset{\text{(97\% yield)}}{\overset{\displaystyle CH_3CH_2CH_2 \quad\quad H}{\underset{\displaystyle H \quad\quad CH_2CH_2CH_3}{C{=}C}}} \quad + \; 2Na^+ \; ^-\ddot{N}H_2 \quad (14.14)$$

The reduction of alkynes with sodium in liquid ammonia is complementary to catalytic hydrogenation of alkynes, which is used to prepare *cis*-alkenes (Sec. 14.6A).

$$R{-}C{\equiv}C{-}R$$

$$\overset{\displaystyle Na/NH_3}{\diagup} \qquad\qquad \overset{\displaystyle H_2/\text{poisoned catalyst}}{\underset{\displaystyle (\text{Sec. 14.6A})}{\diagdown}}$$

$$\textit{trans}\text{-alkene}\quad \underset{\displaystyle H \quad\quad R}{\overset{\displaystyle R \quad\quad H}{C{=}C}} \qquad\qquad \underset{\displaystyle H \quad\quad H}{\overset{\displaystyle R \quad\quad R}{C{=}C}} \quad \textit{cis}\text{-alkene} \qquad (14.15)$$

The stereochemistry of the Na/NH_3 reduction follows from its mechanism. If sodium or other alkali metals are dissolved in pure liquid ammonia, a deep blue solution forms that contains electrons complexed to ammonia (*solvated electrons*).

$$Na{\cdot} + nNH_3 \text{ (liq)} \longrightarrow Na^+ + e^-(NH_3)_n \qquad (14.16)$$

solvated electron

The solvated electron can be thought of as the simplest free radical. Remember that free radicals add to triple bonds (Eq. 14.2). The reaction of solvated electrons with alkynes begins with addition of an electron to the triple bond. The resulting species is an example of a **radical anion**: a species that has both an unpaired electron and a negative charge.

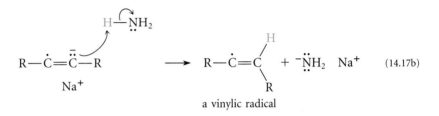

$$R-C\equiv C-R \;\rightleftharpoons\; R-\overset{\displaystyle .}{C}=\overset{\displaystyle ..}{\overset{..}{C}}-R \;\; Na^+ \qquad (14.17a)$$

a radical anion

The radical anion is such a strong base that it readily removes a proton from ammonia to give a *vinylic radical*—a radical in which the unpaired electron is associated with one carbon of a double bond. The destruction of the radical anion in this manner pulls the unfavorable equilibrium in Eq. 14.17a to the right:

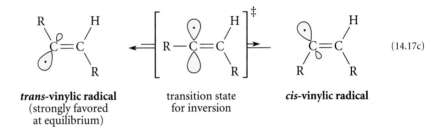

a vinylic radical

The vinylic radical, like the unshared electron pair of an amine (Sec. 6.10B), rapidly undergoes inversion, and the equilibrium between the *cis* and *trans* radicals favors the *trans* radical for the same reason that *trans*-alkenes are more stable than *cis*-alkenes: repulsions between the R groups are reduced.

$$\qquad (14.17c)$$

trans-vinylic radical   transition state   **cis-vinylic radical**
(strongly favored   for inversion
at equilibrium)

The *cis* and *trans* stereoisomers of this radical probably react at about the same rate in the subsequent steps of the mechanism. However, since there is much more of the *trans* radical, the ultimate product of the reaction is the one derived from this radical—the *trans*-alkene.

   Two more mechanistic steps give the alkene product. First, the vinylic radical accepts an electron to form an anion:

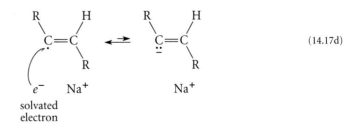

$$\qquad (14.17d)$$

solvated
electron

This anion is also more basic than the solvent, and therefore removes a proton from ammonia to complete the addition.

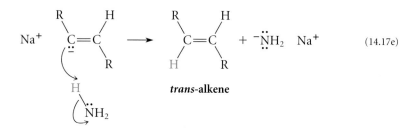

Because ordinary alkenes do not react with the solvated electron (the initial equilibrium analogous to Eq. 14.17a is too unfavorable), the reaction stops at the *trans*-alkene stage.

14.15   What product is obtained in each case when 3-hexyne is treated in each of the following ways? (*Hint:* The products of the two reactions are stereoisomers.)
  *(a)   with sodium in liquid ammonia and the product of that reaction with D_2 over Pd/C
  (b)   with H_2 over Pd/C and quinoline and the product of that reaction with D_2 over Pd/C

14.7   Acidity of 1-Alkynes

A.   Acetylenic Anions

Hydrocarbons are not usually regarded as acids. Nevertheless it is possible to envision the removal of a proton from a hydrocarbon by a very strong base B:⁻.

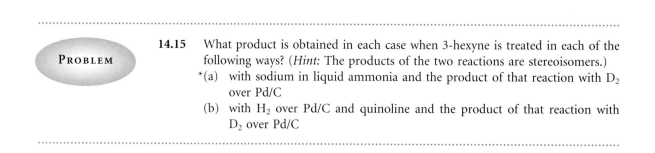

In this equation the hydrocarbon acts as a Brønsted acid. Its conjugate base, a species with an unshared electron pair and a negative charge on carbon, is a carbon anion, or *carbanion* (see Sec. 8.7B).

The conjugate base of an alkane, called generally an *alkyl anion*, has an electron pair in an sp^3 orbital. An example of such an ion is the 2-propanide anion:

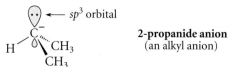

The conjugate base of an alkene, called generally a *vinylic anion*, has an electron pair in an sp^2 orbital. An example of this type of carbanion is the 1-propenide anion:

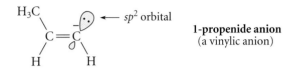

1-propenide anion
(a vinylic anion)

The anion derived from the ionization of a 1-alkyne, called generally an *acetylenic anion*, has an electron pair in an sp orbital. An example of this type of anion is the 1-propynide anion:

sp orbital

$$CH_3C \equiv \overset{-}{C} : $$

1-propynide anion
(an acetylenic anion)

The approximate acidities of the different types of aliphatic hydrocarbons have been measured or estimated:

$$R_3C—H \qquad \overset{\diagdown}{\underset{\diagup}{C}} = \overset{\diagup}{\underset{\diagdown}{C}}—H \qquad —C \equiv C—H \qquad (14.19)$$

type of hydrocarbon	alkane	alkene	alkyne
approximate pK_a	$\geq$55	42	25

These data show, first, that carbanions are extremely strong bases (that is, hydrocarbons are very weak acids); and second, that alkynes are the most acidic of the aliphatic hydrocarbons.

Alkyl anions and vinylic anions are seldom if ever formed by proton removal from the corresponding hydrocarbons; the hydrocarbons are simply not acidic enough. However, alkynes are sufficiently acidic that their conjugate-base acetylenic anions can be formed with strong bases. One base commonly used for this purpose is sodium amide, or sodamide, $Na^+ \ ^-:NH_2$, dissolved in its conjugate acid, liquid ammonia. The amide ion, $^-:NH_2$, is the conjugate base of ammonia, which, as an *acid*, has a pK_a of about 35.

☞
STUDY GUIDE LINK:
✓14.2
*Ammonia, Solvated
Electrons, and Amide
Anion*

$$B:^- \ + \ :NH_3 \ \rightleftharpoons \ ^-:\overset{..}{N}H_2 \ + \ B—H \qquad (14.20)$$
$$pK_a = 35$$
amide ion

Because the amide ion is a much stronger base than an acetylenic anion, the equilibrium for removal of the acetylenic proton by amide ion is very favorable:

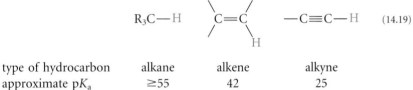

In fact, the sodium salt of an alkyne can be formed from a 1-alkyne quantitatively with NaNH$_2$. Since the amide ion is a much *weaker* base than either a vinylic anion or an alkyl anion, these ions *cannot* be prepared using sodium amide (Problem 14.18).

The relative acidity of alkynes is also reflected in the method usually used to prepare *acetylenic Grignard reagents*: the reaction of an alkylmagnesium halide with an alkyne. (Acetylenic Grignard reagents are reagents of the form R—C$\equiv$C—MgBr.)

$$CH_3CH_2CH_2CH_2\text{—}C\equiv C\text{—}H + CH_3CH_2\text{—}MgBr \longrightarrow$$

$$CH_3CH_2CH_2CH_2\text{—}C\equiv C\text{—}MgBr + CH_3CH_3 \qquad (14.22)$$

an acetylenic Grignard reagent

$$H\text{—}C\equiv C\text{—}H + CH_3CH_2\text{—}MgBr \xrightarrow{\text{THF}} H\text{—}C\equiv C\text{—}MgBr + CH_3CH_3 \qquad (14.23)$$

ethynylmagnesium bromide

This reaction, called **transmetallation**, is really another Brønsted acid-base reaction:

conjugate acid-base pair

$$BrMg\text{—}CH_2CH_3 \quad H\text{—}C\equiv C\text{—}R \longrightarrow CH_3CH_3 + BrMg^+ \quad :\bar{C}\equiv C\text{—}R \qquad (14.24)$$

conjugate base-acid pair

This reaction is similar in principle to the reaction of a Grignard reagent with water or alcohols (Eq. 8.25). Like all Brønsted acid-base equilibria, this one favors formation of the weaker base. The release of ethane gas in the reaction with ethylmagnesium bromide makes the reaction irreversible and is also a useful test for 1-alkynes. Alkynes with an internal triple bond do not react because they lack an acidic acetylenic hydrogen.

What is the reason for the relative acidities of the hydrocarbons? Sec. 3.6A discussed two important factors that affect the acidity of an acid A—H. These factors are the A—H bond strength, and the electronegativity of the group A. Bond dissociation energies show that acetylenic C—H bonds are the strongest of all the C—H bonds in the aliphatic hydrocarbons:

$$\equiv C\text{—}H \quad > \quad =\overset{|}{C}\text{—}H \quad > \quad \overset{|}{\underset{|}{\text{—}C}}\text{—}H \qquad (14.25)$$

acetylenic C—H	vinylic C—H	alkyl C—H
(548 kJ mol,	(460 kJ/mol,	(402–418 kJ/mol,
131 kcal/mol)	110 kcal/mol)	96–100 kcal/mol)

If bond strength were the major factor controlling hydrocarbon acidity, then alkynes would be the *least* acidic hydrocarbons. Since they are in fact the most acidic hydrocarbons, the electronegativities of the *carbons themselves* must govern acidity. Thus, the relative electronegativities of carbon atoms increase in the order $sp^3 < sp^2 < sp$.

This trend in electronegativity with hybridization can be explained in the following way. The electrons in sp-hybridized orbitals are closer to the nucleus, on the average, than sp^2 electrons, which in turn are closer than sp^3 electrons (Sec. 14.2). In other words, electrons in orbitals with larger amounts of s-character are drawn closer to the nucleus. This is a stabilizing effect because the interaction energy between particles of opposite charge (electrons and nuclei) decreases as the distance between them decreases (the electrostatic law; Eq. 3.40). Thus, the stabilization of unshared electron pairs is in the order $sp^3 < sp^2 < sp$. In other words, *unshared electron pairs have lower energy when they are in orbitals of greater s-character.*

PROBLEMS

*14.16   Each of the following compounds protonates on nitrogen. Draw the conjugate acid of each. Rank the compounds in order of their basicities, least basic first, and explain your reasoning.

$$CH_3—CH=\ddot{N}H \qquad CH_3—C≡N: \qquad CH_3—\ddot{N}H_2$$

$$A \qquad\qquad\qquad B \qquad\qquad C$$

14.17   Rank the two compounds below in order of their N—H acidities, least acidic first. Explain your reasoning.

$$CH_3—CH=\ddot{N}H \qquad CH_3—\ddot{N}H_2$$

$$A \qquad\qquad\qquad B$$

14.18   (a)   Using the pK_a values of the hydrocarbons and ammonia, estimate the equilibrium constant for *(1) the reaction in Eq. 14.21 and (2) the analogous reaction of an alkane with amide ion. (*Hint:* See Study Problem 3.5, Sec. 3.4D.)

   *(b)   Use your calculation to explain why sodium amide cannot be used to form alkyl anions from alkanes.

B. Acetylenic Anions as Nucleophiles

Although acetylenic anions are the weakest bases of the simple hydrocarbon anions, they are nevertheless strong bases—much stronger, for example, than hydroxide or alkoxides. They undergo many of the characteristic reactions of strong bases, such as their reactions as nucleophiles in S_N2 reactions with alkyl halides or alkyl sulfonates (Secs. 9.4, 10.3A). Thus, acetylenic anions can be used as nucleophiles in S_N2 reactions to prepare other alkynes.

$$CH_3CH_2CH_2CH_2—Br + Na^+ :\bar{C}≡CH \xrightarrow{NH_3 \text{ (liq)}}$$

1-bromobutane   **sodium acetylide**

$$CH_3CH_2CH_2CH_2—C≡CH + Na^+ Br^- \qquad (14.26)$$

1-hexyne
(64% yield)

$$CH_3CH_2CH_2CH_2—C≡\bar{C}: Na^+ + CH_3—Br \longrightarrow$$

$$CH_3CH_2CH_2CH_2—C≡C—CH_3 + Na^+ Br^- \qquad (14.27)$$

The acetylenic anions in these reactions are formed by the reactions of the appropriate 1-alkynes with $NaNH_2$ in liquid ammonia (Sec. 14.7A). The alkyl halides and sulfonates, as in most other S_N2 reactions, must be primary or unbranched secondary compounds (Why? See Sec. 9.5F.)

The reaction of acetylenic anions with alkyl halides or sulfonates is important because *it is another method of carbon-carbon bond formation.* Let's review the methods covered so far:

1. Cyclopropane formation by addition of carbenes to alkenes (Sec. 9.8)

2. Reaction of Grignard reagents with ethylene oxide (Sec. 11.4C)

3. Reaction of acetylenic anions with alkyl halides or sulfonates (this section).

PROBLEMS

14.19   Give the structures of the products in each of the following reactions.

*(a)   butyl tosylate + Ph—C≡C̄: Na⁺ $\longrightarrow$

(b) $CH_3C≡\bar{C}$: Na⁺ + CH_3CH_2—I $\longrightarrow$

*(c) $CH_3C≡C$—MgBr + ethylene oxide $\xrightarrow{\quad H_3O^+ \quad}$

(d)   Br—$(CH_2)_5$—Br + HC≡C̄: Na⁺ (excess) $\longrightarrow$

*14.20   Explain why graduate student Choke Fumely, in attempting to synthesize 4,4-dimethyl-2-pentyne using the reaction of CH_3—C≡C:⁻ Na⁺ with *tert*-butyl bromide, obtained none of the desired product.

*14.21   Outline two different preparations of 2-pentyne that involve an alkyne and an alkyl halide.

14.22   Suggest another pair of starting materials that could be used to prepare the alkyne product of Eq. 14.27.

14.23   Outline a preparation for each of the following compounds from an alkyne and an alkyl halide.
*(a)   3-heptyne   (b)   4,4-dimethyl-2-pentyne (See Problem 14.20.)

14.8   Organic Synthesis Using Alkynes

Let's tie together what you've learned about alkyne reactions and organic synthesis. The solution to the following study problem requires all of the fundamental operations of organic synthesis: formation of carbon-carbon bonds, transformation of functional groups, and establishment of stereochemistry (Sec. 11.8).

Notice that this problem stipulates the use of starting materials containing five or fewer carbons. This stipulation is made because such compounds are readily available from commercial sources and are relatively inexpensive.

Outline a synthesis of the following compound from acetylene and any other compounds containing no more than five carbons:

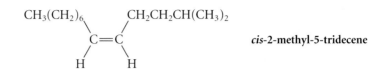

cis-2-methyl-5-tridecene

Solution   Working backward, notice the stereochemistry of the target molecule: it is a *cis*-alkene. We've covered only one method of preparing *cis*-alkenes: hydrogenation of alkynes (Sec. 14.6A). This reaction, then, is used in the last step of the synthesis:

$$CH_3(CH_2)_6—C\equiv C—CH_2CH_2CH(CH_3)_2 \xrightarrow[\text{Lindlar catalyst}]{H_2}$$

2-methyl-5-tridecyne

$$
\begin{array}{cc}
CH_3(CH_2)_6 & CH_2CH_2CH(CH_3)_2 \\
\diagdown & \diagup \\
C&=&C \\
\diagup & \diagdown \\
H & H
\end{array}
$$
   (14.28a)

(target molecule)

The next task is to prepare the alkyne used as the starting material in Eq. 14.28a. Since the desired alkyne contains fourteen carbons and the problem stipulates the use of compounds with five or fewer carbons, we'll have to use several reactions that form carbon-carbon bonds. There are two primary alkyl groups on the triple bond; the order in which they are introduced is arbitrary. Let's introduce the five-carbon fragment on the right-hand side of this alkyne in the last step of the alkyne synthesis. This is accomplished by forming the conjugate-base acetylenic anion of 1-nonyne and allowing it to react with the appropriate alkyl halide, 1-bromo-3-methylbutane (Sec. 14.7B):

$$CH_3(CH_2)_6—C\equiv C—H \xrightarrow[\text{NH}_3\,(\text{liq})]{\text{NaNH}_2} CH_3(CH_2)_6C\equiv \overset{-}{C}: \xrightarrow[\text{1-bromo-3-methylbutane}]{\overset{\displaystyle Br—CH_2CH_2CH(CH_3)_2}{}}$$

1-nonyne

$$CH_3(CH_2)_6—C\equiv C—CH_2CH_2CH(CH_3)_2 \quad (14.28b)$$

The starting material for this reaction, 1-nonyne, is prepared by reaction of 1-bromo-heptane with the sodium salt of acetylene itself.

$$H—C\equiv C—H \xrightarrow[\text{NH}_3\,(\text{liq})]{\text{NaNH}_2} H—C\equiv \overset{-}{C}: Na^+ \xrightarrow[]{Br—CH_2CH_2CH_2CH_2CH_2CH_2CH_3}$$

(large excess
relative to NaNH₂)

$$H—C\equiv C—CH_2CH_2CH_2CH_2CH_2CH_2CH_3 \quad (14.28c)$$

The large excess of acetylene relative to sodium amide is required to ensure formation of the *monoanion*. If there were more sodium amide than acetylene, some *dianion*

:C≡C: could form, and other reactions would occur. (What are they?) Since acetylene is cheap and is easily separated from the products (it is a gas), use of a large excess presents no practical problem.

Since the 1-bromoheptane used in Eq. 14.28c has more than five carbons, it must be prepared as well. The following sequence of reactions will accomplish this objective. (See also Problem 14.24a.)

$$CH_3CH_2CH_2CH_2CH_2{-}Br \xrightarrow[\text{ether}]{Mg} \xrightarrow{H_2C{-}CH_2 \text{ (O)}} \xrightarrow{H_3O^+}$$

1-bromopentane

$$CH_3CH_2CH_2CH_2CH_2CH_2CH_2{-}OH \xrightarrow{\text{conc. HBr}} CH_3CH_2CH_2CH_2CH_2CH_2CH_2{-}Br \qquad (14.28d)$$

1-heptanol   **1-bromoheptane**

The synthesis is now complete. To summarize:

$$HC{\equiv}CH \text{ (excess)} \xrightarrow[\text{NH}_3 \text{ (liq)}]{NaNH_2} \xrightarrow{CH_3(CH_2)_6Br \text{ (Eq. 14.28d)}} CH_3(CH_2)_6C{\equiv}CH \xrightarrow[\text{NH}_3 \text{ (liq)}]{NaNH_2}$$

$$\xrightarrow{(CH_3)_2CHCH_2CH_2Br} CH_3(CH_2)_6C{\equiv}CCH_2CH_2CH(CH_3)_2 \xrightarrow[\substack{\text{Lindlar}\\\text{catalyst}}]{H_2} \underset{\substack{\\ H \qquad\qquad H}}{\overset{\substack{CH_3(CH_2)_6 \qquad CH_2CH_2CH(CH_3)_2}}{C{=}C}}$$

$$(14.28e)$$

PROBLEM

14.24 Outline a synthesis for each of the following compounds from acetylene and any other compounds containing five or fewer carbons. Your synthesis for (a) should be different from that shown in Eq. 14.28d.

*(a)   1-heptanol   (b)

$$CH_3{-}\overset{\overset{\displaystyle O}{\|}}{C}{-}(CH_2)_8CH_3$$

14.9   Pheromones

As Problem 14.25 on p. 670 illustrates, the chemistry of alkynes can be applied to the synthesis of a number of **pheromones**—chemical substances used in nature for communication or signaling. An example of a pheromone is a compound or group of compounds that the female of an insect species secretes to signal her readiness for mating. The sex attractant of the grape berry moth is such a compound; it is a mixture of *cis* and *trans* isomers of the following alkene:

$$CH_3CH_2{-}CH{=}CH{-}CH_2CH_2CH_2CH_2CH_2CH_2CH_2{-}O{-}\overset{\overset{\displaystyle O}{\|}}{C}{-}CH_3$$

(96% *cis*, 4% *trans*)

Pheromones are also used for defense (Fig. 14.6), to mark trails, and for many other purposes. It was discovered not long ago that the traditional use of sows in France to discover buried truffles owes its success to the fact that truffles contain a steroid that happens to be identical to a sex attractant secreted in the saliva of boars during premating behavior!

About three decades ago, scientists became intrigued with the idea that pheromones might be used as a species-specific form of insect control. The thinking was that a sex attractant, for example, might be used to attract and trap the male of an insect species selectively without affecting other insect populations. Alternatively, the males of a species might become confused by a blanket of sex attractant and not be able to locate a suitable female. When used successfully, this strategy would break the reproductive cycle of the insect. The harmful environmental impact of the usual pesticides (for example, DDT) stimulated interest in such highly specific methods.

Research along these lines has shown, unfortunately, that pheromones are not effective for broad control of insect populations. They are, however, very useful for trapping target insects, thereby providing an early warning for insect infestations. When this approach is used, conventional pesticides need be applied only when the target insects appear in the traps. This strategy has brought about reductions in the use of conventional pesticides by as much as 70% in many parts of the United States.

Figure 14.6 *Example of a pheromone used for defense. A whip scorpion is ejecting its spray toward an appendage pinched with forceps. The pattern of the spray is visible on acid-sensitive indicator paper. The secretion is 84% acetic acid (CH_3CO_2H), 5% octanoic acid ($CH_3(CH_2)_6CO_2H$), and 11% water.*

PROBLEM

14.25   In the course of the synthesis of the sex attractant of the grape berry moth, both the *cis* and *trans* isomers of the following alkene were needed.

$$CH_3CH_2—CH＝CH—CH_2CH_2CH_2CH_2CH_2CH_2CH_2CH_2—O$$

*(a)   Outline a synthesis for the *cis* isomer of this alkene from the following alkyl halide and any other organic compounds.

$$Br—CH_2CH_2CH_2CH_2CH_2CH_2CH_2CH_2—O$$

(b)   Outline a synthesis of the *trans* isomer of the same alkene from the same alkyl halide and any other organic compounds.

14.10   Occurrence and Use of Alkynes

Naturally occurring alkynes are relatively rare. Alkynes do not occur as constituents of petroleum, but instead are synthesized from other compounds.

Acetylene itself comes from two common sources. Acetylene can be produced by heating coke (carbon from coal) with calcium oxide in an electric furnace to yield calcium carbide, CaC_2.

$$CaO \ + \ 3C \ \xrightarrow{\text{heat}} \ CaC_2 \ + \ CO \tag{14.29}$$

  calcium   calcium   carbon
  oxide   carbide   monoxide

Calcium carbide is an organometallic compound that can be conceptually regarded as the calcium salt of the acetylene dianion:

$$Ca^{2+} \quad :\bar{C}\!\equiv\!\bar{C}: \quad \textbf{calcium carbide}$$

Like any other acetylenic anion, calcium carbide reacts vigorously with water to yield the hydrocarbon; the calcium oxide by-product of this reaction can be recycled in Eq. 14.29.

The second process for manufacture of acetylene is the thermal cracking (Sec. 5.7B) of ethylene at temperatures above 1200° to give acetylene and H_2. (This process is thermodynamically unfavorable at lower temperatures.) The choice between the carbide process and cracking as the source of acetylene depends on various economic factors.

The most important general use of acetylene is for a chemical feedstock, as illustrated by the following examples:

$$HC\!\equiv\!CH \ + \ HCl \ \xrightarrow{\text{HgCl}_2} \ CH_2\!＝\!CHCl \tag{14.30}$$

  acetylene   **vinyl chloride**
                                   (a monomer used in the manufacture
                                    of poly(vinyl chloride), PVC)

$$2HC\equiv CH \xrightarrow{\text{catalyst}} HC\equiv C-CH=CH_2 \xrightarrow{HCl} CH_2=C-CH=CH_2 \quad (14.31)$$

acetylene · · · · · · · · · · · · vinylacetylene · $\overset{|}{Cl}$

chloroprene
(can be polymerized to give
neoprene rubber; Sec. 15.5.)

Oxygen-acetylene welding is an important use of acetylene, although it accounts for a relatively small percentage of acetylene consumption. The acetylene used for this purpose is supplied in cylinders, but it is hazardous because, at concentrations of 2.5–80% in air, it is explosive. Furthermore, since gaseous acetylene at even moderate pressures is unstable, this substance is not sold simply as a compressed gas. Acetylene cylinders contain a porous material saturated with a solvent such as acetone. Acetylene is so soluble in acetone that most of it actually dissolves. As acetylene gas is drawn off, more of the material escapes from solution as the gas is needed—another example of LeChatelier's principle in action!

Key Ideas in Chapter 14

⚗ Alkynes are compounds containing carbon-carbon triple bonds. The carbons of the triple bond are *sp*-hybridized. Electrons in *sp* orbitals are held somewhat closer to the nucleus than those in sp^2 or sp^3 orbitals.

⚗ The carbon-carbon triple bond in an alkyne consists of one σ bond and two mutually perpendicular π bonds. The electron density associated with the π bonds resides in a cylinder surrounding the triple bond. The induced circulation of these π electrons in a magnetic field causes the unique chemical shifts of acetylenic protons as well as those of acetylenic and propargylic carbons observed in NMR spectra.

⚗ The *sp* hybridization state is less stable than the sp^2 or sp^3 state. For this reason, alkynes have greater heats of formation than isomeric alkenes.

⚗ Alkynes have two general types of reactivity:
1. addition to the triple bond;
2. reactions at the acetylenic $\equiv C-H$ bond.

⚗ Useful additions to the triple bond include Hg^{2+}-catalyzed hydration, hydroboration, catalytic hydrogenation, and reduction with sodium in liquid ammonia.

⚗ Both hydration and hydroboration-oxidation of alkynes yield enols, which spontaneously form the isomeric aldehydes or ketones.

(continues)

Catalytic hydrogenation of alkynes gives *cis*-alkenes when a poisoned catalyst is used. The reduction of alkynes with alkali metals in liquid ammonia, a reaction that involves radical-anion intermediates, gives the corresponding *trans*-alkenes.

1-Alkynes, with pK_a values near 25, are the most acidic of the aliphatic hydrocarbons. Acetylenic anions are formed by the reactions of 1-alkynes with the strong base sodium amide. In a related reaction, acetylenic Grignard reagents can be formed in the reactions of 1-alkynes with alkylmagnesium halides.

Acetylenic anions are good nucleophiles and react with alkyl halides and sulfonates in S_N2 reactions to form new carbon-carbon bonds.

ADDITIONAL PROBLEMS

*14.26   Give the principal product(s) expected when 1-hexyne (or the other compounds indicated) is treated with each of the following reagents.
(a) HBr   (b) H_2, Pd/C   (c) H_2, Pd/C, quinoline
(d) product of (c) + O_3, then $(CH_3)_2S$
(e) product of (c) + BH_3 in THF, then H_2O_2/OH^-
(f) Br_2 (one equivalent)   (g) NaNH$_2$ in liquid ammonia
(h) Hg^{2+}, H_2SO_4, H_2O
(i) Na in liquid ammonia
(j) product of (i), then $Hg(OAc)_2$ in H_2O, then $NaBH_4$

14.27   Give the principal product(s), if any, expected when 4-octyne (or the other compounds indicated) is treated with each of the reagents in Problem 14.26.

14.28   In its latest catalog, Blarneystyne, Inc., a chemical company of dubious reputation specializing in alkynes, has offered some compounds for sale under the following names. Although each name unambiguously specifies a structure, all are incorrect. Propose a correct name for each compound.
*(a) 2-hexyn-4-ol   (b) 6-methoxy-1,5-hexadiyne
*(c) 1-butyn-3-ene   (d) 5-hexyne

14.29   Using only structures containing carbon and hydrogen, draw a structure for each of the following compounds.
*(a) a stable alkyne of five carbons containing a ring
(b) a chiral alkyne of six carbon atoms
*(c) an alkyne of six carbon atoms that gives the *same single* product in its reaction either with BH_3 in THF followed by H_2O_2/OH^- or with $H_2O/Hg^{2+}/H_3O^+$
(d) an alkyne of six carbon atoms that gives the *same single* product in its reaction either with Na in liquid ammonia or with H_2 over Pd/C in the presence of pyridine
*(e) a six-carbon alkyne that can exist as diastereomers

14.30 On the basis of the hybrid orbitals involved in the bonds, arrange the bonds in each of the following sets in order of increasing length.

*(a) C—H bonds of ethylene; C—H bonds of ethane; C—H bonds of acetylene

(b) C—C single bond of propane; C—C single bond of propyne; C—C single bond of propene

14.31 Rank the anions within each series in order of increasing basicity, lowest first. Explain.

*(a) $CH_3CH_2\overset{..}{\underset{..}{O}}:^-$, $HC\equiv\bar{C}:$, $:\overset{..}{\underset{..}{F}}:^-$

(b) $CH_3(CH_2)_3—C\equiv\bar{C}:$, $CH_3(CH_2)_4\overset{-}{C}H_2$, $CH_3(CH_2)_3—CH=\overset{..}{\underset{}{C}}H$

14.32 Using simple observations or tests with readily observable results, show how you would distinguish between the compounds in each of the following pairs.

*(a) *cis*-2-hexene and 1-hexyne   (b) 1-hexyne and 2-hexyne

*(c) 4,4-dimethyl-2-hexyne and 3,3-dimethylhexane

(d) propyne and 1-decyne

14.33 Outline a preparation of each of the following compounds from acetylene and any other reagents.

*(a) 　　　　OH　　　　　　　　(b) 1-hexyne
　　　　　　　|
　　　$CH_3CH_2CHCH_2CH_2CH_3$

*(c) $CH_3CH_2CD_2CD_2CH_2CH_3$　(d) 1-hexene

*(e) 　　　　　　　O　　　　　　(f) $(CH_3)_2CHCH_2CH_2CH_2CH=O$
　　　　　　　　　||
　　　$CH_3(CH_2)_7—C—OH$

*(g) *cis*-2-pentene　　　　　　(h) *trans*-3-decene

*(i) *meso*-4,5-octanediol　　　(j) (*Z*)-3-hexen-1-ol

*14.34 Account with a detailed curved-arrow mechanism for the following known reaction of phenylacetylene.

$$Ph—C\equiv C—H \xrightarrow[\text{THF}]{\text{NaOD, D}_2\text{O (large excess)}} Ph—C\equiv C—D$$

14.35 Give the product of the following sequence of reactions.

$$CH_3CH_2CH_2CH_2C\equiv CH \xrightarrow{CH_3CH_2MgBr} \xrightarrow{D_2O}$$

14.36 Using 1-butyne as the only source of carbon in the products, propose a synthesis for each of the following compounds.

*(a) $CH_3CH_2C\equiv C—D$　　　(b) $CH_3CH_2CD_2CD_3$

*(c) $CH_3CH_2CH_2—CO_2H$　　(d) 1-methoxybutane

(*Problem 14.36 continues*)

*(e)   the racemate of (3*R*,4*S*)-CH₃CH₂CHCHCH₂CH₂CH₂CH₃
                                                    | |
                                                    D   D

(f)   octane   *(g)   O
                                                         ‖
                                          CH₃CH₂CH₂CCH₂CH₃

*14.37   A box labeled "C_6H_{10} isomers" contains samples of three compounds, *A*, *B*, and *C*. Along with the compounds are the IR spectra of *A* and *B*, shown in Fig. 14.7. Fragmentary data in a laboratory notebook suggest that the compounds are 1-hexyne, 2-hexyne, and 3-methyl-1,4-pentadiene. Identify the three compounds.

*14.38   You have just been hired by Triple Bond, Inc., a company that specializes in the manufacture of alkynes containing five or fewer carbons. The President, Mr. Al Kyne, needs an outlet for the company's products. You have been asked to develop a synthesis of the housefly sex pheromone, *muscalure*, with the stipulation that all the carbon in the product must come only from the company's alkynes. The muscalure will then be used in an experimental fly trap. You will be equipped

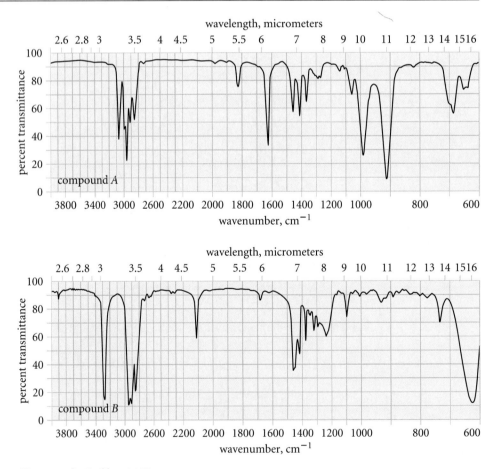

Figure 14.7   *IR spectra for Problem 14.37.*

with a laboratory containing all the company's alkynes, requisition forms for other reagents, and one gross of flyswatters in case you are successful. Outline a preparation of muscalure that meets the company's needs.

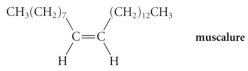

14.39 Outline a preparation of *disparlure*, a pheromone of the Gypsy moth, from acetylene and any other compounds containing not more than five carbon atoms.

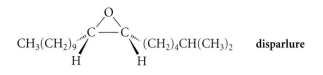

14.40 In the preparation of ethynylmagnesium bromide by Eq. 14.23, ethylmagnesium bromide is added to a large excess of acetylene in THF solution. Two side-reactions that can occur in this procedure are the following:

1. $H—C≡C—MgBr + CH_2CH_3—MgBr \longrightarrow CH_3CH_3 + BrMg—C≡C—MgBr$

2. $2H—C≡C—MgBr \rightleftharpoons H—C≡C—H + BrMg—C≡C—MgBr$

*(a) Suggest a mechanism for Reaction (1), and explain why an excess of acetylene is important for avoiding this reaction.

(b) Suggest a mechanism for Reaction (2), and explain why an excess of acetylene is important for avoiding this reaction.

*(c) Tetrahydrofuran (THF) is used as a solvent because the undesired by-product $BrMg—C≡C—MgBr$ is relatively soluble in this solvent. Explain why it is important for this by-product to be soluble if both side-reactions are to be minimized.

14.41 Identify the following compounds from their IR and proton NMR spectra.
*(a) $C_6H_{10}O$:
NMR: δ 3.31 (3H, s); δ 2.41 (6H, s); δ 1.43 (1H, s)
IR: 2110, 3300 cm^{-1} (sharp)
(b) C_4H_6O:
liberates a gas when treated with C_2H_5MgBr
NMR: δ 2.43 (1H, t, J = 2 Hz); δ 3.41 (3H, s); δ 4.10 (2H, d, J = 2 Hz)
IR: 2125, 3300 cm^{-1}
*(c) C_4H_6O:
NMR in Fig. 14.8, p. 676
IR: 2100, 3300 cm^{-1} (sharp), superimposed on a broad, strong band at 3350 cm^{-1}
(d) C_5H_6O (an alkyne):
NMR: δ 3.10 (1H, d, J = 2 Hz); δ 3.79 (3H, s); δ 4.52 (1H, doublet of doublets, J = 6 Hz and 2 Hz); δ 6.38 (1H, d, J = 6 Hz)

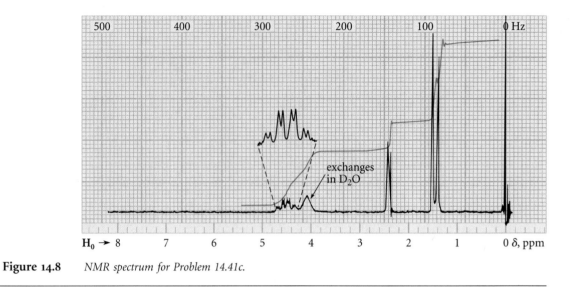

Figure 14.8 *NMR spectrum for Problem 14.41c.*

14.42 *(a) Identify the compound C_6H_{10} that shows IR absorptions at 3300 cm^{-1} and 2100 cm^{-1} and has the following CMR spectrum: δ 27.3, 31.0, 66.7, 92.8.

(b) Explain how you could distinguish between 1-hexyne and 4-methyl-2-pentyne by CMR.

*14.43 Complete the following reactions using intuition developed from this or previous chapters.

(a) $CH_3CH_2CH_2CH_2 - C\equiv C - H +$

$$CH_3CH_2CH_2CH_2 - Li \longrightarrow \xrightarrow{(CH_3)_3SiCl}$$

(*Hint:* Tertiary silyl halides, unlike tertiary alkyl halides, undergo nucleophilic substitution reactions that are not complicated by competing elimination reactions.)

(b)

$$CH_3(CH_2)_6 - C\equiv CH \xrightarrow{NaNH_2} \xrightarrow{\text{diethyl sulfate}}$$

where diethyl sulfate is:

$$CH_3CH_2 - O - \overset{\overset{O}{\|}}{\underset{\underset{O}{\|}}{S}} - O - CH_2CH_3$$

(*Hint:* See Sec. 10.3C.)

(c) $Li - C\equiv CH + F - (CH_2)_5 - Cl \longrightarrow$
(1 equivalent)

(d) $Ph - CH = CH_2 + Br_2 \longrightarrow \xrightarrow{NaNH_2}$ (a hydrocarbon not containing bromine)

(*Hint:* See Sec. 9.5.)

(e) $Ph - C\equiv C - Ph + HCCl_3 \xrightarrow{K^+(CH_3)_3CO^-}$

(*Hint:* See Sec. 9.8A.)

*14.44   Propose mechanisms for each of the following known transformations; use the curved-arrow formalism where possible.

(a)

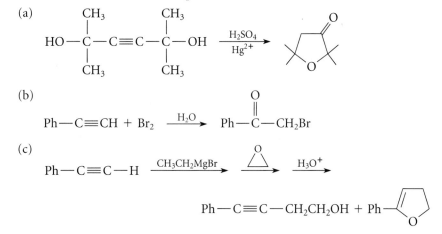

(b)

$$Ph—C{\equiv}CH + Br_2 \xrightarrow{H_2O} Ph—\overset{\overset{\displaystyle O}{\|}}{C}—CH_2Br$$

(c)

$$Ph—C{\equiv}C—H \xrightarrow{CH_3CH_2MgBr} \triangle \overset{O}{} \xrightarrow{H_3O^+}$$

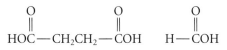

$$Ph—C{\equiv}C—CH_2CH_2OH + Ph—\!\!\!\!$$

*14.45   An optically active alkyne *A* has the following elemental analysis: 89.52% C; 10.48% H. Compound *A* can be catalytically hydrogenated to butylcyclohexane. Treatment of *A* with C_2H_5MgBr liberates no gas. Catalytic hydrogenation of *A* over Pd/C in the presence of quinoline, and treatment of the product with O_3, then H_2O_2, gives an *optically active* tricarboxylic acid $C_8H_{12}O_6$. (A tricarboxylic acid is a compound with three —CO_2H groups.) Give the structure of *A* and account for all observations.

14.46   A compound *A* (C_6H_6) undergoes catalytic hydrogenation over Lindlar catalyst to give a compound *B*, which in turn undergoes ozonolysis followed by workup with aqueous H_2O_2 to yield succinic acid and two equivalents of formic acid. In the absence of a catalyst poison, hydrogenation of *A* gives hexane; and reduction of *A* by sodium in liquid ammonia also gives *B*. Propose a structure for compound *A*.

$$\underset{\text{succinic acid}}{HO\overset{\overset{\displaystyle O}{\|}}{C}—CH_2CH_2—\overset{\overset{\displaystyle O}{\|}}{C}OH} \qquad \underset{\text{formic acid}}{H—\overset{\overset{\displaystyle O}{\|}}{C}OH}$$

15

Dienes, Resonance, and Aromaticity

Dienes are compounds with two carbon-carbon double bonds. Their nomenclature was discussed along with the nomenclature of other alkenes (Sec. 4.2A). Dienes are classified according to the relationship of their double bonds. In **conjugated dienes**, two double bonds are separated by one single bond. These double bonds are called **conjugated double bonds**.

$$CH_2{=\!=}CH{-}CH{=\!=}CH_2 \qquad \textbf{1,3-butadiene}$$
$$\text{(a conjugated diene)}$$

conjugated double bonds

Cumulenes are compounds in which one carbon participates in two carbon-carbon double bonds; these double bonds are called **cumulated double bonds**. Allene (propadiene) is the simplest cumulene. The term *allene* is also sometimes used as a family name for compounds containing only two cumulated double bonds.

$$CH_2{=\!=}C{=\!=}CH_2 \qquad \textbf{allene (propadiene)}$$

one carbon involved in two double bonds

Conjugated dienes and allenes have unique structures and chemical properties that are the basis for much of the discussion in this chapter. On the other hand, dienes in which the double bonds are separated by two or more single bonds have structures and chemical properties more or less like those of simple alkenes and do not require special discussion. These dienes will be called "ordinary" dienes.

$$CH_2{=\!=}CHCH_2CH_2CH_2CH{=\!=}CH_2 \qquad \textbf{1,6-heptadiene}$$
$$\text{(an ordinary diene)}$$

would be lost. The *s-trans* conformation of the molecule is approximately 12.5 kJ/mol (3 kcal/mol) more stable than the *s-cis* conformation. The internal rotation that interconverts these two conformations is very rapid at room temperature.

*15.1   Although the *s-cis* and *s-trans* conformations of 1,3-butadiene interconvert rapidly, the energy barrier to this interconversion is about 31 kJ/mol (7.5 kcal/mol). Suggest a reason why an energy barrier should exist.

15.2   (a) Why does the *s-cis* conformation of 1,3-butadiene have higher energy than the *s-trans* conformation? Use models if necessary.
       (b) It has been suggested that the *s-cis* conformation may be skewed away from planarity by about 15°. Why should this be energetically advantageous to the molecule?

B. Structure and Stability of Cumulated Dienes

The structure of allene is shown in Fig. 15.1. Since the central carbon of allene is bound to two groups, the carbon skeleton of this molecule is linear (Sec. 1.3B). A carbon atom with 180° bond angles is *sp*-hybridized (Sec. 14.2). Therefore, the central carbon of allene, like the carbons in an alkyne triple bond, is *sp*-hybridized. The two remaining carbons are *sp²*-hybridized and have trigonal geometry. Notice from Fig. 15.1 that the bond length of each cumulated double bond is somewhat less than that of an ordinary alkene double bond (see Problem 15.3).

The two bonding π molecular orbitals in allenes are mutually perpendicular, as required by the *sp* hybridization of the central carbon atom (Fig. 15.2). Consequently, *the H—C—H plane at one end of the allene molecule is perpendicular to the H—C—H plane at the other end*, as shown by the Newman projection in Fig. 15.1. Note carefully the difference in the bonding arrangements in allene and the conjugated diene 1,3-butadiene. In the conjugated diene, the π-electron systems of the two double bonds are coplanar and can overlap; all atoms are *sp²*-hybridized. In contrast, allene contains two mutually perpendicular π systems, each spanning two carbons; the central carbon is part of both. Because these two π systems are perpendicular, they do *not* overlap.

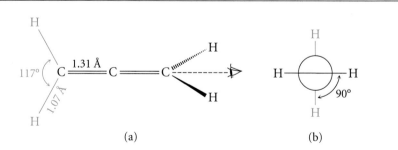

(a)   (b)

Figure 15.1   *Structure of allene, the simplest cumulated diene. (a) Lewis structure showing the bond angles and bond lengths. (b) A Newman projection along the carbon-carbon double bonds as seen by the eye. Notice that the CH₂ groups at opposite ends of the molecule lie in perpendicular planes.*

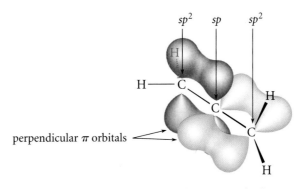

Figure 15.2   *Electronic structure of allene. Note that the two π orbitals are perpendicular.*

Because of their geometries, some allenes are chiral even though they do not contain an asymmetric carbon atom. The following molecule, 2,3-pentadiene, is an example of a chiral allene.

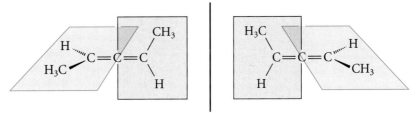

enantiomers

(Using models if necessary, verify the chirality of 2,3-pentadiene by showing that these two structures are not congruent.) Notice that the two sp^2-hybridized carbons are stereocenters. Thus, the enantiomers of 2,3-pentadiene differ by an internal rotation about either double bond. Since internal rotation about a double bond for all practical purposes does not occur, enantiomeric allenes can be isolated. (See Sec. 6.9.)

The sp hybridization of allenes is reflected in their C=C stretching absorptions in the infrared spectrum. This absorption occurs near 1950 cm^{-1}, not far from the C≡C stretching absorption of alkynes.

The data in Table 15.1 show that allenes have greater heats of formation than other types of isomeric dienes. For example, 1,2-pentadiene is considerably less stable than 1,3-pentadiene or 1,4-pentadiene. Thus, the cumulated arrangement is the least stable arrangement of two double bonds. A comparison of the heats of formation of 2-pentyne and 2,3-pentadiene shows that allenes are somewhat less stable than isomeric alkynes as well. In fact, a common reaction of allenes is isomerization to alkynes.

Although a few naturally occurring allenes are known, allenes are relatively rare in nature.

PROBLEMS

*15.3   Explain why cumulated double bonds are somewhat shorter than ordinary double bonds.

*15.4   Explain why there is a larger *difference* between the heats of formation of *trans*-1,3-pentadiene and 1,4-pentadiene (29.3 kJ/mol or 7.1 kcal/mol) than between *trans*-1,3-hexadiene and *trans*-1,4-hexadiene (19.7 kJ/mol or 4.7 kcal/mol).

*15.5   (a)   Draw the two enantiomers of the following allene.

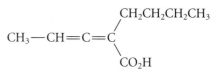

(b)   One enantiomer of this compound has a specific rotation of $-30.7°$. What is the specific rotation of the other?

15.2   Ultraviolet Spectroscopy

The IR and NMR spectra of conjugated dienes are very similar to the spectra of ordinary alkenes. However, another type of spectroscopy can be used to identify organic compounds containing conjugated π-electron systems. In this type of spectroscopy, called **ultraviolet-visible spectroscopy**, the absorption of radiation in the ultraviolet or visible region of the spectrum is recorded as a function of wavelength. The part of the ultraviolet spectrum of greatest interest to organic chemists is the *near ultraviolet* (wavelength range 200×10^{-9} to 400×10^{-9} meter). Visible light, as the name implies, is electromagnetic radiation visible to the human eye (wavelengths from 400×10^{-9} to 750×10^{-9} meter). Because there is a common physical basis for the absorption of both ultraviolet and visible radiation by chemical compounds, both ultraviolet and visible spectroscopy are considered together as one type of spectroscopy, often called simply **UV spectroscopy**.

A.   The UV Spectrum

Like any other absorption spectrum, the UV spectrum of a substance is the graph of radiation absorption by a substance *vs.* the wavelength of radiation. The instrument used to measure a UV spectrum is called a **UV spectrometer**. Except for the fact that it is designed to operate in a different part of the electromagnetic spectrum, it is conceptually much like an IR spectrometer (Figs. 12.3 and 12.12).

A typical UV spectrum, that of 2-methyl-1,3-butadiene (isoprene), is shown in Fig. 15.3. Since isoprene does not absorb visible light, only the ultraviolet region of the spectrum is shown. On the horizontal axis of the UV spectrum is plotted the wavelength λ of the radiation. In UV spectroscopy, the conventional unit of wavelength is the **nanometer** (abbreviated nm). One nanometer equals 10^{-9} meter. (In older literature, the term **millimicron**, abbreviated mμ, was used; one millimicron equals one nanometer.) The relationship between energy of the electromagnetic radiation and its frequency or wavelength should be reviewed again (Sec. 12.1A).

The vertical axis of a UV spectrum shows the **absorbance**. (Absorbance is sometimes called **optical density**, abbreviated O. D.) The absorbance is a measure of the amount of radiant energy absorbed. Suppose the radiation entering a sample has intensity I_0, and the light emerging from the sample has intensity I. The absorbance A is defined as the logarithm of the ratio I_0/I:

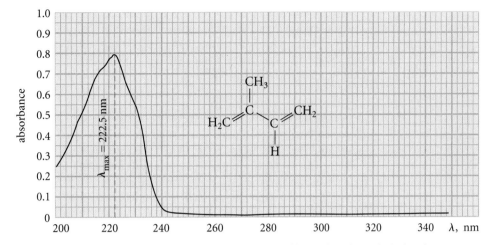

Figure 15.3   *Ultraviolet spectrum of isoprene in methanol. The λ_{max} is the wavelength at which the absorption maximum occurs.*

$$A = \log{(I_0/I)} \qquad (15.2)$$

The more radiant energy is absorbed, the larger is the ratio I_0/I, and the greater is the absorbance.

PROBLEMS

15.6   What is the energy of light (in kJ/mol or kcal/mol) with a wavelength of
*(a)   450 nm?   (b)   250 nm?

15.7   *(a)   What is the absorbance of a sample that transmits one-half of the incident radiation intensity?
(b)   What percent of the incident radiation is transmitted by a sample when its absorbance is 1.0? Zero?

***15.8**   A thin piece of red glass held up to white light appears brighter to the eye than a piece of the same glass that is twice as thick. Which piece has the greater absorbance?

In the UV spectra used in this text, absorbance increases from bottom to top of the spectrum. Therefore, absorption maxima occur as high points or peaks in the spectrum. Notice the difference in how UV and IR spectra are presented. (Absorptions in IR spectra increase from top to bottom, since IR spectra are conventionally presented as plots of *transmittance*, or percentage of light transmitted.) In the UV spectrum shown in Fig. 15.3, the absorbance maximum occurs at a wavelength of 222.5 nm. The wavelength at the maximum of an absorption peak is called the $\boldsymbol{\lambda_{max}}$ (read "lambda-max"). Some compounds have several absorption peaks and a corresponding number of λ_{max} values. Absorption peaks in the UV spectra of compounds in solution are generally quite broad. That is, peak widths span a considerable range of wavelength, typically 50 nm or more.

UV spectroscopy is frequently used for quantitative analysis because the absorbance at a given wavelength depends on the number of molecules in the light path. If a sample is contained in a vessel with a thickness along the light path of l cm, and the absorbing compound is present at a concentration of c moles per liter, then the absorbance is proportional to the product lc.

$$A = \epsilon lc \tag{15.3}$$

This equation is called the *Beer-Lambert law* or simply **Beer's law**. The constant of proportionality ϵ is called the **molar extinction coefficient** or **molar absorptivity**. Although the units of ϵ are $M^{-1}\,cm^{-1}$, these units are typically omitted when values of ϵ are cited. Each absorption in a given spectrum has a unique extinction coefficient that depends on wavelength, solvent, and temperature. The larger is ϵ, the greater is the light absorption at a given concentration c and path length l. For example, isoprene (Fig. 15.3) has an extinction coefficient at 222.5 nm of 10,750 in methanol solvent at 25 °C; its extinction coefficient in alkane solvents is nearly twice as large.

Extinction coefficients of 10^4 to 10^5 are common for molecules with conjugated π-electron systems. This means that strong absorptions can be obtained from very dilute solutions—solutions with concentrations on the order of 10^{-4} to 10^{-6} M. Because of its intrinsic sensitivity, UV spectroscopy was one of the earliest forms of spectroscopy to be used routinely in the laboratory; adequate spectra could be obtained on even the most primitive spectrometers.

Some UV spectra are presented in abbreviated form by citing the λ_{max} values of their principal peaks, the solvent used, and the extinction coefficients. For example, the spectrum in Fig. 15.3 is summarized as follows:

$$\lambda_{max}(CH_3OH) = 222.5 \text{ nm } (\epsilon = 10,750)$$

or $\qquad\lambda_{max}(CH_3OH) = 222.5$ nm $(\log \epsilon = 4.03)$.

PROBLEM

15.9 *(a) From the extinction coefficient of isoprene and its observed absorbance at 222.5 nm (Fig. 15.3), calculate the concentration of isoprene in moles/liter (assume a 1-cm light path).

(b) From the results of part (a) and Fig. 15.3, calculate the extinction coefficient of isoprene at 235 nm.

B. Physical Basis of UV Spectroscopy

What determines whether an organic compound will absorb UV or visible radiation? Ultraviolet and/or visible radiation is absorbed by the π electrons and, in some cases, by the unshared electron pairs in organic compounds. For this reason, UV and visible spectra are sometimes called *electronic spectra*. (The electrons of σ bonds absorb at much lower wavelengths, in the far ultraviolet.) Absorptions by compounds containing only single bonds and unshared electron pairs are generally quite weak (that is, their extinction coefficients are small). However, intense absorption of UV radiation occurs when a compound contains π electrons. Extinction coefficients in such cases are in the range 10^4 to 10^5. The simplest hydrocarbon containing π electrons, ethylene, absorbs UV radiation at $\lambda_{max} = 165$ nm $(\epsilon = 15,000)$. Although this is a strong absorption, the λ_{max}

of ethylene and other simple alkenes is below the usual working range of most conventional UV spectrometers; the lower end of this range is about 200 nm. However, molecules with *conjugated* double or triple bonds (for example, isoprene, Fig. 15.3), have λ_{max} values greater than 200 nm. Therefore, *UV spectroscopy is especially useful for the diagnosis of conjugated double or triple bonds.*

$$CH_2{=}CHCH_2CH_2CH{=}CH_2$$ double bonds not conjugated; no λ_{max} above 200 nm

conjugated double bonds; this compound has a λ_{max} above 200 nm:
$$\lambda_{max} = 227 \text{ nm } (\epsilon = 14{,}200)$$

The structural feature of a molecule responsible for its UV or visible absorption is called a **chromophore**. Thus, the chromophore in isoprene (Fig. 15.3) is the system of conjugated double bonds. Because many important compounds do not contain conjugated double bonds or other chromophores, UV spectroscopy, compared to NMR and IR spectroscopy, has limited utility in structure determination. However, the technique is widely used for quantitative analysis in both chemistry and biology; and, when compounds do contain conjugated multiple bonds, the UV spectrum can be an important element in a structure proof.

The physical phenomenon responsible for the absorption of energy in the UV spectroscopy experiment can be understood from a consideration of what happens when ethylene absorbs UV radiation at 165 nm. The π-electron structure of ethylene is shown in Fig. 4.6, Chapter 4. In the normal state of the ethylene molecule, called the *ground state*, the two π electrons occupy a *bonding π molecular orbital* (Fig. 15.4). When ethylene absorbs energy from light, one π electron is elevated from this bonding molecular orbital to the *antibonding* or π^* molecular orbital. This means that the electron assumes the more energetic wave motion characteristic of the π^* orbital, which includes a node between the two carbon atoms. The resulting state of the ethylene molecule, in which there is one electron in each molecular orbital, is called an *excited state*. To a useful

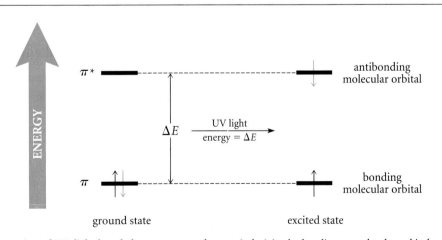

Figure 15.4   *Absorption of UV light by ethylene causes an electron (color) in the bonding π molecular orbital to be promoted to the antibonding π^* molecular orbital.*

A third variable that affects λ_{max} in a less dramatic but yet predictable way is the presence of *substituent groups on the double bond*. For example, each alkyl group on a conjugated double bond adds about 5 nm to the λ_{max} of a conjugated alkene. Thus, the two methyl groups of 2,3-dimethyl-1,3-butadiene add $(2 \times 5) = 10$ nm to the λ_{max} of 1,3-butadiene, which is 217 nm (Table 15.2). The predicted λ_{max} is $(217 + 10) = 227$ nm; the observed value is 226 nm.

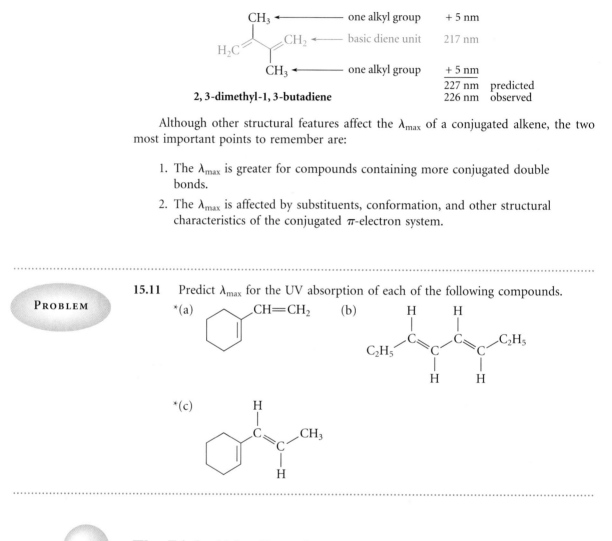

one alkyl group	+ 5 nm
basic diene unit	217 nm
one alkyl group	+ 5 nm
	227 nm predicted
	226 nm observed

2, 3-dimethyl-1, 3-butadiene

Although other structural features affect the λ_{max} of a conjugated alkene, the two most important points to remember are:

1. The λ_{max} is greater for compounds containing more conjugated double bonds.
2. The λ_{max} is affected by substituents, conformation, and other structural characteristics of the conjugated π-electron system.

PROBLEM

15.11   Predict λ_{max} for the UV absorption of each of the following compounds.

*(a)

(b)

*(c)

15.3   The Diels-Alder Reaction

A.   Reactions of Conjugated Dienes with Alkenes

Conjugated dienes undergo several unique reactions. One of these was discovered in 1928, when Otto Diels and Kurt Alder showed that many conjugated dienes undergo addition reactions with certain alkenes or alkynes. The following reaction is typical:

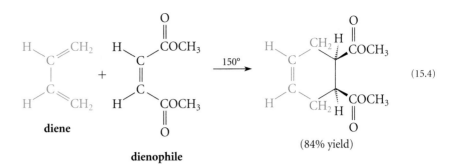

(15.4)

diene

dienophile

(84% yield)

This type of reaction between a conjugated diene and an alkene is called the **Diels-Alder reaction**. For their extensive work on this reaction Diels and Alder shared the 1950 Nobel Prize in chemistry.

When used in a Diels-Alder reaction, the conjugated diene is referred to simply as the *diene*, and the alkene with which it reacts as the *dienophile* (literally, "diene-loving molecule"). The Diels-Alder reaction is an example of a **cycloaddition reaction**—an addition reaction that results in the formation of a ring. Indeed, *the Diels-Alder reaction is an important way of making rings*, as the example in Eq. 15.4 illustrates.

Mechanistically, the Diels-Alder reaction occurs in a single step involving a cyclic flow of electrons. The curved arrows for this mechanism can be drawn in either a clockwise or counterclockwise direction.

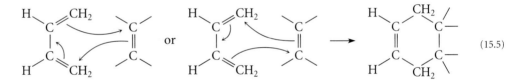

(15.5)

STUDY GUIDE LINK:
✓15.1
*A Terminology
Review*

A concerted reaction that involves a cyclic flow of electrons is called a **pericyclic reaction**. The Diels-Alder reaction is a pericyclic reaction, as is hydroboration of alkenes (Sec. 5.3B). However, hydroboration is not a cycloaddition, because no ring is formed.

Some of the dienophiles that react most rapidly in the Diels-Alder reaction, as in Eq. 15.4, bear substituent groups such as esters ($-CO_2R$), nitriles ($-C\equiv N$), or certain other unsaturated, electronegative groups. However, these substituents are not strictly necessary, as the reactions of many other alkenes can be promoted by heat or pressure. Some alkynes can also serve as dienophiles.

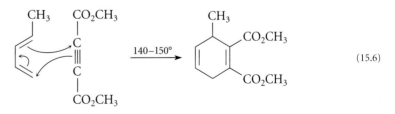

(15.6)

When the diene is cyclic, bicyclic products are obtained in the Diels-Alder reaction. In fact, this reaction is one of the best ways to prepare certain *bicyclic compounds* (Sec. 7.6).

(b) 1,3-butadiene and ethylene

15.13   Give the diene and dienophile that would react in a Diels-Alder reaction to give
each of the following products.

*(a)   (b)   *(c)

15.14   *(a)   Explain why two constitutional isomers are formed in the following Diels-
Alder reaction:

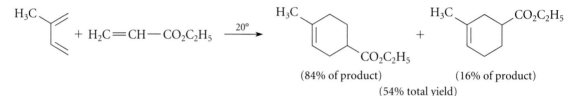

(84% of product)   (16% of product)

(54% total yield)

(b) What two constitutional isomers could be formed in the following Diels-
Alder reaction?

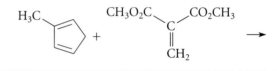

B. Effect of Diene Conformation on the Diels-Alder Reaction

Figure 15.5 shows a diagram of the Diels-Alder transition state, that is, the way in which
the diene and the dienophile are believed to be oriented as they react. This diagram
implies that the diene component is in the *s-cis*, or cisoid, conformation. The experimental
evidence supporting this view is that dienes "locked" into *s-trans*, or transoid, conformations are unreactive in Diels-Alder reactions:

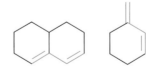

"locked" *s-trans* dienes;
unreactive in Diels-Alder reactions

Conversely, dienes "locked" into *s-cis* conformations are unusually reactive and in many
cases are much more reactive than corresponding noncyclic dienes:

"locked" *s-cis* dienes;
all reactive in the Diels-Alder reaction

**1,3-butadiene in
s-cis conformation**

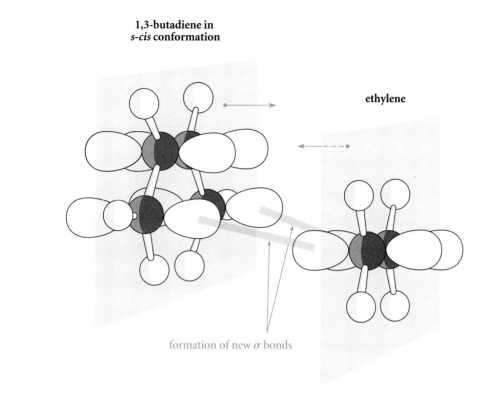

ethylene

formation of new σ bonds

Figure 15.5   *Transition state for the Diels-Alder reaction, shown with 1,3-butadiene as the diene and ethylene as the dienophile. Notice that the diene is in an s-cis, or cisoid, conformation, and that the two mole-cules approach in parallel planes so that the p orbitals of the alkene can overlap with the p orbitals on carbons 1 and 4 of the diene to become the new σ bonds.*

For example, 1,3-cyclopentadiene, which is locked in an *s-cis* conformation, reacts with typical dienophiles hundreds of times as rapidly as 1,3-butadiene, which exists primarily in an *s-trans* conformation.

This same effect of conformation can be observed in some noncyclic dienes. Thus, the *trans* isomer of 1,3-pentadiene reacts about 12,600 times as rapidly as the *cis* isomer with tetracyanoethylene (TCNE), a very reactive dienophile:

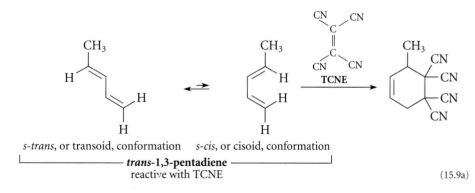

s-trans, or transoid, conformation   *s-cis*, or cisoid, conformation

⎣———————— *trans*-1,3-pentadiene ————————⎦
reactive with TCNE

(15.9a)

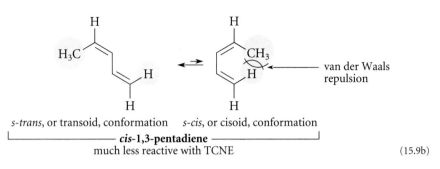

s-trans, or transoid, conformation   *s-cis*, or cisoid, conformation

cis-1,3-pentadiene
much less reactive with TCNE

(15.9b)

The cisoid conformation of the *cis*-diene is destabilized by a severe van der Waals repulsion between the methyl group and a diene hydrogen. The transition states for the Diels-Alder reactions of this diene, which require a cisoid conformation, are destabilized by the same effect. Consequently, the Diels-Alder reactions of the *cis*-diene are much slower than the corresponding reactions of the *trans*-diene, which does not have the destabilizing repulsion in its cisoid conformation.

PROBLEMS

*15.15   A mixture of 0.1 mole of (2E,4E)-2,4-hexadiene and 0.1 mole of (2E,4Z)-2,4-hexadiene was allowed to react with 0.1 mole of TCNE. After the reaction, the unreacted diene was found to consist of only one of the starting 2,4-hexadiene isomers. Which isomer did not react? Explain.

15.16   Complete the following Diels-Alder reaction.

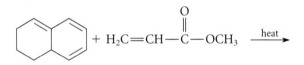

C. Stereochemistry of the Diels-Alder Reaction

When a diene and dienophile react in a Diels-Alder reaction, they approach each other in parallel planes (Fig. 15.5). This type of approach allows the π-electron clouds of the two components to overlap and form the bonds of the product. According to this picture, the diene undergoes a *syn* addition to the dienophile; likewise the dienophile undergoes a 1,4-*syn* addition to the diene. Each component adds to the other at *one face*.

The *syn* addition to the dienophile is illustrated by the following reactions. Groups that are *cis* in the alkene starting material are also *cis* in the product.

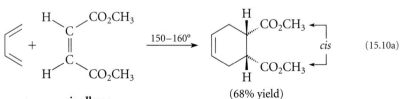

cis-alkene

(68% yield)

(15.10a)

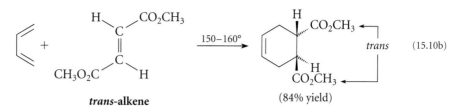

This is, of course, the same stereochemical result observed in other *syn* additions, such as hydroboration and catalytic hydrogenation (Sec. 7.9).

Syn addition to the diene is apparent if the terminal carbons of the diene unit are stereocenters. To assist in the analysis of stereochemistry, let's classify the substituents at the terminal carbons as inner or outer substituents:

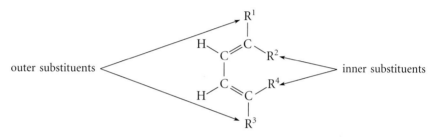

A *syn* addition requires that in the Diels-Alder product, the two inner substituents always have a *cis* relationship; the two outer substituents are also *cis*; and an inner substituent on one carbon is always *trans* to an outer substituent on the other. The following reactions of the stereoisomeric 2,4-hexadienes with the dienophile maleic anhydride demonstrate these points.

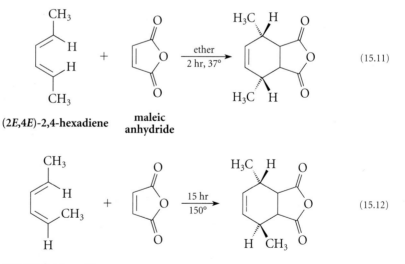

Notice, in Eq. 15.11, that the methyl groups in the diene are both outer substituents, and they are *cis* in the product. In Eq. 15.12, one methyl group in the diene is outer and the other is inner; consequently, they are *trans* in the product.

One further point of stereochemistry in the Diels-Alder reaction becomes important when *both* the terminal carbons of the diene unit *and* the carbons of the dienophile double bond are stereocenters. For a given diene and dienophile, two diastereomeric *syn*-addition products are possible. Each dienophile substituent can be either *cis* or *trans* to each diene substituent. The two possible *syn*-addition products arise from the two distinguishable orientations of the dienophile relative to the diene. The most common situation of this type, illustrated by Eqs. 15.13a and b, involves dienophiles that are *cis* alkenes.

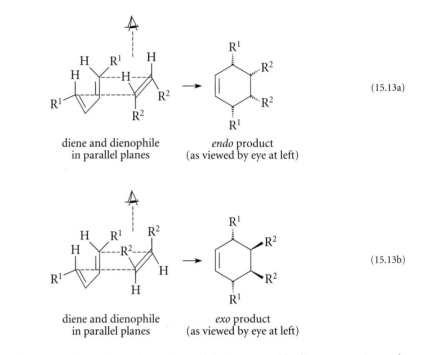

diene and dienophile
in parallel planes

endo product
(as viewed by eye at left)

(15.13a)

diene and dienophile
in parallel planes

exo product
(as viewed by eye at left)

(15.13b)

In such cases, the stereochemical configurations of the two possible diastereomeric products are abbreviated with the terms *endo* and *exo*, respectively. In the *endo* product, the dienophile substituents (R^2) are *cis* to the outer diene substituents (R^1); in the *exo* product, they are *trans*. Although exceptions occur, the *endo* product in many cases is the major one, particularly when cyclic dienes such as 1,3-cyclopentadiene are used.

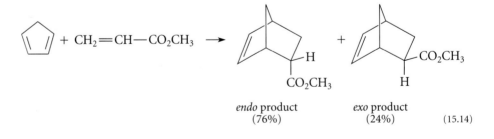

$$+ \; CH_2\!\!=\!\!CH\!-\!CO_2CH_3 \;\longrightarrow$$

endo product
(76%)

exo product
(24%)

(15.14)

....

PROBLEMS

15.17   Give the products formed when each of the following pairs reacts in a Diels-Alder reaction; show the relative stereochemistry of the substituent groups where appropriate.

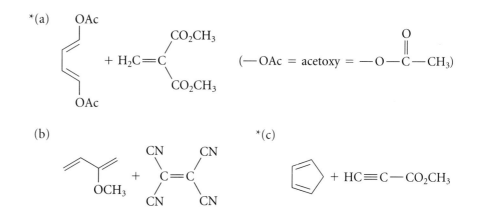

*(a)

(b)   *(c)

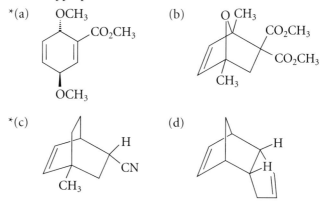

15.18   Give the structures of the starting materials that would yield each of the compounds below in Diels-Alder reactions. Pay careful attention to stereochemistry, where appropriate.

*(a)

(b)

*(c)

(d)

*15.19   In Eq. 15.11, the stereochemistry at the ring fusion is not specified. Show this stereochemistry, assuming that the Diels-Alder reaction gives the *endo* product.

15.4   Addition of Hydrogen Halides to Conjugated Dienes

A.   1,2- and 1,4-Additions

Conjugated dienes, like ordinary alkenes (Sec. 4.7), react with hydrogen halides; however, conjugated dienes give two types of addition products:

$$CH_3—CH=CH—CH=CH—CH_3 + HBr \xrightarrow{-20°}$$

$$CH_3—\underset{\underset{H}{|}}{CH}—\underset{\underset{Br}{|}}{CH}—CH=CH—CH_3 + CH_3—\underset{\underset{H}{|}}{CH}—CH=CH—\underset{\underset{Br}{|}}{CH}—CH_3$$

1,2-addition product   1,4-addition product
(~65%)   (~35%)   (15.15)

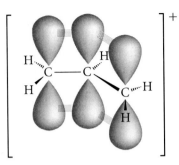

Figure 15.6   *Overlap of p orbitals in the allyl cation, the simplest allylic carbocation. Notice that the carbons on the ends are equivalent, as implied by the resonance structures in Eq. 15.20. Both the positive charge and the π electrons are delocalized.*

this delocalization is not adequately shown by a single Lewis structure, allylic cations are represented as resonance hybrids:

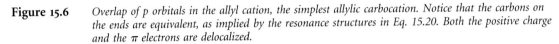

(15.20)

Recall that molecules which are resonance hybrids are more stable than any of their individual contributing structures; such molecules are said to be *resonance stabilized* (Sec. 1.4). The relationship between resonance structures and bonding in allylic carbocations shows *why* resonance hybrids are more stable. *Resonance hybrids have additional bonding associated with electron delocalization; this additional bonding is a stabilizing effect.*

PROBLEMS

15.20   Predict the products of each of the following reactions, and give the mechanisms for their formation.

*(a)   3-chloro-1-methylcyclohexene + C_2H_5OH $\longrightarrow$ (an S_N1 solvolysis)

(b)   1,3-butadiene + HCl $\longrightarrow$

15.21   Suggest a mechanism for each of the following reactions that accounts for both products.

*(a) $(CH_3)_2C{=}CH{-}\underset{\underset{Cl}{|}}{CH}{-}CH_3$ $\xrightarrow{\text{H}_2\text{O/acetone}}$

$(CH_3)_2C{=}CH{-}\underset{\underset{OH}{|}}{CH}{-}CH_3$ + $(CH_3)_2\underset{\underset{OH}{|}}{C}{-}CH{=}CH{-}CH_3$ + HCl

(b) $CH_3{-}CH{=}CH{-}CH_2{-}OH$ + conc. HBr $\xrightarrow[-15°]{\text{H}_2\text{SO}_4}$

$\underset{\text{(84\%)}}{CH_3{-}CH{=}CH{-}CH_2{-}Br}$ + $\underset{\underset{\underset{\text{(16\%)}}{Br}}{|}}{CH_3{-}CH{-}CH{=}CH_2}$ + H_2O

C. Kinetic and Thermodynamic Control

When a conjugated diene reacts with a hydrogen halide to give a mixture of 1,2- and 1,4-addition products, the 1-2-addition product predominates at low temperature:

$$CH_2=CH-CH=CH_2 + HCl \xrightarrow{-80°}$$

$$\underset{\text{(75-80\%)}}{CH_3-\overset{\overset{\textstyle Cl}{|}}{CH}-CH=CH_2} + \underset{\text{(20-25\%)}}{CH_3-CH=CH-CH_2-Cl} \qquad (15.21)$$

Yet alkenes with internal double bonds are more stable than their isomers with terminal double bonds, because the internal double bonds have more alkyl branches (Sec. 4.5B). Hence, in Eq. 15.21, *the major product is the less stable one.* This can be demonstrated experimentally by bringing the two alkyl halide products to equilibrium with heat and Lewis acids:

$$\underset{\text{minor product at equilibrium}}{CH_3-\overset{\overset{\textstyle}{|}}{\underset{\underset{\textstyle Cl}{|}}{CH}}-CH=CH_2} \underset{}{\overset{\text{heat, FeCl}_3}{\rightleftharpoons}} \underset{\text{major product at equilibrium}}{CH_3-CH=CH-CH_2-Cl} \qquad (15.22)$$

Because the more stable isomer always predominates in an equilibrium (Sec. 3.5), the 1,4-addition product is more stable than the 1,2-addition product.

When the less stable product of a reaction is the major product, then two things must be true. First, the less stable product *must be formed more rapidly* than the other products. Remember that a reaction in which two products are formed from the same starting material is in reality two competing reactions (Sec. 4.8). Consequently, the reaction that forms the less stable product has a larger rate. Second, the products *must not come to equilibrium* under the reaction conditions, because otherwise the more stable compound would be present in larger amount. Thus, in the addition of HCl to conjugated dienes, the predominance of the less stable product (Eq. 15.21) shows that 1,2-addition, which gives the less stable product, is faster than 1,4-addition:

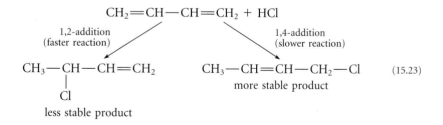

When the product distribution in a reaction differs substantially from the product distribution that would be observed if the products were at equilibrium, the reaction is said to be **kinetically controlled**. Thus, addition of hydrogen halides to conjugated dienes is a kinetically controlled reaction. On the other hand, if the products of a reaction come to equilibrium under the reaction conditions, the product distribution is said to be **thermodynamically controlled**.

In hydrogen halide addition to a conjugated diene, the first and rate-limiting step in the formation of both 1,2- and 1,4-addition products is the same—protonation of the double bond. Consequently, the product distribution must be determined by the relative rates of the *product-determining steps* (Sec. 9.6B): attack of the halide ion on one or the other of the electron-deficient carbons of the allylic carbocation intermediate.

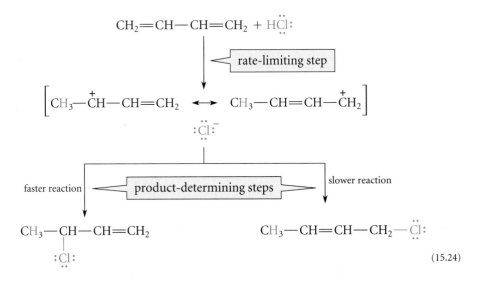

$$CH_2{=}CH{-}CH{=}CH_2 + H\ddot{C}l\!:$$

rate-limiting step

$$\left[CH_3{-}\overset{+}{C}H{-}CH{=}CH_2 \quad\longleftrightarrow\quad CH_3{-}CH{=}CH{-}\overset{+}{C}H_2\right]$$

$$:\!\ddot{C}l\!:^-$$

faster reaction ◁ product-determining steps ▷ slower reaction

$$CH_3{-}CH{-}CH{=}CH_2 \qquad\qquad CH_3{-}CH{=}CH{-}CH_2{-}\ddot{C}l\!:$$
$$\quad\;\;\;|$$
$$\quad\;\;:\!\ddot{C}l\!:$$

(15.24)

When the products are allowed to stand in solution for long periods of time, or when they are treated with heat or Lewis acids (Eq. 15.22), they can come to equilibrium because *the product-determining step for formation of the 1,2-addition product is reversible.* Thus, the alkyl halide products can equilibrate because the 1,2-addition product can undergo S_N1-like ionization to the allylic carbocation and chloride ion, from which both addition products can be re-formed.

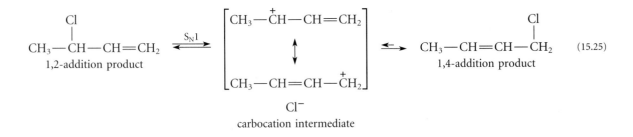

$$\begin{matrix} Cl \\ | \\ CH_3{-}CH{-}CH{=}CH_2 \end{matrix} \underset{}{\overset{S_N1}{\rightleftharpoons}} \left[\begin{matrix} CH_3{-}\overset{+}{C}H{-}CH{=}CH_2 \\ \updownarrow \\ CH_3{-}CH{=}CH{-}\overset{+}{C}H_2 \end{matrix}\right] \longleftrightarrow \begin{matrix} Cl \\ | \\ CH_3{-}CH{=}CH{-}CH_2 \end{matrix}$$

1,2-addition product

$$Cl^-$$
carbocation intermediate

1,4-addition product   (15.25)

Once the 1,4-addition product is formed, it does not re-ionize as rapidly because it is more stable. This point is shown diagrammatically in Fig. 15.7. In other words, formation of the 1,2-addition product is faster but reversible; formation of the 1,4-addition product is *slower but virtually irreversible.*

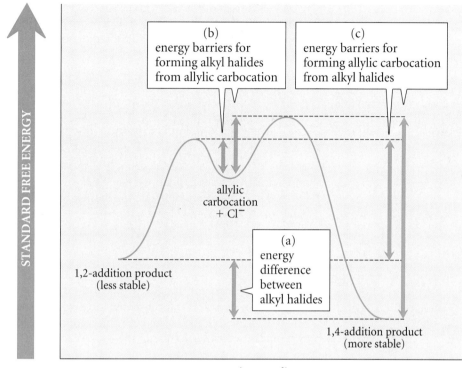

Figure 15.7   *Equilibration of the 1,2- and 1,4-addition products from the reaction of HCl with 1,3-butadiene. (a) The 1,2-addition product is less stable (has higher free energy) than the 1,4-addition product. (b) The 1,2-addition product is formed more rapidly from the allylic carbocation (smaller energy barrier) than the 1,4-addition product. (c) It then follows that reversion of the 1,2-addition product to the allylic carbocation is faster than the same reaction of the 1,4-addition product. Hence, formation of the 1,2-addition product is more reversible. When enough time is allowed for equilibration (or when equilibration is promoted by heat or catalysts), the 1,4-addition product predominates.*

AN ANALOGY FOR KINETIC *VS.* THERMODYNAMIC CONTROL

Imagine a very disoriented person stumbling randomly around a pasture in which is a shallow watering hole and a deep well with a high fence around it. Our disoriented friend is likely to fall into the hole several times; but, because the hole is shallow, he can easily climb out of it and continue staggering around the pasture. He frequently tries to climb the high fence, but doesn't succeed and falls back into the pasture. After a long while, however, he makes it over the fence and falls into the well; once in the well, he is there to stay. If you now imagine a large number of similarly disoriented people staggering around the same (very large) pasture, you should get a reasonably good image of kinetic and thermodynamic control. Initially, a large number of people fall into the shallow hole. If we wait long enough, however, most

(continues)

of them will end up in the deep well. The frequent occurrence—falling into the shallow hole—is reversible, but the rare occurrence—climbing the fence and falling into the well—is irreversible.

Similarly, the allylic carbocation is rapidly attacked by halide ion to give the 1,2-addition product, which is the less stable product—it is in a relatively shallow "energy hole" and is the kinetically controlled product. This reaction can reverse rapidly to give back the carbocation, which can react again. Occasionally, the carbocation is attacked by halide ion to give the more stable 1,4-addition product. This product is in a deeper "energy well" and does not as readily ionize to the carbocation; it is the thermodynamically controlled product. If the reaction mixture is analyzed before the products have a chance to come to equilibrium, then the distribution of products reflects a predominance of 1,2-addition, or kinetic control.

Why is the 1,2-addition product formed more rapidly? When the diene and HCl react, the carbocation and its chloride counterion, when first formed, exist as an *ion pair* (Fig. 8.2). That is, the chloride ion and the carbocation are closely associated. The chloride ion simply finds itself closer to one of the positively charged carbons than to the other. Addition is completed, therefore, at the nearer site of positive charge, giving the 1,2-addition product.

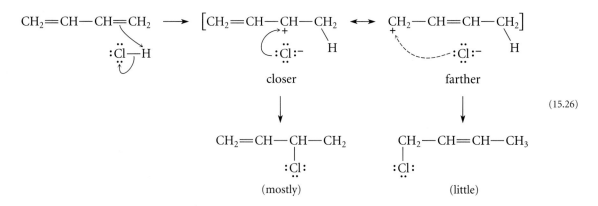

(15.26)

The precise reason for kinetic control, of course, varies from reaction to reaction. Whatever the reason, the relative amounts of products in a kinetically controlled reaction are determined by the relative free energies of the *transition states* for each of the product-determining steps, and *not* by the relative free energies of the products.

<hr />

PROBLEMS

*15.22   Suggest structures for the two constitutional isomers formed when 1,3-butadiene reacts with one equivalent of Br_2. (Ignore any stereochemical issues.) Which of these products would predominate if the two were allowed to come to equilibrium?

15.23 How many product(s) should form when one equivalent of HCl reacts with 1,3-cyclohexadiene? Give their structure(s).

15.5 Diene Polymers

1,3-Butadiene is one of the most important raw materials of the synthetic rubber industry. Annual production of 1,3-butadiene in the United States reaches 3.2 billion pounds.

The polymerization of 1,3-butadiene gives polybutadiene, an important synthetic rubber used in the manufacture of tires:

$$\left[CH_2 - CH = CH - CH_2 \right]_n$$

polybutadiene
(contains both *cis* and *trans* double bonds)

More than one billion pounds of this polymer is manufactured in the United States annually. Polybutadiene is referred to as a *diene polymer*, because it comes from polymerization of a diene monomer. It has only one double bond per unit, because one double bond is lost through the addition that takes place in the polymerization process.

The free-radical polymerization of dienes starts with the addition of an initiating radical (R· in the equations below) to the 1,3-diene unit. In the resulting radical, the unpaired electron is delocalized by resonance:

$$R \cdot \quad CH_2 = CH - CH = CH_2 \longrightarrow$$
$$\left[R - CH_2 - \dot{C}H - CH = CH_2 \quad \longleftrightarrow \quad R - CH_2 - CH = CH - \dot{C}H_2 \right] \quad (15.27a)$$

Addition of this radical to another molecule of butadiene, and repetition of this process many times, yields the final polymer:

$$R - CH_2 - CH = CH - CH_2 \cdot \quad CH_2 = CH - CH = CH_2 \longrightarrow$$
$$R - CH_2 - CH = CH - CH_2 - CH_2 - CH = CH - \dot{C}H_2 \xrightarrow[\text{times}]{\text{repeat many}}$$
$$\left[CH_2 - CH = CH - CH_2 \right]_n \quad \text{(1,4-addition polymer)} \quad (15.27b)$$

Although the product above is shown as the result of 1,4-addition, a small amount of 1,2-addition can occur as well.

1,3-Butadiene can also be polymerized along with styrene (Ph—CH=CH₂), usually in about a 3:1 ratio, to give another type of synthetic rubber called *styrene-butadiene rubber (SBR)*, most of which is used for tires and tread rubber.

$$\left[CH_2 - CH = CH - CH_2 - CH_2 - CH = CH - CH_2 - CH_2 - CH = CH - CH_2 - CH_2 - CH \right]_n$$
$$\underbrace{\hspace{8cm}}_{\text{three butadiene units}} \qquad \underset{\underset{\text{styrene unit}}{|}}{Ph}$$

styrene-butadiene rubber (SBR)

(This structure is oversimplified, because both 1,2- and 1,4-additions to butadiene take place, both *cis* and *trans* double bonds are present, and the order of addition of the butadiene and styrene units is random.) About 1.7 billion pounds of SBR is produced annually in the United States.

SBR is an example of a **copolymer**: a polymer produced by the simultaneous polymerization of two or more monomers.

Natural rubber is (*Z*)-polyisoprene, another diene polymer:

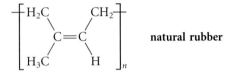

natural rubber

Although it is conceptually a diene polymer, natural rubber is *not* made in nature from isoprene. (The biosynthesis of naturally occurring isoprene derivatives is discussed in Sec. 17.6B.) Rubber hydrocarbon is obtained as a 40% aqueous emulsion from the rubber tree. After isolation, the polymer is subjected to a process called *vulcanization*. In this process, discovered in 1840 by Charles Goodyear, the rubber is kneaded and heated with sulfur. The sulfur forms crosslinks between the polymer chains, which can be represented schematically as follows:

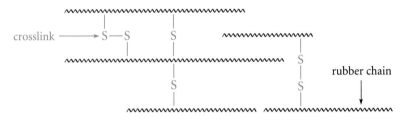

The crosslinks increase the rigidity and strength of the rubber at the cost of some flexibility. Although polyisoprene can be made synthetically, the natural material is generally preferred for economic reasons. Chemists and botanists are investigating the possibility of cultivating other hydrocarbon-producing plants that could become hydrocarbon sources of the future.

PROBLEMS

*15.24   Write a mechanism for the free-radical copolymerization of 1,3-butadiene and styrene to give styrene-butadiene rubber.

15.25   What would be the structure of polybutadiene if every other unit of the polymer resulted from 1,2-addition?

15.6   Resonance

Recall that resonance structures are used to describe a molecule when a single Lewis structure is inadequate (Sec. 1.4). Molecules that can be represented as resonance hybrids are said to be *resonance-stabilized*—that is, they are more stable than any one of their contributing structures. The reason for this additional stability is the additional bonding that results from the delocalization of electrons and additional orbital overlap (Sec. 15.4B). Resonance structures can be derived by the curved-arrow formalism (Sec. 3.3B).

The derivation and use of resonance structures is important for understanding both *molecular structure* and *molecular stability*. This section examines resonance in more detail by reviewing the concepts of resonance and the techniques for writing resonance structures. You'll learn to assess whether a resonance structure is important enough to be considered seriously as an important contributor to a molecular structure. You'll also learn how resonance structures can be used to make deductions about molecular stability.

A.   Writing Resonance Structures

Resonance structures show *the delocalization of electrons*. Resonance structures can be drawn when bonds, unshared electron pairs, or single electrons can be delocalized (moved) by the curved-arrow formalism *without moving any atoms*. Resonance structures are usually placed in brackets to emphasize the fact that they are being used to describe a *single species*. The following are all valid examples of resonance structures:

$$\left[CH_2\!=\!CH\!-\!\overset{+}{C}H_2 \quad \longleftrightarrow \quad {}^+CH_2\!-\!CH\!=\!CH_2 \right] \qquad \qquad (15.28a)$$

delocalization of a bond using electron-pair arrow formalism

$$\left[CH_2\!=\!CH\!-\!\ddot{\underset{..}{O}}\!:^- \quad \longleftrightarrow \quad {}^-\!:\!CH_2\!-\!CH\!=\!\ddot{O}\!: \right] \qquad \qquad (15.28b)$$

delocalization of a bond and an un-shared electron pair

$$\left[CH_2\!=\!CH\!-\!CH_2\!\cdot \quad \longleftrightarrow \quad \cdot CH_2\!-\!CH\!=\!CH_2 \right] \qquad \qquad (15.28c)$$

delocalization of a single electron and a bond by the fishhook formalism (Sec. 5.6B)

As these examples illustrate, some of the most common situations in which resonance structures are used involve the interaction of double or triple bonds with electron-deficient atoms, unshared electron pairs, or unpaired electrons. Notice also in these examples that the delocalization of electrons by resonance can also result in the delocalization of charge (Eqs. 15.28a and b) or the delocalization of an unpaired electron (Eq. 15.28c).

The following structures, although reasonable Lewis structures, are *not* resonance structures, because the movement of an atom takes place; the location of the chlorine is different in the two structures:

$$CH_2\!-\!CH\!=\!CH\!-\!CH_3 \quad \overset{\times}{\longleftrightarrow} \quad CH_2\!=\!CH\!-\!CH\!-\!CH_3 \qquad \qquad (15.29)$$
$$\underset{Cl}{|} \qquad\qquad\qquad\qquad\quad \underset{Cl}{|}$$

In fact, these are separate compounds:

$$CH_2\!-\!CH\!=\!CH\!-\!CH_3 \quad \rightleftharpoons \quad CH_2\!=\!CH\!-\!CH\!-\!CH_3 \qquad \qquad (15.30)$$
$$\underset{Cl}{|} \qquad\qquad\qquad\qquad\quad \underset{Cl}{|}$$

If two structures can exist as different compounds, they cannot be resonance structures. Resonance structures, in contrast, are used to describe a *single* molecule; the molecule is an average of its resonance structures. That is, resonance contributors are *fictitious* structures used to help us understand the structures of *real molecules* for which single Lewis structures are inadequate. Notice that the equilibrium double arrows $\rightleftarrows$ and the resonance double-headed arrow $\leftrightarrow$ have quite different meanings. The equilibrium arrows mean that two *real* compounds are in equilibrium; the resonance arrow means that two or more *fictitious* structures are being used to represent a compound for which a single Lewis structure is inadequate. *You must be careful not to use one symbol in a situation in which the other is appropriate.*

B. Relative Importance of Resonance Structures

In many cases all resonance structures are not of equal importance; that is, the structure of a molecule is most closely approximated by its most important resonance structures. This section shows how to assess the *relative importance* of resonance structures.

To evaluate the relative importance of resonance structures, compare the stabilities of all the resonance structures for a given molecule *as if each structure were a separate molecule.* That is, *imagine* that each structure is real. Then use the relative stabilities of the different structures to determine their relative importance. *The most stable structures are the most important ones.* The following guidelines for evaluating resonance structures emerge from this type of analysis:

1. *Identical structures are equally important descriptions of a molecule.*

Example:

$$\left[^+CH_2\overset{\frown}{-}CH\!=\!CH_2 \quad \leftrightarrow \quad CH_2\!=\!CH\!-\!\overset{+}{C}H_2 \right] \tag{15.31}$$

Because these structures of the allyl carbocation are indistinguishable, they are equally important in describing this species.

2. *Structures with the greater number of bonds are more important.*

Example:

$$\left[\overset{\frown}{C}H_2\!=\!CH\overset{\frown}{-}CH\!=\!\overset{\frown}{C}H_2 \quad \leftrightarrow \quad \overset{\cdot}{C}H_2\!-\!CH\!=\!CH\!-\!\overset{\cdot}{C}H_2 \right] \tag{15.32}$$

Because bonding is energetically favorable, it follows that the more bonds a structure has, the more stable it is. Thus, the structure on the left is more important.

3. *Structures that require the separation of opposite charges are less important than those that do not.*

Example:

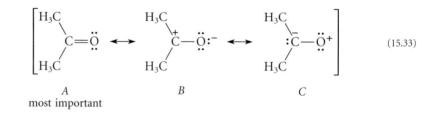

$$\tag{15.33}$$

    A   *B*   *C*
  most important

Since energy is required to separate charge (electrostatic law; Eq. 3.40), structures *B* and *C* are less stable and thus less important. (They also have fewer bonds, and so are minor contributors on two counts.)

Be sure that you understand the difference between "delocalization of charge" and "separation of charge." When a charge is *delocalized* by resonance, charge of a given type is moved to different locations within a molecule, as in Eq. 15.31. When charge is *separated*, two opposite charges are moved away from each other, as in Eq. 15.33.

4. *Structures in which charges and/or electron deficiency are assigned to atoms of appropriate electronegativity are more important.*

Example: In Eq. 15.33, structure *B* is more important than structure *C* because structure *B* assigns an electron deficiency to the more electropositive atom (carbon) and negative charge to the more electronegative atom (oxygen). In applying this guideline, it is important not to confuse *electron deficiency* with *positive charge*. Numerous situations occur in which positively charged atoms are not electron-deficient. Consider, for example, the resonance structures of the following cation:

$$\left[H_2\overset{+}{C} \overset{\frown}{:O} — H \quad \longleftrightarrow \quad H_2C = \overset{+}{O} — H \right] \tag{15.34}$$

In the structure on the left, the carbon atom is electron-deficient *and* positively charged. In the structure on the right, the oxygen, although positively charged, is *not* electron-deficient, because it has a complete octet. Contrast this with structure *C* in Eq. 15.33, in which the oxygen is two electrons short of an octet, and therefore electron-deficient. Because of the importance of the octet rule, an electron-deficient oxygen or nitrogen is a much less stable situation than a positively charged oxygen or nitrogen that has a complete electronic octet. Thus, the structure on the right of Eq. 15.34 is important because it has more bonds (Guideline 2), even though positive charge resides on an electronegative atom. The structure on the left is also important because a less electronegative atom (carbon) bears the positive charge (Guideline 4).

5. *If the orbital overlap symbolized by a resonance structure is not possible, the resonance structure is not important.*

Example:

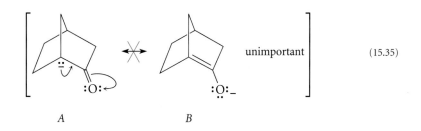

Structure *B* is unimportant because the orbital containing the unshared electron pair lies at a right angle to the π orbital of the C=O group, and therefore cannot overlap with it.

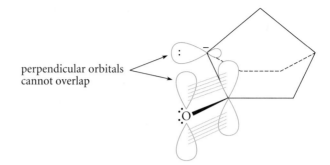

perpendicular orbitals
cannot overlap

Only an energetically costly distortion of the molecule would permit overlap of all the *p* orbitals; but resonance structures cannot involve atomic motion. Notice that in other cases, when orbital overlap is possible, a similar resonance structure *is* important.

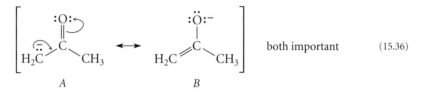

both important   (15.36)

In this example, the ion can easily adopt a conformation in which the carbon orbital containing the unshared electron pair is coplanar with the *p* orbitals of the C=O group.

overlap of *p* orbitals

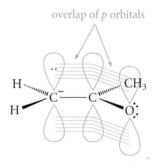

Guideline 5 means that it is not enough to derive a resonance structure correctly with the curved-arrow formalism. We must also keep in mind the *meaning* of the structure in terms of the orbital overlap involved.

C. Use of Resonance Structures

Although resonance does have a mathematical basis in quantum theory, organic chemists generally use resonance arguments in a qualitative way to compare the stabilities of molecules. To make this comparison, the following principle is applied: *All other things being equal, the molecule with the greater number of resonance structures is more stable.* The qualifier "all other things being equal" means that the other aspects of the molecules that affect their stabilities should be about the same.

A comparison of the stabilities of the two carbocations below illustrates the use of resonance arguments. (These are the two carbocations that were compared in the addition of HBr in Eqs. 15.16 and 15.18.) Both carbocations are secondary, and in fact they are isomeric. The two carbocations differ mainly in the number of resonance structures that can be written for each. The more stable carbocation has two resonance structures. The other carbocation has only one Lewis structure, and is less stable.

$$\left[CH_3—CH_2—\overset{+}{C}H—CH=CH—CH_3 \quad \longleftrightarrow \quad CH_3—CH_2—CH=CH—\overset{+}{C}H—CH_3 \right] \quad \text{more stable carbocation}$$

$$CH_3—\overset{+}{C}H—CH_2—CH=CH—CH_3 \quad \text{less stable carbocation}$$

Notice that application of this principle broadens the meaning of the term *resonance stabilization*. Resonance-stabilized molecules are not only more stable than their individual contributing structures; they are also more stable than *other molecules* that have only one Lewis contributor (other things being equal).

The reason that the number of resonance structures is related to the stability of a molecule follows from the electronic basis of resonance itself. Molecules with many resonance structures have extensive electron delocalization and extensive additional bonding that result from the overlap of orbitals. This additional bonding is a stabilizing effect.

STUDY GUIDE LINK:
✓15.2
Resonance Structures

STUDY
PROBLEM
15.2

Which of the following carbocations is more stable?

$$CH_3\overset{..}{\underset{..}{O}}—CH=CH—\overset{+}{C}H_2 \quad \text{or} \quad CH_2=\overset{\overset{\displaystyle :\overset{..}{O}CH_3}{|}}{C}—\overset{+}{C}H_2$$

$$A \qquad\qquad\qquad B$$

Solution   The solution to this problem involves determining which carbocation has the greater number of important resonance structures. Carbocation A has the following important resonance structures:

$$\left[CH_3\overset{..}{\underset{..}{O}}—CH=CH—\overset{+}{C}H_2 \quad \longleftrightarrow \quad CH_3\overset{..}{\underset{..}{O}}—\overset{+}{C}H—CH=CH_2 \quad \longleftrightarrow \quad CH_3\overset{+}{\underset{..}{O}}=CH—CH=CH_2 \right]$$

Carbocation B has only two reasonable structures. In particular, the charge can be delocalized onto the oxygen in ion A but not in ion B:

$$\left[CH_2=\overset{\overset{\displaystyle :\overset{..}{O}CH_3}{|}}{C}—\overset{+}{C}H_2 \quad \longleftrightarrow \quad \overset{+}{C}H_2—\overset{\overset{\displaystyle :\overset{..}{O}CH_3}{|}}{C}=CH_2 \right]$$

Thus, ion A is more stable because it has the larger number of important resonance structures.

....

15.26 In each of the following sets, show by the curved-arrow or fishhook formalism how each structure is derived from any other one, and indicate which structure(s) are most important and why.

*(a) $\left[H_3C-CH_3 \quad \longleftrightarrow \quad H_3C\cdot \quad \cdot CH_3 \quad \longleftrightarrow \right.$

$\left. H_3C:^- \; ^+CH_3 \quad \longleftrightarrow \quad H_3C^+ \; ^-:CH_3 \right]$

(b) $\left[CH_3-\overset{+}{\overset{\cdot\cdot}{C}}=\ddot{N}H \quad \longleftrightarrow \quad CH_3-C\equiv\overset{+}{N}H \quad \longleftrightarrow \quad CH_3-\overset{2+}{C}-\overset{\cdot\cdot}{\underset{\cdot\cdot}{N}}H \right]$

*(c) $\left[CH_3-CH=CH-\overset{+}{C}H_2 \quad \longleftrightarrow \quad CH_3-\overset{+}{C}H-CH=CH_2 \right]$

(d)
$$\left[\begin{array}{cc} \begin{array}{c} H_3C \\ \diagdown \\ \diagup \\ H_3C \end{array} C=CH-\overset{+}{C}H_2 & \longleftrightarrow & \begin{array}{c} H_3C \\ \diagdown \\ \diagup \\ H_3C \end{array} \overset{+}{C}-CH=CH_2 \end{array} \right]$$

15.27 Using resonance arguments, rank the ions or radicals within each set in order of increasing stability, least stable first. Explain.

*(a)
$$\ddot{O}=CH-\overset{\cdot\cdot}{\underset{}{C}}H-C\equiv C-CH_3 \quad \text{or} \quad HC\equiv C-\overset{\overset{\textstyle :\ddot{O}:^-}{|}}{C}=CH-CH_3$$

(b)
$$CH_3-\overset{\overset{\textstyle :\ddot{O}:^-}{|}}{C}H-CH=CH_2 \quad \text{or} \quad CH_3-\overset{\overset{\textstyle :\ddot{O}:^-}{|}}{C}=CH-CH_3$$

*(c)
$$CH_2=CH-CH=CH-\overset{\cdot}{C}H_2 \quad \text{or} \quad \cdot CH_2-\overset{\overset{\textstyle CH_2}{\|}}{C}-CH=CH_2$$

*(d)
$$CH_2=\overset{\overset{\textstyle CH_3}{|}}{C}-\overset{+}{C}H_2 \quad \text{or} \quad CH_3-CH=CH-\overset{+}{C}H_2$$

15.28 In each set below, the two compounds do not differ greatly in stability. Predict which compound within each set should react more rapidly in a S_N1 solvolysis reaction in aqueous acetone. (*Hint:* Assess the relative stabilities of the carbocation intermediates, and apply Hammond's postulate.)

*(a) $CH_3\ddot{O}-CH=CH-CH_2-Cl \qquad CH_3\ddot{O}-\overset{\overset{\textstyle }{}}{C}-CH_2-Cl$

$\qquad\qquad\qquad A \qquad\qquad\qquad\qquad\qquad \overset{\|}{CH_2}$

$\qquad\qquad\qquad\qquad\qquad\qquad\qquad\qquad\qquad B$

(b)

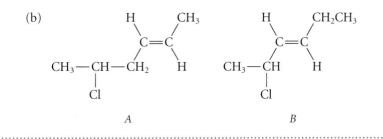

A   *B*

Introduction to Aromatic Compounds

A. Benzene, a Puzzling "Alkene"

Benzene and its derivatives constitute the class of organic substances called **aromatic compounds**.

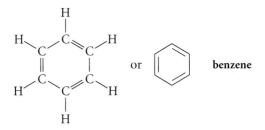 or   benzene

The reason for the term *aromatic* is historical: many fragrant compounds known from earliest times, such as the ones below, proved to be derivatives of benzene.

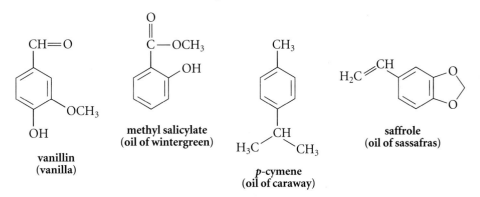

vanillin
(vanilla)

methyl salicylate
(oil of wintergreen)

p-cymene
(oil of caraway)

saffrole
(oil of sassafras)

Although it is known today that many benzene derivatives are not distinguished by unique odors, the word *aromatic* continues to be used as a family name for all benzene derivatives and certain related compounds.

The structure today for benzene was proposed in 1865 by August Kekulé, who claimed in 1890 that it came to him in a dream. (This claim has been disputed by some modern historians.) To nineteenth-century chemists, the problem with the Kekulé structure was that it portrays benzene as a cyclic, conjugated triene. Yet benzene does not undergo any of the addition reactions that are associated with either conjugated dienes or ordinary alkenes. Benzene itself, as well as benzene rings in other compounds, are inert to the usual conditions of halogen addition, hydroboration, hydration, or ozonolysis. This property of the benzene ring is illustrated by the addition of bromine to styrene, a compound that contains both a benzene ring and one additional double bond:

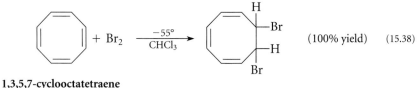

$$(15.37)$$

styrene

The noncyclic double bond in styrene rapidly adds bromine, but the benzene ring remains unaffected.

We might speculate that benzene's lack of reactivity has something to do with its cyclic structure; yet cyclohexene also adds bromine readily. Perhaps, then, it is the cyclic structure and the conjugated double bonds that *together* account for the unusual behavior of benzene. However, 1,3,5,7-cyclooctatetraene (abbreviated in this text as COT) adds bromine smoothly even at low temperature.

(100% yield)   (15.38)

1,3,5,7-cyclooctatetraene
(COT)

Thus, the Kekulé structure clearly had difficulties that could not be easily explained away. In 1869, A. Ladenburg proposed a structure for benzene, called both *Ladenburg benzene* and *prismane*, that seemed to overcome these objections.

Ladenburg benzene or prismane

Although Ladenburg benzene is recognized today as a highly strained molecule (it has been described as a "caged tiger"), an attractive feature of this structure to nineteenth-century chemists was its lack of double bonds.

Several facts, however, ultimately led to the adoption of the Kekulé structure. One of the most compelling arguments was that all efforts to prepare the alkene 1,3,5-cyclo-hexatriene using standard alkene syntheses led to benzene. The argument was, then, that benzene and 1,3,5-cyclohexatriene must be one and the same compound. Synthetic routes involving the same kinds of reactions were also used to prepare COT, which, as Eq. 15.38 illustrates, has the reactivity of an ordinary alkene.

Although the Ladenburg-benzene structure had been discarded for all practical purposes decades earlier, its final refutation came in 1973 with its synthesis by Professor Thomas J. Katz and his colleagues at Columbia University. These chemists found that Ladenburg benzene is an explosive liquid with properties quite different from those of benzene.

How can the Kekulé "cyclic triene" structure for benzene be reconciled with the fact that benzene is inert to the usual reactions of alkenes? The answer to this question will occupy our attention in the next three parts of this section.

B. Structure of Benzene

The structure of benzene (Fig. 15.8) shows that it has *one* type of carbon-carbon bond with a bond length (1.395 Å) intermediate between the lengths of single bonds (1.54 Å) and double bonds (1.33 Å). All atoms in the benzene molecule lie in one plane. The Kekulé structure for benzene shows *two* types of carbon-carbon bonds: single bonds and double bonds. This inadequacy of the Kekulé structure can be remedied, however, by depicting benzene as *the hybrid of two equally contributing resonance structures*:

(15.39)

Benzene is an average of these two structures; it is *one* compound with *one* type of carbon-carbon bond that is neither a single bond nor a double bond, but something in between. Occasionally a benzene ring is represented with either of the following structures, which show the "smearing out" of double-bond character:

It is interesting to compare the structures of benzene and 1,3,5,7-cyclooctatetraene (COT) in view of their greatly different chemical reactivities (Eqs. 15.37 and 15.38). Their structures are remarkably different (Fig. 15.9). First, COT has alternating single and double bonds, which have almost the same lengths as the single and double bonds in 1,3-butadiene. Second, COT is not planar like benzene, but instead is tub-shaped.

The π bonds of benzene and COT are also different (Fig. 15.10). The Kekulé structures for benzene suggest that each carbon atom should be trigonal, and therefore sp^2-hybridized. This means that there is a p orbital on each carbon atom. Since the molecule is planar, and the axes of all six p orbitals of benzene are parallel, these p orbitals overlap

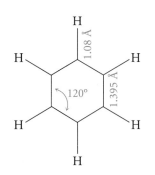

Figure 15.8   *Structure of benzene. (The double bonds are not shown.) Notice that the carbon skeleton has the shape of a planar hexagon, and all carbon-carbon bonds are equivalent.*

Figure 15.9   *Structure of 1,3,5,7-cyclooctatetraene (COT). Compare this structure with that of benzene in Fig. 15.8. Notice the two different types of bonds, single bonds and double bonds. Also notice that the molecule is not planar, but tub-shaped.*

to form a continuous bonding π molecular orbital (Fig. 15.10a). That is, the π-electron density in benzene lies in doughnut-shaped regions both above and below the plane of the ring. This overlap is symbolized by the resonance structures of benzene. In contrast, the carbon atoms of COT are not all coplanar, but they are nevertheless all trigonal. This means that there is a p orbital on each carbon atom of COT, but these orbitals do not overlap to form a continuous π molecular orbital like the one in benzene. Instead (Fig. 15.10b), COT contains four π-electron systems of two carbons each; the π bonds are

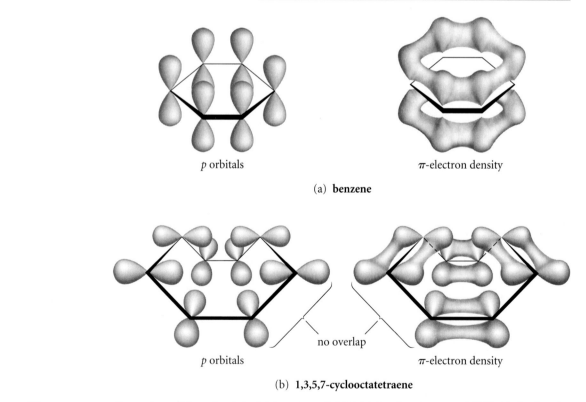

Figure 15.10   *Comparison of the π bonds in (a) benzene and (b) 1,3,5,7-cyclooctatetraene (COT). Notice that the p orbitals in benzene can overlap to form a continuous bonding π molecular orbital; the p orbitals of COT cannot overlap continuously, but are confined to two-carbon units.*

mutually perpendicular, and therefore they do *not* overlap. As far as the π electrons are concerned, COT looks like four isolated ethylene molecules. Because there is no electronic overlap between the π orbitals of adjacent double bonds, *COT does not have resonance structures analogous to those of benzene* (Sec. 15.6B, Guideline 5).

(15.40)

To summarize: resonance structures can be written for benzene, because the p orbitals of benzene overlap to provide additional bonding and additional stability. Resonance structures *cannot* be written for COT because there is no overlap between p orbitals on adjacent double bonds.

C. Stability of Benzene

Chemists of the nineteenth century considered benzene to be unusually stable because it is inert to reagents that react with ordinary alkenes. However, a more precise measure of stability is ΔH_f°, the standard heat of formation. Because benzene and COT have the same empirical formula (CH), we can compare their heats of formation per CH group. The ΔH_f° of benzene is 82.93 kJ/mol (19.82 kcal/mol) or 82.93/6 = 13.8 kJ/mol (3.3 kcal/mol) per CH group. The ΔH_f° of COT is +298.0 kJ/mol (71.23 kcal/mol), or 298.0/8 = 37.3 kJ/mol (8.9 kcal/mol) per CH group. Thus benzene, per CH group, is (37.3 − 13.8) = 23.5 kJ/mol (5.6 kcal/mol) more stable than COT. It follows that benzene is 23.5 × 6 = 141 kJ/mol (33.6 kcal/mol) more stable than a hypothetical six-carbon cyclic conjugated triene with the same stability as COT.

This energy difference of about 141 kJ/mol or 34 kcal/mol is called the **empirical resonance energy** of benzene. This figure is an estimate of just how much special stability is implied by the resonance structures for benzene—thus the name "resonance energy."

> Notice that the resonance energy is the energy by which benzene is *stabilized*; it is therefore an energy that benzene "doesn't have." The empirical resonance energy of benzene has been estimated in several different ways; these estimates range from 126 to 172 kJ/mol (30 to 41 kcal/mol). The important point, however, is not the exact value of this number, but the fact that it is *large*: benzene is a very stable compound!

D. Aromaticity and the Hückel 4*n* + 2 Rule

The unusual stability of benzene is called **aromaticity**. (Once again, this has *nothing* to do with its aroma.) Benzene and its derivatives are not the only compounds with aromatic stability; many other aromatic compounds are known. To be aromatic, a compound must conform to *all* of the following criteria.

Criteria for aromaticity:

1. Aromatic compounds contain one or more rings that have a *cyclic* arrangement of *p* orbitals. Thus, aromaticity is a property of certain *cyclic* compounds.

2. *Every* atom of an aromatic ring has a *p* orbital.

3. Aromatic rings are *planar*.

4. The cyclic arrangement of *p* orbitals in an aromatic compound must contain $4n + 2$ π electrons, where *n* is any positive integer (0, 1, 2, . . .). In other words, an aromatic ring must contain 2, 6, 10, . . . , π electrons.

These criteria for aromatic behavior were first recognized in 1931 by Erich Hückel, a German chemical physicist. They are often called collectively the **Hückel $4n + 2$ rule** or simply the **$4n + 2$ rule**.

The basis of the $4n + 2$ rule lies in the molecular orbital theory of *cyclic π-electron systems*. The theory holds that aromatic stability is observed with *continuous cycles* of *p* orbitals—thus Criteria 1 and 2. The theory also requires that the *p* orbitals must overlap to form π molecular orbitals. This overlap requires that an aromatic ring must be planar; *p* orbitals cannot overlap in rings significantly distorted from planarity—thus Criterion 3. The last criterion has to do with the number of π molecular orbitals and the number of electrons they contain. For example, the overlap of the six carbon *p* orbitals in benzene results in six π molecular orbitals. Three of these are *bonding molecular orbitals* and three are *antibonding molecular orbitals*. Quantum mechanical calculations show that *the bonding molecular orbitals of cyclic π-electron systems have particularly low energy*. Recall (Sec. 1.8A) that each electron in a bonding molecular orbital lowers the energy of a molecule. Thus, a compound has the lowest energy when its bonding molecular orbitals are filled. Because each molecular orbital accommodates two electrons, it takes six π electrons to fill the three bonding molecular orbitals of benzene. (Notice that $4n + 2 = 6$ when $n = 1$.) Molecular orbital theory shows that it always takes $4n + 2$ electrons to fill exactly the bonding molecular orbitals of cyclic π-electron systems—thus Criterion 4.

Application of the $4n + 2$ rule is illustrated in the following Study Problem.

· ·

STUDY PROBLEM 15.3

Decide whether each of the following compounds is aromatic.

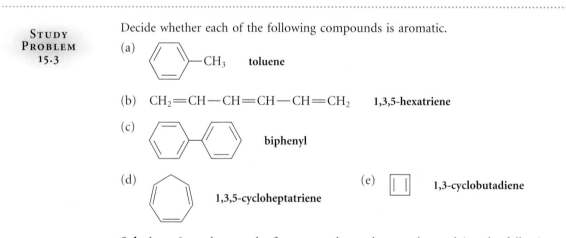

(a)   —CH₃   toluene

(b) $CH_2{=}CH{-}CH{=}CH{-}CH{=}CH_2$ **1,3,5-hexatriene**

(c)   **biphenyl**

(d)   **1,3,5-cycloheptatriene**

(e)   **1,3-cyclobutadiene**

Solution   In each example, first count the π electrons by applying the following rule: *Each double bond contributes two π electrons*. Then apply *all* of the criteria for aromaticity.

(a)   The ring in toluene, like the ring in benzene, is a continuous planar cycle of six π electrons. Hence, the ring in toluene is aromatic. The methyl group is a substituent group on the ring and is not part of the ring system. Because toluene contains an aromatic ring, it is considered to be an aromatic compound. This example shows that *parts of molecules* can be aromatic.

(b)   Although 1,3,5-hexatriene contains six π electrons, it is not aromatic, because it fails Criterion 1 for aromaticity: it is not cyclic. Aromatic species must be cyclic.

(c)   Biphenyl has two rings, each of which is separately aromatic. Hence, biphenyl is an aromatic compound.

(d)   Although 1,3,5-cycloheptatriene has six π electrons, it is not aromatic, because it fails Criterion 2 for aromaticity: one carbon of the ring does not have a p orbital. In other words, the π electron system is not continuous, but is interrupted by the sp^3-hybridized CH_2 group.

(e)   1,3-Cyclobutadiene is not aromatic. Even though it is a continuous cyclic system of p orbitals, it fails Criterion 4 for aromaticity: it does not have $4n + 2$ π electrons.

Aromatic Heterocycles   Aromaticity is not confined solely to hydrocarbons. Some *heterocyclic compounds* (Sec. 8.1C) are aromatic; for example, pyridine and pyrrole are both aromatic nitrogen-containing heterocycles.

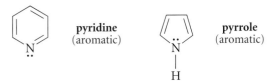

Except for the nitrogen in the ring, the structure of pyridine closely resembles that of benzene. Each atom in the ring, including the nitrogen, is part of a double bond and therefore contributes one π electron. What about the extra electron pair on nitrogen? How does this electron pair figure in the π-electron count? This electron pair resides in an sp^2 orbital in the plane of the ring (Fig. 15.11a). (It has the same relationship to the ring as a hydrogen of benzene.) Because it does not overlap with the ring π-electron system, it is not included in the π-electron count. Thus *vinylic electrons (electrons on doubly bonded atoms) are not counted as π electrons*.

In pyrrole, on the other hand, the electron pair on nitrogen is *allylic* (Fig. 15.11b). The nitrogen has a trigonal geometry and sp^2 hybridization that allow its electron pair to occupy a p orbital and contribute to the π-electron count. The N—H hydrogen lies in the plane of the ring. In general, *allylic electrons are counted as π electrons when they*

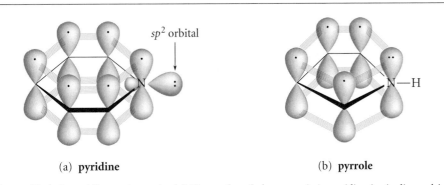

(a) **pyridine**   (b) **pyrrole**

Figure 15.11   *The p orbitals in pyridine and pyrrole. (a) The unshared electron pair in pyridine is vinylic, and is therefore in an sp² orbital and is not part of the aromatic π-electron system. (b) The unshared electron pair in pyrrole is allylic, and can occupy a p orbital that is part of the aromatic π-electron system.*

reside in orbitals that are parallel to the other p orbitals in the molecule. Therefore, pyrrole has six π electrons—four from the double bonds and two from the nitrogen—and is aromatic.

Note carefully the different ways in which we handle the electron pairs on the nitrogens of pyridine and pyrrole. The nitrogen in pyridine is part of a double bond, and the electron pair *is not* part of the π-electron system. The nitrogen in pyrrole is allylic and its electron pair *is* part of the π-electron system.

Aromatic Ions Aromaticity is not restricted to neutral molecules; a number of ions are aromatic. One of the best characterized aromatic ions is the cyclopentadienyl anion:

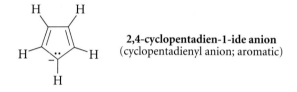

2,4-cyclopentadien-1-ide anion
(cyclopentadienyl anion; aromatic)

This ion resembles pyrrole; however, because the atom bearing the allylic electron pair is carbon rather than nitrogen, its charge is -1. One way to form this ion is by the reaction of sodium with the conjugate acid hydrocarbon, 1,3-cyclopentadiene; notice the analogy to the reaction of Na with H_2O.

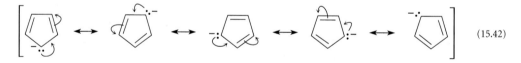

The cyclopentadienyl anion has five equivalent resonance structures; the negative charge can be delocalized to every carbon atom:

$$\left[\text{...} \right] \quad (15.42)$$

These structures show that all carbon atoms of the cyclopentadienyl anion are equivalent. Because of the stability of this anion, its conjugate acid, 1,3-cyclopentadiene, is an unusually strong hydrocarbon acid. With a pK_a of 15, this compound is 10^{10} times as acidic as a 1-alkyne, and about as acidic as water!

Cations, too, may be aromatic.

$$\text{Cl} + SbCl_5 \longrightarrow \left[\text{...} \right] SbCl_6^- \quad (15.43)$$

(a Lewis acid)

cyclopropenyl cation
(aromatic)

Atoms with empty p orbitals are part of the π-electron system, but they contribute no electrons to the π-electron count. Since this cation has two π electrons, it is aromatic ($4n + 2 =$

2 for $n = 0$). The stability of the cyclopropenyl cation, despite its considerable angle strain, is a particularly strong testament to the stabilizing effect of aromaticity.

Counting π electrons accurately is obviously crucial for successful application of the $4n + 2$ rule. Let's summarize the rules for π-electron counting.

1. Each atom that is part of a double bond contributes one π electron.
2. Vinylic electrons do not contribute to the π-electron count.
3. Allylic electrons contribute to the π-electron count if they occupy an orbital parallel to the other p orbitals in the molecule.
4. An atom with an empty p orbital can be part of a continuous aromatic π-electron system, but contributes no π electrons.

Aromatic Polycyclic Compounds   The Hückel $4n + 2$ rule applies strictly to single rings, that is, to monocyclic compounds. However, a number of well-known fused bicyclic and polycyclic compounds are also aromatic. Naphthalene and quinoline are two examples:

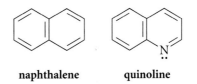

naphthalene   **quinoline**

Although rules have been devised to predict aromaticity of fused-ring compounds, these rules are rather complex, and we need not be concerned with them. It is certainly not difficult to see the resemblance of the two compounds above to benzene, the best-known aromatic compound.

Aromatic Organometallic Compounds   Some remarkable organometallic compounds have aromatic character. For example, the cyclopentadienyl anion, discussed previously in this section as one example of an aromatic anion, forms stable complexes with a number of transition-metal cations. One of the best known of these complexes is *ferrocene*, which is synthesized by the reaction of two equivalents of cyclopentadienyl anion with one equivalent of ferrous ion (Fe^{2+}).

$$2 \; \langle\!\rangle \; + \; FeCl_2 \; \longrightarrow \; Fe^{2+} \left(\ddot{:}\langle\!\rangle \right)_2 \; + \; 2NaCl \qquad (15.44)$$

$$\ddot{:} \; Na^+$$

ferrocene
(90% yield)

Although this synthesis resembles a metathesis (exchange) reaction in which two salts are formed from two other salts, ferrocene is not a salt, but is a remarkable "molecular sandwich" in which a ferrous ion is imbedded between two cyclopentadienyl anions.

PROBLEMS

*15.32   Using the theory of aromaticity, explain the finding that *A* and *B* are different compounds, but *C* and *D* are identical.

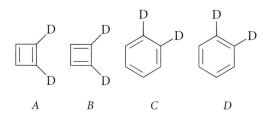

(That *A* and *B* are different molecules was established by Prof. Barry Carpenter and his students at Cornell University in 1980.)

15.33   Which of the compounds in Problem 15.31 are likely to be antiaromatic? Explain.

KEY IDEAS IN CHAPTER 15

🧪   Molecules containing conjugated double or triple bonds have additional stability, relative to unconjugated isomers, that can be attributed to the continuous overlap of their *p* orbitals and to strong sp^2-sp^2 single bonds.

🧪   A cumulene is a compound in which one or more carbon atoms are each part of two double bonds. An allene is a cumulene with two cumulated double bonds. Adjacent π bonds in a cumulene are mutually perpendicular; appropriately substituted allenes are chiral.

🧪   Heats of formation are generally in the order: conjugated dienes < ordinary dienes < alkynes < cumulenes.

🧪   Compounds with conjugated double or triple bonds have UV or visible absorptions at λ_{max} > 200 nm.

🧪   Each conjugated double or triple bond in a molecule contributes 30–50 nm to its λ_{max}. When a compound contains many conjugated double or triple bonds, it absorbs visible light and appears colored.

🧪   The intensity of the UV absorption of a compound is proportional to its concentration (Beer's Law). The constant of proportionality ϵ, called the molar extinction coefficient, is the intrinsic intensity of an absorption.

The Diels-Alder reaction is a pericyclic reaction that involves the cyclo-addition of a conjugated diene and a dienophile (usually an alkene). When the diene is cyclic, bicyclic products are produced.

The diene assumes an *s-cis*, or cisoid, conformation in the transition state of the Diels-Alder reaction; dienes that are locked into *s-trans*, or transoid, conformations are unreactive.

Each component of the Diels-Alder reaction undergoes a *syn* addition to the other. In many cases the *endo* mode of addition is favored over the *exo* mode.

Conjugated dienes react with hydrogen halides to give mixtures of 1,2- and 1,4-addition (conjugate addition) products. Such a mixture of products is accounted for by the formation of a resonance-stabilized allylic carbocation intermediate, which can be attacked by halide ion at either of two positively charged carbons.

When a reaction gives a mixture of products substantially different from that which would be obtained if the products were at equilibrium, the product distribution is said to be kinetically controlled. If the products come to equilibrium under the reaction conditions, the product distribution is said to be thermodynamically controlled. The predominance of the 1,2-addition product in the reaction of hydrogen halides with conjugated alkenes is an example of kinetic control.

Resonance structures are derived by the curved-arrow formalism. A molecule is the weighted average of its resonance structures. That is, the structure of the molecule is most accurately approximated by its most important resonance structures.

Other things being equal, the species with the greatest number of important resonance structures is most stable.

Benzene has special stability that is termed *aromaticity*. All aromatic compounds contain $4n + 2$ π electrons in a continuous, planar, cyclic array.

Compounds that contain $4n$ π electrons in a continuous, planar, cyclic array are antiaromatic and are especially unstable.

ADDITIONAL PROBLEMS

15.34 Use the curved-arrow or fishhook formalism to derive the major resonance structures for each of the following species. Determine which if any structure is the most important one in each case.

(Problem 15.34 continues)

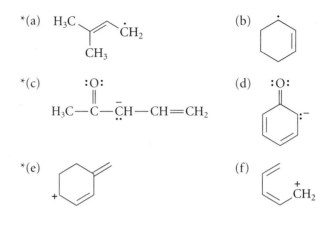

*15.35   Give the principal product(s) expected, if any, when *trans*-1,3-pentadiene reacts under the following conditions. Assume one equivalent of each reagent reacts unless noted otherwise.
(a) Br_2 (dark), CH_2Cl_2 (b) HBr   (c) H_2 (2 equivalents), Pd/C
(d) H_2O, H_3O^+ (e) Na^+ $C_2H_5O^-$ in C_2H_5OH
(f) maleic anhydride (Eq. 15.11, p. 697) heat

15.36   Repeat Problem 15.35 for 3-methylenecycloheptene.

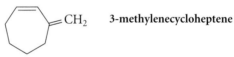   **3-methylenecycloheptene**

15.37   *(a) What five-carbon conjugated diene would give the same *single* product from either 1,2- or 1,4-addition of HCl?
(b) What six-carbon conjugated diene would give the same *single* product from either 1,2- or 1,4-addition of HBr?

15.38   Explain each of the following observations.
*(a) The allene 2,3-heptadiene can be resolved into enantiomers, but the cumulene 2,3,4-heptatriene cannot.
(b) The cumulene in (a) can exist as diastereomers, but the allene in (a) cannot.

15.39   Using the Hückel $4n + 2$ rule, determine whether each of the following compounds is likely to be aromatic. Explain how you arrived at the π-electron count in each case.

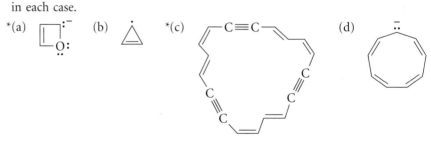

*15.40   Which of the following molecules is likely to be planar and which nonplanar? Explain.

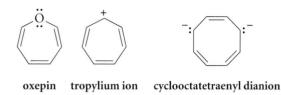

oxepin   tropylium ion   cyclooctatetraenyl dianion

15.41   The following compound appears to meet all the criteria for aromaticity. Yet it is very unstable because it cannot adopt a planar conformation. Using models if necessary, explain what prevents this compound from becoming planar and aromatic.

15.42   Rank the compounds within each set in order of increasing heat of formation (lowest first).

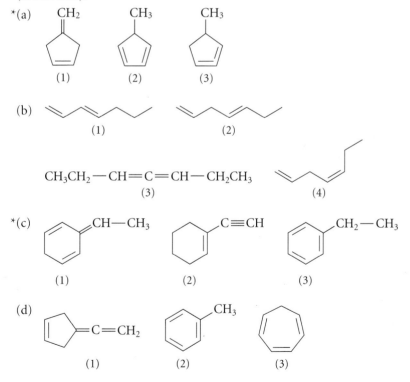

15.43   Assume you have unlabeled samples of the compounds within each set below. Explain how UV spectroscopy could be used to identify each compound.

(Problem 15.43 continues)

*(a)   1,4-cyclohexadiene and 1,3-cyclohexadiene

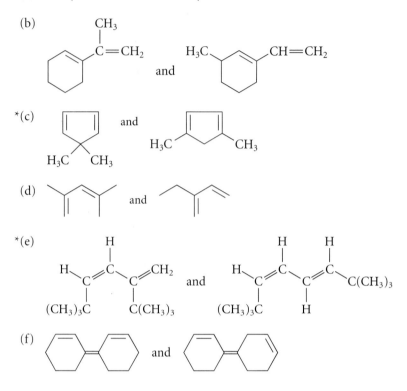

(b)

(c) and

(d) and

(e) and

(f) and

*15.44   A colleague, Ima Hack, has subjected isoprene (Fig. 15.3) to catalytic hydrogenation to give isopentane. Hack has inadvertently stopped the hydrogenation prematurely and wants to know how much unreacted isoprene remains in the sample. The mixture of isoprene and 2-methylbutane (75 mg total) is diluted to one liter with pure methanol and found to have an absorption at 222.5 nm (1-cm path length) of 0.356. Given an extinction coefficient of 10,750 at this wavelength, what mass percent of the sample is unreacted isoprene?

*15.45   A chemist, I. M. Shoddy, has just purchased some compounds in a going-out-of-business sale from Pybond, Inc., a cut-rate chemical supply house. The company, whose motto is "You get what you pay for," has sent Shoddy a compound *A* at a bargain price in a bottle labeled only "C_6H_{10}." Unfortunately, Shoddy has lost his purchase order and cannot remember what he ordered, and he has come to ask your help in identifying the compound. Compound *A* is optically active, and has IR absorption at 2083 cm^{-1}. Partial hydrogenation of *A* with 0.2 equivalent of H_2 over a catalyst gives, in addition to recovered *A*, a mixture of *cis*-2-hexene and *cis*-3-hexene. Identify compound *A* and explain your reasoning.

15.46   Account for the fact that the antibiotic *mycomycin* is optically active.

$$HC{\equiv}C{-}C{\equiv}C{-}CH{=}C{=}CH{-}CH{=}CH{-}CH{=}CH{-}CH_2{-}CO_2H \qquad \textbf{mycomycin}$$

*15.47   Explain the fact that 2,3-dimethyl-1,3-butadiene and maleic anhydride (structure in Eq. 15.11, p. 697) readily react to give a Diels-Alder adduct, but 2,3-di-*tert*-butyl-1,3-butadiene and maleic anhydride do not.

15.48   The following natural product readily gives a Diels-Alder adduct with maleic anhydride (structure in Eq. 15.11, p. 697). What is the most likely configuration of its two double bonds (*cis* or *trans*)? (*Hint:* See Eq. 15.9.)

$$CH_3(CH_2)_5CH=CH-CH=CH(CH_2)_7CO_2H$$

*15.49   Knowing that conjugated dienes react in the Diels-Alder reaction, a student, M. T. Brainpan, has come to you, a noted authority on the Diels-Alder reaction, with an original research idea: to use conjugated alkynes as the diene component in the Diels-Alder reaction (such as the following one). Would Brainpan's idea work? Explain.

$$CH_3-C\equiv C-C\equiv C-CH_3 + \text{maleic anhydride (Eq. 15.11)} \longrightarrow$$

*15.50   (a) Predict the products expected from the addition of one equivalent of HBr to (1) isoprene; (2) *trans*-1,3,5-hexatriene.
         (b) In each case, which are likely to be the kinetically controlled products and which the thermodynamically controlled ones? Explain.

15.51   (a) What two products would be formed when the following diene reacts with one equivalent of DCl? (D = deuterium, the isotope of hydrogen with atomic mass = 2.)

$$H_2C=CH-CH=CH-CH_3 + DCl \longrightarrow$$

         (b) Which product would be formed in greater amount under conditions of kinetic control?

*15.52   The 1,2-addition of one equivalent of HCl to the triple bond of vinylacetylene, $HC\equiv C-CH=CH_2$, gives a chlorine-containing conjugated diene called *chloroprene*. Chloroprene can be polymerized to give *neoprene*, valued for its resistance to oils, oxidative breakdown, and other deterioration. Give the structures of chloroprene and neoprene.

*15.53   When an excess of 1,3-butadiene reacts with Cl_2 in chloroform solvent, two compounds, A and B, both with the formula $C_4H_6Cl_2$, are formed. Compound B reacts with more Cl_2 to form compound D, $C_4H_6Cl_4$, which proves to be a *meso* compound. Compound A reacts with more Cl_2 to form both D and a stereoisomer C. Propose structures for A, B, C, and D.

(Notice that the formation of an aromatic ring in the product helps ensure that the Diels-Alder reaction is driven to completion.) In each of the cases below, use the structure of the Diels-Alder product to deduce the structure of the reactive species formed in the reaction. Show by the curved-arrow formalism how the reactive species is formed, and explain what makes it particularly unstable.

*(a)

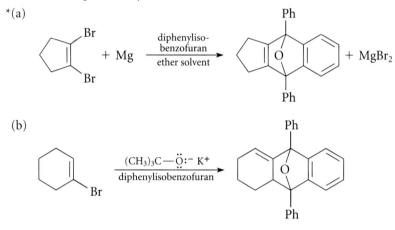

(b)

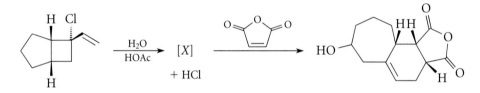

*15.59   Use the structure of the Diels-Alder adduct to deduce the structure of the product X in the reaction below. Then give a mechanism for the formation of X.

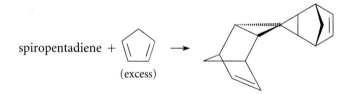

15.60   In 1991, chemists at Rice University reported that they had trapped an unstable compound called *spiropentadiene* using its Diels-Alder reaction with excess 1,3-cyclopentadiene, giving the product shown below. Use the structure of this product to deduce the structure of spiropentadiene.

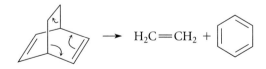

*15.61   The Diels-Alder reaction is an equilibrium that, in some cases, favors the decomposition of the Diels-Alder adduct:

(a) Suggest two reasons why this reaction proceeds in the direction shown.

(b) The compound α-phellandrene (C₁₀H₁₆) adds H₂ in the presence of a catalyst to give 1-isopropyl-4-methylcyclohexane and undergoes the following reaction. Deduce the structure of α-phellandrene.

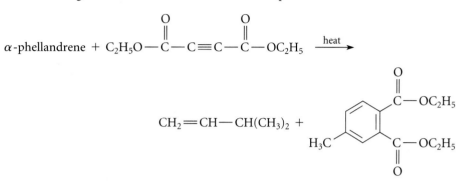

*15.62 When the compound shown below is treated with a strong Brønsted acid, a stable carbocation *A* is formed.

(a) Propose a structure for this carbocation and draw its resonance structures.

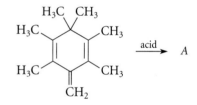

(b) The NMR spectrum of carbocation *A* at −10° consists of four singlets at δ 1.54, δ 2.36, δ 2.63, and δ 2.82 (relative integral 2:2:2:1). Explain why the structure of *A* is consistent with this spectrum by assigning each resonance.

(c) Explain why the NMR spectrum of *A* becomes a single broad line when the temperature is raised to 113°. (*Hint:* See Sec. 13.7.)

*15.63 Dextropimaric acid, isolated from an exudate resin of the cluster pine, is converted by acid and heat into abietic acid. Give a mechanism for this reaction using the curved-arrow formalism.

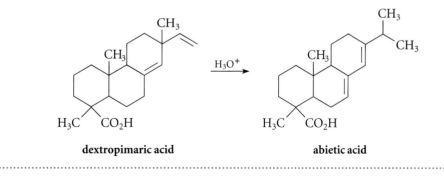

dextropimaric acid   **abietic acid**

16

Chemistry of Benzene and Its Derivatives

Because benzene and its derivatives are *aromatic* compounds, they do not undergo most of the usual addition reactions of alkenes. Instead, they undergo reactions in which a ring hydrogen is *substituted* by another group. Such substitution reactions can be used to prepare a variety of substituted benzenes from benzene itself.

Most of this chapter is concerned with the substitution reactions of benzene and its derivatives. The following two chapters consider the effect of a benzene ring on the reactivity of adjacent functional groups.

16.1 Nomenclature of Benzene Derivatives

The nomenclature of benzene derivatives follows the same rules used for other substituted hydrocarbons:

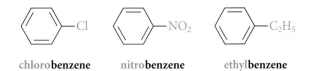

chloro**benzene**   nitro**benzene**   ethyl**benzene**

The *nitro group*, abbreviated —NO_2, which is a part of the nitrobenzene structure above, may be less familiar than the other substituent groups. The nitro group can be represented in more detail as a resonance hybrid of two equivalent dipolar structures:

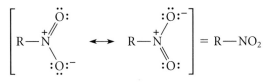

Some monosubstituted benzene derivatives have well-established common names that should be learned.

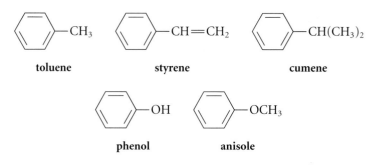

toluene   styrene   cumene

phenol   anisole

The positions of substituent groups in disubstituted benzenes can be designated in two ways. Modern substitutive nomenclature utilizes numerical designations in the same manner used for other compound classes. However, an older system, which is still used, employs the special letter prefixes *o* (for *ortho*), *m* (for *meta*), or *p* (for *para*) for disubstituted benzenes in which the two substituents have a 1,2-, 1,3-, or 1,4-relationship, respectively.

o-dichlorobenzene   *m*-bromonitrobenzene   *p*-fluoroiodobenzene
1,2-dichlorobenzene   1-bromo-3-nitrobenzene   1-fluoro-4-iodobenzene

As these examples illustrate, when none of the substituents qualify as principal groups, they are cited and numbered in alphabetical order. If a substituent is eligible for citation as a principal group, it is assumed to be at position-1 of the ring.

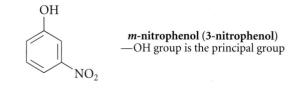

m-nitrophenol (3-nitrophenol)
—OH group is the principal group

Some disubstituted benzene derivatives also have time-honored common names. The dimethylbenzenes are called *xylenes*, and the methylphenols are called *cresols*.

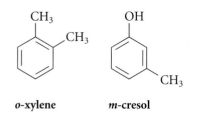

o-xylene   *m*-cresol

The hydroxyphenols also have important common names.

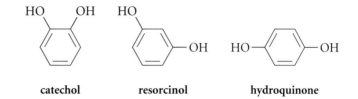

catechol   resorcinol   hydroquinone

When a benzene derivative contains more than two substituents on the ring, the *o*, *m*, and *p* designations are not appropriate; only numbers may be used to designate the positions of substituents. The usual nomenclature rules are followed (Secs. 2.4C, 4.2A, 8.1).

alphabetical citation: ***bromodifluoro***
numbering: **1,2,3**
name: **1-bromo-2,3-difluorobenzene**

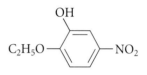

2-ethoxy-5-nitrophenol

Sometimes it is simpler to name a benzene ring as a substituent group. A benzene ring or substituted benzene ring cited as a substituent is referred to generally as an **aryl group**; this term is analogous to *alkyl group* in nonaromatic compounds (Sec. 2.9). When an unsubstituted benzene ring is a substituent, it is called a **phenyl group**. This group can be abbreviated Ph—. It is also sometimes abbreviated by its group formula, C_6H_5—.

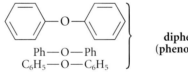

Ph—O—Ph
C_6H_5—O—C_6H_5

diphenyl ether
(phenoxybenzene)

The Ph—CH_2— group is called the **benzyl group**.

Ph—CH_2—Cl   benzyl **chloride**
or **(chloromethyl)benzene**

Be sure to notice the difference between the *phenyl* group, Ph—, and the *benzyl* group, Ph—CH_2—.

PROBLEMS

16.1   Name the following compounds.

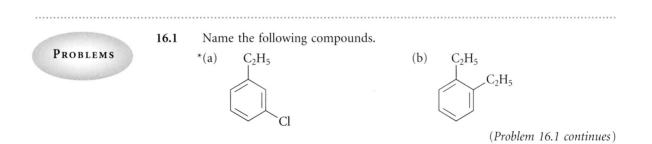

*(a) C_2H_5

(b) C_2H_5
       C_2H_5

(*Problem 16.1 continues*)

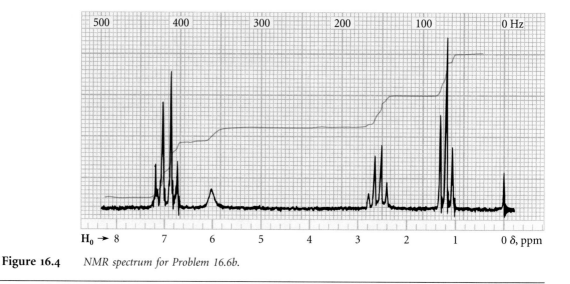

Figure 16.4   *NMR spectrum for Problem 16.6b.*

The chemical shifts of benzylic carbons are in the δ 18–30 region—not appreciably different from the chemical shifts of ordinary alkyl carbons. The ^{13}C chemical shifts observed for ethylbenzene are typical.

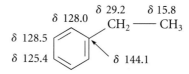

PROBLEMS

*16.7   A benzene derivative known to be a methyl ether with the formula $C_7H_6OCl_2$ has five lines in its CMR spectrum. Propose two possible structures for this compound that fit these facts.

16.8   How would you distinguish mesitylene (1,3,5-trimethylbenzene) from *p*-cymene (*p*-isopropyltoluene) by CMR spectroscopy?

D. UV Spectroscopy

Simple aromatic hydrocarbons have two absorption bands in their UV spectra: a relatively strong band near 210 nm and a much weaker one near 260 nm. The spectrum of ethylbenzene in methanol solvent (Fig. 16.5) is typical: λ_{max} = 208 nm (ϵ = 7520); 261 nm (ϵ = 200). Substituent groups on the ring alter both the λ_{max} values and the intensities of both peaks, particularly if the substituent has an unshared electron pair or p orbitals that can overlap with the π-electron system of the aromatic ring. As is the case in alkenes, more extensive conjugation is associated with an increase in both λ_{max} and intensity. For example, 1-ethyl-4-methoxybenzene (*p*-ethylanisole) in methanol solvent has absorptions at λ_{max} = 224 nm (ϵ = 10,100) and 276 nm (ϵ = 1,930); both absorptions

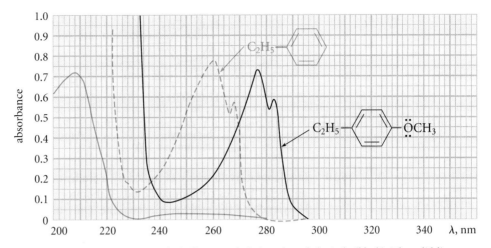

Figure 16.5   *Comparison of the UV spectra of ethylbenzene (color) and p-ethylanisole (black). The solid lines are spectra taken at the same concentrations. Notice that the spectrum of p-ethylanisole is much more intense. The dashed line is the spectrum of ethylbenzene at fiftyfold higher concentration. Notice that the peak in the p-ethylanisole spectrum occurs at higher wavelength.*

occur at higher wavelengths and have greater intensities than the analogous absorptions of ethylbenzene (Fig. 16.5) because the —OCH$_3$ group has electron pairs in orbitals that overlap with the *p* orbitals of the benzene ring.

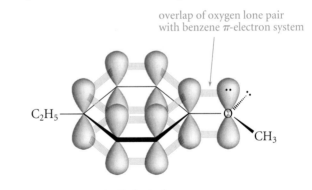

p-ethylanisole

PROBLEMS

*16.9   (a) Explain why compound *A* has a UV spectrum with considerably greater λ$_{max}$ values and intensities than observed for ethylbenzene.

λ$_{max}$ = 256 nm (ε = 20,000)
           283 nm (ε = 5,100)

A

(b) In view of your answer to (a), explain why the UV spectra of compounds *B* and *C* are virtually identical.

(*Problem 16.9 continues*)

By losing a β-proton instead (Eq. 16.6), the carbocation can form a stable aromatic compound, bromobenzene.

PROBLEMS

*16.11   A small amount of a by-product, *p*-dibromobenzene, is also formed in the bromination of benzene shown in Eq. 16.1. Write a stepwise mechanism for formation of this compound.

16.12   Using the curved-arrow formalism, write the mechanism for chlorination of benzene in the presence of $FeCl_3$.

B. Electrophilic Aromatic Substitution

Halogenation of benzene is one of many related reactions that are very typical of benzene and other aromatic compounds. These are called **electrophilic aromatic substitution** reactions. The bromination reaction, for example, is a *substitution* because hydrogen is replaced by another group (bromine). The reaction is *electrophilic* because it involves the reaction of an electrophile, or Lewis acid, with the benzene π electrons. In bromination the Lewis acid is "Br^+" in the complex of bromine and the $FeBr_3$ catalyst.

We've considered two other types of substitution reactions: *nucleophilic substitution* (the S_N2 and S_N1 reactions, Secs. 9.4 and 9.6) and *free-radical substitution* (halogenation of alkanes, Sec. 8.8A). In a nucleophilic substitution reaction, the substituting group acts as a nucleophile, or Lewis base; and in free-radical substitution, free-radical intermediates are involved. In electrophilic substitution, the substituting group reacts as an electrophile, or Lewis acid.

Electrophilic aromatic substitution is the most typical reaction of benzene and its derivatives. As you learn about other electrophilic substitution reactions, it will help you to understand them if you can identify in each reaction the following three mechanistic steps:

Step 1   *Generation of an electrophile.* The electrophile in bromination is the complex of bromine with $FeBr_3$, formed as shown in Eq. 16.3.

Step 2   *Attack of the π electrons of the aromatic ring on the electrophile and formation of a resonance-stabilized carbocation.* This step in the bromination reaction is shown in Eq. 16.5. As shown in Fig. 16.6, the electrophile approaches the π-electron cloud of the aromatic compound above or below the plane of the molecule, and the carbon at the point of attack becomes tetrahedral.

Step 3   *Loss of a proton from the carbocation intermediate to form the substituted aromatic compound.* This step in the bromination reaction is shown in Eq. 16.6.

STUDY PROBLEM 16.1

Give a detailed mechanism for the following electrophilic substitution reaction.

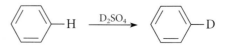

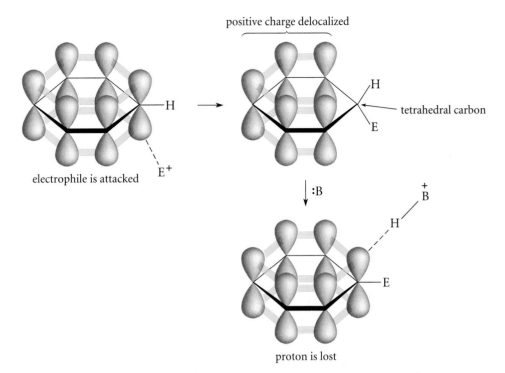

positive charge delocalized

electrophile is attacked E^+

H
tetrahedral carbon
E

:B

H   B

H

E

proton is lost

Figure 16.6   *Electrophilic aromatic substitution. E^+ is the electrophile, and :B is its corresponding base. Notice that attack on the electrophile and loss of the proton occur at opposite faces of the aromatic ring and that the carbocation intermediate has tetrahedral geometry at the site of substitution.*

Solution   Construct the mechanism in terms of the three steps given above.

Step 1   In this reaction, a hydrogen of the benzene ring has been replaced by an isotope D, which must come from the D_2SO_4. Since protons (in the form of Brønsted acids) are good electrophiles, the D_2SO_4 itself can serve as the electrophile.

Step 2   Attack of the benzene π electrons on the electrophile involves protonation of the benzene ring by the isotopically substituted acid:

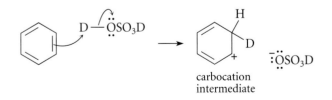

$D-\ddot{\text{O}}SO_3D$

H
D

carbocation
intermediate

$:\ddot{\text{O}}SO_3D$

(If you're asking where that "extra" hydrogen in the carbocation came from, don't forget that each carbon of the benzene ring has a single hydrogen that is not shown explicitly in the skeletal structure. One of these is shown in the carbocation because it is involved in the next step.) You should draw the resonance structures of the carbocation intermediate.

Step 3 Removal of the proton gives the final product:

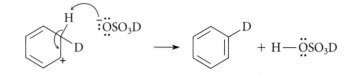

C. Nitration of Benzene

Benzene reacts with concentrated nitric acid, usually in the presence of a sulfuric acid catalyst, to form nitrobenzene. In this reaction, called *nitration*, the nitro group, —NO_2, is introduced into the benzene ring by electrophilic substitution.

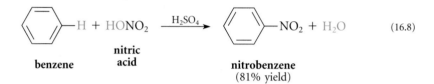

benzene **nitric acid** **nitrobenzene** (81% yield) (16.8)

This reaction fits the mechanistic pattern of the electrophilic aromatic substitution reaction outlined in the previous section:

Step 1 *Generation of the electrophile:* In nitration, the electrophile is $^{+}NO_2$, the *nitronium ion.* This ion is formed by the acid-catalyzed removal of the elements of water from HNO_3.

(16.9a)

nitronium ion (16.9b)

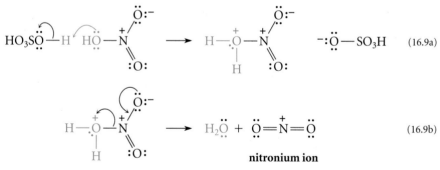

Step 2 *Attack of the benzene π electrons on the electrophile to form a carbocation intermediate.*

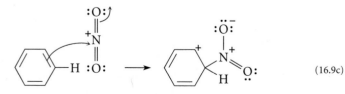

(16.9c)

(Notice that either of the oxygens can accept the electron pair.)

Step 3 *Loss of a proton from the carbocation to give a new aromatic compound.*

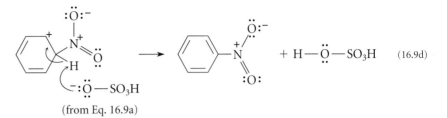

(from Eq. 16.9a)

Nitration is the usual way that nitro groups are introduced into aromatic rings.

D. Sulfonation of Benzene

Another electrophilic substitution reaction of benzene is its conversion into benzene-sulfonic acid by a solution of sulfur trioxide in H_2SO_4.

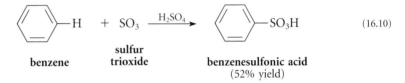

In this reaction, called *sulfonation*, the electrophile is the neutral compound sulfur trioxide, SO_3. Sulfur trioxide is a fuming liquid that reacts violently with water to give H_2SO_4. The source of SO_3 for sulfonation is usually a solution called *fuming sulfuric acid* or *oleum*, which is a solution of SO_3 in concentrated H_2SO_4; this material is one of the most acidic Brønsted acids available commercially.

When sulfur trioxide is attacked by the benzene ring π electrons, an oxygen accepts the electron pair displaced from sulfur.

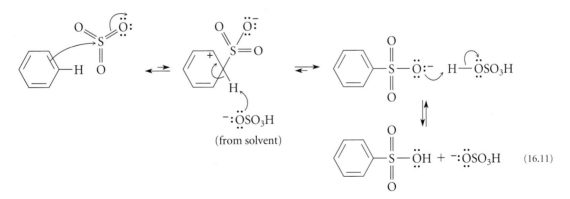

Sulfonic acids such as benzenesulfonic acid are rather strong acids. (Notice the last equilibrium in Eq. 16.11 and the resemblance of benzenesulfonic acid to sulfuric acid.) Many sulfonic acids are isolated from sulfonation reactions as their sodium salts.

Sulfonation, unlike many electrophilic aromatic substitution reactions, is reversible. The —SO_3H (sulfonic acid) group is replaced by a hydrogen when sulfonic acids are heated with steam (Problem 16.13).

PROBLEMS

*16.13 Write a curved-arrow mechanism for the conversion of benzenesulfonic acid to benzene and H_2SO_4 with hot water (steam). (*Hint:* Because sulfonic acids are strong acids, H_3O^+ is present.)

16.14 A compound called *p*-toluenesulfonic acid is formed when toluene is sulfonated at the *para* position. Draw the structure of this compound, and give the mechanism for its formation.

E. Friedel-Crafts Acylation of Benzene

When benzene reacts with an acid chloride in the presence of a Lewis acid catalyst such as aluminum trichloride, $AlCl_3$, a ketone is formed.

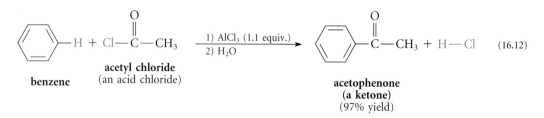

(16.12)

benzene **acetyl chloride** (an acid chloride) **acetophenone** **(a ketone)** (97% yield)

This reaction is an example of a **Friedel-Crafts acylation** (pronounced AY-suh-LAY-shun). In an acylation reaction, an *acyl group* is transferred from one group to another. In the Friedel-Crafts acylation, an acyl group, typically derived from an acid chloride, is introduced into an aromatic ring in the presence of a Lewis acid catalyst.

an acid chloride **an acyl group**

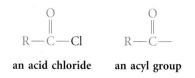

The electrophile in the Friedel-Crafts acylation reaction is a carbocation called an *acylium ion*. This ion is formed when the acid chloride reacts with the Lewis acid $AlCl_3$. (See again Study Guide Link 16.2.)

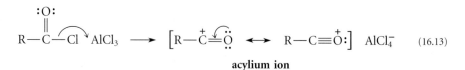

(16.13)

acylium ion

Weaker Lewis acids, such as $FeCl_3$ and $ZnCl_2$, are also used to form acylium ions in Friedel-Crafts acylations of aromatic compounds that are more reactive than benzene.

The acylation reaction is completed by the usual steps of electrophilic aromatic substitution (Sec. 16.4B):

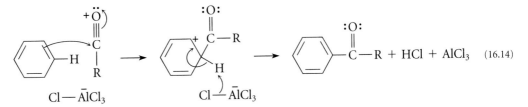

Attack of the benzene π electrons on the acylium ion represents another in the list of carbocation reactions, which now includes the following:

1. Reaction with nucleophiles

2. Rearrangement to other carbocations

3. Loss of a β-proton to give an alkene

4. Reaction with the π electrons of a double bond (or aromatic ring)

The ketone product of the Friedel-Crafts acylation reacts with the Lewis acid catalyst to form a complex that is catalytically inactive. This reaction has two consequences. First, slightly more than one equivalent of the catalyst must be used: one equivalent to react with the product, and an additional catalytic amount to ensure the presence of catalyst throughout the reaction. (Notice, for example, that 1.1 equivalent of $AlCl_3$ is used in Eq. 16.12.) Second, the complex must be destroyed before the ketone product can be isolated. This is usually accomplished by pouring the reaction mixture into ice water.

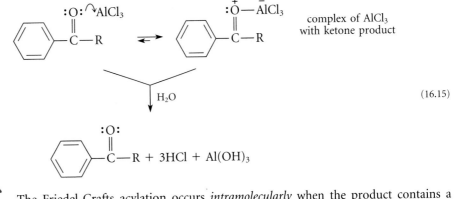

The Friedel-Crafts acylation occurs *intramolecularly* when the product contains a five- or six-membered ring.

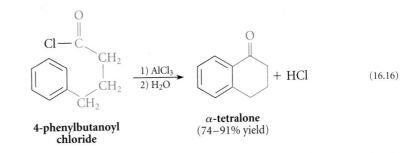

In this reaction, the phenyl ring "bites back" on the acylium ion within the same molecule to form a bicyclic compound. This type of reaction can only occur at an adjacent *ortho*

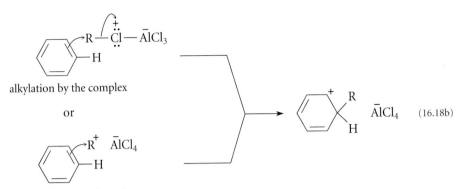

alkylation by the complex

or

alkylation by the carbocation

Loss of a proton to chloride ion completes the alkylation.

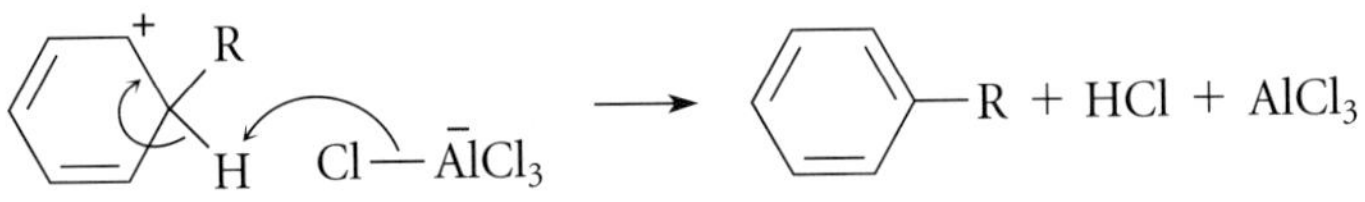

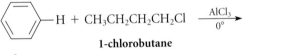

$$\text{(16.18c)}$$

Because some carbocations can rearrange, it is not surprising that rearrangements of alkyl groups are observed in some Friedel-Crafts alkylations:

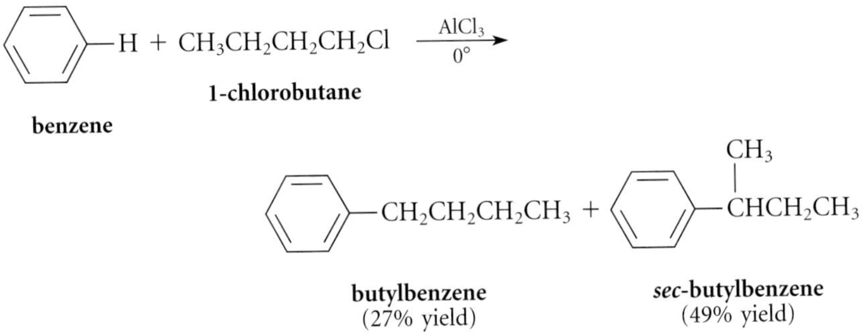

—H + CH$_3$CH$_2$CH$_2$CH$_2$Cl $\xrightarrow[0°]{\text{AlCl}_3}$

benzene   **1-chlorobutane**

—CH$_2$CH$_2$CH$_2$CH$_3$ +
$$\begin{array}{c}\text{CH}_3\\|\\\text{—CHCH}_2\text{CH}_3\end{array}$$
   (16.19)

butylbenzene   ***sec*-butylbenzene**
(27% yield)   (49% yield)

In this example, the alkyl group in the *sec*-butylbenzene product is rearranged. Because primary carbocations are probably too unstable to be involved as intermediates, it is probably the complex of the alkyl halide and AlCl$_3$ that rearranges:

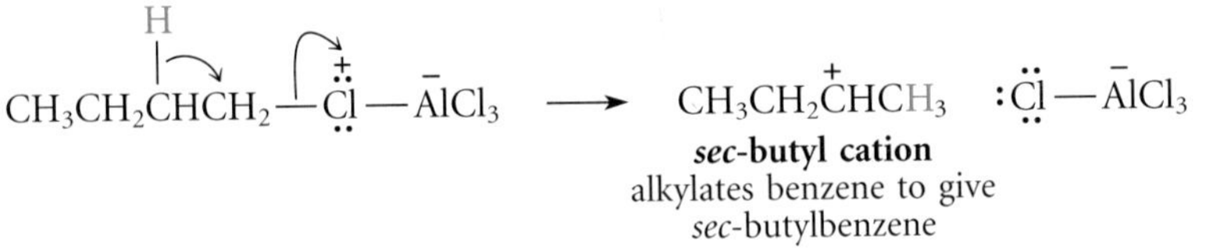

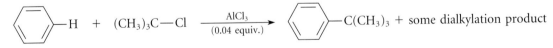

CH$_3$CH$_2$CHCH$_2$—Cl—AlCl$_3$ ⟶ CH$_3$CH$_2$$\overset{+}{\text{C}}HCH_3$:Cl—AlCl$_3$ (16.20)

***sec*-butyl cation**
alkylates benzene to give
sec-butylbenzene

Of course, rearrangement in the Friedel-Crafts alkylation is not observed if the carbocation intermediate is not prone to rearrangement.

—H + (CH$_3$)$_3$C—Cl $\xrightarrow[\text{(0.04 equiv.)}]{\text{AlCl}_3}$ —C(CH$_3$)$_3$ + some dialkylation product

benzene   ***tert*-butyl chloride**   ***tert*-butylbenzene**
(threefold excess)   (66% yield)

(16.21)

In this example, the alkylating cation is the *tert*-butyl cation; because it is tertiary, this carbocation does not rearrange.

Alkylbenzenes such as butylbenzene (Eq. 16.19) that are derived from rearrangement-prone alkyl halides are generally not prepared by the Friedel-Crafts alkylation, but by the Clemmensen or Wolff-Kishner reductions of aryl ketones (Sec. 19.12).

Another complication in Friedel-Crafts alkylation is that the alkylbenzene products are more reactive than benzene itself (for reasons discussed in Sec. 16.5B). This means that the product can undergo further alkylation, and mixtures of products alkylated to different extents are observed.

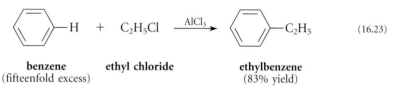

$$\text{(equimolar amounts)} \qquad \text{(16.22)}$$

(This is not a problem in acylation, since the ketone products of acylation are much less reactive than the benzene starting material.) However, a monoalkylation product can be obtained in good yield if a large excess of the starting material is used. For example, in the equation below, the fifteenfold molar excess of benzene ensures that a molecule of alkylating agent is much more likely to encounter a molecule of benzene in the reaction mixture than a molecule of the ethylbenzene product.

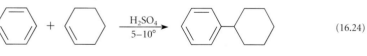

(16.23)

benzene   **ethyl chloride**   **ethylbenzene**
(fifteenfold excess)   (83% yield)

(Notice also the use of excess starting material in Eqs. 16.17 and 16.21.) This strategy is practical only if the starting material is cheap, and if it can be readily separated from the product.

STUDY GUIDE LINK:
✓16.3
*Different Sources of
the Same Reactive
Intermediate*

Alkenes can also be used as the alkylating agents in Friedel-Crafts alkylation reactions. The carbocation electrophiles in such reactions are generated from alkenes by protonation. (Recall that carbocation intermediates are formed in the protonation of alkenes; Secs. 4.7B, 4.9B.)

$$\text{benzene} + \text{cyclohexene} \xrightarrow[5-10°]{H_2SO_4} \text{cyclohexylbenzene} \qquad (16.24)$$

benzene   cyclohexene   **cyclohexylbenzene**
(65–68% yield)

PROBLEMS

*16.18   (a)   Give a curved-arrow mechanism for the reaction shown in Eq. 16.24.
            (b)   Explain why the same product is formed if cyclohexanol is used instead of cyclohexene in this reaction.

16.19   What product is formed when 2-methylpropene is added to a large excess of benzene containing HF and the Lewis acid BF_3? By what mechanism is it formed?

16.20 Predict the products of each of the following reactions, and give mechanisms for their formation. (*Hint:* See Eq. 16.16.)

*(a)

$$H_3C-\text{⟨⟩}-CH_2CH_2CH_2-Cl \xrightarrow{AlCl_3} \quad \text{(a compound } C_{10}H_{12} + HCl)$$

(b)

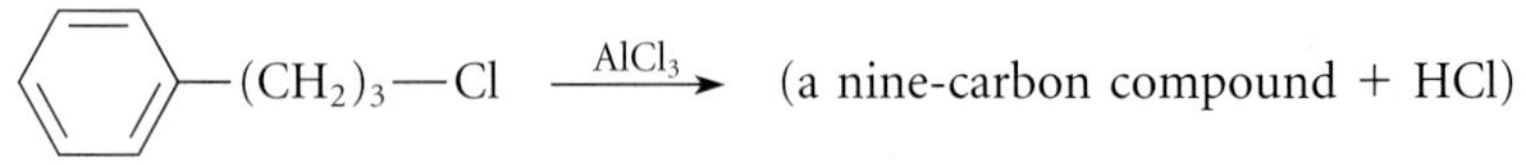

$$\text{⟨⟩}-(CH_2)_3-Cl \xrightarrow{AlCl_3} \quad \text{(a nine-carbon compound + HCl)}$$

16.5 Electrophilic Aromatic Substitution Reactions of Substituted Benzenes

A. Directing Effects of Substituents

When a monosubstituted benzene undergoes an electrophilic aromatic substitution reaction, three possible disubstitution products might be obtained. For example, nitration of bromobenzene could in principle give *ortho-*, *meta-*, or *para*-bromonitrobenzene. If substitution were totally random, an *ortho:meta:para* product ratio of 2:2:1 would be expected (why?). It is found experimentally that the second substitution is *not* random, but is *regioselective*.

Further substitution on a substituted benzene ring occurs in one of two ways. Some substituted benzenes react to give mostly a mixture of *ortho-* and *para*-disubstitution products (with the *para* product predominating in many cases). Bromobenzene, for example, is nitrated to give mostly *o-* and *p*-bromonitrobenzene and almost none of the *meta* isomer.

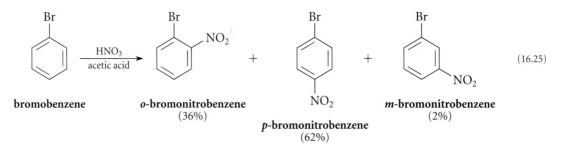

bromobenzene   *o*-**bromonitrobenzene** (36%)   *p*-**bromonitrobenzene** (62%)   *m*-**bromonitrobenzene** (2%)   (16.25)

Other electrophilic substitution reactions of bromobenzene also give mostly *ortho* and *para* isomers. If a substituted benzene undergoes further substitution at the *ortho* and *para* positions, the original substituent is called an **ortho, para-directing group**. Thus, bromine is an *ortho, para*-directing group, because all electrophilic substitution reactions of bromobenzene occur at the *ortho* and *para* positions.

In constrast, some substituted benzenes react in electrophilic aromatic substitution to give mostly the *meta* disubstitution product. For example, the bromination of nitrobenzene gives only the *meta* isomer.

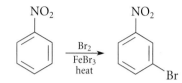

$$(16.26)$$

nitrobenzene   ***m*-bromonitrobenzene**
                          (only product observed)

Other electrophilic substitution reactions of nitrobenzene also give mostly the *meta* isomers. If a substituted benzene undergoes further substitution at the *meta* position, the original substituent group is called a ***meta*-directing group**. Thus, the nitro group is a *meta*-directing group because all electrophilic substitution reactions of nitrobenzene occur at the *meta* position.

A substituent group is either an *ortho, para*-directing group or a *meta*-directing group in *all* electrophilic aromatic substitution reactions; no substituent is *ortho, para* directing in one reaction and *meta* directing in another. A summary of the directing effects of common substituent groups is given in the third column of Table 16.2.

PROBLEM

16.21   Using the information in Table 16.2, predict the product(s) of
*(a)   Friedel-Crafts acylation of anisole with acetyl chloride (structure in Eq. 16.12) in the presence of AlCl$_3$.
(b)   Friedel-Crafts alkylation of a large excess of ethylbenzene with chloromethane in the presence of AlCl$_3$.

What is the reason for these directing effects? These effects occur because *electrophilic substitution reactions at one position of a benzene derivative are much faster than the same reactions at another position.* That is, the substitution reactions at the different ring positions are *in competition.* For example, in Eq. 16.25, *o*- and *p*-bromonitrobenzenes are the major products because the rate of nitration is greater at the *ortho* and *para* positions of bromobenzene than it is at the *meta* position. Understanding these effects thus requires an understanding of the factors that control the rates of aromatic substitution at each position.

***Ortho, Para*-Directing Groups**   Notice that all of the *ortho, para*-directing substituents in Table 16.2 are either *alkyl groups* or *groups that have unshared electron pairs on atoms directly attached to the benzene ring.* Although other types of *ortho, para*-directing groups are known, the principles on which *ortho, para*-directing effects are based can be understood by considering electrophilic substitution reactions of benzene derivatives containing these types of substituents.

Table 16.2 **Summary of Directing and Activating or Deactivating Effects of Some Common Functional Groups**

(Groups are listed in decreasing order of activation.)

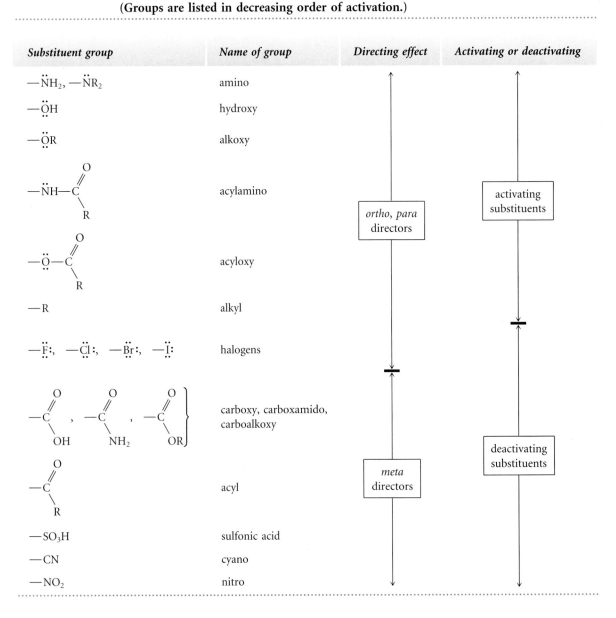

Substituent group	Name of group	Directing effect	Activating or deactivating
—N̈H₂, —N̈R₂	amino		
—ÖH	hydroxy		
—ÖR	alkoxy		
—N̈H—C(=O)R	acylamino	ortho, para directors	activating substituents
—Ö—C(=O)R	acyloxy		
—R	alkyl		
—F̈:, —C̈l:, —B̈r:, —Ï:	halogens		
—C(=O)OH, —C(=O)NH₂, —C(=O)OR	carboxy, carboxamido, carboalkoxy		
—C(=O)R	acyl	meta directors	deactivating substituents
—SO₃H	sulfonic acid		
—CN	cyano		
—NO₂	nitro		

First imagine the reaction of a general electrophile E⁺ with anisole (methoxybenzene). Notice that the atom directly attached to the benzene ring (the oxygen of the methoxy group) has unshared electron pairs. Reaction of E⁺ at the *para* position of anisole gives a carbocation intermediate with the following four important resonance structures:

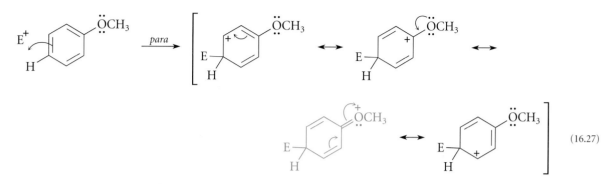

(16.27)

The colored structure shows that the unshared electron pair of the methoxy group can delocalize the positive charge on the carbocation. This is an especially important structure because it contains more bonds than the others, and every atom has an octet.

PROBLEM

*16.22   Draw the carbocation that results from the reaction of the electrophile at the *ortho* position of anisole; show that this ion also has four resonance structures.

If the electrophile reacts with anisole at the *meta* position, the carbocation intermediate that is formed has fewer resonance structures than the ion in Eq. 16.27. In particular, *the charge cannot be delocalized onto the —OCH₃ group when reaction occurs at the meta position.* There is no structure corresponding to the colored structure in Eq. 16.27.

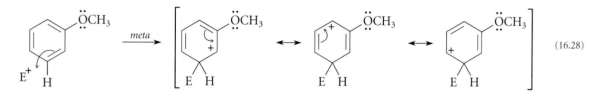

(16.28)

In order for the oxygen to delocalize the charge, it must be adjacent to an electron-deficient carbon, as in Eq. 16.27. The resonance structures show that the positive charge is shared on *alternate* carbons of the ring. When *meta* substitution occurs, the positive charge is not shared by the carbon adjacent to the oxygen.

A comparison of Eq. 16.27 and Problem 16.22 with Eq. 16.28 shows that the reaction of an electrophile at either the *ortho* or *para* positions of anisole gives a carbocation with more resonance structures, that is, a more stable carbocation. The rate-limiting step in many electrophilic aromatic substitution reactions is *formation of the carbocation intermediate*. Hammond's postulate (Sec. 4.8C) suggests that the more stable carbocation should be formed more rapidly. Hence, the products derived from the more rapidly formed carbocation—the more stable carbocation—are the ones observed. Since attack of the electrophile at an *ortho* or *para* position of anisole gives a more stable carbocation than attack at a *meta* position, the products of *ortho, para* substitution are formed more rapidly,

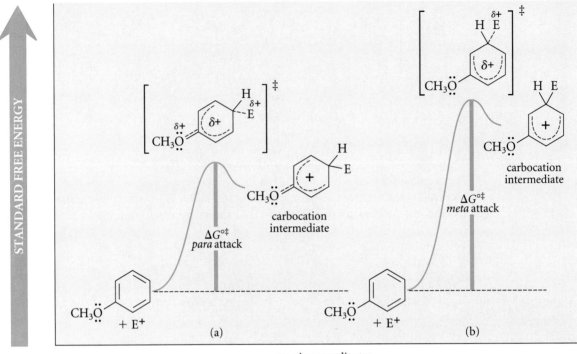

reaction coordinates

Figure 16.7 *Basis of the directing effect of the methoxy group in electrophilic aromatic substitution reactions of anisole. Substitution of anisole by an electrophile E^+ occurs more rapidly at (a) the para position than at (b) the meta position because a more stable carbocation intermediate is involved in para substitution. The dashed lines within the structures symbolize the delocalization of electrons.*

and are thus the products observed (Fig. 16.7). This is why the $—OCH_3$ group is an *ortho, para*-directing group.

To summarize: Substituents containing atoms with unshared electron pairs adjacent to the benzene ring are *ortho, para* directors in electrophilic aromatic substitution reactions because their electron pairs can be involved in the resonance stabilization of the carbocation intermediates.

Now imagine the reaction of an electrophile E^+ with an alkyl-substituted benzene such as toluene. Alkyl groups such as a methyl group have no unshared electrons, but the explanation for the directing effects of these groups is similar. Reaction of E^+ at a position that is *ortho* or *para* to an alkyl group gives an ion that has one tertiary carbocation resonance structure (colored structure in the following equation).

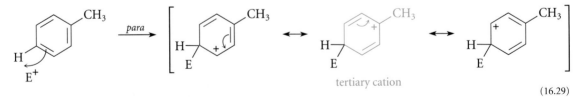

(16.29)

Reaction of the electrophile *meta* to the alkyl group also gives an ion with three resonance structures, but all resonance forms are secondary carbocations.

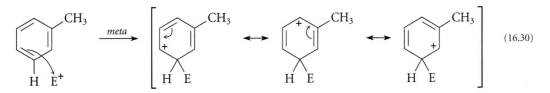

(16.30)

Because reaction at the *ortho* or *para* position gives the more stable carbocation, alkyl groups are *ortho*, *para*-directing groups.

Meta-Directing Groups

The *meta*-directing groups in Table 16.2 are all *electronegative groups that do not have an unshared electron pair on an atom adjacent to the benzene ring*. The directing effect of these groups can be understood by considering as an example the reactions of a general electrophile E⁺ with nitrobenzene at the *meta* and *para* positions.

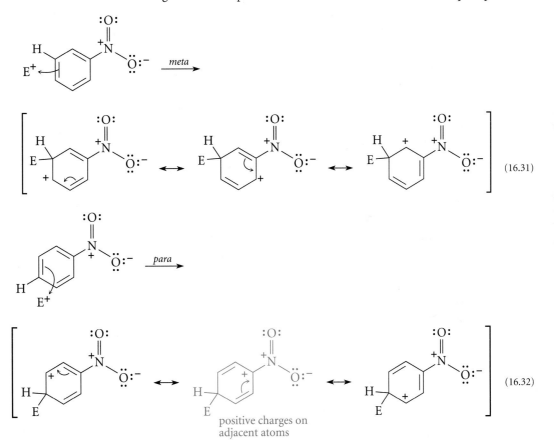

Both reactions give carbocations that have three resonance structures, but attack at the *para* position gives an ion with one particularly unfavorable structure (color). In this structure positive charges are situated on adjacent atoms. Since repulsion between two like charges, and consequently their energy of interaction, increases with decreasing separation, the colored resonance structure in Eq. 16.32 is less important than the others. Thus, the carbocation in Eq. 16.31, with the greater separation of like charges, is more stable than the carbocation in Eq. 16.32. Hammond's postulate (Sec. 4.8C) suggests that the more stable carbocation intermediate should be formed more rapidly. Consequently, the nitro group is a *meta* director because the ion that results from *meta* substitution (Eq. 16.31) is more stable than the one that results from *para* substitution (Eq. 16.32).

In summary, substituents that have positive charges adjacent to the aromatic ring are *meta* directors because *meta* substitution gives the carbocation intermediate in which like charges are farther apart. Notice that not all *meta*-directing groups have full positive charges like the nitro group, but all of them have bond dipoles that place a substantial amount of positive charge next to the benzene ring.

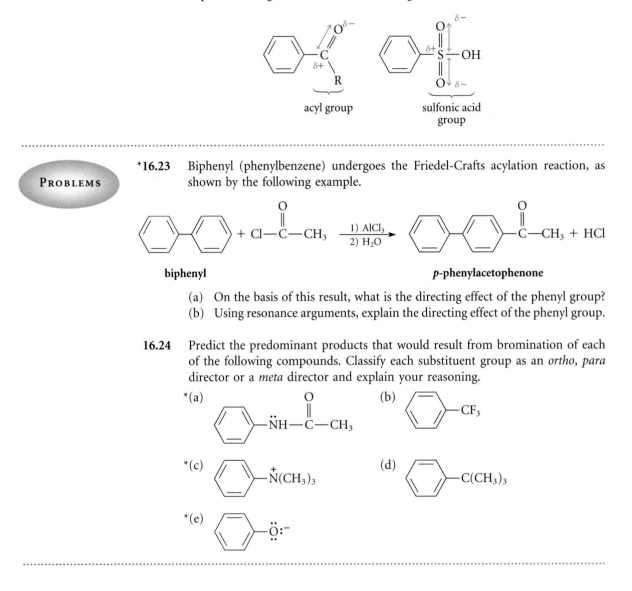

acyl group   sulfonic acid
                         group

PROBLEMS

*16.23   Biphenyl (phenylbenzene) undergoes the Friedel-Crafts acylation reaction, as shown by the following example.

biphenyl   *p*-phenylacetophenone

(a)   On the basis of this result, what is the directing effect of the phenyl group?

(b)   Using resonance arguments, explain the directing effect of the phenyl group.

16.24   Predict the predominant products that would result from bromination of each of the following compounds. Classify each substituent group as an *ortho, para* director or a *meta* director and explain your reasoning.

*(a)

(b)

*(c)

(d)

*(e)

The *Ortho, Para* Ratio   An aromatic substitution reaction of a benzene derivative bearing an *ortho, para*-directing group would give twice as much *ortho* as *para* product if substitution were completely random, because there are two *ortho* positions and only one *para* position available for substitution. However, this situation is rarely observed in practice: it is often found that the *para* substitution product is the major one in the reaction mixture. In some cases this result can be explained by the spatial demands of the electrophile. For example, Friedel-Crafts acylation of toluene gives essentially all *para* substitution product and almost no *ortho* product. The electrophile cannot react at the *ortho* position without developing van der Waals repulsions with the methyl group that is already on the ring. Consequently, reaction occurs at the *para* position, where such repulsions cannot occur.

On the other hand, van der Waals repulsions cannot account for the results in every case. For example, nitration of toluene gives twice as much *o*-nitrobenzene as *p*-nitrobenzene, the result expected on a random basis. In contrast, nitration of fluorobenzene yields mostly the *para* isomer, even though fluorine is much smaller than a methyl group. If van der Waals repulsions determined the outcome, more *ortho* nitration product should be obtained from fluorobenzene than from toluene. The reasons for the *ortho*, *para* ratio vary from case to case, and in some cases are not well understood.

Whatever the reasons for the *ortho*, *para* ratio, if an electrophilic aromatic substitution reaction yields a mixture of *ortho* and *para* isomers, a problem of isomer separation arises that must be solved if the reaction is to be useful. Usually syntheses that give mixtures of isomers are avoided because, in many cases, isomers are difficult to separate. However, the *ortho* and *para* isomers obtained in many electrophilic aromatic substitution reactions have sufficiently different physical properties that they are readily separated (Sec. 16.2). For example, the boiling points of *o*- and *p*-nitrotoluene, 220° and 238°, respectively, are sufficiently different that these isomers can be separated by careful fractional distillation. The melting points of *o*- and *p*-chloronitrobenzene, 34° and 84°, respectively, are so different that the *para* isomer can be selectively crystallized. Most aromatic substitution reactions are so simple and inexpensive to run that when the separation of isomeric products is not difficult, these reactions are useful for organic synthesis despite the product mixtures that are obtained. Thus, you may assume in working problems involving electrophilic aromatic substitution on compounds containing *ortho*, *para*-directing groups that the *para* isomer can be isolated in useful amounts.

B. Activating and Deactivating Effects of Substituents

Different benzene derivatives have greatly different reactivities in electrophilic aromatic substitution reactions. If a substituted benzene derivative reacts more rapidly than benzene itself, then the substituent group is said to be an **activating group**. The Friedel-Crafts acylation of anisole (methoxybenzene), for example, is 300,000 times as fast as the same reaction of benzene under comparable conditions. Furthermore, anisole shows a similar enhanced reactivity relative to benzene in all other electrophilic substitution reactions. Thus, the methoxy group is an *activating group*. On the other hand, if the derivative reacts more slowly than benzene itself, then the substituent is called a **deactivating group**. For example, the rate of the bromination of nitrobenzene is less than 10^{-5} times the rate of the bromination of benzene, and nitrobenzene reacts much more slowly than benzene in all other electrophilic substitution reactions as well. Thus, the nitro group is a *deactivating group*.

A given substituent group is either activating in all electrophilic aromatic substitution reactions or deactivating in all such reactions. Whether a substituent is activating or deactivating is shown in the last column of Table 16.2. In this table the most activating substituent groups are near the top of the table. Three generalizations emerge from examining this table.

1. All *meta*-directing groups are deactivating groups.

2. All *ortho, para*-directing groups except for the halogens are activating groups.

3. The halogens are deactivating groups.

Thus, except for the halogens, there appears to be a correlation between the activating and directing effects of substituents.

In view of this correlation, it is not surprising that the explanation of activating and deactivating effects is closely related to the explanation for directing effects. A key to understanding these effects is the realization that directing effects are concerned with the relative rates of substitution at different positions of the *same* compound, while activating or deactivating effects are concerned with the relative rates of substitution of *different* compounds—a substituted benzene compared with benzene itself. As in the discussion of directing effects, we consider the effect of the substituent on the stability of the intermediate carbocation, and we then assume that the stability of this carbocation is related to the stability of the transition state for its formation, as suggested by Hammond's postulate.

Two properties of substituents must be considered in order to understand activating and deactivating effects. First is the **resonance effect** of the substituent. This is the ability of the substituent to stabilize the carbocation intermediate in electrophilic substitution by delocalization of electrons from the substituent into the ring. The resonance effect is the same effect that is responsible for the *ortho, para*-directing effects of substituents with unshared electron pairs, such as —OCH_3 and halogen (colored structure in Eq. 16.27).

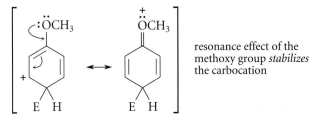

resonance effect of the methoxy group *stabilizes* the carbocation

The second property is the **polar effect** of the substituent. This is the tendency of the substituent group, by virtue of its electronegativity, to pull electrons away from the ring. This is the same effect discussed in connection with substituent effects on acidity (Sec. 3.6B). When a ring substituent is electronegative, it pulls the electrons of the ring toward itself and creates a slight electron deficiency, or positive charge, in the ring. In the carbocation intermediate of an electrophilic substitution reaction, the positive end of the bond dipole interacts repulsively with the positive charge in the ring, thus raising the energy of the ion:

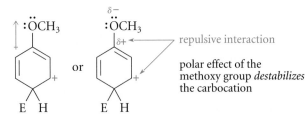

repulsive interaction

polar effect of the methoxy group *destabilizes* the carbocation

Thus, the electron-donating resonance effect of a substituent group with unshared electron pairs *stabilizes* positive charge and *activates* further substitution. If such a group is electronegative, its electron-withdrawing polar effect *destabilizes* positive charge and *deactivates* further substitution. These two effects operate simultaneously. *Whether a substituted derivative of benzene is activated or deactivated toward further substitution depends on the balance of the resonance and polar effects* of the substituent group.

Anisole (methoxybenzene) undergoes electrophilic substitution much more rapidly than benzene because the resonance effect of the methoxy group far outweighs its polar effect. The benzene molecule, in contrast, has no substituent to help stabilize the carbocation intermediate by resonance. Hence, the carbocation intermediate (and the transition state) derived from substitution of anisole is more stable relative to starting materials than the carbocation (and transition state) derived from substitution of benzene. Thus, in a given reaction, *para* substitution of anisole is faster than substitution of benzene. In other words, the methoxy group activates the benzene ring toward *para* substitution.

There is also an important subtlety here. Although the *ortho* and *para* positions of anisole are highly activated toward substitution, the *meta* position is deactivated. When substitution occurs in the *meta* position, the methoxy group cannot exert its resonance effect (Eq. 16.28), and only its rate-retarding polar effect is operative. Thus, whether a group activates or deactivates further substitution really depends on the *position* on the ring that is being considered. Thus, the methoxy group activates *ortho, para* substitution and deactivates *meta* substitution. But this is just another way of saying that the methoxy group is an *ortho, para* director. Because *ortho, para* substitution is the *observed* mode of substitution, the methoxy group is considered to be an activating group. These ideas are summarized in the reaction free-energy diagrams shown in Fig. 16.8.

The deactivating effects of halogen substituents reflect a different balance of resonance and polar effects. Consider the chloro group, for example. Since chlorine and oxygen have similar electronegativities, the polar effects of the chloro and methoxy groups are similar. However, the resonance interaction of chlorine electron pairs with the ring is much less effective than the interaction of oxygen electron pairs because the chlorine valence electrons reside in orbitals with higher quantum numbers. Because these orbitals and the carbon $2p$ orbitals of the benzene ring have *different sizes* and *different numbers of nodes*, they do not overlap so effectively (Fig. 16.9). Since this overlap is the basis of the resonance effect, the resonance effect of chlorine is weak. With a weak rate-enhancing resonance effect and a strong rate-retarding polar effect, chlorine is a deactivating group. Bromine and iodine exert weaker polar effects than chlorine, but their resonance effects are also weaker (why?). Hence, these groups, too, are deactivating groups. Fluorine, as a first-row element, has a stronger resonance effect than the other halogens, but, as the most electronegative element, it has a stronger polar effect as well. Fluorine is also a deactivating group.

The deactivating, rate-retarding polar effects of the halogens are similar at all ring positions, but are offset somewhat by their resonance effects when substitution occurs *para* to the halogen. However, the resonance effect of a halogen cannot come into play at all when substitution occurs at the *meta* position of a halobenzene (why?). Hence, *meta* substitution in halobenzenes is deactivated even more than *para* substitution is. This is another way of saying that halogens are *ortho, para*-directing groups.

Alkyl substituents such as the methyl group have no resonance effect, but the polar effect of any alkyl group toward electron-deficient carbons is an electropositive, stabilizing effect (Sec. 4.7C). It follows that alkyl substituents on a benzene ring stabilize carbocation intermediates in electrophilic substitution, and for this reason, they are activating groups.

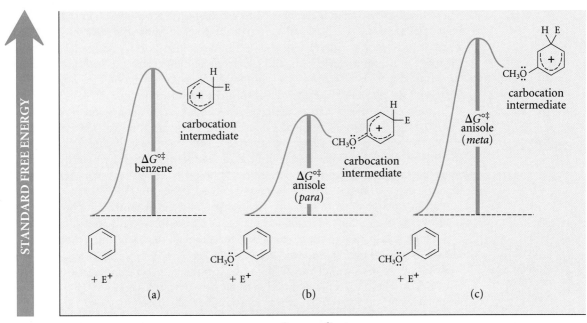

Figure 16.8   *Basis of the activating effect of the methoxy group on electrophilic aromatic substitution in anisole. (a) The energy barrier for substitution of benzene by an electrophile E^+. (b) The energy barrier for substitution of anisole by E^+ at the para position. (c) The energy barrier for substitution of anisole by E^+ at the meta position. The substitution of anisole at the para position is faster than substitution of benzene; the substitution of anisole at the meta position is slower than the substitution of benzene. The methoxy group is an activating group because the observed reaction of anisole, substitution at the para position, is faster than substitution of benzene.*

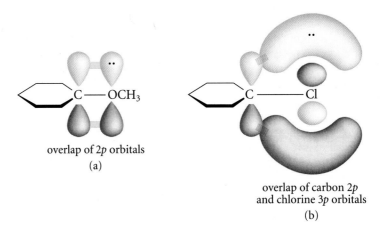

Figure 16.9   *Overlap of carbon and oxygen 2p orbitals (a) is more effective than the overlap of carbon 2p and chlorine 3p orbitals (b), because the orbitals with different quantum numbers have different sizes and different numbers of nodes. The colored and gray parts of the orbitals represent wave peaks and wave troughs, respectively. Bonding overlap occurs only when peaks overlap with peaks and troughs with troughs.*

It turns out that alkyl groups activate substitution at all ring positions, but they are *ortho, para* directors because they activate *ortho, para* substitution more than they activate *meta* substitution (Eqs. 16.29 and 16.30).

Finally, consider the deactivating effects of *meta*-directing groups such as the nitro group. Because a nitro group has no appreciable electron-donating resonance effect, the polar effect of this electronegative group destabilizes the carbocation intermediate and retards electrophilic substitution at *all* positions of the ring. The nitro group is a *meta*-directing group because substitution is retarded more at the *ortho* and *para* positions than at the *meta* positions (Eqs. 16.31 and 16.32). In other words, the *meta*-directing effect of the nitro group is not due to selective activation of the *meta* positions, but rather to greater *deactivation* of the *ortho* and *para* positions. For this reason, the nitro group and the other *meta*-directing groups might be called *meta-allowing groups*.

PROBLEMS

16.25   Draw reaction-free energy profiles analogous to that in Fig. 16.8 in which substitution on benzene by a general electrophile E$^+$ is compared with substitution at the *para* and *meta* positions of *(a) chlorobenzene; (b) nitrobenzene.

16.26   Which should be faster:
  *(a)   nitration of anisole or nitration of thioanisole under the same conditions?
  (b)   bromination of *N,N*-dimethylaniline or bromination of benzene under the same conditions?
  Explain your answers carefully.

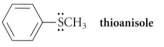

C. Use of Electrophilic Aromatic Substitution in Organic Synthesis

Both activating/deactivating and directing effects of substituents can come into play in planning an organic synthesis that involves electrophilic substitution reactions. The importance of directing effects is illustrated in the following study problem.

STUDY PROBLEM 16.2

Outline a synthesis of *p*-bromonitrobenzene from benzene.

Solution   The key to this problem is whether the bromine or the nitro group should be the first ring substituent introduced. Introduction of the bromine first takes advantage of its directing effect in the subsequent nitration reaction:

(16.33)

benzene   bromobenzene   *p*-nitrobromobenzene

Introduction of the nitro group first followed by bromination would give instead *m*-bromonitrobenzene, because the nitro group is a *meta*-directing group.

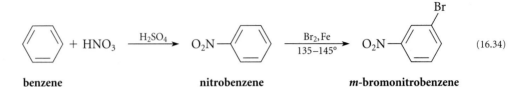

benzene   **nitrobenzene**   **m-bromonitrobenzene**

(16.34)

Hence, to prepare the desired compound, brominate first and *then* nitrate the resulting bromobenzene, as shown in Eq. 16.33.

··

When an electrophilic substitution reaction is carried out on a benzene derivative with more than one substituent, the activating and directing effects are roughly the sum of the effects of the separate substituents. First, let's consider directing effects. In the Friedel-Crafts acylation of *m*-xylene, for example, both methyl groups direct the substitution to the same positions.

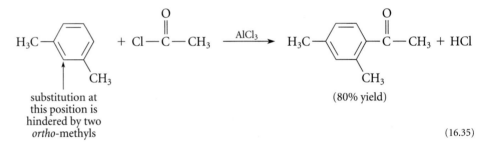

substitution at this position is hindered by two *ortho*-methyls   (80% yield)

(16.35)

Methyl groups are *ortho, para* directors. Substitution at the position *ortho* to both methyl groups is difficult because van der Waals repulsions between both methyls and the electrophile would be present in the transition state. Consequently, substitution occurs at a ring position that is *para* to one methyl and, of necessity, *ortho* to the other.

Two *meta*-directing groups on a ring, such as the carboxylic acid (—CO_2H) groups in the following example, direct further substitution to the remaining open *meta* position:

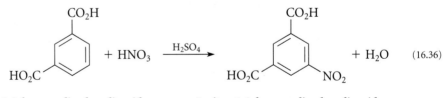

1,3-benzenedicarboxylic acid   **5-nitro-1,3-benzenedicarboxylic acid**
             (96% of product)

In each of the last two examples, both substituents direct the incoming group to the same position. What happens when the directing effects of the two groups are in conflict? If one group is much more strongly activating than the other, the directing effect of the more powerful activating group generally predominates. For example, the —OH group is such a powerful activating group that phenol can be brominated three times. (Notice that the —OH group is near the top of Table 16.2.)

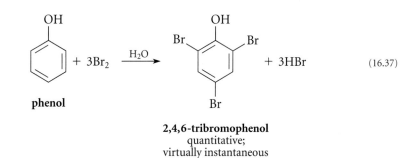

(16.37)

2,4,6-tribromophenol
quantitative;
virtually instantaneous

After the first bromination, the —OH and —Br groups direct subsequent brominations to different positions. The strong activating and directing effect of the —OH group at the *ortho* and *para* positions overrides the weaker directing effect of the —Br group.

In other cases, mixtures of isomers are typically obtained.

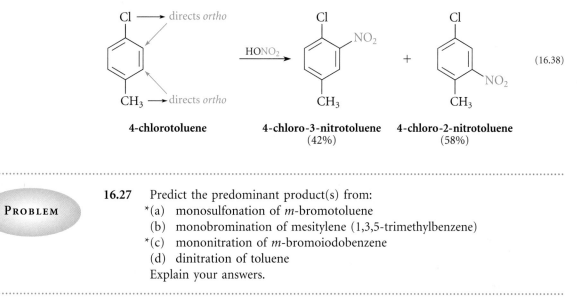

(16.38)

4-chlorotoluene   **4-chloro-3-nitrotoluene**   **4-chloro-2-nitrotoluene**
                                      (42%)   (58%)

PROBLEM

16.27   Predict the predominant product(s) from:
   *(a)   monosulfonation of *m*-bromotoluene
    (b)   monobromination of mesitylene (1,3,5-trimethylbenzene)
   *(c)   mononitration of *m*-bromoiodobenzene
    (d)   dinitration of toluene
    Explain your answers.

STUDY GUIDE LINK:
    ✓16.4
*Reaction Conditions
and Reaction Rate*

The activating or deactivating effects of substituents in an aromatic compound determine the conditions that must be used in an electrophilic substitution reaction. The bromination of nitrobenzene, for example (Eq. 16.26), requires relatively harsh conditions of heat and a Lewis acid catalyst because the nitro group deactivates the ring toward electrophilic substitution. In contrast, mesitylene can be brominated under very mild conditions, because the ring is activated by three methyl groups; a Lewis acid catalyst is unnecessary.

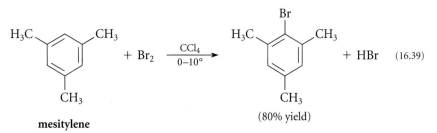

(16.39)

mesitylene   (80% yield)

A similar contrast is apparent in the conditions required to sulfonate benzene and toluene. Sulfonation of benzene requires fuming sulfuric acid (Eq. 16.10). However, because toluene is more reactive than benzene, toluene can be sulfonated with concentrated sulfuric acid, a milder reagent than fuming sulfuric acid.

$$H_3C\!-\!\langle\bigcirc\rangle + H_2SO_4 \longrightarrow H_3C\!-\!\langle\bigcirc\rangle\!-\!SO_3H + H_2O \qquad (16.40)$$

Another important consequence of activating and deactivating effects is that when a deactivating group—for example, a nitro group—is being introduced by an electrophilic substitution reaction, it is easy to introduce one group at a time. Thus, toluene can be nitrated only once because the nitro group that is introduced retards a second nitration on the same ring. Notice that additional nitrations require increasingly harsh conditions.

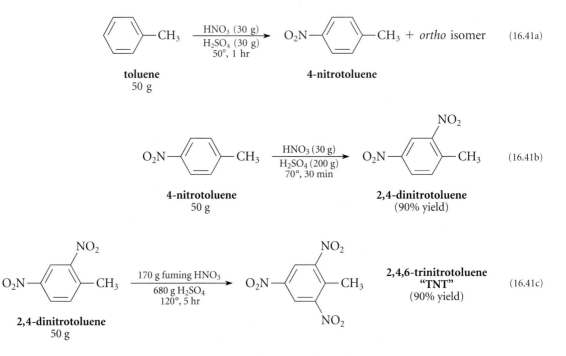

Fuming nitric acid (Eq. 16.41c) is an especially concentrated form of nitric acid. Ordinary nitric acid contains 68% by weight of nitric acid; fuming nitric acid is 95% by weight nitric acid. It owes its name to the layer of colored fumes usually present in the bottle of the commercial product.

Deactivating substituents retard some reactions to the point that they are not useful. For example, Friedel-Crafts *acylation* (Sec. 16.4E) does not occur on a benzene ring substituted *solely* with one or more *meta*-directing groups. In fact, nitrobenzene is so unreactive in the Friedel-Crafts acylation that it can be used as the solvent in the acylation of other aromatic compounds! Similarly, the Friedel-Crafts *alkylation* (Sec. 16.4F) is generally not useful on compounds that are more deactivated than benzene itself.

When an activating group is introduced by electrophilic substitution, additional substitutions can occur easily under the conditions of the first substitution, and as a result, mixtures of products are obtained. This is the situation in Friedel-Crafts alkylation. As noted in the discussion of Eq. 16.22, one way to avoid multiple substitution in such cases is to use a large excess of the starting material.

PROBLEMS

16.28   In each of the following sets, rank the compounds in order of increasing harshness of the reaction conditions required to accomplish the indicated reaction.
   *(a)   sulfonation of benzene, *m*-xylene, or *p*-dichlorobenzene
   (b)   Friedel-Crafts acylation of chlorobenzene, anisole, or toluene.

*16.29   Outline a synthesis of *m*-nitroacetophenone from benzene; explain your reasoning.

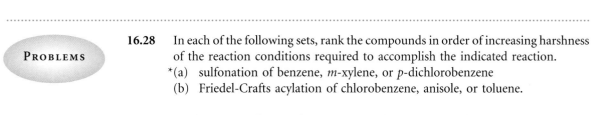

                m-nitroacetophenone

16.6   Hydrogenation of Benzene Derivatives

Because of its aromatic stability, the benzene ring is resistant to conditions used to hydrogenate ordinary double bonds.

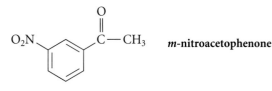

       stilbene   **(2-phenylethyl)benzene**
      (*cis* or *trans*)   (95% yield)

Nevertheless, aromatic rings can be hydrogenated under more extreme conditions of temperature and/or pressure. Typical conditions for carrying out the hydrogenation of benzene derivatives include Rh or Pt catalysts at 5–10 atm of hydrogen pressure and 50–100°, or Ni or Pd catalysts at 100–200 atm and 100–200°.

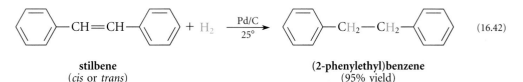

         ethylbenzene   **ethylcyclohexane**
                                      (93% yield)

As this example illustrates, a good way to prepare a substituted cyclohexane in many cases is to prepare the corresponding benzene derivative and then hydrogenate it.

     Catalytic hydrogenation of benzene derivatives gives the corresponding cyclohexanes, and cannot be stopped at the cyclohexadiene or cyclohexene stage. The reason follows from the enthalpies of hydrogenation of benzene, 1,3-cyclohexadiene, and cyclohexene.

$$\text{C}_6\text{H}_6 + \text{H}_2 \longrightarrow \text{C}_6\text{H}_8 \qquad \Delta H° = +24.3 \text{ kJ/mol } (+5.8 \text{ kcal/mol}) \qquad \text{(16.44a)}$$

$$\text{C}_6\text{H}_8 + \text{H}_2 \longrightarrow \text{C}_6\text{H}_{10} \qquad \Delta H° = -111 \text{ kJ/mol } (-26.5 \text{ kcal/mol}) \qquad \text{(16.44b)}$$

$$\text{C}_6\text{H}_{10} + \text{H}_2 \longrightarrow \text{C}_6\text{H}_{12} \qquad \Delta H° = -118 \text{ kJ/mol } (-28.3 \text{ kcal/mol}) \qquad \text{(16.44c)}$$

The hydrogenation of most ordinary alkenes is *exothermic* by 133–126 kJ/mol (27–30 kcal/mol); yet the reaction in Eq. 16.44a is *endothermic*. The unusual $\Delta H°$ of this reaction reflects the aromatic stability of benzene. Because this reaction is endothermic, energy must be added for it to take place—thus the harsh conditions required for the hydrogenation of benzene derivatives. The hydrogenations of 1,3-cyclohexadiene and cyclohexene proceed so rapidly under these vigorous conditions that once these compounds are formed in the hydrogenation of benzene, they react instantaneously.

PROBLEM

16.30   Using benzene and any other reagents, outline a synthesis of each of the following compounds.

   *(a)   cyclohexylcyclohexane   (b)   *tert*-butylcyclohexane

16.7   Source and Industrial Use of Aromatic Hydrocarbons

The most common source of aromatic hydrocarbons is petroleum. Some petroleum sources are relatively rich in aromatic hydrocarbons, and aromatic hydrocarbons can be obtained by catalytic reforming of the hydrocarbons from other sources. Another potentially important, but currently minor, source of aromatic hydrocarbons is *coal tar*, the tarry residue obtained when coal is heated in the absence of oxygen. Once a major source of aromatic hydrocarbons, coal tar may increase in importance as a source of aromatic compounds as the use of coal increases.

Benzene itself is obtained by separation from petroleum fractions, and by demethylation of toluene. Annual production of benzene in the United States is about 1.6 billion gallons. Benzene serves as a principal source of ethylbenzene, styrene, and cumene (see Eq. 16.45), and as one of the sources of cyclohexane. Because cyclohexane is an important intermediate in the production of nylon, benzene has substantial importance to the nylon industry.

Toluene is also obtained by separation from reformates, the products of hydrocarbon interconversion over certain catalysts. As previously noted, some toluene is used in the production of benzene. Toluene is also used as an octane booster for gasoline and as a starting material in the polyurethane industry.

Ethylbenzene and cumene are obtained by the alkylation of benzene with ethylene and propene, respectively, in the presence of acid catalysts (Friedel-Crafts alkylation).

$$(16.45)$$

Cumene is an important intermediate in the manufacture of phenol and acetone (Sec. 18.10). The major use of ethylbenzene is dehydrogenation to styrene, one of the most commercially important aromatic hydrocarbons. Its principal uses are in the manufacture of polystyrene (Sec. 5.7A) and styrene-butadiene rubber (Sec. 15.5). About 1.5 billion gallons of ethylbenzene and 1.2 billion gallons of styrene are produced annually in the United States.

The xylenes are obtained by separation from petroleum and by reforming C_8 petroleum fractions. Of the xylenes, *p*-xylene is the most important commercially. Virtually the entire production of *p*-xylene is used for oxidation to terephthalic acid (Eq. 16.46), an important intermediate in polyester synthesis (for example, Dacron; Sec. 21.12A). (Oxidation of alkylbenzenes is discussed in Sec. 17.5.)

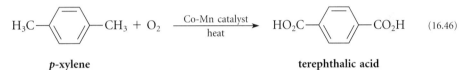

$$(16.46)$$

The relationships of many of the compounds discussed here to the chemical industry as a whole are shown in Fig. 5.3.

AROMATIC COMPOUNDS AND CANCER

Most people appreciate that certain chemicals are hazardous. Among the most worrisome chemical hazards is *carcinogenicity*—the proclivity of a substance to cause cancer. Certain aromatic compounds are **carcinogens**, or cancer-causing chemicals. Both the historical aspects of this finding and the reasons underlying it are interesting.

After the great fire of London in 1666, Londoners began the practice of building homes with long and tortuous chimneys. The use of coal for heating resulted in deposits of black soot that had to be periodically removed from these chimneys, but the only people who could negotiate these narrow passages were small boys, called "sweeping boys." It was common for these boys to contract a disease that we now know is cancer of the scrotum. In 1775 Percivall Pott, a surgeon at London's St. Bartholomew's, Hospital, identified coal dust as the source of "this noisome, painful, and fatal disease," and Pott's findings subsequently led to substantial reform in the child-labor statutes in England. In 1892 Henry T. Butlin, also of St. Bartholomew's, pointed out that the disease did not occur in countries in which the chimney sweeps washed thoroughly after each day's work.

In 1933, the compound benzo[*a*]pyrene was isolated from coal tar; this compound and 7,12-dimethylbenz[*a*]anthracene, both polycyclic aromatic hydrocarbons, are two of the most potent carcinogens known.

(continues)

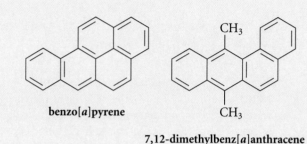

benzo[a]pyrene

7,12-dimethylbenz[a]anthracene
(7,12-DMBA)

These and related compounds are the active carcinogens in soot. Materials such as these are also found in cigarette smoke and in the exhaust of internal combustion engines (that is, automobile pollution). About three thousand tons of benzo[a]pyrene per year are released as particulates into the environment.

Benzene has also been found to be carcinogenic, and it has been supplanted for many uses by toluene, which is not carcinogenic. (Not all aromatic compounds are carcinogens.) However, benzene is much less carcinogenic than the polycyclic hydrocarbons above, and continues to be used with due caution in applications for which it cannot be readily replaced.

Organic chemists and biochemists have learned that the *ultimate carcinogens* (true carcinogens) are not aromatic hydrocarbons themselves, but rather certain epoxide derivatives. The diol-epoxide shown in Eq. 16.47, formed when living cells attempt to metabolize benzo[a]pyrene, is the ultimate carcinogen derived from benzo[a]pyrene.

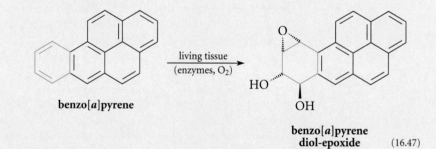

benzo[a]pyrene

$$\xrightarrow[\text{(enzymes, O}_2\text{)}]{\text{living tissue}}$$

benzo[a]pyrene
diol-epoxide   (16.47)

This transformation is particularly remarkable in view of the usual resistance of benzene rings to addition reactions. This type of oxidation is normally employed as a detoxification mechanism; the oxidized derivatives of many hydrocarbons, but not the hydrocarbons themselves, can be secreted and thus removed from the cell. Benzo[a]pyrene diol-epoxide, however, survives long enough to find its way into the nuclei of cells, where the epoxide group reacts with nucleophilic groups on DNA, the molecule that contains the genetic code. (This reaction is discussed in Sec. 27.12B.) It is strongly suspected, although not conclusively proved, that this is a key event in the transformation of cells to a cancerous state caused by benzo[a]pyrene.

Key Ideas in Chapter 16

- Benzene derivatives are distinguished by the NMR absorptions of their ring protons, which occur at lower field than the absorptions of vinylic protons. The unusual chemical shifts of aromatic protons are caused by the ring-current effect. The CMR absorptions of ring carbons are observed in about the same part of the spectrum as the absorptions of the vinylic carbons of alkenes.

- The most characteristic reaction of aromatic compounds is electrophilic aromatic substitution. In this type of reaction, an electrophile is attacked by the π electrons of a benzene ring to form a resonance-stabilized carbocation. Loss of a proton from this ion gives a new aromatic compound.

- Examples of electrophilic aromatic substitution reactions discussed in this chapter are halogenation, used to prepare halobenzenes; nitration, used to prepare nitrobenzene derivatives; sulfonation, used to prepare benzenesulfonic acid derivatives; Friedel-Crafts acylation, used to prepare aryl ketones; and Friedel-Crafts alkylation, used to prepare alkylbenzenes.

- Derivatives containing substituted benzene rings can undergo further substitution either at the *ortho* and *para* positions or at the *meta* position, depending on the ring substituent.

- Benzene rings with alkyl substituents or substituent groups that delocalize positive charge by resonance undergo further substitution at the *ortho* and *para* positions; these substituent groups are called *ortho*, *para*-directing groups.

- Benzene rings with substituents that cannot stabilize carbocations or delocalize positive charge by resonance undergo further substitution at the *meta* position. These substituents are called *meta*-directing groups.

- Whether a substituted benzene undergoes substitution more rapidly or more slowly than benzene itself is determined by the balance of resonance and polar effects of the substituents. Benzene rings containing an *ortho*, *para*-directing group other than halogen react more rapidly in electrophilic aromatic substitution than benzene itself. Benzene rings containing a halogen substituent or any *meta*-directing group react more slowly in electrophilic aromatic substitution than benzene itself.

- The activating and directing effects of substituent groups must be taken into account when planning a synthesis.

- Alkene double bonds can generally be hydrogenated without affecting the benzene ring. Benzene derivatives, however, can be hydrogenated under relatively harsh conditions to cyclohexane derivatives.

*16.31 Give the products expected (if any) when ethylbenzene reacts under the following conditions.
(a) Br_2 in CCl_4 (dark)   (b) HNO_3, H_2SO_4 (c) conc. H_2SO_4

(d)
$$C_2H_5-\overset{\overset{\displaystyle O}{\|}}{C}-Cl, \text{ AlCl}_3 \text{ (1.1 equivalent), then } H_2O$$

(e) CH_3Br, $AlCl_3$ (f) Br_2, $FeBr_3$

16.32 Give the products expected (if any) when nitrobenzene reacts under the following conditions.
(a) Cl_2, $FeCl_3$, heat   (b) fuming HNO_3, H_2SO_4

(c)
$$CH_3-\overset{\overset{\displaystyle O}{\|}}{C}-Cl, \text{ AlCl}_3 \text{ (1.1 equivalent), then } H_2O$$

16.33 Show how you would distinguish the compounds within each of the following pairs using only simple chemical or physical tests with readily observable results.
*(a) phenylacetylene and styrene   (b) ethylbenzene and styrene
*(c) benzene and ethylbenzene

*16.34 Which of the following compounds *cannot* contain a benzene ring? How do you know?
(1) $C_{10}H_{16}$ (2) $C_8H_6Cl_2$ (3) C_5H_4 (4) $C_{10}H_{15}N$

*16.35 (a) Arrange the three isomeric dichlorobenzenes in order of increasing dipole moment (smallest first).
(b) Assuming that the dipole moment is the principal factor governing their relative boiling points, arrange the compounds from part (a) in order of increasing boiling point (smallest first). Explain your reasoning.

*16.36 Explain how you would distinguish each of the following compounds from the others using NMR spectroscopy. Be explicit.

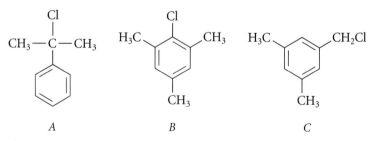

A   *B*   *C*

16.37 Explain how you would distinguish between ethylbenzene, *p*-xylene, and styrene using only NMR spectra.

16.38 Explain why
*(a) the NMR spectrum of the sodium salt of cyclopentadiene consists of a singlet.

 **sodium salt of cyclopentadiene**

(b) the methyl group in the following compound has an unusual chemical shift of δ (−1.67), about four ppm higher field than a typical allylic methyl group.

*16.39 Show how resonance interaction of the electron pairs on the oxygen with the ring π electrons can account for the fact that the chemical shift of protons *a* in *p*-methoxytoluene is smaller than that of protons *b* in spite of the fact that the oxygen has a greater electronegativity than the methyl carbon.

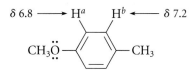

16.40 Outline laboratory syntheses of each of the following compounds, starting with benzene and any other reagents. (The references to equations will assist you with nomenclature.)
*(a) *p*-nitrotoluene   (b) *p*-dibromobenzene
*(c) *p*-chloroacetophenone (Eq. 16.12)
(d) *m*-nitrobenzenesulfonic acid (Eq. 16.10)
*(e) *p*-chloronitrobenzene   (f) 1,3,5-trinitrobenzene
*(g) 2,6-dibromo-4-nitrotoluene   (h) 2,4-dibromo-6-nitrotoluene
*(i) 4-ethyl-3-nitroacetophenone (Eq. 16.12)
*(j) cyclopentylbenzene   (k) methylcyclohexane

16.41 Arrange the following compounds in order of increasing reactivity toward HNO_3 in H_2SO_4. (The references to equations will assist you with nomenclature.)
*(a) mesitylene (Eq. 16.39), toluene, 1,2,4-trimethylbenzene
(b) chlorobenzene, benzene, nitrobenzene
*(c) *m*-chloroanisole, *p*-chloroanisole, anisole
(d) acetophenone (Eq. 16.12), *p*-methoxyacetophenone, *p*-bromoacetophenone

*16.42 Indicate whether each of the following compounds should be nitrated more rapidly or more slowly than benzene, and give the structure of the principal mononitration product in each case. Explain your reasoning.

(Problem 16.42 continues)

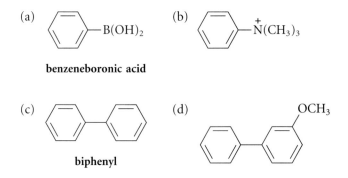

(a) benzeneboronic acid

(b)

(c) biphenyl

(d)

16.43 Rank the following compounds in order of increasing reactivity in bromination. In each case, indicate whether the principal monobromination products will be the *ortho* and *para* isomers or the *meta* isomer, and whether the compound will be more or less reactive than benzene. Explain carefully the points that cause any uncertainty.

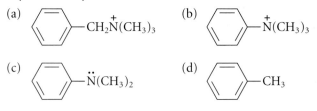

*16.44 Two alcohols, *A* and *B*, have the same molecular formula $C_9H_{10}O$ and react with sulfuric acid to give the same hydrocarbon *C*. Compound *A* is optically active and compound *B* is not. Catalytic hydrogenation of *C* gives a hydrocarbon *D*, C_9H_{10}, which gives two and only two products when nitrated once with HNO_3 in H_2SO_4. Give the structures of *A*, *B*, *C*, and *D*.

16.45 Give the structures of all the hydrocarbons $C_{10}H_{10}$ that would undergo catalytic hydrogenation to give *p*-diethylbenzene.

*16.46 (a) Give a curved-arrow mechanism for the following transformation, which is an example of an intramolecular Friedel-Crafts alkylation.

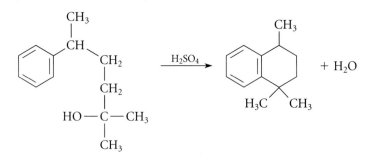

(b) In your mechanism, identify the three basic steps of electrophilic aromatic substitution discussed in Sec. 16.4B.

16.47 When the following compound is treated with H_2SO_4, the product of the resulting reaction has the formula $C_{15}H_{20}$ and does not decolorize Br_2 in CCl_4. Suggest a structure for this product and give a curved-arrow mechanism for its formation.

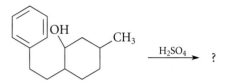

$\xrightarrow{H_2SO_4}$?

*16.48 Celestolide, a perfuming agent with a musk odor, is prepared by the following sequence of reactions.

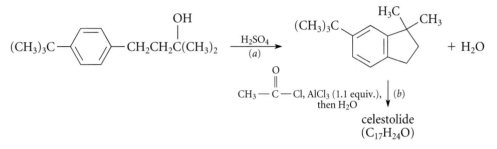

+ H_2O

celestolide
($C_{17}H_{24}O$)

(a) Give a curved-arrow mechanism for reaction (a).

(b) From your knowledge of the reaction used in step (b), give the structure of celestolide. (*Hint:* Consider the steric effect of the *tert*-butyl group.)

*16.49 An optically active compound A ($C_9H_{11}Br$) reacts with sodium ethoxide in ethanol to give an optically inactive hydrocarbon B (NMR spectrum in Fig. 16.10). Compound B undergoes hydrogenation over a Pd/C catalyst at room temperature to give a compound C, which has the formula C_9H_{12}. Give the structures of A, B, and C.

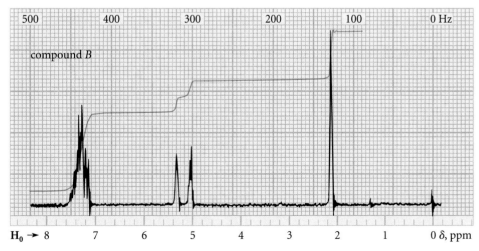

Figure 16.10 *NMR spectrum for Problem 16.49.*

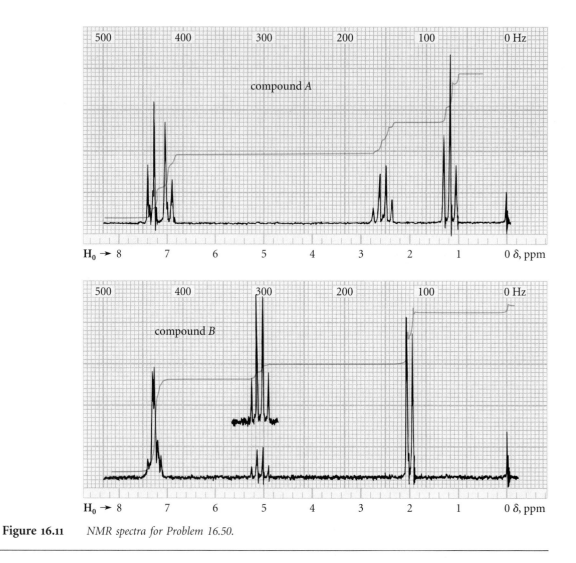

Figure 16.11 *NMR spectra for Problem 16.50.*

16.50 Identify the two compounds, *A* and *B*, both colorless liquids with the formula C_8H_9Br and NMR spectra given in Fig. 16.11.

*16.51 Identify each of the following compounds.
  (a) Compound *A*: IR 1605 cm^{-1}, no O—H stretch
    NMR: δ 3.72 (3*H*, s); δ 6.72 (2*H*, apparent doublet, *J* = 9 Hz); δ 7.15 (2*H*, apparent doublet, *J* = 9 Hz).
    Mass spectrum in Fig. 16.12a. (*Hint:* Notice the M + 2 peak.)
  (b) Compound *B*: IR 1600, 1640 cm^{-1}, no O—H stretch
    UV: λ_{max} = 207 nm (ϵ = 20,400), 258 nm (ϵ = 19,000), and 291 nm (ϵ = 4450).
    NMR spectrum in Fig. 16.12b.
    CMR (attached protons in parentheses): δ 54.9 (3), δ 111.3 (2), δ 114.0 (1), δ 127.4 (1), δ 130.5 (0), δ 136.5 (1), δ 159.7 (0)

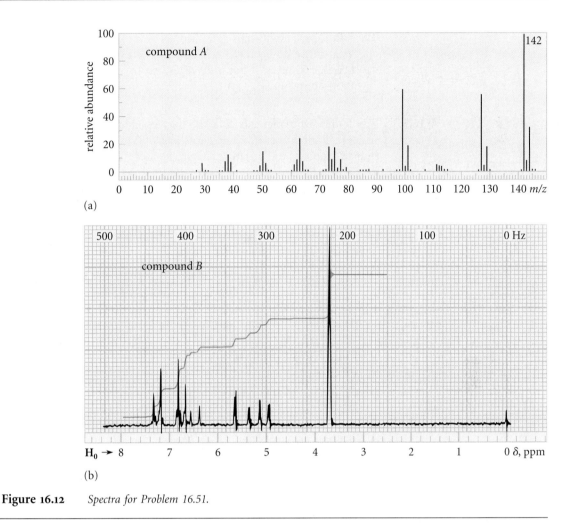

Figure 16.12   *Spectra for Problem 16.51.*

16.52   A method for determining the structures of disubstituted benzene derivatives was proposed in 1874 by Wilhelm Körner of the University of Milan. Körner had in hand three dibromobenzenes, *A*, *B*, and *C*, with melting points of 89°, 6.7°, and −6.5°, respectively. He nitrated each isomer in turn and meticulously isolated *all* of the mononitro derivatives of each. Compound *A* gave one mononitro derivative; compound *B* gave two mononitro derivatives; and compound *C* gave three. These experiments gave him enough information to assign the structures of *o*-, *m*-, and *p*-dibromobenzene.

*(a)   Assuming the correctness of the Kekulé structure for benzene, assign the structures of the dibromobenzene derivatives.

*(b)   Körner had no way of knowing whether the Kekulé or Ladenburg-benzene structure (Sec. 15.7A) was correct. Assuming the correctness of the Ladenburg-benzene structure, assign the structures of the dibromobenzene derivatives.

(c)   It is a testament to Körner's experimental skill that he could isolate all the mononitration products. Of all the mononitration products that he isolated, which one(s) were formed in smallest amount? Explain.

*(d) Jack Körner, Wilhelm's grandnephew twice removed, has decided to repeat great-uncle Wilhelm's experiment by using CMR to identify the dibromobenzene isomers. What differences can he expect in the CMR spectra of these compounds?

*16.53 It is found experimentally that the heat liberated upon hydrogenation of an alkene is about the same for all alkenes with the same number of alkyl substituents at their double bonds. Thus, if benzene were an ordinary alkene, it should have about the same heat of hydrogenation ($\Delta H°$ of hydrogenation) as three cyclohexenes.

(a) Using the data in Eq. 16.44a–c, calculate the heat liberated when benzene is hydrogenated to cyclohexane.

(b) Calculate the heat liberated in the hydrogenation of three moles of cyclohexene to three moles of cyclohexane.

(c) Calculate the discrepancy between the quantities calculated in (a) and (b). This number has been used as another estimate of the *empirical resonance energy* of benzene (Sec. 15.7C).

16.54 Give the structures of the principal organic product(s) expected in each of the following reactions, and explain your reasoning.

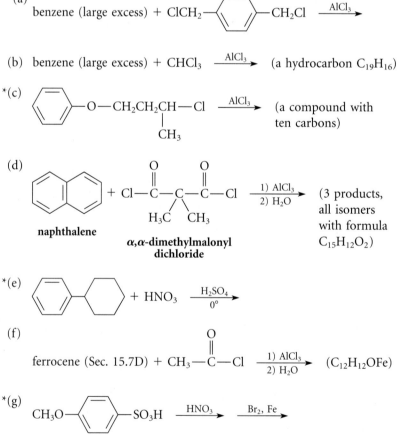

*(a)
benzene (large excess) + ClCH₂—⟨ ⟩—CH₂Cl $\xrightarrow{AlCl_3}$

(b) benzene (large excess) + CHCl₃ $\xrightarrow{AlCl_3}$ (a hydrocarbon $C_{19}H_{16}$)

*(c)
⟨ ⟩—O—CH₂CH₂CH—Cl $\xrightarrow{AlCl_3}$ (a compound with ten carbons)
                    |
                    CH₃

(d)
naphthalene + Cl—C(=O)—C(CH₃)(H₃C)—C(=O)—Cl $\xrightarrow[\text{2) H}_2\text{O}]{\text{1) AlCl}_3}$ (3 products, all isomers with formula $C_{15}H_{12}O_2$)

α,α-dimethylmalonyl dichloride

*(e)
⟨ ⟩—⟨ ⟩ + HNO₃ $\xrightarrow[0°]{H_2SO_4}$

(f)
ferrocene (Sec. 15.7D) + CH₃—C(=O)—Cl $\xrightarrow[\text{2) H}_2\text{O}]{\text{1) AlCl}_3}$ ($C_{12}H_{12}OFe$)

*(g)
CH₃O—⟨ ⟩—SO₃H $\xrightarrow{HNO_3}$ $\xrightarrow{Br_2, Fe}$

16.55   *(a)   Draw all important resonance structures for the carbocation intermediate formed in the nitration of naphthalene at carbon-1.

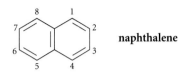

naphthalene

(b)   Draw all important resonance structures for the carbocation intermediate formed in the nitration of naphthalene at carbon-2.

*(c)   On the basis of your answers to parts (a) and (b), at which position should naphthalene nitrate more rapidly?

16.56   Would 1-methoxynaphthalene nitrate more rapidly or more slowly than naphthalene at *(a) carbon-4; (b) carbon-5; *(c) carbon-6?

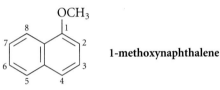

1-methoxynaphthalene

*16.57   Furan is an aromatic heterocyclic compound that undergoes electrophilic aromatic substitution. By drawing resonance structures for the carbocation intermediates involved, deduce whether furan should undergo Friedel-Crafts acylation more rapidly at carbon-2 or carbon-3.

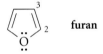

furan

*16.58   Given that anisole protonates primarily on oxygen in concentrated H_2SO_4, explain why 1,3,5-trimethoxybenzene protonates primarily on a carbon of the ring. As part of your reasoning draw the structure of each conjugate acid.

*16.59   A Diels-Alder reaction of 2,5-dimethylfuran and maleic anhydride gives a compound *A* that undergoes acid-catalyzed dehydration to give 3,6-dimethylphthalic anhydride.

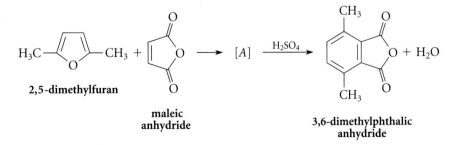

2,5-dimethylfuran   maleic anhydride   3,6-dimethylphthalic anhydride

(*Problem 16.59 continues*)

(a) Deduce the structure of *A*.

(b) Using the curved-arrow formalism, give a mechanism for the conversion of *A* into 3,6-dimethylphthalic anhydride.

*16.60   Using the curved-arrow formalism, propose a mechanism for the following reaction. (*Hint:* Modify Step 3 of the usual aromatic substitution mechanism in Sec. 16.4B).

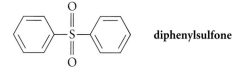

16.61   Diphenylsulfone is a by-product that is formed in the sulfonation of benzene. Give a mechanism for its formation.

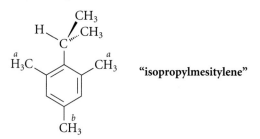

diphenylsulfone

*16.62   At 36° the NMR resonances for the ring methyl groups of "isopropylmesitylene" (protons *a* and *b* in the following structure) are two singlets at δ 2.25 and δ 2.13 with a 2 : 1 intensity ratio, respectively. When the spectrum is taken at −60°, however, it shows three singlets of equal intensity for these groups at δ 2.25, δ 2.17, and δ 2.11. Explain these results.

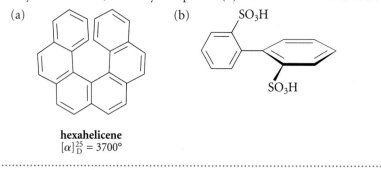

"isopropylmesitylene"

*16.63   Each of the following compounds can be resolved into enantiomers. Explain why each is chiral, and why compound (b) racemizes when it is heated.

(a)

(b)   SO₃H

hexahelicene
$[\alpha]_D^{25} = 3700°$

17

Allylic and Benzylic Reactivity

An **allylic group** is a group on a carbon adjacent to a double bond. A **benzylic group** is a group on a carbon adjacent to a benzene ring or substituted benzene ring.

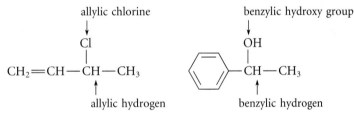

allylic chlorine

allylic hydrogen

benzylic hydroxy group

benzylic hydrogen

In many situations *allylic and benzylic groups are unusually reactive.* This chapter examines what happens when some familiar reactions occur at allylic and benzylic positions, and discusses the reasons for allylic and benzylic reactivity. Sec. 17.6 will show that allylic reactivity is also important in some chemistry that occurs in nature.

17.1 Reactions Involving Allylic and Benzylic Carbocations

Recall that allylic carbocations are resonance-stabilized (Sec. 15.4B).

$$\left[CH_2\!\!=\!\!CH\!\!-\!\!\overset{+}{C}H_2 \quad \longleftrightarrow \quad \overset{+}{C}H_2\!\!-\!\!CH\!\!=\!\!CH_2 \right] \qquad (17.1)$$

resonance structures of the allyl cation

These resonance structures symbolize the delocalization of electrons and positive charge that results from the overlap of *p* orbitals, as shown in Fig. 15.6.

Benzylic carbocations are also resonance-stabilized.

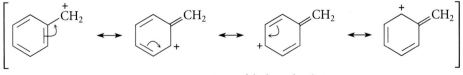

resonance structures of the benzyl cation

(17.2)

789

Table 17.1   Comparison of S_N1 Solvolysis Rates of Allylic and Nonallylic Alkyl Halides

$$R—Cl + C_2H_5OH + H_2O \xrightarrow{44.6°}$$

$$R—OC_2H_5 + R—OH + HCl \text{ (50\% aqueous ethanol)}$$

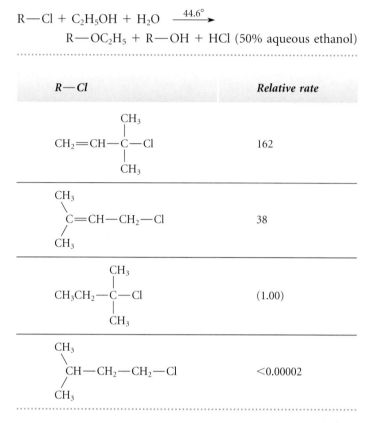

R—Cl	Relative rate
CH$_2$=CH—C(CH$_3$)(CH$_3$)—Cl	162
(CH$_3$)(CH$_3$)C=CH—CH$_2$—Cl	38
CH$_3$CH$_2$—C(CH$_3$)(CH$_3$)—Cl	(1.00)
(CH$_3$)(CH$_3$)CH—CH$_2$—CH$_2$—Cl	<0.00002

The charge on a benzylic carbocation is shared not only by the benzylic carbon, but also by alternate carbons of the ring.

The structures and stabilities of allylic and benzylic carbocations have important consequences for reactions in which they are involved as reactive intermediates. First, *reactions involving benzylic or allylic carbocations as intermediates are generally considerably faster than analogous reactions involving comparably substituted nonallylic or nonbenzylic carbocations.* This point is illustrated by the relative rates of S_N1 solvolysis reactions, shown in Tables 17.1 and 17.2. For example, the tertiary allylic alkyl halide in the first entry of Table 17.1 is more than one hundred times as reactive as the tertiary nonallylic alkyl halide in the third entry. A comparison of the first and third entries of Table 17.2 shows the effect of benzylic substitution. *Tert*-cumyl chloride, the third entry, is more than six hundred times as reactive as *tert*-butyl chloride, the first entry.

The greater reactivities of allylic and benzylic halides are due to the stabilities of the carbocation intermediates that are formed when they react. For example, *tert*-cumyl chloride ionizes to a carbocation with four important resonance structures:

Table 17.2   Comparison of S_N1 Solvolysis Rates of Benzylic and Nonbenzylic Alkyl Halides

$$R—Cl + H_2O \xrightarrow{25°} R—OH + HCl \text{ (90\% aqueous acetone)}$$

Compound	Common name	Relative rate
$(CH_3)_3C—Cl$	*tert*-butyl chloride	1.0

Ph—CH—Cl
    |
    CH_3
	α-phenethyl chloride	1.0

CH_3 | Ph—C—Cl | CH_3	*tert*-cumyl chloride	620
$Ph_2CH—Cl$	benzhydryl chloride	200[a]
$Ph_3C—Cl$	trityl chloride	>600,000

[a] In 80% aqueous ethanol.

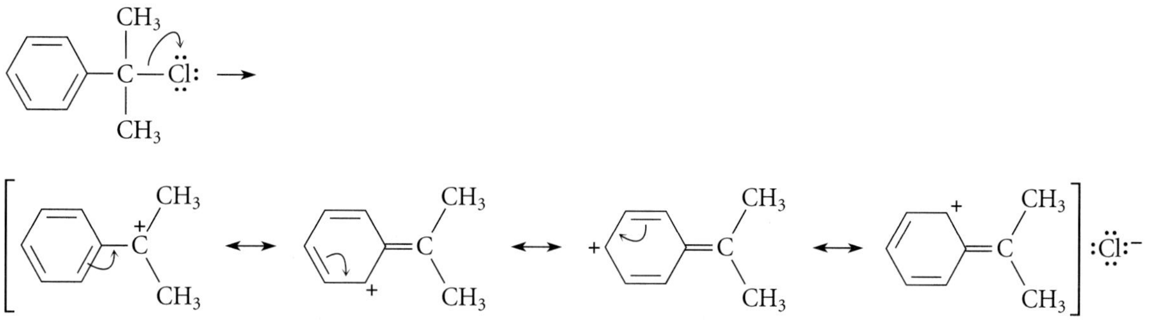

resonance-stabilized carbocation

(17.3)

Ionization of *tert*-butyl chloride, on the other hand, gives the *tert*-butyl cation, a carbocation with only one important contributing structure.

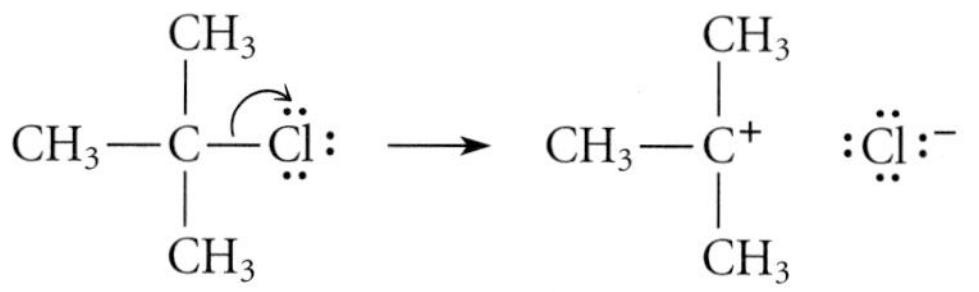

(17.4)

The benzylic cation is more stable relative to its alkyl halide starting material than is the *tert*-butyl cation, and Hammond's postulate says that the more stable carbocation should be formed more rapidly. A similar analysis explains the reactivity of allylic alkyl halides.

Because of the possibility of resonance, *ortho* and *para* substituent groups on the benzene ring that activate electrophilic aromatic substitution further accelerate S_N1 reactions at the benzylic position:

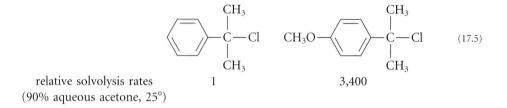

(17.5)

relative solvolysis rates
(90% aqueous acetone, 25°)

$\qquad\qquad$ 1 $\qquad\qquad\qquad$ 3,400

The carbocation derived from the ionization of the *p*-methoxy derivative above not only has the same type of resonance structures as the unsubstituted compound, shown in Eq. 17.3, but also an additional structure (color) in which *charge can be delocalized onto the substituent group itself*:

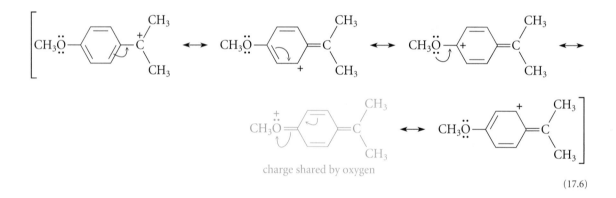

charge shared by oxygen

(17.6)

Other reactions that involve carbocation intermediates are accelerated when the carbocations are allylic or benzylic. Thus, dehydration of alcohols (Sec. 10.1) and reaction of alcohols with hydrogen halides (Sec. 10.2) are also faster when the alcohol is allylic or benzylic. For example, most alcohols require forcing conditions or Lewis acid catalysts to react with HCl to give alkyl chlorides, but such conditions are unnecessary when benzylic alcohols react with HCl. The addition of hydrogen halides to conjugated dienes also reflects the stability of allylic carbocations. Recall that protonation of a conjugated diene gives the allylic carbocation rather than its nonallylic isomer because the allylic carbocation is formed more rapidly (Sec. 15.4A).

A second consequence of the involvement of allylic carbocations as reactive intermediates is that in many cases *more than one product can be formed*. More than one product is possible because the positive charge (and electron deficiency) is shared between two carbons. Nucleophiles can attack either electron-deficient carbon atom and, if the two carbons are not equivalent, two different products result.

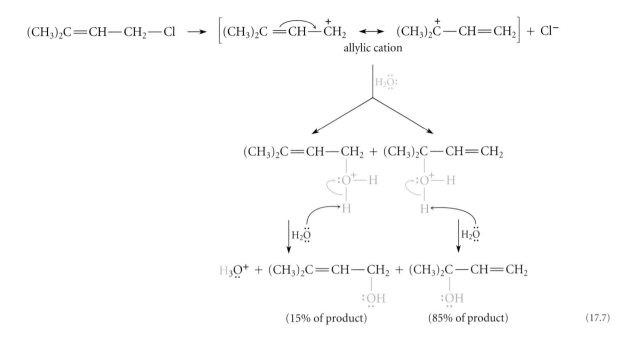

$$\text{(17.7)}$$

The two products are derived from *one* allylic cation that has two resonance forms. Recall that similar reasoning explains why a mixture of products (1,2- and 1,4-addition products) is obtained in the reactions of hydrogen halides with conjugated alkenes (Sec. 15.4A).

Several substitution products in the S_N1 reactions of benzylic alkyl halides might be expected for the same reason.

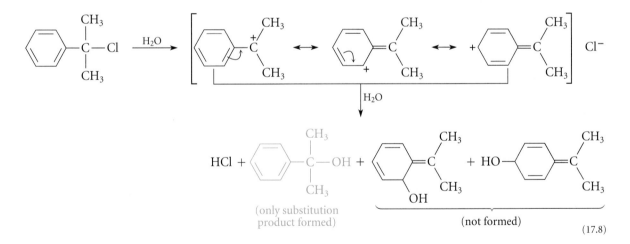

$$\text{(17.8)}$$

But as Eq. 17.8 shows, the products derived from attack of water on the ring are not formed. The reason is that these products are not aromatic, and thus lack the stability associated with the aromatic ring. Aromaticity is such an important stabilizing factor that only the aromatic product (color) is formed.

17.1 Predict the order of relative reactivities of the compounds within each series in S_N1 solvolysis reactions, and explain your answers carefully.

*(a)

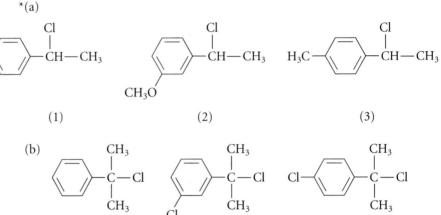

(1)   (2)   (3)

(b)

(1)   (2)   (3)

*17.2 Give the structure of an isomer of the allylic halide reactant in Eq. 17.7 that would react with water in an S_N1 solvolysis reaction to give the same two products. Explain your reasoning.

17.3 Give the structure of an allylic chloride C_5H_9Cl that gives only one S_N1 solvolysis product with water (not counting stereoisomers).

*17.4 Why is trityl chloride much more reactive than the other alkyl halides in Table 17.2?

17.2 Reactions Involving Allylic and Benzylic Radicals

An **allylic radical** has an unpaired electron at an allylic position. Allylic radicals are resonance-stabilized and are more stable than comparably substituted nonallylic radicals.

$$\left[\overset{\frown}{CH_2}=CH\overset{\curvearrowleft}{-}\dot{C}H_2 \quad \longleftrightarrow \quad \dot{C}H_2-CH=CH_2 \right] \tag{17.9}$$

resonance structures of the allyl radical

Similarly, a **benzylic radical**, which has an unpaired electron at a benzylic position, is also resonance-stabilized.

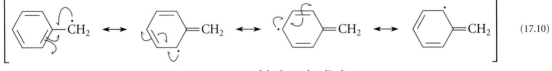

$$\tag{17.10}$$

resonance structures of the benzyl radical

These resonance structures symbolize the delocalization (sharing) of the unpaired electron that results from overlap of p orbitals.

The enhanced stabilities of allylic and benzylic radicals can be experimentally demonstrated with bond dissociation energies. Compare the bond dissociation energies of the two types of —CH_3 hydrogens in 2-pentene:

$$H^a—CH_2—CH{=}CH—CH_2—CH_2—H^b$$

<table>
<tr><td>360 kJ/mol
(86 kcal/mol)</td><td></td><td>418 kJ/mol
(100 kcal/mol)</td><td>(17.11)</td></tr>
</table>

$$H\cdot + \cdot CH_2—CH{=}CH—CH_2—CH_3 \qquad\qquad CH_3—CH{=}CH—CH_2—\dot{C}H_2 + H\cdot$$

allylic radical   alkyl radical

One set of methyl hydrogens is allylic and the other is not. It takes 58 kJ/mol (14 kcal/mol) less energy to remove the allylic hydrogen H^a than the nonallylic one H^b. As Fig. 17.1 shows, the difference in bond dissociation energies is a direct measure of the relative energies of the two radicals. The allylic radical is stabilized by 58 kJ/mol (14 kcal/mol) relative to the nonallylic radical. A similar comparison suggests about the same relative stability for a benzylic radical.

Because allylic and benzylic radicals are especially stable, they are more readily formed as reactive intermediates than ordinary alkyl radicals. Consider, for example, the bromination of cumene:

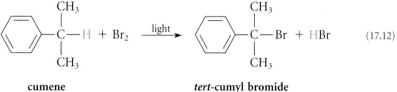

(17.12)

cumene   *tert*-**cumyl bromide**
(nearly quantitative)

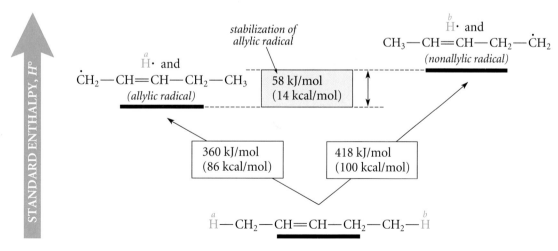

Figure 17.1   *Use of bond dissociation energies to determine the stabilization of an allylic radical. The stabilization of the allylic radical results from the lower energy required (58 kJ/mol, 14 kcal/mol) to remove an allylic hydrogen (a) compared to a nonallylic one (b).*

This is a free-radical chain reaction (Secs. 5.6C, 8.8A). Notice that only the benzylic hydrogen is substituted.

The initiation step in this reaction is dissociation of molecular bromine into bromine atoms; this reaction is promoted by heat or light.

$$:\ddot{Br}\!\!-\!\!\ddot{Br}: \longrightarrow 2 :\ddot{Br}\cdot \tag{17.13a}$$

In the first propagation step, a bromine atom abstracts the one benzylic hydrogen in preference to the six nonbenzylic hydrogens. It is in this propagation step that the selectivity for substitution of the benzylic hydrogen occurs.

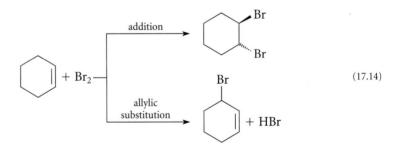 (17.13b)

The reason for this selectivity is the greater stability of the benzylic radical that is formed.

In the second propagation step, the benzylic radical reacts with another molecule of bromine to generate a molecule of product as well as another bromine atom, which can react again in Eq. 17.13b.

(17.13c)

Recall that free-radical halogenation is used to halogenate alkanes industrially (Sec. 8.8A). Because free-radical halogenation of alkanes with different types of hydrogens gives mixtures of products, this reaction is ordinarily not very useful in the laboratory. (It can be used industrially because industry has developed efficient fractional distillation methods that can separate liquids of very similar boiling points.) However, when a benzylic hydrogen is present, it undergoes substitution so much more rapidly than an ordinary hydrogen that a single product is obtained. Consequently, free-radical halogenation can be used for the laboratory preparation of benzylic halides.

Because the allylic radical is also relatively stable, a similar substitution occurs preferentially at the allylic positions of an alkene. But there is a competing reaction in the case of an alkene that is not observed with benzylic substitution: addition of halogen to the alkene double bond (Sec. 5.1A).

(17.14)

(Why doesn't bromine add to the benzene ring in Eq. 17.12?)

Can one reaction be promoted over the other? The answer is yes, if the reaction conditions are chosen carefully. *Addition* of bromine is the predominant reaction if (1) free-radical substitution is suppressed by avoiding conditions that promote free-radical reactions (heat, light, or free-radical initiators); and if (2) the reaction is carried out in solvents of even slight polarity that promote the ionic mechanism for bromine addition. Thus, addition is observed at 25° if the reaction is run in the dark in methylene chloride, CH_2Cl_2. On the other hand, free-radical *substitution* occurs when the reaction is promoted by heat, light, or free-radical initiators, an apolar solvent such as CCl_4 is used, and *the bromine is added slowly so that its concentration remains very low*. To summarize:

Addition:

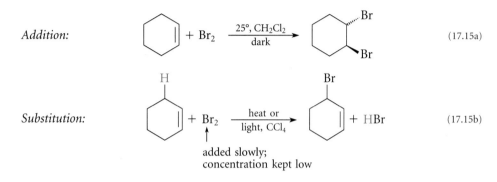

(17.15a)

Substitution:

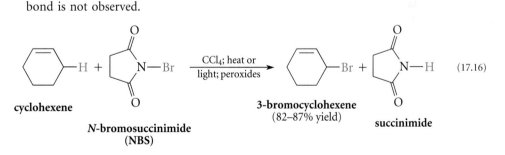

(17.15b)

STUDY GUIDE LINK:
17.1
*Addition vs.
Substitution with
Bromine*

The reason for the effect of bromine concentration has to do with the rate laws for the competing reactions. Addition has a higher kinetic order in $[Br_2]$ than substitution. Hence, the rate of addition is decreased more by lowering the bromine concentration than the rate of substitution.

Adding bromine to a reaction so slowly that it remains at very low concentration is experimentally inconvenient, but a very useful reagent can be used to accomplish the same objective: *N*-bromosuccinimide (abbreviated NBS). When a compound with allylic hydrogens is treated with *N*-bromosuccinimide in CCl_4 under free-radical conditions (heat, light, and/or peroxides), allylic bromination takes place, and addition to the double bond is not observed.

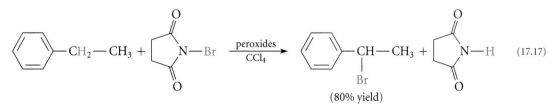

(17.16)

cyclohexene   *N*-bromosuccinimide
                              (NBS)

          3-bromocyclohexene   **succinimide**
              (82–87% yield)

N-Bromosuccinimide can also be used to effect benzylic bromination.

$$\text{Ph}-CH_2-CH_3 + \text{NBS} \xrightarrow[\text{CCl}_4]{\text{peroxides}} \text{Ph}-\underset{\underset{\text{Br}}{|}}{CH}-CH_3 + \text{succinimide}$$

(17.17)

(80% yield)

The initiation step in allylic and benzylic bromination with NBS is the formation of a bromine atom by homolytic cleavage of the N—Br bond in NBS itself. The ensuing substitution reaction has three propagation steps. First, the bromine atom abstracts an allylic hydrogen from the alkene molecule:

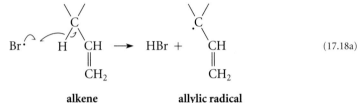

(17.18a)

alkene   **allylic radical**

The HBr thus formed reacts with the NBS in the second propagation step (by an ionic mechanism) to produce a Br_2 molecule.

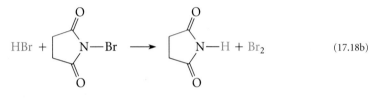

(17.18b)

NBS

The last propagation step is the reaction of this bromine molecule with the radical formed in Eq. 17.18a. A new bromine atom is produced that can begin the cycle anew.

(17.18c)

The first and last propagation steps are identical to those for free-radical substitution with Br_2 itself (Eq. 17.13b,c). The unique role of NBS is to maintain the very low concentration of bromine by reacting with HBr in Eq. 17.18b. The Br_2 concentration remains low because it can be generated no faster than an HBr molecule and an allylic radical are generated in Eq. 17.18a. Thus every time a bromine molecule is formed, an allylic radical is also formed with which the bromine can react.

The low solubility of NBS in CCl_4 (≤ 0.005 M) is crucial to the success of allylic bromination with NBS. When solvents that dissolve NBS are used, different reactions are observed. Hence CCl_4 *must* be used as the solvent in allylic or benzylic bromination with NBS. During the reaction, the insoluble NBS, which is more dense than CCl_4, disappears from the bottom of the flask and the less dense by-product succinimide (Eq. 17.16) forms a layer on the surface of the CCl_4. Equation 17.18b, and possibly other steps of the mechanism, occur at the surface of the insoluble NBS.

STUDY PROBLEM 17.1

What products are expected in the reaction of $CH_2{=}CHCH_2CH_2CH_2CH_3$ (1-hexene) with NBS in CCl_4? Explain your answer.

Solution   Work through the NBS mechanism with 1-hexene. In the step corresponding to Eq. 17.18a the following resonance-stabilized allylic free radical is formed as an intermediate:

$$\left[CH_2{=}CH{-}\overset{\cdot}{C}H{-}CH_2CH_2CH_3 \quad \longleftrightarrow \quad \overset{\cdot}{C}H_2{-}CH{=}CH{-}CH_2CH_2CH_3 \right]$$

$$A \qquad\qquad\qquad\qquad\qquad\qquad B$$

Because the unpaired electron is shared by *two different carbons*, this radical can react in the final propagation step to give *two different products*. Reaction of Br_2 at the radical site shown in structure *A* gives product (1), and reaction at the radical site shown in structure *B* gives product (2):

$$CH_2{=}CH{-}\underset{\underset{Br}{|}}{CH}{-}CH_2CH_2CH_3 \qquad CH_2{-}CH{=}CH{-}CH_2CH_2CH_3$$

$$\overset{}{\underset{Br}{|}}$$

$$(1) \qquad\qquad\qquad\qquad\qquad (2)$$

Product (1) is chiral, and product (2) can exist as both *cis* and *trans* stereoisomers. Hence, bromination of 1-hexene gives racemic (1) as well as *cis-* and *trans-*(2).

PROBLEM

17.5   What product(s) are expected when each of the following compounds reacts with one equivalent of NBS in CCl_4 in the presence of light and/or peroxides? Explain your answers.

*(a)   *trans*-2-pentene   (b)   1-methylcyclopentene

*(c)   3,3-dimethylcyclohexene

(d)   4-*tert*-butyltoluene

17.3   Reactions Involving Allylic and Benzylic Anions

Allylic and benzylic anions are about 59 kJ/mol (14 kcal/mol) more stable than their nonallylic and nonbenzylic counterparts.

$$\left[\overset{\frown}{CH_2}{=}CH{-}\overset{\cdot\cdot}{C}H_2 \quad \longleftrightarrow \quad \overset{\cdot\cdot}{C}H_2{-}CH{=}CH_2 \right] \qquad (17.19)$$

allyl anion

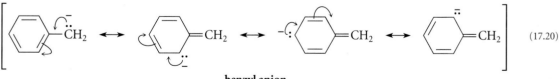

benzyl anion

There are two reasons for the stabilities of these anions. The first is resonance stabilization, as indicated by the resonance structures above. The second reason is the *polar effect* (Sec. 3.6B) of the double bond (in the allyl anion) or the phenyl ring (in the benzyl anion). The polar effect of both groups stabilizes anions. (Opinions differ about the relative importance of resonance and polar effects.)

STUDY GUIDE LINK:
17.2
Polar Effect of
Double Bonds

The enhanced stability of allylic and benzylic anions is reflected in the pK_a values of propene and toluene ($B:^- =$ a base):

$$CH_2{=}CH{-}CH_2{-}H + B:^- \;\rightleftharpoons\; CH_2{=}CH{-}\overset{..}{\underset{..}{C}}H_2 + B{-}H \qquad (17.21)$$

propene
$pK_a \approx 43$

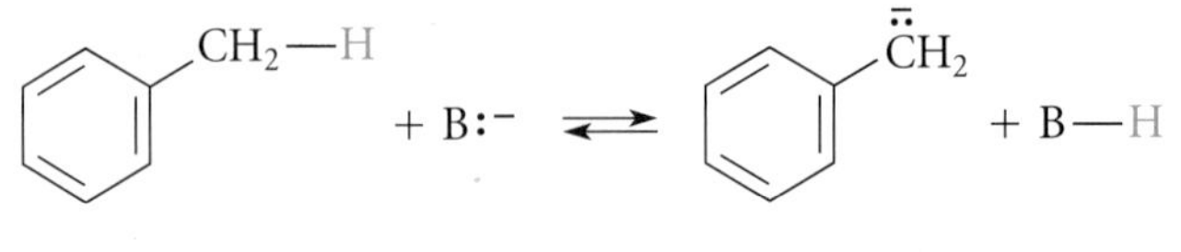

$$+ B:^- \;\rightleftharpoons\; + B{-}H \qquad (17.22)$$

toluene
$pK_a \approx 41$

Although these compounds are very weak acids, their acidities are much greater than the acidities of alkanes that do not contain allylic or benzylic hydrogens. The latter compounds have pK_a values in the range of 55–60.

Relatively few reactions involve free carbanions as reactive intermediates. However, a number of reactions involve species that have *carbanion character*. Two of these are the reactions of Grignard and related organometallic reagents, and E2 eliminations. The following sections show how these reactions are affected when carbanion character occurs at benzylic or allylic positions.

A. Allylic Grignard Reagents

Recall that Grignard reagents have many of the properties expected of *carbanions* (Sec. 8.7B). Thus, allylic Grignard reagents resemble allylic carbanions.

$$CH_2{=}CH{-}CH_2{-}MgBr \quad \text{resembles} \quad CH_2{=}CH{-}\overset{..}{\underset{..}{C}}H_2 \;\; \overset{+}{MgBr} \qquad (17.23)$$

Allylic Grignard reagents undergo a rapid equilibration in which the —MgBr group moves back and forth between the two partially negative carbons at a rate of about 1000 times per second.

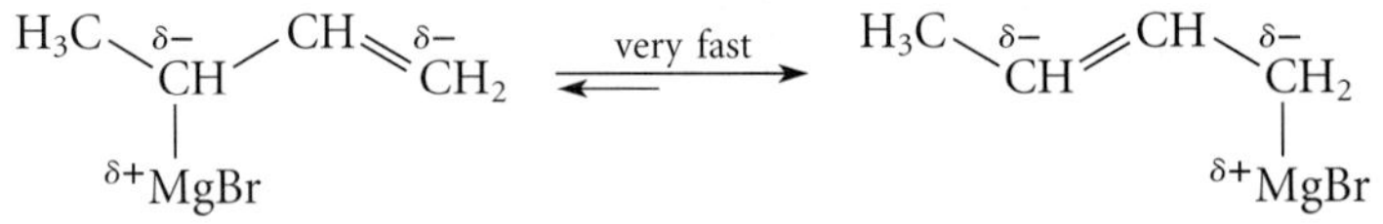

$$\qquad \xrightarrow{\text{very fast}} \qquad (17.24)$$

The transition state for this reaction can be envisioned as an ion pair consisting of an allylic carbanion and a ^+MgBr cation.

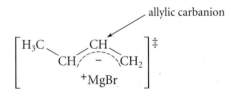

allylic carbanion

Because the allylic carbanion is resonance-stabilized, this transition state has relatively low energy, and consequently the equilibration occurs rapidly.

The equilibration in Eq. 17.24 is an example of an **allylic rearrangement**. An allylic rearrangement involves the simultaneous movement of a group G and a double bond so that one allylic isomer is converted into another.

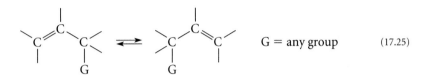

$$G = any\ group \qquad (17.25)$$

Notice that these two structures are *not* resonance structures; they are two *distinct* species in rapid equilibrium.

The rapid allylic rearrangement of an unsymmetrical Grignard reagent, such as the one shown in Eq. 17.24, means that the reagent is actually a mixture of two different reagents. This has two consequences. First, the same mixture of reagents can be obtained from either of two allylically related alkyl halides:

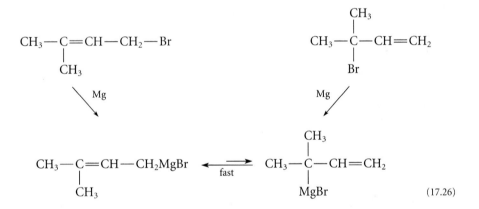

Second, the same mixture of products is obtained when the Grignard reagent is prepared from *either* allylic halide. The following example illustrates this point for a protonolysis reaction:

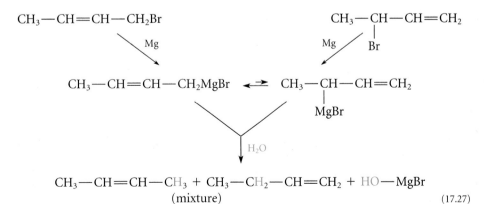

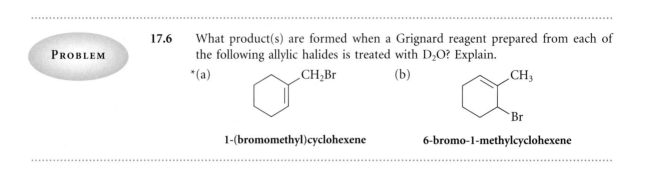

PROBLEM

17.6   What product(s) are formed when a Grignard reagent prepared from each of the following allylic halides is treated with D_2O? Explain.

*(a)

1-(bromomethyl)cyclohexene

(b)

6-bromo-1-methylcyclohexene

B. E2 Eliminations Involving Allylic or Benzylic Hydrogens

Recall that the S_N2 (bimolecular substitution) and E2 (bimolecular elimination) reactions of alkyl halides are *competing reactions,* and that the structure of the alkyl halide is one of the major factors that determine which reaction is the dominant one (Sec. 9.5F). A structural effect in the alkyl halide that tends to promote a greater fraction of elimination is the *acidity of the β-hydrogens.* It is found that a greater ratio of elimination to substitution is observed when the β-hydrogens of the alkyl halide have higher than normal acidity. One example of this situation occurs when the β-hydrogens are allylic or benzylic. (Recall from the introduction to this section that allylic and benzylic hydrogens are more acidic than ordinary alkyl hydrogens.) For example, the E2 reaction of the alkyl bromide in Eq. 17.28 is more than 100 times as fast as the E2 reaction of isopentyl bromide $[(CH_3)_2CHCH_2CH_2Br]$, a comparably branched alkyl halide.

benzylic hydrogens

$$\text{—CH}_2\text{CH}_2\text{—Br} \xrightarrow[\text{C}_2\text{H}_5\text{OH}]{\text{Na}^+ \text{C}_2\text{H}_5\text{O}^-} \text{—CH}=\text{CH}_2 + \text{—CH}_2\text{CH}_2\text{—OC}_2\text{H}_5 \qquad (17.28)$$

(95% elimination)   (5% substitution)

(Elimination predominates because the E2 reaction is particularly fast; the S_N2 component of the competition occurs at a normal rate.)

Why should an acidic β-hydrogen increase the rate of an E2 reaction? In the transition state of the E2 reaction, the base is removing a β-proton, and the transition state of the reaction has *carbanion character* at the β-carbon atom.

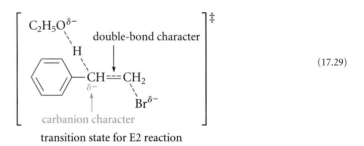

transition state for E2 reaction

(17.29)

This partially formed carbanion is stabilized in the same way that a fully formed carbanion is; a more stable transition state means a faster reaction. Another reason that benzylic E2 reactions are faster is that the alkene double bond which is partially formed in the transition state is conjugated with the benzene ring; recall that conjugated double bonds are particularly stable (Sec. 15.1A).

Let's summarize the structural characteristics of alkyl halides or sulfonate esters that favor E2 reactions over S$_N$2 reactions. Elimination reactions are favored by:

1. Branching at the α-carbon (Sec. 9.5F)

2. Branching at the β-carbon (Sec. 9.5F)

3. Greater acidity of the β-hydrogens (this section).

PROBLEM

17.7 Predict the major product that is obtained when each of the following alkyl halides is treated with potassium *tert*-butoxide. Explain your reasoning.

*(a) [structure: cyclohexene ring with —Br]

(b) [structure: benzene ring —CH(OCH$_3$)—CH$_2$I]

17.4 Allylic and Benzylic S$_N$2 Reactions

S$_N$2 reactions of allylic and benzylic halides are relatively fast even though they do not involve reactive intermediates. The following data for allyl chloride are typical:

relative rate

$$CH_2{=}CH{-}CH_2{-}Cl + I^- \xrightarrow[50°]{\text{acetone}} CH_2{=}CH{-}CH_2{-}I + Cl^- \qquad 73 \qquad (17.30a)$$

$$CH_3{-}CH_2{-}CH_2{-}Cl + I^- \xrightarrow[50°]{\text{acetone}} CH_3{-}CH_2{-}CH_2{-}I + Cl^- \qquad 1 \qquad (17.30b)$$

An even greater acceleration is observed for benzylic halides.

relative rate

[structure: benzene ring —CH$_2$—Cl] $+ I^- \xrightarrow[60°]{\text{acetone}}$ [structure: benzene ring —CH$_2$—I] $+ Cl^-$ $\approx 100,000$ (17.31a)

[structure: (CH$_3$)$_2$CH—CH$_2$—Cl] $+ I^- \xrightarrow[60°]{\text{acetone}}$ [structure: (CH$_3$)$_2$CH—CH$_2$—I] $+ Cl^-$ 1 (17.31b)

Allylic and benzylic S$_N$2 reactions are accelerated because the energies of their transition states are reduced by p-orbital overlap, shown in Fig. 17.2 for an allylic S$_N$2 reaction. In the transition state of the S$_N$2 reaction, the carbon at which substitution occurs is sp^2-hybridized (Fig. 9.2); the incoming nucleophile and the departing leaving group are partially bonded to a p orbital on this carbon. Overlap of this p orbital with the p orbitals of an adjacent double bond or phenyl ring provides additional bonding that lowers the energy of the transition state and accelerates the reaction.

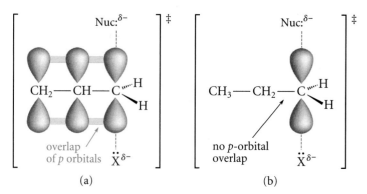

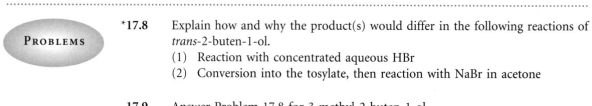

Figure 17.2   *Transition states for S_N2 reactions at (a) allylic and (b) non-allylic carbons. Nuc:⁻ and X:⁻ are the nucleophile and leaving group, respectively. The allylic substitution is faster because the transition state is stabilized by overlap of the p orbital at the site of substitution with the adjacent π bond.*

PROBLEMS

*17.8   Explain how and why the product(s) would differ in the following reactions of *trans*-2-buten-1-ol.
(1) Reaction with concentrated aqueous HBr
(2) Conversion into the tosylate, then reaction with NaBr in acetone

17.9   Answer Problem 17.8 for 3-methyl-2-buten-1-ol.

17.5   Benzylic Oxidation of Alkylbenzenes

Treatment of alkylbenzene derivatives with strong oxidizing agents under vigorous conditions converts the alkyl side chain into a carboxylic acid group. Oxidants commonly used for this purpose are Cr(VI) derivatives, such as $Na_2Cr_2O_7$ (sodium dichromate) or CrO_3; the Mn(VII) reagent $KMnO_4$ (potassium permanganate); or O_2 and special catalysts, a procedure that is used industrially (Eq. 16.46).

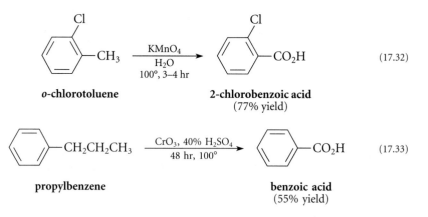

(17.32)

o-chlorotoluene

2-chlorobenzoic acid
(77% yield)

(17.33)

propylbenzene

benzoic acid
(55% yield)

Notice that the benzene ring is left intact, and notice from Eq. 17.33 that the alkyl side-chain, *regardless of length*, is converted into a carboxylic acid group. This reaction is useful for the preparation of some carboxylic acids from alkylbenzenes.

Oxidation of alkyl side chains requires the presence of a benzylic hydrogen. Consequently, *tert*-butylbenzene, which has no benzylic hydrogen, is resistant to benzylic oxidation.

The conditions for this side-chain oxidation are generally vigorous: heat, high concentrations of oxidant, and/or long reaction times. It is also possible to effect less extensive oxidations of side-chain groups. Thus 1-phenylethanol is readily oxidized to acetophenone under milder conditions—the normal oxidation of secondary alcohols to ketones (Sec. 10.6A)—but it is converted into benzoic acid under more vigorous conditions.

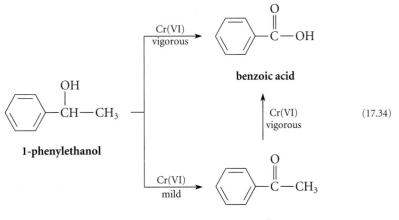

(17.34)

You do not need to be concerned with learning the exact conditions for these reactions; rather, it is important simply to be aware that it is usually possible to find appropriate conditions for each type of oxidation. (See Study Guide Link 16.4.)

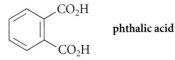

PROBLEMS

17.10   Give the products of vigorous $KMnO_4$ oxidation of each of the following compounds.
  *(a) 1-butyl-4-*tert*-butylbenzene   (b) *p*-nitrobenzyl alcohol

17.11   *(a) A compound *A* has the formula C_8H_{10}. After vigorous oxidation, it yields phthalic acid. What is the structure of *A*?

phthalic acid

  (b) A compound *B* has the formula C_8H_{10}. After vigorous oxidation, it yields benzoic acid (structure in Eq. 17.34). What is the structure of *B*?

STUDY GUIDE LINK:
✓17.3
Synthetic Equivalence

17.6   Terpenes

A. The Isoprene Rule

People have long been fascinated with the pleasant-smelling substances found in nature—for example, the perfume of a rose—and have been curious to learn more about these materials, which have come to be called **essential oils.**

Study Guide Link:
17.4
Essential Oils

Essential oils, particularly oil of turpentine, were known to the ancient Egyptians. However, not until early in the nineteenth century was an effort made to determine the chemical constitution of the essential oils. In 1818, it was found that the C:H ratio in oil of turpentine was 5:8. This same ratio was subsequently found for a wide variety of natural products. These related natural products became known collectively as **terpenes,** a name coined by August Kekulé. The similarity in the atomic compositions of the many terpenes led to the idea that they might possess some unifying structural element.

In 1887, Otto Wallach, a German chemist, pointed out the common structural feature of the terpenes: they all consist of repeating units that have the same carbon skeleton as the five-carbon diene isoprene. This generalization subsequently became known as the **isoprene rule.**

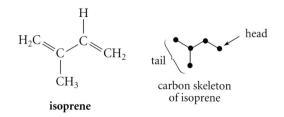

isoprene

For example, citronellol (from oil of rose and other sources) incorporates two isoprene units:

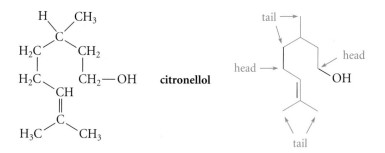

Because of this relationship to isoprene, terpenes are also called **isoprenoids.** Notice carefully that the basis of the terpene or isoprenoid classification is only *the connectivity of the carbon skeleton.* The presence or the positions of double bonds and other functional groups, or the configurations of double bonds and asymmetric carbons, have nothing to do with the terpene classification.

Study Guide Link:
✓17.5
Skeletal Structures

In many terpenes, the isoprene units are connected in a "head-to-tail" arrangement.

Because this arrangement is so common, Wallach assumed the generality of the head-to-tail connectivity in his original statement of the isoprene rule. However, many examples are now known in which the isoprene units have a "head-to-head" connectivity. Furthermore, some compounds are derived from the conventional terpene structures by skeletal rearrangements. Although these compounds do not have the exact terpene connectivity, they are nevertheless classified as terpenes. For our purposes, though, it will be sufficient to recognize terpenes by two criteria.

1. A multiple of five carbon atoms in the main carbon skeleton
2. The carbon connectivity of the isoprene carbon skeleton within each five-carbon unit

Because terpenes are assembled from five-carbon units, their carbon skeletons contain multiples of five carbon atoms ($10, 15, 20, \ldots, 5n$). Terpenes with ten carbon atoms in their carbon chains are classified as **monoterpenes**, those with fifteen carbons **sesquiterpenes**, those with twenty carbons **diterpenes**, and so on. Some examples of terpenes are given in Fig. 17.3 on p. 809. Many of these compounds are familiar natural flavorings or fragrances.

STUDY PROBLEM 17.2

Determine whether the following compound, isolated from the frontal gland secretion of a termite soldier, is a terpene.

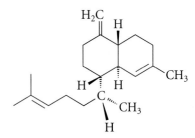

Solution   Because stereochemistry is not an issue, delete all stereochemical details for simplicity. First, count the number of carbons. Because the compound has a multiple of five carbon atoms, it could be a terpene. To check for terpene connectivity, look first for a methyl branch. One is at the end of the long side chain. Identify within this group a chain of four carbons with a methyl branch at the second carbon:

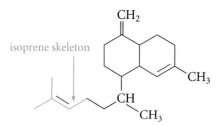

Starting at the next carbon, look for the same pattern. Remember that a bond must connect each isoprene skeleton.

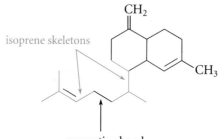

(We arbitrarily chose to proceed clockwise around the ring; you should convince yourself that in this case a counterclockwise path also works.) Continue in this fashion until either the pattern is broken or, as in this case, all carbons are included:

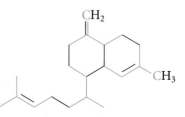

This compound incorporates four isoprene skeletons and is therefore a diterpene.

PROBLEM

17.12   Show the isoprene skeletons within the following compounds of Fig. 17.3.
*(a)   caryophyllene   (b)   vitamin A

B. Biosynthesis of Terpenes

How are terpenes synthesized in nature? What is responsible for the regular repetition of isoprene skeletons? To answer this question, chemists have studied the biosynthesis of terpenes. **Biosynthesis** is the synthesis of chemical compounds by living organisms. The study of biosynthesis is an active area of research that lies at the interface of chemistry and biochemistry. Terpene biosynthesis shows how nature takes advantage of allylic reactivity.

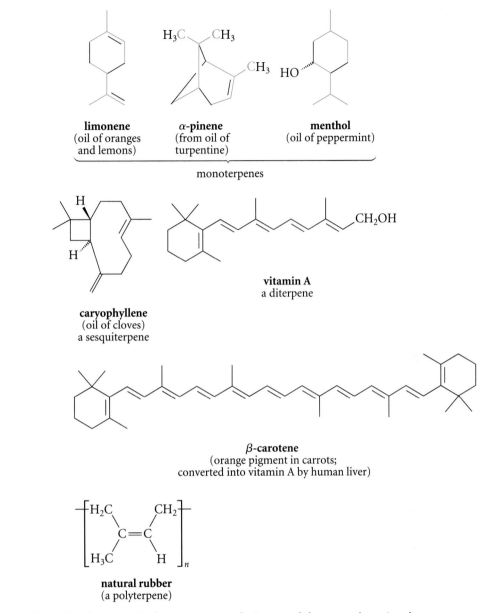

limonene
(oil of oranges
and lemons)

α-pinene
(from oil of
turpentine)

menthol
(oil of peppermint)

monoterpenes

caryophyllene
(oil of cloves)
a sesquiterpene

vitamin A
a diterpene

β-carotene
(orange pigment in carrots;
converted into vitamin A by human liver)

natural rubber
(a polyterpene)

Figure 17.3   *Examples of terpenes. In the monoterpenes, the isoprene skeletons are shown in color.*

The repetitive isoprene skeleton in all terpenes has a common origin in two simple five-carbon compounds:

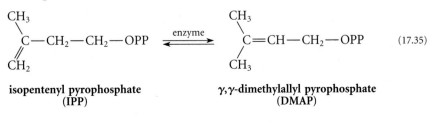

(17.35)

isopentenyl pyrophosphate
(IPP)

γ,γ-dimethylallyl pyrophosphate
(DMAP)

The —OPP in these structures is an abbreviation for the *pyrophosphate group* (color in the structure below), which, in nature, is usually complexed to a metal ion such as Mg^{2+} or Mn^{2+}.

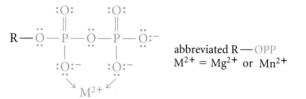

abbreviated R—OPP
$M^{2+} = Mg^{2+}$ or Mn^{2+}

a metal-complexed alkyl pyrophosphate

Alkyl pyrophosphates are esters of the inorganic acid *pyrophosphoric acid* (Sec. 10.3C).

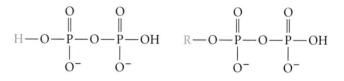

pyrophosphoric acid   **an** alkyl **pyrophosphate**

Pyrophosphate and phosphate are "nature's leaving groups." Just as alkyl halides or alkyl tosylates are used in the laboratory as starting materials for nucleophilic substitution reactions, alkyl pyrophosphates are used by living organisms.

Because IPP and DMAP are readily interconverted in living systems by the reaction of Eq. 17.35, the presence of one ensures the presence of the other. Like all biochemical reactions, including the ones discussed below, this reaction does not occur freely in solution, but is catalyzed by an enzyme. The involvement of enzyme catalysts does not alter the fact that the chemical reactions of living systems are reasonable and understandable in terms of familiar laboratory reactions.

The biosynthesis of the simple monoterpene *geraniol* illustrates the general pattern of terpene biosynthesis. (Geraniol is the fragrant compound in oil of geraniums.) In the first step of geraniol biosynthesis, IPP and DMAP (Eq. 17.35) are bound to the enzyme *prenyl transferase*. The DMAP loses its pyrophosphate leaving group in an S_N1-like process.

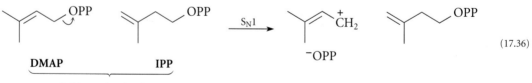

(17.36)

The carbocation formed in Eq. 17.36 is a relatively stable allylic cation (Sec 17.1). Carbocations, like other electrophiles, can be attacked by the π electrons of a double bond. (This same type of reaction is involved in Friedel-Crafts acylations and alkylations; see Secs. 16.4E and F.) The reaction of this carbocation with the double bond of IPP gives a new carbocation. Loss of a proton from a β-carbon of this carbocation gives the monoterpene geranyl pyrophosphate. (B: and BH$^+$ are basic and acidic groups, respectively, of the enzyme catalyst.)

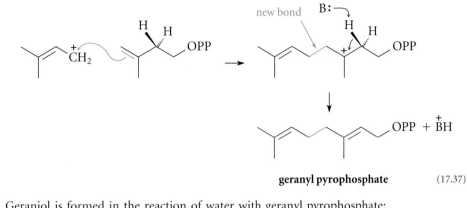

geranyl pyrophosphate   (17.37)

Geraniol is formed in the reaction of water with geranyl pyrophosphate:

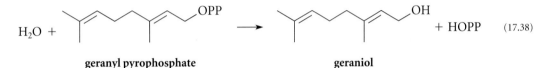

(17.38)

geranyl pyrophosphate   **geraniol**

All head-to-tail terpenes are formed by reactions analogous to the ones shown above. As these examples illustrate, the biosynthesis of terpenes can be understood in terms of carbocation intermediates that are like those involved in laboratory chemistry. As we have noted previously (Secs. 10.7, 11.5B), the organic chemistry of living systems is understandable in terms of laboratory analogies.

The biosynthesis of terpenes also illustrates the economy of nature: a remarkable array of substances is generated from a common starting material. This economy is evident also in other families of natural products. For example, terpenes also serve as the starting point for the biosynthesis of *steroids* (Sec. 7.6D). The isoprene rule is one of the unifying elements that underlie the chemical diversity of nature.

PROBLEMS

*17.13   (a) Give a biosynthetic mechanism for formation of the cyclic terpene limonene (Fig. 17.3) beginning with an intramolecular reaction of the following carbocation. (Assume acids and bases are present as necessary.)

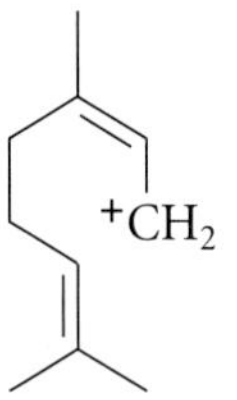

$^+CH_2$

(b) Give a curved-arrow mechanism for the biosynthesis of the carbocation intermediate in (a) from geranyl pyrophosphate.

17.14   Propose a biosynthetic pathway for each of the following natural products. Assume acids and bases are present as necessary.

*(a)

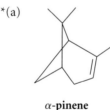

α-pinene

(*Hint:* Start with the carbocation intermediate in Problem 17.13a.)

(b)

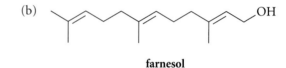

farnesol

(*Hint:* Start with geranyl pyrophosphate, Eq. 17.38.)

KEY IDEAS IN CHAPTER 17

- Functional groups at allylic and benzylic positions are in many cases unusually reactive.

- Addition of hydrogen halides to dienes, and solvolysis of allylic and benzylic alkyl halides, are reactions that involve allylic or benzylic carbocations. The acceleration of reactions that involve these carbocations can be attributed to their resonance stabilization.

- A mixture of isomeric products is typically obtained from a reaction involving an unsymmetrical allylic carbocation as a reactive intermediate because charge is shared by more than one carbon in the ion, and a nucleophile can attack each charged carbon. A reaction involving a benzylic carbocation, in contrast, typically gives only the product derived from attack of a nucleophile at the benzylic position because only in this product is the aromaticity of the ring not disrupted.

- Free-radical halogenation is selective for allylic and benzylic hydrogens because of the stability of the allylic or benzylic free-radical intermediates that are involved.

- *N*-Bromosuccinimide (NBS) in CCl$_4$ solution is used to carry out allylic and benzylic brominations. Benzylic bromination can also be carried out with bromine and light.

- Because the unpaired electron in allylic radicals is shared on different carbons, some reactions that involve allylic radicals give more than one product.

⚗ Allylic and benzylic anions are stabilized by resonance and by the polar effect of double bonds.

⚗ Allylic Grignard and organolithium reagents undergo rapid allylic rearrangements.

⚗ A β-elimination reaction involving allylic or benzylic β-hydrogens is accelerated because the anionic character in the transition state is stabilized in the same manner as an allylic or benzylic anion, and because the developing double bond in the transition state is conjugated.

⚗ S_N2 reactions at allylic and benzylic positions are accelerated because their transition states are stabilized by overlap of *p* orbitals.

⚗ In aromatic compounds, alkyl side chains that contain benzylic hydrogens can be oxidized to carboxylic acid groups under vigorous conditions.

⚗ Terpenes, or isoprenoids, are natural products with carbon skeletons characterized by repetition of the five-carbon isoprene pattern.

⚗ Terpenes are synthesized in nature by enzyme-catalyzed processes involving the reaction of allylic carbocations with double bonds. The biosynthetic precursor of all terpenes is isopentenyl pyrophosphate (IPP).

ADDITIONAL PROBLEMS

17.15 Identify the allylic carbons in each of the following structures.

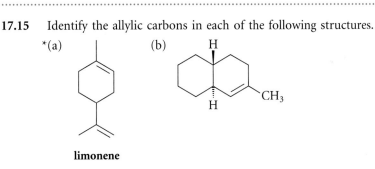

*(a)

(b)

limonene

17.16 Identify the benzylic carbons in each of the following structures.

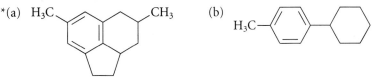

*(a) H₃C ... CH₃

(b) H₃C—

*17.17 Give the principal organic product(s) expected when 4-methylcyclohexene or other compound indicated reacts under the following conditions. Assume one equivalent of each reagent reacts in each case.

(*Problem 17.17 continues*)

(a) Br$_2$, CH$_2$Cl$_2$, dark

(b) *N*-bromosuccinimide in CCl$_4$, light

(c) product(s) of (b), solvolysis in aqueous acetone

(d) product(s) of (b) + Mg in ether   (e) product(s) of (d) + D$_2$O

17.18   Give the principal product(s) expected when *trans*-2-butene or other compound indicated reacts under the conditions in Problem 17.17.

*17.19   Which of the following compounds, all known in nature, can be classified as terpenes? Show the isoprene skeletons in each terpene.

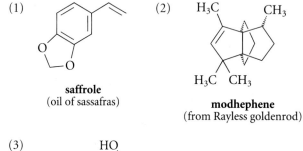

(1)

(2)

saffrole
(oil of sassafras)

modhephene
(from Rayless goldenrod)

(3)

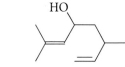

ipsdienol
(one component of the pheromone
of the Norwegian spruce beetle)

(4)

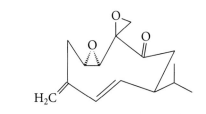

periplanone B
(pheromone of the female American cockroach)

(5)

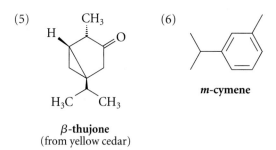

(6)

m-cymene

β-thujone
(from yellow cedar)

17.20   Determine whether the following compound (zoapatanol, used as a fertility-regulating agent in Mexican folk medicine) is a terpene.

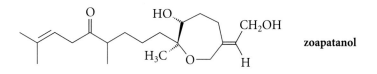

zoapatanol

*17.21 Explain why two products are formed in the first ether synthesis, but only one in the second.

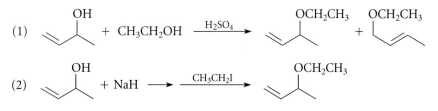

17.22 A student Al Lillich has prepared a pure sample of 3-bromo-1-butene (*A*). Several weeks later he finds that the sample is contaminated with an isomer *B* formed by allylic rearrangement.
(a) Give the structure of *B*.
(b) Give a curved-arrow mechanism for the formation of *B* from *A*.
(c) Which should be the major isomer at equilibrium, *A* or *B*? Explain.

17.23 Outline a synthesis of each of the following compounds from the indicated starting materials and any other reagents.
*(a) 3-phenyl-1-propanol from toluene
(b) benzyl methyl ether from toluene
*(c) (*Z*)-1,4-nonadiene from 1-hexyne

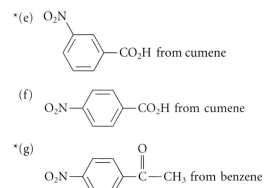

*17.24 Rank the following compounds in order of increasing reactivity (least reactive first) in an S_N1 solvolysis reaction in aqueous acetone. Explain your answers. (The structure of *tert*-cumyl chloride is shown in Table 17.2.)
(1) *m*-nitro-*tert*-cumyl chloride   (2) *p*-methoxy-*tert*-cumyl chloride
(3) *p*-fluoro-*tert*-cumyl chloride   (4) *p*-nitro-*tert*-cumyl chloride

17.25   Arrange the following alcohols according to increasing rates of their acid-catalyzed dehydration to alkene (smallest rate first), and explain your reasoning.

(1)

(2)

(3)

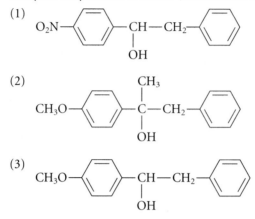

*17.26**   Explain why compound A reacts faster than compound B when each undergoes solvolysis in aqueous acetone.

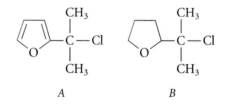

A   B

*17.27**   A hydrocarbon A, C_9H_{10}, is treated with N-bromosuccinimide to give a single monobromo compound B. When B is dissolved in aqueous acetone it reacts to give two compounds, C and D. Catalytic hydrogenation of D gives back A, and C can be separated into enantiomers. When optically active C is oxidized with CrO_3 and pyridine, an optically inactive ketone E is obtained. Vigorous oxidation of A with $KMnO_4$ affords phthalic acid (structure in Problem 17.11). Propose structures for compounds A through E, and explain your reasoning.

17.28   A hydrocarbon A, C_9H_{12}, is treated with N-bromosuccinimide in CCl_4 in the presence of peroxides to give a compound B, $C_9H_{11}Br$. Compound B undergoes rapid solvolysis in aqueous acetone to give an alcohol C, $C_9H_{12}O$, which cannot be oxidized with CrO_3 in pyridine. Vigorous oxidation of compound A with hot chromic acid gives benzoic acid, $Ph—CO_2H$. Identify compounds A–C.

*17.29**   Predict which of the following compounds should undergo the more rapid reaction with K^+ $(CH_3)_3C—O^-$, and give the product of the reaction.

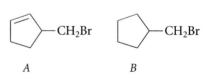

A   B

17.30 Complete the following reactions by proposing structures for the major organic products.

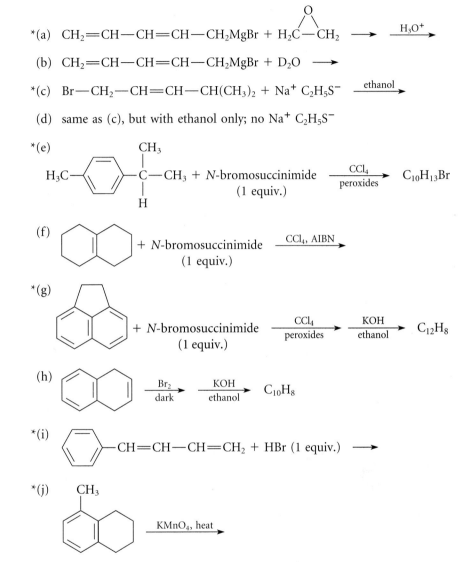

*(a) $CH_2{=}CH{-}CH{=}CH{-}CH_2MgBr$ + $H_2C{-}CH_2$ (with epoxide O) $\longrightarrow$ $\xrightarrow{H_3O^+}$

(b) $CH_2{=}CH{-}CH{=}CH{-}CH_2MgBr$ + D_2O $\longrightarrow$

*(c) $Br{-}CH_2{-}CH{=}CH{-}CH(CH_3)_2$ + $Na^+\ C_2H_5S^-$ $\xrightarrow{ethanol}$

(d) same as (c), but with ethanol only; no $Na^+\ C_2H_5S^-$

*(e)

$H_3C{-}\langle\text{ring}\rangle{-}\overset{\overset{\displaystyle CH_3}{|}}{\underset{\underset{\displaystyle H}{|}}{C}}{-}CH_3$ + *N*-bromosuccinimide (1 equiv.) $\xrightarrow[\text{peroxides}]{CCl_4}$ $C_{10}H_{13}Br$

(f) + *N*-bromosuccinimide (1 equiv.) $\xrightarrow{CCl_4,\ AIBN}$

*(g) + *N*-bromosuccinimide (1 equiv.) $\xrightarrow[\text{peroxides}]{CCl_4}$ $\xrightarrow[\text{ethanol}]{KOH}$ $C_{12}H_8$

(h) $\xrightarrow[\text{dark}]{Br_2}$ $\xrightarrow[\text{ethanol}]{KOH}$ $C_{10}H_8$

*(i) $\langle\text{ring}\rangle{-}CH{=}CH{-}CH{=}CH_2$ + HBr (1 equiv.) $\longrightarrow$

*(j) (1-methyltetralin structure) $\xrightarrow{KMnO_4,\ heat}$

*17.31 When benzyl alcohol (λ_{max} = 258 nm, ϵ = 520) is dissolved in H_2SO_4, a colored solution is obtained that has a different UV spectrum: λ_{max} = 442 nm, ϵ = 53,000. When this solution is added to cold NaOH, the original spectrum of benzyl alcohol is restored. Account for these observations.

17.32 Starting with isopentenyl pyrophosphate, propose a mechanism for the biosynthesis of *(a) zingiberene, compound *A*, obtained from oil of ginger; (b) eudesmol, compound *B*, obtained from eucalyptus.

(*Problem 17.32 continues*)

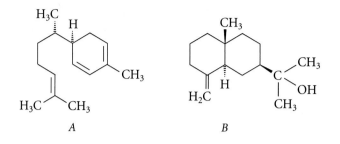

A   B

17.33   Propose a curved-arrow mechanism for each of the following reactions.

*(a)   1-penten-4-yne + Na⁺ ⁻:C≡CH; then allyl bromide ⟶

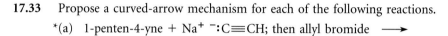

$$CH_2=CH-CH-C≡CH$$
$$|$$
$$CH_2-CH=CH_2$$

(b) $CH_3(CH_2)_3-C≡C-CH_2-Br + Mg \xrightarrow{ether} \xrightarrow{H_2O}$

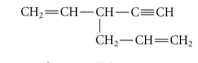

$$CH_3(CH_2)_3-CH=C=CH_2 + CH_3(CH_2)_3-C≡C-CH_3$$

*(c)

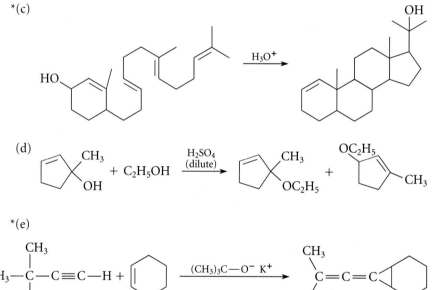

(d)

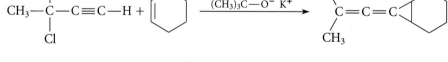

*(e)

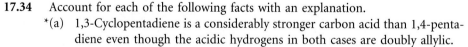

17.34   Account for each of the following facts with an explanation.

*(a)   1,3-Cyclopentadiene is a considerably stronger carbon acid than 1,4-penta-
diene even though the acidic hydrogens in both cases are doubly allylic.

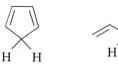

1,3-cyclopentadiene   **1,4-pentadiene**

(b) 3-Bromo-1,4-pentadiene undergoes solvolysis readily in protic solvents, but 5-bromo-1,3-cyclopentadiene is virtually inert.

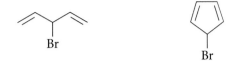

3-bromo-1,4-pentadiene **5-bromo-1,3-cyclopentadiene**

*__17.35__ When 2-hexyne is treated with certain very strong bases, it undergoes the following reaction, an example of the "acetylene zipper" reaction. Give a curved-arrow mechanism for this reaction, and explain why it is irreversible.

$$CH_3{-}C{\equiv}C{-}CH_2CH_2CH_3 + B\!:^- \longrightarrow\ :^-C{\equiv}C{-}CH_2CH_2CH_2CH_3 \;+\; BH \quad (pK_a = 35)$$

$(B\!:^- = \;:^-\ddot{N}H{-}CH_2CH_2CH_2{-}\ddot{N}H_2)$

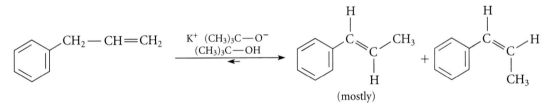

17.36 Propose a mechanism for the following reaction, and give at least two structural reasons why the equilibrium lies to the right.

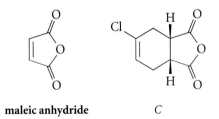

(mostly)

*__17.37__ When 1-buten-3-yne undergoes HCl addition, two compounds *A* and *B* are formed in a ratio of 2.2:1. Neither compound shows a C≡C stretching absorption in its IR spectrum. Compound *B* reacts with maleic anhydride to give compound *C* below, and compound *A* undergoes allylic rearrangement to compound *B* on heating. Propose structures for compounds *A* and *B* and explain your reasoning.

maleic anhydride *C*

*__17.38__ Around 1900, Moses Gomberg, a pioneer in free-radical chemistry, prepared the triphenylmethyl radical, $Ph_3C\cdot$, sometimes called the trityl radical.

(a) Explain why the trityl radical is an unusually stable radical.

(b) The trityl radical is known to exist in equilibrium with a dimer which, for many years, was assumed to be hexaphenylethane. Show how hexaphenylethane could be formed from the trityl radical.

(*Problem 17.38 continues*)

(c) In 1968 the structure of this dimer was investigated using modern methods and found not to be hexaphenylethane, but rather the following compound. Using the fishhook-arrow formalism, show how this compound is formed from the trityl radical, and explain why this compound might form instead of hexaphenylethane. (*Hint:* Can you think of any reason why hexaphenylethane might be unstable?)

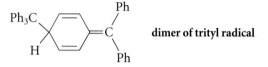

dimer of trityl radical

*17.39 (a) Triphenylmethane (structure below) has a pK_a of 31.5 and, although an alkane, it is almost as acidic as a 1-alkyne. (Most alkanes have p$K_a \geq 55$.) By considering the structure of its conjugate base, suggest a reason why triphenylmethane is such a strong hydrocarbon acid.

(b) Fluoradene is structurally very similar to triphenylmethane, except that the three aromatic rings are "tied together" with bonds. Fluoradene has a pK_a of 11! Suggest a reason why fluoradene is much more acidic than triphenylmethane.

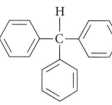   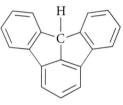

triphenylmethane   **fluoradene**

*17.40 The amount of *anti* addition in the chlorination of alkenes varies with the structure of the alkene, as shown in the following table. (See Sec. 7.9C).

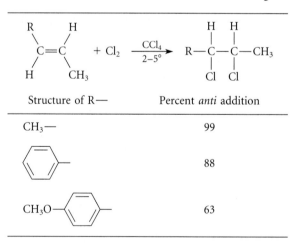

Structure of R—	Percent *anti* addition
CH$_3$—	99
(phenyl)	88
CH$_3$O—(phenyl)	63

Suggest a reason for the variation in the stereochemistry of addition as the alkene structure is varied. (*Hint:* What types of reactive intermediate(s) could account for the stereochemical observations?)

*17.41   In the late 1970s a graduate student at a major west-coast university began synthesizing new classes of drugs and testing them on himself. After being asked to leave the university he began making his living by illegally selling synthetic meperidine (Demerol) to heroin addicts. After shortening his synthetic procedure and self-injecting his product, he developed severe symptoms of Parkinson's disease, as did several of his young clients; one person died. Chemists found that his synthetic meperidine contained two by-products, alcohol *B* and another compound MPTP ($C_{12}H_{15}N$), which, when independently prepared and injected into animals, caused the same symptoms. (Ironically, this has been one of the most significant advances in Parkinsonism research.)

The illicit chemistry is outlined below. Provide the structure of compound *A*, suggest a structure for MPTP, and show how all products are formed.

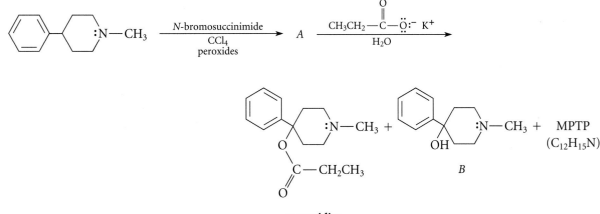

meperidine

*17.42   (a)   For each of the following two reactions suggest a mechanism that is consistent with all of the experimental facts given.

Reaction 1: $CH_3—CH{=}CH—CH_2—Cl + (C_2H_5)_2\overset{..}{N}H$
                (*trans*)

Reaction 2: $CH_3—CH—CH{=}CH_2 + (C_2H_5)_2\overset{..}{N}H$
                         |
                         Cl

$CH_3—CH{=}CH—CH_2—\overset{+}{N}H(C_2H_5)_2$ Cl^-
(82–85% yield)

Experimental observations:

(1)   Both reactions conform to the following rate law, although the rate constants for each reaction are different.

rate = k[alkyl halide]$[(C_2H_5)_2\overset{..}{N}H]$

(2)   The alkyl chloride starting materials do *not* interconvert under the reaction conditions.

(*Problem 17.42 continues*)

(3) The compound below, prepared separately, is *not* converted into the observed product under the reaction conditions.

$$\underset{\overset{|}{CH_3}}{CH_2{=}CH{-}CH{-}\overset{+}{N}H(C_2H_5)_2}\quad Cl^-$$

In particular, explain the importance of facts (2) and (3) in understanding the mechanism.

(b) The mechanism of Reaction 2 is called the S_N2' mechanism. Suggest a reason why this reaction occurs by the S_N2' mechanism and Reaction 1 does not.

*17.43 Is the following S_N2' reaction *syn* or *anti* with respect to the plane of the alkene? How does this result contrast with the stereochemistry of the S_N2 reaction?

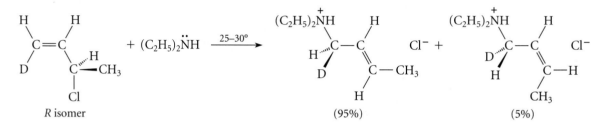

18

Chemistry of Aryl Halides, Vinylic Halides, and Phenols

An **aryl halide** is a compound in which a halogen is bound to the carbon of a benzene ring (or other aromatic ring).

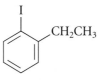

1-ethyl-2-iodobenzene
(an aryl iodide)

not an aryl halide;
halogen not attached directly
to benzene ring

In a **vinylic halide**, a halogen is bound to a carbon of a double bond.

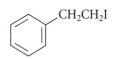

CH_2=CH—Cl

vinyl chloride
(chloroethylene)

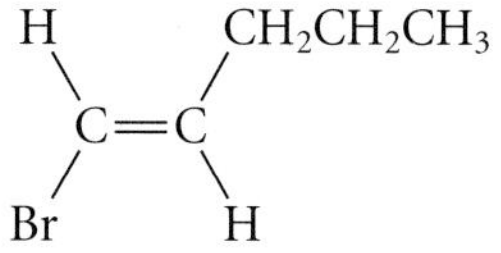

(E)-1-bromo-1-pentene
(a vinylic bromide)

In a **phenol**, a hydroxy (—OH) group is bound to an aromatic ring.

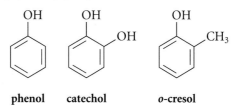

phenol   **catechol**   *o*-cresol

Be sure to differentiate carefully between *vinylic* and *allylic* halides (Chapter 17, Introduction). *Allylic* groups are on a carbon *adjacent* to the double bond. Likewise, be sure that the distinction between *aryl* and *benzylic* halides is clear. *Benzylic* groups are on a carbon *adjacent* to an aromatic ring.

The reactivity of aryl and vinylic halides is quite different from that of ordinary alkyl halides. Although phenols and alcohols have some reactions in common, there are also important differences in the chemical behavior of these two functional groups.

The nomenclature and spectroscopy of aryl halides and phenols were discussed in Secs. 16.1 and 16.3, respectively. The nomenclature of vinylic halides follows the principles of alkene nomenclature (Sec. 4.2A), and the spectroscopy of vinylic halides, except for minor differences due to the halogen, is also similar to that of alkenes.

18.1   Lack of Reactivity of Vinylic and Aryl Halides under S_N2 Conditions

One of the most important differences between vinylic or aryl halides and alkyl halides is their reactivity in nucleophilic substitution reactions. The two most important mechanisms for nucleophilic substitution reactions of alkyl halides are the S_N2 (bimolecular backside attack) mechanism, and the S_N1 (unimolecular carbocation) mechanism (Secs. 9.4, 9.6). What happens to vinylic and aryl halides under the conditions used for S_N1 or S_N2 reactions of alkyl halides?

Consider first the S_N2 reaction. One of the most dramatic contrasts between vinylic or aryl halides and alkyl halides is that simple *vinylic and aryl halides are inert under S_N2 conditions.* For example, when ethyl bromide is allowed to react with Na^+ $C_2H_5O^-$ in C_2H_5OH solvent at 55°, the following S_N2 reaction proceeds to completion in about an hour with excellent yield:

$$CH_3CH_2\!-\!Br + Na^+\ CH_3CH_2\!-\!\ddot{O}\!:^- \xrightarrow[CH_3CH_2OH]{55°} CH_3CH_2\!-\!\ddot{O}\!-\!CH_2CH_3 + Na^+\ Br^- \qquad (18.1)$$

Yet when vinyl bromide or bromobenzene is subjected to the same conditions, nothing happens!

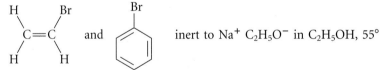

inert to Na^+ $C_2H_5O^-$ in C_2H_5OH, 55°

Why don't aryl halides undergo S_N2 reactions? First, in reaching the transition state, the carbon in the carbon-halogen bond must be rehybridized from sp^2 to sp.

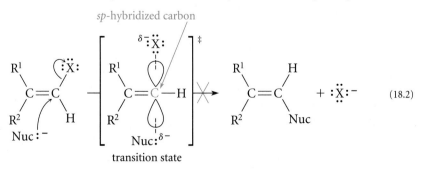

transition state

Contrast this hybridization change with that involved in the S$_N$2 reaction of an *alkyl* halide: *sp^3* to *sp^2* (Fig. 9.2). The *sp* hybridization state has such high energy (Sec. 14.2) that conversion of an *sp^2*-hybridized carbon into an *sp*-hybridized carbon requires about 21 kJ/mol (5 kcal/mol) more energy than is required for an *sp^3* to *sp^2* hybridization change. The relatively high energy of the transition state caused by *sp* hybridization reduces the rate of S$_N$2 reactions of vinylic halides.

A second reason that vinylic halides are unreactive in the S$_N$2 reaction is that the attacking nucleophile (Nuc:$^-$ in Eq. 18.2) must approach the vinylic halide at the backside of the halogen-bearing carbon *and in the plane of the alkene*. This requires that the nucleophile and leaving group approach the alkene substituents R^2 and R^1, respectively, within their van der Waals radii.

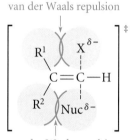

The resulting van der Waals repulsions raise the energy of the transition state and lower the reaction rate.

In summary, both hybridization and van der Waals repulsions (steric effects) within the transition state retard the S$_N$2 reactions of vinylic halides to such an extent that they do not occur.

S$_N$2 reactions of aryl halides have the same problems as those of vinylic halides and two others as well. First, backside attack on the carbon of the carbon-halogen bond would place the nucleophile on a path that goes through the plane of the benzene ring—an obvious impossibility. Furthermore, since the carbon that is attacked undergoes stereochemical inversion, the reaction would necessarily yield a benzene derivative containing a twisted and highly strained double bond.

$$\text{Nuc:}^- \quad \overset{}{\underset{}{\bigcirc}} \!\!-\! \ddot{\text{X}}\text{:} \quad \times\!\!\longrightarrow \quad \underset{\text{Nuc}}{\overset{\text{twisted double bond}}{\bigcirc}} + \text{:}\ddot{\text{X}}\text{:}^- \qquad (18.3)$$

If the impossibility of this result is not clear, try to build a model of the product—but don't break your models!

PROBLEM

18.1   Within each set, rank the compounds in order of increasing rates of their S$_N$2 reactions. Explain your reasoning.

 *(a)   benzyl bromide, (3-bromopropyl)benzene, *p*-bromotoluene

  (b)   1-bromocyclohexene, bromocyclohexane, 1-(bromomethyl)cyclohexene

18.2   Elimination Reactions of Vinylic Halides

Although S_N2 reactions of vinylic halides are completely unknown, base-promoted β-elimination reactions of vinylic halides do occur and can be useful in the synthesis of alkynes.

$$Ph—CH{=}CH—Br \; + \; KOH \xrightarrow{\;200°\;} Ph—C{\equiv}C—H + K^+ \, Br^- + H_2O \qquad (18.4)$$
$$\text{(molten)} \qquad\qquad \text{(distills from}$$
$$\text{reaction mixture;}$$
$$\text{67\% yield)}$$

$$\underset{\displaystyle \overset{\textstyle |}{Ph—CH}}{\overset{\textstyle Br}{\,}}\;\underset{\displaystyle \overset{\textstyle |}{—CH—Ph}}{\overset{\textstyle Br}{\,}} + 2KOH \xrightarrow{\;C_2H_5OH\;} Ph—C{\equiv}C—Ph + 2K^+ \, Br^- + 2H_2O \qquad (18.5)$$
$$\text{(67\% yield)}$$

In Eq. 18.5, two successive eliminations take place. The first gives a vinylic halide and the second gives the alkyne.

Although these reactions involve the energetically unfavorable conversion of the sp^2-hybridized alkene carbons of the starting material into the sp-hybridized alkyne carbons of the product, these eliminations do not involve the van der Waals repulsions associated with vinylic S_N2 reactions, because the base involved attacks a hydrogen rather than the backside of a carbon atom. However, the vinylic eliminations require rather harsh conditions, and some of the more useful examples of this reaction involve elimination of β-hydrogens with enhanced acidity. Notice, for example, that the hydrogens which are eliminated in Eqs. 18.4 and 18.5 are benzylic (Sec. 17.3B).

PROBLEM

*18.2   Arrange the following compounds according to increasing rate of elimination with $NaOC_2H_5$ in C_2H_5OH. What is the product in each case?

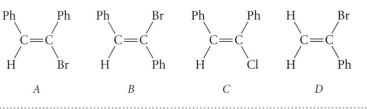

18.3   Lack of Reactivity of Vinylic and Aryl Halides under S_N1 Conditions

Recall that tertiary and some secondary alkyl halides undergo nucleophilic substitution and elimination reactions by the S_N1 and $E1$ mechanisms (Sec. 9.6). Thus, *tert*-butyl bromide undergoes a rapid solvolysis in ethanol to give both substitution and elimination products.

$$CH_3-\underset{\underset{CH_3}{|}}{\overset{\overset{CH_3}{|}}{C}}-\ddot{B}r: \xrightleftharpoons[55°]{C_2H_5OH} \left[CH_3-\underset{\underset{CH_3}{|}}{\overset{\overset{CH_3}{|}}{C^+}} \quad :\ddot{B}r:^- \right] \xrightarrow{C_2H_5OH}$$

$$CH_3-\underset{\underset{CH_3}{|}}{\overset{\overset{CH_3}{|}}{C}}-OC_2H_5 \; + \; \underset{\underset{CH_3}{/}}{\overset{\overset{CH_3}{\backslash}}{C}}=CH_2 \; + \; H\ddot{B}r: \qquad (18.6)$$

$$\text{(72\%)} \qquad\qquad \text{(28\%)}$$

Vinylic and aryl halides, however, are virtually inert to the conditions that promote S_N1 or E1 reactions of alkyl halides. Certain vinylic halides can be forced to react by the S_N1-E1 mechanism under extreme conditions, but such reactions are relatively uncommon.

$$CH_2=\overset{\overset{CH_3}{|}}{C}-Br \; + \; C_2H_5OH \xrightarrow{55°} \text{ no reaction} \qquad (18.7)$$

$$\langle\!\!\bigcirc\!\!\rangle{-}Br \; + \; C_2H_5OH \xrightarrow{55°} \text{ no reaction} \qquad (18.8)$$

To understand why vinylic and aryl halides are inert under S_N1 conditions, consider what would happen if they *were* to undergo the S_N1 reaction. If a vinylic halide undergoes an S_N1 reaction, it must ionize to form a *vinylic cation*.

$$\underset{:\ddot{B}r:}{\overset{R}{\backslash}}C=CH_2 \longrightarrow R-\overset{+}{C}=CH_2 \quad :\ddot{B}r:^- \qquad (18.9)$$

$$\text{a vinylic cation}$$

STUDY GUIDE LINK:
18.1
Vinylic Cations

A **vinylic cation** is a carbocation in which the electron-deficient carbon is also part of a double bond. An orbital diagram of a vinylic cation is shown in Fig. 18.1a. Notice that the electron-deficient carbon is connected to two groups, the R group and the other carbon of the double bond. Hence, the geometry at this carbon is *linear* (Sec. 1.3B) and the electron-deficient carbon is therefore *sp*-hybridized. Notice also that the vacant *p* orbital is not conjugated with the π-electron system of the double bond; in order to be conjugated, it would have to be coplanar with the double-bond π system. Vinylic cations are considerably less stable than alkyl carbocations because of their *sp* hybridization (the same reason that alkynes are less stable than isomeric dienes), and because of the electron-withdrawing polar effect of the double bond. Hence, one reason that vinylic halides do not undergo the S_N1 reaction is the *instability of the vinylic cations* that would necessarily be involved as reactive intermediates.

The second reason that vinylic halides do not undergo the S_N1 reaction is that carbon-halogen bonds are stronger in vinylic halides than they are in alkyl halides. A vinylic carbon-halogen bond involves an sp^2 carbon orbital, whereas an alkyl carbon-halogen bond involves an sp^3 carbon orbital. Hence, a vinylic carbon-halogen bond has more *s* character. Recall that bonds with more *s* character are stronger (Eq. 14.25). Consequently, it takes more energy to break the carbon-halogen bond of a vinylic halide. This additional energy is reflected in a smaller rate of ionization.

However, nucleophilic aromatic substitution reactions do not involve a one-step backside attack for the reasons given in Sec. 18.1. Two clues about the reaction mechanism come from the reactivities of different aryl halides. First, the reaction is faster when there are more nitro groups *ortho* and *para* to the halogen leaving group:

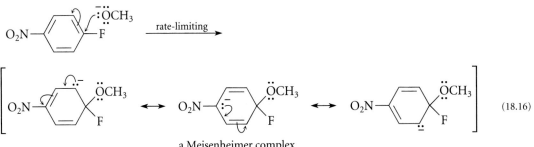

$$O_2N\text{—}\underset{NO_2}{\bigcirc}\text{—}Cl + {}^-OCH_3 \longrightarrow O_2N\text{—}\underset{NO_2}{\bigcirc}\text{—}OCH_3 + Cl^- \qquad 10^5\text{–}10^6 \qquad (18.14a)$$

$$O_2N\text{—}\bigcirc\text{—}Cl + {}^-OCH_3 \longrightarrow O_2N\text{—}\bigcirc\text{—}OCH_3 + Cl^- \qquad 1 \qquad (18.14b)$$

$$\bigcirc\text{—}Cl + {}^-OCH_3 \longrightarrow \text{no reaction} \qquad {\sim}0 \qquad (18.14c)$$

relative rate

Second, the effect of the halogen on the rate of this type of reaction is quite different from that in the S_N1 or S_N2 reaction of alkyl halides. In nucleophilic aromatic substitution reactions, aryl fluorides are most reactive.

Reactivities of aryl halides:

$$\text{Ar—F} \gg \text{Ar—Cl} \approx \text{Ar—Br} \approx \text{Ar—I} \qquad (18.15)$$

In S_N2 and S_N1 reactions of *alkyl* halides, the reactivity order is exactly the reverse: alkyl fluorides are the least reactive alkyl halides (Secs. 9.4E, 9.6C).

These data are consistent with a reaction mechanism in which the nucleophile attacks the π-electron system below (or above) the plane of the aromatic ring to yield a resonance-stabilized anion called a *Meisenheimer complex*. In this anion the negative charge is delocalized throughout the π-electron system of the ring. Formation of this anion is the rate-determining step in many nucleophilic aromatic substitution reactions.

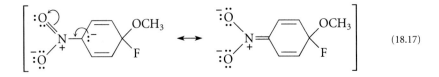

a Meisenheimer complex

The negative charge in this complex is also delocalized into the nitro group.

$$(18.17)$$

The Meisenheimer complex breaks down to products by loss of the halide ion.

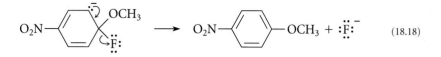

(18.18)

Let's see how this mechanism fits the experimental facts. *Ortho* and *para* nitro groups accelerate the reaction because the transition state resembles the Meisenheimer complex, and the nitro groups stabilize this complex by resonance. Fluorine also stabilizes the negative charge by its electron-withdrawing *polar effect*, which is greater than the polar effect of the other halogens. Because the *loss of halide is not rate-limiting*, the basicity of the halide, or equivalently, the strength of the carbon-halogen bond, is not important in determining the reaction rate.

Notice how the nucleophilic aromatic substitution reaction differs from the S_N2 reaction of alkyl halides. First, there is an actual intermediate in the nucleophilic aromatic substitution reaction—the Meisenheimer complex. (In some cases this is sufficiently stable that it can be directly observed.) There is no evidence for an intermediate in any S_N2 reaction. Second, the nucleophilic aromatic substitution reaction is a frontside displacement; it requires no inversion of configuration. The S_N2 reaction of an alkyl halide, in contrast, is a backside displacement with inversion of configuration. Finally, the effect of the halogen on the reaction rate (Eq. 18.15) is different in the two reactions. Aryl fluorides react most rapidly in nucleophilic aromatic substitution, whereas alkyl fluorides react most slowly in S_N2 reactions.

Study Guide Link:
✓18.2
Contrast of Aromatic Substitution Reactions

PROBLEMS

18.4   Complete the following reactions. (*No reaction* may be the correct response.)

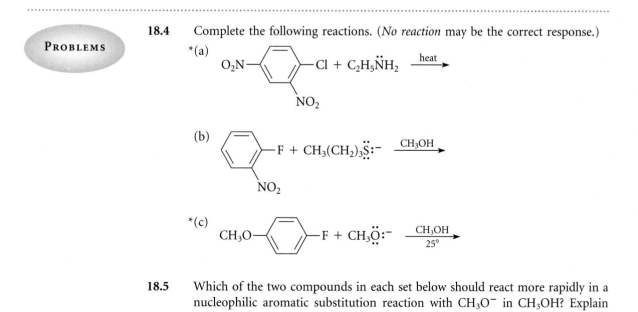

18.5   Which of the two compounds in each set below should react more rapidly in a nucleophilic aromatic substitution reaction with CH_3O^- in CH_3OH? Explain your answers.

(Problem 18.5 continues)

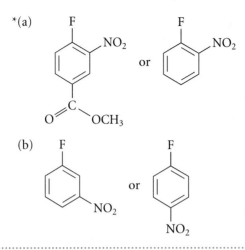

B. Substitution by Elimination-Addition: Benzyne

Recall that vinylic halides undergo β-elimination reactions under vigorous conditions (Sec. 18.2). Imagine if an aryl halide were to undergo an analogous reaction. β-Elimination of an aryl halide would give an interesting "alkyne" called **benzyne**.

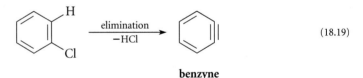

(18.19)

benzyne

An orbital picture of benzyne is shown in Fig. 18.2. Notice that one of the two π bonds in the triple bond of benzyne is perpendicular to the π-electron system of the aromatic ring. This extra π bond is unusual, because the orbitals from which it is formed are more like sp^2 orbitals than p orbitals.

Because alkynes require linear geometry, it is difficult to incorporate them into six-membered rings. Therefore benzyne is highly *strained* (Problem 18.36) and, although it

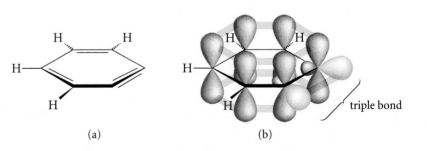

(a)   (b)

Figure 18.2   *Benzyne. (a) A Lewis structure; (b) the corresponding orbital diagram. Notice that the orbitals of one π bond (color) are perpendicular to the aromatic π-electron system of the ring. As a result, one π bond of the triple bond is not stabilized by aromaticity. Benzyne is highly strained, because the ring prevents the triple bond from achieving the preferred linear geometry.*

is a neutral molecule, it is very unstable. Indeed, benzyne has proven to be too reactive to isolate except at temperatures near absolute zero.

However, despite its instability, benzyne is a reactive intermediate in certain reactions of aryl halides. For example, when chlorobenzene is treated with very strong bases such as potassium amide (KNH_2), a substitution reaction takes place.

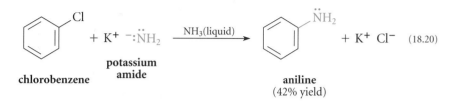

(18.20)

This reaction cannot occur by the S_N1 or S_N2 mechanism for the reasons given in the previous section. Furthermore, since the aryl halide is not substituted with a nitro or other electron-withdrawing group, the aryl halide does not appear to be one that should undergo nucleophilic aromatic substitution.

Evidence for a mechanism involving a benzyne intermediate was obtained in a very clever isotopic labeling experiment conducted by Professor John D. Roberts and his colleagues (of the California Institute of Technology). They carried out the reaction of Eq. 18.20 on a chlorobenzene sample in which the carbon bearing the chlorine (and *only* that carbon) was labeled with the radioactive isotope ^{14}C.

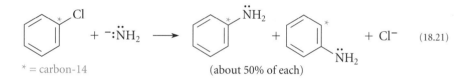

(18.21)

Analysis of the radioactivity in the product showed that in about half of the product, the radioactive carbon is *adjacent* to the substituted carbon, and in the other half, the radioactive carbon is the substituted carbon itself. Now, radioactivity cannot actually move from one carbon to the other. The only way to account for this result is that somehow *carbon-1 and carbon-2 have become equivalent and indistinguishable at some point in the reaction. As a result, subsequent reactions occur at both the labeled and unlabeled carbons with identical rates.* It is also important to notice that only carbon-1 and carbon-2 have become equivalent; the radioactivity is not found at any other carbons.

Let's see how a benzyne intermediate can account for these results. The first step in the mechanism of this reaction is formation of an anion at the *ortho* position. Because benzene derivatives are only weakly acidic (about as acidic as the corresponding alkenes; Sec. 14.7A), this requires a strong base.

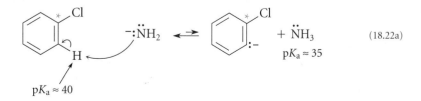

(18.22a)

This anion expels chloride ion, thus completing an elimination and forming benzyne.

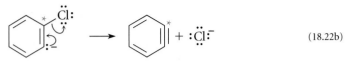

$$(18.22b)$$

benzyne

Because benzyne is very unstable, it undergoes a reaction that is not typical of ordinary alkynes: *it is attacked by the amide ion to give a new anion.* Since benzyne is symmetric (except for its label), the carbons of the triple bond are indistinguishable. Hence, attack of $^-NH_2$ occurs with equal probability at each carbon of the triple bond. As a result, the —NH_2 group in half of the product molecules is bound to a carbon *adjacent to* the radioactive carbon.

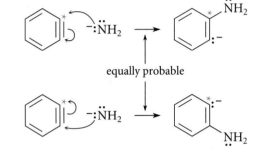

$$(18.22c)$$

Protonation of each anion gives the mixture shown in Eq. 18.21.

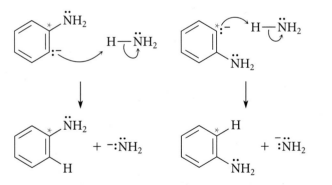

$$(18.22d)$$

The reaction that gives benzyne (Eqs. 18.22a and b) is a *β-elimination reaction.* It somewhat resembles an E2 elimination, except that the reaction occurs in two steps. (Recall that the E2 elimination takes place in a single step; Sec. 9.5A.) Benzyne is subsequently consumed in an *addition reaction* (Eqs. 18.22c and d). Therefore, the overall substitution shown in Eq. 18.21 is really an *elimination-addition* process.

The benzyne mechanism accounts for the fact that, when an aryl halide bears ring substituents, more than one product is observed.

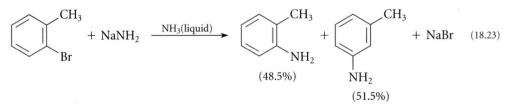

$$(18.23)$$

In Eq. 18.23, the first product results from attack of the amide ion at the carbon of benzyne that was originally bound to the halide ion, and the second product results from attack at the adjacent carbon of benzyne.

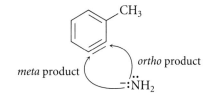

Substitution at a site different from the one occupied by the leaving group is called **cine substitution** (*cine*, from Greek *kinēma*, "movement"; pronounced sin′ ē). Thus, the benzyne mechanism predicts a mixture of direct and *cine*-substitution products.

Reactions involving benzyne intermediates are rarely important as preparative procedures. However, it is important to understand the benzyne mechanism because it accounts for some of the side reactions that occur when aryl halides are subjected to *strongly basic conditions.*

PROBLEM

18.6 According to the benzyne mechanism, what product(s), if any, are expected when each of the following compounds reacts with potassium amide in liquid ammonia?

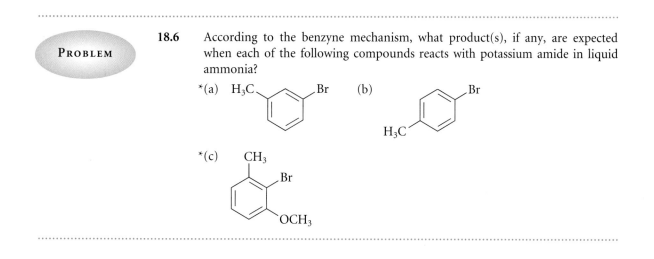

C. Summary: Nucleophilic Substitution Reactions of Aryl Halides

Aryl halides undergo several different types of nucleophilic substitution reactions. Each reaction takes place under a specific set of conditions.

1. Aryl halides substituted with *ortho*- or *para*-nitro groups (or other conjugated, electron-attracting groups) react with Lewis bases in *nucleophilic aromatic substitution* reactions (Sec. 18.4A). This type of reaction involves resonance-stabilized anionic intermediates (Meisenheimer complexes) resulting from nucleophilic attack on the aromatic π-electron system, followed by loss of the halide from this ion.

2. Ordinary alkyl halides undergo substitution by the *benzyne* mechanism (Sec. 18.4B) only in the presence of *very strong bases* such as alkali metal amides and organolithium reagents, or with somewhat weaker bases under *very harsh conditions* (high temperature, long reaction times). Except for the benzyne reaction, ordinary aryl halides do not undergo nucleophilic substitution reactions.

18.5 Aryl and Vinylic Grignard Reagents

Preparation of Grignard and organolithium reagents from aryl and vinylic halides is analogous to the corresponding reactions of alkyl halides.

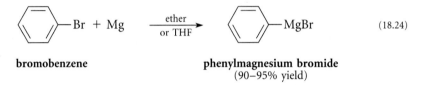

$$\qquad\qquad(18.24)$$

bromobenzene   **phenylmagnesium bromide**
                                    (90–95% yield)

Arylmagnesium bromides or iodides can be prepared from the corresponding bromo- or iodobenzenes in either tetrahydrofuran (THF) or diethyl ether solvent (see Sec. 8.7A). However, THF rather than ether *must* be used as the solvent when Grignard reagents are formed from chlorobenzenes.

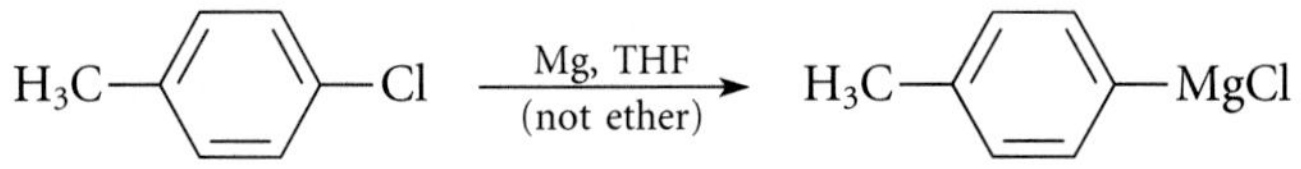

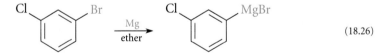

$$\qquad\qquad(18.25)$$

(96% yield)

Because of this solvent effect, the following selective reaction is possible.

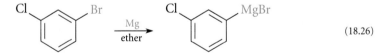

$$\qquad\qquad(18.26)$$

The preparation of vinylic Grignard reagents also requires THF as a solvent.

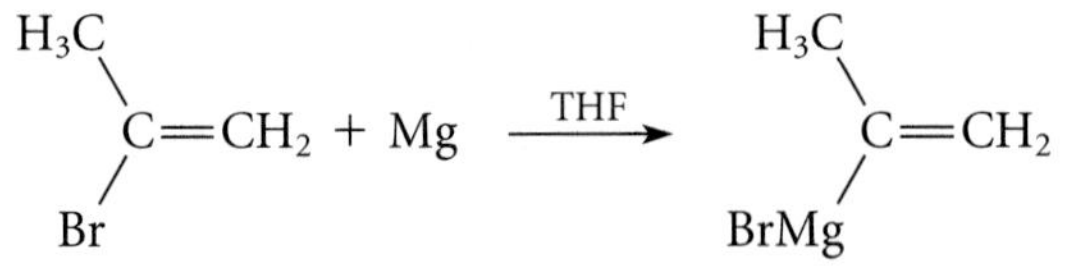

$$\qquad\qquad(18.27)$$

The reasons for these solvent effects are not clearly understood.

PROBLEM

18.7   Outline a synthesis of each of the following compounds from benzene and any other reagents.

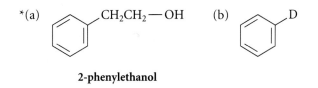

*(a) ... CH₂CH₂—OH

(b) ... D

2-phenylethanol

18.6   Acidity of Phenols

A.   Resonance Effects on Acidity

Phenols, like alcohols, can ionize.

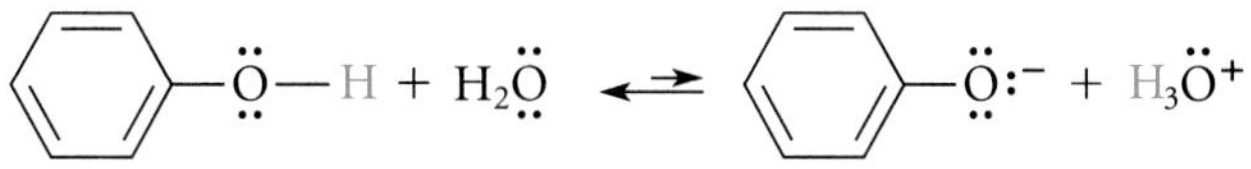

$$\text{phenol} + H_2\ddot{O} \;\rightleftharpoons\; \text{phenoxide ion} + H_3\ddot{O}^+ \tag{18.28}$$

phenol

phenoxide ion (phenolate ion)

The conjugate base of a phenol is named, using common nomenclature, as a *phenoxide ion* or, using substitutive nomenclature, as a *phenolate ion*. Thus, the sodium salt of phenol is called sodium phenoxide or sodium phenolate; the potassium salt of *p*-chlorophenol is called potassium *p*-chlorophenoxide or potassium 4-chlorophenolate.

Phenols are considerably more acidic than alcohols. For example, the pK_a of phenol is 9.95, but that of cyclohexanol is about 17. Thus phenol is approximately 10^7 times as acidic as an alcohol of similar size and shape.

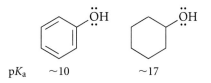

pK_a ∼10   ∼17

Recall from Sec. 3.6B, Fig. 3.2, that the pK_a of an acid is decreased by stabilizing its conjugate base. *The enhanced acidity of phenol is due to stabilization of its conjugate-base anion.*

What is the source of this enhanced stability? First, the phenolate anion is stabilized by resonance:

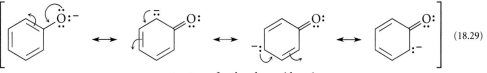

$$\tag{18.29}$$

resonance structures for the phenoxide anion

Study Guide Link:
18.3
Resonance Effects on Phenol Acidity

Second, the polar effect of the benzene ring stabilizes the negative charge. Both resonance stabilization and polar effects are the same effects that stabilize benzylic carbanions (Sec. 17.3). Of course, a phenoxide anion *is* a benzylic anion in which the benzylic group is an oxygen instead of a carbon.

Alkoxides are stabilized neither by resonance nor by the polar effect of benzene rings or double bonds.

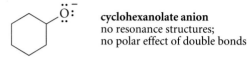

cyclohexanolate anion
no resonance structures;
no polar effect of double bonds

Because phenoxide ions are stabilized by both resonance and polar effects, less energy is required to form phenoxides from phenols than is required to form alkoxides from alcohols. *Since pK_a is directly proportional to free energy of ionization* (Eq. 3.38), phenols have lower pK_a values, and are thus more acidic, than alcohols.

Substituent groups can also affect phenol acidity by both polar and resonance effects. For example, the relative acidities of phenol, *m*-nitrophenol and *p*-nitrophenol reflect the operation of both effects.

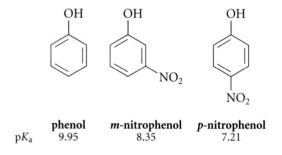

	phenol	***m*-nitrophenol**	***p*-nitrophenol**
pK_a	9.95	8.35	7.21

m-Nitrophenol is more acidic than phenol because the nitro group is very electronegative. The polar effect of the nitro substituent stabilizes the conjugate-base anion for the same reason that it would stabilize the conjugate base of an alcohol (Sec. 8.5B). Yet *p*-nitrophenol is more acidic than *m*-nitrophenol by more than one pK_a unit, even though the *p*-nitro group is farther from the phenol oxygen. This cannot be a polar effect, for polar effects on acidity *decrease* as the distance between the substituent and the acidic group increases. The reason for the increased acidity of *p*-nitrophenol is that the *p*-nitro group stabilizes the conjugate-base anion by resonance (colored structure).

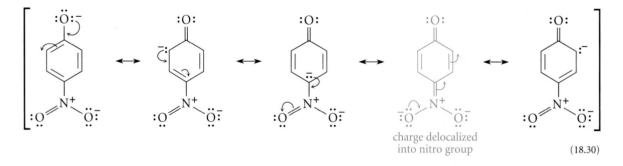

charge delocalized
into nitro group

(18.30)

The colored structure is especially important because it places charge on the electronegative oxygen atom. In *m*-nitrophenol, however, it is not possible to draw resonance structures that delocalize the negative charge into the nitro group.

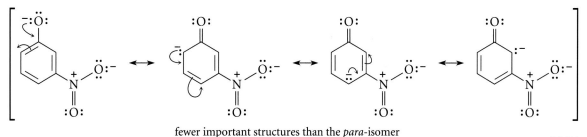

fewer important structures than the *para*-isomer

(18.31)

Since *p*-nitrophenoxide has more resonance structures, it is more stable relative to its corresponding phenol than *m*-nitrophenoxide. Hence *p*-nitrophenol is the more acidic of the two nitrophenols. The acid-strengthening resonance effect of *ortho*- and *para*-nitro groups is so large that 2,4,6-trinitrophenol (picric acid) is actually a strong acid.

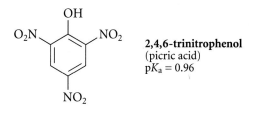

2,4,6-trinitrophenol
(picric acid)
$pK_a = 0.96$

Let's summarize the factors that govern acidity as we've seen them in operation so far:

1. *Element effects:* Other things being equal, compounds are more acidic when the element to which the acidic hydrogen is bound has higher atomic number within either a row or group of the periodic table.
   a. The effect within a row (period) of the periodic table is dominated by relative electronegativities (or electron affinities). Thus, water is much more acidic than methane, and phenol is much more acidic than toluene, because oxygen (higher atomic number) is more electronegative than carbon (lower atomic number).
   b. The effect within a column of the periodic table is dominated by relative bond energies. Thus, thiols are more acidic than alcohols because the S—H bond is weaker than the O—H bond.

2. *Resonance effects:* enhanced delocalization of electrons in the conjugate base enhances acidity.

3. *Polar effects:* stabilization of charge in the conjugate base enhances acidity.

PROBLEM

18.8 Which of the two phenols in each set is more acidic? Explain.
  *(a) 2,5-dinitrophenol or 2,4-dinitrophenol
  (b) phenol or *m*-chlorophenol

(Problem 18.8 continues)

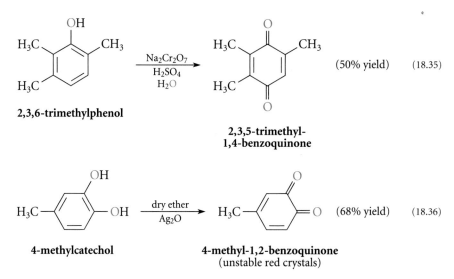

2,3,6-trimethylphenol $\xrightarrow[\substack{H_2SO_4 \\ H_2O}]{Na_2Cr_2O_7}$ **2,3,5-trimethyl-1,4-benzoquinone**   (50% yield)   (18.35)

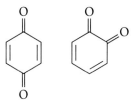

4-methylcatechol $\xrightarrow[Ag_2O]{dry~ether}$ **4-methyl-1,2-benzoquinone**
(unstable red crystals)   (68% yield)   (18.36)

Notice that *p*-hydroxyphenols (hydroquinones), *o*-hydroxyphenols (catechols), and phenols with an unsubstituted position *para* to the hydroxy group are oxidized to quinones.

As the examples above suggest, a **quinone** is any compound containing either of the following structural units.

a *p*-quinone   an *o*-quinone

If the quinone oxygens have a 1,4 (*para*) relationship, the quinone is called a *para*-quinone; if the oxygens are in a 1,2 (*ortho*) arrangement, the quinone is called an *ortho*-quinone. The following compounds are typical quinones.

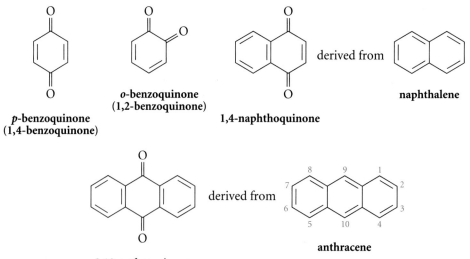

p-benzoquinone
(1,4-benzoquinone)

***o*-benzoquinone**
(1,2-benzoquinone)

1,4-naphthoquinone   derived from   **naphthalene**

9,10-anthraquinone   derived from   **anthracene**

As indicated above, the names of quinones are derived from the names of the corresponding aromatic hydrocarbons: *benzo*quinone is derived from benzene, *naphtho*quinone from naphthalene, and so on.

Ortho-quinones, particularly *ortho*-benzoquinones, are considerably less stable than the isomeric *para*-quinones. One reason for this difference is that in *ortho*-quinones, the C=O bond dipoles are nearly aligned, and therefore have a repulsive, destabilizing interaction. In *para*-quinones these dipoles are further apart.

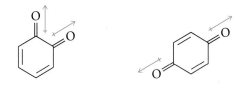

o-quinone
bond dipoles nearly aligned

p-quinone
bond dipoles further apart

A number of quinones occur in nature. *Juglone* occurs in walnut shells. *Doxorubicin* (adriamycin), isolated from a microorganism, is an important antitumor drug.

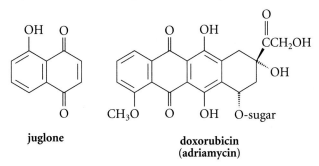

juglone

doxorubicin
(adriamycin)

The oxidation of phenols by air (O_2) to colored, quinone-containing products is the reaction responsible for the darkening that is observed when some phenols are stored for long periods of time. The oxidation of hydroquinone and its derivatives to the corresponding *p*-benzoquinones can also be carried out reversibly in an electrochemical cell. Oxidation potentials of a number of phenols with respect to standard electrodes are well known.

Practical Applications of Phenol Oxidation The oxidation of phenols has several important practical applications. For example, phenols are sometimes used to inhibit free-radical reactions that result in the oxidation of other compounds (Sec. 5.6C). The basis of this effect is that many free radicals (R· in Eq. 18.37a) abstract a hydrogen from hydroquinone to form a very stable radical called a *semiquinone*.

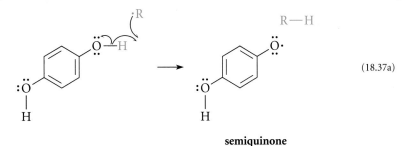

(18.37a)

semiquinone

(The semiquinone radical is resonance-stabilized; draw its resonance structures.) A second free radical can react with the semiquinone to complete its oxidation to quinone.

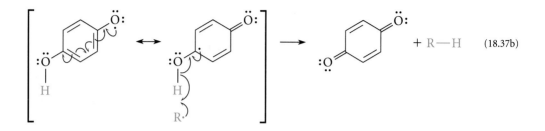

$$+ R\!-\!H \qquad (18.37b)$$

Hydroquinone thus terminates free-radical chain reactions by intercepting free-radical intermediates R· and reducing them to R—H.

The effectiveness of several widely used food preservatives is based on reactions such as these. Examples of such preservatives are "butylated hydroxytoluene" (BHT) and "butylated hydroxyanisole" (BHA).

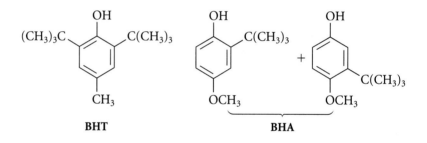

Oxidation involving free-radical processes is one way that foods discolor and spoil. A preservative such as BHT inhibits these processes by donating its O—H hydrogen atom to free radicals in the food (as in Eq. 18.37a). The BHT is thus transformed into a phenoxy radical, which is too stable and unreactive to propagate radical chain reactions. Although the use of BHT and BHA as food additives has generated some controversy because of their potential side effects, without such additives foods could not be stored for reasonable periods of time and transported over long distances.

Recent research indicates that vitamin E, a phenol, is the major compound in the blood responsible for preventing oxidative damage by free radicals. Vitamin E acts by terminating radical chains in the manner shown in Eq. 18.37a.

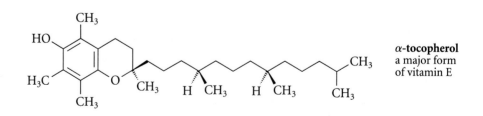

α-tocopherol
a major form
of vitamin E

PHOTOGRAPHY AND PHENOL OXIDATION

The oxidation of hydroquinone lies at the heart of the photographic process. When photographic film is exposed to light, grains of silver bromide in the photographic emulsion on the film absorb light and are activated or *sensitized*.

$$AgBr + light \longrightarrow [AgBr]^* \qquad (18.38)$$
$$\text{sensitized}$$
$$\text{silver bromide}$$

Because silver bromide is trapped in the photographic emulsion, it is immobile. Thus sensitized silver bromide molecules provide a faithful record of the positions on the film that have been struck by light. Now, sensitized silver bromide is a much better oxidizing agent than silver bromide that has not been exposed to light. When exposed film is treated with a solution of hydroquinone (a common photographic developer), [AgBr]* oxidizes hydroquinone to *p*-benzoquinone (which is subsequently washed away), and the silver(I) is reduced to finely divided silver metal, which remains trapped in the photographic emulsion.

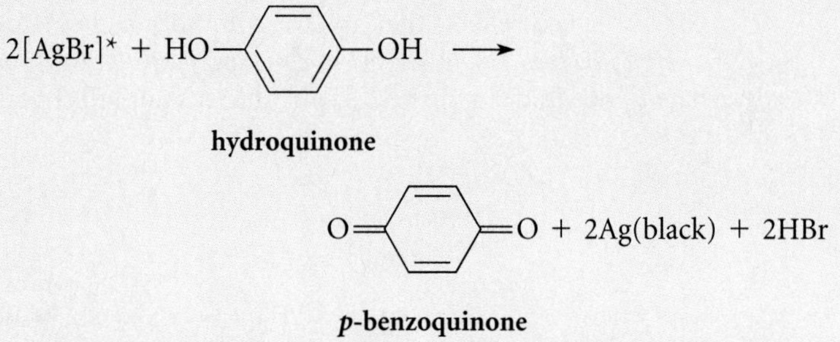

$$2[AgBr]^* + HO\text{---}\bigcirc\text{---}OH \longrightarrow$$

hydroquinone

$$O{=}\bigcirc{=}O + 2Ag(black) + 2HBr \qquad (18.39)$$

_p_-benzoquinone

Because unactivated AgBr oxidizes hydroquinone much more slowly, silver metal forms only where light has impinged on the film. This precipitated silver is the black part of a black-and-white negative.

PROBLEMS

18.11 Given the structure of phenanthrene, below, draw structures of
*(a) 9,10-phenanthraquinone   (b) 1,4-phenanthraquinone
Indicate whether each is an *o*- or a *p*-quinone.

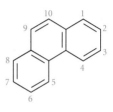

phenanthrene

18.12   Complete the following reactions.

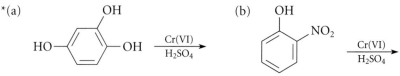

18.13   Draw the important resonance structures of the radicals formed when each of the following reacts with R·, a general free radical.
*(a)   vitamin E   (b)   BHT

 18.8   ## Electrophilic Aromatic Substitution Reactions of Phenols

Phenols are aromatic compounds, and they undergo electrophilic aromatic substitution reactions like those described in Chapter 16. In some of these reactions, the —OH group has special effects that are not common to ther substituent groups.

Because the —OH group is a strongly activating substituent, phenol can be halogenated once under mild conditions that do not affect benzene itself.

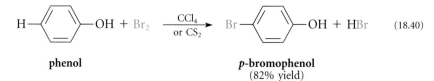

phenol   **p-bromophenol**
                                              (82% yield)

Notice the mild conditions of this reaction. A Lewis acid such as $FeBr_3$ is not required. (A solution of Br_2 in CCl_4 is the reagent usually used for adding bromine to alkenes.) But when phenol reacts with Br_2 in H_2O (bromine water), more extensive bromination occurs and 2,4,6-tribromophenol is obtained.

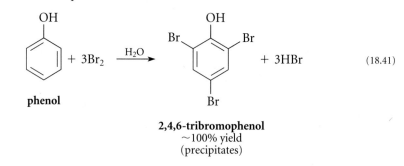

phenol

2,4,6-tribromophenol
~100% yield
(precipitates)

This more extensive bromination occurs for two reasons. First, bromine reacts with water to give protonated hypobromous acid, a more potent electrophile than bromine itself.

$$Br_2 + H_2\overset{..}{O}: \quad \rightleftarrows \quad Br^- \quad + \quad Br-\overset{+}{O}H_2 \quad \rightleftarrows \quad HBr \ + \ Br-\overset{..}{\underset{..}{O}}H \qquad (18.42)$$

protonated   hypobromous
hypobromous acid   acid

Second, in aqueous solutions near neutrality, phenol partially ionizes to its conjugate-base phenoxide anion. Although only a small amount of this anion is present, it is very reactive and brominates instantly, thereby pulling the phenol-phenolate equilibrium to the right.

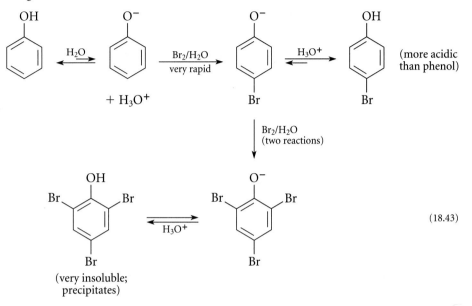

(18.43)

Phenoxide ion is much more reactive than phenol because the reactive intermediate is not a carbocation, but is instead a more stable neutral molecule (colored structure).

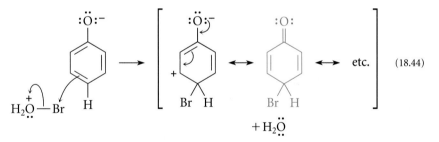

(18.44)

p-Bromophenol is also in equilibrium with its conjugate base p-bromophenoxide anion, which brominates again until all *ortho* and *para* positions have been substituted. Notice in Eq. 18.43 that in the second and third substitutions the powerful *ortho*, *para* directing and activating effects of the —Ö:⁻ group override the weaker deactivating and directing effects of the bromine substituents. In strongly acidic solution, in which formation of the phenolate anion is suppressed, bromination can be stopped at the 2,4-dibromophenol stage.

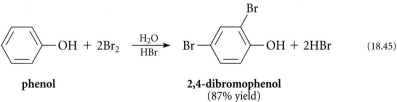

(18.45)

phenol   **2,4-dibromophenol**
                                         (87% yield)

Phenol is also very reactive in other electrophilic substitution reactions, such as nitration. Phenol can be nitrated once under mild conditions. (Notice that H_2SO_4 is not present as it is in the nitration of benzene; Eq. 16.8.)

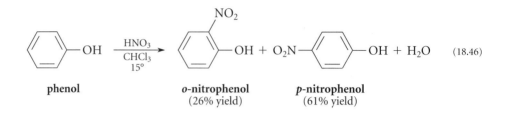

phenol   *o*-nitrophenol   *p*-nitrophenol
                                              (26% yield)   (61% yield)

(18.46)

Because phenol is activated toward electrophilic substitution, it is also possible to nitrate phenol two and three times. However, direct nitration is *not* the preferred method for synthesis of di- and trinitrophenol, because the concentrated HNO_3 required for multiple nitrations is also an oxidizing agent, and phenols are easily oxidized (Sec. 18.7). Thus 2,4-dinitrophenol is synthesized instead by the nucleophilic aromatic substitution reaction of 1-chloro-2,4-dinitrobenzene with ^-OH (Sec. 18.4A).

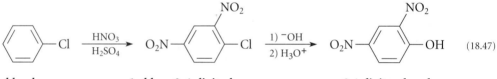

chlorobenzene   **1-chloro-2, 4-dinitrobenzene**   **2,4-dinitrophenol**

(18.47)

The great reactivity of phenol in electrophilic aromatic substitution does not extend to the Friedel-Crafts acylation reaction, because phenol reacts rapidly with the $AlCl_3$ catalyst.

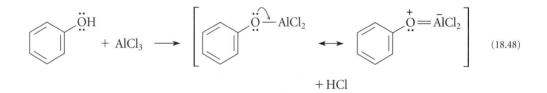

(18.48)

The adduct of phenol and $AlCl_3$ is much less reactive than phenol itself in electrophilic aromatic substitution reactions because, as shown in Eq. 18.48, the oxygen electrons are delocalized onto the electron-deficient aluminum. Because of their delocalization away from the benzene ring, these electrons are less available for resonance stabilization of the carbocation intermediate formed within the ring during Friedel-Crafts acylation (Eq. 16.14). Thus, Friedel-Crafts acylation of phenol occurs slowly, but can be carried out successfully at elevated temperatures. Because it is not highly activated, the ring is acylated only once.

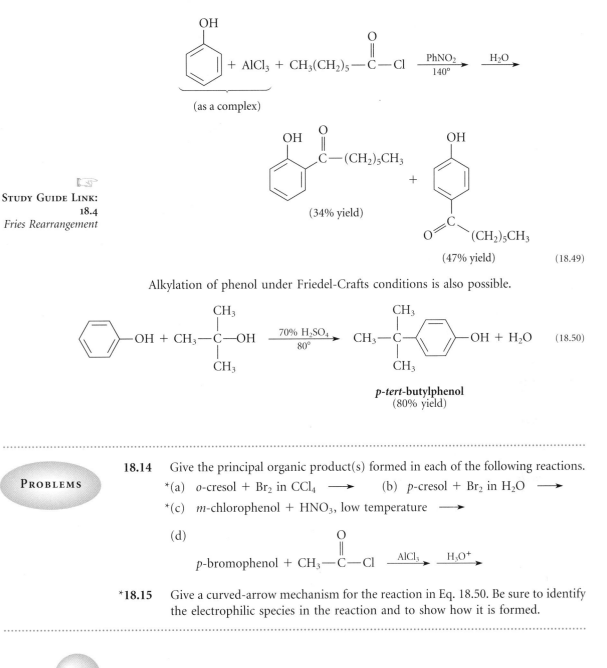

(18.49)

Alkylation of phenol under Friedel-Crafts conditions is also possible.

p-tert-**butylphenol**
(80% yield)

(18.50)

PROBLEMS

18.14 Give the principal organic product(s) formed in each of the following reactions.

*(a)   *o*-cresol + Br$_2$ in CCl$_4$ $\longrightarrow$ (b)   *p*-cresol + Br$_2$ in H$_2$O $\longrightarrow$

*(c)   *m*-chlorophenol + HNO$_3$, low temperature $\longrightarrow$

(d)

$$\text{*p*-bromophenol} + \text{CH}_3\text{—C—Cl} \xrightarrow{\text{AlCl}_3} \xrightarrow{\text{H}_3\text{O}^+}$$

*18.15 Give a curved-arrow mechanism for the reaction in Eq. 18.50. Be sure to identify the electrophilic species in the reaction and to show how it is formed.

18.9   Lack of Reactivity of the Aryl-Oxygen Bond

Just as the reactions of alcohols that break the carbon-oxygen bond have close analogy to the reactions of alkyl halides that break the carbon-halogen bond, the carbon-oxygen reactivity of phenols follows the carbon-halogen reactivity of *aryl* halides. (That is, we can think of phenols as "aryl alcohols.") Recall that aryl halides do not undergo S$_N$1 or

E1 reactions (Sec. 18.3); for the same reasons, phenols and phenyl ethers also do not react under conditions used for the S_N1 or E1 reactions of alcohols. Thus phenols do *not* form aryl bromides with concentrated HBr; they do *not* dehydrate with concentrated H_2SO_4. (Notice, however, that they do undergo sulfonation; see Sec. 16.4D.) The reasons for these observations are exactly the same as those that explain the lack of reactivity of aryl halides. (What are those reasons? See Secs. 18.1 and 18.3.)

More generally, *any* derivative of the form

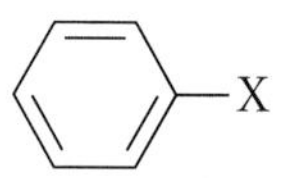

in which X is a good leaving group, such as tosylate, mesylate, or even $\overset{+}{-}OH_2$, has the same lack of reactivity toward S_N1 and S_N2 conditions as aryl halides—and for the same reasons.

The lack of reactivity of the aryl-oxygen bond can be put to good use in the cleavage of aryl ethers. Recall that when ethers cleave—depending on the particular ether and the mechanism—products resulting from cleavage at either of the carbon-oxygen bonds are possible (Sec. 11.3). In the case of aryl ethers, cleavage occurs *only* at the alkyl-oxygen bond; consequently, only one set of products is formed:

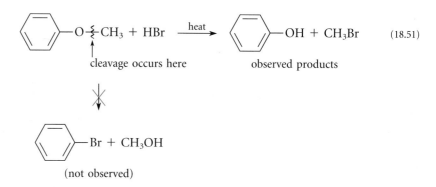

In this example, ether cleavage gives phenol and methyl bromide rather than bromobenzene and methanol because S_N2 attack of bromide ion can only occur *at the methyl group* of the protonated ether:

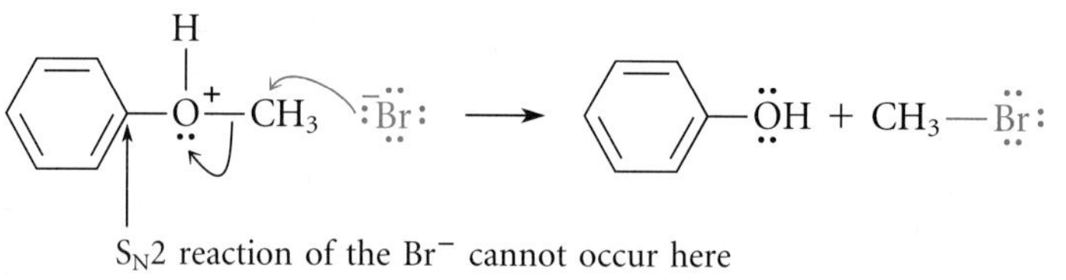

18.16   Within each set, identify the ether that would *not* readily cleave with concentrated HBr and heat, and explain. Then give the products of ether cleavage and the mechanisms of their formation, for the other ether(s) in the set.

PROBLEM

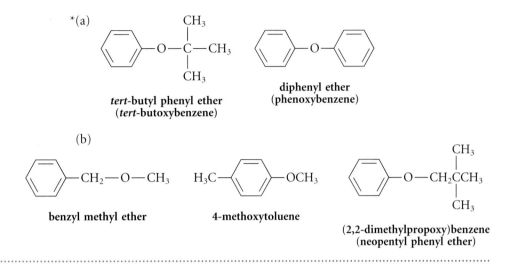

*(a)

tert-butyl phenyl ether
(*tert*-butoxybenzene)

diphenyl ether
(phenoxybenzene)

(b)

benzyl methyl ether

4-methoxytoluene

(2,2-dimethylpropoxy)benzene
(neopentyl phenyl ether)

18.10   Industrial Preparation and Use of Phenol

Historically, phenol has been made in a variety of ways, but the principal method used today is an elegant example of a process that gives two industrially important compounds, phenol and acetone, $(CH_3)_2C{=}O$, from a single starting material. The starting material for the manufacture of phenol is cumene (isopropylbenzene), which comes from benzene and propylene, two compounds obtained from petroleum (Sec. 16.7). The production of phenol and acetone is a two-stage process. In the first stage, cumene undergoes an *autoxidation* to form cumene hydroperoxide. (An **autoxidation** is a reaction with atmospheric oxygen that gives peroxides).

Autoxidation:

$$Ph{-}\underset{\underset{H}{|}}{\overset{\overset{CH_3}{|}}{C}}{-}CH_3 + O_2 \longrightarrow Ph{-}\underset{\underset{O{-}O{-}H}{|}}{\overset{\overset{CH_3}{|}}{C}}{-}CH_3 \qquad (18.53a)$$

cumene   cumene hydroperoxide

In the second stage, cumene hydroperoxide undergoes an acid-catalyzed rearrangement that yields both acetone and phenol.

Rearrangement:

$$Ph{-}\underset{\underset{O{-}O{-}H}{|}}{\overset{\overset{CH_3}{|}}{C}}{-}CH_3 \xrightarrow[\text{H}_2\text{O}]{\overset{5{-}25\%}{\text{H}_2\text{SO}_4}} Ph{-}OH + CH_3{-}\overset{\overset{O}{\|}}{C}{-}CH_3$$

phenol   acetone

$$(18.53b)$$

Phenol is a very important article of commerce. The annual U.S. production of phenol is about 3.7 billion pounds. Phenol is a starting material for the production of phenol-formaldehyde resins (Sec. 19.15), which are polymers that have a variety of uses, including plywood adhesives, glass fiber (Fiberglass) insulation, molded phenolic plastics used in automobiles and appliances, and many others. Fig. 5.3 shows how the manufacture of phenol and acetone fits into the overall chemical economy.

<table>
<tr><td>**PROBLEM**</td></tr>
</table>

*18.17   Compound *A* below is a by-product of the autoxidation of cumene, and compound *B* is a by-product of the acid-catalyzed conversion of cumene hydroperoxide to phenol and acetone.

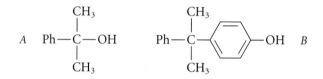

Give a curved-arrow mechanism that shows how compound *A* can react with phenol under the conditions of Eq. 18.53b to give compound *B*.

KEY IDEAS IN CHAPTER 18

⚗ Neither aryl halides, vinylic halides, nor phenols undergo S_N2 reactions because of the *sp* hybridization of the S_N2 transition states and because backside attack of nucleophiles is sterically blocked.

⚗ Neither aryl halides, vinylic halides, nor phenols undergo S_N1 reactions because of the instability of the carbocation intermediates that would be involved in such reactions.

⚗ Because the aryl-oxygen bond is unreactive in both S_N1 and S_N2 reactions, the cleavage of aryl alkyl ethers occurs exclusively at the alkyl-oxygen bond to give phenols.

⚗ Vinylic halides undergo β-elimination reactions in base under vigorous conditions to give alkynes. Strong bases can also promote β-elimination in aryl halides to give benzyne intermediates, which are rapidly consumed by addition reactions with the bases.

⚗ Aryl halides substituted with *ortho-* or *para*-electron-attracting groups, particularly nitro groups, undergo nucleophilic aromatic substitution. This type of reaction involves attack of the nucleophile to form a resonance-stabilized anion (Meisenheimer complex), which then expels the halide leaving group.

⚗ Aryl and vinylic halides form Grignard reagents. However, formation of Grignard reagents from chlorobenzene, and formation of most vinylic Grignard reagents, require tetrahydrofuran (THF) rather than ether as a solvent.

⚗ Phenols are considerably more acidic than alcohols because phenoxide ions, the conjugate bases of phenols, are stabilized both by resonance and by the electron-attracting polar effect of the aromatic ring. Phenols containing substituent groups that stabilize negative charge by resonance and/or polar effects are even more acidic.

⚗ Phenols dissolve in dilute hydroxide solution because they are converted into salts of their conjugate-base anions, which are water-soluble ionic compounds.

⚗ Phenols are oxidized to quinones.

⚗ Some electrophilic aromatic substitution reactions of phenols show unusual effects attributable to the —OH group. Thus phenol brominates three times in bromine water because the —OH group can ionize; and phenol is rather unreactive in Friedel-Crafts acylation reactions because the —OH group reacts with the $AlCl_3$ catalyst.

⚗ The preparation of phenol and acetone by the autoxidation of cumene and rearrangement of the resulting hydroperoxide is an important industrial process.

ADDITIONAL PROBLEMS

*18.18 What reaction (if any) takes place when *p*-iodotoluene is subjected to each of the following conditions?
   (a) CH_3OH, 25°
   (b) CH_3O^- in CH_3OH, 25°
   (c) CH_3O^-, pressure, heat
   (d) Mg in THF
   (e) Li in hexane
   (f) product of (d) + D_2O

*18.19 Give the product(s) expected (if any) when *m*-cresol is subjected to each of the following conditions.
   (a) concentrated H_2SO_4
   (b) Br_2 in CCl_4 (dark)
   (c) Br_2 (excess) in CCl_4, light
   (d) dilute HCl
   (e) 0.1 *M* NaOH solution
   (f) HNO_3, cold

   (g)   O
               ‖
        C_2H_5—C—Cl, $AlCl_3$, heat; then H_2O

   (h) $Na_2Cr_2O_7$ in H_2SO_4

18.20 Arrange the compounds within each set in order of increasing acidity, and explain your reasoning.
   *(a) cyclohexyl mercaptan, cyclohexanol, benzenethiol
   (b) cyclohexanol, phenol, benzyl alcohol

   (Problem 18.20 continues)

*(c)

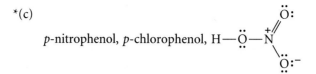

p-nitrophenol, p-chlorophenol, $H-\overset{..}{\underset{..}{O}}-\overset{+}{N}$

(d) 4-nitrobenzenethiol, 4-nitrophenol, phenol

*(e) 2,4-dimethylphenol, 2,6-di(*tert*-butyl)phenol, 2,6-dichlorophenol

*(f)

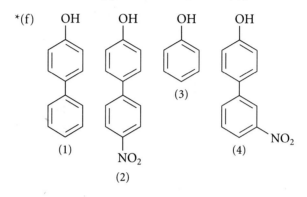

(1)   (2)   (3)   (4)

*18.21   Although enols are unstable compounds (Sec. 14.5A), suppose that the acidity of an enol could be measured. Which would be more acidic: enol *A* or alcohol *B*? Why?

$$A \quad CH_3-\overset{\overset{\displaystyle OH}{|}}{C}=CH_2 \qquad CH_3-\overset{\overset{\displaystyle OH}{|}}{CH}-CH_3 \quad B$$

*18.22   Identify compounds *A*, *B*, and *C* from the following information.

(a) Compound *A*, $C_8H_{10}O$, is insoluble in water but soluble in aqueous NaOH solution, and yields 3,5-dimethylcyclohexanol when hydrogenated over a nickel catalyst at high pressure.

(b) Aromatic compound *B*, $C_8H_{10}O$, is insoluble in both water and aqueous NaOH solution. When treated successively with concentrated HBr, then Mg in THF, then water, it gives *p*-xylene.

(c) Compound *C*, $C_9H_{12}O$, is insoluble in water and in NaOH solution, but reacts with concentrated HBr and heat to give *m*-cresol and a volatile alkyl bromide.

18.23   Complete the following reaction sequence by giving the major organic product.

$$CH_3CH_2-\!\!\left\langle\!\!\bigcirc\!\!\right\rangle\!\!-OH \xrightarrow{\text{NaOH solution}} \xrightarrow{\text{dimethyl sulfate}} \xrightarrow[\text{high pressure}]{H_2,\ \text{catalyst}} \quad ?$$

*18.24   Contrast the reactivities of cyclohexanol and phenol with each of the following reagents, and explain.

(a) aqueous NaOH solution

(b) *p*-toluenesulfonyl chloride in pyridine

(c) NaH in THF
(d) concentrated aqueous HBr, H_2SO_4 catalyst
(e) Br_2 in CCl_4 (f) $Na_2Cr_2O_7$ in H_2SO_4 (g) H_2SO_4, heat

18.25 (a) Give the products (if any) when each of the following compounds reacts with HBr and heat.

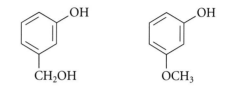

3-(hydroxymethyl)phenol   **3-methoxyphenol**

(b) Give the products (if any) when each of the following compounds reacts with Br_2 in CCl_4 in the dark.

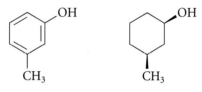

m-cresol   **cis-3-methylcyclohexanol**

*18.26 Choose the one compound within each set that meets the indicated criterion, and explain your choice.
(a) The compound that reacts with alcoholic KOH to liberate fluoride ion:

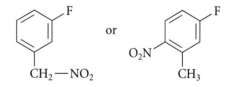

(b) The compound that *cannot* be prepared by a Williamson ether synthesis:

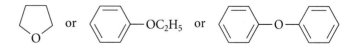

(c) The compound that gives an acidic solution when allowed to stand in aqueous ethanol:

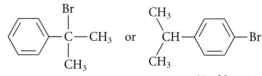

(*Problem 18.26 continues*)

(d)   The compound that reacts with magnesium in ether:

p-chlorotoluene   or   benzyl chloride

(e)   The ether that would cleave more rapidly in HI:

phenyl cyclohexyl ether   or   diphenyl ether

(f)   The compound that gives two products when it reacts with KNH$_2$ in liquid ammonia:

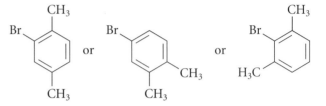

*18.27   Explain the following observations, which were recorded in the chemical literature, concerning the reaction between *tert*-butyl bromide and potassium benzenethiolate (the potassium salt of benzenethiol): "The attempts to prepare phenyl *tert*-butyl sulfide by this route failed. If the reactants were kept at room temperature KBr was formed, but the benzenethiol was recovered unchanged."

18.28   What products (if any) are formed when 3,5-dimethylbenzenethiol is treated first with one equivalent of Na$^+$ C$_2$H$_5$O$^-$ in ethanol, and then with each of the following?
(a)   allyl bromide   (b)   bromobenzene
(c)   2-bromo-2-methylbutane and heat

*18.29   A mixture of *p*-cresol, pK_a = 10.2, and 2,4-dinitrophenol, pK_a = 4.11, is dissolved in ether. The ether solution is then vigorously shaken with one of the following aqueous solutions. Which solution effects the best separation of the two phenols by dissolving one in the water layer and leaving the other in the ether solution? Explain. (*Hint:* Apply Eq. 3.14.)
(1)   a 0.1 *M* aqueous HCl solution
(2)   a solution that contains a large excess of pH = 4 buffer
(3)   a solution that contains a large excess of pH = 7 buffer
(4)   0.1 *M* NaOH solution

*18.30   Phenols, like alcohols, are Brønsted bases.
(a)   Write the reaction in which phenol reacts as a base with the acid H$_2$SO$_4$.
(b)   On the basis of resonance and polar effects, decide whether phenol or cyclohexanol should be the stronger base.

*18.31   The UV spectrum of *p*-nitrophenol in aqueous solution is shown in Fig. 18.3 (trace *A*). When a few drops of concentrated NaOH are added, the solution turns yellow and the spectrum changes (trace *B*). On addition of a few drops of concentrated acid, the color disappears and the spectrum shown in trace *A* is restored. Explain these observations.

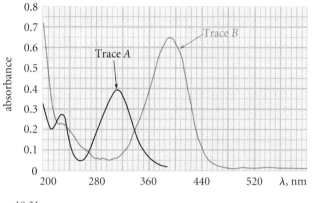

Figure 18.3   *UV spectra for Problem 18.31.*

18.32   Vanillin is the active component of the natural flavoring vanilla.

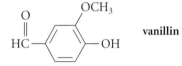   **vanillin**

When a few drops of vanillin are added to an aqueous NaOH solution, the characteristic vanilla odor is not present. Upon acidification of the solution, a strong vanilla odor is present. Explain.

*18.33   It has been suggested that the solvolysis of 2-(bromomethyl)-5-nitrophenol at alkaline pH values involves the intermediate shown in brackets below.

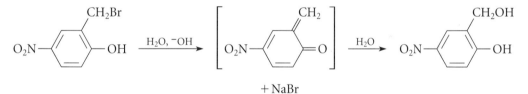

+ NaBr

Give a curved-arrow mechanism for the formation of this intermediate and for its reaction with water to give the final product.

18.34   When a suspension of 2,4,6-tribromophenol is treated with an excess of bromine water, the white precipitate of 2,4,6-tribromophenol disappears and is replaced by a precipitate of a yellow compound that has the following structure. Give a curved-arrow mechanism for the formation of this compound.

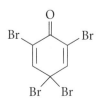

18.35   Complete each reaction by giving the major organic product(s). *No reaction* may be an appropriate response.

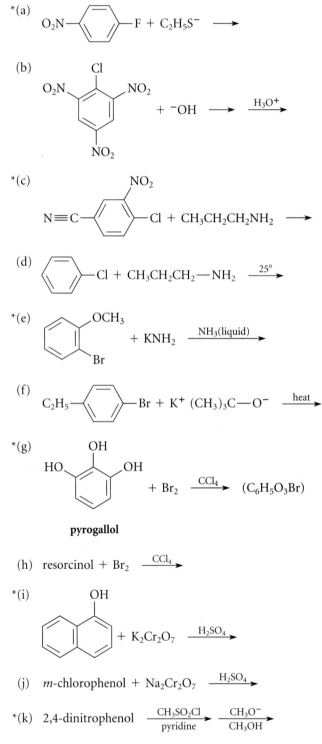

*(a) O_2N—⟨ ⟩—F + $C_2H_5S^-$ ⟶

(b)   ⟨structure⟩ + ^-OH ⟶ $\xrightarrow{H_3O^+}$

*(c)   ⟨structure⟩ + $CH_3CH_2CH_2NH_2$ ⟶

(d)   ⟨ ⟩—Cl + $CH_3CH_2CH_2$—NH_2 $\xrightarrow{25°}$

*(e)   ⟨structure with OCH_3 and Br⟩ + KNH_2 $\xrightarrow{NH_3(liquid)}$

(f) C_2H_5—⟨ ⟩—Br + $K^+ (CH_3)_3C$—O^- $\xrightarrow{heat}$

*(g)   ⟨pyrogallol structure⟩ + Br_2 $\xrightarrow{CCl_4}$ ($C_6H_5O_3Br$)

pyrogallol

(h)   resorcinol + Br_2 $\xrightarrow{CCl_4}$

*(i)   ⟨naphthol structure⟩ + $K_2Cr_2O_7$ $\xrightarrow{H_2SO_4}$

(j)   *m*-chlorophenol + $Na_2Cr_2O_7$ $\xrightarrow{H_2SO_4}$

*(k)   2,4-dinitrophenol $\xrightarrow[\text{pyridine}]{CH_3SO_2Cl}$ $\xrightarrow[\text{CH}_3\text{OH}]{CH_3O^-}$

(l)

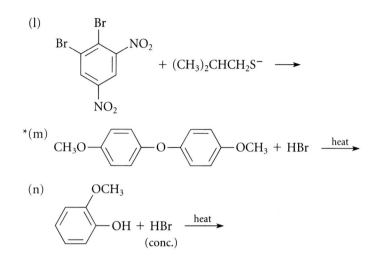

+ (CH₃)₂CHCH₂S⁻ $\longrightarrow$

*(m)

CH₃O—⟨⟩—O—⟨⟩—OCH₃ + HBr $\xrightarrow{\text{heat}}$

(n)

⟨⟩—OH + HBr $\xrightarrow{\text{heat}}$
(conc.)
(with OCH₃ substituent)

*18.36   The heat (or enthalpy) of hydrogenation is the heat liberated when a substance undergoes catalytic hydrogenation. The hydrogenation reactions of all ordinary internal alkynes (alkynes with alkyl substituents at both carbons of the triple bond) to the corresponding *cis*-alkenes give off about the same amount of heat.
(a)   Using the heats of formation below, calculate the approximate heat of hydrogenation of an ordinary internal alkyne.

Compound	ΔH_f° (kJ/mol)	(kcal/mol)	Compound	ΔH_f° (kJ/mol)	(kcal/mol)
2-butyne	146	35.0	benzyne	494	118.0
cis-2-butene	−7.1	−1.7	benzene	83	19.8
2-pentyne	129	30.8			
cis-2-pentene	−28	−6.7			

(b)   Calculate the heat of hydrogenation of benzyne to benzene. Use the discrepancy between this number and the number you calculated in (a) to estimate the energy by which benzyne is destabilized relative to an ordinary alkyne.

18.37   Outline a synthesis for each of the following compounds from the indicated starting material and any other reagents.
*(a)   1-chloro-2,4-dinitrobenzene from benzene
(b)   1-chloro-3,5-dinitrobenzene from benzene

*(c) 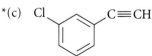   from *m*-bromochlorobenzene

(d)

Ph—C≡C—Ph   from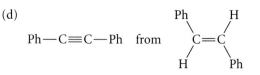

(Problem 18.37 continues)

18.45   Using the curved-arrow formalism where appropriate, give stepwise mechanisms for the following reactions.

*(a)

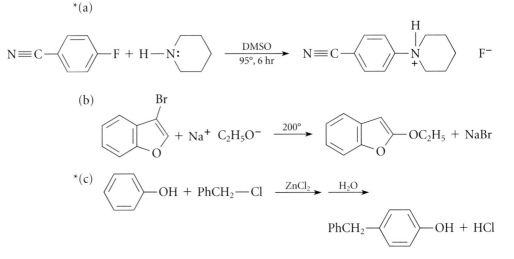

(b)

(c)

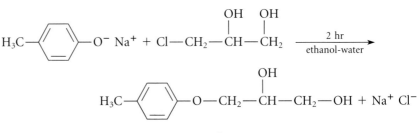

(*Hint:* ZnCl$_2$ is a Lewis acid; see Sec. 16.4F.)

(d)

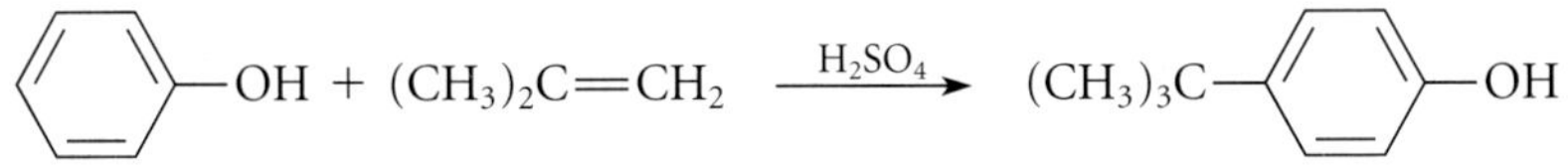

*18.46   The following reaction, used to prepare the drug Mephenesin (a skeletal muscle relaxant), appears to be a simple Williamson ether synthesis.

Mephenesin

During this reaction a precipitate of NaCl forms after only about ten minutes, but a considerably longer reaction time is required to obtain a good yield of Mephenesin. Taking these facts into account, suggest a mechanism for this reaction. (*Hint:* See Sec. 11.6.)

*18.47   Explain why the dipole moment of 4-chloronitrobenzene (2.69 D) is *less* than that of nitrobenzene (3.99 D), and the dipole moment of 4-nitroanisole (1-methoxy-4-nitrobenzene, 4.92 D) is *greater* than that of nitrobenzene, even though the electronegativities of chlorine and oxygen are about the same.

*18.48   In the following conversion the Diels-Alder reaction is used to trap a very interesting intermediate by its reaction with anthracene. From the structure of the product deduce the structure of the intermediate. Then write a mechanism that shows how the intermediate is formed from the starting material.

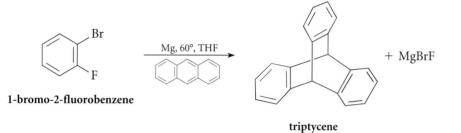

1-bromo-2-fluorobenzene   triptycene

*18.49   Propose a structure for the product *A* obtained in the following oxidation of 2,4,6-trimethylphenol. (Compound *A* is a rather unstable type of compound called generally a *quinone methide*.)

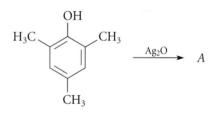

Proton NMR of *A*: δ 1.90 (*6H*), δ 5.49 (*2H*), δ 6.76 (*2H*), all broad singlets.

*18.50   For the reactions below, explain why different products are obtained when different amounts of AlBr$_3$ catalyst are used. (*Hint:* See Sec. 18.8B.)

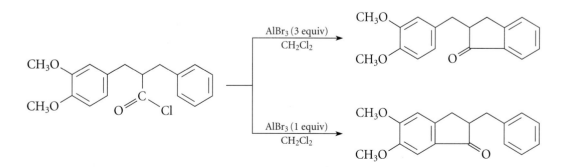

*18.51   When the following unusual epoxide reacts with water, *p*-cresol is the only product formed, and it has the labeling pattern shown. This rearrangement is called the "NIH shift," because it was discovered by chemists at the National Institutes of Health.

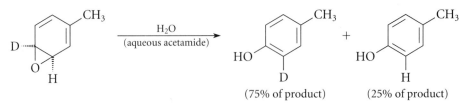

(75% of product)   (25% of product)

(a) Suggest a mechanism for this reaction that accounts for the position of the label, and explain why some of the deuterium is lost.

(b) Explain why only *p*-cresol and no *m*-cresol is formed.

19

Chemistry of Aldehydes and Ketones; Carbonyl-Addition Reactions

T his chapter begins the study of **carbonyl compounds**—compounds containing the **carbonyl group**, C=O. Aldehydes, ketones, carboxylic acids, and the carboxylic acid derivatives—esters, amides, anhydrides, and acid chlorides—are all carbonyl compounds.

This chapter focuses on the nomenclature, properties, and characteristic carbonyl-group reactions of aldehydes and ketones. Chapters 20 and 21 will consider carboxylic acids and carboxylic acid derivatives, respectively. Chapter 22 deals with ionization, enolization, and condensation reactions, which are common to the chemistry of all classes of carbonyl compounds.

Aldehydes and ketones have the following general structures:

Examples:

$$CH_3-\underset{\substack{\|\\O}}{C}-H \qquad CH_3-\underset{\substack{\|\\O}}{C}-CH_3$$

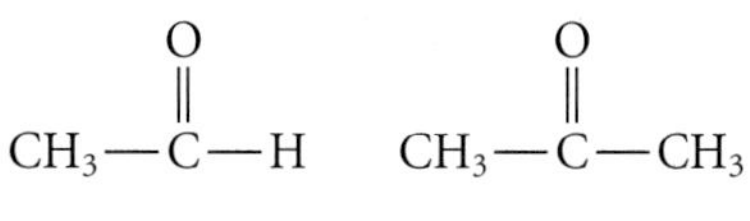

acetaldehyde
(an aldehyde)

acetone
(a ketone)

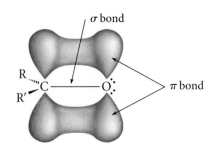

Figure 19.1   *Bonding in a typical carbonyl compound. The carbonyl group and the two R groups lie in the same plane, and there are both a σ bond and a π bond between the carbonyl carbon and oxygen.*

In a **ketone**, the groups bound to the carbonyl carbon (R and R′ in the structures above) are alkyl or aryl groups. In an **aldehyde**, at least one of the groups at the carbonyl carbon atom is a hydrogen, and the other may be alkyl, aryl, or a second hydrogen.

The carbonyl carbon of a typical aldehyde or ketone is sp^2-hybridized with bond angles approximating 120°. The carbon-oxygen double bond consists of a σ bond and a π bond, much like the double bond of an alkene (Fig. 19.1). Just as C—O single bonds are shorter than C—C single bonds, C=O bonds are shorter than C=C bonds (Sec. 1.3B). The structures of some simple aldehydes and ketones are given in Fig. 19.2.

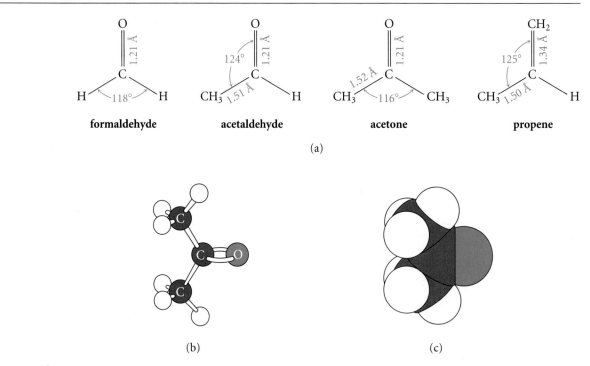

(a)

(b)   (c)

Figure 19.2   *Structures of aldehydes and ketones. (a) The structures of formaldehyde, acetaldehyde, and acetone compared to the structure of propene. Notice that the C=O bonds are shorter than the C=C bond. Notice also that the carbonyl carbon is trigonal planar with bond angles very close to 120°. (b) Ball-and-stick model of acetone. (c) Space-filling model of acetone.*

19.1   Nomenclature of Aldehydes and Ketones

A.   Common Nomenclature

Common names are almost always used for the simplest aldehydes. In common nomenclature the suffix *aldehyde* is added to a prefix that indicates the chain length. A list of prefixes is given in Table 19.1.

$$CH_2\!\!=\!\!O \qquad CH_3\!-\!CH_2\!-\!CH_2\!-\!CH\!\!=\!\!O$$

formaldehyde   butyr + *aldehyde* = **butyraldehyde**

Acetone is the common name for the simplest ketone.

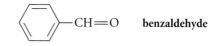

CH$_3$—C—CH$_3$ **acetone**

Benzaldehyde is the simplest aromatic aldehyde.

—CH$=$O   **benzaldehyde**

Certain aromatic ketones are named by attaching the suffix *ophenone* to the appropriate prefix from Table 19.1.

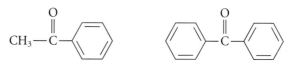

acet + *ophenone* = **acetophenone**   **benzophenone**

The common names of some ketones are constructed by citing the two groups on the carbonyl carbon followed by the word *ketone*.

Table 19.1   Prefixes Used in Common Nomenclature of Carbonyl Compounds

Prefix	R— in R—CH=O or R—CO$_2$H	Prefix	R— in R—CH=O or R—CO$_2$H
form	H—	isobutyr	(CH$_3$)$_2$CH—
acet	CH$_3$—	valer	CH$_3$CH$_2$CH$_2$CH$_2$—
propion, propi[a]	CH$_3$CH$_2$—	isovaler	(CH$_3$)$_2$CHCH$_2$—
butyr	CH$_3$CH$_2$CH$_2$—	benz, benzo[b]	Ph—

[a] Used in phenone nomenclature as discussed in the text.
[b] Used in carboxylic acid nomenclature (Sec. 20.1A).

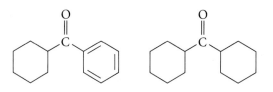

cyclohexyl phenyl ketone   **dicyclohexyl ketone**

Simple substituted aldehydes and ketones can be named in the common system by designating the positions of substituents with Greek letters, beginning at the position *adjacent* to the carbonyl group.

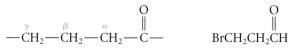

β-bromopropionaldehyde

As suggested by this nomenclature, a carbon *adjacent* to the carbonyl group is termed the **α-carbon**, and the hydrogens on the α-carbon are termed **α-hydrogens**.

Many common carbonyl-containing substituent groups are named by a simple extension of the terminology in Table 19.1: the suffix *yl* is added to the appropriate prefix. The following names are examples:

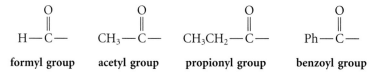

formyl group   **acetyl group**   **propionyl group**   **benzoyl group**

Such groups are called in general **acyl groups**. (This is the source of the term *acylation*, used in Sec. 16.4E.) To be named as an acyl group, a substituent group must be connected to the remainder of the molecule *at its carbonyl carbon*.

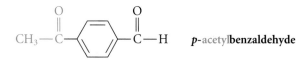   *p*-acetylbenzaldehyde

Be careful not to confuse the *benzoyl* group, an acyl group, with the *benzyl* group, an alkyl group.

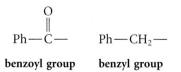

benzoyl group   **benzyl group**

A great many aldehydes and ketones were well known long before any system of nomenclature existed. These are known by traditional names that are illustrated by the following examples:

Ph—CH=CH—CH=O   —CH=O   CH₂=CH—CH=O

$$Ph-CH=CH-CH=O \qquad \text{furfural} \qquad CH_2=CH-CH=O$$

cinnamaldehyde   **furfural**   **acrolein**

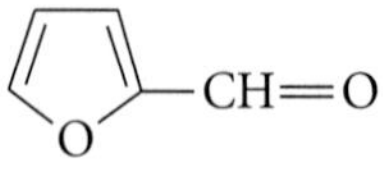

B. Substitutive Nomenclature

The substitutive name of an aldehyde is constructed from a prefix indicating the length of the carbon chain followed by the suffix *al*. The prefix is the name of the corresponding hydrocarbon without the final *e*.

$$CH_3CH_2CH_2CH{=}O \qquad butan\cancel{e} + al = \textbf{butanal}$$

In numbering the carbon chain of an aldehyde, the carbonyl carbon receives the number one.

$$\overset{4}{C}H_3\overset{3}{C}H_2\overset{2}{C}H\overset{1}{C}H{=}O \qquad \textbf{2-methylbutanal}$$
$$\underset{CH_3}{|}$$

Note carefully the difference in chain numbering of aldehydes in common and substitutive nomenclature. In common nomenclature, numbering begins at the carbon *adjacent* to the carbonyl (the *α*-carbon); in substitutive nomenclature, numbering begins at the carbonyl carbon itself.

As with diols, the final *e* is not dropped when the carbon chain has more than one aldehyde group.

$$O{=}CH{-}CH_2CH_2CH_2CH_2{-}CH{=}O \qquad \textbf{hexanedial}$$

When an aldehyde group is attached to a ring, the suffix *carbaldehyde* is appended to the name of the ring. (In older literature, the suffix *carboxaldehyde* was used.)

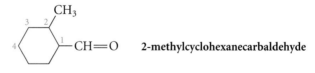

cyclohexanecarbaldehyde

In aldehydes of this type, carbon-1 is not the carbonyl carbon, but rather the ring carbon attached to the carbonyl group.

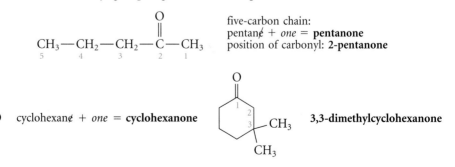

The name *benzaldehyde* (Sec. 19.1A) is used in both common and substitutive nomenclature.

A ketone is named by giving the hydrocarbon name of the longest carbon chain containing the carbonyl group, dropping the final *e*, and adding the suffix *one*. The position of the carbonyl group is given the lowest possible number.

$$\underset{5}{C}H_3{-}\underset{4}{C}H_2{-}\underset{3}{C}H_2{-}\overset{\overset{\textstyle O}{\|}}{\underset{2}{C}}{-}\underset{1}{C}H_3$$

five-carbon chain:
pentan$\cancel{e}$ + *one* = **pentanone**
position of carbonyl: **2-pentanone**

As with diols and dialdehydes, the final *e* of the hydrocarbon name is not dropped in the nomenclature of diones, triones, etc.

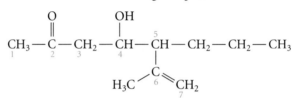

six-carbon chain:
hexane + *dione* = **hexanedione**
positions of carbonyls: **2,4-hexanedione**

Aldehyde and ketone carbonyl groups receive higher priority than —OH or —SH groups for citation as *principal groups* (Sec. 8.1B).

Priority for citation as principal group:

$$\underset{\text{—CH (aldehyde)}}{\overset{\overset{\displaystyle O}{\|}}{}} > \underset{\text{—C— (ketone)}}{\overset{\overset{\displaystyle O}{\|}}{}} > \text{—OH} > \text{—SH} \qquad (19.1)$$

STUDY
PROBLEM
19.1

Provide a substitutive name for the following compound.

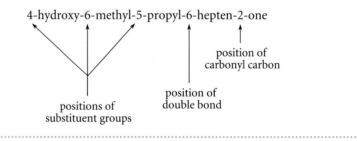

(The numbers are utilized in the solution below.)

Solution   To name this compound use the nomenclature rules in Sec. 8.1B. First, *identify the principal group*. Possible candidates are the carbonyl group at carbon-2 and the hydroxy group at carbon-4. Because ketones have a higher citation priority than hydroxy groups (Eq. 19.1), the compound is named as a ketone with the suffix *one*. Next, *identify the principal chain*. This is the longest carbon chain containing the principal group and the greatest number of double and triple bonds. This chain (numbered in the structure above) contains seven carbons. Notice that the longer carbon chain within the molecule is not the principal chain because the presence of double bonds takes precedence over length. Hence, the compound is named as a heptenone with hydroxy, methyl, and propyl substituents cited in alphabetical order. Finally, *number the principal chain*. Number from the end of the chain so that the principal group—the carbonyl group—receives the lowest possible number. Thus, the carbonyl carbon is carbon-2, the hydroxy carbon is carbon-4, the carbon bearing the propyl group is carbon-5, and the first alkene carbon is carbon-6 (see numbering in the structure above). Cite the substituent groups in alphabetical order. The name is thus:

4-hydroxy-6-methyl-5-propyl-6-hepten-2-one

positions of
substituent groups

position of
double bond

position of
carbonyl carbon

When a ketone carbonyl group is treated as a substituent, its position is designated by the term *oxo*.

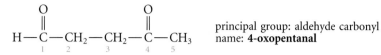

principal group: aldehyde carbonyl
name: **4-oxopentanal**

19.1   Give the structure of the following compounds.

  *(a) isobutyraldehyde   (b) valerophenone
  *(c) *o*-bromoacetophenone   (d) γ-chlorobutyraldehyde
  *(e) 4-(2-chlorobutyryl)benzaldehyde   (f) 3-cyclohexenone
  *(g) 2-oxocyclopentanecarbaldehyde   (h) 3-hydroxy-2-butanone
  *(i) *m*-methoxypropiophenone

19.2   Give the substitutive name for each of the following compounds.

  *(a) diisopropyl ketone   (b) acetone

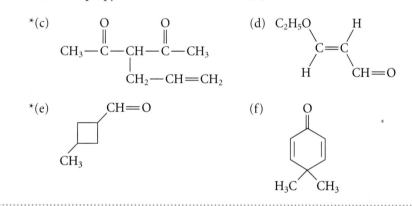

*(e)

(f)

Physical Properties of Aldehydes and Ketones

Most simple aldehydes and ketones are liquids. However, formaldehyde is a gas, and acetaldehyde has a boiling point (20.8°) very near room temperature, although it is usually sold as a liquid.

Aldehydes and ketones are polar molecules because of their C=O bond dipoles.

$$\delta^- \; :\!\ddot{O}: \uparrow$$
$$\underset{\delta^+}{C}$$

Because of their polarities, aldehydes and ketones have higher boiling points than alkenes or alkanes with similar molecular masses and shapes. But because aldehydes and ketones are not hydrogen-bond donors, their boiling points are considerably lower than those of the corresponding alcohols.

	$CH_3CH{=}CH_2$	$CH_3CH{=}O$	CH_3CH_2OH
boiling point	$-47.4°$	$20.8°$	$78.3°$
dipole moment	0.4 D	2.7 D	1.7 D

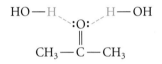

boiling point	$-6.9°$	$56.5°$	$82.3°$
dipole moment	0.5 D	2.7 D	1.7 D

Aldehydes and ketones with four or fewer carbons have considerable solubilities in water because they can accept hydrogen bonds from water at the carbonyl oxygen.

$$HO-H \quad\quad\quad H-OH$$
$$\overset{\displaystyle :O:}{\underset{\displaystyle CH_3-C-CH_3}{\|}}$$

Acetaldehyde and acetone are miscible with water (that is, soluble in all proportions). The water solubility of aldehydes and ketones along a series diminishes rapidly with increasing molecular mass.

Acetone and 2-butanone are especially valued as solvents because they dissolve not only water but also a wide variety of organic compounds. These solvents have sufficiently low boiling points that they can be easily separated from other less volatile compounds. Acetone, with a dielectric constant of 21, is a polar solvent, and is often used as a solvent or co-solvent for nucleophilic substitution reactions.

19.3   Spectroscopy of Aldehydes and Ketones

A. IR Spectroscopy

The principal infrared absorption of aldehydes and ketones is the $C{=}O$ stretching absorption, a strong absorption that occurs in the vicinity of 1700 cm^{-1}. In fact, this is one of the most important of all infrared absorptions. Because the $C{=}O$ bond is stronger than the $C{=}C$ bond, the stretching frequency of the $C{=}O$ bond is greater.

The position of the $C{=}O$ stretching absorption varies predictably for different types of carbonyl compounds. It generally occurs at 1710–1715 cm^{-1} for simple ketones and at 1720–1725 cm^{-1} for simple aldehydes. The carbonyl absorption is clearly evident, for example, in the IR spectrum of butyraldehyde (Fig. 19.3). The stretching absorption of the carbonyl-hydrogen bond of aldehydes near 2710 cm^{-1} is another characteristic absorption; however, NMR spectroscopy provides a more reliable way to diagnose the presence of this type of hydrogen (Sec. 19.3B).

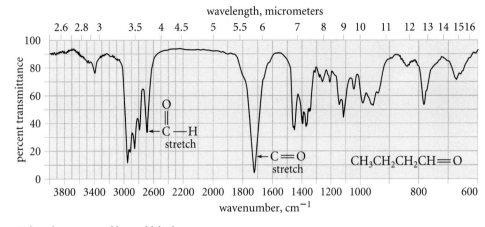

Figure 19.3   *Infrared spectrum of butyraldehyde.*

Compounds in which the carbonyl group is conjugated with aromatic rings, double bonds, or triple bonds have lower carbonyl stretching frequencies than unconjugated carbonyl compounds.

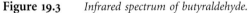

	acetophenone	3-buten-2-one		1-butene	2-butanone
$\diagdown C=O \diagup$	1685 cm^{-1}	1670 cm^{-1}		—	1715 cm^{-1}
$\diagdown C=C \diagup$	1600 cm^{-1}	1613 cm^{-1}		1642 cm^{-1}	—

$$(19.2)$$

Note that the carbon-carbon double-bond stretching frequencies are also lower in the conjugated molecules. These effects can be explained by the resonance structures for these compounds. Since the C=O and C=C bonds have some single-bond character, as indicated by the following resonance structures, they are somewhat weaker than ordinary double bonds, and therefore absorb in the IR at lower frequency.

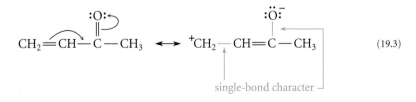

$$(19.3)$$

In cyclic ketones with rings containing fewer than six carbons, the carbonyl absorption frequency increases significantly as the ring size decreases.

STUDY GUIDE LINK:
19.1
*IR Absorptions of
Cyclic Ketones*

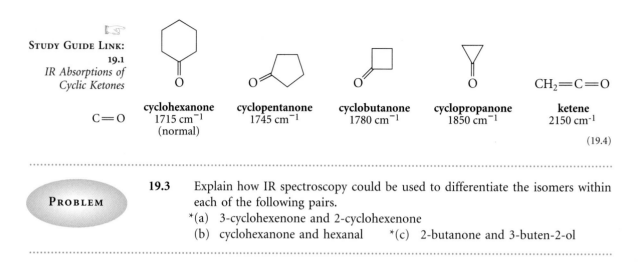

	cyclohexanone	cyclopentanone	cyclobutanone	cyclopropanone	ketene
C=O	$1715\ cm^{-1}$ (normal)	$1745\ cm^{-1}$	$1780\ cm^{-1}$	$1850\ cm^{-1}$	$2150\ cm^{-1}$

(19.4)

PROBLEM

19.3   Explain how IR spectroscopy could be used to differentiate the isomers within each of the following pairs.
*(a)   3-cyclohexenone and 2-cyclohexenone
(b)   cyclohexanone and hexanal   *(c)   2-butanone and 3-buten-2-ol

B. Proton NMR Spectroscopy

The characteristic NMR absorption common to both aldehydes and ketones is that of the protons on the carbons *adjacent* to the carbonyl group—the α-protons. This absorption is in the δ 2.0–2.5 region of the spectrum (see also Table 13.2). This absorption is slightly farther downfield than the absorptions of allylic protons; this makes sense because the C=O group is more electronegative than the C=C group. In addition, the absorption of the aldehydic proton is quite distinctive, occurring in the δ 9–10 region of the NMR spectrum, at lower field than most other NMR absorptions.

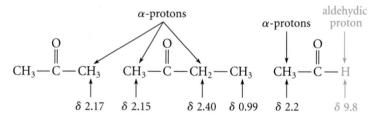

In general, aldehydic protons are significantly deshielded. The reason for this deshielding is the same as that for the deshielding effects of the carbon-carbon double bond (Sec. 13.6A). However, the carbonyl group has a greater deshielding effect because of the electronegativity of the carbonyl oxygen.

PROBLEM

19.4   Deduce the structures of the following compounds.
*(a) C_4H_8O: IR 1720, 2710 cm^{-1}
NMR in Fig. 19.4.
(b) C_4H_8O: IR 1717 cm^{-1}
NMR δ 0.95 (3H, t, J = 8 Hz); δ 2.03 (3H, s); δ 2.38 (2H, q, J = 8 Hz)
*(c) $C_{10}H_{12}O_2$: IR 1690 cm^{-1}, 1612 cm^{-1}
NMR δ 1.4 (3H, t, J = 8 Hz); δ 2.5 (3H, s); δ 4.1 (2H, q, J = 8 Hz);
δ 6.9 (2H, d, J = 9 Hz); δ 7.9 (2H, d, J = 9 Hz)

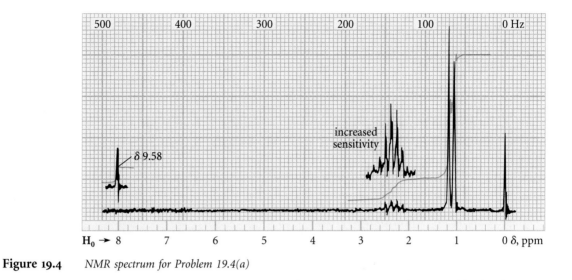

Figure 19.4   *NMR spectrum for Problem 19.4(a)*

C. Carbon NMR Spectroscopy

The most characteristic absorption of aldehydes and ketones in CMR spectroscopy is that of the carbonyl carbon, which occurs typically in the δ 190–220 ppm range. This large downfield shift is due to the induced electron circulation in the π bond, as in alkenes (Fig. 13.15, Sec. 13.6A), and to the additional chemical-shift effect of the electronegative carbonyl oxygen. Because the carbonyl carbon of a ketone bears no hydrogens, its CMR absorption, like that of other quaternary carbons, is characteristically rather weak (Sec. 13.8). This effect is evident in the CMR spectrum of propiophenone (Fig. 19.5).

The α-carbon absorptions of aldehydes and ketones show modest downfield shifts, typically in the δ 30–50 ppm range, with, as usual, greater shifts for more branched carbons. The α-carbon shift of propiophenone, 31.7 ppm (Fig. 19.5, carbon *b*) is typical. Because shifts in this range are also observed for other functional groups, these absorptions are less useful than the carbonyl carbon resonances for identifying aldehydes and ketones.

PROBLEMS

*19.5   Propose a structure for a compound $C_6H_{12}O$ that has IR absorption at 1705 cm^{-1}, no proton NMR absorption at a chemical shift greater than δ 3, and the following CMR spectrum: δ 24.4, δ 26.5, δ 44.2, δ 212.6. The resonances at δ 44.2 and δ 212.6 have very low intensity.

19.6   The CMR spectrum of 2-ethylbutanal consists of the following absorptions: δ 11.5, δ 21.7, δ 55.2, δ 204.7. Draw the structure of this aldehyde, label each chemically nonequivalent set of carbons, and assign each absorption to the appropriate carbon(s).

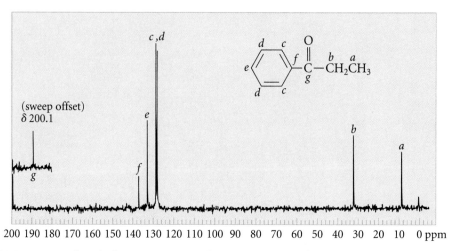

Figure 19.5   *CMR spectrum of propiophenone. Notice two things particularly about the spectrum: the large down-field shift of the carbonyl carbon g, and the small signals for the two carbons (f and g) that bear no protons. Recall (Sec. 13.8) that, although signal intensities in carbon spectra generally do not accurately correspond to numbers of carbons, quaternary carbons generally have weaker signals than proton-bearing carbons.*

D. UV Spectroscopy

The $\pi \rightarrow \pi^*$ absorptions (Sec. 15.2B) of unconjugated aldehydes and ketones occur at about 150 nm, a wavelength well below the operating range of common UV spectrometers. Simple aldehydes and ketones also have another, much weaker, absorption at higher wavelength, in the 260–290 nm region. This absorption is caused by excitation of the unshared electrons on oxygen (sometimes called the *n* electrons). This high-wavelength absorption is usually referred to as an $n \rightarrow \pi^*$ **absorption**.

$$(CH_3)_2 C = \overset{..}{\underset{..}{O}} \qquad n \longrightarrow \pi^* \qquad 271 \text{ nm } (\epsilon = 16) \text{ (in ethanol)}$$

$$\underset{n \text{ electrons}}{\uparrow}$$

This absorption is easily distinguished from a $\pi \rightarrow \pi^*$ absorption because it is only 10^{-2} to 10^{-3} times as strong. However, it is strong enough that aldehydes and ketones cannot be used as solvents for UV spectroscopy.

Like conjugated dienes, the π electrons of compounds in which carbonyl groups are conjugated with double or triple bonds have strong absorption in the UV spectrum. The spectrum of 1-acetylcyclohexene (Fig. 19.6) is typical. The 232-nm peak is due to light absorption by the conjugated π-electron system and is thus a $\pi \rightarrow \pi^*$ absorption. It has a very large extinction coefficient, much like that of a conjugated diene. The weak 308-nm absorption is an $n \rightarrow \pi^*$ absorption.

conjugated
π-electron system

$\lambda_{max} = 232$ nm $(\epsilon = 13{,}200)$
$\lambda_{max} = 308$ nm $(\epsilon = 150)$
(in methanol)

1-acetylcyclohexene
[1-(cyclohexen-1-yl)ethanone]

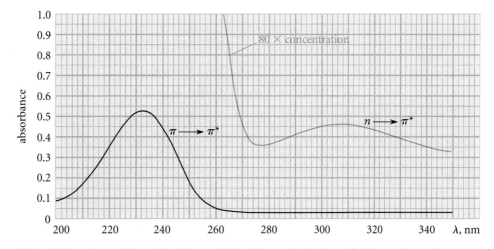

Figure 19.6   *Ultraviolet spectrum of 1-acetylcyclohexene [1-(cyclohexen-1-yl)ethanone]. The spectrum of a more concentrated solution (color) reveals the "forbidden" n → π* absorption, which is so weak that it is not apparent in the spectrum taken on a more dilute solution (black).*

The λ_{max} of a conjugated aldehyde or ketone is governed by the same variables that affect the λ_{max} values of conjugated dienes: the number of conjugated double bonds, substitution on the double bond, and so on. When an aromatic ring is conjugated with a carbonyl group, the typical aromatic absorptions are more intense and shifted to higher wavelengths than those of benzene.

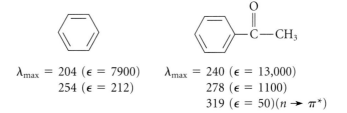

$\lambda_{max} = 204 \ (\epsilon = 7900)$ $\lambda_{max} = 240 \ (\epsilon = 13{,}000)$
$254 \ (\epsilon = 212)$ $278 \ (\epsilon = 1100)$
$319 \ (\epsilon = 50)(n \rightarrow \pi^{*})$

The $\pi \rightarrow \pi^{*}$ absorptions of conjugated carbonyl compounds, like those of conjugated alkenes, arise from the promotion of a π electron from a bonding to an antibonding (π^{*}) molecular orbital (Sec. 15.2B). An $n \rightarrow \pi^{*}$ absorption arises from promotion of one of the n (unshared) electrons on a carbonyl oxygen to a π^{*} molecular orbital. As stated above, $n \rightarrow \pi^{*}$ absorptions are weak. Spectroscopists say that these absorptions are *forbidden*. This term refers to certain physical reasons for the very low intensity of these absorptions. The 254 nm absorption of benzene, which has a very low extinction coefficient of 212, is another example of a "forbidden" absorption.

PROBLEMS

19.7   Explain how the compounds within each set can be distinguished using only UV spectroscopy.

(Problem 19.7 continues)

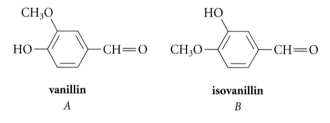

*(a)

and

(b) 2-cyclohexenone and 3-cyclohexenone

*(c) 1-phenyl-2-propanone and *p*-methylacetophenone

*19.8   Which one of the following compounds should have $\pi \rightarrow \pi^*$ UV absorption at the greater λ_{max} when the compound is dissolved in NaOH solution? Explain.

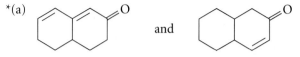

vanillin
A

isovanillin
B

19.9   In neutral alcohol solution, the UV spectra of *p*-hydroxyacetophenone and *p*-methoxyacetophenone are virtually identical. When NaOH is added to the solution, the λ_{max} of *p*-hydroxyacetophenone increases by about 50 nm, but that of *p*-methoxyacetophenone is unaffected. Explain these observations.

E. Mass Spectrometry

Important fragmentations of aldehydes and ketones are illustrated by the mass spectrum of 5-methyl-2-hexanone (Fig. 19.7). The three most important peaks occur at $m/z = 71$, 58, and 43. The peaks at $m/z = 71$ and $m/z = 43$ arise from cleavage of the parent ion at the bond between the carbonyl group and an adjacent carbon atom by two mechanisms that were discussed in Sec. 12.6C: *inductive cleavage* and *α-cleavage*. Inductive cleavage accounts for the $m/z = 71$ peak. In this cleavage the alkyl fragment carries the charge and the carbonyl fragment carries the unpaired electron.

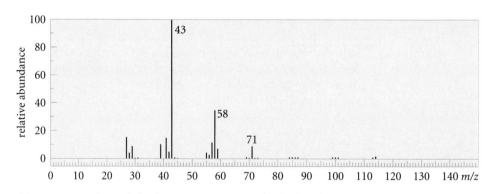

Figure 19.7   *Mass spectrum of 5-methyl-2-hexanone. Notice particularly the odd-electron ion at $m/z = 58$.*

$$\left[(CH_3)_2CHCH_2CH_2-\overset{\overset{+}{:\!\overset{..}{O}:}}{\underset{\|}{C}}-CH_3 \longleftrightarrow (CH_3)_2CHCH_2CH_2-\overset{\overset{+}{:\!\overset{..}{O}:}}{\underset{\cdot}{C}}-CH_3\right] \longrightarrow$$

molecular ion from loss of unshared electron

$$(CH_3)_2CHCH_2\overset{+}{C}H_2 + \cdot\overset{\overset{:O:}{\|}}{C}-CH_3 \qquad (19.5)$$
$$m/z = 71$$

α-Cleavage accounts for the $m/z = 43$ peak. In this case the same molecular ion fragments in such a way that the carbonyl fragment carries the charge and the alkyl fragment carries the unpaired electron:

$$(CH_3)_2CHCH_2CH_2-\overset{\overset{+}{:\!\overset{..}{O}\cdot}}{\underset{\|}{C}}-CH_3 \longrightarrow (CH_3)_2CHCH_2\overset{\cdot}{C}H_2 + :\overset{..}{O}\equiv\overset{+}{C}-CH_3 \qquad (19.6)$$
$$m/z = 43$$

An analogous cleavage at the carbon-hydrogen bond accounts for the fact that many aldehydes show a strong M $-$ 1 peak.

What accounts for the $m/z = 58$ peak? This peak is an odd-electron ion. As discussed in Sec. 12.6C, a common mechanism for formation of odd-electron ions is hydrogen transfer followed by loss of a stable neutral molecule; indeed, exactly such a mechanism is responsible for the $m/z = 58$ peak. The oxygen radical in the molecular ion abstracts a hydrogen atom from a carbon *five atoms away*, and the resulting radical then undergoes α-cleavage.

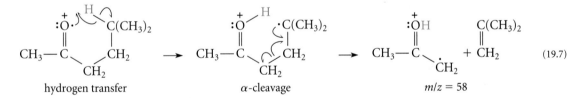

If we count the hydrogen that is transferred, the rearrangement occurs through a transient six-membered ring. This process is called a **McLafferty rearrangement**, after Professor Fred McLafferty, now of Cornell University, who investigated this type of fragmentation extensively. The McLafferty rearrangement and subsequent α-cleavage is a common mechanism for the production of odd-electron fragment ions in the mass spectrometry of carbonyl compounds.

PROBLEMS

*19.10 Explain each of the following observations resulting from a comparison of the mass spectra of 2-hexanone (*A*) and 3,3-dimethyl-2-butanone (*B*).

(a) The $m/z = 57$ fragment peak is much more intense in the spectrum of *B* than it is in the spectrum of *A*.

(b) The spectrum of compound *A* shows a fragment at $m/z = 58$, but that of compound *B* does not.

19.11   Using only mass spectrometry, how would you distinguish 2-heptanone from 3-heptanone?

19.4   Synthesis of Aldehydes and Ketones

Several reactions already presented can be used for the preparation of aldehydes and ketones. The three most important of these are

1. Oxidation of alcohols (Sec. 10.6A). Primary alcohols can be oxidized to aldehydes, and secondary alcohols can be oxidized to ketones.
2. Friedel-Crafts acylation (Sec. 16.4E). This reaction provides a way to synthesize aryl ketones. It also involves the formation of a carbon-carbon bond—the bond between the aryl ring and the carbonyl group.
3. Hydration and hydroboration-oxidation of alkynes (Sec. 14.5)

Other reactions have been discussed that give aldehydes or ketones as products, but these are less important as synthetic methods:

4. Ozonolysis of alkenes (Sec. 5.4)
5. Periodate cleavage of glycols (Sec. 10.6C)

Ozonolysis and periodate cleavage are reactions that break carbon-carbon bonds. Because an important aspect of organic synthesis is the *making* of carbon-carbon bonds, use of these reactions in effect wastes some of the effort that goes into making the alkene or glycol starting materials. Nevertheless, these reactions can be used synthetically in certain cases.

Other important methods of preparing aldehydes and ketones start with carboxylic acid derivatives; these methods are discussed in Chapter 21.

19.5   Introduction to Aldehyde and Ketone Reactions

The reactions of aldehydes and ketones can be conveniently grouped into two categories: (1) reactions of the carbonyl group, which are considered in this chapter; and (2) reactions involving the α-carbon, which are presented in Chapter 22.

The great preponderance of carbonyl-group reactions of aldehydes and ketones fall into three categories:

1. *Reactions with acids.* The carbonyl oxygen is weakly basic and thus reacts with Lewis and Brønsted acids. With E^+ as a general electrophile, this reaction can be represented as follows:

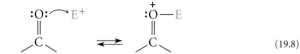

(19.8)

Carbonyl basicity is important because it plays a role in several other carbonyl-group reactions.

2. *Addition reactions.* The most important carbonyl-group reaction is addition to the C=O double bond. With E—Y symbolizing a general reagent, addition can be represented in the following way:

$$:O: \qquad\qquad :\ddot{O}\!-\!E $$
$$ \parallel \qquad\qquad\qquad | $$
$$ C \quad +\ E\!-\!Y \ \longrightarrow \ -C- \qquad (19.9) $$
$$ \qquad\qquad\qquad\qquad\qquad | $$
$$ \qquad\qquad\qquad\qquad\qquad Y $$

Superficially, carbonyl addition is analogous to alkene addition (Sec. 4.6).

Many reactions of aldehydes and ketones are simple additions that conform exactly to the model in Eq. 19.9. Others are multistep processes in which addition is followed by other reactions.

3. *Oxidation of aldehydes.* Aldehydes can be oxidized to carboxylic acids:

$$ \begin{array}{ccc} O & & O \\ \parallel & oxidation & \parallel \\ -C\!-\!H & \xrightarrow{\quad\quad} & -C\!-\!OH \end{array} \qquad (19.10) $$

19.6   Basicity of Aldehydes and Ketones

Aldehydes and ketones are weakly basic and react at the carbonyl oxygen with protons or Lewis acids.

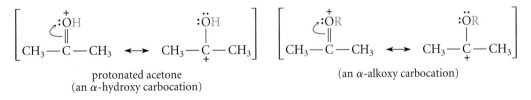

As Eq. 19.11 shows, the protonated form of an adehyde or ketone is resonance-stabilized. The resonance structure on the right shows that the protonated carbonyl compound has carbocation character. In fact, in some cases the conjugate acids of aldehydes and ketones undergo typical carbocation reactions.

Closely related to protonated aldehydes and ketones are *α-alkoxy carbocations:* cations in which the acidic proton of a protonated ketone is replaced by an alkyl group.

$$ \left[\begin{array}{ccc} \overset{+}{:}\!\ddot{O}H & & :\ddot{O}H \\ \parallel & & | \\ CH_3\!-\!C\!-\!CH_3 & \longleftrightarrow & CH_3\!-\!\overset{+}{C}\!-\!CH_3 \end{array} \right] \left[\begin{array}{ccc} \overset{+}{:}\!\ddot{O}R & & :\ddot{O}R \\ \parallel & & | \\ CH_3\!-\!C\!-\!CH_3 & \longleftrightarrow & CH_3\!-\!\overset{+}{C}\!-\!CH_3 \end{array} \right] $$

protonated acetone
(an α-hydroxy carbocation)   (an α-alkoxy carbocation)

α-Hydroxy carbocations and α-alkoxy carbocations are considerably more stable than ordinary carbocations. For example, a comparably substituted α-alkoxy carbocation is about 100 kJ/mol (24 kcal/mol) more stable than an ordinary tertiary carbocation in the gas phase.

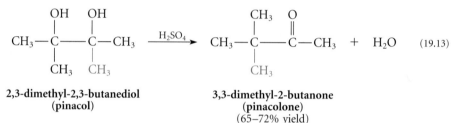

(19.12)

An α-alkoxy carbocation, like a protonated aldehyde or ketone, owes its stability to the resonance interaction of the electron-deficient carbon with the neighboring oxygen. This resonance effect far outweighs the electron-attracting polar effect of the oxygen, which, by itself, would destabilize the carbocation.

STUDY PROBLEM 19.2

Many 1,2-diols, under the acidic conditions used for dehydration of alcohols, undergo a reaction called the *pinacol rearrangement*:

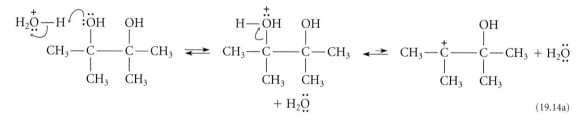

(19.13)

2,3-dimethyl-2,3-butanediol
(pinacol)

3,3-dimethyl-2-butanone
(pinacolone)
(65–72% yield)

Propose a curved-arrow mechanism for this reaction, and explain why the rearrangement step is energetically favorable.

Solution   First, analyze the connectivity changes that take place. A methyl group shifts to an adjacent carbon, and one of the —OH groups is lost as water. The fact that a rearrangement occurs suggests a carbocation intermediate; such a carbocation can be generated (as in the dehydration of any alcohol) by protonation of an —OH group and loss of H_2O:

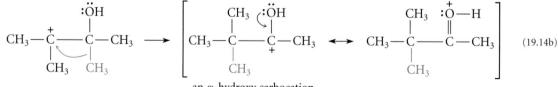

(19.14a)

The rearrangement can now take place. The product of the rearrangement is an α-hydroxy carbocation.

(19.14b)

an α-hydroxy carbocation

You've just learned that such carbocations are especially stable, a fact indicated by their resonance structures. Thus, the rearrangement step is favorable because the α-hydroxy carbocation is more stable than the tertiary carbocation. The second resonance structure in Eq. 19.14b emphasizes the point that an α-hydroxy carbocation is also a protonated ketone. Removal of a proton from the carbonyl oxygen gives the product.

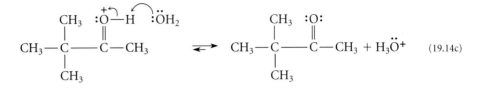

(19.14c)

Aldehydes and ketones in solution are considerably less basic than alcohols (Sec. 8.6). In other words, their conjugate acids are more acidic than those of alcohols.

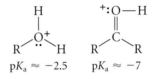

Because protonated aldehydes and ketones are resonance-stabilized and protonated alcohols are not, we might have expected protonated carbonyl compounds to be *more stable* relative to their conjugate bases and therefore *less acidic*. The relative acidity of protonated alcohols and carbonyl compounds is an example of a solvent effect. In the *gas phase*, aldehydes and ketones *are* indeed more basic than alcohols. One reason for the greater basicity of alcohols in solution is that protonated alcohols have more O—H protons to participate in hydrogen bonding to solvent than do protonated aldehydes or ketones.

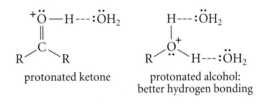

protonated ketone protonated alcohol: better hydrogen bonding

PROBLEMS

19.12 *(a) Write an S_N1 mechanism for the solvolysis of $CH_3—\ddot{O}—CH_2Cl$ [chloro(methoxy)methane] in ethanol; draw appropriate resonance structures for the carbocation intermediate.

(b) Explain why the alkyl halide in part (a) undergoes solvolysis much more rapidly than 1-chlorobutane. (In fact, it reacts in ethanol more than 100 times as rapidly.)

19.13 Predict the product of the pinacol rearrangements of each of the following diols.

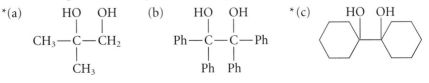

19.14 Use resonance arguments to explain why

*(a) *p*-methoxybenzaldehyde is more basic than *p*-nitrobenzaldehyde

(b) 3-buten-2-one is more basic than 2-butanone

19.7 Reversible Addition Reactions of Aldehydes and Ketones

One of the most typical reactions of aldehydes and ketones is *addition* to the carbon-oxygen double bond. To begin with, let's focus on two simple addition reactions, hydration and addition of hydrogen cyanide (HCN).

Hydration (addition of water):

$$CH_3-\overset{\overset{\displaystyle O}{\|}}{C}-H \; + \; H-OH \; \rightleftharpoons \; CH_3-\overset{\overset{\displaystyle OH}{|}}{\underset{\underset{\displaystyle OH}{|}}{C}}-H \qquad (19.15)$$

acetaldehyde

acetaldehyde hydrate

The product of water addition is called a *hydrate*, or *gem*-diol. (The prefix *gem* stands for *geminal*, from the Latin word for twin, and is used in chemistry when two identical groups are present on the same carbon.)

Addition of HCN:

$$CH_3-\overset{\overset{\displaystyle O}{\|}}{C}-CH_3 \; + \; H-C\equiv N \; \underset{}{\overset{pH\ 9-10}{\rightleftharpoons}} \; CH_3-\overset{\overset{\displaystyle OH}{|}}{\underset{\underset{\displaystyle C\equiv N}{|}}{C}}-CH_3 \qquad (19.16)$$

acetone

acetone cyanohydrin
(77–78% yield)

The product of HCN addition is termed a *cyanohydrin*. Cyanohydrins constitute a special class of *nitriles* (organic cyanides). (The chemistry of nitriles is considered in Chapter 21.) Notice that the preparation of cyanohydrins is another method of forming carbon-carbon bonds.

All carbonyl-addition reactions are regioselective. The more electropositive species (for example, the hydrogen of H—OH in hydration) adds to the carbonyl oxygen, and the more electronegative species (for example, the —OH in hydration) adds to the carbonyl carbon.

A. Mechanisms of Carbonyl-Addition Reactions

Carbonyl-addition reactions occur by two general types of mechanisms. The first occurs under *basic conditions*. In this mechanism, a nucleophile attacks the carbonyl group at the carbonyl carbon, and the carbonyl oxygen becomes negatively charged. In cyanohydrin formation, for example, the cyanide ion, formed by ionization of HCN, is the nucleophile.

$$H-CN \; + \; ^-OH \; \rightleftharpoons \; ^-:CN \; + \; H_2O \qquad (19.17a)$$

$pK_a = 9.4$ \qquad\qquad **cyanide ion**

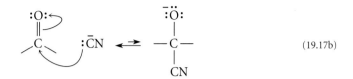

(19.17b)

The negatively charged oxygen—essentially an alkoxide ion—is a relatively strong base, and is protonated by either water or HCN to complete the addition:

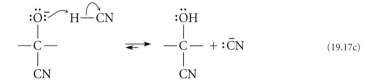

(19.17c)

   This mechanism, called **nucleophilic addition**, *has no analogy in the reactions of ordinary alkenes.* This pathway occurs with aldehydes and ketones because, in the transition state, negative charge is placed on oxygen, an electronegative atom. The same reaction of an alkene would place negative charge on a relatively electropositive carbon atom.

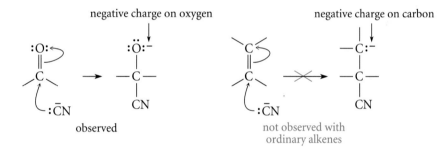

negative charge on oxygen

observed

negative charge on carbon

not observed with ordinary alkenes

Study Guide Link:
✓**19.2**
Why Nucleophiles React at the Carbonyl Carbon

Attack of the nucleophile occurs on the carbon of the carbonyl group rather than on the oxygen for the same reason: negative charge is "pushed" onto the more electronegative atom—oxygen. Notice that this mechanism accounts for the regioselectivity of carbonyl additions under basic conditions. The nucleophile—the species with unshared electron pairs—is typically the more electronegative partner of the groups that add, and *the nucleophile always attacks the carbonyl carbon.*

Study Guide Link:
19.3
Analogy for Nucleophilic Addition

   An orbital picture of nucleophilic addition, along with the geometry of nucleophilic attack on the carbonyl group, is shown in Fig. 19.8.

   The second mechanism for carbonyl addition occurs under *acidic conditions,* and is closely analogous to the mechanism for the addition of acids to alkenes (Secs. 4.7, 4.9B). Acid-catalyzed hydration of aldehydes and ketones is an example of this mechanism. The first step in hydration is protonation of the carbonyl oxygen (Sec. 19.6).

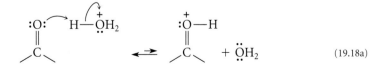

(19.18a)

Protonation of the carbonyl oxygen gives it a positive charge. A positively charged oxygen attracts electrons even more strongly than the oxygen of an unprotonated carbonyl group.

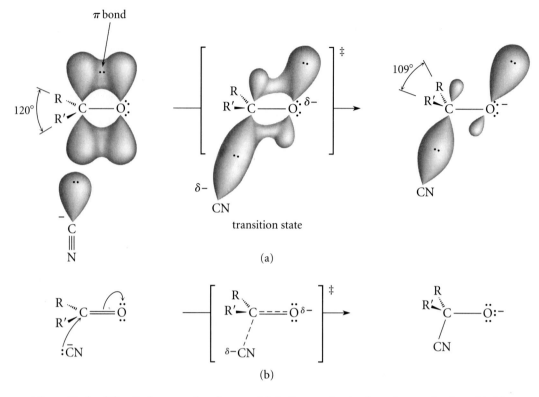

Figure 19.8   *Nucleophilic attack on a carbonyl group. (a) A diagram showing how the π molecular orbital is converted into a σ bond to the nucleophile and a hybrid orbital containing one of the oxygen unshared electron pairs. Notice that the nucleophile approaches the carbonyl carbon from above or below the plane of the molecule. Notice also that the bond angle between the R and R′ groups changes from 120° in the carbonyl compound to 109° in the addition product; thus the R and R′ groups are closer to each other in the product than they are in the starting material. (b) The curved-arrow formalism and Lewis structures for the same reaction.*

In other words, the protonated carbonyl compound is a much stronger *Lewis acid* (electron acceptor) than an unprotonated carbonyl compound. As a result, even the relatively weak base H_2O can react at the carbonyl carbon. Loss of a proton to solvent completes the reaction.

Study Guide Link:
✓**19.4**
Acids and Bases in Reaction Mechanisms

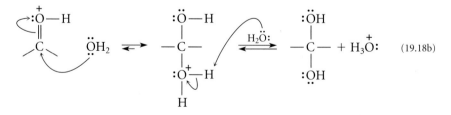

(19.18b)

Hydration of aldehydes and ketones also occurs in neutral and basic solution (Problem 19.15).

PROBLEMS

*19.15   Write a curved-arrow mechanism for the hydroxide-catalyzed hydration of acetaldehyde.

19.16   Write a curved-arrow mechanism for *(a) the acid-catalyzed addition of methanol to benzaldehyde; and for (b) the methoxide-catalyzed addition of methanol to benzaldehyde.

B. Equilibria and Rates in Carbonyl-Addition Reactions

Hydration and cyanohydrin formation are both reversible reactions. (Not all carbonyl additions are reversible.) Whether the equilibrium for a reversible addition favors the addition product or the carbonyl compound *depends strongly on the structure of the carbonyl compound.* For example, cyanohydrin formation favors the cyanohydrin addition product in the case of aldehydes and methyl ketones, but the equilibrium favors the carbonyl compound when aryl ketones are used.

The effect of aldehyde or ketone structure on the addition equilibrium for hydration is illustrated by the data in Table 19.2. Note the following trends in the table.

1. Addition is more favorable for aldehydes than for ketones.

2. Electronegative groups near the carbonyl carbon make carbonyl addition more favorable.

3. Addition is less favorable when groups are present that donate electrons by resonance to the carbonyl carbon.

Table 19.2   Equilibrium Constants for Hydration of Aldehydes and Ketones

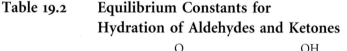

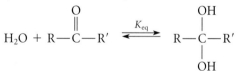

Aldehydes	K_{eq}	Ketones	K_{eq}
$CH_2{=}O$	2.2×10^3	$(CH_3)_2C{=}O$	1.4×10^{-3}
$CH_3CH{=}O$	1.0		
$(CH_3)_2CHCH{=}O$	0.5–1.0	$Ph{-}\overset{O}{\overset{\|}{C}}{-}CH_3$	6.6×10^{-6}
$PhCH{=}O$	8.3×10^{-3}	$Ph_2C{=}O$	1.2×10^{-7}
$ClCH_2CH{=}O$	37	$(ClCH_2)_2C{=}O$	10
$Cl_3CCH{=}O$	2.8×10^4	$(CF_3)_2C{=}O$	too large to measure

The trends in this table and the reasons behind them are important for two reasons. First, the equilibria for all addition reactions show similar effects of structure. Second, and more important, the *rates* of carbonyl addition reactions, that is, the *reactivities* of carbonyl compounds, follow similar trends. In other words, *the more a compound favors addition at equilibrium, the more rapidly it reacts in addition reactions.*

What is the reason for the effect of structure on carbonyl addition? The stability of the carbonyl compound relative to that of the addition product governs the $\Delta G°$ for addition. This point is illustrated in Fig. 19.9. The conclusion from this figure is that *added stability in the carbonyl compound increases the energy change ($\Delta G°$), and hence decreases the equilibrium constant, for formation of an addition product.*

What stabilizes carbonyl compounds? The major effects involved can be understood by considering the resonance structures of the carbonyl group:

$$\left[\begin{array}{ccc} \overset{\displaystyle :O:}{\underset{\displaystyle R—C—R}{\|}} & \longleftrightarrow & \overset{\displaystyle :\overset{..}{O}:^-}{\underset{\displaystyle R—\overset{+}{C}—R}{|}} \end{array} \right] \qquad (19.19)$$

The structure on the right, although not as important a contributor as the one on the left, reflects the polarity of the carbonyl group, and has the characteristics of a carbocation.

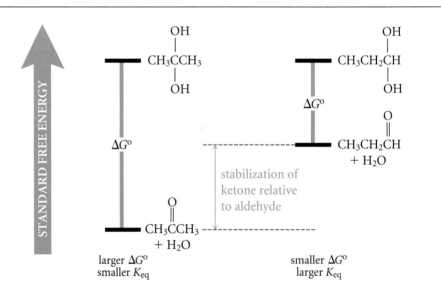

Figure 19.9 *The greater stability of a ketone relative to an aldehyde causes the ketone to have a greater standard free energy of hydration and therefore a smaller equilibrium constant for hydration. (The two hydrates have been placed at the same energy level for comparison purposes.) Because the transition states for addition reactions resemble the addition products, then the hydrates approximate transition states, and the free energies approximate standard free energies of activation. The less stable carbonyl compound—the aldehyde in this example—has the smaller standard free energy of activation, and hence reacts more rapidly in addition reactions.*

Therefore anything that stabilizes carbocations also tends to stabilize carbonyl compounds. Because alkyl groups stabilize carbocations, ketones (R = alkyl) are more stable than aldehydes (R = H). This stability is reflected in the relative heats of formation of aldehydes and ketones. For example, acetone, with $\Delta H_f^\circ = -218$ kJ/mol (-52.0 kcal/mol), is 26 kJ/mol (6.1 kcal/mol) more stable than its isomer propionaldehyde, for which $\Delta H_f^\circ = -192$ kJ/mol (-45.9 kcal/mol). Because alkyl groups stabilize carbonyl compounds, the equilibria for additions to ketones are less favorable than those for additions to aldehydes (Trend 1). Formaldehyde, with *two* hydrogens and no alkyl groups bound to the carbonyl, has a very large equilibrium constant for hydration.

Electronegative groups such as halogens destabilize carbocations by their polar effect, and for the same reason destabilize carbonyl compounds. Thus, halogens make the equilibria for addition more favorable (Trend 2). In fact, chloral hydrate (known in medicine as a hypnotic) is a stable crystalline compound.

$$
\underset{\substack{\text{chloral}\\ \text{(2,2,2-trichloroethanal)}}}{Cl_3C-\overset{\displaystyle O}{\overset{\|}{C}}H} + H_2O \longrightarrow \underset{\substack{\\ \text{chloral hydrate}}}{Cl_3C-\underset{\displaystyle OH}{\overset{\displaystyle OH}{\underset{|}{\overset{|}{C}H}}}} \tag{19.20}
$$

Groups that are conjugated with the carbonyl group, such as the phenyl group of benzaldehyde, stabilize carbocations by resonance, and hence stabilize carbonyl compounds.

$$\tag{19.21}$$

Resonance stabilization cannot occur in the hydrate because the carbonyl group is no longer present. Consequently, aryl aldehydes and ketones have relatively unfavorable hydration equilibria (Trend 3).

The trends in relative rates of addition can be predicted from the trends in equilibrium constants. That is, *compounds with the most favorable addition equilibria tend to react most rapidly in addition reactions.* Thus, *aldehydes are generally more reactive than ketones in addition reactions; formaldehyde is more reactive than many other simple aldehydes.* The reason for the parallel trends in rates and equilibria is that the transition states for addition reactions resemble the addition products. Just as destabilization of aldehydes or ketones decreases the ΔG° for their addition reactions (Fig. 19.9), the same destabilization decreases the *free energy of activation* $\Delta G^{\circ\ddagger}$ for addition and thus increases the rate of addition.

This section has covered two examples of addition to the carbonyl group. Subsequent sections will deal with other addition reactions as well as more complex reactions that have mechanisms in which the initial steps are addition reactions. These addition reactions all have mechanisms similar to the ones discussed in this section. *Addition to the carbonyl group is a common thread that runs throughout most of aldehyde and ketone chemistry.*

Problems

*19.17 The compound *ninhydrin* exists as a hydrate. Which carbonyl group is hydrated? Explain, and give the structure of the hydrate.

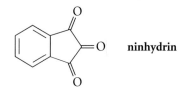

ninhydrin

19.18 Which carbonyl compound should form the greater proportion of cyanohydrin at equilibrium? Draw the structure of the cyanohydrin, and explain your reasoning.

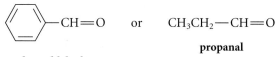

benzaldehyde

or $CH_3CH_2-CH{=}O$

propanal

19.19 Within each set, which compound should be more reactive in carbonyl-addition reactions? Explain your choices.

*(a) $CH_3-\overset{O}{\overset{\|}{C}}-CH_2CH_2Br$ or $CH_3-\overset{O}{\overset{\|}{C}}-CH_2Br$

(b) $CH_3-\overset{O}{\overset{\|}{C}}-\overset{O}{\overset{\|}{C}}-CH_3$ or $CH_3-\overset{O}{\overset{\|}{C}}-CH_2CH_3$

*(c) or

*(d) or

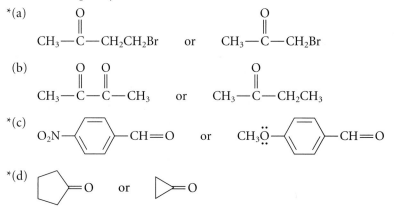

(*Hint:* Note the bond angles in Fig. 19.8a.)

19.8 Reduction of Aldehydes and Ketones to Alcohols

Aldehydes and ketones are reduced to alcohols with either lithium aluminum hydride, $LiAlH_4$, or sodium borohydride, $NaBH_4$. These reactions result in the net *addition* of the elements of H_2 across the C=O bond.

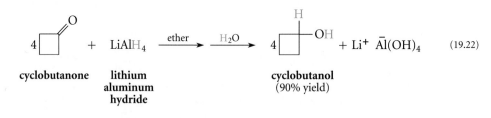

(19.22)

cyclobutanone **lithium aluminum hydride** **cyclobutanol** (90% yield)

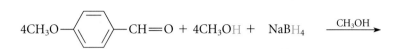

p-methoxybenzaldehyde   **sodium borohydride**

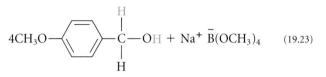

p-methoxybenzyl alcohol
(96% yield)

As these examples illustrate, reduction of an aldehyde gives a primary alcohol, and reduction of a ketone gives a secondary alcohol.

Lithium aluminum hydride is one of the most useful reducing agents in organic chemistry. It serves generally as a source of $H:^-$, the *hydride ion*. This is understandable because hydrogen is more electronegative than aluminum (Table 1.1). Thus, the Al—H bonds of the $^-AlH_4$ ion carry a substantial fraction of the negative charge. In other words,

$$
\begin{matrix} H \\ | \\ H\!-\!Al\!-\!H \\ | \\ H \end{matrix} \quad \text{reacts as if it were} \quad \begin{matrix} H \\ | \\ H\!-\!Al \\ | \\ H \end{matrix} \ H:^- \qquad (19.24)
$$

The hydride ion in $LiAlH_4$ is very basic. For example, $LiAlH_4$ reacts violently with water and therefore must be used in dry solvents such as anhydrous ether and THF.

$$
Li^+ \ \begin{matrix} H \\ | \\ H\!-\!Al\!-\!H \\ | \\ H \end{matrix} \ H\!-\!\ddot{O}H \longrightarrow \begin{matrix} H \\ | \\ H\!-\!Al \\ | \\ H \end{matrix} \ + \ H\!-\!H \ + \ Li^+ \ ^-\!:\!\ddot{O}H \qquad (19.25)
$$

lithium aluminum hydride   (reacts further with water)   hydrogen gas

Like many other strong bases, the hydride ion in $LiAlH_4$ is a good nucleophile. The reaction of $LiAlH_4$ with aldehydes and ketones involves the nucleophilic attack of hydride (delivered from $^-AlH_4$) on the carbonyl carbon. A lithium ion coordinated to the carbonyl oxygen acts as a Lewis-acid catalyst.

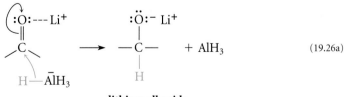

lithium alkoxide

$$(19.26a)$$

The product, an alkoxide salt (which actually exists as a complex with the Lewis acid AlH_3 or other trivalent-aluminum species present in solution), is converted by protonation into the alcohol product. The proton source is water (or an aqueous solution of a weak acid such as ammonium chloride), which is added in a separate step.

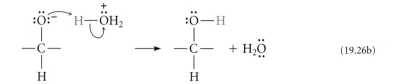

$$(19.26b)$$

As the stoichiometry of Eq. 19.22 indicates, all four hydrides of LiAlH$_4$ are active.

The reaction of sodium borohydride with aldehydes and ketones is conceptually similar to that of LiAlH$_4$. The sodium ion, however, does not form so strong a bond to the carbonyl oxygen as the lithium ion. For this reason, NaBH$_4$ reductions are carried out in protic solvents such as alcohols. Hydrogen bonding between the alcohol solvent and the carbonyl group serves as a weak acid catalysis that activates the carbonyl group. NaBH$_4$ reacts only slowly with alcohols and can even be used in water if the solution is not acidic.

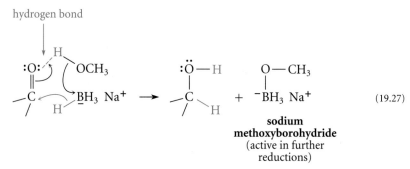

$$(19.27)$$

sodium methoxyborohydride (active in further reductions)

As Eq. 19.23 shows, all four hydride equivalents of NaBH$_4$ are active in the reduction.

Because LiAlH$_4$ and NaBH$_4$ are hydride donors, reductions by these and related reagents are generally referred to as **hydride reductions**. The important mechanistic point about these reactions is that they are further examples of *nucleophilic addition*. Hydride ion from LiAlH$_4$ or NaBH$_4$ is the nucleophile, and the proton is delivered from acid added in a separate step (in the case of LiAlH$_4$ reductions) or solvent (in the case of NaBH$_4$ reductions).

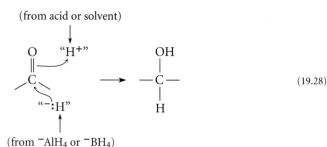

$$(19.28)$$

Unlike the additions discussed in Sec. 19.7, hydride reductions are *not* reversible. Reversal of carbonyl addition would require that the original attacking group, in this case, H:$^-$, be expelled as a leaving group. As in S$_N$1 or S$_N$2 reactions, the best leaving groups are the weakest bases. Hydride ion is such a strong base that it is not easily expelled as a leaving group. Hence, hydride reductions of all aldehydes and ketones are not reversible—they go to completion.

Both LiAlH$_4$ and NaBH$_4$ are highly useful in the reduction of aldehydes and ketones. Lithium aluminum hydride is, however, a much more *reactive* agent than sodium borohydride. A number of functional groups react with LiAlH$_4$ but not NaBH$_4$, for example, alkyl halides, alkyl tosylates, and nitro groups. Sodium borohydride can be used as a reducing agent in the presence of these groups.

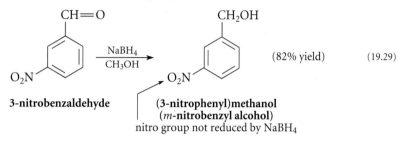

3-nitrobenzaldehyde

(3-nitrophenyl)methanol
(*m*-nitrobenzyl alcohol)
nitro group not reduced by NaBH$_4$

(82% yield)   (19.29)

Sodium borohydride is also a much less hazardous reagent than lithium aluminum hydride. The greater selectivity and safety of NaBH$_4$ make it the preferred reagent in many applications, but either reagent can be used for the reduction of simple aldehydes and ketones. Both are very important in organic chemistry.

DISCOVERY OF NaBH$_4$ REDUCTIONS

The discovery of NaBH$_4$ reductions illustrates that interesting research findings are sometimes obtained by accident. In the early 1940s, the U.S. Army Signal Corps became interested in methods for generation of hydrogen gas in the field. NaBH$_4$ was proposed as a relatively safe, portable source of hydrogen: addition of acidified water to NaBH$_4$ results in the evolution of hydrogen gas at a safe, moderate rate. In order to supply the required quantities of NaBH$_4$, a large-scale synthesis was necessary. The following reaction appeared to be suitable for this purpose.

$$4NaH + B(OCH_3)_3 \longrightarrow NaBH_4 + 3NaOCH_3 \qquad (19.30)$$

The problem with this process was that the sodium borohydride had to be separated from the sodium methoxide by-product. Several solvents were tried in the hope that a significant difference in solubilities could be found. In the course of this investigation, acetone was tried as a recrystallization solvent, and it was found to react with the NaBH$_4$ to yield isopropyl alcohol. Thus was born the use of NaBH$_4$ as a reducing agent for carbonyl compounds.

These investigations, carried out by Herbert C. Brown (1912–   ), now Professor Emeritus of Chemistry at Purdue University, were part of what was to become a major research program in the boron hydrides, shortly thereafter leading to the discovery of hydroboration (Sec. 5.3B). Brown even describes his interest in the field of boron chemistry as something of an accident, because it sprung from his reading a book about boron and silicon hydrides that was given to him by his girlfriend (now his wife) as a graduation present.

(continues)

> Mrs. Brown observes that the choice of this particular book was dictated by
> the fact that it was among the least expensive chemical titles in the bookstore;
> in the depression era students had to be careful how they spent their money!
> For his work in organic chemistry Brown shared the Nobel Prize in Chemistry
> in 1979 with Georg Wittig (Sec. 19.13).

Aldehydes and ketones can also be reduced to alcohols by catalytic hydrogenation. This reaction is analogous to the catalytic hydrogenation of alkenes (Sec. 4.9A).

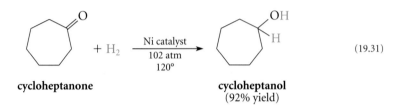

$$\text{cycloheptanone} + H_2 \xrightarrow[\substack{102\ atm\\120°}]{\text{Ni catalyst}} \text{cycloheptanol (92\% yield)} \tag{19.31}$$

Catalytic hydrogenation is less important for the reduction of carbonyl groups than it once was because of the modern use of hydride reagents.

It is usually possible to use catalytic hydrogenation for the selective reduction of an alkene double bond in the presence of a carbonyl group. Palladium catalysts are particularly effective for this purpose.

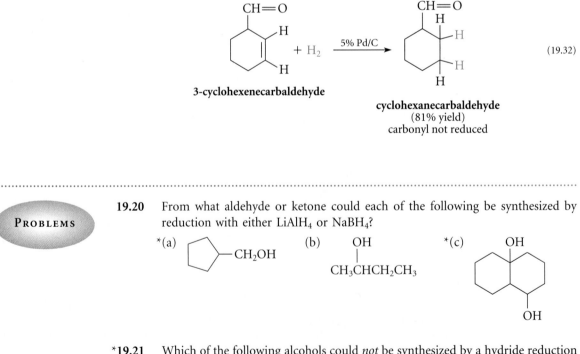

(19.32)

3-cyclohexenecarbaldehyde + H₂ → 5% Pd/C → **cyclohexanecarbaldehyde** (81% yield) carbonyl not reduced

PROBLEMS

19.20   From what aldehyde or ketone could each of the following be synthesized by reduction with either LiAlH₄ or NaBH₄?

*(a) cyclopentyl—CH₂OH   (b) CH₃CHCH₂CH₃ with OH   *(c) decalin with OH, OH

*19.21   Which of the following alcohols could *not* be synthesized by a hydride reduction of an aldehyde or ketone? Explain.

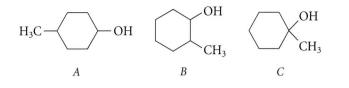

A B C

19.9   Reactions of Aldehydes and Ketones with Grignard and Related Reagents

The reaction of Grignard reagents with carbonyl groups is the most important application of the Grignard reagent in organic chemistry. Addition of Grignard reagents to aldehydes and ketones in an ether solvent, followed by protonolysis, gives alcohols.

$$(CH_3)_2CHCH + BrMg-CH_2CH_3 \xrightarrow{\text{ether}} \xrightarrow{H_3O^+} (CH_3)_2CHCHCH_2CH_3 \qquad (19.33)$$

2-methylpropanal   **ethylmagnesium bromide**   **2-methyl-3-pentanol** (68% yield)

$$CH_3-\overset{O}{\underset{}{C}}-CH_3 + CH_3CH_2CH_2-MgBr \xrightarrow{\text{ether}} \xrightarrow{H_3O^+} CH_3-\overset{OH}{\underset{CH_2CH_2CH_3}{C}}-CH_3 \qquad (19.34)$$

acetone   **propylmagnesium bromide**   **2-methyl-2-pentanol** (68% yield)

The reaction of Grignard reagents with aldehydes and ketones is another example of *carbonyl addition*. In this reaction, the magnesium of the Grignard reagent, a Lewis acid, bonds to the carbonyl oxygen. This bonding, much like protonation in acid-catalyzed hydration, makes the carbonyl carbon more electrophilic (that is, makes it more reactive toward nucleophiles) by making the carbonyl oxygen a better acceptor of electrons. The carbon group of the Grignard reagent attacks the carbonyl carbon. Recall that this group is a strong base that behaves much like a *carbanion* (Secs. 8.7B, 11.4C).

☞
Study Guide Link:
✓19.5
Lewis Acid Catalysis

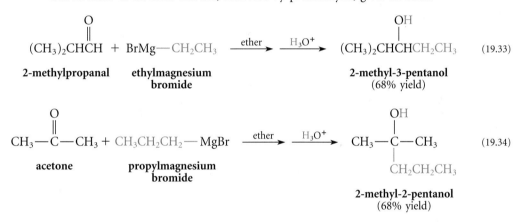

$$(19.35a)$$

a bromomagnesium alkoxide

The product of this addition, a bromomagnesium alkoxide, is essentially the magnesium salt of an alcohol. Addition of dilute acid to the reaction mixture gives an alcohol.

$$(19.35b)$$

Because of the great basicity of Grignard reagents, this addition, like hydride reductions, is not reversible, and works with just about any aldehyde or ketone.

The reactions of organolithium and sodium acetylide reagents with aldehydes and ketones are fundamentally similar to the Grignard reaction.

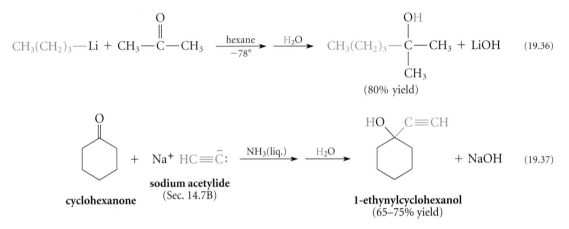

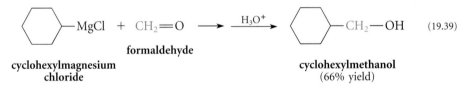

The reaction of Grignard and related reagents with aldehydes and ketones is important not only because it can be used to convert aldehydes or ketones into alcohols, but also because it is an excellent method of *carbon-carbon bond formation.*

STUDY GUIDE LINK:
✓19.6
Reactions that Form
Carbon-Carbon
Bonds

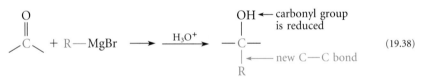

The possibilities for alcohol synthesis with the Grignard reaction are almost endless. Primary alcohols are synthesized by the addition of a Grignard reagent to formaldehyde.

Because Grignard reagents are made from alkyl halides, which in many cases can be synthesized from alcohols, this reaction can be used as a net one-carbon chain extension of an alcohol:

$$R\!-\!OH \longrightarrow R\!-\!Br \xrightarrow[\text{ether}]{\text{Mg}} R\!-\!MgBr \xrightarrow[\text{ether}]{CH_2\!=\!O} \xrightarrow{H_3O^+} R\!-\!CH_2\!-\!OH \qquad (19.40)$$

net one-carbon
chain extension

Addition of a Grignard reagent to an aldehyde other than formaldehyde gives a secondary alcohol (Eq. 19.33), and addition to a ketone gives a tertiary alcohol (Eq. 19.34). The Grignard synthesis of a tertiary alcohol or, in some cases, a secondary alcohol can also be turned into an alkene synthesis by dehydration of the alcohol with strong acid during the protonolysis step.

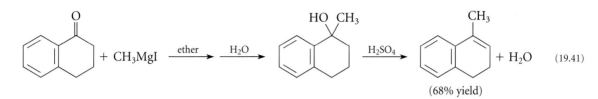

$$(19.41)$$

(68% yield)

When you are asked to prepare an alcohol, you can determine whether it can be synthesized by the reaction of a Grignard reagent with an aldehyde or ketone if you understand that the *net effect* of the Grignard reaction, followed by protonolysis, is addition of R—H (R = an alkyl or aryl group) across the C=O double bond:

$$(19.42)$$

STUDY GUIDE LINK:
✓**19.7**
Alcohol Syntheses

Once you grasp this relationship, you can determine the starting materials for a particular synthesis by mentally subtracting R and H from the target alcohol. This approach is illustrated in the following Study Problem.

STUDY PROBLEM 19.3

Propose a synthesis of 2-butanol by the reaction of a Grignard reagent with an aldehyde or ketone.

Solution   The carbonyl carbon of the starting material becomes the α-carbon of the alcohol. Consequently, any alkyl group bound to this carbon in the product can be derived from a Grignard reagent. The O—H proton is derived from the water or acid used in the protonolysis step. Thus, one possible analysis of the required synthesis is as follows:

$$(19.43)$$

(The double arrow is read, "Implies as starting materials.") Another possibility for a Grignard synthesis of 2-butanol can be found by a similar analysis; what is it?

PROBLEMS

19.22   Show how ethyl bromide can be used as a starting material in the preparation of each of the following compounds. (*Hint:* How are Grignard reagents prepared?)

*(a)   OH
       |
   PhCHCH$_2$CH$_3$

(b)   OH
       |
   (CH$_3$CH$_2$)$_2$CCH$_3$

*(c)   1-butanol

(d)   1-propanol

*(e)   Ph
            |
   Ph—C=CH—CH$_3$

(f)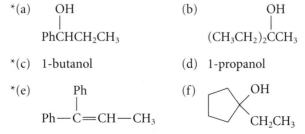

*(g)   CH$_3$CH$_2$CH=O

19.23   Outline two different Grignard syntheses for each of the following compounds.

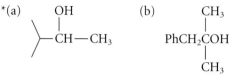

19.10 Acetals and Their Use as Protecting Groups

The preceding sections dealt with simple carbonyl-addition reactions—first, reversible additions (hydration and cyanohydrin formation); then, irreversible additions (hydride reduction and addition of Grignard reagents). This and the following sections consider some reactions that begin as additions but incorporate other types of mechanistic steps.

A. Preparation and Hydrolysis of Acetals

When an aldehyde or ketone reacts with a large excess of an alcohol in the presence of a trace of strong acid, an *acetal* is formed.

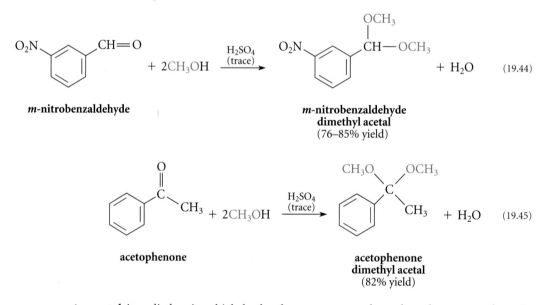

An **acetal** is a diether in which both ether oxygens are bound to the same carbon. In other words, acetals are the ethers of *gem*-diols (Sec. 19.7). (Acetals derived from ketones were once called *ketals*, but this name is no longer used.)

Notice that two equivalents of alcohol are consumed in each of the reactions above. However, 1,2- and 1,3-diols contain two —OH groups within the same molecule. Hence, one equivalent of a 1,2- or 1,3-diol can react to form a cyclic acetal, in which the acetal group is part of a five- or six-membered ring, respectively.

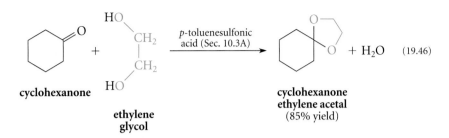

cyclohexanone   **ethylene glycol**   **cyclohexanone ethylene acetal** (85% yield)

The formation of acetals is reversible. The reaction is driven to the right by applying *LeChatelier's principle* (Sec. 9.2). In acetal formation this is accomplished either by the use of excess alcohol as the solvent, by removal of the water by-product, or both. In Eq. 19.46, for example, the water can be removed as an *azeotrope* with benzene. (The benzene-water azeotrope is a mixture of benzene and water that has a lower boiling point than either benzene or water alone.)

The first step in the mechanism of acetal formation is acid-catalyzed *addition* of the alcohol to the carbonyl group to give a **hemiacetal**—a compound with an —OR and —OH group on the same carbon (*hemi* = half; *hemiacetal* = half acetal).

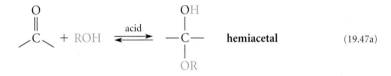

Hemiacetal formation is completely analogous to acid-catalyzed hydration. (Write the stepwise mechanism of this reaction; see Problem 19.16a.)

STUDY GUIDE LINK:
✓19.8
Hemiacetal Protonation

The hemiacetal reacts further when the —OH group is protonated and water is lost to give a relatively stable carbocation, an α-alkoxy carbocation (Sec. 19.6).

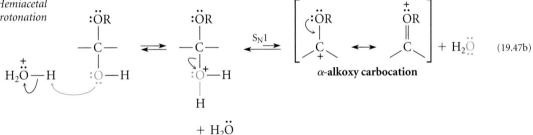

Loss of water from the hemiacetal is an S_N1 reaction analogous to the loss of water in the dehydration of an ordinary alcohol (Eq. 10.2b). Attack of an alcohol molecule on the cation and deprotonation of the attacking oxygen complete the mechanism.

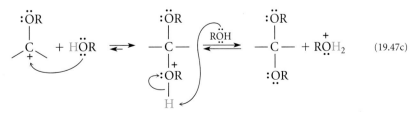

The mechanism for acetal formation is really a combination of other familiar mechanisms. It involves an *acid-catalyzed carbonyl addition* followed by a *substitution* that occurs by the S_N1 mechanism.

Because the formation of acetals is reversible, acetals in the presence of acid and excess water are transformed rapidly back into the corresponding carbonyl compounds and alcohols; this process is called **acetal hydrolysis**. (A *hydrolysis* is a cleavage reaction involving water.) As expected from the principle of microscopic reversibility, the mechanism of acetal hydrolysis is the reverse of the mechanism of acetal formation. Hence, acetal hydrolysis, like hemiacetal formation, is acid-catalyzed.

The formation of *hemiacetals* is catalyzed not only by acids, but by bases as well (Problem 19.16b). However, the conversion of hemiacetals into acetals is catalyzed *only* by acids (Eqs. 19.47b and c). This is why acetal formation, which is a combination of the two reactions, is catalyzed by acids but not by bases.

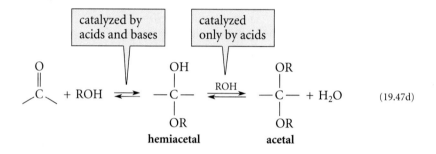

(19.47d)

As expected from the principle of microscopic reversibility, the hydrolysis of hemiacetals to aldehydes and ketones is also catalyzed by bases, but the hydrolysis of acetals to hemiacetals is catalyzed *only* by acids. Hence, *acetals are stable in basic and neutral solution.*

Hemiacetals, the intermediates in acetal formation (Eq. 19.47a), in most cases cannot be isolated because they react further to yield acetals (in alcohol solution) or decompose to aldehydes or ketones plus water. Simple aldehydes, however, form appreciable amounts of hemiacetals in alcohol solution, just as they form appreciable amounts of hydrates in water (see Table 19.2).

$$CH_3—CH{=}O + \underset{\text{solvent}}{C_2H_5OH} \; \rightleftarrows \; CH_3—\overset{\displaystyle OH}{\underset{\displaystyle OC_2H_5}{C}H} \quad \text{(97\% at equilibrium)} \qquad (19.48)$$

Five- and six-membered *cyclic* hemiacetals form spontaneously from the corresponding hydroxy aldehydes, and most are stable, isolable compounds.

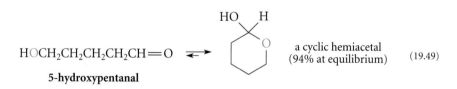

(19.49)

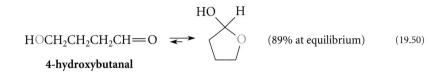

$$\text{HOCH}_2\text{CH}_2\text{CH}_2\text{CH}{=}\text{O} \ \rightleftharpoons \qquad \text{(89\% at equilibrium)} \qquad (19.50)$$

4-hydroxybutanal

The five- and six-carbon sugars are important biological examples of cyclic hemiacetals.

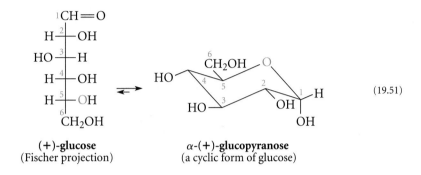

(19.51)

(+)-glucose
(Fischer projection)

α-(+)-glucopyranose
(a cyclic form of glucose)

STORAGE OF ALDEHYDES AS ACETALS

Some aldehydes are stored as acetals. Acetaldehyde, when treated with a trace of acid, readily forms a cyclic acetal called *paraldehyde*. Each molecule of paraldehyde is formed from three molecules of acetaldehyde. (Notice that an alcohol is not required for formation of paraldehyde.) Paraldehyde, with a boiling point of 125°, is a particularly convenient way to store acetaldehyde, which itself boils near room temperature. Upon heating with a trace of acid, acetaldehyde can be distilled from a sample of paraldehyde.

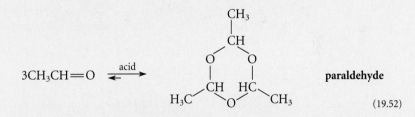

(19.52)

Formaldehyde can be stored as the acetal polymer *paraformaldehyde*, which precipitates from concentrated formaldehyde solutions.

$$\text{HO}{-}\!\!\left[\text{CH}_2{-}\text{O}\right]_{\!n}\!\!{-}\text{H}$$

paraformaldehyde

(An alcohol is not involved in paraformaldehyde formation.) Because it is a solid, paraformaldehyde is a useful form in which to store formaldehyde, itself a gas. Formaldehyde is liberated from paraformaldehyde by heating.

19.24   Write the structure of the product formed in each of the following reactions.

*(a)

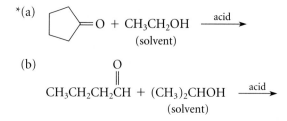

(b)

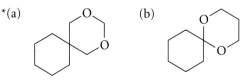

19.25   Propose syntheses of each of the following acetals from carbonyl compounds and alcohols.

*(a)   (b)

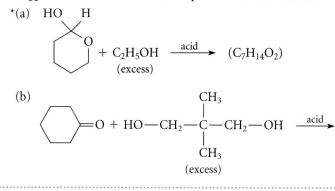

19.26   Suggest a structure for the acetal product of each reaction.

*(a)   HO   H

(b)

B. Protecting Groups

A common tactic of organic synthesis is the use of *protecting groups*. The method is illustrated by the following analogy. Suppose you and a friend are both unwelcome at a party, but are determined to attend it anyway. To avoid recognition and confrontation you wear a disguise, which might be a wig, a false mustache, or even more drastic accoutrements. Your friend doesn't bother with such deception. The host recognizes your friend and throws him out of the party, but, because you are not recognized, you remain and enjoy the evening, removing your disguise only after the party is over. Now, suppose two groups in a molecule, *A* and *B*, are both known to react with a certain reagent, but we want to let only group *A* react and leave group *B* unaffected. The solution to this problem is to *disguise*, or *protect*, group *B* in such a way that it cannot react. After group *A* is allowed to react, the disguise of group *B* is removed. The "chemical disguise" used with group *B* is called a **protecting group**. The following study problem illustrates the use of a protecting group.

Propose a method for carrying out the following conversion.

$$Br-\underset{}{\bigcirc}-\overset{\overset{O}{\parallel}}{C}-CH_3 \xrightarrow{?} HO-CH_2CH_2-\underset{}{\bigcirc}-\overset{\overset{O}{\parallel}}{C}-CH_3 \qquad (19.53)$$

Solution   It might seem that the way to effect this conversion would be to convert the starting halide into the corresponding Grignard reagent, and then allow this reagent to react with ethylene oxide, followed by dilute aqueous acid (Sec. 11.4C). However, Grignard reagents react with ketones (Sec. 19.9). Hence, the Grignard reagent derived from one molecule of the starting material would react with the carbonyl group of another molecule, and thus the ketone group would not survive this reaction. However, the ketone can be *protected* as an acetal, which does *not* react with Grignard reagents. (Acetals are really ethers, and ethers are unaffected by Grignard reagents.) The following synthesis incorporates this strategy.

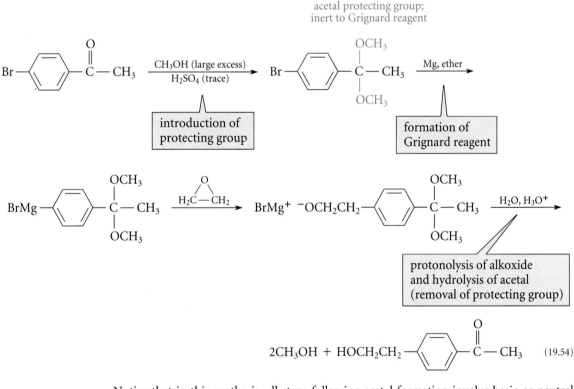

$$2CH_3OH + HOCH_2CH_2-\underset{}{\bigcirc}-\overset{\overset{O}{\parallel}}{C}-CH_3 \qquad (19.54)$$

Notice that in this synthesis, all steps following acetal formation involve basic or neutral conditions. Acid can be used only when destruction of the acetal is desired.

Carbonyl groups react with a number of reagents that react with other functional groups. Acetals are commonly used to protect the carbonyl groups of aldehydes and ketones from basic, nucleophilic reagents. Once the protection is no longer needed, the protecting group is easily removed, and the carbonyl group re-exposed, by treatment with dilute aqueous acid. Because acetals are unstable in acid, they do *not* protect carbonyl groups under acidic conditions.

19.27   Outline a synthesis of each of the following compounds from *p*-bromobenzaldehyde and any other reagents.

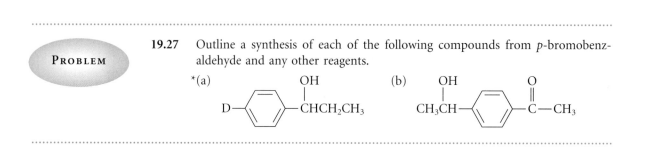

19.11   Reactions of Aldehydes and Ketones with Amines

A.   Reaction with Primary Amines and Other Monosubstituted Derivatives of Ammonia

A **primary amine** is an organic derivative of ammonia in which only one ammonia hydrogen is replaced by an alkyl or aryl group. An **imine** is a nitrogen analog of an aldehyde or ketone in which the C=O group is replaced by a C=N—R group, where R = alkyl, aryl, or H.

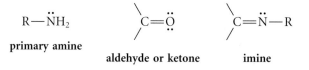

(Imines are sometimes called **Schiff bases**.) Imines are prepared by the reaction of aldehydes or ketones with primary amines.

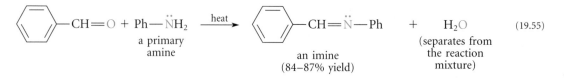

Formation of imines is reversible, and generally takes place with acid or base catalysis, or with heat. Imine formation is typically driven to completion by precipitation of the imine, removal of water, or both.

The mechanism of imine formation begins as a nucleophilic addition to the carbonyl group. In this case, the nucleophile is the amine, which reacts with the aldehyde or ketone to give an unstable addition product called a **carbinolamine**. A carbinolamine is a compound with an amine group (—NH₂, —NHR, or —NR₂) and a hydroxy group on the same carbon.

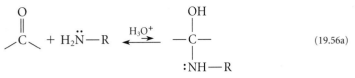

$$ \text{carbinolamine} \tag{19.56a} $$

Study Guide Link:
✓19.9
Mechanism of Carbinolamine Formation

(You should write the detailed mechanism, which is analogous to the mechanism of other reversible additions.) Carbinolamines are not isolated, but undergo acid-catalyzed dehydration to form imines. This reaction is essentially an alcohol dehydration (Sec. 10.1), except that it is typically much faster than dehydration of an ordinary alcohol.

$$ \text{carbinolamine} \xrightarrow{\text{acid}} \text{imine} + H_2O \quad \text{(acid-catalyzed dehydration)} \tag{19.56b} $$

Study Guide Link:
✓19.10
Dehydration of Carbinolamines

(Write the mechanism of this reaction as well.)

Typically the dehydration of the carbinolamine is the rate-limiting step of imine formation. This is why imine formation is catalyzed by acids. Yet the acid concentration cannot be too high becuase amines are basic compounds.

$$ R\ddot{N}H_2 + H_3\ddot{O}^+ \rightleftarrows R\overset{+}{N}H_3 + H_2\ddot{O}\colon \tag{19.57} $$

Protonation of the amine pulls the equilibrium in Eq. 19.56a to the left and carbinolamine formation cannot occur. For this reason, many imine syntheses are carried out in very dilute acid.

To summarize: imine formation is a sequence of two reactions that have close analogies to familiar reactions: *carbonyl addition* followed by *β-elimination*.

Imines are used to prepare amines (Section 23.7B). Before NMR spectroscopy assumed a central role in structure elucidation, certain types of *N*-substituted imines were especially important in organic chemistry. These derivatives and the amines from which they are derived are given in Table 19.3 on p. 906. For example, the 2,4-DNP derivative of acetone is prepared as follows:

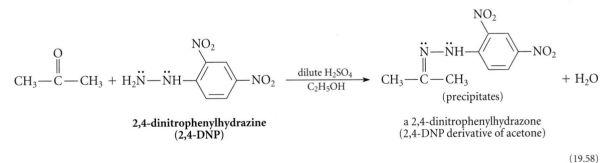

$$ \tag{19.58} $$

When a new compound was synthesized, it was typically characterized by conversion into two crystalline derivatives. These derivatives served as the basis for subsequent identification of the new compound when it was isolated from another source or from a different reaction. The reason it was important to prepare derivatives is that they

Table 19.3　Some *N*-Substituted Imine Derivatives of Aldehydes and Ketones

$$R_2C{=}O + H_2\ddot{N}{-}R' \longrightarrow R_2C{=}\ddot{N}{-}R' + H_2O$$

Amine	Name	Carbonyl Derivative	Name
$H_2\ddot{N}{-}\ddot{O}H$	hydroxylamine	$R_2C{=}\ddot{N}{-}\ddot{O}H$	oxime
$H_2\ddot{N}{-}\ddot{N}H_2$	hydrazine	$R_2C{=}\ddot{N}{-}\ddot{N}H_2$	hydrazone
$H_2\ddot{N}{-}\ddot{N}H{-}\bigcirc$	phenylhydrazine	$R_2C{=}\ddot{N}{-}\ddot{N}H{-}\bigcirc$	phenylhydrazone
$H_2\ddot{N}{-}\ddot{N}H{-}\bigcirc({-}NO_2)({-}NO_2)$	2,4-dinitrophenyl-hydrazine (2,4-DNP)	$R_2C{=}\ddot{N}{-}\ddot{N}H{-}\bigcirc({-}NO_2)({-}NO_2)$	2,4-dinitrophenyl-hydrazone (2,4-DNP derivative)
$H_2\ddot{N}{-}\ddot{N}H{-}\overset{O}{\overset{\|}{C}}{-}\ddot{N}H_2$	semicarbazide	$R_2C{=}\ddot{N}{-}\ddot{N}H{-}\overset{O}{\overset{\|}{C}}{-}\ddot{N}H_2$	semicarbazone

eliminate the ambiguity that can arise if two different compounds have the same melting points or boiling points. It almost never happens that two compounds with the same melting or boiling points give two crystalline derivatives that also have the same melting and boiling points. The derivatives in Table 19.3, because they are almost always crystalline solids, were widely used to characterize aldehydes and ketones.

To illustrate how such derivatives might be used in structure verification, suppose that a chemist has isolated a liquid that could be either 6-methyl-2-cyclohexenone or 2-methyl-2-cyclohexenone. The boiling points of these compounds are too similar for an unambiguous identification. Yet the melting point of either a 2,4-DNP derivative or a semicarbazone (see Table 19.3) would quickly establish which compound has been isolated.

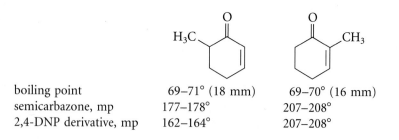

boiling point	69–71° (18 mm)	69–70° (16 mm)
semicarbazone, mp	177–178°	207–208°
2,4-DNP derivative, mp	162–164°	207–208°

Of course, the identity of the compound could be readily established today by spectroscopy. (Explain how.)

It is important to be familiar with the imine derivatives in Table 19.3 because references to the use of such derivatives are commonplace in the older literature of chemistry.

19.28   Draw the structure of

*(a)   the semicarbazone of cyclohexanone

(b)   the 2,4-DNP derivative of 2-methylpropanal

*(c)   the imine formed in the reaction between 2-methylhexanal and ethylamine ($C_2H_5NH_2$).

*19.29   Write a curved-arrow mechanism for the acid-catalyzed formation of the hydrazone of acetaldehyde.

19.30   Write a curved-arrow mechanism for the acid-catalyzed hydrolysis of the imine derived from benzaldehyde and ethylamine ($CH_3CH_2\overset{..}{N}H_2$). Use the principle of microscopic reversibility (Sec. 10.1) to guide you.

B.   Reaction with Secondary Amines

A **secondary amine** has the general structure $R_2\overset{..}{N}H$, in which two ammonia hydrogens are replaced by alkyl or aryl groups. An **enamine** has the following general structure:

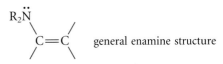

general enamine structure

Formation of an enamine occurs when a secondary amine reacts with an aldehyde or ketone, provided that the carbonyl compound has an α-hydrogen.

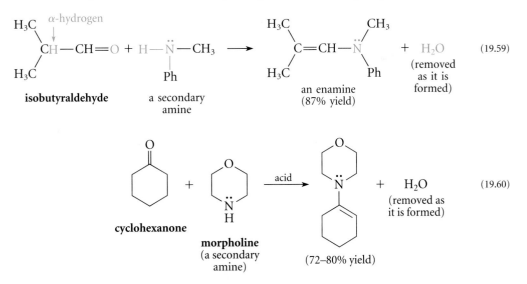

Notice that the two alkyl groups of a secondary amine, as in Eq. 19.60, may be part of a ring.

Like imine formation, enamine formation is reversible, and must be driven to completion by the removal of one of the reaction products (usually water; see Eq. 19.60).

Enamines, like imines, revert to the corresponding carbonyl compounds and amines in aqueous acid.

The mechanism of enamine formation begins like the mechanism of imine formation, as a nucleophilic addition to give a carbinolamine intermediate. (Write the mechanism of the this reaction.)

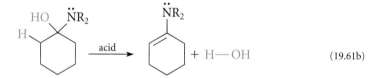

$$(19.61a)$$

Because no hydrogen remains on the nitrogen of this carbinolamine, imine formation cannot occur. Instead, dehydration of the carbinolamine involves loss of a hydrogen from an adjacent *carbon*.

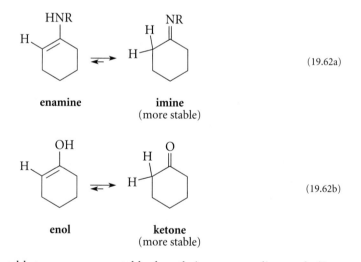

$$(19.61b)$$

Why don't primary amines react with aldehydes or ketones to form enamines rather than imines? The answer is the enamines bear the same relationship to imines that *enols* bear to ketones.

$$(19.62a)$$

enamine   **imine**
(more stable)

$$(19.62b)$$

enol   **ketone**
(more stable)

Just as most aldehydes and ketones are more stable than their corresponding enols (Sec. 14.5A), most imines are more stable than their corresponding enamines. Because secondary amines *cannot* form imines, they form enamines instead.

To summarize: aldehydes and ketones react with primary amines (RNH_2) to give imines, and with secondary amines (R_2NH) to give enamines. In a third type of amine, a **tertiary amine** (R_3N), all hydrogens of ammonia are replaced by alkyl or aryl groups. *Tertiary amines do not react with aldehydes and ketones to form stable derivatives.* Although most tertiary amines are good nucleophiles, they have no N—H hydrogens, and therefore cannot even form carbinolamines. Their adducts with aldehydes and ketones are unstable and can only break down to starting materials.

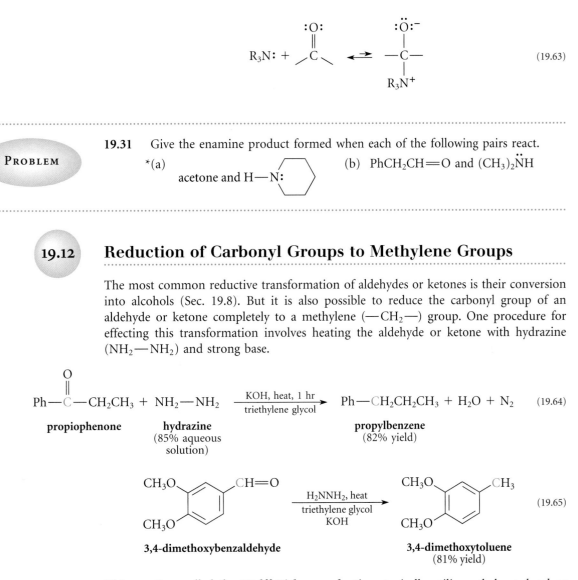

$$R_3N: + \quad \overset{\displaystyle :\overset{\displaystyle \ddot{O}:}{\|}}{\underset{\displaystyle}{C}} \quad \rightleftharpoons \quad \overset{\displaystyle :\ddot{O}:^{-}}{\underset{\displaystyle \underset{\displaystyle R_3N^{+}}{|}}{\overset{\displaystyle |}{C}}} \tag{19.63}$$

PROBLEM

19.31   Give the enamine product formed when each of the following pairs react.

*(a)   acetone and H—N: ⬡

(b) $PhCH_2CH{=}O$ and $(CH_3)_2\ddot{N}H$

19.12   Reduction of Carbonyl Groups to Methylene Groups

The most common reductive transformation of aldehydes or ketones is their conversion into alcohols (Sec. 19.8). But it is also possible to reduce the carbonyl group of an aldehyde or ketone completely to a methylene ($—CH_2—$) group. One procedure for effecting this transformation involves heating the aldehyde or ketone with hydrazine ($NH_2—NH_2$) and strong base.

$$\underset{\textbf{propiophenone}}{Ph{-}\overset{\displaystyle \overset{\displaystyle O}{\|}}{C}{-}CH_2CH_3} \ + \ \underset{\substack{\textbf{hydrazine}\\ \textbf{(85\% aqueous}\\ \textbf{solution)}}}{NH_2{-}NH_2} \ \xrightarrow[\text{triethylene glycol}]{\text{KOH, heat, 1 hr}} \ \underset{\substack{\textbf{propylbenzene}\\ \textbf{(82\% yield)}}}{Ph{-}CH_2CH_2CH_3} + H_2O + N_2 \tag{19.64}$$

$$\underset{\textbf{3,4-dimethoxybenzaldehyde}}{\begin{array}{c}CH_3O\\ \\CH_3O\end{array}\!\!\!\bigcirc\!\!\!\begin{array}{c}CH{=}O\\ \\ \\ \end{array}} \ \xrightarrow[\substack{\text{triethylene glycol}\\ \text{KOH}}]{H_2NNH_2,\ \text{heat}} \ \underset{\substack{\textbf{3,4-dimethoxytoluene}\\ \textbf{(81\% yield)}}}{\begin{array}{c}CH_3O\\ \\CH_3O\end{array}\!\!\!\bigcirc\!\!\!\begin{array}{c}CH_3\\ \\ \\ \end{array}} \tag{19.65}$$

This reaction, called the **Wolff-Kishner reduction**, typically utilizes ethylene glycol or similar compounds (triethylene glycol in the above cases) as co-solvents. The high boiling points of these solvents allow the reaction mixtures to reach the high temperatures required for the reduction to take place at a reasonable rate.

The Wolff-Kishner reduction is an extension of imine formation (Sec. 19.11A) because a *hydrazone* (Table 19.3) is an intermediate in the reaction. A series of Brønsted acid-base reactions (not discussed here) lead ultimately to expulsion of nitrogen and formation of the product.

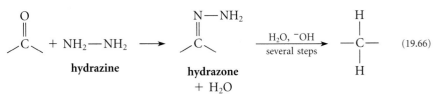

$$\overset{\displaystyle \overset{\displaystyle O}{\|}}{C} \ + \ \underset{\textbf{hydrazine}}{NH_2{-}NH_2} \ \longrightarrow \ \underset{\substack{\textbf{hydrazone}\\ +\ H_2O}}{\overset{\displaystyle \overset{\displaystyle N{-}NH_2}{\|}}{C}} \ \xrightarrow[\text{several steps}]{H_2O,\ ^-OH} \ \overset{\displaystyle \overset{\displaystyle H}{|}}{\underset{\displaystyle \underset{\displaystyle H}{|}}{C}} \tag{19.66}$$

The Wolff-Kishner reduction takes place under strongly basic conditions. The same overall transformation can be achieved under acidic conditions by a reaction called the **Clemmensen reduction**. In this reaction, an aldehyde or ketone is reduced with zinc amalgam (a solution of zinc metal in mercury) in the presence of HCl.

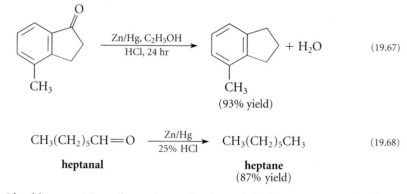

$$CH_3(CH_2)_5CH{=}O \xrightarrow[25\% \text{ HCl}]{Zn/Hg} CH_3(CH_2)_5CH_3 \qquad (19.68)$$

heptanal   **heptane**
                                 (87% yield)

There is considerable uncertainty about the mechanism of the Clemmensen reduction.

One of the most useful applications of the Wolff-Kishner or Clemmensen reductions is the introduction of alkyl substituents into benzene rings. This is illustrated in the following Study Problem.

....

STUDY PROBLEM 19.5

Outline a synthesis of butylbenzene from benzene and any other reagents.

Solution When you are asked to prepare an alkylbenzene from benzene, Friedel-Crafts alkylation (Sec. 16.4F) should come to mind. Indeed, this reaction is useful for introducing groups that do not rearrange, such as methyl groups, ethyl groups, and *tert*-butyl groups, into benzene rings. But when this reaction is used to prepare butylbenzene from benzene and 1-chlorobutane, a major amount of rearranged product is observed:

$$\langle\text{benzene}\rangle{-}H + CH_3CH_2CH_2CH_2{-}Cl \xrightarrow{AlCl_3}$$

benzene

$$\langle\rangle{-}\underset{\underset{CH_3}{|}}{CH}CH_2CH_3 + \langle\rangle{-}CH_2CH_2CH_2CH_3 + HCl \qquad (19.69a)$$

sec-**butylbenzene**   **butylbenzene**
(65%)   (35%)

Butylbenzene can be easily prepared free of isomers, however, by the Wolff-Kishner reduction of butyrophenone:

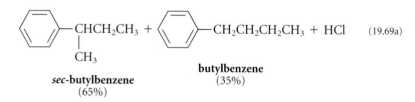

In turn, butyrophenone is readily prepared by Friedel-Crafts *acylation* (Sec. 16.4E), which is not plagued by the rearrangement problems associated with the *alkylation*.

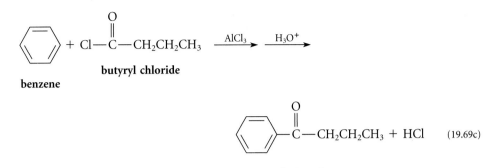

(19.69c)

butyrophenone

PROBLEMS

*19.32   Outline a synthesis of 1,4-dimethoxy-2-propylbenzene from hydroquinone and any other reagents.

19.33   Draw the structures of all aldehydes or ketones that could in principle give the following product after application of either the Wolff-Kishner or Clemmensen reduction.

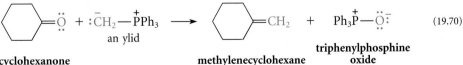

19.13   The Wittig Alkene Synthesis

Our tour through aldehyde and ketone chemistry started with simple additions; then addition followed by substitution (acetal formation); then additions followed by elimination (imine and enamine formation). Another addition-elimination reaction, called the **Wittig reaction**, is an important method for preparing alkenes from aldehydes and ketones.

An example of the Wittig reaction is the preparation of methylenecyclohexane from cyclohexanone.

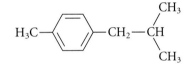

(19.70)

cyclohexanone   methylenecyclohexane   triphenylphosphine oxide

The Wittig reaction is especially important because it gives alkenes in which the *position* of the double bond is unambiguous; in other words, the Wittig reaction is completely *regioselective*. It can be used for the preparation of alkenes that would be difficult to prepare by other reactions. For example, methylenecyclohexane, which is readily prepared by the Wittig reaction (Eq. 19.70), cannot be prepared by dehydration of 1-methylcyclohexanol; 1-methylcyclohexene is obtained instead, because alcohol dehydration gives the alkene isomer(s) in which the double bond has the greatest number of alkyl substituents (Sec. 10.1).

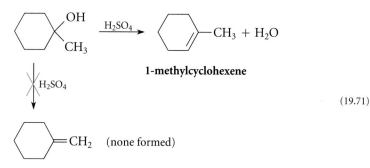

1-methylcyclohexene

(19.71)

methylenecyclohexane

The nucleophile in the Wittig reaction is a type of ylid. An **ylid** is any compound with opposite charges on adjacent, covalently bound atoms, each of which has an electronic octet.

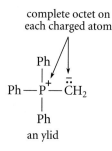

Because phosphorus, like sulfur (Sec. 10.9), can accommodate more than eight valence electrons, a phosphorus ylid has an uncharged resonance structure.

$$\left[\text{Ph}_3\overset{+}{\text{P}} \overset{\frown}{} \overset{..}{\overset{-}{\text{C}}}\text{H}_2 \quad \longleftrightarrow \quad \text{Ph}_3\text{P}=\text{CH}_2 \right]$$

ten electrons around
phosphorus

(19.72)

Although the structures of phosphorus ylids are sometimes written with phosphorus-carbon double bonds, the charged structures, in which each atom has an octet of electrons, are very important contributors.

The mechanism of the Wittig reaction, like the mechanisms of other carbonyl reactions, involves attack of a nucleophile on the carbonyl carbon. The nucleophile in the Wittig reaction is the anionic carbon of the ylid. The anionic oxygen in the resulting species reacts with phosphorus to form an *oxaphosphetane* intermediate. (An oxaphosphetane is a saturated four-membered ring containing both oxygen and phosphorus as ring atoms.)

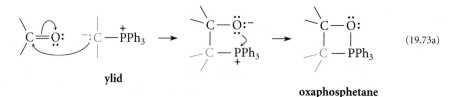

ylid

(19.73a)

oxaphosphetane

Under the usual reaction conditions, the oxaphosphetane spontaneously decomposes to the alkene and the by-product triphenylphosphine oxide.

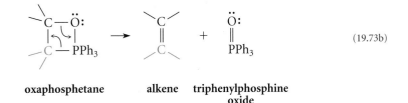

                    oxaphosphetane   alkene   triphenylphosphine
                                                                        oxide

The ylid starting material in the Wittig reaction is prepared by the reaction of an alkyl halide with triphenylphosphine (Ph_3P) in an S_N2 reaction to give a *phosphonium salt*.

$$Ph_3P\colon\; + \quad CH_3 - \ddot{Br}\colon \xrightarrow[\text{2 days}]{\text{benzene}} Ph_3\overset{+}{P} - CH_3 \quad \colon\!\ddot{Br}\colon^{-} \qquad (19.74a)$$

**triphenylphosphine   methyl bromide   methyltriphenyl-
                                                     phosphonium bromide
                                                     (a phosphonium salt;
                                                          99% yield)**

The phosphonium salt can be converted into its conjugate base, the ylid, by reaction with a strong base such as an organolithium reagent.

$$Ph_3\overset{+}{P} - CH_2 \;\; Br^{-} \longrightarrow Ph_3\overset{+}{P} - \overset{\cdot\cdot}{C}H_2 + H - CH_2CH_2CH_2CH_3 + LiBr \qquad (19.74b)$$

$$\overset{|}{H}$$

$$Li - CH_2CH_2CH_2CH_3$$

an ylid

butyllithium

To plan the synthesis of an alkene by the Wittig reaction, consider the origin of each part of the product, and then reason deductively. Thus, one carbon of the alkene double bond originates from the alkyl halide used to prepare the ylid; the other is the carbonyl carbon of the aldehyde or ketone:

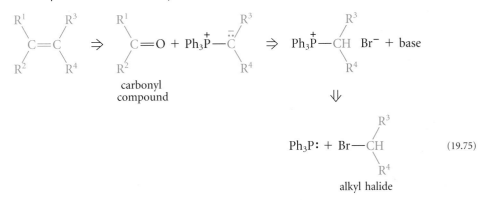

                  carbonyl
                  compound

                                                                    alkyl halide

(Again, the double arrows used in this deductive analysis are read "implies as starting material.") This analysis also shows that there are, in principle, two possible Wittig syntheses for any given alkene; in the other possibility, the R^1 and R^2 groups could originate from the alkyl halide and the R^3 and R^4 groups from the aldehyde or ketone. However, remember that the reaction used to form the phosphonium salt is an S_N2 reaction; consequently, this reaction is fastest with methyl and primary alkyl halides. In other words, most Wittig syntheses are planned so that the most reactive alkyl halide can be used as one of the starting materials.

STUDY PROBLEM 19.6

Outline two Wittig alkene syntheses of 1-hexene. Is one synthesis preferred over the other? Why?

Solution   The analysis in Eq. 19.75 suggests that the "right-hand" part of the alkene can be derived from the aldehyde pentanal:

$$CH_2{=}CHCH_2CH_2CH_2CH_3 \quad \Rightarrow \quad Ph_3\overset{+}{P}{-}\overset{..}{\overset{-}{C}}H_2 + O{=}CHCH_2CH_2CH_2CH_3$$

1-hexene $\Downarrow$ **pentanal**

$$Ph_3P \quad + \quad CH_3Br$$

methyl bromide   (19.76)

Another possibility, however, is that the "left-hand" part of the alkene is derived from formaldehyde:

$$CH_2{=}CHCH_2CH_2CH_2CH_3 \quad \Rightarrow \quad H_2C{=}O + Ph_3\overset{+}{P}{-}\overset{..}{\overset{-}{C}}HCH_2CH_2CH_2CH_3$$

1-hexene $\Downarrow$

$$Ph_3P + BrCH_2CH_2CH_2CH_2CH_3$$

1-bromopentane   (19.77)

Both syntheses are reasonable. However, because methyl bromide is more reactive than 1-bromopentane in the S_N2 reaction (why? Sec. 9.4C), the first reaction is preferred.

One problem with the Wittig reaction is that it gives mixtures of E and Z isomers.

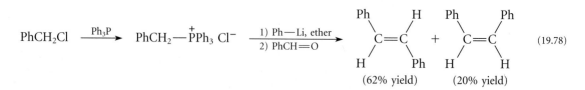

Although certain modifications of the Wittig reaction that avoid this problem have been developed, these are outside the scope of our discussion.

DISCOVERY OF THE WITTIG REACTION

The Wittig alkene synthesis is named for Georg Wittig (1897–1987), who was Professor of Chemistry at the University of Heidelberg. Wittig and his coworkers discovered the alkene synthesis in the course of other work in phosphorus chemistry; they had not set out to develop this reaction explicitly. Once the significance of the reaction was recognized, it was widely exploited. Wittig shared the Nobel Prize for Chemistry with H. C. Brown (Sec. 19.8) in 1979.

The Wittig reaction is not only important as a laboratory reaction; it has also been industrially useful. For example, it is an important reaction in the industrial synthesis of vitamin A derivatives.

PROBLEMS

19.34 Give the structure of the alkene(s) formed in each of the following reactions.

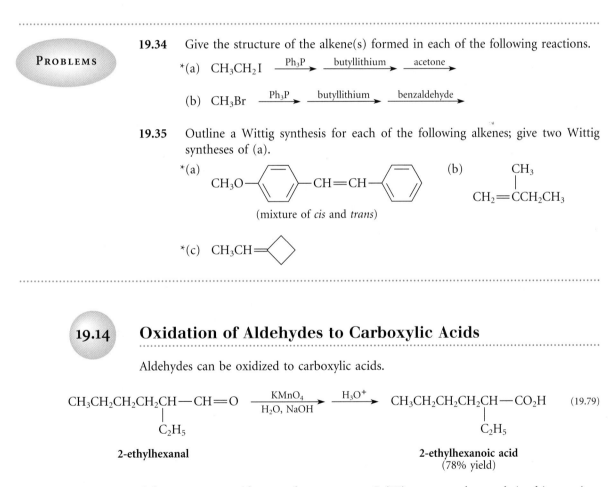

*(a) $CH_3CH_2I \xrightarrow{Ph_3P} \xrightarrow{butyllithium} \xrightarrow{acetone}$

(b) $CH_3Br \xrightarrow{Ph_3P} \xrightarrow{butyllithium} \xrightarrow{benzaldehyde}$

19.35 Outline a Wittig synthesis for each of the following alkenes; give two Wittig syntheses of (a).

*(a)
$$CH_3O-\langle\!\!\!\!\!\bigcirc\!\!\!\!\!\rangle-CH=CH-\langle\!\!\!\!\!\bigcirc\!\!\!\!\!\rangle$$
(mixture of *cis* and *trans*)

(b)
$$\underset{\overset{|}{CH_2=CCH_2CH_3}}{\overset{CH_3}{}}$$

*(c) $CH_3CH=\langle\!\!\!\diamondsuit\!\!\!\rangle$

19.14 Oxidation of Aldehydes to Carboxylic Acids

Aldehydes can be oxidized to carboxylic acids.

$$CH_3CH_2CH_2CH_2CH\!\!-\!\!CH=\!\!O \xrightarrow[H_2O,\ NaOH]{KMnO_4} \xrightarrow{H_3O^+} CH_3CH_2CH_2CH_2CH\!\!-\!\!CO_2H \quad (19.79)$$
$$\underset{C_2H_5}{\overset{|}{}} \qquad\qquad\qquad\qquad\qquad \underset{C_2H_5}{\overset{|}{}}$$

2-ethylhexanal   **2-ethylhexanoic acid**
                                     (78% yield)

Other common oxidants such as aqueous Cr(VI) reagents also work in this reaction. These oxidizing agents are the same ones used for oxidizing alcohols (Sec. 10.6A).

Some aldehyde oxidations begin as addition reactions. For example, in the oxidation of aldehydes by Cr(VI) reagents, the *hydrate*, not the aldehyde, is actually the species oxidized. (See Eq. 10.37, p. 468.)

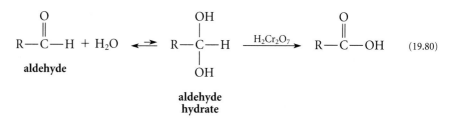

$$(19.80)$$

That is, the "aldehyde" oxidation is really an "alcohol" oxidation, the "alcohol" being the hydrate formed by addition of water to the aldehyde carbonyl group. For this reason, some water should be present in solution in order that aldehyde oxidations with Cr(VI) occur at a reasonable rate.

In the laboratory, aldehydes can be conveniently oxidized to carboxylic acids with Ag(I) reagents.

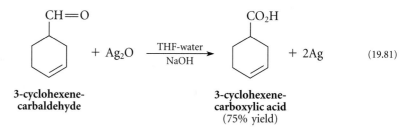

$$(19.81)$$

The expense of silver limits its use to small-scale reactions, as a rule. However, the Ag_2O oxidation is especially handy when the aldehyde to be oxidized contains double bonds or alcohol —OH groups, functional groups that react with other oxidizing reagents but do not react with Ag_2O.

Sometimes, as in Eq. 19.81, the Ag(I) is used as a slurry of brown Ag_2O, which changes to a black precipitate of silver metal as the reaction proceeds. If the silver ion is solubilized as its ammonia complex, $^+Ag(NH_3)_2$, oxidation of the aldehyde is accompanied by the deposition of a metallic silver mirror on the walls of the reaction vessel. This observation can be used as a convenient test for aldehydes, known as **Tollens' test**.

Many aldehydes are oxidized by the oxygen in air upon standing for long periods of time. This process, another example of *autoxidation* (Sec. 18.10), is responsible for the contamination of some aldehyde samples with appreciable amounts of carboxylic acids.

Ketones cannot be oxidized without breaking carbon-carbon bonds (see Table 10.1, p. 466). Ketones are resistant to mild oxidation with Cr(VI) reagents, and acetone can even be used as a solvent for oxidations with such reagents. Potassium permanganate, however, oxidizes ketones, and it is therefore not useful as an oxidizing reagent in the presence of ketones.

PROBLEMS

*19.36 Give the structure of an aldehyde $C_8H_8O_2$ that would be oxidized to terephthalic acid by $KMnO_4$.

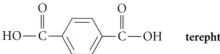 **terephthalic acid**

19.37   What product is formed when the following compound is treated with Ag_2O?

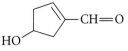

Manufacture and Use of Aldehydes and Ketones

The most important commercial aldehyde is formaldehyde, which is manufactured by the oxidation of methanol over a silver catalyst.

$$CH_3-OH \xrightarrow[O_2,\ 600-650°]{Ag\ catalyst} CH_2=O \qquad (19.82)$$

About 7 billion pounds of formaldehyde, valued at approximately \$625 million, is produced annually in the United States.

The single most important use of formaldehyde is in the synthesis of a class of polymers known as *phenol-formaldehyde resins*. (A **resin** is a polymer with a rigid three-dimensional network of repeating units.) Although the exact structure and properties of a phenol-formaldehyde resin depend on the conditions of the reaction used to prepare it, a typical segment of such a resin can be represented schematically as follows:

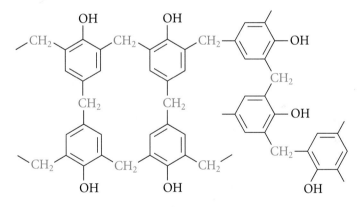

Phenol-formaldehyde resins are produced by heating phenol and formaldehyde with acidic or basic catalysts. The $-CH_2-$ groups (color) are derived from the formaldehyde, which in some cases is supplied in the form of its addition product with ammonia. Different formulations of these resins are used for telephones, adhesives in exterior-grade plywood, and heat-stable bondings for brake linings. A phenol-formaldehyde resin called *Bakelite*, patented in 1909 by the Belgian immigrant Leo H. Baekeland, was the first useful synthetic polymer.

The simplest ketone, acetone, is co-produced with phenol by the autoxidation-rearrangement of cumene (Sec. 18.10). About 2.4 billion pounds of acetone, valued at \$528 million, is produced annually in the United States. Acetone is used both as a solvent and as a starting material for polymers.

KEY IDEAS IN CHAPTER 19

- The functional group in aldehydes and ketones is the carbonyl group.

- Aldehydes and ketones are polar molecules. Simple aldehydes and ketones have boiling points higher than those of hydrocarbons but lower than those of alcohols. The aldehydes and ketones of low molecular mass have substantial solubilities in water.

- The carbonyl stretching absorption is the most important infrared absorption of aldehydes and ketones. The proton NMR spectra of aldehydes have distinctive low-field absorptions at δ 9–11 for the aldehydic protons; and α-protons of both aldehydes and ketones absorb near δ 2.5. The most characteristic absorptions in the CMR spectra of aldehydes and ketones are the carbonyl carbon resonances at δ 190–220. Aldehydes and ketones have weak $n \to \pi^*$ UV absorptions, and compounds that contain double bonds conjugated with the carbonyl group have $\pi \to \pi^*$ absorptions. α-Cleavage, inductive cleavage, and the McLafferty rearrangement are the important fragmentation modes observed in the mass spectra of aldehydes and ketones.

- Aldehydes and ketones are weak bases, and are protonated on their carbonyl oxygens to give α-hydroxy carbocations. The interaction of a proton or Lewis acid with a carbonyl group activates it toward addition reactions.

- The most characteristic carbonyl-group reactions of aldehydes and ketones are carbonyl-addition reactions, which, depending on the reaction, occur under acidic or basic conditions or both. Hydration and cyanohydrin formation are examples of simple reversible carbonyl additions. Hydride reductions and Grignard reactions are examples of simple additions that are not reversible.

- Acetal formation is an example of addition to the carbonyl group followed by substitution.

- Imine formation, enamine formation, and the Wittig alkene synthesis are examples of addition to the carbonyl group followed by elimination.

- The carbonyl group of an aldehyde or ketone can be reduced to a methylene group by either the Wolff-Kishner or Clemmensen reduction.

- Aldehydes can be oxidized to carboxylic acids; ketones cannot be oxidized without breaking carbon-carbon bonds.

- Aldehydes and ketones can be converted into alcohols (Grignard reactions, hydrogenation, and hydride reductions), alkenes (Wittig reaction), and alkanes (Wolff-Kishner and Clemmensen reductions).

*19.38   Give the products expected (if any) when acetone reacts with each of the following reagents.

(a) H_3O^+ (b) $NaBH_4$ in CH_3OH, then H_2O
(c) CrO_3, pyridine   (d) NaCN, pH 10, H_2O

(e) CH_3OH (excess),   (f) 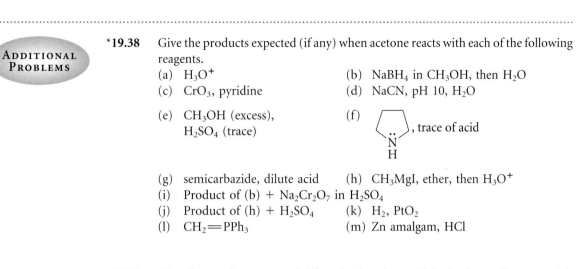, trace of acid
     H_2SO_4 (trace)

(g) semicarbazide, dilute acid   (h) CH_3MgI, ether, then H_3O^+
(i) Product of (b) + $Na_2Cr_2O_7$ in H_2SO_4
(j) Product of (h) + H_2SO_4 (k) H_2, PtO_2
(l) $CH_2{=}PPh_3$ (m) Zn amalgam, HCl

19.39   Give the product expected (if any) when butyraldehyde (butanal) reacts with each of the following reagents.

(a) PhMgBr, then H_3O^+ (b) $LiAlH_4$ in ether, then H_3O^+
(c) alkaline $KMnO_4$, then H_3O^+ (d) aqueous $H_2Cr_2O_7$
(e) NH_2OH, pH = 5   (f) Ag_2O
(g) Zn amalgam, HCl

19.40   Sodium bisulfite adds reversibly to aldehydes and a few ketones to give *bisulfite addition products*.

$$\left[Na^+\ H{-}\overset{\overset{\displaystyle :O:}{\|}}{\underset{\underset{\displaystyle :O:}{\|}}{S}}{-}\ddot{\underset{\cdot\cdot}{O}}:^- \rightleftharpoons Na^+\ ^-:\overset{\overset{\displaystyle :O:}{\|}}{\underset{\underset{\displaystyle :O:}{\|}}{S}}{-}\ddot{\underset{\cdot\cdot}{O}}H \right] + R{-}\overset{\overset{\displaystyle :O:}{\|}}{C}H \rightleftharpoons R{-}\overset{\overset{\displaystyle :\ddot{O}H}{|}}{\underset{\underset{\displaystyle :\ddot{O}:^-\ Na^+}{|}}{\underset{\underset{\displaystyle \ddot{O}{=}S{=}\ddot{O}}{|}}{C}}H$$

sodium bisulfite

*(a) Write a curved-arrow mechanism for this addition reaction; assume water is the solvent.

*(b) The reaction can be reversed by adding either H_3O^+ or ^-OH. Explain this observation using LeChatelier's principle and your knowledge of sodium bisulfite reactions from inorganic chemistry.

(c) Deduce the structure of the bisulfite addition product of 2-methylpentanal.

19.41   Each of the following reactions gives a mixture of two separable isomers. What are the two isomers formed in each case?

*(a)

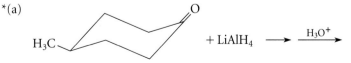 + $LiAlH_4$ $\longrightarrow$ $\xrightarrow{H_3O^+}$

(b) $PhCH{=}O + CH_3{-}CH{=}PPh_3 \longrightarrow$

*19.42   Give the structures of the four *separable* isomers $C_9H_{18}O_3$ formed in the reaction of hexanal with glycerol (1,2,3-propanetriol) in the presence of an acid catalyst.

19.43   (a) What are the two structurally isomeric cyclic acetals that could in principle be formed in the acid-catalyzed reaction of acetone and glycerol (1,2,3-propanetriol)?
   (b) Only one of the two compounds is actually formed. Given that it can be resolved into enantiomers, which isomer in (a) is the one that is produced?

*19.44   Give the structures of the two separable isomers formed in the following reaction.

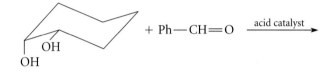

19.45   Complete the following reactions by giving the principal organic product(s).

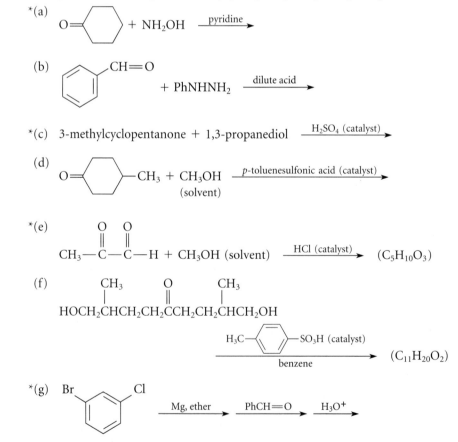

(h) propiophenone + Ph—MgBr $\xrightarrow{\text{ether}}$ $\xrightarrow{\text{H}_3\text{O}^+}$

*(i)

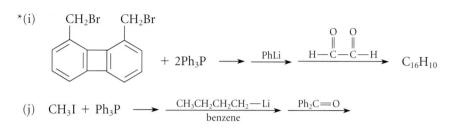

(j) CH₃I + Ph₃P $\longrightarrow$ $\xrightarrow[\text{benzene}]{\text{CH}_3\text{CH}_2\text{CH}_2\text{CH}_2\text{—Li}}$ $\xrightarrow{\text{Ph}_2\text{C=O}}$

19.46 Using known reactions and mechanisms discussed in the text, complete the following reactions.

*(a)

$$\underset{\text{CH}_3\text{CH}_2\text{CH}_2\text{CH}}{\overset{\overset{\text{O}}{\underset{|}{\text{Ph}_2\text{P}}}}{}}-\underset{\overset{\|}{\text{C}}}{\overset{\text{O}}{}}-\text{CH}_2\text{CH}_3 \xrightarrow[\text{MeOH}]{\text{NaBH}_4} \xrightarrow{\text{NaH}}$$

(*Hint:* See Eqs. 19.73a and b.)

(b)

$$\underset{\text{CH}_3\text{CH}_2\text{CH}_2\text{CCH}_3}{\overset{\overset{\text{O}}{\|}}{}} + \text{H}_2\ddot{\text{N}}-\text{Ph} \longrightarrow \xrightarrow{\text{NaBH}_4}$$

(*Hint:* The C=N bond undergoes addition much like the C=O bond.)

*19.47 Compound *A*, C₈H₈O, when treated with Zn amalgam and HCl, gives a xylene isomer that in turn gives only one ring monobromination product with Br₂ and Fe. Propose a structure for *A*.

19.48 Complete the following series of reactions by giving the major organic product.

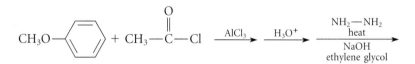

19.49 Suggest routes by which each of the following compounds could be synthesized from the indicated starting materials and any other reagents.
*(a) 1-phenyl-1-butanone (butyrophenone) from butyraldehyde
(b) 2-cyclohexyl-2-propanol from cyclohexanone
*(c) cyclohexyl methyl ether from cyclohexanone
(d) (CH₃)₂CHCH=CH₂ from isobutyraldehyde
*(e) 2,3-dimethyl-2-hexene from 3-methyl-2-hexanone
(f) 2,3-dimethyl-1-hexene from 3-methyl-2-hexanone

*(g)

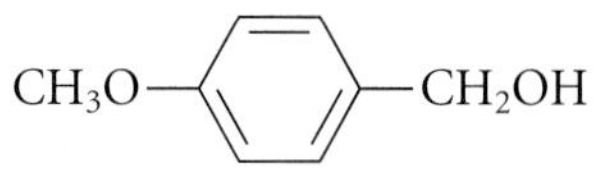

CH₃O—⟨⟩—CH₂OH   from *p*-bromoanisole

(*Problem 19.49 continues*)

(h)   3-methyl-3-hexanol from propyl bromide

*(i)   1,6-hexanediol from cyclohexene

*(j)   1-butyl-4-propylbenzene from benzene

*(k)

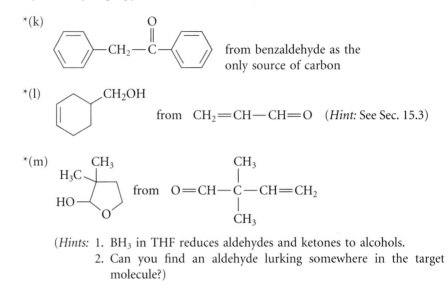

from benzaldehyde as the only source of carbon

*(l)

from $CH_2{=}CH{-}CH{=}O$ (*Hint:* See Sec. 15.3)

*(m)

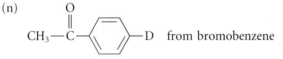

from $O{=}CH{-}\underset{\underset{CH_3}{|}}{\overset{\overset{CH_3}{|}}{C}}{-}CH{=}CH_2$

(*Hints:* 1. BH_3 in THF reduces aldehydes and ketones to alcohols.
2. Can you find an aldehyde lurking somewhere in the target molecule?)

(n)

   from bromobenzene

CH_3—C(=O)—⟨ ⟩—D

*19.50   (a) The following compound is unstable and spontaneously decomposes to acetophenone and HBr. Give a mechanism for this transformation.

$$Ph{-}\underset{\underset{Br}{|}}{\overset{\overset{OH}{|}}{C}}{-}CH_3$$

(b) Use the information in (a) to complete the following reaction:

$$CH_2{=}CH{-}Br + OsO_4 \xrightarrow{H_2O}$$

*19.51   The product *A* of the following reaction hydrolyzes in dilute aqueous acid to give acetophenone. Identify *A* and give a mechanism for its formation that accounts for the regioselectivity of the reaction.

$$Ph{-}C{\equiv}C{-}H + CH_3OH \underset{\text{(solvent)}}{\xrightarrow[\text{(catalyst)}]{H_2SO_4}} \quad [A]\ (C_{10}H_{14}O_2)$$

19.52   What are the starting materials for synthesis of each of the following imines?

*(a)

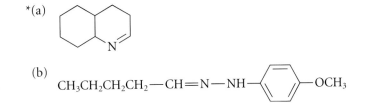

(b)

$CH_3CH_2CH_2CH_2 - CH = N - NH -$ 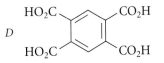 $-OCH_3$

*19.53   Compound A, $C_{11}H_{12}O$, which gave no Tollens' test, was treated with $LiAlH_4$, followed by dilute acid, to give compound B, which could be resolved into enantiomers. When optically active B was treated with CrO_3 in pyridine, an optically inactive sample of A was obtained. Heating A with hydrazine in base gave hydrocarbon C, which, when heated with alkaline $KMnO_4$, gave carboxylic acid D. Identify all compounds and explain your reasoning.

D

$$HO_2C \diagdown \diagup CO_2H$$
$$HO_2C \diagup \diagdown CO_2H$$

19.54   Compound A, $C_6H_{12}O_2$, was found to be optically active, and was slowly oxidized to an optically active carboxylic acid B, $C_6H_{12}O_3$, by $^+Ag(NH_3)_2$. Oxidation of A by anhydrous CrO_3 gave an optically inactive compound that reacted with Zn amalgam/HCl to give 3-methylpentane. With aqueous H_2CrO_4, compound A was oxidized to an optically inactive dicarboxylic acid C, $C_6H_{10}O_4$. Give structures for compounds A, B, and C.

19.55   Remembering that a protonated aldehyde or ketone is a type of carbocation, and that carbocations are electrophiles, propose curved-arrow mechanisms for the following electrophilic aromatic substitution reactions. (See Sec. 16.4B.)

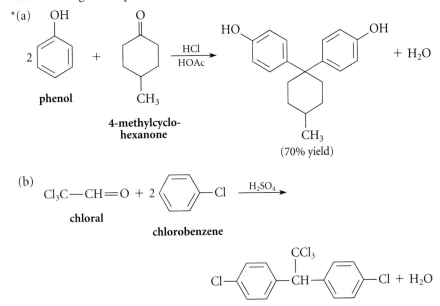

*(a)

phenol   +   4-methylcyclo-hexanone $\xrightarrow[HOAc]{HCl}$ (70% yield)   + H_2O

(b) $Cl_3C - CH = O + 2$ chlorobenzene $\xrightarrow{H_2SO_4}$

DDT (an insecticide)   + H_2O

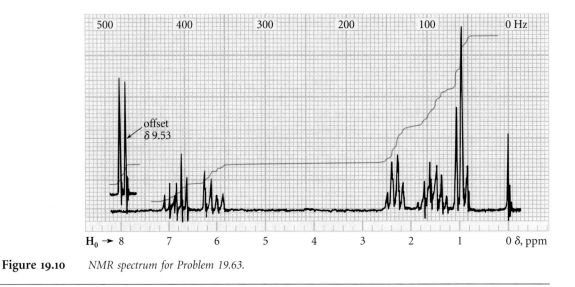

Figure 19.10 *NMR spectrum for Problem 19.63.*

(b) When acetaldehyde is mixed with a tenfold excess of ethanethiol, its $n \rightarrow \pi^*$ absorption at 280 nm is nearly eliminated.

(c) The following compound gives a Tollens' test much more slowly than benzaldehyde.

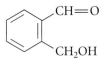

19.63 Identify the following compounds.

*(a) $C_{10}H_{10}O_2$ NMR: δ 2.82 (6H, s), δ 8.13 (4H, s)
IR: 1681 cm^{-1}, no O—H stretch

(b) $C_5H_{10}O$ NMR: δ 9.8 (1H, s), δ 1.1 (9H, s)

*(c) $C_6H_{10}O$ NMR in Fig. 19.10
IR: 1701 cm^{-1}, 970 cm^{-1}
UV: λ_{max} = 215 nm (ϵ = 17,400), 329 nm (ϵ = 26)
This compound gives a positive Tollens' test.

*19.64 Identify the compound with the mass spectrum and proton NMR spectrum given in Fig. 19.11. This compound has IR absorptions at 1678 and 1600 cm^{-1}.

19.65 Identify the compound that has the following spectroscopic properties:
IR: 1686, 1600 cm^{-1}
Mass spectrum: Molecular ion and base peak at m/z = 82
NMR spectrum in Fig. 19.12.

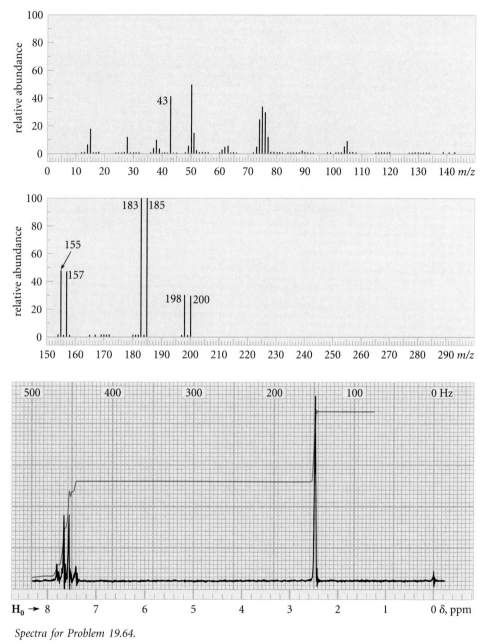

Figure 19.11 *Spectra for Problem 19.64.*

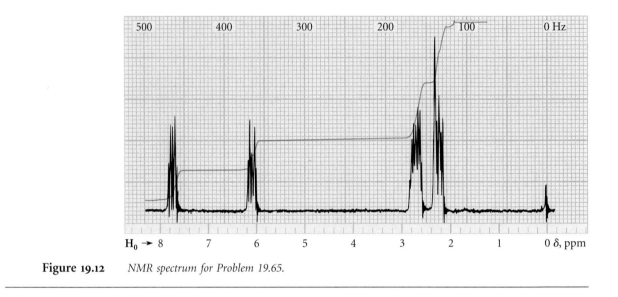

Figure 19.12   *NMR spectrum for Problem 19.65.*

*19.66   Identify the compound *A* ($C_6H_{12}O_3$) that has IR absorption at 1710 cm^{-1} (no absorption in the 3200–3400 cm^{-1} region), as well as the following CMR-DEPT spectrum (attached hydrogens in parentheses): δ 30.6 (3), δ 47.2 (2), δ 53.5 (3), δ 101.7 (1), δ 204.9 (0). One of the proton NMR absorptions of compound *A* is a singlet at δ 2.1.

20

Chemistry of Carboxylic Acids

The characteristic functional group in a **carboxylic acid** is the **carboxy group**.

acetic acid
(a simple carboxylic acid)

condensed structure

Carboxylic acids and their derivatives rank with aldehydes and ketones among the most important organic compounds because they occur widely in nature, and because they serve important roles in organic synthesis. This chapter is concerned with the structures, properties, acidities, and carbonyl-group reactions of carboxylic acids themselves. Chapter 21 is devoted to a study of carboxylic acid derivatives.

This chapter also surveys briefly some of the chemistry of **sulfonic acids**.

condensed structure

methanesulfonic acid
(a simple sulfonic acid)

20.1 Nomenclature of Carboxylic Acids

A. Common Nomenclature

Common nomenclature is widely used for the simpler carboxylic acids. A carboxylic acid is named by adding the suffix *ic* and the word *acid* to the prefix for the appropriate group given in Table 19.1.

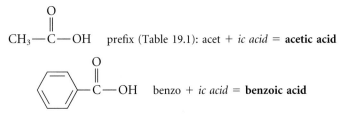

CH$_3$—C—OH   prefix (Table 19.1): acet + *ic acid* = **acetic acid**

—C—OH   benzo + *ic acid* = **benzoic acid**

Some of these names owe their origin to the natural source of the acid. For example, formic acid occurs in the venom of the red ant (Latin, *formica*, ant); acetic acid is the acidic component of vinegar (Latin, *acetus*, vinegar); and butyric acid is the foul-smelling component of rancid butter (Latin, *butyrum*, butter). The common names of carboxylic acids, given in Table 20.1, are used as much or more than the substitutive names.

As with aldehydes and ketones, substitution in the common system is denoted with Greek letters rather than numbers. The position *adjacent* to the carboxy group is designated as α.

$$\overset{\gamma}{CH_3}-\overset{\beta}{CH_2}-\overset{\alpha}{CH}-\overset{\overset{\displaystyle O}{\|}}{C}-OH \qquad \textbf{\textit{α}-bromobutyric acid}$$
$$|$$
$$Br$$

Carboxylic acids with two carboxy groups are called **dicarboxylic acids**. The unbranched dicarboxylic acids are particularly important, and are invariably known by their common names. Some important dicarboxylic acids are also listed in Table 20.1.

$$HO_2C-CH_2CH_2-CO_2H \qquad \textbf{succinic acid}$$

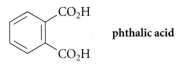

$$HO_2C-CH_2\overset{\overset{\displaystyle CH_3}{|}}{\underset{\underset{\displaystyle CH_3}{|}}{C}}CH_2-CO_2H \qquad \textbf{\textit{β,β}-dimethylglutaric acid}$$

A mnemonic device used by generations of organic chemistry students for remembering the names of the dicarboxylic acids is the phrase, "*Oh, My, Such Good Apple Pie*," in which the first letter of each word corresponds to the name of successive dicarboxylic acids: oxalic, malonic, succinic, glutaric, adipic, and pimelic acids.

Phthalic acid is an important aromatic dicarboxylic acid.

phthalic acid

Many carboxylic acids were known long before any system of nomenclature existed, and their time-honored traditional names are widely used. The following are examples of these.

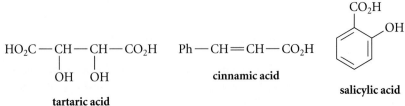

$$HO_2C-\underset{\underset{\displaystyle OH}{|}}{CH}-\underset{\underset{\displaystyle OH}{|}}{CH}-CO_2H \qquad Ph-CH=CH-CO_2H$$

tartaric acid   **cinnamic acid**

salicylic acid

Table 20.1   Names and Structures of Some Carboxylic Acids

Systematic name	Common name	Structure
methanoic[a] acid	formic acid	HCO_2H
ethanoic[a] acid	acetic acid	CH_3CO_2H
propanoic acid	propionic acid	$C_2H_5CO_2H$
butanoic acid	butyric acid	$CH_3CH_2CH_2CO_2H$
2-methylpropanoic acid	isobutyric acid	$(CH_3)_2CHCO_2H$
pentanoic acid	valeric acid	$CH_3(CH_2)_3CO_2H$
3-methylbutanoic acid	isovaleric acid	$(CH_3)_2CHCH_2CO_2H$
2,2-dimethylpropanoic acid	pivalic acid	$(CH_3)_3CCO_2H$
hexanoic acid	caproic acid	$CH_3(CH_2)_4CO_2H$
octanoic acid	caprylic acid	$CH_3(CH_2)_6CO_2H$
decanoic acid	capric acid	$CH_3(CH_2)_8CO_2H$
dodecanoic acid	lauric acid	$CH_3(CH_2)_{10}CO_2H$
tetradecanoic acid	myristic acid	$CH_3(CH_2)_{12}CO_2H$
hexadecanoic acid	palmitic acid	$CH_3(CH_2)_{14}CO_2H$
octadecanoic acid	stearic acid	$CH_3(CH_2)_{16}CO_2H$
2-propenoic[a] acid	acrylic acid	$CH_2{=}CHCO_2H$
2-butenoic[a] acid	crotonic acid	$CH_3CH{=}CHCO_2H$
benzoic acid	benzoic acid	$PhCO_2H$
Dicarboxylic acids		
ethanedioic[a] acid	oxalic acid	$HO_2C{-}CO_2H$
propanedioic[a] acid	malonic acid	$HO_2CCH_2CO_2H$
butanedioic[a] acid	succinic acid	$HO_2C(CH_2)_2CO_2H$
pentanedioic[a] acid	glutaric acid	$HO_2C(CH_2)_3CO_2H$
hexanedioic[a] acid	adipic acid	$HO_2C(CH_2)_4CO_2H$
heptanedioic[a] acid	pimelic acid	$HO_2C(CH_2)_5CO_2H$
1,2-benzenedicarboxylic[a] acid	phthalic acid	
(Z)-2-butenedioic[a] acid	maleic acid	
(E)-2-butenedioic[a] acid	fumaric acid	

[a] Substitutive name is almost never used.

B. Substitutive Nomenclature

A carboxylic acid is named systematically by dropping the final *e* from the name of the hydrocarbon with the same number of carbon atoms and adding the suffix *oic* and the word *acid*.

$$CH_3CH_2\!-\!\overset{\displaystyle O}{\overset{\|}{C}}\!-\!OH \quad propan\cancel{e} + \textit{oic acid} = \textbf{propanoic acid}$$

The final *e* is not dropped in the name of dicarboxylic acids.

$$HO_2C\!-\!CH_2CH_2CH_2CH_2CH_2CH_2\!-\!CO_2H \quad \text{octanedioic acid}$$

When a carboxylic acid is derived from a cyclic hydrocarbon, the suffix *carboxylic* and the word *acid* are added to the name of the hydrocarbon. (This nomenclature is similar to that for the corresponding aldehydes; Sec. 19.1B.)

cyclohexanecarboxylic acid　　**1,2,4-benzenetricarboxylic acid**

One exception to this nomenclature is benzoic acid (Sec. 20.1A), for which the IUPAC recognizes the common name.

The principal chain in substituted carboxylic acids is numbered, as in aldehydes, by assigning the number 1 to the carbonyl carbon.

$$\underset{5}{CH_3}\!-\!\underset{4}{CH_2}\!-\!\underset{3}{\overset{\displaystyle \overset{\textstyle CH_3}{|}}{CH}}\!-\!\underset{2}{CH_2}\!-\!\underset{1}{CO_2H} \qquad \textbf{3-methylpentanoic acid}$$

This numbering scheme should be contrasted with that used in the common system, in which numbering begins with the Greek letter α at carbon-2.

In carboxylic acids derived from cyclic hydrocarbons, numbering begins at the ring carbon bearing the carboxy group.

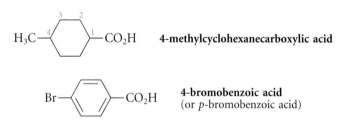

When carboxylic acids contain other functional groups, the carboxy groups receive priority over aldehyde and ketone carbonyl groups, over hydroxy groups, and over mercapto groups for citation as the principal group.

Priority for citation as principal group:

$$\underset{}{\overset{\displaystyle O}{\overset{\|}{-C}}}\!-\!OH > \overset{\displaystyle O}{\overset{\|}{-C}}\!-\!H > \overset{\displaystyle O}{\overset{\|}{-C}}\!- \; > \; -OH \; > \; -SH \qquad (20.1)$$

Provide a substitutive name for the following compound.

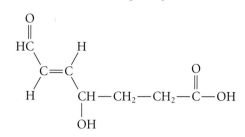

Solution   First, decide on the principal group. From the order in Eq. 20.1, the carboxy group has highest priority. The aldehyde oxygen and the hydroxy group are treated as substituents. The structure has seven carbons and one double bond, and hence is a heptenoic acid. The carboxy group is given the number 1; hence, the double bond is at carbon-5, and the molecule is a 5-heptenoic acid. The —OH group is named as a 4-hydroxy substituent, and the aldehyde oxygen as a 7-oxo substituent. Application of the priority rules for double-bond stereochemistry (Sec. 4.2B) shows that the double bond has *E* stereochemistry. The name is therefore (*E*)-4-hydroxy-7-oxo-5-heptenoic acid. Although carbon-5 is an asymmetric carbon, its stereochemistry is not specified in the structure, and is therefore omitted in the name.

The carboxy group is sometimes named as a substitutent:

carboxymethyl group

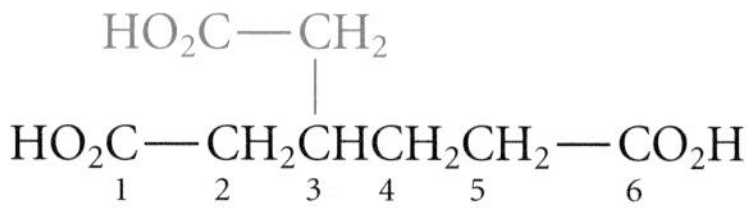

$$HO_2C—CH_2$$
$$HO_2C—CH_2CHCH_2CH_2—CO_2H$$
$$1\quad 2\quad 3\quad 4\quad 5\quad 6$$ **3-(carboxymethyl)hexanedioic acid**

A complete list of nomenclature priorities for all of the functional groups covered in this text is given in Appendix I.

20.1   Give the structure of each of the following compounds.
*(a) γ-hydroxybutyric acid
(b) β,β-dichloropropionic acid
*(c) (*Z*)-3-hexenoic acid
(d) 4-methylhexanoic acid
*(e) 1,4-cyclohexanedicarboxylic acid
(f) *p*-methoxybenzoic acid
*(g) α,α-dichloroadipic acid
(h) oxalic acid

20.2   Name each of the following compounds. Use a common name for at least one compound.

*(a)

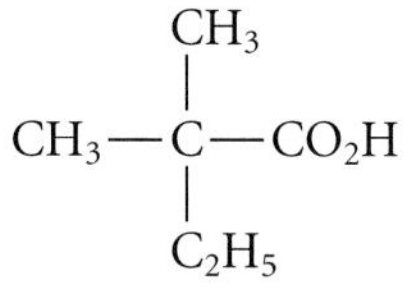

$$CH_3$$
$$CH_3—C—CO_2H$$
$$C_2H_5$$

(b) $HO_2C(CH_2)_7CH(CH_3)_2$

(Problem 20.2 continues)

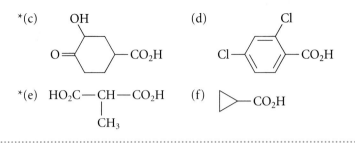

*(c)   OH / O= / —CO₂H

(d)   Cl / Cl— —CO₂H

*(e)   HO₂C—CH—CO₂H
                |
               CH₃

(f)   ▷—CO₂H

20.2 Structure and Physical Properties of Carboxylic Acids

The structure of a simple carboxylic acid, acetic acid, is compared with the structures of other oxygen-containing compounds in Figure 20.1. Carboxylic acids, like aldehydes and ketones, have trigonal geometry at their carbonyl carbons. Notice that the two oxygens of a carboxylic acid are quite different. One, the **carbonyl oxygen**, is part of the C=O double bond. Fig. 20.1 demonstrates that the C=O bonds of aldehydes, ketones, and carboxylic acids have the same length. The other oxygen, called the **carboxylate oxygen**, is part of the C—O single bond. Notice from Fig. 20.1 that the C—O bond in a carboxylic acid is considerably shorter than the C—O bond in an alcohol or ether (1.36 Å vs. 1.42 Å). One reason for this is that the C—O bond in an acid is an sp^2-sp^3 single bond, whereas the C—O bond in an alcohol or ether is an sp^3-sp^3 single bond. Another reason is that a carboxylic acid has a resonance structure in which the C—O bond has some double-bond character.

$$\left[\begin{array}{c} :\!O\!: \\ \| \\ R\!-\!C\!-\!\ddot{O}H \end{array} \quad \longleftrightarrow \quad \begin{array}{c} :\!\ddot{O}\!:^{-} \\ | \\ R\!-\!C\!\!=\!\!\overset{+}{\ddot{O}}H \end{array} \right] \tag{20.2}$$

double-bond character

The carboxylic acids of lower molecular mass are high-boiling liquids with acrid, piercing odors. They have considerably higher boiling points than many other organic compounds of about the same molecular mass and shape:

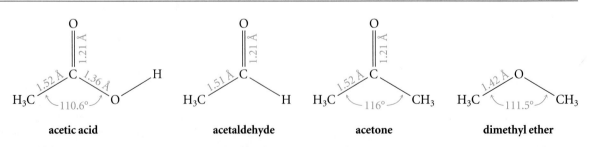

acetic acid   acetaldehyde   acetone   dimethyl ether

Figure 20.1   *Comparison of the structures of acetic acid and other oxygen-containing compounds. Notice that the carbonyl compounds have identical C=O bond lengths, and that the C—O bond in a carboxylic acid is shorter than that in an ether.*

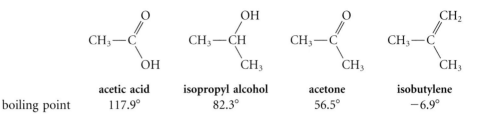

	acetic acid	isopropyl alcohol	acetone	isobutylene
boiling point	117.9°	82.3°	56.5°	−6.9°

The high boiling points of carboxylic acids can be attributed not only to their polarity, but also to the fact that they form very strong hydrogen bonds. In the solid state, and under some conditions in both the gas phase and solution, carboxylic acids exist as hydrogen-bonded dimers. (A **dimer** is any structure derived from two identical smaller units.)

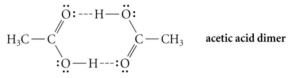

acetic acid dimer

The equilibrium constants for the formation of such dimers in solution are on the order of 10^6 to 10^7 M^{-1}. (The equilibrium constant for dimerization of ethanol, in contrast, is 11 M^{-1}.)

Many aromatic and dicarboxylic acids are solids. For example, the melting points of benzoic acid and succinic acid are 122° and 188°, respectively.

The simpler carboxylic acids have substantial solubilities in water, as expected from their hydrogen-bonding capabilities; the unbranched carboxylic acids below pentanoic acid are miscible with water. Many dicarboxylic acids also have significant water solubilities.

PROBLEM

*20.3   At a given concentration of acetic acid, in which solvent would you expect the amount of acetic acid dimer to be greater: CCl_4 or water? Explain.

20.3 Spectroscopy of Carboxylic Acids

A. IR Spectroscopy

Two important absorptions are found in the infrared spectrum of a typical carboxylic acid. One is the C=O stretching absorption, which occurs near 1710 cm^{-1} for carboxylic acid dimers. (The IR spectra of carboxylic acids are nearly always run under conditions such that they are in the dimer form. The carbonyl absorptions of carboxylic acid monomers occur near 1760 cm^{-1} but are rarely observed.) The other important carboxylic acid absorption is the O—H stretching absorption. This absorption is much broader than the O—H stretching absorption of an alcohol or phenol, and covers a very wide region of the spectrum—typically 2400–3600 cm^{-1}. (In many cases this absorption obliterates the C—H stretching absorption of the acid.) The carbonyl absorption and

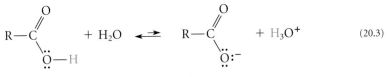

$$\text{(20.3)}$$

carboxylic acid   carboxylate ion

The conjugate bases of carboxylic acids are called generally **carboxylate ions**. Carboxylate salts are named by replacing the *ic* in the name of the acid (in any system of nomenclature) with the suffix *ate*.

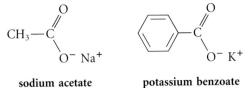

sodium acetate   **potassium benzoate**
(acet~~ic~~ + *ate* = **acetate**)

STUDY GUIDE LINK:
✓20.2
*Bases that React with
Carboxylic Acids*

Carboxylic acids are among the most acidic organic compounds; acetic acid, for example, has a pK_a of 4.76. This pK_a is low enough that an aqueous solution of acetic acid gives an acid reaction with litmus or pH paper.

Carboxylic acids are more acidic than alcohols or phenols, other compounds with O—H bonds.

——— increasing acidity ———→

C_2H_5—O—H   ⟨benzene⟩—O—H CH_3—$\overset{\overset{O}{\|}}{C}$—O—H

pK_a 15.9   9.95   4.76

The acidity of carboxylic acids is due to two factors. First is the *polar effect* of the carbonyl group. The carbonyl group, because of its sp^2-hybridized atoms and the presence of oxygen, is a very electronegative group. The carbonyl group is much more electronegative than the phenyl ring of a phenol or the alkyl group of an alcohol. The polar effect of the carbonyl group stabilizes charge in the carboxylate ion. Remember that *stabilization of a conjugate base enhances acidity* (Sec. 3.6B).

STUDY GUIDE LINK:
✓20.3
*Resonance Effect on
Carboxylic Acid
Acidity*

The second factor that accounts for the acidity of carboxylic acids is the resonance stabilization of their conjugate-base carboxylate ions.

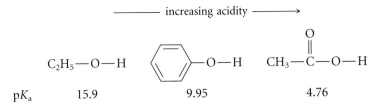

$$\text{(20.4)}$$

resonance structures of acetate ion

Although typical carboxylic acids have pK_a values in the 4–5 range, the acidities of carboxylic acids vary with structure. Recall, for example (Sec. 3.6B), that halogen substitution within the alkyl group of a carboxylic acid enhances acidity by a polar effect.

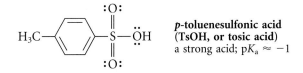

$$\text{pK}_a \quad\quad 4.76 \quad\quad\quad\quad 2.66 \quad\quad\quad\quad\quad 1.24 \quad\quad\quad\quad\quad 0.23$$

Trifluoroacetic acid, commonly abbreviated TFA, is such a strong acid that it is often used in place of HCl and H_2SO_4 when an acid of moderate strength is required.

The pK_a values of some carboxylic acids are given in Table 20.2, and the pK_a values of the simple dicarboxylic acids in Table 20.3. The data in these tables give some idea of the range over which the acidities of carboxylic acids vary.

Sulfonic acids are much stronger than comparably substituted carboxylic acids.

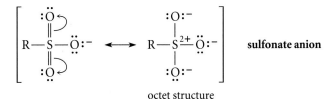

One reason that sulfonic acids are more acidic than carboxylic acids is the high oxidation state of sulfur. The octet structure for a sulfonate anion indicates that sulfur has considerable positive charge. This positive charge stabilizes the negative charge on the oxygens.

octet structure

Table 20.2 pK_a Values of Some Carboxylic Acids

Acid[a]	pK_a
formic	3.75
acetic	4.76
propionic	4.87
2,2-dimethylpropanoic (pivalic)	5.05
acrylic	4.26
chloroacetic	2.85
phenylacetic	4.31
benzoic	4.18
p-methylbenzoic (p-toluic)	4.37
p-nitrobenzoic	3.43
p-chlorobenzoic	3.98
p-methoxybenzoic (p-anisic)	4.47
2,4,6-trinitrobenzoic	0.65

[a] See Table 20.1 for structures.

Table 20.3 pK_a Values of Some Dicarboxylic Acids

Acid[a]	First pK_a	Second pK_a
carbonic	3.58	6.35
oxalic	1.27	4.27
malonic	2.86	5.70
succinic	4.21	5.64
glutaric	4.34	5.27
adipic	4.41	5.28
phthalic	2.95	5.41

[a] See Table 20.1 for structures.

Sulfonic acids are useful as acid catalysts in organic solvents because they are more soluble than most inorganic acids. For example, *p*-toluenesulfonic acid is moderately soluble in benzene and toluene and can be used as a strong acid catalyst in those solvents. (Sulfuric acid, in contrast, is completely insoluble in benzene and toluene.)

Many carboxylic acids of moderate molecular mass are not soluble in water. Their alkali metal salts, however, are ionic compounds, and in many cases are much more soluble in water. Therefore many water-insoluble carboxylic acids dissolve in solutions of alkali metal hydroxides (NaOH, KOH) because the insoluble acids are converted completely into their soluble salts.

$$\underset{\substack{\Vert \\ \text{R}-\text{C}-\overset{\cdot\cdot}{\underset{\cdot\cdot}{\text{O}}}\text{H}}}{\overset{\text{O}}{}} + \text{NaOH} \longrightarrow \underset{\substack{\Vert \\ \text{R}-\text{C}-\overset{\cdot\cdot}{\underset{\cdot\cdot}{\text{O}}}\text{:}^- \text{ Na}^+}}{\overset{\text{O}}{}} + \text{H}_2\text{O} \tag{20.6}$$

Even 5% sodium bicarbonate ($NaHCO_3$) solution is basic enough (pH $\approx$ 8.5) to dissolve a carboxylic acid. This can be understood from the equilibrium expression for ionization of a carboxylic acid RCO_2H with a dissociation constant K_a.

$$K_a = \frac{[\text{RCO}_2^-][\text{H}_3\text{O}^+]}{[\text{RCO}_2\text{H}]} \tag{20.7a}$$

or

$$\frac{K_a}{[\text{H}_3\text{O}^+]} = \frac{[\text{RCO}_2^-]}{[\text{RCO}_2\text{H}]} \tag{20.7b}$$

In order for the carboxylic acid to dissolve in water, it has to be mostly ionized and in its soluble conjugate-base form RCO_2^-; that is, in Eq. 20.7b, the ratio $[\text{RCO}_2^-]/[\text{RCO}_2\text{H}]$ has to be *large*. As a practical matter, we can say that when this ratio is 100 or greater, the acid has been completely converted into its anion. (Of course, there is no pH at which the acid exists *completely* as its anion; but when this ratio is $\geq$100, the concentration of acid is negligible.) Since K_a is a constant, Eq. 20.7b shows that this ratio can be made large by making the hydrogen-ion concentration $[\text{H}_3\text{O}^+]$ small. In other words, $[\text{H}_3\text{O}^+]$ *must be small in comparison with the K_a of the acid.* Taking negative logarithms of Eq. 20.7b gives the following result:

$$\text{pH} - \text{p}K_a = \log \frac{[\text{RCO}_2^-]}{[\text{RCO}_2\text{H}]} \tag{20.7c}$$

If $[\text{RCO}_2^-]/[\text{RCO}_2\text{H}]$ is $\geq$100, then the logarithm of this ratio is $\geq$2. Equation 20.7c shows that, for conversion of an acid into its anion, the pH of the solution must be two or more units greater than the pK_a of the acid. Since 5% sodium bicarbonate solution has a pH of about 8.5, and most carboxylic acids have pK_a values in the range 4–5, sodium bicarbonate solution is more than basic enough to dissolve a typical acid, provided of course that enough bicarbonate is present that it is not completely consumed by reaction with the acid.

A typical carboxylic acid, then, can be separated from mixtures with other water-insoluble, nonacidic substances by extraction with NaOH, Na_2CO_3, or $NaHCO_3$ solution. The acid dissolves in the basic aqueous solution, but nonacidic compounds do not. After separating the basic aqueous solution, it can be acidified with a strong acid to yield the carboxylic acid, which may be isolated by filtration or extraction with organic solvents. (A similar idea was used in the separation of phenols; Sec. 18.6B.) Carboxylic acids can also be separated from phenols by extraction with 5% $NaHCO_3$ if the phenol is not one that is unusually acidic. Since the pK_a of a typical phenol is about 10, it remains largely un-ionized and thus insoluble in a solution with a pH of 8.5. (This conclusion follows from an equation for phenol ionization analogous to Eq. 20.7c.)

PROBLEMS

*20.7   (a) Write the equations for the first and second ionizations of succinic acid. Label each with the appropriate pK_a values from Table 20.3.
        (b) Why is the first pK_a value of succinic acid lower than the second pK_a value?

20.8   Explain why the first and second pK_a values of the dicarboxylic acids become closer together as the lengths of their carbon chains increase (Table 20.3).

*20.9   Imagine that you have just carried out a conversion of p-bromotoluene into p-bromobenzoic acid and wish to separate the product from the unreacted starting material. Design a separation of these two substances that would enable you to isolate the purified acid.

B. Basicity of Carboxylic Acids

Although we think of carboxylic acids primarily as acids, the carbonyl oxygens of acids, like those of aldehydes or ketones, are weakly basic.

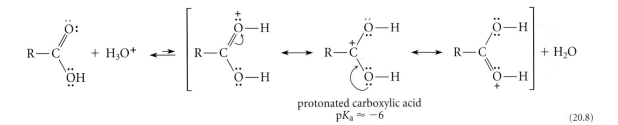

protonated carboxylic acid
$pK_a \approx -6$

(20.8)

The basicity of carboxylic acids plays a very important role in many of their reactions.

Protonation of an acid on the carbonyl oxygen occurs because, as Eq. 20.8 shows, a resonance-stabilized cation is formed. Protonation on the carboxylate oxygen is much less favorable because it does not give a resonance-stabilized cation, and because the positive charge on oxygen is destabilized by the polar effect of the carbonyl group.

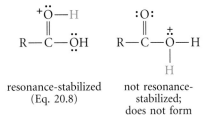

resonance-stabilized   not resonance-
(Eq. 20.8)   stabilized;
                           does not form

20.5   Fatty Acids, Soaps, and Detergents

Carboxylic acids with long, unbranched carbon chains are called **fatty acids** because many of them are liberated from fats and oils by a hydrolytic process called *saponification* (Sec. 21.7A). Some fatty acids contain carbon-carbon double bonds. Fatty acids with *cis*-double bonds occur widely in nature, but those with *trans*-double bonds are rare. The following compounds are examples of common fatty acids:

$$CH_3(CH_2)_{14}CO_2H \quad \text{or} \quad CH_3CH_2CH_2CH_2CH_2CH_2CH_2CH_2CH_2CH_2CH_2CH_2CH_2CH_2CH_2CO_2H$$

palmitic acid
(from palm oil)

$$CH_3(CH_2)_{16}CO_2H$$

stearic acid
(Greek, *stear*, "tallow," or "beef fat")

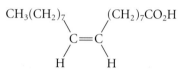

oleic acid

The sodium and potassium salts of fatty acids, called **soaps**, are the major ingredients of commercial soap.

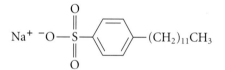

sodium stearate: a soap

Soaps are one type of detergent. A **detergent** is any substance used for cleaning a solid object (such as cloth) by immersing it in a liquid solution. Closely related to soaps are synthetic detergents. The following compound, the sodium salt of a sulfonic acid, is used in household laundry detergent formulations.

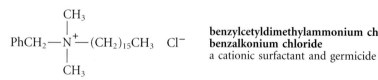

Many soaps and detergents have not only cleansing properties, but also germicidal characteristics.

Soaps and synthetic detergents are two examples of a larger class of molecules known as **surfactants**. Surfactants are molecules with two structural parts that interact with water in opposing ways: a *polar head group*, which is readily solvated by water, and a *hydrocarbon tail*, which, like a long alkane, is not well solvated by water. In a soap, the polar head group is the carboxylate anion, and the hydrocarbon tail is obviously the carbon chain. The soap and the detergent shown above are examples of *anionic surfactants*, that is, surfactants with an anionic polar head group. *Cationic surfactants* are also known:

$$PhCH_2 - \overset{\overset{\displaystyle CH_3}{\displaystyle |}}{\underset{\underset{\displaystyle CH_3}{\displaystyle |}}{N^+}} - (CH_2)_{15}CH_3 \quad Cl^-$$

**benzylcetyldimethylammonium chloride
benzalkonium chloride**
a cationic surfactant and germicide

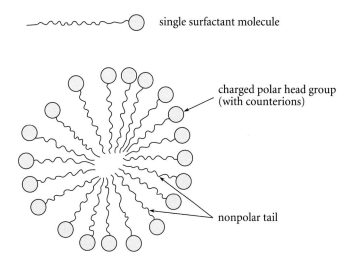

Figure 20.3   *Schematic diagram of micelle structure. Each micelle contains 50–150 molecules and is approximately spherical. Notice that the polar head groups within the micelle are directed outward toward the solvent water, and that the nonpolar tails interact with each other and are isolated from water within the interior of the micelle.*

Although small amounts of surfactant molecules dissolve in water, when the surfactant concentration is raised above a certain value, called the **critical micelle concentration (CMC)**, the surfactant molecules spontaneously form **micelles**, which are approximately spherical aggregates of 50–150 surfactant molecules (Fig. 20.3). Think of a micelle as a large ball in which the polar head groups are exposed on the outside and the nonpolar tails are buried on the inside. The micellar structure satisfies the solvation requirements of both the polar head groups, which are close to water, and the "greasy groups"—the nonpolar tails—which associate with each other on the inside of the micelle. (Recall "like-dissolves-like"; Sec. 8.4B.)

The detergent properties of surfactants are easy to understand once the rationale for the formation of micelles is clear (Fig. 20.4). When a fabric with greasy dirt is exposed

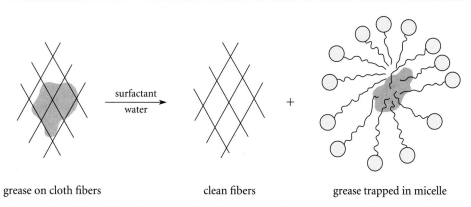

grease on cloth fibers   clean fibers   grease trapped in micelle

Figure 20.4   *Schematic description of the detergent action of a surfactant. (The size of the surfactant micelle is greatly enlarged.) Dirt (grease) is extracted from the fibers into the "greasy" interior of the micelle.*

to an aqueous solution containing micelles of a soap or detergent, the dirt associates with the "greasy" hydrocarbon chains on the interior of the micelle and is incorporated into the micellar aggregate. Then the dirt is lifted away from the surface of the fabric and carried into solution. The antiseptic action of some surfactants owes its success to a similar phenomenon. A cell membrane—the envelope that surrounds the contents of the cell—is made up of molecules, called *phospholipids*, which are also surfactants (Sec. 21.12B). When the bacterial cell is exposed to a solution containing a surfactant, phospholipids of the cell membrane tend to associate with the surfactant. In some cases this disrupts the membrane enough that the cell can no longer function, and it dies.

☞
Study Guide Link:
20.4
More on Surfactants

So-called *hard water* disrupts the cleaning action of detergents, and causes the formation of a scum when mixed with soaps. Hard water contains Ca^{2+} and Mg^{2+} ions. Hard-water scum ("bathtub ring") is a precipitate of the calcium or magnesium salts of fatty acids, which (unlike the sodium and potassium salts) do not form micelles and are insoluble in water. These offending ions can be solubilized and removed by complexation with phosphates. Phosphates, however, have been found to cause excessive growth of algae in rivers and streams, and their use has been curtailed (thus, "low-phosphate detergents"). Unfortunately, no completely acceptable substitute for phosphates has yet been found.

Surfactants are used not only in "soap-and-water"-type cleaning operations. They are also extremely important as components of fuels and lubricating oils. For example, detergents in engine oils assist in keeping deposits suspended in the oil and thus prevent them from building up on engine surfaces.

20.6   Synthesis of Carboxylic Acids

Two reactions covered in previous chapters are important for the preparation of carboxylic acids:

1. Oxidation of primary alcohols and aldehydes (Secs. 10.6B, 19.14)
2. Side-chain oxidation of alkylbenzenes (Sec. 17.5)

Ozonolysis (Sec. 5.4) can also be used to prepare carboxylic acids, although it is less important because it breaks carbon-carbon bonds.

Another important method for preparation of carboxylic acids is the reaction of Grignard reagents with carbon dioxide, followed by protonolysis. Typically the reaction is run by pouring an ether solution of the Grignard reagent over crushed dry ice.

$$\underset{CH_3CH_2CHMgBr}{\overset{CH_3}{|}} + CO_2 \longrightarrow \xrightarrow{H_3O^+} \underset{CH_3CH_2CH}{\overset{CH_3}{|}}\overset{O}{\overset{||}{-C}}-OH \qquad (20.9)$$
(76–86% yield)

Carbon dioxide is itself a carbonyl compound. The mechanism of this reaction is much like that for Grignard additions to other carbonyl compounds (Sec. 19.9).

Addition of the Grignard reagent to carbon dioxide gives the bromomagnesium salt of a carboxylic acid. When aqueous acid is added to the reaction mixture in a separate reaction step, the free carboxylic acid is formed.

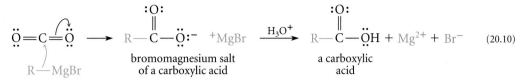

Notice that the reaction of Grignard reagents with CO_2, unlike the other reactions listed at the beginning of this section, is another method for forming carbon-carbon bonds. (Be sure to review the others; Appendix V.)

All of the methods used for preparing carboxylic acids are summarized in Appendix IV. A number of these are discussed in the next two chapters.

PROBLEM

20.10   Outline a synthetic scheme for each of the following transformations.
  *(a)   cyclopentanecarboxylic acid from cyclopentanol
  (b)   octanoic acid from 1-heptene

20.7   Introduction to Carboxylic Acid Reactions

The reactions of carboxylic acids can be categorized into four types.

1. Reactions at the carbonyl group
2. Reactions at the carboxylate oxygen
3. Loss of the carboxy group as CO_2 (decarboxylation)
4. Reactions involving the α-carbon

The most typical reaction at the carbonyl group is *substitution at the carbonyl carbon.* Let E—Y be a general reagent in which E is an electrophilic group (for example, a hydrogen) and Y is a nucleophilic group. Typically the —OH of the carboxy group is substituted by the group —Y.

$$\underset{R-C-OH}{\overset{O}{\parallel}} + E-Y \;\rightleftharpoons\; \underset{R-C-Y}{\overset{O}{\parallel}} + HO-E \qquad (20.11)$$

Another reaction at the carbonyl group of carboxylic acids and their derivatives is reaction of the carbonyl oxygen with an electrophile (Lewis acid or Brønsted acid), that is, the reaction of the carbonyl oxygen as a *base*:

$$\underset{R-C-\ddot{O}H}{\overset{:\ddot{O}:}{\parallel}} \;\rightleftharpoons\; \left[\underset{R-C-\ddot{O}H}{\overset{:\overset{+}{\ddot{O}}-E}{\parallel}} \;\longleftrightarrow\; \underset{R-C=\overset{+}{\ddot{O}}H}{\overset{:\ddot{O}-E}{\vert}} \right] + Y:^{-} \qquad (20.12)$$

One such reaction, protonation of the carbonyl oxygen, was discussed in Sec. 20.4B. Many substitution reactions at the carbonyl carbon are acid-catalyzed; that is, the reactions of nucleophiles at the carbonyl *carbon* are catalyzed by the reactions of acids at the carbonyl *oxygen*.

You've already studied one *reaction at the carboxylate oxygen*: the ionization of carboxylic acids (Sec. 20.4A):

$$R—\overset{\overset{\displaystyle :O:}{\|}}{C}—\overset{..}{\underset{..}{O}}—H + H_2\overset{..}{O}: \rightleftharpoons \left[R—\overset{\overset{\displaystyle :O:}{\|}}{C}—\overset{..}{\underset{..}{O}}:^- \longleftrightarrow R—\overset{\overset{\displaystyle :\overset{..}{O}:^-}{|}}{C}=\overset{..}{O}: \right] + H_3\overset{+}{\overset{..}{O}}: \quad (20.13)$$

Another general reaction involves reaction of the carboxylate oxygen as a nucleophile ($Y:^-$ = halide, sulfonate ester, or other leaving group).

$$R—\overset{\overset{\displaystyle O}{\|}}{C}—\overset{..}{\underset{..}{O}}:^- + E—Y \longrightarrow R—\overset{\overset{\displaystyle O}{\|}}{C}—\overset{..}{\underset{..}{O}}—E + Y:^- \quad (20.14)$$

Decarboxylation is loss of the carboxy group as CO_2.

$$R—\overset{\overset{\displaystyle O}{\|}}{C}—O—H \longrightarrow R—H + O=C=O \quad (20.15)$$

This reaction is more important for some types of carboxylic acids than for others (Sec. 20.11).

This chapter concentrates on the first three types of reactions—reactions at the carbonyl group, reactions at the carboxylate oxygen, and decarboxylation. Chapter 21 considers for the most part reactions at the carbonyl group of carboxylic acid *derivatives*. Chapter 22 takes up the fourth type of reaction—reactions involving the α-carbon—for both carboxylic acids and their derivatives.

20.8   Conversion of Carboxylic Acids into Esters

A. Acid-Catalyzed Esterification

Esters are carboxylic acid derivatives with the following general structure:

$$R—\overset{\overset{\displaystyle O}{\|}}{C}—O—R' \quad (R = H, \text{alkyl, or aryl}; R' = \text{alkyl or aryl})$$

When a carboxylic acid is treated with a large excess of an alcohol in the presence of a strong acid catalyst, an ester is formed.

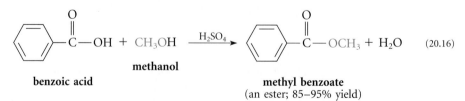

benzoic acid   **methanol**   **methyl benzoate**
(an ester; 85–95% yield)

This reaction is called **acid-catalyzed esterification**, or sometimes *Fischer esterification*, after the great German chemist Emil Fischer (1852–1919).

The equilibrium constants for esterifications with most primary alcohols are near unity; for example, the equilibrium constant for the esterification of acetic acid with ethyl alcohol is 3.38. The reaction is driven to completion by applying *LeChatelier's principle*. Use of the reactant alcohol as the solvent, which ensures that it is present in large excess, drives the equilibrium toward the ester product.

Acid-catalyzed esterification *cannot* be applied to the synthesis of esters from phenols or tertiary alcohols. Tertiary alcohols undergo dehydration (Sec. 10.1) and other reactions under the acidic conditions of the reaction, and the equilibrium constants for esterification of phenols are much less favorable than those for esterification of alcohols by a factor of about 10^4. Although it is possible in principle to drive the esterification of phenols to completion, there are simpler ways for preparing esters of both phenols and tertiary alcohols that are discussed in Chapter 21.

A very important question relevant to the mechanism of acid-catalyzed esterification is whether the oxygen of the water liberated in the reaction comes from the acid or the alcohol.

$$Ph-\overset{\overset{\text{O}}{\|}}{C}-OH + CH_3-OH \quad \overset{?}{\longrightarrow} \quad \begin{cases} \longrightarrow Ph-\overset{\overset{\text{O}}{\|}}{C}-OCH_3 + H_2O & (20.17a) \\ \text{or} \\ \xrightarrow{\times} Ph-\overset{\overset{\text{O}}{\|}}{C}-OCH_3 + H_2O & (20.17b) \end{cases}$$

This question was answered in 1938, when it was found, using the ^{18}O isotope to label the alcohol oxygen, that the OH of the water produced comes exclusively from the carboxylic acid. Acid-catalyzed esterification is therefore a *substitution of* —OH *at the carbonyl group of the acid by the oxygen of the alcohol* (Eq. 20.17a). Thus, acid-catalyzed esterification is an example of *substitution at a carbonyl carbon*.

The mechanism of acid-catalyzed esterification is very important, for it serves as a model for the mechanisms of other acid-catalyzed reactions of carboxylic acids and their derivatives. In the mechanism that follows, the formation of a methyl ester in the solvent methanol is shown for concreteness. The first step of the mechanism is protonation of the carbonyl oxygen (Sec. 20.4B):

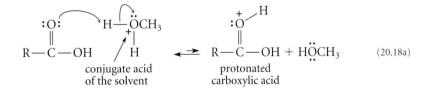

$$\text{(20.18a)}$$

The catalyzing acid is the conjugate acid of the solvent. Notice that this is the actual acid present when a strong acid such as H_2SO_4 is dissolved in methanol.

Recall (Sec. 19.7A) that *protonation of a carbonyl oxygen makes the carbonyl carbon more electrophilic* because the carbonyl oxygen becomes a better electron acceptor. The carbonyl carbon of a protonated carbonyl group is electrophilic enough to react with the weakly basic methanol molecule. Attack of methanol on the carbonyl carbon, followed by loss of a proton, gives a **tetrahedral addition intermediate**.

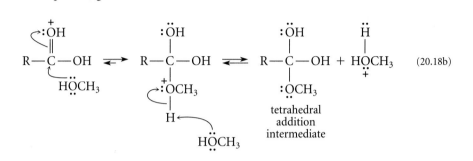

(20.18b)

tetrahedral addition intermediate

Formation of the tetrahedral addition intermediate is essentially the same reaction as the acid-catalyzed reaction of an alcohol with a protonated aldehyde or ketone to form a hemiacetal (Sec. 19.10A). The tetrahedral addition intermediate, after protonation, loses water to give the conjugate acid of the ester:

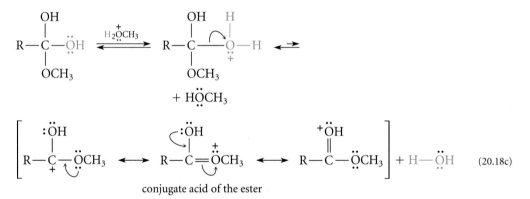

conjugate acid of the ester

(20.18c)

Loss of a proton gives the ester product.

$$R—C—OCH_3 \quad H \;\rightleftharpoons\; R—C—OCH_3 + HOCH_3$$

The acid catalyst is also regenerated in this step.

Notice that *the mechanism of esterification is an extension of the mechanism of carbonyl addition.* In esterification, a nucleophile attacks the carbonyl carbon, just as in carbonyl-addition reactions (Sec. 19.7A, Fig. 19.8), and an addition compound is formed. In esterification, however, the addition compound—the tetrahedral addition intermediate—reacts further; the —OH group from the carboxylic acid, after protonation, is expelled from the addition compound, and a carboxylic acid derivative, an ester, is formed.

Esterification illustrates a very general mechanistic pattern that occurs in many substitution reactions of carboxylic acids and their derivatives. An addition intermediate is formed, and a leaving group —X is expelled from this intermediate to give a new carboxylic acid derivative.

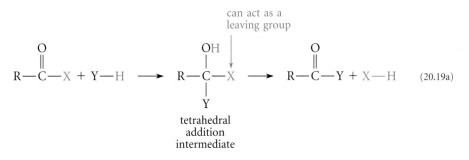

$$\text{(20.19a)}$$

In other words, substitution at a carbonyl carbon is really a sequence of two processes: *addition* to the carbonyl group followed by *elimination* to regenerate the carbonyl group.

Note that although esterification is catalyzed only by acids, a number of other carbonyl-substitution reactions, like carbonyl-addition reactions, are base-catalyzed. Several reactions of this type are discussed in Chapters 21 and 22.

Why don't aldehydes and ketones undergo substitution at their carbonyl carbons? When a nucleophile attacks the carbonyl carbon of an aldehyde or ketone, *neither of the groups attached to the carbonyl carbon can act as a leaving group.*

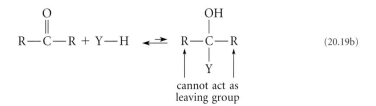

$$\text{(20.19b)}$$

The reason is that the H— or the R— group of an aldehyde or ketone, in order to act as a leaving group, would be expelled as H:⁻ or R:⁻, respectively, either of which is a very *strong* base. This cannot occur because in carbonyl-substitution reactions, as in S_N2 and S_N1 reactions, the best leaving groups are generally the *weakest* bases (Secs. 9.4E, 9.6C). In contrast, one of the groups attached to the carbonyl carbon of a carboxylic acid derivative is either a weak base, or (as in the case of esterification, Eq. 20.18c), is converted by protonation into a weak base; in either case, such a group can act as a good leaving group, and substitution results.

Study Guide Link:
20.5
Orthoesters

Problems

20.11   Give the structure of the product formed when
*(a) adipic acid is heated in a large excess of 1-propanol (as solvent) with a sulfuric acid catalyst.
(b) 3-methylhexanoic acid is heated with a large excess of ethanol (as solvent) with a sulfuric acid catalyst.

*20.12   (a) Using the principle of microscopic reversibility, give a detailed mechanism for the acid-catalyzed hydrolysis of methyl benzoate (structure in Eq. 20.16) to benzoic acid.
(b) How would you change the reaction conditions of acid-catalyzed esterification to favor ester *hydrolysis* rather than ester *formation*?

(Problem 20.12 continues)

(c) The hydrolysis of methyl benzoate is also promoted by ¯OH. Write a mechanism for the hydrolysis of methyl benzoate in NaOH solution.

(d) Methyl benzoate *cannot* be prepared from benzoic acid and methanol containing sodium methoxide (Na^+ CH_3O^-) because the acid reacts with methoxide in another way. What is this reaction?

B. Esterification by Alkylation

The esterification discussed in the previous section involves attack of a nucleophile at the carbonyl carbon. This section considers a different method of forming esters that illustrates another mode of carboxylic acid reactivity: nucleophilic reactivity of the *carboxylate oxygen*.

When a carboxylic acid is treated with diazomethane in ether solution, it is rapidly converted into its methyl ester.

$$CH_3(CH_2)_4CH=CH-\overset{\overset{\textstyle O}{\|}}{C}-OH \ + \ :\bar{C}H_2-\overset{+}{N}\equiv N: \ \xrightarrow{\text{ether}}$$

2-octenoic acid   **diazomethane**

$$CH_3(CH_2)_4CH=CH-\overset{\overset{\textstyle O}{\|}}{C}-OCH_3 \ + \ :N\equiv N: \qquad (20.20)$$

methyl 2-octenoate
(91% yield)

Diazomethane, a toxic yellow gas (bp $= -23°$), is usually generated chemically as it is needed and is codistilled with ether into a flask containing the carboxylic acid to be esterified. Diazomethane is both explosive and allergenic, and is therefore only used in small quantities. Nevertheless, esterification with diazomethane is so mild and free of side reactions that in many cases it is the method of choice for the synthesis of methyl esters, particularly in small-scale reactions.

The acidity of the carboxylic acid is important in the mechanism of this reaction. Protonation of diazomethane by the carboxylic acid gives the methyldiazonium ion.

$$R-\overset{\overset{\textstyle O}{\|}}{C}-\overset{..}{\underset{..}{O}}-H \quad :CH_2-\overset{+}{N}\equiv N: \ \rightleftarrows \ R-\overset{\overset{\textstyle O}{\|}}{C}-\overset{..}{\underset{..}{O}}:^- \ + \ CH_3-\overset{+}{N}\equiv N: \qquad (20.21a)$$

methyldiazonium ion

This ion has one of the best leaving groups, molecular nitrogen. An S_N2 reaction of the methyldiazonium ion with the carboxylate oxygen results in the displacement of N_2 and formation of the ester.

$$R-\overset{\overset{\textstyle O}{\|}}{C}-\overset{..}{\underset{..}{O}}:^- \ + CH_3-\overset{+}{N}\equiv N: \ \longrightarrow \ R-\overset{\overset{\textstyle O}{\|}}{C}-\overset{..}{\underset{..}{O}}-CH_3 \ + \ :N\equiv N: \qquad (20.21b)$$

Carboxylate ions are less basic, and therefore less nucleophilic, than alkoxides or phenoxides, but they do react with especially reactive alkylating agents. The methyl-diazonium ion formed by the protonation of diazomethane is one of the most reactive alkylating agents known. Notice, though, that it is the *acid* that reacts with diazomethane, because the acid is required to protonate diazomethane in the first step of the reaction.

The nucleophilic reactivity of carboxylates is also illustrated by the reaction of certain alkyl halides with carboxylate ions.

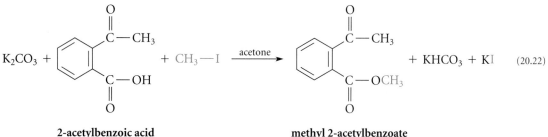

2-acetylbenzoic acid   methyl 2-acetylbenzoate
                                              (65% yield)

This is an S_N2 reaction in which the carboxylate ion, formed by the acid-base reaction of the acid and K_2CO_3, acts as the nucleophile that attacks the alkyl halide. Because carboxylate ions are such weak nucleophiles, this reaction works best on alkyl halides that are especially reactive in S_N2 reactions, such as methyl iodide and benzylic or allylic halides (Sec. 17.4).

Let's contrast the esterification reactions above with acid-catalyzed esterification in Sec. 20.8A. In all of the reactions discussed in this section, *the carboxylate oxygen of the acid acts as a nucleophile.* This oxygen is *alkylated* by an alkyl halide or diazomethane. In acid-catalyzed esterification, *the carbonyl carbon, after protonation of the carbonyl oxygen, acts as an electrophile* (a Lewis acid). The nucleophile in acid-catalyzed esterification is the oxygen atom of the solvent alcohol molecule.

PROBLEMS

20.13   Give the structure of the ester formed when
     *(a)   isobutyric acid reacts with diazomethane in ether;
      (b)   succinic acid reacts with a large excess of diazomethane in ether;
     *(c)   isobutyric acid reacts with benzyl bromide and K_2CO_3 in acetone;
      (d)   benzoic acid reacts with allyl bromide and K_2CO_3 in acetone.

*20.14   *Tert*-butyl esters can be prepared by the acid-catalyzed reaction of 2-methylpro-pene (isobutylene) with carboxylic acids.

$$CH_3-\overset{\overset{O}{\|}}{C}-OH \; + \; \overset{H_2C}{\underset{H_3C}{\diagdown}}C-CH_3 \; \xrightarrow[\text{(pressure)}]{H_2SO_4} \; CH_3-\overset{\overset{O}{\|}}{C}-O-\overset{\overset{CH_3}{|}}{\underset{CH_3}{\underset{|}{C}}}-CH_3$$

acetic acid   2-methylpropene   *tert*-butyl acetate
                          (isobutylene)   (85% yield)

                                          (*Problem 20.14 continues*)

(a) Suggest a mechanism for this reaction that accounts for the role of the acid catalyst. (*Hint:* See Sec. 4.7B.)

(b) Would you classify this esterification reaction mechanistically as one that is related to acid-catalyzed esterification, or one that is related to the reactions in Sec. 20.8B? Explain.

20.9   Conversion of Carboxylic Acids into Acid Chlorides and Anhydrides

A.   Synthesis of Acid Chlorides

Acid chlorides are carboxylic acid derivatives with the following general structure:

$$
\begin{array}{c}
\text{O} \\
\parallel \\
\text{R}-\text{C}-\text{Cl}
\end{array}
\quad (\text{R} = \text{H, alkyl, or aryl})
$$

Acid chlorides are invariably prepared from carboxylic acids. Two reagents used for this purpose are thionyl chloride, $SOCl_2$, and phosphorus pentachloride, PCl_5.

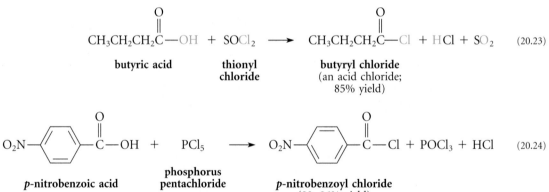

$$
\begin{array}{c}
\text{O} \\
\parallel \\
\text{CH}_3\text{CH}_2\text{CH}_2\text{C}-\text{OH}
\end{array}
+ \text{SOCl}_2 \longrightarrow
\begin{array}{c}
\text{O} \\
\parallel \\
\text{CH}_3\text{CH}_2\text{CH}_2\text{C}-\text{Cl}
\end{array}
+ \text{HCl} + \text{SO}_2 \qquad (20.23)
$$

butyric acid   **thionyl chloride**   **butyryl chloride** (an acid chloride; 85% yield)

$$
\text{O}_2\text{N}-\!\!\!\left\langle\;\;\right\rangle\!\!\!-
\begin{array}{c}
\text{O} \\
\parallel \\
\text{C}-\text{OH}
\end{array}
+ \quad \text{PCl}_5 \longrightarrow
\text{O}_2\text{N}-\!\!\!\left\langle\;\;\right\rangle\!\!\!-
\begin{array}{c}
\text{O} \\
\parallel \\
\text{C}-\text{Cl}
\end{array}
+ \text{POCl}_3 + \text{HCl} \qquad (20.24)
$$

***p*-nitrobenzoic acid**   **phosphorus pentachloride**   ***p*-nitrobenzoyl chloride** (90–96% yield)

☞
STUDY GUIDE LINK:
✓**20.6**
*Mechanism of Acid
Chloride Formation*

Notice that acid chloride synthesis fits the general pattern of *substitution at a carbonyl group*; in this case, —OH is substituted by —Cl.

$$
\begin{array}{c}
\text{O} \\
\parallel \\
\text{R}-\text{C}-\text{OH}
\end{array}
\longrightarrow
\begin{array}{c}
\text{O} \\
\parallel \\
\text{R}-\text{C}-\text{Cl}
\end{array}
\qquad (20.25)
$$

Notice also that thionyl chloride is the same reagent used for making alkyl chlorides from alcohols (Sec. 10.3D), a reaction in which —OH is replaced by —Cl at the carbon of an *alkyl group*.

As discussed in Chapter 21, acid chlorides are very reactive and for this reason are very useful for the synthesis of other carbonyl compounds.

☞
STUDY GUIDE LINK:
✓**20.7**
*More on Synthetic
Equivalents*

Sulfonyl chlorides, the acid chlorides of sulfonic acids, are prepared by treatment of sulfonic acids or their sodium salts with PCl_5.

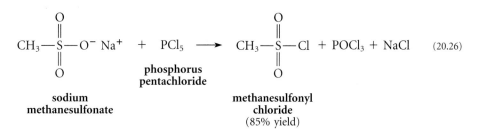

sodium
methanesulfonate

phosphorus
pentachloride

methanesulfonyl
chloride
(85% yield)

Aromatic sulfonyl chlorides can be prepared directly by the reaction of aromatic compounds with chlorosulfonic acid.

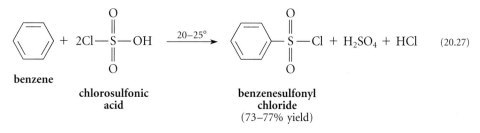

benzene

chlorosulfonic
acid

benzenesulfonyl
chloride
(73–77% yield)

This reaction is a variation of aromatic sulfonation, an electrophilic aromatic substitution reaction (Sec. 16.4D). Chlorosulfonic acid, the acid chloride of sulfuric acid, acts as an electrophile in this reaction just as SO_3 does in sulfonation.

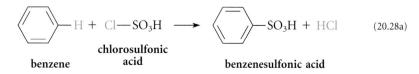

benzene

chlorosulfonic
acid

benzenesulfonic acid

The sulfonic acid produced in the reaction is converted into the sulfonyl chloride by reaction with another equivalent of chlorosulfonic acid.

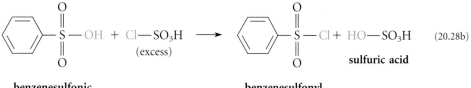

benzenesulfonic
acid

sulfuric acid

benzenesulfonyl
chloride

This part of the reaction is analogous to the reaction of a carboxylic acid with thionyl chloride (Eq. 20.23).

PROBLEMS

20.15 Draw the structures of the acid chlorides derived from *(a) 2-methylbutanoic acid; (b) *p*-methoxybenzoic acid; *(c) 1-propanesulfonic acid.

20.16 Give the structure of the acid chloride formed in each of the following transformations.
  *(a) sodium ethanesulfonate + PCl_5 ⟶
  (b) benzoic acid + $SOCl_2$ ⟶

(Problem 20.16 continues)

*(c)   toluene + excess chlorosulfonic acid ⟶

(d)   chlorobenzene + excess chlorosulfonic acid ⟶

*20.17   Outline a synthesis of the following compound from benzoic acid and any other reagents.

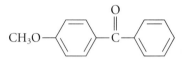

(*Hint:* See Study Guide Link 20.7.)

B. Synthesis of Anhydrides

Carboxylic acid *anhydrides* have the following general structure:

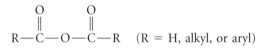

The name *anhydride*, which means "without water," comes from the fact that an anhydride reacts with water to give two equivalents of a carboxylic acid.

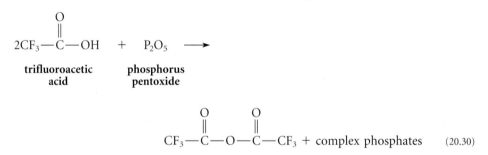

The name *anhydride* also graphically describes the way that anhydrides are prepared: treatment of carboxylic acids with strong dehydrating agents.

$$2CF_3-\overset{\overset{\textstyle O}{\|}}{C}-OH \;+\; P_2O_5 \longrightarrow$$

| **trifluoroacetic** | **phosphorus** |
| **acid** | **pentoxide** |

$$CF_3-\overset{\overset{\textstyle O}{\|}}{C}-O-\overset{\overset{\textstyle O}{\|}}{C}-CF_3 \;+\; \text{complex phosphates} \qquad (20.30)$$

trifluoroacetic anhydride
(74% yield)

Phosphorus pentoxide (actual formula P_4O_{10}) is an extremely hygroscopic white powder that reacts violently with water. It is also used as a potent desiccant. This compound is a complex anhydride of phosphoric acid, because it gives phosphoric acid when it reacts with water.

Most anhydrides may themselves be used to form other anhydrides. (As noted above, P_2O_5 is an inorganic anhydride that is used to form anhydrides of carboxylic acids.) In the following example, a dicarboxylic acid reacts with acetic anhydride to form a *cyclic anhydride*—a compound in which the anhydride group is part of a ring:

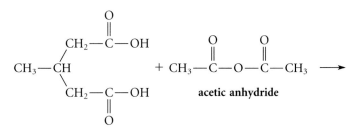

β-methylglutaric acid

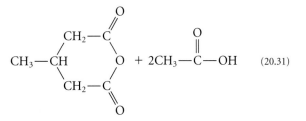

β-methylglutaric anhydride
(>90% yield; a cyclic anhydride)

Phosphorus oxychloride ($POCl_3$) and P_2O_5 (Eq. 20.30) can also be used for the formation of cyclic anhydrides. Cyclic anhydrides containing five- and six-membered anhydride rings are readily prepared from their corresponding dicarboxylic acids. Compounds containing either larger or smaller anhydride rings generally cannot be prepared this way. Formation of cyclic anhydrides with five- and six-membered rings is so facile that in some cases it occurs on heating the dicarboxylic acid.

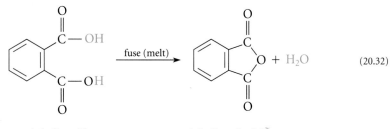

phthalic acid   **phthalic anhydride**

    This reaction, like many other carboxylic acid reactions that have been discussed, fits the pattern of substitution at the carbonyl carbon: the —OH of one carboxylic acid molecule is substituted by the *acyloxy group* (color) of another.

STUDY GUIDE LINK:
✓20.8
Mechanism of
Anhydride Formation

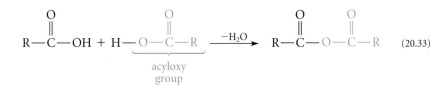

Anhydrides, like acid chlorides, are used in the synthesis of other carboxylic acid derivatives (Sec. 21.8B).

PROBLEMS

20.18   Complete each of the following reactions by giving the structure of the organic product.

*(a)   α-chloroacetic acid + P_2O_5 ⟶

(b)   *p*-chlorobenzoic acid + P_2O_5 ⟶

20.19   *(a)   Fumaric and maleic acids (Table 20.1) are *E,Z*-isomers. One forms a cyclic anhydride on heating and one does not. Which one forms the cyclic anhydride? Explain.

(b)   Which one of the following compounds forms a cyclic anhydride on heating: α-methylmalonic acid or 2,3-dimethylbutanedioic acid?

20.10   Reduction of Carboxylic Acids to Primary Alcohols

When a carboxylic acid is treated with lithium aluminum hydride, $LiAlH_4$, then dilute acid, a primary alcohol is formed.

$$2CH_3CH_2\overset{\displaystyle O}{\overset{\displaystyle \|}{CH\!\!-\!\!C}}\!\!-\!\!OH + LiAlH_4 \xrightarrow{\text{ether}} \xrightarrow{H_3O^+} 2CH_3CH_2CHCH_2\!\!-\!\!OH \qquad (20.34)$$
$$\underset{\displaystyle CH_3}{} \qquad\qquad\qquad\qquad \underset{\displaystyle CH_3}{}$$

2-methylbutanoic acid   **2-methyl-1-butanol**
                                 (83% yield)

This is an important method for the preparation of primary alcohols.

Before the reduction itself takes place, $LiAlH_4$, a source of the very strongly basic hydride ion ($H:^-$), reacts with the acidic hydrogen of the carboxylic acid to give the lithium salt of the carboxylic acid and one equivalent of hydrogen gas.

$$R\!\!-\!\!\overset{\displaystyle \overset{..}{O}:\ Li^+}{\overset{\displaystyle \|}{C}}\!\!-\!\!O\!\!-\!\!H\ \ H\!\!-\!\!\bar{A}lH_3 \longrightarrow R\!\!-\!\!\overset{\displaystyle :\overset{..}{O}\!\!:^-\ Li^+}{C}\!\!=\!\!O + H_2 + AlH_3 \qquad (20.35a)$$

The lithium salt of the carboxylic acid is the species that is actually reduced.

The reduction occurs in two stages. In the first stage, the AlH_3 formed in Eq. 20.35a reduces the carboxylate ion to an aldehyde. The aldehyde is rapidly reduced further to give, after protonolysis, the primary alcohol (Sec. 19.8).

$$R\!\!-\!\!\overset{\displaystyle O}{\overset{\displaystyle \|}{C}}\!\!-\!\!O^-\ Li^+ \xrightarrow[\text{Eq. 20.35a}]{AlH_3\ (\text{from}} R\!\!-\!\!\overset{\displaystyle O}{\overset{\displaystyle \|}{C}}\!\!-\!\!H \xrightarrow{LiAlH_4} \xrightarrow{H_3O^+} R\!\!-\!\!CH_2OH \qquad (20.35b)$$

☞

STUDY GUIDE LINK:
20.9
Mechanism of the
LiAlH₄ *Reduction of*
Carboxylic Acids

Because the aldehyde is more reactive than the carboxylate salt, it *cannot* be isolated.

Notice that the $LiAlH_4$ reduction of a carboxylic acid incorporates two different types of carbonyl reactions. The first is a net *substitution* at the carbonyl carbon to give the aldehyde intermediate. The second is an *addition* to the aldehyde.

$$R-\overset{\overset{\displaystyle O}{\parallel}}{C}-OH \xrightarrow{\text{(substitution)}} R-\overset{\overset{\displaystyle O}{\parallel}}{C}-H \xrightarrow{\text{(addition)}} R-\overset{\overset{\displaystyle OH}{|}}{\underset{\underset{\displaystyle H}{|}}{C}}-H \quad (20.36)$$

Many of the reactions of carboxylic acid derivatives discussed in Chapter 21 also fit the same pattern of addition followed by substitution.

Note that sodium borohydride, $NaBH_4$, another important hydride reducing agent (Sec. 19.8), does *not* reduce carboxylic acids, although it does react with the acidic hydrogens of acids in a manner analogous to Eq. 20.35a.

The $LiAlH_4$ reduction of carboxylic acids can be combined with the Grignard synthesis of carboxylic acids (Sec. 20.6) to provide a one-carbon chain extension of carboxylic acids, as illustrated by the following study problem.

STUDY PROBLEM 20.2

Fatty acids containing an even number of carbon atoms are readily obtained from natural sources, but those containing an odd number of carbons are relatively rare. Outline a synthesis of the rare tridecanoic acid, $CH_3(CH_2)_{10}CH_2CO_2H$, from the readily available lauric acid (see Table 20.1).

Solution   The problem requires the synthesis of a carboxylic acid with the addition of one carbon atom to a carbon chain. The Grignard synthesis of carboxylic acids will accomplish this objective:

$$CH_3(CH_2)_{10}CH_2Br \xrightarrow[\text{2) } CO_2,\text{ then } H_3O^+]{\text{1) Mg, ether}} CH_3(CH_2)_{10}CH_2CO_2H$$

1-bromododecane   **tridecanoic acid**

The required alkyl bromide, 1-bromododecane, comes from treatment of the corresponding alcohol with concentrated HBr; and the alcohol in turn comes from the $LiAlH_4$ reduction of lauric acid:

$$CH_3(CH_2)_{10}CO_2H \xrightarrow[\text{2) } H_3O^+]{\text{1) } LiAlH_4 \text{ in ether}} CH_3(CH_2)_{10}CH_2OH \xrightarrow{\text{conc. HBr}} CH_3(CH_2)_{10}CH_2Br$$

lauric acid   **1-dodecanol**   **1-bromododecane**

PROBLEMS

20.20   Give the structure of a compound with the indicated formula that would give the following diol in a $LiAlH_4$ reduction followed by protonolysis.

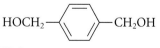

*(a) $C_8H_6O_3$ (b) $C_8H_6O_4$

20.21   Propose reaction sequences for each of the following conversions:
   *(a) benzoic acid into 3-phenyl-1-propanoic acid
   (b) benzoic acid into phenylacetic acid, $PhCH_2CO_2H$

20.11 Decarboxylation of Carboxylic Acids

The loss of carbon dioxide from a carboxylic acid is called **decarboxylation**.

$$R—\overset{\overset{\displaystyle O}{\|}}{C}—O—H \longrightarrow R—H + O{=}C{=}O \qquad (20.37)$$

Although decarboxylation is not an important reaction for most ordinary carboxylic acids, certain types of carboxylic acids are readily decarboxylated. Among these are

1. β-keto acids
2. malonic acid derivatives
3. carbonic acid derivatives

β-Keto acids—carboxylic acids with a keto group in the β-position—readily decarboxylate at room temperature in *acidic* solution.

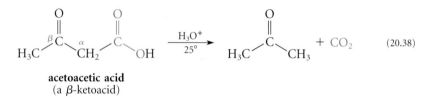

acetoacetic acid
(a β-ketoacid)

Decarboxylation of a β-keto acid involves an *enol intermediate* that is formed by an internal proton transfer from the carboxylic acid group to the carbonyl oxygen atom of the ketone. The enol is transformed spontaneously into the corresponding ketone (Sec. 14.5A).

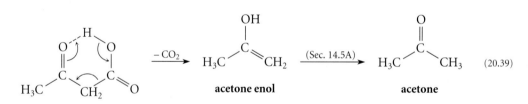

acetone enol   acetone

The *acid* form of the β-keto acid decarboxylates more readily than the conjugate-base carboxylate form because the latter has no acidic proton that can be donated to the β-carbonyl oxygen. In effect, the carboxy group promotes its own removal!

Malonic acid and its derivatives readily decarboxylate upon heating in acidic solution.

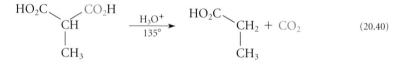

α-methylmalonic acid   propionic acid

This reaction, which also does not occur in base, bears a close resemblance to the decarboxylation of β-keto acids, because both types of acids have a carbonyl group β to the carboxy group.

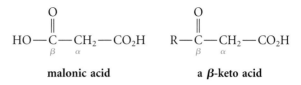

malonic acid   **a β-keto acid**

Because decarboxylation of malonic acid and its derivatives requires heating, the acids themselves can be isolated at room temperature.

Carbonic acid is unstable and decarboxylates spontaneously in acidic solution to carbon dioxide and water. (Carbonic acid is formed reversibly when CO_2 is bubbled into water; carbonic acid gives carbonated beverages their acidity, and CO_2 gives them their "fizz.")

$$\underset{\text{carbonic acid}}{\overset{\displaystyle O}{\underset{HO}{\parallel}}{\overset{}{C}}{\underset{OH}{}}} \quad \rightleftharpoons \quad CO_2 + H_2O \qquad (20.41)$$

Similarly, any carbonic acid derivative with a free carboxylic acid group will also decarboxylate under acidic conditions.

$$\underset{\text{methyl carbonate}}{CH_3O-\overset{O}{\overset{\parallel}{C}}-OH} \quad \underset{\longleftarrow}{\overset{H_3O^+}{\longrightarrow}} \quad CH_3OH + CO_2 \qquad (20.42)$$

$$\underset{\text{carbamic acid}}{H_2N-\overset{O}{\overset{\parallel}{C}}-OH} \quad \underset{\longleftarrow}{\overset{H_3O^+}{\longrightarrow}} \quad CO_2 + NH_3 \quad \underset{\longleftarrow}{\overset{H_3O^+}{\longrightarrow}} \quad \overset{+}{N}H_4 \qquad (20.43)$$

Under basic conditions, carbonic acid is converted into its salts and does not decarboxylate. These salts—sodium bicarbonate ($NaHCO_3$) and sodium carbonate (Na_2CO_3)—are familiar stable compounds.

Carbonic acid derivatives in which *both* carboxylate oxygens are involved in ester or amide formation are stable. Dimethyl carbonate (a diester of carbonic acid) and urea (the diamide of carbonic acid) are examples of such stable compounds. Likewise, the acid chloride phosgene is also stable.

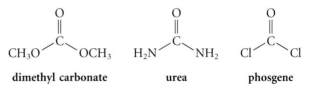

dimethyl carbonate   **urea**   **phosgene**

PROBLEMS

*20.22   Give the product expected when each of the following compounds is treated with acid.

(*Problem 20.22 continues*)

(a)

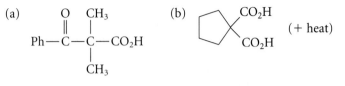

$$\text{Ph}-\overset{\overset{\displaystyle O}{\|}}{\text{C}}-\overset{\overset{\displaystyle CH_3}{|}}{\underset{\underset{\displaystyle CH_3}{|}}{\text{C}}}-\text{CO}_2\text{H}$$

(b) (+ heat)

(c) $\text{C}_2\text{H}_5-\text{NH}-\text{CO}_2^-\ \text{Na}^+$

20.23   Give the structures of all the β-keto acids that will decarboxylate to yield 2-methylcyclohexanone.

*20.24   One piece of evidence supporting the enol mechanism in Eq. 20.39 is that β-keto acids which *cannot* form enols are stable to decarboxylation. For example, the following β-keto acid can be distilled at 310° without decomposition. Attempt to construct a model of the enol that would be formed when this compound decarboxylates. Use your model to explain why this β-keto acid resists decarboxylation. (*Hint:* See Sec. 7.6C.)

KEY IDEAS IN CHAPTER 20

○ The carboxy group is the characteristic functional group of carboxylic acids.

○ Carboxylic acids are solids or high-boiling liquids, and carboxylic acids of relatively low molecular mass have substantial solubilities in water.

○ Carbonyl and O—H absorptions are the most important infrared absorptions of carboxylic acids. In proton NMR spectra the α-hydrogens of carboxylic acids absorb in the δ 2.0–2.5 region, and the O—H protons, which can be exchanged with D_2O, absorb in the δ 9–13 region. The carbonyl carbon resonances in the CMR spectra of carboxylic acids occur at δ 170–180, about 20 ppm higher field than the carbonyl carbon resonances of aldehydes and ketones.

○ Typical carboxylic acids have pK_a values between 4 and 5, although pK_a values are influenced significantly by polar effects. Sulfonic acids are even more acidic. The acidity of carboxylic acids is due to a combination of polar and resonance effects. The conjugate base of a carboxylic acid is a carboxylate ion.

◻ Carboxylic acids with long unbranched carbon chains are called fatty acids. The alkali metal salts of fatty acids are soaps. Soaps and other detergents are surfactants; they form micelles in aqueous solution.

◻ Because of their acidities, carboxylic acids dissolve not only in aqueous NaOH, but also in aqueous solutions of weaker bases such as sodium bicarbonate.

◻ The reaction of Grignard reagents with CO_2 is both a synthesis of carboxylic acids and a method of carbon-carbon bond formation.

◻ The reactivity of the carbonyl carbon toward nucleophiles plays an important role in many reactions of carboxylic acids and their derivatives. In these reactions, a nucleophile attacks the carbonyl carbon to form a tetrahedral addition intermediate, which then breaks down by loss of a leaving group. The result is a net substitution reaction at the carbonyl carbon. Acid-catalyzed esterification and lithium aluminum hydride reduction are two examples of nucleophilic carbonyl substitution. In lithium aluminum hydride reduction the substitution product, an aldehyde, reacts further in an addition reaction.

◻ The nucleophilic reactivity of the carboxylate oxygen is important in some reactions of carboxylic acids, such as ester formation by alkylation with diazomethane or alkyl halides.

◻ Carboxylic acids are readily converted into acid chlorides with phosphorus pentachloride or thionyl chloride, and into anhydrides with P_2O_5. Cyclic anhydrides containing five- and six-membered rings are readily formed on heating the corresponding dicarboxylic acids.

◻ β-Keto acids as well as derivatives of malonic acid and carbonic acid decarboxylate in acidic solution; most malonic acid derivatives require heating. The decarboxylation reactions of β-keto acids, and probably malonic acids as well, involve enol intermediates.

ADDITIONAL PROBLEMS

*20.25   Give the product expected when butyric acid (or other compound indicated) reacts with each of the following reagents.
(a) ethanol (solvent), H_2SO_4 catalyst   (b) NaOH solution
(c) $LiAlH_4$ (excess), then H_3O^+ (d) heat
(e) $SOCl_2$ (f) diazomethane in ether
(g) product of (c) (excess) + $CH_3CH{=}O$, HCl (catalyst)
(h) product of (e), $AlCl_3$, benzene, then H_2O
(i) product of (h), H_2NNH_2, base, heat

20.26   Give the product expected when benzoic acid reacts with each of the following reagents.
(a) CH_3I, K_2CO_3
(b) concentrated HNO_3, H_2SO_4
(c) PCl_5
(d) P_2O_5, heat
(e) CH_3MgBr (1 equivalent) in ether

*20.27   Draw the structures and give the names of all the dicarboxylic acids with the formula $C_6H_{10}O_4$. Indicate which are chiral, which would readily form cyclic anhydrides on heating, and which would decarboxylate on heating.

20.28   *(a) What is the molecular mass of a carboxylic acid containing a single carboxylic acid group if 8.61 mL of 0.1 M NaOH solution are required to neutralize 100 mg of the acid?
(b) How many milliliters of 0.1 M NaOH solution are required to neutralize 100 mg of succinic acid?

20.29   Give the product(s) formed and give the curved-arrow formalism for the reaction of acetic acid with one equivalent each of the following reagents.
*(a) Na^+ CH_3O^-
(b) Cs^+ ^-OH
*(c) CH_3CH_2—$MgBr$
(d) CH_3—Li
*(e) $:NH_3$
(f) NaH

*(g)
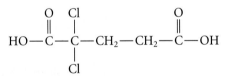

*20.30   Draw the structure of the major species present in solution when 0.01 mole of the following acid in aqueous solution is treated with 0.01 mole of NaOH. Explain.

20.31   Rank the following compounds in order of increasing acidity. Explain your answers.

| | CO₂H | CO₂H | CO₂H |

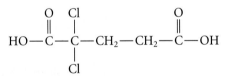

A   *B*   *C*

*20.32   Common pH paper turns red at pH values below about 3. Show that a 0.1 M solution of acetic acid (pK_a = 4.76) will turn litmus red, and that a 0.1 M solution of phenol (pK_a = 9.95) will not.

20.33   What is the pH of a solution containing an acetic acid-sodium acetate buffer in which the actual [acetic acid]/[sodium acetate] ratio is (a) 1/3? (b) 3? (c) 1?

20.34   *(a)   Explain why the most acidic species in any solvent is the conjugate acid of the solvent.

(b)   Show why HBr is a stronger acid in acetic acid solvent than it is in water.

*20.35   Explain why all efforts to synthesize a carboxylic acid containing the isotope oxygen-18 at *only* the carbonyl oxygen fail and yield instead a carboxylic acid in which the labeled oxygen is distributed equally between both oxygens of the carboxy group. (*O = ^{18}O)

$$R-\overset{\overset{*O}{\|}}{C}-OH \longrightarrow R-\overset{\overset{*O}{\|}}{C}-OH + R-\overset{\overset{O}{\|}}{C}-\overset{*}{O}H$$

(50% of label on each oxygen)

20.36   Outline a synthesis of each of the following compounds from isobutyric acid (2-methylpropanoic acid) and any other reagents.

*(a) $(CH_3)_2CH-\overset{\overset{O}{\|}}{C}-OCH_3$

(b) $(CH_3)_2CH-\overset{\overset{O}{\|}}{C}-OCH_2CH_2CH_3$

*(c) $(CH_3)_2CHCH_2-\overset{\overset{O}{\|}}{C}-OCH_3$

(d) $(CH_3)_2CHCH_2OH$

*(e)   3-methyl-2-butanone

(f)   isobutyrophenone

*(g) $(CH_3)_2CHCH=CH_2$

*20.37   A graduate student, Al Kane, has been given by his professor a very precious sample of $(-)$-3-methylhexane, along with optically active samples of both enantiomers of 4-methylhexanoic acid, each of known absolute configuration. Kane has been instructed to determine the absolute configuration of $(-)$-3-methylhexane. Kane has come to you for assistance. Show what he should do to deduce the configuration of the optically active hydrocarbon from the acids of known configuration. Be specific.

20.38   The sodium salt of *valproic acid* is a drug that has been used in the treatment of epilepsy. (Valproic acid is a name used in medicine.)

$$\begin{array}{c} CH_3CH_2CH_2 \\ \diagdown \\ CH-CO_2H \qquad \textbf{valproic acid} \\ \diagup \\ CH_3CH_2CH_2 \end{array}$$

(a)   Give the substitutive name of valproic acid.

(b)   Give the common name of valproic acid.

(c)   Outline a synthesis of valproic acid from carbon sources containing fewer than five carbons and any other reagents.

*20.39   Because the radioactive isotope carbon-14 is used at very low ("tracer") levels, its presence cannot be detected by spectroscopy. It is generally detected by counting its radioactive decay in a device called a scintillation counter. The location of carbon-14 in a chemical compound must be determined by carrying out chemical degradations, isolating the resulting fragments that represent different carbons in the molecule, and counting them.

A well-known biologist, Fizzi O. Logicle, has purchased a sample of phenyl-acetic acid advertised to be labeled with the radioactive isotope carbon-14 *only* at the carbonyl carbon. Before using this compound in experiments designed to test a promising theory of biosynthesis, she has wisely decided to be sure that the radiolabel is located only at the carbonyl carbon as claimed. Knowing your expertise in organic chemistry, she has asked that you devise a way to determine what fraction of the ^{14}C is at the carbonyl carbon and what fraction is elsewhere in the molecule. Outline a reaction scheme that could be used to make this determination.

20.40   You have been employed by a biochemist, Fungus P. Gildersleeve, who has given you a very expensive sample of benzoic acid labeled equally in both oxygens with ^{18}O. He asks you to prepare methyl benzoate (structure in Eq. 20.16), preserving as much ^{18}O label in the ester as possible. Which method(s) of ester synthesis would you use to carry out this assignment? Why?

*20.41   You are a chemist for Chlorganics, Inc., a company specializing in chlorinated organic compounds. A process engineer, Turner Switchback, has accidentally mixed the contents of four vats containing, respectively, *p*-chlorophenol, 4-chlorocyclohexanol, *p*-chlorobenzoic acid, and chlorocyclohexane. The president of the company, Hal Ogen (green with anger) has ordered you to design an expeditious separation of these four compounds. Success guarantees you a promotion; accommodate him.

20.42   Penicillin-G is a widely used member of the penicillin family of drugs. In which fluid would you expect penicillin-G to be more soluble: stomach acid (pH = 2) or the bloodstream (pH 7.4)? Explain.

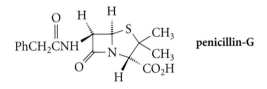   **penicillin-G**

*20.43   (a)   The relatively stable carbocation *crystal violet* has a deep blue-violet color in aqueous solution. When NaOH is added to the solution the blue color fades because the carbocation reacts with sodium hydroxide in about 1–2 minutes to give a colorless product. Show the reaction of crystal violet with NaOH.

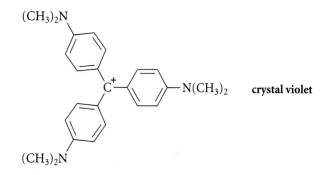

crystal violet

(b) When the detergent sodium dodecyl sulfate (SDS) is present in solution above its critical micelle concentration, the bleaching of crystal violet with NaOH takes several days. Account for the effect of SDS on the rate of this reaction.

$$CH_3CH_2CH_2CH_2CH_2CH_2CH_2CH_2CH_2CH_2CH_2CH_2OSO_3^- \ Na^+$$

sodium dodecyl sulfate (SDS)

20.44 Draw the structure of the cyclic anhydride that forms when each of the following acids is heated.

*(a)

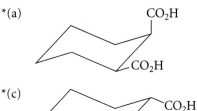

(b) *meso*-α,β-dimethylsuccinic acid

*(c)

(*Hint:* Don't forget about the chair flip.)

*(d)

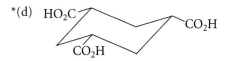

20.45 Propose a synthesis of each of the following compounds from the indicated starting material(s) and any other reagents.

*(a) 2-pentanol from propionic acid

(b)

$$CH_3CH_2-\overset{\overset{\displaystyle O}{\|}}{C}-O-CH_2CH=CH_2$$ from allyl alcohol as the only carbon source

*(c) 2-methylpentane from 4-methylpentanoic acid

(d) 2-methylheptane from pentanoic acid

*(e) *m*-nitrobenzoic acid from toluene

(*Problem 20.45 continues*)

(f) *p*-nitrobenzoic acid from toluene

*(g) 1,3-diphenylpropane from benzoic acid

*(h)

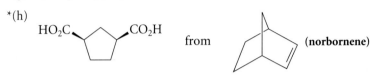

*(i) 5-oxohexanoic acid from 5-bromo-2-pentanone. (*Hint:* Use a protecting group.)

20.46 Complete each of the following reactions by giving the principal organic product(s). Give the reasons for your answers.

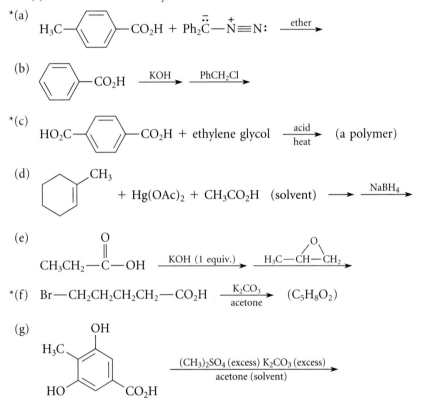

*(a)

H_3C—⟨benzene⟩—CO_2H + $Ph_2\overset{..}{C}$—$\overset{+}{N}$≡N: $\xrightarrow{\text{ether}}$

(b) ⟨benzene⟩—CO_2H $\xrightarrow{\text{KOH}}$ $\xrightarrow{\text{PhCH}_2\text{Cl}}$

*(c) HO_2C—⟨benzene⟩—CO_2H + ethylene glycol $\xrightarrow[\text{heat}]{\text{acid}}$ (a polymer)

(d) ⟨cyclohexene with CH_3⟩ + $Hg(OAc)_2$ + CH_3CO_2H (solvent) $\longrightarrow$ $\xrightarrow{\text{NaBH}_4}$

(e) CH_3CH_2—$\overset{\overset{\text{O}}{\|}}{C}$—$OH$ $\xrightarrow{\text{KOH (1 equiv.)}}$ $\xrightarrow{H_3C-CH-CH_2 \text{ (epoxide)}}$

*(f) Br—$CH_2CH_2CH_2CH_2$—CO_2H $\xrightarrow[\text{acetone}]{\text{K}_2\text{CO}_3}$ ($C_5H_8O_2$)

(g)

H₃C, OH, HO, CO_2H substituted benzene $\xrightarrow[\text{acetone (solvent)}]{(CH_3)_2SO_4 \text{ (excess) } K_2CO_3 \text{ (excess)}}$

20.47 *(a) Decarboxylation of compound *A* below gives *two* separable products; draw their structures and explain.

(b) How many products are formed when compound *B* below is decarboxylated?

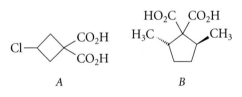

A   *B*

20.48 *(a) Squaric acid (structure below) has pK_a values of about 1 and 3.5. Draw the reactions corresponding to the two successive ionizations of squaric acid and label each with the appropriate pK_a value.

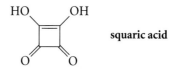

squaric acid

*(b) Most enols have pK_a values in the 10–12 range. Using polar and resonance effects, explain why squaric acid is much more acidic.

(c) Given that squaric acid behaves like a dicarboxylic acid, draw structures for the products formed when it reacts with excess $SOCl_2$; with ethanol solvent in the presence of an acid catalyst.

*20.49 Organolithium reagents such as methyllithium (CH_3Li) react with carboxylic acids to give ketones:

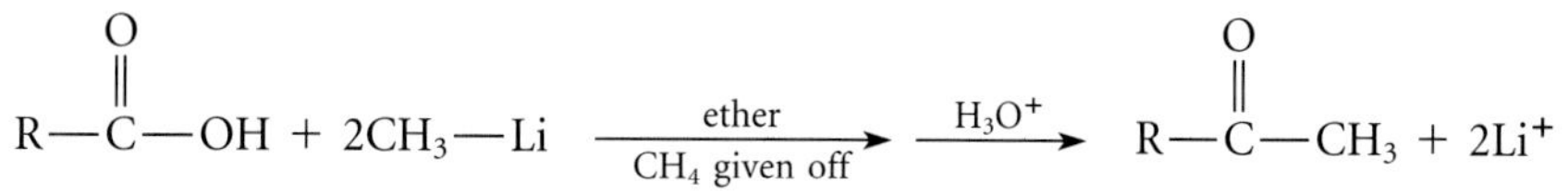

$$R{-}\overset{\overset{\displaystyle O}{\|}}{C}{-}OH + 2CH_3{-}Li \xrightarrow[\text{CH}_4 \text{ given off}]{\text{ether}} \xrightarrow{H_3O^+} R{-}\overset{\overset{\displaystyle O}{\|}}{C}{-}CH_3 + 2Li^+$$

Notice that *two* equivalents of the lithium reagent are required, and that the ketone does not react further. Suggest a mechanism for this reaction that accounts for these facts. (*Hint:* Start by looking at Problem 20.29d.)

20.50 Using the results from the previous problem, give the products of the following reaction and the mechanism by which they are formed.

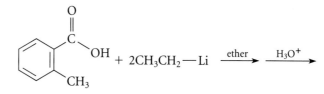

20.51 Give a curved-arrow mechanism for each of the following known reactions.

*(a)

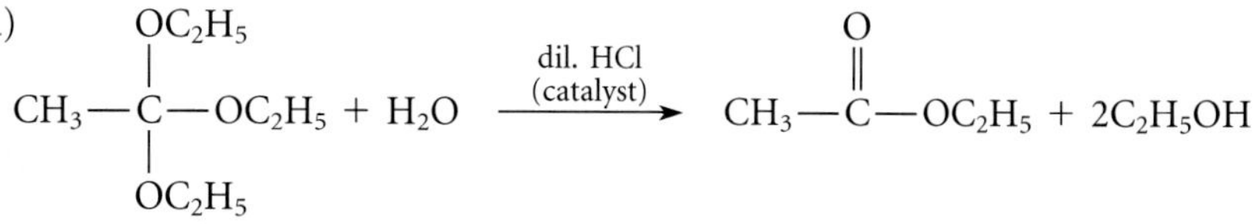

$$CH_3{-}\overset{\overset{\displaystyle OC_2H_5}{|}}{\underset{\underset{\displaystyle OC_2H_5}{|}}{C}}{-}OC_2H_5 + H_2O \xrightarrow[\text{(catalyst)}]{\text{dil. HCl}} CH_3{-}\overset{\overset{\displaystyle O}{\|}}{C}{-}OC_2H_5 + 2C_2H_5OH$$

an orthoester

(b)

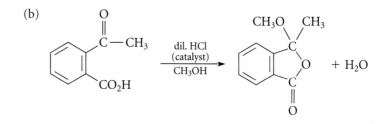

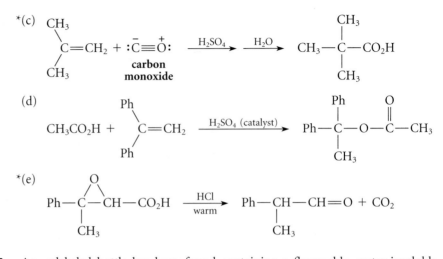

*20.52   An unlabeled bottle has been found containing a flammable, water-insoluble substance *A* that decolorizes Br_2 in CH_2Cl_2 and has the following elemental analysis: 87.7% carbon, 12.3% hydrogen. The base peak in the mass spectrum occurs at $m/z = 67$. The proton NMR of *A* is complex, but integration shows that about 30% of the protons have chemical shifts in the δ 1.8–2.2 region of the spectrum. Treatment of *A* successively with OsO_4, then periodic acid, and finally with Ag_2O, gives a single dicarboxylic acid *B* that can be resolved into enantiomers. Neutralization of a solution containing 100 mg of *B* requires 13.7 mL of 0.1 *M* NaOH solution (see Problem 20.28). Compound *B*, when treated with $POCl_3$, forms a cyclic anhydride. Give the structures of *A* and *B*.

20.53   Give the structure of each of the following compounds.
    *(a) $C_9H_{10}O_3$: IR 2300–3200, 1710, 1600 cm^{-1}
        NMR spectrum in Fig. 20.5a
    (b) $C_9H_{10}O_3$: IR 2400–3200, 1700, 1630 cm^{-1}
        NMR spectrum in Fig. 20.5b
    *(c)   The compound with IR absorptions at 2300–3400, 1710, 1670, and 974 cm^{-1}, UV absorption at 208 nm ($\epsilon = 12{,}300$), peaks in the mass spectrum at $m/z = 86$ (molecular ion) and 41, and the NMR spectrum in Fig. 20.5c.

20.54   Propose reasonable fragmentation mechanisms that explain why
    *(a)   the mass spectrum of 2-methylpentanoic acid has a strong peak at $m/z = 74$.
    (b)   the mass spectrum of benzoic acid shows major peaks at $m/z = 105$ and $m/z = 77$.

20.55   *(a)   Give a structure for compound *A*, mp 121°, $C_6H_{10}O_4$, that can be resolved into enantiomers and has the following NMR spectra:
        CMR: δ 13.5, δ 41.2, δ 177.9
        proton NMR: δ 1.13 (6*H*, d, *J* = 7 Hz); δ 2.65 (2*H*, quintet, *J* = 7 Hz);
            δ 9.9 (2*H*, broad s, exchanges with D_2O)
    (b)   Give a structure for compound *B*, an isomer of *A*, that has a melting point of 208° and both CMR and proton NMR spectra that are virtually identical to those of *A*.

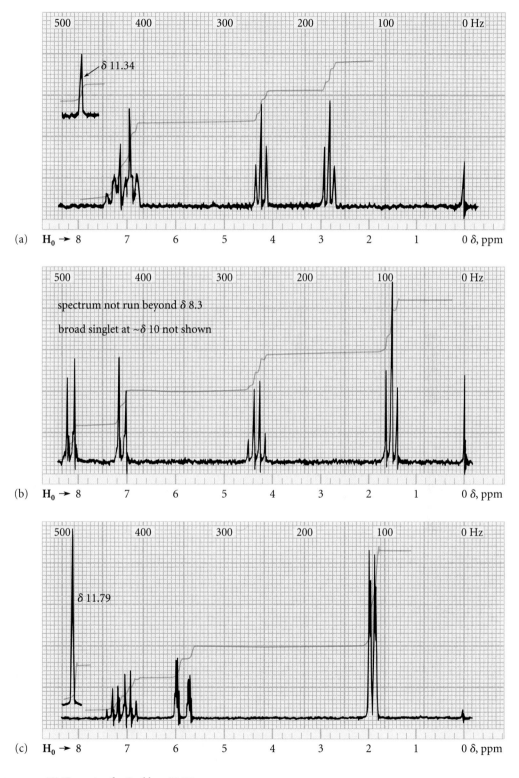

(a) $H_0 \rightarrow$

(b) $H_0 \rightarrow$

(c) $H_0 \rightarrow$

Figure 20.5   *NMR spectra for Problem 20.53.*

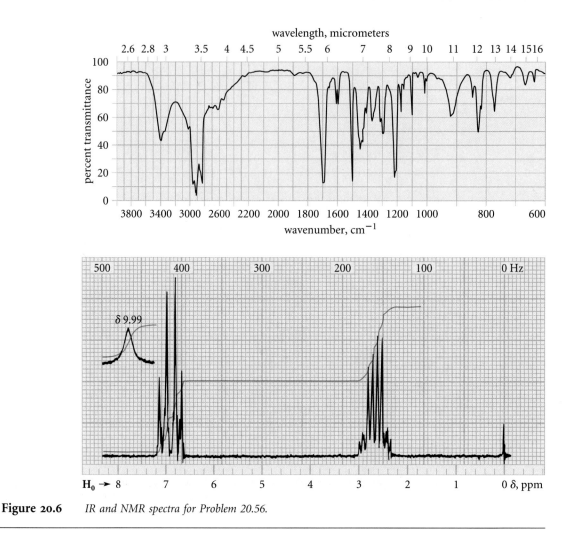

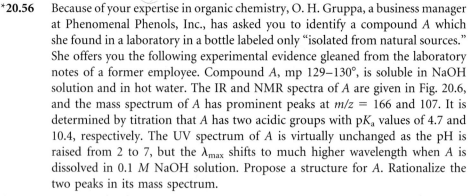

Figure 20.6   *IR and NMR spectra for Problem 20.56.*

*20.56   Because of your expertise in organic chemistry, O. H. Gruppa, a business manager at Phenomenal Phenols, Inc., has asked you to identify a compound *A* which she found in a laboratory in a bottle labeled only "isolated from natural sources." She offers you the following experimental evidence gleaned from the laboratory notes of a former employee. Compound *A*, mp 129–130°, is soluble in NaOH solution and in hot water. The IR and NMR spectra of *A* are given in Fig. 20.6, and the mass spectrum of *A* has prominent peaks at $m/z = 166$ and 107. It is determined by titration that *A* has two acidic groups with pK_a values of 4.7 and 10.4, respectively. The UV spectrum of *A* is virtually unchanged as the pH is raised from 2 to 7, but the λ_{max} shifts to much higher wavelength when *A* is dissolved in 0.1 *M* NaOH solution. Propose a structure for *A*. Rationalize the two peaks in its mass spectrum.

21

Chemistry of Carboxylic Acid Derivatives

Carboxylic acid derivatives are compounds that can be hydrolyzed under acidic or basic conditions to give a related carboxylic acid. All of them can be conceptually derived by replacing a small part of the carboxylic acid structure with other groups, as shown in Table 21.1 on page 972.

Carboxylic acids and their derivatives have not only structural similarities but also close relationships in their chemistry. With the exception of nitriles, all carboxylic acid derivatives contain a *carbonyl group*. Many important reactions of these compounds occur at the carbonyl group, and the $-C \equiv N$ (cyano) group of nitriles has reactivity that resembles that of a carbonyl group. Thus, the chemistry of carboxylic acid derivatives, like that of aldehydes, ketones, and carboxylic acids, involves the chemistry of the carbonyl group.

21.1 Nomenclature and Classification of Carboxylic Acid Derivatives

A. Esters and Lactones

Esters are named as derivatives of their parent carboxylic acids by applying a variation of the system used in naming carboxylate salts (Sec. 20.4A). The group attached to the carboxylate oxygen is named first as a simple alkyl or aryl group. This name is followed by the name of the parent carboxylate, which, as you have learned, is constructed by dropping the final *ic* from the name of the acid and adding the suffix *ate*. This procedure is used in both common and substitutive nomenclature.

Table 21.1   Structures of Carboxylic Acid Derivatives

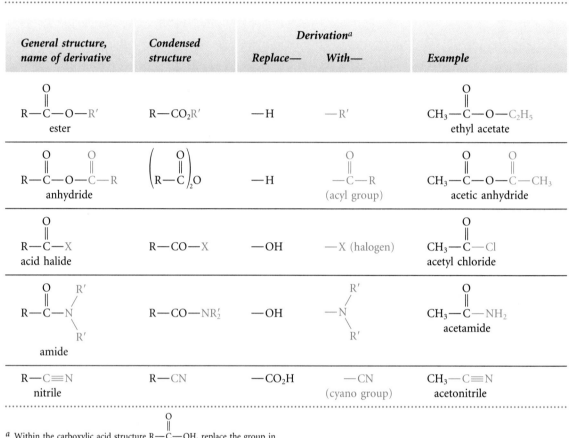

General structure, name of derivative	Condensed structure	Derivation[a]		Example
		Replace—	With—	
R—C(=O)—O—R' ester	R—CO₂R'	—H	—R'	CH₃—C(=O)—O—C₂H₅ ethyl acetate
R—C(=O)—O—C(=O)—R anhydride	(R—C(=O))₂O	—H	—C(=O)—R (acyl group)	CH₃—C(=O)—O—C(=O)—CH₃ acetic anhydride
R—C(=O)—X acid halide	R—CO—X	—OH	—X (halogen)	CH₃—C(=O)—Cl acetyl chloride
R—C(=O)—N(R')(R') amide	R—CO—NR'₂	—OH	—N(R')(R')	CH₃—C(=O)—NH₂ acetamide
R—C≡N nitrile	R—CN	—CO₂H	—CN (cyano group)	CH₃—C≡N acetonitrile

[a] Within the carboxylic acid structure R—C(=O)—OH, replace the group in column 3 with the group in column 4 to obtain the derivative. (Note that this shows the relationship of structures, but not necessarily how they are interconverted chemically.)

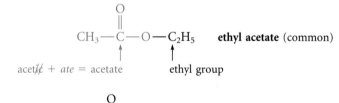

$$CH_3-C(=O)-O-C_2H_5 \qquad \textbf{ethyl acetate } (common)$$

acet~~ic~~ + *ate* = acetate   ethyl group

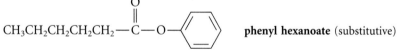

$$CH_3CH_2CH_2CH_2CH_2-C(=O)-O-\bigcirc \qquad \textbf{phenyl hexanoate } (substitutive)$$

Substitution is indicated by numbering the acid portion of the ester as in carboxylic acid nomenclature, beginning with the carbonyl as carbon-1 (substitutive nomenclature), or with the adjacent carbon as the α-position (common nomenclature). The alkyl or aryl group is numbered (using numbers in substitutive nomenclature, Greek letters in common nomenclature) from the point of attachment to the carboxylate oxygen.

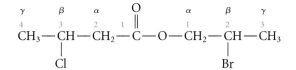

common system numbering
substitutive numbering

common name: **β-bromopropyl β-chlorobutyrate**
substitutive name: **2-bromopropyl 3-chlorobutanoate**

Esters of other acids are named by analogous extensions of acid nomenclature.

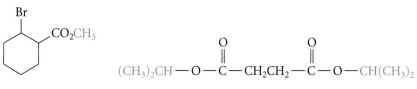

methyl **2-bromocyclohex-anecarboxylate**

diisopropyl **succinate**

Cyclic esters are called **lactones**.

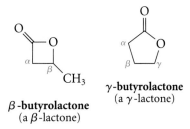

β-**butyrolactone**
(a β-lactone)

γ-**butyrolactone**
(a γ-lactone)

In common nomenclature, illustrated in the examples above, the *name* of a lactone is derived from the acid with the same number of carbons in its principal chain; the *ring size* is denoted by a Greek letter corresponding to the point of attachment of the lactone ring oxygen to the carbon chain. Thus, in a β-lactone, the ring oxygen is attached at the β-carbon to form a four-membered ring.

The substitutive nomenclature of lactones is a specialized extension of heterocyclic nomenclature that will not be considered.

B. Acid Halides

Acid halides are named in any system of nomenclature by replacing the *ic* ending of the acid with the suffix *yl*, followed by the name of the halide.

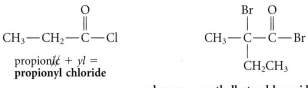

propion~~ic~~ + *yl* =
propionyl chloride

α-bromo-α-methylbutyryl bromide (common)
2-bromo-2-methylbutanoyl bromide (substitutive)

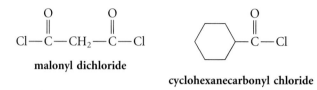

malonyl dichloride

cyclohexanecarbonyl chloride

Notice in the last example the special nomenclature required when the acid halide group is attached to a ring: the compound is named as an alkanecarbonyl halide.

C. Anhydrides

To name an anhydride, the name of the parent acid is followed by the word *anhydride*.

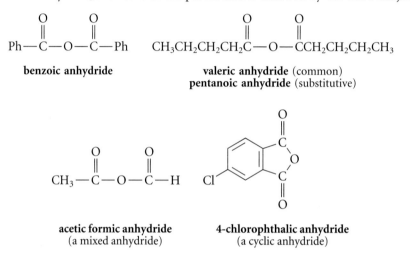

benzoic anhydride

valeric anhydride (common)
pentanoic anhydride (substitutive)

acetic formic anhydride
(a mixed anhydride)

4-chlorophthalic anhydride
(a cyclic anhydride)

Acetic formic anhydride (above) is an example of a **mixed anhydride**, an anhydride derived from two different acids. Mixed anhydrides are named by citing the two parent acids in alphabetical order.

D. Nitriles

Nitriles are named in the common system by dropping the *ic* or *oic* from the name of the acid *with the same number of carbon atoms* (counting the nitrile carbon) and adding the suffix *onitrile*. In substitutive nomenclature, the suffix *nitrile* is added to the name of the hydrocarbon with the same number of carbon atoms.

Ph—C≡N: **benzonitrile** (benz~~oic~~ + *onitrile*)

CH₃—C≡N: **acetonitrile** (acet~~ic~~ + *onitrile*)

CH₃—CH—CH₂—C≡N: **isovaleronitrile** (common)
| **3-methylbutanenitrile** (substitutive)
CH₃

:N≡C—CH₂—CH₂—C≡N: **succinonitrile** (common)
                              butanedinitrile (substitutive)

The name of the three-carbon nitrile is shortened in common nomenclature:

$CH_3CH_2—CN$ **propionitrile** (not propiononitrile)

When the nitrile group is attached to a ring, a special *carbonitrile* nomenclature is used.

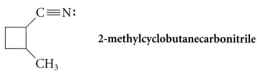

2-methylcyclobutanecarbonitrile

E. Amides, Lactams, and Imides

Simple amides are named in any system by replacing the *ic* or *oic* suffix of the acid name with the suffix *amide*.

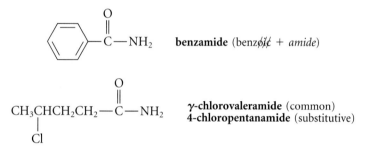

benzamide (benzoic + *amide*)

γ-chlorovaleramide (common)
4-chloropentanamide (substitutive)

When the amide functional group is attached to a ring, the suffix *carboxamide* is used.

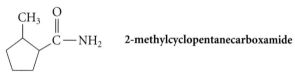

2-methylcyclopentanecarboxamide

Amides are classified as **primary**, **secondary**, or **tertiary** according to the number of hydrogens on the amide nitrogen.

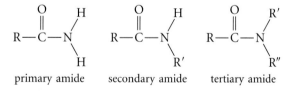

primary amide   secondary amide   tertiary amide

Notice that this classification, unlike that of alkyl halides and alcohols, refers to substitution *at nitrogen* rather than substitution at carbon. Thus, the following compound is a *secondary amide*, even though there is a tertiary alkyl group bound to nitrogen.

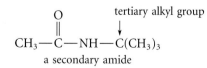

a secondary amide

Substitution on nitrogen in secondary and tertiary amides is designated with the letter *N* (italicized or underlined).

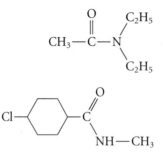

N,N-diethylacetamide
(double *N* designation shows that
both ethyl groups are on nitrogen)

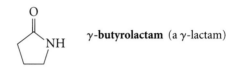

N-methyl-4-chlorocyclohexanecarboxamide

Cyclic amides are called **lactams**, and the common nomenclature of the simple lactams is analogous to that of lactones. Lactams, like lactones, are classified by ring size as γ-lactams (5-membered lactam ring), β-lactams (4-membered lactam ring), and so on.

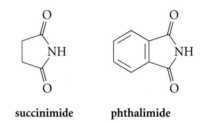

γ-butyrolactam (a γ-lactam)

Imides are formally the nitrogen analogs of anhydrides. Cyclic imides, of which the two compounds below are examples, are of greater importance than open-chain imides, although the latter are also known compounds.

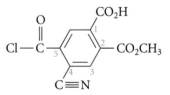

succinimide   **phthalimide**

F. Nomenclature of Substituent Groups

The priorities for citing principal groups in a carboxylic acid derivative are as follows:

$$\text{acid} > \text{anhydride} > \text{ester} > \text{acid halide} > \text{amide} > \text{nitrile} \qquad (21.1)$$

All of these groups have citation priority over aldehydes and ketones, as well as the other functional groups considered in previous chapters. (A complete list of group priorities is given in Appendix I.) The names used for citing these groups as substituents are given in Table 21.2. The following compounds illustrate the use of these names:

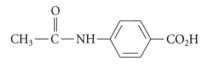

p-acetamidobenzoic acid
4-(acetylamino)benzoic acid

**5-chloroformyl-4-cyano-2-meth-
oxycarbonylbenzoic acid**

Table 21.2   Names of Carboxylic Acid Derivatives When Used as Substituent Groups

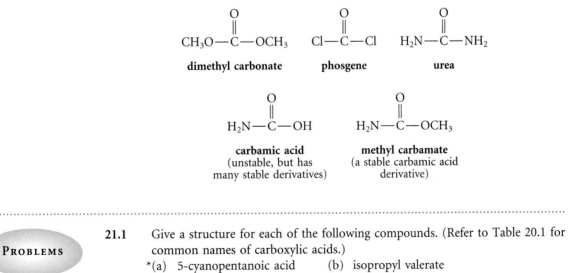

Group	Name	Group	Name
—C(=O)—OH	carboxy	—C(=O)—Cl	chloroformyl
—C(=O)—OCH₃	methoxycarbonyl	—C(=O)—NH₂	carbamoyl
—C(=O)—OC₂H₅	ethoxycarbonyl	—NH—C(=O)—CH₃	acetamido or acetylamino[a]
—CH₂—C(=O)—OH	carboxymethyl	—C≡N	cyano
—O—C(=O)—CH₃	acetoxy or acetyloxy[a]		

[a] Used by Chemical Abstracts.

G. Carbonic Acid Derivatives

Esters of carbonic acid are named like any other ester, but other important carbonic acid derivatives have special names that should be learned.

$$CH_3O-\overset{\overset{\displaystyle O}{\|}}{C}-OCH_3 \qquad Cl-\overset{\overset{\displaystyle O}{\|}}{C}-Cl \qquad H_2N-\overset{\overset{\displaystyle O}{\|}}{C}-NH_2$$

dimethyl carbonate   **phosgene**   **urea**

$$H_2N-\overset{\overset{\displaystyle O}{\|}}{C}-OH \qquad H_2N-\overset{\overset{\displaystyle O}{\|}}{C}-OCH_3$$

carbamic acid
(unstable, but has
many stable derivatives)

methyl carbamate
(a stable carbamic acid
derivative)

PROBLEMS

21.1   Give a structure for each of the following compounds. (Refer to Table 20.1 for common names of carboxylic acids.)

*(a) 5-cyanopentanoic acid   (b) isopropyl valerate

*(c) ethyl methyl malonate   (d) cyclohexyl acetate

(*Problem 21.1 continues*)

*(e) *N*-methylmaleimide   (f) *N,N*-dimethylformamide
*(g) γ-valerolactone   (h) α-chloroisobutyryl chloride
*(i) glutarimide   *(j) 3-ethoxycarbonylhexanedioic acid

21.2   Name the following compounds.

*(a)   (b) CH₃CH₂CH₂CN   *(c)

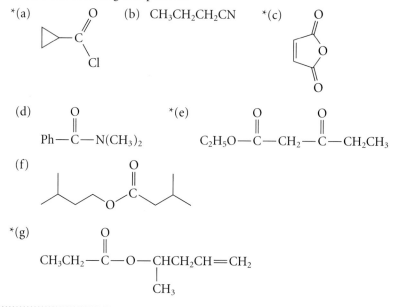

(d)

Ph—C—N(CH₃)₂

*(e)   C₂H₅O—C—CH₂—C—CH₂CH₃

(f)

*(g)

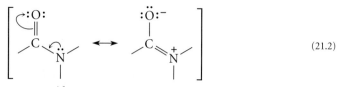

- -

21.2   Structures of Carboxylic Acid Derivatives

The structures of many carboxylic acid derivatives are very similar to what would be expected from the structures of other carbonyl compounds. For example, the C=O bond length is about 1.21 Å, and the carbonyl group and its two attached atoms are planar. The nitrile C≡N bond length, 1.16 Å, is significantly shorter than the acetylene C≡C bond length, 1.20 Å. This is another example of the shortening of bonds to smaller atoms (Sec. 1.3B).

In an amide, not only the carbonyl carbon, but also the amide nitrogen, have essentially trigonal planar bonding patterns (Fig. 21.1). The trigonal planar geometry at the amide nitrogen can be understood on the basis of the following resonance structures, which show that the bond between the nitrogen and the carbonyl carbon has considerable double-bond character.

$$\left[\begin{array}{ccc} & & \\ \end{array}\right]$$

(21.2)

amide resonance structures

Because of this trigonal planar geometry at nitrogen, secondary and tertiary amides can exist in both *E* and *Z* conformations about the carbonyl-nitrogen bond; the *Z* conformation

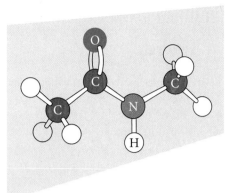

Figure 21.1   *Ball-and-stick model of N-methylacetamide. The labeled atoms all lie in the same plane.*

predominates in most secondary amides because, in this form, van der Waals repulsions between the largest groups are avoided.

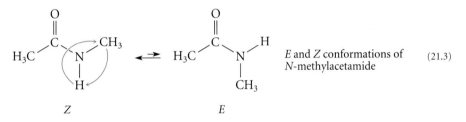

The interconversion of *E* and *Z* forms of amides is too rapid at room temperature to permit their separate isolation, but is very slow compared with rotation about ordinary carbon-carbon single bonds. A typical energy barrier for rotation about the carbonyl-nitrogen bond of an amide is 71 kJ/mol (17 kcal/mol), which results in an internal rotation rate of about ten times per second. (In contrast, an internal rotation in butane occurs about 10^{11} times per second.) The relatively low rate of internal rotation is caused by the significant double-bond character in the carbon-nitrogen bond; recall that rotation about double bonds is much slower than rotation about single bonds.

Study Guide Link:
21.1
NMR Evidence for Internal Rotation in Amides

Problems

*21.3   Draw the structure of an amide that *must* exist in an *E* conformation about the carbonyl-nitrogen bond.

21.4   Draw and label the *E* and *Z* conformations of the amino acid derivative *N*-acetylproline.

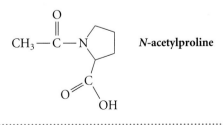

21.3 Physical Properties of Carboxylic Acid Derivatives

A. Esters

Esters are polar molecules, but they lack the capability of acids to donate hydrogen bonds. The lower esters are typically volatile, fragrant liquids that have lower densities than water. Most esters are not soluble in water. The low boiling point of a typical ester (color) is illustrated by the following comparison:

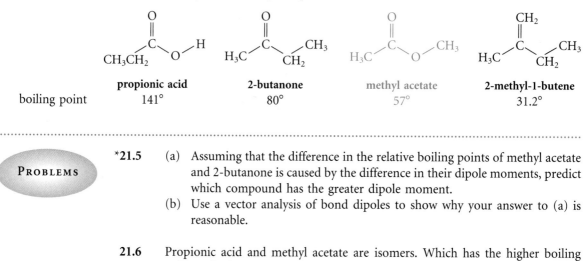

	propionic acid	2-butanone	methyl acetate	2-methyl-1-butene
boiling point	141°	80°	57°	31.2°

PROBLEMS

*21.5 (a) Assuming that the difference in the relative boiling points of methyl acetate and 2-butanone is caused by the difference in their dipole moments, predict which compound has the greater dipole moment.

(b) Use a vector analysis of bond dipoles to show why your answer to (a) is reasonable.

21.6 Propionic acid and methyl acetate are isomers. Which has the higher boiling point and why?

B. Anhydrides and Acid Chlorides

Most of the lower anhydrides and acid chlorides are dense, water-insoluble liquids with acrid, piercing odors. Their boiling points are not very different from those of other polar molecules of about the same molecular mass and shape.

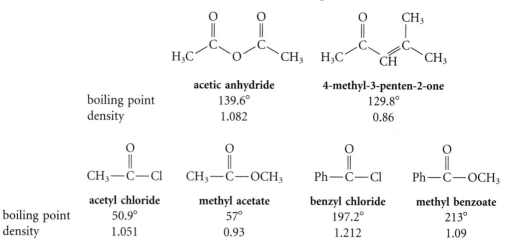

	acetic anhydride	4-methyl-3-penten-2-one
boiling point	139.6°	129.8°
density	1.082	0.86

	acetyl chloride	methyl acetate	benzyl chloride	methyl benzoate
boiling point	50.9°	57°	197.2°	213°
density	1.051	0.93	1.212	1.09

The simplest anhydride, formic anhydride, and the simplest acid chloride, formyl chloride, are unstable and cannot be isolated under ordinary conditions.

C. Nitriles

Nitriles are among the most polar organic compounds. Acetonitrile, for example, has a dipole moment of 3.4 D. The polarity of nitriles is reflected in their boiling points, which are rather high despite the absence of hydrogen bonding.

$$CH_3-C\equiv N\colon \qquad CH_3CH_2-C\equiv N\colon \qquad CH_3-C\equiv C-H$$

	acetonitrile	propionitrile	propyne
boiling point	81.6°	97.4°	−23.3°

Although nitriles are very poor hydrogen-bond acceptors (because they are very weak bases; see Sec. 21.5), acetonitrile is miscible with water and propionitrile has a moderate solubility in water. Higher nitriles are insoluble in water. Acetonitrile is a particularly valuable polar aprotic solvent because of its moderate boiling point and its relatively high dielectric constant of 38.

D. Amides

The lower amides are water-soluble, polar molecules with high boiling points. Primary and secondary amides, like carboxylic acids (Sec. 20.2), tend to associate into hydrogen-bonded dimers or higher aggregates in the solid state, in the pure liquid state, or in solvents that do not form hydrogen bonds. This association has a noticeable effect on the properties of amides and is of substantial biological importance in the structures of proteins (Sec. 26.8C). For example, simple amides have very high boiling points; many are solids.

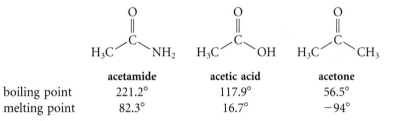

	acetamide	acetic acid	acetone
boiling point	221.2°	117.9°	56.5°
melting point	82.3°	16.7°	−94°

Primary amides have two hydrogens on the amide nitrogen that can form hydrogen bonds. Along a series in which these hydrogens are replaced by methyl groups, the capacity for hydrogen bonding is reduced, and boiling points decrease in spite of the increase in molecular mass.

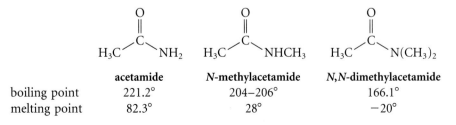

	acetamide	N-methylacetamide	N,N-dimethylacetamide
boiling point	221.2°	204–206°	166.1°
melting point	82.3°	28°	−20°

A number of amides have high dielectric constants (see Table 8.2). *N,N*-Dimethyl-formamide (DMF), for example, dissolves a number of inorganic salts, and is widely used as a polar aprotic solvent, despite its high boiling point.

21.4   Spectroscopy of Carboxylic Acid Derivatives

A. IR Spectroscopy

The most important feature in the IR spectra of most carboxylic acid derivatives is the C=O stretching absorption. For nitriles, the most important feature in the IR spectrum is the C≡N stretching absorption. These absorptions are summarized in Table 21.3, along with the absorptions of other carbonyl compounds. Some of the noteworthy trends in this table are the following:

1. Esters are readily differentiated from carboxylic acids, aldehydes, or ketones by the unique ester carbonyl absorption at 1735 cm^{-1}.

2. Lactones, lactams, and cyclic anhydrides, like cyclic ketones, have carbonyl absorption frequencies that increase significantly as the ring size decreases. (See Study Guide Link 19.1.)

3. Anhydrides and some acid chlorides have two carbonyl absorptions. The two carbonyl absorptions of anhydrides are due to the symmetrical and unsymmetrical stretching vibrations of the carbonyl group (Fig. 12.7). (The reason for the double absorption of acid chlorides is more obscure.)

4. The carbonyl absorptions of amides occur at much lower frequencies than those of other carbonyl compounds.

5. The C≡N stretching absorptions of nitriles generally occur in the triple-bond region of the spectrum. These absorptions are stronger, and occur at higher frequencies, than the C≡C absorptions of alkynes. (Why? Sec. 12.3B.)

The IR spectra of some carboxylic acid derivatives are shown in Fig. 21.2a–c on p. 984.

Other useful absorptions in the IR spectra of carboxylic acid derivatives are also summarized in Table 21.3. For example, primary and secondary amides show an N—H stretching absorption in the 3200–3400 cm^{-1} region of the spectrum. Many primary amides show two N—H absorptions, and secondary amides show a single strong N—H absorption. In addition, a strong N—H bending absorption occurs in the vicinity of 1640 cm^{-1}, typically appearing as a shoulder on the low-frequency side of the amide carbonyl absorption. Obviously, tertiary amides lack both of these N—H vibrations. The presence of these absorptions in a primary amide and their absence in a tertiary amide are evident in the comparison of the two spectra in Fig. 21.3 on p. 985.

B. NMR Spectroscopy

The α-proton resonances of all carboxylic acid derivatives are observed in the δ 1.9–3 region of the proton NMR spectrum (see Table 13.2). In esters, the chemical shifts of protons on the alkyl carbon adjacent to the carboxylate oxygen occur at about 0.6 ppm lower field than the analogous protons in alcohols and ethers. This shift is attributable to the electron-attracting character of the carbonyl group.

Table 21.3   Important Infrared Absorptions of Carbonyl Compounds and Nitriles

Compound	Carbonyl absorption, cm^{-1}	Other absorptions, cm^{-1}
ketone	1710–1715	
α,β-unsaturated ketone	1670–1680	
aryl ketone	1680–1690	
cyclopentanone	1745	
cyclobutanone	1780	
aldehyde	1720–1725	aldehydic C—H stretch at 2720
α,β-unsaturated aldehyde	1680–1690	
aryl aldehyde	1700	
carboxylic acid (dimer)	1710	OH stretch at 2400–3000 (strong, broad); C—O stretch at 1200–1300
aryl carboxylic acid	1680–1690	
ester or 6-membered		
lactone (δ-lactone)	1735	C—O stretch at 1000–1300
α,β-unsaturated ester	1720–1725	
5-membered lactone (γ-lactone)	1770	
4-membered lactone (β-lactone)	1840	
acid chloride	1800	a second weaker band is sometimes observed at 1700–1750
anhydride	1760, 1820 (two absorptions)	C—O stretch as in ester
6-membered cyclic anhydride	1750, 1800	
5-membered cyclic anhydride	1785, 1865	
amide	1650–1655	N—H bend at 1640
		N—H stretch at 3200–3400; double absorption for primary amide
6-membered lactam (δ-lactam)	1670	
5-membered lactam (γ-lactam)	1700	
4-membered lactam (β-lactam)	1745	
nitrile		C≡N stretch at 2200–2250

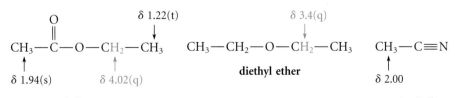

ethyl acetate   diethyl ether   acetonitrile

The *N*-alkyl protons of amides have chemical shifts in the δ 2.6–3 chemical shift region, and the N—H proton resonances of primary and secondary amides are observed

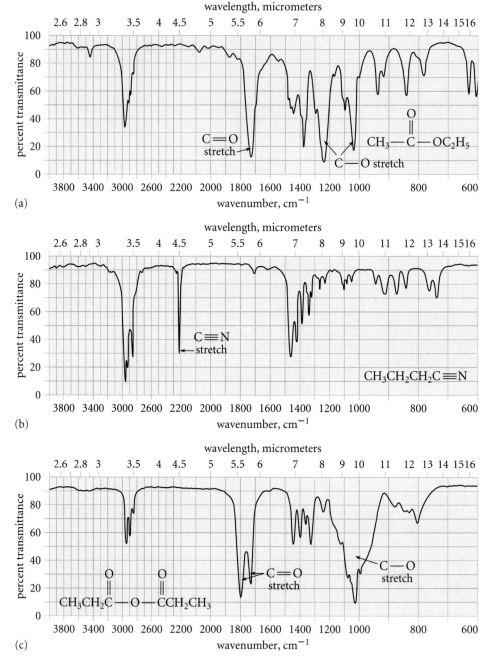

Figure 21.2 *Infrared spectra of some carboxylic acid derivatives: (a) ethyl acetate; (b) butyronitrile; (c) propionic anhydride.*

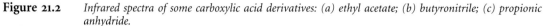

in the δ 7.5–8.5 region. The resonances for these protons, like those of carboxylic acid O—H protons, are sometimes broad. This broadening is caused by a slow chemical exchange with the protons of other protic substances (such as traces of moisture) and

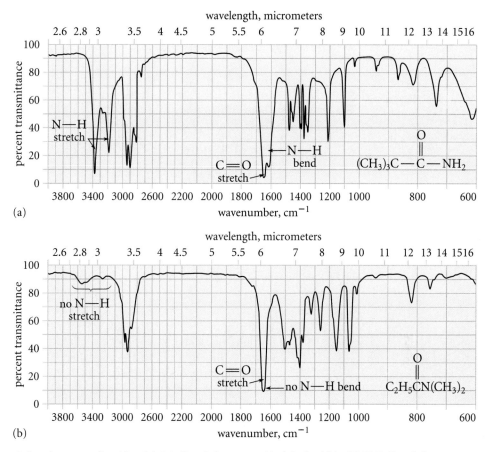

Figure 21.3   *Infrared spectra of amides: (a) 2,2-dimethylpropanamide (pivalamide); (b) N,N-dimethylpropan-amide. Notice that the N—H stretching and bending absorptions seen in (a) are absent in (b).*

by unresolved splitting with ^{14}N, which has a nuclear spin. Amide N—H resonances, like the O—H signals of acids and alcohols, can be washed out by exchange with D_2O (Sec. 13.6D).

STUDY GUIDE LINK:
✓ **21.2**
Solving Structure Problems Involving Nitrogen-Containing Compounds

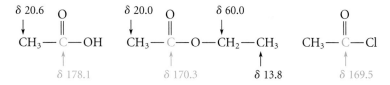

N-methylacetamide

In carbon NMR (CMR) spectra, the carbonyl chemical shifts of carboxylic acid derivatives are in the range δ 165–180, very much like those of carboxylic acids.

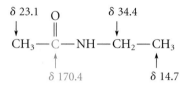

The chemical shifts of nitrile carbons are considerably smaller, occurring in the δ 115–120 range. These shifts are much greater, however, than those of acetylenic carbons.

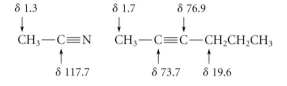

PROBLEMS

21.7   How would you differentiate between the compounds in each of the following pairs?
*(a) $CH_3—O—CH_2CH_2—O—CH_2CH=CH_2$ and ethyl butyrate by IR spectroscopy
(b)   *p*-ethylbenzoic acid and ethyl benzoate by IR spectroscopy
*(c)   2,4-dimethylbenzonitrile and *N*-methylbenzamide by NMR spectroscopy
(d)   methyl propionate and ethyl acetate by NMR spectroscopy
*(e)   ethyl butyrate and ethyl isobutyrate by CMR spectroscopy

*21.8   Identify the compound with the proton NMR spectrum given in Fig. 21.4. This compound has molecular mass = 87 and absorptions at 3300 and 1650 cm^{-1} in its IR spectrum.

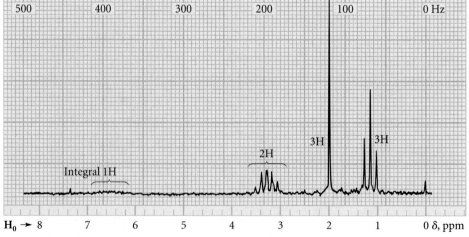

Figure 21.4   *NMR spectrum for Problem 21.8.*

21.5   Basicity of Carboxylic Acid Derivatives

Like carboxylic acids themselves, carboxylic acid derivatives are weakly basic at the carbonyl oxygen, and nitriles are weakly basic at nitrogen. These basicities are particularly important in some of the acid-catalyzed reactions of esters, amides, and nitriles.

The basicity of an ester is about the same as the basicity of the corresponding carboxylic acid.

$$
CH_3-\overset{\overset{\displaystyle :O:}{\|}}{C}-\ddot{O}CH_3 + H_3O^+ \;\rightleftharpoons
$$

$$
\left[\; CH_3-\overset{\overset{\displaystyle +\ddot{O}-H}{\|}}{C}-\ddot{O}CH_3 \;\longleftrightarrow\; CH_3-\overset{\overset{\displaystyle :\ddot{O}-H}{|}}{\underset{+}{C}}-\ddot{O}CH_3 \;\longleftrightarrow\; CH_3-\overset{\overset{\displaystyle :\ddot{O}-H}{|}}{C}=\underset{+}{\ddot{O}}CH_3 \;\right] + H_2O \qquad (21.4a)
$$

protonated ester;
$pK_a \approx -6$

Amides are considerably more basic than other carboxylic acid derivatives. This basicity, relative to esters, is a reflection of the reduced electronegativity of nitrogen relative to oxygen. That is, the resonance structures in which positive charge is shared on nitrogen are particularly important for a protonated amide.

$$
CH_3-\overset{\overset{\displaystyle :O:}{\|}}{C}-\ddot{N}H_2 + H_3O^+ \;\rightleftharpoons
$$

$$
\left[\; CH_3-\overset{\overset{\displaystyle +\ddot{O}-H}{\|}}{C}-\ddot{N}H_2 \;\longleftrightarrow\; CH_3-\overset{\overset{\displaystyle :\ddot{O}-H}{|}}{\underset{+}{C}}-\ddot{N}H_2 \;\longleftrightarrow\; CH_3-\overset{\overset{\displaystyle :\ddot{O}-H}{|}}{C}=\underset{+}{N}H_2 \;\right] + H_2O \qquad (21.4b)
$$

protonated amide;
$pK_a \approx -0.5$ to -1

Notice carefully that both esters and amides, like carboxylic acids (Sec. 20.4B), protonate on the *carbonyl oxygen*. Protonation of esters on the carboxylate oxygen, or amides on the nitrogen, would give a cation that is *not* resonance-stabilized, and additionally, that is destabilized by the electron-attracting polar effect of the carbonyl group. The site of protonation of amides was for many years a subject of controversy, because ammonia and amines (R_3N:) are protonated on nitrogen. It has been estimated, however, that nitrogen protonation of an amide is less favorable than carbonyl protonation by about 8 pK_a units.

Nitriles are very weak bases; protonated nitriles have a pK_a of about -10.

Study Guide Link:
✓**21.3**
Basicity of Nitriles

$$
CH_3-C\equiv N: + H_3O^+ \;\rightleftharpoons\; \left[\; CH_3-C\equiv \overset{+}{N}-H \;\longleftrightarrow\; CH_3-\overset{+}{C}=\ddot{N}-H \;\right] + H_2O \qquad (21.5)
$$

protonated nitrile;
$pK_a \approx -10$

21.9 Which of the two compounds in each set below should have the greater basicity at the carbonyl oxygen? Explain.

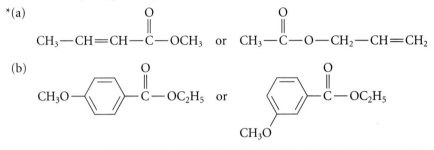

21.6 Introduction to Reactions of Carboxylic Acid Derivatives

The reactions of carboxylic acid derivatives can be categorized as follows:

1. Reactions at the carbonyl group (or cyano group of a nitrile)
   a. Reactions at the carbonyl oxygen or cyano nitrogen
   b. Reactions at the carbonyl carbon or cyano carbon
2. Reactions involving the α-carbon
3. Reactions at the nitrogen of amides

One *carbonyl-group reaction* of carboxylic acid derivatives is the reaction of the carbonyl oxygen—and, by analogy, the nitrile nitrogen—as a base. This type of reaction, discussed in the previous section, often serves as the first step in acid-catalyzed reactions of carboxylic acid derivatives.

As with carboxylic acids, the major carbonyl-group reaction of carboxylic acid derivatives is *substitution at the carbonyl carbon*, also called **acyl substitution**. Acyl substitution can be represented generally as follows, with E = an electrophilic group and Y = a nucleophilic group:

$$
\underset{\substack{\text{carboxylic acid}\\\text{derivative}}}{R-\overset{\overset{\displaystyle O}{\|}}{C}-X} \ + \ E-Y \ \longrightarrow \ \underset{\substack{\text{another}\\\text{carboxylic acid}\\\text{derivative}}}{R-\overset{\overset{\displaystyle O}{\|}}{C}-Y} \ + \ E-X \tag{21.6}
$$

$$
\text{an acyl group} \longrightarrow \boxed{R-\overset{\overset{\displaystyle O}{\|}}{C}\text{—}X}
$$

The term *acyl substitution* comes from the fact that substitution occurs at the carbonyl carbon of an *acyl group*. In other words, an acyl group is transferred, in the general example above, between an —X and a —Y group. The group —X might be the —Cl of an acid chloride, the —OR of an ester, and so on; this group is substituted by another group —Y. This is precisely the same type of reaction as esterification of carboxylic acids (—X = —OH, E—Y = H—OCH_3; Sec. 20.8A). Acyl substitution reactions of carboxylic acid derivatives are the major focus of this chapter.

Although nitriles are not carbonyl compounds, the C≡N bond behaves chemically much like a carbonyl group. For example, a typical reaction of nitriles is *addition*.

$$R—C≡N\colon + E—Y \longrightarrow R—\overset{\overset{\displaystyle Y}{\displaystyle |}}{C}=\overset{\cdot\cdot}{N}—E \qquad (21.7)$$

(Compare this reaction with addition to the carbonyl group of an aldehyde or ketone.) Although the resulting addition products are stable in some cases, in most situations they react further.

Like aldehydes and ketones, carboxylic acid derivatives undergo certain reactions involving the α-carbon. The α-carbon reactions of all carbonyl compounds are grouped together in Chapter 22. The reactivity of amides at nitrogen is discussed in Sec. 23.11C.

21.7 Hydrolysis of Carboxylic Acid Derivatives

All carboxylic acid derivatives have in common the fact that they undergo *hydrolysis* (a cleavage reaction with water) to yield carboxylic acids.

A. Hydrolysis of Esters

Saponification of Esters   One of the most important reactions of esters is the cleavage reaction with hydroxide ion to yield a carboxylate salts and an alcohol. The carboxylic acid itself is formed when a strong acid is subsequently added to the reaction mixture.

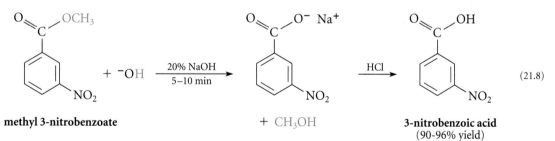

methyl 3-nitrobenzoate   + CH_3OH 3-nitrobenzoic acid
                                                                 (90-96% yield)

Ester hydrolysis in aqueous hydroxide is called **saponification** because it is used in the production of soaps from fats (Sec. 21.12B). Despite its association with fatty-acid esters, the term *saponification* is sometimes used to refer to hydrolysis in base of any acid derivative.

The mechanism of ester saponification involves attack by the nucleophilic hydroxide anion to give a tetrahedral addition intermediate from which an alkoxide ion is expelled.

$$R-\overset{\overset{\displaystyle :O:}{\|}}{\underset{\underset{\displaystyle -:\ddot{O}H}{|}}{C}}-\ddot{O}CH_3 \;\rightleftharpoons\; \left[\,R-\overset{\overset{\displaystyle :\ddot{O}:^{-}}{|}}{\underset{\underset{\displaystyle :OH}{|}}{C}}-\ddot{O}CH_3\,\right] \;\rightleftharpoons\; R-\overset{\overset{\displaystyle :O:}{\|}}{C}-\ddot{O}H \;+\; -:\ddot{O}CH_3 \qquad (21.9a)$$

tetrahedral
addition intermediate

The alkoxide ion thus formed (methoxide in the above example) reacts with the acid to give the carboxylate salt and the alcohol.

$$R-\overset{\overset{\displaystyle :O:}{\|}}{C}-\underset{\underset{\displaystyle pK_a = 4.5}{}}{\ddot{O}-H} \;+\; -:\ddot{O}CH_3 \;\rightleftharpoons\; R-\overset{\overset{\displaystyle :O:}{\|}}{C}-\ddot{O}:^{-} \;+\; \underset{\underset{\displaystyle pK_a = 15}{}}{H-\ddot{O}CH_3} \qquad (21.9b)$$

The equilibrium in this reaction lies far to the right because the carboxylic acid is a much stronger acid than methanol. LeChatelier's principle operates: the reaction in Eq. 21.9b removes the carboxylic acid from the equilibrium in Eq. 21.9a as its salt and thus drives the hydrolysis to completion. Hence, *saponification is effectively irreversible*. Although an excess of hydroxide ion is often used as a matter of convenience, many esters can be saponified with just one equivalent of ⁻OH. Saponification can also be carried out in an alcohol solvent, even though an alcohol is one of the products of the reaction. If saponification were reversible, an alcohol could not be used as the solvent, because the equilibrium would be driven towards starting materials.

Acid-Catalyzed Ester Hydrolysis   Because esterification of an acid with an alcohol is a reversible reaction (Sec. 20.8A), esters can be hydrolyzed to carboxylic acids in aqueous solutions of strong acids. In most cases this reaction is slow and must be carried out with an excess of water, in which most esters are insoluble. Saponification, followed by acidification, is a much more convenient method for hydrolysis of most esters because it is faster, it is irreversible, and it can be carried out not only in water, but also in a variety of solvents—even alcohols.

As expected from the principle of microscopic reversibility (Sec. 10.1), the mechanism of acid-catalyzed hydrolysis is the exact reverse of the mechanism of acid-catalyzed esterification (Sec. 20.8A). The ester is first protonated by the acid catalyst:

$$\overset{\overset{\displaystyle :O:}{\|}}{R-C-OCH_3} \quad \overset{H-\overset{+}{\ddot{O}}H_2}{} \;\rightleftharpoons\; R-\overset{\overset{\displaystyle :O}{\|}}{C}-OCH_3 \;+\; H_2\ddot{O}: \qquad (21.10a)$$

As in other acid-catalyzed reactions at the carbonyl group, protonation makes the carbonyl carbon more electrophilic by making the carbonyl oxygen a better acceptor of electrons. Water, acting as a nucleophile, attacks the carbonyl carbon and then loses a proton to give the tetrahedral addition intermediate:

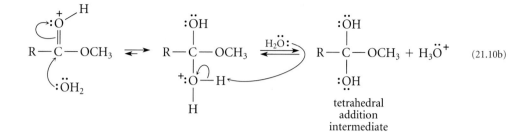

$$(21.10b)$$

Protonation of the leaving oxygen converts it into a better leaving group. Loss of this group gives a protonated carboxylic acid, from which a proton is removed to give the carboxylic acid itself.

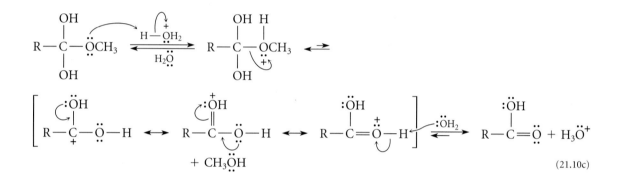

$$(21.10c)$$

STUDY GUIDE LINK:
✓21.4
Mechanism of Ester Hydrolysis

STUDY GUIDE LINK:
21.5
Cleavage of Tertiary Esters and Carbonless Carbon Paper

Let's summarize the important differences between acid-catalyzed ester hydrolysis and ester saponification. First, in acid-catalyzed hydrolysis, the carbonyl carbon can be attacked by the relatively weak nucleophile water because the carbonyl oxygen is protonated. In base, the carbonyl oxygen is not protonated; hence, a much stronger base than water, namely, hydroxide ion, is required to attack the carbonyl carbon. Second, acid *catalyzes* ester hydrolysis, but base *is not a catalyst* because it is consumed by the reaction in Eq. 21.9b. Finally, acid-catalyzed ester hydrolysis is reversible, but saponification is irreversible, again because of the ionization in Eq. 21.9b.

Ester hydrolysis and saponification are both examples of *acyl substitution* (Sec. 21.6). Specifically, the mechanisms of these reactions are classified as **nucleophilic acyl substitution** mechanisms. In a nucleophilic acyl substitution reaction, the substituting group attacks the carbonyl carbon as a nucleophile. This nucleophile is ⁻OH in saponification, and H_2O in acid-catalyzed hydrolysis; each group displaces, or substitutes for, the —OR group of the ester. With the exception of the reactions of nitriles, most of the reactions in the remainder of this chapter are nucleophilic acyl substitution reactions.

Hydrolysis and Formation of Lactones   Because lactones are cyclic esters, they undergo many of the reactions of esters, including saponification. Saponification converts a lactone completely into the salt of the corresponding hydroxy acid.

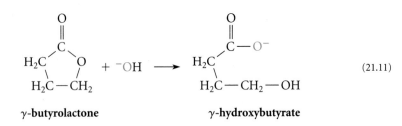

$$\gamma\text{-butyrolactone} \qquad\qquad\qquad \gamma\text{-hydroxybutyrate}$$

(21.11)

Upon acidification, the hydroxy acid forms. However, *if a hydroxy acid is allowed to stand in acidic solution, it comes to equilibrium with the corresponding lactone.* The formation of a lactone from a hydroxy acid is nothing more than an *intramolecular* esterification (an esterification within the same molecule) and, like esterification, the lactonization equilibrium is acid-catalyzed.

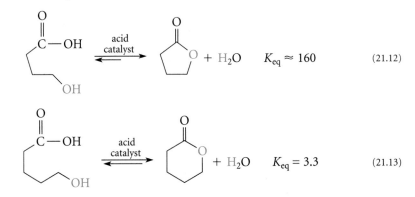

(21.12)

(21.13)

As the examples in Eqs. 21.12 and 21.13 illustrate, lactones containing five- and six-membered rings are favored at equilibrium over their corresponding hydroxy acids. Although lactones with ring sizes smaller than five or larger than six are well known, they are less stable than their corresponding hydroxy acids. Consequently, the lactonization equilibria for these compounds favor instead the hydroxy acids.

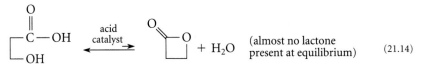

(21.14)

B. Hydrolysis of Amides

Amides can be hydrolyzed to carboxylic acids and ammonia or amines by heating them in acidic or basic solution.

$$\underset{\substack{\text{2-phenylbutanamide}}}{\overset{\substack{\text{Ph O}\\ | \ \ ||}}{CH_3CH_2CHC}}-NH_2 + H_2O \ \xrightarrow[\text{heat, 2 hr}]{55 \text{ wt \% } H_2SO_4}\ \underset{\substack{\text{2-phenylbutanoic acid}\\(88\text{–}90\% \text{ yield})}}{\overset{\substack{\text{Ph O}\\ | \ \ ||}}{CH_3CH_2CHC}}-OH + \overset{+}{N}H_4 \ HSO_4^-$$

(21.15)

In acid, protonation of the ammonia or amine by-product drives the hydrolysis equilibrium to completion. The amine can be isolated, if desired, by addition of base to the reaction mixture following hydrolysis, as in the following example.

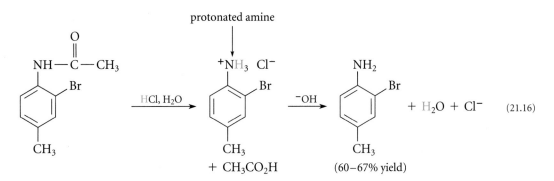

Hydrolysis of amides in base is analogous to saponification of esters. In base, the reaction is driven to completion by formation of the carboxylic acid salt.

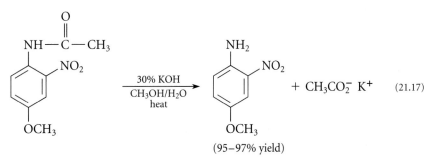

The conditions for both acid- and base-promoted amide hydrolysis are considerably more severe than the corresponding reactions of esters. That is, amides are considerably *less reactive* than esters. The relative reactivities of carboxylic acid derivatives are discussed in Sec. 21.7E.

The mechanisms of amide hydrolysis are typical nucleophilic acyl substitution mechanisms; you are asked to explore this point in Problem 21.10.

PROBLEMS

21.10 Show in detail the hydrolysis mechanism of *N*-methylbenzamide *(a) in acidic solution; (b) in aqueous NaOH. Assume that each mechanism involves a tetrahedral addition intermediate.

21.11 Give the structures of the hydrolysis products that result from each of the following reactions.

*(a)

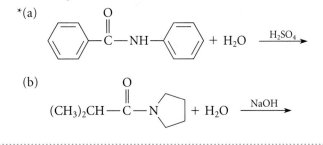

(b)

C. Hydrolysis of Nitriles

Nitriles are hydrolyzed to carboxylic acids and ammonia by heating them in acidic or basic solution.

$$PhCH_2\!-\!C\!\equiv\!N + 2H_2O + H_2SO_4 \xrightarrow[\text{3 hr}]{\text{heat}} PhCH_2\!-\!CO_2H + NH_4^+\ HSO_4^-$$ (21.18)

phenylacetonitrile   (57 wt %)   **phenylacetic acid**
                                                             (78% yield)

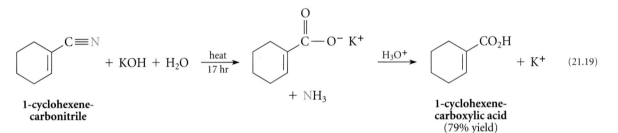

$$+\ KOH + H_2O \xrightarrow[\text{17 hr}]{\text{heat}}$$

**1-cyclohexene-
carbonitrile**

$$+\ NH_3$$

$$\xrightarrow{H_3O^+}$$

$$+\ K^+$$ (21.19)

**1-cyclohexene-
carboxylic acid**
(79% yield)

Nitriles hydrolyze more slowly than esters and amides. Consequently, the conditions required for the hydrolysis of nitriles are correspondingly more severe.

The mechanism of nitrile hydrolysis in acidic solution involves, first, protonation of the nitrogen (Sec. 21.5):

$$R\!-\!C\!\equiv\!N\colon\quad H\!-\!\overset{+}{O}H_2 \rightleftharpoons R\!-\!C\!\equiv\!\overset{+}{N}\!-\!H + \colon\!\ddot{O}H_2$$ (21.20a)

This protonation makes the nitrile carbon much more electrophilic, just as protonation of a carbonyl oxygen makes a carbonyl carbon more electrophilic. Attack of the nucleophile water on the nitrile carbon and loss of a proton gives an intermediate called an *imidic acid*.

$$R\!-\!\overset{\colon\ddot{O}H_2}{\underset{}{C}}\!=\!\overset{+}{N}\!-\!H \rightleftharpoons R\!-\!\overset{H\ddot{O}\!-\!H}{\underset{}{C}}\!=\!\ddot{N}\!-\!H \xrightarrow{\colon\ddot{O}H_2} R\!-\!\overset{H\ddot{O}\colon}{\underset{}{C}}\!=\!N\!-\!H + H_3\overset{+}{O}$$ (21.20b)

an imidic acid

The imidic acid is unstable and is converted under the reaction conditions into an amide:

$$R\!-\!\overset{\colon\ddot{O}H}{\underset{}{C}}\!=\!\ddot{N}\!-\!H \xrightleftharpoons[\,]{H\!-\!\overset{+}{O}H_2} \left[\, R\!-\!\overset{\colon\ddot{O}\!-\!H}{\underset{}{C}}\!=\!\overset{+}{N}H_2 \leftrightarrow R\!-\!\overset{\colon\overset{+}{O}\!-\!H}{\underset{}{C}}\!-\!NH_2 \,\right] \xrightleftharpoons[\,]{\colon\ddot{O}H_2}$$

$$R\!-\!\overset{\colon O\colon}{\underset{}{C}}\!-\!\ddot{N}H_2 + H_3\overset{+}{O}$$ (21.20c)

Because amide hydrolysis is faster than nitrile hydrolysis, the amide formed in Eq. 21.20c does not survive under the vigorous conditions of nitrile hydrolysis, and is hydrolyzed to a carboxylic acid and ammonium ion, as discussed in Sec. 21.7B. Thus, the ultimate product of nitrile hydrolysis in acid is a carboxylic acid.

Notice that nitriles behave mechanistically much like carbonyl compounds. Compare, for example, the mechanism of acid-promoted nitrile hydrolysis in Eqs. 21.20a and b with that for the acid-catalyzed hydration of an aldehyde or ketone (Sec. 19.7A). In both mechanisms, an electronegative atom is protonated (nitrogen of the C≡N bond, or oxygen of the C=O bond), and water attacks the carbon of the resulting cation.

The parallel between nitrile and carbonyl chemistry is further illustrated by the hydrolysis of nitriles in base. The nitrile group, like a carbonyl group, is attacked by basic nucleophiles and, as a result, the electronegative nitrogen assumes a negative charge. Proton transfer gives an imidic acid (which, like a carboxylic acid, ionizes in base).

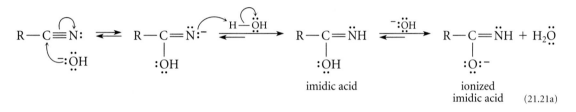

$$\text{(21.21a)}$$

As in acid-promoted hydrolysis, the imidic acid reacts further to give the corresponding amide, which, in turn, hydrolyzes under the reaction conditions to the carboxylate salt of the corresponding carboxylic acid (Sec. 21.7B).

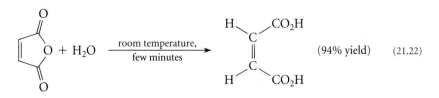

$$\text{(21.21b)}$$

D. Hydrolysis of Acid Chlorides and Anhydrides

Acid chlorides and anhydrides react *rapidly* with water, even in the absence of acid or base catalysts.

$$\text{(94\% yield)} \qquad \text{(21.22)}$$

$$\underset{Ph}{\overset{Ph}{\diagdown}}C=CH-\overset{\overset{\displaystyle O}{\|}}{C}-Cl + H_2O \xrightarrow[0°]{\overset{1)\ Na_2CO_3}{2)H_3O^+}} \underset{Ph}{\overset{Ph}{\diagdown}}C=CH-\overset{\overset{\displaystyle O}{\|}}{C}-OH \qquad (21.23)$$

(>95% yield)

However, the hydrolysis reactions of acid chlorides and anhydrides are almost never used for the preparation of carboxylic acids, because these derivatives are themselves usually prepared from acids (Sec. 20.9). Rather, these reactions serve as reminders that if samples of acid chlorides and anhydrides are allowed to come into contact with moisture they will rapidly become contaminated with the corresponding carboxylic acids.

E. Mechanisms and Reactivity in Nucleophilic Acyl Substitution Reactions

As you've seen, all carboxylic acid derivatives can be hydrolyzed to carboxylic acids; however, the *conditions* under which the different derivatives are hydrolyzed differ considerably. Hydrolysis reactions of amides and nitriles require heat as well as acid or base; hydrolysis reactions of esters require acid or base, but require heating only briefly, if at all; and hydrolysis reactions of acid chlorides and anhydrides occur rapidly at room temperature even in the absence of acid and base. These trends in reactivity, which are observed not only in hydrolysis, but in *all* nucleophilic acyl substitution reactions, can be summarized as follows:

Reactivities of carboxylic acid derivatives in nucleophilic acyl substitution reactions:

acid chlorides > anhydrides >> esters, acids > amides > nitriles   (21.24)

(The reactions of nitriles are additions, not substitutions, but are included for comparison.)

The practical significance of this reactivity order is that selective reactions are possible. In other words, an ester can be hydrolyzed under conditions that will leave an amide in the same molecule unaffected; likewise, nucleophilic substitution reactions on an acid chloride can be carried out under conditions that will leave an ester group unaffected.

Understanding the trends in relative reactivity requires, first, an understanding of the mechanisms by which nucleophilic acyl substitution reactions take place. (The reactivity of nitriles is considered later.) For the sake of simplicity, imagine a reaction of a nucleophile, $^-$:Nuc, with a carboxylic acid derivative containing a leaving group X under neutral or basic conditions. The substitution reaction involves formation of a tetrahedral addition intermediate.

$$R-\overset{\overset{\displaystyle :O:}{\|}}{\underset{:Nuc}{C}}-X \rightleftarrows R-\overset{\overset{\displaystyle :O:^-}{|}}{\underset{\underset{Nuc}{|}}{C}}-X \rightleftarrows R-\overset{\overset{\displaystyle :O:}{\|}}{C}-Nuc + {}^-:X \qquad (21.25)$$

tetrahedral addition
intermediate

Which step is rate-limiting depends on the carbonyl compound and the nucleophile.

Recall that reactivity is governed by the *standard free energy of activation* $\Delta G^{\circ\ddagger}$, the difference in the standard free energies of the transition state and the reactants (Sec. 4.8A). Also recall that reactions with larger $\Delta G^{\circ\ddagger}$ are slower than reactions with smaller $\Delta G^{\circ\ddagger}$. The first thing that affects the relative reactivities of the different carboxylic acid derivatives is their relative stabilities; a similar effect was discussed in Sec. 19.7B. If the carbonyl compound starting material is stabilized relative to the transition state, the standard free energy of the starting material is lowered relative to that of the transition state and the reactivity of the carbonyl compound is *decreased* (Fig. 21.5). This effect is important regardless of the step that is rate-limiting.

The major factor affecting the stability of the carbonyl compound is the *resonance interaction of the unshared electron pairs of the group* X *with the carbonyl π electrons*. For example, this interaction is depicted for an amide, in which $-X = -NH_2$, as follows:

$$\left[\begin{array}{ccc} & \ddot{\text{O}}: & \\ & \parallel & \\ R - C - \ddot{N}H_2 & \longleftrightarrow & R - C = \overset{+}{N}H_2 \end{array} \quad \begin{array}{c} :\ddot{\text{O}}:^- \\ | \\ R - C = \overset{+}{N}H_2 \end{array} \right] \qquad (21.26)$$

Esters, anhydrides, and acid chlorides all have analogous resonance structures. In an ester, such a resonance interaction places a positive charge on oxygen.

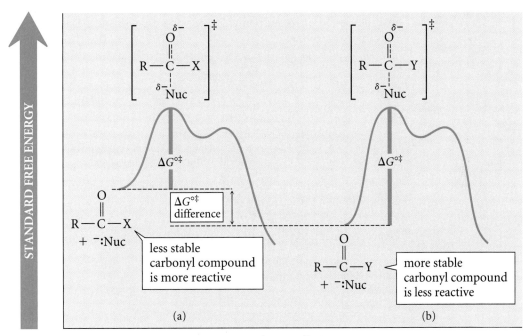

Figure 21.5   *How stability of the carbonyl compound affects reactivity in nucleophilic acyl substitution. The two transition states have been arbitrarily placed at the same energy for comparison purposes. Although the first step, formation of the tetrahedral addition intermediate, is rate-limiting in this diagram, this effect is present regardless of the rate-limiting step. (a) The less stable carbonyl compound is more reactive. (b) The more stable carbonyl compound is less reactive.*

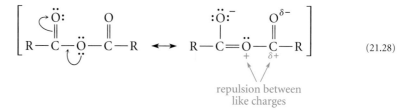

$$(21.27)$$

Because oxygen is more electronegative than nitrogen, this resonance interaction is less important in an ester than in an amide; hence, amides are stabilized by resonance more than esters. An additional unfavorable interaction occurs in the resonance structure of an anhydride: the repulsion between the positive charge on the carboxylate oxygen and the partial positive charge on the carbonyl carbon.

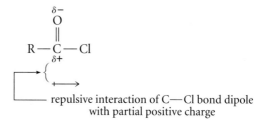

$$(21.28)$$

Consequently, anhydrides are stabilized less by resonance than esters and amides. Finally, the analogous resonance interaction in an acid chloride requires the overlap of a chlorine $3p$ orbital with a carbon $2p$ orbital. Because this overlap is very ineffective (Fig. 16.9), acid chlorides are stabilized even less by resonance than anhydrides. What's more, the polar effect of the chlorine also destabilizes the carbonyl compound through an unfavorable interaction of the carbon-chlorine bond dipole with the partial positive charge on the carbonyl carbon:

$$
\begin{array}{c}
\overset{\delta-}{O} \\
\parallel \\
R\!-\!\underset{\delta+}{C}\!-\!Cl
\end{array}
$$

repulsive interaction of C—Cl bond dipole
with partial positive charge

To summarize: the less important the resonance stabilization of the carbonyl compound, the more reactive the compound is.

Resonance stabilization of carbonyl compounds (X = leaving group in Eq. 21.25):

$$
-X = -Cl < -X = \overset{\overset{\text{O}}{\parallel}}{-O-C-R} < -X = -OR < -X = -NH_2 \qquad (21.29)
$$

$$\longleftarrow \text{decreasing reactivity} \longrightarrow$$

Notice that this is the observed order of reactivity given in Eq. 21.24.

When loss of the leaving group from the tetrahedral addition intermediate (the second step of Eq. 21.25) is rate-limiting, not only is the stabilization of the carbonyl compound important in determining reactivity, but a second effect is important as well: stabilization of the transition state (Fig. 21.6). The major factor accounting for differences in transition-state stability is *the relative base strengths of the different leaving groups* —X. As in S_N2, S_N1, and other reactions, the best leaving groups are the weakest bases.

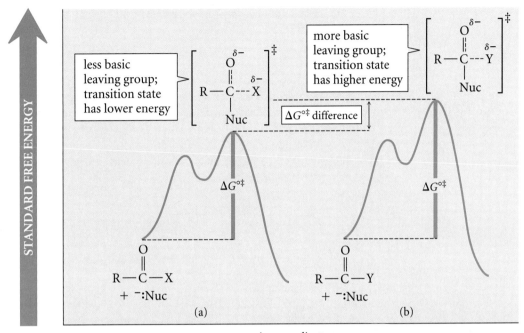

Figure 21.6   *Effects of the leaving group on reactivity in nucleophilic acyl substitution. This effect is most important in reactions for which the second step, loss of the leaving group from the tetrahedral addition intermediate, is rate-limiting as indicated in this diagram. The two carbonyl compounds are arbitrarily placed at the same energy to show the transition-state effects. (a) The carbonyl compound with the better (less basic) leaving group X has the transition state of lower energy, and is therefore more reactive. (b) The carbonyl compound with the poorer (more basic) leaving group Y has the transition state of higher energy, and is therefore less reactive.*

Relative basicities of leaving groups:

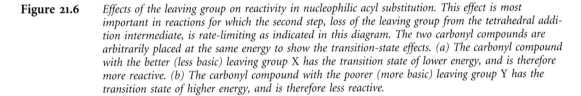

$$-X = -Cl < -X = -O-\overset{\overset{\textstyle O}{\|}}{C}-R < -X = -OR < -X = -NH_2 \qquad (21.30)$$

—————————————— decreasing reactivity —————————————→

Again, notice that this is the observed reactivity order in Eq. 21.24.

Now let's put everything together. *Regardless of which step is rate-limiting*, and regardless of whether stabilization of the carbonyl compound or basicity of the leaving group is the major effect on reactivity, we reach the same conclusion: the reactivity order is that shown in Eq. 21.24.

Although this detailed analysis has been carried out for reactions that involve a negatively charged nucleophile, the same conclusions are obtained from an analysis of acid-catalyzed reactions.

What about nitriles? Reactions of nitriles in base are slower than those of other acid derivatives because nitrogen is less electronegative than oxygen and accepts additional electrons less readily. Reactions of nitriles in acid are slower because of their extremely low basicities. It is the protonated form of a nitrile that reacts with nucleophiles in acid solution; but so little of this form is present (Sec. 21.5) that the rate of the reaction is very small.

*21.12   Use an analysis of resonance effects and leaving-group basicities to explain why acid-catalyzed hydrolysis of esters is faster than acid-catalyzed hydrolysis of amides.

21.13   Which should be faster: hydrolysis of an acid fluoride or hydrolysis of an acid chloride? Explain your reasoning.

21.14   Complete the following reactions.

*(a)

$$N\equiv C-CH_2-\overset{\overset{\displaystyle O}{\|}}{C}-OCH_3 + {}^-OH \text{ (1 equiv)} \xrightarrow[CH_3OH]{H_2O}$$

(b)

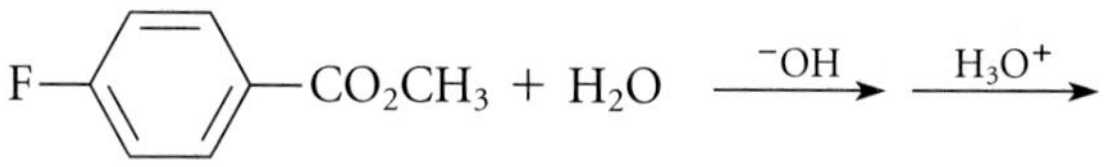

$$F-\underset{}{\bigcirc}-CO_2CH_3 + H_2O \xrightarrow{{}^-OH} \xrightarrow{H_3O^+}$$

*(c)

$$H_2N-\overset{\overset{\displaystyle O}{\|}}{C}-NH_2 + H_2O \xrightarrow[\text{heat}]{H_3O^+}$$

(d)

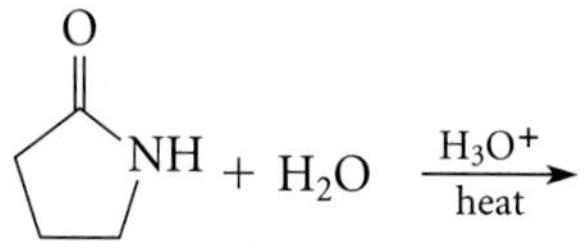

$$\underset{}{\overset{O}{\|}}\text{(cyclic)}NH + H_2O \xrightarrow[\text{heat}]{H_3O^+}$$

21.8   Reactions of Carboxylic Acid Derivatives with Nucleophiles

The previous section showed that all carboxylic acid derivatives hydrolyze to carboxylic acids. Water and hydroxide ion are only two of the nucleophiles that react with carboxylic acid derivatives. This section shows how the reactions of other nucleophiles with carboxylic acid derivatives can be used to prepare other carboxylic acid derivatives. As you proceed through this section, notice how all of the reactions fit the pattern of nucleophilic acyl substitution.

A.   Reactions of Acid Chlorides with Nucleophiles

Among the most useful ways of preparing carboxylic acid derivatives are the reactions of acid chlorides with various nucleophiles. Because of the great reactivity of acid chlorides, such reactions are typically very rapid and can be carried out under mild conditions.

Reactions of Acid Chlorides with Ammonia and Amines

Acid chlorides react rapidly and irreversibly with ammonia or amines to give amides. Reaction of an acid chloride with *ammonia* yields a primary amide:

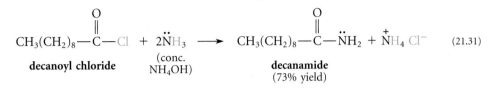

$$ (21.31) $$

decanoyl chloride   (conc. NH₄OH)   **decanamide** (73% yield)

Reaction of an acid chloride with a *primary amine* (an amine of the form R—NH₂; Sec. 19.11A) gives a secondary amide:

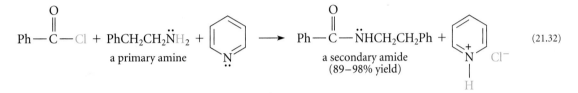

a primary amine   a secondary amide (89–98% yield)

$$ (21.32) $$

Reaction of an acid chloride with a *secondary amine* (an amine of the form R₂NH; Sec. 19.11B) gives a tertiary amide:

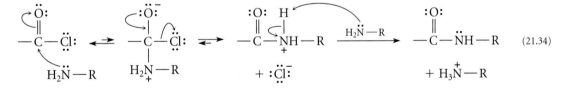

a secondary amine   a tertiary amide (77–81% yield)

$$ (21.33) $$

These reactions are all additional examples of nucleophilic acyl substitution.

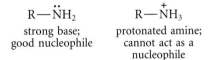

$$ (21.34) $$

Notice that a proton is removed from the amide nitrogen in the last step of the mechanism. Unless another base is added to the reaction mixture, *the starting amine acts as the base in this step.* Hence, for each equivalent of amide that is formed, an equivalent of amine is protonated. When the amine is protonated, its electron pair is taken "out of action," and the amine is no longer nucleophilic.

R—N̈H₂   R—N⁺H₃
strong base;   protonated amine;
good nucleophile   cannot act as a
                      nucleophile

Hence, if the only base present is the amine nucleophile, then at least *two* equivalents must be used: one equivalent as the nucleophile and one as the base in the final proton-transfer step.

The use of excess amine is practical when the amine is cheap and readily available. Another alternative is to use a *tertiary amine* (an amine of the form R_3N:) such as triethylamine or pyridine as the base (Eq. 21.32).

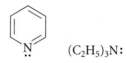

$(C_2H_5)_3N$:

pyridine **triethylamine**

STUDY GUIDE LINK: 21.6 *Reaction of Tertiary Amines with Acid Chlorides*

The presence of a tertiary amine does not interfere with amide formation by another amine because a tertiary amine itself cannot form an amide (why?). The use of a tertiary amine is particularly appropriate if the amine used to form the amide is expensive and cannot be used in excess.

Yet another alternative is to use the *Schotten-Baumann* technique. In this method, the reaction is run with a water-insoluble acid chloride in a separate layer (either alone or in a solvent) over an aqueous solution of NaOH (Eq. 21.33). Hydrolysis of the acid chloride by NaOH is avoided because the water-insoluble acid chloride is not in contact with the water-soluble hydroxide ion. The amine, which is soluble in the acid chloride solution, reacts to yield an amide. The aqueous NaOH extracts and neutralizes the protonated amine that is formed.

(21.35)

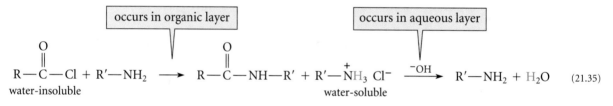

The important point about all the methods for preparing amides is that either two equivalents of amine must be used, or an equivalent of base must be added to effect the final neutralization.

Reaction of Acid Chlorides with Alcohols and Phenols Esters are formed rapidly when acid chlorides react with alcohols or phenols. In principle, the HCl liberated in the reaction need not be neutralized, since alcohols and phenols are not basic enough to be extensively protonated by the acid. However, some esters (such as *tert*-butyl esters; see Study Guide Link 21.5) and alcohols (such as tertiary alcohols; Secs. 10.1, 10.2), are sensitive to acid. In practice, a tertiary amine like pyridine is added to the reaction mixture or is even used as the solvent to neutralize the HCl.

(21.36a)

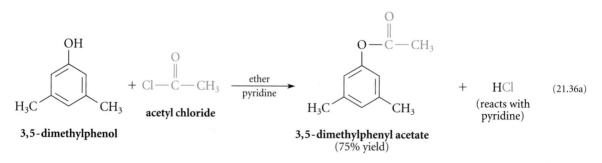

benzoyl chloride ***tert*-butyl alcohol** ***tert*-butyl benzoate** (71–76% yield) (reacts with quinoline) (21.36b)

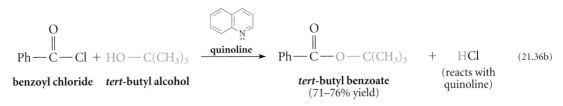

As these examples illustrate, esters of tertiary alcohols and phenols, which cannot be prepared by acid-catalyzed esterification, can be prepared by this method.

Sulfonate esters (esters of sulfonic acids) are prepared by the analogous reactions of sulfonyl chlorides (the acid chlorides of sulfonic acids) with alcohols. This reaction was introduced in Sec. 10.3A.

1-butanol

***p*-toluenesulfonyl chloride (tosyl chloride)**

(reacts with pyridine) (21.37)

butyl *p*-toluenesulfonate (butyl tosylate)
(88–90% yield)

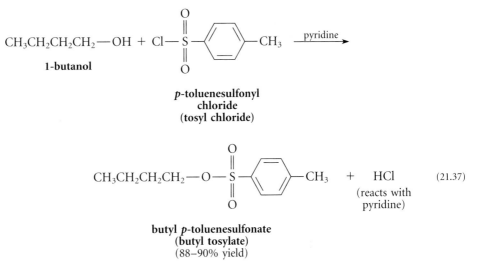

Reaction of Acid Chlorides with Carboxylate Salts Even though carboxylate salts are weak nucleophiles, acid chlorides are reactive enough to be attacked by carboxylate salts to give anhydrides.

propionyl chloride **sodium acetate** (excess) **acetic propionic anhydride** (60% yield) (21.38)

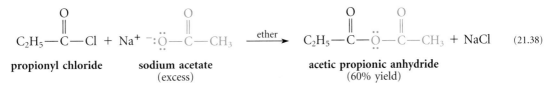

This is a second general method for the synthesis of anhydrides. Unlike the anhydride synthesis discussed in Sec. 20.9B, the reactions of acid chlorides with carboxylate salts can be used to prepare mixed anhydrides, as the example in Eq. 21.38 illustrates.

STUDY GUIDE LINK: ✓21.7 *Another Look at Friedel-Crafts Acylation*

Summary: Use of Acid Chlorides in Organic Synthesis One of the most important general methods for converting a carboxylic acid into an ester, amide, or anhydride is first to convert the carboxylic acid into its acid chloride (Sec. 20.9A) and then use one of the acid chloride reactions discussed in this section to form the desired carboxylic acid derivative. To summarize:

21.15   Using an acid chloride synthesis as a first step, outline a conversion of hexanoic acid into each of the following compounds.
*(a)   ethyl hexanoate   (b)   N-methylhexanamide

21.16   Complete the following reactions by giving the major organic products.

*(a) $CH_3CH_2CO_2H$ $\xrightarrow[\text{(excess)}]{SOCl_2}$ $\xrightarrow[\text{(excess)}]{(CH_3)_2NH}$

(b)

$$\triangleright\!\!-\!\!\overset{\overset{\textstyle O}{\|}}{C}\!-\!Cl + HO\!-\!\triangleleft \xrightarrow[\text{ether}]{\text{pyridine}}$$

*(c)

$$PhCH_2\!-\!\overset{\overset{\textstyle O}{\|}}{C}\!-\!Cl + CH_3CH_2SH \longrightarrow$$

(d)

$$CH_3CH_2CH_2\!-\!\overset{\overset{\textstyle O}{\|}}{C}\!-\!Cl + Na^+\ {}^-O\!-\!\overset{\overset{\textstyle O}{\|}}{C}\!-\!CH_3 \longrightarrow$$

*(e)

$$Cl\!-\!\overset{\overset{\textstyle O}{\|}}{C}\!-\!Cl \text{ (excess)} + CH_3OH \longrightarrow$$

(f)

$$Cl\!-\!\overset{\overset{\textstyle O}{\|}}{C}\!-\!Cl + CH_3OH \text{ (excess)} \longrightarrow$$

*(g)

$$C_2H_5O\!-\!\overset{\overset{\textstyle O}{\|}}{C}\!-\!OC_2H_5 + HO\!-\!CH_2CH_2\!-\!OH \xrightarrow[\text{heat}]{\text{acid catalyst}} (C_3H_4O_3)$$

(h)   phthalic anhydride + $CH_3OH \longrightarrow$

STUDY GUIDE LINK:
✓21.8
*Esters and
Nucleophiles*

*21.17   Contrast the location of ^{18}O in the products of the following two reactions, and explain.

(a)

$$PhCH_2\!-\!O\!-\!\overset{\overset{\textstyle O}{\|}}{\underset{\underset{\textstyle O}{\|}}{S}}\!-\!CH_3 + {}^{18}OH^- \longrightarrow$$

(b)

$$PhCH_2\!-\!O\!-\!\overset{\overset{\textstyle O}{\|}}{C}\!-\!CH_3 + {}^{18}OH^- \longrightarrow$$

21.18   How would you synthesize each of the following compounds from an acid chloride?

*(a)

$$\underset{\displaystyle CH_3CHOSO_2}{\overset{\displaystyle Ph}{|}}\!-\!\!\left\langle\!\!\bigcirc\!\!\right\rangle\!\!-\!CH_3$$

(b)

$$CH_3\!-\!\overset{\overset{\textstyle O}{\|}}{C}\!-\!O\!-\!\!\left\langle\!\!\bigcirc\!\!\right\rangle\!\!-\!NO_2$$

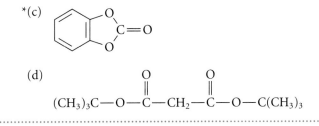

(d)

$$(CH_3)_3C-O-\overset{\overset{\displaystyle O}{\|}}{C}-CH_2-\overset{\overset{\displaystyle O}{\|}}{C}-O-C(CH_3)_3$$

21.9   Reduction of Carboxylic Acid Derivatives

Most of the reactions discussed in the previous sections have been straightforward nucleophilic acyl substitution reactions. In this and the following section, some reactions are considered that involve nucleophilic acyl substitution steps followed by other reactions.

A. Reduction of Esters to Primary Alcohols

Lithium aluminum hydride reduces all carboxylic acid derivatives. Reduction of esters with this reagent, like reduction of carboxylic acids, gives primary alcohols.

$$2CH_3CH_2-\overset{\overset{\displaystyle}{\underset{\underset{\displaystyle CH_3}{|}}{CH}}}{}-\overset{\overset{\displaystyle O}{\|}}{C}-OC_2H_5 \;+\; LiAlH_4 \xrightarrow{\text{ether}} \xrightarrow{\text{H}_3\text{O}^+}$$

ethyl 2-methylbutanoate   lithium aluminum hydride

$$2CH_3CH_2-\underset{\underset{\displaystyle CH_3}{|}}{CH}-CH_2-OH \;+\; 2C_2H_5OH \;+\; Li^+,\, Al^{3+}\ salts \qquad (21.47)$$

ethanol

2-methyl-1-butanol
(91% yield)

Notice that *two* alcohols are formed in this reaction, one derived from the *acyl group* of the ester (2-methyl-1-butanol in Eq. 21.47), and one derived from the alkoxy group (ethanol in Eq. 21.47). In most cases, a methyl or ethyl ester is used in this reaction, and the by-product methanol or ethanol is discarded; the alcohol from the acyl portion of the ester is the product of interest.

As noted several times (Sec. 20.10), the active nucleophile in LiAlH$_4$ reductions is the *hydride ion* (H:$^-$) delivered from $^-$AlH$_4$, and this reduction is no exception. Hydride replaces alkoxide at the carbonyl group of the ester to give an aldehyde. (Write the mechanism of this reaction, another example of nucleophilic acyl substitution.)

$$Li^+ \ {}^-AlH_4 \;+\; R-\overset{\overset{\displaystyle O}{\|}}{C}-OC_2H_5 \;\longrightarrow\; R-\overset{\overset{\displaystyle O}{\|}}{C}-H \;+\; Li^+\ C_2H_5O^- \;+\; AlH_3 \qquad (21.48a)$$

The aldehyde reacts rapidly with LiAlH$_4$ to give, after protonolysis, the alcohol (Sec. 19.8).

$$R-\overset{\overset{\displaystyle O}{\|}}{C}-H \xrightarrow{\text{LiAlH}_4} \xrightarrow{\text{H}_3\text{O}^+} R-\underset{\underset{\displaystyle H}{|}}{\overset{\overset{\displaystyle OH}{|}}{C}}-H \qquad (21.48b)$$

The reduction of esters to alcohols thus involves a *nucleophilic acyl substitution* reaction followed by a *carbonyl addition* reaction.

Sodium borohydride, another useful hydride reducing agent, is much less reactive than lithium aluminum hydride. It reduces aldehydes and ketones, but reacts very sluggishly with most esters; in fact, NaBH$_4$ can be used to reduce aldehydes and ketones selectively in the presence of esters.

Acid chlorides and anhydrides also react with LiAlH$_4$ to give primary alcohols. However, since acid chlorides and anhydrides are usually prepared from carboxylic acids, and since carboxylic acids themselves can be reduced to alcohols with LiAlH$_4$, the reduction of acid chlorides and anhydrides is seldom used.

B. Reduction of Amides to Amines

Amines are formed when amides are reduced with LiAlH$_4$.

$$\text{LiAlH}_4 + 2\text{Ph}-\overset{\overset{\displaystyle O}{\|}}{C}-\text{NH}_2 \longrightarrow \xrightarrow{\text{H}_3\text{O}^+} 2\text{Ph}-\text{CH}_2-\text{NH}_2 + \text{Li}^+, \text{Al}^{3+} \text{ salts} + 2\text{H}_2 \qquad (21.49)$$

lithium   **benzamide**   **benzylamine**
aluminum   (80% yield)
hydride

Amide reduction can be used not only to prepare primary amines from primary amides, but also secondary and tertiary amines from secondary and tertiary amides, respectively.

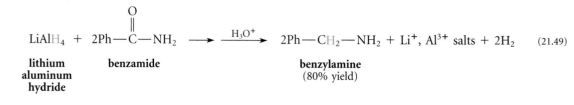

$$(88\% \text{ yield})$$

Notice carefully that the reaction of LiAlH$_4$ with an amide differs from its reaction with an ester. In the reduction of an ester, the *carboxylate oxygen* is lost as a leaving group. If amide reduction were strictly analogous to ester reduction, the nitrogen would be lost, and a primary alcohol would be formed; clearly, this is not the case. Instead, it is the *carbonyl oxygen* that is lost in amide reduction.

Ester reduction:

$$R-\overset{\overset{\displaystyle O}{\|}}{C}-OR' \xrightarrow{\text{LiAlH}_4} \xrightarrow{\text{H}_3\text{O}^+} R-\text{CH}_2\overset{\uparrow}{\text{O}}\text{H} + R'\text{OH} \qquad (21.51a)$$

$$\text{(carbonyl oxygen retained)}$$

Amide reduction:

$$R\overset{\overset{\displaystyle O}{\|}}{-C}-NR_2' \xrightarrow{\text{LiAlH}_4} \xrightarrow{\text{H}_3\text{O}^+} R-CH_2NR_2' \quad \text{(carbonyl oxygen lost)} \quad (21.51b)$$

Let's consider the reason for this difference, using as a case study the reduction of a secondary amide. (The mechanisms of reduction of primary and tertiary amides are somewhat different, but have the same result.)

In the first step of the mechanism, the weakly acidic amide proton reacts with an equivalent of hydride, a strong base, to give hydrogen gas, AlH_3, and the lithium salt of the amide.

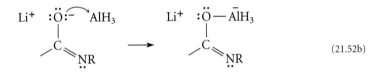

$$(21.52a)$$

The lithium salt of the amide, a Lewis base, reacts with the Lewis acid AlH_3.

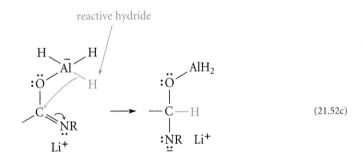

$$(21.52b)$$

The resulting species is an active hydride reagent conceptually much like LiAlH_4, and can deliver hydride to the C=N double bond.

reactive hydride

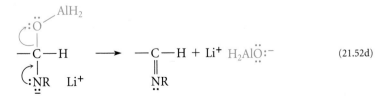

$$(21.52c)$$

Here is why oxygen rather than nitrogen is lost from the amide. If nitrogen were lost from the tetrahedral addition intermediate, it would have to assume a second negative charge. On the other hand, loss of oxygen requires expulsion of $^-\text{OAlH}_2$, which is actually a fairly good leaving group. Loss of this group gives an *imine* (Sec. 19.11).

$$(21.52d)$$

Hence, loss of the carbonyl oxygen occurs because it is converted into a better leaving group than the nitrogen.

The C=N of the imine, like the C=O of an aldehyde, undergoes nucleophilic addition with "H:⁻" from ⁻AlH₄ or from one of the other hydride-containing species in the reaction mixture. Addition of water or acid to the reaction mixture converts the addition intermediate into an amine by protonolysis.

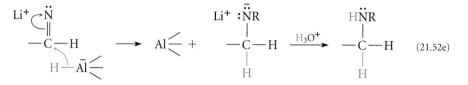

$$(21.52e)$$

C. Reduction of Nitriles to Primary Amines

Nitriles are reduced to primary amines by reaction with LiAlH₄, followed by the usual protonolysis step.

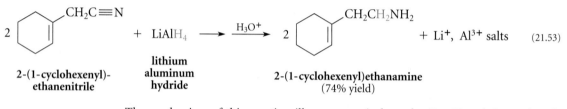

$$(21.53)$$

2-(1-cyclohexenyl)-ethanenitrile   **lithium aluminum hydride**   **2-(1-cyclohexenyl)ethanamine** (74% yield)

The mechanism of this reaction illustrates again how the C≡N and C=O bonds react in similar ways. This reaction probably occurs as two successive *nucleophilic additions*.

$$R—C{\equiv}N\colon \longrightarrow R—C{=}N\colon^{Li} + AlH_3 \qquad (21.54a)$$
$$H—\bar{A}lH_3 \qquad\qquad H$$
$$\text{imine salt}$$

In the second addition, the imine salt reacts in a similar manner with AlH₃ (or another equivalent of AlH₄⁻).

$$R—CH{=}N\colon^{Li} \longrightarrow \left[R—CH_2—N\colon^{Li}_{AlH_2} \longleftrightarrow R—CH_2—\overset{+}{N}{<}^{Li}_{\underset{=}{A}lH_2} \right] \qquad (21.54b)$$
$$H—\bar{A}lH_2$$

In the resulting derivative, both the N—Li and the N—Al bonds are very polar, and the nitrogen has a great deal of anionic character. Both bonds are susceptible to protonolysis. Hence, an amine is formed when water is added to the reaction mixture.

$$R—CH_2—N\colon^{Li}_{Al} \xrightarrow{H_3O^+} R—CH_2—\ddot{N}H_2 + Li^+,\ Al^{3+}\ \text{salts} \qquad (21.54c)$$

Nitriles are also reduced to primary amines by catalytic hydrogenation.

$$CH_3(CH_2)_4C{\equiv}N + 2H_2 \xrightarrow[\substack{2000 \text{ psi} \\ 120-130°}]{\text{Raney Ni}} CH_3(CH_2)_4CH_2NH_2 \qquad (21.54d)$$

hexanenitrile   **1-hexanamine**

An intermediate in the reaction is the imine, which is not isolated but is hydrogenated to the amine product. (See also Problem 21.22.)

$$R{-}C{\equiv}N \xrightarrow{H_2,\text{ catalyst}} [R{-}CH{=}NH] \xrightarrow{H_2,\text{ catalyst}} R{-}CH_2{-}NH_2 \qquad (21.55)$$

                                        imine

The reductions discussed in this and the previous section allow the formation of the *amine* functional group from amides and nitriles, the nitrogen-containing carboxylic acid derivatives. Hence, any synthesis of a carboxylic acid can be used as part of an amine synthesis, provided that the amine has the following form:

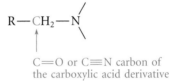

C=O or C≡N carbon of
the carboxylic acid derivative

As indicated, the carbon of the carbonyl group or cyano group in the carboxylic acid derivative ends up as the —CH₂— group adjacent to the amine nitrogen.

. . . .

**STUDY
PROBLEM
21.1**

Outline a synthesis of (cyclohexylmethyl)methylamine from cyclohexanecarboxylic acid.

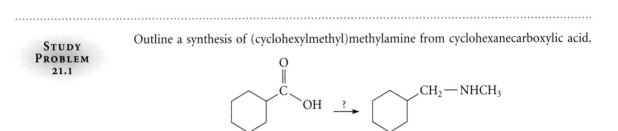

cyclohexanecarboxylic acid   **(cyclohexylmethyl)methylamine**

Solution   Any carboxylic acid derivative used to prepare the amine must contain nitrogen; the two such derivatives are amides and nitriles. However, notice that the only type of amine that can be prepared directly by nitrile reduction is an amine of the form —CH₂NH₂. Since the desired product is not of this form, the reduction of nitriles must be rejected as an approach to this target.

The amide that could be reduced to the desired amine is *N*-methylcyclohexanecarboxamide:

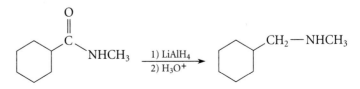

***N*-methylcyclohexanecarboxamide**

This amide can be prepared, in turn, by reaction of the appropriate amine, in this case methylamine, with an acid chloride:

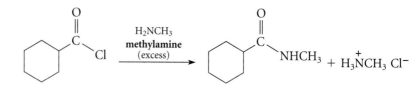

cyclohexanecarbonyl chloride

Finally, the acid chloride is prepared from the carboxylic acid (Sec. 20.9A).

D. Reduction of Acid Chlorides to Aldehydes

Acid chlorides can be reduced to aldehydes by either of two procedures. In the first, the acid chloride is hydrogenated over a catalyst that has been deactivated, or *poisoned*, with an amine, such as quinoline, that has been heated with sulfur. (Amines and sulfides are catalyst poisons.) This reaction is called the **Rosenmund reduction.**

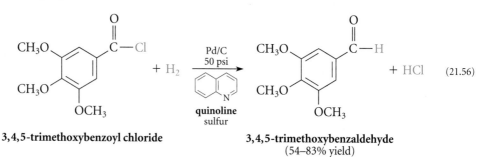

(21.56)

3,4,5-trimethoxybenzoyl chloride   **3,4,5-trimethoxybenzaldehyde**
                                                     (54–83% yield)

The poisoning of the catalyst prevents further reduction of the aldehyde product.

A second, more recent, method of converting acid chlorides into aldehydes is the reaction of an acid chloride at low temperature with a "cousin" of LiAlH₄, lithium tri(*tert*-butoxy)aluminum hydride.

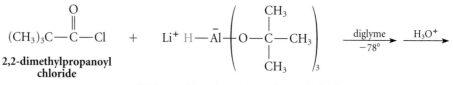

2,2-dimethylpropanoyl   **lithium tri(*tert*-butoxy)aluminum hydride**
chloride

$$(CH_3)_3C-\overset{O}{\overset{\|}{C}}-H + 3CH_3-\overset{CH_3}{\underset{CH_3}{\overset{|}{\underset{|}{C}}}}-OH + LiCl + Al^{3+}\ salts \qquad (21.57)$$

2,2-dimethylpropanal

The hydride reagent used in this reduction is derived by the replacement of three hydrogens of lithium aluminum hydride by *tert*-butoxy groups. As the hydrides of LiAlH$_4$ are replaced successively with alkoxy groups, less reactive reagents are obtained. (Can you think of a reason why this should be so?) In fact, the preparation of LiAlH[O—C(CH$_3$)$_3$]$_3$ owes its success to the poor reactivity of its hydride: the reaction of LiAlH$_4$ with *tert*-butyl alcohol stops after three moles of alcohol have been consumed.

$$\text{Li}^+ \ ^-\text{AlH}_4 + 3(\text{CH}_3)_3\text{C—O—H} \longrightarrow \text{Li}^+ \ \text{H—}\overline{\text{Al}}[\text{O—C(CH}_3)_3]_3 + 3\text{H}_2 \qquad (21.58)$$

The one remaining hydride reduces only the most reactive functional groups. Because *acid chlorides are more reactive than aldehydes toward nucleophiles*, the reagent reacts preferentially with the acid chloride reactant rather than with the product aldehyde. In contrast, lithium aluminum hydride is so reactive that it fails to discriminate to a useful degree between the aldehyde and acid chloride groups, and reduces acid chlorides to primary alcohols.

The reduction of acid chlorides adds another synthesis of aldehydes and ketones to those given in Sec. 19.4. A complete list of methods for preparing aldehydes and ketones is given in Appendix IV.

....

PROBLEMS

21.19   Show how benzoyl chloride can be converted into each of the following compounds.

*(a)   benzyl alcohol   (b) benzaldehyde

*(c)

    PhCH$_2$—N⟨   ⟩

(d) PhCH$_2$—NH$_2$

***21.20**   Complete the following reactions by giving the principal organic product(s).

(a)

$$\text{C}_2\text{H}_5\text{O}-\overset{\overset{\textstyle O}{\|}}{\text{C}}-\text{CH}_2-\text{CN} \quad \xrightarrow[\text{(excess)}]{\text{LiAlH}_4} \quad \xrightarrow{\text{H}_2\text{O}}$$

(b) PhCH$_2$C≡N + H$_2$ $\xrightarrow[\text{heat}]{\text{Raney Ni}}$

(c)

$$\underset{\text{Ph—CH—CO}_2\text{C}_2\text{H}_5}{\overset{\overset{\textstyle O}{\|}}{\underset{|}{\text{O—C—CH}_3}}} + \text{LiAlH}_4 \text{ (excess)} \longrightarrow \xrightarrow{\text{H}_3\text{O}^+}$$

21.21   Give the structures of two compounds that would give the following product after LiAlH$_4$ reduction.

$$(\text{CH}_3)_2\text{CHCH}_2\text{CH}_2\text{CH}_2\text{NH}_2$$

*21.22   (a)   In the catalytic hydrogenation of some nitriles to primary amines, secondary amines are obtained as by-products:

$$R—C\equiv N \xrightarrow{\text{H}_2 \text{ (catalyst)}} RCH_2NH_2 \ + \ (RCH_2)_2NH$$

secondary amine

Suggest a mechanism for the formation of this by-product. (*Hint:* What is the intermediate in the reduction? How can this intermediate react with an amine?)

(b)   Explain why ammonia added to the reaction mixture prevents the formation of this by-product.

E. Relative Reactivities of Carbonyl Compounds

Recall that the reaction of lithium aluminum hydride with a carboxylic acid (Sec. 20.10) or ester (Sec. 21.9A) involves an aldehyde intermediate. But the product of such a reaction is a primary alcohol, not an aldehyde, because *the aldehyde intermediate is more reactive than the acid or ester.* The instant a small amount of aldehyde is formed, it is in competition with the remaining acid or ester for the LiAlH$_4$ reagent. Because it is more reactive, the aldehyde reacts faster than the remaining ester does. Hence, the aldehyde cannot be isolated under such circumstances. On the other hand, the lithium tri(*tert*-butoxy)aluminum hydride reduction of acid chlorides can be stopped at the aldehyde because acid chlorides are more reactive than aldehydes. When the aldehyde is formed as a product, it is in competition with the remaining acid chloride for the hydride reagent. Because the acid chloride is more reactive, it is consumed before the aldehyde has a chance to react.

These examples show that the outcome of many reactions of carboxylic acid derivatives is determined by the *relative reactivity of carbonyl compounds* toward nucleophilic reagents, which can be summarized as follows. (Nitriles are included as "honorary carbonyl compounds.")

Relative reactivities toward nucleophiles:

acid chlorides > aldehydes > ketones >> esters, acids > amides > nitriles   (21.59)

The explanation of this reactivity order is the same one used in Sec. 21.7E. Relative reactivity is determined by the stability of each type of carbonyl compound relative to its transition state for addition or substitution. *The more a compound is stabilized, the less reactive it is; the more a transition state for nucleophilic addition or substitution is stabilized, the more reactive the compound is* (Figs. 21.5 and 21.6). For example, esters are stabilized by resonance (Eq. 21.27) in a way that aldehydes and ketones are not. Hence, esters are less reactive than aldehydes. In contrast, resonance stabilization of acid chlorides is much less important, and acid chlorides are destabilized by the electron-attracting polar effect of the chlorine. Moreover, the transition-state energies for nucleophilic substitution reactions of acid chlorides are lowered by favorable leaving-group properties of chlorine. For these reasons, acid chlorides are more reactive than aldehydes, in which these effects of the chlorine are absent.

21.10

Reactions of Carboxylic Acid Derivatives with Organometallic Reagents

A. Reaction of Esters with Grignard Reagents

Most carboxylic acid derivatives react with Grignard or organolithium reagents. One of the most important reactions of this type is the reaction of esters with Grignard reagents. In this reaction, a tertiary alcohol is formed after protonolysis. (Secondary alcohols are formed from esters of formic acid; see Problem 21.24.)

$$(CH_3)_2CH-\overset{\overset{\displaystyle O}{\|}}{C}-OC_2H_5 \ + \ 2CH_3MgI \ \xrightarrow{\text{ether}} \ \xrightarrow{H_3O^+}$$

ethyl 2-methylpropanoate

$$(CH_3)_2CH-\underset{\underset{\displaystyle CH_3}{|}}{\overset{\overset{\displaystyle OH}{|}}{C}}-CH_3 \ + \ C_2H_5OH \ + \ Mg^{2+} \text{ salts} \qquad (21.60)$$

2,3-dimethyl-2-butanol
(92% yield)

$$\overset{\overset{\displaystyle O}{\|}}{C}-OCH_3 \ + \ 2\,Li-(CH_2)_3CH_3 \ \xrightarrow{\text{THF}} \ \xrightarrow{H_3O^+} \ \underset{\underset{\displaystyle (CH_2)_3CH_3}{|}}{\overset{\overset{\displaystyle OH}{|}}{C}}-(CH_2)_3CH_3 \ + \ CH_3OH \ + \ Li^+$$

methyl
2-methylpropenoate   **3-butyl-2-methyl-1-hepten-3-ol**   (21.61)

Notice that two equivalents of organometallic reagent react per mole of ester. Notice also that a second alcohol is produced in the reaction (ethanol and methanol in Eqs. 21.60 and 21.61, respectively). Recall that a similar situation occurs in the LiAlH₄ reduction of esters (Sec. 21.9A). This alcohol is generally not the one of interest and is discarded as a by-product.

Like the LiAlH₄ reduction of esters, this reaction is a nucleophilic acyl substitution followed by an addition. A ketone is formed in the substitution step. (Fill in the details of the mechanism.)

$$R-\overset{\overset{\displaystyle O}{\|}}{C}-OC_2H_5 \ + \ CH_3-MgI \ \longrightarrow \ R-\overset{\overset{\displaystyle O}{\|}}{C}-CH_3 \ + \ I-Mg-OC_2H_5 \qquad (21.62a)$$

The ketone intermediate is not isolated because *ketones are more reactive than esters toward nucleophilic reagents* (Eq. 21.59). The ketone therefore reacts with a second equivalent of the Grignard reagent to form a magnesium alkoxide, which, after protonolysis, gives the alcohol (Sec. 19.9).

$$R-\overset{\overset{\displaystyle O}{\|}}{C}-CH_3 \ + \ CH_3-MgI \ \xrightarrow{[\text{Sec. 19.9}]} \ \xrightarrow[H_2O]{H_3O^+} \ R-\underset{\underset{\displaystyle CH_3}{|}}{\overset{\overset{\displaystyle OH}{|}}{C}}-CH_3 \qquad (21.62b)$$

From what ester and Grignard reagent could 3-methyl-3-pentanol be prepared by a single reaction, followed by protonolysis?

Solution The problem can be rephrased in terms of structures:

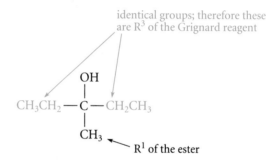

In other words, what choices should be made for R^1, R^2, and R^3? There are two keys to solving this problem. First, the carbonyl carbon of the ester starting material becomes the α-carbon of the target alcohol; hence, R^1 must therefore be attached to this carbon. Second, the two *identical* groups on the α-carbon of the target alcohol *must* correspond to group R^3 of the Grignard reagent.

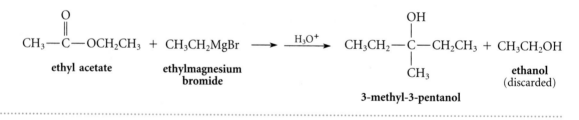

These deductions follow from the examples in the text or from the mechanism of the reaction. What about the group R^2 in the ester? It doesn't matter, because $-OR^2$ is the leaving group that becomes the by-product alcohol which is discarded. Since methyl or ethyl esters are common and relatively inexpensive, R^2 = methyl or ethyl is a good choice. Hence, the reaction required to prepare the desired alcohol is

As Study Problem 21.2 demonstrates, the reaction of a Grignard reagent with an ester is an important way to prepare alcohols in which at least two of the groups on the α-carbon are identical. (A complete list of methods for preparing alcohols is found in Appendix IV.)

B. Reaction of Acid Chlorides with Lithium Dialkylcuprates

Because acid chlorides are more reactive than ketones, the reaction of an acid chloride with a Grignard reagent can in principle be stopped at the ketone.

$$\underset{}{R-\overset{\overset{\displaystyle O}{\|}}{C}-Cl} + R'-MgBr \longrightarrow R-\overset{\overset{\displaystyle O}{\|}}{C}-R' + Cl-Mg-Br \qquad (21.63)$$

However, Grignard reagents are so reactive that this transformation is difficult to achieve in practice without careful control of the reaction conditions; that is, it is hard to prevent the reaction of the product ketone with the Grignard reagent to give an alcohol.

Another type of organometallic reagent, called a *lithium dialkylcuprate*, can be used to effect this transformation cleanly. **Lithium dialkylcuprate** reagents are prepared by the reaction of two equivalents of an organolithium reagent with one equivalent of cuprous chloride, CuCl. The first equivalent forms an alkylcopper compound:

$$R-Li + Cu-Cl \longrightarrow R-Cu + LiCl \qquad (21.64a)$$

The driving force for this reaction is the preference of lithium, the more electropositive metal, to exist as an ionic compound (LiCl). Because the copper of an alkylcopper reagent is a Lewis acid, it reacts accordingly with "alkyl anion" from a second equivalent of the organolithium reagent.

$$Li^+ \; R\overset{\frown}{\colon} \quad Cu-R \longrightarrow R-\overset{-}{C}u-R \;\; Li^+ \qquad (21.64b)$$
$$\text{lithium dialkylcuprate}$$

The product of this reaction is the lithium dialkylcuprate. Although the copper bears a negative charge, it is electropositive relative to carbon, and the reagent can be conceptualized as an alkyl anion complexed to copper:

$$R-Cu\text{-}\text{-}\text{-}\text{-}\overline{\colon}R \;\; Li^+$$

Lithium dialkylcuprates react in some ways like Grignard or lithium reagents; however, because the "alkyl anion" is complexed by copper, a less electropositive element than lithium, the alkyl-copper bond has more covalent character, and these reagents are less reactive. They typically react with acid chlorides and aldehydes, very slowly with ketones, and not at all with esters. The reaction of lithium dialkylcuprates with acid chlorides gives ketones in excellent yield.

$$\underset{\substack{\text{hexanoyl} \\ \text{chloride}}}{CH_3(CH_2)_4\overset{\overset{\displaystyle O}{\|}}{C}-Cl} + \underset{\substack{\text{lithium} \\ \text{dimethylcuprate}}}{(CH_3)_2Cu^-\,Li^+} \xrightarrow[\text{THF}]{\substack{-78° \\ 15\,\text{min}}} \xrightarrow{H_2O} \underset{\substack{\text{2-heptanone} \\ (81\%\ \text{yield})}}{CH_3(CH_2)_4\overset{\overset{\displaystyle O}{\|}}{C}-CH_3} + \underset{\substack{\text{(reacts} \\ \text{with } H_2O)}}{CH_3-Cu} + Li^+\,Cl^- \qquad (21.65)$$

Because ketones are much less reactive than acid chlorides toward lithium dialkylcuprates, they do not react further.

The reaction of acid chlorides with lithium dialkylcuprates is one of several excellent methods for the preparation of ketones. Be sure to review the others in Appendix IV.

Both this reaction and the reaction of Grignard reagents with esters also provide additional methods for the formation of carbon-carbon bonds. You should also review the other reactions used for this purpose, found in Appendix V.

PROBLEMS

21.23   Suggest a sequence of reactions for carrying out each of the following conversions.
   *(a)   benzoic acid to triphenylmethanol
   (b)   butyric acid to 3-methyl-3-hexanol
   *(c)   isobutyronitrile to 2,3-dimethyl-2-butanol (2 ways)
   (d)   propionic acid to 3-pentanone

21.24   *(a)   What is the general structure of the alcohols obtained by the reaction of Grignard reagents of the form R—MgBr with ethyl formate, followed by protonolysis?
   (b)   Outline a synthesis of 3-pentanol from a Grignard reagent and ethyl formate.

21.25   *(a)   What is the general structure of the alcohols obtained by the reaction of Grignard reagents of the form R—MgBr with diethyl carbonate (Sec. 21.1G), followed by protonolysis?
   (b)   Outline a synthesis of 3-isopropyl-2,4-dimethyl-3-pentanol from diethyl carbonate and a Grignard reagent.

21.26   Predict the product when each of the following compounds reacts with one equivalent of lithium dimethylcuprate, followed by protonolysis. Explain.

   *(a)   (b)
$$N{\equiv}C(CH_2)_{10}{-}\overset{\displaystyle O}{\overset{\|}{C}}{-}Cl \qquad CH_3(CH_2)_3O{-}\overset{\displaystyle O}{\overset{\|}{C}}(CH_2)_4\overset{\displaystyle O}{\overset{\|}{C}}{-}Cl$$

21.11   ## Synthesis of Carboxylic Acid Derivatives

Reactions in this and the previous chapter demonstrate that many syntheses of carboxylic acid derivatives begin with other carboxylic acid derivatives. Let's review the methods that have been covered:

Synthesis of esters:

   1. Acid-catalyzed esterification of carboxylic acids (Sec. 20.8A)
   2. Alkylation of carboxylic acids or carboxylates (Sec. 20.8B)
   3. Reaction of acid chlorides and anhydrides with alcohols or phenols (Sec. 21.8A)
   4. Transesterification of other esters (Sec. 21.8C)

Synthesis of acid chlorides:

   1. Reaction of carboxylic acids with $SOCl_3$ or PCl_5 (Sec. 20.9A)

Synthesis of anhydrides:

1. Reaction of carboxylic acids with dehydrating agents (Sec. 20.9B)
2. Reaction of acid chlorides with carboxylate salts (Sec. 21.8A)

Synthesis of amides:

1. Reaction of acid chlorides, anhydrides, or esters with amines (Sec. 21.8A, C)

Synthesis of nitriles:

The synthesis of nitriles is an important exception to the generalization that carboxylic acid derivatives are usually prepared from other carboxylic acid derivatives. Two syntheses of nitriles are:

1. Cyanohydrin formation (Sec. 19.7A)
2. S_N2 reaction of cyanide ion with alkyl halides or sulfonate esters

An S_N2 reaction of cyanide ion, like all S_N2 reactions, requires a primary or unbranched secondary alkyl halide or sulfonate ester.

$$PhCH_2Cl \;+\; Na^+ \; ^-{:}CN \xrightarrow{\text{EtOH/H}_2\text{O}} PhCH_2CN \;+\; Na^+ \, Cl^- \qquad (21.66)$$

benzyl chloride α**-phenylacetonitrile**
                                        (80–90% yield)

$$Br(CH_2)_3Br \;+\; 2Na^+ \, ^-CN \xrightarrow{\text{EtOH/H}_2\text{O}} NC(CH_2)_3CN + 2Na^+ \, Br^- \qquad (21.67)$$

1,3-dibromopropane   **glutaronitrile**
                                        (77–86% yield)

Both this reaction and the synthesis of cyanohydrins are noteworthy because they provide additional ways to form carbon-carbon bonds. You should see how many reactions you can list that form carbon-carbon bonds; check yourself against the summary in Appendix V.

Because carboxylic acid derivatives can be hydrolyzed to carboxylic acids, any synthesis of a carboxylic acid derivative can be used as part of the synthesis of a carboxylic acid itself. Because nitriles are prepared from compounds other than carboxylic acid derivatives, the preparation of a nitrile can be particularly useful as an intermediate step in the preparation of a carboxylic acid. The following study problem illustrates this approach.

STUDY PROBLEM 21.3

Outline a synthesis of pentanoic acid (valeric acid) from 1-butanol.

$$CH_3CH_2CH_2CH_2\!-\!OH \xrightarrow{\;?\;} CH_3CH_2CH_2CH_2\!-\!\overset{\displaystyle O}{\overset{\displaystyle \|}{C}}\!-\!OH$$

1-butanol   **pentanoic acid**

Solution   Notice that a new carbon-carbon bond must be formed at some point in this synthesis. Because nitriles can be hydrolyzed to carboxylic acids, the immediate precursor of the carboxylic acid can be the corresponding nitrile.

$$CH_3CH_2CH_2CH_2-C\equiv N \xrightarrow[\text{heat}]{H_2O, H_3O^+} CH_3CH_2CH_2CH_2-\overset{\displaystyle O}{\overset{\displaystyle \|}{C}}-OH$$

The nitrile, in turn, can be prepared from the corresponding alkyl halide:

$$CH_3CH_2CH_2CH_2-Br \xrightarrow{^-CN} CH_3CH_2CH_2CH_2-C\equiv N$$

Notice the formation of the new carbon-carbon bond at this point. The alkyl halide is formed from the alcohol by any of the methods summarized in Sec. 10.4, for example, by reaction with concentrated HBr:

$$CH_3CH_2CH_2CH_2-OH \xrightarrow{HBr} CH_3CH_2CH_2CH_2-Br$$

Note that this alkyl halide could also be converted into the target carboxylic acid by the method discussed in Sec. 20.6.

PROBLEMS

*21.27   Outline two methods for the preparation of 5-methylhexanoic acid from 1-bromo-4-methylpentane. (See Sec. 20.6.)

21.28   Outline a preparation of pentanoic acid from 1-butanol different from the one shown in Study Problem 21.3.

*21.29   (a) From what nitrile can 2-hydroxypropanoic acid (lactic acid) be prepared?
        (b) How can this nitrile be prepared from acetaldehyde? (*Hint:* see Sec. 19.7A.)

21.12   Use and Occurrence of Carboxylic Acids and Their Derivatives

A. Nylon and Polyesters

Two of the most important polymers produced on an industrial scale are *nylon* and *polyesters*. The chemistry of carboxylic acids and their derivatives plays an important role in the synthesis of these polymers.

Nylon is the general name given to a group of polymeric amides. The two most widely used are nylon-6,6 and nylon-6.

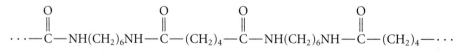

nylon-6,6

$$\text{or} \quad \left[NH(CH_2)_6NH\overset{\displaystyle O}{\overset{\displaystyle \|}{C}}(CH_2)_4\overset{\displaystyle O}{\overset{\displaystyle \|}{C}} \right]_n$$

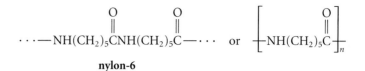

nylon-6

About 2.6 billion pounds of nylon are produced annually in the United States. Nylon is used in tire cord, carpet, and apparel.

The starting material for the industrial synthesis of nylon-6,6 is adipic acid. In one process, adipic acid is converted into its dinitrile and then into 1,6-hexanediamine (hexamethylenediamine).

$$HO_2C—(CH_2)_4—CO_2H \xrightarrow{\text{several steps}} N≡C—(CH_2)_4—C≡N \xrightarrow[\text{cat}]{H_2}$$

$$H_2N—(CH_2)_6—NH_2 \qquad (21.68)$$

hexamethylenediamine

When hexamethylenediamine and adipic acid are mixed, they form a salt. Heating the salt forms the polymeric amide.

$$\underset{\text{salt}}{\overset{O}{\underset{\|}{-C-O^- \overset{+}{H_3}N-}}} \rightleftharpoons \underset{\text{acid}}{\overset{O}{\underset{\|}{-C-OH}}} + \underset{\text{amine}}{H_2N-} \xrightarrow{-H_2O} \underset{\substack{\text{nylon} \\ \text{(an amide)}}}{\overset{O}{\underset{\|}{-C-NH-}}} \qquad (21.69)$$

The reaction of an amine with an acid to form an amide is analogous to the reaction of an amine with an ester (Sec. 21.8C). However, much more vigorous conditions are required because the equilibrium on the left of Eq. 21.69 strongly favors the salt. In the salt, the amine is protonated and therefore not nucleophilic, and the carboxylate ion is very unreactive toward nucleophiles (why?). The small amount of amine and carboxylic acid in equilibrium with the salt react when the salt is heated, pulling the equilibrium to the right.

Nylon is an example of a **condensation polymer**, a polymer formed in a reaction that liberates a small molecule, in this case H_2O. This is in contrast to an *addition polymer*, which is formed without the liberation of a small molecule (compare Sec. 5.7A).

Adipic acid, from which nylon is made, is prepared industrially in several steps from petroleum. This is a classic example of the dependence of an important segment of the chemical economy on petroleum feedstocks. About 1.8 billion pounds of adipic acid are synthesized annually in the United States.

Polyesters are condensation polymers derived from the reaction of diols and dicarboxylic acids. One widely used polyester, poly(ethylene terephthalate), can be produced by the esterification of ethylene glycol and terephthalic acid.

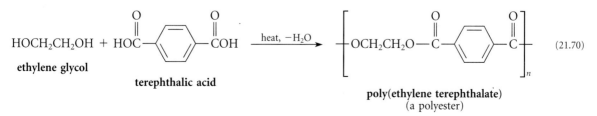

Certain familiar polyester fibers and films are sold under the trade names Dacron® and Mylar®, respectively. About 3.6 billion pounds of polyester are produced annually in the United States. Polyester production also depends on raw materials produced from petroleum.

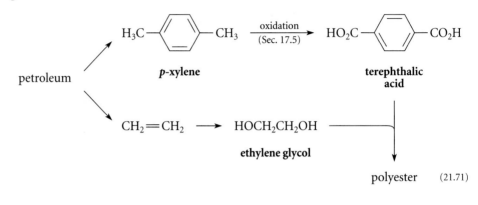

polyester   (21.71)

21.30 *(a) What would happen to the polyester in Eq. 21.70 if it were allowed to react with LiAlH₄, followed by dilute acid?

(b) Which polymer should be more resistant to strong base: nylon-6,6 or the polyester in Eq. 21.70? Explain.

*21.31   One interesting process for making nylon-6,6 demonstrates the potential of using biomass as an industrial starting material. The raw material for this process, outlined below, is the aldehyde furfural, obtained from sugars found in oat hulls. Suggest conditions for carrying out each of the steps in this process indicated by italicized letters.

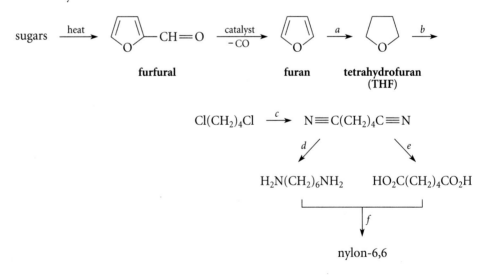

nylon-6,6

*21.32   ε-Caprolactam is polymerized to nylon-6 when it is heated with a *catalytic amount* of water.

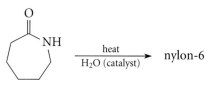

ϵ-caprolactam

Give a mechanism for the polymerization, showing clearly the role of the water.

B. Waxes, Fats, and Phospholipids

Waxes, fats, and phospholipids are all important naturally occurring ester derivatives of fatty acids. A **wax** is an ester of a fatty acid and a "fatty alcohol," an alcohol with a long unbranched carbon chain. For example, carnauba wax, obtained from the leaves of the Brazilian carnauba palm, and valued for its hard, brittle characteristics, consists of about 80% of esters derived from C_{24}, C_{26}, and C_{28} fatty acids and C_{30}, C_{32}, and C_{34} alcohols. The following compound is a typical constituent of carnauba wax.

$$CH_3(CH_2)_{24}-\overset{\overset{O}{\|}}{C}-O-(CH_2)_{31}CH_3 \quad \text{constituent of carnauba wax}$$

A **fat** is an ester derived from a molecule of glycerol and three molecules of fatty acid.

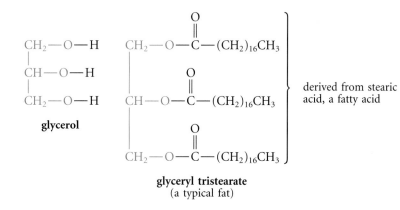

glyceryl tristearate
(a typical fat)

The three acyl groups in a fat may be the same, as in a glyceryl tristearate, or different, and they may contain unsaturation, which is typically in the form of one or more *cis* double bonds. Fats with no double bonds, termed **saturated fats**, are typically solids; lard is an example of a saturated fat. Fats containing double bonds, termed **unsaturated fats**, are in many cases oily liquids; olive oil is an example of an unsaturated fat.

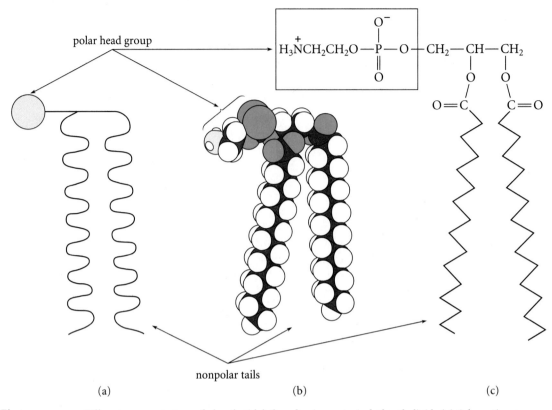

Figure 21.7 *Different representations of phosphatidylethanolamine, a typical phospholipid. (a) Schematic representation. (b) Space-filling model. (c) Lewis structure.*

Treatment of a fat with NaOH yields glycerol and the sodium salts of the fatty acids (*soaps*; Sec. 20.5). This is the reaction from which *saponification* (Sec. 21.7A) gets its name.

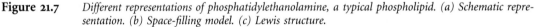

$$\begin{array}{l} CH_2 - O_2C(CH_2)_{16}CH_3 \\ | \\ CH - O_2C(CH_2)_{16}CH_3 \\ | \\ CH_2 - O_2C(CH_2)_{16}CH_3 \end{array} + 3NaOH \longrightarrow \begin{array}{l} CH_2OH \\ | \\ CHOH \\ | \\ CH_2OH \end{array} + 3CH_3(CH_2)_{16}CO_2^- \ Na^+ \qquad (21.72)$$

a soap

Fats, which are stored in highly concentrated form in the body, serve as the biological storehouse of energy reserves.

Phospholipids are closely related to fats, because they too are esters of glycerol. Phospholipids differ from fats in that one of the terminal oxygens of glycerol in a phospholipid is esterified to a special type of organic phosphate derivative, forming a polar head group in the molecule. The remaining two oxygens of glycerol are esterified to fatty acids, which serve as "nonpolar tails" (Fig. 21.7).

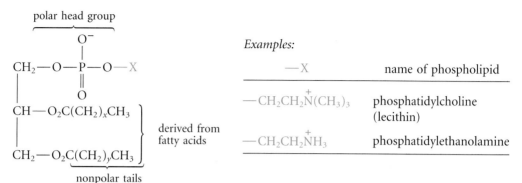

Examples:

—X	name of phospholipid
$-CH_2CH_2\overset{+}{N}(CH_3)_3$	phosphatidylcholine (lecithin)
$-CH_2CH_2\overset{+}{N}H_3$	phosphatidylethanolamine

Phospholipids closely resemble soaps (Sec. 20.5), because both types of molecules are **amphipathic**—that is, they have both polar and nonpolar ends.

Phospholipids, along with cholesterol (a steroid) and imbedded proteins, are major constituents of *cell membranes*, the envelopes that surround living cells. The phospholipids give membranes a *bilayer* structure in which the polar head groups interact with aqueous solution, and the nonpolar tails interact with each other, isolated from the aqueous solution (Fig. 21.8). This *phospholipid bilayer* forms a barrier that is impermeable to most ions and to many organic molecules, and allows the cell to control "what gets in" and "what gets out." (Look again at Sec. 20.5; can you see why detergents can disrupt cell membranes?)

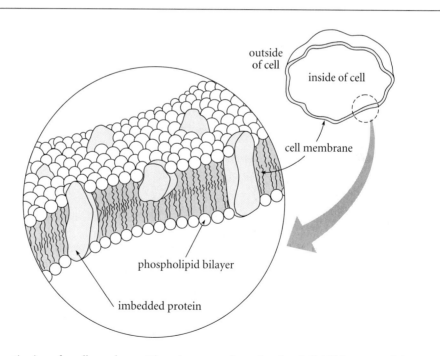

Figure 21.8   *Schematic view of a cell membrane. The enlargement shows the phospholipid bilayer containing imbedded proteins. In this view the phospholipids are represented schematically as shown in Fig. 21.7a, with the polar head groups as circles and the nonpolar tails as "squiggles." Note that the polar head groups are exposed to water, and the nonpolar tails form a hydrocarbonlike region on the interior of the membrane, isolated from the aqueous solvent.*

The most important occurrence of amides in nature is in *proteins*, which are polymers in which α-amino acid units are connected by amide linkages. Chapter 26 is devoted to a discussion of amino acids and proteins.

KEY IDEAS IN CHAPTER 21

Carboxylic acid derivatives are polar molecules; except for amides, they have low water solubilities and low-to-moderate boiling points. Because of their capacity for hydrogen bonding, amides have high boiling points (many are solids) and moderate solubilities in water.

The carbonyl absorptions (and the $C\equiv N$ absorption of nitriles) are the most important absorptions in the infrared spectra of carboxylic acid derivatives (Table 21.3). Proton NMR spectra show typical δ 2–3 absorptions for the α-protons, as well as other characteristic absorptions: δ 3.5–4.5 for the *O*-alkyl groups of esters, δ 2.6–3 for the *N*-alkyl groups of amides and δ 7.5–8.5 for the N—H protons of amides.

In CMR spectra, the chemical shifts of carbonyl carbons are in the δ 170 region, and those of nitrile carbons are in the δ 115–120 range.

Carboxylic acid derivatives, like carboxylic acids themselves, are weak bases that can be protonated by strong acids on the carbonyl carbon or the nitrile nitrogen.

The most characteristic reaction of carboxylic acid derivatives is nucleophilic acyl substitution. Hydrolysis reactions and the reactions of acid chlorides, anhydrides, and esters with nucleophiles are examples of this type of reactivity. Nucleophilic acyl substitution typically involves addition of a nucleophile at the carbonyl carbon to form a tetrahedral addition intermediate, which then loses a leaving group to form product.

In some reactions of carboxylic acid derivatives, nucleophilic acyl substitution is followed by addition. This is the case, for example, in the LiAlH$_4$ reductions and Grignard reactions of esters, in which the aldehyde and ketone intermediates, respectively, react further.

The relative reactivities of carbonyl compounds and nitriles with nucleophiles are in the order: acid chlorides > aldehydes > ketones >> esters > amides > nitriles.

Nitriles react by addition, much like aldehydes and ketones. In some reactions, the product of addition undergoes a second addition (as in reduction to primary amines); and in other cases it undergoes a substitution reaction (as in hydrolysis).

Polyesters and polyamides (for example, nylon) are important commercial examples of carboxylic acid derivatives. In nature, waxes, fats, and phospholipids are important natural examples of esters. Phospholipids spontaneously form bilayers in water; such bilayers are a major constituent of cell membranes.

ADDITIONAL PROBLEMS

*21.33   Give the principal organic product(s) expected when ethyl benzoate or other compound indicated reacts with each of the following reagents.
(a) H_2O, heat, acid catalyst   (b) NaOH, H_2O
(c) aqueous NH_3, heat   (d) $LiAlH_4$, then H_2O
(e) excess $CH_3CH_2CH_2MgBr$, then H_2O
(f) product of (e) + acetyl chloride, pyridine, 0°
(g) product of (e) + benzenesulfonyl chloride
(h) NaOEt, EtOH

21.34   Give the principal organic product(s) expected when propionyl chloride reacts with each of the following reagents.
(a) H_2O (b) ethanethiol, pyridine, 0°
(c) $(CH_3)_3COH$, pyridine   (d) $(CH_3)_2CuLi$, −78°, then H_2O
(e) H_2, Pd catalyst (quinoline/S poison)
(f) $AlCl_3$, toluene, then H_2O (g) $(CH_3)_2CHNH_2$ (2 equiv.)
(h) sodium benzoate   (i) p-cresol, pyridine

21.35   Give the structure of a compound that satisfies each of the following criteria.
*(a) A compound C_3H_5N that liberates ammonia on treatment with hot aqueous KOH
(b) A compound C_3H_7ON that liberates ammonia on treatment with hot aqueous KOH
*(c) A compound that gives 1-butanol and 2-methyl-2-propanol on treatment with excess CH_3MgBr followed by protonolysis
(d) A compound that gives equal amounts of 1-hexanol and 2-hexanol on treatment with $LiAlH_4$ followed by protonolysis
*(e) A compound $C_3H_2N_2$ that gives CO_2, $^+NH_4$, and acetic acid after boiling in concentrated aqueous HCl

*21.36   Complete the following diagram by filling in all missing reagents or intermediates.

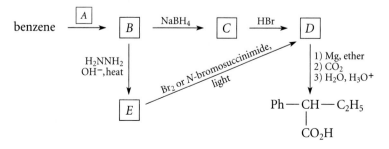

*21.37   When (*R*)-(−)-mandelic acid (*α*-hydroxy-*α*-phenylacetic acid) is treated with CH₃OH and H₂SO₄, and the resulting compound is treated with excess LiAlH₄ in ether, then H₂O, a levorotatory product is obtained that reacts with periodic acid. Give the structure, name, and absolute configuration of this product.

21.38   Contrast the results to be expected when levulinic acid (4-oxopentanoic acid) is treated with (a) excess LiAlH₄, then H₃O⁺ and (b) excess NaBH₄ in methanol, then H₃O⁺.

*21.39   In clinical studies with atherosclerotic patients it was found that one of the metabolites of the hyperlipidemia drug (*Z*)-3-methyl-4-phenyl-3-butenamide (*A*, below) is a compound *B* which has the formula $C_{11}H_{15}NO_2$. When compound *B* is heated in aqueous acid, lactone *C* is formed along with the ammonium ion (NH₄⁺).

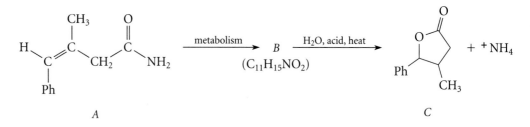

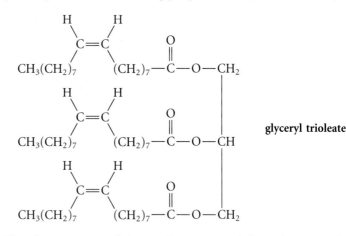

A   *C*

(a) Propose a structure for compound *B*, and explain your reasoning.
(b) Give a curved-arrow mechanism for the conversion of *B* into *C*.

21.40   A major component of olive oil is glyceryl trioleate (structure below).

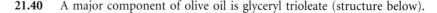

glyceryl trioleate

(a) Give the structures of the products expected from the saponification of glyceryl trioleate with excess aqueous NaOH.
(b) Give the structure of the saturated fat glyceryl tristearate, produced by the hydrogenation of glyceryl trioleate.

21.41   Propose a synthesis for each of the following compounds from the indicated starting materials and any other reagents.

*(a) Compound *A* from 2-bromobenzoic acid.

(b) Compound *B*, the active ingredient in some insect repellants, from 3-methylbenzaldehyde.

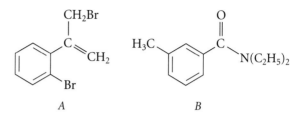

A   *B*

21.42 Propose a synthesis for each of the following compounds from butyric acid and any other reagents.

  *(a) 4-methyl-4-heptanol   (b) 2-methyl-2-pentanol
  *(c) 4-heptanol   *(d) $CH_3CH_2CH_2CH_2CH_2CH_2NH_2$
  *(e) $CH_3CH_2CH_2CH_2CH_2NH_2$ (f) $CH_3CH_2CH_2CH_2NH_2$

***21.43** (a) Draw the structures of all the products that would be obtained when $(\pm)$-α-phenylglutaric anhydride is treated with $(\pm)$-1-phenylethanol. Which products should be separable without optical resolution? Which products should be obtained in equal amounts? In different amounts?

(b) Answer the same question for the reaction of the *S* enantiomer of the same alcohol with $(\pm)$-α-phenylglutaric anhydride.

21.44 Treatment of acetic propionic anhydride with ethanol gives a mixture of two esters consisting of 36% of the higher boiling one, *A*, and 64% of the lower boiling one, *B*. Identify *A* and *B*.

***21.45** Predict the relative reactivity of thiol esters (see structure below) toward hydrolysis with aqueous NaOH.

(1) Much more reactive than acid chlorides

(2) Less reactive than acid chlorides, but more reactive than esters

(3) Less reactive than esters

Explain your choice.

$$\underset{\text{general thiol ester structure}}{R\!-\!\overset{\displaystyle O}{\overset{\|}{C}}\!-\!SR'}$$

21.46 Explain why carboxylate salts are much less reactive than esters in nucleophilic acyl substitution reactions.

21.47 You are employed by Fibers Unlimited, a company specializing in the manufacture of specialty polymers. The vice-president for research, Strong Fishlein, has asked you to design laboratory preparations of the following polymers. You are to use as starting materials the company's extensive stock of dicarboxylic acids containing six or fewer carbon atoms. Accommodate him.

(*Problem 21.47 continues*)

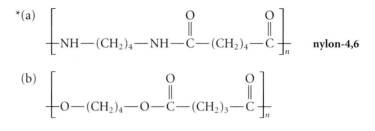

*(a)   nylon-4,6

(b)

*21.48   A compound *A* has prominent infrared absorptions at 1050, 1786, and 1852 cm^{-1}, and shows a single absorption in the proton NMR at δ 3.00. When heated gently with methanol, compound *B*, $C_5H_8O_4$, is obtained. Compound *B* has IR absorption at 2500–3000 (broad), 1730, and 1701 cm^{-1}, and its proton NMR spectrum in D_2O consists of two singlets at δ 2.7 and δ 3.7 in the intensity ratio 4:3. Identify *A* and *B*, and explain your reasoning.

21.49   Propose a structure for a compound *A* which has an infrared absorption at 1820 cm^{-1} and a single proton NMR absorption at δ 1.5. Compound *A* reacts with water to give α,α-dimethylmalonic acid and with methanol to give the monomethyl ester of the same acid. (Compound *A* was unknown until its preparation in 1978 by chemists at the University of California, San Diego.)

21.50   Klutz McFingers, a graduate student in his ninth year of study, has suggested the following synthetic procedures, and has come to you in the hope that you can explain why none of them works very well (or not at all).

*(a)   Noting the fact that primary alcohols + HBr give alkyl halides, Klutz has proposed by analogy the following nitrile synthesis:

$$Ph—CH_2—OH + HCN \xrightarrow{H_2O} Ph—CH_2—CN + H_2O$$

(*Hint:* HCN is a weak acid, pK_a = 9.4.)

*(b)   Klutz has proposed the following synthesis for the half-ester of adipic acid:

$$HO_2C—(CH_2)_4—CO_2H + CH_3OH \xrightarrow{H_2SO_4}$$
$$\text{(1.0 mole)} \qquad \text{(1.0 mole)}$$
$$HO_2C—(CH_2)_4—CO_2CH_3 + H_2O$$

(c)   Klutz has proposed the following synthesis of acetic benzoic anhydride:

$$CH_3—\overset{O}{\overset{||}{C}}—OH + Ph—\overset{O}{\overset{||}{C}}—OH \xrightarrow{P_2O_5} CH_3—\overset{O}{\overset{||}{C}}—O—\overset{O}{\overset{||}{C}}—Ph$$

*(d)   Noting correctly that methyl benzoate is completely saponified by one molar equivalent of NaOH, Klutz has suggested that methyl salicylate should also undergo saponification with one equivalent of NaOH.

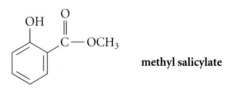

methyl salicylate

*(e) Klutz, finally able to secure a position with a pharmaceutical company working on β-lactam antibiotics, has proposed the following reaction for deamidation of a cephalosporin derivative:

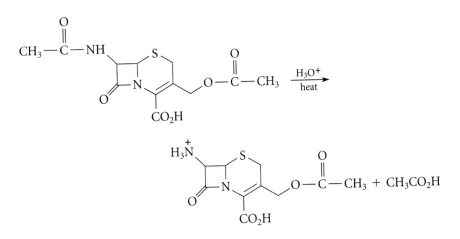

*21.51 You are a chemist working for the Imahot Pepper Company, and have been asked to provide some information about *capsaicin*, the active ingredient of hot peppers.

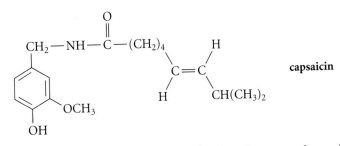

How should capsaicin or other compound indicated react under each of the following conditions?

(a) Br$_2$/CCl$_4$ (b) 5% aqueous NaOH
(c) 5% aqueous HCl   (d) H$_2$, catalyst
(e) product of (d) + 6 *M* HCl, heat   (f) product of (b) + CH$_3$I
(g) product of (d) + concentrated aqueous HBr

*21.52 Exactly 2.00 g of an ester *A* containing only C, H, and O was saponified with 15.00 mL of a 1.00 *M* NaOH solution. Following the saponification the solution required 5.30 mL of 1.00 *M* HCl to titrate the unused NaOH. Ester *A*, as well as its acid and alcohol saponification products *B* and *C*, respectively, were all optically active. Compound *A* was not oxidized by K$_2$Cr$_2$O$_7$, nor did compound *A* decolorize Br$_2$ in CCl$_4$. Alcohol *C* was oxidized to acetophenone by K$_2$Cr$_2$O$_7$. When acetophenone was reduced with NaBH$_4$, a compound *D* was formed that reacted with the acid chloride derived from *B* to give two optically active compounds: *A* (identical with starting ester) and *E*. Propose a structure for each compound that is consistent with the data. (Note that the absolute stereochemical configurations of chiral substances cannot be determined from the data.)

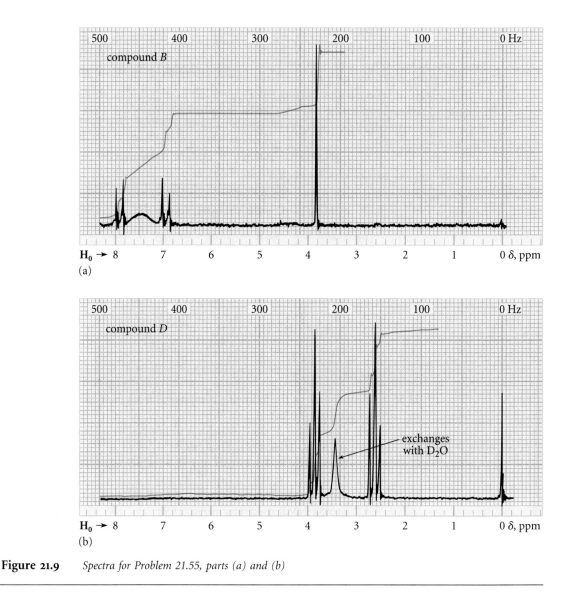

Figure 21.9   *Spectra for Problem 21.55, parts (a) and (b)*

21.56   Outline a synthesis of each of the following compounds from the indicated starting materials and any other reagents.

*(a)   *o*-methylbutyrophenone from *o*-bromotoluene

(b)   1-cyclohexyl-2-methyl-2-propanol from bromocyclohexane

*(c)    NH from phthalic acid

(d)   PhNHCH$_2$CH$_2$CH(CH$_3$)$_2$ from isovaleric acid

(*Problem 21.56 continues on p. 1036*)

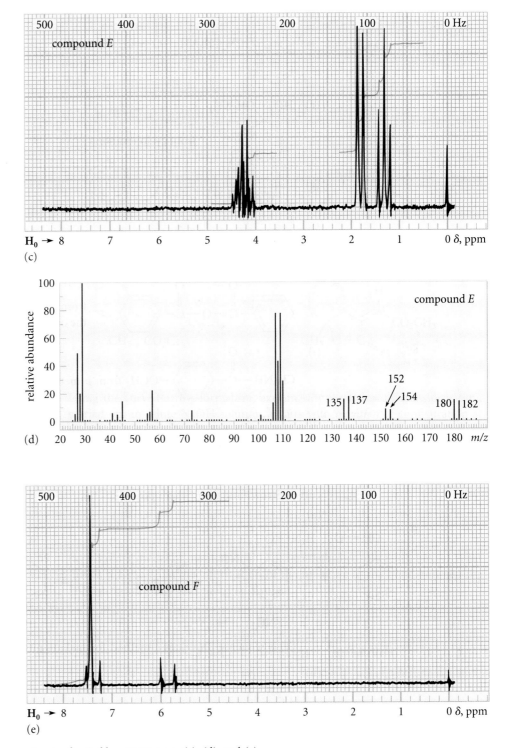

Figure 21.9 *Spectra for Problem 21.55, parts (c), (d), and (e)*

Similarly, carboxylic acid O—H hydrogens ($pK_a \approx 4$–5) are also α-hydrogens. We can think of amide conjugate-base anions as "nitrogen analogs" of enolate ions and carboxylate anions as "oxygen analogs" of enolate anions. Notice that the acidity order carboxylic acids > amides > (aldehydes, ketones) corresponds to the relative electronegativities of the atoms to which the acidic hydrogens are bound—oxygen, nitrogen, and carbon, respectively (*element effect*; Sec. 3.6A).

*22.1   Explain why diethyl malonate ($pK_a = 12.9$) and ethyl acetoacetate (ethyl 3-oxobutanoate, $pK_a = 10.7$) are much more acidic than ordinary esters. (To answer this question, you first have to identify the acidic hydrogen in each of these compounds.)

22.2   Which is more acidic: the diamide of succinic acid or the imide succinimide (Sec. 21.1E)? Why?

B. Introduction to Reactions of Enolate Ions

The acidities of aldehydes, ketones, and esters are particularly important because enolate ions are key reactive intermediates in many important reactions of carbonyl compounds. Let's consider the types of reactivity we can expect to observe with enolate ions.

First, enolate ions are Brønsted bases, and they react with Brønsted acids. (The reaction of an enolate ion with an acid is the reverse of Eq. 22.1 or 22.2.) The formation of enolate ions and their reactions with Brønsted acids have two simple but important consequences. First, the α-hydrogens of an aldehyde or ketone—and no others—can be exchanged for deuterium by treating the carbonyl compound with a base in D_2O.

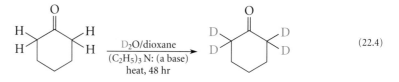

$$(22.4)$$

*22.3   Write a mechanism involving an enolate ion intermediate for the reaction shown in Eq. 22.4. Explain why *only* the α-hydrogens are replaced by deuterium.

22.4   Explain how the proton NMR spectrum of 2-butanone would change if the compound were treated with D_2O and a base.

The second consequence of enolate-ion formation and protonation is that, if an optically active aldehyde or ketone owes its chirality solely to an asymmetric α-carbon, and if this carbon bears a hydrogen, the compound will be racemized by base.

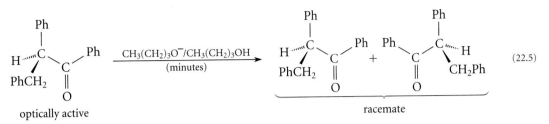

$$\text{(22.5)}$$

optically active   racemate

The reason racemization occurs is that the enolate ion, which forms in base, is *achiral* because of the sp^2 hybridization at its anionic carbon (Fig. 22.1). That is, the ionic α-carbon and its attached groups lie in one plane. The anion can be reprotonated at either face to give either enantiomer with equal probability. Although not very much enolate ion is present at any one time, the reactions involved in the ionization equilibrium are relatively fast, and racemization occurs relatively quickly if the carbonyl compound is left in contact with base.

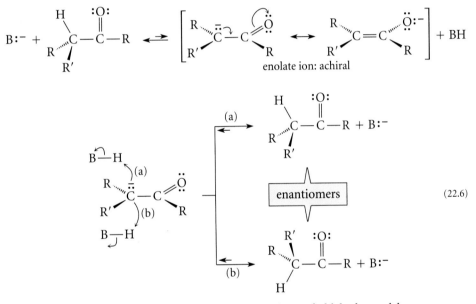

enolate ion: achiral

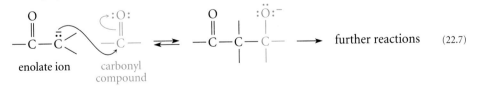

$$\text{(22.6)}$$

α-Hydrogen exchange and racemization reactions of aldehydes and ketones occur much more readily than those of esters, because aldehydes and ketones are more acidic, and therefore form enolate ions more rapidly and under milder conditions.

Enolate ions are not only Brønsted bases, but Lewis bases as well. Hence, enolate ions react as *nucleophiles*. Like other nucleophiles, enolate ions attack the carbons of carbonyl groups:

$$\text{(22.7)}$$

enolate ion   carbonyl
                   compound

This type of process is the first step of a variety of *carbonyl addition* reactions and *nucleophilic acyl substitution* reactions involving enolate ions as nucleophiles. Much of this chapter will be devoted to a study of such reactions.

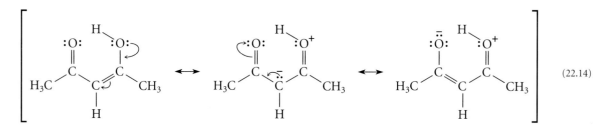

$$(22.14)$$

The second stabilizing effect is the intramolecular hydrogen bond present in each of these enols. This provides another source of increased bonding and hence, increased stabilization.

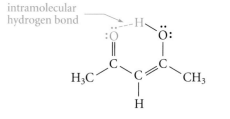

- -

PROBLEMS

22.7   Draw all enol tautomers of the following compounds. If there are none, explain why.
  *(a) 2-butanone   (b) 2-methylcyclohexanone
  *(c) 2-methylpentanoic acid   (d) benzaldehyde
  *(e) *N*,*N*-dimethylacetamide

22.8   Draw the "enol" tautomers of the following compounds.

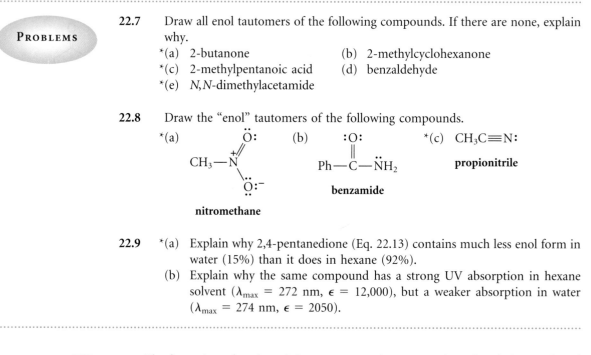

22.9   *(a) Explain why 2,4-pentanedione (Eq. 22.13) contains much less enol form in water (15%) than it does in hexane (92%).
    (b) Explain why the same compound has a strong UV absorption in hexane solvent (λ_{max} = 272 nm, ϵ = 12,000), but a weaker absorption in water (λ_{max} = 274 nm, ϵ = 2050).

- -

STUDY GUIDE LINK:
✓**22.2**
Kinetic vs.
Thermodynamic
Stability of Enols

The formation of enols and the reverse reaction, conversion of enols into carbonyl compounds, are catalyzed by both acids and bases. The rapid conversion of enols into carbonyl compounds accounts for the fact that enols are difficult to isolate and observe under ordinary circumstances.

Base-catalyzed enolization involves the intermediacy of an *enolate ion*, and is thus a consequence of the acidity of the α-hydrogen.

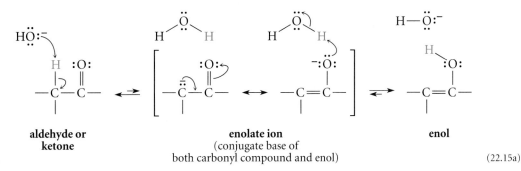

aldehyde or ketone   **enolate ion**
(conjugate base of
both carbonyl compound and enol)

enol

(22.15a)

Protonation of the enolate anion by water on the α-carbon gives back the carbonyl compound; protonation on oxygen gives the enol. Notice that the enolate ion is the conjugate base of not only the carbonyl compound, but also the enol—this is why it is called an *enolate ion*!

Acid-catalyzed enolization involves the conjugate acid of the carbonyl compound. Recall that this ion has carbocation characteristics (Sec. 19.6). Loss of the proton from oxygen gives back the starting carbonyl compound; loss of the proton from the α-carbon gives the enol. Notice that an enol and its carbonyl tautomer have the same conjugate acid.

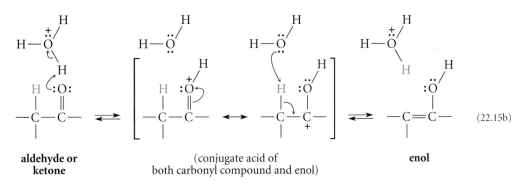

aldehyde or ketone   (conjugate acid of
both carbonyl compound and enol)

enol

(22.15b)

Exchange of α-hydrogens for deuterium as well as racemization at the α-carbon are catalyzed not only by bases (Sec. 22.1B) but also by acids.

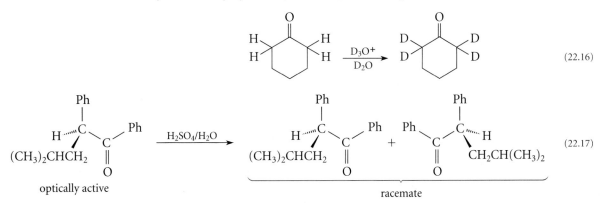

optically active   racemate

(22.16)

(22.17)

Both acid-catalyzed processes can be explained by the intermediacy of enols. As Eq. 22.15b shows, formation of a carbonyl compound from an enol introduces hydrogen from solvent at the α-carbon; this fact accounts for the observed isotope exchange. This carbon of an enol, like that of an enolate ion, is not asymmetric. The absence of chirality in the enol accounts for the racemization observed in acid.

PROBLEM

22.10 Using the curved-arrow formalism, give the mechanism for *(a) the racemization shown in Eq. 22.17; (b) the exchange shown in Eq. 22.16.

22.3 α-Halogenation of Carbonyl Compounds

This section begins a survey of reactions that involve enols and enolate ions as reactive intermediates.

A. Acid-Catalyzed α-Halogenation of Aldehydes and Ketones

Halogenation of an aldehyde or ketone in *acidic* solution usually results in the replacement of one α-hydrogen by halogen.

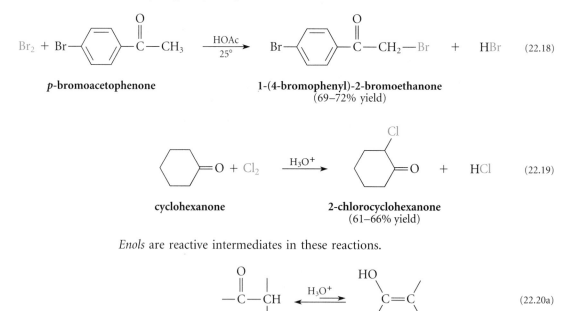

p-bromoacetophenone

1-(4-bromophenyl)-2-bromoethanone
(69–72% yield)

cyclohexanone

2-chlorocyclohexanone
(61–66% yield)

aldehyde or ketone

enol

Enols are reactive intermediates in these reactions.

Like other "alkenes," enols react with halogens; but unlike ordinary alkenes, enols add only one halogen atom. After addition of the first halogen to the double bond, the resulting carbocation intermediate loses a proton instead of adding the second halogen.

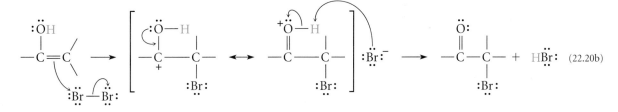

(22.20b)

Acid-catalyzed halogenation provides a particularly instructive case study that shows the importance of the rate law in determining the mechanism of a reaction. Under the usual reaction conditions, the rate law for acid-catalyzed halogenation is found to be

$$\text{rate} = k[\text{ketone}][H_3O^+] \tag{22.21}$$

This rate law implies that even though the reaction is a halogenation, *the rate is independent of the halogen concentration.* This rate law means that halogens *cannot* be involved in the transition state for the rate-limiting step of the reaction (Sec. 9.3B). From this observation and others, it was deduced that *enol formation* (Eq. 22.20a) *is the rate-limiting process in acid-catalyzed halogenation of aldehydes and ketones.* Since halogen is not involved in enol formation, it does not appear in the rate law.

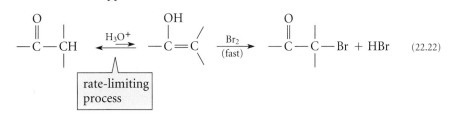

(22.22)

Because only one halogen can be introduced at a given α-carbon in acidic solution, it follows that introduction of a second halogen is much slower than introduction of the first. The reason is not known with certainty, but it may be related to the stability of the carbocation intermediate that is formed by attack of the halogen on the halogenated enol. This carbocation is destabilized by the electron-attracting polar effect of *two* halogens:

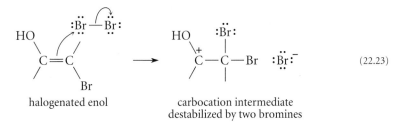

(22.23)

halogenated enol

carbocation intermediate
destabilized by two bromines

If the rate-limiting transition state resembles this carbocation, then the transition state should have very high energy and the rate should be small.

*22.11   (a) Sketch a reaction-free energy diagram for acid-catalyzed enol formation using the mechanism in Eq. 22.15b as your guide. Assume that the second step, proton removal from the α-carbon, is rate-limiting.

(b) Incorporating the results of part (a), sketch a reaction-free energy diagram for acid-catalyzed halogenation of an aldehyde or ketone.

22.12   Explain why:

*(a)   the rate of iodination of optically active 1-phenyl-2-methyl-1-butanone in acetic acid/HNO_3 is identical to its rate of racemization under the same conditions

(b)   the rates of bromination and iodination of acetophenone are identical at a given acid concentration.

B.   Halogenation of Aldehydes and Ketones in Base: the Haloform Reaction

Halogenation of aldehydes and ketones with α-hydrogens also occurs in base. In this reaction, *all* α-hydrogens are substituted by halogen.

$$(CH_3)_3C—\overset{O}{\overset{\|}{C}}—CH_3 + Br_2 \xrightarrow[\substack{H_2O/dioxane \\ 0°}]{NaOH} (CH_3)_3C—\overset{O}{\overset{\|}{C}}—CBr_3 \qquad (22.24a)$$

α-hydrogens

no α-hydrogens

When the aldehyde or ketone starting material is either acetaldehyde or a methyl ketone (as in Eq. 22.24a), the trihalo carbonyl compound is not stable under the reaction conditions, and it reacts further to give, after acidification of the reaction mixture, a carboxylic acid and a haloform. (Recall from Sec. 8.1A that a *haloform* is a trihalomethane, that is, a compound of the form HCX_3, where X = halogen.)

$$(CH_3)_3C—\overset{O}{\overset{\|}{C}}—CBr_3 \xrightarrow{\;OH^-\;} \xrightarrow{\;H_3O^+\;} (CH_3)_3C—\overset{O}{\overset{\|}{C}}—OH \;+\; HCBr_3 \qquad (22.24b)$$

(71–74% yield)

bromoform

Eqs. 22.4a and b constitute an example of the **haloform reaction**. Notice that, in a haloform reaction, a carbon-carbon bond is broken.

The mechanism of the haloform reaction involves the formation of an *enolate ion* as a reactive intermediate.

$$R—\overset{O}{\overset{\|}{C}}—CH_3 + OH^- \;\rightleftharpoons\; R—\overset{O}{\overset{\|}{C}}—\overset{..}{\overset{-}{C}}H_2 + H_2O \qquad (22.25a)$$

enolate ion

The enolate ion reacts as a nucleophile with halogen to give an α-halo carbonyl compound.

$$R—\overset{O}{\overset{\|}{C}}—\overset{..}{\overset{-}{C}}H_2 + :\overset{..}{Br}—\overset{..}{Br}: \longrightarrow R—\overset{O}{\overset{\|}{C}}—CH_2\overset{..}{Br}: + :\overset{..}{\overset{..}{Br}}:^- \qquad (22.25b)$$

However, halogenation does not stop here, because the enolate ion of the α-halo ketone is formed even more rapidly than the enolate ion of the starting ketone. The reason is that the inductive effect of the halogen stabilizes the enolate ion and, by Hammond's postulate, the transition state for its formation. Consequently, a second bromination occurs.

$$R-\overset{\overset{\displaystyle O}{\|}}{C}-\overset{\overset{\displaystyle H}{|}}{\underset{\underset{\displaystyle H}{|}}{C}}-Br \longrightarrow R-\overset{\overset{\displaystyle O}{\|}}{C}-\overset{\overset{\displaystyle H}{|}}{\underset{\displaystyle \ddot{}}{C}}-Br \xrightarrow{Br-Br} R-\overset{\overset{\displaystyle O}{\|}}{C}-CHBr_2 + Br^- \qquad (22.25c)$$

$$\underset{\displaystyle =:\!\ddot{O}H}{} \qquad\qquad + H_2\ddot{O}:$$

The dihalo carbonyl compound brominates again (why?).

$$R-\overset{\overset{\displaystyle O}{\|}}{C}-CHBr_2 \xrightarrow[OH^-]{Br_2} R-\overset{\overset{\displaystyle O}{\|}}{C}-CBr_3 + Br^- + H_2O \qquad (22.25d)$$

A carbon-carbon bond is broken when the trihalo carbonyl compound undergoes a *nucleophilic acyl substitution reaction*.

$$R-\overset{\overset{\displaystyle :O:}{\|}}{C}-CBr_3 \underset{\displaystyle =:\!\ddot{O}H}{\rightleftarrows} R-\overset{\overset{\displaystyle :\ddot{O}:^-}{|}}{\underset{\displaystyle :OH}{C}}-CBr_3 \rightleftarrows R-\overset{\overset{\displaystyle :O:}{\|}}{C}-\ddot{O}-H + {}^-:CBr_3 \longrightarrow$$

$$\underset{\displaystyle \text{a trihalomethyl anion}}{}$$

$$R-\overset{\overset{\displaystyle :O:}{\|}}{C}-\ddot{O}:^- + H-CBr_3 \qquad (22.25e)$$

The leaving group in this reaction is a trihalomethyl anion. Usually, carbanions are too basic to serve as leaving groups; but trihalomethyl anions are much less basic than ordinary carbanions (why?). However, the basicity of trihalomethyl anions, while low enough for them to act as leaving groups, is high enough for them to react irreversibly with the carboxylic acid by-product, as shown in the last part of Eq. 22.25e. This acid-base reaction drives the overall haloform reaction to completion. (This is analogous to saponification, which is also driven to completion by ionization of the carboxylic acid product; Sec. 21.7A.) The carboxylic acid itself can be isolated by acidifying the reaction mixture.

$$R-\overset{\overset{\displaystyle O}{\|}}{C}-\ddot{O}:^- + H_3O^+ \longrightarrow R-\overset{\overset{\displaystyle O}{\|}}{C}-\ddot{O}H + H_2O \qquad (22.25f)$$

Occasionally, the haloform reaction is used to prepare carboxylic acids from readily available methyl ketones. More often, however, this reaction is used as a qualitative test for methyl ketones, called the **iodoform test**. In this test, a compound of unknown structure is mixed with alkaline I_2. A yellow precipitate of iodoform (HCI_3) is taken as evidence for a methyl ketone (or acetaldehyde, the "methyl aldehyde"). Alcohols of the form shown in Eq. 22.26 also give a positive iodoform test because they are oxidized to methyl ketones (or to acetaldehyde, in the case of ethanol) by the basic iodine solution.

$$R-\overset{\overset{\displaystyle OH}{|}}{CH}-CH_3 \xrightarrow[\text{base}]{I_2} R-\overset{\overset{\displaystyle O}{\|}}{C}-CH_3 \qquad (22.26)$$

$$\underset{\displaystyle \text{undergoes iodoform reaction}}{}$$

22.13   Give the products expected (if any) when each of the following compounds reacts with Br_2 in NaOH.

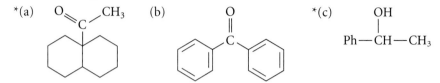

    *(a)   (b)   *(c)

*22.14   Give the structure of a compound $C_6H_{10}O_2$ that gives succinic acid and iodoform on treatment with a solution of I_2 in aqueous NaOH, followed by acidification.

C. α-Bromination of Carboxylic Acids

Carboxylic acids can be brominated at their α-carbons. A bromine is substituted for an α-hydrogen when a carboxylic acid is treated with Br_2 and a catalytic amount of red phosphorus or PBr_3. (The actual catalyst is PBr_3; phosphorus can be used because it reacts with Br_2 to give PBr_3.)

$$CH_3CH_2CH_2CH_2CH_2CO_2H + Br_2 \xrightarrow{\text{P or PBr}_3} CH_3CH_2CH_2CH_2\underset{\underset{Br}{|}}{C}HCO_2H + HBr \qquad (22.27)$$

hexanoic acid

2-bromohexanoic acid
(83–89% yield)

This reaction is called the **Hell-Volhard-Zelinsky reaction** after its discoverers, and is sometimes nicknamed the *HVZ reaction.*

      The first stage in the mechanism of this reaction is the conversion of the carboxylic acid into a small amount of acid bromide by the catalyst PBr_3 (Sec. 20.9A).

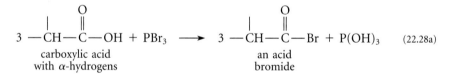

$$3 \;-\underset{|}{\overset{|}{C}}H-\overset{\overset{O}{\|}}{C}-OH + PBr_3 \longrightarrow 3\;-\underset{|}{\overset{|}{C}}H-\overset{\overset{O}{\|}}{C}-Br + P(OH)_3 \qquad (22.28a)$$

carboxylic acid   an acid
with α-hydrogens   bromide

From this point the mechanism closely resembles that for the acid-catalyzed bromination of ketones. The *enol* of the acid bromide is the species that actually brominates.

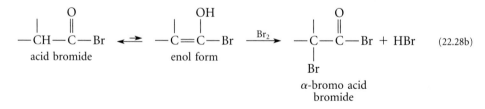

acid bromide   enol form   α-bromo acid
bromide

When *a small amount* of PBr_3 catalyst is used, the α-bromo acid bromide reacts with the carboxylic acid to form more acid bromide, which is then brominated as shown above.

$$-\overset{|}{\underset{Br}{CH}}-\overset{O}{\overset{\|}{C}}-OH \;+\; -\overset{|}{\underset{|}{C}}-\overset{O}{\overset{\|}{C}}-Br \;\rightleftharpoons\; -\overset{|}{\underset{Br}{CH}}-\overset{O}{\overset{\|}{C}}-Br \;+\; -\overset{|}{\underset{Br}{C}}-\overset{O}{\overset{\|}{C}}-OH \qquad (22.29)$$

enters bromination
sequence at Eq. 22.28b

α-bromo acid

Thus, when a catalytic amount of PBr$_3$ is used, the reaction product is the α-bromo acid.

If *one full equivalent* of PBr$_3$ catalyst is used, the α-bromo acid bromide is the reaction product; this can be used in many of the reactions of acid halides discussed in Sec. 21.8A. For example, the reaction mixture can be quenched with an alcohol to give an α-bromo ester:

$$CH_3CH_2-\overset{O}{\overset{\|}{C}}-OH \xrightarrow[Br_2]{P\ (1\ equiv.)} CH_3-\overset{|}{\underset{Br}{CH}}-\overset{O}{\overset{\|}{C}}-Br \xrightarrow[\substack{(CH_3)_2\ddot{N}-Ph \\ (a\ base)}]{(CH_3)_3COH}$$

propanoic acid

$$CH_3-\overset{|}{\underset{Br}{CH}}-\overset{O}{\overset{\|}{C}}-OC(CH_3)_3 \qquad (22.30)$$

tert-**butyl 2-bromopropanoate**

D. Reactions of α-Halo Carbonyl Compounds

Most α-halo carbonyl compounds are very reactive in S$_N$2 reactions, and can be used to prepare other α-substituted carbonyl compounds.

$$(CH_3)_2\ddot{S}: \quad :\ddot{Br}-CH_2-\overset{O}{\overset{\|}{C}}-Ph \xrightarrow[25°,\ 30\ min]{acetone/H_2O} CH_3-\overset{CH_3}{\overset{|}{\underset{+}{S}}}-CH_2-\overset{O}{\overset{\|}{C}}-Ph \quad :\ddot{Br}:^- \qquad (22.31)$$

(85% yield)

In the case of α-halo ketones, nucleophiles used in these reactions must not be too basic. For example, dimethyl sulfide, used in Eq. 22.31, is a very weak base but a fairly strong nucleophile. (Stronger bases promote enolate-ion formation; and the enolate ions of α-halo ketones undergo other reactions.) More basic nucleophiles can be used with α-halo acids because, under basic conditions, α-halo acids are ionized to form their carboxylate conjugate-base anions; a second ionization to give an enolate ion, which would introduce a second negative charge into the molecule, does not occur.

$$Cl-\overset{Cl}{\underset{}{\langle\ \rangle}}-OH + Cl-CH_2-CO_2H \xrightarrow[2)\ H_3O^+]{\substack{1)\ NaOH \\ (2\ equiv.)}} Cl-\overset{Cl}{\underset{}{\langle\ \rangle}}-O-CH_2-CO_2H + Cl^- \qquad (22.32)$$

2,4-dichlorophenol

2,4-dichlorophenoxyacetic acid
(2,4-D, a selective herbicide;
87% yield)

$$^-:C\equiv N: \ + \ Cl-CH_2-CO_2^- \ \xrightarrow{\ H_3O^+\ } \ :N\equiv C-CH_2-CO_2H \ + \ Cl^- \quad (22.33)$$

chloroacetate anion   **cyanoacetic acid** (77–80% yield)

The following comparison gives a quantitative measure of the S_N2 reactivity of α-halo carbonyl compounds:

$$Cl-CH_2-\overset{\overset{\displaystyle O}{\parallel}}{C}-CH_3 + KI \ \xrightarrow{\text{acetone}} \ I-CH_2-\overset{\overset{\displaystyle O}{\parallel}}{C}-CH_3 + KCl \qquad \begin{array}{c} \textit{relative rate:} \\ 35{,}000 \end{array} \quad (22.34a)$$

$$Cl-CH_2CH_2CH_3 + KI \ \xrightarrow{\text{acetone}} \ I-CH_2CH_2CH_3 + KCl \qquad 1 \quad (22.34b)$$

The explanation for the enhanced reactivity is probably similar to that for the increased reactivity of allylic alkyl halides in S_N2 displacements (Sec. 17.4).

In contrast, α-halo carbonyl compounds react so slowly by the S_N1 mechanism that this reaction is not useful.

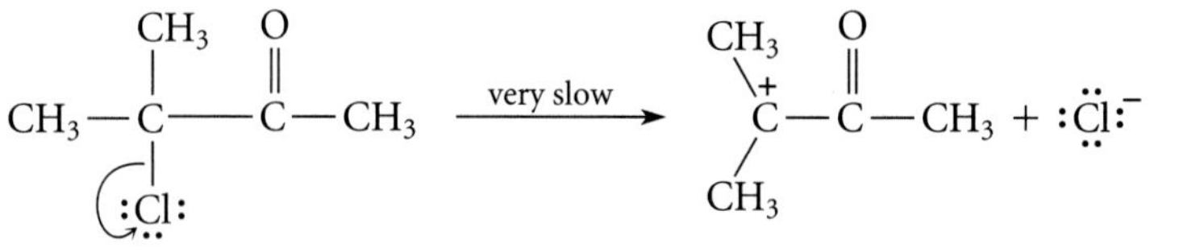

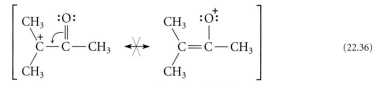

$$\quad (22.35)$$

In fact, *reactions that require the formation of carbocations alpha to carbonyl groups generally do not occur.* Although it might seem that an α-carbonyl carbocation should be resonance-stabilized, its resonance structure is not important (why?).

$$\left[\ \begin{array}{c} CH_3 \quad :O: \\ \diagdown \overset{+}{C}{-}\overset{\parallel}{C}{-}CH_3 \\ \diagup \\ CH_3 \end{array} \quad \overset{\times}{\longleftrightarrow} \quad \begin{array}{c} CH_3 \quad :\overset{+}{O}: \\ \diagdown C{=}C{-}CH_3 \\ \diagup \\ CH_3 \end{array} \ \right] \quad (22.36)$$

not an important structure

PROBLEMS

22.15   What product is formed when
*(a)   propionic acid is treated first with Br_2 and one equivalent of PBr_3, then with a large excess of ammonia?
(b)   α-phenylacetic acid is treated first with Br_2 and one equivalent of PBr_3, then with a large excess of ethanol?

22.16   Give the product of each of the following reactions.
*(a)

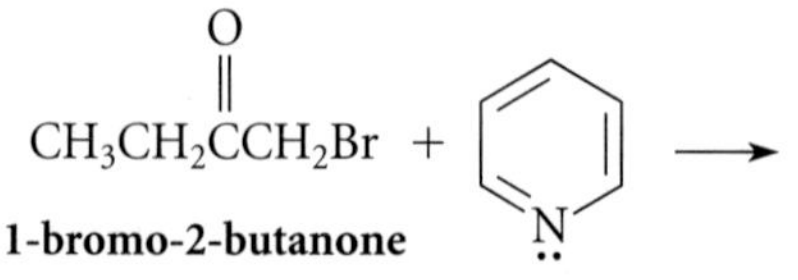

$$CH_3CH_2\overset{\overset{\displaystyle O}{\parallel}}{C}CH_2Br \ + \ \underset{\underset{\displaystyle H}{\overset{\displaystyle}{N}}}{\bigcirc} \ \longrightarrow$$

1-bromo-2-butanone

pyridine

(b)

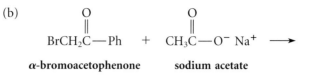

α-bromoacetophenone   **sodium acetate**

*22.17   Give the mechanism of the reaction in Eq. 22.32. Your mechanism should show why *two equivalents* of NaOH must be used.

22.4   Aldol Addition and Aldol Condensation

A.   Base-Catalyzed Aldol Reactions

In aqueous base, acetaldehyde undergoes a reaction called the **aldol addition**.

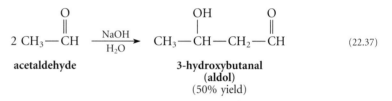

acetaldehyde   **3-hydroxybutanal**
                                       (aldol)
                                      (50% yield)

The term *aldol* is both a trivial name for 3-hydroxybutanal and a generic name for β-hydroxy aldehydes. The aldol addition is a very important and general reaction of aldehydes and ketones that have α-hydrogens. Notice that this reaction is another method of forming carbon-carbon bonds.

The base-catalyzed aldol addition involves an *enolate ion* as an intermediate. In this reaction, an enolate ion, formed by reaction of acetaldehyde with aqueous NaOH, adds to a second molecule of acetaldehyde.

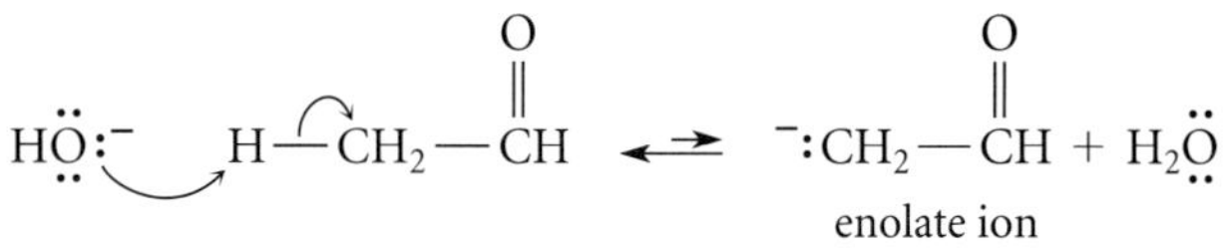

$$HO:^- \quad H-CH_2-CH \overset{O}{\underset{\text{enolate ion}}{\rightleftharpoons}} \quad ^-:CH_2-CH + H_2O \quad (22.38a)$$

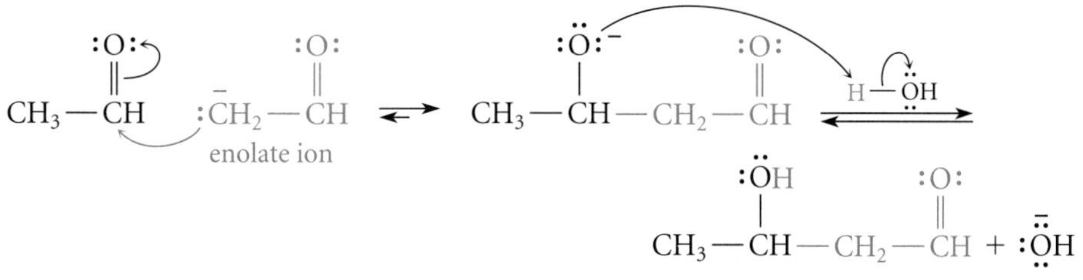

enolate ion

$$CH_3-CH \quad :CH_2-CH \rightleftharpoons CH_3-CH-CH_2-CH \quad \underset{H-OH}{\rightleftharpoons}$$

enolate ion

$$CH_3-CH-CH_2-CH + :OH \quad (22.38b)$$

Notice that the aldol addition is another nucleophilic addition to a carbonyl group. In this reaction, the nucleophile is an enolate ion. The reaction may *look* more complicated than some additions because of the number of carbon atoms in the product. However, it is not conceptually different from other nucleophilic additions, for example, cyanohydrin formation.

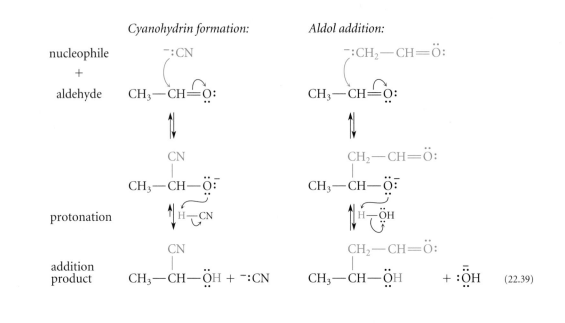

Cyanohydrin formation: *Aldol addition:*

(22.39)

....

PROBLEM

22.18 Use the reaction mechanism to deduce the product of the aldol addition reaction of *(a) α-phenylacetaldehyde; (b) propionaldehyde.

....

The aldol addition is reversible. Like many other carbonyl addition reactions (Sec. 19.7B), the equilibrium for the aldol condensation is more favorable for aldehydes than for ketones.

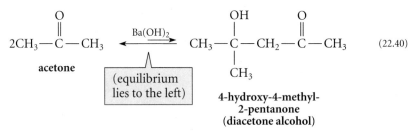

(22.40)

In this aldol addition of acetone, the equilibrium favors the ketone reactant rather than the addition product diacetone alcohol. This product can be isolated in good yield only if an apparatus is used that allows the product to be removed from the base catalyst as it is formed.

Under more severe conditions (higher base concentration and/or heat), the product of aldol addition undergoes a dehydration reaction.

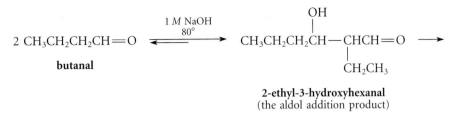

2-ethyl-3-hydroxyhexanal
(the aldol addition product)

$$CH_3CH_2CH_2CH=CCH=O + H_2O \quad (22.41)$$
$$| \atop CH_2CH_3$$

2-ethyl-2-hexenal
(86% yield)

The sequence of reactions consisting of the aldol addition followed by dehydration, as in Eq. 21.41, is called the **aldol condensation**. (A **condensation** is a reaction in which two molecules combine to form a larger molecule with the elimination of a small molecule, in many cases water.)

> The term *aldol condensation* has been used historically to refer to the aldol addition reaction as well as to the addition and dehydration reactions together. To eliminate ambiguity, *aldol condensation* is used in this text only for the addition-dehydration sequence. The term *aldol reactions* is used to refer generically to both addition and condensation reactions.

The dehydration part of the aldol condensation is catalyzed by base, and occurs in two distinct steps through a *carbanion intermediate*.

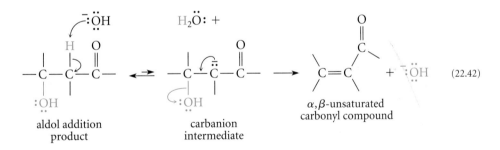

aldol addition   carbanion α,β-unsaturated
product   intermediate   carbonyl compound

Notice that, although this is a β-elimination, it is *not* concerted. In this respect it differs from the E2 reaction.

A base-catalyzed dehydration reaction of alcohols is unusual; ordinary alcohols do *not* dehydrate in base. However, β-hydroxy aldehydes and β-hydroxy ketones do for two reasons. First, their α-hydrogens are relatively acidic. Recall that β-eliminations are particularly rapid when acidic hydrogens are involved (Sec. 17.3B). Second, the product is conjugated and therefore is particularly stable.

The product of the aldol condensation is an *α,β-unsaturated carbonyl compound*. The aldol condensation is an important method for the preparation of certain α,β-unsaturated carbonyl compounds. Whether the aldol addition product or the condensation product is formed depends on reaction conditions, which must be worked out on a case-by-case basis. You can assume for purposes of problem-solving, unless stated otherwise, that either the addition product or the condensation product can be prepared.

> ### MUSICAL HISTORY OF THE ALDOL CONDENSATION
>
> Discovery of the aldol condensation is usually attributed solely to Charles Adolphe Wurtz, a French chemist who trained Friedel and Crafts. However, the reaction was first investigated during the period 1864–1873 by Aleksandr Borodin, a Russian chemist who was also a self-taught and proficient composer of music. (Borodin's musical themes were used as the basis of songs in the musical *Kismet*.) Borodin found it difficult to compete with Wurtz's large, modern, well-funded laboratory. Borodin also lamented that his professional duties so burdened him with "examinations and commissions" that he could only compose when he was at home ill. Knowing this, his musical friends used to greet him, "Aleksandr, I hope you are ill today!"

B. Acid-Catalyzed Aldol Condensation

Aldol condensations are also catalyzed by acid.

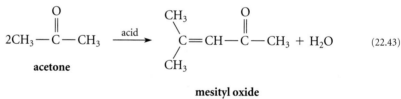

$$2CH_3 - \overset{\overset{\textstyle O}{\|}}{C} - CH_3 \quad \xrightarrow{\text{acid}} \quad \underset{CH_3}{\overset{CH_3}{\diagdown}} C = CH - \overset{\overset{\textstyle O}{\|}}{C} - CH_3 + H_2O \qquad (22.43)$$

acetone

mesityl oxide
(79% yield)

Acid-catalyzed aldol condensations, as in this example, generally give α,β-unsaturated carbonyl compounds as products; addition products cannot be isolated.

In acid-catalyzed aldol condensations, the conjugate acid of the aldehyde or ketone is a key reactive intermediate.

$$CH_3 - \overset{\overset{\textstyle :O:}{\|}}{C} - CH_3 \quad \overset{\overset{\displaystyle H - \overset{+}{O}H_2}{\frown}}{\rightleftharpoons} \quad CH_3 - \overset{\overset{\textstyle \overset{+}{:O} \diagup H}{\|}}{C} - CH_3 + \overset{\cdot\cdot}{O}H_2 \qquad (22.44a)$$

This protonated ketone serves two roles. First, it is transformed into a small amount of the *enol* (Eq. 22.15b). Second, the protonated ketone is the electrophilic species in the reaction. It is attacked by the π electrons of the enol to give an α-hydroxy carbocation, which is the conjugate acid of the addition product:

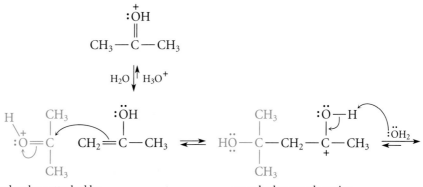

carbonyl carbon attacked by
enol π electrons

an α-hydroxy carbocation
(protonated ketone; Sec. 19.6)

$$\overset{\cdot\cdot}{HO}-\underset{\underset{CH_3}{|}}{\overset{\overset{CH_3}{|}}{C}}-CH_2-\overset{\overset{:O:}{\|}}{C}-CH_3 + H_3\overset{\cdot\cdot}{O}^+ \qquad (22.44b)$$

As the second part of Eq. 22.44b shows, the α-hydroxy carbocation loses a proton to give the β-hydroxy ketone product. Under the acidic conditions, this material spontaneously undergoes acid-catalyzed dehydration to give an α,β-unsaturated carbonyl compound:

STUDY GUIDE LINK:
✓22.3
*Dehydration of
β-Hydroxy Carbonyl
Compounds*

$$HO-\underset{\underset{CH_3}{|}}{\overset{\overset{CH_3}{|}}{C}}-CH_2-\overset{\overset{O}{\|}}{C}-CH_3 \xrightarrow{H_3O^+} \underset{\underset{CH_3}{\diagdown}}{\overset{\overset{CH_3}{\diagup}}{C}}=CH-\overset{\overset{O}{\|}}{C}-CH_3 + H_2O \qquad (22.44c)$$

This dehydration drives the aldol condensation to completion. (Recall that without this dehydration, the aldol condensation of ketones is unfavorable; Eq. 22.40).

Let's contrast the species involved in the acid- and base-catalyzed aldol reactions. An *enol*, not an enolate ion, is the nucleophilic species in an acid-catalyzed aldol condensation. Enolate ions are too basic to exist in acidic solution. Although an enol is much less nucleophilic than an enolate ion, it reacts at a useful rate because it attacks a potent electrophile, a protonated carbonyl compound (an α-hydroxy carbocation). In a base-catalyzed aldol reaction, an *enolate ion* is the nucleophile. A protonated carbonyl compound is *not* an intermediate because it is too acidic to exist in basic solution. The electrophile attacked by the enolate ion is a *neutral* carbonyl compound. To summarize:

Reaction:	Nucleophile:	Electrophile:
Base-catalyzed aldol reaction	enolate ion	neutral carbonyl compound
Acid-catalyzed aldol condensation	enol	protonated carbonyl compound

C. Special Types of Aldol Reactions

Crossed Aldol Reactions The discussion above considered only aldol reactions between two molecules of the same aldehyde or ketone. When two *different* carbonyl compounds are used, the reaction is termed a **crossed aldol reaction**. In many cases, the result of a crossed aldol reaction is a difficult-to-separate mixture, as the following study problem illustrates.

Study Problem 22.1

Give the structures of the aldol addition products expected from the base-catalyzed reaction of acetaldehyde and propionaldehyde.

Solution Such a reaction involves four different species: acetaldehyde (*A*) and its enolate ion (*A'*), as well as propionaldehyde (*P*) and its enolate ion (*P'*):

$$
\underset{A}{CH_3-\overset{\displaystyle O}{\overset{\|}{C}}-H} \qquad \underset{A'}{:\!\overset{-}{C}H_2-\overset{\displaystyle O}{\overset{\|}{C}}-H} \qquad \underset{P}{CH_3CH_2-\overset{\displaystyle O}{\overset{\|}{C}}-H} \qquad \underset{P'}{CH_3\overset{\cdot\cdot}{C}H-\overset{\displaystyle O}{\overset{\|}{C}}-H}
$$

Four possible addition products can arise from the attack of each enolate ion on each aldehyde:

$$
\underset{A\,+\,A'}{\overset{\displaystyle OH}{\overset{|}{CH_3CHCH_2CH}}=O} \qquad \underset{P\,+\,A'}{\overset{\displaystyle OH}{\overset{|}{CH_3CH_2CHCH_2CH}}=O} \qquad \underset{\underset{A\,+\,P'}{CH_3}}{\overset{\displaystyle OH}{\overset{|}{CH_3CHCHCH}}=O} \qquad \underset{\underset{P\,+\,P'}{CH_3}}{\overset{\displaystyle OH}{\overset{|}{CH_3CH_2CHCHCH}}=O}
$$

(Be sure you see how each product is formed; write a mechanism for each, if necessary.) Notice also a further complication: diastereomers are possible for the last two products because each has two asymmetric carbons.

Crossed aldol reactions that provide complex mixtures, such as the one in Study Problem 22.1, are not very useful, because the product of interest is not formed in very high yield, and because isolation of one product from a complex mixture is in most cases extremely tedious. Although conditions that favor one product or another in crossed aldol reactions have been worked out in specific cases, under the usual conditions (aqueous or alcoholic acid or base), useful crossed aldol reactions as a practical matter are limited to situations in which *a ketone with α-hydrogens is condensed with an aldehyde that has no α-hydrogens*. An important example of this type is the **Claisen-Schmidt condensation**. In this reaction, a ketone with α-hydrogens—acetone in the following example—is condensed with an aromatic aldehyde that has no α-hydrogens—benzaldehyde in this case.

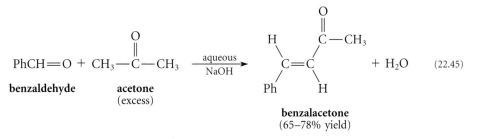

(22.45)

benzaldehyde   **acetone**
                         (excess)

benzalacetone
(65–78% yield)

Notice that the addition product cannot be isolated in this reaction; the highly conjugated condensation product is formed as its most stable stereoisomer—the *trans* isomer in Eq. 22.45.

In view of the complex mixture obtained in the example used in Study Problem 22.1 it is reasonable to ask why only one product is obtained from the crossed aldol condensation in Eq. 22.45. First, because the aldehyde in the Claisen-Schmidt reaction has no α-hydrogens, it cannot act as the enolate component of the aldol condensation; consequently, two of the four possible products cannot form. The other possible side reaction is the aldol addition reaction of the ketone with itself, as in Eq. 22.40; why doesn't this reaction occur? The enolate ion from acetone can react either with another molecule of acetone or with benzaldehyde. Recall that addition to a ketone occurs more slowly than addition to an aldehyde (Sec. 19.7B). Furthermore, even if addition to acetone does occur, the aldol addition reaction of two ketones is reversible (Eq. 22.40) and addition to an aldehyde has a more favorable equilibrium constant than addition to a ketone (Sec. 19.7B). Thus, in Eq. 22.45, both the rate and equilibrium for addition to benzaldehyde are more favorable than they are for addition to a second molecule of acetone. Thus, the product shown in Eq. 22.45 is the only one formed.

The Claisen-Schmidt condensation, like other aldol condensations, can also be catalyzed by acid.

☞
Study Guide Link:
*✓22.4
Understanding
Condensation
Reactions*

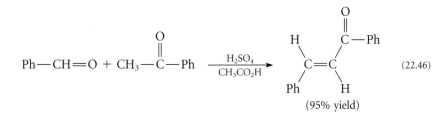

(22.46)

(95% yield)

Intramolecular Aldol Condensation   When a molecule contains more than one aldehyde or ketone group, an *intramolecular* reaction (a reaction within the same molecule) is possible. In such a case the aldol condensation results in formation of a ring. Intramolecular aldol condensations are particularly favorable when five- and six-membered rings can be formed.

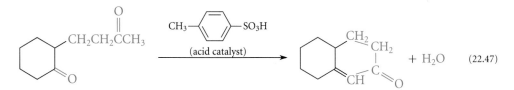

(22.47)

PROBLEM

22.19 Predict the product(s) in each of the following aldol condensations.

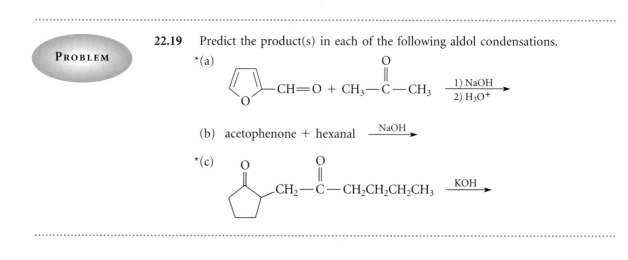

*(a)

(b) acetophenone + hexanal $\xrightarrow{\text{NaOH}}$

*(c)

D. Synthesis with the Aldol Condensation

The aldol condensation can be applied to the synthesis of a wide variety of α,β-unsaturated aldehydes and ketones. Notice that it also represents another method for the formation of carbon-carbon bonds. (See the complete list in Appendix V.) If you desire to prepare a particular α,β-unsaturated aldehyde or ketone by the aldol condensation, you must ask two questions: (1) What starting materials are required in the aldol condensation? (2) With these starting materials, is the aldol condensation of these compounds a feasible one?

The starting materials for an aldol condensation can be determined by mentally "splitting" the α,β-unsaturated carbonyl compound at the double bond:

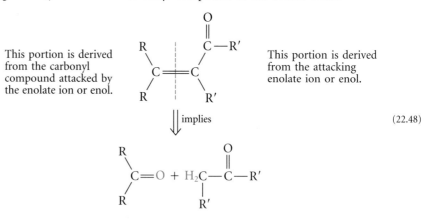

This portion is derived from the carbonyl compound attacked by the enolate ion or enol.

This portion is derived from the attacking enolate ion or enol.

(22.48)

That is, work backward from the desired synthetic objective by replacing the double bond on the carbonyl side by two hydrogens and on the other side by a carbonyl oxygen (=O) to obtain the structures of the starting materials in the aldol condensation.

Knowing the potential starting materials for an aldol condensation is not enough; you must also know whether the condensation is one that works, or whether instead it is one that is likely to give troublesome mixtures(see Study Guide Link 4.5). In other words, you can't make every conceivable α,β-unsaturated aldehyde or ketone by the aldol condensation—only certain ones. This point is illustrated in the following study problem.

Study Problem 22.2

Determine whether the following α,β-unsaturated ketone can be prepared by an aldol condensation.

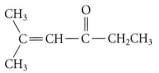

Solution Following the procedure in Eq. 22.48, analyze the desired product as follows:

acetone **2-butanone**

required starting materials

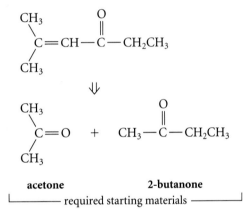

The desired product requires a crossed condensation between two similar ketones. It would not be a useful reaction because a number of possible constitutionally isomeric products could form (see Study Problem 22.1). As noted above, such a mixture should be avoided if possible.

Problems

22.20 Some of the following molecules can be synthesized in good yield using an aldol condensation. Identify these and give the structures of the required starting materials. Others cannot be synthesized in good yield by an aldol condensation. Identify these, and explain why the required aldol condensation would not be likely to succeed.

*(a)

(b)

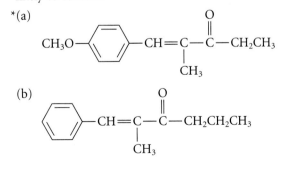

(Problem 22.20 continues)

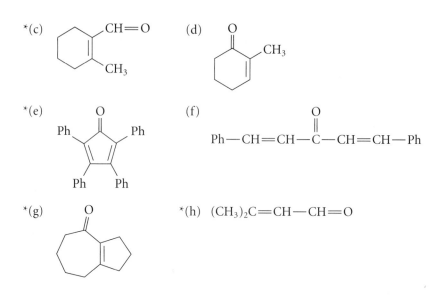

*22.21   Analyze the aldol condensation in Eq. 22.47 using the method given in Eq. 22.48. Show that there are four possible aldol condensation products that might in principle result from the starting material. Argue why the observed product is the most reasonable one.

22.5 Condensation Reactions Involving Ester Enolate Ions

This section begins the use of more compact abbreviations for several commonly occurring organic groups. These abbreviations, shown in Table 22.1, not only save space, but also

Table 22.1   Abbreviations of Some Common Organic Groups

Group	Structure	Abbreviation
methyl	CH_3—	Me
ethyl	CH_3CH_2—	Et
propyl	$CH_3CH_2CH_2$—	Pr
isopropyl	$(CH_3)_2CH$—	*i*-Pr
butyl	$CH_3CH_2CH_2CH_2$—	Bu
isobutyl	$(CH_3)_2CHCH_2$—	*i*-Bu
tert-butyl	$(CH_3)_3C$—	*t*-Bu
acetyl	$CH_3-\overset{\overset{\displaystyle O}{\|\|}}{C}-$	Ac
acetate (or acetoxy)	$CH_3-\overset{\overset{\displaystyle O}{\|\|}}{C}-O-$	AcO

make the structures of large molecules less cluttered and more easily read. Just as Ph—
is used to symbolize the phenyl ring, Me— can be used for methyl, Et— for ethyl,
Pr— for propyl, and so on. Thus, ethyl acetate is abbreviated EtOAc; sodium ethoxide
(NaOC$_2$H$_5$) is simply written as NaOEt; and methanol is abbreviated as MeOH.

PROBLEM

22.22   Write the structure that corresponds to each of the following abbreviations.
   *(a) Et$_3$C—OH   (b) *i*-Pr—Ph   *(c) *t*-BuOAc
   (d) Pr—OH   *(e) Ac$_2$O   (f) Ac—Ph

A. Claisen Condensation

The aldol reactions discussed in the last section involve the enols or enolates of *aldehydes*
and *ketones*. This section discusses condensation reactions that involve the enolate ions
of *esters*.

   Ethyl acetate undergoes a condensation reaction in the presence of one equivalent
of sodium ethoxide in ethanol to give a compound known commonly as ethyl acetoacetate.

$$2\ CH_3\overset{\overset{\textstyle O}{\|}}{-C}-OEt \quad \xrightarrow[\text{EtOH}]{\underset{\text{(1 equiv.)}}{\text{NaOEt}}} \quad \xrightarrow{\ H_3O^+\ } \quad CH_3\overset{\overset{\textstyle O}{\|}}{-C}-CH_2\overset{\overset{\textstyle O}{\|}}{-C}-OEt\ +\ EtOH \qquad (22.49)$$

ethyl acetate   **ethyl acetoacetate**
                                          (75–76% yield)

This is the best-known example of the **Claisen condensation**, named for Ludwig Claisen
(1851–1930), who was a Professor at the University of Kiel. (Don't confuse this reaction
with the Claisen-Schmidt condensation in the previous section—same Claisen, different
reaction!) The product of this reaction, ethyl acetoacetate, is an example of a **β-keto
ester**: a compound with a ketone carbonyl group β to an ester carbonyl group.

$$CH_3\overset{\overset{\textstyle O}{\|}}{\underset{\beta}{-C}}-\underset{\alpha}{CH_2}\overset{\overset{\textstyle O}{\|}}{-C}-OEt$$

   The first step in the mechanism of the Claisen condensation is formation of an
enolate ion by the reaction of the ester with the ethoxide base.

$$Et\ddot{\underset{\cdot\cdot}{O}}\!:^- + \overset{\frown}{H}\!-\!CH_2\overset{\overset{\textstyle O}{\|}}{-C}-OEt \;\rightleftharpoons\; {}^-\!:CH_2\overset{\overset{\textstyle O}{\|}}{-C}-OEt + Et\ddot{\underset{\cdot\cdot}{O}}H \qquad (22.50a)$$

                              enolate ion
                              of ethyl acetate

Because ethoxide ion is a nucleophile, you might also ask whether it can also attack the
carbonyl group of the ester to give the usual nucleophilic acyl substitution reaction. This
reaction undoubtedly takes place, but the products are the same as the reactants! This is
why ethoxide ion is used as a base with ethyl esters in the Claisen condensation (see
Study Guide Link 22.1).

Although the ester enolate ion is formed in low concentration, it is a strong base and good nucleophile, and it undergoes a *nucleophilic acyl substitution reaction* with a second molecule of ester (Eq. 22.50b).

$$CH_3-\overset{\overset{\displaystyle :O:}{\|}}{C}-OEt \quad {}^-\!:CH_2-\overset{\overset{\displaystyle O}{\|}}{C}-OEt \;\rightleftharpoons\; CH_3-\overset{\overset{\displaystyle :\ddot{O}:^-}{|}}{\underset{\underset{\displaystyle :OEt}{|}}{C}}-CH_2-\overset{\overset{\displaystyle O}{\|}}{C}-OEt \;\rightleftharpoons$$

tetrahedral addition intermediate

$$CH_3-\overset{\overset{\displaystyle :O:}{\|}}{C}-CH_2-\overset{\overset{\displaystyle O}{\|}}{C}-OEt \;+\; Et\ddot{O}:^- \qquad (22.50b)$$

The overall equilibrium as written in Eq. 22.50b favors the reactants; that is, *all β-keto esters are less stable than the esters from which they are derived.* For this reason, the Claisen condensation has to be driven to completion by applying LeChatelier's principle. The most common technique is to use one equivalent of ethoxide catalyst. In the β-keto ester product, the hydrogens on the carbon adjacent to both carbonyl groups (color in Eq. 22.50c) are especially acidic (why?), and the ethoxide removes one of these protons to form quantitatively the conjugate base of the product.

$$CH_3-\overset{\overset{\displaystyle O}{\|}}{C}-CH_2-\overset{\overset{\displaystyle O}{\|}}{C}-OEt \;+\; Na^+\;EtO^- \;\rightleftharpoons$$

$$\underset{pK_a\,=\,10.7}{}$$

$$CH_3-\overset{\overset{\displaystyle O}{\|}}{C}-\underset{\underset{\displaystyle Na^+}{}}{\overset{\displaystyle \ddot{C}H}{}}-\overset{\overset{\displaystyle O}{\|}}{C}-OEt \quad + \quad EtOH \qquad (22.50c)$$

$$\underset{pK_a\,=\,15\text{--}16}{}$$

The un-ionized β-keto ester product in Eq. 22.49 is formed when acid is added subsequently to the reaction mixture.

Notice that ethoxide ion is a catalyst for the reactions in Eqs. 22.50a and b, but it is consumed in Eq. 22.50c. Thus, ethoxide is a reactant rather than a catalyst in the overall reaction, and for this reason *one full equivalent* of ethoxide must be used in the Claisen condensation.

The removal of a product by ionization is the same strategy employed to drive ester saponification to completion (Sec. 21.7A). The importance of this strategy in the success of the Claisen condensation is evident if the condensation is attempted with an ester that has only one α-hydrogen: *little or no condensation product is formed.* In this case, the desired condensation product has a quaternary α-carbon, and therefore it has no α-hydrogens acidic enough to react completely with ethoxide.

$$2\;(CH_3)_2CH-\overset{\overset{\displaystyle O}{\|}}{C}-OEt \;\xrightleftharpoons[EtOH]{^-OEt}\; (CH_3)_2CH-\overset{\overset{\displaystyle O}{\|}}{C}-\overset{\overset{\displaystyle CH_3}{|}}{\underset{\underset{\displaystyle CH_3}{|}}{C}}-CO_2Et \qquad (22.51)$$

no acidic hydrogen here

(no product observed)

Furthermore, if the desired product of Eq. 22.51 (prepared by another method) is subjected to the conditions of the Claisen condensation, it readily decomposes back to starting materials because of the reversibility of the Claisen condensation.

The Claisen condensation is another example of *nucleophilic substitution*. In this reaction, the nucleophile is an enolate ion derived from an ester. Although the reaction may seem complex because of the number of carbon atoms in the product, it is not conceptually different from other nucleophilic acyl substitutions, for example, ester saponification:

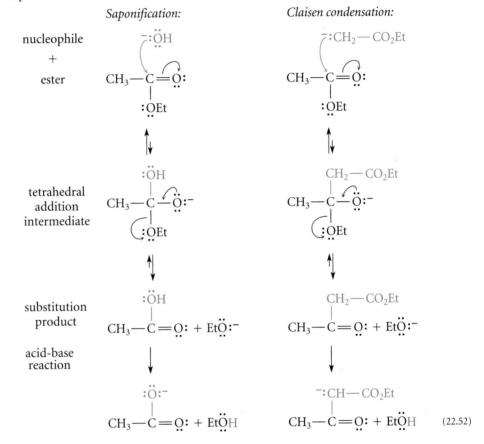

$$(22.52)$$

You have now studied two types of condensation reaction: the aldol condensation and the Claisen condensation. These condensations are quite different and should not be confused. To compare:

1. The aldol condensation is an *addition* reaction of an enolate ion or an enol with an aldehyde or ketone followed by a *dehydration*. The Claisen condensation is a *nucleophilic acyl substitution* reaction of an enolate ion with an ester group.

2. The aldol condensation is catalyzed by both base and acid. The Claisen condensation requires a full equivalent of base, and is *not* catalyzed by acid.

3. The aldol addition requires only one α-hydrogen. A second α-hydrogen is required, however, for the dehydration step of the aldol condensation. In

the Claisen condensation, the ester starting material must have at least *two* α-hydrogens, one for each of the ionizations shown in Eqs. 22.50a and 22.50c.

 PROBLEMS

22.23 Give the Claisen condensation product formed in the reaction of each of the following esters with one equivalent of NaOEt.
*(a) ethyl α-phenylacetate   (b) ethyl butyrate

*22.24 Hydroxide ion is about as basic as ethoxide ion. Would NaOH be a suitable base for the Claisen condensation of ethyl acetate? Explain. (*Hint:* See Study Guide Link 22.1.)

B. Dieckmann Condensation

Intramolecular Claisen condensations, like intramolecular aldol condensations, take place readily when five- or six-membered rings can be formed. The intramolecular Claisen condensation reaction is called the **Dieckmann condensation**.

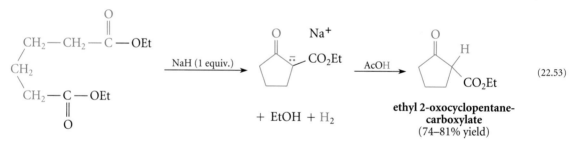

(22.53)

diethyl adipate

ethyl 2-oxocyclopentane-carboxylate
(74–81% yield)

The base used in this reaction, sodium hydride, was discussed in Sec. 8.5A. It removes an α-proton from the ester starting material. The ethoxide ion displaced as a leaving group in the nucleophilic acyl substitution reaction serves as the base that removes the α-proton from the product, thus driving the reaction to completion. (In principle, one equivalent of sodium ethoxide can be used as the base in this reaction instead of sodium hydride.)

 PROBLEM

22.25 *(a) Explain why compound *A*, when treated with one equivalent of NaOEt, followed by acidification, is completely converted into compound *B*.
   (b) Give the structure of the only product formed when diethyl α-methyladipate (*C*) reacts in the Dieckmann condensation. Explain your reasoning.

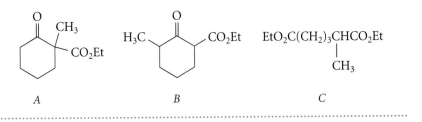

A   *B*   *C*

C. Crossed Claisen Condensation

The Claisen condensation of two *different* esters is called a **crossed Claisen condensation**. The crossed Claisen condensation of two esters that both have α-hydrogens gives a mixture of four compounds that are typically difficult to separate. Such reactions in most cases are not synthetically useful.

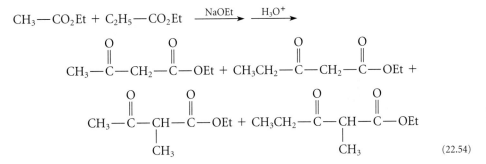

This problem is conceptually similar to the problem with crossed aldol reactions, discussed in Study Problem 21.1, Sec. 22.4C.

Crossed Claisen condensations are useful, however, if one ester is especially reactive or has no α-hydrogens. For example, formyl groups ($-CH=O$) are readily introduced with esters of formic acid such as ethyl formate:

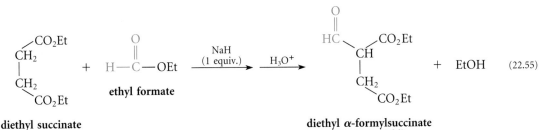

Formate esters fulfill both of the criteria for a crossed Claisen condensation. First, they have no α-hydrogens; second, their carbonyl reactivity is considerably greater than that of other esters. The reason for their greater reactivity is that the carbonyl group in a formate ester is "part aldehyde," and aldehydes are particularly reactive toward nucleophiles (Eq. 21.59).

A less reactive ester without α-hydrogens can be used if it is present in excess. For example, an ethoxycarbonyl group can be introduced with diethyl carbonate.

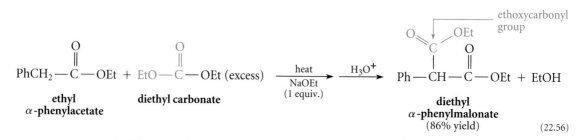

$$\text{(22.56)}$$

In this example, the enolate ion of ethyl α-phenylacetate condenses preferentially with diethyl carbonate rather than with another molecule of itself, because of the much higher concentration of diethyl carbonate. Of course, the excess diethyl carbonate must then be separated from the product.

Another type of crossed Claisen condensation is the reaction of ketones with esters. In this type of reaction the enolate ion of a ketone attacks the carbonyl group of an ester.

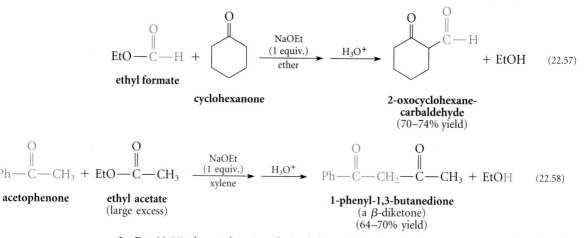

$$\text{(22.57)}$$

$$\text{(22.58)}$$

In Eq. 22.57, the enolate ion derived from the ketone cyclohexanone is acylated by the ester ethyl formate. In Eq. 22.58, the enolate ion of the ketone acetophenone is acylated by the ester ethyl acetate. In these reactions, several side reactions are possible in principle but in fact do not interfere. In Eq. 22.57, a possible side reaction is the aldol addition of cyclohexanone with itself. However, the equilibrium for the aldol addition of two ketones favors the reactants, whereas the Claisen condensation is irreversible because *one equivalent of base is used to form the enolate ion of the product.* Because the ester has no α-hydrogens, it cannot condense with itself.

The ester in Eq. 22.58, however, *does* have α-hydrogens and is known to condense with itself (Eq. 22.49). Why is such a condensation not an interfering side reaction? The answer is that ketones are far more acidic than esters (by about 5–7 pK_a units). Thus the enolate ion of the ketone is formed in much greater concentration than the enolate ion of the ester. The ketone enolate ion can react with another molecule of ketone—an unfavorable equilibrium—or it can be intercepted by the excess of ethyl acetate to give the observed product, which is a β-diketone. Like a β-keto ester, a β-diketone is especially acidic and is ionized completely by the one equivalent of NaOEt. (Be sure to identify the acidic hydrogens of the product in Eq. 22.58). Hence, β-diketone formation is observed because ionization makes this an irreversible reaction.

These examples illustrate that the crossed Claisen condensation can be used for the synthesis of a wide variety of β-dicarbonyl compounds.

PROBLEM

22.26   Complete the following reactions. Assume that one equivalent of NaOEt is present in each case.

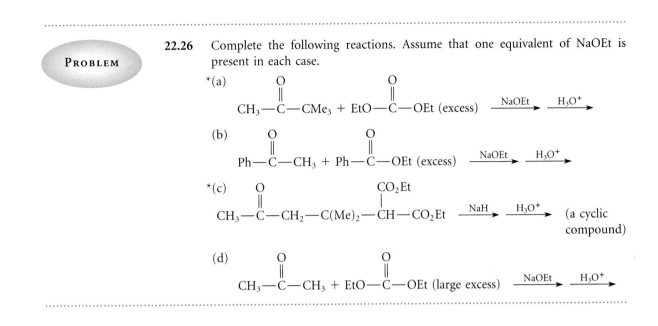

*(a)

$$CH_3-\overset{\overset{\displaystyle O}{\|}}{C}-CMe_3 + EtO-\overset{\overset{\displaystyle O}{\|}}{C}-OEt \text{ (excess)} \xrightarrow{\text{NaOEt}} \xrightarrow{\text{H}_3\text{O}^+}$$

(b)

$$Ph-\overset{\overset{\displaystyle O}{\|}}{C}-CH_3 + Ph-\overset{\overset{\displaystyle O}{\|}}{C}-OEt \text{ (excess)} \xrightarrow{\text{NaOEt}} \xrightarrow{\text{H}_3\text{O}^+}$$

*(c)

$$CH_3-\overset{\overset{\displaystyle O}{\|}}{C}-CH_2-C(Me)_2-\overset{\overset{\displaystyle CO_2Et}{|}}{CH}-CO_2Et \xrightarrow{\text{NaH}} \xrightarrow{\text{H}_3\text{O}^+} \text{(a cyclic compound)}$$

(d)

$$CH_3-\overset{\overset{\displaystyle O}{\|}}{C}-CH_3 + EtO-\overset{\overset{\displaystyle O}{\|}}{C}-OEt \text{ (large excess)} \xrightarrow{\text{NaOEt}} \xrightarrow{\text{H}_3\text{O}^+}$$

D. Synthesis with the Claisen Condensation

As the examples in the previous sections have shown, the Claisen condensation and related reactions can be used for the synthesis of β-dicarbonyl compounds: β-keto esters, β-diketones, and the like. Compare these types of compounds with those prepared by the aldol condensation and note the differences carefully.

In planning the synthesis of a β-dicarbonyl compound, you should adopt the usual two-step strategy: examine the target molecule and work backward to reasonable starting materials. Then don't forget to analyze the reaction of these starting materials to see whether the desired reaction is reasonable or whether other reactions will occur instead.

To determine the starting materials for a Claisen condensation, mentally reverse the condensation by adding the elements of ethanol (or other alcohol) across either of the carbon-carbon bonds *between* the carbonyl groups. Because there are two such bonds, you will generally find two possible sets of starting materials by this procedure.

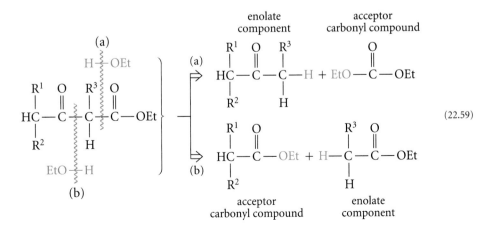

(22.59)

for the preparation of laboratory samples, or for preparation of compounds not available from the malonic ester synthesis, the preparation and alkylation of enolate ions with amide bases is particularly valuable.

The possibility of the Claisen condensation as a side reaction was noted in the discussion of Eq. 22.64. The use of a very strong amide base avoids the Claisen condensation for the following reason. The reaction is run by *adding the ester to the base*. When a molecule of ester enters the solution, either it can react with the strong base to form an enolate ion, or it can react with a molecule of already-formed enolate ion in the Claisen condensation. The reaction of esters with strong amide bases is so much faster, even at −78°, than the Claisen condensation that the enolate ion is formed instantly and never has a chance to undergo the Claisen condensation. In other words, the Claisen condensation is avoided because the ester and its enolate ion are never present simultaneously (except for an instant) in the reaction flask.

Another potential side reaction is attack of the amide base (or even its conjugate acid amine, which is, after all, still a base) on the ester. Because amines react with esters to give products of aminolysis (Sec. 21.8C), it would not be unreasonable to expect the *conjugate bases* of amines—very strong bases indeed—to react even more rapidly with esters. That this does not happen is once again the result of a competition. When an amide base reacts with the ester, it can either remove a proton, or it can attack the carbonyl carbon. Attack on the carbonyl carbon is retarded by van der Waals repulsions between groups on the carbonyl compound and the large branched groups on the bases. (These van der Waals repulsions have been aptly termed *F-strain*, or "front strain.") For such a branched amide base to attack the carbonyl carbon is somewhat like trying to put a dinner plate into the coin slot of a vending machine. If the amide base could be in contact with the ester long enough, it would eventually react at the carbonyl carbon; but the base instead reacts more rapidly a different way: it abstracts an α-proton. Reaction with a tiny hydrogen does not cause the van der Waals repulsions that would occur if the base were to attack the carbonyl carbon. Hence, the amide base takes the path of least resistance: it forms the enolate ion. Notice that van der Waals repulsions are used productively in this example—to avoid an *undesired* reaction.

PROBLEMS

22.31   Outline a synthesis of each of the following compounds from either diethyl malonate or ethyl acetate. Since the branched amide bases are relatively expensive, you may use them in only one reaction.

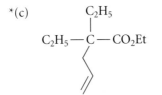

*(a)   (structure: CH—CO$_2$H with CH$_3$ and propenyl group)

(b)   CH$_3$CH$_2$CH$_2$ and CH$_3$CH$_2$CH$_2$ attached to CH—CO$_2$H

valproic acid
(used in treatment of epilepsy)

*(c)   C$_2$H$_5$—C(—CO$_2$Et)(C$_2$H$_5$) with alkyne group

22.32   The reactions of ester enolate ions are not restricted to alkylation. With this in mind, suggest the structure of the product formed when the enolate ion formed by the reaction of *tert*-butyl acetate with LCHIA reacts with each of the following compounds at −78° followed by dilute HCl.
*(a)   acetone   (b)   benzaldehyde

C. Acetoacetic Ester Synthesis

Recall that β-keto esters, like malonic esters, are substantially more acidic than ordinary esters (Eq. 22.50c), and are completely ionized by alkoxide bases.

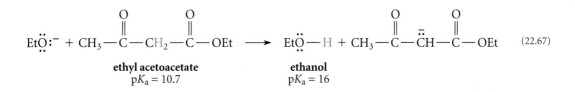

The enolate ions derived from β-keto esters, like those from malonate ester derivatives, can be alkylated by alkyl halides or sulfonate esters.

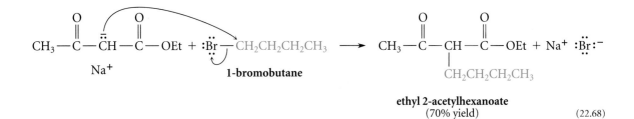

Dialkylation of β-keto esters is also possible.

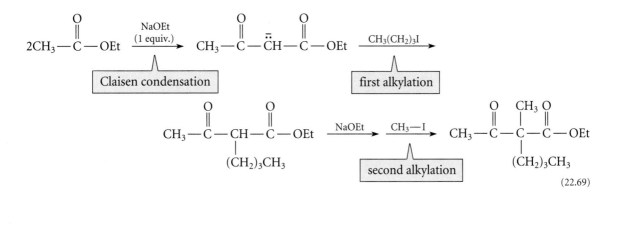

Alkylation of a Dieckmann condensation product is the same type of reaction:

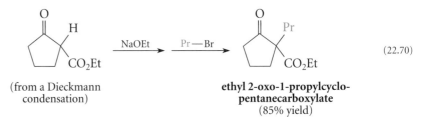

(from a Dieckmann condensation)

ethyl 2-oxo-1-propylcyclo-pentanecarboxylate
(85% yield)

(22.70)

Like esters of substituted malonic acids, the alkylated derivatives of ethyl acetoacetate can be hydrolyzed and decarboxylated to give ketones. Ester saponification and protonation gives a substituted β-keto acid; and β-keto acids spontaneously decarboxylate at room temperature (Sec. 20.11). This series of reactions is illustrated with the product of Eq. 22.68:

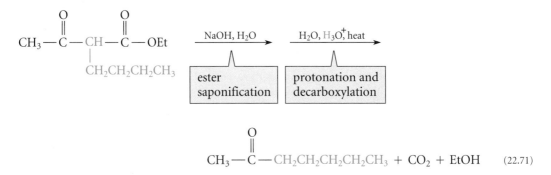

The alkylation of ethyl acetoacetate followed by saponification, protonation, and decarboxylation to give a ketone is called the **acetoacetic ester synthesis**. The alkylation part of this sequence, like the alkylation of diethyl malonate, involves the construction of new carbon-carbon bonds.

Whether a target ketone can be prepared by the acetoacetic ester synthesis can be determined by mentally reversing the synthesis.

This analysis involves replacing an α-hydrogen of the target ketone with a $-CO_2Et$ group. This process unveils the β-keto ester required for the synthesis. The β-keto ester, in turn, can either be prepared directly by a Claisen condensation or can be prepared from other β-keto esters by alkylation or di-alkylation with appropriate alkyl halides, as indicated by the possibilities in Eq. 22.72.

Outline a preparation of 2-methyl-3-pentanone by a reaction sequence that involves at least one Claisen condensation.

Solution   The discussion above leads to the following analysis:

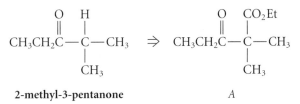

2-methyl-3-pentanone   *A*

The β-keto ester *A* cannot be prepared directly by a Claisen condensation because it would require a crossed Claisen condensation (see Eq. 22.59), and because the reaction could not be made irreversible by deprotonation. A second option is to provide one of the methyl groups by alkylation of the enolate ion derived from β-keto ester *B*:

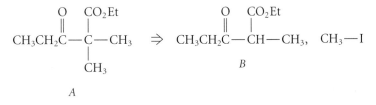

A   *B*

The enolate ion of compound *B*, in turn, can be prepared directly by the Claisen condensation of ethyl propionate. (This follows from the analysis in Eq. 22.59, Sec. 22.5D.)

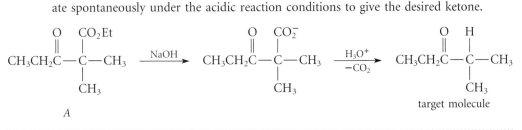

Saponifying *A* and acidifying the solution will give the β-keto acid, which will decarboxylate spontaneously under the acidic reaction conditions to give the desired ketone.

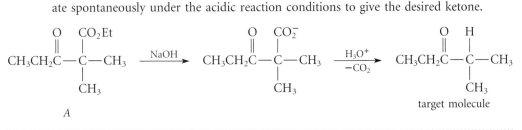

Do not let the large number of reactions in this chapter obscure a very important central theme: *enolate ions are nucleophiles,* and they do many of the things that other nucleophiles do: addition to carbonyl groups, nucleophilic acyl substitution, S_N2 reactions with alkyl halides, and so on. The reactions of enolate ions presented here are only a small fraction of those that are known. Yet if you grasp the central idea that enolate ions are nucleophiles, and if you understand the other reactions of nucleophiles, you should have little difficulty understanding (and perhaps even predicting) other reactions of enolate ions.

PROBLEMS

22.33 Outline a synthesis of each of the following compounds from ethyl acetoacetate and any other reagents.

*(a) 5-methyl-2-hexanone (b) 4-phenyl-2-butanone

22.34 Outline a synthesis of each of the following compounds from a β-keto ester; then show how the β-keto ester itself can be prepared.

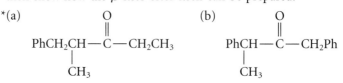

STUDY GUIDE LINK:
✓22.8
Further Analysis of
the Claisen
Condensation

*22.35 Predict the outcome of the following reaction by identifying *A*, then *B*, then the final product.

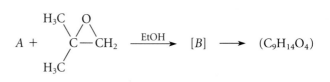

22.7 **Biosynthesis of Fatty Acids**

The utility of the Claisen condensation and the aldol reactions is not confined to the laboratory; these reactions are also important in the biological world. The biosynthesis of *fatty acids* (Sec. 20.5) illustrates how nature uses a reaction very similar to the Claisen condensation to build long carbon chains.

The starting material for the biosynthesis of fatty acids is a thiol ester of acetic acid called *acetyl-CoA*.

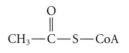

acetyl-CoA

The abbreviation acetyl-CoA stands for **acetyl-coenzyme A**, the complete structure of which is shown in Fig. 22.3. The complex functionality in this molecule is required for its recognition by enzymes. However, this complexity has no direct role in its chemical transformations and can be ignored for our purposes.

Acetyl-CoA is first converted into malonyl-CoA by carboxylation of the α-carbon.

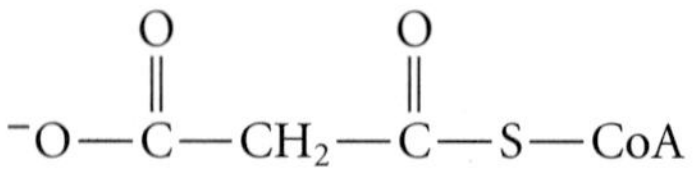

$$^-O-\overset{\overset{\displaystyle O}{\|}}{C}-CH_2-\overset{\overset{\displaystyle O}{\|}}{C}-S-CoA$$

malonyl-CoA

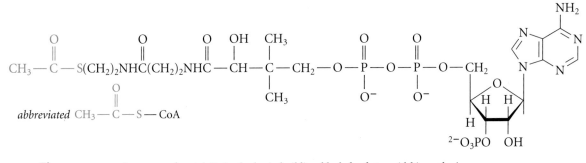

Figure 22.3 *Structure of acetyl-CoA, the basic building block for fatty acid biosynthesis.*

The —SCoA group in both acetyl- and malonyl-CoA is then replaced by a different group —SR, called the *acyl carrier protein*. Although this is an important aspect of the biochemistry, it makes no difference in the chemical transformations involved. In a reaction closely resembling the Claisen condensation, these malonyl and acetyl thiol esters react in an enzyme-catalyzed reaction to give an acetoacetyl thiol ester. (In this equation, BH⁺ and B: are acidic and basic groups, respectively, that are part of the enzyme catalyst.)

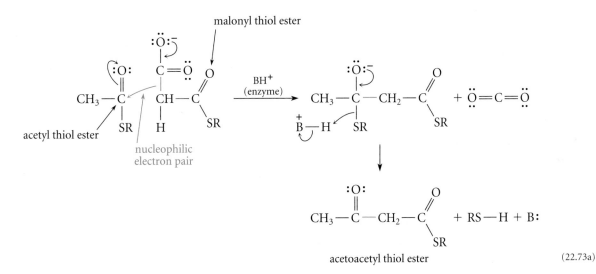

acetoacetyl thiol ester   (22.73a)

The nucleophilic electron pair (color in Eq. 22.73a) is made available not by proton abstraction, but by loss of CO_2 from malonyl-CoA. The loss of CO_2 as a gaseous by-product also serves another role: to drive the Claisen condensation to completion. Recall that in the laboratory, a Claisen condensation is driven to completion by ionization of the product with a strong base like ethoxide. Such a strong base cannot be used within living cells, in which all reactions must occur near neutral pH.

The product of Eq. 22.73a, an acetoacetyl thiol ester, then undergoes successively a carbonyl reduction, a dehydration, and a double-bond reduction, each catalyzed by an enzyme.

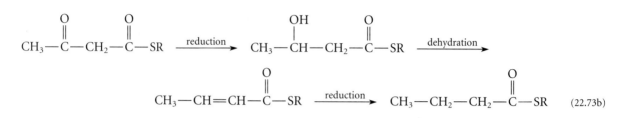

The net result of Eqs. 22.73a and b is that the acetyl thiol ester is converted into a thiol ester with *two additional carbons*. This sequence of reactions is then repeated, thus adding yet another two carbons to the chain.

$$CH_3CH_2CH_2C{-}SR \ + \ {}^-O{-}C{-}CH_2{-}C{-}SR \ \xrightarrow{H_3O^+}$$

$$CH_3CH_2CH_2C{-}CH_2{-}C{-}SR \ + \ R{-}SH \ + \ CO_2 \qquad (22.73c)$$

$$CH_3CH_2CH_2C{-}CH_2{-}C{-}SR \ \xrightarrow[\text{of Eq. 22.73b}]{\text{three reactions}} \ CH_3CH_2CH_2CH_2CH_2{-}C{-}SR \qquad (22.73d)$$

These four reactions are repeated with the addition of two carbons to the carbon chain at each cycle until a fatty acid with the proper chain length is obtained. The fatty acid thiol ester is then transesterified by glycerol to form fats and phospholipids (Sec. 21.12B).

The biosynthetic mechanism outlined above shows clearly why the common fatty acids have an *even number of carbon atoms*: they are formed from the successive addition of two-carbon acetate units. Fatty acids with an odd number of carbon atoms, although known, are relatively rare.

Fatty acids are not the only compounds in nature synthesized from acetyl-CoA. Isopentenyl pyrophosphate (the basic building block of isoprenoids and steroids; Sec. 17.6B) as well as a number of aromatic compounds found in nature are also ultimately derived from acetyl-CoA. Claisen condensations and aldol reactions play significant roles in the synthesis of these complex natural products.

This text has presented a number of illustrations of how chemistry is carried out in nature. *All of these processes have close analogies in laboratory chemistry.* With benefit of hindsight, it might seem obvious that natural chemistry and laboratory chemistry should be closely related. However, this point was far from obvious to early chemists. The serendipitous synthesis of urea by Friedrich Wöhler (Sec. 1.1B) signaled the beginning of an age in which the chemistry of living systems and laboratory chemistry are regarded as branches of the same basic science.

The "traditional" way of learning biochemistry is to memorize the many pathways and try to understand the relationships between them. The better way to learn biochemical pathways is to see them as logical sequences of transformations that make sense in terms of the organic chemistry involved. The student who brings an understanding of the fundamental mechanisms of organic chemistry to his or her study of biochemistry is empowered to take this more logical, and certainly less tedious, approach.

PROBLEM

22.36   Outline the biosynthetic reactions by which *(a) the thiol ester of hexanoic acid is converted into the thiol ester of octanoic acid; (b) the thiol ester of octanoic acid is converted into the thiol ester of decanoic acid.

22.8   Conjugate-Addition Reactions

A. Conjugate Addition to α,β-Unsaturated Carbonyl Compounds

The conjugated arrangement of C=C and C=O bonds endows α,β-unsaturated carbonyl compounds with unique reactivity, which is illustrated by the reaction of an α,β-unsaturated ketone with HCN.

$$
\underset{}{\text{Ph—CH=CH—}\overset{\overset{\text{O}}{\|}}{\text{C}}\text{—Ph}} + \text{NaCN} \xrightarrow[\text{EtOH}]{\overset{35°}{\text{HOAc}}} \underset{\underset{\text{CN}}{|}}{\text{Ph—CH—CH}_2\text{—}\overset{\overset{\text{O}}{\|}}{\text{C}}\text{—Ph}} \qquad (22.74)
$$

(93–96% yield)

In this reaction, the elements of HCN appear to have added across the C=C bond. Yet this is not a reaction of ordinary double bonds:

$$
\text{CH}_3\text{CH=CH}_2 + \text{NaCN} \xrightarrow{\text{HOAc}} \text{no reaction} \qquad (22.75)
$$

Nucleophilic addition to the double bond in an α,β-unsaturated carbonyl compound occurs because it gives a resonance-stabilized enolate ion intermediate:

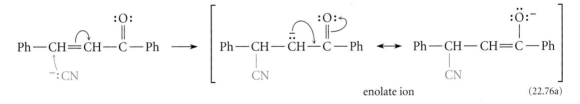

enolate ion   (22.76a)

(Nucleophilic addition to the alkene in Eq. 22.75, in contrast, would give a very unstable alkyl anion.) The enolate ion can be protonated on either oxygen or carbon. In either case a carbonyl group is eventually regenerated. The *overall* result of the reaction is net addition to the double bond.

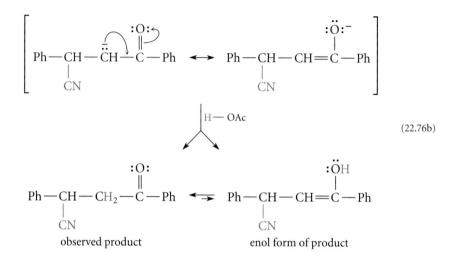

(22.76b)

Ph—CH—CH₂—C—Ph   ⇌   Ph—CH—CH=C—Ph
  | |
  CN   CN
observed product   enol form of product

Nucleophilic addition to the carbon-carbon double bonds of α,β-unsaturated aldehydes, ketones, esters, and nitriles is a rather general reaction that can be observed with a variety of nucleophiles. Some additional examples follow; try to write the mechanisms of these reactions.

$$CH_3—CH{=}CH—CO_2Et + NaCN \xrightarrow[H_2O]{EtOH} \xrightarrow{H_3O^+}$$

ethyl crotonate

$$CH_3—\underset{\underset{CN}{|}}{CH}—CH_2—CO_2Et \xrightarrow[heat]{H_2O,\ H_3O^+} CH_3—\underset{\underset{CO_2H}{|}}{CH}—CH_2—CO_2H \quad (22.77)$$

ethyl β-cyanobutyrate   **α-methylsuccinic acid**
                                   (66–70% yield)

$$\text{—SH} + CH_2{=}CH—CO_2Me \xrightarrow[MeOH]{NaOMe} \text{—S—CH}_2—CH_2—CO_2Me \quad (22.78)$$

2-propanethiol   **methyl acrylate**

methyl 3-(isopropylthio)propanoate
(97% yield)

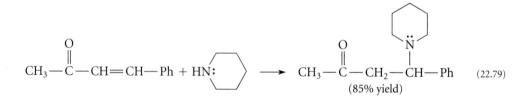

(22.79)
(85% yield)

$$CH_3SH + CH_2{=}CH—CN \xrightarrow[MeOH]{NaOMe} CH_3S—CH_2—CH_2—CN \quad (22.80)$$

methanethiol   **acrylonitrile**   **3-(methylthio)propanenitrile**
                                         (91% yield)

Notice that the addition of cyanide in Eq. 22.77 forms a new carbon-carbon bond, and that the nitrile group can then be converted into a carboxylic acid group by acid hydrolysis. The addition of a nucleophile to acrylonitrile (as in Eq. 22.80) is a useful reaction called **cyanoethylation**.

Quinones (Sec. 18.7) are α,β-unsaturated carbonyl compounds, and they also undergo similar addition reactions.

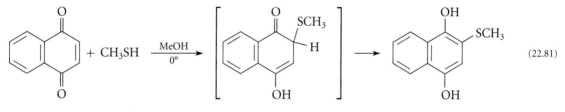

$$(22.81)$$

The examples above occur in base, but acid-catalyzed additions to the carbon-carbon double bonds of α,β-unsaturated carbonyl compounds are also known.

$$CH_2{=}CH{-}CO_2Me + HBr \xrightarrow{\ Et_2O\ } Br{-}CH_2CH_2{-}CO_2Me \qquad (22.82)$$

methyl acrylate   **methyl β-bromopropionate**
                                       (80–84% yield)

$$CH_2{=}CH{-}CH{=}O + HCl \xrightarrow{\ -15°\ } Cl{-}CH_2CH_2{-}CH{=}O \qquad (22.83)$$

Although such reactions appear to be nothing more than simple additions to the carbon-carbon double bond, this is not the case. The more basic position of an α,β-unsaturated carbonyl compounds is not the double bond, but rather the carbonyl oxygen (why?). Protonation on the carbonyl oxygen is followed by the attack of halide ion. The electrophilic oxygen can accept electrons as a result of nucleophilic attack either at the carbonyl carbon or, because of the conjugated arrangement of π bonds, at the β-carbon:

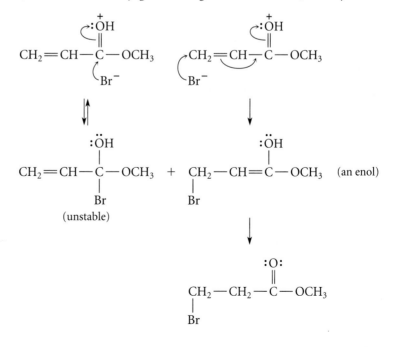

$$(22.84)$$

Attack of Br⁻ at the carbonyl carbon yields an unstable tetrahedral addition intermediate; attack at the β-carbon yields an enol, which rapidly reverts to the observed carbonyl product.

An addition to the double bond of an α,β-unsaturated carbonyl compound is an example of *conjugate addition*. The mechanism of the conjugate addition of HBr shown in Eq. 22.84 is similar to the conjugate addition of HBr to 1,3-butadiene (Sec. 15.4A); both involve carbocation intermediates. However, the *nucleophilic conjugate addition*, such as the addition of cyanide in Eq. 22.74, has no parallel in the reactions of simple conjugated dienes.

B. Conjugate Addition *vs.* Carbonyl-Group Reactions

Conjugate addition *competes* with carbonyl-group reactions. In the case of aldehydes and ketones, conjugate addition competes with *addition to the carbonyl group*. (Nuc = nucleophile; for example, in cyanide addition, H—Nuc = H—CN.)

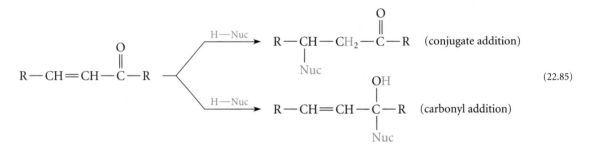

(22.85)

In the case of esters, conjugate addition competes with *nucleophilic acyl substitution*.

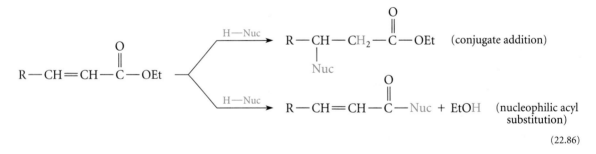

(22.86)

When can we expect to observe conjugate addition, and when can we expect reactions at the carbonyl carbon?

Consider first the reactions of aldehydes and ketones. Relatively weak bases that give *reversible* carbonyl-addition reactions with ordinary aldehydes and ketones tend to give conjugate addition with α,β-unsaturated aldehydes and ketones. Among the relatively weak bases in this category are cyanide ion, amines, thiolate ions, and enolate ions derived from β-dicarbonyl compounds. Conjugate addition is observed with these nucleophiles because, in most cases, it is *irreversible*. In other words, *conjugate-addition products are more stable than carbonyl-addition products.* If carbonyl addition is reversible—even if it occurs more rapidly—then conjugate addition can drain the carbonyl compound from the addition equilibrium, and the conjugate-addition product is formed ultimately.

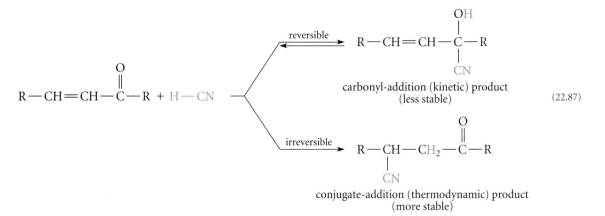

$$\text{(22.87)}$$

This, then, is another case of *kinetic vs. thermodynamic control of a reaction* (Sec. 15.4C). The conjugate-addition product is the thermodynamic (more stable) product of the reaction.

Why is the conjugate-addition product more stable? The answer lies in a simple bond energy argument. Conjugate addition retains a carbonyl group at the expense of a carbon-carbon double bond. Carbonyl addition retains a carbon-carbon double bond at the expense of a carbonyl group. Because a $C{=}O$ bond is considerably stronger than a $C{=}C$ bond (Table 5.3), conjugate addition gives a more stable product. (Of course, other bonds are broken and formed as well, but the major effect is the relative strengths of the two kinds of double bonds.) These same factors are reflected in the relative heats of formation of the isomers allyl alcohol and propionaldehyde:

$$
\begin{array}{ccc}
& CH_2{=}CH{-}CH_2{-}OH & CH_3{-}CH_2{-}CH{=}O \\
\Delta H_f^\circ & -132 \text{ kJ/mol} & -192 \text{ kJ/mol} \\
& (-31.6 \text{ kcal/mol}) & (-45.9 \text{ kcal/mol})
\end{array}
$$

As Eq. 22.87 suggests, carbonyl addition is in many cases the kinetically favored process, that is, it is faster than conjugate addition. When nucleophiles are used that undergo *irreversible* carbonyl additions, then the carbonyl addition product is observed rather than the conjugate addition product. This is exactly what happens with very powerful nucleophiles such as $LiAlH_4$ and organolithium reagents: these species add irreversibly to carbonyl groups, and form carbonyl-addition products whether the reactant carbonyl compound is α,β-unsaturated or not. (These reactions are discussed further in Secs. 22.9 and 22.10A.)

Many of the same nucleophiles that undergo conjugate addition with aldehydes and ketones also undergo conjugate addition with esters. In contrast, stronger bases that react irreversibly at the carbonyl carbon react with esters to give nucleophilic acyl substitution products. Thus, hydroxide ion reacts with an α,β-unsaturated ester to give products of saponification, a nucleophilic acyl substitution reaction, because saponification is not reversible. Likewise, $LiAlH_4$ reduces α,β-unsaturated esters at the carbonyl group, because attack of hydride ion on the carbonyl group is irreversible.

To summarize: Conjugate addition usually occurs with nucleophiles that are relatively weak bases. Stronger bases give irreversible carbonyl addition or nucleophilic acyl substitution reactions.

22.37 Give the product expected when methyl methacrylate (methyl 2-methylpropenoate) reacts with each of the following reagents.
*(a) ⁻CN in acidic EtOH   (b) C₂H₅SH and NaOEt (1 equiv) in EtOH
*(c) HBr   (d) NaOH

22.38 Give a mechanism for each of the following reactions.
*(a) MeN̈H₂ + 2CH₂=CH—CO₂Me ⟶

$$\text{MeO}_2\text{C—CH}_2\text{CH}_2—\overset{\displaystyle\cdot\cdot}{\underset{\displaystyle|}{\text{N}}}—\text{CH}_2\text{CH}_2—\text{CO}_2\text{Me}$$
$$\text{Me}$$

(b)

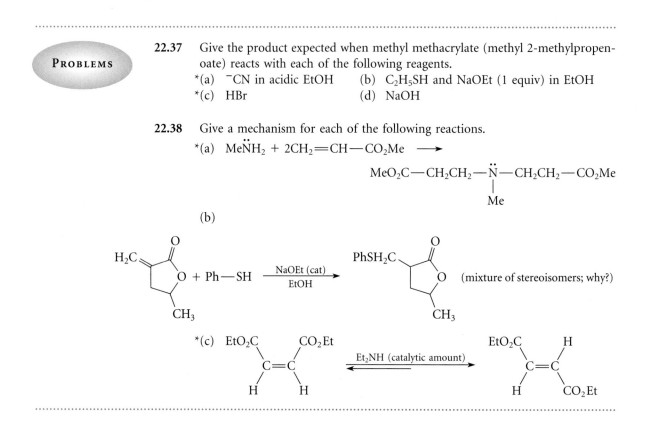

(mixture of stereoisomers; why?)

C. Conjugate Addition of Enolate Ions

Enolate ions derived from malonic ester derivatives, β-keto esters, and the like undergo conjugate-addition reactions with α,β-unsaturated carbonyl compounds, as in the following example:

$$\underset{\text{O}}{\overset{\text{O}}{\text{CH}_3—\overset{\|}{\text{C}}—\text{CH}=\text{CH}_2}} + \text{CH}_2(\text{CO}_2\text{Et})_2 \xrightarrow[\text{EtOH}]{\text{NaOEt (catalyst)}} \text{CH}_3—\overset{\text{O}}{\overset{\|}{\text{C}}}—\text{CH}_2\text{CH}_2\text{CH}(\text{CO}_2\text{Et})_2 \qquad (22.88)$$
$$\text{(65–71\% yield)}$$

The mechanism of this reaction follows exactly the same pattern established for other nucleophilic conjugate additions; the nucleophile is the enolate ion formed in the reaction of ethoxide with diethyl malonate (Eq. 22.61a, p. 1074). Notice that, in contrast to the Claisen ester condensation (Sec. 22.5A), this reaction requires only a catalytic amount of base. The reaction does *not* rely on ionization of the product to drive it to completion. It goes to completion because a carbon-carbon π bond in the starting α,β-unsaturated carbonyl compound is replaced by a stronger carbon-carbon σ bond.

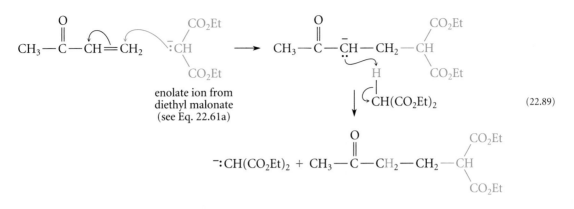

enolate ion from
diethyl malonate
(see Eq. 22.61a)

(22.89)

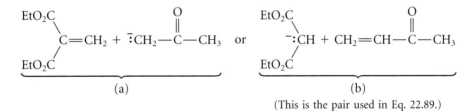

Conjugate additions of carbanions to α,β-unsaturated carbonyl compounds are called **Michael additions**, after Arthur Michael (1853–1942), a Harvard professor who investigated these reactions extensively.

Proper planning is needed to utilize the Michael addition in a synthesis. The product of many Michael additions could originate from either of two pairs of reactants. For example, in the reaction shown in Eq. 22.89, the same product (in principle) might be obtained by the Michael addition reaction of either of the following pairs of reactants (convince yourself of this point):

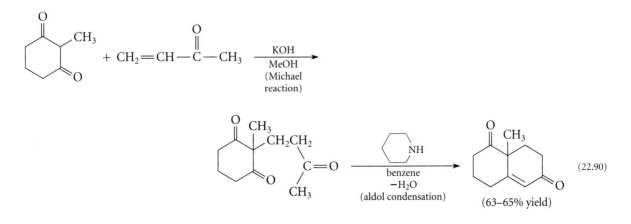

(This is the pair used in Eq. 22.89.)

Which pair of reactants should be used? To answer this question, use the result in Sec. 22.8B: Weaker bases tend to give conjugate addition, and stronger bases tend to give carbonyl-group reactions. Hence, to maximize conjugate addition, *choose the pair of reactants with the less basic enolate ion*—pair (b) in the case above.

In one useful variation of the Michael addition, the immediate product of the reaction can be subjected to an aldol condensation that closes a ring.

(22.90)

(Write the detailed mechanisms of these reactions.) This sequence is an example of a **Robinson annulation**, named for Sir Robert Robinson, a British chemist who pioneered its use. (An annulation is a ring-forming reaction, from Latin *annulus*, "ring.")

**STUDY
PROBLEM
22.6**

Outline a synthesis of tricarballylic acid from diethyl fumarate and any other reagents.

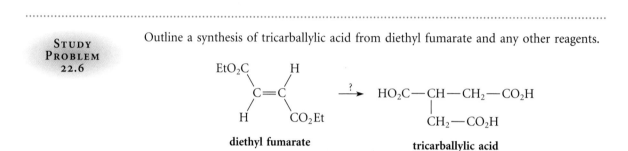

diethyl fumarate   **tricarballylic acid**

Solution   Two of the carboxylic acid groups required in the target are already in place as the ester units in diethyl fumarate. A Michael addition of some species that could be converted into a —CH₂CO₂H group is required. Notice that the desired product is conceptually a substituted acetic acid:

STUDY GUIDE LINK:
✓22.9
*Synthetic Equivalents
in Conjugate
Addition*

$$HO_2C—CH—CH_2—CO_2H$$
$$\quad\quad\quad |$$
$$\quad CH_2—CO_2H \quad\quad \text{substituted acetic acid}$$

Recall that one way of preparing substituted acetic acids is the malonic ester synthesis (Sec. 22.6A). The malonic ester synthesis can be employed if a Michael reaction with diethyl fumarate as the alkylating agent is used instead of an S_N2 reaction with an alkyl halide.

$$CH_2(CO_2Et)_2 \xrightarrow{\text{NaOEt}} {}^-:CH(CO_2Et)_2 \xrightarrow[\substack{\text{EtOH} \\ \text{Michael reaction}}]{} EtO_2C—CH—CH_2—CO_2Et$$
$$\quad\quad\quad\quad\quad\quad\quad\quad\quad\quad\quad\quad\quad\quad\quad\quad | $$
$$\quad\quad\quad\quad\quad\quad\quad\quad\quad\quad\quad\quad\quad\quad\quad CH(CO_2Et)_2$$

Saponification of all four ester groups, protonation, and decarboxylation would yield the desired tricarboxylic acid:

$$EtO_2C—CH—CH_2—CO_2Et \xrightarrow[\substack{\text{2) H}_3\text{O}^+}]{\substack{\text{1) NaOH} \\ \text{(saponification)}}}$$
$$\quad\quad | $$
$$\quad CH—CO_2Et$$
$$\quad\quad | $$
$$\quad CO_2Et$$

$$HO_2C—CH—CH_2—CO_2H \xrightarrow{\text{heat}} HO_2C—CH—CH_2—CO_2H + CO_2$$
$$\quad\quad | \quad\quad\quad\quad\quad\quad\quad\quad\quad\quad\quad\quad | $$
$$\quad CH—CO_2H \quad\quad\quad\quad\quad\quad\quad CH_2—CO_2H$$
$$\quad\quad | $$
$$\quad CO_2H$$

PROBLEMS

22.39   Provide structures for the missing nucleophiles that could be used in the following transformations.

*(a)

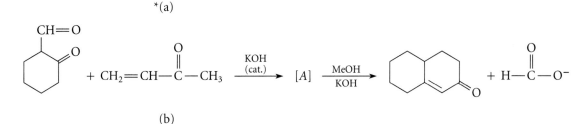

(b) X + CH₂=C(CH₃)—CO₂Et $\xrightarrow[\text{EtOH}]{\text{NaOEt}}$ $\xrightarrow[\text{heat}]{\text{H}_3\text{O}^+}$ HO₂CCH₂CH₂CHCO₂H (with CH₃ substituent)

22.40   Give a curved-arrow mechanism for each of the following reactions. In each reaction identify the bracketed intermediate.

*(a)

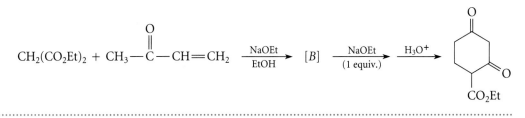

(b)

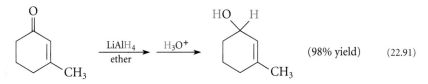

Reduction of α,β-Unsaturated Carbonyl Compounds

22.9

The carbonyl group of an α,β-unsaturated aldehyde or ketone, like that of an ordinary aldehyde or ketone (Sec. 19.8), is reduced to an alcohol with lithium aluminum hydride.

$$\text{(cyclohexenone with CH}_3\text{)} \xrightarrow[\text{ether}]{\text{LiAlH}_4} \xrightarrow{\text{H}_3\text{O}^+} \text{(cyclohexenol with CH}_3\text{)} \quad \text{(98\% yield)} \quad (22.91)$$

This reaction, like other LiAlH₄ reductions, involves the attack of hydride at the carbonyl carbon and is therefore a carbonyl addition.

   Why is carbonyl addition, rather than conjugate addition, observed in this case? The answer follows from the discussion in Sec. 22.8B. Carbonyl addition is not only *faster* than conjugate addition but, in this case, is also *irreversible*. It is irreversible because hydride is a very poor leaving group. Because carbonyl addition of LiAlH₄ is irreversible, conjugate addition never has a chance to occur, and is therefore not observed.

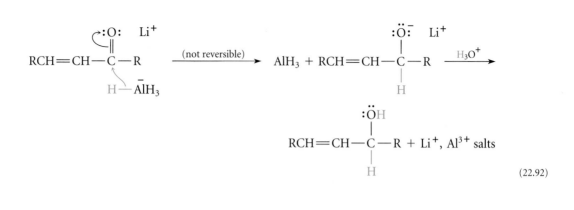

$$(22.92)$$

In other words, reduction of the carbonyl group with LiAlH$_4$ is a *kinetically controlled* reaction.

Many α,β-unsaturated carbonyl compounds are reduced by NaBH$_4$ to give mixtures of both carbonyl-addition products and conjugate-addition products. Because mixtures are obtained, NaBH$_4$ reductions of α,β-unsaturated ketones are not useful. Why conjugate addition is observed with NaBH$_4$ is not well understood. Although some cases of conjugate addition with LiAlH$_4$ are known, this reagent usually reduces carbonyl groups, including the carbonyl groups of esters, without affecting double bonds.

The carbon-carbon double bond of an α,β-unsaturated carbonyl compound can in most cases be reduced selectively by catalytic hydrogenation. (See also Eq. 19.32, p. 894.)

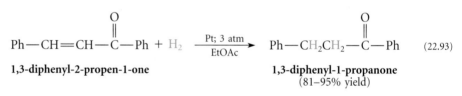

$$(22.93)$$

1,3-diphenyl-2-propen-1-one   **1,3-diphenyl-1-propanone**
(81–95% yield)

PROBLEM

22.41   Show how ethyl 2-butenoate can be used as a starting material to prepare
*(a) ethyl butanoate; (b) 2-buten-1-ol.

22.10 Reactions of α,β-Unsaturated Carbonyl Compounds with Organometallic Reagents

A. Addition of Organolithium Reagents to the Carbonyl Group

Organolithium reagents react with α,β-unsaturated carbonyl compounds to yield products of carbonyl addition.

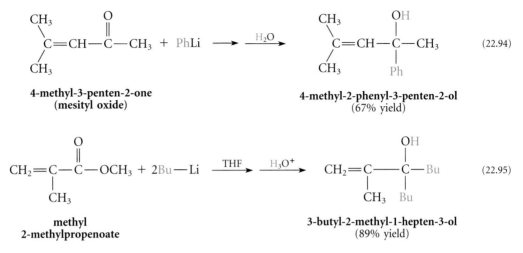

4-methyl-3-penten-2-one
(mesityl oxide)

4-methyl-2-phenyl-3-penten-2-ol
(67% yield)

(22.94)

methyl
2-methylpropenoate

3-butyl-2-methyl-1-hepten-3-ol
(89% yield)

(22.95)

The reason for carbonyl addition rather than conjugate addition is the same as in the case of LiAlH₄ reduction (Sec. 22.9): carbonyl addition is more rapid than conjugate addition and it is also irreversible.

Because Grignard and organolithium reagents undergo many of the same types of reactions, it is reasonable to ask whether Grignard reagents also undergo carbonyl addition. Grignard reagents in many cases give mixtures of conjugate addition and carbonyl addition. For this reason, organolithium reagents are used with α,β-unsaturated carbonyl compounds when carbonyl addition is the desired reaction.

B. Conjugate Addition of Lithium Dialkylcuprate Reagents

Lithium dialkylcuprate reagents (Sec. 21.10B) give exclusively products of *conjugate addition* when they react with α,β-unsaturated esters and ketones.

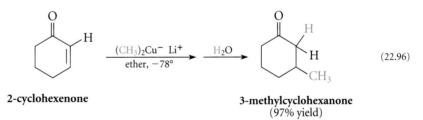

2-cyclohexenone

3-methylcyclohexanone
(97% yield)

(22.96)

Even α,β-unsaturated aldehydes, which are normally very reactive at the carbonyl group, give all or mostly products of conjugate addition, especially at low temperature.

$$Et_2C{=}CH{-}CH{=}O \xrightarrow[\substack{ether \\ -50°}]{(CH_3)_2CuLi} \xrightarrow{H_3O^+} Et_2C{-}CH_2{-}CH{=}O + Et_2C{=}CH{-}\overset{OH}{\underset{CH_3}{CH}}$$

(22.97)

(95% of product)   (5% of product)
70% total yield

*(a), (b)   3,4-dimethyl-2-hexanone (2 ways)

*(c)

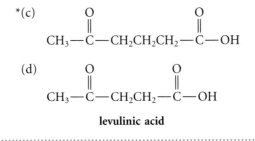

(d)

$$CH_3-\overset{\overset{\displaystyle O}{\|}}{C}-CH_2CH_2-\overset{\overset{\displaystyle O}{\|}}{C}-OH$$

levulinic acid

It is important to realize that many of the reactions discussed in this chapter can be used to form carbon-carbon bonds.

1. Aldol addition and condensation reactions (Sec. 22.4)
2. Claisen and Dieckmann condensations (Sec. 22.5)
3. Malonic ester synthesis (Sec. 22.6A)
4. Alkylation of ester enolates with amide bases and alkyl halides (Sec. 22.6B)
5. Acetoacetic ester synthesis (Sec. 22.6C)
6. Addition of cyanide ions (Sec. 22.8A) and enolate ions (Sec. 22.8C) to α,β-unsaturated carbonyl compounds
7. Reaction of lithium dialkylcuprates with α,β-unsaturated carbonyl compounds (Sec. 22.10B)

(A complete list of methods for forming carbon-carbon bonds is given in Appendix V.) Their utility for carbon-carbon bond formation accounts in large measure for the importance of these reactions in organic chemistry.

KEY IDEAS IN CHAPTER 22

Hydrogens on carbon atoms α to carbonyl groups and cyano groups are acidic. Ionization of these hydrogens gives enolate ions.

Most carbonyl compounds with α-hydrogens are in equilibrium with small amounts of enol (vinylic alcohol) tautomers. Generally, carbonyl-enol equilibria favor carbonyl tautomers, but there are important exceptions, such as phenols and β-diketones, in which enol tautomers are the major forms.

Enolization is catalyzed by acids and bases.

Enolate ions can act as nucleophiles in a number of reactions. Enolate ions can
1. undergo α-halogenation (haloform reaction)
2. add to carbonyl groups (aldol addition and condensation)
3. substitute at carbonyl groups (Claisen and Dieckmann condensations)
4. react with alkyl halides (enolate alkylation)
5. add to α,β-unsaturated carbonyl compounds (Michael addition).

Enols undergo α-halogenation and aldol condensation reactions in acidic solution.

The aldol addition and the Claisen condensation are reversible reactions. The aldol addition is generally favorable for aldehydes, but unfavorable for most ketones. It can be driven to completion by dehydration of the β-hydroxy carbonyl compound. The Claisen condensation is generally driven to completion by ionization of the product.

Two types of addition to α,β-unsaturated carbonyl compounds are possible: addition to the carbonyl group and addition to the double bond (conjugate addition, or 1,4-addition). When carbonyl addition is reversible, conjugate addition is observed because it gives the more stable product. When carbonyl addition is irreversible, it is observed instead because it is faster.

Reactions closely resembling enolate condensations are observed in nature. Many such reactions employ acetyl-CoA as a starting material.

ADDITIONAL PROBLEMS

*22.45   Give the principal organic product expected when 3-buten-2-one (methyl vinyl ketone) reacts with each of the following reagents.
(a) HBr   (b) Br_2 in CCl_4
(c) $LiAlH_4$, then H_2O (d) HCN in water, pH 10
(e) $Et_2Cu^- Li^+$, then H_3O^+ (f) diethyl malonate and NaOEt
(g) ethylene glycol, HCl (cat.)   (h) 1,3-butadiene

22.46   Give the principal organic products expected when ethyl *trans*-2-butenoate (ethyl crotonate) reacts with each of the following reagents.
(a) $^-$CN in water, then H_2O/H_3O^+, heat   (b) Me_2NH, room temperature
(c) NaOH, H_2O, heat   (d) CH_3Li (excess), then H_3O^+
(e) H_2, catalyst   (f) 1,3-cyclopentadiene

22.47   Give the structure of a compound that meets each criterion.
*(a) an optically active compound $C_6H_{12}O$ that racemizes in base
(b) an achiral compound $C_6H_{12}O$ that does not give a positive Tollens' test (Sec. 19.14)

(*Problem 22.47 continues*)

22.58   When compound *A* is treated with NaOCH₃ in CH₃OH, isomerization to compound *B* occurs.

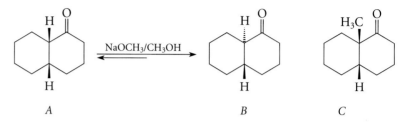

A   *B*   *C*

*(a)   Give a mechanism for this reaction and explain why the equilibrium favors compound *B*.

(b)   Explain why, when compound *C* is subjected to the same conditions, no isomerization occurs.

22.59   Indicate which hydrogens are replaced by deuterium when each of the following compounds is treated with dilute NaOD in a large excess of CH₃OD.

*(a)   (b)

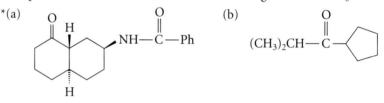

22.60   In either acid or base, 3-cyclohexenone comes to equilibrium with 2-cyclohexenone:

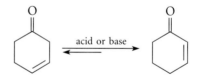

(a)   Explain why the equilibrium favors the α,β-unsaturated ketone over its β,γ-unsaturated isomer.

(b)   Give a mechanism for this reaction in aqueous NaOH.

(c)   Give a mechanism for the same reaction in dilute aqueous H₂SO₄. (*Hint:* The enol 1,3-cyclohexadienol is an intermediate in the acid-catalyzed reaction.)

(d)   Is the equilibrium constant for the analogous reaction of 4-methyl-3-cyclohexenone expected to be greater or smaller? Explain.

22.61   In 3-methyl-2-cyclohexenone the eight hydrogens H^a, H^b, H^c, and H^d can be exchanged for deuterium in CH₃O⁻/CH₃OD.

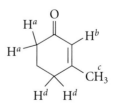

(a) Write a mechanism for the base-catalyzed exchange of hydrogens H^a, H^c, and H^d.

(b) Explain why hydrogen H^b is much less acidic than hydrogens H^a, H^c, and H^d, even though it is an α-hydrogen.

(c) Although hydrogen H^b is not unusually acidic, it nevertheless exchanges readily in base. Write a mechanism for the exchange of H^b. (*Hint:* Notice the equilibrium in Problem 22.60.)

22.62   In the following compound, identify the hydrogens that are exchanged for deuterium in CH_3O^-/CH_3OD.

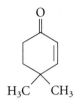

22.63   *(a) Compound *A* below, γ-pyrone, has a conjugate acid with an unusually high pK_a of -0.4. The pK_a of the conjugate acid of compound *B*, in contrast, is about -3.

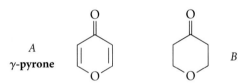

Draw the structures of the conjugate acids of both molecules, and explain why *A* is more basic than *B*.

(b) Tropone reacts with one equivalent of HBr to give a stable crystalline conjugate acid salt with a pK_a of -0.6, which is greater (i.e., less negative) than the pK_a values of most protonated α,β-unsaturated ketones. Give the structure of the conjugate acid of tropone, and explain why tropone is unusually basic.

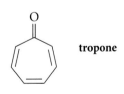

tropone

*22.64   (a) The following resonance structures can be written for an α,β-unsaturated carboxylic acid:

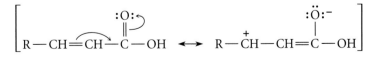

Would this type of resonance interaction increase or diminish the acidity of a carboxylic acid relative to that of an ordinary carboxylic acid?

(Problem 22.64 continues)

(b) Consider the following pK_a data:

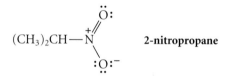

$$CH_2=CH-CH_2-CO_2H$$
$$pK_a = 4.37$$

$$CH_3CH_2CH_2-CO_2H$$
$$pK_a = 4.87$$

$$pK_a = 4.70$$

Show how these data are consistent with the polar effect of the double bond and with the resonance effect in part (a).

*22.65   (a) The pK_a of 2-nitropropane is 10. Give the structure of its conjugate base, and suggest reason(s) why 2-nitropropane has a particularly acidic C—H bond.

$$(CH_3)_2CH-\overset{+}{N}\underset{:\ddot{O}:^-}{\overset{\ddot{O}:}{\Big\Vert}}$$ **2-nitropropane**

(b) When the conjugate base of 2-nitropropane is protonated, an isomer of 2-nitropropane is formed, which, on standing, is slowly converted into 2-nitropropane itself. Give the structure of this isomer.

(c) What product forms when 2-nitropropane reacts with ethyl acrylate $(CH_2=CH-CO_2Et)$ in the presence of NaOEt in EtOH?

22.66   Cringe Labrack, a graduate student in his tenth year of study, has suggested each of the following synthetic procedures. Explain why each one cannot be expected to work.

*(a) $CH_3CH_2CO_2Et \xrightarrow[\text{EtOH}]{\text{NaOEt}} \xrightarrow{CH_3I} (CH_3)_2CHCO_2Et$

(b) $CH_2(CO_2Et)_2 \xrightarrow[\text{EtOH}]{\text{NaOEt}} \xrightarrow{PhBr} Ph-CH(CO_2Et)_2$

*(c)
$$\overset{\overset{\text{OH}}{|}}{CH_3CHCO_2Et} \xrightarrow[\text{heat}]{H_3O^+} CH_2=CH-CO_2Et$$

(d)
$$CH_3CH_2CO_2H \xrightarrow[\text{Br}_2]{\text{PBr}_3 \text{ (cat.)}} \xrightarrow[\text{ether}]{\text{Mg}} \xrightarrow[\text{2) H}_3O^+]{\text{1) CH}_3\text{CH}=\text{O}} CH_3-\overset{\overset{\text{OH}}{|}}{CH}-\overset{\overset{\text{CH}_3}{|}}{CH}-CO_2H$$

*(e)
[structure: acetophenone] $+ Br_2 \xrightarrow{\text{AlBr}_3 \text{ (cat)}}$ [structure: 3-bromoacetophenone]

(f)

$$C_2H_5 \overset{\overset{\displaystyle O}{\|}}{-}C-CH_3 + CH_3-\overset{\overset{\displaystyle O}{\|}}{CH} \xrightarrow{\ OH^-\ } CH_3-CH=CH-\overset{\overset{\displaystyle O}{\|}}{C}-C_2H_5$$

22.67 Crossed aldol condensations can be carried out if one of the carbonyl compounds is unusually acidic.

*(a) Give the structure of the α,β-unsaturated carbonyl compound that results from the crossed aldol condensation of diethyl malonate and acetone in NaOEt/EtOH.

(b) Explain why the aldol condensation of acetone with itself does not compete with the crossed aldol condensation in (a).

*(c) Outline a sequence of reactions for the conversion of the product from (a) into 3,3-dimethylbutanoic acid.

*22.68 Identify the intermediates A and B in the following transformation, and show how they are formed.

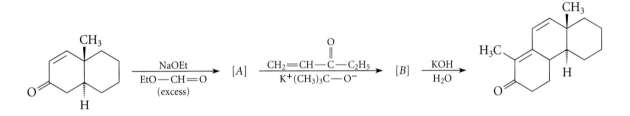

22.69 Explain the following findings.

(a) One full equivalent of base must be used in the Claisen or Dieckmann condensation.

(b) Ethyl acetate readily undergoes a Claisen condensation in the presence of one equivalent of sodium ethoxide, but phenyl acetate does *not* undergo a Claisen condensation in the presence of one equivalent of sodium phenoxide.

*22.70 When 2,4-pentanedione in ether is treated with one equivalent of NaH, a gas is evolved, and a species A is formed.

(a) Give the structure of A. Which atoms of A should be nucleophilic? Explain.

(b) When A reacts with CH_3I, three isomeric compounds, B, C, and D $(C_6H_{10}O_2)$, are formed. Suggest structures for these compounds.

22.71 Propose syntheses of each of the following compounds from the indicated starting materials and any other reagents.

*(a) 3-ethylcyclopentanol from 2-cyclopentenone

(b) 1,3,3-trimethylcyclohexanol from 3-methyl-2-cyclohexenone

*(c) 2-ethyl-1,3-hexanediol from butyric acid

(d) 2-benzylcyclohexanone from $EtO_2C(CH_2)_5CO_2Et$

(*Problem 22.71 continues*)

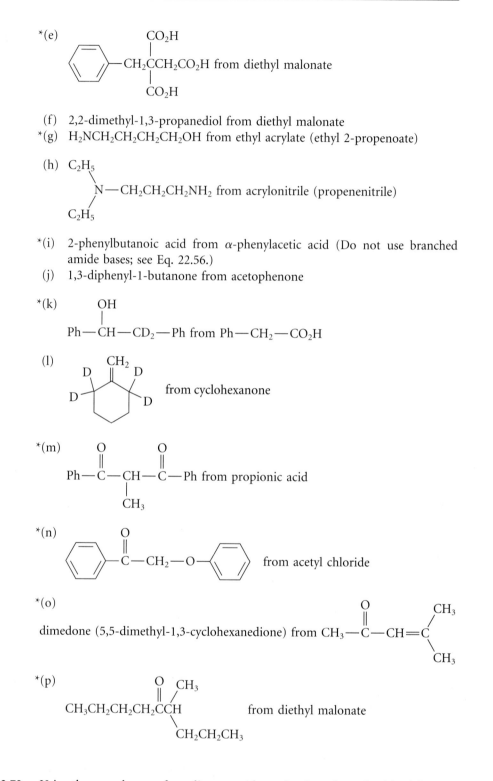

*(e)

$CH_2CCH_2CO_2H$ from diethyl malonate (with CO_2H groups and phenyl)

(f) 2,2-dimethyl-1,3-propanediol from diethyl malonate

*(g) $H_2NCH_2CH_2CH_2CH_2OH$ from ethyl acrylate (ethyl 2-propenoate)

(h) C_2H_5
N—$CH_2CH_2CH_2NH_2$ from acrylonitrile (propenenitrile)
C_2H_5

*(i) 2-phenylbutanoic acid from α-phenylacetic acid (Do not use branched amide bases; see Eq. 22.56.)

(j) 1,3-diphenyl-1-butanone from acetophenone

*(k)

Ph—CH—CD_2—Ph from Ph—CH_2—CO_2H (with OH)

(l)

from cyclohexanone

*(m)

Ph—C—CH—C—Ph from propionic acid (with CH$_3$)

*(n)

C—CH_2—O from acetyl chloride

*(o)

dimedone (5,5-dimethyl-1,3-cyclohexanedione) from CH_3—C—CH=C with CH$_3$ groups

*(p)

$CH_3CH_2CH_2CH_2CCH$ with CH$_3$ and $CH_2CH_2CH_3$ from diethyl malonate

22.72   Using the curved-arrow formalism, provide mechanisms for each of the following reactions.

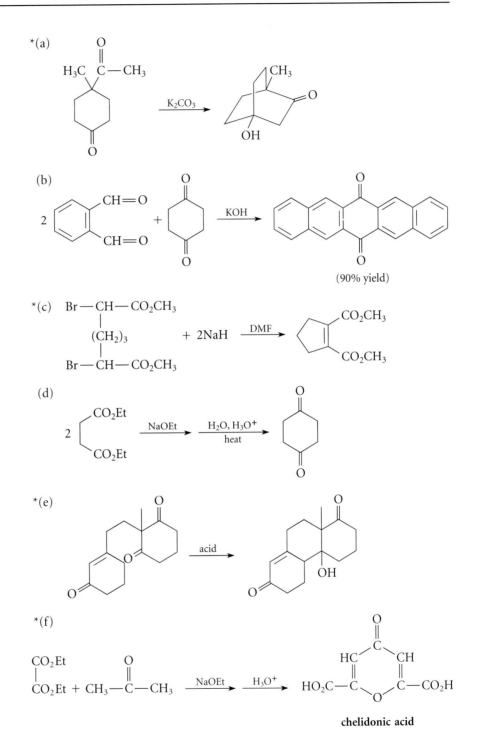

(90% yield)

chelidonic acid

*22.73   When the diester of a malonic acid derivative is treated with sodium ethoxide and urea, a *barbiturate* is formed. (Barbiturates are hypnotic drugs.)

(*Problem 22.73 continues*)

22.81 Using the reaction discussed in the previous problem, predict the final product of the following Wittig reaction-hydrolysis sequence:

$$CH_3OCH_2Cl + Ph_3P \xrightarrow{} \xrightarrow{Bu-Li} \xrightarrow{cyclohexanone} \xrightarrow{H_3O^+, H_2O}$$

22.82 Complete the following reactions by giving the major organic products. Explain your reasoning.

*(a)

$$CH_3-\overset{O}{\overset{||}{C}}-CH_2-\overset{O}{\overset{||}{C}}-OEt + Br(CH_2)_4Br \xrightarrow[EtOH]{NaOEt \text{ (excess)}} \xrightarrow[heat]{H_3O^+} (C_7H_{12}O)$$

(b) γ-butyrolactone $\xrightarrow{Li^+ [(CH_3)_2CH]_2\ddot{N}:^-} \xrightarrow{CH_3I}$

*(c)

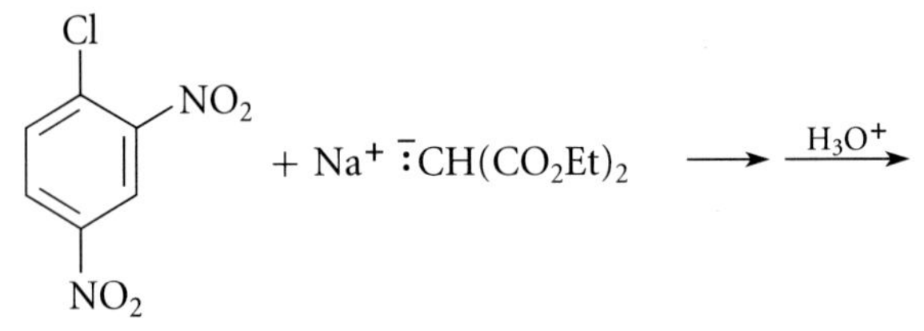

 + Na$^+$ $:$CH(CO$_2$Et)$_2$ $\xrightarrow{H_3O^+}$

(d)

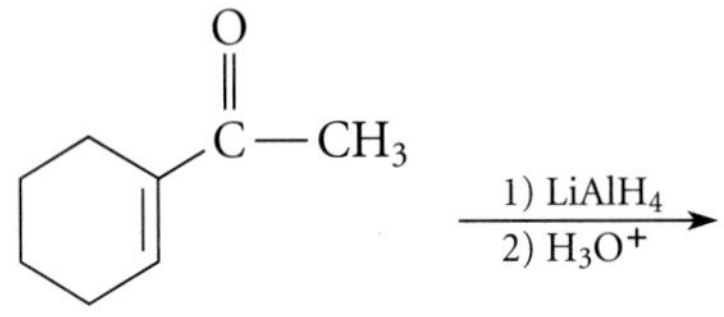

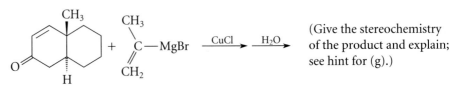

1) LiAlH$_4$
2) H$_3$O$^+$

*(e)

CH$_3$ ... CO$_2$Et + CH$_3$CH(CO$_2$Et)$_2$ $\xrightarrow[EtOH]{NaOEt}$

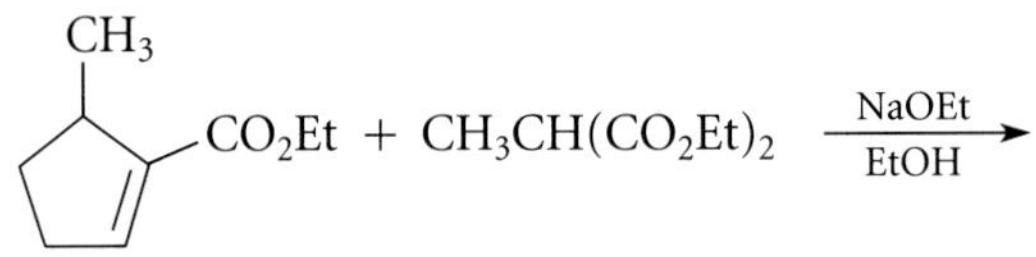

(f)

$$CH_3-\overset{O}{\overset{||}{C}}-CH_2-CO_2Et + CH_2{=}CH-CO_2Et \xrightarrow{EtO^-} \xrightarrow{H_3O^+}$$

*(g)

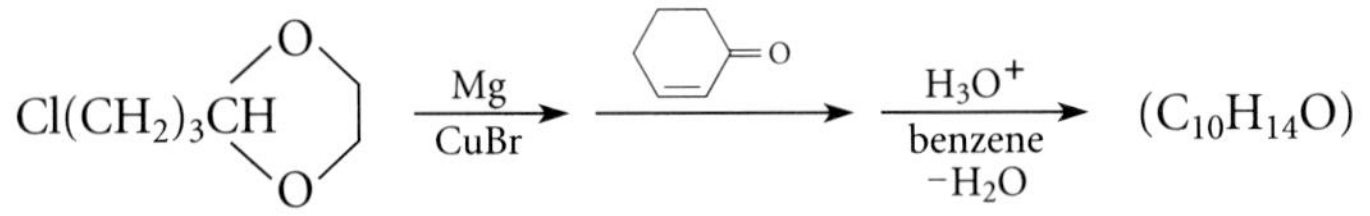

Cl(CH$_2$)$_3$CH ... $\xrightarrow[CuBr]{Mg}$ ... $\xrightarrow[\substack{benzene \\ -H_2O}]{H_3O^+}$ (C$_{10}$H$_{14}$O)

(*Hint:* Grignard reagents treated with CuBr or CuCl react like lithium dialkylcuprate reagents.)

(h)

... + $\overset{CH_3}{\underset{CH_2}{\overset{|}{C}}}$—MgBr $\xrightarrow{CuCl} \xrightarrow{H_2O}$

(Give the stereochemistry of the product and explain; see hint for (g).)

*(i)

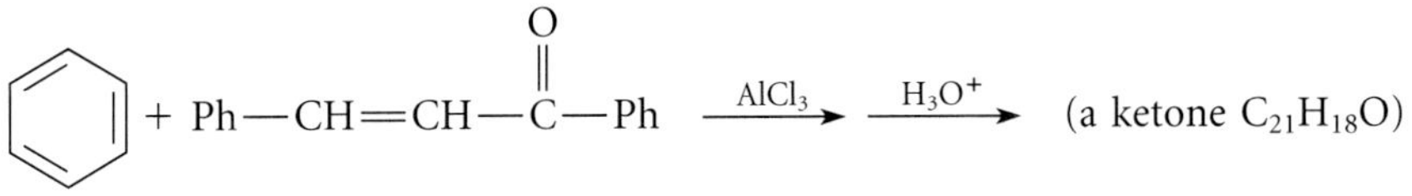

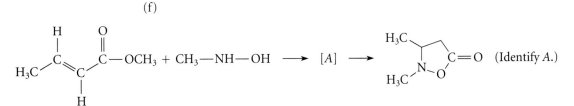

(a ketone $C_{21}H_{18}O$)

22.83   Using the curved-arrow formalism, provide mechanisms for each of the following reactions.

*(a)

$$CH_3-\overset{\overset{\displaystyle O}{\|}}{C}-CH_2-CO_2Et + PhCO_2Et \xrightarrow{\text{NaOEt}} \xrightarrow{H_3O^+}$$

$$Ph\overset{\overset{\displaystyle O}{\|}}{C}-CH_2-CO_2Et + CH_3CO_2Et \quad \text{(removed as it is formed)}$$

(b)

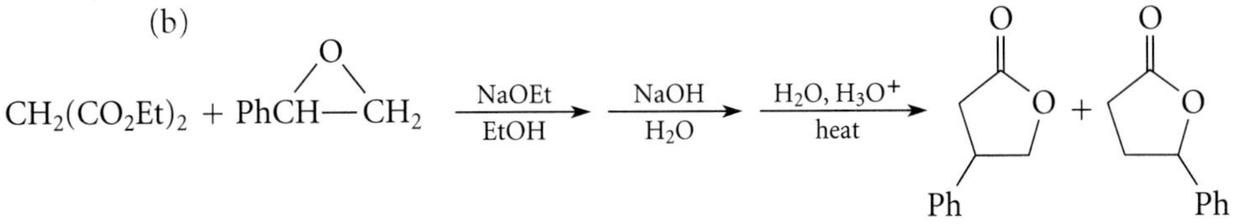

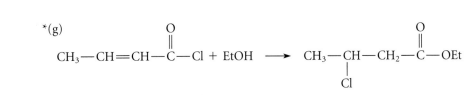

*(c)

$$PhO-CH=CH-\overset{\overset{\displaystyle O}{\|}}{C}-CH_3 \xrightarrow{H_2O, H_3O^+} O=CH-CH_2-\overset{\overset{\displaystyle O}{\|}}{C}-CH_3 + PhOH$$

(d)

$$PhO-CH=CH-\overset{\overset{\displaystyle O}{\|}}{C}-CH_3 \xrightarrow[\text{2) } H_3O^+]{\text{1) } H_2O, \, ^-OH} O=CH-CH_2-\overset{\overset{\displaystyle O}{\|}}{C}-CH_3 + PhOH$$

*(e)

$$Ph-\overset{\overset{\displaystyle OH}{|}}{CH}-CH=CH-CH_3 \xrightarrow[\text{EtOH/H}_2\text{O}]{\text{KOH}} Ph-\overset{\overset{\displaystyle O}{\|}}{C}-CH_2CH_2CH_3$$

(f)

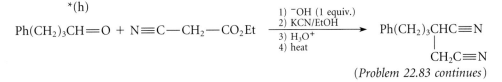

*(g)

$$CH_3-CH=CH-\overset{\overset{\displaystyle O}{\|}}{C}-Cl + EtOH \longrightarrow CH_3-\underset{\overset{|}{Cl}}{CH}-CH_2-\overset{\overset{\displaystyle O}{\|}}{C}-OEt$$

*(h)

$$Ph(CH_2)_3CH=O + N\equiv C-CH_2-CO_2Et \xrightarrow[\substack{\text{3) } H_3O^+ \\ \text{4) heat}}]{\substack{\text{1) } ^-OH \text{ (1 equiv.)} \\ \text{2) KCN/EtOH}}} Ph(CH_2)_3\underset{\overset{|}{CH_2C\equiv N}}{CHC\equiv N}$$

(*Problem 22.83 continues*)

23.1   Nomenclature of Amines

A. Common Nomenclature

In common nomenclature an amine is named by appending the suffix *amine* to the name of the alkyl group; the name of the amine is written as one word.

$$C_2H_5NH_2 \qquad (CH_3)_3N$$

ethylamine   trimethylamine

When two or more alkyl groups in a secondary or tertiary amine are different, the compound is named as an *N*-substituted derivative of the larger group.

$$(CH_3)_2N\!-\!CH_2CH_2CH_2CH_3 \qquad C_2H_5\!-\!NH\!-$$

N,N-dimethylbutylamine

N-ethylcyclohexylamine

This type of notation is required to show that the substituents are on the amine nitrogen and not on an alkyl group carbon.

Aromatic amines are named as derivatives of aniline.

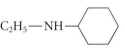

aniline   **3-nitroaniline**   **N-ethylaniline**
                  (***m*-nitroaniline**)

B. Substitutive Nomenclature

Because the IUPAC system for amine nomenclature is not logically consistent with IUPAC nomenclature of other organic compounds, the most widely used system of substitutive amine nomenclature is that of *Chemical Abstracts*, a comprehensive index to the world's chemical literature. In this system an amine is named in much the same way as the analogous alcohol, except that the suffix *amine* is used.

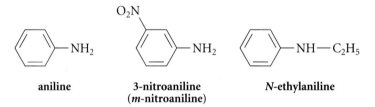

$$\begin{array}{cc} \text{OH} & \text{NH}_2 \\ | & | \\ CH_3CHCH_2CH_2CH_3 & CH_3CHCH_2CH_2CH_3 \end{array}$$

2-pentanol   **2-pentanamine**

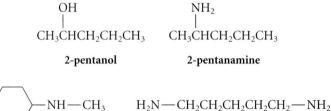

$$-NH\!-\!CH_3 \qquad H_2N\!-\!CH_2CH_2CH_2CH_2CH_2\!-\!NH_2$$

1,5-pentanediamine

N-methylcyclohexanamine

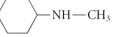

$$CH_3\!-\!NH\!-\!CH_2CH_2CH_2\!-\!NH\!-\!C_2H_5$$

N-ethyl-N′-methyl-1,3-propanediamine

Notice that in diamine nomenclature, as in diol nomenclature, the final *e* of the hydrocarbon name is retained. In the last example, the prime is used to show that the ethyl and methyl groups are on different nitrogens.

The priority of citation of amine groups as principal groups is just below that of alcohols:

$$\underset{\substack{\text{and derivatives)}}}{\overset{\displaystyle O}{\overset{\|}{-C}}-OH}\ \text{(carboxylic acid} > \overset{\displaystyle O}{\overset{\|}{-C}}-H\ \text{(aldehyde)} -\overset{\displaystyle O}{\overset{\|}{C}}-\ \text{(ketone)} > -OH > -NR_2 \qquad (23.1)$$

(A complete list of group priorities is given in Appendix I.) When cited as a substituent, the $-NH_2$ group is called the **amino** group.

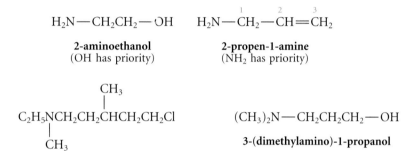

$$H_2N-CH_2CH_2-OH \qquad \overset{1}{H_2N}-\overset{2}{CH_2}-\overset{3}{CH}=CH_2$$

2-aminoethanol   **2-propen-1-amine**
(OH has priority)   (NH_2 has priority)

$$\underset{\displaystyle \underset{\displaystyle CH_3}{|}}{\overset{\displaystyle \overset{\displaystyle CH_3}{|}}{C_2H_5NCH_2CH_2CHCH_2CH_2Cl}} \qquad\qquad (CH_3)_2N-CH_2CH_2CH_2-OH$$

3-(dimethylamino)-1-propanol

5-chloro-*N*-ethyl-*N*,3-dimethyl-1-pentanamine

An *N* designation in the last example is unnecessary because the position of the methyl groups is clear from the parentheses.

Although *Chemical Abstracts* calls aniline *benzenamine*, the more common practice is to use the common name *aniline* in substitutive nomenclature.

The nomenclature of *heterocyclic compounds* was introduced in Sec. 8.1C in the discussion of ether nomenclature. Many important nitrogen-containing heterocyclic compounds are known by specific names that should be learned. Some important saturated heterocyclic amines are the following:

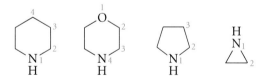

piperidine   morpholine   pyrrolidine   aziridine

As in the oxygen heterocyclics, numbering generally begins with the heteroatom. The following are examples of substituted derivatives:

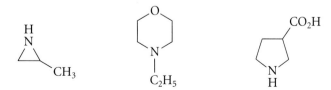

2-methylaziridine   *N*-ethylmorpholine   3-pyrrolidinecarboxylic acid
(4-ethylmorpholine)
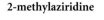

The fact that primary and secondary amines can both donate and accept hydrogen bonds also accounts for the fact that they have higher boiling points than ethers. On the other hand, alcohols are better hydrogen-bond donors than amines because they are more acidic. Therefore, alcohols have higher boiling points than amines.

$(C_2H_5)_2NH$ $(C_2H_5)_2O$ $(C_2H_5)_2CH_2$

diethylamine   **diethyl ether**   **pentane**

boiling point: 56.3°   37.5°   36.1°

$CH_3CH_2CH_2CH_2OH$ $CH_3CH_2CH_2CH_2NH_2$

1-butanol   **1-butanamine**

boiling point: 117.3°   77.8°

The water miscibility of most primary and secondary amines with four or fewer carbons, as well as trimethylamine, is consistent with their hydrogen-bonding abilities. Amines with large carbon groups have little or no water solubility.

23.4   Spectroscopy of Amines

A. IR Spectroscopy

The most important absorptions in the infrared spectra of primary amines are the N—H stretching absorptions, which usually occur as two or more peaks at 3200–3375 cm^{-1}. Also characteristic of primary amines is an NH$_2$ scissoring absorption (see Fig. 12.7) near 1600 cm^{-1}. These absorptions are illustrated in the IR spectrum of butylamine (Fig. 23.1). Most secondary amines show a single N—H stretching absorption rather than the multiple peaks observed for primary amines, and the absorptions associated with the various NH$_2$ bending vibrations of primary amines are not present. For example, diethylamine lacks the NH$_2$ scissoring absorption present in the butylamine spectrum. Tertiary amines obviously show no absorptions associated with N—H vibrations. The C—N stretching absorptions of amines, which occur in the same general part of the spectrum as C—O stretching absorptions (1050–1225 cm^{-1}), are not very useful.

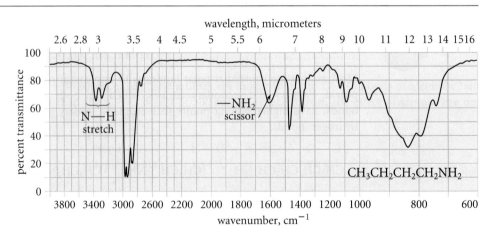

Figure 23.1   *IR spectrum of butylamine.*

B. NMR Spectroscopy

The characteristic resonances in the proton NMR spectra of amines are those of the protons adjacent to the nitrogen (the α-protons) and the N—H protons. In alkylamines, the α-protons are observed in the δ 2.5–3.0 region of the spectrum. In aromatic amines, the α-protons of N-alkyl groups are somewhat further downfield (why?), near δ 3. The following chemical shifts are typical:

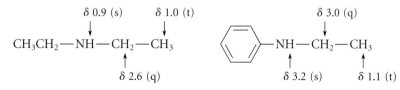

The chemical shift of the N—H proton, like that of the O—H proton in an alcohol, depends on the concentration of the amine, and on other conditions of the NMR experiment. In alkylamines, this resonance typically occurs at rather high field—typically around δ 1. In aromatic amines, this resonance is at lower field, as in the example above.

Like the OH protons of alcohols, phenols, and carboxylic acids, the NH protons of amines under most conditions undergo rapid exchange (Secs. 13.6D and 13.7). For this reason, splitting between the amine N—H and adjacent C—H groups is usually not observed. Thus, in the NMR spectrum of diethylamine the N—H resonance is a singlet rather than the triplet expected from splitting by the adjacent —CH₂— protons. In some amine samples the N—H resonance is broadened and, like the O—H proton of alcohols, it can be obliterated from the spectrum by exchange with D₂O.

The characteristic CMR absorptions of amines are those of the α-carbons—the carbons attached directly to the nitrogen. These absorptions occur in the δ 30–50 chemical-shift range. As expected from the relative electronegativities of oxygen and nitrogen, these shifts are somewhat less than the α-carbon shifts of ethers.

C. Mass Spectrometry

α-Cleavage (Sec. 12.6C) is a particularly important fragmentation mode of amines observed in mass spectrometry. For example, in the mass spectrum of butylamine, the only significant peaks are the molecular ion at *m/z* = 73 and the base peak at *m/z* = 30.

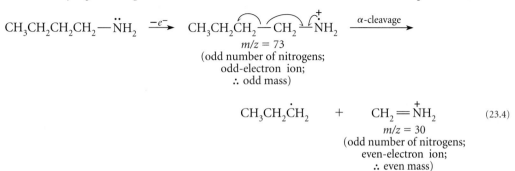

It is helpful to realize that compounds containing an odd number of nitrogens have odd molecular masses. (See Study Guide Link 21.2.) It follows that in the mass spectrum of an amine, the molecular ion occurs at an *odd mass* if the amine contains an *odd* number of nitrogens. Like the molecular ion, odd-electron ions containing an odd number of nitrogens are observed at odd mass, but even-electron ions containing an odd number of nitrogens are observed at even mass. These points are illustrated by the fragment ions shown in Eq. 23.4.

PROBLEMS

*23.4   Identify the compound that has the following spectra.
IR spectrum: 3279 cm^{-1}
NMR spectrum: δ 0.91 (1*H*, s), δ 1.07 (3*H*, t, *J* = 7 Hz),
δ 2.60 (2*H*, q, *J* = 7 Hz), δ 3.70 (2*H*, s), δ 7.18 (5*H*, apparent s)
Mass spectrum: *m/z* = 135 (M, 17%), 120 (40%), 91 (base peak)

23.5   A compound has IR absorptions at 3400–3500 cm^{-1} and the following NMR spectrum: δ 2.07 (6*H*, s), δ 2.16 (3*H*, s), δ 3.19 (broad, exchanges with D$_2$O), δ 6.63 (2*H*, s). To which one of the following compounds do these spectra belong? Explain.
(1) 2,4-dimethylbenzylamine   (2) 2,4,6-trimethylaniline
(3) *N*,*N*-dimethyl-*p*-methylaniline   (4) 3,5-dimethyl-*N*-methylaniline
(5) 4-ethyl-2,6-dimethylaniline

*23.6   An amine *A* has a mass spectrum with a base peak at *m/z* = 72. An amine *B* has a mass spectrum with a base peak at *m/z* = 58. One amine is 2-methyl-2-heptanamine, and the other is *N*-ethyl-4-methyl-2-pentanamine. Which is which?

23.7   Explain how you could distinguish between the two compounds in each of the following sets using *only* CMR spectroscopy.
*(a) 2,2-dimethyl-1-propanamine and 2-methyl-2-butanamine
(b) *trans*-1,2-cyclohexanediamine and *trans*-1,4-cyclohexanediamine

23.5 Basicity and Acidity of Amines

A. Basicity of Amines

Amines, like ammonia, are strong enough bases that they are completely protonated in dilute acid solutions.

$$CH_3\overset{..}{N}H_2 \;+\; H—Cl \;\rightleftharpoons\; CH_3—\overset{+}{N}H_3 \; Cl^- \;+\; H_2O \qquad (23.5)$$

methylamine   **methylammonium
chloride**

The salts of protonated amines are called **ammonium salts**. The ammonium salts of simple alkylamines are named as substituted derivatives of the ammonium ion. Other ammonium salts are named by replacing the final *e* in the name of the amine with the suffix *ium*.

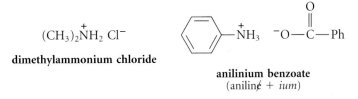

dimethylammonium chloride

anilinium benzoate
(anilin*e* + *ium*)

Always remember that ammonium salts are fully ionic compounds. Although ammonium chloride is often written as NH_4Cl, the structure is more properly represented as $^+NH_4$ Cl^-. Although the N—H bonds are covalent, there is no covalent bond between the nitrogen and the chlorine. (A covalent bond would violate the octet rule.)

Recall that the basicity of any compound, including an amine, is expressed in terms of the pK_a of its conjugate acid (Sec. 3.4C). The higher the pK_a of an ammonium ion, the more basic is its conjugate-base amine. (A discussion of the relationship between basicity constants K_b and dissociation constants K_a is given in Sec. 3.4C.)

B. Substituent Effects on Amine Basicity

The pK_a values for the conjugate acids of some representative amines are given in Table 23.1. As this table shows, the exact basicity of an amine depends on its structure.

Table 23.1   **Basicities of Some Amines**

(Each pK_a value is for the dissociation of the corresponding conjugate-acid ammonium ion.)

Amine	pK_a	Amine	pK_a	Amine	pK_a
CH_3NH_2	10.62	$(CH_3)_2NH$	10.64	$(CH_3)_3N$	9.76
$C_2H_5NH_2$	10.63	$(C_2H_5)_2NH$	10.98	$(C_2H_5)_3N$	10.65
$PhCH_2NH_2$	9.34				
$PhNH_2$	4.62	$PhNHCH_3$	4.85	$PhN(CH_3)_2$	5.06
O_2N—⟨⟩—NH_2	~1.0	O_2N-⟨⟩—NH_2	2.45		
Cl—⟨⟩—NH_2	3.81	Cl-⟨⟩—NH_2	3.32	Cl-⟨⟩—NH_2	2.62
H_3C—⟨⟩—NH_2	5.07	H_3C-⟨⟩—NH_2	4.67	CH_3-⟨⟩—NH_2	4.38

Three types of effects influence the basicity of amines—the same effects that influence the acid-base properties of other compounds. They are:

1. the effect of alkyl substitution
2. the polar effect
3. the resonance effect.

Recall that the pK_a of an ammonium ion, like that of any other acid, is directly related to the standard free-energy difference $\Delta G°$ between it and its conjugate base by the following equation (Sec. 3.6B).

$$\Delta G° = 2.3RT \, (\text{p}K_a) \tag{23.6}$$

The effect of a substituent group on pK_a can be analyzed in terms of how it affects the energy of either an ammonium ion or its conjugate-base amine, as shown in Fig. 23.2. For example, if a substituent stabilizes an amine more than it stabilizes the conjugate-acid ammonium ion (Fig. 23.2a), the standard free energy of the amine is lowered, $\Delta G°$ is decreased, and the pK_a of the ammonium ion is reduced; that is, the amine is less basic than the amine without the substituent. If a substituent stabilizes the ammonium ion more than its conjugate-base amine (Fig. 23.2b), the opposite effect is observed: the pK_a is increased, and the amine basicity is also increased.

Consider first the effect of alkyl substitution. Most common alkylamines are somewhat more basic than ammonia in aqueous solution:

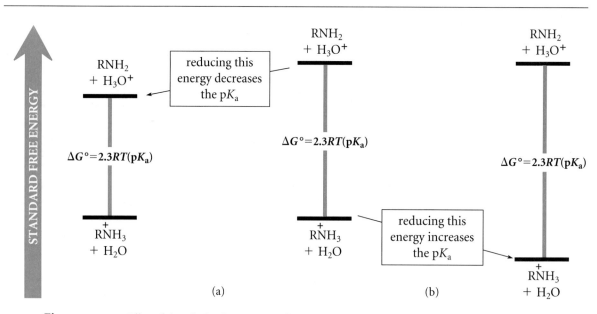

(a)   (b)

Figure 23.2   *Effect of the relative free energies of ammonium ions and amines on the pK_a values of the ammonium ions. (a) Reducing the energy of an amine decreases the pK_a of its conjugate-base ammonium ion and thus reduces the basicity of the amine. (b) Reducing the energy of an ammonium ion increases its pK_a and thus increases the basicity of its conjugate-base amine.*

pK_a *in aqueous solution:*

$$\overset{+}{N}H_4 \quad Et\overset{+}{N}H_3 \quad Et_2\overset{+}{N}H_2 \quad Et_3\overset{+}{N}H \qquad (23.7)$$
$$9.21 \quad\quad 10.63 \quad\quad 10.98 \quad\quad 10.65$$

These pK_a values show that the secondary amine is the most basic. Two opposing factors are actually at work here. The first is the tendency of alkyl groups to stabilize charge through a *polarization* effect. The electron clouds of the alkyl groups distort so as to create a net attraction between them and the positive charge of the ammonium ion:

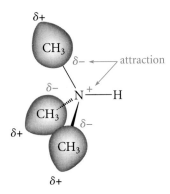

Because the ammonium ion is stabilized by this effect, its pK_a is increased (Fig. 23.2b). This effect is evident in the *gas-phase* basicities of amines. In the gas phase, the acidity of ammonium ions decreases regularly with increasing alkyl substitution:

Gas-phase acidity:

$$\overset{+}{N}H_4 > C_2H_5\overset{+}{N}H_3 > (C_2H_5)_2\overset{+}{N}H_2 > (C_2H_5)_3\overset{+}{N}H \qquad (23.8)$$

The same polarization effect also operates in the gas-phase acidity of alcohols (Sec. 8.5C), except in the opposite direction. In other words, the polarization of alkyl groups can act to stabilize *either* positive or negative charge. In the presence of a positive charge, as in ammonium ions, electrons polarize *toward* the charge; in the presence of a negative charge, as in alkoxide ions, they polarize *away from* the charge. We might say that electron clouds are like some politicians: they polarize in whatever way is necessary to create the most favorable situation.

The second factor involved in the effect of alkyl substitution on amine basicity must be a *solvent effect*, because the basicity order of amines in the gas phase (Eq. 23.8) is different from that in aqueous solution (Eq. 23.7). In other words, the solvent water must play an important role in the solution basicity of amines. An explanation of this solvent effect is that ammonium ions in solution are stabilized not only by alkyl groups, but also by hydrogen-bond donation to the solvent:

$$
\begin{array}{c}
H\text{---}:\!\ddot{O}H_2 \\
| \\
CH_3\!-\!\overset{+}{N}\!-\!H\text{---}:\!\ddot{O}H_2 \\
| \\
H\text{---}:\!\ddot{O}H_2
\end{array}
$$

Primary ammonium salts have three hydrogens that can be donated to form hydrogen bonds, but a tertiary ammonium salt has only one. Thus, primary ammonium ions are stabilized by hydrogen bonding more than tertiary ones.

The pK_a values of alkylammonium salts reflect the operation of both hydrogen bonding and alkyl-group polarization. Because these effects work in opposite directions, the basicity in Eq. 23.7 maximizes at the secondary amine.

Ammonium-ion pK_a values, like the pK_a values of other acids, are also sensitive to the *polar effects* of substituents.

$$\overset{+}{Et_2NH}CH_2C\equiv N \qquad \overset{+}{Et_2NH}CH_2CH_2C\equiv N \qquad \overset{+}{Et_2NH}(CH_2)_4C\equiv N \qquad \overset{+}{Et_3NH} \qquad (23.9)$$

$$pK_a \qquad 4.55 \qquad\qquad 7.65 \qquad\qquad 10.08 \qquad\qquad 10.65$$

An electronegative (electron-withdrawing) group such as halogen or cyano destabilizes an ammonium ion because of a repulsive electrostatic interaction between the positive charge on the ammonium ion and the positive end of the substituent bond dipole.

$$\overset{\delta-}{N}\equiv\overset{\delta+}{C}-CH_2-CH_2-\overset{+}{NH_3} \quad \text{repulsive interaction}$$

Notice that the polar effects of substituent groups operate largely on the *conjugate acid* of the amine—the alkylammonium ion—because this cation is the charged species in the acid-base equilibrium. (Recall from Sec. 3.6B that polar effects on stability are greatest on the charged species in a chemical equilibrium.) In the case of a carboxylic acid, alcohol, or phenol, the *conjugate-base anion* is the charged species; consequently, the polar effects of substituents are reflected primarily in their effects on the stabilities of these anions.

The data in Eq. 23.9 show that the base-weakening effect of electron-withdrawing substituents, like all polar effects, decreases rapidly with distance between the substituent and the charged atom.

Resonance effects on amine basicity are illustrated by the difference between the pK_a values of the conjugate acids of aniline and cyclohexylamine, two primary amines of almost the same shape and molecular mass.

$$\langle\!\!\!\!\bigcirc\!\!\!\!\rangle-\overset{+}{NH_3} \qquad \bigcirc-\overset{+}{NH_3} \qquad\qquad (23.10)$$

$$pK_a \qquad 4.62 \qquad\qquad 10.64$$

(Notice that the pK_a values of the substituted anilinium ions in Table 23.1 are considerably lower than the pK_a values of the alkylammonium ions.) Aniline is stabilized by resonance interaction of the unshared electron pair on nitrogen with the aromatic ring (Eq. 23.3). When aniline is protonated, this resonance stabilization is no longer present, because the unshared pair is bound to a proton and is "out of circulation." The stabilization of aniline relative to its conjugate acid reduces its basicity (Eq. 23.6 and Fig. 23.2). In other words, the resonance stabilization of aniline lowers the energy required for its formation from its conjugate acid, and thus lowers its basicity relative to that of cyclohexylamine, in which the resonance effect is absent.

The electron-withdrawing polar effect of the aromatic ring also contributes significantly to the reduced basicity of aromatic amines as it does to the increased acidities of phenols relative to alcohols.

STUDY
PROBLEM
23.1

Arrange the following three amines in order of increasing basicity.

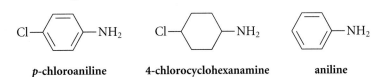

| *p*-chloroaniline | 4-chlorocyclohexanamine | aniline |

Solution Because the chloro substituent is an electron-withdrawing group, it reduces the basicity of an aniline. Hence, *p*-chloroaniline is less basic than aniline. Because of the resonance effect, *p*-chloroaniline is also less basic than 4-chlorocyclohexanamine. But what about the relative basicity of aniline and 4-chlorocyclohexanamine? The resonance and polar effects of the aromatic ring and the polar effect of the chlorine are all base-weakening effects. The problem is to decide whether the effect of the aromatic ring or the effect of the chlorine is more important in reducing basicity. To make this decision, reason *by analogy*. Examine the effect of an electron-withdrawing group on the basicity of an amine in which the polar effect is the only effect that can operate. For example, the series in Eq. 23.9 shows that an electronegative group four carbons away from the amine nitrogen has a very modest effect. On the other hand, the comparison in Eq. 23.10 shows that the resonance and polar effects of an aromatic ring change the pK_a of an amine by about six units. Consequently, the base-weakening effect of "changing" a cyclohexane ring to a phenyl ring is much more important than the base-weakening effect of "replacing" a hydrogen with a 4-chloro group in cyclohexanamine. Hence, the basicity order is:

$$\text{\textit{p}-chloroaniline} < \text{aniline} \ll \text{4-chlorocyclohexanamine}$$

PROBLEMS

*23.8 Arrange the amines within each set in order of increasing basicity, least basic first.
(a) propylamine, ammonia, dipropylamine
(b) methyl 3-aminopropanoate, *sec*-butylamine, $\overset{+}{H_3N}CH_2CH_2NH_2$
(c) aniline, methyl *m*-aminobenzoate, methyl *p*-aminobenzoate
(d) benzylamine, *p*-nitrobenzylamine, cyclohexylamine, aniline

27.9 Explain the basicity order of the following three amines:

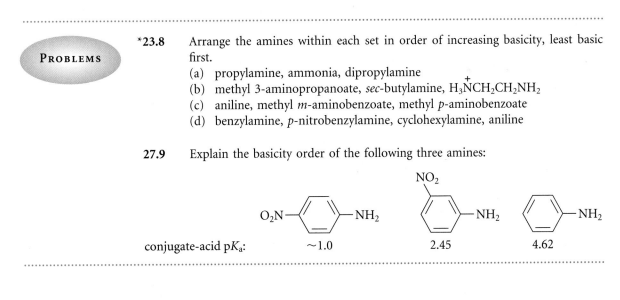

conjugate-acid pK_a: ~1.0   2.45   4.62

C. Separations Using Amine Basicity

Because ammonium salts are ionic compounds, many have appreciable water solubilities. Hence, when a water-insoluble amine is treated with dilute aqueous acid, for example,

5% HCl solution, the amine dissolves as its ammonium salt. Upon treatment with base, the ammonium salt is converted back into the amine. These observations can be used to design separations of amines from other compounds, as the following study problem illustrates.

STUDY PROBLEM 23.2

A chemist has treated *p*-chloroaniline with acetic anhydride, and wants to separate the amide product from any unreacted amine. Design a separation based on the basicities of the two compounds.

Solution   If the mixture is treated with 5% aqueous HCl, the amine will form the hydrochloride salt and dissolve in the aqueous solution. The amide, however, is not basic enough to be protonated in 5% HCl, and therefore does not dissolve.

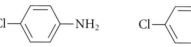

p-chloroaniline (water-insoluble)	**N-(p-chlorophenyl)acetamide** (water-insoluble)

5% aqueous HCl	5% aqueous HCl

                                            no appreciable reaction

soluble in aqueous solution

(23.11)

The water-insoluble amide can be filtered or extracted away from the aqueous solution of the ammonium salt. Then the aqueous solution of the ammonium salt can be treated with NaOH to liberate the free amine.

Amine basicities can play a key role in the design of enantiomeric resolutions. The enantiomeric resolution discussed in Sec. 6.8 should be reviewed with this idea in mind.

PROBLEMS

*23.10   Using solubilities in acidic or basic solution as a basis, design a separation of a mixture containing *p*-chlorobenzoic acid, *p*-chloroaniline, and *p*-chlorotoluene into its pure components.

23.11   Design an enantiomeric resolution of racemic 2-phenylpropanoic acid using a pure enantiomer of 1-phenylethanamine as resolving agent. (Assume that a solvent can be found in which the diasteromeric salts have different solubilities.) Discuss the importance of amine basicity to the success of this scheme.

D. Acidity of Amines

Although amines are normally considered to be bases, primary and secondary amines are also very weakly acidic. The conjugate base of an amine is called an **amide** (not to be confused with amide derivatives of carboxylic acids).

The amide conjugate base of ammonia itself is usually prepared by dissolving an alkali metal such as sodium in liquid ammonia in the presence of a trace of ferric ion. When sodium is used, the resulting base is called *sodium amide* (or *sodamide*).

$$2Na + 2\overset{\cdot\cdot}{N}H_3 \xrightarrow{\ Fe^{3+}\ } 2Na^+ \ ^-{:}\overset{\cdot\cdot}{N}H_2 + H_2 \tag{23.12}$$

sodium amide
(sodamide)

The conjugate bases of alkylamines are prepared by treating the amine with butyllithium in an ether solvent such as THF.

$$(Me_2CH)_2\overset{\cdot\cdot}{N}-H + CH_3CH_2CH_2CH_2-Li \longrightarrow (Me_2CH)_2\overset{\cdot\cdot}{N}{:}^- \ Li^+ + CH_3CH_2CH_2CH_3 \tag{23.13}$$

diisopropylamine   **butyllithium**   **lithium
diisopropylamide**

STUDY GUIDE LINK:
23.1
Structures of Amide
Bases

The pK_a of a typical amine is about 35. Thus, amides are *very* strong bases. This is why they can be used to form acetylide or enolate ions (Secs. 14.7A, 22.6B).

E. Summary of Acidity and Basicity

We have now surveyed the acidity and basicity of the most important organic functional groups. This information is summarized in the tables in Appendix VI. Although the values given in these tables are typical values, remember that acidity and basicity are affected by alkyl substitution, polar effects, and resonance effects. The acid-base properties of organic compounds are important not only in predicting many of their chemical properties, but also in their industrial and medicinal applications.

23.6 Quaternary Ammonium Salts

Closely related to ammonium salts are compounds in which all four hydrogens of $^+NH_4$ are replaced by alkyl or aryl groups. Such compounds are called **quaternary ammonium salts**. The following compounds are examples:

$(CH_3)_4N^+ \ Cl^-$ $Ph\overset{+}{N}Et_3 \ Br^-$

tetramethylammonium chloride   ***N,N,N*-triethylanilinium bromide**

$PhCH_2\overset{+}{N}(CH_3)_3 \ ^-OH$

benzyltrimethylammonium hydroxide
(Triton B)

$$CH_3(CH_2)_{15}-\overset{\overset{\displaystyle CH_3}{|}}{\underset{\underset{\displaystyle CH_3}{|}}{N^+}}-CH_2-Ph \ Cl^-$$

"benzalkonium chloride"
a cationic surfactant (Sec. 20.5)

Like the corresponding ammonium ions, quaternary ammonium salts are fully ionic compounds. Many quaternary ammonium salts containing large organic groups are soluble in nonaqueous solvents. Triton B, for example, is used as a source of hydroxide ion that is soluble in organic solvents. Benzalkonium chloride, a common antiseptic, acts as a surfactant in water (Sec. 20.5) and is also soluble in several organic solvents. The quaternary ammonium ions in such compounds can be conceptualized as "positive charges surrounded by greasy groups."

PROBLEMS

23.12   Draw a structure of each of the following quaternary ammonium salts.
 *(a) tetraethylammonium fluoride
 (b) dibenzyldimethylammonium bromide

*23.13 Explain why compound *A* can be isolated in optically active form, but compound *B* cannot.

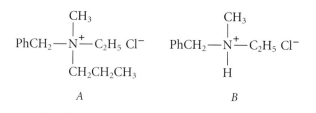

23.7 Alkylation and Acylation Reactions of Amines

The previous section showed that amines are *Brønsted bases*. Amines, like many other Brønsted bases, are also *nucleophiles* (Lewis bases). Three reactions of nucleophiles are:

 1. S_N2 reaction with alkyl halides, sulfonate esters, or epoxides (Secs. 9.1, 9.4, 10.3, 11.4)

 2. Addition to aldehydes, ketones, and α,β-unsaturated carbonyl compounds (Secs. 19.7, 19.11, 22.8A)

 3. Nucleophilic acyl substitution at the carbonyl groups of carboxylic acid derivatives (Sec. 21.8)

This section covers or reviews reactions of amines that fit into each of these categories.

A. Direct Alkylation of Amines

Treatment of ammonia or an amine with an alkyl halide or other alkylating agent results in alkylation of the nitrogen.

$$R_3N: + CH_3{-}I \longrightarrow R_3\overset{+}{N}{-}CH_3 \ I^- \tag{23.14}$$

This process is an example of an S_N2 reaction in which the amine acts as the nucleophile.

The product of the reaction shown in Eq. 23.14 is an alkylammonium ion. If this ammonium ion has N—H bonds, further alklylations can take place to give a complex product mixture, as in the following example:

$$\ddot{N}H_3 + CH_3I \longrightarrow CH_3\overset{+}{N}H_3\ I^- + (CH_3)_2\overset{+}{N}H_2\ I^- + (CH_3)_3\overset{+}{N}H\ I^- + (CH_3)_4N^+\ I^- \quad (23.15)$$

A mixture of products is formed because the methylammonium ion produced initially is partially deprotonated by the ammonia starting material. Because the resulting methylamine is also a good nucleophile, it too reacts with methyl iodide.

$$\ddot{N}H_3 + CH_3 - I \longrightarrow H_3\overset{+}{N} - CH_3\ I^- \quad (23.16a)$$

$$\ddot{N}H_3 + H - \overset{+}{N}H_2 - CH_3\ I^- \rightleftharpoons \overset{+}{N}H_4\ I^- + H_2\ddot{N} - CH_3 \quad (23.16b)$$

$$CH_3 - \ddot{N}H_2 + CH_3 - I \longrightarrow (CH_3)_2\overset{+}{N}H_2\ I^- \quad (23.16c)$$

Analogous deprotonation-alkylation reactions give the other products of the mixture shown in Eq. 23.15 (See Problem 23.18).

Epoxides, as well as α,β-unsaturated carbonyl compounds and α,β-unsaturated nitriles, also react with amines and ammonia. As the following results show, multiple alkylation can occur with these alkylating agents as well.

$$t\text{-Bu}-NH_2 + H_2C\overset{O}{\underset{}{-}}CH_2 \xrightarrow{H_2O} t\text{-Bu}-NH-CH_2CH_2-OH + t\text{-Bu}-N(CH_2CH_2OH)_2 \quad (23.17)$$

$$NH_3 \text{ (excess)} + CH_2{=}CH-CN \longrightarrow \underset{\text{(32\% yield)}}{H_2N-CH_2CH_2CN} + \underset{\text{(57\% yield)}}{HN(CH_2CH_2CN)_2} \quad (23.18)$$

In an alkylation reaction, the exact amount of each product obtained depends on the precise reaction conditions and on the relative amounts of starting amine and alkyl halide. Because a mixture of products results, the utility of alkylation as a preparative method for amines is limited, although in specific cases, conditions have been worked out to favor particular products. Sec. 23.11 discusses other methods that are more useful for the preparation of amines.

Quaternization of Amines   Amines can be converted into quaternary ammonium salts with excess alkyl halide under forcing conditions. This process, called **quaternization**, is one of the most important synthetic applications of amine alkylation. The reaction is particularly useful when especially reactive alkyl halides, such as methyl iodide or benzylic halides, are used.

$$\underset{\textbf{benzyldimethylamine}}{PhCH_2\ddot{N}Me_2} + MeI \xrightarrow{EtOH} \underset{\substack{\textbf{benzyltrimethylammonium}\\\textbf{iodide}\\(94–99\% \text{ yield})}}{PhCH_2\overset{+}{N}Me_3\ I^-} \quad (23.19)$$

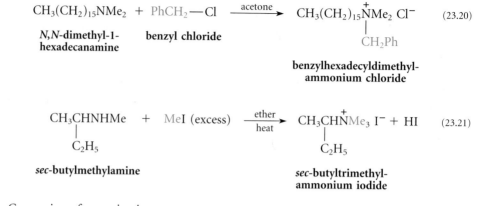

Conversion of an amine into a quaternary ammonium salt with excess methyl iodide (as in Eq. 23.21) is called **exhaustive methylation**.

B. Reductive Amination

When primary and secondary amines react with either aldehydes or ketones, they form imines and enamines, respectively (Sec. 19.11). In the presence of a reducing agent, imines and enamines are reduced to amines.

$$EtNH_2 + CH_3\overset{\overset{\displaystyle O}{\|}}{-C}-CH_3 \xrightarrow{-H_2O} \left[CH_3\overset{\overset{\displaystyle NEt}{\|}}{-C}-CH_3 \right] \xrightarrow[\substack{30\ psi\\ EtOH}]{H_2,\ Pt} CH_3\overset{\overset{\displaystyle NH-Et}{|}}{-CH}-CH_3 \quad (23.22)$$

an imine
(not isolated)

Reduction of the C=N double bond is analogous to reduction of the C=O double bond (Sec. 19.8). Notice that the imine or enamine does not have to be isolated, but is reduced within the reaction mixture as it forms. Because imines and enamines are reduced much more rapidly than carbonyl compounds, reduction of the carbonyl compound is not a competing reaction.

The formation of an amine from the reaction of an aldehyde or ketone with another amine and a reducing agent is called **reductive amination**. Sodium borohydride, NaBH$_4$, and a related compound, sodium cyanoborohydride, NaBH$_3$CN, find frequent use as reducing agents in reductive amination.

$$PhCH=O + Ph-NH_2 \xrightarrow[EtOH]{NaBH_4} PhCH_2-NH-Ph \quad (23.23)$$

(83% yield)

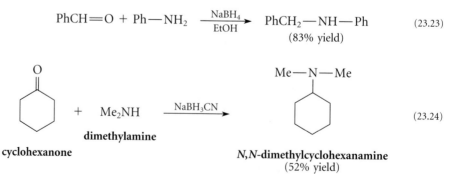

(23.24)

(Sodium cyanoborohydride is supplied as an easily handled, commercially available powder that is stable even in aqueous solution at pH values above 3.)

In the following synthesis, the secondary amine benzylethylamine is reductively aminated with formaldehyde.

$$Et{-}NH{-}CH_2Ph \; + \; H_2C{=}O \quad \xrightarrow{\; NaBH_3CN \;} \quad Et{-}\overset{\displaystyle CH_3}{\underset{|}{N}}{-}CH_2Ph \qquad (23.25)$$

benzylethylamine   **formaldehyde**   **benzylethylmethylamine**

Neither an imine nor an enamine can be an intermediate in this reaction (why?). In this case a small amount of a cationic intermediate, an *imminium ion*, is formed in solution by ionization of the carbinolamine intermediate. The imminium ion is rapidly and irreversibly reduced.

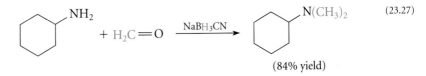

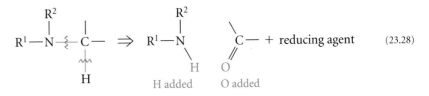

an imminium ion   (23.26)

As Eq. 23.26 and the following equation show, the reaction of an amine with an excess of formaldehyde is a useful way to introduce methyl groups to the level of a tertiary amine.

$$\text{(cyclohexyl)}{-}NH_2 \; + \; H_2C{=}O \quad \xrightarrow{\; NaBH_3CN \;} \quad \text{(cyclohexyl)}{-}N(CH_3)_2 \qquad (23.27)$$

(84% yield)

(Quaternization does not occur in this reaction; why?)

Suppose you want to prepare a given amine and want to determine whether reductive amination would be a suitable preparative method. How do you determine the required starting materials? Adopt the usual strategy for analyzing a synthesis: start with the target molecule and mentally reverse the reductive amination process. Mentally break one of the C—N bonds and replace it on the nitrogen side with an N—H bond. On the carbon side, drop a hydrogen from the carbon and add a carbonyl oxygen.

$$R^1{-}\overset{\displaystyle R^2}{\underset{|}{N}}\overset{\xi}{-}\underset{|}{\overset{|}{C}}{-} \quad \Rightarrow \quad R^1{-}\overset{\displaystyle R^2}{\underset{\underset{H}{\diagdown}}{N}} \qquad \overset{O}{\underset{\diagup}{\overset{\parallel}{C}}}{-} \; + \; \text{reducing agent} \qquad (23.28)$$

H added   O added

As this analysis shows, a hydrogen must be on the "disconnected" carbon. This process is applied in the following study problem.

STUDY
PROBLEM
23.3
Outline a preparation of *N*-ethyl-*N*-methylaniline from suitable starting materials using a reductive amination sequence.

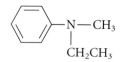

N-ethyl-N-methylaniline

Solution   Either the *N*-methyl or *N*-ethyl bond can be used for analysis. (The *N*-phenyl bond cannot be used because the carbon in the C—N bond has no hydrogen.) We arbitrarily choose the N—CH₃ bond and make the appropriate replacements to reveal the following starting materials:

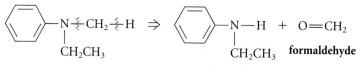

N-ethylaniline

Thus, treatment of *N*-ethylaniline with formaldehyde and NaBH₃CN should give the desired amine.

C. Acylation of Amines

Recall that amines can be converted into amides by reaction with acid chlorides, anhydrides, or esters (Sec. 21.8).

$$\underset{\substack{\| \\ O}}{R'{-}C}{-}Cl + 2R_2NH \longrightarrow \underset{\substack{\| \\ O}}{R'{-}C}{-}NR_2 + R_2\overset{+}{N}H_2\ Cl^- \tag{23.29}$$

$$\underset{\substack{\| \\ O}}{R'{-}C}{-}O{-}\underset{\substack{\| \\ O}}{C{-}R'} + 2R_2NH \longrightarrow \underset{\substack{\| \\ O}}{R'{-}C}{-}NR_2 + \underset{\substack{\| \\ O}}{R'{-}C}{-}O^- + R_2\overset{+}{N}H_2 \tag{23.30}$$

$$\underset{\substack{\| \\ O}}{R'{-}C}{-}OR'' + R_2NH \longrightarrow \underset{\substack{\| \\ O}}{R'{-}C}{-}NR_2 + R''OH \tag{23.31}$$

In this type of reaction, a bond is formed between the amine and a carbonyl carbon. These are all examples of *acylation*: a reaction involving the transfer of an *acyl group*.

Recall that the reaction of an amine with an acid chloride or an anhydride requires either *two equivalents* of the amine or one equivlaent of the amine and an additional equivalent of another base such as a tertiary amine or hydroxide ion. These and other aspects of amine acylation should be reviewed in Sec. 21.8.

PROBLEMS

*23.14   Suggest two syntheses of *N*-ethylcyclohexanamine by reductive amination.

23.15   Outline a second synthesis of *N*-ethyl-*N*-methylaniline (the target molecule in Study Problem 23.3) by reductive amination.

*23.16   Outline a synthesis of the quaternary ammonium salt $(CH_3)_3\overset{+}{N}CH_2Ph\ Br^-$ from each of the following combinations of starting materials.
  *(a) dimethylamine and any other reagents
  (b) benzylamine and any other reagents

*23.17   A chemist Caleb J. Cookbook heated ammonia with bromobenzene expecting to form tetraphenylammonium bromide. Can Caleb expect this reaction to succeed? Explain.

23.18   Continue the sequence of reactions in Eq. 23.16a–c to show how trimethyl-ammonium iodide is formed as one of the products in Eq. 23.15.

23.19   Outline a preparation of each of the following from an amine and an acid chloride.
  *(a) *N*-phenylbenzamide   (b) *N*-benzyl-*N*-ethylpropanamide

23.8 Hofmann Elimination of Quaternary Ammonium Hydroxides

The previous section discussed ways to *make* carbon-nitrogen bonds. In these reactions, amines react as *nucleophiles*. The subject of this section is an elimination reaction used to *break* carbon-nitrogen bonds. In this reaction, which involves *quaternary ammonium hydroxides* ($R_4N^+\ {}^-OH$) as starting materials, amines act as *leaving groups*.

When a quaternary ammonium hydroxide is heated, a β-elimination reaction takes place to give an alkene, which distills from the reaction mixture.

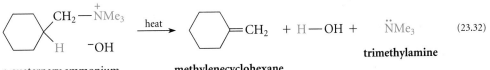

$$\text{(23.32)}$$

a quaternary ammonium   methylenecyclohexane   **trimethylamine**
hydroxide   (74% yield)

This type of elimination reaction is called a **Hofmann elimination**, after August Wilhelm Hofmann (1818–1895), a German chemist who became professor at the Royal College of Chemistry in London. Hofmann was particularly noted for his work on amines.

A quaternary ammonium hydroxide used as the starting material in Hofmann elimi-nations is formed by treating a quaternary ammonium salt with silver hydroxide (AgOH), which, in turn, is formed from water and silver oxide (Ag_2O).

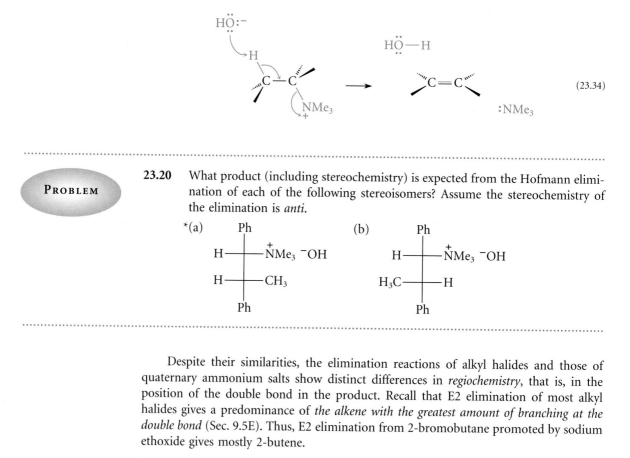

Thus, alkenes can be formed from amines by a three-step process: exhaustive methylation (see Eq. 23.21), conversion of the ammonium salt to the hydroxide (Eq. 23.33), and Hofmann elimination (Eq. 23.32).

The Hofmann elimination is conceptually analogous to the E2 reaction of alkyl halides (Sec. 9.5), in which a proton and a halide ion are eliminated; in the Hofmann elimination, a proton and a tertiary amine are eliminated. Because the amine leaving group is very basic, and therefore a relatively poor leaving group, the conditions of the Hofmann elimination are typically harsh.

Like the analogous E2 reaction of alkyl halides, the Hofmann elimination generally occurs as an *anti* elimination (Sec. 9.5D).

PROBLEM

23.20   What product (including stereochemistry) is expected from the Hofmann elimination of each of the following stereoisomers? Assume the stereochemistry of the elimination is *anti*.

Despite their similarities, the elimination reactions of alkyl halides and those of quaternary ammonium salts show distinct differences in *regiochemistry*, that is, in the position of the double bond in the product. Recall that E2 elimination of most alkyl halides gives a predominance of *the alkene with the greatest amount of branching at the double bond* (Sec. 9.5E). Thus, E2 elimination from 2-bromobutane promoted by sodium ethoxide gives mostly 2-butene.

$$CH_3CH_2CHCH_3 \xrightarrow{\text{NaOEt}} CH_3CH=CHCH_3 + CH_3CH_2CH=CH_2 + CH_3CH_2CHCH_3 \quad (23.35)$$

(81% of alkene formed; mostly *trans*)   (19% of alkene formed)

with Br below the first reactant and OEt below the third product.

In contrast, Hofmann elimination of the corresponding trimethylammonium salt gives mostly 1-butene.

$$CH_3CH_2CHCH_3 \xrightarrow{\text{heat}} CH_3CH_2CH{=}CH_2 + CH_3CH{=}CHCH_3 \qquad (23.36)$$

with $+NMe_3$ ^-OH below, and (95%) under the first product, (5%; *cis* and *trans*) under the second.

In general, *elimination of a trialkylammonium hydroxide occurs so that the base abstracts a proton from the β-carbon with the least branching.*

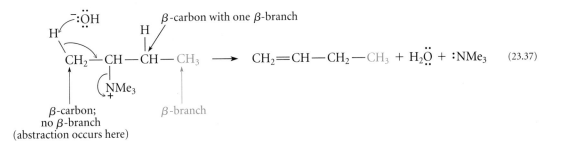

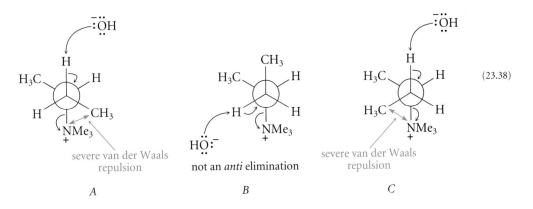

$$(23.38)$$

An energetically unfavorable feature can be found in each of these transition-state conformations. Although transition states *A* and *C* are *anti* eliminations, they contain severe van der Waals repulsions because of the *gauche* relationship of the trimethylammonium group and the methyl group. (A trimethylammonium group is about the same size as a *tert*-butyl group.) Although transition state *B* avoids the *gauche* relationship between the trimethylammonium group and the methyl group, it is not an *anti* elimination. Thus no transition state for elimination across the C2-C3 bond incorporates both *anti* elimination and the absence of significant van der Waals repulsions. The relatively higher energy of these transition states causes the corresponding reactions to be relatively slow. In contrast, because elimination at the C1-C2 bond—Hofmann elimination—can occur with *anti* stereochemistry and without severe van der Waals repulsions, this reaction is faster, and is therefore the major one observed.

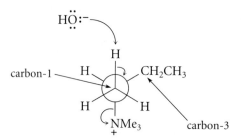

In the case of elimination from 2-bromobutane (Eq. 23.35), the transition states for *anti* elimination require a *gauche* relationship of a methyl and a bromine. (A bromine is about the same size as a methyl group.) The resulting van der Waals repulsions are relatively modest—about like those in *gauche*-butane—and not great enough to offset the energetic advantage of forming the more stable alkene. Consequently, *anti* elimination occurs to give mostly the more stable butene isomer.

Especially acidic β-hydrogens tend to be eliminated even if they are on a more highly branched carbon. Recall that the elimination of acidic hydrogens is also observed in other elimination reactions (Sec. 17.3B). In the following example, a hydrogen on the carbon adjacent to the phenyl group is eliminated because it is more acidic than a methyl hydrogen.

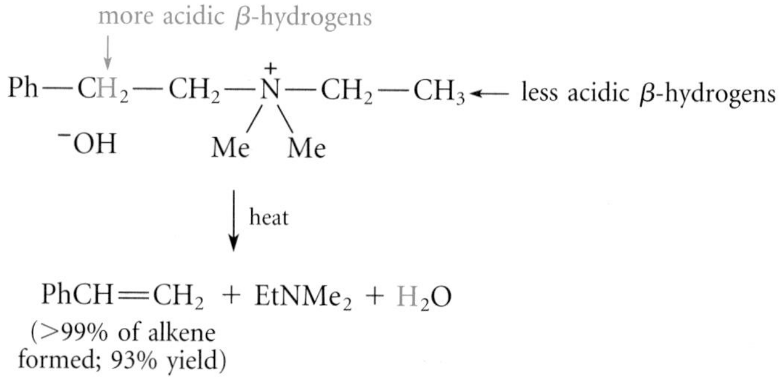

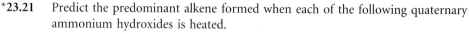

(23.39)

$$PhCH{=}CH_2 \; + \; EtNMe_2 \; + \; H_2O$$
(>99% of alkene
formed; 93% yield)

Hydrogens on carbons adjacent to carbonyl groups, which are also unusually acidic (Sec. 22.1A) are also preferentially abstracted in the Hofmann elimination (see Problem 23.21b).

The Hofmann elimination in conjunction with exhaustive methylation played a particularly important role in determining the structures of amine-containing natural products in the older literature (Problem 23.23).

....

PROBLEMS

*23.21   Predict the predominant alkene formed when each of the following quaternary ammonium hydroxides is heated.

(a)   (b)

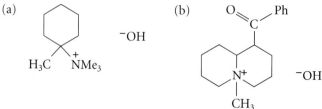

23.22   Give the structures of three alkenes that could in principle form when the following compound is heated. Rank these products in order of the relative amounts produced, greatest first.

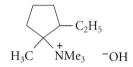

*23.23   *Coniine* is the toxic component of hemlock, the poisonous plant believed to have killed Socrates. Coniine is a secondary amine containing a piperidine ring. Exhaustive methylation of coniine yields a salt that, when treated with Ag_2O, then heat, gives a mixture of alkenes from which a compound A ($C_{10}H_{21}N$) is isolated. Exhaustive methylation of A and treatment with Ag_2O and heat give a mixture of 1,4-octadiene and 1,5-octadiene. Propose a structure for coniine.

23.9   Aromatic Substitution Reactions of Aniline Derivatives

Aromatic amines can undergo *electrophilic aromatic substitution* reactions on the aromatic ring (Sec. 16.4). The amino group is one of the most powerful *ortho, para*-directing groups in electrophilic substitution. If the conditions of the reaction are not too acidic, aniline and its derivatives undergo rapid ring substitution. For example, aniline, like phenol, brominates three times under mild conditions.

(23.40)

The conditions of most aromatic substitution reactions involve strong Brønsted or Lewis acids. For example, under the strongly acidic conditions of nitration, the amino group is protonated, and a significant amount of *m*-nitroaniline is formed.

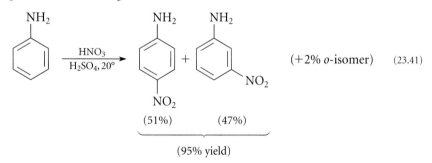

(+2% *o*-isomer)   (23.41)

A *protonated* amino group does not have the unshared electron pair on nitrogen that gives rise through resonance to the activating, *ortho, para*-directing effect of the free amino group. Ammonium salts are largely *meta*-directing groups. In view of the *meta*-directing effect of ammonium groups, it is perhaps somewhat surprising that a significant

STUDY GUIDE LINK:
✓23.2
Nitration of Aniline

amount of *para* isomer is formed in the nitration of Eq. 23.41. It is likely that the *p*-nitroaniline product arises by nitration of the minuscule amount of highly reactive unprotonated aniline in the reaction mixture. As the unprotonated aniline reacts, it is replenished, by LeChatelier's principle.

Aniline can be nitrated regioselectively at the *para* position if the nitrogen is first *protected* from protonation. (The general idea of a *protecting group* was introduced in Sec. 19.10B.) This strategy is used in the solution to the following study problem.

STUDY PROBLEM 23.4

Outline a preparation of *p*-nitroaniline from aniline and any other reagents.

Solution   An amide group is much less basic than an amino group. Hence, acylation of the amino group of aniline with acetyl chloride to give *N*-phenylacetamide (acetanilide) will protect the nitrogen from protonation. The acetamido group, although much less activating than a free amino group, is nevertheless an activating, *ortho, para*-directing group in aromatic substitution (Table 16.2). Following nitration of acetanilide, the acetyl group is removed to give *p*-nitroaniline, the target compound.

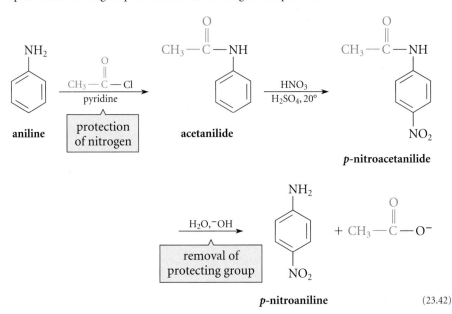

Because the acetamido group is considerably less basic than an amino group, it is only partially protonated under the acidic reaction conditions of nitration. Because the acetamido group is less activating than a free amino group (why?), nitration occurs only once.

PROBLEMS

*23.24   Outline a preparation of sulfanilamide, a sulfa drug, from aniline and any other reagents. (*Hint:* See Sec. 20.9A, Eq. 20.27; also note that sulfonyl chlorides react with amines to form amides in much the same manner as carboxylic acid chlorides.)

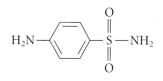

sulfanilamide

23.25   Outline a preparation of each of the following compounds from aniline and any other reagents.
  *(a)   2,4-dinitroaniline
   (b)   sulfathiazole, a sulfa drug. (*Hint:* 2-Aminothiazole is a readily available amine.)

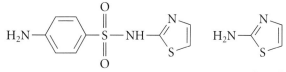

sulfathiazole

2-aminothiazole

23.10   Diazotization; Reactions of Diazonium Ions

A.   Formation and Substitution Reactions of Diazonium Salts

The reactions considered in previous sections show that the chemistry of the amino group vaguely resembles that of the hydroxy group. Thus, amino groups, like hydroxy groups, can both donate and accept hydrogen bonds. Amino groups, like hydroxy groups, are basic and nucleophilic (only more so). Amino groups, like hydroxy groups (with suitable activation), can serve as leaving groups. And amino groups, like hydroxy groups, activate aromatic rings toward electrophilic aromatic substitution.

In contrast, when it comes to oxidation reactions, there is no parallel between amines and alcohols or phenols. Oxidation of amines generally occurs at the amino nitrogen, whereas oxidation of alcohols and phenols occurs at the α-carbon. An important oxidation reaction of amines that illustrates this point is called **diazotization**: the reaction of primary amines with nitrous acid (HNO_2) to form diazonium salts. A **diazonium salt** is a compound of the form $R{-}\overset{+}{N}{\equiv}N{:}\, X^-$, in which X^- is a typical anion (chloride, bromide, sulfate, etc.). Because nitrous acid is unstable, it is usually generated as needed by the reaction of sodium nitrite ($NaNO_2$) with a strong acid such as HCl or H_2SO_4. Both aliphatic and aromatic primary amines are readily diazotized:

☞
Study Guide Link:
    23.3
Mechanism of Diazotization

$$ \underset{\substack{\textbf{2-butanamine} \\ (\textbf{\textit{sec}-butylamine})}}{\underset{\underset{\text{NH}_2}{|}}{\text{CH}_3\text{CHCH}_2\text{CH}_3}} \xrightarrow[\text{H}_2\text{O}]{\text{NaNO}_2,\ \text{HCl}} \underset{\substack{\textbf{2-butanediazonium chloride} \\ \text{(an aliphatic diazonium salt)} \\ \text{unstable; cannot be isolated}}}{\left[\underset{\underset{+}{\overset{|}{\text{N}{\equiv}\text{N}{:}\ \text{Cl}^-}}}{\text{CH}_3\text{CHCH}_2\text{CH}_3}\right]} \qquad (23.43) $$

$$Ph—NH_2 \xrightarrow[\text{H}_2\text{O}]{\text{NaNO}_2,\ \text{HCl}} Ph—\overset{+}{N}\equiv N\!:\ Cl^- \qquad (23.44)$$

aniline

benzenediazonium chloride
(an aromatic diazonium salt)
can be isolated

Notice that diazonium salts incorporate one of the very best leaving groups—molecular nitrogen (color in Eq. 23.43). For this reason, *aliphatic* diazonium salts react immediately as they are formed by S_N1, E1, and/or S_N2 mechanisms to give substitution and elimination products along with nitrogen gas.

$$CH_3CHCH_2CH_3 \underset{NH_2}{\quad} \xrightarrow[\text{H}_2\text{O}]{\text{NaNO}_2,\text{H}_2\text{SO}_4} \left[\begin{matrix} CH_3CHCH_2CH_3 \\ \underset{+}{N}\equiv N\!: \end{matrix} \right] \longrightarrow \left[CH_3\overset{+}{C}HCH_2CH_2 \right] + :N\equiv N\!:\ \uparrow$$
nitrogen gas

S_N1 E1

OH_2

$$H_3O^+ + CH_3CHCH_2CH_2 + CH_2\!=\!CHCH_2CH_3 + CH_3CH\!=\!CHCH_3$$
$$\underset{OH}{\mid}$$
(9% of product)   (31% of product;
(60% of product)   10% *cis* and 21% *trans*)

$$(23.45)$$

(The rapid liberation of nitrogen gas on treatment with nitrous acid is a qualitative test for primary alkylamines.) Because of the complex mixture of products that results, the reactions of aliphatic diazonium salts are not generally useful in organic synthesis.

Recall that benzene rings bearing good leaving groups do not readily undergo S_N1 or S_N2 reactions (Secs. 18.1, 18.3). For this reason, *aromatic* diazonium salts may be isolated and used in a variety of reactions. In practice, though, they are usually prepared in solution at 0–5 °C and used without isolation, because they lose nitrogen on heating and they are explosive in the dry state.

Among the most important reactions of aryldiazonium salts are substitution reactions with cuprous halides; in these reactions the diazonium group is replaced by a halogen.

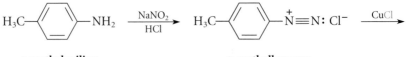

***p*-methylaniline**
(***p*-toluidine**)   ***p*-methylbenzene-**
diazonium chloride

$$H_3C-\!\!\left\langle\!\!\bigcirc\!\!\right\rangle\!\!-Cl\ +\ N_2 \qquad (23.46)$$

***p*-chlorotoluene**
(70–71% yield)

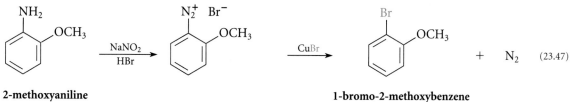

2-methoxyaniline

1-bromo-2-methoxybenzene
(*o*-bromoanisole)
(88–93% yield)

(23.47)

An analogous reaction occurs with cuprous cyanide, CuCN.

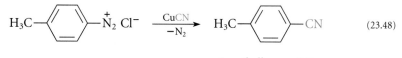

4-methylbenzonitrile
(67% yield)

(23.48)

This reaction is another way of forming a carbon-carbon bond, in this case to an aromatic ring (see Appendix V). The resulting nitrile can be converted by hydrolysis into a carboxylic acid, which can, in turn, serve as the starting material for a variety of other types of compounds. The reaction of an aryldiazonium ion with a cuprous salt is called the **Sandmeyer reaction**. This reaction is an important method for the synthesis of aryl halides and nitriles.

Aryl iodides can also be made by the reaction of diazonium salts with the potassium salt KI.

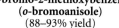

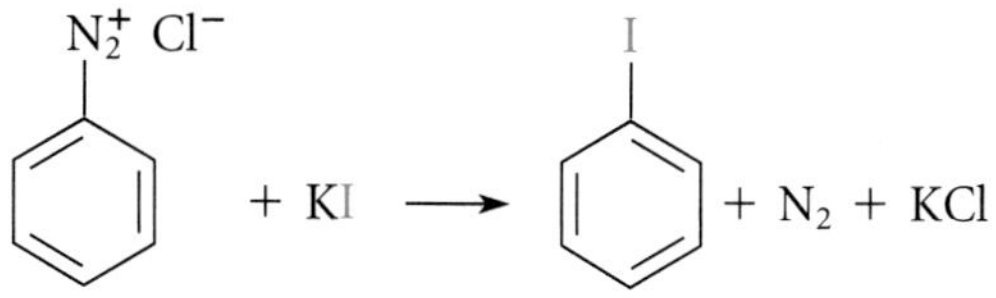

(23.49)

Notice that the analogous reactions with KBr and KCl do not work; cuprous salts are required.

Aryldiazonium salts can be hydrolyzed to phenols by heating them in water. A relatively recent variation of this reaction reminiscent of the Sandmeyer reaction is the use of cuprous oxide (Cu$_2$O) and an excess of aqueous cupric nitrate [Cu(NO$_3$)$_2$] at room temperature.

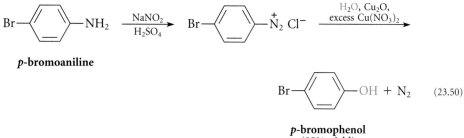

***p*-bromoaniline**

***p*-bromophenol**
(95% yield)

(23.50)

Finally, the diazonium group is replaced by hydrogen when the diazonium salt is treated with hypophosphorous acid, H_3PO_2.

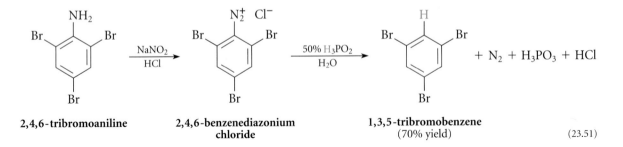

2,4,6-tribromoaniline   **2,4,6-benzenediazonium**   **1,3,5-tribromobenzene**
                                              chloride   (70% yield)   (23.51)

Notice that 1,3,5-tribromobenzene, the product of Eq. 23.51, cannot be prepared by the direct bromination of benzene (why?). Recall that the starting material, 2,4,6-tribromo-aniline, is prepared by the bromination of aniline (Eq. 23.40). In this bromination reaction, the positions of the bromines are determined by the powerful directing effect of the amino nitrogen. Once the amino group has fulfilled its role as an activating and directing group, it can be removed using the reaction in Eq. 23.51.

The diazonium salt reactions shown in Eqs. 23.46–23.51 are all substitution reactions, but none are S_N2 or S_N1 reactions because aromatic rings do *not* undergo substitution by these mechanisms (Secs. 18.1, 18.3). It turns out that the Sandmeyer and related reactions occur by radical-like mechanisms mediated by the copper. The reaction of diazonium salts with KI (Eq. 23.49), although not involving copper, probably occurs by a similar mechanism. The reaction of diazonium salts with H_3PO_2 (Eq. 23.51) has been shown definitively to be a free-radical chain reaction.

PROBLEMS

23.26   Outline a synthesis for each of the following compounds from the indicated starting materials using a reaction sequence involving a diazonium salt.
*(a) 2-bromobenzoic acid from *o*-toluidine
(b) 2,4,6-tribromobenzoic acid from aniline

*23.27   When (*R*)-1-deuterio-1-butanamine (structure below) is diazotized with nitrous acid in water, the alcohol product formed has the *S* configuration. (D = ^2H, the isotope of hydrogen with atomic mass = 2.)

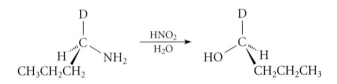

(*R*)-1-deuterio-1-butanamine

(a) Give the stereochemical configuration of the diazonium ion formed as an intermediate in this reaction. Draw its structure.

(b) What mechanism for reaction of the diazonium ion with water is consistent with the stereochemical result in the equation above?

B. Aromatic Substitution with Diazonium Ions

Aryldiazonium ions react with aromatic compounds containing strongly activating substituent groups, such as amines and phenols, to give substituted *azobenzenes*. (Azobenzene itself is Ph—N=N—Ph.)

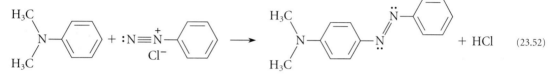

$$(23.52)$$

butter yellow
(an azobenzene)

This is an electrophilic aromatic substitution reaction in which the terminal nitrogen of the diazonium ion is the electrophile. The mechanism follows the usual pattern of electrophilic aromatic substitution (Sec. 16.4B). First, the electrophile is attacked by the π electrons of the aromatic compound to give a resonance-stabilized carbocation:

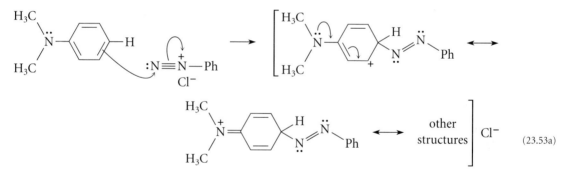

$$(23.53a)$$

This carbocation then loses a proton to give the substitution product.

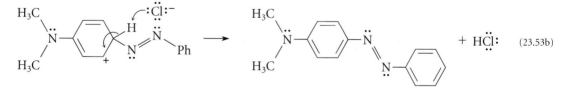

$$(23.53b)$$

(Why does substitution occur at the *para* position? See Sec. 16.5A.)

The azobenzene derivatives formed in these reactions have extensive conjugated π-electron systems, and most of them are colored (Sec. 15.2C). Some of these compounds are used as dyes and indicators; as a class they are known as **azo dyes**. (An azo dye is a colored derivative of azobenzene.) For example, the azo dye methyl orange is a well-known acid-base indicator.

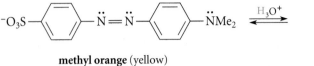

methyl orange (yellow)
an azo dye

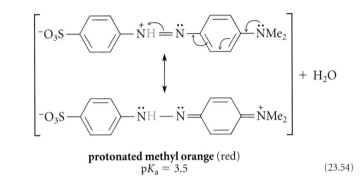

protonated methyl orange (red)
$pK_a = 3.5$

(23.54)

Because methyl orange changes color when it protonates, it can be used as an acid-based indicator at pH values near its pK_a of 3.5. Some azo dyes are used in dyeing fabrics, foodstuffs, and cosmetics. For example, FD & C #6 (FD & C = food, drug, and cosmetic) is a yellow compound used to color gelatin desserts, ice cream, beverages, candy, and so on.

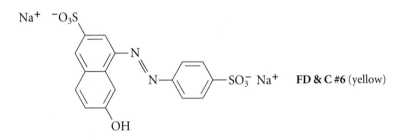

FD & C #6 (yellow)

C. Reactions of Secondary and Tertiary Amines with Nitrous Acid

Secondary amines react with nitrous acid to give *N*-nitrosoamines, compounds of the form $R_2N-N=O$, usually called simply **nitrosamines**.

$$Me_2NH + HNO_2 \longrightarrow Me_2N-N=O + H_2O \qquad (23.55)$$

***N,N*-dimethyl-
nitrosamine**
(89–90% yield)

$$Ph-NH-CH_3 \xrightarrow[\text{HCl}]{\text{NaNO}_2} Ph-N-CH_3 \qquad (23.56)$$
$$\begin{array}{c} | \\ N=O \end{array}$$

***N*-methyl-*N*-nitrosoaniline**

NITROSAMINES AND CANCER

The fact that many nitrosamines are known to be potent carcinogens has created a conflict over the use of sodium nitrite ($NaNO_2$) as a meat preservative. The meat-packing industry has argued that sodium nitrite is important in preventing the botulism that results from meat spoilage. But because sodium nitrite is, in combination with acid, a diazotizing reagent, it has the capacity for producing nitrosamines. For example, the frying of bacon generates nitrosamines that concentrate in the fat. (Well-drained bacon contains fewer nitrosamines, and nitrosamines are destroyed by ascorbic acid (vitamin C), which is present in fruit and vegetable juices. Perhaps this is a good reason for drinking orange juice when having bacon for breakfast!) The potential hazards of sodium nitrite led to a long campaign by consumer groups to have it banned as a meat preservative; the campaign was fought by the meat-packing industry. Then researchers found that nitrite is produced by the bacteria in the normal human intestine. It became questionable whether the risk from nitrite in meat is any greater than the risk faced all along from normal intestinal flora. These findings caused the Food and Drug Administration in 1980 to back away from banning sodium nitrite as a preservative, recommending only that it be kept to a minimum. Here again (see Sec. 8.8B) is a situation in which risk (risk of cancer from nitrosamines) is weighed against benefit (ability to preserve foods; freedom from toxic effects of food spoilage) over which reasonable persons may differ.

The nitrogen of tertiary amines does not react under the strongly acidic conditions used in diazotization reactions. However, *N,N*-disubstituted aromatic amines undergo electrophilic aromatic substitution on the benzene ring. The electrophile is the nitrosyl cation, $^+\ddot{N}{=}\ddot{O}$, which is generated from nitrous acid under acidic conditions.

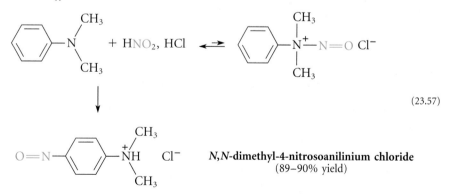

(23.57)

N,N-dimethyl-4-nitrosoanilinium chloride
(89–90% yield)

PROBLEMS

*23.28   Design a synthesis of methyl orange (Eq. 23.54) using aniline as the only aromatic starting material.

23.29   What two compounds would react in a diazo coupling reaction to form FD & C dye #6?

23.30 *(a) Using the curved-arrow formalism, show how the nitrosyl cation, $^+\ddot{N}=\ddot{O}$, is generated from HNO_2 under acidic conditions.

(b) Give a curved-arrow mechanism for the electrophilic aromatic substitution reaction shown in Eq. 23.57.

23.11 Synthesis of Amines

A. Gabriel Synthesis of Primary Amines

Recall that direct alkylation of ammonia is generally not a good synthetic method for the preparation of amines because multiple alkylation takes place (Sec. 23.7A). This problem can be avoided by protecting the amine nitrogen so that it can react only once with alkylating reagents. One approach of this sort begins with the imide *phthalimide*. Because the pK_a of phthalimide is about 9, its conjugate-base anion is easily formed with KOH or NaOH. This anion is a good nucleophile, and is alkylated by alkyl halides or sulfonate esters in S_N2 reactions.

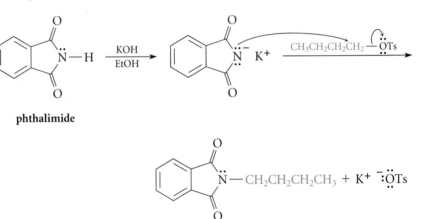

phthalimide

N-butylphthalimide   (23.58a)

The alkyl halides and sulfonates used in this reaction are primary or unbranched secondary (why?). Since the *N*-alkylated phthalimide formed in this reaction is really a double amide, it can be converted into the free amine by amide hydrolysis in either strong acid or base.

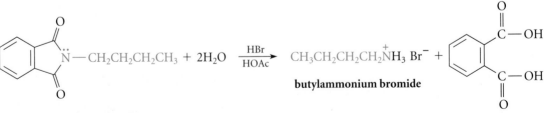

butylammonium bromide

N-butylphthalimide

(23.58b)

In this example, acidic hydrolysis gives the ammonium salt, which can be converted into the free amine by neutralization with base.

The alkylation of phthalimide anion followed by hydrolysis of the alkylated derivative to the primary amine is called the **Gabriel synthesis**. Notice that the nitrogen in phthalimide has only one acidic hydrogen, and thus can be alkylated only once. Although *N*-alkylphthalimides also have a pair of unshared electrons on nitrogen, they do not alkylate further, because neutral imides are *much* less basic, and therefore less nucleophilic, than the phthalimide anion. Hence, multiple alkylation, which occurs in the direct alkylation of ammonia, does not occur in the Gabriel synthesis.

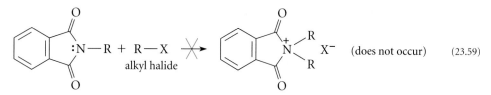

B. Reduction of Nitro Compounds

Nitro compounds can be reduced to amines under a variety of conditions. The nitro group is generally reduced easily by catalytic hydrogenation:

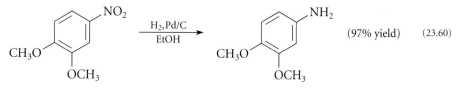

1,2-dimethoxy-4-nitrobenzene   **3,4-dimethoxyaniline**

An older but effective method for reducing the nitro group is the reduction of aromatic nitro compounds to primary amines with finely divided metal powders and HCl; iron or tin powder is frequently used.

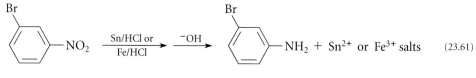

1-bromo-3-nitrobenzene   *m*-**bromoaniline**
(80% yield)

In this reaction the nitro compound is reduced *at nitrogen*, and the metal, which is oxidized to a metal salt, is the reducing agent. Although the methods shown in both Eqs. 23.60 and 23.61 also work with aliphatic nitro compounds, they are particularly important with aromatic nitro compounds as methods for introducing an amino group into an aromatic ring.

In view of the utility of lithium aluminum hydride ($LiAlH_4$) and sodium borohydride ($NaBH_4$) as reducing agents for other compounds, it is reasonable to ask what happens when nitro compounds are treated with these reagents. Aromatic nitro compounds do react with $LiAlH_4$, but the reduction products are azobenzenes (Sec. 23.10B), not amines:

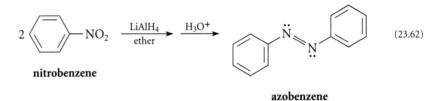

nitrobenzene

azobenzene (23.62)

Nitro groups do not react at all with sodium borohydride under the usual conditions.

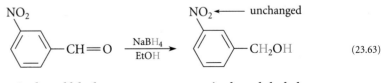

m-nitrobenzaldehyde → *m*-nitrobenzyl alcohol (23.63)

Hence, $LiAlH_4$ and $NaBH_4$ are *not* useful in forming aromatic amines from nitro compounds.

PROBLEM

23.31 Outline syntheses of the following compounds from the indicated starting materials.

*(a) *p*-iodoanisole from phenol and any other reagents

(b) *m*-bromoiodobenzene from nitrobenzene

C. Curtius and Hofmann Rearrangements

A very useful synthesis of amines starts with a class of compounds called *acyl azides*. An **acyl azide** has the following general structure:

$$\underbrace{R-\overset{\overset{\displaystyle O}{\|}}{C}}_{\text{acyl group}}-\overbrace{\ddot{\underset{..}{N}}^{-}-\overset{+}{N}\equiv N:}^{\text{azide group}} \quad \text{or} \quad R-\overset{\overset{\displaystyle O}{\|}}{C}-N_3 \quad \text{an acyl azide} \tag{23.64}$$

When an acyl azide is heated in an inert solvent such as benzene or toluene, it is transformed with loss of nitrogen into an **isocyanate**, a compound of the general structure R—N═C═O.

$$\underset{\textbf{dodecanoyl azide}}{CH_3(CH_2)_{10}-\overset{\overset{\displaystyle O}{\|}}{C}-N_3} \xrightarrow[\text{benzene}]{\text{heat}} \underset{\substack{\textbf{undecyl isocyanate}\\ (81\text{–}86\%\text{ yield})}}{CH_3(CH_2)_{10}-N=C=O} + N_2 \tag{23.65}$$

This reaction, called the **Curtius rearrangement**, is a concerted reaction that can be represented as follows:

$$\overset{:O:}{\underset{R}{\overset{\|}{C}}}\overset{\frown}{\underset{N}{\overset{+}{N}}} \equiv N: \longrightarrow \ddot{O}=C=\ddot{N}-R + :N\equiv N: \qquad (23.66)$$

STUDY GUIDE LINK:
✓ **23.4**
Mechanism of the Curtius Rearrangement

The isocyanate product of a Curtius rearrangement can be transformed into an amine by hydration in either acid or base. Hydration involves, first, addition of water across the C=N bond to give a carbamic acid:

$$H_2O + R-N=C=O \xrightarrow{H_3O^+} \overset{O}{\underset{}{R-NH-\overset{\|}{C}-OH}} \qquad (23.67)$$

an isocyanate   a carbamic acid

Carbamic acids are among those types of carboxylic acids that spontaneously decarboxylate (see Eq. 20.43). Decarboxylation gives the amine, which is protonated under the acidic conditions of the reaction. The free amine is obtained by neutralization:

$$\overset{O}{\underset{}{R-NH-\overset{\|}{C}-OH}} \xrightarrow{H_3O^+} R-\overset{+}{N}H_3 \xrightarrow{^-OH} R-NH_2 + H_2O \qquad (23.68)$$
$$+ CO_2$$

Notice carefully the overall transformation that occurs as a result of the Curtius rearrangement followed by hydration: the carbonyl carbon of the acyl azide is removed as CO_2.

$$\overset{O}{\underset{}{R-\overset{\|}{C}-N_3}} \xrightarrow[-N_2]{\text{heat}} R-N=C=O \xrightarrow[H_3O^+]{H_2O} \left[\overset{O}{\underset{}{R-NH-\overset{\|}{C}-OH}}\right] \longrightarrow R\overset{+}{N}H_3 + CO_2\uparrow$$

acyl azide   **isocyanate**   **carbamic acid** $\downarrow ^-OH$ $\qquad (23.69)$
(unstable)

$$RNH_2 + H_2O$$

amine

STUDY GUIDE LINK:
✓ **23.5**
Formation and Decarboxylation of Carbamic Acids

An important use of the Curtius rearrangement is for the preparation of carbamic acid derivatives (see Sec. 21.1G). Such derivatives are produced by allowing the isocyanate products to react with nucleophiles other than water. Reaction of isocyanates with alcohols or phenols yields carbamate esters; and reaction with amines yields ureas.

$$R-N=C=O \begin{cases} \xrightarrow[\substack{\text{(alcohol or} \\ \text{phenol)}}]{R'OH} \overset{O}{\underset{}{R-NH-\overset{\|}{C}-OR'}} & \text{a carbamate ester} \\[2em] \xrightarrow{H_2O} CO_2 + R-NH_2 & \text{an amine} \qquad (23.70) \\[2em] \xrightarrow[\text{(amine)}]{R'NH_2} \overset{O}{\underset{}{R-NH-\overset{\|}{C}-NH-R'}} & \text{a urea} \end{cases}$$

(from Curtius rearrangement)

How can we prepare the acyl azides used in the Curtius rearrangement? The key is to recognize that these compounds are carboxylic acid derivatives. The most straightforward preparation is the reaction of an acid chloride with sodium azide.

$$\text{Ph—CH}_2\text{—}\overset{\overset{\displaystyle O}{\|}}{C}\text{—Cl} + \text{NaN}_3 \longrightarrow \text{Ph—CH}_2\text{—}\overset{\overset{\displaystyle O}{\|}}{C}\text{—N}_3 + \text{NaCl} \qquad (23.71)$$

α-phenylacetyl chloride   **α-phenylacetyl azide**
(an acyl azide)

Another widely used method is to convert an ethyl ester into an acyl derivative of hydrazine ($\text{NH}_2\text{—NH}_2$) by aminolysis (Sec. 21.8C). The resulting amide, an *acyl hydrazide*, is then diazotized with nitrous acid to give the acyl azide.

$$\text{Ph—CH}_2\text{—}\overset{\overset{\displaystyle O}{\|}}{C}\text{—OEt} + \text{NH}_2\text{NH}_2 \xrightarrow{-\text{EtOH}} \text{Ph—CH}_2\text{—}\overset{\overset{\displaystyle O}{\|}}{C}\text{—NHNH}_2 \xrightarrow[\substack{\text{HCl} \\ -10°}]{\text{NaNO}_2}$$

ethyl α-phenylacetate   **hydrazine**   **α-phenylacetyl hydrazide**
(an acyl hydrazide)
(80–100% yield)

$$\text{PhCH}_2\text{—}\overset{\overset{\displaystyle O}{\|}}{C}\text{—N}_3 \qquad (23.72)$$

α-phenylacetyl azide

Notice the similarity of this diazotization to the diazotization of alkylamines:

$$\text{R—CH}_2\text{—NH}_2 + \text{HONO} \longrightarrow \text{R—CH}_2\text{—}\overset{+}{\text{N}}\equiv\text{N:}$$

$$\text{R—}\overset{\overset{\displaystyle O}{\|}}{C}\text{—NH—NH}_2 + \text{HONO} \longrightarrow \text{R—}\overset{\overset{\displaystyle O}{\|}}{C}\text{—}\overset{\overset{\displaystyle H}{|}}{N}\text{—}\overset{+}{\text{N}}\equiv\text{N:} \xrightarrow{\text{H}_2\text{O}} \text{R—}\overset{\overset{\displaystyle O}{\|}}{C}\text{—}\overset{..}{\text{N}}\text{—}\overset{+}{\text{N}}\equiv\text{N:} \qquad (23.73)$$

conjugate acid $+ \text{H}_3\text{O}^+$
of the acyl azide

Because the conjugate acid of the acyl azide is quite acidic (why?), it loses a proton from the adjacent nitrogen to the give the neutral acyl azide.

A reaction closely related to the Curtius rearrangement is the **Hofmann rearrangement** or **Hofmann hypobromite reaction**. The starting material for this reaction is a primary amide rather than an acyl azide. Treatment of an amide with bromine in base gives rise to a rearrangement.

$$\text{Br}_2 + 2\text{NaOH} + (\text{CH}_3)_3\text{CCH}_2\text{—}\overset{\overset{\displaystyle O}{\|}}{C}\text{—NH}_2 \longrightarrow$$

3,3-dimethylbutanamide

$$(\text{CH}_3)_3\text{CCH}_2\text{—NH}_2 + \text{O}=\text{C}=\text{O} + 2\text{NaBr} + \text{H}_2\text{O} \qquad (23.74)$$

2,2-dimethyl-1-propanamine
(neopentylamine)

The first step in the mechanism of the Hofmann rearrangement is ionization of the amide N—H (Sec. 22.1A); the resulting anion is then brominated.

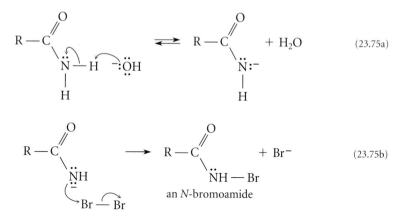

(23.75a)

(23.75b)

an N-bromoamide

(This reaction is analogous to α-bromination of a ketone in base; Sec. 22.3B.) The N-bromoamide product is even more acidic than the amide starting material (why?), and it too ionizes.

$$R—\overset{\overset{O}{\|}}{C}—\overset{\overset{..}{N}}{\underset{Br}{|}}—H \quad \overset{..}{:}\overset{..}{O}H \quad \rightleftharpoons \quad R—\overset{\overset{O}{\|}}{C}—\overset{=}{\overset{..}{N}}: \quad + \quad H_2\overset{..}{\underset{..}{O}} \qquad \text{(23.75c)}$$

The N-bromo anion does not brominate again, because it undergoes an even faster reaction: rearrangement to an isocyanate.

$$R\overset{:O:}{\underset{\underset{..}{N}}{\overset{\|}{C}}}\overset{..}{Br}: \quad \longrightarrow \quad \overset{..}{O}=C=N—R \quad + \quad :\overset{..}{Br}:^- \qquad \text{(23.75d)}$$

an isocyanate

Notice that the rearrangement steps of the Hofmann and Curtius reactions are conceptually identical; the only difference is the leaving group.

Hofmann:

Curtius:

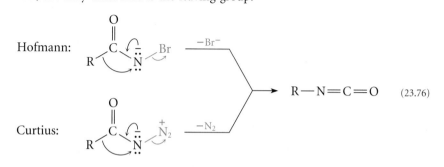

$$R—N=C=O \qquad \text{(23.76)}$$

Because the Hofmann rearrangement is carried out in aqueous base, the isocyanate cannot be isolated as it is in the Curtius rearrangement. It spontaneously hydrates to a carbamate ion, which then decarboxylates to the amine product under the strongly basic reaction conditions.

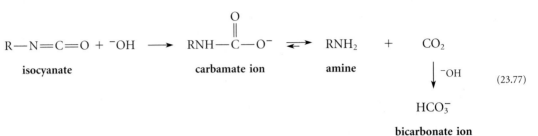

$$\text{bicarbonate ion}$$ (23.77)

Although the reaction of amines with CO_2 is reversible, formation of the amine in the Hofmann rearrangement is driven to completion by the reaction of hydroxide ion with CO_2 to form bicarbonate ion (or carbonate ion) under the strongly basic conditions of the reaction.

An interesting and very useful aspect of both the Hofmann and Curtius rearrangements is that they take place with complete *retention of stereochemical configuration* in the migrating alkyl group:

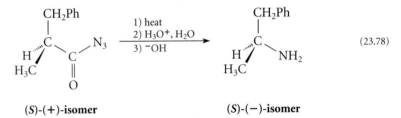

(23.78)

Hence, optically active carboxylic acid derivatives can be used to prepare optically active amines of known stereochemical configuration.

The advantage of the Curtius rearrangement over the Hofmann rearrangement is that the Curtius reaction can be run under mild, neutral conditions, and the isocyanate can be isolated if desired. The disadvantage is that some acyl azides in the pure state can detonate without warning, and extreme caution is required in handling them. Amides, in contrast, are stable and easily handled organic compounds.

PROBLEMS

23.32 *(a) Could *tert*-butylamine be prepared by the Gabriel synthesis? If so, write out the synthesis. If not, explain why.

(b) Propose a synthesis of *tert*-butylamine by another route.

23.33 Write a curved-arrow mechanism for each of the following reactions.

*(a) ethyl isocyanate (CH_3CH_2—N=C=O) with ethanol to yield ethyl *N*-ethylcarbamate

(b) ethyl isocyanate with ethylamine to yield *N,N'*-diethylurea.

23.34 What product is formed when 2-methylpropanamide is subjected to the conditions of the Hofmann rearrangement *(a) in ethanol solvent? (b) in aqueous NaOH?

*23.35 When hexanamide is subjected to the conditions of the Hofmann rearrangement, pentanamine (*A*) is obtained as expected. However, a significant by-product is

N,N′-dipentylurea (*B*). Explain the origin of *B*. (*Hint:* Neither pentyl isocyanate nor pentanamine has appreciable solubility in aqueous base.)

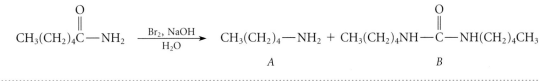

$$CH_3(CH_2)_4\overset{\overset{\displaystyle O}{\|}}{C}-NH_2 \quad \xrightarrow[H_2O]{Br_2,\ NaOH} \quad CH_3(CH_2)_4-NH_2 \ + \ CH_3(CH_2)_4NH-\overset{\overset{\displaystyle O}{\|}}{C}-NH(CH_2)_4CH_3$$

$$\qquad\qquad\qquad\qquad\qquad\qquad\qquad\qquad\qquad A \qquad\qquad\qquad\qquad\qquad\qquad B$$

D. Synthesis of Amines: Summary

The following amine syntheses have been covered in this and previous sections:

1. Reduction of amides and nitriles with LiAlH$_4$ (Secs. 21.9B, 21.9C)
2. Direct alkylation of amines (Sec. 23.7A). This reaction is of limited utility, but is useful for preparing quaternary ammonium salts.
3. Reductive amination (Sec. 23.7B)
4. Aromatic substitution reactions of anilines (Sec. 23.9)
5. Gabriel synthesis of primary amines (Sec. 23.11A)
6. Reduction of nitro compounds (Sec. 23.11B)
7. Hofmann and Curtius rearrangements (Sec. 23.11C)

Methods 2, 3, and 4 represent methods of preparing amines from other amines. To the extent that an amide used in method 1 can be prepared from an amine, this method, too, represents a method for obtaining one amine from another. Methods 5–7 are limited to the preparation of primary amines, and methods 1, 3, 6, and 7 can be used for obtaining amines from other functional groups.

PROBLEM

23.36   Show how 2-cyclopentyl-*N,N*-dimethylethanamine could be synthesized from each of the following starting materials.

*(a)   [cyclopentyl]—CH$_2$—CO$_2$H

(b)   [cyclopentyl]—CH$_2$—CN

*(c)   [cyclopentyl]—CH$_2$CH$_2$—CO$_2$H

(d)   [cyclopentyl]—CH$_2$—CH=O   (two ways)

23.12   Use and Occurrence of Amines

A. Industrial Use of Amines and Ammonia

Among the relatively few industrially important amines is hexamethylenediamine, H$_2$N(CH$_2$)$_6$NH$_2$, used in the synthesis of nylon-6,6 (Sec. 21.12A). Ammonia is also an important "amine," and is a key source of nitrogen in a number of manufacturing processes. In agricultural chemistry, for example, liquid ammonia itself and urea, which is made from ammonia and CO_2, are important nitrogen fertilizers. Ammonia is manufactured by the hydrogenation of N_2. Although it might not seem that the industrial synthesis of ammonia has anything to do with organic chemistry, the hydrogen used in its manufacture in fact comes from the cracking of alkanes (Sec. 5.7B). Thus, the availability of ammonia is tied inexorably to the availability of hydrocarbons.

B. Naturally Occurring Amines

Alkaloids   Among the many types of naturally occurring amines are the **alkaloids**: nitrogen-containing bases that occur naturally in plants. This simple definition encompasses a highly diverse group of compounds; the structures of a few alkaloids are shown in Figure 23.3. Because amines are the most common organic bases, it is not surprising that most alkaloids are amines, including heterocyclic amines. It is believed that the first alkaloid ever isolated and studied is morphine, discovered in 1805. Many alkaloids have biological activity (Fig. 23.3); others have no known activity, and their functions within the plants from which they come are, in many cases, obscure. Investigations dealing with

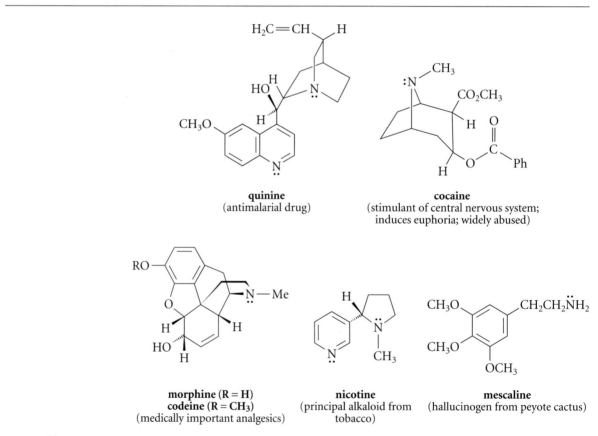

quinine
(antimalarial drug)

cocaine
(stimulant of central nervous system; induces euphoria; widely abused)

morphine (R = H)
codeine (R = CH₃)
(medically important analgesics)

nicotine
(principal alkaloid from tobacco)

mescaline
(hallucinogen from peyote cactus)

Figure 23.3   *Structures of some alkaloids. Notice that each compound has at least one basic amine group.*

the isolation, structure, and medicinal properties of alkaloids continue to be major research activities in organic chemistry.

PROBLEM

23.37   Show that *(a) morphine and (b) mescaline (Fig. 23.3) are Brønsted bases by giving the structures of their conjugate acids.

Hormones and Neurotransmitters   Epinephrine (adrenaline) is an amine secreted by both the adrenal medulla and sympathetic nerve endings; it is an example of a **hormone**—a compound that regulates the biochemistry of multicellular organisms, particularly vertebrates.

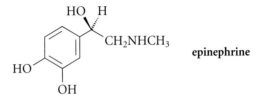

epinephrine

Epinephrine, for example, is associated with the "fight or flight" response to external stimuli; you might feel the effects of epinephrine secretion when you walk unprepared into your organic chemistry class and your instructor says, "Pop quiz today." The mechanisms by which hormones exert their effects are important research areas in contemporary biochemistry.

Norepinephrine, another amine, and acetylcholine, a quaternary ammonium ion, are examples of *neurotransmitters.*

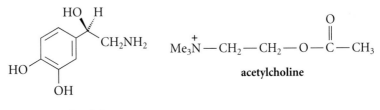

acetylcholine

norepinephrine

Neurotransmitters are molecules that are important in the communication between nerve cells, or between nerve cells and their target organs. This communication occurs at cellular junctions called *synapses.* A nerve impulse is transmitted when a neurotransmitter is released from a nerve cell on one side of the synapse, moves by diffusion across the synapse, and binds to a protein receptor molecule of another nerve cell or a target organ on the other side. This binding triggers either the transmission of the impulse down the nerve cell to the next synapse or a response of the target organ. Different neurotransmitters are involved in different parts of the nervous system.

Significant advances have recently occurred in understanding the chemistry that takes place in the human brain (*neurochemistry*). These advances are being made by teams of molecular biologists, biochemists, and organic chemists. It is conceivable that an understanding of neurochemistry will lead to treatments for such widespread and tragic afflictions as Parkinson's disease and Alzheimer's disease. Sigmund Freud perhaps anticipated these developments when he wrote in 1930, "The hope of the future lies in organic chemistry. . . ."

KEY IDEAS IN CHAPTER 23

▽ Amines are classified as primary, secondary, or tertiary. Quaternary ammonium salts are compounds in which all four hydrogens of the ammonium ion are formally replaced with alkyl or aryl groups.

▽ Most amines undergo rapid inversion at nitrogen. This inversion interconverts an amine and its mirror image.

▽ Simple amines are liquids with unpleasant odors. Amines of low molecular mass are miscible with water.

▽ The N—H stretching absorption is the most important infrared absorption of primary and secondary amines. In the NMR spectra of amines, the α-protons have chemical shifts in the δ 2.5–3.0 range. The CMR chemical shifts of α-carbons are in the δ 30–50 range.

▽ Basicity is one of the most important chemical properties of amines. The basicity of an amine is expressed by the pK_a of its conjugate-acid ammonium ion. Ammonium-ion pK_a values are affected by alkyl substitution on the nitrogen, the polar effects of nearby substituent groups, and resonance interaction of the amine's unshared electrons with an adjacent aromatic ring.

▽ Because amines are basic, they are also nucleophilic. The nucleophilicity of amines is important in many of their reactions, for example, alkylation, imine formation, and acylation.

▽ Amines can serve as leaving groups. Thus, quaternary ammonium hydroxides when heated undergo the Hofmann elimination, in which an amine is lost from the α-carbon and a proton is lost from the least branched β-carbon atom.

▽ The amino group is *ortho*, *para*-directing in electrophilic aromatic substitution reactions. However, in many reactions of this type, such as nitration and sulfonation, protection of the nitrogen as an amide is required to prevent protonation of the amine under the acidic reaction conditions.

Treatment of primary amines with nitrous acid gives diazonium ions. Aliphatic diazonium ions decompose under the diazotization conditions to give a complex mixture of products. Aryldiazonium ions, however, can be used in a number of substitution reactions: the Sandmeyer reaction (substitution of N_2 by halide or cyanide groups) or hydrolysis (substitution with —OH). Because aryldiazonium ions are electrophiles, they react with activated aromatic compounds, such as aromatic amines and phenols, to give substituted azobenzenes, some of which are dyes. Secondary amines react with nitrous acid to give nitrosamines. Under acidic conditions, tertiary amines do not react. (Aromatic tertiary amines give ring-nitrosated products.)

Amines can be synthesized from amides, nitriles, nitro compounds, and other amines, as summarized in Sec. 23.11D. The Curtius and Hofmann rearrangements yield amines with one fewer carbon atom. The Curtius rearrangement can also be used to prepare isocyanates, which react with nucleophiles by addition across the C=N bond to give carbamic acid derivatives.

ADDITIONAL PROBLEMS

*23.38   Give the principal organic product(s) expected when *p*-chloroaniline or other compound indicated reacts with each of the following reagents.
(a)   dilute HBr   (b) C_2H_5MgBr (c) $NaNO_2$, HCl, 0°
(d)   *p*-toluenesulfonyl chloride
(e)   product of (c) with H_2O, Cu_2O, and excess $Cu(NO_3)_2$
(f)   product of (c) with CuBr   (g)   product of (c) with H_3PO_2
(h)   product of (c) with CuCN   (i)   product of (d) + NaOH, 25°

23.39   Give the principal organic product(s) expected when *N*-methylaniline reacts with each of the following reagents.
(a) Br_2 (b)   benzoyl chloride
(c)   benzyl chloride, then dilute ⁻OH   (d)   *p*-toluenesulfonic acid
(e) $NaNO_2$, HCl   (f)   excess CH_3I, heat, then Ag_2O
(g) $CH_3CH{=}O$, $NaBH_3CN$

*23.40   Give the principal organic product(s), if any, expected when isopropylamine or other compound indicated reacts with each of the following reagents.
(a)   dilute H_2SO_4 (b)   dilute NaOH solution
(c)   butyllithium in THF, −78°   (d)   acetyl chloride, pyridine
(e) $NaNO_2$, aqueous HBr, 0°   (f)   acetone, H_2/Pd/C
(g)   excess CH_3I, heat   (h)   benzoic acid, 25°
(i)   formaldehyde, $NaBH_4$ (j)   product of (g) + Ag_2O, then heat
(k)   product of (d) with $LiAlH_4$, then H_3O^+

23.41   Give the structure of a compound that fits each description. (There may be more than one correct answer for each.)

*(a)   a chiral primary amine C_4H_7N with no triple bonds

(b)   a chiral primary amine $C_4H_{11}N$

*(c)   two secondary amines, which, when treated with CH_3I, then Ag_2O and heat, give propene and *N*,*N*-dimethylaniline

*(d)   a compound C_4H_9N that reacts with $NaBH_4$ to give *N*-methyl-2-propanamine

23.42   Explain how you would distinguish the compounds within each set by a simple chemical test with readily observable results, such as solubility in acid or base, evolution of a gas, etc.

*(a)   *N*-methylhexanamide; 1-octanamine; *N*,*N*-dimethyl-1-hexanamine

(b)   *p*-methylaniline, benzylamine, *p*-cresol, anisole

*23.43   On the package insert for the drug *labetalol*, used in the control of blood pressure and hypertension, is given the following structure:

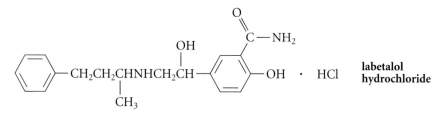

labetalol
hydrochloride

A pharmacist, Prunella Pillpeddler, knowing your expertise in chemistry, has come to you with some questions about the chemical behavior of labetalol, which you should answer.

(a)   Labetalol is claimed to be a salt. Explain by giving a more detailed structure.

(b)   What happens to labetalol·HCl when treated with one equivalent of NaOH at room temperature?

(c)   What happens to labetalol when it is treated with an excess of aqueous NaOH and heat?

(d)   What are the products formed when labetalol is treated with 6 *M* aqueous HCl and heat?

23.44   (a)   Give the structure of cocaine (Fig. 23.3) as it would exist in 1 *M* aqueous HCl solution.

(b)   What products would form if cocaine were treated with an excess of aqueous NaOH and heat?

(c)   What products would form if cocaine were treated with an excess of concentrated aqueous HCl and heat?

*23.45   Design a separation of a mixture containing the following four compounds into its pure components. Describe exactly what you would do and what you would expect to observe.

nitrobenzene, aniline, *p*-chlorophenol, and *p*-nitrobenzoic acid

23.46   How would the basicity of trifluralin, a widely used herbicide, compare with that of N,N-diethylaniline: much greater, about the same, or much less? Explain.

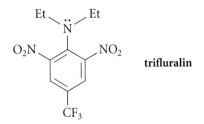

trifluralin

*23.47**   (a)   When anthranilic acid (below) is treated with $NaNO_2$ in aqueous HCl solution, and the resulting solution is treated with N,N-dimethylaniline, a dye called *methyl red* is formed. Give the structure of methyl red.

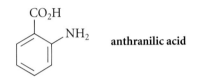

anthranilic acid

(b)   When an acidic solution of methyl red is titrated with base, the dye behaves as a diprotic acid with pK_a values of 2.3 and 5.0. The color of the methyl red solution changes very little as the pH is raised past 2.3, but as the pH is raised past 5.0, the color of the solution changes dramatically from red to yellow. Explain.

23.48   Alizarin yellow R is an azo dye that changes color from yellow to red between pH 10.2 and 12.2.

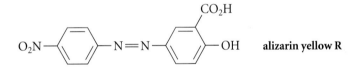

alizarin yellow R

(a)   Outline a synthesis of alizarin yellow R from aniline, salicylic acid (o-hydroxybenzoic acid), and any other reagents.
(b)   Draw the structure of alizarin yellow R as it exists in its yellow form at pH = 9
(c)   Draw the structure of alizarin yellow R as it exists at pH > 12. Why does it change color?

23.49   Amanda Amine, an organic chemistry student, has proposed the following reactions. Indicate in each case why the reaction would not succeed as written.

*(a)

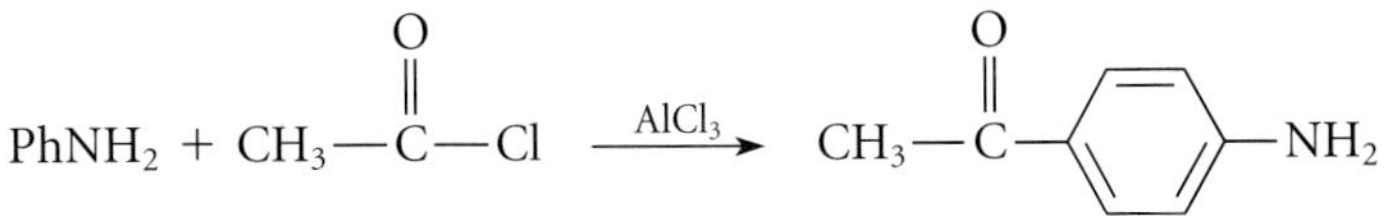

$$PhNH_2 + CH_3 - \overset{O}{\underset{\|}{C}} - Cl \xrightarrow{AlCl_3} CH_3 - \overset{O}{\underset{\|}{C}} - \underset{}{\underset{}{\bigcirc}} - NH_2$$

(Problem 23.49 continues)

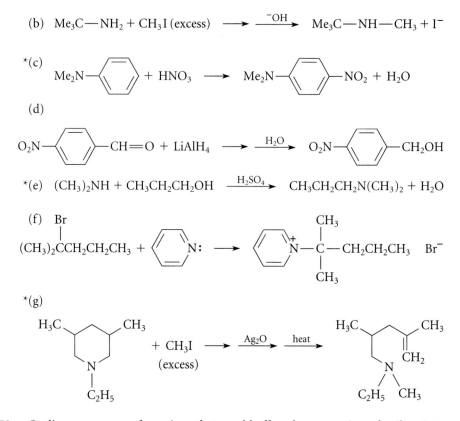

(b) $Me_3C—NH_2 + CH_3I$ (excess) $\longrightarrow$ $\xrightarrow{^-OH}$ $Me_3C—NH—CH_3 + I^-$

*(c)

(d)

*(e) $(CH_3)_2NH + CH_3CH_2CH_2OH$ $\xrightarrow{H_2SO_4}$ $CH_3CH_2CH_2N(CH_3)_2 + H_2O$

(f)

*(g)

23.50   Outline a sequence of reactions that would effect the conversion of aniline into each of the following compounds.
*(a) benzylamine   (b) benzyl alcohol
*(c) 2-phenylethanamine   (d) N-phenyl-2-butanamine
*(e) p-chlorobenzoic acid

*23.51   When 1,5-dibromopentane reacts with ammonia, among several products isolated is a water-soluble compound *A* that rapidly gives a precipitate of AgBr with acidic AgNO₃ solution. Compound *A* is unchanged when treated with dilute base, but treatment of *A* with concentrated NaOH and heat gives a new compound *B* ($C_{10}H_{19}N$) that decolorizes Br_2 in CCl_4. Compound *B* is identical to the product obtained from the following reaction sequence:

4-pentenoic acid $\xrightarrow{SOCl_2}$ $\xrightarrow{\text{piperidine}}$ $\xrightarrow[\text{2) } H_3O^+]{\text{1) } LiAlH_4}$ *B*

Identify *A* and explain your reasoning.

23.52   When p-aminophenol reacts with one molar equivalent of acetic anhydride, a compound acetaminophen (*A*, $C_8H_9NO_2$) is formed that dissolves in dilute NaOH. When *A* is treated with one equivalent of NaOH followed by ethyl iodide, an ethyl ether *B* is formed. What is the structure of acetaminophen? Explain your reasoning.

23.53   Give an explanation for each of the following facts.

*(a) The barrier to rotation about the *N*-phenyl bond in *N*-methyl-*p*-nitroaniline is considerably higher (42–46 kJ/mol, or 10–11 kcal/mol) than that in *N*-methylaniline itself (about 25 kJ/mol, or 6 kcal/mol).

(b) *Cis* and *trans*-1,3-dimethylpyrrolidine rapidly interconvert.

*(c) CH_3NH—CH_2—$NHCH_3$ is unstable in aqueous solution.

(d) The following compound exists as the enamine tautomer shown rather than as an imine:

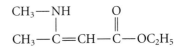

*(e) Diazotization of 2,4-cyclopentadien-1-amine gives a diazonium salt which, unlike most aliphatic diazonium ions, does not decompose to a carbocation.

23.54 Imagine that you have been given a sample of racemic 2-phenylbutanoic acid. Outline steps that would allow you to obtain pure samples of each of the following compounds from this starting material and any other reagents. (Note that optical resolutions are time-consuming. One resolution that would serve all four syntheses would be most efficient.)

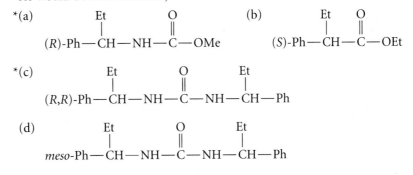

*23.55 Outline the preparation of the following compounds from 3-methylbenzoic acid and any other reagents.

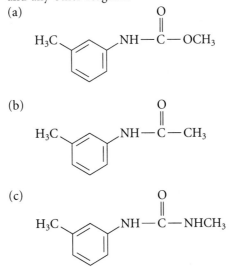

23.56 Show how the insecticide *carbaryl* can be prepared from methyl isocyanate, $CH_3—N=C=O$.

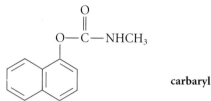

carbaryl

*23.57 A compound A ($C_{22}H_{27}NO$) is insoluble in acid and base but reacts with concentrated aqueous HCl and heat to give a clear aqueous solution from which, on cooling, benzoic acid precipitates. When the supernatant solution is made basic, a liquid B separates. Compound B is achiral. Treatment of B with benzoyl chloride in pyridine gives back A. Evolution of gas is not observed when B is treated with an aqueous solution of $NaNO_2$ and HCl. Treatment of B with excess CH_3I, then Ag_2O and heat, gives a compound C, $C_9H_{19}N$, plus styrene, $Ph—CH=CH_2$. Compound C, when treated with excess CH_3I, then Ag_2O and heat, gives a single alkene D that is identical to the compound obtained when cyclohexanone is treated with the ylid $:\overset{-}{C}H_2—\overset{+}{P}(Ph)_3$. Give the structure of A and explain your reasoning.

23.58 Three bottles A, B, and C have been found, each of which contains a liquid and is labeled "amine $C_8H_{11}N$." As an expert in amine chemistry, you have been hired as a consultant and asked to identify each compound. Compounds A and B give off a gas when they react with $NaNO_2$ and HCl at 0°; C does not. However, when the aqueous reaction mixture from the diazotization of C is warmed, a gas is evolved. Compound A is optically inactive, but when it reacts with (+)-tartaric acid, two isomeric salts with different physical properties are obtained. Titration of C with aqueous HCl reveals that its conjugate acid has a $pK_a = 5.1$. Oxidation of C with H_2O_2 (a reagent known to oxidize amino groups to nitro groups), followed by vigorous oxidation with $KMnO_4$ gives *p*-nitrobenzoic acid. Oxidation of B in a similar manner yields 1,4-benzenedicarboxylic acid (terephthalic acid), and oxidation of A yields benzoic acid. Identify compounds A, B, and C.

*23.59 You have been hired as a consultant by the Chicago police, who have brought to you a compound A confiscated in an important drug raid. Compound A is a water-soluble white solid that rapidly gives a precipitate of AgCl with acidic $AgNO_3$ solution. Treatment of the solid with 5% NaOH solution liberates a water-insoluble liquid B that is found to be optically active and dextrorotatory; a gas is evolved when B is treated with $NaNO_2$ and HCl at 0°. The elemental analysis laboratory returns the following analysis for B: 79.9% carbon, 9.7% hydrogen, and 10.4% nitrogen, and the mass spectrum of B suggests a molecular mass of 135. Treatment of B successively with excess CH_3I, Ag_2O, and then heat gives C, a mixture of alkenes. Catalytic hydrogenation of C yields a single compound D that is identical to the compound obtained when propiophenone is heated strongly in ethylene glycol with hydrazine (NH_2NH_2) and KOH. The NMR spectrum of A is complex, but contains a clean doublet at δ 1.1 ($J = 7.5$

Hz). Identify compound *A*. (Do not try to determine its absolute configuration.) Conviction of a notorious drug syndicate may hinge on your successful analysis.

23.60 Complete the following reactions by giving the structure(s) of the major product(s). Explain how you arrived at your answers.

*(a) $NH_3 + HNO_2 \longrightarrow$

*(b)

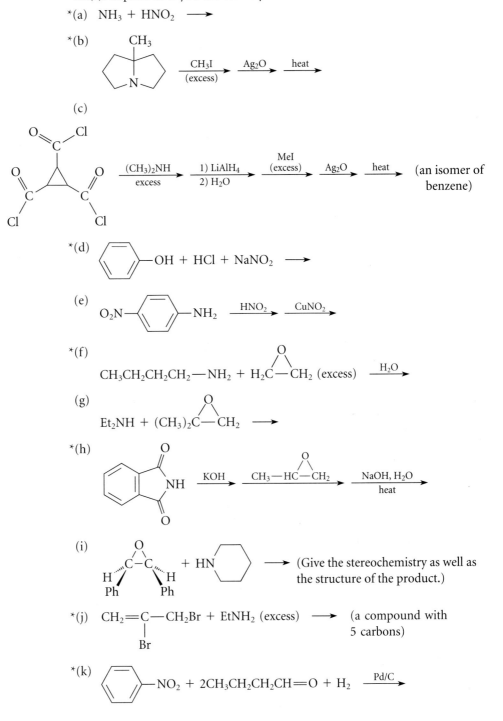

(c)

*(d)

(e)

*(f) $CH_3CH_2CH_2CH_2-NH_2 + H_2C-CH_2$ (excess) $\xrightarrow{H_2O}$

(g) $Et_2NH + (CH_3)_2C-CH_2 \longrightarrow$

*(h)

(i)   + HN $\langle \rangle$ $\longrightarrow$ (Give the stereochemistry as well as the structure of the product.)

*(j) $CH_2=C-CH_2Br + EtNH_2$ (excess) $\longrightarrow$ (a compound with 5 carbons)
$\quad\quad\quad |$
$\quad\quad\quad Br$

*(k) $-NO_2 + 2CH_3CH_2CH_2CH=O + H_2 \xrightarrow{Pd/C}$

23.61   Outline a synthesis for each of the following compounds from the indicated starting materials and any other reagents. The starting material for the compounds in parts (a) through (e) is pentanoic acid.

*(a) *N*-methyl-1-hexanamine   (b) pentylamine
*(c) *N,N*-dimethyl-1-pentanamine   (d) butylamine
*(e) hexylamine

(f)

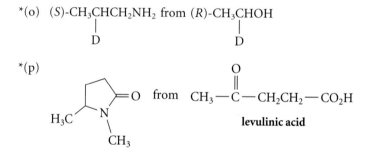

$$PhCH_2-\overset{\overset{\textstyle CH_3}{|}}{\underset{\underset{\textstyle CH_3}{|}}{N^+}}-CH_2CH_2CH_2CH_3 \ Br^- \text{ from butyraldehyde}$$

*(g) *N*-ethyl-3-phenyl-1-propanamine from toluene
*(h) 2-pentanamine from diethyl malonate
*(i) isobutylamine from acetone   (j) isopentylamine from acetone
*(k) *m*-chlorobromobenzene from nitrobenzene
 (l) *p*-chlorobromobenzene from nitrobenzene
*(m) *p*-methoxybenzonitrile from phenol
 (n) *p*-bromoaniline from *p*-nitroaniline

*(o) (*S*)-CH$_3$CHCH$_2$NH$_2$ from (*R*)-CH$_3$CHOH
              | |
              D   D

*(p)

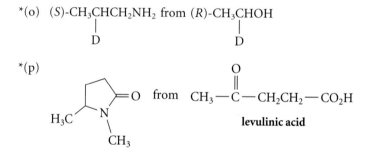

             from   CH$_3$—C—CH$_2$CH$_2$—CO$_2$H
                          ‖
                          O

                    levulinic acid

*23.62   Rank the following compounds according to increasing relative rates of Hofmann elimination, and explain your answers carefully.

$$R-CH_2-CH_2-\overset{+}{N}Me_3 \ {}^-OH$$

compound *A*   R = —C(CH$_3$)$_3$
compound *B*   R = —H
compound *C*   R = —CH$_3$

*23.63   In the NMR spectrum of a concentrated (4.5 *M*) aqueous solution of methylamine, the methyl group appears as a quartet when the solution pH is 1. At intermediate pH, the methyl group appears as a broad line. At pH = 9 the methyl group is observed as a single sharp line. Explain these observations.

*23.64   Aniline has a UV spectrum with peaks at λ_{max} = 230 nm (ϵ = 8600) and 280 nm (ϵ = 1430). In the presence of dilute HCl, the spectrum of aniline changes dramatically: λ_{max} = 203 (ϵ = 7500) and 254 (ϵ = 160). This spectrum is nearly identical to the UV spectrum of benzene. Account for the effect of acid on the UV spectrum of aniline.

23.65   In the warehouse of the company Tumany Amines, Inc., two unidentified compounds have been found. The president of the company, Wotta Stench, has hired you to identify them from their spectra:
*(a) $C_{10}H_{15}NO$: IR spectrum 3355, 1618 cm^{-1}, no carbonyl absorption. NMR spectrum in Fig. 23.4a (p. 1166).
(b) C_3H_9NO: NMR and IR spectra in Fig. 23.4b and c (p. 1166)

*23.66   Propose a structure for the compound A ($C_6H_{15}O_2N$) that is unstable in aqueous acid and has the following NMR spectra:
Proton NMR: δ 2.30 (6H, s); δ 2.45 (2H, d, J = 6 Hz); δ 3.27 (6H, s); δ 4.50 (1H, t, J = 6 Hz)
Carbon NMR: δ 46.3, δ 53.2, δ 68.8, δ 102.4

23.67   *(a) Propose a structure for an amine A (C_4H_9N), which liberates a gas when treated with $NaNO_2$ and HCl. The CMR spectrum of A is as follows, with attached protons in parentheses: δ 14(2), δ 34.3 (2), δ 50.0(1)
(b) Propose a structure for an amine B (C_4H_9N), which does *not* liberate a gas when treated with $NaNO_2$ and HCl, and has IR absorptions at 917 cm^{-1}, 990 cm^{-1}, and 1640 cm^{-1}, as well as N—H absorption at 3300 cm^{-1}. The CMR spectrum of B is as follows: δ 36.0, δ 54.4, δ 115.8, δ 136.7.

23.68   Give a mechanism for each of the following rearrangement reactions.
*(a)

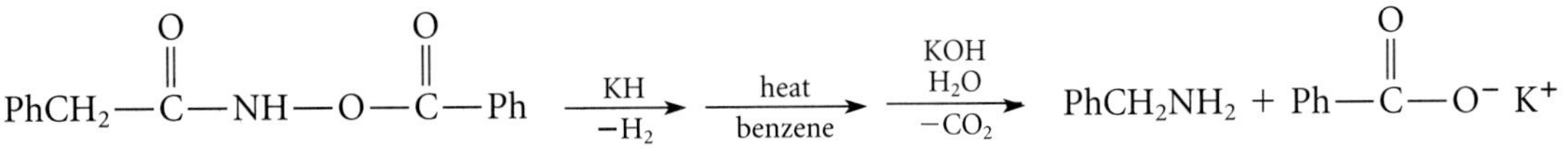

(b)

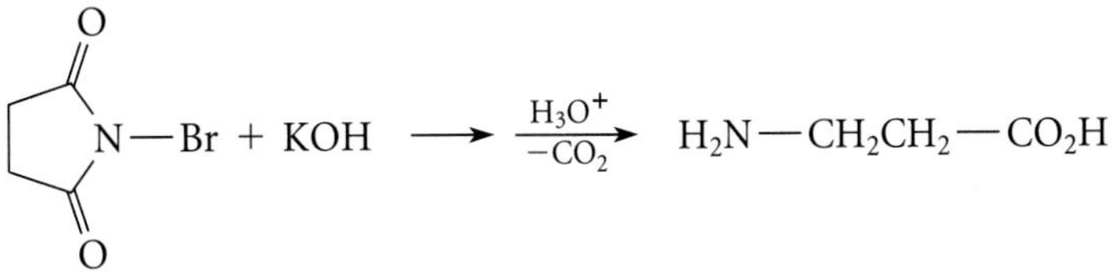

*(c)

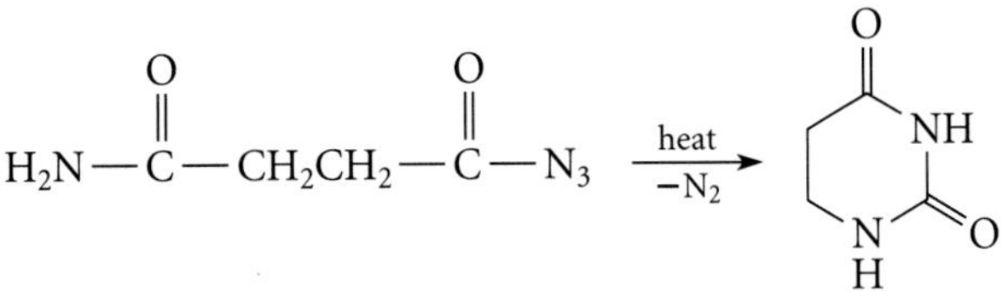

*23.69   Suggest an explanation for each of the following reactions.
(a)

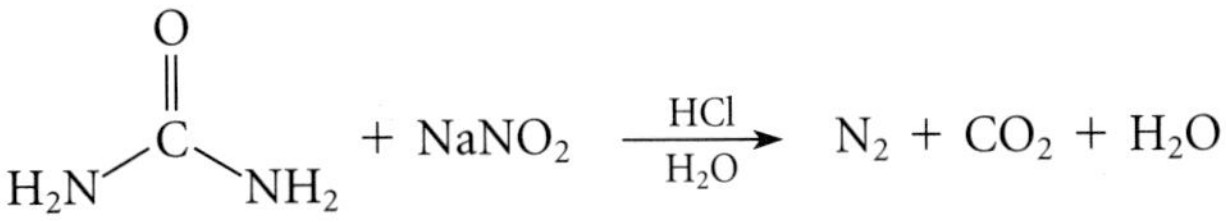

(This reaction can be used to scavenge unwanted nitrous acid.)

(*Problem 23.69 continues on p. 1167*)

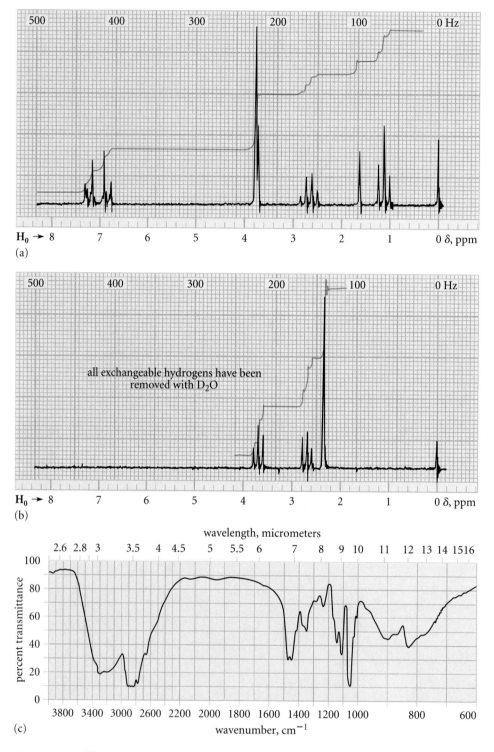

Figure 23.4 *Spectra for Problem 23.65.*

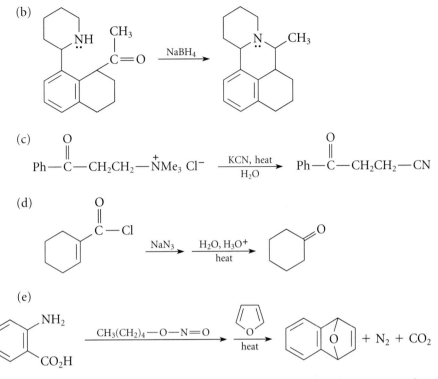

(Hint: $CH_3(CH_2)_4-O-N=O$, amyl nitrite, causes the same transformation of primary amines as nitrous acid.)

23.70 A chemist, Mada Meens, treated ammonia with pentanal in the presence of hydrogen gas and a catalyst in the expectation of obtaining 1-pentanamine by reductive amination. However, in addition to 1-pentanamine, she also obtained dipentylamine and tripentylamine. Explain how these by-products are formed.

$$CH_3(CH_2)_3CH=O + NH_3 + H_2 \xrightarrow{\text{catalyst}}$$
$$CH_3(CH_2)_3CH_2NH_2 + [CH_3(CH_2)_3CH_2]_2NH + [CH_3(CH_2)_3CH_2]_3N$$

*23.71 Explain the fact that the following amines, despite obvious similarities in structure, have considerably different basicities. (Hint: Make a model of compound B. Look at the relationship of the nitrogen unshared electron pair to the p orbitals of the benzene ring.)

(Problem 23.71 continues)

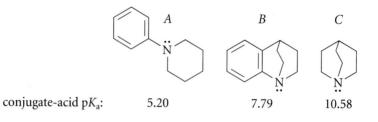

conjugate-acid pK_a: 5.20   7.79   10.58

23.72   Amide *A* below, δ-valerolactam, is a typical amide with a conjugate-acid pK_a of 0.8. The two cyclic tertiary amines *B* and *C* also have typical conjugate-acid pK_a values. In contrast, the conjugate-acid pK_a of amide *D* is unusually high for an amide, and it hydrolyzes much more rapidly than other amides. Draw the structure of the conjugate acid of amide *D*, and suggest a reason for both its unusual pK_a and its rapid hydrolysis.

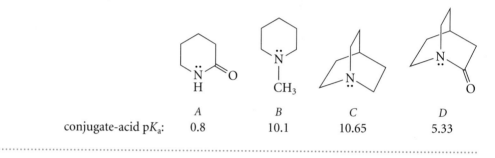

conjugate-acid pK_a: *A*   *B*   *C*   *D*
                          0.8   10.1   10.65   5.33

24

Chemistry of Naphthalene and the Aromatic Heterocycles

Our discussion of aromatic compounds has focused mainly on derivatives of benzene. However, some aromatic compounds contain two or more fused rings. (The aromaticity of such compounds is not covered by the Hückel $4n + 2$ rule, which applies only to *monocyclic* aromatic compounds; Sec. 15.7D.) Aromatic hydrocarbons that contain two or more fused rings are called **polycyclic aromatic hydrocarbons**. The simplest of the polycyclic aromatic hydrocarbons is *naphthalene*.

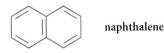

naphthalene

This chapter considers the chemistry of naphthalene, and then takes up the chemistry of some important aromatic heterocyclic compounds. In particular, the chemistry of both naphthalene and the heterocycles is contrasted with the chemistry of benzene.

The structures of some other polycyclic aromatic hydrocarbons are shown in Fig. 24.1. Although the Lewis structures of these compounds consist of alternating single and double bonds, the π electrons are extensively delocalized within π molecular orbitals.

One form of carbon itself has a polycyclic aromatic structure. *Graphite* is a carbon polymer that consists of layers of fused benzene rings (Fig. 24.2) The softness of graphite and its ability to act as a lubricant can be attributed to the ease with which its layers slide past each other. Even though it is not an ionic compound, graphite is an excellent electrical conductor because of the ease with which π electrons can be delocalized across its structure.

Another remarkable polycyclic aromatic form of carbon with the formula C_{60} was discovered recently, first in clouds of interstellar gas and then in soot. In 1985, Harold W. Kroto of the University of Sussex, Brighton, U.K., and Richard E. Smalley of Rice

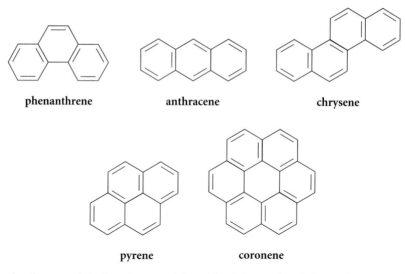

Figure 24.1 *Some polycyclic aromatic hydrocarbons consisting of fused six-membered rings with conjugated double bonds.*

University in Houston, announced their proposal for the structure of this species, shown in Fig. 24.3. This structure, which corresponds to the seams on a modern soccerball, was named *buckminsterfullerene* because of its resemblance to a geodesic dome, designed by Buckminster Fuller. All carbons in this structure are equivalent. Indeed, the CMR of buckminsterfullerene consists of a single resonance at δ 143. A number of different fullerenes (as such ball-shaped structures have been nicknamed) of different sizes have been documented.

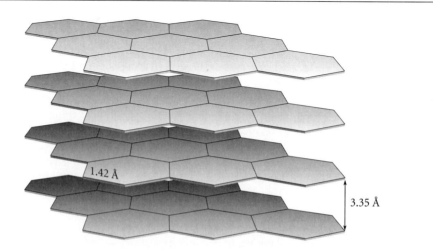

Figure 24.2 *Structure of graphite. Graphite consists of fused aromatic rings in layers separated by 3.35 Å. (Only four of many layers are shown, and the double bonds within the aromatic rings are not shown.)*

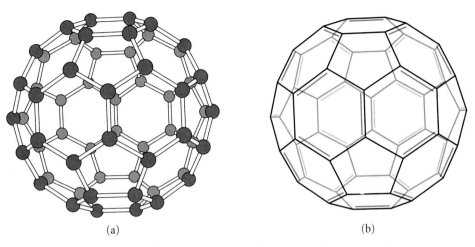

(a)   (b)

Figure 24.3   *(a) Ball-and-stick model and (b) a Lewis structure of buckminsterfullerene. In the Lewis structure, double bonds are shown in color.*

24.1   Chemistry of Naphthalene

A. Physical Properties and Structure

Naphthalene is a solid (mp 80.5°), with a high vapor pressure. The familiar odor of moth balls is due to naphthalene.

Naphthalene can be represented by three resonance structures; two are equivalent, and one is unique.

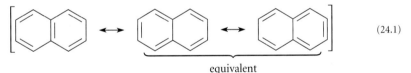

(24.1)

equivalent

The carbon-carbon bond lengths in naphthalene (1.36 Å, 1.41 Å, and 1.42 Å) are very similar to those of benzene (1.395 Å) and graphite (Fig. 24.2), and the C—C—C bond angles are very close to 120°, the value expected for a planar aromatic arrangement of six-membered rings.

PROBLEM

*24.1   Using the resonance structures for naphthalene (Eq. 24.1) to guide you, predict which carbon-carbon bond should be the shortest.

B. Nomenclature

In substitutive nomenclature, carbon-1 of naphthalene is the carbon adjacent to a *bridgehead* carbon (a vertex at which the rings are fused). Substituents are given the lowest numbers consistent with this scheme and their relative priorities. The following examples illustrate this idea:

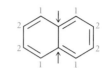

bridgehead carbons (→)

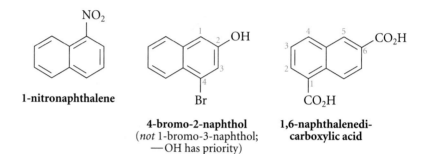

1-nitronaphthalene

4-bromo-2-naphthol
(*not* 1-bromo-3-naphthol;
—OH has priority)

**1,6-naphthalenedi-
carboxylic acid**

Naphthalene also has a common nomenclature that uses Greek letters. In this system the 1-position is designated as α and the 2-position as β. Common and substitutive nomenclature should not be mixed.

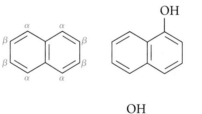

substitutive: 1-naphthol
common: α-naphthol

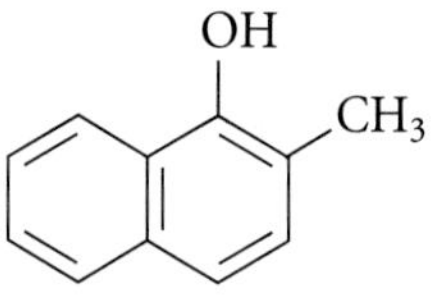

correct: β-methyl-α-naphthol,
or 2-methyl-1-naphthol
incorrect: 2-methyl-α-naphthol
(mixes common and substitutive
nomenclature)

The naphthalene ring can also be named as a substituent group, the *naphthyl* group. The term *naphthyl* for a naphthalene ring is analogous to the term *phenyl* for a benzene ring.

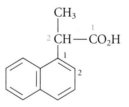

2-(1-naphthyl)propanoic acid

position of attachment on
naphthalene ring

position of substitution on the
parent propanoic acid

PROBLEM

24.2   Name the following compounds.

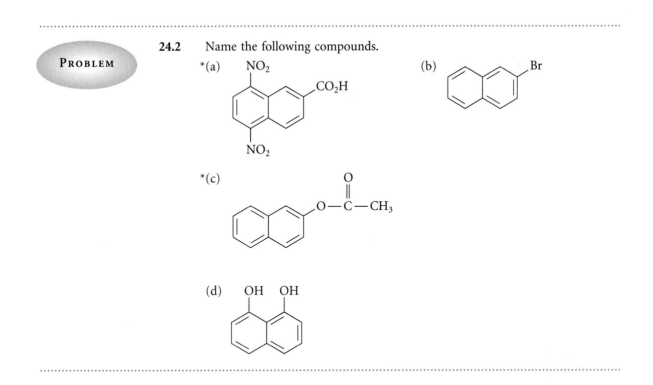

*(a)

*(c)

(d)

C. Electrophilic Substitution Reactions

Electrophilic Aromatic Substitution Reactions of Naphthalene   Because naphthalene is an aromatic compound, it should not be surprising that it undergoes electrophilic aromatic substitution reactions much like those of benzene. (You should review the electrophilic aromatic substitution reactions of benzene found in Secs. 16.4, 16.5, 18.8, and 23.9.)

An understanding of electrophilic substitution in naphthalene hinges on the answers to two questions: (1) At what position does naphthalene undergo substitution? (2) How reactive is naphthalene in electrophilic aromatic substitution reactions?

Electrophilic aromatic substitution of naphthalene generally occurs at the 1-position.

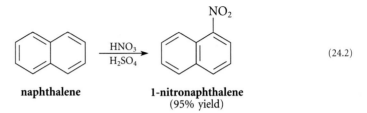

naphthalene   **1-nitronaphthalene**
                                    (95% yield)

(24.2)

Why 1-substitution rather than 2-substitution occurs is revealed by comparing resonance structures for the carbocation intermediates in the two processes. Seven resonance structures can be drawn for the carbocation intermediate in electrophilic substitution at the 1-position.

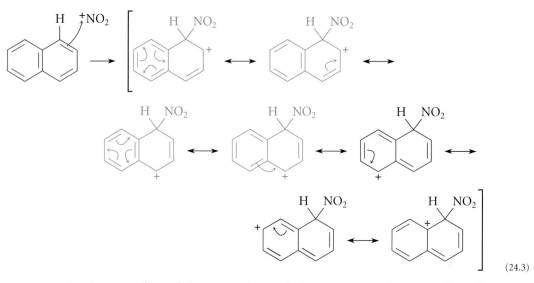

(24.3)

Furthermore, four of these (color) contain intact benzene rings. This is an important point because *structures in which benzene rings are left intact are more important than those in which the formal double bonds are moved out of the ring.* The reason this is so is that structures lacking the intact benzene rings are not aromatic, and are thus less stable. (Recall from Sec. 15.6B that the importance of resonance structures is determined by imagining they are separate molecules, even though they really aren't.) Substitution in the 2-position, in contrast, gives a carbocation intermediate for which there are six resonance structures (draw these!). Of these six, only two contain intact benzene rings (identify these). Therefore substitution at the 1-position gives the more stable carbocation intermediate and, by Hammond's postulate (Sec. 4.8C), occurs more rapidly. Notice that there is nothing wrong with 2-substitution; 1-substitution is simply more favorable.

An interesting result occurs in the sulfonation of naphthalene, a result which hinges on the fact that *aromatic sulfonation is a reversible reaction.* This reversibility is apparent if any arylsulfonic acid derivative is heated in the presence of acid: the sulfonic acid group is replaced by hydrogen.

$$H_2O + \langle\!\!\!\bigcirc\!\!\!\rangle\!-\!SO_3H \xrightarrow{\text{acid, heat}} \langle\!\!\!\bigcirc\!\!\!\rangle\!-\!H + H_2SO_4 \qquad (24.4)$$

Sulfonation of naphthalene under mild conditions gives mostly 1-naphthalenesulfonic acid, a result expected from the outcome of other electrophilic substitutions of naphthalene. However, under more vigorous conditions, sulfonation yields mostly 2-naphthalenesulfonic acid.

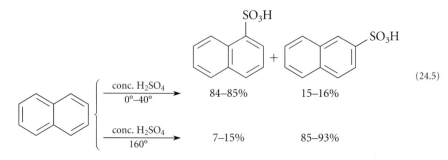

(24.5)

	84–85%	15–16%
conc. H$_2$SO$_4$, 0°–40°		
conc. H$_2$SO$_4$, 160°	7–15%	85–93%

This is another case of *kinetic vs. thermodynamic control* of a reaction (Sec. 15.4C). At low temperature, substitution at the 1-position is observed because it is faster. At higher temperature, formation of 1-naphthalenesulfonic acid is reversible, and the *more stable*, but more slowly formed, 2-naphthalenesulfonic acid is observed. This reversibility is confirmed when 1-naphthalenesulfonic acid is subjected to high-temperature conditions.

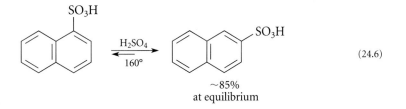

$$(24.6)$$

Why is the 2-isomer more stable? The answer lies in the van der Waals repulsions that occur between groups in the 1- and 8-positions. The sulfonic acid group is large—about as large as a *tert*-butyl group. The unfavorable interaction between this group and the hydrogen in the 8-position, called a *peri* interaction, destabilizes the 1-isomer.

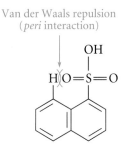

A *peri* interaction is much more severe than the interaction of the same two groups in *ortho* positions because bonds in *ortho* positions diverge from each other, whereas the bonds in *peri* positions are parallel. Thus, 1-naphthalenesulfonic acid, if allowed to equilibrate, is converted into the 2-isomer to avoid this unfavorable steric interaction. In contrast, such an equilibrium is not observed in other electrophilic substitution reactions that are not reversible.

Now let's consider the reactivity of naphthalene. *Naphthalene is considerably more reactive than benzene in electrophilic aromatic substitution.* For example, recall that bromination of benzene requires a Lewis acid catalyst such as FeBr$_3$ (Sec. 16.4A). In contrast, naphthalene is readily brominated in CCl$_4$ without catalysts.

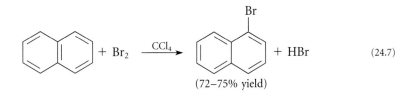

$$(24.7)$$

The greater reactivity of naphthalene in electrophilic aromatic substitution reactions reflects the considerable resonance stabilization of the carbocation intermediate (see Eq. 24.3).

Electrophilic Aromatic Substitution Reactions of Substituted Naphthalenes

Recall that in electrophilic aromatic substitution reactions of substituted benzenes, substituents on the ring either *accelerate* or *retard* a substitution reaction relative to benzene itself; that is, substituents may *activate* or *deactivate* further substitution. Recall also that substituents exert *directing effects* in aromatic substitution reactions; that is, the nature of the substituent determines the ring position(s) at which further substitution takes place (Sec. 16.5). The same effects are observed in naphthalene chemistry. However, an additional question arises in naphthalene chemistry that has no counterpart in the chemistry of benzene: does the second substitution occur on the ring that is already substituted, or does it occur on the unsubstituted ring?

The following trends are observed in most cases:

1. When one ring of naphthalene is substituted with deactivating groups (such as $-NO_2$ or $-SO_3H$), further substitution occurs on the *unsubstituted* ring at an open α-position (if available).

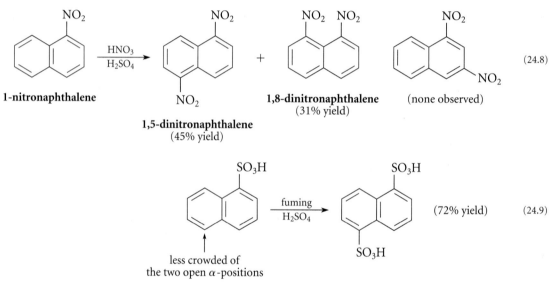

2. When one ring of naphthalene is substituted with activating groups (such as $-CH_3$ or $-OCH_3$), further substitution occurs in the substituted ring at the *ortho* or *para* positions.

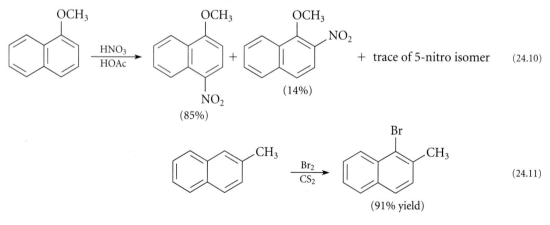

In the last example, substitution occurs at an *ortho* position because no *para* position is open. Of the two open *ortho* positions (identify them), substitution at an α-position is preferred to substitution at a β-position (why?).

PROBLEMS

24.3   Complete the following reactions by giving the structure(s) of the major organic product(s). Explain your answers.

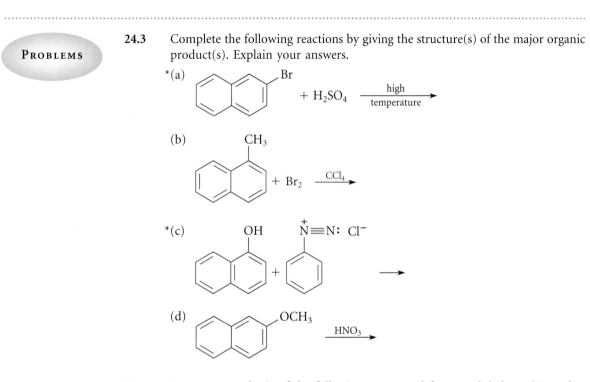

*24.4   Propose a synthesis of the following compound from naphthalene. (Note that the Friedel-Crafts acylation reaction cannot be used because it gives a mixture of 1- and 2-acetylnaphthalene that is difficult to separate.)

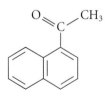

24.2

Introduction to the Aromatic Heterocycles

Heterocyclic compounds are compounds with rings that contain more than one type of atom. The heterocyclic compounds of greatest interest to organic chemists have carbon rings containing one or two **heteroatoms**—atoms other than carbon. Although the chemistry of many *saturated* heterocyclic compounds is analogous to that of their noncyclic counterparts, a significant number of *unsaturated* heterocyclic compounds exhibit aromatic behavior. Some of these are shown in Fig. 24.4. The remainder of this chapter focuses primarily on the unique chemistry of a few of these aromatic heterocycles. The

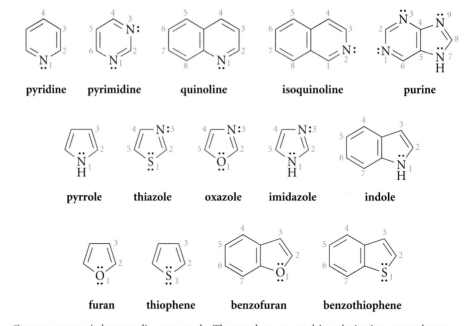

Figure 24.4   *Common aromatic heterocyclic compounds. The numbers are used in substitutive nomenclature.*

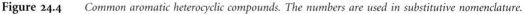

principles that emerge should enable you to understand the chemistry and properties of other heterocyclic compounds that you may encounter.

A. Nomenclature

The names and structures of some common aromatic heterocyclic compounds are given in Fig. 24.4. This figure also shows how the rings are numbered in substitutive nomenclature. In all but a few cases, a heteroatom is given the number 1. (Isoquinoline is an exception.) Notice in thiazole and oxazole that oxygen and sulfur are given a lower number than nitrogen when a choice exists. Substituent groups are given the lowest number consistent with this scheme. (These are the same rules used in numbering and naming saturated heterocyclic compounds; see Secs. 8.1C and 23.1B.)

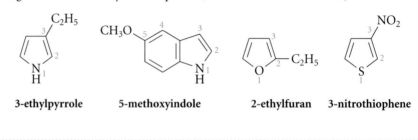

3-ethylpyrrole   **5-methoxyindole**   **2-ethylfuran**   **3-nitrothiophene**

PROBLEMS

24.5   Draw the structure of *(a) 4-(dimethylamino)pyridine;
(b) 4-ethyl-2-nitroimidazole.

24.6   Name the following compounds.

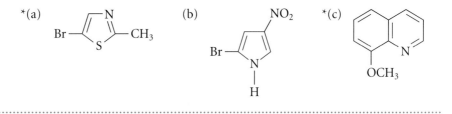

B. Structure and Aromaticity

The aromatic heterocyclic compounds furan, thiophene, and pyrrole can be written as resonance hybrids, illustrated here for furan.

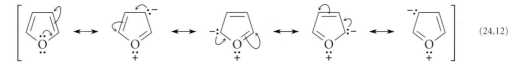

(24.12)

Because separation of charge is present in all but the first structure, the first structure is considerably more important than the others. Nevertheless, the importance of the other structures is evident in a comparison of the dipole moments of furan and tetrahydrofuran, a saturated heterocyclic ether.

	tetrahydrofuran	furan
dipole moment:	1.7 D	0.7 D
boiling point:	67°	31.4°

The dipole moment of tetrahydrofuran is attributable mostly to the bond dipoles of its polar C—O single bonds. That is, electrons in the σ bonds are pulled toward the oxygen because of its electronegativity. This same effect is present in furan, but in addition there is a second effect: the resonance delocalization of the oxygen unshared electrons into the ring shown in Eq. 24.12. This tends to push electrons away from oxygen into the π-electron system of the ring.

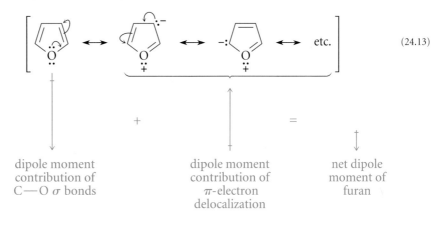

(24.13)

Because these two effects in furan nearly cancel, furan has a very small dipole moment. The relative boiling points of tetrahydrofuran and furan reflect the difference in their dipole moments.

Pyridine, like benzene, can be represented by two equivalent neutral resonance structures. Three additional structures of less importance reflect the relative electronegativity of nitrogen.

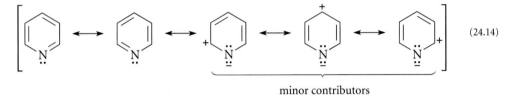

(24.14)

minor contributors

The aromaticity of some heterocyclic compounds was considered in the discussion of the Hückel $4n + 2$ rule (Sec. 15.7D). It is important to understand which unshared electron pairs in a heterocyclic compound are part of the $4n + 2$ aromatic π-electron system, and which are not (Fig. 24.5). Heteroatoms involved in double bonds of the Lewis structure—such as the nitrogen of pyridine—contribute one π electron to the six π-electron aromatic system, just like each of the carbon atoms in the π system. The orbital containing the unshared electron pair of the pyridine nitrogen is perpendicular to the p orbitals of the ring and is therefore not involved in π bonding. An unshared electron pair on a heteroatom in an allylic position—such as the unshared pair on the nitrogen of pyrrole—is part of the aromatic π system. Thus, the hydrogen of pyrrole lies

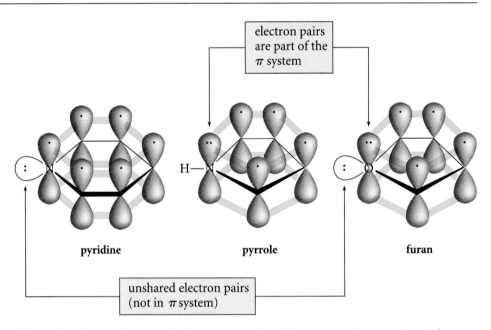

pyridine   pyrrole   furan

Figure 24.5   *The configurations of the unshared electron pairs and π electrons in pyridine, pyrrole, and furan. The orbitals in each $4n + 2$-electron π system are shown in gray; π interactions are shown in color. Unshared electron pairs not in the π system are shown in white. Notice that the unshared electron pair on the nitrogen of pyridine and one unshared pair in furan are not part of the aromatic π-electron system. In contrast, the unshared electron pair of pyrrole is part of the aromatic system.*

Table 24.1   Empirical Resonance Energies of Some Aromatic Compounds

Compound	Resonance energy		Compound	Resonance energy	
	kJ/mol	kcal/mol		kJ/mol	kcal/mol
benzene	138–151	33–36	furan	67	16
pyridine	96–117	23–28	pyrrole	89–92	21–22
naphthalene	255	61	thiophene	121	29

in the plane of the ring. The oxygen of furan contributes one unshared electron pair to the aromatic π-electron system, and the other unshared electron pair occupies a position analogous to the hydrogen of pyrrole—in the ring plane, perpendicular to the p orbitals of the ring.

How much stability does each heterocyclic compound owe to its aromatic character? Recall that the *empirical resonance energy* can be used to estimate this stability (Sec. 15.7C). (Remember that this is the energy a compound "doesn't have" because of its aromaticity, that is, its aromatic stability.) The empirical resonance energies of benzene, naphthalene, and some heterocyclic compounds are given in Table 24.1. To the extent that resonance energy is a measure of aromatic character, furan has the least aromatic character of the heterocyclic compounds in the table.

PROBLEMS

*24.7   Draw the important resonance structures for pyrrole.

24.8   *(a) The dipole moments of pyrrole and pyrrolidine are similar in magnitude but have opposite directions. Explain, indicating the direction of the dipole moment in each compound.

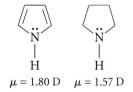

$$\mu = 1.80\,D \quad \mu = 1.57\,D$$

*(b) Explain why the dipole moments of furan and pyrrole have opposite directions.
(c) Should the dipole moment of 3,4-dichloropyrrole be greater or less than that of pyrrole? Explain.

*24.9   Each of the following NMR chemical shifts goes with a proton at carbon-2 of either pyridine, pyrrolidine, or pyrrole. Match each chemical shift with the appropriate heterocyclic compound, and explain your answer. δ 8.51; δ 6.41; and δ 2.82.

C. Basicity and Acidity of the Nitrogen Heterocycles

Basicity   Pyridine and quinoline act as ordinary amine bases.

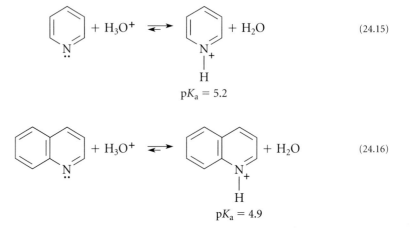

$$pK_a = 5.2$$

$$pK_a = 4.9$$

However, pyridine and quinoline are less basic than aliphatic tertiary amines because of the sp^2 hybridization of their nitrogen unshared electron pairs. (Recall that the basicity of an unshared electron pair decreases with increasing *s* character; Sec. 14.7A.)

Because pyrrole and indole look like amines, it may come as a surprise that neither of these two heterocycles has appreciable basicity. These compounds are protonated only in strong acid, and protonation occurs on carbon, not nitrogen.

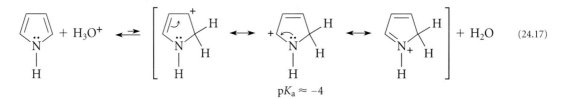

$$pK_a \approx -4$$

The marked contrast between the basicities of pyridine and pyrrole can be understood by considering the role of nitrogen's unshared electron pair in the aromaticity of each compound (Fig. 24.5). Protonation of the pyrrole nitrogen would disrupt the aromatic six π-electron system by taking the nitrogen's unshared pair "out of circulation."

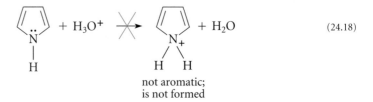

not aromatic;
is not formed

Although protonation of the carbon of pyrrole (Eq. 24.17) also disrupts the aromatic π-electron system, at least the resulting cation is resonance-stabilized. On the other hand, protonation of the pyridine unshared electron pair occurs easily because this electron pair is *not* part of the π-electron system. Hence protonation of this electron pair does not destroy aromaticity.

Imidazole is a base; the pK_a of its conjugate acid is 6.95. On which nitrogen does imidazole protonate?

Solution   Imidazole has two nitrogens: one has the electronic configuration of pyridine, but the other is like the nitrogen of pyrrole. Consequently, protonation occurs on the pyridinelike nitrogen—the nitrogen whose unshared electron pair is *not* part of the aromatic sextet.

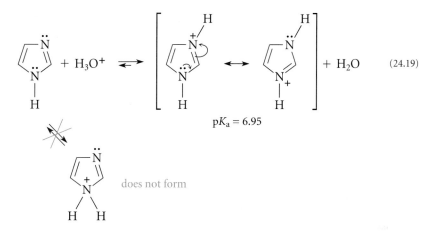

$$\text{(24.19)}$$

Notice that, according to the resonance structures in Eq. 24.19, the two nitrogens of protonated imidazole are equivalent; consequently, deprotonation to give imidazole can occur at either nitrogen.

Acidity   Pyrrole and indole are weak acids.

$$\text{(24.20)}$$

With pK_a values of about 17.5, pyrrole and indole are about as acidic as alcohols and about 15–17 pK_a units more acidic than amines (Sec. 23.5D). The greater acidity of pyrroles and indoles is a consequence of the resonance stabilization of their conjugate-base anions (Eq. 24.20; draw the three missing resonance structures in this equation.) Pyrrole and indole are acidic enough to behave as acids toward basic organometallic compounds such as Grignard or organolithium reagents.

$$\text{(24.21)}$$

24.10   *(a) Suggest a reason why pyridine is miscible with water, whereas pyrrole has little water solubility.

(b) Indicate whether you would expect imidazole to have high or low water solubility, and why.

24.11   *(a) The compound 4-(dimethylamino)pyridine protonates to give a conjugate acid with a pK_a value of 9.9. This compound is thus 4.7 pK_a units more basic than pyridine itself. Draw the structure of the conjugate acid of 4-(dimethylamino)pyridine, and explain why 4-(dimethylamino)pyridine is much more basic than pyridine.

(b) What product is expected when 4-(dimethylamino)pyridine reacts with CH_3I?

*24.12   Protonation of aniline causes a dramatic shift of its UV spectrum to lower wavelengths, but protonation of pyridine has almost no effect on its UV spectrum. Explain the difference.

24.3   Chemistry of Furan, Pyrrole, and Thiophene

A. Electrophilic Aromatic Substitution

Furan, thiophene, and pyrrole, like benzene and naphthalene, undergo electrophilic aromatic substitution reactions. At which position—carbon-2 or carbon-3—does substitution occur in these compounds? Let's try to predict the experimental result by examining the carbocation intermediates involved in the substitution reactions at the two different positions.

STUDY
PROBLEM
24.2

Using the nitration of pyrrole as an example, predict whether electrophilic aromatic substitution occurs predominantly at carbon-2 or carbon-3.

Solution   Recall (Sec. 16.4C) that the electrophile in nitration is the nitronium ion, $^+NO_2$. Substitution at the two different positions of pyrrole by the nitronium ion gives different carbocation intermediates:

Substitution at carbon-2:

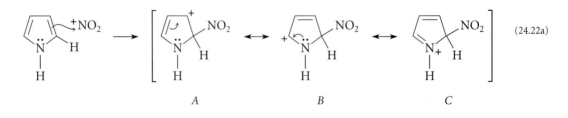

(24.22a)

Substitution at carbon-3:

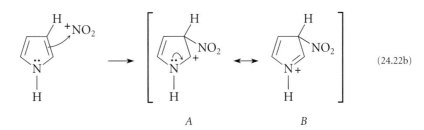

(24.22b)

$$A \qquad\qquad B$$

The carbocation resulting from substitution at carbon-2 has more important resonance structures, and is therefore more stable, than the carbocation resulting from substitution at carbon-3. Hammond's postulate suggests that the reaction involving the more stable intermediate should be the faster reaction. Consequently, the prediction is that nitration occurs at carbon-2. The experimental facts are as follows:

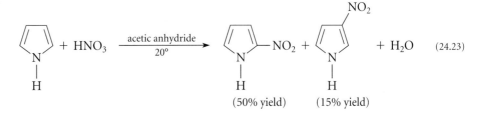

(24.23)

(50% yield)   (15% yield)

2-Nitropyrrole is the major nitration product of pyrrole, as predicted. Notice that there is nothing wrong with substitution at carbon-3; substitution at carbon-2 is simply more favorable. (Notice that some 3-nitropyrrole is obtained in the reaction.)

As the study problem suggests, electrophilic substitution of pyrrole occurs predominantly at the 2-position. Similar results are observed with furan and thiophene:

$$\text{furan} + CH_3-\overset{O}{\overset{\|}{C}}-O-\overset{O}{\overset{\|}{C}}-CH_3 \xrightarrow[\text{CH}_3\text{CO}_2\text{H}]{\text{BF}_3} \text{furan}-\overset{O}{\overset{\|}{C}}-CH_3 + CH_3-\overset{O}{\overset{\|}{C}}-OH \quad (24.24)$$

(a Friedel-Crafts reaction)   (75–92% yield)

$$\text{thiophene} + HNO_3 \xrightarrow{\text{acetic anhydride}} \text{thiophene}-NO_2 + \text{thiophene-}NO_2 + H_2O \quad (24.25)$$

(70% yield)   (5% yield)

How do pyrrole, furan, thiophene, and benzene compare in their relative reactivities in electrophilic aromatic substitution? Furan, pyrrole, and thiophene are all much more reactive than benzene. Although precise reactivity ratios depend on the particular reaction, the relative rates of bromination are typical:

$$\begin{array}{ccccccc} \text{pyrrole} & > & \text{furan} & > & \text{thiophene} & > & \text{benzene} \\ 3 \times 10^{18} & & 6 \times 10^{11} & & 5 \times 10^{9} & & 1 \end{array} \quad (24.26)$$

Milder reaction conditions must be used with more reactive compounds. (Reaction conditions that are too vigorous in many cases bring about polymerization and tar formation.) For example, a less reactive acylating reagent is used in the acylation of furan than in the acylation of benzene. (Recall that anhydrides are less reactive than acid chlorides; Sec. 21.7E.)

$$
\bigcirc + CH_3-\overset{\overset{\displaystyle O}{\|}}{C}-Cl \xrightarrow[\text{2) H}_2\text{O}]{\text{1) AlCl}_3} \bigcirc-\overset{\overset{\displaystyle O}{\|}}{C}-CH_3 + HCl \quad (24.27a)
$$

(97% yield)

$$
\bigcirc_O + CH_3-\overset{\overset{\displaystyle O}{\|}}{C}-O-\overset{\overset{\displaystyle O}{\|}}{C}-CH_3 \xrightarrow[\text{AcOH}]{\text{BF}_3} \bigcirc_O-\overset{\overset{\displaystyle O}{\|}}{C}-CH_3 + CH_3CO_2H \quad (24.27b)
$$

(75–92% yield)

The reactivity order of the heterocycles (Eq. 24.26) is a consequence of the relative abilities of the heteroatoms to stabilize positive charge in the intermediate carbocations (for example, structure *C* in Eq. 24.22a). Both pyrrole and furan have heteroatoms from the first row of the periodic table. Because nitrogen is better than oxygen at delocalizing positive charge (nitrogen is less electronegative), pyrrole is more reactive than furan. The sulfur of thiophene is a second row element and, although it is less electronegative than oxygen, its *3p* orbitals overlap less efficiently with the *2p* orbitals of the aromatic π-electron system (see Fig. 16.9). In fact, the reactivity *order* of the heterocycles in aromatic substitution parallels the reactivity order of the correspondingly substituted benzene derivatives:

Relative reactivities:

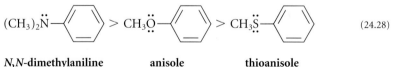

N,N-dimethylaniline   **anisole**   **thioanisole**

What about the activating and directing effects of substituents in furan, pyrrole, and thiophene rings? As might be expected from benzene and naphthalene chemistry, the usual activating and directing effects of substituents in aromatic substitution apply (see Table 16.2). Superimposed on these effects is the normal effect of the heterocyclic atom in directing substitution to the 2-position. The following example illustrates these effects:

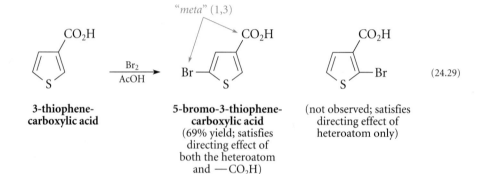

3-thiophene-carboxylic acid

5-bromo-3-thiophene-carboxylic acid
(69% yield; satisfies directing effect of both the heteroatom and —CO₂H)

(not observed; satisfies directing effect of heteroatom only)

(24.29)

In this example, the —CO_2H group directs the second substituent into a *"meta"* (1,3) relationship; the thiophene ring tends to substitute at the 2-position. The observed product satisfies both of these directing effects. (Notice that we count around the *carbon* framework of the heterocyclic compound, not through the heteroatom, when using this *ortho, meta, para* analogy.) In the following example, the chloro group is an *ortho, para*-directing group. Since the position *"para"* to the chloro group is also a 2-position, both the sulfur of the ring and the chloro group direct the incoming nitro group to the same position.

(24.30)

2-chlorothiophene   **2-chloro-5-nitrothiophene**

When the directing effects of substituents and the ring compete, it is not unusual to observe mixtures of products.

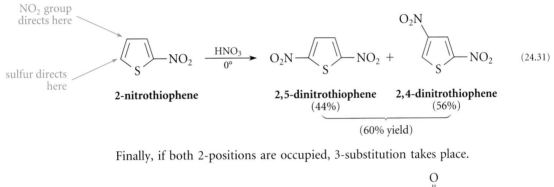

(24.31)

2-nitrothiophene   **2,5-dinitrothiophene**   **2,4-dinitrothiophene**
                              (44%)   (56%)

(60% yield)

Finally, if both 2-positions are occupied, 3-substitution takes place.

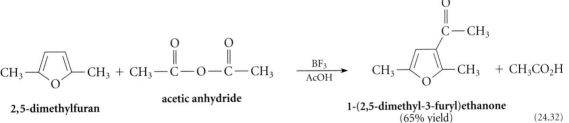

(24.32)

2,5-dimethylfuran   **acetic anhydride**   **1-(2,5-dimethyl-3-furyl)ethanone**
                                                         (65% yield)

B. Addition Reactions of Furan

The previous sections focused on the aromatic character of furan, pyrrole, and thiophene. A furan, pyrrole, or thiophene could, however, be viewed as a 1,3-butadiene with its terminal carbons "tied down" by a heteroatom bridge.

"butadiene" unit within furan

Do the heterocycles ever behave chemically as if they are conjugated dienes? The answer is yes. Of the three heterocyclic compounds furan, pyrrole, and thiophene, furan

A. Fischer Indole Synthesis

One of the best-known methods for preparing indoles is the **Fischer indole synthesis**, named for the great German chemist Emil Fischer (see Sec. 27.9A). In this reaction, an aldehyde or ketone with at least two α-hydrogens is reacted with a phenylhydrazine derivative in the presence of an acid catalyst and/or heat.

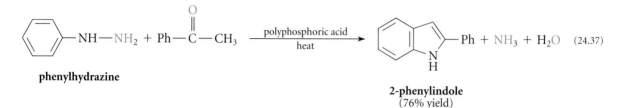

phenylhydrazine

2-phenylindole
(76% yield)

A variety of Brønsted or Lewis acid catalysts can be used: H_2SO_4, BF_3, $ZnCl_2$, and others. (The acid used in Eq. 24.37, *polyphosphoric acid*, is a syrupy mixture of P_2O_5 and phosphoric acid.) The reaction also works with many different substituted phenylhydrazines and carbonyl compounds. However, acetaldehyde, which could in principle be used to give indole itself, does not work in this reaction, probably because it polymerizes under the reaction conditions.

   The mechanism of the Fischer indole synthesis begins with a familiar reaction: conversion of the carbonyl compound into a *phenylhydrazone*, a type of imine (Table 19.3).

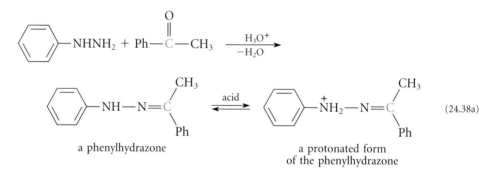

a phenylhydrazone

a protonated form
of the phenylhydrazone

(24.38a)

The protonated phenylhydrazone is in equilibrium with a small amount of a protonated enamine tautomer.

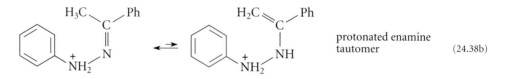

protonated enamine
tautomer   (24.38b)

Study Guide Link:
✓**24.2**
*Fischer Indole
Synthesis*

The latter species undergoes a pericyclic reaction involving *three electron pairs* (six electrons) to give a new intermediate in which the N—N bond of the phenylhydrazone has been broken.

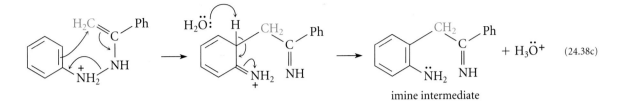

The intermediate formed in Eq. 24.38c is an imine (Sec. 19.11A). Imines react like ketones. Thus, the imine, after protonation on the imine nitrogen, undergoes nucleophilic addition with the amino group in the same molecule. The resulting "enamine" derivative is the product indole.

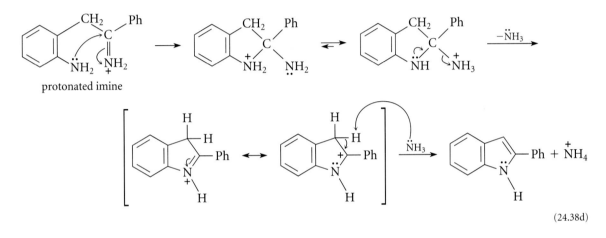

(24.38d)

Although substituted phenylhydrazines work in the Fischer indole synthesis, some are difficult to prepare. For this reason, the Fischer synthesis is most often used with phenylhydrazine itself, that is, to prepare indoles that are substituted at the 2- or 3-position rather than in the phenyl ring.

PROBLEMS

24.15   What starting materials are required for the synthesis of each of the following compounds by the Fischer indole synthesis?

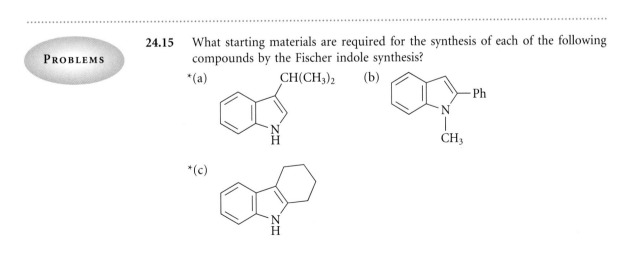

*24.16   When phenylhydrazine is reacted with 2-butanone under conditions of the Fischer indole synthesis, a mixture of two isomeric indoles is formed. Give their structures and explain.

B. Reissert Indole Synthesis

The Fischer indole synthesis occurs under acidic conditions. The Reissert indole synthesis, in contrast, occurs under basic conditions. The key starting materials for this indole synthesis are diethyl oxalate and *o*-nitrotoluene or a substituted derivative.

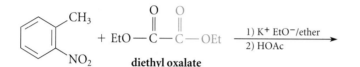

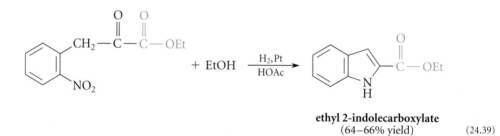

ethyl 2-indolecarboxylate
(64–66% yield)   (24.39)

The *o*-nitro group is an essential element in the success of this reaction because its presence makes the benzylic methyl hydrogens acidic enough to be removed by ethoxide; the resulting anion is resonance-stabilized.

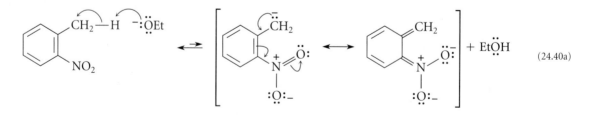

(24.40a)

In a variation of the Claisen condensation (Sec. 22.5A), this nucleophilic anion attacks a carbonyl group of diethyl oxalate, displacing ethanol. Like the Claisen condensation, this reaction is driven to completion by ionization of the product. For this reason, at least one equivalent of the base must be used.

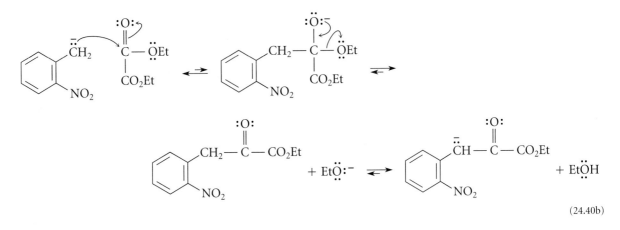

(24.40b)

The anion is neutralized by protonation in acetic acid, and the nitro group is converted into an amino group in a separate reduction step. (Catalytic hydrogenation is the reduction method used in Eq. 24.39.) The amino group thus formed reacts with the neighboring ketone to yield, after acid-base equilibria, an "enamine," that is, the aromatic indole. (Fill in the mechanistic details for the formation of the product from the amine.)

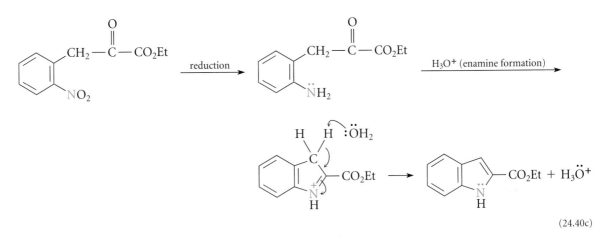

(24.40c)

2-Indolecarboxylic acid, like other heterocyclic carboxylic acids, can be decarboxylated (see Eq. 24.36). Consequently, the Reissert reaction followed by decarboxylation can be used to prepare indole itself.

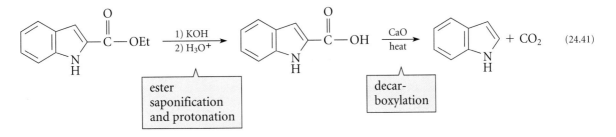

(24.41)

If substituted nitrotoluenes are used in the Reissert reaction, this reaction, in conjunction with the final decarboxylation step, can be used to prepare indoles that are substituted in the benzene ring and unsubstituted at the 2- and 3-positions. In this sense, the Reissert synthesis is complementary to the Fischer synthesis. (Recall that the Fischer synthesis is most often used to prepare indoles that are substituted at the 2- or 3-position.)

PROBLEM

24.17 Outline Reissert syntheses of the following indole derivatives from the indicated starting materials and any other reagents.
  *(a) 5-bromoindole from *m*-toluidine (3-methylaniline)
  (b) 6-indolecarbonitrile from *p*-toluidine (4-methylaniline)

24.5 Chemistry of Pyridine and Quinoline

A. Electrophilic Aromatic Substitution

In general, it is difficult to prepare monosubstituted pyridines by electrophilic aromatic substitution because pyridine has a very low reactivity; it is much less reactive than benzene. An important reason for this low reactivity is that pyridine is protonated under the very acidic conditions of most electrophilic aromatic substitution reactions (Eq. 24.15). The resulting positive charge on nitrogen makes it difficult to form a carbocation intermediate, which would place a second positive charge within the same ring.

Fortunately, a number of monosubstituted pyridines are available from natural sources. Among these are the methylpyridines, or *picolines:*

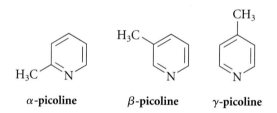

α-**picoline**   β-**picoline**   γ-**picoline**

The picolines (and other methylated pyridines) are obtained from *coal tar* (Sec. 16.7). Another very useful monosubstituted derivative of pyridine is *nicotinic acid* (pyridine-3-carboxylic acid), which is conveniently prepared in a number of ways, one of which is side-chain oxidation of nicotine, an alkaloid present in tobacco (Fig. 23.3).

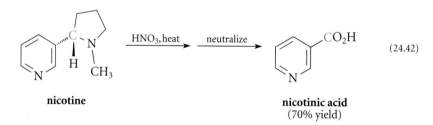

nicotine   **nicotinic acid**
                                                (70% yield)

$$(24.42)$$

(Nitric acid in this reaction is used as an oxidizing agent.)

Although electrophilic aromatic substitution reactions are not very useful for introducing substituents into pyridine itself, pyridine rings substituted with activating groups such as methyl groups do undergo electrophilic aromatic substitution reactions.

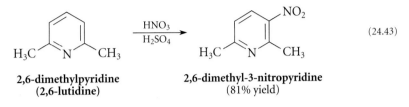

(24.43)

2,6-dimethylpyridine
(2,6-lutidine)

2,6-dimethyl-3-nitropyridine
(81% yield)

As this example illustrates, substitution in pyridine generally takes place in the 3-position. Although the methyl groups in Eq. 24.43 also direct substitution to the 3-position, the tendency of pyridine to undergo 3-substitution is general even in the absence of such directing groups. As with other electrophilic substitutions, an understanding of this directing effect comes from an examination of the carbocation intermediates formed in substitution at different positions. Substitution in the 3-position gives a carbocation with three different resonance structures:

3-Substitution:

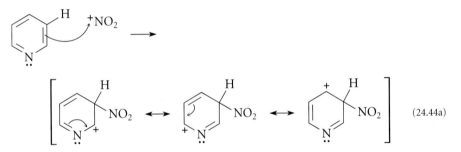

(24.44a)

Substitution at the 4-position also involves a carbocation intermediate with three resonance structures, but the one shown in color is particularly unfavorable because *the nitrogen, an electronegative atom, is electron-deficient.*

4-Substitution:

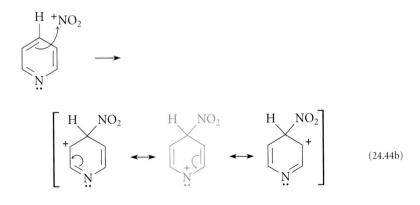

(24.44b)

You must be sure to understand that the nitrogen in the colored structure is very different from the nitrogen in pyrrole (Eq. 24.22a, structure *C*). The pyrrole nitrogen is also positively charged, but it is not electron-deficient because it has a complete octet. In contrast, an *electron-deficient* electronegative atom such as the one in Eq. 24.44b is very unfavorable energetically.

If electrophilic substitution in pyridine occurs at the 3-position, how can we obtain pyridine derivatives substituted at other positions? One compound used to obtain 4-substituted pyridines is pyridine-*N*-oxide, formed by oxidation of pyridine with 30% hydrogen peroxide.

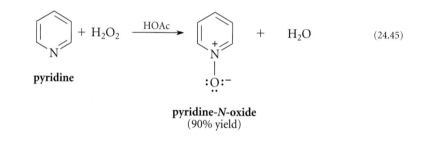

$$\text{(24.45)}$$

An analogy to pyridine-*N*-oxide from benzene chemistry is phenoxide, the conjugate base of phenol. Just as phenol or phenoxide is much more reactive in electrophilic aromatic substitution than benzene (Sec. 18.8), pyridine-*N*-oxide is much more reactive than pyridine. Of course, since there is a positive charge on the nitrogen of pyridine-*N*-oxide, this compound is *much* less reactive than phenol or phenoxide. Nevertheless, pyridine-*N*-oxide undergoes useful aromatic substitution reactions, and substitution occurs in the 4-position.

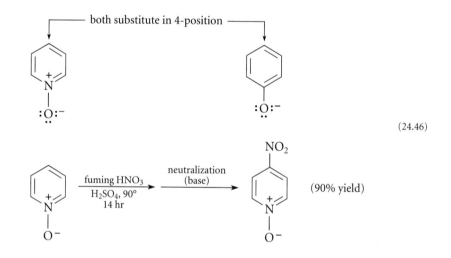

$$\text{(24.46)}$$

Once the *N*-oxide function is no longer needed, it can be removed by catalytic hydrogenation; this procedure also reduces the nitro group. Reaction with trivalent phosphorus compounds, such as PCl_3, removes the *N*-oxide function without reducing the nitro group.

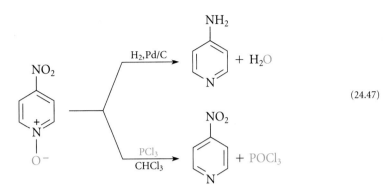

(24.47)

Similar reactions are possible with quinoline.

*24.18   Which should be more reactive in nitration: β-picoline or α-picoline? Explain using resonance structures, and give the major nitration product(s) in each case.

24.19   By drawing resonance structures for the carbocation intermediates, show why aromatic substitution in pyridine-N-oxide occurs at the 4-position rather than at the 3-position.

*24.20   When quinoline is nitrated, two mononitration products are formed. Give their structures and explain. (*Hint:* Use what you know about substitution in naphthalene.)

B.   Nucleophilic Aromatic Substitution

In contrast to its low reactivity in *electrophilic* aromatic substitution, the pyridine ring readily undergoes *nucleophilic* aromatic substitution. A rather unusual reaction of this type, called the **Chichibabin reaction**, can be used to prepare 2-aminopyridine. In this reaction, treatment of a pyridine derivative with the strong base sodium amide (Na^+ $^-NH_2$; Sec. 23.5D) brings about the direct substitution of an amino group for a ring hydrogen.

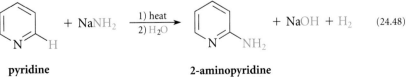

(24.48)

pyridine   **2-aminopyridine**
                                (66–76% yield)

In the first step of the mechanism, the amide ion, acting as a nucleophile, attacks the 2-position of the ring to form a *tetrahedral addition intermediate.*

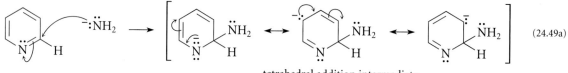

$$(24.49a)$$

tetrahedral addition intermediate

This step of the mechanism can be understood by recognizing that the C=N linkage of the pyridine ring is somewhat analogous to a carbonyl group; that is, carbon at the 2-position has some of the character of a carbonyl carbon, and can be attacked by nucleophiles. The C=N group of pyridine is, of course, *much* less reactive than a carbonyl group because it is part of an aromatic system.

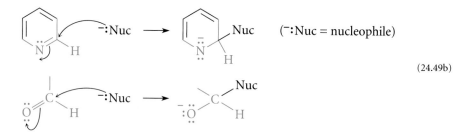

$$(24.49b)$$

In the second step of the mechanism, the leaving group, a *hydride ion*, is lost.

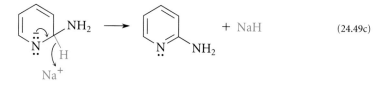

$$(24.49c)$$

Hydride ion is a very poor, and thus very unusual, leaving group because it is very basic. There are two reasons why this reaction occurs. First, the aromatic pyridine ring is reformed; aromaticity lost in the formation of the tetrahedral addition intermediate is regained when the leaving group departs. Second, the basic hydride produced in the reaction reacts with the —NH$_2$ group irreversibly to form hydrogen gas and the resonance-stabilized conjugate-base anion of 2-aminopyridine.

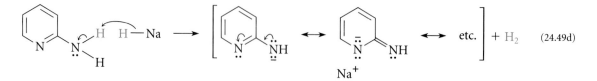

$$(24.49d)$$

The neutral 2-aminopyridine is formed when water is added in a separate step.

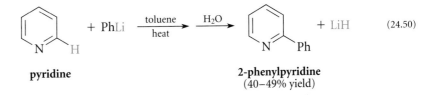

A reaction similar to the Chichibabin reaction occurs with organolithium reagents.

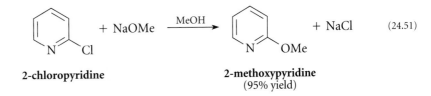

When pyridine is substituted with a better leaving group than hydride at the 2-position, it reacts more rapidly with nucleophiles. The 2-halopyridines, for example, readily undergo substitution of the halogen by other nucleophiles under conditions milder than those used in the Chichibabin reaction.

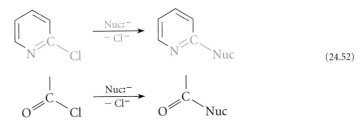

This nucleophilic substitution can also be related to the analogous reaction of a carbonyl compound. This reaction of a 2-chloropyridine resembles the nucleophilic acyl substitution reaction of an acid chloride—except that acid chlorides are *much* more reactive than 2-halopyridines.

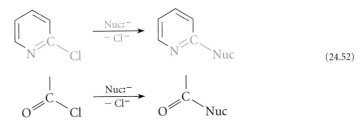

The nucleophilic substitution reactions of pyridines can be classified as *nucleophilic aromatic substitution* reactions. Recall that aryl halides undergo nucleophilic aromatic substitution when the benzene ring is substituted with electron-withdrawing groups (Sec. 18.4A). The "electron-withdrawing group" in the reactions of pyridines is the pyridine nitrogen itself. The tetrahedral addition intermediate (Eq. 24.49a) is analogous to the Meisenheimer complex of nucleophilic aromatic substitution (Eq. 18.16). Thus, there is a mechanistic parallel between three types of reactions: (1) nucleophilic acyl substitution, a typical reaction of carboxylic acid derivatives; (2) nucleophilic aromatic substitution; and (3) nucleophilic substitution on the pyridine ring.

The 2-aminopyridines formed in the Chichibabin reaction serve as starting materials for a variety of other 2-substituted pyridines. For example, diazotization of 2-aminopyridine gives a diazonium ion that can undergo substitution reactions (see Sec. 23.10A).

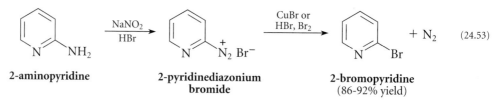

2-aminopyridine   2-pyridinediazonium   2-bromopyridine
                              bromide   (86-92% yield)

(24.53)

When the diazonium salt reacts with water, it is hydrolyzed to 2-hydroxypyridine, which in most solvents exists in its carbonyl form, 2-pyridone, despite the aromaticity of the 2-hydroxy form. In water the ratio of 2-hydroxypyridine to 2-pyridone is about 1:340.

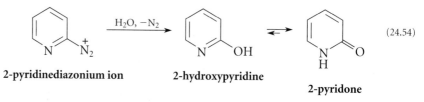

2-pyridinediazonium ion   2-hydroxypyridine

2-pyridone

(24.54)

Although 2-pyridone exists largely in the carbonyl form, it nevertheless undergoes some reactions reminiscent of hydroxy compounds. For example, treatment of 2-pyridone with PCl_5 gives 2-chloropyridine.

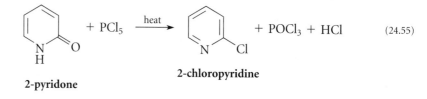

2-pyridone   2-chloropyridine

(24.55)

If you think of 2-pyridone in terms of its 2-hydroxypyridine tautomer, this reaction is similar to the preparation of acid chlorides from carboxylic acids.

similar to

(24.56)

Notice again the analogy between pyridine chemistry and carbonyl chemistry.

Pyridines with leaving groups in the 4-position also undergo nucleophilic substitution reactions.

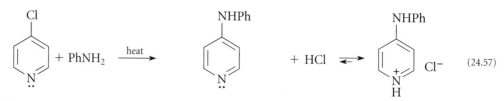

4-chloropyridine   4-(*N*-phenylamino)pyridine

(24.57)

As the examples in this section suggest, nucleophilic substitution reactions at the 2- and 4-positions of a pyridine ring are particularly common. The reason is clear from the mechanism of this type of reaction: negative charge in the addition intermediate is delocalized onto the electronegative pyridine nitrogen.

Substitution at carbon-2: (Y = leaving group, ⁻:Nuc = nucleophile)

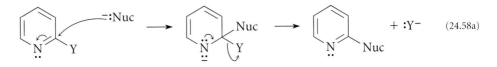

(24.58a)

Substitution at carbon-4:

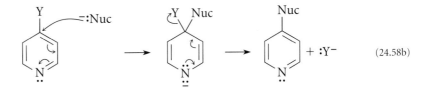

(24.58b)

What about substitution at carbon-3? 3-Substituted pyridines are *not* reactive in nucleophilic substitution because negative charge in the addition intermediate *cannot* be delocalized onto the electronegative nitrogen:

Substitution at carbon-3:

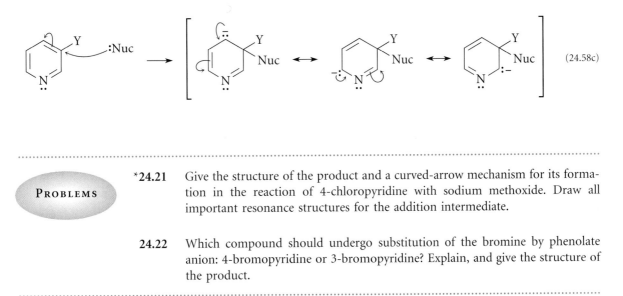

(24.58c)

PROBLEMS

*24.21   Give the structure of the product and a curved-arrow mechanism for its formation in the reaction of 4-chloropyridine with sodium methoxide. Draw all important resonance structures for the addition intermediate.

24.22   Which compound should undergo substitution of the bromine by phenolate anion: 4-bromopyridine or 3-bromopyridine? Explain, and give the structure of the product.

C. Pyridinium Salts and Their Reactions

Pyridine, like many Lewis bases, is a nucleophile. When pyridines react in S$_N$2 reactions with alkyl halides or sulfonate esters, quaternary ammonium salts, called *pyridinium salts*, are formed.

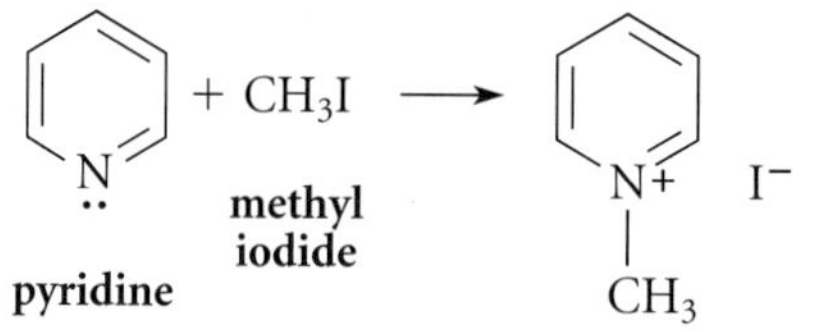

pyridine **methyl iodide** **1-methylpyridinium iodide** (a pyridinium salt) (nearly quantitative) (24.59)

Pyridinium salts are activated toward nucleophilic reactions at the 2- and 4-positions of the ring much more than pyridines themselves because the positively charged nitrogen is more electronegative, and is therefore a better electron acceptor, than the neutral nitrogen of a pyridine. When the nucleophiles in such displacement reactions are anions, charge is neutralized. In the following reaction, for example, the pyridinium salt is attacked at the 2-position by hydroxide ion; the resulting hydroxy compound is then oxidized by potassium ferricyanide [$K_3Fe(CN)_6$] present in the reaction mixture.

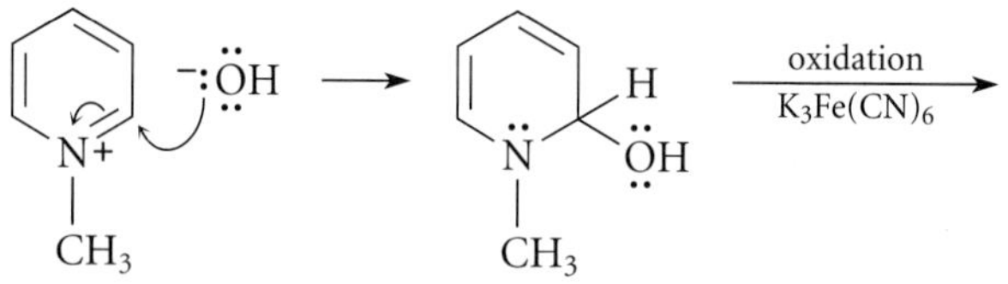

(24.60)

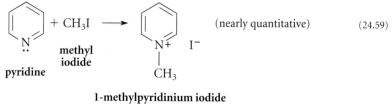

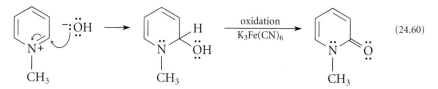

A biological example of nucleophilic addition to the 4-position of a pyridinium ring is found in biological oxidations with NAD^+ (Sec. 10.7).

Pyridine-*N*-oxides are in one sense pyridinium ions, and they react with nucleophiles in much the same way as quaternary pyridinium salts:

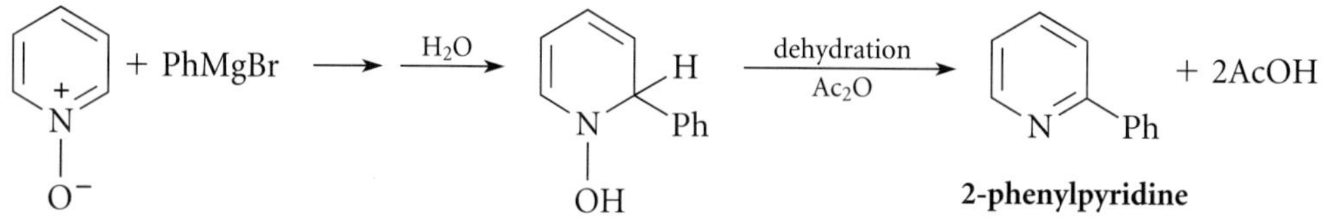

(24.61)

pyridine *N*-oxide **2-phenylpyridine**

D. Side-Chain Reactions of Pyridine Derivatives

The "benzylic" hydrogens of an alkyl group at the 2- or 4-position of a pyridine ring are more acidic than ordinary benzylic hydrogens because the electron pair (and charge) in the conjugate-base anion is delocalized onto the electronegative pyridine nitrogen.

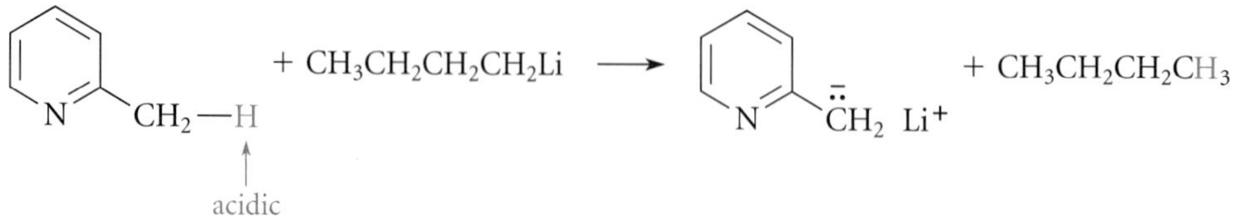

(24.62)

(Write the resonance structures of this ion and verify that charge is delocalized onto the pyridine nitrogen.) As the example in Eq. 24.62 illustrates, strongly basic reagents such as organolithium reagents or $NaNH_2$ abstract a "benzylic" hydrogen from 2- or 4-alkylpyridines. The anion formed in this way has a reactivity much like that of other organo-lithium reagents. In Eq. 24.63, for example, it adds to the carbonyl group of an aldehyde to give an alcohol (Secs. 19.9, 22.10A).

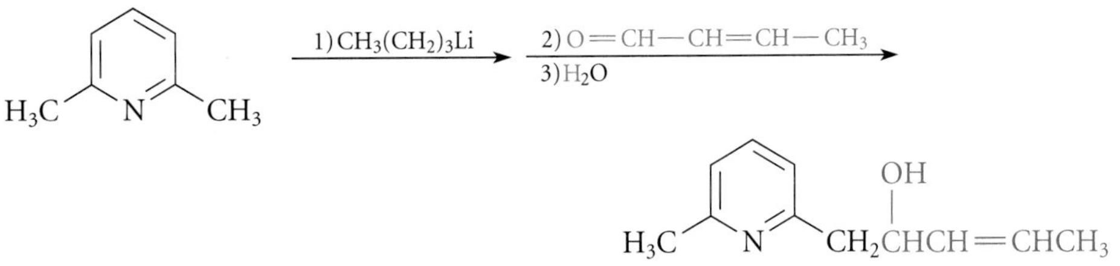

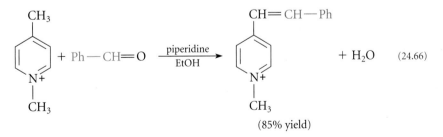

$$\text{(24.63)}$$

In this example, notice the analogy between pyridine chemistry and carbonyl chemistry. If the C=N linkage of a pyridine ring is analogous to a carbonyl group, then the "benzylic" anion is analogous to an enolate anion.

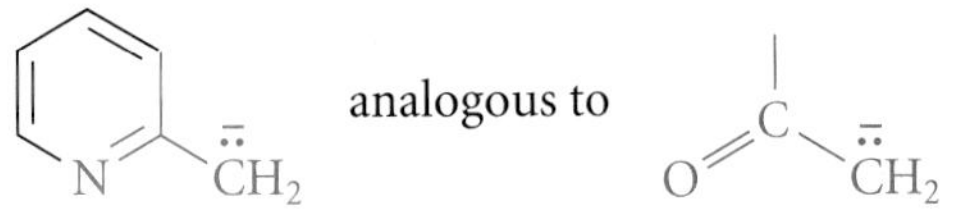

$$\text{analogous to} \qquad \text{(24.64)}$$

On the basis of this analogy, then it is reasonable that these anions should undergo some of the reactions of enolate anions, such as the aldol addition in Eq. 24.63.

The "benzylic" hydrogens of 2- or 4-alkylpyridinium salts are much more acidic than those of the analogous pyridines because the conjugate-base "anion" is actually a neutral compound, as the following resonance structures show:

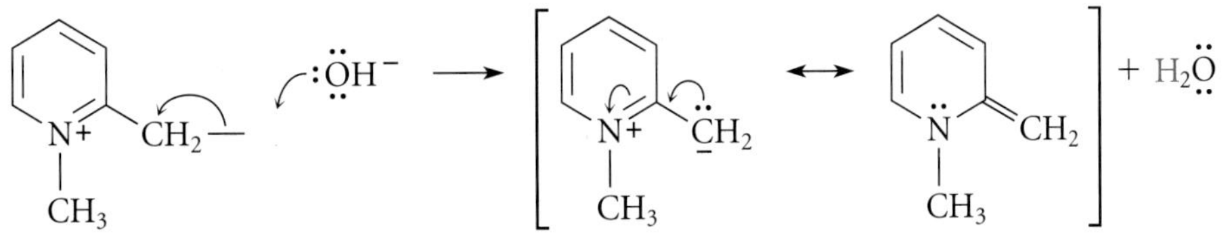

$$\text{(24.65)}$$

The "benzylic" hydrogens of 2- or 4-alkylpyridinium salts are acidic enough that the conjugate-base "anions" can be formed in useful concentrations by aqueous NaOH or amines. In the following reaction, which exploits this acidity, the conjugate base of a pyridinium salt is used as the "enolate" component in a variation of the Claisen-Schmidt condensation (Sec. 22.5C).

$$\text{(24.66)}$$

(85% yield)

Many side-chain reactions of pyridines are analogous to those of the corresponding benzene derivatives. For example, side-chain oxidation (Sec. 17.5) is a useful reaction of both alkylbenzenes and alkylpyridines. The oxidation of nicotine to nicotinic acid (Eq. 24.42) is an example of such a reaction.

PROBLEMS

24.23 Give the principal organic product in the reaction of quinoline with each of the following reagents. (*Hint:* Consider the similar reactions of pyridine.)
*(a) 30% H_2O_2 (b) $NaNH_2$, heat; then H_2O
*(c) product of (a), then HNO_3, H_2SO_4

24.24 Outline a synthesis for each of the following compounds from the indicated starting material and any other reagents. (Recall that a *picoline* is a methyl-pyridine.)
*(a) 3-methyl-4-nitropyridine from 3-picoline
(b) 4-methyl-3-nitropyridine from 4-picoline

*(c)

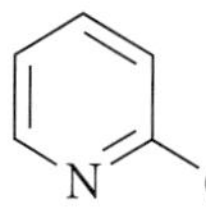

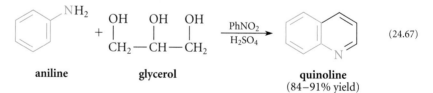

CH_2—CO_2H from 2-picoline (*Hint:* See Sec. 20.6.)

*(d) 3-aminopyridine from 3-picoline

24.25 Predict the predominant product in each of the following reactions. Explain your answer.
*(a) 3,4-dimethylpyridine + butyllithium (1 equiv.), then CH_3I ➤ ($C_8H_{11}N$)
(b) 3,4-dibromopyridine + NH_3, heat ➤ ($C_5H_5BrN_2$)

E. Skraup Synthesis of Quinolines

A number of reasonably versatile syntheses of quinolines from acyclic compounds are known. This is fortunate, since many direct substitution reactions of the quinoline nucleus give mixtures (Problem 24.20). One of the best known syntheses of quinolines is the **Skraup synthesis**, an acid-catalyzed reaction of glycerol with aniline or its derivatives.

$$
\text{aniline} + \underset{\text{glycerol}}{\overset{\text{OH OH OH}}{\underset{CH_2-CH-CH_2}{|\quad|\quad|}}} \xrightarrow[\text{H}_2\text{SO}_4]{\text{PhNO}_2} \text{quinoline} \qquad (24.67)
$$

aniline   **glycerol**   **quinoline**
(84–91% yield)

In this reaction, glycerol undergoes an acid-catalyzed dehydration to provide a small but continuously replenished amount of acrolein, an α,β-unsaturated aldehyde. (If acrolein itself were used as a reactant at high concentration, it would polymerize.)

STUDY GUIDE LINK:
24.3
Dehydration of Glycerol

$$
\underset{\text{glycerol}}{\overset{\text{OH OH OH}}{\underset{CH_2-CH-CH_2}{|\quad|\quad|}}} \xrightarrow[-H_2O]{H_2SO_4} O{=}CH-CH_2-CH_2OH \xrightarrow[-H_2O]{H_2SO_4}
$$

$$
O{=}CH-CH{=}CH_2 \qquad (24.68a)
$$

acrolein

Aniline undergoes a conjugate addition (Sec. 22.8A) with the acrolein (Reaction *a* below). Next, the resulting aldehyde is protonated (Reaction *b*). Because the protonated aldehyde has carbocation character, it acts as an electrophile in an intramolecular electrophilic aromatic substitution reaction (Reaction *c*). Dehydration of the resulting alcohol yields 1,2-dihydroquinoline. (You should fill in the details of the mechanism outlined in Eq. 24.68a and b.)

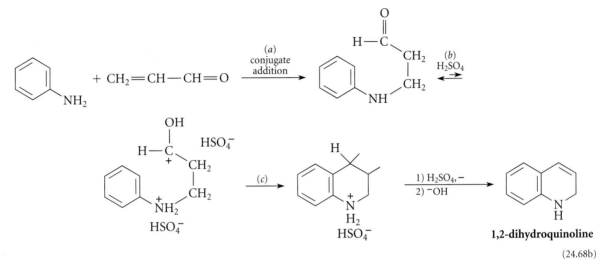

(24.68b)

The 1,2-dihydroquinoline product differs from quinoline by only one degree of unsaturation, and it is readily oxidized to the aromatic quinoline by mild oxidants. Nitrobenzene, As_2O_5, or Fe^{3+} are commonly used oxidants in the Skraup synthesis; these are included in the reaction mixture.

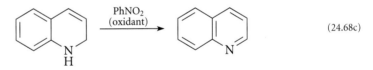

(24.68c)

α,β-Unsaturated aldehydes and ketones that are less prone to polymerize than acrolein can be used instead of glycerol in the Skraup synthesis to give substituted quinolines.

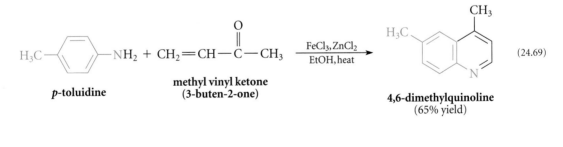

(24.69)

p-toluidine

methyl vinyl ketone
(3-buten-2-one)

4,6-dimethylquinoline
(65% yield)

·············

24.26 What product is expected when *p*-methoxyaniline (*p*-anisidine) reacts with each of the following compounds under the conditions of the Skraup synthesis?
*(a) 1-phenyl-2-buten-1-one (b) glycerol

*24.27 What reactants are required for a Skraup synthesis of 6-chloro-3,4-dimethyl-quinoline?

24.28   When 3-methylaniline (*m*-toluidine) reacts with glycerol, nitrobenzene, and H$_2$SO$_4$, two isomeric quinolines are obtained. Give their structures and explain.

24.6   Occurrence of Heterocyclic Compounds

Nitrogen heterocycles occur widely in nature. Sec. 23.12B discussed the *alkaloids* (Fig. 23.3), many of which contain heterocyclic ring systems. The naturally occurring amino acids proline, histidine, and tryptophan, which are covered in Chapter 26, contain respectively a pyrrolidine, imidazole, and indole ring (Fig. 24.6). A number of vitamins are heterocyclic compounds; without these compounds, many important metabolic processes could not take place. For example, vitamin B$_6$ is a substituted pyridine; NAD$^+$ (Sec. 10.7) is a pyridinium ion; and vitamin B$_1$ (thiamin) contains a pyrimidine ring and an *N*-substituted thiazolium salt. The nucleic acids, which carry and transmit genetic information in the cell, contain purine and pyrimidine rings (Fig. 24.4) in combined form (Chapter 27).

Heterocyclic compounds are involved in some of the colors of nature that have intrigued mankind from the earliest times. Why is blood red? Why is grass green? The color of blood is due to an iron complex of heme, a heterocycle composed of pyrrole units. This type of heterocycle is called a **porphyrin**.

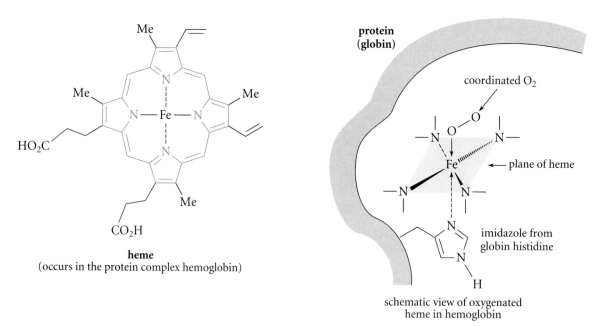

heme
(occurs in the protein complex hemoglobin)

schematic view of oxygenated
heme in hemoglobin

Heme is an aromatic heterocycle that is found in red blood cells as a tight complex with a protein called *globin*; the complex is called **hemoglobin**. The iron, held in position by coordination with the nitrogens of heme and an imidazole of globin, complexes reversibly

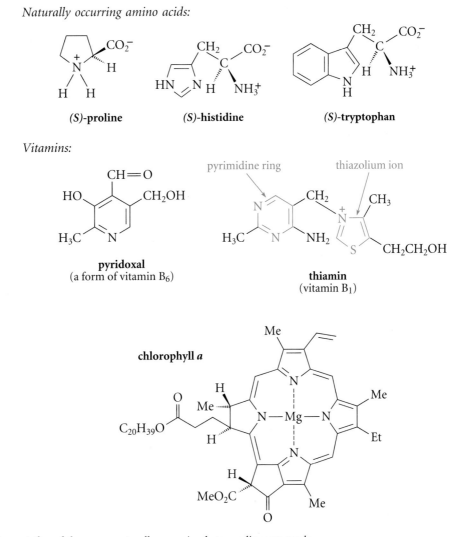

Naturally occurring amino acids:

(S)-proline *(S)*-histidine *(S)*-tryptophan

Vitamins:

pyridoxal
(a form of vitamin B$_6$)

pyrimidine ring   thiazolium ion

thiamin
(vitamin B$_1$)

chlorophyll *a*

Figure 24.6 *A few of the many naturally occurring heterocyclic compounds.*

with oxygen. Thus, hemoglobin is the oxygen carrier of blood, and the red color of blood is due to oxygenated hemoglobin. Carbon monoxide and cyanide, two well-known respiratory poisons, also complex with the iron in hemoglobin as well as with iron in the heme groups of other respiratory proteins.

The green color of plants is caused by *chlorophyll*, a class of compounds closely related to the porphyrins (Fig. 24.6). The absorption of sunlight by chlorophylls is the first step in the conversion of sunlight into usable energy by plants. Thus the chlorophylls are nature's "solar energy collectors."

KEY IDEAS IN CHAPTER 24

◊ Electrophilic aromatic substitution reactions of naphthalene generally occur at the 1-position. Sulfonation occurs most rapidly at the 1-position, but the most stable sulfonation product, formed at higher temperature, comes from sulfonation at the 2-position.

◊ Naphthalene derivatives undergo electrophilic aromatic substitution at the more activated ring. If a naphthalene derivative contains electron-donating, activating substituents, further substitution occurs on the substituted ring. If the substituted ring bears deactivating substituents, further substitution occurs on the unsubstituted ring.

◊ The aromatic heterocycles containing nitrogen atoms as part of a double bond, for example, imidazole, pyridine, and quinoline, are good Brønsted bases. Those with a nitrogen in which the unshared electron pair is part of the π-electron system, for example pyrrole and indole, are not basic, because protonation of the nitrogen would disrupt the aromatic π-electron system.

◊ Pyrrole, furan, and thiophene are all more reactive than benzene in electrophilic aromatic substitution, and undergo substitution predominantly at the 2-position. The reactivity order is pyrrole > furan > thiophene.

◊ Because furan has a relatively small empirical resonance energy, it undergoes some conjugate-addition reactions, such as the Diels-Alder reaction.

◊ Pyridine reacts very slowly in electrophilic aromatic substitution. Pyridine and its derivatives undergo substitution at the 3-position. Electrophilic substitution reactions of pyridine-*N*-oxides, however, occur at the 4-position.

◊ Many side-chain reactions of heterocyclic compounds proceed normally without disrupting the heterocyclic rings.

◊ Pyridine derivatives undergo nucleophilic aromatic substitution reactions at the 2- and 4-positions. Thus pyridines react in the Chichibabin and related reactions; 2- and 4-chloropyridines undergo nucleophilic aromatic substitution reactions. Pyridinium salts are even more reactive than pyridines in these reactions. The chemistry of the pyridine C=N linkage has some similarity to that of the carbonyl group.

◊ The "benzylic" hydrogens of pyridines and especially pyridinium salts are acidic enough to be removed by bases. The resulting anions can act as nucleophiles in substitution and condensation reactions.

🧪 Indoles are prepared by the reactions of arylhydrazines with aldehydes or ketones that have at least two α-hydrogens (the Fischer indole synthesis), or by the reaction of o-nitrotoluene and its derivatives in base with diethyl oxalate (the Reissert synthesis).

🧪 The reactions of α,β-unsaturated aldehydes or ketones with aniline or its derivatives give 1,2-dihydroquinolines, which are oxidized to the corresponding quinolines by mild oxidants present in the reaction mixture (Skraup synthesis).

ADDITIONAL PROBLEMS

24.29   Give the principal organic product(s) expected when 1-methylnaphthalene reacts with each of the following reagents.
*(a)   concentrated HNO_3 (b) H_2SO_4, 40°
*(c)   N-bromosuccinimide, CCl_4, light   (d) Br_2, CCl_4

24.30   Give the principal organic product(s) expected when 2-methylthiophene or other compound indicated reacts with each of the following reagents.
*(a)   acetic anhydride, BF_3, acid   (b) HNO_3
*(c)   N-bromosuccinimide, CCl_4, light   *(d)   dilute aqueous HCl
(e)   dilute aqueous NaOH
*(f)   product of (c) + Mg/ether, then CO_2, then H_3O^+
(g)   product of (a) + Ph—CH=O and NaOH

24.31   Give the principal organic product(s) expected when 2-methylpyridine or other compound indicated reacts with each of the following reagents.
*(a)   dilute aqueous HCl   (b)   dilute aqueous NaOH
*(c) $CH_3CH_2CH_2CH_2$—Li   (d) $NaNH_2$, heat, then H_2O
*(e) HNO_3, H_2SO_4, heat; then ^-OH *(f)   30% H_2O_2
(g) CH_3I
*(h)   product of (c) + PhCH=O, then H_3O^+
(i)   product of (e) + H_2, catalyst
*(j)   product of (d) + $NaNO_2$/HCl, then H_2O

24.32   Rank the compounds within each of the following series in order of increasing reactivity toward nitration with HNO_3/H_2SO_4 and explain your choices.
*(a)   naphthalene, pyridine, quinoline
(b)   thiophene, benzene, 3-methylthiophene

*24.33   Rank each of the following compounds in order of increasing S_N1 solvolysis reactivity in ethanol, and explain your choices.

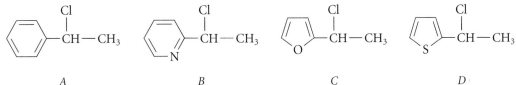

A   *B*   *C*   *D*

24.34   Draw the structure, including all important resonance structures, of the carbocation intermediate involved in each of the following reactions.
    *(a)   Friedel-Crafts acetylation of benzofuran (Fig. 24.4) at carbon-2
    (b)   nitration of benzothiophene (Fig. 24.4) at carbon-3

24.35   Draw the carbonyl tautomers of the compounds below. Which compound within each set contains the greatest percentage of carbonyl tautomer? Explain.
    *(a)   2-hydroxyfuran or 2-hydroxypyrrole
    (b)   phenol or 2-hydroxypyridine

***24.36**   Bromination of 1,6-dimethylnaphthalene gives a mixture of three isomeric monobromo derivatives, all brominated in the naphthalene ring. Two of these are formed in major amount, and one in very small amount. Give the structures and names of the three isomers, and indicate which are the major products. Explain your reasoning.

24.37   Sulfonation of 1-naphthalenesulfonic acid with fuming H_2SO_4 at 40° gives one major product, but sulfonation at 180° gives two. Give the structures of the product(s) formed in each case and explain your reasoning.

24.38   Rank the compounds within each of the following sets in order of increasing basicity, and explain your reasoning.
    *(a)   pyridine, 4-methoxypyridine, 5-methoxyindole, 3-methoxypyridine
    (b)   pyridine, 3-nitropyridine, 3-chloropyridine

    *(c)

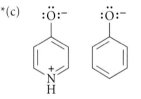

    *(d)   imidazole and oxazole   (e)   imidazole and thiazole

***24.39**   The following compound is a very strong base; its conjugate acid has a pK_a of about 13.5. Give the structure of its conjugate acid and show that it is stabilized by resonance.

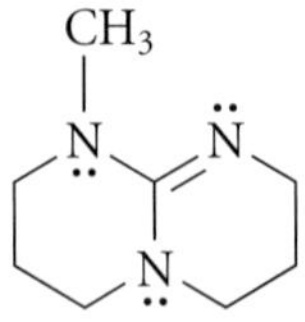

24.40   Draw the structure of the major form of each of the following compounds present in an aqueous solution containing initially one molar equivalent of 1 M HCl.
    *(a)   quinine (Fig. 23.3)   (b)   nicotine (Eq. 24.42 or Fig. 23.3)

*(c)

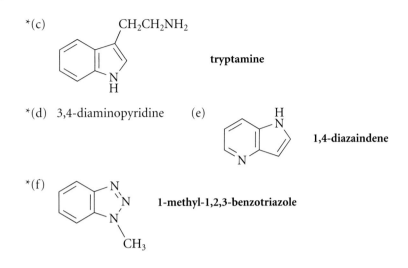

tryptamine

*(d) 3,4-diaminopyridine   (e)

1,4-diazaindene

*(f)

1-methyl-1,2,3-benzotriazole

24.41   Complete the following reactions by giving the major organic product(s).

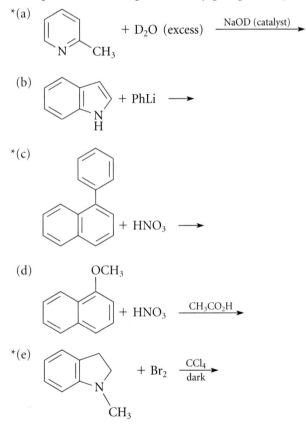

*(a)

+ D₂O (excess) $\xrightarrow{\text{NaOD (catalyst)}}$

(b)

+ PhLi $\longrightarrow$

*(c)

+ HNO₃ $\longrightarrow$

(d)

+ HNO₃ $\xrightarrow{\text{CH}_3\text{CO}_2\text{H}}$

*(e)

+ Br₂ $\xrightarrow[\text{dark}]{\text{CCl}_4}$

(*Note:* The starting material has no double bond between positions 2 and 3; that is, it is not an indole.)

(f)

+ H₂ $\xrightarrow[25°]{\text{Pt/C}}$ (C₇H₉N)

(*Problem 24.41 continues*)

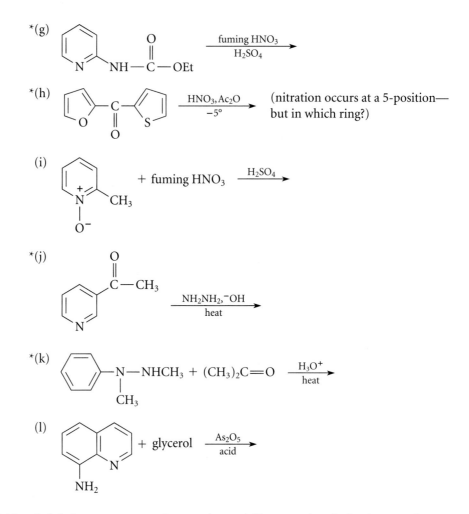

*(g)

*(h)   (nitration occurs at a 5-position—
but in which ring?)

(i)

*(j)

*(k)

(l)

*24.42   Indole in many cases undergoes electrophilic aromatic substitution at carbon-3. Using this observation, give the structure of the azo dye formed in the following reaction.

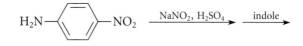

*24.43   Anthracene (structure below), like furan, behaves as a conjugated diene in Diels-Alder reactions, in which additions occur to the 9,10-diene unit.

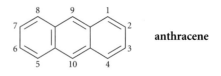

anthracene

*(a)   Give the structure of triptycene, the Diels-Alder adduct formed when anthracene reacts with benzyne (Sec. 18.4B).

(b)   Give the structure of the Diels-Alder adduct of furan and benzyne.

*24.44 By considering the resonance structures of the carbocation intermediates, predict the ring position at which bromination of anthracene occurs. (The structure of anthracene is given in Problem 24.43.)

24.45 Outline a synthesis of each of the following compounds from naphthalene and any other reagents.
*(a) 1-chloro-4-nitronaphthalene   (b) 1-bromonaphthalene
*(c) 1-naphthalenecarboxylic acid   (d) 2-(1-naphthyl)ethanol
*(e) 1-naphthyl acetate   (f) 1-bromo-4-chloronaphthalene
*(g) 1,4-naphthalenediamine
*(h) 5-amino-1-naphthalenesulfonic acid

*24.46 Doreen Dimwhistle has proposed the following variations on the Chichibabin reaction:
(a) indole + NaNH₂ → 2-aminoindole
(b) 2-chloropyridine + NaNH₂ → 2-amino-6-chloropyridine
She is shocked to find that neither of these reactions works as planned and has come to you for an explanation. Explain what reaction, if any, occurs instead in each case.

24.47 The following compound is isolated as a by-product in the Chichibabin reaction of pyridine and sodium amide. Give a curved-arrow mechanism for its formation.

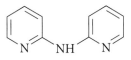

*24.48 When 3-methyl-2-butanone reacts with phenylhydrazine and a Lewis acid, the following compound, an example of an *indolenine*, is formed. Give a curved-arrow mechanism for the formation of this compound.

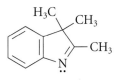

*24.49 When pyrrole is treated with 5.5 *M* HCl at 0° for 30 sec., a crystalline product *B* is obtained. A likely intermediate in this reaction is *A*.

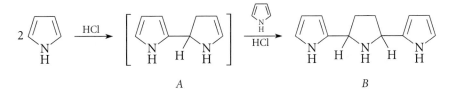

*(a) Draw a curved-arrow mechanism for the formation of *A*.
(b) Draw a curved-arrow mechanism for the formation of *B* from *A*, pyrrole, and HCl.

24.50   Outline a synthesis for each of the following compounds from the indicated starting material and any other reagents:

*(a)

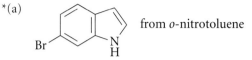

from *o*-nitrotoluene

(b)   2-ethyl-3,5-dimethylindole from 3-pentanone

*(c)

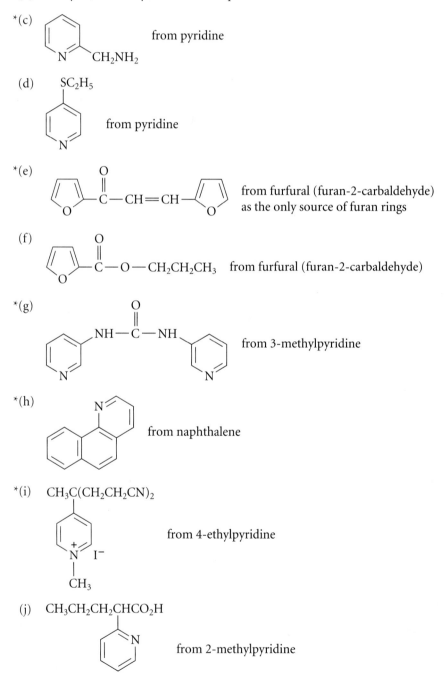

from pyridine

(d)   SC$_2$H$_5$

    from pyridine

*(e)   from furfural (furan-2-carbaldehyde) as the only source of furan rings

(f)   from furfural (furan-2-carbaldehyde)

*(g)   from 3-methylpyridine

*(h)   from naphthalene

*(i)   CH$_3$C(CH$_2$CH$_2$CN)$_2$

    from 4-ethylpyridine

(j)   CH$_3$CH$_2$CH$_2$CHCO$_2$H

    from 2-methylpyridine

*24.51 A compound *A*, $C_8H_{11}NO$, smells as if it might have been isolated from an extract of dirty socks. This compound can be resolved into enantiomers and it dissolves in 5% aqueous HCl. Oxidation of *A* with concentrated HNO_3 and heat gives nicotinic acid (3-pyridinecarboxylic acid). (See Eq. 24.42.) When *A* reacts with CrO_3 in pyridine, a compound *B* (C_8H_9NO) is obtained. Compound *B*, when treated with dilute NaOD in D_2O, incorporates five deuterium atoms per molecule. Identify *A* and explain your reasoning.

24.52 Many furan derivatives are unstable in strong acid. Hydrolysis of 2,5-dimethyl-furan in aqueous acid gives a compound *A*, $C_6H_{10}O_2$, that has a proton NMR spectrum consisting entirely of two singlets at δ 2.1 and δ 2.6 in the ratio 3:2, respectively. On treatment of compound *A* with very dilute NaOD in D_2O, both NMR signals virtually disappear. Treatment of *A* with zinc amalgam and HCl gives hexane. Propose a structure for *A* and give a mechanism for its formation.

*24.53 (a) Identify *A*, *B*, and *C* in the following scheme.

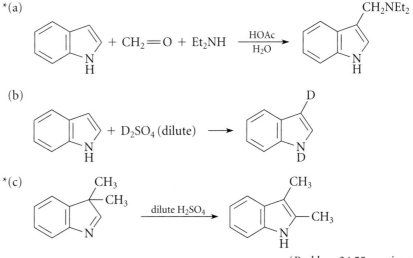

(b) Explain why *C* cannot be synthesized in one step from thiophene.

24.54 Identify the bracketed compounds in the reaction sequence below. Note that there are two reasonable possibilities for compound *D*. The correct structure can be deduced from the fact that the dipole moment of compound *D* is zero.

3-methylpyridine $\xrightarrow{\text{KMnO}_4,\ \text{heat}}$ [*A*] $\xrightarrow{\text{SOCl}_2}$ $\xrightarrow{\text{NH}_3}$

[*B*] $\xrightarrow{\text{Br}_2,\ \text{NaOH}}$ [*C*] $\xrightarrow{\text{glycerol, acid, As}_2\text{O}_5}$ [*D*]

24.55 Outline rational mechanisms for each of the following reactions. Give the structure for the bracketed intermediates in parts (d) and (f).

*(a)

*(b)

*(c)

(Problem 24.55 continues)

*(d)

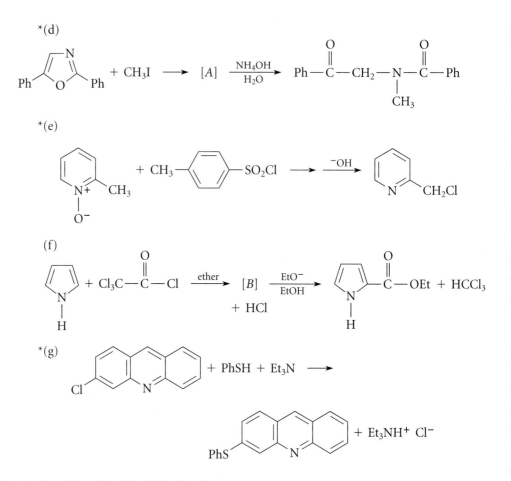

*(e)

*(f)

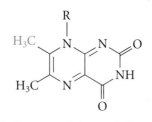

*(g)

24.56 Explain each of the following facts.

*(a) The hydrogens of the methyl group shown in color, as well as the imide proton, in the compound below are readily exchanged for deuterium by dilute NaOD in D₂O, but those of the other methyl group are not.

(b) In the ion below, the hydrogens of the methyl group shown in color are most acidic, even though the other methyl group is directly attached to the positively charged ion.

*(c) The following reaction takes place in aqueous base:

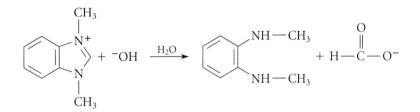

*(d) The compound 2-pyridone does not hydrolyze in aqueous NaOH using conditions that bring about the rapid hydrolysis of δ-butyrolactam.

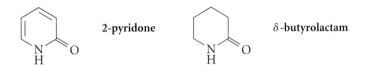

(e) Treatment of 4-chloropyridine with ammonia gives 4-aminopyridine, but treatment of 3-aminopyridine under the same conditions gives no reaction.

*(f) Treatment of 3-chloropyridine with sodamide (NaNH$_2$) gives a mixture of 3- and 4-aminopyridines.

24.57 Each of the following reactions is an example of a heterocyclic ring synthesis that was not discussed explicitly in the text. Using the curved-arrow formalism, give a mechanism for each reaction. Remember to begin by analyzing the relationship of atoms in the reactants and products.

*(a)

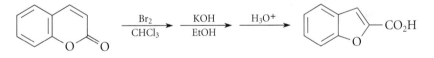

(b) Hinsberg thiophene synthesis:

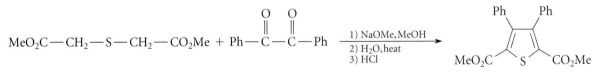

*(c) Friedlander quinoline synthesis:

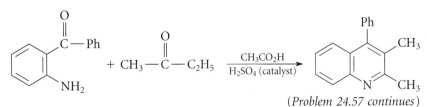

(*Problem 24.57 continues*)

(d) Combes quinoline synthesis:

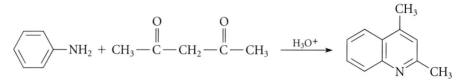

*(e) Hantzsch dihydropyridine synthesis:

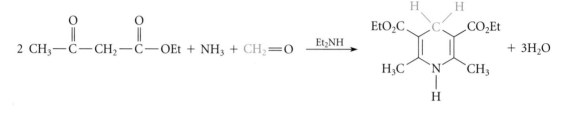

*24.58 You work for a pharmaceutical company whose management has decided to produce synthetic vitamin B_6. The company is in possession of some fragmentary notes from Strong E. Nuff, one of their early chemists, that outline the following synthesis of pyridoxine (a form of vitamin B_6). Unfortunately, reagents for each of the numbered steps have been omitted. They have hired you as a consultant; suggest the reagents that would accomplish each step.

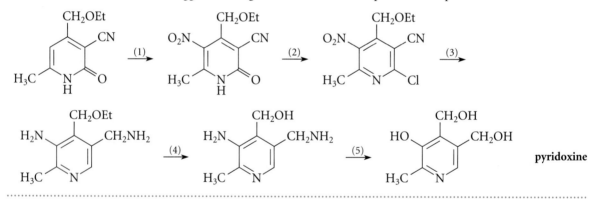

25

Pericyclic Reactions

Pericyclic reactions are reactions that occur by a concerted cyclic shift of electrons. This definition states two key elements. First, a pericyclic reaction is *concerted*. In a *concerted reaction*, reactant bonds are broken and product bonds are formed at the same time, without intermediates. Second, a pericyclic reaction involves a *cyclic shift of electrons*. (The word *pericyclic* means "around the circle.") The Diels-Alder reaction (Sec. 15.3) and the S_N2 reaction (Sec. 9.4) are both concerted reactions, but only the Diels-Alder reaction occurs by a *cyclic electron shift*. Hence the Diels-Alder reaction is a pericyclic reaction, but the S_N2 reaction is not.

This chapter is concerned with three major types of pericyclic reactions, although there are others. The first type is the **electrocyclic reaction**: an intramolecular reaction of an acyclic π-electron system in which a ring is formed with a new σ bond, and the product has one fewer π bonds than the starting material.

$$(25.1)$$

The second type of reaction is the **cycloaddition**: a reaction of two separate π-electron systems in which a ring is formed with two new σ bonds, and the product has two fewer π bonds than the reactants.

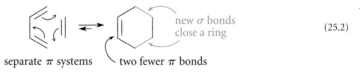

$$(25.2)$$

The third type of reaction is the **sigmatropic reaction**: a reaction in which an allylic σ bond at one end of a π-electron system appears to migrate to the other end of the π-electron system. The π bonds change positions in the process, and their total number is unchanged.

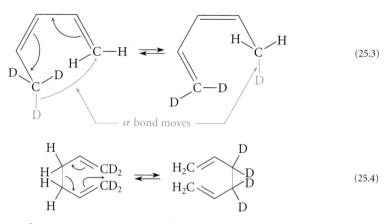

(25.3)

(25.4)

Three features of any given type of pericyclic reaction are intimately related:

1. The way the reaction is activated (heat or light)
2. The number of electrons involved in the reaction
3. The stereochemistry of the reaction

Before illustrating these points, let's clarify the first two terms in this list. Point 1 refers to the fact that many pericyclic reactions require no catalysts or reagents other than the reacting partners. Such reactions take place either on heating or on irradiation with ultraviolet light. Many reactions activated by heat are not activated by light, and vice-versa. Recall, for example, that many Diels-Alder reactions occur merely on heating the diene and dienophile together (Sec. 15.3). These reactions are not activated by light.

The number of electrons involved in a pericyclic reaction (Point 2) is twice the number of curved arrows required to write the reaction mechanism in the curved-arrow notation. For example:

(25.5)

three curved arrows;
six electrons

Note that the direction of "electron flow" in pericyclic reactions indicated by the curved arrows is arbitrary. Although it is clockwise in Eq. 25.5, it could be written counterclockwise and be equally correct.

Specifying any two of the features in the foregoing list for a particular type of reaction specifies the third. To illustrate, consider the following electrocyclic reactions:

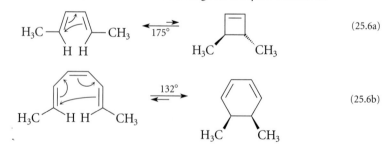

(25.6a)

(25.6b)

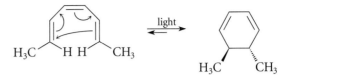

(25.6c)

First compare Eq. 25.6a and 25.6b. Both are activated by heat; however, the former reaction, involving four electrons, gives only the *trans*-disubstituted isomer of the cyclic product, whereas the latter reaction, involving six electrons, gives only the *cis* isomer.

Next compare Eqs. 25.6b and 25.6c. Both reactions involve six electrons. When the starting material is heated, only the *cis*-disubstituted isomer of the cyclic product is obtained. When the starting material is irradiated with ultraviolet light, the only product obtained is the *trans* isomer.

Correlations such as these had been observed for many years, but the reasons for them were not understood. In 1965 a theory that clearly explained these observations and successfully predicted many new ones was put forth by Robert B. Woodward (1917–1979), then a professor of chemistry at Harvard University, and Roald Hoffmann, at the time a Junior Fellow at Harvard and presently Professor of Chemistry at Cornell University. For this theory, called *Conservation of Orbital Symmetry*, Hoffmann received the 1981 Nobel Prize in Chemistry. He shared the prize with Kenichi Fukui, a professor of chemistry at Kyoto University in Japan, who had advanced a related theory, called *Frontier-Orbital Theory*. (The two theories make the same predictions; they are alternative ways of looking at the same reactions.) Woodward undoubtedly would have also shared the Nobel Prize had he not died prior to its announcement. (The terms of Nobel's bequest require that the prize be awarded only to living scientists.) Woodward had, however, received an earlier Nobel Prize for his work in organic synthesis. This chapter presents elements of the Woodward-Hoffmann-Fukui theory that will enable you to understand and predict the outcome of pericyclic reactions.

PROBLEM

25.1   Classify each of the following pericyclic reactions as an electrocyclic, cycloaddition, or sigmatropic reaction. Give the curved-arrow formalism for each.

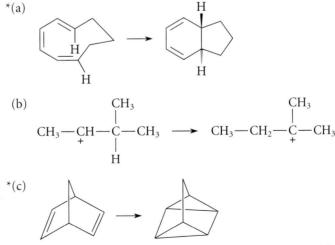

*(a)

(b)

*(c)

(Problem 25.1 continues)

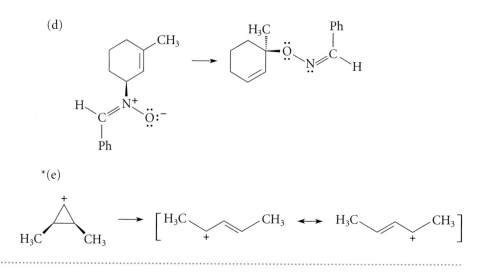

25.1   Molecular Orbitals of Conjugated π-Electron Systems

Understanding the theory of pericyclic reactions requires an understanding of some rudiments of *molecular orbital theory*, particularly as it applies to molecules containing π electrons. Molecular orbital theory was introduced in Secs. 1.8 and 4.1A; these sections should be reviewed carefully.

A. Molecular Orbitals of Conjugated Alkenes

The overlap of p orbitals to give π molecular orbitals is described by the mathematics of quantum theory. However, the mathematical aspects of this theory are not required to appreciate the results. This section considers the molecular orbital theory of ethylene and conjugated alkenes. The π molecular orbitals for such molecules can be constructed according to the following generalizations, which are applied to ethylene and 1,3-butadiene in Figs. 25.1 and 25.2, respectively, on pp. 1223 and 1224.

Be sure you understand how *each generalization* applies to *each example*.

1. When a number (say m) of p orbitals interact, the resulting π-electron system contains the same number m of molecular orbitals (MOs), all with different energies.

Because two p orbitals contribute to the π-electron system of ethylene, this molecule has the same number—two—of π MOs, which are designated as ψ_1 and ψ_2. Similarly, the four p orbitals of 1,3-butadiene combine to form four MOs, ψ_1, ψ_2, ψ_3, and ψ_4.

2. Half of the molecular orbitals have lower energies than the isolated p orbitals. These are called **bonding molecular orbitals**. The other half have higher energies than the isolated p orbitals. These are called **antibonding molecular orbitals**.

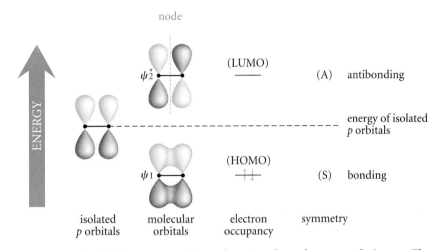

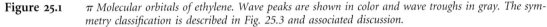

Figure 25.1   *π Molecular orbitals of ethylene. Wave peaks are shown in color and wave troughs in gray. The symmetry classification is described in Fig. 25.3 and associated discussion.*

To emphasize this distinction, antibonding MOs will be indicated with asterisks. Thus, ethylene has one bonding MO (ψ_1) and one antibonding MO ($\psi_2^\star$); 1,3-butadiene has two bonding MOs (ψ_1 and ψ_2) and two antibonding MOs ($\psi_3^\star$ and $\psi_4^\star$).

3. The bonding molecular orbital of lowest energy, ψ_1, has no nodes (except, of course, the plane of the molecule, which is a node of the component p orbitals). Each molecular orbital of increasingly higher energy has one additional node.

Recall from Sec. 1.6B that a *node* is a plane at which any wave, including an electron wave (orbital), is zero; that is, when an electron is in a given MO there is zero *probability of finding the electron*, or zero *electron density*, at the node. A particularly important feature of the node for understanding pericyclic reactions is that the electron wave has a *peak* on one side of the node (color in Figs. 25.1 and 25.2) and a *trough* on the other side. (See also Fig. 1.9, p. 25.) It is said that the orbital *changes phase* at the node.

Thus ψ_1 of ethylene has no nodes, and $\psi_2^\star$ has one node. In 1,3-butadiene, ψ_1 has no nodes, ψ_2 has one node, $\psi_3^\star$ has two, and $\psi_4^\star$ has three.

4. The nodes occur *between* atoms and are arranged symmetrically with respect to the center of the π-electron system.

The node in $\psi_2^\star$ of ethylene is between the two carbon atoms, in the center of the π system. The node in ψ_2 of 1,3-butadiene is also symmetrically placed in the center of the π system. The two nodes in $\psi_3^\star$ are placed between carbons 1 and 2, and between carbons 3 and 4, respectively—equidistant from the center of the π system. Each of the three nodes in $\psi_4^\star$, the orbital of highest energy, must occur between carbon atoms.

The next generalization relates to the *symmetry* of the molecular orbitals.

5. Odd-numbered MOs (ψ_1, ψ_3, ψ_5 ...) are symmetric with respect to an imaginary *reference plane* at the center of the π-electron system and perpendicular to the plane of the molecule. Even-numbered MOs (ψ_2, ψ_4, ψ_6 ...) are antisymmetric with respect to this plane.

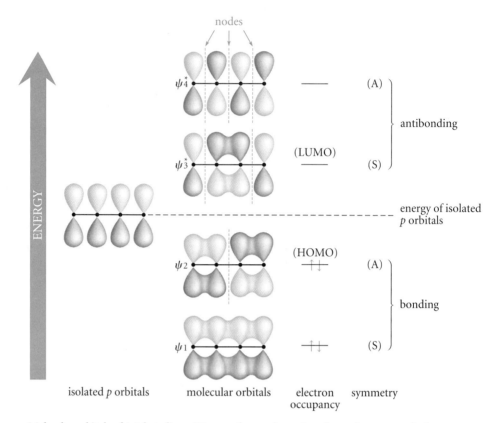

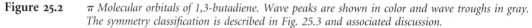

Figure 25.2   *π Molecular orbitals of 1,3-butadiene. Wave peaks are shown in color and wave troughs in gray. The symmetry classification is described in Fig. 25.3 and associated discussion.*

The *reference plane* in this generalization is shown for the 1,3-butadiene molecule in Fig. 25.3. A **symmetric MO** is one in which peaks reflect across the reference plane into peaks and troughs into troughs, as shown for $\psi_3^\star$ of 1,3-butadiene in Fig. 25.3. An **antisymmetric MO** is one in which peaks reflect into troughs, as shown for ψ_2 of 1,3-butadiene in Fig. 25.3. Of particular importance for the analysis of pericyclic reactions is the relative phase of each MO at its *terminal carbons*. Notice that within any symmetric MO, such as ψ_1 or $\psi_3^\star$ of 1,3-butadiene, the phase (that is, the relative orientation of peaks and troughs) at the two terminal carbons is *the same*; within any antisymmetric MO, such as ψ_2 and $\psi_4^\star$ of 1,3-butadiene, the phase at the two terminal carbons is *different*. Be sure to verify for yourself that the MOs in Figs. 25.1 and 25.2 fit this pattern.

The last generalization deals with the distribution of the available π electrons within the MOs.

6. Electrons are placed pairwise into each molecular orbital, beginning with the orbital of lowest energy (Aufbau principle).

This point is illustrated in Figs. 25.1 and 25.2 in the column labeled "electron occupancy." An alkene has the same number of π electrons as it has p orbitals. Thus ethylene, with two p orbitals, has two π electrons. These are both placed (with opposite spin) into ψ_1 (Fig. 25.1). 1,3-Butadiene, with four p orbitals, has four π electrons. Two are placed in

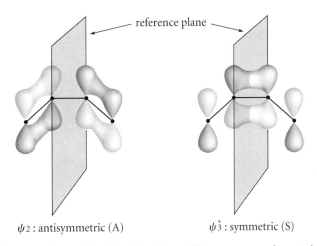

ψ_2 : antisymmetric (A) ψ_3^* : symmetric (S)

Figure 25.3   *Symmetry classification of ψ_2 and ψ_3^* of 1,3-butadiene with respect to a reference plane through the center of the molecule and perpendicular to the plane of the molecule. The symmetry classifications of the MOs in Figs. 25.1 and 25.2 are indicated by the abbreviations (A) and (S).*

ψ_1 and two in ψ_2 (Fig. 25.2). These examples show that the bonding MOs are fully filled in conjugated alkenes and that the antibonding MOs are empty.

The presence of unconjugated substituents (for example, alkyl groups), to a useful approximation, does not alter the π molecular orbital structure of a conjugated alkene. For example, the molecular orbital structures of 1,3-butadiene and 1,3-pentadiene are essentially the same.

$$CH_2\!\!=\!\!CH\!-\!CH\!\!=\!\!CH_2 \qquad\qquad CH_2\!\!=\!\!CH\!-\!CH\!\!=\!\!CH\!-\!CH_3 \qquad (25.7)$$

1,3-butadiene   **1,3-pentadiene**

same π molecular orbital structures

The π-electron contribution to the energy of a molecule is determined by the energies of its *occupied* MOs. Because bonding MOs have lower energies than isolated p orbitals, there is an energetic advantage to π-molecular orbital formation; this is why π bonds exist.

Two MOs are of particular importance in understanding pericyclic reactions. One is the occupied molecular orbital of highest energy, termed the **highest occupied molecular orbital (HOMO)**. The other is the unoccupied molecular orbital of lowest energy, termed the **lowest unoccupied molecular orbital (LUMO)**. These are labeled in Figs. 25.1 and 25.2. In ethylene, ψ_1 is the HOMO and ψ_2^* the LUMO; in 1,3-butadiene, ψ_2 is the HOMO and ψ_3^* the LUMO. Notice that *the HOMO and LUMO of a conjugated alkene have opposite symmetries*. Also notice that *the HOMO has a lower energy than the LUMO*.

The HOMO and LUMO are sometimes collectively termed **frontier orbitals** because they are the molecular orbitals at the energy extremes: the HOMO is the occupied molecular orbital of *highest energy*, and the LUMO is the unoccupied molecular orbital of *lowest energy*. The analysis of pericyclic reactions focuses heavily on the symmetries of frontier orbitals.

PROBLEMS

*25.2   Answer the following questions for 1,3,5-hexatriene, the conjugated triene containing six carbons.

(a) How many π molecular orbitals are there?

(b) Classify each MO as symmetric or antisymmetric.

(c) Which MOs are bonding? Which are antibonding?

(d) Which MOs are the frontier molecular orbitals?

(e) Within the HOMO, is the phase at the terminal carbons the same or different?

(f) Within the LUMO, is the phase at the terminal carbons the same or different?

25.3   State whether the π-molecular orbital ψ_6 in 1,3,5,7,9-decapentaene (a ten-carbon conjugated alkene) is symmetric or antisymmetric with respect to the reference plane; is bonding or antibonding; is a frontier MO; and if so, is a HOMO or a LUMO.

B. Molecular Orbitals of Conjugated Ions and Radicals

Conjugated unbranched ions and radicals have an odd number of carbon atoms. For example, the allyl cation has three carbon atoms and three p orbitals, hence, three MOs.

$$\left[CH_2{=}CH{-}\overset{+}{C}H_2 \quad \longleftrightarrow \quad \overset{+}{C}H_2{-}CH{=}CH_2\right] \quad \textbf{allyl cation}$$

The MOs of such species follow many of the same patterns as those of conjugated alkenes. The MOs for the allyl and 2,4-pentadienyl systems are shown in Figs. 25.4 and 25.5, respectively. These figures show two important differences between these MOs and those of conjugated alkenes. First, in each case one MO is neither bonding nor antibonding, but has the same energy as the isolated p orbitals; this MO is called a **nonbonding molecular orbital**. The nonbonding MO in the allyl system is ψ_2. The remaining orbitals are either bonding or antibonding, and there are an equal number of each type. Second, in some of the MOs, nodes pass through carbon atoms. For example, in the allyl system, there is a node on the central carbon of ψ_2. This means that electrons in ψ_2 have no electron density on the central carbon. This is why, for example, charge in the allyl anion resides only on the terminal carbons, a point deduced from resonance arguments:

$$\left[CH_2{=}CH{-}\overset{\cdot\cdot}{C}H_2 \quad \longleftrightarrow \quad \overset{\cdot\cdot}{C}H_2{-}CH{=}CH_2\right] \quad \textbf{allyl anion}$$

no charge here

Just as the charge in an atomic anion is associated with an excess of valence electrons, the charge in a conjugated carbanion can be associated with the electrons in its HOMO.

Notice that cations, radicals, and anions involving the same π system have the same molecular orbitals. For example, the MOs of the allyl system apply equally well to the allyl cation, allyl radical, and allyl anion, because all three species contain the same p orbitals. These species differ only in the *number* of π electrons, as shown in the "electron occupancy" column of Fig. 25.4. Thus, the HOMO of the allyl cation is ψ_1 and the LUMO is ψ_2. In contrast, the HOMO of the allyl anion is ψ_2, and the LUMO is ψ_3^*.

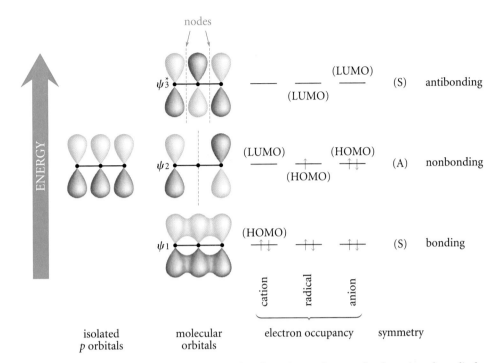

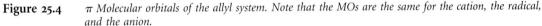

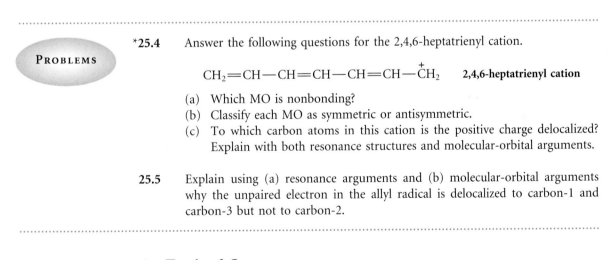

isolated   molecular   electron occupancy   symmetry
p orbitals   orbitals

Figure 25.4   *π Molecular orbitals of the allyl system. Note that the MOs are the same for the cation, the radical, and the anion.*

PROBLEMS

*25.4   Answer the following questions for the 2,4,6-heptatrienyl cation.

$$CH_2{=}CH{-}CH{=}CH{-}CH{=}CH{-}\overset{+}{C}H_2$$ **2,4,6-heptatrienyl cation**

(a) Which MO is nonbonding?
(b) Classify each MO as symmetric or antisymmetric.
(c) To which carbon atoms in this cation is the positive charge delocalized? Explain with both resonance structures and molecular-orbital arguments.

25.5   Explain using (a) resonance arguments and (b) molecular-orbital arguments why the unpaired electron in the allyl radical is delocalized to carbon-1 and carbon-3 but not to carbon-2.

C. Excited States

The molecules and ions we have been discussing can absorb energy from light of certain wavelengths. This process, which is also responsible for the UV spectra of these species (Sec. 15.2A), is shown schematically in Fig. 25.6 on p. 1228 for 1,3-butadiene. The normal electronic configuration of any molecule is called the **ground state**. Energy from absorbed light is used to promote an electron from the HOMO of ground-state 1,3-butadiene (ψ_2) into the LUMO ($\psi_3^{\star}$). A species with a promoted electron is called an **excited state**. In

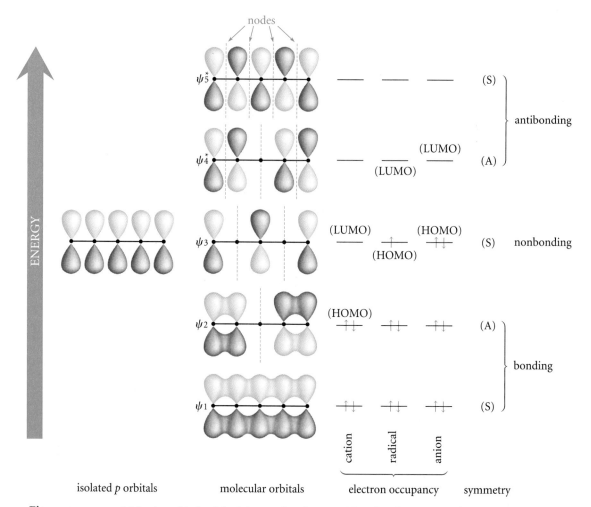

Figure 25.5 π *Molecular orbitals of the 2,4-pentadienyl system. Note that the MOs are the same for the cation, the radical, and the anion.*

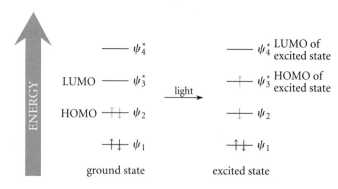

Figure 25.6   *Light absorption by 1,3-butadiene promotes an electron from the HOMO to the LUMO and produces an excited state.*

the excited state of 1,3-butadiene, ψ_3^* is the HOMO. Notice that *the HOMOs of the ground state and the excited state have opposite symmetries.*

25.2   Electrocyclic Reactions

A. Thermal Electrocyclic Reactions

This section begins the application of MO theory to pericyclic reactions with a discussion of *thermal* electrocyclic reactions, that is, electrocyclic reactions activated by heat. Sec. 25.2B considers electrocyclic reactions activated by light. (See Eq. 25.1 for the definition of electrocyclic reactions.)

When an electrocyclic reaction takes place, the carbons at each end of the conjugated π system must turn in a concerted fashion so that the p orbitals can overlap (and rehybridize) to form the σ bond that closes the ring. To illustrate, consider the reaction shown in Eq. 25.6a, the electrocyclic closure of (2E,4E)-2,4-hexadiene to give 3,4-dimethyl-cyclobutene. This turning can occur in two stereochemically distinct ways. In a **conrotatory** closure the two carbon atoms turn in the same direction. (The colored arrows show the direction of motion, not electron flow.)

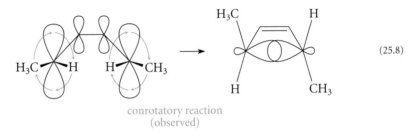

$$(25.8)$$

conrotatory reaction
(observed)

(There are, of course, two conrotatory modes, clockwise and counterclockwise; the clockwise mode is shown, but the counterclockwise mode in this case is equally probable.) In the second mode of ring closure, called a **disrotatory** mode, the carbon atoms turn in opposite directions.

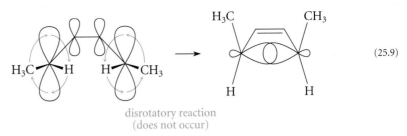

$$(25.9)$$

disrotatory reaction
(does not occur)

The two modes of ring closure can be distinguished by the stereochemistry of the product. As noted in Eq. 25.8, *trans-*, not *cis*-3,4-dimethylcyclobutene is the observed product. Hence, *the mode of ring closure is conrotatory.*

Molecular orbital theory explains this result. A simple way to look at the reaction is to focus on the *HOMO of the diene.* This molecular orbital contains the π electrons of highest energy. These π electrons are to a molecule as valence electrons are to an atom. Just as the atomic valence electrons are the ones involved in most chemical reactions, the electrons in the HOMO are the ones that govern the course of pericyclic reactions.

When the ring closure takes place, the two *p* orbitals on the ends of the π system must overlap. But simple overlap is not enough: they must overlap *in phase*. That is, the wave peak on one carbon must overlap with the wave peak on the other, or a wave trough must overlap with a wave trough. If a peak were to overlap with a trough, the electron waves would cancel and no bond would form.

Let's see what it takes to provide the required bonding overlap. First, the diene must assume the *s-cis* conformation; only in this conformation are the terminal carbons of the π-electron system close enough to each other that their *p* orbitals can overlap. Next, recall that alkyl substituents, to a useful approximation, do not affect the π molecular orbital structure of a conjugated alkene (Sec. 25.1A). Consequently, the π molecular orbital structure of 2,4-hexadiene is more or less the same as that of 1,3-butadiene (Fig. 25.2). In other words, the methyl groups at each end of the molecule can be largely ignored when considering the MOs of the system. An examination of the HOMO of a conjugated diene (ψ_2 in Fig. 25.2) reveals that because of the antisymmetric nature of ψ_2, conrotatory ring closure is required for in-phase, or bonding, overlap, as observed:

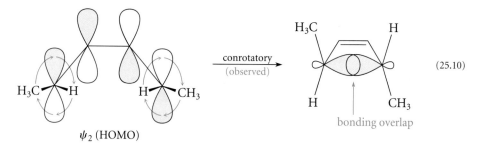

(25.10)

In contrast, disrotatory ring closure gives out-of-phase overlap, an antibonding (and hence unstable) situation:

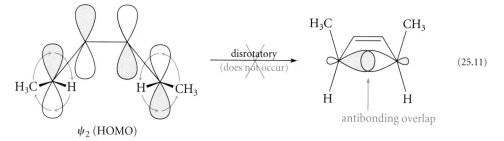

(25.11)

Thus, it is the relative orbital phase at the terminal carbon atoms of the HOMO—the *orbital symmetry*—that determines whether the reaction is conrotatory or disrotatory. This observation suggests that *all* conjugated polyenes with *antisymmetric* HOMOs should undergo conrotatory ring closure, and indeed, such is the case. The electrocyclic reactions of other conjugated alkenes can be predicted by a similar analysis, as the following study problem illustrates.

....

STUDY PROBLEM 25.1

Predict the stereochemistry of the thermal electrocyclic ring closure of (2E,4Z,6E)-2,4,6-octatriene to 5,6-dimethyl-1,3-cyclohexadiene.

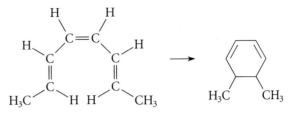

Solution   First, ignore the substituent groups and examine the HOMO of the simpler triene, 1,3,5-hexatriene (Problem 25.2). Because the HOMO of this triene (ψ_3) is *symmetric*, the HOMO has the *same phase* at each end of the π system. Hence, bonding overlap can occur only if the ring closure is *disrotatory*.

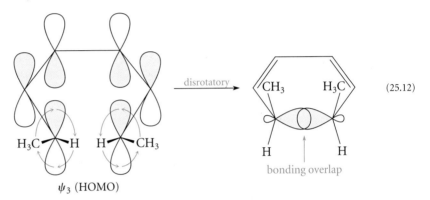

$$\psi_3 \text{ (HOMO)}$$

(25.12)

The disrotatory motion, as Eq. 25.12 shows, requires that the methyl groups have a *cis* relationship in the product. As Eq. 25.6b shows, this is indeed the observed stereochemistry of the reaction.

To summarize: electrocyclic closure of a conjugated diene is conrotatory, and that of a conjugated triene is disrotatory. The reason for the difference is the phase relationships within the HOMO at the terminal carbons of the π system. In the diene the HOMO has opposite phase at these two carbons; in the triene the HOMO has the same phase. A different type of rotation is thus required in each case for bonding overlap.

This result can be generalized. Conjugated alkenes with $4n$ π electrons (n = any integer) have antisymmetric HOMOs, and undergo conrotatory ring closure; those with $4n + 2$ π electrons have symmmetric HOMOs and undergo disrotatory ring closure. That is, conrotatory ring closure is *allowed* for systems with $4n$ π electrons; it is *forbidden* for systems with $4n + 2$ π electrons. Conversely, disrotatory ring closure is *allowed* for systems with $4n + 2$ π electrons; it is *forbidden* for systems with $4n$ π electrons.

B. Excited-State (Photochemical) Electrocyclic Reactions

When a molecule absorbs light, it reacts through its *excited state* (Sec. 25.1C). The HOMO of the excited state is different from the HOMO of the ground state, and has different symmetry. For example, as Eq. 25.6c shows, the *photochemical* ring closure of (2E,4Z,6E)-2,4,6-octatriene is *conrotatory*. This is understandable in terms of the symmetry of ψ_4^*, the HOMO of the excited state.

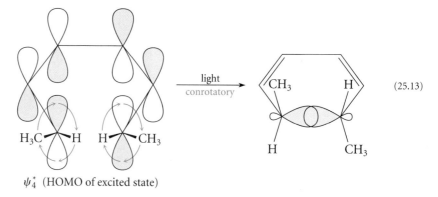

$$\xrightarrow[\text{conrotatory}]{\text{light}}$$ (25.13)

ψ_4^* (HOMO of excited state)

Contrast this result with that of the ground-state reaction in Eq. 25.12. The stereochemical result is different because the symmetry of the HOMO is different.

To generalize this result: the mode of ring closure in *photochemical* electrocyclic reactions—reactions that occur through electronically excited states—differs from that of thermal electrocyclic reactions, which occur through electronic ground states. These results can be summarized with a series of *selection rules* for electrocyclic reactions, given in Table 25.1.

C. Selection Rules and Microscopic Reversibility

The selection rules in Table 25.1 are based on the orbital symmetry of the open-chain (conjugated alkene) reactant. However, it is important to understand that these rules (as well as others to be considered) refer to the *rates* of pericyclic reactions, but have nothing to say about the *positions of the equilibria* involved. Thus the electrocyclic reaction of the diene in Eq. 25.6a to give a cyclobutene favors the diene at equilibrium because of the strain in the cyclobutene, but the electrocyclic reaction of the conjugated triene in Eq. 25.6b favors the cyclic compound because σ bonds are stronger than π bonds, and because six-membered rings are relatively stable.

It is also common for a photochemical reaction to favor the less stable isomer of an equilibrium because the energy of light is harnessed to drive the equilibrium energetically "uphill." For example, in the following reaction, the conjugated alkene absorbs UV light, but the bicyclic compound does not; hence, the photochemical reaction favors the latter.

Table 25.1   Selection Rules for Electrocyclic Reactions

Number of electrons[a]	Mode of activation	Allowed stereochemistry
4n	thermal photochemical	conrotatory disrotatory
4n + 2	thermal photochemical	disrotatory conrotatory

[a] n = an integer

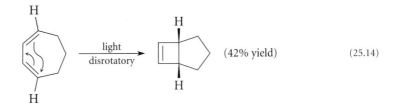

$$(25.14)$$

Thus, the selection rules do not indicate which component of an equilibrium will be favored—only whether the equilibrium will be established at a reasonable rate.

The *principle of microscopic reversibility* (Sec. 10.1) assures us that selection rules apply equally well to the forward and reverse of any pericyclic reaction, because the reaction in both directions must proceed through the same transition state. Hence, an electrocyclic ring *opening* must follow the same selection rules as its reverse, an electrocyclic ring *closure*. Thus, the thermal ring-opening reaction of the cyclobutene in Eq. 25.15, like the reverse ring-closure reaction, must be a conrotatory process (Table 25.1).

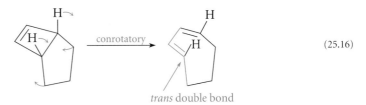

$$(25.15)$$

In the following electrocyclic ring-opening reaction, the allowed thermal conrotatory process would give a highly strained molecule containing a *trans* double bond within a small ring.

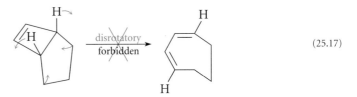

$$(25.16)$$

Although the selection rules suggest that the reaction could occur, it does not because of the strain in the product. In other words, "allowed" reactions are sometimes prevented from occurring for reasons having nothing to do with the selection rules. Concerted ring opening to the relatively unstrained all-*cis* diene also does not occur, because this would be a disrotatory process—a process "forbidden" by the selection rules in Table 25.1 for a concerted 4*n*-electron reaction.

$$(25.17)$$

Hence, the bicyclic compound is effectively "trapped into existence"; that is, there is no concerted thermal pathway by which it can reopen. (Note that it is formed *photochemically* from the all-*cis* diene; Eq. 25.14.)

Two rather spectacular examples of the effect of the selection rules for pericyclic reactions are the benzene isomers *prismane* (Sec. 15.7A) and *Dewar benzene*:

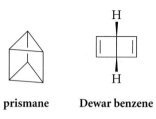

prismane   **Dewar benzene**

Despite the tremendous amount of strain in both molecules and the aromatic stability of benzene, neither prismane nor Dewar benzene is spontaneously transformed to benzene because, in each case, such a transformation violates pericyclic selection rules (see Problem 25.49). Because no simple concerted pathway exists for its conversion into the much more stable isomer benzene, prismane has been referred to in the literature as a "caged tiger."

PROBLEMS

*25.6   Which one of the following electrocyclic reactions should occur readily by a concerted mechanism?

(a)

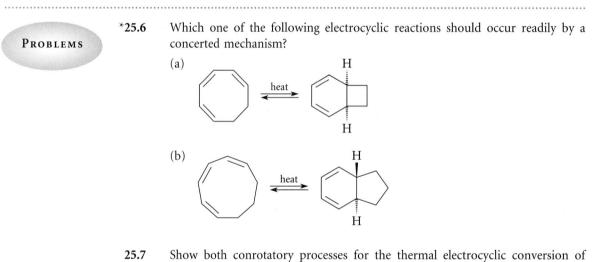

(b)

25.7   Show both conrotatory processes for the thermal electrocyclic conversion of (2E,4E)-2,4-hexadiene into 3,4-dimethylcyclobutene (Eq. 25.8). Explain why the two processes are equally likely.

*25.8   In the thermal ring opening of *trans*-3,4-dimethylcyclobutene, *two* products could be formed by a conrotatory mechanism, but only one is observed. Give the two possible products. Which one is observed and why?

25.9   After heating to 200°, the following compound is converted in 95% yield into an isomer *A* that can be hydrogenated to cyclodecane. Give the structure of *A*, including its stereochemistry.

25.3 Cycloaddition Reactions

A *cycloaddition reaction* (Eq. 25.2) is classified, first, by the number of electrons involved in the reaction. The reaction in Eq. 25.18a is a [4 + 2] cycloaddition because the reaction involves four electrons from one reacting component and two electrons from the other. The reaction in Eq. 25.18b is a [2 + 2] cycloaddition.

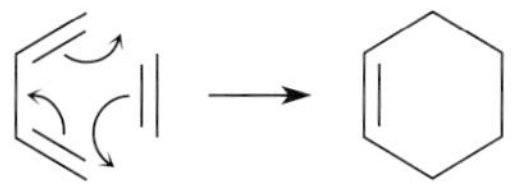

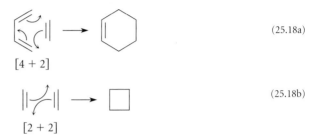

(25.18a)

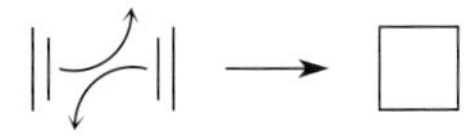

(25.18b)

As in electrocyclic reactions, the number of electrons involved is determined by writing the reaction mechanism in the curved-arrow notation. The number of electrons contributed by a given reactant is equal to twice the number of curved arrows originating from that component (two electrons per arrow).

A cycloaddition reaction is also classified by its stereochemistry with respect to *the plane of each reacting molecule.* (Recall that the carbons and their attached atoms in π-electron systems are coplanar.) This classification is shown for a [4 + 2] cycloaddition in Fig. 25.7. A cycloaddition may in principle occur either across the same face, or across opposite faces, of the planes in each reacting component. If the reaction occurs across the same face of a π system, the reaction is said to be **suprafacial** with respect to that

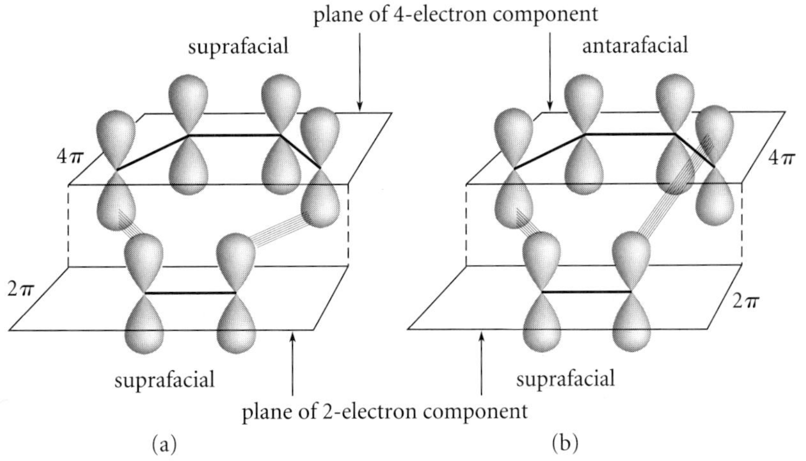

Figure 25.7 *Classification of cycloaddition reactions, illustrated for (a) a [4s + 2s] cycloaddition and (b) a [4a + 2s] cycloaddition. The s and a designations refer to the stereochemistry of the cycloaddition (suprafacial or antarafacial) with respect to the planes of the reacting components. For example, in (b), the addition, indicated by the colored lines, occurs below the plane of the 4π-electron component at one end and above the plane at the other, and is therefore classified as 4a, or antarafacial on the 4π-electron component. The addition occurs on the 2π-electron component on the same side of its plane at both ends, and is therefore classified as 2s, or suprafacial on the 2π-electron component. This reaction is therefore a [4a + 2s] cycloaddition.*

π system. A suprafacial addition is nothing more than a *syn* addition (Sec. 7.9A) that occurs in a single mechanistic step. If the reaction occurs at opposite faces of a π system, it is said to be **antarafacial**. An antarafacial addition is just an *anti* addition that occurs in one mechanistic step. Thus a [4s + 2s] cycloaddition is one that occurs suprafacially (or *syn*) on both the 4π component and the 2π component. A [4a + 2s] cycloaddition occurs antarafacially (or *anti*) on the 4π component, but suprafacially on the 2π component.

In order for a cycloaddition to occur, bonding overlap must take place between the *p* orbitals at the terminal carbons of each π-electron system, because these are the carbons at which new bonds are formed. This bonding overlap begins when the HOMO of one component interacts with the LUMO of the other. The electrons in the HOMO of one component are analogous to the valence electrons in an atom: they are the reacting electrons. The LUMO of the other component is the empty orbital of lowest energy into which the electrons from the HOMO must flow. It doesn't matter whether we consider the HOMO from the 4π-electron component and the LUMO from the 2π-electron component, or vice-versa. The important point is that the two frontier MOs involved in the interaction must have *matching phases* if bonding overlap is to be achieved.

STUDY GUIDE LINK:
✓25.2
Frontier Orbitals

This phase match is achieved when a [4 + 2] cycloaddition occurs *suprafacially* on each component, that is, when the cycloaddition is a [4s + 2s] process. (A [4a + 2a] process is also theoretically allowed, but is geometrically impossible.) Using the HOMO from the 4π-electron component and the LUMO from the 2π-electron component, this overlap can be represented as follows:

ψ_2 (HOMO of 4-electron component)

ψ_2^* (LUMO of 2-electron component)

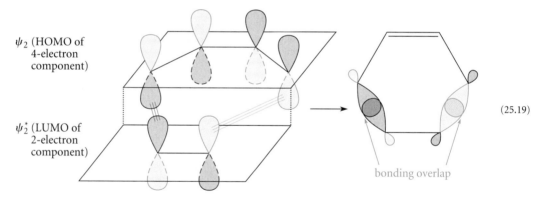

bonding overlap

(25.19)

Recall that the Diels-Alder reaction, the most important example of a [4s + 2s] cycloaddition, indeed occurs suprafacially on each component (Sec. 15.3C). This can be seen from the retention of stereochemistry observed in both the diene and dienophile in the following example:

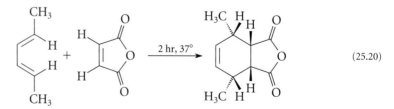

(25.20)

You should convince yourself that the [4s + 2a] and [4a + 2s] modes of cycloaddition do *not* provide bonding overlap at both ends of the π-electron systems. You should also convince yourself that it doesn't matter which component provides the HOMO and which provides the LUMO (Problem 25.10).

The situation is different in a [2 + 2] cycloaddition. Again we use the HOMO of one component and the LUMO of the other. Notice that the orbital symmetries do *not* accommodate a cycloaddition that is suprafacial on both components:

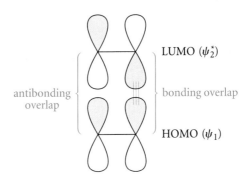

However, an addition that is suprafacial on one component but antarafacial on the other is allowed by orbital symmetry, but is geometrically more difficult. For this reason, the thermal [2 + 2] cycloaddition is a much less common reaction than the Diels-Alder reaction. Many of the known [2 + 2] additions occur by nonconcerted mechanisms, and therefore do not fall under the purview of the rules for pericyclic reactions.

Although the [2s + 2s] cycloaddition is forbidden by orbital symmetry under thermal conditions, it is allowed under photochemical conditions. Under these conditions it is the *excited state* of one alkene that reacts with the other alkene. The HOMO of the excited state has the proper symmetry to interact in a bonding way with the LUMO of the reacting partner:

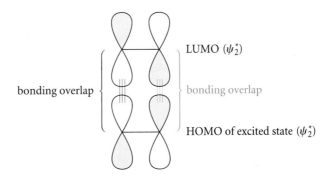

Indeed, many examples of photochemical [2s + 2s] cycloadditions are known. Such processes are widely used for making cyclobutanes.

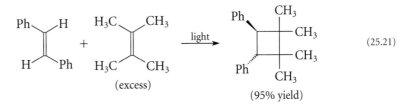

$$(25.21)$$

Table 25.2 Selection Rules for Cycloaddition Reactions

Number of electrons[a]	Mode of activation	Allowed stereochemistry[b]
$4n$	thermal	supra-antara antara-supra
	photochemical	supra-supra antara-antara
$4n + 2$	thermal	supra-supra antara-antara
	photochemical	supra-antara antara-supra

[a] n = an integer
[b] supra = suprafacial; antara = antarafacial

The results of this section can be generalized to the cycloaddition selection rules shown in Table 25.2. Notice that all-suprafacial cycloadditions are allowed thermally for systems in which the total number of reacting electrons is $4n + 2$, and they are allowed photochemically for systems in which the number is $4n$.

A consequence of the selection rules is that suprafacial cycloadditions should be allowed for certain systems with more than six π electrons. For example, the following all-suprafacial cycloaddition is a $[6s + 4s]$ process involving ten electrons (five curved arrows). Notice that $4n + 2$ equals 10 for $n = 2$.

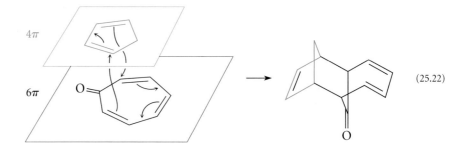

(25.22)

PROBLEMS

*25.10 Show that using the HOMO from the 2π-electron component and the LUMO from the 4π-electron component also gives bonding overlap in a $[4s + 2s]$ cycloaddition.

25.11 Show by a frontier orbital analysis that the $[4a + 2s]$ and $[4s + 2a]$ modes of cycloaddition are not allowed.

*25.12   Give the product of the following reaction, which involves an $[8s + 2s]$ cycloaddition:

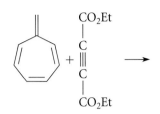

*25.13   The photochemical cycloaddition of two molecules of *cis*-2-butene gives a mixture of two products, *A* and *B*. The analogous photochemical cycloaddition of *trans*-2-butene also gives a mixture of two products, *B* and *C*. The photochemical reaction of a *mixture* of *cis*- and *trans*-2-butene gives a mixture of *A*, *B*, and *C*, along with a fourth product, *D*. Propose structures for all four compounds.

25.4   Sigmatropic Reactions

A.   Classification and Stereochemistry

Sigmatropic reactions (Eqs. 25.3 and 25.4) are classified by using bracketed numbers to indicate the number of atoms over which a σ bond appears to migrate. In some reactions, both ends of a σ bond migrate. In the following reaction, for example, each end of a σ bond migrates over three atoms. (Count the point of original attachment as atom #1.) The following reaction is therefore a [3,3] sigmatropic reaction.

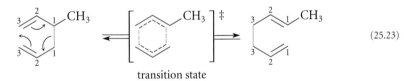

(25.23)

transition state

In other reactions, one end of a σ bond remains fixed to the same group and the other end migrates. For example, the following reaction is a [1,5] sigmatropic reaction because one end of the bond "moves" from atom #1 to atom #1 (that is, it doesn't move), and the other end moves over five atoms.

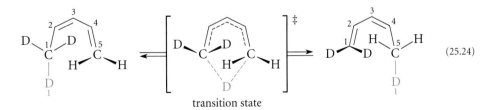

(25.24)

transition state

Sigmatropic reactions, like other pericyclic reactions, are also classified by their stereochemistry. This classification is based on whether the migrating bond moves over the same face, or between opposite faces, of the π-electron system. If the migrating bond moves across one face of the π system, the reaction is said to be *suprafacial*. For example, if the [1,5] sigmatropic reaction of Eq. 25.24 were suprafacial, it would occur in the following manner:

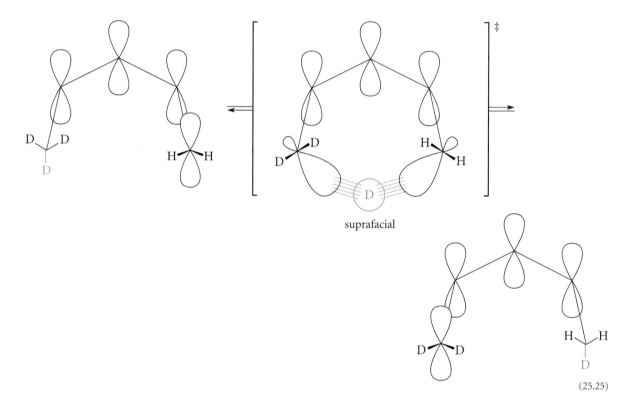

suprafacial

(25.25)

If the reaction were antarafacial, it would occur instead as shown in Eq. 25.26:

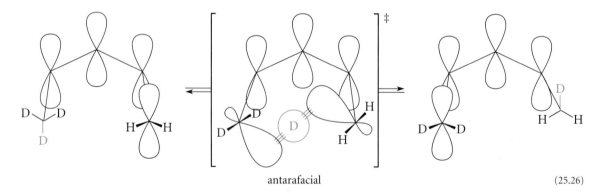

antarafacial   (25.26)

When both ends of a σ bond migrate, the reaction can be suprafacial or antarafacial with respect to either π system. For example, if the [3,3] sigmatropic reaction in Eq. 25.23 were suprafacial on both π systems, it could occur as follows:

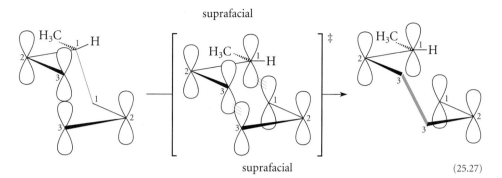

suprafacial

suprafacial   (25.27)

The stereochemistry of a sigmatropic reaction is revealed experimentally only if the molecules involved have stereocenters at the appropriate carbons. This point is illustrated in the following study problem.

STUDY PROBLEM 25.2

Classify the following sigmatropic reaction by giving its bracketed-number designation and its stereochemistry with respect to the plane of the π-electron system.

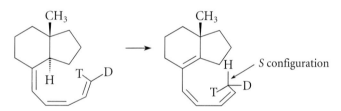

Solution   First identify the bond that is migrating. Because the hydrogen atom migrates, one end of the migrating bond remains fixed; in other words, this is a [1,?] sigmatropic reaction in which we have to determine "?" by counting the carbons over which the migration takes place:

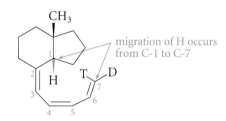

migration of H occurs from C-1 to C-7

Notice that the original point of attachment is counted as carbon-1. Consequently, this is a [1,7] sigmatropic reaction. To determine the stereochemistry, imagine that carbons 1 through 7 all lie in the same plane. As the molecule is depicted above, the migrating hydrogen is *below* that plane (dashed wedge). Because rotation about the C6-C7 bond cannot occur until after the reaction is over (it is a double bond), depict the T and D in the same relative orientations in the starting material and product (as they are depicted in the problem). This reveals that the hydrogen is *above* the plane of the π-electron system in the product. Consequently, the hydrogen has migrated from the lower to the upper face of the π-electron system. Therefore the reaction is a [1,7] *antarafacial* sigmatropic reaction.

25.14 *(a) Refer to Study Problem 25.2 and, assuming an antarafacial migration, give the structure of a starting material that would give a stereoisomer of the product with the *R* configuration at the isotopically substituted carbon.

(b) Give the structure of *another* starting material that would give the same stereoisomer as in (a).

25.15 Classify the following sigmatropic reactions with bracketed numbers.

*(a)

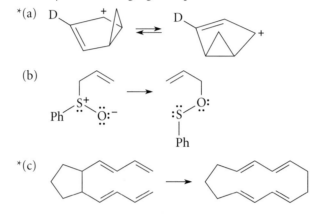

(b)

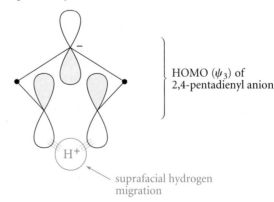

*(c)

Molecular orbital theory provides the connection between the type of sigmatropic reaction and its stereochemistry. Consider, for example, a [1,5] sigmatropic migration of hydrogen across a π-electron system. Think of this reaction as the migration of a proton from one end of the 2,4-pentadienyl anion (Fig. 25.5) to the other:

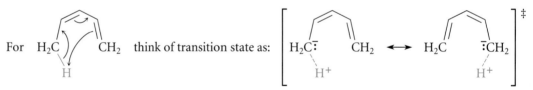

STUDY GUIDE LINK:
✓25.3
Orbital Analysis of Sigmatropic Reactions

The interaction of the proton LUMO—an empty 1*s* orbital—with the HOMO of the π system controls the stereochemistry of the reaction. For the 2,4-pentadienyl anion (Fig. 25.5), the HOMO is symmetric. This means that bonding overlap can occur if the migration occurs suprafacially:

HOMO (ψ_3) of
2,4-pentadienyl anion

H$^+$

suprafacial hydrogen
migration

An ingenious experiment published in 1970 by Wolfgang Roth and his collaborators at the University of Cologne revealed the stereochemistry of the [1,5] sigmatropic hydrogen shift. In the isotopically labeled, optically active alkene shown in Eq. 25.28, a suprafacial [1,5] hydrogen shift is possible from each of the conformations shown:

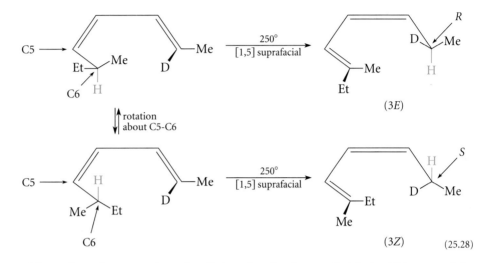

(3E)

rotation about C5-C6

(3Z)   (25.28)

It follows from these equations that if the migration is suprafacial, the *3E* isomer of the product must have the *R* configuration, and the *3Z* isomer of the product must have the *S* configuration. The suprafacial migration shown in Eq. 25.28 was observed.

The [1,3] hydrogen shift involves the HOMO of the allyl anion, an antisymmetric orbital:

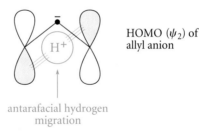

HOMO (ψ_2) of allyl anion

antarafacial hydrogen migration

In order for a [1,3] hydrogen shift to occur, the migrating hydrogen must pass from one face of the allyl π system to the other. Despite the fact that this reaction is "allowed" by MO theory, it requires that the migrating proton bridge too great a distance for adequate bonding. Alternatively, the terminal lobes of the allyl π system could twist; but then a new problem would arise: these lobes would not overlap with the *p* orbital of the central carbon. The resulting loss of conjugation would raise the energy of the transition state. As these arguments suggest, the concerted sigmatropic [1,3] hydrogen shift is virtually nonexistent in organic chemistry. (See Problem 25.16.)

Several interesting experiments have been conducted in which a molecule could in principle undergo both [1,5] and [1,3] hydrogen shifts. In one such experiment—an experiment of elegant simplicity—carried out in 1964 also by Roth, 1,3,5-cyclooctatriene was labeled at carbons 7 and 8 with deuterium and then allowed to undergo many hydrogen shifts for a long period of time. When the molecule undergoes [1,5] hydrogen shifts, the D should migrate part of the time, and the H part of the time. However, after a long time, the D should eventually scramble to all positions that have a 1,5-relationship. In such a case, only carbons 3, 4, 7, and 8 would be partially deuterated:

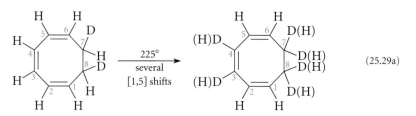

$$(25.29a)$$

(You should write a series of steps for this transformation to convince yourself that it is the predicted result.) On the other hand, if the molecule undergoes successive [1,3] hydrogen (or deuterium) shifts, the deuterium should be scrambled eventually to all positions.

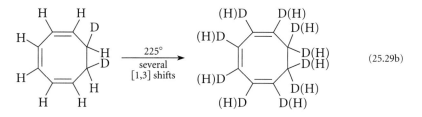

$$(25.29b)$$

The experimental result was that even after very long reaction times, deuterium appeared only in the positions predicted by the [1,5] shift.

Although the suprafacial [1,3] shift of a hydrogen is *not* allowed, the corresponding shift of a carbon atom *is* allowed, provided that certain stereochemical conditions are met. Suppose that an alkyl group (suitably substituted so that its stereochemistry can be traced) were to undergo a suprafacial [1,3] sigmatropic shift. This shift could occur in two stereochemically distinct ways. In the first way, the carbon migrates with *retention* of configuration.

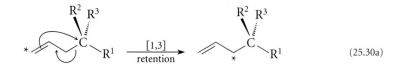

$$(25.30a)$$

(In this and the following equation, one carbon of the allyl group is marked with an asterisk so that its fate can be traced.) In the second way, the carbon migrates with *inversion* of configuration.

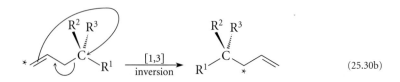

$$(25.30b)$$

Consider the orbital symmetry relationships in these two modes of reaction. Think of the migrating group as an alkyl cation migrating between the ends of an allyl anion.

The LUMO of the alkyl cation—an empty p orbital—interacts with the HOMO of an allyl anion (ψ_2 in Fig. 25.4). In the case of migration with *retention*, the phase relationships between the orbitals involved lead to antibonding overlap:

Retention:

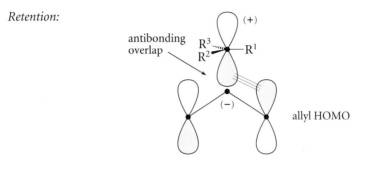

Hence, suprafacial carbon migration with retention is forbidden by orbital symmetry, in the same sense that hydrogen migration is forbidden. If migration occurs with *inversion*, however, there is bonding overlap in the transition state.

Inversion:

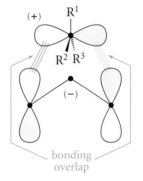

Thus, carbon migration with inversion is allowed by orbital symmetry.

This analysis shows that it is the node in the p orbital that makes the [1,3] suprafacial migration of carbon possible; each of the two lobes of the p orbital, which have opposite phase, can overlap with each end of the allyl π system. Because a bond is broken at one side of the migrating carbon and formed at the other side, inversion of configuration is observed. In the migration of a hydrogen, the orbital involved is a $1s$ orbital, which has no nodes. Hence, [1,3] suprafacial migration of hydrogen is not allowed.

Orbital symmetry, then, makes a very straightforward prediction: The suprafacial [1,3] sigmatropic migration of carbon must occur with inversion of configuration. The following result confirms this prediction:

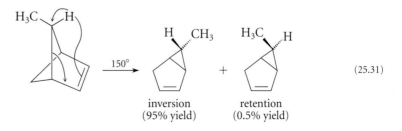

(25.31)

Migration with retention of configuration might have been expected to be the most straightforward, least contorted pathway that the rearrangement could take; yet the theory of orbital symmetry predicts otherwise. One of the remarkable things about the theory is that it correctly predicts so many reactions that otherwise would have appeared unlikely.

As might be expected, orbital symmetry dictates the opposite stereochemistry for [1,5] migrations. Carbon, like hydrogen, undergoes suprafacial [1,5] migrations with retention of configuration (Problem 25.18).

PROBLEMS

*25.16 Explain why the hydrogen migration shown in reaction (1) occurs readily and why the very similar migration shown in (2) does not take place even under forcing conditions. (The asterisked carbons indicate a carbon isotope present so that the rearrangement can be observed.)

(1)

(2)

25.17 Predict the result that would have been expected in the experiment described by Eq. 25.28 for an antarafacial migration.

25.18 *(a) Carry out an orbital symmetry analysis to show that suprafacial [1,5] carbon migrations should occur with retention of configuration in the migrating group.
  (b) Indicate what type of sigmatropic reactions are involved in the following transformation. Is the stereochemistry of the first step in accord with the predictions of orbital symmetry?

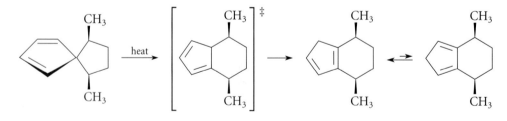

B. [3,3] Sigmatropic Rearrangements

Let's now examine a sigmatropic reaction in which both ends of a σ bond change positions. One of the most common and useful examples of this type of reaction is the [3,3] sigmatropic rearrangement. Using the same logic as before, the transition state for this rearrangement can be visualized as the interaction of two allylic systems, one a cation and one an anion.

The frontier orbitals involved are the HOMO of the anion and the LUMO of the cation, which are the same orbital (ψ_2) of the allyl system (Fig. 25.4).

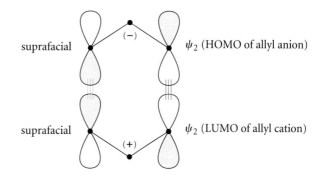

The two MOs involved achieve bonding overlap when the [3,3] sigmatropic rearrangement occurs suprafacially on both components. (You should convince yourself that a reaction that is antarafacial on both π systems is also allowed by orbital symmetry, but one that is suprafacial on one component and antarafacial on the other is forbidden.)

One of the best known types of [3,3] sigmatropic rearrangement is the **Cope rearrangement**, which was extensively investigated by Arthur C. Cope of the Massachusetts Institute of Technology long before the principles of orbital symmetry were known. The Cope rearrangement is simply a 1,5-diene isomerization:

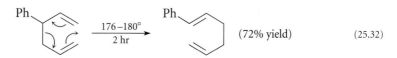

An interesting variation of the Cope rearrangement is the "oxyCope" reaction. In this type of reaction, an enol is formed initially; tautomerization of the enol (Sec. 22.2) into the corresponding carbonyl compound is a very favorable equilibrium that drives the reaction to completion.

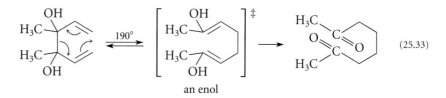

In the **Claisen rearrangement**, an ether that is both allylic and vinylic or an allylic aryl ether undergoes a [3,3] sigmatropic rearrangement.

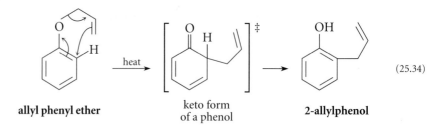

allyl phenyl ether keto form
of a phenol **2-allylphenol** (25.34)

If both *ortho* positions are blocked by substituent groups, the *para*-substituted derivative
is obtained:

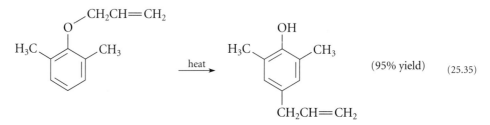

(95% yield) (25.35)

This reaction occurs by a sequence of two Claisen rearrangements, followed by tautomerization of the product to the phenol:

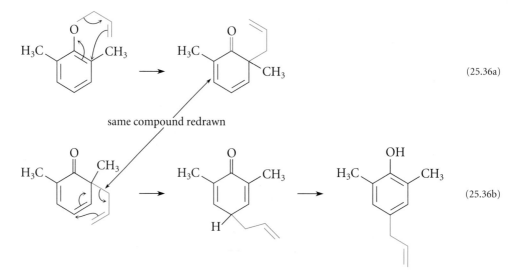

(25.36a)

same compound redrawn

(25.36b)

Problem 25.20 considers the Claisen rearrangement of an aliphatic ether.

C. Summary: Selection Rules for Sigmatropic Reactions

The stereochemistry of sigmatropic reactions is a function of the number of electrons
involved. (As with other pericyclic reactions, the number of electrons involved is determined from the curved-arrow formalism: Count the curved arrows and multiply by two.)
All-suprafacial sigmatropic reactions occur when $4n + 2$ electrons are involved in the
reaction, that is, an odd number of electron pairs or curved arrows. In contrast, a
sigmatropic reaction must be antarafacial on one component and suprafacial on the other

Table 25.3   Selection Rules for Thermal Sigmatropic Reactions

Number of electrons[b]	Allowed stereochemistry[a]	
	Generalized stereochemistry	Stereochemistry of single-atom migrations
4n	supra-antara antara-supra	supra-inversion antara-retention
4n + 2	supra-supra antara-antara	supra-retention antara-inversion

[a] supra = suprafacial; antara = antarafacial
[b] n = an integer

when 4n electrons (an even number of electron pairs or curved arrows) are involved. When a single carbon migrates, the term "suprafacial" is taken to mean "retention of configuration," and the term "antarafacial" is taken to mean "inversion of configuration."

These generalizations are summarized in Table 25.3 as selection rules for sigmatropic reactions. These selection rules hold for thermal reactions—reactions that occur from electronic ground states. As with other pericyclic reactions, there are selection rules for photochemical sigmatropic reactions—reactions that occur from excited states. However, because these are rather uncommon, they are not listed in Table 25.3.

PROBLEMS

25.19 *(a) What allowed and reasonable sigmatropic reaction(s) can account for the following transformation?

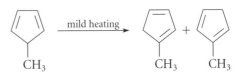

(b) What product(s) are expected from a similar reaction of 2,3-dimethyl-1,3-cyclopentadiene?

25.20 *(a) Aliphatic allylic vinylic ethers undergo the Claisen rearrangement. Complete the following reaction:

$$(CH_3)_2C\!=\!CH\!-\!CH_2\!-\!O\!-\!CH\!=\!CH_2 \xrightarrow{\text{heat}}$$

(b) What starting material would give the following compound in an aliphatic Claisen rearrangement?

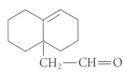

*25.21 Show how the transition state for a [3,3] sigmatropic reaction can be analyzed as the interaction of two allylic radicals, and that the same stereochemical outcome is predicted. (See Study Guide Link 25.3.)

25.22 Show by an orbital symmetry analysis that a [3,3] sigmatropic reaction which is antarafacial on both components is allowed. Would you expect such a reaction to be very common? Why?

25.5 Summary of the Pericyclic Selection Rules

It is always possible to derive the allowed stereochemistry of a pericyclic reaction by using the phase relationships within the molecular orbitals involved. However, a convenient way to remember the selection rules without re-deriving them each time they are needed is summarized in Table 25.4. This involves assigning either a +1 or a −1 to each of the following aspects of the reaction: thermal or photochemical activation; $4n + 2$ or $4n$ reacting electrons; and stereochemistry. Apply Table 25.4 using the following steps.

1. Assign +1 to thermal reactions and −1 to photochemical reactions.
2. Assign +1 to systems with $4n + 2$ reacting electrons and −1 to systems with $4n$ reacting electrons. Remember that the number of reacting electrons is twice the number of arrows needed to describe the reaction in conventional curved-arrow notation.
3. Assign +1 to disrotatory, suprafacial, or retention stereochemistry, and −1 to conrotatory, antarafacial, or inversion stereochemistry. The stereochemical assignment is made for *each component* of the reaction. Thus, for a sigmatropic reaction or a cycloaddition, two such assignments must be made.
4. Multiply together the resulting numbers. If the product is +1, the reaction is allowed; if the product is a −1, the reaction is forbidden.

The following study problem illustrates these steps.

Table 25.4 An Aid for Applying the Pericyclic Selection Rules

For			*Assign*
Mode of activation	**Stereochemistry**	**Number of electrons**[a]	
Thermal	Disrotatory/suprafacial/ retention	$4n + 2$	+1
Photochemical	Conrotatory/antarafacial/ inversion	$4n$	−1

[a] n = an integer

STUDY
PROBLEM
25.3

Use Table 25.4 to predict whether each of the following reactions is allowed.
  (a) Thermal disrotatory electrocyclic ring closure of (Z)-1,3,5-hexatriene
  (b) A photochemical [4s + 2s] cycloaddition
  (c) The thermal [1,3] suprafacial migration of carbon with inversion of config-
       uration in the migrating group

Solution   For (a), assign a +1 for the thermal mode, a +1 for disrotatory stereochemistry, and a +1 for $4n + 2$ electrons. Result: $(+1)(+1)(+1) = +1$: the reaction is allowed.
   For (b), assign a −1 for the photochemical mode, a +1 for $4n + 2$ electrons, and
+1 for the suprafacial stereochemistry on each component of the addition. (Recall that
this is the meaning of the 4s and 2s notation.) Result: $(-1)(+1)(+1)(+1) = -1$: the
reaction is forbidden.
   For (c), assign a +1 for the thermal mode, +1 for suprafacial stereochemistry on
one component, −1 for inversion stereochemistry on the other, and a −1 for $4n$ electrons.
Result: $(+1)(+1)(-1)(-1) = +1$: the reaction is allowed.

The introduction to this chapter stated that three aspects of a given pericyclic reaction
were interrelated: the mode of activation (heat or light), the number of reacting electrons,
and the stereochemistry. Table 25.4 and Study Problem 25.3 succinctly summarize the
relationships between these reaction characteristics. The underlying basis of this table is
orbital symmetry—the symmetry characteristics of molecular orbitals.

PROBLEM

25.23 Without consulting Tables 25.1–25.3, classify the stereochemistry (+1 or −1)
of each of the following reactions. Indicate whether each is allowed or forbidden.
  *(a) [4s + 4a] thermal cycloaddition
  (b) suprafacial [1,7] thermal hydrogen migration
  *(c) disrotatory photochemical electrocyclic ring closure reaction of
        (2E,4Z,6Z,8E)-2,4,6,8-decatetraene

25.6   Fluxional Molecules

A number of compounds continually undergo rapid sigmatropic rearrangements at room
temperature. One such compound is *bullvalene*, which was first prepared in 1963.

bullvalene

The [3,3] sigmatropic rearrangements in bullvalene rapidly interconvert *identical forms
of the molecule*:

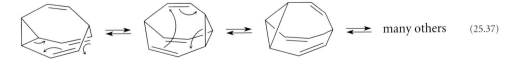

many others   (25.37)

Vitamin D_2, sometimes called "irradiated ergosterol," is the form of vitamin D that is commonly added to milk and other foods as a dietary supplement.

*25.25   When previtamin D_2 is isolated and irradiated, ergosterol is obtained along with a stereoisomer, *lumisterol*. Explain mechanistically the origin of lumisterol.

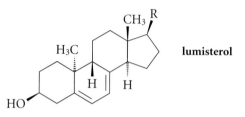

lumisterol

25.26   When previtamin D_2 is *heated*, two compounds *A* and *B* are obtained that are stereoisomers of both ergosterol and lumisterol. Suggest structures for these compounds and explain mechanistically how they are formed.

*25.27   When the compounds *A* and *B* in the previous problem are *irradiated*, two stereoisomeric compounds *C* and *D*, respectively, are obtained, each of which contains a cyclobutene ring. Suggest structures for *C* and *D* and explain mechanistically how they are formed. Explain why irradiation of either *A* or *B* does *not* give back previtamin D_2.

KEY IDEAS IN CHAPTER 25

🧪   Pericyclic reactions are concerted reactions that occur by cyclic electron shifts. Electrocyclic reactions, cycloadditions, and sigmatropic rearrangements are important pericyclic reactions.

🧪   Electrocyclic reactions are stereochemically classified as conrotatory or disrotatory; cycloadditions and sigmatropic rearrangements are classified as suprafacial or antarafacial.

🧪   The stereochemical course of a pericyclic reaction is governed largely by the symmetry of the reactant HOMO (highest occupied molecular orbital), or, if there are two reacting components, by the relative symmetries of the HOMO of one component and the LUMO (lowest unoccupied molecular orbital) of the other.

⌂   Considerations of orbital symmetry lead to selection rules for pericyclic reactions. Whether a pericyclic reaction is allowed or forbidden depends on the number of electrons involved, the mode of activation (thermal or photochemical), and the stereochemical course of the reaction. The selection rules are summarized in Tables 25.1–25.4 and Sec. 25.5.

ADDITIONAL PROBLEMS

25.28 Without consulting tables or figures, answer the following questions:

  *(a) Is a thermal disrotatory electrocyclic reaction involving twelve electrons allowed or forbidden?

  (b) Is a [8s + 4s] photochemical cycloaddition allowed?

  *(c) Is the HOMO of (Z)-3,4-dimethyl-1,3,5-hexatriene symmetric or antisymmetric?

25.29 What do the pericyclic selection rules have to say about the *position of equilibrium* in each of the following reactions? Which side of each equilibrium is favored and why?

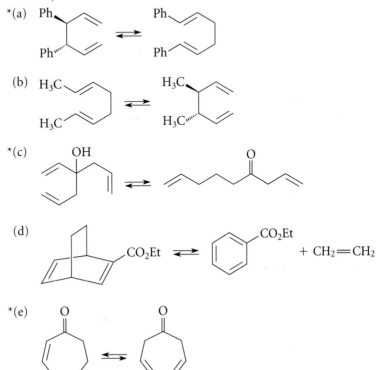

  *(a)

  (b)

  *(c)

  (d)

  *(e)

25.30 *(a) Predict the stereochemistry of compounds *B* and *C*.

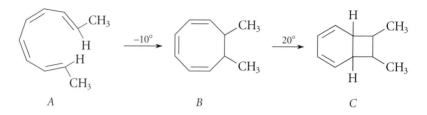

A   *B*   *C*

(b) What stereoisomer of *A* also gives compound *C* on heating?

25.31 *(a) Classify the following pericyclic reaction. (More than one classification is possible.)

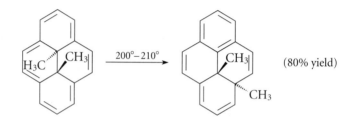

(b) Suppose the migrating methyl group in (a) were labeled with the hydrogen isotopes deuterium (D) and tritium (T) so that it is a —CHDT group with the *S* configuration. What would be the configuration of this group in the product? Explain your reasoning.

*25.32 Complete the following reactions by giving the major organic product(s).

(a)

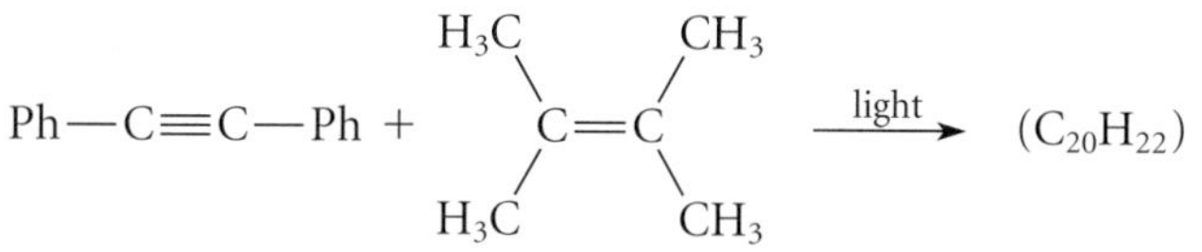

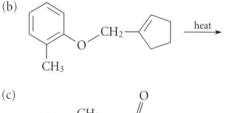

(b)

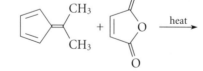

(c)

25.33 When compound *A* is irradiated with ultraviolet light for 115 hours in pentane, an isomeric compound *B* is obtained that decolorizes bromine in CCl_4 and reacts with ozone to give, after the usual workup, a compound *C*.

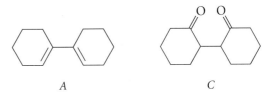

A   C

*(a) Give the structure of *B* and the stereochemistry of both *B* and *C*.

(b) On heating to 90°, compound *D*, a stereoisomer of *B*, is converted into *A*, but compound *B* is virtually inert under the same conditions. Identify compound *D* and account for these observations.

*25.34 *Heptafulvalene* undergoes a thermal reaction with tetracyanoethylene (TCNE) to give the adduct shown below. What is the stereochemistry of this adduct? Explain.

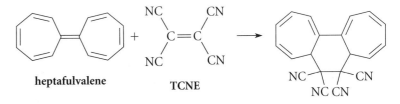

heptafulvalene   **TCNE**

25.35 Suggest a mechanism for each of the following transformations. Some involve pericyclic reactions only; others involve pericyclic reactions as well as other steps. Invoke the appropriate selection rules to explain any stereochemical features observed.

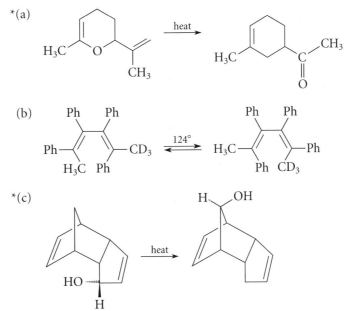

*(a)

(b)

*(c)

(*Problem 25.36 continues*)

(d)

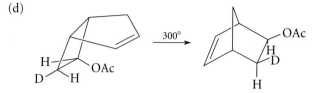

*25.36   When 1,3,5-cyclooctatriene, *A*, is heated to 80–100°, it comes to equilibrium with an isomeric compound *B*. Treatment of the mixture of *A* and *B* with $CH_3O_2C—C≡C—CO_2CH_3$ gives a compound *C*, which, when heated to 200° for twenty minutes, gives dimethyl phthalate and cyclobutene. Identify compounds *B* and *C*, and explain what reactions have occurred.

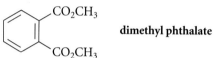

dimethyl phthalate

25.37   The following reaction occurs as a sequence of two pericyclic reactions. Identify the intermediate *A* and describe the two reactions.

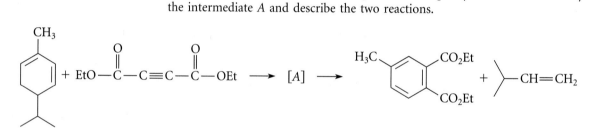

α-phellandrene

*25.38   Black Hemptra bugs, generally observed in the tropical regions of India immediately after the rainy season, give off a characteristic nauseating smell whenever they are disturbed or crushed. Substance *A*, the compound causing the odor, can be obtained either by extracting the bugs with petroleum ether (which no doubt disturbs them greatly), or it can be prepared by heating the compound below at 170–180° for a short time. Give the structure of compound *A*.

$$CH_2=CH—\overset{\overset{\displaystyle C_2H_5}{|}}{CH}—O—\underset{\underset{\displaystyle OC_2H_5}{|}}{C}=CH_2$$

25.39   When 2-methyl-2-propenal is treated with allylmagnesium chloride ($CH_2=CH—CH_2—MgCl$) in ether, then with dilute aqueous acid, a compound *A* is obtained, which, when heated strongly, yields an aldehyde *B*. Give the structures of compounds *A* and *B*.

*25.40 Compound A (C$_{11}$H$_{14}$O$_3$) is insoluble in base and gives an isomeric compound B when heated strongly. Compound B gives a sodium salt when treated with NaOH. Treatment of the sodium salt of B with dimethyl sulfate gives a new compound C (C$_{12}$H$_{16}$O$_3$) that is identical in all respects to a natural product *elemicin*. Ozonolysis of elemicin followed by oxidation gives the carboxylic acid D below. Propose structures for compounds A, B, and C.

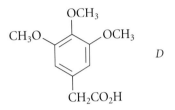

D

25.41 Using phenol and any other reagents as starting materials, outline a synthesis of each of the following compounds.

*(a) 1-ethoxy-2-propylbenzene   (b)

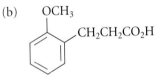

25.42 Each of the following reactions involves a sequence of two pericyclic reactions. Identify the bracketed intermediate X or Y involved in each reaction, and describe the pericyclic reactions involved.

*(a)

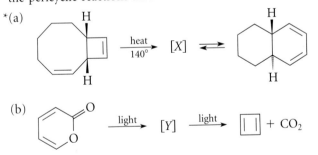

(b)

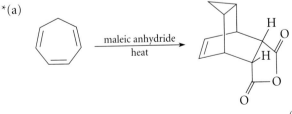

25.43 When each of the compounds below is heated in the presence of maleic anhydride, an intermediate is trapped as a Diels-Alder adduct. What is the intermediate formed in each reaction, and how is it formed from the starting material?

*(a)

(Problem 25.43 continues)

25.48   Anticipating the isolation of the potentially aromatic hydrocarbon *B*, a group of chemists irradiated compound *A* below with ultraviolet light. Compound *C* was obtained as a product instead of *B*.

*(a)   Explain why compound *B* might be expected to be unstable in spite of its cyclic array of $4n + 2$ π electrons.

(b)   Explain why the formation of compound *B* is allowed by the pericyclic selection rules.

*(c)   Account for the formation of the observed product *C*.

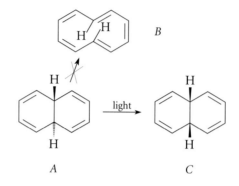

25.49   *(a)   What type of pericyclic reaction is required to form benzene from Dewar benzene?

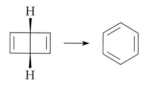

Dewar benzene   **benzene**

(b)   Explain why Dewar benzene, although a very unstable molecule, is not spontaneously transformed to benzene. (Although Dewar benzene forms benzene when heated, this reaction requires a surprisingly high temperature and is believed not to be concerted.)

*25.50   (a)   Identify the hydrocarbon *B* and the intermediate *A* (both with the formulas $C_{11}H_{10}$) in the reaction sequence below. Compound *B* is formed spontaneously from *A* in a pericyclic reaction.

$$\text{———} \xrightarrow{2Br_2} \xrightarrow[\text{EtOH}]{\text{KOH}} [A] \longrightarrow [B]$$

(b)   The proton NMR spectrum of *B* consists of a complex absorption at δ 7.1 (8*H*) and a singlet at δ (−0.5) (2*H*). Account for the absorption at a negative chemical shift; that is, to which protons does this absorption correspond, and why do they absorb at a negative chemical shift?

26

Amino Acids, Peptides, and Proteins

Amino acids, as the name implies, are compounds that contain both an amino group and a carboxylic acid group.

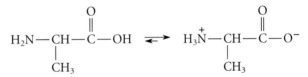

alanine
(an α-amino acid)

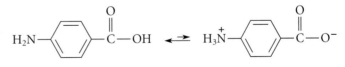

p-aminobenzoic acid
(PABA, a component
of folic acid, a vitamin)

☞
STUDY GUIDE LINK:
✓ **26.1**
Neutral Amino Acids

As these structures show, a neutral amino acid—an amino acid with an *overall* charge of zero—can contain within the same molecule two groups of opposite charge. Molecules containing oppositely charged groups are known as **zwitterions** (German "hybrid ion"). A zwitterionic structure is possible because the basic amino group can accept a proton and the acidic carboxylic acid group can lose a proton. Each of the **α-amino acids**, of which alanine above is an example, has an amino group on the α-carbon—the carbon adjacent to the carboxylic acid group.

Peptides are biologically important polymers in which α-amino acids are joined into chains through amide bonds, called **peptide bonds**. A peptide bond is derived from the amino group of one amino acid and the carboxylic acid group of another.

1263

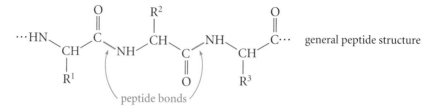

general peptide structure

Proteins are very large peptides, and some proteins are aggregates of more than one peptide. The name *protein* (Greek "of first rank") is particularly apt because peptides and proteins serve many important roles in biology. For example, enzymes (biological catalysts) and some hormones are peptides or proteins.

26.1   Nomenclature of Amino Acids and Peptides

A.   Nomenclature of Amino Acids

Some amino acids are named substitutively as carboxylic acids with amino substituents.

$$\overset{+}{H_3N}-CH_2CH_2CH_2-CO_2^- \;\rightleftharpoons\; H_2N-CH_2CH_2CH_2-CO_2H$$

4-aminobutanoic acid
(γ-aminobutyric acid)

—CO₂H

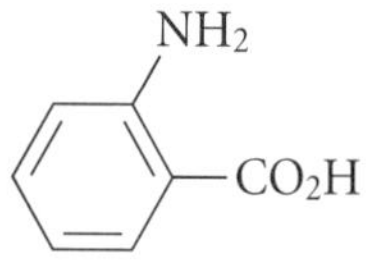

2-aminobenzoic acid
(*o*-**aminobenzoic acid or**
anthranilic acid)

$$(CH_3)_2\overset{+}{N}H-CH_2CH_2-\overset{O}{\overset{\|}{C}}-O^-$$

3-(dimethylamino)propanoic acid

Notice that even if they exist as zwitterions, amino acids are named as uncharged compounds.

Twenty α-amino acids are known by widely accepted traditional names. These are the amino acids that occur commonly as constituents of proteins. The names and structures of these amino acids are given in Table 26.1 on pp. 1266–1267.

Two points about the structures of the α-amino acids will help you to remember them. First, with the exception of proline, all α-amino acids have the same general structure, differing only in the identity of the side chain R.

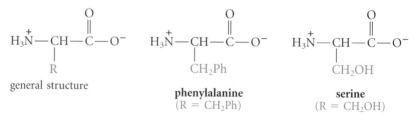

general structure

phenylalanine
(R = CH₂Ph)

serine
(R = CH₂OH)

Proline is the only naturally occurring amino acid with a secondary amino group. In proline the —NH— and the side chain are "tied together" in a ring.

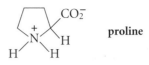

proline

Second, as Table 26.1 shows, the amino acids can be organized into six groups according to the nature of their side chains.

1. Amino acids with —H or aliphatic hydrocarbon side chains
2. Amino acids with side chains containing aromatic groups
3. Amino acids with side chains containing —SH, —SCH$_3$, or alcohol —OH groups
4. Amino acids with side chains containing carboxylic acid or amide groups
5. Amino acids with basic side chains
6. Proline

☞
STUDY GUIDE LINK:
✓26.2
*Names of the
Amino Acids*

The α-amino acids are often designated by either three-letter or single-letter abbreviations, which are given in Table 26.1.

B. Nomenclature of Peptides

The terminology and nomenclature associated with peptides are best illustrated by an example. Consider the following peptide formed from the three amino acids alanine, valine, and lysine.

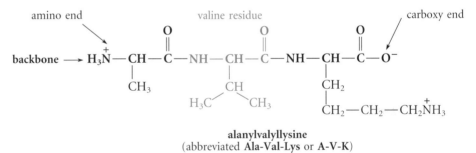

alanylvalyllysine
(abbreviated **Ala-Val-Lys** or **A-V-K**)

The **peptide backbone** is the repeating sequence of nitrogen, α-carbon, and carbonyl groups shown in boldface type in the structure above. The characteristic amino acid side chains are attached to the peptide backbone at the respective α-carbon atoms. Each amino acid unit in the peptide is called a **residue**. For example, the part of the peptide derived from valine, the *valine residue*, is shown in color. The ends of a peptide are labeled as the **amino end** or **amino terminus** and the **carboxy end** or **carboxy terminus**. A peptide can be characterized by the number of residues it contains. For example, the peptide above is a **tripeptide** because it contains three amino acid residues. A peptide containing two, three, or five amino acids would be termed a **dipeptide**, **tripeptide**, or **pentapeptide**, respectively. A relatively short peptide of unspecified length containing a few amino acids is sometimes referred to as an **oligopeptide** (from a Greek root meaning "scant" or "few").

A peptide is conventionally named by giving successively the names of the amino acid residues, *starting at the amino end*. The names of all but the carboxy-terminal residue

Table 26.1 **Names, Structures, Abbreviations, and Properties of the Twenty Common Naturally Occurring Amino Acids**

General structure:

$$H_3\overset{+}{N}-\underset{\underset{R}{|}}{CH}-\overset{\overset{O}{\|}}{C}-O^-$$

Name and abbreviations	R^a	Optical rotation of L enantiomer in H_2O	pK_{a1}	pK_{a2}	pK_{a3}	Isoelectric point, pI	Water solubility, wt % at 25 °C
Amino Acids with Simple Aliphatic Side Chains							
glycine, Gly, G	—H		2.34	9.60	—	5.97	20
alanine, Ala, A	—CH_3	(+)	2.35	9.69	—	6.02	14
valine, Val, V	—$CH(CH_3)_2$	(+)	2.32	9.62	—	5.97	6.5
leucine, Leu, L	—$CH_2CH(CH_3)_2$	(−)	2.36	9.60	—	5.98	2.2
isoleucine, Ile, I	—$CH-C_2H_5$ $\underset{}{\overset{}{\underset{CH_3}{|}}}$ [S configuration]	(+)	2.36	9.68	—	6.02	3.9
Amino Acids with Aromatic Side Chains							
phenylalanine, Phe, F	—CH_2Ph	(−)	1.83	9.13	—	5.48	2.9
tryptophan, Trp, W		(−)	2.38	9.39	—	5.88	1.1
tyrosine, Tyr, Y		(−)	2.20	9.11	10.07	5.65	0.05
histidine, His, H^b		(−)	1.82	6.00	9.17	7.58	7.1

a Side chains are shown in their uncharged form.
b Histidine is a weakly basic amino acid.

Table 26.1 (continued)

Name and abbreviations	R^a	Optical rotation of L enantiomer in H_2O	pK_{a1}	pK_{a2}	pK_{a3}	Isoelectric point, pI	Water solubility, wt % at 25 °C
Amino Acids with Side-Chain Alcohol —OH, —SH, and —SCH₃ Groups							
serine, Ser, S	—CH₂—OH	(−)	2.21	9.15	—	5.68	4.8
threonine, Thr, T	—CH—OH | CH₃ [R configuration]	(−)	2.71	9.62	—	5.16	17
methionine, Met, M	—CH₂CH₂—SCH₃	(−)	2.28	9.21	—	5.75	3.4
cysteine, Cys, C^c	—CH₂—SH	(−)	1.71	8.18	10.28	5.02	very sol.
Amino Acids with Side Chains Containing Carboxylic Acid or Amide Groups							
aspartic acid, Asp, D	—CH₂—CO₂H	(+)	1.88	3.65	9.60	2.76	0.50
glutamic acid, Glu, E	—CH₂CH₂—CO₂H	(+)	2.16	4.32	9.67	3.24	0.84
asparagine, Asn, N	—CH₂—CO—NH₂	(−)	2.02	8.80	—	5.41	3.0
glutamine, Gln, Q	—CH₂CH₂—CO—NH₂	(+)	2.17	9.13	—	5.65	3.5
Amino Acids with Side Chains Containing Strongly Basic Groups							
lysine, Lys, K	—(CH₂)₄—NH₂	(+)	2.18	9.12	10.53	9.82	very sol.
arginine, Arg, R	—(CH₂)₃NH—C(=NH)—NH₂	(+)	2.17	9.04	12.48	10.76	13
Cyclic (Secondary) Amino Acid							
proline, Pro, P		(−)	1.99	10.60	—	6.10	63

^{a} Side chains are shown in their uncharged form.

^{c} Cysteine often occurs in proteins as a disulfide dimer, called cystine: H_2N—CH—CH₂—S—S—CH₂—CH—NH_2. For this reason, cysteine is sometimes called **half-cystine** and abbreviated Cys/2.
  |CO₂H   CO₂H|

are formed by dropping the final ending (*ine*, *ic*, or *an*) and replacing it with *yl*. Thus, the peptide above is named alanylvalyllysine. In practice, this type of nomenclature is cumbersome for all but the smallest peptides. A simpler way of naming peptides is to connect with hyphens the three-letter (or one-letter) abbreviations of the component amino acid residues beginning with the amino-terminal residue. Thus, the peptide above is also written as Ala-Val-Lys or A-V-K.

Large peptides of biological importance are known by common names. Thus, *insulin* is an important peptide hormone that contains fifty-one amino acid residues; *ribonuclease*, an enzyme, is a protein containing 124 amino acid residues (and a rather small protein at that!).

PROBLEMS

26.1 Draw the structures of the following peptides.
*(a) tryptophylglycylisoleucylaspartic acid
(b) Glu-Gln-Phe-Arg (or E-Q-F-R)

*26.2 Using three-letter abbreviations for the amino acid residues, name the following peptide.

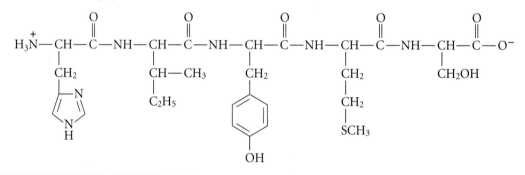

26.2 Stereochemistry of the α-Amino Acids

With the exception of glycine, all common naturally occurring α-amino acids have an asymmetric α-carbon atom, and are chiral molecules. The chiral amino acids in Table 26.1 are found within naturally occurring proteins in only one enantiomeric form, which has the following configuration:

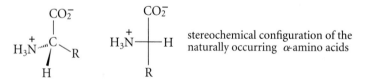

This configuration is *S* in all cases except for cysteine (see Problem 26.4).

The stereochemistry of α-amino acids is often specified with an older system, the **D,L-system**. An L amino acid *by definition* has the amino group on the left and the hydrogen on the right when the carboxylic acid group is up and the side chain is down in a Fischer projection of the α-carbon. Therefore, the naturally occurring amino acids have the L-configuration. Thus, (*S*)-serine can also be called L-serine; its enantiomer is D-serine.

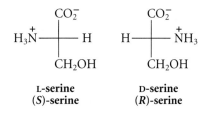

L-serine   **D-serine**
(*S*)-serine   (*R*)-serine

Note that the correspondence between *S* and L *is not general* (see Problem 26.4).

The D or L designation for an α-amino acid refers to the configuration of the α-carbon *regardless of the number of asymmetric carbons in the molecule*. Thus, L-threonine, which has two asymmetric carbons, is the (2*S*,3*R*) stereoisomer. Its enantiomer, D-threonine, is the (2*R*,3*S*) stereoisomer.

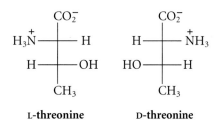

L-threonine   **D-threonine**

Because threonine has two asymmetric carbons, it also has two other stereoisomers—its diastereomers. In the D,L system, *diastereomers are given different names*. The diastereomers of threonine are called *allothreonine*. Thus, L-allothreonine is the (2*S*,3*S*) stereoisomer of threonine, and D-allothreonine is its enantiomer. We need not dwell further on this older system except to be aware that it is still used for amino acids and sugars.

PROBLEMS

26.3 *(a) L-Isoleucine has two asymmetric carbons and has the (2*S*,3*S*) configuration. Draw a Fischer projection of L-isoleucine.
  (b) Alloisoleucine is the diastereomer of isoleucine. Draw Fischer projections of L-alloisoleucine and D-alloisoleucine.

26.4 *(a) What is the α-carbon configuration of L-cysteine in the *R*,*S* system?
  (b) Explain why it is that L-cysteine and L-serine have different configurations in the *R*,*S* system.

26.3 **Acid-Base Properties of Amino Acids and Peptides**

A. Zwitterionic Structures of Amino Acids and Peptides

As suggested in the introduction to this chapter, the neutral forms of the α-amino acids are *zwitterions*. How do we know this is so? Some of the evidence is as follows:

1. Amino acids are insoluble in apolar aprotic solvents such as ether. On the other hand, most unprotonated amines and un-ionized carboxylic acids dissolve in ether. Although the water solubilities of the different amino acids vary from case to case, all are more soluble in water than they are in ether.

2. Amino acids have very high melting points. For example, glycine melts at 262° (with decomposition), and tyrosine melts at 310° (also with decomposition). Hippuric acid, a much larger molecule than glycine, and glycinamide, the amide of glycine, have much lower melting points. The former compound lacks the amino group, the latter lacks the carboxylic acid group, and neither can exist as a zwitterion.

$$Ph-\overset{\overset{\displaystyle O}{\|}}{C}-NH-CH_2-CO_2H \qquad H_2N-CH_2-\overset{\overset{\displaystyle O}{\|}}{C}-NH_2 \qquad \overset{+}{H_3N}-CH_2-\overset{\overset{\displaystyle O}{\|}}{C}-O^-$$

N-benzoylglycine (hippuric acid)	**glycinamide**	**glycine**
mp 190°	mp 67–68°	mp 262° (d)

The high melting points and greater solubilities in water than in ether are characteristics expected of salts, not uncharged organic compounds. These saltlike characteristics are, however, what *would* be expected of a zwitterionic compound. The strong forces in the solid states of the amino acids that result from the attraction of full positive and negative charges on *different molecules* are much like those between the ions in a salt. These attractions stabilize the solid state and resist conversion of the solid into a liquid—whether a pure liquid melt or a solution. Water is the best solvent for most amino acids because it solvates ionic groups much as it solvates the ions of a salt (Sec. 8.4B).

3. The dipole moments of the amino acids are very large—much larger than those of similar-sized molecules with only one amine or carboxylic acid group.

$$\overset{+}{H_3N}-CH_2-CO_2^- \qquad CH_3-CH_2-CO_2H \qquad CH_3CH_2CH_2CH_2-NH_2$$

glycine	**propanoic acid**	**butylamine**
$\mu \sim 14$ D	$\mu = 1.7$ D	$\mu = 1.4$ D

A large dipole moment is expected for molecules that contain a great deal of separated charge (Sec. 1.2D and Study Guide Link 1.2).

4. The pK_a values for amino acids are what would be expected for the zwitterionic forms of the neutral molecules.

Suppose a neutral amino acid is titrated with acid. When one equivalent of acid is added, the *basic* group of the amino acid will have been protonated. When this experiment is carried out with glycine, the pK_a of the basic group is found to be 2.3. If glycine is indeed a zwitterion, this basic group can only be the carboxylate ion. If glycine is not a zwitterion, this basic group has to be the amine.

$$H_3O^+ + \overset{+}{H_3N}-CH_2-\overset{\overset{\displaystyle O}{\|}}{C}-\overset{..}{\underset{..}{O}}:^- \rightleftharpoons \overset{+}{H_3N}-CH_2-\overset{\overset{\displaystyle O}{\|}}{C}-\overset{..}{\underset{..}{O}}H + H_2O \qquad (26.1a)$$

basic group of the zwitterionic form

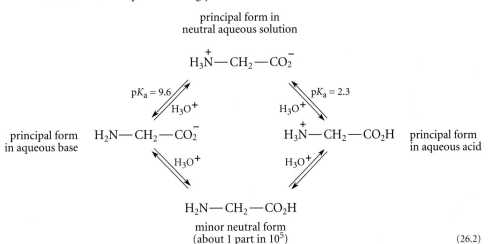

$$H_3O^+ + H_2\overset{..}{N}-CH_2-\overset{\overset{\displaystyle O}{\|}}{C}-OH \quad \rightleftharpoons \quad H_3\overset{+}{N}-CH_2-\overset{\overset{\displaystyle O}{\|}}{C}-OH + H_2O \quad (26.1b)$$

basic group of the nonzwitterionic form

Which is the correct description of the titration? The pK_a of 2.3 is that expected of a carboxylic acid in a molecule containing a nearby electron-withdrawing group (in this case, the $H_3\overset{+}{N}$— group). In contrast, the conjugate acids of amines have pK_a values in the 8–10 range. This analysis suggests that the zwitterion, not the uncharged form, is being titrated.

Along the same lines, if NaOH is added to neutral glycine, a group is titrated with $pK_a = 9.6$. This is a reasonable pK_a value for an alkylammonium ion, but would be very unusual for a carboxylic acid. This comparison also suggests that the neutral form of glycine is a zwitterion.

The acid-base equilibria for glycine can be summarized as follows:

principal form in
neutral aqueous solution

$$H_3\overset{+}{N}-CH_2-CO_2^-$$

$pK_a = 9.6$ $pK_a = 2.3$

H_3O^+ H_3O^+

principal form $H_2N-CH_2-CO_2^-$ $H_3\overset{+}{N}-CH_2-CO_2H$ principal form
in aqueous base   in aqueous acid

H_3O^+ H_3O^+

$$H_2N-CH_2-CO_2H$$

minor neutral form
(about 1 part in 10^5)

(26.2)

The major neutral form of any α-amino acid is the zwitterion. In fact, it can be estimated that the ratio of the uncharged form of an α-amino acid to the zwitterion form is about one part in 10^5, as shown for glycine in Eq. 26.2.

Peptides also exist as zwitterions; that is, at pH values near 7, amino groups are protonated and carboxylic acid groups are ionized.

PROBLEM

26.5   Draw the structure of the major neutral form of each of the following peptides.
     *(a) A-K-V-E-M   (b) G-D-G-L-F

B. Isoelectric Points of Amino Acids and Peptides

An important measure of the acidity or basicity of an amino acid is its **isoelectric point** or **isoelectric pH**. This is the pH of a dilute aqueous solution of the amino acid at which the total charge on all molecules of the amino acid is zero. At the isoelectric point two conditions are met. First, the concentration of conjugate acid molecules A equals the concentration of conjugate base molecules B. Because the conjugate acid A is positively charged and the conjugate base B is negatively charged, the equality of the two concentrations means that the total charge on all molecules of the amino acid is zero. Second, the relative concentration of the neutral form N is greater than at any other pH.

To illustrate, consider the ionization equilibria of the amino acid alanine.

$$\overset{+}{H_3N}\text{—CH—CO}_2H \underset{H_3O^+}{\overset{H_2O}{\rightleftharpoons}} \overset{+}{H_3N}\text{—CH—CO}_2^- \underset{H_3O^+}{\overset{H_2O}{\rightleftharpoons}} H_2N\text{—CH—CO}_2^- \qquad (26.3)$$
$$\underset{CH_3}{|} \qquad\qquad \underset{CH_3}{|} \qquad\qquad \underset{CH_3}{|}$$
$$A \qquad\qquad\qquad N \qquad\qquad\qquad B$$

The isoelectric point of alanine is 6.0. Thus, in a solution of alanine in which the pH has been adjusted to 6.0, a very small amount of alanine is in form A, an exactly equal amount is in form B, and most of the alanine is in neutral (zwitterionic) form N. The isoelectric points of the twenty common amino acids are listed in Table 26.1.

The isoelectric pH has a simple relationship to the pK_a values of an amino acid. Let K_{a1} be the dissociation constant of the carboxylic acid group of form A, and K_{a2} the dissociation constant of the ammonium group in form N (Eq. 26.3).

$$K_{a1} = \frac{[N][H_3O^+]}{[A]} \qquad K_{a2} = \frac{[B][H_3O^+]}{[N]} \qquad (26.4)$$

Dividing both equations through by $[H_3O^+]$,

$$\frac{K_{a1}}{[H_3O^+]} = \frac{[N]}{[A]} \quad \text{and} \quad \frac{K_{a2}}{[H_3O^+]} = \frac{[B]}{[N]} \qquad (26.5)$$

Letting $pK_{a1} = -\log K_{a1}$, and $pK_{a2} = -\log K_{a2}$,

$$pK_{a1} - pH = \log\left(\frac{[A]}{[N]}\right) \qquad (26.6a)$$

and

$$pH - pK_{a2} = \log\left(\frac{[B]}{[N]}\right) \qquad (26.6b)$$

These two equations show that the relative concentrations of A, B, and N depend on the relationship between the pH of the solution and the respective pK_a values. This point is illustrated in Fig. 26.1, which is a plot of the relative amounts of each form *vs.* pH. At low pH (pH $\ll pK_{a1}$), form A predominates, as shown by Eq. 26.6a. As the pH is raised, the concentration of A falls and that of N grows. As the pH is raised further, the concentration of form N falls and that of B grows. At high pH, form B predominates, as required by Eq. 26.6b. As Fig. 26.1 demonstrates, form N has its maximum concentration at a pH value higher than pK_{a1} and lower than pK_{a2}. In fact, the pH at which form N has a maximum concentration—the isoelectric point, p*I*—is the average of the two pK_a values:

$$\text{isoelectric point} = pI = \frac{pK_{a1} + pK_{a2}}{2} \qquad (26.7)$$

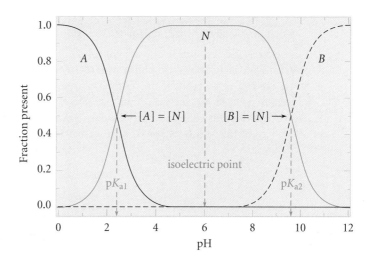

Figure 26.1   *Variation in the concentrations of the three forms of alanine shown in Eq. 26.3 as a function of pH. ——, concentration of the conjugate acid A; ----, concentration of the conjugate base B; ——, concentration of the neutral zwitterionic form N. The dissociation constant of A is K_{a1}, and that of B is K_{a2} (Eq. 26.4). Notice that the concentration of the neutral form N is a maximum at a pH that is halfway between pK_{a1} and pK_{a2}; this pH is the isoelectric point.*

(See Problem 26.9.) Thus, alanine, with pK_a values of 2.3 and 9.7 (Table 26.1), has an isoelectric point of $(2.3 + 9.7)/2 = 6.0$.

The significance of the isoelectric point is that it indicates not only the pH value at which a solution of the amino acid contains the greatest amount of neutral form N but also the sign of the net charge on the amino acid at *any* pH. For example, at a pH value less than the isoelectric point, more molecules of an amino acid are in form A than in form B; in this situation, the amino acid is said to be *positively charged*. At a pH value greater than the isoelectric point, more molecules of an amino acid are in form B than in form A. In this situation, the amino acid is said to be *negatively charged*. Sec. 26.3C will show why a knowledge of the net charge can be useful.

When an amino acid has a side chain containing an acidic or basic group, the isoelectric point is markedly changed. The amino acid lysine (Lys), for example, has a basic side-chain amino group as well as its α-amino and carboxy groups.

$$\overset{+}{H_3N}\!-\!CH\!-\!CO_2^- \underset{H_3O^+}{\overset{H_2O}{\rightleftarrows}} H_2N\!-\!CH\!-\!CO_2^- \underset{H_3O^+}{\overset{H_2O}{\rightleftarrows}} H_2N\!-\!CH\!-\!CO_2^- \qquad (26.8)$$

$$\underset{\overset{|}{^+NH_3}}{\underset{|}{(CH_2)_4}} \qquad\qquad \underset{\overset{|}{^+NH_3}}{\underset{|}{(CH_2)_4}} \qquad\qquad \underset{\overset{|}{NH_2}}{\underset{|}{(CH_2)_4}}$$

$$\qquad A \qquad\qquad\qquad\qquad N \qquad\qquad\qquad\qquad B$$

The isoelectric point of lysine is 9.82, which is the average of its two *highest* pK_a values of 9.12 and 10.53. At the isoelectric point of lysine, equal amounts of forms A and B of lysine are present, and form N has its maximum concentration. The lowest pK_a of lysine—the pK_a of the carboxylic acid group—doesn't enter the picture because neither of the equilibria involving the neutral form N involves ionization of the carboxylic acid group.

Let's compare the charge state of alanine and lysine at pH 6. Because pH 6 is the isoelectric point of alanine, its charge is zero. Because pH 6 is much lower than the isoelectric point of lysine, the net charge on lysine molecules is positive at this pH. Amino acids with high isoelectric points are classified as *basic amino acids*. Lysine and arginine are the two most basic of the common naturally occurring amino acids (Table 26.1). As indicated by its isoelectric point, arginine is the more basic of the two. Its side chain carries the basic guanidino group, the conjugate acid of which has a pK_a of 12.5. The basicity of this group is a consequence of the fact that its conjugate acid is resonance-stabilized.

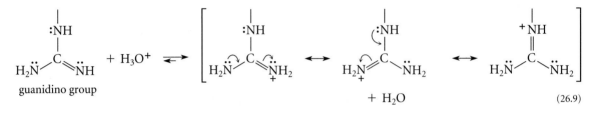

guanidino group

$$(26.9)$$

The amino acids aspartic acid (Asp) and glutamic acid (Glu) have carboxylic acid groups on their side chains and have low isoelectric points. The isoelectric point of aspartic acid, for example, is 2.76, the average of its two *lowest* pK_a values. (You should show why this is reasonable.) Amino acids with low isoelectric points are classified as *acidic amino acids*. Molecules of an acidic amino acid carry a net negative charge at pH 6. Aspartic acid and glutamic acid are the two most acidic of the common naturally occurring amino acids.

Amino acids with isoelectric points near 6, such as glycine or alanine, are classified as *neutral amino acids*. Of course, pH 6 is not *exactly* neutral; "neutral" amino acids are actually slightly acidic because the carboxylic acid group is somewhat more acidic than the amino group is basic.

Let's summarize the charge situation in basic, neutral, and acidic amino acids at pH 6:

Principal forms at pH 6:

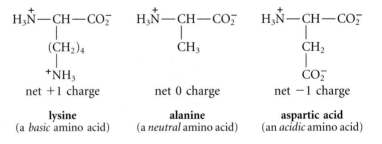

lysine	alanine	aspartic acid
(a *basic* amino acid)	(a *neutral* amino acid)	(an *acidic* amino acid)

Peptides with both acidic and basic groups also have isoelectric points. We can tell by inspection whether a peptide is acidic, basic, or neutral by examining the number of acidic and basic groups that it contains. A peptide with more amino and guanidino groups than carboxylic acid groups, for example, will have a high isoelectric point. Conversely, a peptide with more carboxylic acid groups than amino or guanidino groups will have a low isoelectric point.

PROBLEMS

*26.6   (a) Point out the ionizable groups of the amino acid *tyrosine* (Table 26.1).
         (b) What is the net charge on tyrosine at pH 6? How do you know?
         (c) Draw the structure of the major form(s) of tyrosine present at this pH.

26.7   (a) Point out the ionizable groups of the amino acid *histidine* (Table 26.1).
         (b) Draw all the acid-base equilibria for histidine.
         (c) What is the net charge on histidine at pH 6? How do you know?
         (d) Of the forms you drew in (b), which are the major one(s) present at this pH?

26.8   Classify the following peptides as acidic, basic, or neutral. What is the net charge on each peptide at pH = 6?
         *(a) Gly-Leu-Val
         (b) Leu-Trp-Lys-Gly-Lys
         *(c) *N*-acetyl-Asp-Val-Ser-Arg-Arg (*N*-acetyl means that the terminal amino group of the peptide is acetylated.)
         (d) Glu-Lys-Asp-Ala-Phe-Ile

*26.9   By definition the isoelectric point of an amino acid is that pH at which $[A] = [B]$ (Eq. 26.3). Use this condition, along with Eq. 26.5, to derive Eq. 26.7, the formula for the isoelectric point of a neutral amino acid.

C. Separations of Amino Acids and Peptides Using Acid-Base Properties

Isoelectric points are often used to design separations of amino acids and peptides. Consider, for example, the water solubilities of amino acids and peptides. Most peptides and amino acids, like carboxylic acids and amines, are most soluble when they carry a net charge and least soluble in their neutral forms. Thus, some peptides, proteins, and amino acids precipitate from water when the pH is adjusted to their isoelectric points. These same compounds are more soluble at pH values far from their isoelectric points, because they carry a net charge at these pH values.

A separation technique used a great deal in amino acid and peptide chemistry is **ion-exchange chromatography**. This method, too, depends on the isoelectric points of amino acids and peptides. In this technique, a hollow tube or column is filled with a buffer solution in which is suspended a finely powdered, insoluble polymer called an *ion-exchange resin*. The resin bears acidic or basic groups. One popular resin, for example, is a sulfonated polystyrene—a polystyrene in which the phenyl rings contain strongly acidic sulfonic acid groups. If the pH of the buffer is such that the sulfonic acid groups are ionized, the resin bears a negative charge. This charge is the key to ion-exchange separations.

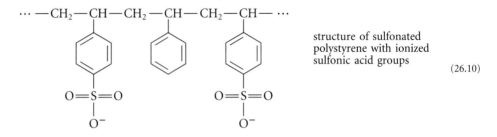

structure of sulfonated
polystyrene with ionized
sulfonic acid groups

(26.10)

The way ion-exchange chromatography works is shown in Fig. 26.2. Suppose the buffer in the column has a pH of 6, and a solution containing a mixture of the two amino acids Val and Lys in the same buffer is added to the top of the column. Buffer is then allowed to flow through the column; a frit (a porous glass plate) keeps the resin from washing out. Since valine has zero charge at this pH, it is not attracted by the ionic groups on the column, and is washed through the column with a relatively small volume of buffer. Lysine, on the other hand, bears a net positive charge at pH 6, and is strongly attracted to the negatively charged resin. Because of this attraction, lysine is retained on the column and emerges only after a considerably larger amount of buffer has passed through the column. The two amino acids are thus separated. Thus, whether an amino acid or peptide is adsorbed by the column depends on its charge—which, in turn, depends on the relationship of its isoelectric point to the pH of the buffer.

In the experiment shown in Fig. 26.2, the ion-exchange resin is negatively charged and adsorbs cations; it is therefore called a *cation-exchange resin*. Resins that bear positively charged pendant groups adsorb anions, and are called *anion-exchange resins*.

Ion Exchange and Water Softeners

Ion exchange has very important commercial applications, for example, in water treatment. Commercial water softeners contain cation-exchange resins much like the one used in this example, which adsorb the more highly charged calcium and magnesium ions in hard water and replace them with sodium ions with which the column is supplied. When the supply of sodium ions is exhausted, the column has to be flushed extensively, or regenerated, with concentrated NaCl solution to replace the adsorbed calcium and magnesium ions with sodium ions.

Problems

*26.10   (a) How might the structure of the resin in Eq. 26.10 be altered to make the resin an anion exchanger (that is, an anion-binding resin)? (*Hint:* What type of organic functional group can carry a positive charge?)

(b) Predict the order of elution of the following peptides from an anion-exchange resin at pH 6: A-V-G, D-E-E-G, D-N-N-G. Explain your reasoning.

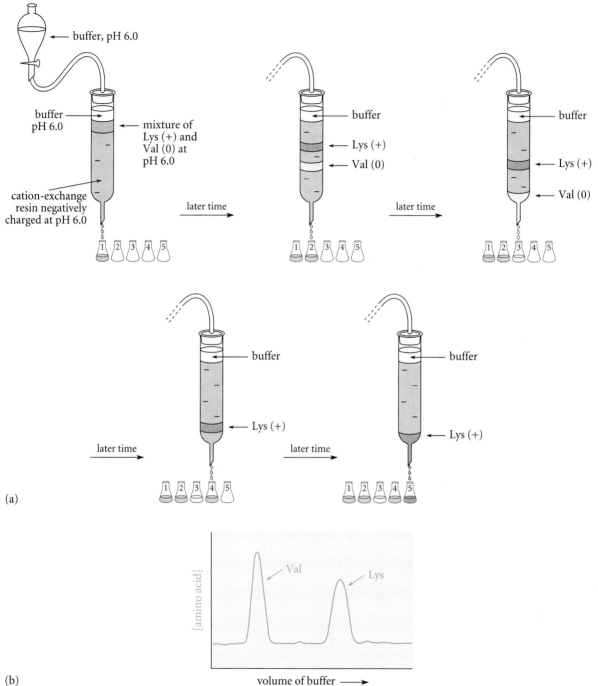

(a)

(b)

Figure 26.2   *(a) Diagram of the cation-exchange separation of valine and lysine. (b) Amino acid concentration as a function of the volume of buffer passed through the cation-exchange column. Lysine, which carries a positive charge at the pH of the buffer, is attracted to the negatively charged resin and moves through the column more slowly than valine, which carries zero charge.*

26.11 A mixture of *N*-acetyl-Leu-Gly, Lys-Gly-Arg, and Lys-Gly-Leu is applied to a sulfonated polystyrene cation-exchange column at a buffer pH of 6.0. Predict the order in which these three peptides will elute from the column and explain your reasoning. (See Problem 26.8c for an explanation of the *N*-acetyl nomenclature.)

26.4 Synthesis and Enantiomeric Resolution of α-Amino Acids

A. Alkylation of Ammonia

Some α-amino acids can be prepared by alkylation of ammonia with α-bromo carboxylic acids.

$$CH_3CH_2\overset{\underset{\displaystyle |}{CH_3}}{CH}-\overset{\underset{\displaystyle |}{Br}}{CH}-CO_2H + 2NH_3 \xrightarrow{\text{(large excess)}} CH_3CH_2\overset{\underset{\displaystyle |}{CH_3}}{CH}-\overset{\underset{\displaystyle |}{^+NH_3}}{CH}-CO_2^- + \overset{+}{N}H_4\,Br^- \qquad (26.11)$$

isoleucine
(49% yield)

This is an S_N2 reaction in which ammonia acts as the nucleophile. (Recall from Sec. 22.3D that α-halo carbonyl compounds are very reactive in S_N2 reactions.) Alkylation of ammonia is generally not an acceptable method for preparing primary amines because ammonia can be alkylated more than once to give complex mixtures (Sec. 23.7A). However, the use of a large excess of ammonia in this synthesis favors monoalkylation. Furthermore, amino acids are less reactive toward alkylating agents than simple alkylamines because the amino groups of amino acids are less basic, and therefore less nucleophilic, than ammonia and simple alkylamines, and because branching in amino acids retards further alkylation.

B. Alkylation of Aminomalonate Derivatives

One of the most widely used methods for preparing α-amino acids is a variation of the malonic ester synthesis (Sec. 22.6A). The malonic ester derivative used is one in which a protected amino group is already in place: diethyl α-acetamidomalonate. This derivative is treated with sodium ethoxide in ethanol to form the enolate ion, which is then alkylated with an alkyl halide (benzyl chloride in the following example):

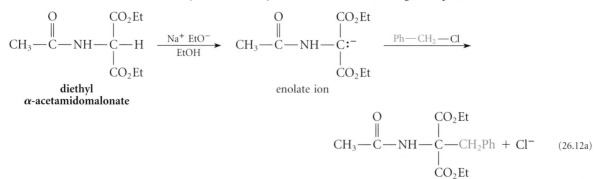

The resulting compound is then treated with hot aqueous HCl or HBr. This treatment accomplishes three things: First, the ester groups are hydrolyzed to carboxylic acids (Sec. 21.7A), yielding a substituted malonic acid (draw it!). Second, the malonic acid derivative decarboxylates under the reaction conditions (Sec. 20.11). Third, the acetamido group, an amide, is also hydrolyzed (Sec. 21.7B). Neutralization gives the α-amino acid.

$$
\underset{\displaystyle \begin{matrix} \text{O} \\ \parallel \\ \text{CH}_3\!-\!\text{C}\!-\!\text{NH}\!-\!\underset{\displaystyle \text{CO}_2\text{Et}}{\overset{\displaystyle \text{CO}_2\text{Et}}{\text{C}}}\!-\!\text{CH}_2\text{Ph} \end{matrix}}{} \xrightarrow[\text{heat}]{\text{H}_2\text{O, HBr}} \text{CH}_3\text{CO}_2\text{H} + 2\text{EtOH} + \text{CO}_2 + \underset{\displaystyle \text{CO}_2\text{H}}{\overset{\displaystyle +}{\text{H}_3\text{N}}}\!-\!\text{CH}\!-\!\text{CH}_2\text{Ph} \;\; \text{Br}^-
$$

phenylalanine hydrobromide
(65% yield after
neutralization)

(26.12b)

C. Strecker Synthesis

An important method for the synthesis of carboxylic acids is the hydrolysis of nitriles (Sec. 21.7C, 21.11). Thus, α-amino nitriles can be hydrolyzed to give α-amino acids. α-Amino nitriles, in turn, are prepared by treatment of aldehydes with ammonia in the presence of cyanide ion.

$$
\text{CH}_3\text{CH}\!=\!\text{O} + \overset{+}{\text{N}}\text{H}_4\,\text{Cl}^- + \text{Na}^+\,\text{CN}^- \longrightarrow \underset{\substack{\text{2-aminopropanenitrile} \\ (\text{an } \alpha\text{-amino nitrile})}}{\text{CH}_3\!-\!\overset{\displaystyle \text{NH}_2}{\underset{\displaystyle \text{CN}}{\text{CH}}}} + \text{NaCl} + \text{H}_2\text{O} \xrightarrow[\text{H}_2\text{O}]{\substack{\text{heat} \\ \text{HCl}}} \xrightarrow{^-\text{OH}}
$$

acetaldehyde

$$
\text{NH}_3 + \text{CH}_3\!-\!\underset{\displaystyle ^+\text{NH}_3}{\text{CH}}\!-\!\overset{\displaystyle \text{O}}{\overset{\displaystyle \parallel}{\text{C}}}\!-\!\text{O}^- \qquad (26.13)
$$

alanine
(52–60% yield)

This preparation of α-amino acids is called the **Strecker synthesis.**

The mechanism of α-amino nitrile formation probably involves an imine intermediate.

$$
\text{CH}_3\!-\!\text{CH}\!=\!\text{O} + \overset{..}{\text{N}}\text{H}_3 \xrightarrow{-\text{H}_2\text{O}} \underset{\text{an imine}}{\text{CH}_3\!-\!\text{CH}\!=\!\overset{..}{\text{N}}\text{H}} \rightleftharpoons \text{CH}_3\!-\!\text{CH}\!=\!\overset{+}{\text{N}}\text{H}_2 + \text{H}_2\text{O} \qquad (26.14a)
$$

The conjugate acid of the imine reacts with cyanide under the conditions of the reaction to give the α-amino nitrile.

$$CH_3—CH{=}\overset{+}{N}H_2 + :\bar{C}N \longrightarrow CH_3—\underset{\underset{CN}{|}}{CH}—\overset{..}{N}H_2 \qquad (26.14b)$$

The addition of cyanide to an imine is analogous to the formation of a cyanohydrin from an aldehyde or ketone (Sec. 19.7A).

$$HCN + CH_3—CH{=}O \longrightarrow CH_3—\underset{\underset{CN}{|}}{CH}—OH \qquad (26.15)$$

PROBLEM

26.12   Indicate which of the methods in this section could be used to prepare each of the following amino acids. For each method that can be used, give an equation. For each case in which a method would not work, give a reason.
   *(a) α-phenylglycine   (b) leucine

D. Enantiomeric Resolution of α-Amino Acids

Amino acids synthesized by common laboratory methods are *racemic*. Since many applications require pure enantiomers, the racemic mixtures must be resolved. As useful as the diastereomeric salt method is (Sec. 6.8), it can be tedious and time-consuming. An alternative approach to the preparation of enantiomerically pure amino acids, and one that is used industrially, is the synthesis of amino acids by microbiological fermentation. Some cultures of microorganisms can be used to produce industrial quantities of certain amino acids in the natural L form.

Certain enzymes—biological catalysts—can be used to resolve racemic amino acids into enantiomers. For example, a preparation of the enzyme *acylase* from hog kidney selectively catalyzes the hydrolysis of *N*-acetyl-L-amino acids and leaves the corresponding D isomers unaffected. Thus, treatment of the *N*-acetylated racemate with this enzyme affords the free L-amino acid only:

$$H_2O + CH_3—\overset{\overset{O}{\|}}{C}—NH—\underset{\underset{CH_3}{|}}{CH}—CO_2H \xrightarrow{\text{hog-kidney acylase}}$$

N-acetyl-D,L-alanine
(racemate)

$$\overset{+}{H_3N}\overset{\overset{H}{|}}{\underset{\underset{CO_2^-}{|}}{\text{———}}}CH_3 + CH_3\overset{\overset{H}{|}}{\underset{\underset{CO_2H}{|}}{\text{———}}}NH—\overset{\overset{O}{\|}}{C}—CH_3 + CH_3CO_2H \qquad (26.16)$$

L-(+)-alanine   *N*-acetyl-D-alanine
(insoluble in EtOH)   (soluble in EtOH)

In this example the liberated L-alanine is precipitated from ethanol; the *N*-acetyl-D-alanine remains in solution, from which it can be recovered and hydrolyzed in aqueous acid to D-alanine.

The enzyme differentiates between the two enantiomers of *N*-acetylalanine because it is an *optically pure chiral* compound. Recall that enantiomers have different reactivities with chiral reagents (Sec. 7.7A).

26.5   Acylation and Esterification Reactions of Amino Acids

Amino acids undergo many of the reactions characteristic of both amines and carboxylic acids. *Acylation* is an amine reaction that is very important in amino acid chemistry. Acylation by acetic anhydride is shown in Eq. 26.17.

$$\overset{+}{H_3}N-CH-CO_2^- + CH_3-\overset{\displaystyle O}{\overset{\|}{C}}-O-\overset{\displaystyle O}{\overset{\|}{C}}-CH_3 \xrightarrow{CH_3CO_2H} \xrightarrow{H_2O}$$

$$\underset{\displaystyle \overset{|}{CH_2}}{\underset{\displaystyle \overset{|}{CH(CH_3)_2}}{}}$$

acetic anhydride

leucine

$$CH_3-\overset{\displaystyle O}{\overset{\|}{C}}-NH-CH-CO_2H \qquad (26.17)$$
$$\underset{\displaystyle \overset{|}{CH_2}}{}$$
$$\underset{\displaystyle \overset{|}{CH(CH_3)_2}}{}$$

N-acetylleucine
(85–95% yield)

Acylation by acid chlorides is also a useful reaction (Sec. 21.8A).

Amino acids, like ordinary carboxylic acids, are easily esterified by heating with an alcohol and a mineral acid catalyst (acid-catalyzed esterification; Sec. 20.8A).

$$H_2N-\!\!\left\langle\!\!\bigcirc\!\!\right\rangle\!\!-\overset{\displaystyle O}{\overset{\|}{C}}-OH + C_2H_5OH \xrightarrow[\text{heat}]{H_2SO_4} \xrightarrow{NaHCO_3} H_2N-\!\!\left\langle\!\!\bigcirc\!\!\right\rangle\!\!-\overset{\displaystyle O}{\overset{\|}{C}}-OC_2H_5 + H_2O$$

p-aminobenzoic acid (PABA)   ethyl *p*-aminobenzoate
                                        (benzocaine, a local anesthetic)   (26.18)

PROBLEMS

26.13   Give the major product expected when
    *(a)   leucine is treated with *p*-toluenesulfonyl chloride (tosyl chloride).
    (b)   alanine is heated in methanol solvent with HCl catalyst.

*26.14   If the hydrochloride salt of glycine methyl ester is neutralized and allowed to stand in solution, a polymer forms. If the hydrochloride itself is allowed to stand, nothing happens. Explain these observations.

26.6 Determination of Peptide Structure

A. Hydrolysis of Peptides; Amino Acid Analysis

This section covers several reactions that are used to determine the structures of unknown peptides. An important reaction used for this purpose is hydrolysis of the peptide (amide) bonds of a peptide to give its constituent amino acids (Sec. 21.7B). This hydrolysis is typically carried to completion in 6 M aqueous HCl for 20–24 hours at 110 °C.

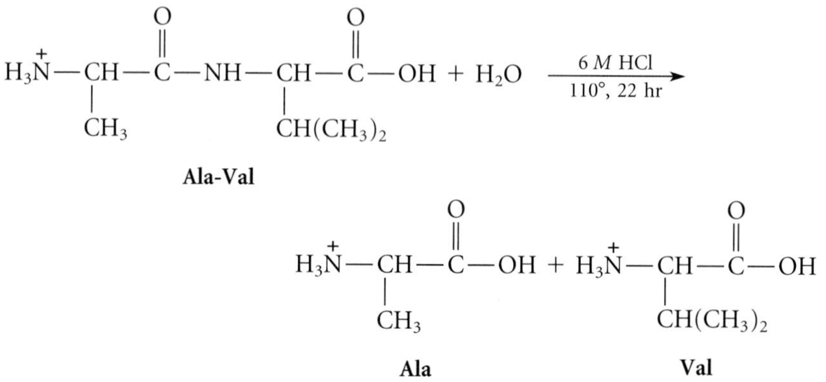

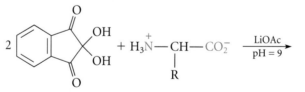

Ala-Val

$$H_3\overset{+}{N}-CH-\overset{\displaystyle O}{\overset{\|}{C}}-OH + H_3\overset{+}{N}-CH-\overset{\displaystyle O}{\overset{\|}{C}}-OH \qquad (26.19)$$

Ala   **Val**

When a peptide or protein is hydrolyzed, the product amino acids can be separated, identified, and quantitated by a technique called **amino acid analysis**. In one variation of this method, the amino acids in the hydrolyzed mixture are separated by passing them through a cation-exchange column (see Fig. 26.2) under very carefully defined conditions. The time at which each amino acid emerges from the column is accurately known from standards. As each amino acid emerges, it is mixed with *ninhydrin*, a reagent that reacts with primary amines and α-amino acids to give a dye called *Ruhemann's purple*, which has an intense blue-violet color.

☞
Study Guide Link:
26.3
Reaction of α-Amino Acids with Ninhydrin

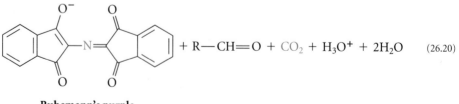

ninhydrin

Ruhemann's purple
$\lambda_{max} = 570$ nm

The intensity of the resulting color is proportional to the amount of the amino acid present. The color intensity, and therefore the amount of each amino acid, is recorded

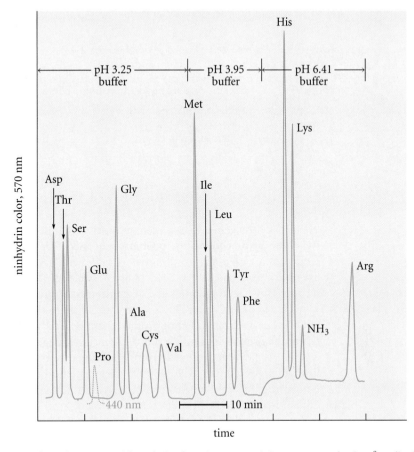

Figure 26.3   *Chart trace from the amino acid analysis of a mixture containing 10 nanomoles (10^{-8} mol) of each amino acid. The pH values at the top of the chart show the pH of the buffers used to elute (wash) the amino acids from the cation-exchange column. The peaks are from the color resulting from mixing the column effluent with ninhydrin.*

STUDY GUIDE LINK:
26.4
Amino Acid Analysis

as a function of time. Figure 26.3 shows the recorder trace obtained from a mixture of 10 nanomoles (10^{-8} mole) of each α-amino acid. The trace obtained from the analysis of a peptide of unknown composition contains a peak corresponding to each amino acid present; the area of the peak is proportional to the amount of the amino acid. Thus, by hydrolysis, reaction with ninhydrin, and quantitation of the color produced, the *identities* and *relative amounts* of the amino acid residues in a peptide can be determined.

For example, imagine that a hypothetical peptide *P* has been hydrolyzed and subjected to amino acid analysis, and that the results are as follows:

P: Ala_3,Arg,(Asp or Asn),Gly_2,His,Ile,Leu,Lys,Met,NH_3,Phe,Pro,Trp,Tyr,Val

According to this analysis, the peptide contains three times as much Ala and twice as much Gly as Arg, His, Lys, or the other amino acids present. The absolute number of each amino acid residue is not known unless, in addition, the molecular mass of the peptide is known. Notice also that the relative order of the amino acid residues within the peptide is also not known. In this sense, amino acid analysis is to the amino acid composition of a peptide as elemental analysis is to the molecular formula of an organic compound.

26.15 *(a) Explain why acid hydrolysis followed by amino acid analysis does not distinguish between Asp and Asn. (*Hint:* Why is ammonia present in the peptide *P* above?)

  (b) What other pair of amino acids are not differentiated by amino acid analysis?

B. Sequential Degradation of Peptides

The actual arrangement, or sequential order, of amino acid residues in a peptide is called its **amino acid sequence** or **primary sequence**. There are a large number of possible sequences for even a relatively small peptide with a given amino acid composition. For example, more than 10^{14} sequences are possible for peptide *P* described in the previous section! How can the amino acid sequence of a peptide or protein be determined?

This apparently complex problem is actually solved rather easily. It is possible to remove one residue at a time from the amino end of a peptide, identify it, and then repeat the process sequentially on the remaining peptide.

The standard technique for implementing this strategy is called the **Edman degradation**, after Per Edman, a Swedish biochemist who devised the method in 1952. In an Edman degradation, the peptide is treated with *phenyl isothiocyanate* (often called the **Edman reagent**). The peptide reacts with the Edman reagent at its amino groups to give thiourea derivatives. Although reaction with the Edman reagent also occurs at the side-chain amino groups of lysine residues (see Problem 26.30), only the reaction at the terminal amino group is relevant to the degradation. (In this and subsequent equations, the abbreviation Pep^N is used for the amino-terminal part of a peptide and the abbreviation Pep^C for the carboxy-terminal part.)

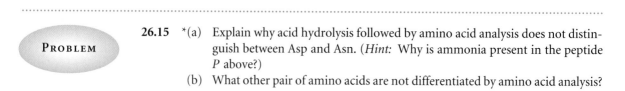

(26.21a)

thiourea derivative

This reaction is exactly analogous to the reaction of amines with *isocyanates*, the oxygen analogs of *isothiocyanates* (Eq. 23.70, p. 1149). Any remaining phenyl isothiocyanate is removed, and the modified peptide is then treated with anhydrous trifluoroacetic acid. As a result of this treatment, the sulfur of the thiourea, which is nucleophilic, displaces the amino group of the adjacent residue to yield a five-membered heterocycle called a *thiazolinone*; the other product of the reaction is *a peptide that is one residue shorter*.

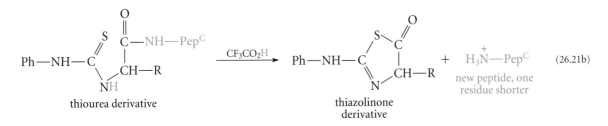

thiourea derivative   CF$_3$CO$_2$H   thiazolinone derivative   +   H$_3$N—Pep^C (26.21b)

new peptide, one residue shorter

When treated subsequently with aqueous acid, the thiazolinone derivative forms an isomer called a **phenylthiohydantoin**. This probably occurs by reopening of the thiazolinone to the thiourea, followed by ring formation involving the thiourea nitrogen. Notice in this and the previous equation the intramolecular formation of five-membered rings.

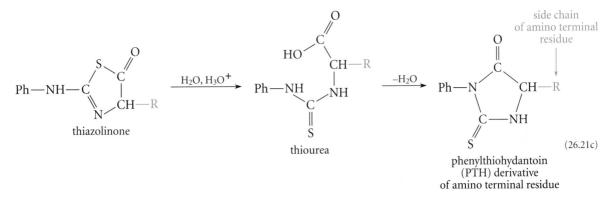

thiazolinone   H$_2$O, H$_3$O$^+$ thiourea   –H$_2$O   side chain of amino terminal residue

phenylthiohydantoin (PTH) derivative of amino terminal residue   (26.21c)

(You should fill in the details of these reactions using the curved-arrow formalism.)

Because the phenylthiohydantoin (PTH) derivative carries the characteristic side chain of the amino-terminal residue, identification of the PTH identifies the amino acid residue that was removed. Methods for identifying PTH derivatives by chromatography are well established. The peptide liberated in Eq. 26.21b can be subjected in turn to the Edman degradation again to yield the PTH derivative of the next amino acid and a new peptide that is shorter by yet another residue.

In principle, the Edman degradation can be continued indefinitely for as many residues as necessary to define completely the sequence of the peptide. In practice, because the yields at each step are not perfectly quantitative, an increasingly complex mixture of peptides is formed with each successive step in the cleavage, and after a number of such steps the results become ambiguous. Hence, the number of residues in a sequence that can be determined by the Edman method is limited. Nevertheless, instruments are now in use that can apply Edman chemistry to structure determination of peptides in a highly standardized, automated, and reproducible form. In such instruments the sequential degradation of twenty residues is common, and the degradation of as many as sixty or seventy amino acid residues is sometimes possible. The application of the Edman degradation to the determination of peptide structure is illustrated in the following study problem.

STUDY PROBLEM 26.1

A pentapeptide *A* contains the following amino acids: Gly, Ile, Leu, Lys, Phe. Successive cycles of the Edman degradation give PTH derivatives with the R groups shown in the table below. Propose a structure for *A*.

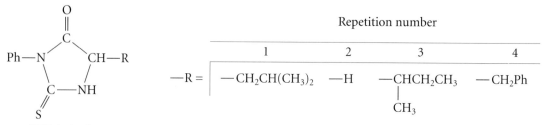

PTH derivatives

	Repetition number			
	1	2	3	4
$-R =$	$-CH_2CH(CH_3)_2$	$-H$	$-CHCH_2CH_3$	$-CH_2Ph$
			$\quad\quad\;\; CH_3$	

Solution Each side chain R in the table above corresponds to the side chain of the amino acid liberated in the given cycle of the Edman degradation, beginning at the amino terminus. R for cycle 1 corresponds to the side chain of Leu; cycle 2, Gly; cycle 3, Ile; and cycle 4, Phe. Hence, the sequence of *A* is Leu-Gly-Ile-Phe-Lys. Note that Lys is inferred to be at the carboxy terminus because it is the only amino acid not liberated in the Edman degradation.

PROBLEMS

26.16 Using the curved-arrow formalism, write in detail the mechanisms for the reactions in *(a) Eq. 26.21a; (b) Eq. 26.21b; *(c) Eq. 26.21c.

*26.17 Some peptides found in nature have an amino-terminal acetyl group (color):

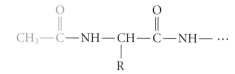

Can these peptides undergo the Edman degradation? Explain.

C. Specific Cleavage of Peptides

Most common proteins contain hundreds of amino acid residues. Hence the Edman chemistry, which is limited by yield to 20–60 consecutive residues, cannot be used to determine the structure of such proteins in a single set of sequential degradations. The amino acid sequence of most large proteins is determined by breaking the protein into smaller peptides and sequencing these peptides individually. Then the sequence of the protein is reconstructed from the sequences of the peptides. In other words, the large sequencing problem is divided into a series of smaller sequencing problems. When breaking a larger protein into smaller peptides, it is desirable to use reactions that cleave the protein in high yield at well-defined points so that a relatively small number of peptides are obtained. (Cleavages at random points in the protein chain would give complex, difficult-to-separate mixtures.) This section describes two of the most common methods used by protein chemists to cleave peptides at specific amino acid residues into smaller fragments.

Peptide Cleavage at Methionine with Cyanogen Bromide When a peptide reacts with *cyanogen bromide* (Br—C≡N) in aqueous HCl, a peptide bond is cleaved specifically at the carboxy side of each methionine residue.

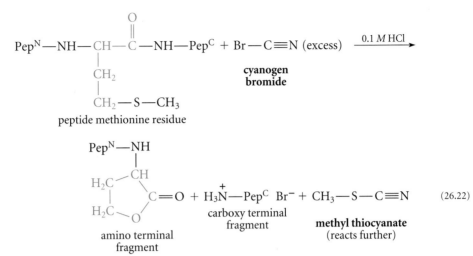

Note that the amino terminal fragment from the cleavage shown in Eq. 26.22 has a carboxy terminal *homoserine lactone* residue instead of the starting methionine.

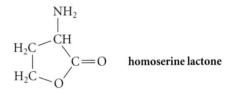

Methionine is a relatively rare amino acid; hence, when a typical protein is cleaved with BrCN, relatively few cleavage peptides are obtained, and all of them are derived from cleavage at methionine residues.

Why does this cleavage work? Cyanogen bromide has the character of an acid halide. Although it can in principle react with any nucleophilic amino acid side chain, under the conditions of the reaction only its reaction at methionine leads to a peptide cleavage, for reasons that are apparent in the mechanism outlined below. (Fill in the details of each step using the curved-arrow formalism.)

The sulfur in the methionine side chain acts as a nucleophile, displacing bromide from cyanogen bromide to give a type of *sulfonium ion* (Sec. 11.5A).

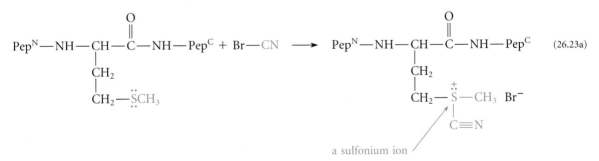

The sulfonium ion, with its electron-withdrawing cyanide, is an excellent leaving group, and is displaced by the oxygen of the neighboring amide bond to form a five-membered ring. (The amide oxygen is normally not a very good nucleophile, but reactions that occur with the formation of five- or six-membered rings are particularly rapid; Sec. 11.6.)

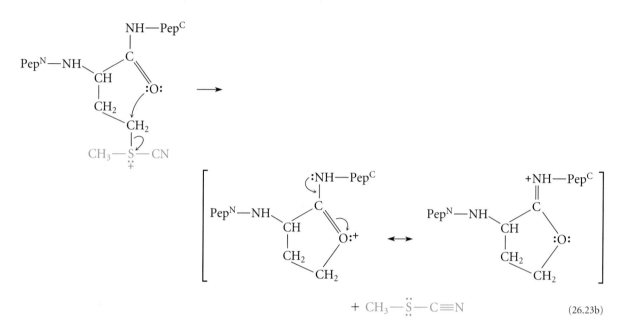

$$(26.23b)$$

This step shows why the cleavage is specific for methionine: only methionine has a side chain that can form a five-membered ring by such a mechanism.

The last mechanistic step in the cleavage is hydrolysis of the ion formed in Eq. 26.23b. It is in this step that the peptide bond is actually broken:

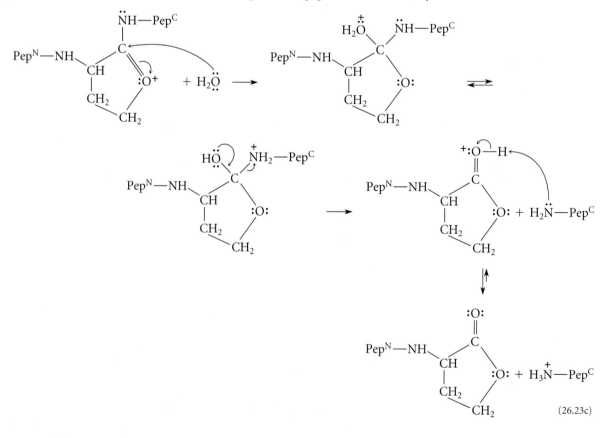

$$(26.23c)$$

Peptide Cleavage with Proteolytic Enzymes The cyanogen bromide cleavage is one of relatively few nonenzymatic cleavages of peptides. In contrast, a number of enzymes catalyze the hydrolysis of peptide bonds at specific points in an amino acid sequence. Such peptide-hydrolyzing enzymes are called **proteases**, **peptidases**, or **proteolytic enzymes**. One of the most widely used proteases is the enzyme *trypsin*. This enzyme catalyzes the hydrolysis of peptides or proteins at the carbonyl group of arginine or lysine residues, provided that these residues (a) are not at the amino end of the protein, and (b) are not followed by a proline residue.

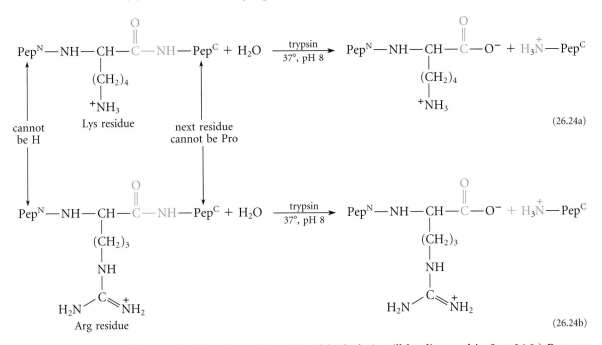

(26.24a)

(26.24b)

(The mechanism of trypsin-catalyzed hydrolysis will be discussed in Sec. 26.9.) Because trypsin catalyzes the hydrolysis of peptides at internal rather than terminal residues, it is called an **endopeptidase**. (Enzymes that cleave peptides only at terminal residues are termed **exopeptidases**.)

The use of both cyanogen bromide and trypsin-catalyzed hydrolysis in the determination of peptide structure is illustrated in the following study problem.

....

STUDY PROBLEM 26.2

Treatment of the peptide *P* (page 1283) with cyanogen bromide gives two new peptides, *A* and *B*, with the following compositions determined by amino acid analysis:

composition of A: Ala,Gly$_2$,His,Ile,Leu,Lys,Phe,Pro,Tyr,homoserine lactone
composition of B: Ala$_2$,Arg,(Asp or Asn),Val,Trp

Treatment of peptide *A* with trypsin gives two new peptides, *C* and *D*.

composition of C: Gly,Ile,Leu,Phe,Tyr,Lys
composition of D: Ala,Gly,His,Pro,homoserine lactone

Treatment of peptide *B* with trypsin also gives two new peptides, *E* and *F*.

composition of E: Ala,Trp,Arg
composition of F: Ala,(Asp or Asn),Val

Deduce as much information as you can from these data about the primary sequence of the original peptide *P*.

Solution   First, the results of the cyanogen bromide cleavage suggest this partial structure:

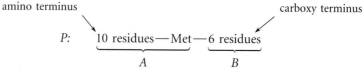

Peptide *A* must precede peptide *B* in the sequence because the homoserine lactone residue produced in the cyanogen bromide cleavage must occur at the *carboxy terminus* of the peptide at the "left" (amino-terminal side) of the cleavage point (Eq. 26.22).

Now consider the cleavage with trypsin. Because *C* and *D* both come from *A*, *D* must be derived from the carboxy terminus of *A* because both *A* and *D* contain homoserine lactone. This means that *C* must come from the amino terminal side of *A*. This is also reasonable because *C* contains the Lys residue; trypsin cleaves at the "right" (carboxy-terminal) side of Lys (Eq. 26.24a). The partial structure of *P* so far is:

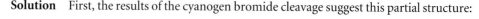

Peptide *E* comes from the "left" (amino-terminal) side of *B*, because this peptide contains an Arg residue; presumably the carboxy-terminal side of Arg is the cleavage point. It therefore follows that Arg is at the carboxy-terminal end of *E*. If peptide *E* comes from the "left" side of *B*, then *F* must come from the "right" side. Therefore, the partial structure of *P* can be refined to the following. (Cleavage points with cyanogen bromide and trypsin are shown.)

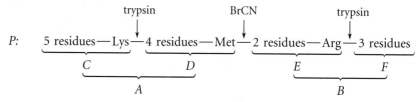

This is the most detail that can be obtained from the data given. In principle the sequence could be completed by applying the Edman method to the short peptides *C*–*F*.

Several enzymes besides trypsin are also used to cleave peptides. Chymotrypsin, a protein related to trypsin, is used to cleave peptides at amino acid residues with aromatic side chains and, to a lesser extent, residues with large hydrocarbon side chains. Thus, chymotrypsin cleaves peptides at Phe, Trp, Tyr, and occasionally Leu and Ile residues. Chymotrypsin and trypsin are mammalian digestive enzymes; their biological role, understandably, is to catalyze the hydrolytic breakdown of dietary proteins in the intestine. An important endopeptidase from a microorganism, *Staphylococcus aureus*, catalyzes the hydrolysis of peptides at glutamic acid residues. Thus, biochemists have an arsenal of different proteases that can be used to cleave proteins into peptides at specific sites.

*26.18   (a) What product would you expect from the reaction of cyanogen bromide with free amino acid groups in a protein, such as the side-chain amino group of lysine?

   (b) What conditions of the reaction with BrCN prevent such a reaction? Explain. (See Eq. 26.22.)

*26.19   A peptide Q has the following amino acid composition.

$$Q: \quad Arg_2, Ile_2, Glu, Gly_2, Leu, Lys, Phe, Pro, Ser, Trp$$

When Q was subjected to a cycle of the Edman degradation, the PTH derivative of leucine was formed, along with a new peptide. Treatment of Q with trypsin gave the following peptides. (Their individual sequences were determined by the Edman degradation.)

$$Q \xrightarrow{\text{trypsin}} \underset{A}{\text{Gly-Arg,}} \; \underset{B}{\text{Ile-Trp-Phe-Pro-Gly-Arg,}} \; \underset{C}{\text{Leu-Lys,}} \; \underset{D}{\text{Ser-Glu-Ile}}$$

Cleavage of Q with chymotrypsin gave the following peptides:

$$Q \xrightarrow{\text{chymotrypsin}} \begin{array}{l} E: \quad \text{partial sequence Leu-Lys-Gly} \dots ; \quad \text{and} \\ F: \quad \text{partial sequence Phe-Pro-Gly-Arg-Ser} \dots \end{array}$$

From these data construct the amino acid sequence of Q. Explain why the additional cleavage data from chymotrypsin are necessary to define the sequence. (This and the following problem illustrate the use of *overlapping peptides*, a technique frequently used in the sequencing of large proteins.)

26.20   A peptide R is cleaved with BrCN into two new peptides, A and B. Peptide A, when treated with trypsin, gives two peptides, C and D, with the following compositions: C: (Arg,Gly); D: (Ala_2,Leu,Trp). Peptide B is also hydrolyzed in the presence of trypsin to give two peptides, E and F, with the following compositions: E: (Gly,Lys); F: (Asp,homoserine lactone.) Peptide R is cleaved with the enzyme chymotrypsin to give a mixture of four peptides G, H, I, and J with the following compositions: G: (Ala_2,Leu); H: (Arg,Asp,Gly_2,Lys,Met,Trp); I: (Arg,Asp,Gly_2,Leu,Lys,Met,Trp); J: (contains only Ala). Deduce the complete sequence of peptide R from these data.

●

Solid-Phase Peptide Synthesis

Of the many procedures available for the synthesis of peptides, the most widely used are variations of an ingenious method called **solid-phase peptide synthesis**. In this method the carboxy-terminal amino acid is covalently anchored to an *insoluble* polymer, and the peptide is "grown" by adding one residue at a time to this polymer. Solutions containing the appropriate reagents are allowed to contact the polymer with shaking. At the conclusion of each step, the polymer containing the peptide is simply filtered away from the solution, which contains soluble by-products and impurities. The completed peptide is removed from the polymer by a reaction that cleaves its bond to the resin, just as a plant

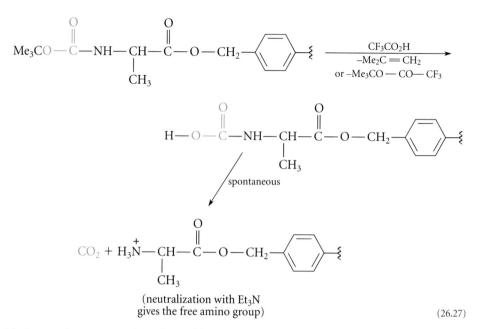

$$(26.27)$$

Notice two important points about this reaction. First, the carboxylic acid group exposed in this reaction (color) is of the carbamic acid type, and carbamic acids decarboxylate under acidic conditions (Eq. 20.43). Second, this deprotection step, after neutralization of the ammonium salt, exposes the free amino group of the resin-bound amino acid, which is used as a nucleophile in the next reaction.

Next comes the coupling of a second residue to the resin-bound alanine. Coupling of Boc-glycine to the free amino group of the resin-bound Ala is effected by the reagent *N,N'*-dicyclohexylcarbodiimide, or DCC. The result of this reaction is that the resin-bound Ala is acylated by the Boc-Gly, and a new peptide bond is formed.

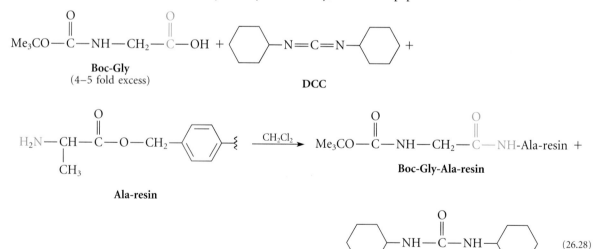

$$(26.28)$$

***N,N*-dicyclohexylurea (DCU)**

What is the role of DCC in this reaction? Addition of the amino acid carboxylic acid group to a double bond of DCC gives a derivative called an *O*-acylisourea.

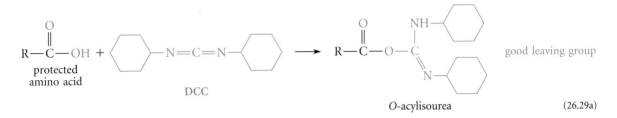

protected
amino acid

DCC

O-acylisourea

good leaving group

(26.29a)

This derivative behaves somewhat like an anhydride, and is an excellent acylating agent. It reacts in either of two ways. First, it can react directly with the amino group of the resin-bound amino acid to form the peptide bond. This reaction is an example of ester or anhydride aminolysis (Sec. 21.8B,C).

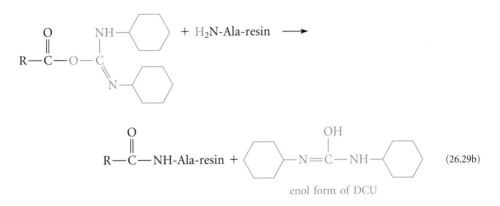

enol form of DCU

(26.29b)

The by-product of this reaction, the enol form of DCU, is transformed into DCU itself under the reaction conditions. (You should write the mechanistic details of each of these reactions using the curved-arrow formalism.)

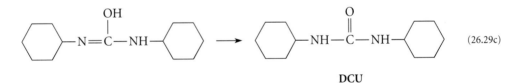

DCU

(26.29c)

The second way in which the *O*-acylisourea can react is with the carboxy group of another equivalent of the protected amino acid, giving an anhydride:

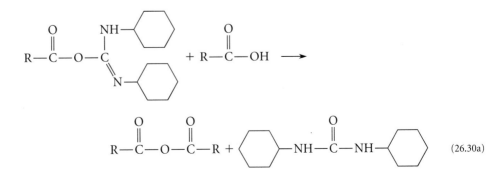

(26.30a)

This anhydride can also acylate the amino group of the resin-bound amino acid to complete the formation of the peptide bond.

$$R - \overset{\overset{\displaystyle O}{\|}}{C} - O - \overset{\overset{\displaystyle O}{\|}}{C} - R + H_2N\text{-Ala-resin} \longrightarrow$$

$$R - \overset{\overset{\displaystyle O}{\|}}{C} - NH - \text{Ala-resin} + R - \overset{\overset{\displaystyle O}{\|}}{C} - OH \qquad (26.30b)$$

Which of these acylation reactions actually occurs depends on the conditions actually used in the solid-phase peptide synthesis. In either case, the desired peptide bond is formed. The by-product DCU and the excess of the Boc-amino acid are simply washed away from the peptide because they are not attached covalently to the resin.

Completion of the peptide synthesis requires deprotection of the resin-bound dipeptide in the usual way and a final coupling step with Boc-Phe and DCC:

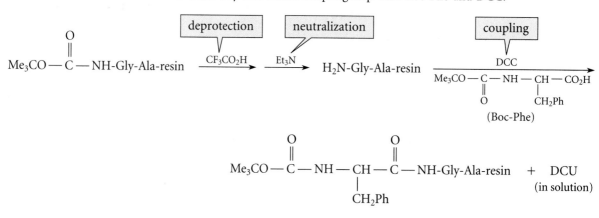

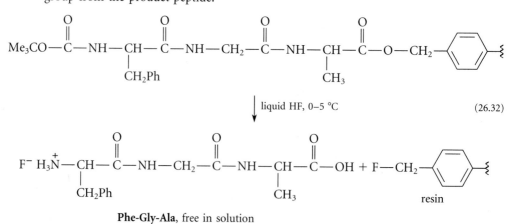

Boc-Phe-Gly-Ala-resin   (26.31)

Once all the peptide bonds in the desired tripeptide are assembled, the completed peptide must be removed from the resin. The ester linkage that connects the peptide to the resin, like most esters, is more easily cleaved than the peptide (amide) bonds (Sec. 21.7E), and is typically broken by liquid hydrogen fluoride, which also removes the Boc group from the product peptide.

Phe-Gly-Ala, free in solution

The method of solid-phase peptide synthesis discussed above, which employs the Boc group as the amino-terminal protecting group, is one of two major methods in common use today. The other important method involves a conceptually similar stepwise approach but employs a different protecting group, the (9-fluorenylmethyloxy)carbonyl (Fmoc) group, that can be removed by a mild base; recall that the Boc group is removed by acid. A piperidine solution in DMF is usually used for the removal of this group.

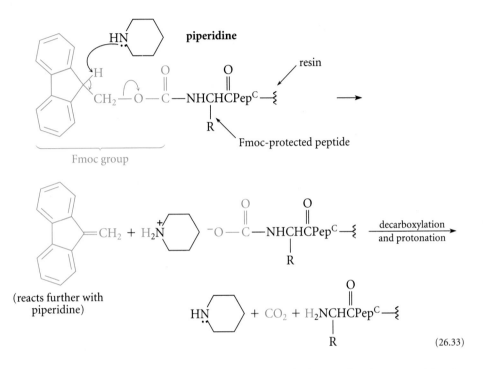

$$(26.33)$$

This β-elimination reaction is very rapid because the hydrogen that is removed by the piperidine is very acidic (Sec. 17.3B). (The conjugate-base anion of the Fmoc group is an aromatic species.) Because the protecting group can be removed by base, the linkage of the peptide to the resin does not have to be stable to acid, as it is in the Boc procedure. Consequently, a number of peptide-resin linkages have been developed that can be removed by dilute trifluoroacetic acid. For this reason, the very hazardous liquid HF reagent can be avoided.

The same reagents used for solid-phase peptide synthesis can also be used for peptide synthesis in solution, but removal of the DCU from the product peptide is sometimes difficult. The advantage of the solid-phase method, then, is the ease with which dissolved impurities and by-products are removed from the resin-bound peptide by simple filtration.

Despite its advantages, solid-phase peptide synthesis has one unique problem. Suppose, for example, that a coupling reaction is incomplete, or that other side reactions take place to give impurities that remain covalently bound to the resin. These are then carried along to the end of the synthesis, when they are also removed from the resin and must be separated (in some cases tediously) from the desired peptide product. (This situation is something like what might occur if a flight attendant on a flight from San Francisco to New York discovers over Denver a passenger without a ticket; the offending party cannot be removed until the end of the line.) In order to avoid impurities, then, each step in the solid-phase synthesis must occur with virtually 100% yield. Remarkably, this ideal is often approached closely in practice (Problem 26.21).

---

*26.21   Calculate the average yield of each of the 369 steps in the synthesis of ribonuclease by the solid-phase method discussed above, assuming the reported overall yield of 17%.

26.22   What average yield per amino acid would have to be obtained in order to synthesize a protein containing 100 amino acids in 50% overall yield?

26.23   *(a)   An aspiring peptide chemist, Mo Bonds, has decided to attempt the synthesis of the peptide Gly-Lys-Ala using the solid-phase method. To the Ala-resin he couples the following derivative of lysine:

$$Boc-NH-CH-CO_2H \qquad \alpha,\epsilon\text{-diBoc-lysine}$$
$$|$$
$$(CH_2)_4$$
$$|$$
$$NH-Boc$$

Why are *two* Boc groups necessary for the protection of lysine?

*(b)   After the coupling, he deprotects his resin-bound peptide with anhydrous CF_3CO_2H, and then completes the synthesis in the usual way by coupling Boc-Gly, deprotecting the peptide and removing it from the resin. He is shocked to find a mixture of several peptide products. Two of them give the amino acid analysis (Ala,Gly,Lys), and one gives the amino acid analysis (Ala,Gly$_2$,Lys). Suggest a structure for each product and explain what happened.

(c)   On the suggestion of a colleague, Uri Thane, Mo Bonds repeats the chemistry in (b) using the lysine derivative below, and obtains a good yield of the desired tripeptide. Explain.

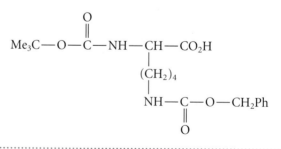

Problem 26.23 shows that certain amino acid side chains can also react under the conditions of peptide synthesis. Special *protecting groups* (Sec. 19.10B) must be introduced on these side chains. These protecting groups must survive the entire synthesis, including

the removal of the amino-protecting group at each stage, yet themselves be removable at the end of the synthesis. The design of protecting groups that can meet these exacting requirements is an important aspect of research in peptide synthesis.

26.8   Structures of Peptides and Proteins

A.   Primary Structure

The structures of molecules as large as peptides can be described at different levels of complexity. The simplest description of a peptide or protein structure is its covalent structure or **primary structure**. The most important aspect of any primary structure is the amino acid sequence (Sec. 26.6). However, peptide bonds are not the only covalent bonds that connect amino acid residues. In addition, **disulfide bonds** (Sec. 10.9) link the cysteine residues in different parts of a sequence.

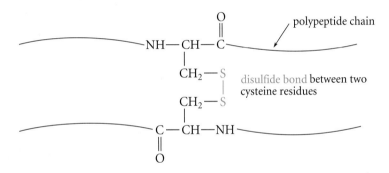

Disulfide bonds thus serve as crosslinks between different parts of a peptide chain. A number of proteins contain several peptide chains; disulfide bonds hold these chains together. The primary structure of a peptide or protein, then, includes its amino acid sequence and its disulfide bonds. The primary structure of *lysozyme*, a small enzyme that is abundant in hen egg white, is shown in Fig. 26.4. Lysozyme is a single polypeptide chain of 129 amino acids that includes eight cysteine residues linked together into four disulfide bonds.

The disulfide bonds of a protein are readily reduced to free cysteine thiols by other thiols. Two commonly used thiol reagents are 2-mercaptoethanol ($HSCH_2CH_2OH$), and dithiothreitol (known to biochemists as DTT, or Cleland's reagent).

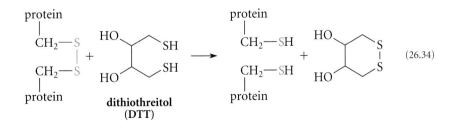

(26.34)

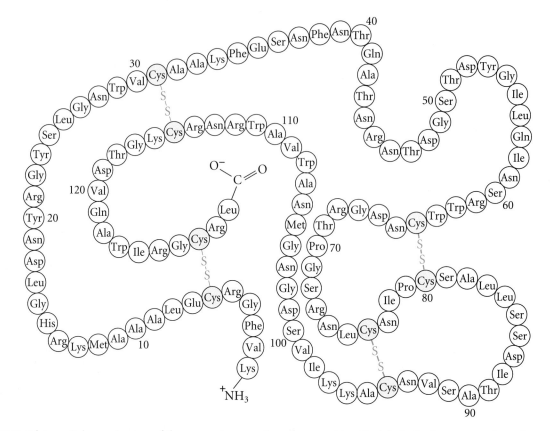

Figure 26.4 *Primary structure of the enzyme lysozyme from hen egg white. Physiologically, lysozyme catalyzes the hydrolysis of bacterial cell walls. Different variants of this enzyme are found in tears, nasal mucus, and even viruses—anywhere antibacterial action is important. Lysozyme is one of the smallest known enzymes. Individual amino acid residues, connected by peptide bonds, are numbered from the amino terminus. The disulfide bonds are shown in color.*

This reaction is simply a biological example of the thiol-disulfide equilibrium shown in Eq. 10.59, p. 485. Typically, when the extraneous thiols are removed, the thiols of the protein spontaneously reoxidize in air back to disulfides.

A PRACTICAL EXAMPLE OF DISULFIDE-BOND REDUCTION

An interesting example of the biological effects of disulfide-bond reduction occurs in the ordinary hair permanent. Hair (protein) is treated with a thiol solution; this solution is responsible for the unpleasant smell of permanents. The thiol solution reduces the disulfide bonds in the hair. With the hair in curlers, the disulfides are allowed to reoxidize. The hair is thus set by disulfide-bond reformation into the conformation dictated by the curlers. Only after a long period of time do the disulfide bonds rescramble to their normal configuration, when another permanent becomes necessary.

An industrial example of the use of disulfide bonds is the process of *vulcanization* (Sec. 15.5), which introduces disulfide bonds into synthetic polymers. Vulcanization provides a polymer with greater rigidity.

B. Secondary Structure

The description of the primary structure of a protein gives no indication of how the molecule might actually appear in three dimensions. The structural characteristics of a typical peptide bond are shown in Fig. 26.5. With few exceptions, the amide units in most peptides are planar. Recall that rotation about the carbonyl-nitrogen bond of most amides is relatively slow, and that the preferred conformation about this bond is Z (Sec. 21.2); the same is true for the amide bonds in a peptide. There are two other single bonds in a typical peptide residue: the two bonds to the α-carbon. In principle, rotation about these bonds should occur more rapidly. Because a protein contains many such bonds, it might seem that a very large number of conformations could occur in a protein or large peptide. However, it turns out that only three conformations are found most commonly.

In one type of peptide conformation, the peptide backbone adopts the conformation of a **right-handed alpha (α)-helix**, shown in Fig. 26.6 on p. 1302. "Right-handed" means that the helix turns in a clockwise fashion along the helical axis. In this conformation, the side-chain groups are positioned on the outside of the helix, and the helix is stabilized by *hydrogen bonds* between the amide N—H of one residue and the carbonyl oxygen four residues further along the helix. The alpha (α) terminology refers to a characteristic X-ray diffraction pattern that was observed for certain proteins before chemists fully

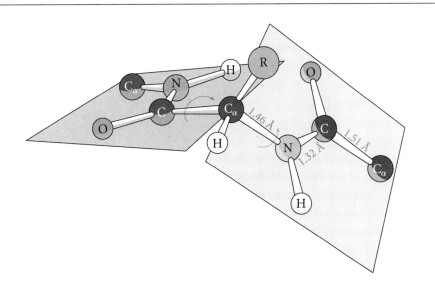

Figure 26.5   *Typical dimensions of a peptide bond. The two planes are those of the adjacent amide groups, and the amino acid side chain is represented by R. In principle, rotations about the bonds to the α-carbons (marked with arrows) are possible.*

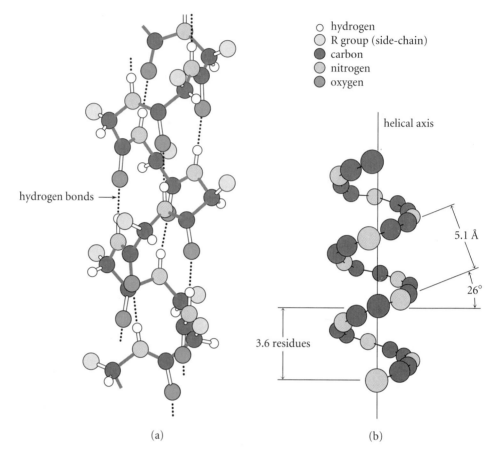

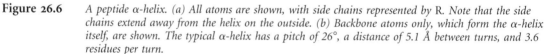

○ hydrogen
○ R group (side-chain)
● carbon
○ nitrogen
● oxygen

helical axis

hydrogen bonds →

5.1 Å

26°

3.6 residues

(a)   (b)

Figure 26.6   *A peptide α-helix. (a) All atoms are shown, with side chains represented by R. Note that the side chains extend away from the helix on the outside. (b) Backbone atoms only, which form the α-helix itself, are shown. The typical α-helix has a pitch of 26°, a distance of 5.1 Å between turns, and 3.6 residues per turn.*

understood their structures. The α type of pattern was eventually shown to be associated with the right-handed helix, thus the name α-helix.

Another commonly occurring X-ray diffraction pattern, called a β pattern, was eventually found to be characteristic of a second commonly occurring peptide conformation, called **beta (β)-structure** or **pleated sheet**. In this type of structure, a peptide chain adopts an open, zigzag conformation, and is engaged in hydrogen bonding with another peptide chain (or a different part of the same chain) in a similar conformation. The successive hydrogen-bonded chains can run (in the amino terminal to carboxy terminal sense) in the same direction (**parallel pleated sheet**) or in opposite directions (**antiparallel pleated sheet**). The antiparallel pleated sheet structure is shown in Fig. 26.7. The name "pleated sheet" is derived from the fanlike surface described by the aggregate of several hydrogen-bonded chains (Fig. 26.7b). Notice that the side-chain R-groups alternate between positions above and below the sheet.

A third peptide conformation is the **random coil**. As the name implies, peptides that adopt a random coil show no discernible pattern in their conformations. An apt

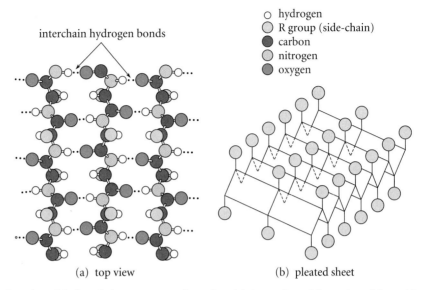

interchain hydrogen bonds

○ hydrogen
◯ R group (side-chain)
● carbon
◯ nitrogen
◯ oxygen

(a) top view   (b) pleated sheet

Figure 26.7   *The β-antiparallel pleated sheet structure of proteins. (a) A top view of the antiparallel peptide chains. Notice the hydrogen bonds between chains (dotted lines). (b) A schematic view of the pleated-sheet surface formed by the backbone atoms.*

analogy for the random coil is the appearance of a tangled ball of yarn after an hour's encounter with a playful house cat.

Although other conformations are known in peptides, the α-helix, β-structure, and random coil are the major ones. Some peptides and proteins exist entirely in one of these three conformations. For example, the α-keratins, major proteins of hair and wool, exist in the α-helical conformation. In these proteins, several α-helices are coiled about each other to form "molecular ropes." These structures have considerable physical strength. In contrast, silk fibroin, the fiber secreted by the silkworm, adopts the β-antiparallel pleated sheet conformation.

The description of peptide or protein structure in terms of α-helix, β-structure, and random-coil conformations is called **secondary structure**. Despite the examples above, proteins that contain a single type of secondary structure are relatively rare. Rather, *most proteins contain different types of secondary structure* in different parts of their peptide chains.

C. Tertiary and Quaternary Structures

The complete three-dimensional description of protein structure at the atomic level is called **tertiary structure**. The tertiary structures of proteins are determined by X-ray crystallography; each crystallographic structure analysis requires significant effort. Nevertheless, since the first protein crystallographic structure was determined in 1960, hundreds of protein structures have been elucidated, and more structures are continually appearing.

The tertiary structure of any given protein is an aggregate of α-helix, β-sheet, random coil, and other structural elements. In recent years it has become evident that in many proteins certain higher-order structural motifs are common. For example, a common motif is a bundle of four helices (called a *four-helix bundle*), each running approximately antiparallel to the next and separated by short turns in the peptide chain. Another common structural motif is the *beta barrel*, literally a bag consisting of β-sheets connected by short turns. Several such motifs can occur within a given protein, so that a protein might consist of several smaller, relatively ordered structures connected by short random loops. These ordered sub-structures are sometimes termed **domains**. For example, the enzyme lysozyme (Fig. 26.8) contains two domains, one on either side of the plane indicated by the dashed line.

In general, the tertiary structures of proteins are determined by the *noncovalent* interactions between groups within protein molecules, and between groups of the protein with the surrounding solvent. The following three general types of interactions are illustrated in Fig. 26.9.

1. Hydrogen bonds

2. Van der Waals attractions

3. Electrostatic interactions

Recall that *hydrogen bonds* stabilize both α-helices and β-sheets (Figs. 26.6, 26.7). All protein structures appear to contain many stable hydrogen bonds, not only within regions of helix or β-sheet, but also in other regions. Protein conformations are also stabilized in part by hydrogen bonding of certain groups to solvent water.

Van der Waals interactions, or dispersion forces, are the same interactions that provide the cohesive force in a liquid hydrocarbon (Sec. 2.6A and Fig. 2.8), and can be regarded as examples of the "like-dissolves-like" phenomenon. For example, we know that benzene does not dissolve in water, but dissolves readily in hexane. In the same sense, the benzene ring of a phenylalanine side chain energetically prefers to be near other aromatic or hydrocarbon side chains rather than the "waterlike" side chain of a serine, the charged, polar side chain of an aspartic acid, or solvent water on the outside of a protein. (This does *not* mean that *all* phenylalanine residues are buried next to other hydrocarbonlike residues; it does mean, however, that phenylalanine residues, more often

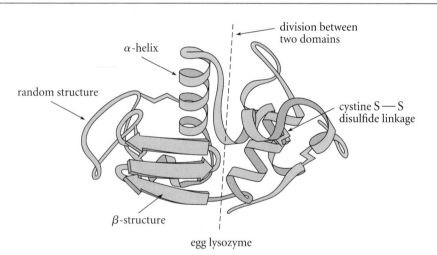

Figure 26.8   *Conformation of lysozyme from hens' eggs. In this figure the peptide backbone is traced as a ribbon that shows the regions of α-helix, β-sheet, and random structure. The two domains of lysozyme occur on either side of the plane indicated by the dashed line. (The primary structure of lysozyme is shown in Fig. 26.4.)*

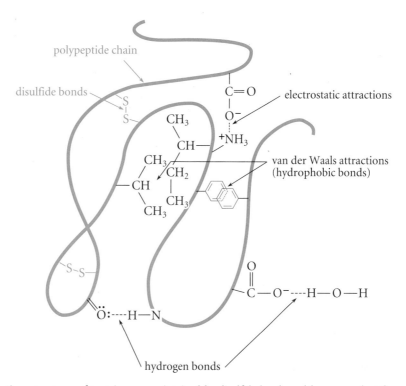

Figure 26.9   *The tertiary structures of proteins are maintained by disulfide bonds and by noncovalent forces: van der Waals attractions (hydrophobic bonds), hydrogen bonds, and electrostatic attractions and repulsions.*

than not, will be found in hydrocarbonlike environments.) The van der Waals interactions of this type between hydrocarbonlike residues are sometimes called *hydrophobic bonds*, because the hydrocarbon groups energetically prefer each other to water. Residues such as the side chain of phenylalanine or isoleucine are sometimes called *hydrophobic residues*. In contrast, polar residues, such as the carboxy groups of aspartic acid and glutamic acid, or the amino groups of lysine, are often found to interact with other polar residues or with the aqueous solvent. Such residues are sometimes termed *hydrophilic residues*.

Electrostatic interactions are noncovalent interactions between charged groups governed by the electrostatic law (Eq. 3.40). A typical stabilizing electrostatic interaction is the attraction of a protonated, positively charged amino group and a nearby ionized, negatively charged carboxylate ion.

A protein adopts a tertiary structure in which favorable interactions are maximized and unfavorable interactions are minimized. Although there are exceptions, most soluble proteins are globular and compact rather than extended. For example, lysozyme (Fig. 26.8) is a globular protein. Globular, nearly spherical, shapes are a consequence of the fact that proteins expose to aqueous solvent as small a surface as possible. (A sphere is the geometrical object with the minimum surface-to-volume ratio.) The reason for minimizing the exposed surface is that the majority of the residues in most proteins are hydrophobic, and the interaction of hydrophobic side chains with solvent water is unfavorable. The conformations of proteins are probably as much a consequence of these *unfavorable* interactions with water as the *favorable* interactions of the amino acid residues with each other. Indeed, to a first approximation, water-soluble proteins are large "grease balls" (with a few charged or polar groups on their surfaces) floating about in aqueous solution. In this respect, proteins somewhat resemble micelles (Sec. 20.5).

Suppose we were to synthesize a protein. Would the finished protein automatically "know" what conformation to assume, or is some external agent required to direct the protein into its naturally occurring conformation? This question was first answered by two elegant experiments with the enzyme ribonuclease. First of all, synthetic ribonuclease was prepared by the solid-phase method and found to be an active enzyme. Because the enzyme must have its natural, or native, conformation in order to be active, it follows that ribonuclease, once synthesized, spontaneously folds into this conformation.

The second type of experiment involved **denaturation** of ribonuclease. Denaturation is illustrated schematically in Fig. 26.10. When a protein is denatured, it is converted entirely into a random-coil structure. (A common example of *irreversible* protein denaturation occurs when an egg is fried; the denaturation and precipitation of the proteins in egg white are responsible for the change in appearance of the white as it is cooked.) Some proteins, including ribonuclease, can be denatured *reversibly* by chemical agents. Typically, a protein is denatured by breaking its disulfide bonds with thiols, such as DTT or 2-mercaptoethanol (Eq. 26.34), and by treating it with 8 *M* urea, detergents, or heat. Ribonuclease was denatured by treatment with 2-mercaptoethanol and 8 *M* urea. After the urea was removed, and the cysteine —SH groups allowed to reoxidize back to disulfides, the protein spontaneously reassumed its original, native conformation. This experiment shows that *the amino acid sequence of ribonuclease specifies its conformation; that is, the native structure is the most stable structure.* If this were not so, another, more stable structure would have formed when the protein was allowed to refold after the urea was removed.

There is a great deal of evidence that the ribonuclease result is fairly general: many proteins spontaneously assume their native conformations at the time of their biosynthesis. That is, *primary structure dictates tertiary structure.*

To summarize:

1. Many proteins appear to fold spontaneously into their native conformations.

2. The noncovalent forces that determine conformation are: (a) hydrogen bonds; (b) van der Waals forces (hydrophobic bonds); and (c) electrostatic forces.

Figure 26.10   *When a protein is denatured, its disulfide bonds are broken and it is converted entirely into a random coil.*

3. Most soluble proteins appear to be compact, globular structures. Although there are exceptions, hydrocarbonlike amino acid residues tend to be found on the interior of a protein, away from solvent water, and the polar residues tend to be on the exterior of a protein, where they can form hydrogen bonds with water.

Some (but not all) proteins are aggregates of other proteins. The best-known example of such proteins is *hemoglobin*, which transports oxygen in the bloodstream. Hemoglobin (Fig. 26.11) is an aggregate of four smaller proteins, or *subunits*, two of one type (called alpha subunits) and two of another (called beta subunits). (This terminology has nothing to do with α- and β-structure.) The alpha and beta subunits are similar, but differ somewhat in their primary structures. These subunits are held together solely by noncovalent forces—hydrogen bonds, electrostatic interactions, and van der Waals interactions. Notice in Fig. 26.11 that the individual subunits lie more or less at the vertices of a regular tetrahedron. This shape is the most compact arrangement that can be assumed by four objects. Many important proteins are aggregates of individual polypeptide subunits. In some proteins the subunits are identical; in other cases they are different. The description of subunit arrangement in a protein is called **quaternary structure**.

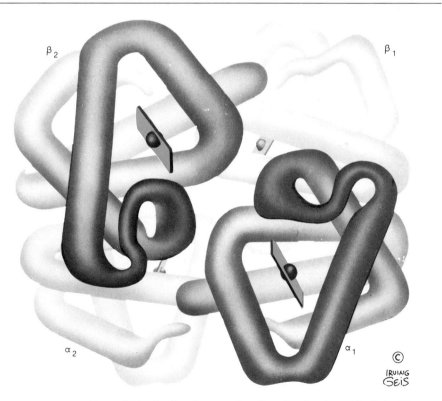

IRVING GEIS

Figure 26.11   *Quaternary structure of hemoglobin showing the general outline of each polypeptide chain. The rectangles represent the heme groups (Sec. 24.6) and the spheres represent the iron at which oxygen is bound to hemoglobin. Notice the tetrahedral orientation of the subunits.*

PROBLEM

*26.24   What would you expect to happen when hemoglobin is treated with a denaturant such as 8 *M* urea? Explain.

26.9   Enzymes: Biological Catalysts

Many of the proteins that occur naturally are **enzymes**. Enzymes are the catalysts for biological reactions (Sec. 4.9C). Through detailed studies of certain enzymes, chemists have come to understand some of the reasons why these proteins are efficient catalysts. To illustrate enzyme catalysis, let's consider the mechanism by which the enzyme *trypsin* catalyzes the hydrolysis of peptide bonds. Trypsin, it will be recalled, is the enzyme used in the sequencing of proteins (Sec. 26.6C). With a molecular weight of about 24,000, trypsin is an enzyme of modest size. It is a globular protein containing three polypeptide chains held together by disulfide bonds. The following comparison provides some idea of the catalytic effectiveness of trypsin. Peptides in the presence of trypsin are rapidly hydrolyzed at 37° and pH = 8. In the absence of trypsin, the same peptides are indefinitely stable under the same conditions, and hydrolysis requires boiling them in 6 *M* HCl for several hours. However trypsin, in contrast to hot HCl solution, does not catalyze hydrolysis at just *any* peptide bonds. It is specific for hydrolysis of the peptide bonds at lysine and arginine residues (Eq. 26.24). These two aspects of trypsin catalysis—*catalytic efficiency* and *specificity*—are characteristic of catalysis by all enzymes. Understanding these phenomena is the basis for understanding enzyme catalysis in general.

If an enzyme catalyzes a reaction of a certain compound, the compound is said to be a **substrate** for the enzyme. Enzymes act on their substrates in at least three stages, shown schematically in the following equation.

$$\underset{\substack{\text{enzyme}}}{E} + \underset{\substack{\text{substrate}}}{S} \rightleftharpoons \underset{\substack{\text{noncovalent}\\\text{enzyme-substrate}\\\text{complex}}}{E \cdot S} \rightleftharpoons \underset{\substack{\text{noncovalent}\\\text{enzyme-product}\\\text{complex}}}{E \cdot P} \rightleftharpoons \underset{\substack{\text{enzyme}}}{E} + \underset{\substack{\text{product}}}{P} \qquad (26.35)$$

First, the substrate binds to the enzyme in a noncovalent **enzyme-substrate complex**. The binding occurs at a part of the enzyne called the **active site**. Within the active site are groups that attract the substrate by interacting favorably with it. The noncovalent interactions that cause a substrate to bind to an enzyme are typically the same ones that stabilize protein conformations: electrostatic interactions, hydrogen bonding, and van der Waals interactions, or hydrophobic bonds.

In the second stage of enzyme catalysis, the enzyme promotes the appropriate chemical reaction(s) on the bound substrate to give an enzyme-product complex. The necessary chemical transformations are brought about by groups in the active site of the enzyme. In most enzymes, these groups are particular amino acid side chains of the enzyme itself. However, in some cases, other molecules, called **coenzymes**, are also required. (An example of a coenzyme is NAD^+; Sec. 10.7. Most vitamins are coenzymes.)

In the last stage of enzyme catalysis, the product(s) depart from the active site, leaving the enzyme ready to repeat the process on a new substrate molecule.

Enzymes are true catalysts; their concentrations are typically much lower than the concentrations of their substrates. They do not affect the equilibrium constants of the reactions they catalyze. They catalyze equally both forward and reverse reactions of an equilibrium.

Let's see how the trypsin-catalyzed hydrolysis of peptide bonds fits this general picture of enzyme catalysis. The active site of trypsin containing a bound peptide substrate is shown in Fig. 26.12 on p. 1310. The active site of trypsin consists of a cavity, or "pocket," that just accommodates the amino acid side chain of a lysine or arginine residue from the substrate; an arginine is shown in Fig. 26.12. Several hydrophobic residues line this cavity. At the bottom of the cavity is the side-chain carboxylic acid group of an aspartic acid residue (Asp-189 in the trypsin sequence). This group is ionized, and therefore *negatively charged*, at neutral pH. The amino group of a lysine side chain and the guanidino group of an arginine side chain are both protonated, and therefore *positively charged*, at neutral pH. The favorable electrostatic attraction between the ionized Asp-189 side chain of the enzyme and the positively charged side chain of the substrate helps stabilize the enzyme-substrate complex. This complex is also stabilized by the van der Waals interactions between the —CH_2— groups of the substrate side chain and the hydrophobic residues that line the cavity. Here, then, can be seen some of the reasons for the *specificity* of trypsin. The active site just "fits" the substrate (or vice-versa), and it contains groups that are noncovalently attracted to groups on the substrate.

Near the mouth of the active site are two amino acid residues that serve a critical catalytic function: a serine (Ser-195) and a histidine (His-57). The way that these residues act to catalyze peptide bond hydrolysis is shown in Fig. 26.13 on p. 1312. The —OH group of the serine side chain acts as a nucleophile to displace the peptide leaving group from the carbonyl group of the substrate. The resulting product is an *acyl-enzyme*; in this covalent complex, the residual peptide substrate is actually esterified to the enzyme! The imidazole group of histidine-57 serves as a base catalyst to remove the proton from the attacking serine. When water enters the active site, it too is deprotonated by the histidine as it attacks the carbonyl group of the acyl-enzyme, to give the free carboxy group of the substrate and regenerate the enzyme. After the product leaves the active site, the enzyme is ready for a new substrate molecule.

The *catalytic efficiency* of trypsin, as well as that of other enzymes, is attributable mostly to the fact that all of the necessary reactive groups are positioned in proximity within the enzyme-substrate complex: the substrate carbonyl, a nucleophile (the serine —OH group), and an acid-base catalyst (the imidazole of the histidine). These groups do not have to "find" each other by random collision, as they would if they were all free in solution. Notice that the reactions shown in Fig. 26.13 are not particularly unusual; enzyme-catalyzed transformations find close analogy in common organic reactions. As this example shows, understanding the chemistry of life does not require the idea of a "vital force" so prevalent in Wöhler's day; it is simply good organic chemistry! We hasten to add that the rationality of this chemistry makes it no less remarkable and elegant.

The discussion in this chapter shows conceptually why most enzyme-catalyzed reactions are stereoselective, that is, why an enzyme catalyzes the reaction of only one enantiomer of a chiral substrate (see, for example, Eq. 26.16). Enzymes are polymers of L-amino acids; that is, *enzymes are enantiomerically pure chiral molecules*. Each enzyme occurs naturally in only one enantiomeric form. When an enzyme reacts with its substrate, the enzyme-substrate complex is formed. The enzyme-substrate complex derived from the *enantiomer* of the substrate is the *diastereomer* of the complex formed from the substrate itself.

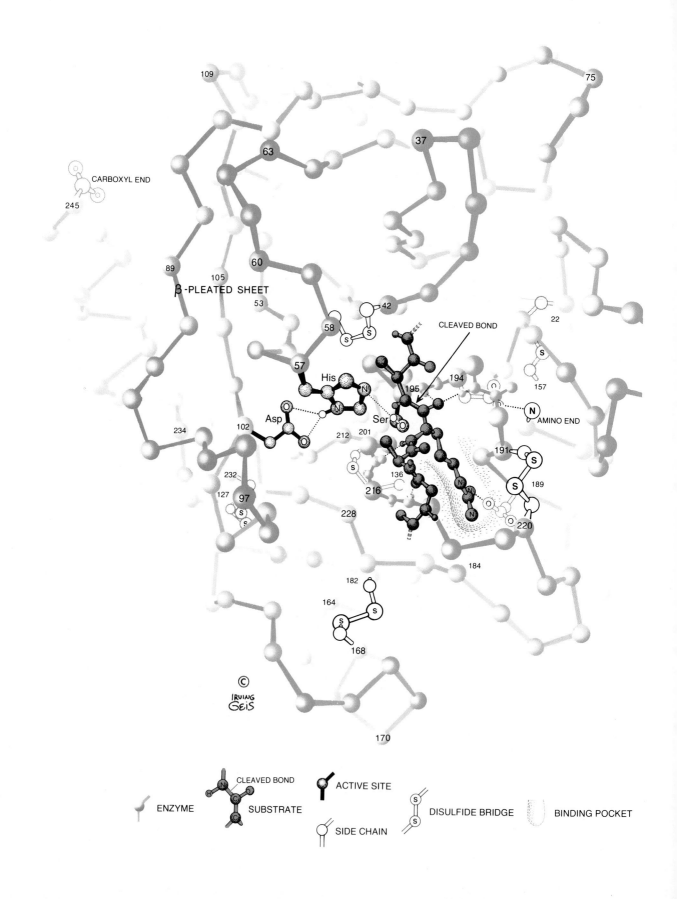

109

75

CARBOXYL END

245

63

37

60

89

105

β-PLEATED SHEET

53

42

58

S   S

57

22

S

157

His

N

CLEAVED BOND

194

N   AMINO END

Asp   O   O   N

Ser   195   O

102

234

212   201

191   S

S

232   136

S   189

127   97   216   N   N

S   S   N   O   O   220

228

184

182

164   S

170   S

168

© IRVING GEIS

ENZYME   CLEAVED BOND   N   H   C   C   SUBSTRATE   ACTIVE SITE   S   S   DISULFIDE BRIDGE   BINDING POCKET

SIDE CHAIN

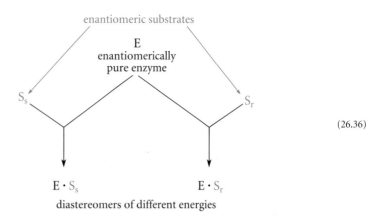

enantiomeric substrates

E
enantiomerically
pure enzyme

S_s

S_r

(26.36)

$E \cdot S_s$ $E \cdot S_r$

diastereomers of different energies

Because the two complexes are diastereomers, they have different energies, and one is more stable than the other. For the same reason, the rates at which these complexes are converted into products also differ. This is an important example of the *differentiation of enantiomers by a chiral reagent* (Sec. 7.7A). In fact, most enzymes do not catalyze reactions involving the enantiomers of their natural substrates to any significant extent. Putting a substrate enantiomer into the chiral active site is like putting a right hand into a left-handed glove: things simply don't fit!

Study Guide Link:
✓26.6
An All-D Enzyme

If scientists understand the details of enzyme catalysis, they should be able to design and synthesize artificial enzymes that bind specific compounds and act on them catalytically. Success in this endeavor might yield an arsenal of rationally designed molecules that could catalyze industrially important transformations under mild, environmentally friendly conditions. This sort of activity has yet to meet with general success. The rational synthesis of compounds with the *catalytic efficiency* and *specificity* of enzymes is an area of research that will undoubtedly occupy chemists of the future.

26.10 Occurrence of Peptides and Proteins

The previous section showed that proteins serve an important role as enzymes—biological catalysts. Peptides and proteins also serve other important biological roles. For example, proteins serve as transporters; thus, hemoglobin transports oxygen in the bloodstream. Some proteins have a structural role: collagen, a protein, is the major component of connective tissue. Proteins and peptides act as a line of defense: antibodies, or immunoglobulins, are the proteins that protect higher animals from invasion by foreign substances, including infectious agents. In some cases the active components present in venoms and toxins (for example snake and bee venoms) are proteins. Many hormones are peptides. Two classical examples of peptide hormones are insulin, which, among other things, controls the uptake of glucose into cells, and glucagon, which counterbalances the action of insulin. Gastrin is a peptide that controls the release of stomach acid.

Figure 26.12 *The enzyme trypsin. Only the α-carbons are shown. Disulfide bridges are shown in outline, and a portion of the polypeptide chain is shown in color. A substrate peptide containing an arginine residue is shown in the active site; the positively charged side chain lies in the "specificity pocket," sketched in shading. The negatively charged side-chain carboxylate group of Asp-189 lies at the bottom of this pocket. The catalytically important His-57 and Ser-195 residues are prepared for cleavage of the peptide bond marked with an arrow.*

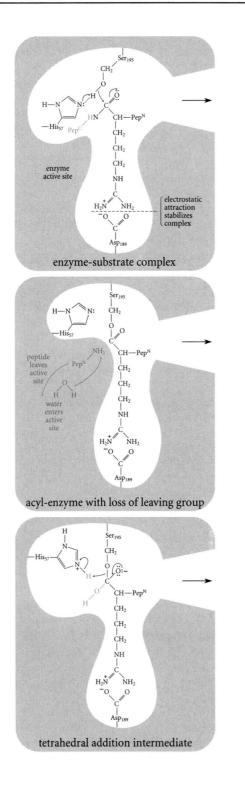

enzyme-substrate complex

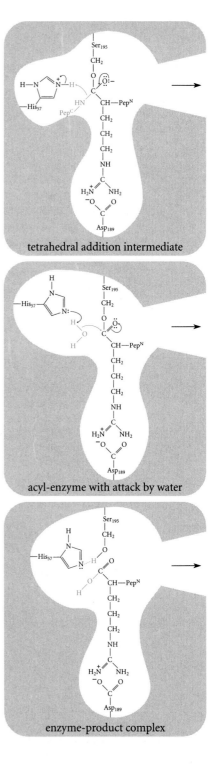

tetrahedral addition intermediate

acyl-enzyme with loss of leaving group

acyl-enzyme with attack by water

tetrahedral addition intermediate

enzyme-product complex

Figure 26.13   *Mechanism of the trypsin-catalyzed hydrolysis of an arginyl-peptide bond, beginning with the enzyme-substrate complex and ending with the enzyme-product complex. Notice that the imidazole in the side chain of the His-57 residue acts alternately as an acid and as a base to facilitate the necessary proton transfers. Pep^N and Pep^C denote the amino- and carboxy-terminal parts of the peptide substrate, respectively.*

An important discovery in the peptide field was prompted by the curiosity of scientists to find out why morphine, a compound that does not occur naturally in the human body, relieves pain. They reasoned that the body must contain another natural substance that might have the same effects as morphine. Certain peptides, called *enkephalins*, and β-endorphin, a longer peptide from which the enkephalins are derived, were discovered and found to have morphinelike effects. These may be the substances ultimately responsible for tolerance to pain. (It has been theorized that "joggers' high," the state of quasi-addiction to long-distance running, may be attributed to the release of these peptides.) It is now clear that peptide hormones control a large number of biological functions. Who knows: perhaps your ability to learn organic chemistry is regulated by one or more peptides!

Chemists have, on occasion, taken a cue for synthetic substances from protein chemistry. Nylon (Sec. 21.12A), with its many amide bonds, might be regarded as "synthetic silk"; silk is a protein fiber. In the 1970s, peptide chemists at G. D. Searle Co. accidently discovered a sweet-tasting peptide, L-aspartyl-L-phenylalanine methyl ester (aspartame), which is now an important artificial sweetener sold under the trade name NutraSweet®.

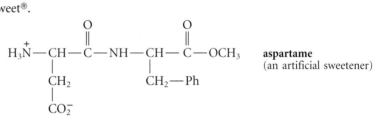

aspartame
(an artificial sweetener)

Peptide chemists are working to develop new, metabolically stable analogs of physiologically important peptides that can be used in medicine. Peptide and protein chemistry is an important branch of organic chemistry that will undoubtedly yield more exciting developments in the future.

KEY IDEAS IN CHAPTER 26

Amino acids are compounds that contain both an amino group and a carboxylic acid group. Peptides and proteins are polymers of the α-amino acids.

(continues)

The common naturally occurring chiral amino acids have the L config-uration, which is the same as the *S* configuration for all amino acids except cysteine.

Amino acids, as well as peptides that contain both acidic and basic groups, exist at neutrality as zwitterions.

The acidic or basic character of an amino acid, peptide, or protein is indicated by its isoelectric point.

Common methods for the synthesis of α-amino acids include the alkyl-ation of ammonia with α-halo acids; alkylation of α-acetamidomalo-nate esters, followed by hydrolysis and decarboxylation; and the Strecker synthesis.

Amino acids react both as amines and carboxylic acids. Thus, the amino group can be acylated; the carboxylic acid group can be esterified.

Peptides can be hydrolyzed to their constituent amino acids by heating them for several hours in aqueous acid or base. Hydrolysis of a pep-tide, separation of the resulting amino acids by ion-exchange chroma-tography, and quantitation give the relative number of each amino acid in the peptide. This procedure is called amino acid analysis.

Peptides and proteins can be cleaved specifically at methionine residues with cyanogen bromide, and at arginine and lysine residues by trypsin-catalyzed hydrolysis.

The primary sequence of a peptide or protein can be determined by the Edman degradation: treatment of the peptide with phenyl isothio-cyanate, followed by acid. Each cycle of this degradation affords a PTH derivative of the amino terminal amino acid plus a new peptide that is one residue shorter.

Peptide synthesis strategically involves attaching amino acids, one at a time, to a peptide chain beginning at the carboxy terminus. Protecting groups such as the Boc group must be used to prevent competing reac-tions that would occur in their absence. In solid-phase peptide synthe-sis, a protected carboxy-terminal amino acid is covalently bound to an insoluble resin; deprotected; condensed with another protected amino acid using DCC; deprotected; and the cycle continued until the peptide chain has been assembled, at which point the peptide is removed from the resin by cleavage with liquid HF or other acidic reagents.

The primary structure of a protein is a description of its covalent bonds, including disulfide bonds. Its secondary structure is a descrip-tion of its content of α-helix, β-structure, and random coil; and its tertiary structure is a complete description of its three-dimensional structure. The manner in which smaller proteins aggregate into larger ones is termed quaternary structure.

⚗   The noncovalent forces that determine the three-dimensional structure of a protein include hydrogen bonds, van der Waals attractions, and electrostatic interactions.

⚗   Enzymes are highly specific and efficient biological catalysts. A substrate is bound noncovalently at the enzyme active site before it is transformed into products.

ADDITIONAL
PROBLEMS

26.25   Give the structures of the products expected when *(1) valine and (2) proline (or other compounds indicated) react with each of the following reagents:
(a) ethanol (solvent) and H_2SO_4 (b) benzoyl chloride, Et_3N
(c) aqueous HCl solution   (d) aqueous NaOH solution
(e) benzaldehyde, heat, NaCN

(f)   O   O
                   ‖   ‖
        Me_3CO—C—O—C—$OCMe_3$, DMF, Et_3N

(g) product of (f) + DCC + glycine *tert*-butyl ester
(h) product of (g) + anhydrous CF_3CO_2H, then neutralize with NaOH
(i) product of (h) + 6 *M* HCl, heat

26.26   Referring to Table 26.1, identify the amino acid(s) that satisfy each of the following criteria.
*(a) the most acidic amino acid   (b) the most basic amino acid
*(c) the amino acids that can exist as diastereomers
(d) the amino acid that has zero optical rotation under all conditions
*(e) the amino acids that are converted into other amino acids on treatment with concentrated aqueous NaOH solution followed by neutralization.

26.27   In repeated attempts to synthesize the dipeptide Val-Leu, aspiring peptide chemist Polly Styreen performs each of the following operations. Explain what, if anything, is wrong with each procedure.
*(a) The cesium salt of leucine is allowed to react with chloromethylated polystyrene (Merrifield resin). The resulting derivative is treated with Boc-Val and DCC, then with liquid HF.
(b) The cesium salt of Boc-Leu is allowed to react with the Merrifield resin. The resulting derivative is then treated with Boc-Val and DCC, then with liquid HF.

*26.28   According to its amino acid composition (Fig. 26.4), lysozyme has an isoelectric point that is (choose one and explain):
(1) ≪6   (2) about 6   (3) ≫6

26.29   Which of the statements below would correctly describe the isoelectric point of *cysteic acid*, an oxidation product of cysteine? Explain your answer.

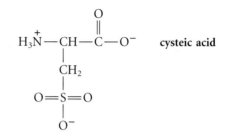

cysteic acid

(1) lower than that of aspartic acid
(2) about the same as that of aspartic acid
(3) about the same as that of cysteine
(4) about the same as that of lysine
(5) higher than that of lysine

*26.30   A peptide was subjected to one cycle of the Edman degradation, and the compound below was obtained. What is the amino-terminal residue of the peptide?

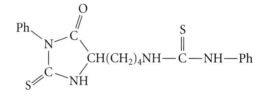

*26.31   *Dansyl chloride* (5-dimethylamino-1-naphthalenesulfonyl chloride) reacts with amino groups to give a fluorescent derivative. After a peptide *P* with the composition (Arg,Asp,Gly,Leu$_2$,Thr,Val) reacts with dansyl chloride at pH 9, it is hydrolyzed in 6 *M* aqueous HCl. The derivative shown in the the following equation, detected by its fluorescence, is isolated after neutralization, along with the free amino acids Arg, Asp, Gly, Leu, and Thr. What conclusion can be drawn about the structure of the peptide from this result?

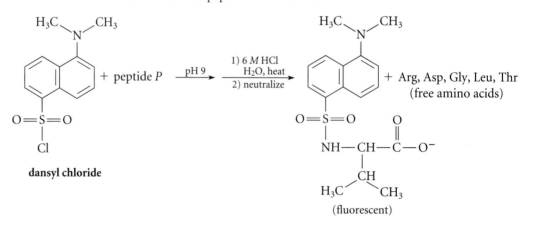

*26.32 A peptide Q has the following composition by amino acid analysis:

$$Q: \quad \text{Ala,Arg,Asp,Gly}_2\text{,Glu,Leu,Val}_2\text{,NH}_3$$

Treatment of Q once with the Edman reagent followed by anhydrous acid gives a new peptide R with the following composition by amino acid analysis:

$$R: \quad \text{Ala,Arg,Asp,Gly}_2\text{,Glu,Val}_2\text{,NH}_3$$

Treatment of Q and R with the enzyme *dipeptidylaminopeptidase* (DPAP) yields a mixture of the following peptides:

$$Q \xrightarrow{\text{DPAP}} \text{Arg-Gly, Gln-Ala, Leu-Val, Val-Asp, Gly}$$

$$R \xrightarrow{\text{DPAP}} \text{Ala-Gly, Asp-Gln, Gly-Val, Val-Arg}$$

What is the amino acid sequence of Q?

26.33 The peptide hormone glucagon has the following amino acid sequence:

> His-Ser-Gln-Gly-Thr-Phe-Thr-Ser-Asp-Tyr-Ser-Lys-Tyr-Leu-
> Asp-Ser-Arg-Arg-Ala-Gln-Asp-Phe-Val-Gln-Trp-Leu-Met-
> Asn-Thr

Give the products that would be obtained when this protein is treated with
(a) trypsin at pH 8   (b) cyanogen bromide in HCl
(c) Ph—N=C=S, then CF_3CO_2H, then aqueous acid

*26.34 A peptide C was found to have a molecular mass of about 1100. Amino acid analysis of C revealed its composition to be $(\text{Ala}_2\text{,Arg,Gly,Ile})$. The peptide was unchanged on treatment with the Edman reagent, then CF_3CO_2H. Treatment of C with trypsin gave a single peptide D with an amino acid analysis identical to that of C. Three cycles of the Edman degradation applied to D revealed the partial sequence Ala-Ile-Gly. Suggest a structure for peptide C.

26.35 When bovine insulin is treated with the Edman reagent followed by anhydrous CF_3CO_2H, then by aqueous acid, the PTH derivatives of *both* glycine and phenylalanine are obtained in nearly equal amounts. What can be deduced about the structure of insulin from this information?

*26.36 An amino acid A, isolated from the acid hydrolysis of a peptide antibiotic, gave a positive test with ninhydrin and had an optical rotation (HCl solution) of $+37.5°$. Compound A was not identical to any of the amino acids in Table 26.1. In amino acid analysis, compound A emerged from the ion-exchange column with the basic amino acids, and it could be prepared by the reaction of L-glutamine with Br_2 in NaOH, followed by neutralization. Suggest a structure for A.

26.37   A previously unknown amino acid, γ-carboxyglutamic acid (Gla), was discovered in the amino acid sequence of the blood-clotting protein prothrombin.

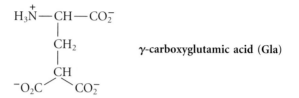

γ-carboxyglutamic acid (Gla)

This amino acid escaped detection for many years because, on acid hydrolysis, it is converted into another common amino acid. Explain.

*26.38   (a)   With what reagent would the Merrifield chloromethylated polystyrene resin react to give the following resin?

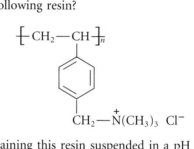

(b)   To a column containing this resin suspended in a pH 6 buffer is added a mixture of the amino acids Arg, Glu, and Leu, and the column is eluted with the same buffer. In what order will the amino acids be washed from the column? Explain.

26.39   In *paper electrophoresis*, amino acids and peptides can be separated by their differential migration in an electric field. To the center of a strip of paper, wet with buffer at pH 6, is applied a mixture of the following three peptides in a single small spot: Gly-Lys, Gly-Asp, and Gly-Ala. A positively charged electrode (anode) is attached to the left side of the paper, and a negatively charged electrode (cathode) to the right side. A voltage is applied across the ends of the paper for a period of time, after which the peptides have separated into three spots: one near the cathode, one near the anode, and one in the center, at the location of the original spot. Which peptide is in each spot? Explain.

*26.40   Explain each of the following observations.
   (a)   The optical rotations of alanine are different in water, 1 *M* HCl, and 1 *M* NaOH.
   (b)   Two mono-*N*-acetyl derivatives of lysine are known.
   (c)   The peptide Gly-Ala-Arg-Ala-Glu is readily cleaved by trypsin in water at pH = 8, but is inert to trypsin in 8 *M* urea at the same pH.
   (d)   After peptides containing cysteine are treated with $HSCH_2CH_2OH$, then with aziridine, they can be cleaved by trypsin at their (modified) cysteine residues.

$$\underset{H_2C-CH_2}{\overset{NH}{\triangle}} \quad \textbf{aziridine}$$

(e) When L-methionine is oxidized with H_2O_2, two separable methionine sulfoxides with the following structure are formed:

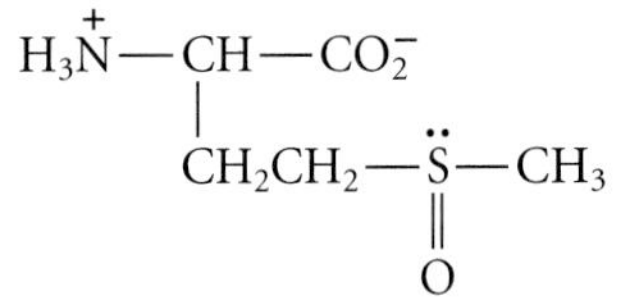

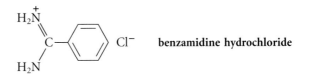

(f) When benzamidine hydrochloride is present in solution, trypsin no longer catalyzes peptide hydrolysis. (Benzamidine hydrochloride is not a denaturant at the concentrations used.)

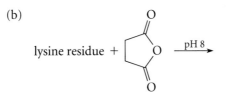

benzamidine hydrochloride

*26.41   Poly-L-lysine (a peptide containing only lysine residues) exists entirely in an α-helical conformation at pH > 11. Below pH 10, however, the peptide assumes a random-coil conformation. Poly-L-glutamic acid, on the other hand, exists in the α-helical conformation at pH < 4, but above pH 5 it assumes a random-coil conformation. Explain the effect of pH on the secondary structure of both polymers. Your explanation should indicate why the helical conformation predominates at high pH in one case and low pH in the other. (*Hint:* Look carefully at the location of the amino acid side chains in Fig. 26.6.)

26.42   Complete the following reactions, assuming the amino acid residue is part of a peptide and is at neither the amino nor the carboxy terminus.

*(a)   lysine residue + $CH_2{=}O$ + $NaBH_4$ $\xrightarrow{\text{pH 9}}$
       (excess of both)

(b)

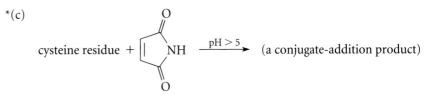

lysine residue +

succinic anhydride $\xrightarrow{\text{pH 8}}$

*(c)

cysteine residue + [maleimide structure] $\xrightarrow{\text{pH > 5}}$ (a conjugate-addition product)

maleimide

(Problem 26.42 continues)

(d)

$$\text{cysteine residue} + \text{I}-CH_2-\overset{\displaystyle O}{\overset{\|}{C}}-O^- \xrightarrow{\text{pH } 8-9}$$

iodoacetate

*(e)

aspartic acid residue +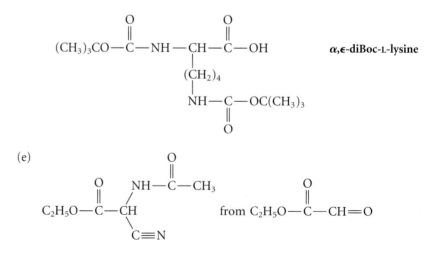

a water-soluble carbodiimide that
reacts much like DCC

$$H_2N-CH_2-CO_2CH_3 \longrightarrow$$

*(f) tyrosine residue + $Ph\overset{+}{N}\equiv N \ Cl^- \xrightarrow{\text{pH } 9}$

*26.43 Outline a synthesis of each of the following compounds from the indicated starting material and any other reagents.

(a)

$$Ph-\overset{\displaystyle O}{\overset{\|}{C}}-NH-\underset{}{\bigcirc}-\overset{\displaystyle O}{\overset{\|}{C}}-OC_2H_5 \text{ from } p\text{-aminobenzoic acid}$$

(b) $Ph-\underset{\underset{+NH_3}{|}}{CD}-CO_2^-$ from benzoic acid

(c) $CD_3-\underset{\underset{+NH_3}{|}}{CH}-CO_2^-$ from $CD_3-CH=O$

(d) L-Lys-L-Ala-L-Pro from L-proline, L-alanine, and the following compound using a solid-phase peptide synthesis:

$$(CH_3)_3CO-\overset{\displaystyle O}{\overset{\|}{C}}-NH-\underset{\underset{\underset{\underset{O}{\|}}{NH-C-OC(CH_3)_3}}{\overset{|}{(CH_2)_4}}}{\overset{|}{CH}}-\overset{\displaystyle O}{\overset{\|}{C}}-OH \qquad \boldsymbol{\alpha,\epsilon\text{-diBoc-L-lysine}}$$

(e)

$$C_2H_5O-\overset{\displaystyle O}{\overset{\|}{C}}-\underset{\underset{C\equiv N}{|}}{\overset{\overset{\displaystyle NH-\overset{\displaystyle O}{\overset{\|}{C}}-CH_3}{\diagup}}{CH}} \qquad \text{from } C_2H_5O-\overset{\displaystyle O}{\overset{\|}{C}}-CH=O$$

(f)

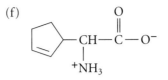

from the product of (e) and cyclopentene. (Cyclopentenyl amino acids are produced by certain plants. *Hint:* Hydrogens α to a cyano group are about as acidic as those α to an ester group.)

(g) The polymer *p*-aramid from terephthalic acid (1,4-benzenedicarboxylic acid). (This polymer is used in tire cord and other applications that require rigidity and strength.)

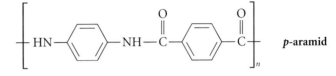

p-aramid

26.44 Show how the acetamidomalonate method can be used to prepare the following unusual amino acids from the indicated starting material and any other reagents.

(a) $(CH_3)_2CDCH_2$—CH—CO_2^- from isobutylene (2-methylpropene)
        |
       $^+NH_3$

(b) Ph—CHD—CH—CO_2^- from benzaldehyde
            |
           $^+NH_3$

(c)

γ-**oxohomotyrosine**
from anisole (methoxybenzene)

*26.45 When peptides containing a 2,3-diaminopropanoic acid (DAPA) residue are treated with the Edman reagent and then with acid, a peptide cleavage occurs in addition to degradation of the amino-terminal residue. Using the curved-arrow formalism to rationalize your answer, propose a structure for X.

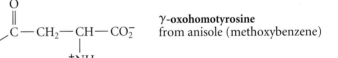

DAPA residue

26.46 The artificial sweetener *aspartame* was withheld from the market for several years because, on storage for extended periods of time in aqueous solution, it forms a *diketopiperazine*.

(*Problem 26.46 continues*)

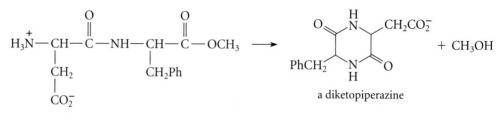

aspartame

(Extensive biological testing was required to show that this by-product was safe for consumers.) Give a curved-arrow mechanism for the formation of the diketopiperazine.

*26.47 Complete the following reactions by giving the structure of the major organic product(s).

(a) ethylamine + Ph—N=C=S ⟶

(b) PhCH=O + KCN + CH_3NH_2 ⟶ $\dfrac{H_3O^+/H_2O}{\text{heat}}$ ⟶

(c) $\overset{+}{H_3N}$—CH—CO_2^- $\dfrac{\text{NaOH}}{\text{(1 equiv.)}}$ $\dfrac{(CH_3)_2CH\text{—CH=O}}{H_2O}$ $\dfrac{\text{NaBH}_4}{}$
　　　　|
　　　　CH_3

(d)
$(CH_3)_3C$—O—C(=O)—NH—CH(CH_3)—C(=O)—OCH_3 + NaOH (1 equiv.) ⟶

(e)
Ph—C(=O)—NH—NH_2 + $NaNO_2$ $\xrightarrow{\text{HCl}}$

(f) product of (e) $\xrightarrow{\text{heat in benzene}}$

(g) product of (f) + valine methyl ester ⟶

(h)
H—C(=O)—NH—$CH(CO_2Et)_2$ $\dfrac{\text{NaOEt}}{\text{EtOH}}$ [with $H_2C=$, C—CH_2Cl, H_3C] $\dfrac{H_2O, H_3O^+}{\text{heat}}$

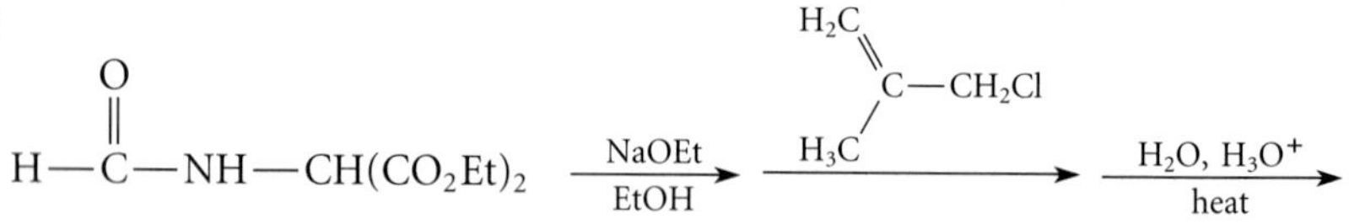

(i)
N≡C—CH(CH(CH_3)_2)—C(=O)—OEt + H_2N—NH_2 ⟶ $\xrightarrow{\text{NaNO}_2,\ \text{HCl}}$ $\xrightarrow{\text{EtOH, heat}}$ $\dfrac{HCl/H_2O}{\text{heat}}$

26.48 Identify each of the bracketed compounds in the following reaction scheme. Explain your answers.

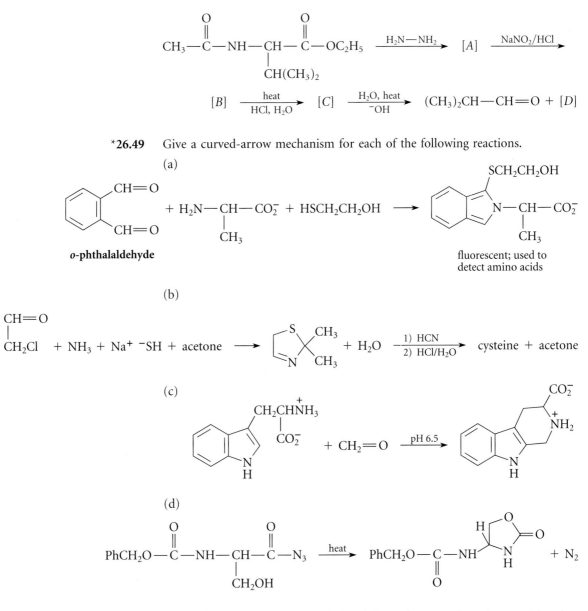

*26.49 Give a curved-arrow mechanism for each of the following reactions.

(a)

o-phthalaldehyde

fluorescent; used to
detect amino acids

(b)

(c)

(d)

*26.50 (a) In most peptides the amide bonds have the Z configuration; explain why.

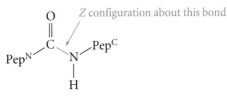

Z configuration about this bond

(b) One particular amino acid residue in the Pep^C position adopts the E configuration in some cases. Which amino acid residue should be most likely to
assume an E configuration and why?

*26.51   When peptides containing the Asn-Gly sequence, such as *H* in the equation below, are stored in aqueous solution at neutral or slightly basic solution, ammonia is liberated and a derivative *I* is formed. On continued storage, species *I* reacts to give two new peptides, *J* and *K*. Peptide *J* is the same as peptide *H* except that Asn is replaced by Asp, and peptide *K* is an isomer of peptide *J*. Propose structures for peptides *J* and *K*, and rationalize their formation using the curved-arrow formalism. (These reactions are believed to be a major source of deterioration associated with aging in naturally occurring peptides and proteins.)

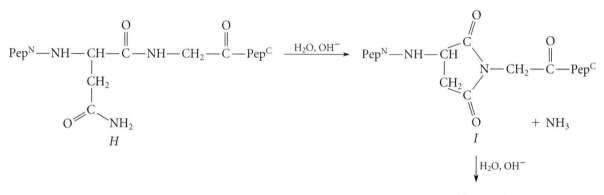

peptides *J* and *K*

*26.52   (a)   Explain why two monomethyl esters of *N*-acetyl-L-aspartic acid are known. Draw their structures.
   (b)   Explain why a mixture of these two compounds can be separated by anion-exchange chromatography at pH = 3.0, but not at pH = 7. (*Hint:* Use the pK_a values of aspartic acid in Table 26.1.)
   (c)   Which of the two compounds would emerge first from an anion-exchange column at pH = 3.0? Explain.

26.53   When *N*-acetyl-L-aspartic acid is treated with acetic anhydride, an optically active compound *A*, $C_6H_7NO_4$, is formed. Treatment of *A* with the amino acid L-valine yields two separable, isomeric peptides, *B* and *C*, that are both converted into a mixture of L-alanine and L-aspartic acid by acid hydrolysis. Suggest structures for *A*, *B*, and *C*.

27

Carbohydrates and Nucleic Acids

Because of their abundance in the natural world and their importance to living things, sugars have been the subject of intense investigation since the earliest days of scientific inquiry. Scientists refer to sugars and their derivatives as *carbohydrates*. As this name implies, most of the common sugars have molecular formulas that fit a "hydrate of carbon" pattern, that is, a formula of the form $C_n(H_2O)_m$. For example, sucrose (table sugar) has the formula $C_{12}(H_2O)_{11}$ or $C_{12}H_{22}O_{11}$, and both glucose and fructose (a sugar prevalent in honey) have the formula $C_6(H_2O)_6$ or $C_6H_{12}O_6$. This "hydrate of carbon" pattern is more than an apparent relationship. Anyone familiar with the conversion of table sugar into carbon by concentrated H_2SO_4 (or anyone who has made caramel sauce, a less extreme example of the same phenomenon) has witnessed in practice the dehydration of carbohydrates:

> A quarter pound of nice white lump sugar put into a breakfast cup with the smallest possible dash of boiling water and then the addition of plenty of oil of vitriol [H_2SO_4] is a truly wonderful spectacle, and more instructive than much reading, to see the white sugar turn black, then boil spontaneously, and now, rising out of the cup in solemn black, it heaves and throbs as the oil of vitriol continues its work in the lower part of the cup, emitting volumes of steam.... [J.W. Pepper, *Scientific Amusements for Young People*, 1863]

As the result of a more modern understanding of their structures, **carbohydrates** are now defined as aldehydes and ketones containing a number of hydroxy groups on an unbranched carbon chain, as well as their chemical derivatives.

Two common carbohydrate structures:

$$O{=}CH-CH-CH-CH-CH-CH_2OH \qquad HOCH_2-C-CH-CH-CH-CH_2OH$$

Less precisely, but more descriptively, carbohydrate chemistry can be regarded as the chemistry of sugars and their derivatives.

Carbohydrates are among the most abundant organic compounds on the earth. In polymerized form as cellulose, carbohydrates account for 50–80% of the dry weight of plants. Carbohydrates are a major source of food; sucrose (table sugar) and lactose (milk sugar) are examples. Even the shells of arthropods such as lobsters consist largely of carbohydrate.

The study of carbohydrates relies heavily on the principles of stereochemistry (Chapter 6) and on the conformational aspects of cyclohexane rings (Chapter 7). Do not hesitate to use molecular models when necessary.

Nucleic acids are so named because they are a principal component of the cell nucleus (although they also occur elsewhere). Nucleic acids are of two types: *deoxyribonucleic acids* (DNA) and *ribonucleic acids* (RNA). DNA is the storehouse of genetic information in the cell. Your hair color, your sex, and the color of your eyes are all determined by the structure of your DNA. RNA serves various roles in translating and processing the information encoded within the structure of DNA. Both DNA and RNA are polymers. Just as proteins are polymers of α-amino acids, RNA and DNA are polymers of fundamental building blocks called *nucleotides*. A study of nucleic acids concludes this chapter. It is appropriate to consider nucleotides and their polymers along with the carbohydrates because nucleotides, as you'll see subsequently, are derivatized carbohydrates.

27.1   Classification and Properties of Carbohydrates

Carbohydrates can be classified in several ways. Certain classifications that are based on structure are illustrated by the following examples.

$$HOCH_2-CH-CH-CH-CH-CH{=}O$$
$$\quad\quad\quad\ \ |\quad\ \ |\quad\ \ |\quad\ \ |$$
$$\quad\quad\quad OH\ \ OH\ \ OH\ \ OH$$

an *aldose* (aldehyde carbonyl group)
a *hexose* (six carbon atoms)
an *aldohexose* (combination of the above classifications)

$$HOCH_2-CH-CH-C-CH_2OH$$
$$\quad\quad\quad\ \ |\quad\ \ |\quad\ \ \|$$
$$\quad\quad\quad OH\ \ OH\ \ O$$

a *ketose* (ketone carbonyl group)
a *pentose* (five carbon atoms)
a *ketopentose* or *pentulose* (combination of the above classifications)

One type of classification is based on the type of carbonyl group in the carbohydrate. A carbohydrate with an aldehyde carbonyl group is called an **aldose**; a carbohydrate with a ketone carbonyl group is called a **ketose**. Carbohydrates can also be classified by the number of carbon atoms they contain. A six-carbon carbohydrate is called a **hexose**, and a five-carbon carbohydrate is called a **pentose**. These two classifications can be combined: an **aldohexose** is an aldose containing six carbon atoms, and a **ketopentose** is a ketose containing five carbon atoms. A ketose can also be indicated with the suffix *ulose*; thus, a five-carbon ketose is also termed a **pentulose**.

Another type of classification scheme is based on the hydrolysis of certain carbohydrates to simpler carbohydrates. **Monosaccharides** cannot be converted into simpler carbohydrates by hydrolysis. Glucose and fructose are examples of monosaccharides. Sucrose, however, is a **disaccharide**—a compound that can be converted by hydrolysis into two monosaccharides.

$$\text{sucrose } (C_{12}H_{22}O_{11}) + H_2O \xrightarrow{\text{acid or certain enzymes}} \text{glucose } (C_6H_{12}O_6) + \text{fructose } (C_6H_{12}O_6) \quad (27.1)$$

a disaccharide

monosaccharides

Likewise, **trisaccharides** can be hydrolyzed to three monosaccharides, **oligosaccharides** to a "few" monosaccharides (in the same sense that *oligopeptides* can be hydrolyzed to a few amino acids), and **polysaccharides** to a very large number of monosaccharides.

Because of their many hydroxy groups, carbohydrates are very soluble in water. The ease with which a large amount of table sugar dissolves in water to make syrup is an example from common experience of carbohydrate solubility. Carbohydrates are virtually insoluble in nonpolar solvents.

27.2 Structures of the Monosaccharides

A. Stereochemistry and Configuration

We'll consider the stereochemistry of carbohydrates by focusing largely on the aldoses with six or fewer carbons. The aldohexoses have four asymmetric carbons and exist as 2^4 or sixteen possible stereoisomers. These can be divided into two enantiomeric sets of eight diastereomers.

$$\text{HOCH}_2-\underset{\overset{|}{\text{OH}}}{\text{CH}}-\underset{\overset{|}{\text{OH}}}{\text{CH}}-\underset{\overset{|}{\text{OH}}}{\text{CH}}-\underset{\overset{|}{\text{OH}}}{\text{CH}}-\text{CH}{=}\text{O}$$

aldohexoses
four asymmetric carbons
$2^4 = 16$ stereoisomers

Similarly, there are two enantiomeric sets of four diastereomers (eight stereoisomers total) in the aldopentose series. Each diastereomer is a *different carbohydrate* with *different properties*, known by a *different name*. The aldoses with six or fewer carbons are given in Fig. 27.1 as Fischer projections. Be sure you understand how to draw and interpret Fischer projections, as they are widely used in carbohydrate chemistry (Sec. 6.11).

Each of the monosaccharides in Fig. 27.1 has an enantiomer. For example, the two enantiomers of glucose have the following structures:

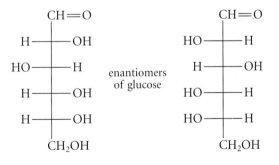

It is important to specify the enantiomers of carbohydrates in a simple way. Suppose you had a model of one of these glucose enantiomers in your hand; how would you explain to someone who cannot see the model (for example, over the telephone) which enantiomer you were holding? You could, of course, use the R,S system to describe the configuration of one or more of the asymmetric carbon atoms. A different system, however, was in use long before the R,S system was established. The **D,L system**, which came from proposals made in 1906 by a New York University chemist, M. A. Rosanoff, is used for this purpose. In this system, the configuration of a carbohydrate enantiomer is specified by applying the following conventions:

1. The naturally occurring stereoisomer of the aldotriose glyceraldehyde (the 2*R* enantiomer) is arbitrarily said to have the D configuration; its enantiomer is then said to have the L configuration.

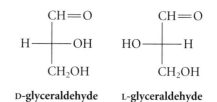

   D-glyceraldehyde   L-glyceraldehyde

2. The other aldoses or ketoses are written in a Fischer projection with their carbon atoms in a straight vertical line, and the carbons are numbered consecutively as they would be in systematic nomenclature, so that the carbonyl carbon receives the lowest number.

3. The *asymmetric carbon of highest number* is designated as a *reference carbon*. If this carbon has the H, OH, and CH_2OH groups in the same relative configuration as the same three groups of D-glyceraldehyde, the carbohydrate is said to have the D configuration. If this carbon has the same configuration as L-glyceraldehyde, the carbohydrate is said to have the L configuration.

The application of these conventions is illustrated in the following study problem.

....

STUDY PROBLEM 27.1

Determine whether the following carbohydrate derivative, shown in Fischer projection, has the D or L configuration.

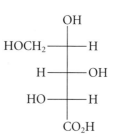

Solution   First redraw the structure so that the carbon with the lowest number in substitutive nomenclature—the carboxylic acid group—is at the top. This can be done by rotating the structure 180° in the plane of the page. Then carry out a cyclic permutation of the three groups at the bottom so that all carbons lie in a vertical line. Recall (Sec. 6.11) that these are allowed manipulations of Fischer projections.

Figure 27.1   *The D family of aldoses. Each compound shown here has an enantiomer in the L family. The colored arrows show how the aldoses are related by the Kiliani-Fischer synthesis (Sec. 27.8).*

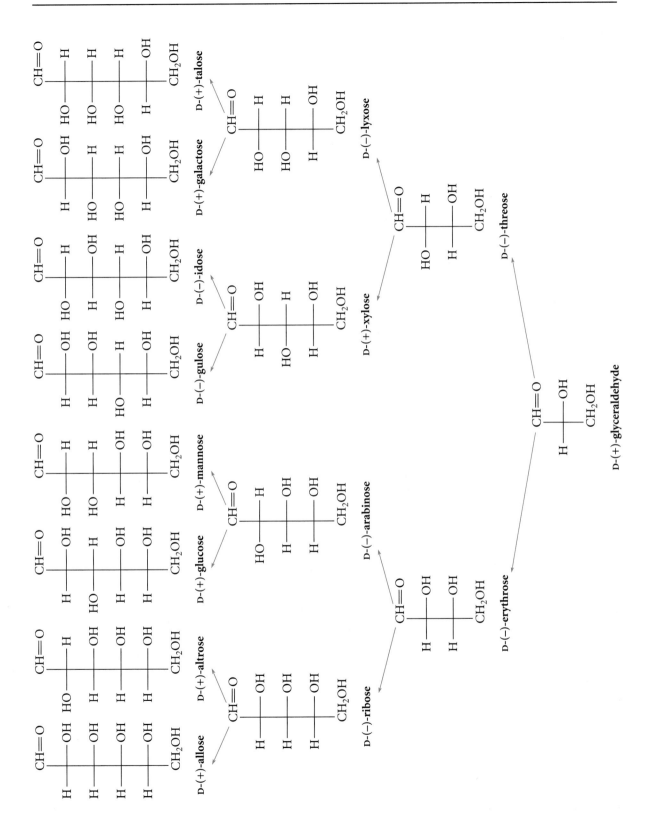

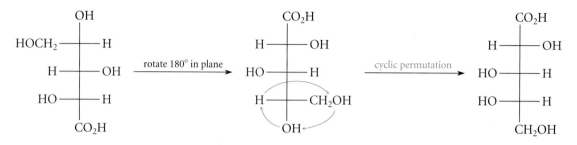

Finally, compare the configuration of the highest numbered carbon with that of D-glyceraldehyde. Because the configuration is different, the molecule has the L-configuration.

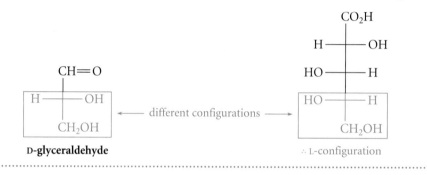

The monosaccharides shown in Fig. 27.1 are the D family of enantiomers. Each of the compounds in this figure has an enantiomer with the L configuration. Recall that *there is no general correspondence between configuration and the sign of the optical rotation* (see Sec. 6.3C). Notice, for example, that some D aldoses have positive rotations, but others have negative rotations. Also, *there is no simple relationship between the D,L system and the R,S system.* The R,S system is used to specify the configuration of *each* asymmetric carbon atom in a molecule, but *the D,L system specifies a particular enantiomer of a molecule that might contain many asymmetric carbons.* Although use of the D,L system is straightforward for carbohydrates and amino acids, it has been virtually abandoned for other compounds.

An annoying aspect of the D,L system is that each diastereomer is given a different name. This is one reason why the D,L system has been generally replaced with the R,S system, which can be used with systematic nomenclature. Nevertheless, the common names of many carbohydrates are so well entrenched that they remain important.

A few of the aldoses in Fig. 27.1 are particularly important, and their structures should be learned. D-Glucose, D-mannose, and D-galactose are the most important aldohexoses because of their wide natural occurrence. The structures of the latter two aldoses are easy to remember once the structure of glucose is learned, because their configurations differ in a simple way from the configuration of glucose. D-Glucose and D-mannose differ in configuration only at carbon-2; D-glucose and D-galactose differ only at carbon-4. Compounds that differ in configuration at only one of several asymmetric carbons are called **epimers**. Hence, D-glucose and D-mannose are epimeric at carbon-2; D-glucose and D-galactose are epimeric at carbon-4.

D-Ribose is a particularly important aldopentose; its structure is easy to remember because all of its —OH groups are on the right in the standard Fischer projection. D-Fructose is an important naturally occurring ketose:

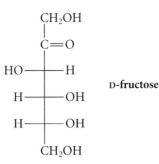

D-fructose

Notice that carbons 3, 4, and 5 of D-fructose have the same stereochemical configuration as carbons 3, 4, and 5 of D-glucose.

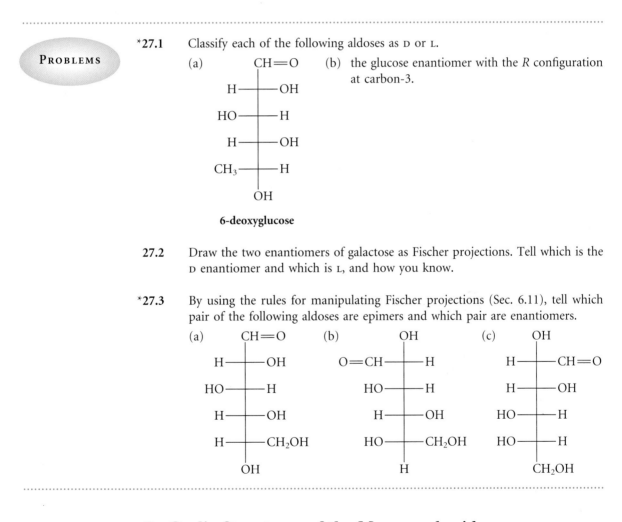

PROBLEMS

*27.1   Classify each of the following aldoses as D or L.

(a)

6-deoxyglucose

(b)   the glucose enantiomer with the *R* configuration at carbon-3.

27.2   Draw the two enantiomers of galactose as Fischer projections. Tell which is the D enantiomer and which is L, and how you know.

*27.3   By using the rules for manipulating Fischer projections (Sec. 6.11), tell which pair of the following aldoses are epimers and which pair are enantiomers.

(a)   (b)   (c)

B.   Cyclic Structures of the Monosaccharides

Recall that γ- or δ-hydroxy aldehydes exist predominantly as *cyclic hemiacetals* (Sec. 19.10A).

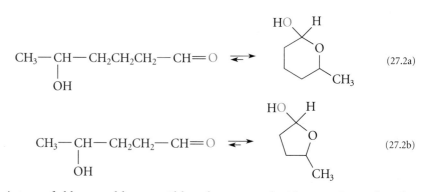

$$CH_3\!-\!CH\!-\!CH_2CH_2CH_2\!-\!CH\!=\!O \;\rightleftharpoons\; \qquad\qquad (27.2a)$$
$$\qquad\quad |$$
$$\qquad\quad OH$$

$$CH_3\!-\!CH\!-\!CH_2CH_2\!-\!CH\!=\!O \;\rightleftharpoons\; \qquad\qquad (27.2b)$$
$$\qquad\quad |$$
$$\qquad\quad OH$$

The same is true of aldoses and ketoses. Although monosaccharides are often written by convention as acylic carbonyl compounds, they exist predominantly as cyclic acetals. For example, in aqueous solution glucose consists of about 0.003% aldehyde and a trace of the hydrate; the rest—more than 99.99%—is cyclic hemiacetals.

In many carbohydrates both five- and six-membered cyclic hemiacetals are possible, depending on which hydroxy group undergoes cyclization.

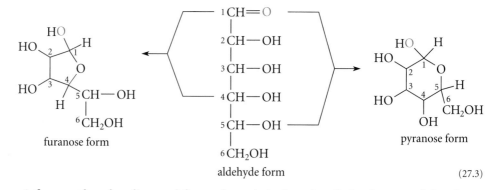

furanose form   aldehyde form   pyranose form

$$(27.3)$$

A five-membered cyclic acetal form of a carbohydrate is called a **furanose** (after furan, a five-membered oxygen heterocycle); a six-membered cyclic acetal form of a carbohydrate is called a **pyranose** (after pyran, a six-membered oxygen heterocycle).

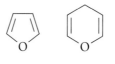

furan   pyran

The aldohexoses and aldopentoses exist predominantly as pyranoses, but the furanose forms of some carbohydrates are important.

A name such as *glucose* is used when referring to any or all forms of the carbohydrate. To name a cyclic hemiacetal form of a carbohydrate, start with a prefix derived from the name of the carbohydrate (for example, *gluco* for glucose, *manno* for mannose) followed by a suffix that indicates the type of hemiacetal ring (*pyranose* for a six-membered ring, *furanose* for a five-membered ring). Thus, a six-membered cyclic hemiacetal form of D-glucose is called D-glucopyranose; a five-membered cyclic hemiacetal form of D-mannose is called D-mannofuranose.

The cyclic hemiacetal structures of aldoses were suspected originally because they do not undergo (or undergo very slowly) certain reactions of aldehydes. This fact suggested that the aldehyde group is somehow masked. Confirmation of these cyclic structures was originally obtained from chemical degradations. However, the cyclic structures of carbohydrates are readily apparent today from NMR spectroscopy. For example, the aldehydic proton resonance of glucose in the δ 9–10 region of its proton NMR spectrum is too weak to detect under ordinary circumstances, yet there is a doublet at δ 5.2 corresponding to a proton α to two oxygens—the proton at carbon-1 of the pyranose structure. Similarly, in the CMR spectrum, the carbonyl carbon absorption in the δ 200 region of the spectrum is not discernible in ordinary spectra.

Anomers   It is important to notice that *the furanose or pyranose form of a carbohydrate has one more asymmetric carbon than the open chain form*—carbon-1. Thus there are two possible diastereomers of D-glucopyranose.

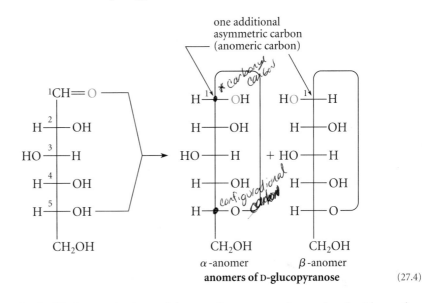

anomers of D-glucopyranose   (27.4)

(The rings in the Fischer projections of these cyclic compounds are closed with a rather strange-looking long bond.) Both of these compounds are forms of D-glucopyranose, and in fact, glucose in solution exists as a mixture of both. They are diastereomers, and are therefore separable compounds with different properties. When two cyclic forms of a carbohydrate differ in configuration only at their hemiacetal carbons, they are said to be **anomers**. Thus, the two forms of D-glucopyranose are anomers of glucose. The hemiacetal carbon (carbon-1 of an aldose) is sometimes referred to as the **anomeric carbon**.

As the structures above illustrate, anomers are named with the Greek letters α and β. This nomenclature refers to the Fischer projection of the cyclic form of a carbohydrate, written with all carbon atoms in a straight vertical line. *In the α-anomer the hemiacetal* —OH *group is on the same side of the Fischer projection as the oxygen at the configurational carbon*. (The configurational carbon is the one used for specifying the D,L designation; that is, carbon-5 for the aldohexoses.) Conversely, in the β-anomer the hemiacetal, —OH group is on the side of the Fischer projection opposite the oxygen at the configurational carbon. The application of these definitions to the nomenclature of the D-glucopyranose anomers is as follows:

STUDY GUIDE LINK:
27.1
Nomenclature of Anomers

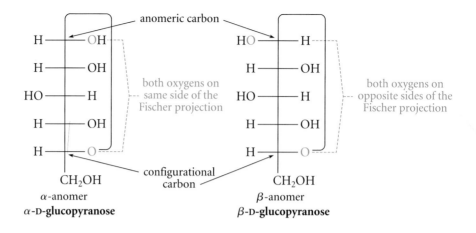

α-anomer
α-D-**glucopyranose**

β-anomer
β-D-**glucopyranose**

Conformational Representations of Pyranoses

Fischer projections of carbohydrates are convenient for specifying their *configurations* at each asymmetric carbon, but Fischer projections contain no information about the *conformations* of carbohydrates. It is important to relate Fischer projections to conformational representations of the carbohydrates. The following study problem shows how to establish this relationship in a systematic manner for the pyranoses.

STUDY PROBLEM 27.2

Convert the Fischer projection of β-D-glucopyranose into a chair conformation.

Solution   First redraw the Fischer projection for β-D-glucopyranose in an equivalent Fischer projection in which the ring oxygen is in a down position. This is done by using a cyclic permutation of the groups on carbon-5, an allowed manipulation of Fischer projections (Sec. 6.11).

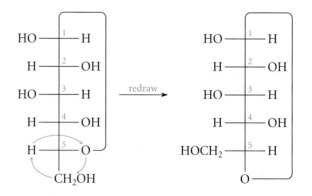

Recall that the carbon backbone of such a Fischer projection is imagined to be folded around a barrel or drum (Fig. 6.18c). Such an interpretation of the Fischer projection of β-D-glucopyranose yields the following structure, in which the ring lies in a plane perpendicular to the page. (The ring hydrogens are not shown.)

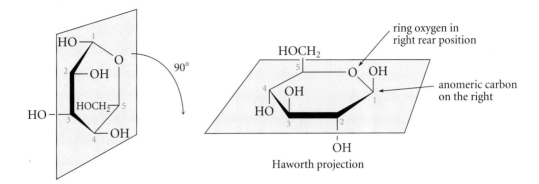

Haworth projection

When the plane of the ring is turned 90° so that the *anomeric carbon is on the right and the ring oxygen is in the rear*, the groups in *up* positions are those that are on the *left* in the Fischer projection; the groups in *down* positions are those that are on the *right* in the Fischer projection. A planar structure of this sort is called a **Haworth projection**. In a Haworth projection, the ring is drawn in a plane at right angles to the page and positions of the substituents are indicated with up or down bonds. The shaded bonds are in front of the page, and the others are in back.

A Haworth projection does not indicate the conformation of the ring. Six-membered carbohydrate rings are like substituted cyclohexanes, and, like substituted cyclohexanes, exist in chair conformations. Thus, to complete the conformational representation of β-D-glucopyranose, draw either one of the two chair conformations in which the anomeric carbon and the ring oxygen are in the same relative positions as they are in the Haworth projection above. Then place *up* and *down* groups in axial or equatorial positions, as appropriate.

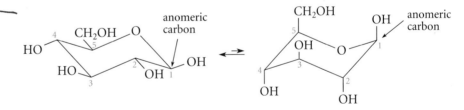

Remember: Although the chair flip changes equatorial groups to axial, and vice-versa, it does not change whether a group is up or down. Consequently, it doesn't matter which of the two possible chair conformations you draw first.

To summarize the conclusions of Study Problem 27.2: When a carbohydrate ring is drawn with the anomeric carbon on the right and the ring oxygen in the rear, substituents that are on the left in the Fischer projection are *up* in either the Haworth projection or the chair structures; groups that are on the right in the Fischer projection are *down* in either the Haworth projection or the chair structures.

Although the five-membered rings of furanoses are nonplanar, they are close enough to planarity that Haworth formulas are good approximations to their actual structures. Haworth projections are frequently used for furanoses for this reason. Thus, a Haworth projection of β-D-ribofuranose is derived as follows:

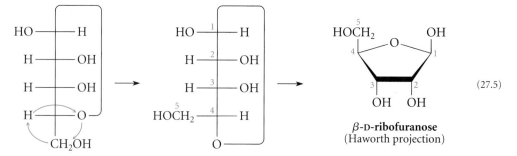

(27.5)

β-D-ribofuranose
(Haworth projection)

The Haworth formula is named for Sir Walter Norman Haworth (1883–1950), a noted British carbohydrate chemist who carried out important research on the cyclic structures of carbohydrates. Haworth received the Nobel Prize in Chemistry in 1937 and was knighted in 1947.

Although the procedure in Study Problem 27.2 can be used for any carbohydrate, in some cases it is sometimes simpler to derive a cyclic structure from its relationship to another cyclic structure. First, notice that the structure of β-D-glucopyranose is easy to remember because, in one conformation, all ring substituents are equatorial (Study Problem 27.2). Suppose, now, that we want to draw the conformation of β-D-galactopyranose. Because D-galactose and D-glucose are epimers at carbon-4, then the conformational representation of β-D-galactopyranose can be quickly derived by interchanging the —H and —OH groups at carbon-4 of β-D-glucopyranose.

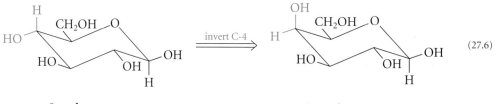

β-D-**glucopyranose**   β-D-**galactopyranose**

(27.6)

Likewise, because mannose and glucose are epimeric at carbon-2, the structure of a D-mannopyranose can be simply derived by interchanging the —H and —OH groups at carbon-2 of the corresponding D-glucopyranose structure.

Sometimes it becomes necessary to draw the conformation of a carbohydrate that either is a mixture of anomers or is of uncertain anomeric composition. In such cases, the configuration at the anomeric carbon is represented by a "squiggly bond."

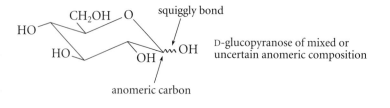

squiggly bond

D-glucopyranose of mixed or uncertain anomeric composition

anomeric carbon

PROBLEMS

26.4 Draw a Fischer projection, a Haworth projection, and, for the pyranoses, a chair structure for each of the following compounds.

*(a)   α-D-fructopyranose   (b)   β-D-mannopyranose

*(c)   β-D-xylofuranose   (d)   α-D-glucopyranose

*(e)   α-L-glucopyranose

(f)   a mixture of the α- and β-anomers of L-glucopyranose.

27.5   Name each of the following aldoses. In (a), work back to the Fischer projection and consult Fig. 27.1. In (b), decide which carbons have configurations epimeric to those of glucose, and which have the same configurations; then use Fig. 27.1.

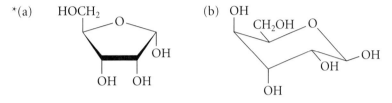

Mutarotation of Carbohydrates

When pure α-D-glucopyranose is dissolved in water, its specific rotation is found to be +112°. With time, however, the rotation of the solution decreases, ultimately reaching a stable value of +52.7° (Fig. 27.2). When pure β-D-glucopyranose is dissolved in water,

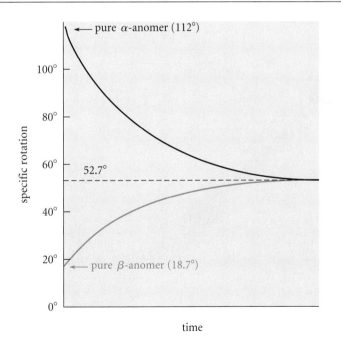

Figure 27.2   *Mutarotation of D-glucose. Equimolar aqueous solutions of pure α- or β-glucopyranose gradually change their specific optical rotations to the same final value that is characteristic of the equilibrium mixture.*

it has a specific rotation of $+18.7°$. The rotation of this solution increases with time, also to $+52.7°$. This change of optical rotation with time is called **mutarotation** (*muta =* change). Mutarotation also occurs when pure anomers of other carbohydrates are dissolved in aqueous solution.

The mutarotation of glucose is caused by the conversion of the α- and β-glucopyranose anomers into an equilibrium mixture of both. The same equilibrium mixture is formed, as it must be, from either pure α-D-glucopyranose or β-D-glucopyranose. Mutarotation is catalyzed by both acid and base, but also occurs even in pure water.

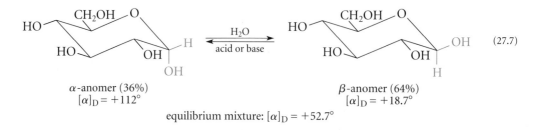

α-anomer (36%)
$[\alpha]_D = +112°$

β-anomer (64%)
$[\alpha]_D = +18.7°$

equilibrium mixture: $[\alpha]_D = +52.7°$

(27.7)

Mutarotation occurs, first, by opening of the pyranose ring to the free aldehyde form. This is nothing more than the reverse of hemiacetal formation (Sec. 19.10A). Then a 180° rotation about the carbon-carbon bond to the carbonyl group permits re-closure of the hemiacetal ring by attack of the hydroxy group on the opposite face of the carbonyl carbon.

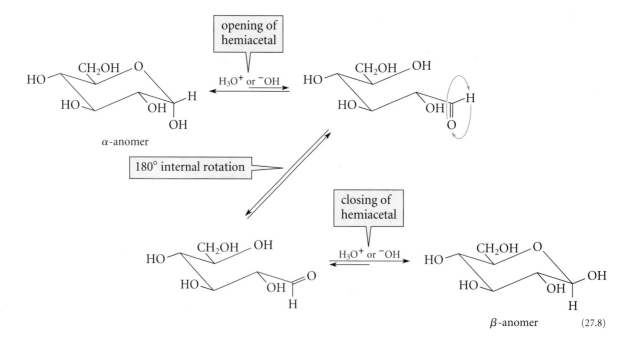

α-anomer

β-anomer   (27.8)

The mutarotation of glucose is due almost entirely to the interconversion of its two pyranose forms. Other carbohydrates undergo more complex mutarotations. For example, the crystalline form of D-fructose, a 2-ketohexose, is β-D-fructopyranose. When crystalline D-fructose is dissolved in aqueous solution, it equilibrates to both pyranose and furanose forms.

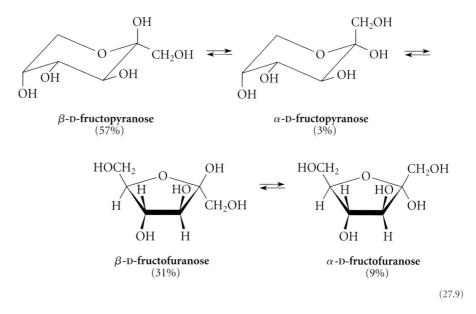

β-D-**fructopyranose**
(57%)

α-D-**fructopyranose**
(3%)

β-D-**fructofuranose**
(31%)

α-D-**fructofuranose**
(9%)

(27.9)

Glucose in solution also contains furanose forms, but these are present in very small amounts—about 0.2% each.

The foregoing discussion shows that a single hexose can exist in at least five forms: the acyclic aldehyde or ketone form, the α- and β-pyranose forms, and the α- and β-furanose forms. Modern techniques, particularly NMR spectroscopy, have enabled chemists to determine for many carbohydrates the amounts of the different forms that are present at equilibrium. The results for some monosaccharides are summarized in Table 27.1 on p. 1340.

Some general conclusions from this table are:

1. Most aldohexoses and aldopentoses exist primarily as pyranoses, although a few have substantial amounts of furanose forms.

2. There are relatively small amounts of noncyclic carbonyl forms of most monosaccharides.

3. Mixtures of α- and β-anomers are usually found, although the exact amounts of each vary from case to case.

The fraction of any form in solution at equilibrium is determined by its stability relative to that of all other forms. To predict the data in Table 27.1 for a given monosaccharide would require an understanding of all the factors that contribute to the stability or instability of *every* one of its isomeric forms in aqueous solution. In some cases, though, the principles of cyclohexane conformational analysis (Sec. 7.3, 7.4) can be applied, as Problem 27.8 illustrates.

Table 27.1   **Compositions of Monosaccharides at Equilibrium in Aqueous Solution at 40°**

| | *pyranose* | | *furanose* | | *aldehyde* |
Sugar	*α*	*β*	*α*	*β*	*or ketone*
D-glucose	36	64	trace[a]		0.003
D-galactose[b]	27–36	64–73	trace		trace
D-mannose	68	32	trace	0	trace
D-allose	18	70	5	7	
D-altrose	27	40	20	13	
D-idose[c]	39	36	11	14	
D-talose	40	29	20	11	
D-arabinose[d]	63	34		3	
D-xylose	37	63			
D-ribose	20	56	6	18	0.02
D-fructose	0–3	57–75	4–9	21–31	0.25

Percent at equilibrium

[a] In 10% aqueous dioxane, glucose contains 0.1–0.2% of each furanose.
[b] At 25 °C, galactose contains 29% α-pyranose, 64% β-pyranose, 3% α-furanose, and 4% β-furanose.
[c] 25 °C
[d] At 25 °C, arabinose contains 60% α-pyranose, 35% β-pyranose, 3% α-furanose, and 2% β-furanose.

PROBLEMS

*27.6   Using the curved-arrow formalism, fill in the details for acid-catalyzed mutarotation of glucopyranose shown in Eq. 27.8. Begin by protonating the ring oxygen.

27.7   Using the curved-arrow formalism, fill in the details for base-catalyzed mutarotation of glucopyranose. Begin by removing a proton from the hydroxy group at carbon-1.

*27.8   Consider the β-D-pyranose forms of glucose and talose. Suggest one reason why talose contains a smaller fraction of β-pyranose form than glucose.

27.9   Draw a conformational representation of:
(a) β-D-allopyranose   (b) α-D-idofuranose.

*27.10   From the specific rotations shown in Fig. 27.2, calculate the amounts of α- and β-D-glucopyranose present at equilibrium. (Assume that the amounts of aldehyde and furanose forms are negligible.) Compare your answer to the data given in Table 27.1.

27.4 Base-Catalyzed Isomerization of Aldoses and Ketoses

In base, aldoses and ketoses rapidly equilibrate to mixtures of other aldoses and ketoses.

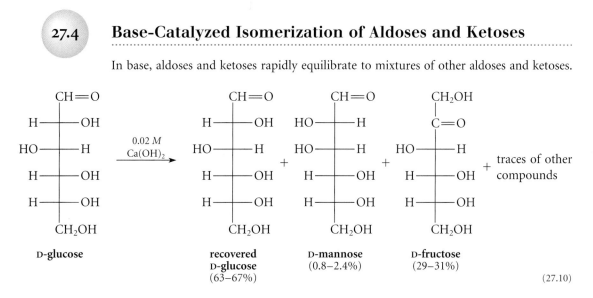

D-glucose → recovered D-glucose (63–67%) + D-mannose (0.8–2.4%) + D-fructose (29–31%) + traces of other compounds

(27.10)

This transformation is an example of the **Lobry de Bruyn-Alberda van Ekenstein reaction,** named for two Dutch chemists, Cornelius Adriaan van Troostenbery Lobry de Bruyn (1857–1904) and Willem Alberda van Ekenstein (1858–1907). Despite its rather formidable name, this reaction is a simple one, and is closely related to processes you have already studied.

Although glucose in solution exists mostly in its cyclic hemiacetal forms, it is also in equilibrium with a small amount of its acyclic aldehyde form. This aldehyde, like other carbonyl compounds with α-hydrogens, ionizes to give small amounts of its enolate ion in base. Protonation of this enolate ion at one face of the double bond gives back glucose; protonation at the other face gives mannose. This is much like the process shown in Eq. 22.6.

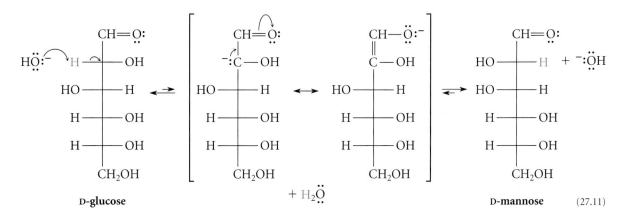

D-glucose ⇌ [... + H₂O] ⇌ D-mannose   (27.11)

The enolate ion can also be protonated on oxygen to give a new enol, called an **enediol.** An enediol contains a hydroxy group at each end of a double bond. The enediol derived from glucose is simultaneously the enol of not only the aldoses glucose and mannose, but also the ketose fructose.

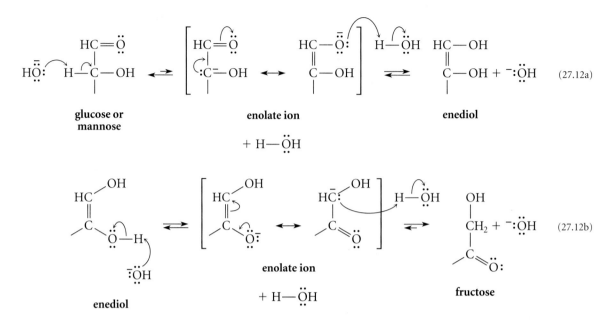

Such base-catalyzed epimerizations and aldose-ketose equilibria need not stop at carbon-2. For example, fructose epimerizes on prolonged treatment with base (why?).

Several transformations of this type are important in metabolism. One such reaction, the conversion of glucose-6-phosphate into fructose-6-phosphate, occurs in the breakdown of glucose (glycolysis), the series of reactions by which glucose is utilized as a food source. Since biochemical reactions occur near pH 7, there is too little hydroxide ion present to catalyze the reaction. Instead, the reaction is catalyzed by an enzyme, glucose-6-phosphate isomerase.

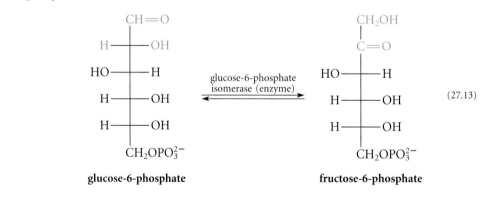

(27.13)

glucose-6-phosphate   **fructose-6-phosphate**

PROBLEM

27.11   Into what other aldose and 2-ketose would each of the following aldoses be transformed on treatment with base? Give the structure and name of the aldose, and the structure of the 2-ketose.

*(a) D-galactose   (b) D-allose

27.5   Glycosides

Most monosaccharides react with alcohols under acidic conditions to yield cyclic acetals.

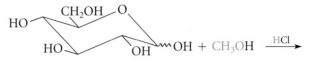

D-glucose

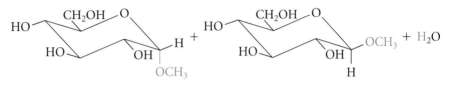

methyl α-D-glucopyranoside   **methyl β-D-glucopyranoside**
(83–85% yield; separated by fractional crystallization)   (27.14)

Such compounds are called **glycosides**. They are special types of acetals in which one of the oxygens of the acetal linkage is the ring oxygen of the pyranose or furanose.

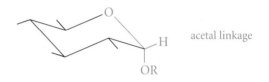

As illustrated in Eq. 27.14, glycosides are named as derivatives of the parent carbohydrate. The term *pyranoside* indicates that the glycoside ring is a six-membered ring. The term *furanoside* is used for a five-membered ring.

Glycoside formation, like acetal formation, is catalyzed by acid and involves an α-alkoxy carbocation intermediate (Sec. 19.6).

STUDY GUIDE LINK:
✓ 27.2
*Acid Catalysis of
Carbohydrate
Reactions*

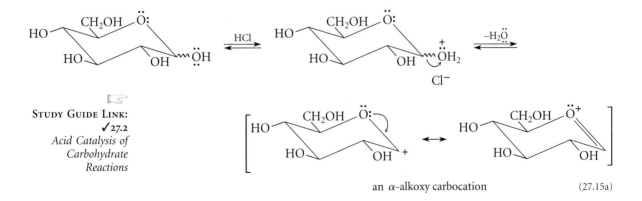

an α-alkoxy carbocation   (27.15a)

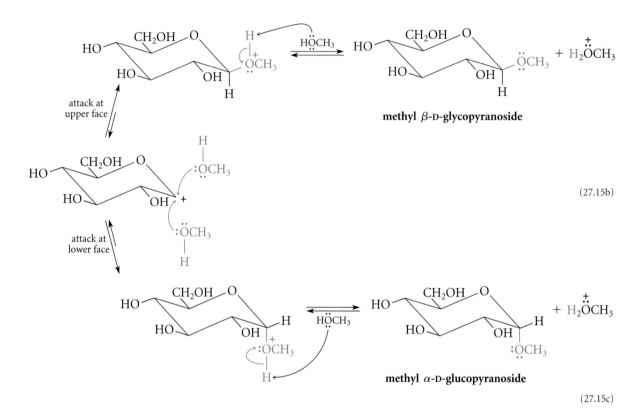

methyl β-D-glycopyranoside

(27.15b)

methyl α-D-glucopyranoside

(27.15c)

It is useful to contrast what happens to a carbohydrate and what happens to an ordinary aldehyde or ketone in the presence of alcohol and acid. When an aldehyde reacts with methanol to form an acetal (Sec. 19.10A), two additional carbon atoms are incorporated into the molecule. When glucose forms a methyl glycoside, only one additional carbon atom is incorporated; the other part of the acetal group is derived from the carbohydrate itself.

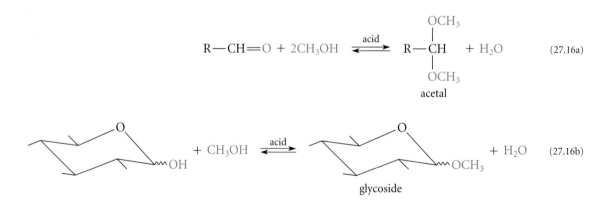

This difference between aldoses and ordinary aldehydes is one piece of evidence that led early chemists to suspect the cyclic (pyranose or furanose) structure of carbohydrates.

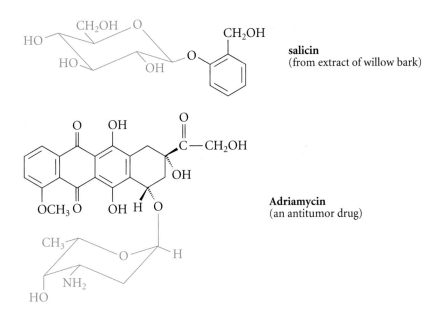

salicin
(from extract of willow bark)

Adriamycin
(an antitumor drug)

Figure 27.3   *Two naturally occurring glycosides of medicinal interest. The carbohydrate part of each glycoside is shown in color.*

Like other acetals, glycosides are *stable to base*, but are hydrolyzed in dilute aqueous acid back to their parent carbohydrates.

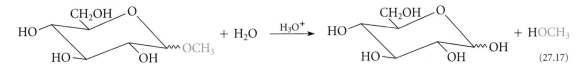

(27.17)

Many compounds occur naturally as glycosides; two examples are shown in Fig. 27.3. In addition, glycoside formation plays an important role in the removal of some chemicals from the body. In this process, a carbohydrate is joined to an —OH group of the substance to be removed. The added carbohydrate group makes the substance more soluble in water and, hence, more easily excreted.

Like simple methyl glycosides, the glycoside of a natural product can be hydrolyzed to its component alcohol or phenol and carbohydrate.

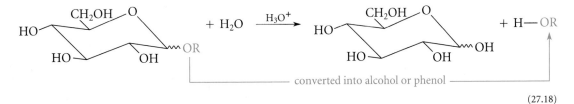

converted into alcohol or phenol

(27.18)

*27.12   (a)   Name the following glycoside.

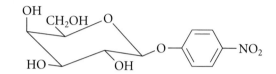

(b)   Into what products will this glycoside be hydrolyzed in aqueous acid?

27.13   Vanillin (the natural vanilla flavoring) occurs in nature as a β-glycoside of glucose. Suggest a structure for this glycoside.

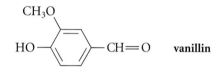

vanillin

27.14   Draw structures for:
     *(a)   methyl β-D-fructofuranoside   (b)   isopropyl α-D-galactopyranoside

27.6   Ether and Ester Derivatives of Carbohydrates

Because carbohydrates contain many —OH groups, it should not be surprising that carbohydrates undergo many of the reactions of alcohols. One such reaction is ether formation. In the presence of concentrated base, carbohydrates are converted into ethers by reactive alkylating agents such as dimethyl sulfate, methyl iodide, or benzyl chloride.

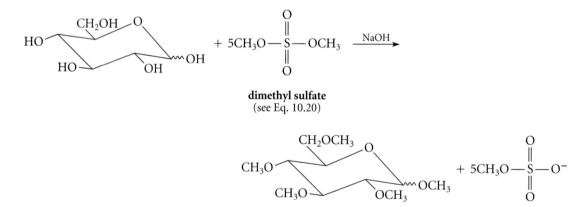

dimethyl sulfate
(see Eq. 10.20)

methyl 2,3,4,6-tetra-
O-methyl-D-glucopyranoside

(27.19)

(Note that the ethers are named as *O*-alkyl derivatives of the carbohydrates.) These reactions are examples of the Williamson ether synthesis (Sec. 11.1A). This synthesis with most alcohols requires a base stronger than ⁻OH to form the conjugate-base alkoxide.

The hydroxy groups of carbohydrates, however, are more acidic ($pK_a \approx 12$) than those of ordinary alcohols. (The higher acidity of carbohydrate hydroxy groups is attributable to the polar effect of the many neighboring oxygens in the molecule.) Consequently, substantial concentrations of their conjugate-base alkoxide ions are formed in concentrated NaOH. A large excess of the alkylating reagent is used because hydroxide itself, present in large excess, also reacts with alkylating agents. It is interesting that little or no base-catalyzed epimerization (Sec. 27.4) is observed in this reaction, despite the strongly basic conditions used. Evidently, alkylation of the hydroxy group at the anomeric carbon is much faster than epimerization. Once this oxygen is alkylated, epimerization can no longer occur (why?).

Other reagents used to form methyl ethers of carbohydrates include CH_3I/Ag_2O, and the strongly basic $NaNH_2$ (sodium amide) in liquid NH_3 followed by CH_3I.

Remember that the alkoxy group at the anomeric carbon is different from the other alkoxy groups in an alkylated carbohydrate because it is part of the glycosidic linkage. Because it is an acetal, it can be hydrolyzed in aqueous acid under mild conditions:

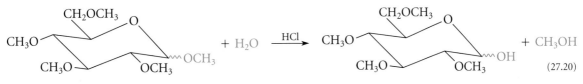

(27.20)

The other alkoxy groups are ordinary ethers and do not hydrolyze under these conditions. They require *much* stronger conditions for cleavage (Sec. 11.3).

Another reaction of alcohols is esterification; indeed, the hydroxy groups of carbohydrates, like those of other alcohols, can be esterified.

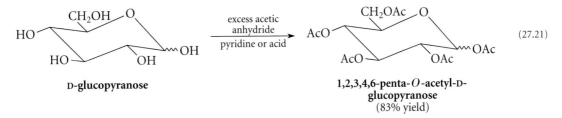

(27.21)

D-glucopyranose

1,2,3,4,6-penta-*O*-acetyl-D-glucopyranose
(83% yield)

Ester derivatives of carbohydrates can be saponified in base or removed by transesterification with an alkoxide such as methoxide:

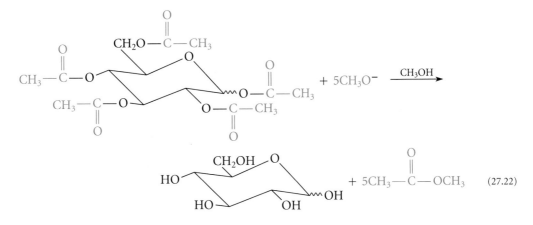

(27.22)

Ethers and esters are used as protecting groups in reactions involving carbohydrates. Because ethers and esters of carbohydrates have broader solubility characteristics and greater volatility than the carbohydrates themselves, they also find use in the characterization of carbohydrates by chromatography and mass spectrometry.

STUDY PROBLEM 27.3

Outline a sequence of reactions by which glucose can be converted into methyl 2,3,4,6-tetra-*O*-acetylglucopyranoside.

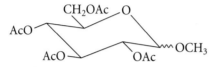 **methyl 2,3,4,6-tetra-*O*-acetylglucopyranoside**

Solution In solving problems of this sort, in which apparently similar hydroxy groups are converted into different derivatives, the key is to recognize that hydroxy groups and alkoxy groups at the anomeric position (carbon-1 in aldoses) behave quite differently than these groups at other positions. As the text discussion above shows, the hydroxy group at carbon-1 of glucose is part of a *hemiacetal* group, and alkoxy groups at the same carbon are part of an *acetal* group. Acetals are formed and hydrolyzed under much milder conditions than ordinary ethers. Consequently, the methyl "ether" (actually, a methyl acetal) at carbon-1 can be formed by treating glucose with methanol and acid. The remaining hydroxy groups can then be esterified with excess acetic anhydride (as in Eq. 27.21) to give the desired product:

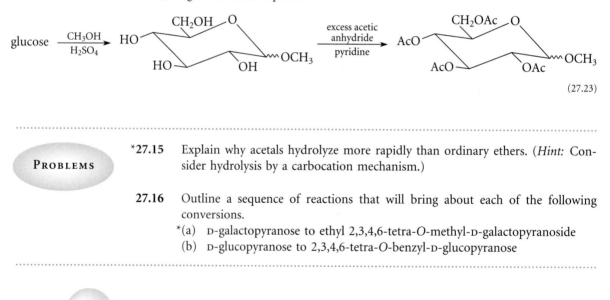

(27.23)

PROBLEMS

*27.15 Explain why acetals hydrolyze more rapidly than ordinary ethers. (*Hint:* Consider hydrolysis by a carbocation mechanism.)

27.16 Outline a sequence of reactions that will bring about each of the following conversions.
  *(a) D-galactopyranose to ethyl 2,3,4,6-tetra-*O*-methyl-D-galactopyranoside
  (b) D-glucopyranose to 2,3,4,6-tetra-*O*-benzyl-D-glucopyranose

27.7

Oxidation and Reduction Reactions of Carbohydrates

Like simpler aldehydes, the aldehyde group of aldoses can be both oxidized and reduced. It is also possible to oxidize selectively the primary alcohol group of an aldose. The structures and names of the common oxidation and reduction products of aldoses are summarized in Table 27.2.

Table 27.2 Structures of Common Oxidation and Reduction Products of Aldoses

General structure:

$$X \left[\begin{array}{c} CH \\ | \\ OH \end{array} \right]_n Y$$

Derivative structure		General name	Example derived from glucose
$X-$ =	$-Y$ =		
$HOCH_2-$	$-CH=O$	aldose	glucose
$HOCH_2-$	$-CO_2H$	aldonic acid	gluconic acid
HO_2C-	$-CO_2H$	aldaric acid	glucaric acid
$HOCH_2-$	$-CH_2OH$	alditol	glucitol
HO_2C-	$-CH=O$	uronic acid	glucuronic acid

PROBLEM

27.17 Using Table 27.2 to assist you, draw a Fischer projection for the structure of *(a) galacturonic acid, the uronic acid derived from galactose; (b) ribitol, the alditol derived from ribose.

A. Oxidation to Aldonic Acids

Treatment of an aldose with bromine water oxidizes the aldehyde group to a carboxylic acid. The oxidation product is an *aldonic acid* (see Table 27.2).

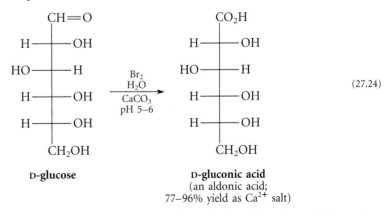

D-glucose

D-gluconic acid
(an aldonic acid;
77–96% yield as Ca^{2+} salt)

(27.24)

Although it is customary to represent aldonic acids in the free carboxylic acid form, they, like other γ- and δ-hydroxy acids (Sec. 20.8A), exist in acidic solution as lactones called *aldonolactones*. The lactones with five-membered rings are somewhat more stable than those with six-membered rings.

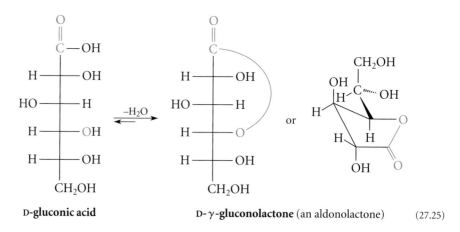

D-gluconic acid   D-γ-gluconolactone (an aldonolactone)   (27.25)

Oxidation with bromine water is a useful test for aldoses. Aldoses can also be oxidized with other reagents, for example, Tollens' reagent ($Ag^+(NH_3)_2$; Sec. 19.14). However, because Tollens' reagent is alkaline and causes base-catalyzed epimerization of aldoses (Sec. 27.4), it is less useful synthetically. Because the alkaline conditions of Tollens' test also promote the equilibration of aldoses and ketoses, ketoses also give positive Tollens' tests. Glycosides are *not* oxidized by bromine water, because the aldehyde carbonyl group is protected as an acetal.

B. Oxidation to Aldaric Acids

Dilute nitric acid oxidizes aldehydes and primary alcohols to carboxylic acids *without affecting secondary alcohols.* Consequently, this is a very useful reagent for converting aldoses (or aldonic acids) into aldaric acids (Table 27.2).

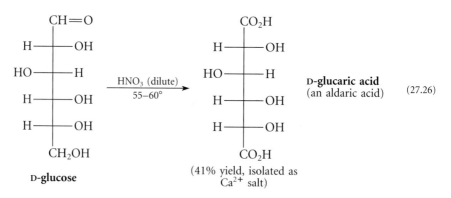

D-glucose   (41% yield, isolated as Ca^{2+} salt)   D-glucaric acid (an aldaric acid)   (27.26)

Like aldonic acids, aldaric acids in acidic solution form lactones. Two different five-membered lactones are possible, depending on which carboxylic acid group undergoes lactonization. Furthermore, under certain conditions, some aldaric acids can be isolated as dilactones, in which both carboxylic acid groups are lactonized.

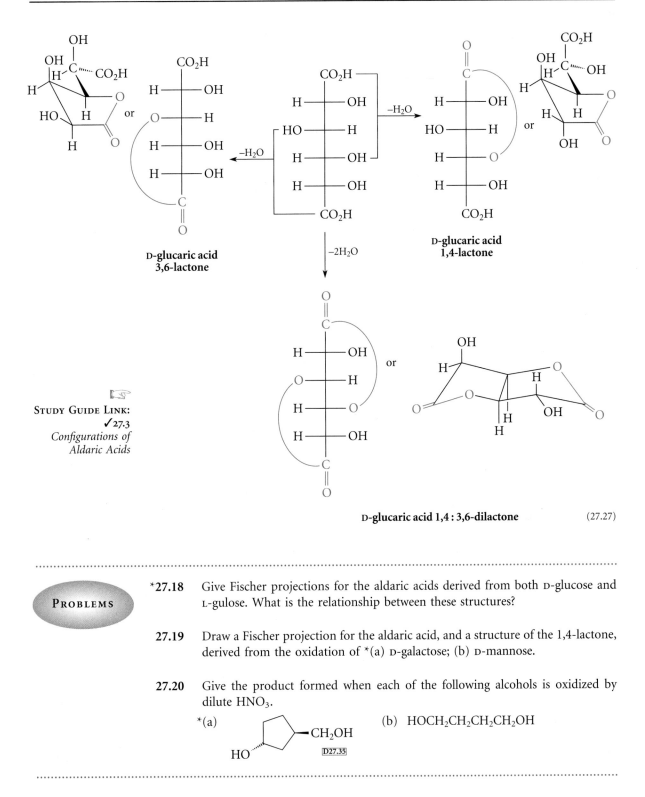

D-glucaric acid 3,6-lactone

D-glucaric acid 1,4-lactone

STUDY GUIDE LINK:
✓27.3
Configurations of Aldaric Acids

D-glucaric acid 1,4 : 3,6-dilactone   (27.27)

....

PROBLEMS

*27.18   Give Fischer projections for the aldaric acids derived from both D-glucose and L-gulose. What is the relationship between these structures?

27.19   Draw a Fischer projection for the aldaric acid, and a structure of the 1,4-lactone, derived from the oxidation of *(a) D-galactose; (b) D-mannose.

27.20   Give the product formed when each of the following alcohols is oxidized by dilute HNO_3.

*(a)   [structure: cyclopentane with —CH₂OH and HO substituents]   D27.35

(b) $HOCH_2CH_2CH_2CH_2OH$

C. Periodate Oxidation

Many carbohydrates contain vicinal glycol units and, like other 1,2-glycols, are oxidized by periodic acid (Sec. 10.6C). A complication arises when, as in many carbohydrates, more than two adjacent carbons bear hydroxy groups. When one of the oxidation products is an α-hydroxy aldehyde, as in the following example, it is oxidized further to formic acid and another aldehyde.

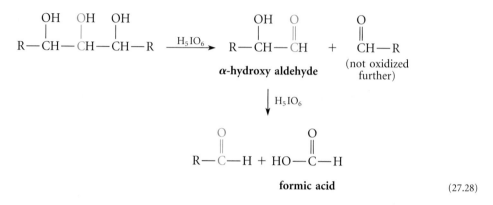

By analogy, an α-hydroxy ketone is oxidized to an aldehyde and a carboxylic acid.

Because it is possible to determine accurately both the amount of periodate consumed and the amount of formic acid produced, periodate oxidation can be used to differentiate between pyranose and furanose structures of saccharide derivatives. For example, periodate oxidation of methyl α-D-glucopyranoside liberates one equivalent of formic acid:

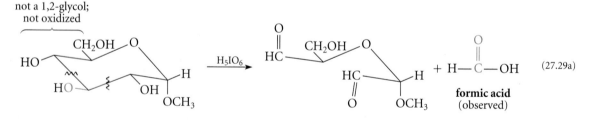

A furanose form of this glycoside, however, would liberate only formaldehyde:

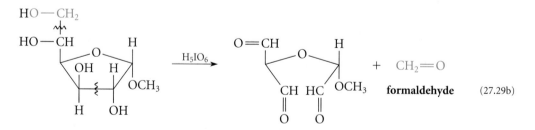

The periodate oxidation of carbohydrates was developed by C. S. Hudson (1881–1952), a noted American carbohydrate chemist. It was used extensively to relate the anomeric configurations of many carbohydrate derivatives. How this was done is suggested by Problems 27.21 and 27.22.

PROBLEMS

*27.21   Explain why the methyl α-D-pyranosides of *all* D-aldohexoses give, in addition to formic acid, the same compound when oxidized by periodate.

27.22   Assuming you knew the properties of the compound obtained in Problem 27.21, including its optical rotation, show how you could use periodate oxidation to distinguish methyl α-D-galactopyranoside from methyl β-D-galactopyranoside.

D. Reduction to Alditols

Aldohexoses, like ordinary aldehydes, undergo many of the usual carbonyl reductions. For example, sodium borohydride ($NaBH_4$, Sec. 19.8) reduces aldoses to alditols (Table 27.2).

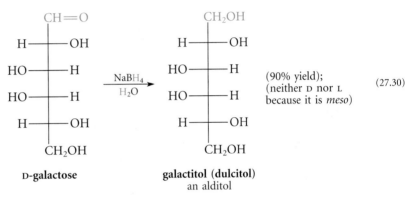

D-galactose   galactitol (dulcitol)
                          an alditol

Catalytic hydrogenation (for example, H_2 with a Raney nickel catalyst in aqueous ethanol) can also be used for the same transformation.

   In the oxidation and reduction reactions discussed in this section aldoses have been depicted in their carbonyl forms rather than in their cyclic hemiacetal forms. Do not lose sight of the fact that all forms are present at equilibrium, and the aldehyde form can react even though it is present in a very small amount. Once it reacts, it is immediately replenished (LeChatelier's principle). Thus, when the aldehyde group reacts with $NaBH_4$ to give alditol, the equilibrium provides more of the aldehyde form:

$$\text{cyclic (hemiacetal) forms} \;\rightleftharpoons\; \text{aldehyde form} \;\xrightarrow[\text{H}_2\text{O}]{\text{NaBH}_4}\; \text{alditol} \qquad (27.31)$$

Furthermore, the acidic or basic conditions of many aldehyde reactions (basic conditions in the case of $NaBH_4$ reduction) catalyze this equilibrium. Not only is more aldehyde formed, but it is formed *rapidly*.

27.8   Kiliani-Fischer Synthesis

Aldoses, like other aldehydes, add hydrogen cyanide to give cyanohydrins (Sec. 19.7). Notice that the following reaction, like several others that have been discussed, involves the aldehyde form of the sugar.

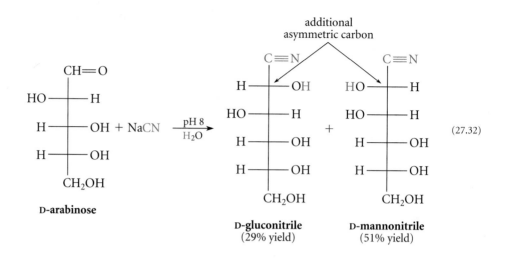

D-arabinose

D-gluconitrile
(29% yield)

D-mannonitrile
(51% yield)

(27.32)

Because the cyanohydrin product has an additional asymmetric carbon, it is formed as a mixture of two epimers. Because these epimers are diastereomers, they are typically formed in different amounts (Sec. 7.8B). Although the exact amount of each is not easily predicted, in most cases significant amounts of both are obtained.

The mixture of cyanohydrins can be converted into a mixture of aldoses by catalytic hydrogenation, and these aldoses can be separated.

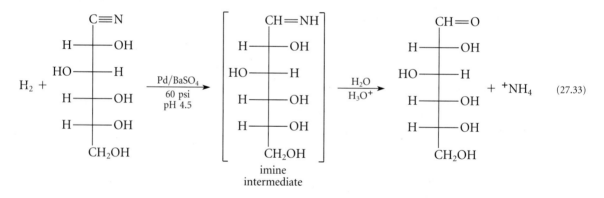

imine
intermediate

(27.33)

As this equation shows, the hydrogenation reaction involves reduction of the nitrile to an imine (or a cyclic carbinolamine derivative of the imine). Under the reaction conditions, the imine hydrolyzes readily to the aldose and ammonium ion.

This example shows that cyanohydrin formation followed by reduction converts an aldose into two epimeric aldoses with one additional carbon. That is, two aldohexoses, epimeric at carbon-2, are formed from an aldopentose. Notice particularly that this synthesis does not affect the stereochemistry of carbons 2, 3, and 4 in the starting material.

The formation of cyanohydrins from aldoses was developed by Heinrich Kiliani (1855–1945), Head of the Medicinal Chemistry Laboratory at the University of Freiburg. Kiliani also showed that the cyanohydrins could be hydrolyzed to aldonic acids. Emil Fischer, whose remarkable accomplishments in carbohydrate chemistry are described in the following section, developed a method to reduce the aldonic acids (as their lactones) to aldoses. The three processes—cyanohydrin formation, hydrolysis, and reduction—provided a way to convert an aldose into two other aldoses with one additional carbon. The overall transformation came to be known as the **Kiliani-Fischer synthesis**. The chemistry shown in Eqs. 27.32–27.33 is a modern variation of the Kiliani-Fischer synthesis developed in the 1970s.

Kiliani, a noted authority on carbohydrates, also proved the structures of several monosaccharides, including the 2-ketose structure of fructose.

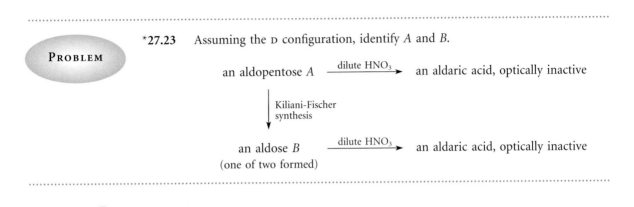

PROBLEM

*27.23   Assuming the D configuration, identify A and B.

an aldopentose A $\xrightarrow{\text{dilute } HNO_3}$ an aldaric acid, optically inactive

Kiliani-Fischer
synthesis

an aldose B $\xrightarrow{\text{dilute } HNO_3}$ an aldaric acid, optically inactive
(one of two formed)

27.9   Proof of Glucose Stereochemistry

The aldohexose structure of (+)-glucose was established around 1870. The van't Hoff-LeBel theory of the tetrahedral carbon atom, published in 1874 (Sec. 6.12), suggested the possibility that glucose and the other aldohexoses could be stereoisomers. The problem to be solved, then, was: Which one of the 2^4 possible stereoisomers is glucose? This problem was solved in two stages.

A.   Which Diastereomer? The Fischer Proof

The first (and major) part of the solution to the problem of glucose stereochemistry was published in 1891 by Emil Fischer, a German chemist who carried out landmark investigations in several fields of organic chemistry. It would be reason enough to study Fischer's proof as one of the most brilliant pieces of reasoning in the history of chemistry. However, it also will serve to sharpen your understanding of stereochemical relationships.

It is important to understand that in Fischer's day there was no way to determine the absolute stereochemical configuration of any chemical compound. Consequently, Fischer arbitrarily *assumed* that carbon-5 (the configurational carbon in the D,L system) of (+)-glucose has the —OH on the right in the standard Fischer projection; that is, Fischer assumed that (+)-glucose has what we now call the D configuration. No one knew whether this assumption was correct; the solution to this problem had to await the development of special physical methods not available in Fischer's day. If Fischer's guess had been wrong, then it would have been necessary to reverse all of his stereochemical assignments. Fischer, then, proved the stereochemistry of (+)-glucose *relative* to an assumed configuration at carbon-5. The remarkable thing about his proof is that it allowed him to assign relative configurations in space using only chemical reactions and optical activity. The logic involved is direct, simple, and elegant, and can be summarized in four steps:

Step 1   (−)-Arabinose is converted into both (+)-glucose and (+)-mannose by a Kiliani-Fischer synthesis. From this fact (see Sec. 27.8), Fischer deduced that (+)-glucose and (+)-mannose are epimeric at carbon-2, and that the configuration of (−)-arabinose at carbons 2, 3, and 4 is the same as that of (+)-glucose and (+)-mannose at carbons 3, 4, and 5, repectively.

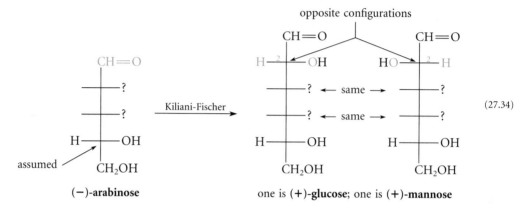

(27.34)

(−)-arabinose   one is **(+)-glucose**; one is **(+)-mannose**

Step 2   (−)-Arabinose can be oxidized by dilute HNO_3 (Sec. 27.7B) to an optically active aldaric acid. From this, Fischer concluded that the —OH group at carbon-2 of arabinose must be on the left. If this —OH group were on the right, then the aldaric acid of arabinose would have to be *meso*, and thus optically inactive, *regardless of the configuration of the —OH group at carbon-3*. (Be sure you see why this is so; if necessary, draw both possible structures for (−)-arabinose to verify this deduction.)

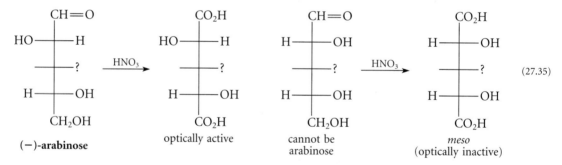

(27.35)

(−)-arabinose   optically active   cannot be arabinose   *meso* (optically inactive)

The relationships among arabinose, glucose, and mannose established in Steps 1 and 2 require the following partial structures for (+)-glucose and (+)-mannose.

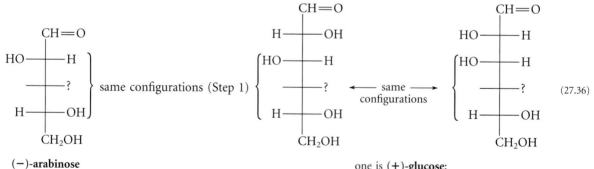

(27.36)

(−)-arabinose   one is **(+)-glucose**; one is **(+)-mannose**

Step 3   Oxidations of *both* (+)-glucose and (+)-mannose with HNO_3 give optically active aldaric acids. From this, Fischer deduced that the —OH group at carbon-4 is on the right in both (+)-glucose and (+)-mannose. Recall that whatever the configuration at carbon-4 in these two aldohexoses, it must be the same in both. Only if the —OH is on the right will *both* structures yield, on oxidation, optically active aldaric acids. If the —OH were on the left, *one* of the two aldohexoses would have given a *meso*, and hence, optically inactive, aldaric acid.

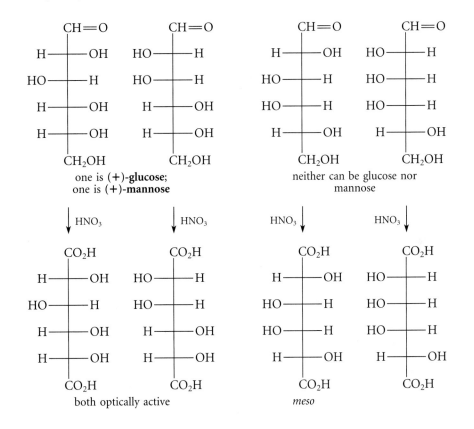

one is (**+**)-**glucose**;
one is (**+**)-**mannose**

neither can be glucose nor
mannose

both optically active

meso

Because the configuration at carbon-4 of (+)-glucose and (+)-mannose is the same as that at carbon-3 of (−)-arabinose (Step 1), at this point Fischer could deduce the complete structure of (−)-arabinose.

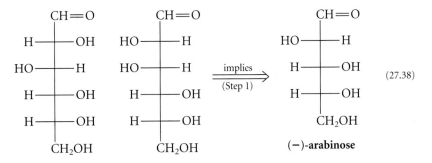

(27.38)

one is (**+**)-**mannose**; one is (**+**)-**glucose**

(−)-**arabinose**

Step 4 The previous steps had established that (+)-glucose had one of the two structures in Eq. 27.38 and (+)-mannose had the other, but Fischer did not yet know which structure goes with which sugar. This point is confusing to some students. Fischer's situation was like that of a young man who has just met two sisters, but he doesn't know their names. So he asks a friend: "What are their names?" The friend says, "Oh, they are Mannose and Glucose; only I don't know which is which!" Just because the young man knows *both* names doesn't mean that he can associate *each* name with *each* face. Similarly, although Fischer knew the structures associated with both (+)-glucose and (+)-mannose, he did not yet know how to correlate *each* aldose with *each* structure.

This problem was solved when Fischer found that another aldose, (+)-gulose, can be oxidized with HNO_3 to the same aldaric acid as (+)-glucose. (Fischer had synthesized (+)-gulose in the course of his research.) How does this fact differentiate between (+)-glucose and (+)-mannose? Two *different* aldoses can give the same aldaric acid only if their —CH=O and —CH$_2$OH groups are *at opposite ends of an otherwise identical molecule* (Problem 27.18). Interchange of the —CH$_2$OH and —CH=O groups in one of the aldohexose structures in Eq. 27.38 gives the *same* aldohexose. (You should verify that these two structures are identical by rotating either one 180° in the plane of the page and comparing it to the other.)

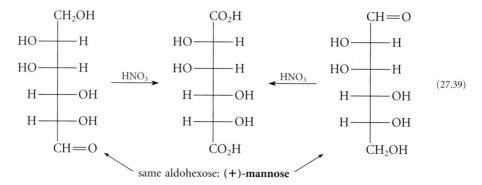

$$(27.39)$$

same aldohexose: **(+)-mannose**

Because there is only one adohexose that can be oxidized to this aldaric acid, that aldohexose cannot be (+)-glucose; therefore it must be (+)-mannose. Interchanging the end groups of the other aldohexose structure in Eq. 27.38 gives a *different* aldose:

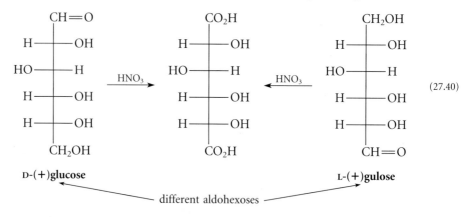

$$(27.40)$$

D-(+)glucose   L-(+)gulose

— different aldohexoses —

Consequently, one of these two structures must be that of (+)-glucose. *Only the structure on the left* is one of the possibilities listed in Eq. 27.38; consequently, this is the structure of (+)-glucose. The structure on the right of Eq. 27.40, then, is that of (+)-gulose. (Note that (+)-gulose has the L-configuration; that is, the —OH group at carbon-5 in the standard Fischer projection is on the *left*. Rotate the (+)-gulose structure 180° in the plane of the page to see this.)

STUDY GUIDE LINK:
27.4
More on the Fischer Proof

EMIL FISCHER

Emil Fischer (1852–1919) studied with Adolph von Baeyer and ultimately became Professor at Berlin in 1892. Fischer carried out important research on sugars, proteins (he devised the first rational syntheses of peptides), and heterocycles (for example, the Fischer indole synthesis). Fischer was a technical advisor to Kaiser Wilhelm. The following story gives some indication of the authority that Fischer commanded in Germany. It is said that one day he and the Kaiser were arguing questions of science policy, and the Kaiser sought to end debate by pounding his fist on the table, shouting, "Ich bin der Kaiser!" (I am the King!) Fischer, not to be silenced, responded in kind: "Ich bin Fischer!" Another story, perhaps apocryphal, attributes an important laboratory function to Fischer's long, flowing beard. It was said that when a student had difficulty crystallizing a sugar derivative (some of which are notoriously difficult to crystallize), Fischer would shake his beard over the flask containing the recalcitrant compound. The accumulated seed crystals in his beard would fall into the flask and bring about the desired crystallization. Fischer was awarded the Nobel Prize in 1902.

PROBLEMS

*27.24 An aldopentose A can be oxidized with dilute HNO_3 to an optically active aldaric acid. A Kiliani-Fischer synthesis starting with A gives two new aldoses B and C. Aldose B can be oxidized to an achiral, and therefore optically inactive, aldaric acid, but aldose C is oxidized to an optically active aldaric acid. Assuming the D configuration, give the structures of A, B, and C.

27.25 An aldohexose A is either D-idose or D-gulose (see Fig. 27.1). It is found that a different aldohexose, L-(−)-glucose, gives the same aldaric acid as A. What is the identity of A?

B. Which Enantiomer? The Absolute Configuration of D-(+)-Glucose

Fischer never learned whether his arbitrary assignment of the absolute configuration of (+)-glucose was correct, that is, whether the —OH at carbon-5 of (+)-glucose was really on the right in its Fischer projection (as assumed) or on the left. The groundwork for solving this problem was laid when the configuration of (+)-glucose was correlated to that of (−)-tartaric acid. (Stereochemical correlation was introduced in Sec. 6.5.) This was done in the following way.

(+)-Glucose was converted into (−)-arabinose by a reaction called the **Ruff degradation**. In this reaction sequence, an aldose is oxidized to its aldonic acid (Sec. 27.7A), and the calcium salt of the aldonic acid is treated with ferric ion and hydrogen peroxide. This treatment decarboxylates the calcium salt and simultaneously oxidizes carbon-2 to an aldehyde.

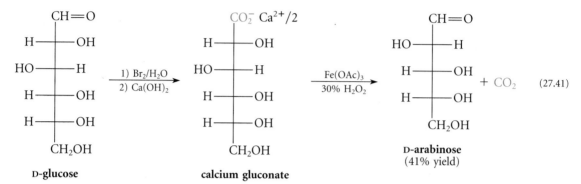

D-glucose	**calcium gluconate**

D-arabinose
(41% yield)

(27.41)

In other words, an aldose is degraded to another aldose with one fewer carbon atom, *its stereochemistry otherwise remaining the same.* Because the relationship between (+)-glucose and (−)-arabinose was already known from the Kiliani-Fischer synthesis (see Step 1 of the Fischer proof in the last section), this reaction served to establish the course of the Ruff degradation. Next, (−)-arabinose was converted into (−)-erythrose by another cycle of the Ruff degradation.

$$(-)\text{-arabinose} \xrightarrow{\text{Ruff degradation}} (-)\text{-erythrose} \qquad (27.42)$$

D-Glyceraldehyde, in turn, was related to (−)-erythrose by a Kiliani-Fischer synthesis:

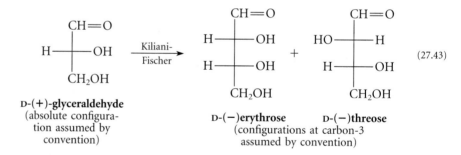

(27.43)

D-(+)-glyceraldehyde
(absolute configura-
tion assumed by
convention)

D-(−)erythrose   **D-(−)threose**
(configurations at carbon-3
assumed by convention)

This sequence of reactions showed that (+)-glucose, (−)-erythrose, (−)-threose, and (+)-glyceraldehyde were all of the same stereochemical series—the D series. Oxidation of D-(−)-threose with dilute HNO_3 gave D-(−)-tartaric acid.

In 1950 the absolute configuration of naturally occurring (+)-tartaric acid (as its potassium rubidium double salt) was determined by a special technique of X-ray crystallography called *anomalous dispersion*. This determination was made by J.M. Bijvoet, A.F. Peerdeman, and A.J. van Bommel, Dutch chemists who worked, appropriately enough, at the van't Hoff laboratory in Utrecht. If Fischer had made the right choice for the D configuration, the assumed structure for D-(−)-tartaric acid and the experimentally determined structure of (+)-tartaric acid determined by the Dutch crystallographers would be enantiomers. If Fischer had guessed incorrectly, the assumed structure for (−)-tartaric acid would be the same as the experimentally determined structure of

(+)-tartaric acid, and would have to be reversed. To quote Bijvoet and his colleagues: *"The result is that Emil Fischer's convention* [for the D configuration] *appears to answer to reality."*

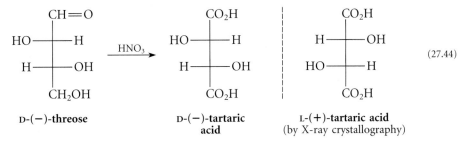

(27.44)

D-(−)-threose   D-(−)-tartaric   L-(+)-tartaric acid
                                  acid   (by X-ray crystallography)

PROBLEMS

*27.26   Given the structure of D-glyceraldehyde, how would you assign a structure to each of the two aldoses obtained from it by Eq. 27.43, assuming that these compounds were previously unknown?

27.27   Imagine that a scientist reexamines the crystallographic work that established the absolute configuration of (+)-tartaric acid and finds that the structure of this compound is the mirror image of the one given in the text above. What changes would have to be made in Fischer's structure of D-(+)-glucose?

27.10   Disaccharides and Polysaccharides

A.   Disaccharides

Disaccharides consist of two monosaccharides connected by a glycosidic linkage. **(+)-Lactose** is an example of a disaccharide. ((+)-Lactose is present to the extent of about 4.5% in cow's milk and 6–7% in human milk.)

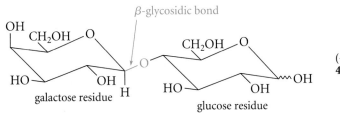

(+)-**lactose** or
4-O-(β-D-**galactopyranosyl**)-D-**glucopyranose**

In (+)-lactose, a D-glucopyranose molecule is linked by its oxygen at carbon-4 to carbon-1 of D-galactopyranose. In effect, (+)-lactose is a glycoside in which galactose is the carbohydrate and glucose is the "alcohol." Recall that the glycosidic linkage is an acetal, and therefore hydrolyzes under acidic conditions (Sec. 27.5). Hence, (+)-lactose can be hydrolyzed in acidic solution to give one equivalent each of D-glucose and D-galactose, in the same sense that a methyl glycoside can be hydrolyzed to give methanol and a carbohydrate.

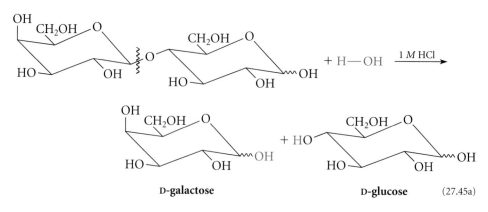

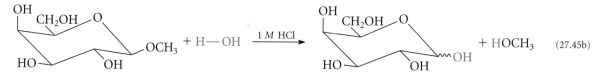

Certain enzymes also catalyze the same reaction at pH values near neutrality. Note that Eq. 27.45a demonstrates the structural basis for the definition of disaccharides presented in Sec. 27.1: A disaccharide is a carbohydrate that can be hydrolyzed to two monosaccharides. Hydrolysis occurs at the glycosidic bond between the two monosaccharide residues.

The stereochemistry of the glycosidic bond in (+)-lactose is β. That is, the stereochemistry of the oxygen linking the two monosaccharide residues in the glycosidic bond corresponds to that in the β-anomer of D-galactopyranose. This stereochemistry is very important in biology, because higher animals possess an enzyme, β-galactosidase, that catalyzes the hydrolysis of this β-glycosidic linkage; this hydrolysis allows lactose to act as a source of glucose. α-Glycosides of galactose are inert to the action of this enzyme.

Because carbon-1 of the galactose residue in (+)-lactose is involved in a glycosidic linkage, it cannot be oxidized. However, carbon-1 of the glucose residue is part of a hemiacetal group, which, like the hemiacetal group of monosaccharides, is in equilibrium with the free aldehyde, and can undergo characteristic aldehyde reactions. Thus, oxidation of (+)-lactose with bromine water (Sec. 27.7A) effects oxidation of the glucose residue:

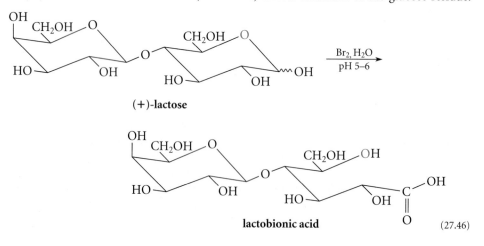

Carbohydrates such as (+)-lactose that can be oxidized in this way are called **reducing sugars**. The glucose residue is said to be at the *reducing end* of the disaccharide, and the galactose residue at the *nonreducing end*. Because of its hemiacetal group, (+)-lactose also undergoes many other reactions of aldose hemiacetals, such as mutarotation.

(+)-**Sucrose**, or table sugar, is another important disaccharide. About ninety million tons of sucrose are produced annually in the world. Sucrose consists of a D-glucopyranose residue and a D-fructofuranose residue connected by a glycosidic bond (color) at the anomeric carbons of *both* monosaccharides.

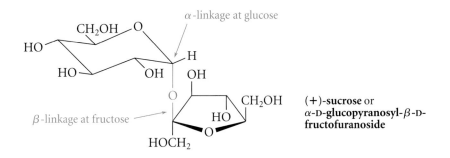

(+)-**sucrose** or
α-D-**glucopyranosyl-β-D-fructofuranoside**

The glycosidic bond in (+)-sucrose is different from the one in lactose. Only one of the residues of lactose—the galactose residue—contains an acetal (glycosidic) carbon. In contrast, *both* residues of (+)-sucrose have an acetal carbon. The glycosidic bond in (+)-sucrose bridges carbon-2 of the fructofuranose residue and carbon-1 of the glucopyranose residue. These are the carbonyl carbons in the noncyclic forms of the individual monosaccharides; remember that the carbonyl carbons become the acetal or hemiacetal carbons in the cyclic forms.

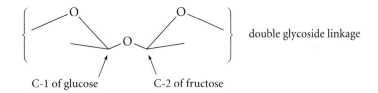

double glycoside linkage

C-1 of glucose   C-2 of fructose

Thus, neither the fructose nor the glucose part of sucrose has a free hemiacetal group. Hence, (+)-sucrose cannot be oxidized by bromine water, nor does it undergo mutarotation. Carbohydrates such as (+)-sucrose that cannot be oxidized by bromine water are classified as **nonreducing sugars**.

Like other glycosides, (+)-sucrose can be hydrolyzed to its component monosaccharides. Sucrose is hydrolyzed by aqueous acid or by enzymes (called *invertases*) to an equimolar mixture of D-glucose and D-fructose. This mixture is sometimes called *invert sugar* because, as hydrolysis of sucrose proceeds, the positive rotation of the solution changes to a negative rotation characteristic of the glucose-fructose mixture. This rotation is negative because the strongly negative rotation (−92°) of fructose (sometimes called *levulose*) has a greater magnitude than the positive rotation (+52.7°) of glucose (sometimes called *dextrose*). Fructose, which is the sweetest of the common sugars (about twice as sweet as sucrose), accounts for the intense sweetness of honey, which is mostly invert sugar.

*27.28 What products are expected from each of the following reactions?
(a) lactobionic acid (Eq. 27.46) + 1 M aqueous HCl →
(b) (+)-lactose + dimethyl sulfate, NaOH →
(c) product of (b) + 1 M aqueous H_2SO_4 →

27.29 The structure of cellobiose, a disaccharide obtained from the hydrolysis of the polysaccharide cellulose, is given below. Into what monosaccharide(s) is cellobiose hydrolyzed by aqueous HCl?

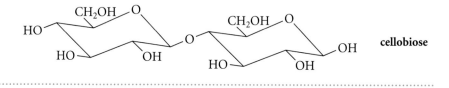

cellobiose

B. Polysaccharides

In principle, any number of monosaccharide residues can be linked together with glycosidic bonds to form chains. When such chains are long, the sugars are called **polysaccharides**. This section surveys a few important polysaccharides.

Cellulose   Cellulose, the principal structural component of plants, is the most abundant organic compound on the earth. Cotton is almost pure cellulose; wood is cellulose combined with a polymer called *lignin*. About 5×10^{14} kg of cellulose are biosynthesized and degraded annually on the earth.

Cellulose is a regular polymer of D-glucopyranose residues connected by β-1,4-glycosidic linkages.

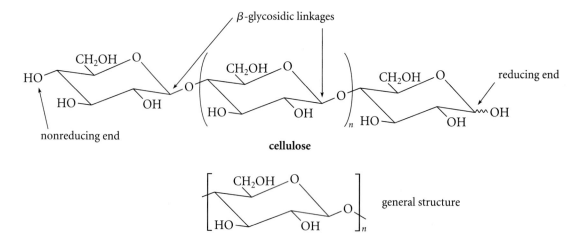

Like disaccharides, polysaccharides can be hydrolyzed to their constituent monosaccharides. Thus, cellulose can be hydrolyzed to D-glucose residues. Mammals lack the enzymes that catalyze the hydrolysis of the β-glycosidic linkages of cellulose; this is why humans cannot digest grasses, which are principally cellulose. Cattle can, of course, derive nourishment from grasses, but this is because the bacteria in their rumens provide the appropriate enzymes that break down plant cellulose to glucose.

Processed cellulose (cellulose that has been specially treated) has many other uses. It can be spun into fibers (rayon) or made into wraps (cellophane). The paper on which this book is printed is largely processed cellulose. Nitration of the cellulose hydroxy groups gives nitrocellulose, a powerful explosive. Cellulose acetate, in which the hydroxy groups of cellulose are esterified with acetic acid, is known by the trade names Celanese, Arnel, etc., and is used in knitting yarn and decorative household articles.

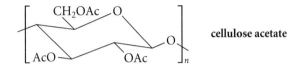

cellulose acetate

Cellulose is potentially important as an alternative energy source. Recall that biomass is largely cellulose, and cellulose is merely polymerized glucose. The glucose derived from hydrolysis of cellulose can be fermented to ethanol, which can be used as a fuel (as in gasohol). And plants obtain the energy to manufacture cellulose from the sun. Thus, the cellulose in plants—the most abundant source of carbon on the earth—can be regarded as a storehouse of solar energy.

Starch   Starch, like cellulose, is also a polymer of glucose. In fact, starch is a mixture of two different types of glucose polymers. In one, **amylose**, the glucose residues are connected by α-1,4-glycosidic linkages. Conceptually, the only chemical difference between amylose and cellulose is the stereochemistry of the glycosidic bond.

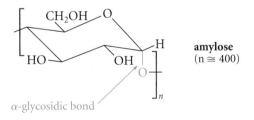

amylose
(n $\cong$ 400)

The other constituent of starch is **amylopectin**, a branched polysaccharide. Amylopectin contains relatively short chains of glucose residues in α-1,4-linkages. In addition, it contains branches that involve α-1,6-glycosidic linkages. Part of a typical amylopectin molecule might look as follows:

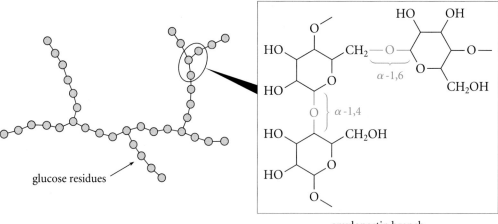

amylopectin branch

Starch is the important storage polysaccharide in corn, potatoes, and other starchy vegetables. Humans have enzymes that catalyze the hydrolysis of the α-glycosidic bonds in starch, and can therefore use starch as a source of glucose.

Chitin   Chitin is a polysaccharide that also occurs widely in nature—notably, in the shells of arthropods (for example, lobsters and crabs). Crab shell is an excellent source of nearly pure chitin.

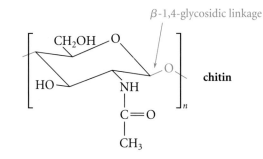

chitin

Chitin is a polymer of *N*-acetyl-D-glucosamine (or, as it is known systematically, 2-acetamido-2-deoxy-D-glucose). Residues of this carbohydrate are connected by β-1,4-glycosidic linkages within the chitin polymer. *N*-Acetyl-D-glucosamine is liberated when chitin is hydrolyzed in aqueous acid. Stronger acid brings about hydrolysis of the amide bond to give D-glucosamine hydrochloride and acetic acid.

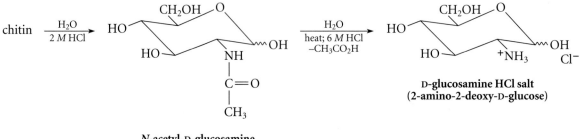

N-acetyl-D-glucosamine
(2-acetamido-2-deoxy-D-glucose)

D-glucosamine HCl salt
(2-amino-2-deoxy-D-glucose)

(27.47)

Glucosamine and *N*-acetylglucosamine are the best-known examples of the **amino sugars**. A number of amino sugars occur widely in nature. Amino sugars linked to proteins (glycoproteins) are found at the outer surfaces of cell membranes, and some of these are responsible for blood-group specificity.

DISCOVERY OF D-GLUCOSAMINE

In 1876 Georg Ledderhose was a premedical student working in the laboratory of his uncle, Friedrich Wöhler (the same chemist who first synthesized urea; Sec. 1.1B). One day, Wöhler had lobster for lunch, and returned to the laboratory carrying the lobster shell. "Find out what this is," he told his nephew. History does not record Ledderhose's thoughts on receiving the refuse from his uncle's lunch, but he proceeded to do what all chemists did with unknown material—he boiled it in concentrated HCl. After hydrolysis of the shell, crystals of the previously unknown D-glucosamine hydrochloride precipitated from the cooled solution (see Eq. 27.47).

Principles of Polysaccharide Structure   Studies of many polysaccharides have revealed the following generalizations about polysaccharide structure:

1. Polysaccharides are mostly long chains with some branches; there are no highly cross-linked, three-dimensional networks. Some cyclic oligosaccharides are known.

2. The linkages between monosaccharide units are in every case glycosidic linkages; thus, monosaccharides can be liberated from all polysaccharides by acid hydrolysis.

3. A given polysaccharide contains only one stereochemical type of glycoside linkage. Thus, the glycoside linkages in cellulose are all β; those in starch are all α.

PROBLEM

*27.30   What product(s) would be obtained when cellulose is treated first exhaustively with dimethyl sulfate/NaOH, then 1 *M* aqueous HCl?

27.11   Nucleosides, Nucleotides, and Nucleic Acids

A.   Nucleosides and Nucleotides

A **ribonucleoside** is a β-glycoside formed between the furanose form of D-ribose and a heterocyclic compound. The heterocyclic group is commonly referred to as the **base**, and the ribose as the **sugar**. A **deoxyribonucleoside** is a β-glycoside of D-2-deoxyribose and a heterocyclic base. The prefix *deoxy* means "without oxygen;" thus, 2-deoxyribose is a ribose that lacks an —OH group at carbon-2.

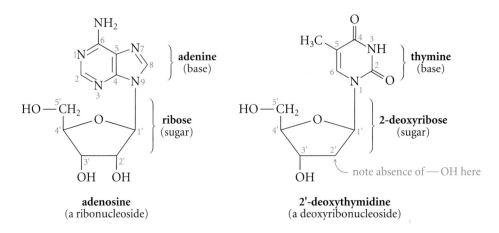

adenosine
(a ribonucleoside)

2'-deoxythymidine
(a deoxyribonucleoside)

Notice that in these structures the sugar ring and the heterocyclic ring are numbered separately. To differentiate the two sets of numbers, primes (′) are used in referring to the atoms of the sugar. For example, the 2′ (pronounced *two-prime*) carbon of adenosine is carbon-2 of the sugar ring.

The bases that occur most frequently in nucleosides are derived from two heterocyclic ring systems: **pyrimidine** and **purine**. Notice particularly the numbering of these rings. Three bases of the pyrimidine type and two of the purine type occur most commonly.

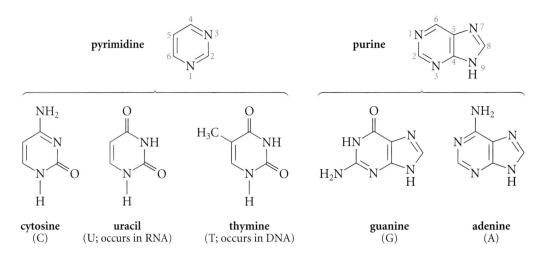

cytosine
(C)

uracil
(U; occurs in RNA)

thymine
(T; occurs in DNA)

guanine
(G)

adenine
(A)

The base is attached to the sugar at N-9 of the purines and N-1 of the pyrimidines, as in the examples above.

In nature, the 5′ —OH group of the ribose in a nucleoside is usually found esterified to a phosphate group. A 5′-phosphorylated nucleoside is called a **nucleotide**. A **ribonucleotide** is derived from the monosaccharide ribose; a **deoxyribonucleotide** is derived from 2′-deoxyribose. Some nucleotides contain a single phosphate group; others contain two or three phosphate groups condensed in phosphoric anhydride linkages.

Table 27.3   **Nomenclature of Nucleic Acid Bases, Nucleosides, and Nucleotides**[a]

Base	Nucleoside	Nucleotide monophosphate	Abbreviation for the monophosphate
adenine (A)	adenosine	adenylic acid	AMP
uracil (U)	uridine	uridylic acid	UMP
thymine (T)	thymidine	thymidylic acid	TMP
cytosine (C)	cytidine	cytidylic acid	CMP
guanine (G)	guanosine	guanidylic acid	GMP

[a] The deoxyribonucleosides and deoxyribonucleotides are named by appending the prefix *deoxy*:

deoxyadenosine   deoxyadenylic acid   *d*AMP

The prefix *deoxy* means 2′-deoxy unless stated otherwise.

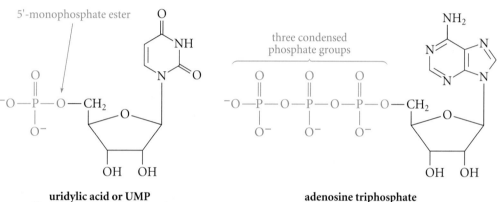

uridylic acid or UMP
(for *u*ridine *m*ono*p*hosphate)

adenosine triphosphate
or **ATP**

Although the ionization state of the phosphate groups depends on pH, these groups are written conventionally in the ionized form.

Nomenclature of the five common bases and their corresponding nucleosides and nucleotides is summarized in Table 27.3. This table gives the names of the ribonucleosides and ribonucleotides. To name the corresponding 2′-deoxy derivatives, the prefix *2′-deoxy* (or simply *deoxy*) is appended to the names of the corresponding ribose derivatives. For example, the 2′-deoxy analog of adenosine is called 2′-deoxyadenosine or simply deoxyadenosine. In addition, the names of the mono-, di-, and triphosphonucleotide derivatives are often abbreviated. Thus, adenylic acid is abbreviated AMP (for adenosine monophosphate); the di- and tri-phosphorylated derivatives are called ADP and ATP, respectively (see example above). The abbreviations for the corresponding deoxy derivatives contain a *d* prefix. Thus, 2′-deoxythymidylic acid can be abbreviated *d*TMP.

Nucleotides are important because they are the building blocks of ribonucleic acid (RNA) and deoxyribonucleic acid (DNA), polymeric molecules that are responsible for the storage and transmission of genetic information. Ribonucleotides also have other important biochemical functions, some of which have already been presented. NAD$^+$, one of nature's important oxidizing agents (Fig. 10.1, Sec. 10.7), and coenzyme A (Fig. 22.3, Sec. 22.7) are both ribonucleotides. One of the most ubiquitous nucleotides is ATP (adenosine triphosphate), which serves as the fundamental energy source for the living cell. The hydrolysis of ATP to ADP, shown in Eq. 27.48, liberates 30.5 kJ/mol (7.3 kcal/mol) of energy at pH 7; living systems harness this energy to drive energy-requiring biochemical processes.

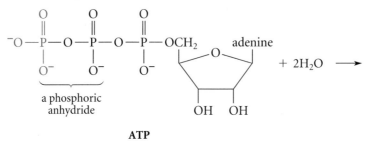

ATP

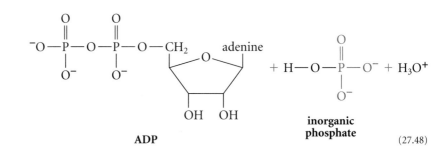

ADP   **inorganic phosphate**   (27.48)

Muscle contraction—obviously an energy-requiring process—is an example of the biological use of ATP hydrolysis to provide energy. Abbreviating inorganic phosphate as P_i, the overall process for muscle contraction can be summarized as follows:

$$\text{energy} + \text{muscle} \longrightarrow \text{contracted muscle}$$
$$\text{ATP} + H_2O \longrightarrow \text{ADP} + P_i + \text{energy} \qquad (27.49)$$
$$sum: \quad \text{ATP} + \text{muscle} + H_2O \longrightarrow \text{ADP} + P_i + \text{contracted muscle}$$

(A human might use 0.5 kilogram of ATP/hr during strenuous exercise!) *How* living organisms use the energy from ATP hydrolysis is a subject that we leave for your study of biochemistry; the important point is that overall ATP hydrolysis is invariably involved in any biological process that requires energy. The energy for making ATP from ADP is ultimately derived from the foods we eat, for example, carbohydrates; and the carbohydrates that we use as foods are produced by plants using solar energy harnessed by the processes of photosynthesis.

PROBLEM

27.31   Draw the structure of *(a) deoxythymidine monophosphate (*d*TMP); (b) GDP.

B. Structures of DNA and RNA

Deoxyribonucleic acid (DNA) is a polymer of deoxyribonucleotides, and is the storehouse of genetic information throughout all of nature (with the exception of certain viruses).

A typical section of DNA is shown in Fig. 27.4. This figure shows that the nucleotide residues in DNA are interconnected by phosphate groups that are esterified both to the 3' —OH group of one ribose and the 5' —OH of another. The DNA polymer incorporates adenine, thymine, guanine, and cytosine as the nucleotide bases. Although only four residues are shown in Fig. 27.4, a typical strand of DNA might be thousands of nucleotides long. Just as each amino acid residue in a peptide is differentiated by its amino acid side chain, *each residue in a polynucleotide is differentiated by the identity of its base*. The DNA polymer is thus a backbone of alternating phosphates and 2'-deoxyribose groups to which are connected bases that differ from residue to residue. The ends of the DNA polymer are labeled 3' or 5', corresponding to the deoxyribose carbon on which the terminal hydroxy group is attached.

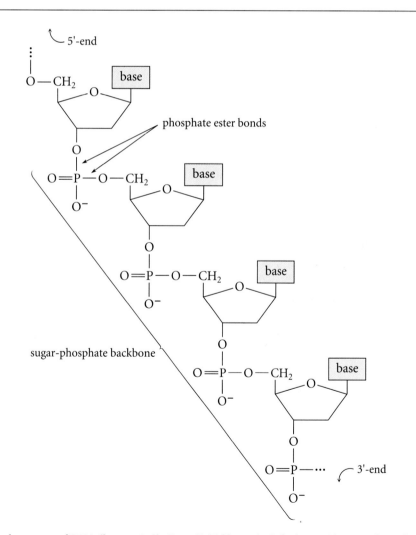

Figure 27.4   *General structure of DNA (base = A, T, G, or C; Table 27.3). Only four residues are shown here; a typical strand of DNA contains thousands of residues.*

Ribonucleic acid (RNA) polymers are conceptually much like DNA polymers, except that ribose, rather than 2′-deoxyribose, is the sugar. RNA also incorporates essentially the same bases as DNA, except that uracil occurs in RNA instead of thymine, and some rare bases (not considered here) are found in certain types of RNA.

It was known for many years before the detailed structure of DNA was determined that DNA carries genetic information. It was also known that DNA is *replicated*, or copied, during cell reproduction. In 1950, Erwin Chargaff of Columbia University showed that the ratios of adenine to thymine, and guanosine to cytosine, in DNA are both 1.0; these observations are called *Chargaff's rules*. How these facts relate to the storage and transmission of genetic information, however, remained a mystery. It became clear to a number of scientists that a knowledge of the three-dimensional structure of DNA would be essential in order to understand how DNA functions as it does. The importance of this problem was sufficiently obvious that several scientists worked feverishly to be the first to determine the three-dimensional structure of DNA. In 1953, James D. Watson and Francis C. Crick, then at Cambridge University, proposed a structure for DNA. Their proposal was based on X-ray diffraction patterns of DNA fibers obtained by their colleagues at the Medical Research Council laboratory in England, Rosalind Franklin and Maurice Wilkins. (For an intriguing account of the race for the DNA structure, see "The Double Helix," by James D. Watson; Atheneum, 1968.) For their work on the structure of DNA, Watson, Wilkins, and Crick received the Nobel Prize for Medicine and Physiology in 1962.

The Watson-Crick structure of DNA is shown in Fig. 27.5. The structure has the following important features:

1. The structure of DNA contains *two* right-handed helical polynucleotide chains that run in opposite directions, coiled around a common axis; the structure is therefore that of a *double helix*. The helix makes a complete turn every ten nucleotide residues. (Other helical conformations of DNA are also known; current research is aimed at elucidating the biological roles of different DNA conformations.)

2. The sugars and phosphates, which are rich in —OH groups and charges, are on the outside of the helix, where they can interact with solvent water; the bases, which are more hydrocarbonlike, are largely buried in the interior of the double helix, away from water.

3. The chains are held together by hydrogen bonds between bases. *Each adenine (A) in one chain hydrogen-bonds to a thymine (T) in the other, and each guanosine (G) in one chain hydrogen-bonds to a cytosine (C) in the other.* Thus, every purine in one chain is hydrogen-bonded to a pyrimidine in the other. For this reason, A is said to be *complementary* to T, and G is *complementary* to C. A closer look at these hydrogen-bonded *Watson-Crick base pairs* is shown in Fig. 27.6, p. 1374. Notice that the A–T pair has about the same spatial dimensions as the G–C pair.

4. The planes of successive complementary base pairs are stacked, one on top of the other, and are perpendicular to the axis of the helix. The distance between each successive base-pair plane is 3.4 Å. Since the helix

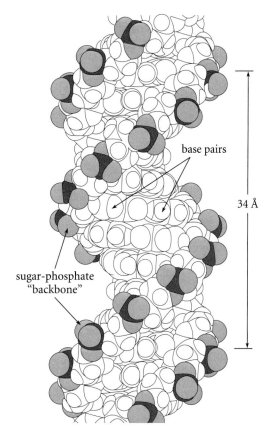

base pairs

34 Å

sugar-phosphate
"backbone"

Figure 27.5   *Space-filling model of the DNA double helix. The phosphate oxygens of the sugar-phosphate back-bones are shown in color and the phosphorus atoms in gray. All other atoms are shown in outline only. The sugar-phosphate backbones trace the outline of the double helix.*

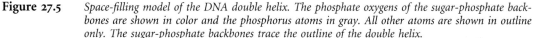

makes a complete turn every ten residues, this means that there is a distance of $10 \times 3.4 = 34$ Å along the helix per complete turn.

5. There is no intrinsic restriction on the sequence of bases in a polynucleotide; however, because of the hydrogen bonding described in Point 3, the sequence of one polynucleotide strand (the "Watson" strand) in the double helix is complementary to that in the other strand (the "Crick" strand). Thus everywhere there is an A in one strand, there is a T in the other; everywhere there is a G in one strand, there is a C in the other.

Hydrogen-bonding complementarity in DNA accounts nicely for Chargaff's rules: if A always hydrogen bonds to T and G always hydrogen bonds to C, then the number of As must equal the number of Ts, and the number of Gs must equal the number of Cs. This structure also suggests a reasonable mechanism for the duplication of DNA during cell division: the two strands can come apart, and a new strand can be grown as

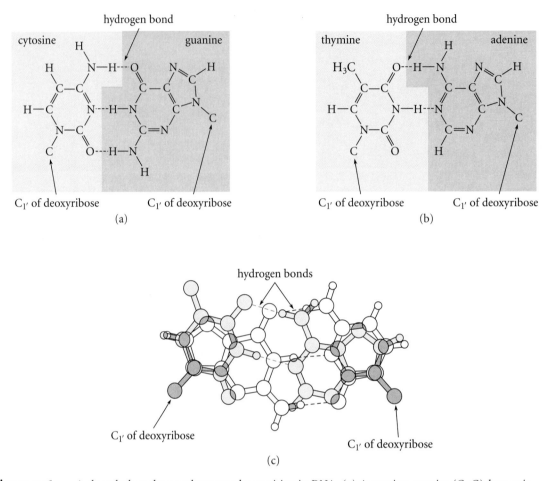

Figure 27.6   *A closer look at the complementary base pairing in DNA. (a) A cytosine–guanine (C–G) base pair involves three hydrogen bonds. (b) A thymine–adenine (T–A) base pair involves two hydrogen bonds. (c) Superposition of the C–G and T–A (color) base pairs shows that the two occupy the same space.*

a complement of each original strand. In other words, *the proper sequence of each new DNA strand during cellular reproduction is assured by hydrogen-bonding complementarity* (Fig. 27.7).

PROBLEMS

*27.32   Draw in detail the structure of a section of RNA four residues long which, from the 5′-end, has the following sequence of bases: A, U, C, G.

27.33   Would you expect Chargaff's rules to apply within an individual strand of DNA? Explain.

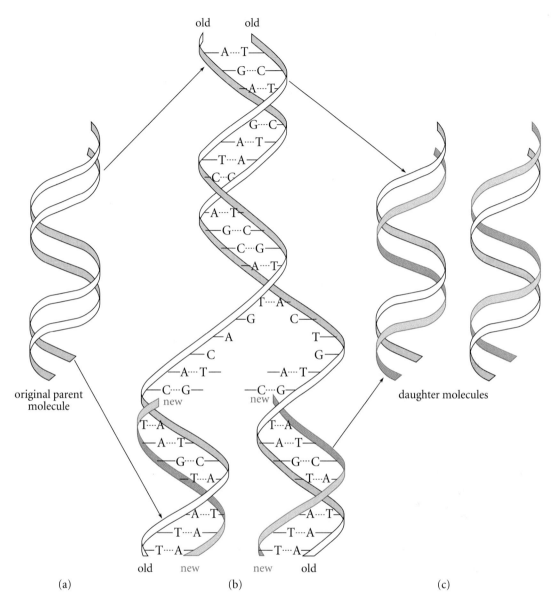

Figure 27.7   *Complementary base pairing in DNA is crucial to its faithful replication. (a) A typical DNA double helix. (b) In the replicating DNA a new strand grows on each of the original strands. (c) Two new molecules of DNA, each containing one old strand and one new strand.*

27.12   DNA, RNA, and the Genetic Code

A. Role of DNA and RNA in Protein Synthesis

The sequence of nucleotides in DNA forms a linear code for every protein and RNA molecule. To understand this point, let us see how the following strand of DNA, which might be imagined as part of a gene in some organism, could be used biologically to direct the synthesis of a section of a specific protein. If this were a DNA strand from a cell, it would be one of the two strands of the double helix; each letter in the sequence identifies a residue of DNA by its particular base.

$$\longleftarrow \quad 3'\text{-end} \qquad\qquad\qquad 5'\text{-end} \quad \longrightarrow$$
$$\ldots\text{–A–A–A–G–A–T–T–C–A–C–C–C–C–T–C–A–T–C–}\ldots$$

STUDY GUIDE LINK:
✓ 27.5
DNA Transcription

First, a strand of DNA directs the synthesis of a complementary strand of RNA. This process is called **transcription**. The ultimate RNA product of transcription—the transcript—is called **messenger RNA (mRNA)**. The sequence of the mRNA transcript is complementary to one DNA strand of the gene. For example, everywhere there is a G in DNA, there is a C (the complementary base) in the mRNA transcript. An adenine (A) in DNA is transcribed into a uracil (U). (Messenger RNA contains uracil rather than thymine (T); uracil is just a thymine without its methyl group.) Thus the gene fragment above would be transcribed as follows:

DNA, a polydeoxyribonucleotide

3'-end   5'-end
...–A–A–A–G–A–T–T–C–A–C–C–C–C–T–C–A–T–C–...
...–U–U–U–C–U–A–A–G–U–G–G–G–G–A–G–U–A–G–...
5'-end   3'-end

mRNA, a polyribonucleotide

Notice that the complementary sequence of mRNA runs in the direction opposite to that of its parent DNA—the 3'-end of RNA matches the 5'-end of DNA, and vice-versa.

Once the mRNA synthesis is complete, the mRNA sequence is used by the cell to direct the synthesis of a specific protein from its component amino acids. This process is called **translation**. Each successive three-residue triplet in the sequence of mRNA is translated as a specific amino acid in the sequence of a protein according to the **genetic code** given in Table 27.4. Thus, the particular stretch of mRNA shown above would be translated into a protein sequence as follows:

mRNA

5'-end   3'-end
...–U–U–U–C–U–A–A–G–U–G–G–G–G–A–G–U–A–G–...
...–Phe – Leu – Ser – Gly – Glu   STOP!
amino end   carboxy end

peptide chain

Just as a sequence of dots and dashes in Morse code can be used to form words, *the precise sequence of bases in DNA (by way of its complementary mRNA transcription product) codes for the successive amino acids of a protein.* Morse code has two coding units—the dot and the dash. In DNA or mRNA, there are four: A, T (U in mRNA), G, and C, the four nucleotide bases. Notice that the sequences of DNA and mRNA contain no "commas." The protein-synthesizing "machinery" of the cell knows where one amino acid code ends and another starts because, as Table 27.4 shows, there is a specific "start" signal—either of the nucleotide sequences AUG or GUG—at the appropriate point in the mRNA. Because mRNA also contains "stop" signals (UAA, UGA, or UAG), protein synthesis is also terminated at the right place.

Some amino acids have multiple codes. For example, Table 27.4 shows that glycine, the most abundant amino acid in proteins, is coded by GGU, GGC, GGA, and GGG.

It is possible for the change of only one base in the DNA (and consequently in the mRNA) of an organism to cause the change of an amino acid in the corresponding protein. A dramatic example of such a change is the genetic disease *sickle-cell anemia*. In this painful disease, the red blood cells take on a peculiar sickle shape that causes them to clog capillaries. The molecular basis for this disease is a single amino acid substitution in hemoglobin, the protein that transports oxygen in the blood (Fig. 26.11). In sickle-cell hemoglobin, glutamic acid at position 6 in one of the protein chains of normal hemoglobin is changed to valine. That is, sickle-cell disease results from a change in but one of the 141 amino acids in this hemoglobin chain! The mRNA genetic code for Glu is GAA and GAG; that for Val is GUA and GUG (among others). In other words, a change of only one nucleotide (A → U) of the (3 × 141), or 423, nucleotides that code for this chain of hemoglobin is responsible for the disease.

Although we've focused on the structure and function of DNA, the structure of RNA is also important. Although a detailed discussion of this topic is beyond the scope of this text, there are many different types of RNA besides messenger RNA, each with a specific function in the cell.

Very powerful methods have been developed for sequencing DNA and RNA. With these methods, DNA and RNA sequences containing several hundred nucleotides can be determined in a relatively short time. These nucleotide sequences can be used to check, or even predict, protein sequences by applying the genetic code. Similarly, effective methods for the synthesis of DNA and RNA fragments also exist. These methods strategically resemble peptide synthesis in the sense that a strand of DNA or RNA is "grown"

Table 27.4   The Genetic Code

5'-OH Terminal base of mRNA	Middle base of mRNA				3'-OH Terminal base of mRNA
	U	C	A	G	
U	Phe	Ser	Tyr	Cys	U
	Phe	Ser	Tyr	Cys	C
	Leu	Ser	(Stop)	(Stop)	A
	Leu	Ser	(Stop)	Trp	G
C	Leu	Pro	His	Arg	U
	Leu	Pro	His	Arg	C
	Leu	Pro	Gln	Arg	A
	Leu	Pro	Gln	Arg	G
A	Ile	Thr	Asn	Ser	U
	Ile	Thr	Asn	Ser	C
	Ile	Thr	Lys	Arg	A
	Met[a]	Thr	Lys	Arg	G
G	Val	Ala	Asp	Gly	U
	Val	Ala	Asp	Gly	C
	Val	Ala	Glu	Gly	A
	Val[a]	Ala	Glu	Gly	G

[a] Sometimes used as "start" codons.

from individual nucleotides by using a series of protection, coupling, and deprotection steps. Molecular biologists have also discovered ways in which foreign DNA can be incorporated into, and expressed by, host organisms. All of these techniques used together have led to new biotechnologies that have been termed collectively "genetic engineering." One major pharmaceutical house employs a lowly bacterium—*Escherichia coli*—for the commercial production of *human* insulin using these techniques. Formerly all insulin used for the treatment of diabetes came from horses and pigs, and shortages of this important hormone occurred. The new process has made available an abundant supply of human insulin, and was one of the first of many such processes that have been developed for the production of complex biological materials for use in human medicine.

B. DNA Modification and Chemical Carcinogenesis

We've shown how the double-helical structure of DNA, DNA replication, and the fidelity of DNA transcription into RNA involve very specific base-pairing complementarity. Other important processes, such as the recognition of the three-base triplet code of mRNA during protein biosynthesis, also involve this type of complementarity. The molecular basis of this complementarity is the specific hydrogen bonding between a pyrimidine and a purine base. You can perhaps imagine that, if this hydrogen bonding were upset, the base-pairing complementarity would also be upset, and with it, some or all of the biological processes that rely on this phenomenon. There is strong circumstantial evidence that chemical damage to DNA can interfere with this hydrogen-bonding complementarity and can trigger the state of uncontrolled cell division known as cancer.

One type of chemical damage to DNA is caused by *alkylating agents* (Sec. 10.3B). Certain types of alkylating agents react with DNA by alkylating one or more of the nucleotide bases. These same alkylating agents are also carcinogens (cancer-causing compounds). A few such compounds are shown below.

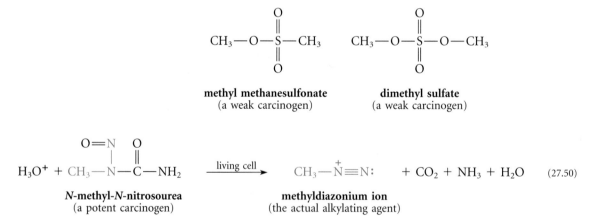

When such alkylating agents (abbreviated CH_3—X below) react with DNA, alkylated guanosines are among the products. The major product is alkylated on N-7 of the guanine base, but an important minor product is alkylated on the oxygen at C-6 (called the O-6 position).

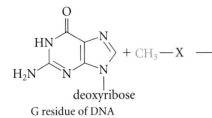 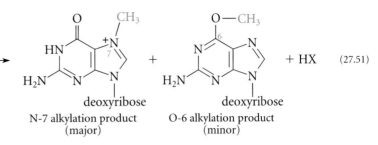

deoxyribose   deoxyribose   deoxyribose

G residue of DNA   N-7 alkylation product   O-6 alkylation product
                         (major)   (minor)   (27.51)

(An analogous alkylation occurs at O-4 of thymine; see Problem 27.34.) Notice that the alkylation at O-6 prevents the N-1 nitrogen from acting as a hydrogen-bond donor in a Watson-Crick base pair (Fig. 27.6) because the hydrogen is lost from this nitrogen as a result of alkylation.

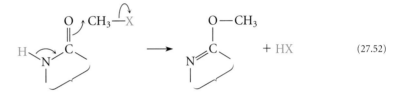

(27.52)

The N-7 alkylation, in contrast, does not directly affect any of the atoms involved in the hydrogen-bonding complementarity. It has been found that the alkylating agents which are the most potent carcinogens also yield the greatest amount of the guanines alkylated at O-6 and thymines alkylated at O-4. Although this correlation does not prove that these alkylations are primary events in carcinogenesis, it provides strong circumstantial evidence in this direction.

The way in which aromatic hydrocarbons are converted into carcinogenic epoxides by enzymes in living systems was discussed in Sec. 16.7. These epoxides have been shown to react with DNA; among the products of this reaction is a guanosine residue alkylated on the nitrogen at carbon-2 of the guanine base.

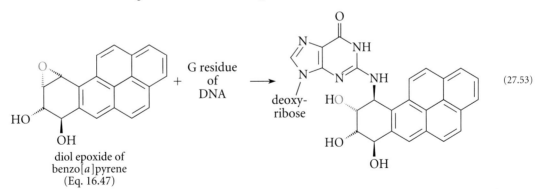

diol epoxide of
benzo[a]pyrene
(Eq. 16.47)

(27.53)

This nitrogen is also involved in the hydrogen-bonding interaction of G with C (Fig. 27.6). Thus, it may be that alkylation by aromatic hydrocarbon epoxides also triggers the onset of cancer by interfering with the base-pairing complementarity.

One last example of DNA damage is caused by ultraviolet radiation. Ultraviolet light promotes the [2s + 2s] cycloaddition (Sec. 25.3) of two pyrimidines when they occur in adjacent positions on a strand of DNA. In the following example, a thymine dimer is formed from two adjacent thymines.

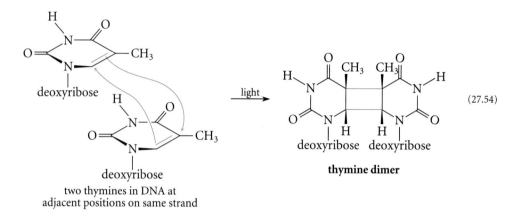

(27.54)

thymine dimer

two thymines in DNA at
adjacent positions on same strand

Most people have a biological repair system that effects the removal of the modified pyrimidines and repairs the DNA. People with a rare skin disease, *xeroderma pigmentosum*, have a genetic deficiency in the enzyme that initiates this repair. Most of these people contract skin cancer and die at an early age. Here, then, is a situation in which the chemical modification of DNA has been clearly associated with the onset of cancer.

As these examples show, it is possible to understand the molecular basis of some diseases. Certainly further progress in human medicine will stem from an understanding of the organic chemistry of the living cell.

PROBLEM

27.34   There is evidence that alkylation at O-4 of thymine, like alkylation at O-6 of guanine, is another mutagenic event that can lead to cancer.
   *(a)   Draw the structure of a thymine residue as it would exist after O-4 methylation.
   (b)   Explain why O-4 alkylation at thymine would disrupt Watson-Crick base pairing.

KEY IDEAS IN CHAPTER 27

⚗   Carbohydrates are aldehydes and ketones that contain a number of hydroxy groups on an unbranched carbon chain, as well as their chemical derivatives.

⚗   The D,L system is an older but widely used method for specifying carbohydrate enantiomers. The D enantiomer is the one in which the asymmetric carbon of highest number has the same configuration as (2R)-glyceraldehyde (D-glyceraldehyde).

Monosaccharides exist in cyclic furanose or pyranose forms in which a hydroxy group and the carbonyl group of the aldehyde or ketone have reacted to form a cyclic hemiacetal.

The cyclic forms of monosaccharides are in equilibrium with small amounts of their respective aldehydes or ketones, and can therefore undergo a number of aldehyde and ketone reactions. These include oxidation (bromine water or dilute nitric acid); reduction with sodium borohydride; cyanohydrin formation (the first step in the Kiliani-Fischer synthesis); and base-catalyzed enolization and enolate-ion formation (the Lobry de Bruyn-Alberda van Eckenstein reaction).

The —OH groups of carbohydrates undergo many typical reactions of alcohols and glycols, such as ether formation, ester formation, and glycol cleavage with periodate.

Because the hemiacetal carbons of monosaccharides are asymmetric, the cyclic forms of monosaccharides exist as diastereomers called anomers. The equilibration of anomers is why carbohydrates undergo mutarotation.

In a glycoside the —OH group at the anomeric carbon of a carbohydrate is substituted with an ether (—OR) group. In disaccharides or polysaccharides, the —OR group is derived from another saccharide residue. The —OR group of glycosides can be replaced with an —OH group by hydrolysis. Thus, higher saccharides can be hydrolyzed to their component monosaccharides in aqueous acid.

Disaccharides, trisaccharides, etc., can be classified as reducing or nonreducing sugars. Reducing sugars have at least one free hemiacetal group. In nonreducing sugars all anomeric carbons are involved in glycosidic linkages.

Ribonucleotides and deoxyribonucleotides, which are phosphorylated derivatives of ribonucleosides and deoxyribonucleosides, are the building blocks of RNA and DNA, respectively. These compounds are β-glycosides of either ribose or 2′-deoxyribose, respectively, and a purine or pyrimidine base. Adenine, guanine, and cytosine are bases in both DNA and RNA; thymine is unique to DNA, and uracil to RNA.

An important conformation of DNA is the double helix, in which two strands of DNA running in opposite directions wrap around a common helical axis. The sugars and phosphate groups lie on the outside of the helix, and the bases are stacked in parallel planes on the inside. The two strands of the double helix are held together by purine-pyrimidine hydrogen bonds between complementary residues. A number of known carcinogens apparently modify DNA in such a way that this complementary hydrogen bonding is disrupted.

*27.35   Give the product(s) expected when D-mannose (or other compound indicated) reacts with each of the following reagents. (Assume that cyclic mannose derivatives are pyranoses.)
(a) $Ag^+(NH_3)_2$ (b) dilute HCl   (c) dilute NaOH
(d) Br_2/H_2O, then H_3O^+ (e) CH_3OH, HCl   (f) acetic anhydride
(g) product of (d) + $Ca(OH)_2$, then $Fe(OAc)_3$, H_2O_2
(h) product of (e) + $PhCH_2Cl$ (excess) and NaOH

27.36   Give the products expected when D-ribose (or other compound indicated) reacts with each of the following reagents.
(a) dilute HNO_3 (b) ^-CN, H_2O (c) product of (b) + $H_2/Pd/BaSO_4$
(d) CH_3OH, HCl (four isomeric compounds; two pyranosides and two furanosides)
(e) products of (d) + $(CH_3)_2SO_4$ (excess) and NaOH

27.37   Draw the indicated type of structure for each compound below.
*(a) CDP (sugar ring in Haworth projection)
(b) α-D-talopyranose (chair)
*(c) propyl β-L-arabinopyranoside (chair)
(d) (+)-lactose (Haworth projection)

27.38   Name the specific form of each aldose shown below.

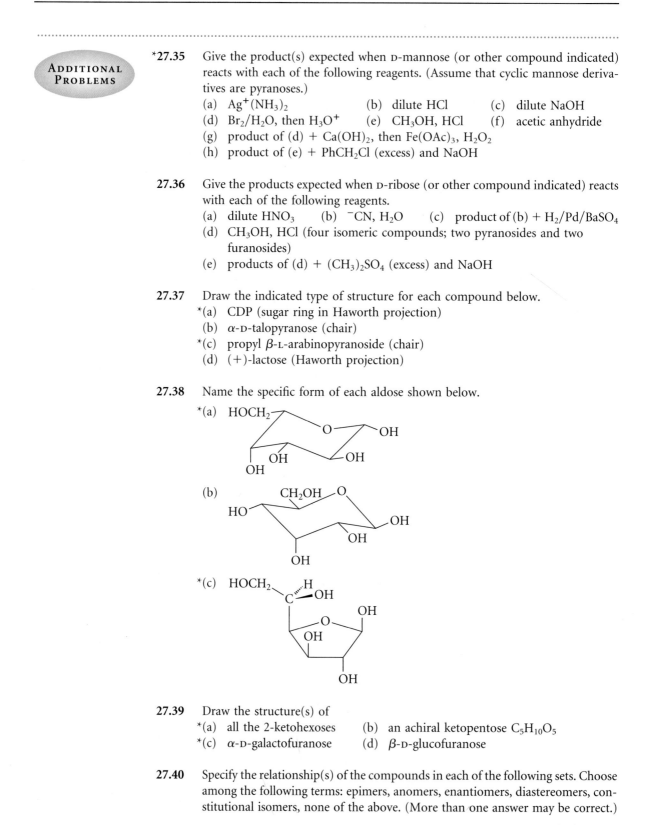

*(a) HOCH₂

(b)

*(c)

27.39   Draw the structure(s) of
*(a) all the 2-ketohexoses   (b) an achiral ketopentose $C_5H_{10}O_5$
*(c) α-D-galactofuranose   (d) β-D-glucofuranose

27.40   Specify the relationship(s) of the compounds in each of the following sets. Choose among the following terms: epimers, anomers, enantiomers, diastereomers, constitutional isomers, none of the above. (More than one answer may be correct.)

*(a) α-D-glucopyranose and β-D-glucopyranose
 (b) α-D-glucopyranose and α-D-mannopyranose
*(c) β-D-mannopyranose and β-L-mannopyranose
 (d) α-D-ribofuranose and α-D-ribopyranose
*(e) aldehyde form of D-glucose and α-D-glucopyranose
 (f) methyl α-D-fructofuranoside and 2-*O*-methyl-α-D-fructofuranose

27.41 Tell whether each structure or term is a correct description of the L-sorbose structure below or a form with which it is in equilibrium.

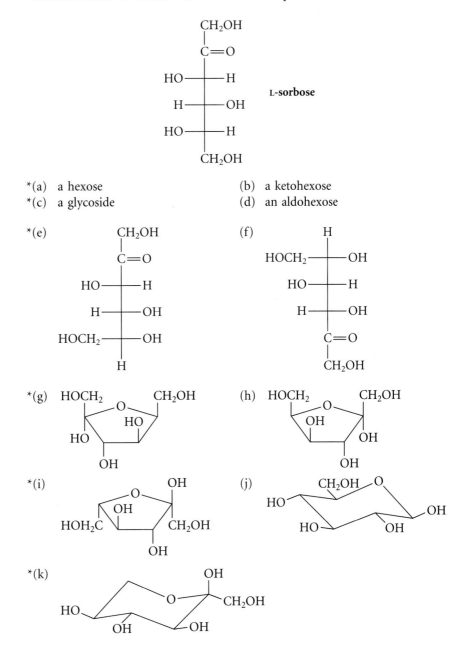

*(a) a hexose   (b) a ketohexose
*(c) a glycoside   (d) an aldohexose

*27.42 Consider the structure of *raffinose*, a trisaccharide found in sugar beets and a number of higher plants.

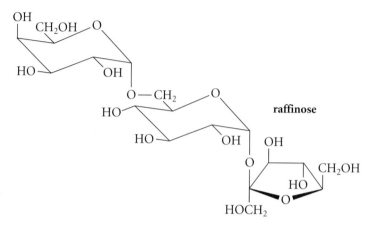

OH
CH2OH
O
HO
OH
O—CH2
O
HO
raffinose
HO
OH
OH
O
CH2OH
HO
O
HOCH2

(a) Classify raffinose as a reducing or nonreducing sugar, and tell how you know.
(b) Identify the glycoside linkages in raffinose, and classify each as either α or β.
(c) Name the monosaccharides formed when raffinose is hydrolyzed in aqueous acid.
(d) What products are formed when raffinose is treated with dimethyl sulfate in NaOH, and then with aqueous acid and heat?

27.43 Draw the structure of 3-*O*-β-D-glucopyranosyl-α-D-arabinofuranose, a disaccharide that is the β-glycoside formed between D-glucopyranose at the nonreducing end and the —OH group at carbon-3 of α-D-arabinofuranose at the reducing end.

27.44 Complete the following reactions by giving the major organic product(s).

*(a) phenyl β-D-glucopyranoside + CH_3OH (solvent) $\xrightarrow{H_2SO_4}$

(b) $HOCH_2CH_2CH_2CH_2CH{=}O$ + CH_3OH (solvent) $\xrightarrow{HCl}$

*(c)

1) OsO4
2) H2O, NaHSO3
H5IO6
NaBH4
CH3OH

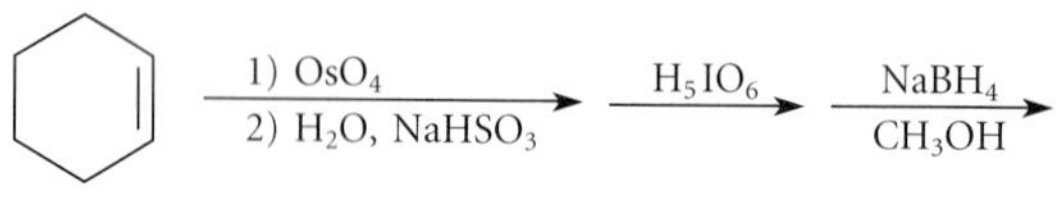

(d)

1) OsO4
2) H2O, NaHSO3
acetone
dil. HCl

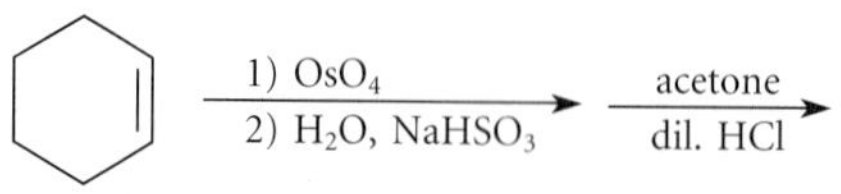

*(e) (+)-sucrose + CH_3I (excess) $\xrightarrow{Ag_2O}$

(f) (+)-lactose + C_2H_5OH $\xrightarrow[heat]{H_2SO_4}$

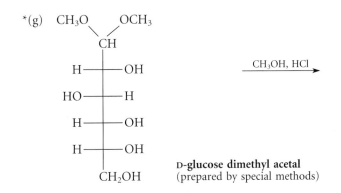

*(g)

D-glucose dimethyl acetal
(prepared by special methods)

$\xrightarrow{\text{CH}_3\text{OH, HCl}}$

27.45 An important reaction used by Emil Fischer in his research on carbohydrate chemistry was the reaction of aldoses and ketoses with phenylhydrazine to give *osazones*, shown below. Osazones, unlike many carbohydrates, form crystalline solids that are useful in characterizing carbohydrates.

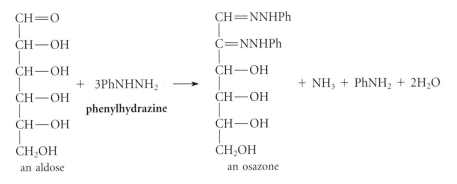

*(a) Glucose and mannose give the same osazone. Given that these two compounds are aldohexoses, what could a scientist who knows nothing about the stereochemistry of these carbohydrates deduce about their stereochemical relationship from this fact?

(b) What aldopentose gives the same osazone as D-arabinose?

27.46 A biologist, Simone Spore, needs the following isotopically labeled aldoses for some feeding experiments. Realizing your expertise in the saccharide field, she has come to you to ask whether you will synthesize these compounds for her. She has agreed to provide an unlimited supply of D-(−)-arabinose as a starting material. (* = ^{14}C, T = ^{3}H = tritium)

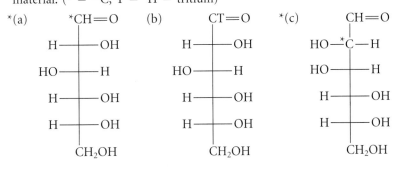

Available commercial sources of isotopes include $Na_2{}^*CO_3$, Na^*CN, $^3H_2$, and ^3H_2O. Outline a synthesis of each isotopically labeled compound.

*27.47   Compound *A*, known to be a monomethyl ether of D-glucose, can be oxidized to a carboxylic acid *B* with bromine water. When the calcium salt of *B* is subjected to ferric acetate and hydrogen peroxide (Ruff degradation), another aldose monomethyl ether is obtained that can also be oxidized with bromine water. When *A* is subjected to the Kiliani-Fischer synthesis, two new methyl ethers are obtained. Both are optically active, and one of them can be oxidized with dilute nitric acid to an optically inactive compound. Suggest a structure for *A*, including its stereochemistry.

27.48   When D-ribose-5-phosphate was treated with an extract of mouse spleen, an optically *inactive* compound *X*, $C_5H_{10}O_5$, was produced. Treatment of *X* with $NaBH_4$ gave a mixture of the alditols ribitol and xylitol. (See Table 27.2.) Treatment of *X* with periodic acid produced two molar equivalents of formaldehyde. Suggest a structure for *X*.

*27.49   The *Wohl degradation* can be used to convert an aldose into another aldose with one fewer carbon. Give the structure of the missing compounds as well as the curved-arrow mechanisms for the conversion of *B* to *C* and *C* to arabinose.

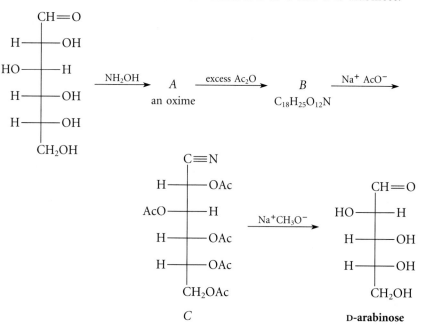

*27.50   The following sequence of reactions, called the *Weerman degradation*, can be used to degrade an aldose to another aldose with one fewer carbon atom.

$$\text{aldose} \xrightarrow{\ Br_2/H_2O\ } \xrightarrow{\ H_3O^+\ } [A] \xrightarrow[\text{heat}]{\ NH_3\ } [B] \xrightarrow[NaOH]{\ Cl_2\ } \text{aldose with one fewer carbon}$$

Using glucose as the aldose, explain what is happening in each step of this sequence. Your explanation should include the identity of each compound in brackets. (*Hint:* Compound *A* is a lactone, and a lactone is a type of ester.)

*27.51   L-Rhamnose is a 6-deoxyaldose with the structure shown below. When a methyl glycoside of L-rhamnose, methyl α-L-rhamnopyranoside, was treated with periodic acid, compound A, $C_6H_{12}O_5$ was obtained that showed no evidence of a carbonyl group in its IR spectrum. Treatment of A with CH_3I/Ag_2O gave a derivative B, $C_8H_{16}O_5$. Treatment of A with H_2/Ni or $NaBH_4$ gave compound C, shown below in Fischer projection. Give the structure of A. Explain why A gives no detectable carbonyl absorption in its IR spectrum, yet reacts with $NaBH_4$.

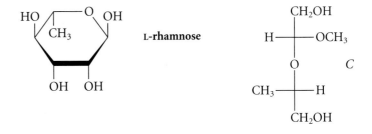

27.52   Oligosaccharides of the following type are obtained from the partial hydrolysis of starch amylopectin.

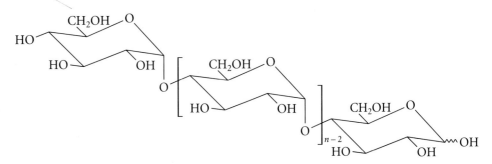

What ratio of erythritol to glycerol would be obtained from successive treatment of a twelve-residue oligosaccharide of the type shown above with periodic acid, then $NaBH_4$, and then hydrolysis in aqueous acid?

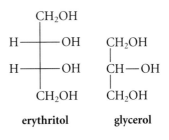

27.53   One theory of genetic mutation postulates that some mutations arise as the result of mis-pairing of bases caused by the existence of relatively rare tautomeric forms of the bases in DNA. Show the hydrogen-bonding complementarity that can result from *(a) the pairing of an imine tautomer of C with A; (b) the pairing of an enol tautomer of T with G.

27.54 The stability of a DNA double helix can be measured by its *melting temperature,* T_m, defined as the temperature at which the helix is 50% dissociated into individual chains.

*(a) Explain why the double helix formed between polydeoxyadenylic acid (polyA) and polydeoxythymidylic acid (polyT) has a considerably lower T_m (68°) than that of the double helix formed between polydeoxyguanidylic acid (polyG) and polydeoxycytidylic acid (polyC) (91°). (*Hint:* See Fig. 27.6.)

(b) Which of the following viruses has the higher ratio of (G + C)/(A + T) in its DNA? Explain.

Viral DNA source	T_m, °C
Human adenovirus I	58.5
Fowl pox	35

*27.55 Maltose is a disaccharide obtained from the hydrolysis of starch. Maltose can be hydrolyzed to two equivalents of glucose, and can be oxidized to an acid, maltobionic acid, with bromine water. Treatment of maltose with dimethyl sulfate and sodium hydroxide, followed by hydrolysis of the product in aqueous acid, yields one equivalent each of 2,3,4,6-tetra-*O*-methyl-D-glucose and 2,3,6,-tri-*O*-methyl-D-glucose. Hydrolysis of maltose is catalyzed by α-amylase, an enzyme known to affect only α-glycosidic linkages. Give *two* structures of maltose consistent with these data, and explain your answers.

Treatment of maltobionic acid with dimethyl sulfate and sodium hydroxide followed by hydrolysis of the product in aqueous acid gives 2,3,4,6-tetra-*O*-methyl-D-glucose and 2,3,5,6-tetra-*O*-methyl-D-gluconic acid. (See Eq. 27.25 for the structure of D-gluconic acid.) Give the structure of maltose.

27.56 Planteose, a carbohydrate isolated from tobacco seeds, can be hydrolyzed in dilute acid to yield one equivalent each of D-fructose, D-glucose, and D-galactose. Almond emulsin (an enzyme preparation that hydrolyzes α-galactosides) catalyzes the hydrolysis of planteose to D-galactose and sucrose. Planteose does not react with bromine water. Treatment of planteose with $(CH_3)_2SO_4/NaOH$, followed by dilute acid hydrolysis, yields, among other compounds, 1,3,4-tri-*O*-methyl-D-fructose. Suggest a structure for planteose.

27.57 *(a) What is the amino acid sequence of a peptide coded by a strand of mRNA with the following base sequence?

5′ AUGAAACAAGAUUUUUAUUGGGGG 3′

(b) What is the sequence of the DNA from which the mRNA is transcribed? Be sure to specify the 3′ and 5′ ends.

*(c) What would be the translation product resulting from a single mutation in which a change occurs from U to A at position 18 from the 5′ end in the mRNA?

*27.58   Outline a mechanism for the following reaction, an example of the *Amadori rearrangement*.

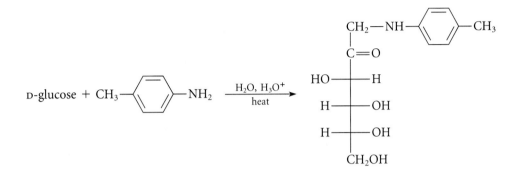

27.59   Explain with a mechanism why treatment of the 2-deoxy-2-amino derivative of D-glucose (D-glucosamine) with aqueous NaOH liberates ammonia.

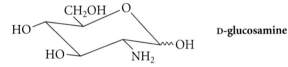

D-glucosamine

*27.60   L-Ascorbic acid (vitamin C) has the following structure:

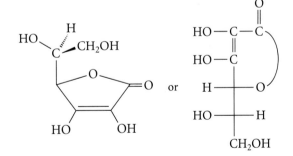

(a) Ascorbic acid has pK$_a$ = 4.21, and is thus about as acidic as a typical carboxylic acid. Identify the acidic hydrogen and explain.

(b) Thousands of tons annually of ascorbic acid are made commercially from D-glucose. In the synthesis below give the structures of the compounds [A]–[C].

(*Problem 27.60 continues*)

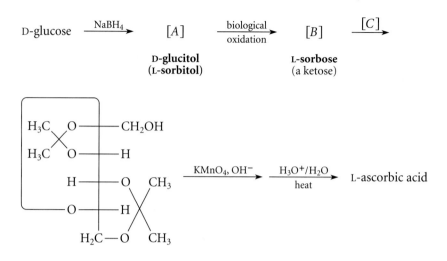

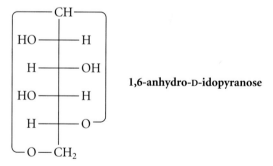

*27.61   When RNA is treated with periodic acid, and the product of that reaction treated with base, only the nucleotide residue at the 3'-end is removed.
(a)   Explain why the degradation is reasonable by showing its chemistry.
(b)   Would the same degradation work with DNA? Explain.

27.62   When DNA is treated with 0.5 M NaOH at 25°, no reaction takes place, but when RNA is subjected to the same conditions, it is rapidly cleaved into mononucleotide 2- and 3-phosphates. Explain. (*Hint:* What is the only structural difference between RNA and DNA? How can this difference promote the observed behavior? See Sec. 11.6.)

*27.63   At 100°, D-idose exists mostly (about 86%) as a 1,6-anhydropyranose:

1,6-anhydro-D-idopyranose

(a)   Draw the chair conformation of this compound.
(b)   Explain why D-idose has more of the anhydro form than D-glucose. (Under the same conditions glucose contains only 0.2% of the 1,6-anhydro form.)

Appendices

Appendix I. Substitutive Nomenclature of Organic Compounds

The substitutive name of an organic compound is based on its *principal group* and *principal chain*.

The *principal group* is *assigned* according to the following priorities:

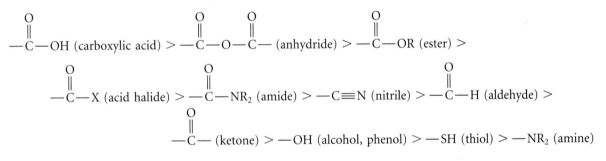

The *principal chain* is *identified* by applying the following criteria in order until a decision can be made.

1. Maximum number of substituents corresponding to the principal group
2. Maximum number of double and triple bonds considered together
3. Maximum length
4. Maximum number of double bonds
5. Maximum number of substituents cited as prefixes

A *principal chain* is *numbered* by applying the following criteria in order until there is no ambiguity. Where multiple numbers are possible, comparisons are made at the first point of difference.

1. Lowest number for the principal group cited as a suffix, that is, the group on which the name is based
2. Lowest numbers for multiple bonds, with double bonds having priority over triple bonds

3. Lowest numbers for other substituents, taking into account the "first point of difference" rule (Sec. 2.4C, Rule 8)

4. Lowest number for the substituent named as a prefix that is cited first in the name

The *name* is *constructed* starting with the hydrocarbon corresponding to the principal chain.

1. Cite the principal group by its suffix and number; its number is the last one cited in the name.

2. If there is no principal group, name the compound as a substituted hydrocarbon.

3. Cite the names and numbers of the other substituents in alphabetical order at the beginning of the name.

These lists cover most of the cases cited in the text. (See Study Problems 8.1 and 8.2 for illustrations.) For a more complete discussion of nomenclature, see, *Nomenclature of Organic Chemistry, 1979 Edition*, by the International Union of Pure and Applied Chemistry, published by Pergamon Press.

II

Appendix II. Important Infrared Absorptions

Type of Absorption	Frequency, cm^{-1} (Intensity)	Comment
	Alkanes	
C—H stretch	2850–3000 (m)	occurs in all compounds with aliphatic C—H bonds
	Alkenes	
C=C stretch		
—CH=CH$_2$	1640 (m)	
$\diagdown$C=CH$_2$ $\diagup$	1655 (m)	
others	1660–1675 (w)	not observed if alkene is symmetrical
=C—H stretch	3000–3100 (m)	
=C—H bend		
—CH=CH$_2$	910–990 (s)	
$\diagdown$C=CH$_2$ $\diagup$	890 (s)	
H$\diagdown$C=C$\diagup$ $\diagup$ $\diagdown$H	960–980 (s)	

Type of Absorption	Frequency, cm^{-1} (Intensity)	Comment
	675–730 (s)	position is highly variable
	800–840 (s)	

Alcohols and Phenols

Type of Absorption	Frequency, cm^{-1} (Intensity)	Comment
O—H stretch	3200–3400 (s)	
C—O stretch	1050–1250 (s)	also present in other compounds with C—O bonds: ethers, esters, etc.

Alkynes

Type of Absorption	Frequency, cm^{-1} (Intensity)	Comment
C≡C stretch	2100–2200 (m)	not present or weak in many internal alkynes
≡C—H stretch	3300 (s)	present in 1-alkynes only

Aromatic Compounds

Type of Absorption	Frequency, cm^{-1} (Intensity)	Comment
C=C stretch	1500, 1600 (s)	two absorptions
C—H bend	650–750 (s)	
overtone	1660–2000 (w)	

Aldehydes

Type of Absorption	Frequency, cm^{-1} (Intensity)	Comment
C=O stretch		
ordinary	1720–1725 (s)	
α,β-unsaturated	1680–1690 (s)	
benzaldehydes	1700 (s)	
C—H stretch	2720 (m)	

Ketones

Type of Absorption	Frequency, cm^{-1} (Intensity)	Comment
C=O stretch		
ordinary	1710–1715 (s)	increases with decreasing ring size (Table 21.3)
α,β-unsaturated	1670–1680 (s)	
aryl ketones	1680–1690 (s)	

Carboxylic Acids

Type of Absorption	Frequency, cm^{-1} (Intensity)	Comment
C=O stretch		
ordinary	1710 (s)	
benzoic acids	1680–1690 (s)	
O—H stretch	2400–3000 (s)	very broad

Esters and Lactones

Type of Absorption	Frequency, cm^{-1} (Intensity)	Comment
C=O stretch	1735 (s)	increases with decreasing ring size (Table 21.3)

Acid Chlorides

Type of Absorption	Frequency, cm^{-1} (Intensity)	Comment
C=O stretch	1800 (s)	second weaker band sometimes observed at 1700–1750

Anhydrides

Type of Absorption	Frequency, cm^{-1} (Intensity)	Comment
C=O stretch	1760, 1820 (s)	two bands; increases with decreasing ring size in cyclic anhydrides

(Table continues)

Appendix II   (continued)

Type of Absorption	Frequency, cm^{-1} (Intensity)	Comment
	Amides and Lactams	
C=O stretch	1650–1655 (s)	increases with decreasing ring size (Table 21.3)
N—H bend	1640 (s)	
N—H stretch	3200–3400 (m)	doublet absorption observed for some primary amides
	Nitriles	
C≡N stretch	2200–2250 (m)	
	Amines	
N—H stretch	3200–3375 (m)	several absorptions sometimes observed, especially for primary amines

Appendix III.   Important Proton NMR Chemical Shifts

A. Protons within Functional Groups

Group	Chemical shift, ppm	Group	Chemical shift, ppm
—C—C—H	0.7–1.5	O‖—C—H	9–11
C=C (H)	4.6–5.7	O‖—C—N—H	7.5–9.5
—O—H	varies with solvent and with acidity of O—H	—C—NH—	0.5–1.5
—C≡C—H	1.7–2.5	⟨benzene⟩—NH—	2.5–3.5
⟨benzene⟩H	6.5–8.5		

B. Alkyl Protons Adjacent to Functional Groups

Group G	δ for CH$_3$—G, ppm	σ$_G$ ^{a}
—H	0.2	—
—CR$_3$ or —CR$_2$— (R = H, alkyl)	0.9	0.6
—F	4.3	3.6
—Cl	3.0	2.5

Group G	δ for CH₃—G, ppm	σ_G ᵃ
—Br	2.7	2.3
—I	2.2	1.8
—CR=CR₂ (R = H, alkyl)	1.8	1.3
—C≡C—R (R = H, alkyl)	2.0	1.4
—OH	3.5	2.6
—OR (R = alkyl)	3.3	2.4
—OR (R = aryl)	3.7	2.9
—SH, —SR	2.4	1.6
(phenyl ring)	2.3	1.8
$\overset{O}{\overset{\|}{-C-R}}$	2.1 (R = aryl) 2.6 (R = aryl)	1.5
$\overset{O}{\overset{\|}{-C-OR}}$ (R = H, alkyl)	2.1	1.5
$\overset{O}{\overset{\|}{-O-C-R}}$	3.6 (R = alkyl) 3.8 (R = aryl)	3.0
$\overset{O}{\overset{\|}{-C-NR_2}}$ (R = H, alkyl)	2.0	1.5
$\overset{O}{\overset{\|}{-NR-C-R}}$ (R = H, alkyl)	2.8	—
—NR₂ (R = H, alkyl)	2.2	1.6
—NR (R = aryl)	2.9	—
—C≡N	2.0	1.7

ᵃ Effective contribution to the chemical shift of the methylene protons in a group $G_1-CH_2-G_2$ according to the equation

$$\delta(-CH_2-) = 0.2 + \sigma_{G_1} + \sigma_{G_2}$$

Some ¹³C chemical shifts are given in Table 13.7, page 623.

IV

Appendix IV.   Summary of Synthetic Methods

The methods below are listed in order of their occurrence in the text; the section reference follows each reaction in parentheses. Thus, a review at any point in the text is possible by considering the methods listed for earlier sections.

Don't forget that in many cases a method can be applied to compounds containing more than one functional group. Thus, catalytic hydrogenation can be used to convert phenols into alcohols, but it is listed under "Synthesis of Alkanes and Aromatic Hydrocarbons" because the actual transformation is the formation of —CH₂—CH₂— groups from —CH=CH— groups; the presence of the —OH group is incidental.

Reaction summaries for each chapter are found in the Study Guide.

Synthesis of Alkanes and Aromatic Hydrocarbons

1. Catalytic hydrogenation of alkenes (4.9A)
2. Protonolysis of Grignard or related reagents (8.7B)
3. Cyclopropane formation by the addition of carbenoids to alkenes (Simmons-Smith reaction; 9.8B)
4. Catalytic hydrogenation of alkynes (14.6A)
5. Friedel-Crafts alkylation of aromatic compounds (16.4F)
6. Catalytic hydrogenation of aromatic compounds (16.6)
7. Wolff-Kishner or Clemmensen reductions of aldehydes or ketones (19.12)
8. Reaction of aryldiazonium salts with hypophosphorous acid (23.10A)

Synthesis of Alkenes

1. β-Elimination reactions of alkyl halides or sulfonates (9.5, 10.3A, 17.3B)
2. Acid-catalyzed dehydration of alcohols (10.1)
3. Catalytic hydrogenation of alkynes (gives *cis*-alkenes when used with internal alkynes; 14.6A)
4. Reduction of alkynes with alkali metals in liquid ammonia (gives *trans*-alkenes when used with internal alkenes; 14.6B)
5. Diels-Alder reaction of dienes and alkenes to give cyclic alkenes (15.3, 25.3)
6. Wittig reaction of aldehydes or ketones (19.13)
7. Aldol condensation reactions of aldehydes or ketones to give α,β-unsaturated aldehydes or ketones (22.4)
8. Hofmann elimination of quaternary ammonium hydroxides (23.8)

Synthesis of Alkynes

1. Alkylation of acetylenic anions with alkyl halides or sulfonates (14.7B)
2. β-Elimination reactions of alkyl dihalides or vinylic halides (18.2)

Synthesis of Alkyl and Aryl Halides

1. Addition of hydrogen halides to alkenes (4.7, 15.4A)
2. Addition of halogens to alkenes (5.1)
3. Peroxide-promoted addition of HBr to alkenes (5.6)
4. Synthesis of dihalocyclopropanes by addition of dihalomethylene to alkenes (9.8A)
5. Reaction of alcohols with HBr or thionyl chloride (10.2, 10.3D, 17.1)
6. Reaction of sulfonate esters or other alkyl halides with halide ions (10.3A, 17.4)
7. Halogenation of aromatic compounds (16.4A)

8. Allylic and benzylic bromination of alkenes or aromatic hydrocarbons (17.2)

9. α-Halogenation of aldehydes, ketones, or carboxylic acids (22.3A,C)

10. Synthesis of aryl halides by the reaction of cuprous chloride, cuprous bromide, or potassium iodide with aryldiazonium salts (Sandmeyer and related reactions; 23.10A)

Synthesis of Grignard Reagents and Related Organometallic Compounds

1. Reaction of alkyl or aryl halides with metals (8.7A, 18.5)

2. Preparation of acetylenic Grignard reagents by metal-hydrogen exchange (14.7A)

3. Preparation of lithium dialkylcuprates by reaction of alkyllithium reagents with cuprous halides (21.10B)

Synthesis of Alcohols and Phenols

(Syntheses apply only to alcohols unless noted otherwise.)

1. Acid-catalyzed hydration of alkenes (used industrially, but generally not a good laboratory method; 4.9B)

2. Synthesis of halohydrins from alkenes (5.1B)

3. Oxymercuration-reduction of alkenes (5.3A)

4. Hydroboration-oxidation of alkenes (5.3B)

5. Ring-opening reactions of epoxides (11.4A,B)

6. Reaction of ethylene oxide with Grignard reagents (11.4C)

7. Reduction of aldehydes or ketones (19.8, 27.7D, 22.9)

8. Reaction of aldehydes or ketones with Grignard or related reagents (19.9, 22.10A)

9. Reduction of carboxylic acids to primary alcohols (20.10)

10. Reduction of esters to primary alcohols (21.9A)

11. Reaction of esters with Grignard or related reagents (21.10A)

12. Aldol addition reactions of aldehydes or some ketones to give β-hydroxy aldehydes or ketones (22.4)

13. Reaction of diazonium salts with water to give phenols (23.10A)

14. Synthesis of phenols by the Claisen rearrangement of allylic aryl ethers (25.4B)

Synthesis of Glycols

1. Reaction of alkenes with osmium tetroxide or alkaline potassium permanganate (5.5)

2. Acid-catalyzed hydrolysis of epoxides (11.4B)

Synthesis of Ethers

1. Alkylation of alkoxides or phenoxides with alkyl halides or alkyl sulfonates (Williamson synthesis; 11.1A, 18.6B, 27.6)
2. Alkoxymercuration-reduction of alkenes (11.1B)
3. Acid-catalyzed dehydration of alcohols (11.1C)
4. Acid-catalyzed addition of alcohols to alkenes (11.1C)
5. Acetal formation by the acid-catalyzed reaction of alcohols with aldehydes or ketones (19.10A, 27.5)

Synthesis of Epoxides

1. Oxidation of alkenes with peroxycarboxylic acids (11.2A)
2. Cyclization of halohydrins (11.2B)

Synthesis of Disulfides

1. Oxidation of thiols (10.9)

Synthesis of Aldehydes

1. Ozonolysis of alkenes (of limited utility because carbon-carbon bonds are broken; 5.4)
2. Oxidation of primary alcohols (10.6A)
3. Oxidative cleavage of glycols (of limited utility because carbon-carbon bonds are broken; 10.6C)
4. Hydroboration-oxidation of alkynes (14.5B)
5. Reduction of acid chlorides (21.9D)
6. Aldol addition reactions of aldehydes to give β-hydroxy aldehydes (22.4)
7. Aldol condensation reactions of aldehydes to give α,β-unsaturated aldehydes (22.4)
8. Synthesis of aldoses from other aldoses by the Kiliani-Fischer synthesis (27.8) and the Ruff degradation (27.9B)

Synthesis of Ketones

1. Ozonolysis of alkenes (of limited utility because carbon-carbon bonds are broken; 5.4)
2. Oxidation of secondary alcohols (10.6A)
3. Oxidative cleavage of glycols (of limited utility because carbon-carbon bonds are broken; 10.6C)

   4. Mercuric-ion catalyzed hydration of alkynes (14.5A)

   5. Friedel-Crafts acylation of aromatic compounds (16.4E)

   6. Oxidation of phenols to quinones (18.7)

   7. Reaction of acid chlorides with lithium dialkylcuprates (21.10B)

   8. Aldol condensation reactions of ketones to give α,β-unsaturated ketones (22.4)

   9. Claisen and Dieckmann condensation reactions of esters to give β-keto esters (22.5A,B)

   10. Crossed Claisen condensation reactions of esters to give β-diketones (22.5C)

   11. Acetoacetic ester synthesis (22.6C)

   12. Conjugate addition reactions of α,β-unsaturated ketones (22.8), including addition of lithium dialkylcuprate reagents (22.10B)

Synthesis of Sulfoxides and Sulfones

   1. Oxidation of sulfides (11.7)

Synthesis of Carboxylic and Sulfonic Acids

(Syntheses apply only to carboxylic acids unless noted otherwise.)

   1. Ozonolysis of alkenes (of limited utility because carbon-carbon bonds are broken; 5.4)

   2. Oxidation of primary alcohols (10.6B)

   3. Oxidation of thiols to sulfonic acids (10.9)

   4. Sulfonation of aromatic compounds to give arylsulfonic acids (16.4D)

   5. Side-chain oxidation of alkylbenzenes (17.5)

   6. Oxidation of aldehydes (19.14)

   7. Reaction of Grignard or related reagents with carbon dioxide (20.6)

   8. Hydrolysis of carboxylic acid derivatives, especially nitriles (21.7, 21.11, 26.4C)

   9. Haloform reaction of methyl ketones (of limited utility, since this reaction breaks carbon-carbon bonds; 22.3B)

   10. Malonic ester synthesis (22.6A)

Synthesis of Esters

   1. Reaction of alcohols with sulfonyl chlorides (for sulfonate esters; 10.3A)

   2. Acid-catalyzed esterification of carboxylic acids with primary or secondary alcohols (20.8A, 26.5)

   3. Alkylation of carboxylic acids with diazomethane (20.8B)

   4. Alkylation of carboxylate salts with alkyl halides (20.8B)

   5. Reaction of acid chlorides, anhydrides, or esters with alcohols (21.8, 27.6)

6. Claisen and Dieckman condensation reactions of esters to give β-keto esters (22.5A,B)

7. Alkylation of ester enolate ions (22.6A,B)

8. Conjugate addition reactions of α,β-unsaturated esters (22.8)

Synthesis of Anhydrides

1. Reaction of carboxylic acids with dehydrating agents (20.9B)

2. Reaction of acid chlorides with carboxylate salts (21.8A)

Synthesis of Acid Chlorides

1. Reaction of carboxylic or sulfonic acids with thionyl chloride, phosphorus pentachloride, or related reagents (20.9A)

2. Synthesis of sulfonyl halides by chlorosulfonation of aromatic compounds (20.9A)

Synthesis of Amides

1. Reaction of acid chlorides, anhydrides, or esters with amines (21.8, 26.5)

2. Condensation of amines and carboxylic acids with dicyclohexylcarbodi-imide (26.7)

Synthesis of Nitriles

1. Formation of cyanohydrins from aldehydes or some ketones (19.7A, 27.8A)

2. Reaction of alkyl halides or sulfonates with cyanide ion (21.11)

3. Conjugate addition of cyanide ion to α,β-unsaturated carbonyl compounds (22.8A)

4. Reaction of cuprous cyanide with aryldiazonium salts (23.10A)

Synthesis of Amines

1. Reduction of amides (21.9B)

2. Reduction of nitriles to primary amines (21.9C)

3. Direct alkylation of ammonia and amines (of limited utility because of the possibility of overalkylation; 23.7A, 26.4A)

4. Reductive amination of aldehydes and ketones (23.7B)

5. Aromatic substitution reactions of aniline derivatives (23.9)

6. Gabriel synthesis of primary amines (23.11A)

7. Reduction of nitro compounds (23.11B)

8. Curtius and Hofmann rearrangements (23.11C)

Synthesis of Nitro Compounds

1. Nitration of aromatic compounds (16.4C)

V   Appendix V.   Reactions Used to Form Carbon-Carbon Bonds

Reactions are listed in the order that they are discussed in the text. The section reference follows each reaction in parentheses.

1. Cyclopropane formation by addition of carbenes or carbenoids to alkenes (9.8)

2. Reaction of Grignard reagents with ethylene oxide (11.4C)

3. Reaction of acetylenic anions with alkyl halides or sulfonates (14.7B)

4. Diels-Alder reactions (15.3, 25.3)

5. Friedel-Crafts acylation (16.4E) and alkylation reactions (16.4F)

6. Cyanohydrin formation (19.7, 26.4C, 27.8A)

7. Reaction of Grignard and related reagents with aldehydes or ketones (19.9)

8. Wittig alkene synthesis (19.13)

9. Reaction of Grignard and related reagents with carbon dioxide (20.6)

10. Reaction of Grignard and related reagents with esters (21.10A)

11. Reaction of lithium dialkylcuprates with acid chlorides (21.10B)

12. Reaction of cyanide ion with alkyl halides or sulfonates (21.11)

13. Aldol addition and condensation reactions (22.4)

14. Claisen and related condensation reactions (22.5)

15. Malonic ester synthesis (22.6A, 26.4B)

16. Alkylation of ester enolate ions with alkyl halides or sulfonates (22.6B)

17. Acetoacetic ester synthesis (22.6C)

18. Conjugate-addition reactions of cyanide ion (22.8A) or enolate ions (22.8C) to α,β-unsaturated carbonyl compounds

19. Conjugate addition of lithium dialkylcuprate reagents to α,β-unsaturated carbonyl compounds (22.10B)

20. Reaction of aryldiazonium salts with cuprous cyanide (23.10A)

21. Fischer and Reissert indole syntheses (24.4)

22. Skraup quinoline synthesis (24.5E)

23. Formation of rings by electrocyclic reactions (25.2)

24. Claisen rearrangement (25.4B)

Appendix VI. Typical Acidities and Basicities of Organic Functional Groups

A. Acidities of Groups That Ionize to Give Anionic Conjugate Bases

Functional group	Structure	Typical pK_a
sulfonic acid[a,b]	R—S(=O)(=O)—O—H[c]	<1 (strong acid)
carboxylic acid[a,b]	R—C(=O)—O—H	3–5
phenol[b]	⬡—O—H	9–11
thiol[b]	R—S—H	9–11
sulfonamide[b]	R—S(=O)(=O)—N(H)—R	10
amide	R—C(=O)—N(H)—R	15–17
alcohol	R—O—H	15–19
aldehyde, ketone	R—C(=O)—CR$_2$(H)	17–20
ester	R$_2$C(H)—C(=O)—OR	20–25
nitrile	R$_2$C(H)—C≡N	25
amine	R$_2$N—H	32–35
alkane	R$_3$C—H	50–60 (?)

[a] Acidic enough to dissolve in aqueous 5% $NaHCO_3$ solution.
[b] Acidic enough to dissolve in aqueous 5% $NaOH$ solution.
[c] Acidic hydrogens in color.

B. Basicities of Groups That Protonate to Give Cationic Conjugate Acids

Functional group	Structure of conjugate acid	Typical pK_a of conjugate acid	Functional group	Structure of conjugate acid	Typical pK_a of conjugate acid
alkylamine[a]	$R_3\overset{+}{N}$—H[b]	9–11			
aromatic amine[a]	Ar—$\overset{+}{N}R_2$—H	4–5	thiol, sulfide (R,R' = alkyl, H)	R—$\overset{+}{S}$—R with H above S	−6 to −7
amide	R—C(=$\overset{+}{O}$—H)—NR$_2$	−1	ester, acid	R—C(=$\overset{+}{O}$—H)—OR	−6
alcohol, ether (R,R' = alkyl, H)	R—$\overset{+}{O}$—R with H above O	−2 to −3	aldehyde, ketone	R—C(=$\overset{+}{O}$—H)—R	−7
phenol, aromatic ether	Ar—$\overset{+}{O}$—R with H above O	−6 to −7	nitrile	R—C≡$\overset{+}{N}$—H	−10

[a] Basic enough to dissolve in 5% aqueous HCl.
[b] Acidic hydrogens in color.

Credits

Photo of author, page xiii, courtesy of Kyle Loudon.

Figure 1.13: Adapted from *University Chemistry*, 3rd Edition by Bruce H. Mahan. © 1975 by Addison-Wesley Publishing Company.

Figure 2.10: Courtesy of The Dow Chemical Company, Texas Operations.

Figure 2.11a: T. J. Beveridge, University of Guelph and G. Sprott, National Research Council of Canada/BPS.

Figure 2.11b: T. J. Beveridge and C. Forsberg, University of Geulph/BPS.

Figure 6.3: From the Nobel Lecture of Vladimir Prelog, *Science*, **193**:17 (1976). Copyright © 1976 The Nobel Foundation.

Figure 6.19: Reprinted with permission from G. B. Kauffman and R. D. Myers, *Journal of Chemical Education*, **52**:777 (1975). Copyright © 1975 by American Chemical Society.

Figure 8.5: Reprinted with permission from R. Ember, P. S. Layman, Wil Lepkowski, and P. S. Zurer, *Chemical and Engineering News*, November 24, 1986. Copyright © 1986 by American Chemical Society.

Figures 12.4, 12.8, 12.9, 12.10, 12.11, 12.20, 12.21, 12.22, and 12.23: Adapted from *Aldrich Library of FT-IR Spectra*, Charles J. Pouchert, editor. © 1985 Aldrich Chemical Company. Used with permission.

Figures 12.14, 12.15, 12.16, 12.17, 12.18, 12.24, and 12.25: From *NPA-NIH Mass Spectral Data Base*, published by the National Bureau of Standards, United States Department of Commerce.

Figure 12.19: Adapted from F. W. McLafferty, *Interpretation of Mass Spectra*. © 1977 The Benjamin/Cummings Publishing Company, Inc.

Figures 13.2, 13.3, 13.4, 13.7, 13.9, 13.12, 13.14, 13.17, 13.21, 13.22, 13.27, 13.29, and 13.30: © Sadtler Research Laboratories, Division of Bio-Rad Laboratories, Inc.

Figures 13.13 and 13.23: Spectra courtesy of John Kozlowski.

Figure 13.18: From Daniel J. Pasto and Carl R. Johnson, *Organic Structure Determination*. © 1969, Prentice-Hall, Inc., pp. 192 and 200. Adapted by permission of Prentice-Hall, Inc., Englewood Cliffs, NJ.

Figure 13.20: Reprinted with permission from F. A. Bovey, *Chemical and Engineering News*, August 30, 1965. © 1965 by the American Chemical Society.

Figure 13.24: Adapted from *Aldrich Library of FT-IR Spectra*, Charles J. Pouchert, editor. © 1985 Aldrich Chemical Company. Used with permission. © Sadtler Research Laboratories, Division of Bio-Rad Laboratories, Inc. From *NPA-NIH Mass Spectral Data Base*, published by the National Bureau of Standards, United States Department of Commerce.

Figure 13.26: Courtesy of Dr. Paul J. Keller, Barrow Neurological Institute.

Figures 13.28 and 13.31: Adapted from *Aldrich Library of FT-IR Spectra*, Charles J. Pouchert, editor. © 1985 Aldrich Chemical Company. Used with permission. © Sadtler Research Laboratories, Division of Bio-Rad Laboratories, Inc.

Figure 13.32: From *NPA-NIH Mass Spectral Data Base*, published by the National Bureau of Standards, United States Department of Commerce. Adapted from *Aldrich Library of FT-IR Spectra*, Charles J. Pouchert, editor. © 1985 Aldrich Chemical Company. Used with permission. © Sadtler Research Laboratories, Division of Bio-Rad Laboratories, Inc.

Figures 14.3, 14.5, and 14.7: Adapted from *Aldrich Library of FT-IR Spectra*, Charles J. Pouchert, editor. © 1985 Aldrich Chemical Company. Used with permission.

Figure 14.6: Photo courtesy of Professor Tom Eisner, Cornell University.

Index

Key: Page numbers in **boldface** give locations of exact definitions. When the page number is follows by *p*, *f*, or *t*, the item is found in a *problem*, *figure*, or *table*, respectively. SGL indicates a Study Guide Link.

A

Abietic acid, 735*p*
Absolute stereochemical configuration, **233**
  relationship to optical rotation, 238
Absorbance, **684**
Absorptions, infrared
  confirmatory, 543
  diagnostic, 543
Absorption spectroscopy, **538**
Absorptivity, *see* Extinction coefficient
Abstraction, of atoms in free-radical reactions, 198
Ac, abbreviation for acetyl group, 1064*t*
Acceptor, hydrogen-bond, **348**
Acebutolol, structure, 85*p*
Acetaldehyde
  aldol addition, 1055
  enolization, 1044
  from biological oxidation of ethanol, 472–475
  structure, 866*f*, 934*f*
  synthesis from acetylene by hydration, 656
  trimerization, 901
Acetals, **898**, 898–902
  as protecting groups, 902–904
  glycosides, 1344
  synthesis, 898
2-Acetamido-2-deoxy-D-glucose, 1366
Acetamido group, 977*t*

Acetanilide, nitration, 1139
Acetate group, abbreviation, 1064*t*
Acetic acid
  in spoiled wine, 474
  pK_a, 108, 939*t*, 939
  solvent properties, 352, 354*t*
  structure, 930, 934*f*
Acetic anhydride
  in acetylation of amino acids, 1281
  in Friedel-Crafts acylation, 1185–1186
  reaction with carbohydrates, 1347
Acetoacetic acid, decarboxylation, 958
Acetoacetic ester synthesis, 1079–1082
Acetone
  aldol reactions, 1056, 1058
  industrial synthesis and use, 851–852, 917
  production from petroleum, 213*f*
  solvent properties, 354*t*
  structure, 866*f*, 934*f*
Acetonitrile, solvent properties, 355*t*
Acetophenone, 867
  acidity, 1039
Acetoxy group, 977*t*
  abbreviation, 1064*t*
Acetylacetone, *see* 2,4-Pentanedione
Acetylamino group, 977*t*
Acetylcholine, 1155
AcetylCoA, 1082, 1083*f*
1-Acetylcyclohexene, UV spectrum, 877
Acetylene
  bonding, 648–649
  hydration to acetaldehyde, 656
  industrial synthesis and use, 670
  structure, 14*f*, 17
Acetylenes, **46**; *see also* Alkynes
Acetylenic anions, 662–665
  as nucleophiles, 665–666
  synthesis, 663
N-Acetyl-D-glucosamine, 1366
Acetyl group, 868
  abbreviation, 1064*t*

  as protecting group, 1138
Acetyloxy group, 977*t*
Achiral, 227
Acid, *see* specific types, *e.g.*, Brønsted, Lewis, carboxylic acid, etc.
Acid anhydrides, *see* Anhydrides
Acid-base reactions, importance to organic chemistry, 87
Acid bromides, in α-bromination of carboxylic acids, 1053
Acid chlorides, *see* Acid halides
Acid halides
  hydrolysis, 995–996
  in Friedel-Crafts acylation reactions, 754
  in organic synthesis, summary, 1003–1004
  IR spectroscopy, 982–983
  NMR and CMR spectroscopy, 982–986
  nomenclature, 973–974, 976–977
  physical properties, 980
  reactions
    reduction to aldehydes, 1012–1013
    with amines and ammonia, 1001–1002
    with alcohols and phenols, 1002–1003
    with carboxylate salts, 1003
    with lithium dialkylcuprates, 1017–1018
  relative carbonyl reactivity, 1014
  relative reactivity in nucleophilic acyl substitution, 996–1000
  synthesis from carboxylic acids, 952–953
    mechanism, SGL 20.6
Acidity
  of alcohols, gas-phase, 365–366
  of alcohols and thiols, 360–366
  of functional groups, summary, A12
  relationship to orbital hybridization, 664

AcO, abbreviation for acetoxy group, 1064*t*

Acrilan, 209*t*

Acrolein, 868
  intermediate in Skraup quinoline synthesis, 1204

Acrylic acid
  pK_a, 939*t*
  structure, 931*t*

Acrylonitrile, conjugate-addition reactions, 1086–1087

Activating group
  in electrophilic aromatic substitution, **767**, 767–771
  in electrophilic aromatic substitution of naphthalene derivatives, 1176

Active site, of enzymes, **1308**

Acylamino group, activating and directing effects in electrophilic aromatic substitution, 762*t*

Acylase, 1280

Acylation, *see also* specific reactions, *e.g.*, Friedel-Crafts reaction
  of amines, 1132
  of α-amino acids, 1281
  of phenols, 848–849

Acyl azides, **1148**
  Curtius rearrangement, 1148–1150
  synthesis, 1150

Acyl carrier protein, 1083

Acyl enzyme, intermediate in trypsin catalysis, 1309, 1312

Acyl group, 988
  introduction in Friedel-Crafts acylations, 754

Acyl hydrazides, **1150**

O-Acylisourea, intermediate in solid-phase peptide synthesis, 1295

Acylium ion, intermediate in Friedel-Crafts acylations, 754

Acyloxy group, activating and directing effects in electrophilic aromatic substitution, 762*t*

Acyl substitution, **988**
  nucleophilic, **991**
    mechanisms and reactivity, 996–1000

1,2-Addition, **700**

1,4-Addition, *see* Conjugate addition

Addition polymer, **208**

Addition reactions
  carbonyl, *see also* Nucleophilic addition
    mechanisms, 884–886
    relationship to nucleophilic acyl substitution, 949
    structural effects on rates and equilibria, 887–890
  of alkenes, introduction, 144
    stereochemistry, 310–311

of alkynes, 653–662

Adenine, 1368

Adenosine, 1369*t*

Adenosine triphosphate, 1369
  hydrolysis, 1370

S-Adenosylmethionine, 517–518
  structure, 518*f*

Adenylic acid, 1369*t*

Adipic acid
  in nylon synthesis, 1021
  pK_a values, 939*t*
  structure, 931*t*

ADP, 1370

Adriamycin, 1345*f*

Adriamycin, 843

AIBN, 198

Alanine
  enantiomeric resolution, 1280
  isoelectric point, 1272–1273
  protonation state vs. pH, 1273*f*
  structure and properties, 1266*f*

Alcohol, *see also* Ethanol, *e.g.*, Ethanol, industrial

Alcohol dehydrogenase, 472
  and stereochemistry of ethanol oxidation, 480–482

Alcohols, 335
  acidity, 360–366
    in gas phase, 365
    polar effects, 364
  addition to aldehydes and ketones, 898–901
  amphoterism, 366
  basicity, 366–367
    role in alcohol dehydration, 444
  boiling points, 348
    role of hydrogen bonding, 350
  by-products in ester reduction, 1007
  classification, 335
  conversion into alkoxides, 362–363
  conversion into alkyl halides, summary, 457–458
  conversion into esters, 946–949
  conversion into sulfonate esters, 451–452
  dehydration, 443–446
  IR spectroscopy, 553–554
  NMR spectroscopy, 616–619
  nomenclature, 338–342
  pK_a, 99*t*, 365*t*
  reactivity in iodoform test, 1051
  reactions
    acetal formation, 898–902
    conversion into ethers by dehydration, 500–503
    dehydration to ethers, 496*p*
    oxidation to aldehydes and ketones, 467–469

oxidation to carboxylic acids, 467–470
  with Grignard reagents, 371
  with acid chlorides, 1002–1003
  with anhydrides, 1004–1005
  with epoxides, 509, 511
  with esters, 1005
  with hydrogen halides, 447–450
    comparison to ether cleavage, 508
  with thionyl chloride, 456–457
  sulfonate ester derivatives, 450–454
  synthesis
    alkene hydration, 163–165
    ester reduction, 1007–1008
    from alkenes, comparison of methods, 187–188
    Grignard reactions of aldehydes and ketones, 895–897,
    Grignard reactions of esters, 1015–1016
    Grignard reactions with ethylene oxide, 515–516
    hydroboration-oxidation of alkenes, 185–187
    oxymercuration-reduction of alkenes, 180–182
    reduction of aldehydes and ketones, 890–894
    reduction of carboxylic acids, 956–957
    reduction of esters
    ring-opening reactions of epoxides, 509–511
    summary, 485–486, A-7

Aldaric acids
  synthesis from aldoses, 1350–1351
  structure, 1349*t*

Aldehydes, *see also* α,β-Unsaturated carbonyl compounds
  acidity, 1039–1042
  base-catalyzed halogenation, 1050–1051
  basicity, 881–883
    in gas phase, 883
  bonding, 866
  CMR spectroscopy, 875
  formation in ozonolysis, 190–191
  formation in periodate cleavage of glycols, 470–472
  α-halo, reactivity, 1053–1054
  α-halogenation, acid-catalyzed, 1048–1049
  hydrates, role in aldehyde oxidation, 468, 916
  α-hydroxy, from periodate oxidation of carbohydrates, 1352
  intermediates in LiAlH$_4$ reduction of esters, 1007–1008
  IR spectroscopy, 872–873

manufacture and use, 917
mass spectrometry, 878–879
NMR spectroscopy, 874
nomenclature, 867–871
physical properties, 871–872
  summary, 880, A-8
reactions
  acetal formation, 898–899
  aldol reactions, 1055–1063
  autoxidation, 916
  bisulfite addition, 919p
  Clemmensen reduction, 910
  enolization, 1044–1048
  exchange of α-hydrogens
    acid-catalyzed, 1047
    base-catalyzed, 1042
  halogenation
    acid-catalyzed, 1048–1049
    base-catalyzed, 1050–1051
  hydration, 884–886
  Fischer indole synthesis, 1190–1191
  Grignard reactions, 895–897
  imine and enamine formation,
    904–909
  oxidation to carboxylic acids, 468,
    915–916
  racemization
    acid-catalyzed, 1047
    base-catalyzed, 1043
  reduction to alcohols, 890–894
  reductive amination, 1130–1132
  Wittig reaction, 911–915
  Wolff-Kishner reduction, 909–911
relative carbonyl reactivity, 1014
stability, 889
structure, 865
synthesis
  aldol condensation to give α,β-unsat-
    urated aldehydes, 1057–1063
  conjugate addition to α,β-unsatu-
    rated aldehydes, 1085–1093
  hydroboration-oxidation of alkynes,
    657–659
  oxidation of primary alcohols,
    467–469
  reduction of acid chlorides,
    1012–1013
UV spectroscopy, 876–877
Alder, Kurt, 690
Alditols
  synthesis from aldoses, 1353
  structure, 1349t
Aldohexose, 1326
Aldol, 1055
Aldol addition, see Aldol reactions
Aldol condensation, see Aldol reactions
Aldol reactions, 1055–1063
  acid-catalyzed, 1058–1063

base-catalyzed, 1055–1058
crossed, 1060–1061
in synthesis, 1062–1063
intramolecular, 1061–1062
of methylpyridines, 1203
of pyridinium salts, 1203
variations, SGL 22.5
Aldonic acids
  structure, 1349t
  synthesis from aldoses, 1349–1350
Aldonolactones, 1350
Aldoses, 1326; see also Monosaccharides
  Amadori rearrangement, 1389p
  isomerization in base, 1341–1342
  Kiliani-Fischer synthesis, 1353–1354
  oxidation to aldonic acids, 1349–1350
  Ruff degradation, 1360
  structures, 1329f
  Weerman degradation, 1386p
  Wohl degradation, 1386p
Alizarin yellow R, 1159p
Alkaloids, 1154
Alkanes, 45–79, 46
  acidity, 663
  benzylic bromination, 795–796
  boiling points, 67–70
  combustion, 72–75
  conformations, 49–55
  cracking, 76
  cyclic, 271–298; see also Cycloalkanes
  halogenation, 371–373
  insolubility in water, 71
  IR spectroscopy, 549–550
  melting points, 70–71
  motor fuels, 77–78
  NMR spectroscopy, 615
  nomenclature, 56–61, SGL 2.2
  occurrence and use, 75–78
  physical properties, 47t, 67–72
  reforming, 76
  synthesis
    from aldehydes and ketones,
      909–911
    from alkynes, 659–660
    summary, A-5
  thermal cracking, 210–212
  unbranched, 47–48
Alkenes, 121–223, 121; see also Dienes
  acidity, 663
  carbon hybridization, 122–126
  elemental analysis, 72–75
  conjugated
    electrocyclic reactions, 1229–1231
    molecular orbitals, 1222–1226
    UV spectroscopy, 688–690
  formation in β-elimination reactions,
    387–389
  heats of formation, 140–142, 680t

industrial source, 210–212
IR spectroscopy, 550–553, 551t
NMR spectroscopy, 612–614
  chemical shift table for alkene pro-
    tons, 612t
nomenclature, 129–136, 646
physical properties, 138–139
reactions
  addition of halogens, 175–176
    stereochemistry, 313–317, 820p
  addition of hydrogen halides,
    145–153
  addition reactions, introduction, 144
  alkoxymercuration-reduction,
    499–500
  allylic bromination, 796–797
  catalytic hydrogenation, 161–162
    stereochemistry, 319
  conversion into dihalocyclopropanes,
    428–429
  conversion into epoxides with peroxy
    acids, 503–505
  conversion into glycols, 192–194
    stereochemistry, 320
  conversion into halohydrins,
    177–178
  Diels-Alder reaction, 690–699
  free-radical addition of HBr,
    194–207
  free-radical polymerization, 207–210
  Friedel-Crafts alkylation reactions,
    759
  hydration, 163–165
  hydroboration-oxidation, 183–187
    stereochemistry, 317–318
  oxymercuration-reduction, 180–182
    stereochemistry, 319–320
  ozonolysis, 188–192
  Simmons-Smith reaction, 430–433
relative stabilities, 139–143
  and regiochemistry in alcohol dehy-
    dration, 445–446
  and regiochemistry in E2 reactions,
    410–411
stereoisomerism, 129
structure and bonding, 121–129
synthesis
  dehydration of alcohols, 443–446
  Diels-Alder-reaction, 690–699
  β-elimination reactions, 411, 414,
    454, 1133–1137
  Grignard reactions, 896
  reduction of alkynes to cis alkenes,
    659–660
  reduction of alkynes to trans alkenes,
    660–662
  summary, A-6
  Wittig reaction, 911–915

Alkoxides, **61**
  base strength, 99*t*
  effect of structure on E2-S$_N$2 competi-
      tion, 413–415
  in transesterification of esters, 1005
  in Williamson ether synthesis, 497–499
  stabilization by polarization of alkyl
      groups, 365
  synthesis
    from alcohols with sodium hydride,
        362
    from alcohols with alkali metals,
        362–363
α-Alkoxy carbocations, 881
Alkoxy group, **343**
  activating and directing effects in elec-
      trophilic aromatic substitution,
      762*t*
Alkylating agent, 454–455, **455**
Alkylation, **454**; *see also* specific reactions,
      e.g., Friedel-Crafts alkylation
  of amines, 1128–1130
  of esters, 1076–1078
  of β-keto esters, 1079–1082
  of malonic ester derivatives, 1074–1075
  of phenols, 849
Alkyl bromides, *see* Alkyl halides
Alkyl chlorides, *see* Alkyl halides
Alkyl groups, **57**
  activating and directing effects in elec-
      trophilic aromatic substitution,
      762*t*
  abbreviation by R, 79
  abbreviations, 1064*t*
Alkyl halides, **335**, 385–441
  allylic, relative S$_N$1 reactivities, 790*t*
  benzylic, relative S$_N$1 reactivities, 791*t*
  boiling points, 346–347
  classification, 335
  conversion into ethers by Williamson
      synthesis, 497–499
  E2-S$_N$2 competition, effect of structure,
      411–415
  IR spectroscopy, 550
  NMR spectroscopy, 616
  nomenclature, 336
  predicting results of S$_N$1-S$_N$2-E1-E2
      competition, 423–427
  reactions
    alkylation of amines, 1128–1130
    alkylation of β-keto esters,
        1079–1082
    aryl ether formation with phenoxide
        ions, 841
    E2 reaction, 404–416; *see also* E2
        reaction
      involving allylic and benzylic
          hydrogens, 802–803

α-elimination reactions, 427–429
β-elimination reactions, 387–389
ester alkylation, 1076–1078
ester formation with carboxylate
    salts, 951
Friedel-Crafts alkylation, 757
Gabriel synthesis of amines,
    1146–1147
Grignard and organolithium reagent
    formation, 368
malonic ester alkylation, 1074–1075
nucleophilic substitution reactions,
    386*t*, 385–387
S$_N$1 and E1 reactions, 416–423
S$_N$2 reactions, 394–404; *see also* S$_N$2
    reactions
Williamson ether synthesis, 497–499
with acetylenic anions, 665–666
with cyanide ion, 1019
Wittig alkene synthesis, 913–914
synthesis
  alkane halogenation, 371–373
  allylic bromination, 796–797
  benzylic bromination, 795–796
  hydrogen halide addition to alkenes,
      145–153
  free-radical addition of HBr to
      alkenes, 195
  free-radical halogenation of alkanes,
      371–373
  from alcohols
    reaction with hydrogen halides,
        447–450
    reaction with thionyl chloride,
        456–457
    reaction with phosphorus tribro-
        mide, 457*p*
  S$_N$2 reactions of sulfonate esters, 453
  summary, A-6
uses, 373–374
Alkyl iodides, *see* Alkyl halides
Alkylthio group, **343**
Alkyne anions, 662
Alkynes, **46**, 645–677
  acidity of 1-alkynes, 662–665
  heats of formation, 859*p*
  IR spectroscopy, 650
  NMR and CMR spectroscopy, 650–652
  nomenclature, 645–647
  occurrence and use, 670–671
  physical properties, 649–650
  reactions
    addition of bromine, 653
    addition of hydrogen halides, 653
    catalytic hydrogenation, 659–660
    Diels-Alder reactions, 691
    formation of acetylenic Grignard
        reagents, 663–664

hydration, 654–657
hydroboration-oxidation, 657–659
reduction with sodium in liquid
    ammonia, 660–662
structure and bonding, 647–649
synthesis
  elimination reactions of vinylic dihal-
      ides, 826
  from other alkynes, 665–666
  test for 1-alkynes, 664
Alkynyl groups, 646
Allenes, 679
  structure and bonding, 682*f*, 683*f*
  chirality, 683
Alloisoleucine, 1269*p*
  physical properties, 244*t*
D-(+)-Allose
  equilibrium composition, 1340*t*
  structure, 1329*f*
Allothreonine, 1269
Allowed pericyclic reactions, **1231**
Allyl alcohol, structure, 338
Allyl anion
  molecular orbitals, 1226–1227
  resonance structures, 96*p*, 799
Allyl cation
  molecular orbitals, 1226–1227
  resonance structures, 96*p*, 789
  structure and bonding, 702
Allyl chloride, structure, 337
Allyl group, **131**, **337**
Allylic, 701
Allylic group, 701, **789**
Allylic protons, chemical shifts in NMR,
    612
Allylic rearrangement, **801**
  of Grignard reagents, 801
Allyl radical, molecular orbitals,
    1226–1227
D-(+)-Altrose
  equilibrium composition, 1340*t*
  structure, 1329*f*
Aluminum trichloride
  catalyst in Friedel-Crafts acylations,
      754–755
  reaction with phenol, 848
Amadori rearrangement, 1389*p*
Amide anion, 663
Amides (carboxylic acid derivatives)
  acidity, 1041–1042
  basicity, 987
  classification, 975
  intermediates in nitrile hydrolysis, 994
  internal rotation, 979, SGL 21.1
  IR spectroscopy, 982–983
  NMR and CMR spectroscopy, 982–986
  nomenclature, 975–977
  physical properties, 981

reactions
  Hofmann rearrangement, 1150–1153
  hydrolysis, 992–993
  reduction to amines, 1008–1010
relative carbonyl reactivity, 1014
relative reactivity in nucleophilic acyl
    substitution, 996–1000
structure, 978–979
synthesis
  from acid chlorides, 1001–1002
  from carboxylic acids and amines,
    1291–1298
  from esters, 1005
  summary, A-10
Amides (conjugate bases of amines and
    ammonia) 1127
  basicity, 99t
  structure, SGL 23.1
Amination, reductive, 1130–1132
Amines, 1113–1168
  acidity, 1127
  as leaving groups, 1133–1137
  basicity, 99t, 1120–1121
    effect of alkyl substitution, 1123
    gas-phase, 1123
    polar effects, 1124
    use in enantiomeric resolutions,
      249–250
    use in separations, 1125–1126
  industrial use, 1153–1154
  inversion, 254–255, 1116–1117
  IR spectroscopy, 1118
  mass spectrometry, 1119–1120
  naturally occurring, 1154–1155
  NMR spectroscopy, 1119
  nomenclature, 1114–1116
  oxidation, 1139
  physical properties, 1117–1118
  primary, 1113
  reactions
    acylation with acid chlorides,
      1001–1002
    acylation with anhydrides,
      1004–1005
    alkylation, 1128–1130
    alkylation by reductive amination,
      1130–1132
    formation of diazonium salts,
      1139–1143
    imine and enamine formation,
      905–909
    quaternization, 1129–1130
    oxidation of pyridines and quino-
      lines, 1196
    summary, 1153
    with acid chlorides, 1001–1002
    with anhydrides, 1004–1005
    with esters, 1005

with nitrous acid, 1144–1145
secondary, 1113
structure, 1116–1117
synthesis
  Curtius and Hofmann
    rearrangements, 1148–1153
  Gabriel synthesis, 1146–1147
  reduction of amides, 1008–1010
  reduction of nitriles, 1010–1012
  reduction of nitro compounds,
    1147–1148
  summary, A-10
tertiary, 1113
  reaction with aldehydes and ketones,
    908–909
Amino radical, electron affinity, 107
Amino acid analysis, 1282–1284
Amino acids, 1263–1324, 1263; see also
    α-amino acids
  acid-base properties, 1269–1278
  acylation, 1281
  dipole moments, 1270
  esterification, 1281
  isoelectric points, 1271–1275
  melting points, 1270
  nomenclature, 1266–1267
  pKa, 1270–1271
  separation using acid-base properties,
    1275–1277
  solubility, 1270
  zwitterionic structures, evidence,
    1269–1271
α-Amino acids, 1263; see also Amino
    acids
  enantiomeric resolution, 1280–1281
  formation in peptide hydrolysis,
    1282–1284
  optical rotations, 1267–1268t
  pKa values, 1267–1268t
  separation, 1283f
  stereochemistry, 1268–1269
  synthesis, 1278–1280
1-Amino-1-cyclopropanecarboxylic acid,
    in biosynthesis of ethylene, 212
2-Amino-2-deoxy-D-glucose, 1366
Amino group, 1115
  activating and directing effects in elec-
    trophilic aromatic substitution,
    762t
  directing effect, effect of protonation,
    1137–1138
2-Aminopyridine
  diazotization, 1200
  synthesis by Chichibabin reaction, 1197
Amino sugars, 1367
Amino terminus, of a peptide, 1267
2-Aminothiazole, 1139p
Ammonia

acidity and pKa, 99t
bond dissociation energy, 106
industrial use and synthesis, 1154
liquid, reagent in alkyne reduction,
    660–662
nitrogen hybridization, 36
reaction with acid chlorides, 1001–1002
reaction with esters, 1005
solvent for forming acetylenic anions,
    663
structure, 14f, 17
Ammonium cyanate, in synthesis
    of urea, 2
Ammonium salts, 1120
  quaternary, 1127–1128
    Hofmann elimination, 1133–1137
AMP, 1369t
Amphipathic, 1025
Amphoteric compounds, 97
  alcohols, 366
  pKa values, 102
Amplitude, of a bond vibration, 542
Amylopectin, 1365–1366
Amylose, 1365
Anchimeric assistance, see Neighboring-
    group participation
Anesthetics, 373
Angle strain, 293
Angular momentum quantum
    number, 23
Anhydrides
  cyclic, 955
  IR spectroscopy, 982–983
  mixed, 947
  nomenclature, 974
  physical properties, 980
  reactions
    hydrolysis, 995–996
    with alcohols and phenols, 1005
    with amines, 1004–1006
  relative reactivity in nucleophilic acyl
    substitution, 996–1000
  synthesis
    from acid chlorides and carboxylate
      salts, 1003
    from carboxylic acids, 954–955
    mechanism, SGL 20.8
    summary, A-10
Anhydropyranose, 1390p
Aniline, 1114
  and derivatives, electrophilic aromatic
    substitution, 1137–1139
  bromination, 1137
  diazotization, 1140
  in Skraup quinoline synthesis, 1204
  nitration, 1137
  structure, 1117
Anion-exchange resin, 1276

Anions, *see also* Carbanions
    solvation, 357
Anisole, structure, 738
Annulation, 1092
Anomalous dispersion, 240
Anomeric carbon, **1333**
Anomers, **1333**, 1333–1334
    configuration, determination by perio-
        date reaction, 1352–1353
Antarafacial, 1235–1236, 1240
Anthracene, 1170f
    addition reactions, 1212p
    resonance structures, 117p
Anthranilic acid, 1159p
9,10-Anthraquinone, 842
Anti
    stereochemistry of addition, **311**
    conformation, *see* Butane, conforma-
        tions; Conformations
Antiaromatic compounds, 725–726
Antibiotic, **360**
    ionophore, 360
Antibonding molecular orbitals, **32**, **1222**
    role in ultraviolet absorption, 687
Anti elimination, **407**, 407–409
Antifreeze, automotive, 377
Antisymmetric molecular orbital, **1224**
Apolar, solvent classification, **352**
Applied frequency, in NMR spectroscopy,
    582
Aprotic, solvent classification, **352**
D-(−)-Arabinose
    equilibrium composition, 1340t
    Fischer proof of stereochemistry,
        1356–1357
    structure, 1329f
    synthesis from glucose, 1360
p-Aramid, 1321p
Arginine, structure and properties, 1267t
Arnel, 1365
Aromatic, *see* Aromaticity
Aromatic compounds, *see also* Benzene
        derivatives
    biosynthesis, 1084
    carcinogenicity, 777–778
    hydrocarbons, source and industrial
        use, 776–777
    polycyclic aromatic hydrocarbons,
        1169, 1169–1171
Aromaticity, 715–726
    4n + 2 rule, 719–726
    evidence from NMR spectra, 743
Aromatic substitution
    contrast of different types, SGL 18.2
    electrophilic, 750–751
        of benzene, 748–760
        of benzene derivatives, 760–775
        of furan, pyrrole, and thiophene,
            1184–1188

of naphthalene and derivatives,
    1176–1177
of pyridine and derivatives,
    1194–1197
ortho, para ratio, 766–767
nucleophilic, 829–832
of pyridine derivatives, 1197–1201
Aryl cation, 828
Aryl group, **79**
Aryl halides, **823**
    elimination to give benzyne, 832–835
    in synthesis of phenols, 848
    nucleophilic substitution reactions,
        829–836
    summary, 835
    S_N1 reactions, 826–829
    S_N2 reactions, 824–825
    synthesis
        electrophilic halogenation, 748–750
        Sandmeyer reaction, 1140–1141
L-(−)-Ascorbic acid, 1389–1390p
Asparagine, structure and properties,
    1267t
Aspartame, 1313
Aspartic acid
    isoelectric point, 1274
    structure and properties, 1267t
Aspirin, 1004
Association reactions, Lewis acid-base, **88**
Asymmetric carbon, **228**
    and chirality, 251
Atomic orbitals, **22**; *see also* Orbitals,
    atomic
ATP, *see* Adenosine triphosphate
Attack, **90**
Aufbau principle
    and atomic orbitals, **28**
    and molecular orbitals, **32**–33,
        1224–1225
Autoxidation, **377**
    of aldehydes, 916
    of ethers, 377–378
    of phenol, 851
Axial, substituent positions in cyclohex-
    anes, **274**, 274–275
Aziridine, 1115
Aziridines, nitrogen inversion, 327p
Azobenzene, **1143**
Azobenzenes, synthesis from nitro com-
    pounds, 1148
Azo dyes, **1143**
Azoisobutyronitrile (AIBN), as free-radical
    initiator, 198

B

Backside attack
    in epoxide ring opening, 510

in halohydrin formation, 506
in S_N2 reactions, 396–397
of nucleophiles, stereochemical conse-
    quences, 316
Bacteria, anaerobic, source of methane, 76
Baeyer test, for unsaturation, 193
Bakelite, 917
Barbier, François Phillipe Antoine,
    367–368
Barbital, 1108p
Barbiturates, 1107–1108p
Barium sulfate, support for hydrogenation
    catalyst, 162
Base, *see also* specific type, *e.g.*, Brønsted,
    Lewis, etc.
    in nucleic acids, 1367–1368
Base pairs, in DNA structure, 1372–1375
Base peak, in mass spectrometry, **558**
Basicity, *see also* Acidity
    and leaving-group effectiveness, 403
    and nucleophilicity, 399
    Brønsted, and nucleophilicity, 401
    constant, 100–101
    gas-phase, 1123
    of aldehydes and ketones, 883
    of alcohols, 366–367
        role in alcohol dehydration to
            alkenes, 444
        role in alcohol dehydration to
            ethers, 501
        role in reactions of alcohols with
            hydrogen halides, 447–450
    of ethers, 366–367
        importance in ether cleavage, 507
    of epoxides, importance in ring-
        opening reactions, 512
    of functional groups, summary, A-13
    of leaving groups, importance in nucleo-
        philic acyl substitution, 998–999
Beer's law, 686
Bending vibration, **546**
Benhydryl chloride, 791t
Benzalacetone, 1061
Benzaldehyde, 867
Benzalkonium chloride, 942, 1127–1128
Benzenamine, *see* Aniline
Benzene, *see also* Benzene derivatives
    evolution of structure, 715–717
    heat of formation, 719
    heat of hydrogenation, 776
    lack of reactivity in addition reac-
        tions, 716
    NMR spectrum, 742
    π-orbitals, 718
    physical properties, 740
    production from petroleum, 213f
    reactions
        conversion into benzenesulfonyl chlo-
            ride, 953

electrophilic aromatic substitution reactions, 748–760
  Friedel-Crafts acylation, 754–757
  Friedel-Crafts alkylation, 757–759
  halogenation, 748–750
  nitration, 752–753
  sulfonation, 753–754
resonance structures, 21p, 96p
resonance energy, 719, 1181t
  estimate by heats of hydrogenation, 786p
solvent properties, 354t
source and industrial use, 776
stability, 719
structure and bonding, 79, 717–719, 717f
Benzene derivatives, *see also* Aryl halides, Phenol
  CMR spectroscopy, 745–746
  IR spectroscopy, 741–742
  NMR spectroscopy, 742–746
  nomenclature, 737–739
  physical properties, 740–741
  reactions
    benzylic bromination, 795–796
    benzylic oxidation, 804–805
    catalytichydrogenation, 775–776
    conversion into sulfonyl chlorides, 953
    electrophilic aromatic substitution, 760–775, 1137–1139
    with diazonium salts, 1143–1144
  source and industrial use, 776–777
  synthesis
    electrophilic aromatic substitution, 771–775
    from diazonium salts, 1140–1143
    Wolff-Kishner and Clemmensen reductions, 910–911
  UV spectroscopy, 746–747
Benzenediazonium chloride, synthesis, 1140
1,2-Benzenedicarboxylic acid, *see* Phthalic acid
Benzenesulfonic acid
  structure, 450
  synthesis from benzene, 753
Benzofuran, structure, 1178f
Benzoic acid
  structure, 930
  pK_a, 941t
Benzophenone, 867
Benzo[a]pyrene
  carcinogenicity, 777–778
  guanosine alkylation in DNA, 1379
1,4-Benzoquinone, 841
p-Benzoquinone, *see* 1,4-Benzoquinone
Benzothiophene, structure, 1178f

Benzoyl group, 868
Benzyl alcohol, structure, 338
Benzylamine, pK_a of conjugate acid, 1121t
Benzyl anion, resonance structures, 799
Benzyl bromide, structure, 337
Benzyl cation, resonance structures, 789
Benzyl chloride, reaction with carbohydrates, 1346
Benzyl group, **337**, **739**
Benzylic group, **789**
Benzylic protons, chemical shifts, 744
Benzyne, 832–835
  heat of formation, 859p
Berzelius, Jöns Jacob, 1
BHA, *see* Butylated hydroxyanisole
BHT, *see* Butylated hydroxytoluene
Biacetyl, 925p
Bicyclic, **294**
Bicyclic compounds, 294–302
  bridged, **295**
    stereochemistry, 333p
  fused, **295**
  nomenclature, 295–296
  synthesis by Diels-Alder reactions, 691–694
Bicyclo[4.4.0]decane, *see* Decalin
Bimesityl, UV absorption, comparison to mesitylene, 748
Bimolecular, 394
Biosynthesis, **808**
Biosynthesis, of fatty acids, 1082–1085
Biot, Jean-Baptiste, 261–262
Bisulfate, basicity, 99t
Boat, conformation of cyclohexane, **276**, 277–279, 278f
Boc group, 1292
Boilermakers, best in Big, 10
Boiling point, **67**; *see also* specific compounds or classes
  principles, 67–70
  summary of factors governing, 350
  trends with structure, 70
Bond angle, **13**, 14–15
Bond dipole, **11**
  effect on acid strength, 109–110
Bond dissociation energies, 204–207
  effect on acid strength, 106–107
  effect on IR absorption positions, 545
  table of values, 205t
Bonding molecular orbital, **32**, **1222**
Bond length, **13**
  rules, 13–14
Bond order, **13**
Bonds
  bent or "banana," 294
  carbon-carbon, formation in organic synthesis, 523
  chemical, theories, 3–37

covalent, 5–12, **6**
  and molecular orbital theory, 30–34
  double, **7**
  ionic, 4–5
  pi (π), 124–125
  polar covalent, **10**, 10–11
  sigma (σ) **33**
  strength, 204
  triple, **7**
Borane
  complexes with Lewis bases, 184
  in hydroboration of alkenes, 183
Borazole, 732p
Boron trifluoride
  complex with diethyl ether, 367
  structure, 16
Boron trifluoride etherate, 367
Breathalyzer test, 492p
Bredt's rule, **299**
Bridgehead carbons, **295**
  in naphthalene, 1172
Bridged bicyclic compound, **295**
Bromide, basicity, 99t
Bromination, *see also* Halogenation
  allylic, 796–797
  benzylic, 795–796
  of aldehydes and ketones, acid-catalyzed, 1048–1049
  of aniline, 1137
  of benzene, 748–750
  of carboxylic acids, 1052–1053
  of naphthalene, 1175
  of phenol, 846–847
Bromine
  addition to alkenes, 175–176
    stereochemistry, 313–317
    as an oxidation, 461
  addition to alkynes, 653
  addition to furan, 1188
  aqueous, uses
    bromination of phenols, 846–847
    halohydrin formation, 177–178
    oxidation of aldoses, 1349
  atom
    electron affinity, 107
    exact isotopic masses and natural abundances, 559t
  group, activating and directing effects in electrophilic aromatic sunstitution, 762t
  oxidation of thiols to disulfides, 484
N-Bromoamide, intermediate in Hofmann rearrangement, 1151
1-Bromo-3-chloropropane, NMR spectrum and splitting analysis, 606
7-Bromo-1,3,5-cycloheptatriene, 732p
5-Bromo-1,3-cyclopentadiene, solvolysis, 819p

1-Bromo-2,2-dimethylpropane, in substitution and elimination reactions, 390, 426
Bromoethane
  NMR spectrum, 594f
  physical properties, 347
Bromoform
  product of haloform reaction of methyl ketones, 1050
  structure, 337
Bromohydrin, **177**
Bromomethane, structure, comparison to other methyl derivatives, 345t
2-Bromo-2-methylpropane, formation by HBr addition to 2-methylpropene, 148
Bromonium ion, **176**
  formation, relationship to epoxide formation from alkenes, 504
  in halohydrin formation, 177–178
  relationship to mercurinium ion, 181
  stereochemical evidence, 315–316
N-Bromosuccinimide
  reaction with HBr, 798
  reagent for allylic bromination, 797–798
Bromotrifluoromethane, industrial use, 373
Brønsted acid-base reactions, 96–102
  equilibrium constants, 101–102, 119p
  relationship to nucleophilic substitution reactions, 389
  relation to hydrogen bonding, 349
Brønsted acids, **96**, 96–102
  free energy of dissociation, 103
  identification, SGL 3.3
  strengths, 98–100, 99t
    element effect, 105–108
    polar effect, 108–111
    trends, 110
Brønsted bases, **96**, 96–102
  identification, SGL 3.3
  strengths, 100–101
Brown, Herbert C., 186, 893–894
Bu, abbreviation for butyl group, 1064t
i-Bu, abbreviation for isobutyl group, 1064t
t-Bu, abbreviation for tert-butyl group, 1064t
Buckminsterfullerene, 1170–1171, 1171f
Bullvalene, 1251–1252
1,3-Butadiene
  conformations, 681
  molecular orbitals, 1223f
    symmetry classification, 1225f
  structure and bonding, 679, 681
  UV absorption, 688t
Butanal, IR spectrum, 873f

1-Butanamine
  boiling point and dipole moment, 1117
  IR spectrum, 1118f
Butane
  conformations, 52–55
    analysis of stereochemistry, 253
  physical properties, 47t
Butanedioic acid, see Succinic acid
2,3-Butanediol, analysis of stereoisomers, 245–247
2,3-Butanedione, 925p
Butanoic acid, pK_a, 108
2-Butanol, analysis of chirality, 226–227
2-Butanone, dielectric constant, 353p
1-Butene, heat of formation, 141p
2-Butene, cis-trans isomerism, 126
cis-2-Butene
  heat of formation, 140
  space-filling model, 142f
  stereochemistry of bromine addition, 313–317
trans-2-Butene
  space-filling model, 142f
  heat of formation, 140
  stereochemistry of bromine addition, 313–317
Butlin, Henry T., 777
tert-Butoxycarbonyl group, see Boc group
tert-Butoxy radical, 196
Butyraldehyde, see Butanal
Butter yellow, 1143
tert-Butyl acetate, synthesis, 951p
tert-Butyl alcohol, see 2-Methyl-2-propanol
Butylated hydroxyanisole (BHA), food preservative, 844
Butylated hydroxytoluene (BHT), food preservative, 844
tert-Butylbenzene, synthesis from benzene, 758
tert-Butyl bromide, see 2-Bromo-2-methylpropane
Butyl cation, heat of formation, 149t
sec-Butyl cation, heat of formation, 149t
tert-Butyl cation, 148
  heat of formation, 149t
  structure, 149–150, 150f
Butyl group, abbreviation, 1064t
tert-Butyl group, abbreviation, 1064t
Butyllithium
  in synthesis of amide bases, 1077, 1127
  in Wittig alkene synthesis, 913
  reaction with 2-picoline, 1202
  synthesis, 369
tert-Butyl methyl ether, see 2-Methoxy-2-methylpropane
4-tert-Butylphenol, synthesis, 849
sec-Butyl radical, heat of formation, 203t

sec-Butyl group, structure, 58t
tert-Butyl group, skeletal structure, 66–67
Butyric acid, structure, 931t
γ-Butyrolactam, 976
γ-Butyrolactone, 975
^{13}C NMR spectroscopy, see CMR spectroscopy

C

Cahn-Ingold-Prelog system, **132**, 231–233; see also E,Z, R,S
Calcium carbide, 670
Camphor, DEPT CMR spectrum, 627f
Cancer
  and aromatic hydrocarbons, 777–778
  and DNA alkylation, 1378–1379
  and nitrosamines, 1145
Capric acid, structure, 931t
Caproic acid, structure, 931t
ε-Caprolactam, polymerization, 1022–1023
Caprylic acid, structure, 931t
Capsaicin, 1031p
Carbaldehyde, suffix in substitutive nomenclature, 869
Carbamate esters, see Esters, of carbamic acids
Carbamic acid, 977
  decarboxylation, 959, 1149, 1152
  from hydration of isocyanates, 1149
Carbamoyl group, 977t
Carbanions, 662–665
  allylic and benzylic, as reactive intermediates, 799–803
Carbaryl, 1162p
Carbene, **428**; see also Methylene
Carbenes, 427–431
Carbenoids, **431**
Carbinolamines
  dehydration, 905, SGL 19.10
  intermediates in enamine formation, 907–908
  intermediates in imine formation, 905–906
  mechanism of formation, SGL 19.9
Carboalkoxy groups, activating and directing effects in electrophilic aromatic substitution, 762t
Carbocations, **147**
  activating and directing effects in electrophilic aromatic substitution, 762–766
  α-alkoxy, 881
  allylic, 701–702

as reactive intermediates, 789–794
aryl, 828
benzylic, as reactive intermediates, 789–794
classification, **149**
generation from different starting materials, SGL 11.2, SGL 16.3
heats of formation, 149*t*
α-hydroxy, intermediates in acetal formation and hydrolysis, 899–900
intermediates
  in acid-catalyzed alkene hydration, 163
  in acid-catalyzed halogenation of aldehydes and ketones, 1049
  in alcohol dehydration, 444
  in biosynthesis of terpenes, 808–812
  in electrophilic aromatic substitution, 750, 1174, 1195
  in ether cleavage, 507–508
  in formation and hydrolysis of glycosides, 1343–1346
  in Friedel-Crafts acylations, 754
  in Friedel-Crafts alkylations, 757–758
  in hydrogen halide addition to alkenes, 146–148
  in hydrogen halide addition to conjugated dienes, 700–702
  in mass spectrometry fragmentations, 561–565
  in pinacol rearrangement, 882
  in reactions of alcohols with hydrogen halides, 449–450
  in S_N1-E1 reactions, 417–419, 421
  in synthesis of ethers from tertiary alcohols, 501
rearrangement, 151–153
  in Friedel-Crafts alkylation, 758
structure and stability, 148–151
summary of fundamental reactions, 422, 755
vinylic, 827, 828*f*
Carbohydrates, 1325–1390, **1325**; *see also* Aldoses, Ketoses
alkylation, 1346–1347
classification and properties, 1326–1327
ether and ester derivatives, 1346–1348
glycosides, 1343–1346
mutarotation, 1337–1340
oxidation and reduction, 1348–1353
periodate oxidation, 1352
Carbon
asymmetric, *see* Asymmetric carbon
catalyst support, 162
electronic configuration, 29
exact isotopic masses and natural abundances, 559*t*
hybridization, 35, 123, 648

effect on rates of S_N2 reactions, 824
α-Carbon
  in alkyl halides, **387**
  in carbonyl compounds, **868**
β-Carbon, 387
Carbon-13, *see* ^{13}C
Carbon dioxide
  environmental effects, 375–376
  reaction with Grignard reagents, 944–945
Carbonic acid
  decarboxylation, 959
  derivatives, 977–978
  pK_a values, 939*t*
Carbonium ions, **147**; *see also* Carbocations
Carbon monoxide
  in commercial methanol production, 376
  production from petroleum, 213*f*
Carbon NMR spectroscopy, *see* CMR spectroscopy
Carbon tetrachloride
  solvent properties, 354*t*
  solvent effect in NBS brominations, 798
Carbonyl-addition reactions, *see* Addition reactions, carbonyl
Carbonyl compounds, **865**; *see also* specific types, *e.g.*, Ketones
Carbonyl group
  bond length, 978
  treatment in *E,Z*-system, 135
Carbonyl oxygen, **934**
Carboxaldehyde, suffix in substitutive nomenclature, 869
Carboxamido group, activating and directing effects in electrophilic aromatic substitution, 762*t*
γ-Carboxyglutamic acid, 1318*p*
Carboxy group, 977*t*
  activating and directing effects in electrophilic aromatic substitution, 762*t*
Carboxylate ions, 108, **938**
  basicity, 99*t*
  resonance structures, 938
Carboxylate oxygen, **934**
Carboxylate salts
  reaction with acid chlorides, 1003
  reactivity, 1029*p*
Carboxylic acids, 929–970
  acidity, 937–941
    polar effects, 108–111
    use in purification, 940–941
  basicity, 941
  α-bromo, synthesis, 1052–1053
  conjugate acids, 941

derivatives, 971–1038; *see also* specific types, *e.g.*, Esters
  summary, 972*t*
fatty acids soaps and detergents, 942–944
formation in ozonolysis, 190–191
α-halo
  reactivity in nucleophilic substitution, 1053–1054
infrared spectroscopy, 935–936
solubility and ionization, 940
NMR and CMR spectroscopy, 935–937
nomenclature, 929–933
physical properties, 934–935
pK_a values, 99*t*
reactions
  α-bromination, 1052–1053
  conversion into acid chlorides, 952–953
  conversion into anhydrides, 954–955
  conversion into esters, 946–949
  conversion into primary alcohols, 956–957
  decarboxylation, 958–959
  dimerization, 935
  α-halogenation, 1052–1053
  reduction with LiAlH$_4$, 956–957
  summary, 945–946
  with NaBH$_4$, 959
relative carbonyl reactivity, 1014
relative reactivity in nucleophilic acyl substitution, 996–1000
resonance structures, 934
structure, 934–935
synthesis
  benzylic oxidation, 804–805
  Grignard reactions of CO$_2$, 944–945
  hydrolysis of carboxylic acid derivatives, 989–996, 1019–1020
  malonic ester synthesis, 1075
  oxidation of aldehydes, 915–916
  oxidation of primary alcohols, 467, 469–470
  summary, A-9
Carboxymethyl group, 933, 977*t*
Carboxy terminus, of a peptide, **1267**
Carnauba wax, 1023
β-Carotene
  structure, 689, 809*f*
  and visible absorption, 689
Carpenter, Barry, 726
Caryophyllene, structure, 809*f*
Catalysis, 161–166
  by enzymes, 165–166
Catalysts, **161**
  and equilibrium constant, 166
  chiral, 306–307
  heterogeneous, **161**

Catalysts (*continued*)
  homogeneous, **161**
  poisons, 161, 659
Catalytic hydrogenation, *see* Hydrogenation
Catechol, 739, 823
Cation-exchange resin, 1276
Cations, solvation, 357
Celanese, 1365
Celestolide, 783*p*
Cellobiose, 1364*p*
Cellulose, 1364–1365
Cellulose acetate, 1365
CFC, *see* Chlorofluorocarbons
Chain reactions, *see* Free-radical chain reactions, 197–200
Chair, conformation of cyclohexane, 272–279
  how to draw, 274
  of cyclohexane derivatives, separation, 282–283
Chair flip
  in cyclohexane, 276–277
  in decalins, 298
  in Assembly Hall, 1985
  role in halohydrin formation, 506
  stereochemical consequences, 289–292
Chair interconversion, *see* Chair flip
Chargaff, Erwin, 1372
Chargaff's rules, 1372–1374
Charge *see also* Formal charge
  delocalization, 711
  separation, 711
Chelidonic acid, 1107*p*
Chemical Abstracts, 62
Chemical equilibrium, *see* Equilibrium
Chemical equivalence, 476–480, **476**
  determination, 584
  relationship to constitutional equivalence, 476
  test, 479–480
Chemical exchange, **617**
  and NMR spectroscopy, 617–618, 620–621
Chemical-ionization mass spectra, **567**
Chemical literature, **62**
Chemical shift, 582–590, 583
  dependence on operating frequency, 583
  differences, and validity of $n + 1$ rule, 608
  effect of hydrogen bonding, 617
  group contributions, 588–590, 589*t*
  in CMR spectroscopy, 623*t*
  of alkene protons, 612*t*
  relationship to structure, 587–591
  use in solving NMR problems, 592–593
Chichibabin reaction, 1197–1201
Chiral, **226**

Chirality, 225–227, **226**
  and symmetry, 229–230
  importance, 227
  of allenes, 683
Chiral molecules, reactions, SGL 7.4
Chitin, 1366–1367
Chloral, hydration, 889
Chloral hydrate, 889
Chlordane
  decomposition in base, 436*p*
  structure and use, 374
Chloride, basicity, 99*t*
Chlorination, *see also* Halogenation
  of aldehydes and ketones, acid-catalyzed, 1048–1049
  of benzene, 748
Chlorine
  addition to alkenes, 175–176
    stereochemistry, 820*p*
  atom, electron affinity, 107
  exact isotopic masses and natural abundances, 559*t*
  group, activating and directing effects in electrophilic aromatic substitution, 762*t*
Chloroacetic acid, pK_a, 939*t*
2-Chloroaniline, pK_a of conjugate acid, 1121*t*
3-Chloroaniline, pK_a of conjugate acid, 1121*t*
4-Chloroaniline, pK_a of conjugate acid, 1121*t*
Chlorobenzene, reaction with potassium amide, 853–854
4-Chlorobenzoic acid, pK_a, 939*t*
2-Chloro-1,3-butadiene (chloroprene), 731*p*
  industrial synthesis, 671
1-Chlorobutane, CMR spectrum, 624*f*
  NMR spectra and analysis, 608*f*, 609*f*
  physical properties, 347
2-Chlorobutanoic acid, pK_a, 108
3-Chlorobutanoic acid, pK_a, 108
4-Chlorobutanoic acid, pK_a, 108
Chlorofluorocarbons (CFCs)
  environmental effects, 373
  reactions with stratospheric ozone, 200
Chloroform
  solvent properties, 354*t*
  structure, 337
  use in generating dichloromethylene, 427
Chloroformyl group, 977*t*
Chlorohydrin, **177**
Chloromethane
  dipole moment, 41*p*
  structure, comparison to other methyl derivatives, 345*t*

m-Chloroperoxybenzoic acid, in conversion of alkenes into epoxides, 503
Chlorophyll *a*, 1207*f*
Chloroprene, *see* 2-Chloro-1,3-butadiene
2-Chloropyridine
  nucleophilic aromatic substitution reactions, 1199–1200
  synthesis, 1200
4-Chloropyridine, nucleophilic aromatic substitution reactions, 1200
Chlorosulfonic acid, in synthesis of sulfonyl chlorides, 953
Cholecalciferol (Vitamin D$_3$), 1253
Cholesterol, structure, 301
Chromate, oxidizing agent for alcohols, 467–469
Chromate ester, intermediate in alcohol oxidation, 468
Chromic anhydride, *see* Chromium trioxide
Chromium trioxide
  in alcohol oxidations, 467–469
  in benzylic oxidations, 804
Chromophore, **687**
Chrysene, 1170*f*
Chymotrypsin, use in peptide hydrolysis, 1290, 1291*p*
CI, *see* Chemical ionization
Cine substitution, 835
Cinnamaldehyde, 868
Cinnamic acid, structure, 930
Cis; see also Cis-trans isomerism
  alkene stereochemistry, 129
  cycloalkane substitution, 283–287
  ring fusion in polycyclic compounds, 297–299
Cisoid conformation, of dienes, 681
Cis-trans isomerism
  in alkenes, 129
  in substituted cyclohexanes, 283–287
  ring fusion in polycyclic compounds, 297–299
Citronellol, 806
Claisen, Ludwig, 1065
Claisen condensation, 1065–1073
  crossed, 1069–1070,
  relationship to Reissert indole synthesis, 1192
  side reaction in ester alkylation, 1078
  use in synthesis, 1071–1073, SGL 22.8
  variations, SGL 22.5
Claisen rearrangement, 1247–1248
Claisen-Schmidt condensation, 1060–1061
  of pyridinium salts, 1203
α-Cleavage, 564
  in mass spectrometry of amines, 1119
  in mass spectrometry of carbonyl compounds, 879

Cleland's reagent, 1299
Clemmensen reduction, 909–911
CMP, 1369*t*
Carbon NMR (CMR) spectroscopy,
        622–628; *see also* specific com-
        pound classes, *e.g.,* Ketones, CMR
        spectroscopy
    DEPT technique, 635, 627*f*
    proton-decoupled spectra, 624
Coal tar
    as source of aromatic compounds,
    as source of pyridine derivatives, 1194
Cocaine, 1154*f*
Codeine, 1154*f*
Coenzyme A, 1082–1083
Coenzymes, **474, 1308**
Cole, Thomas W., 296
Collins reagent, oxidizing agent for alco-
        hols, 467–468
Color, effect of light absorption,
        688–689
Combes synthesis of quinolines, 1218*p*
Combination bands in infrared spectra,
        742
Combustion, **72**
    and respiration, 1
    of alkanes, 72–75
    use in elemental analysis, 72–75
Common nomenclature, 336; *see also* spe-
        cific compound classes, *e.g.,*
        Alkenes, nomenclature
Concerted mechanism, **185,** 189
Concerted reaction, SGL 15.1
Condensation, **1057**
Condensation polymer, 1021
Conformation, **18,** 18–19
    anti, **53**
    detection by NMR spectroscopy,
        619–621
    eclipsed, **50**
    gauche, **53**
    of alkanes, generalizations, 54
    of butane, 52–55
        chirality, 253
    of cyclohexane, 272–279
    of ethane, 49–52
    of 1,3-butadiene, 681
    of dienes
        effect on Diels-Alder reaction,
            694–696
        effect on UV absorption, 689–690
    of proteins, 1301–1303
    staggered, 50
Conformational analysis, 279–287, **282**
    of cyclohexanes, 280–282, 286–287
Conformational diastereomers, **253**
Conformational enantiomers, **253**
Congruence, testing molecules for,
        226–227

Congruent, **225**
Coniine, 1137*p*
Conjugate acid, **97**
Conjugate acid-base pair, **97**
Conjugate addition, **692**
    competition with carbonyl group reac-
        tions, 1088–1090
    of enolate ions, 1090–1093
    of hydrogen halides to conjugated
        dienes, 699–700, 703–706
    of lithium dialkyl cuprates to α,β-unsat-
        urated carbonyl compounds,
        1095–1098
    of α,β-unsaturated carbonyl com-
        pounds, 1085–1093
    use in organic synthesis, 1097–1098
Conjugate base, **97**
Conjugation, **679**
    effect on IR spectra, 873
    effect on UV spectra, 687–690
Connectivity, **47**
    and stereoisomerism, 126
    and principal chain, 56
    and constitutional equivalence, 478
Conrotatory, **1229**
Conservation of orbital symmetry, 1221
Constitutional equivalence, **476**
    and chemical equivalence, 476,
        585–586
Cope rearrangement, 1247
Copolymer, **708**
Coronene, 1170*f*
Cortisone, structure, 301
COT, *see* 1,3,5,7-Cyclooctatetraene
Couper, Archibald Scott, 46
Coupled protons, **595**
Coupling constant, **595**
    and validity of *n* + 1 splitting rule, 608
    of alkene protons, 614*t*
    of aromatic protons, 744*t*
Covalent bond, 5–12, **6;** *see also* Bond,
        covalent
    and molecular orbitals, 30–34
Covalent compounds, structure,
        12–19
Cracking, of alkanes, 76, 210–212
Crafts, James Mason, 756
Cram, Donald J., 360
Cresols, 738
o-Cresol, 823
Crick, Francis C., 1372
Critical micelle concentration, 943
Crosslinks, in rubber, 708
Crotonic acid, structure, 931*t*
[12]-Crown-4, structure, 358
[18]-Crown-6, structure, 358
Crown ethers, 358–359
Crystal violet, 965*p*
Cubane, structure, 296

Cumene
    in phenol synthesis, 851
    production from petroleum, 213*f*
    source and industrial use, 776–777
    structure, 738
Cumene hydroperoxide, intermediate in
        phenol synthesis, 851
Cumulenes, **679**
tert-Cumyl chloride, 791*t*
Cupric nitrate, reaction with diazonium
        salts, 1141
Cuprous bromide, reaction with diazo-
        nium salts, 1141
Cuprous chloride, reaction with diazoni-
        umsalts, 1140–1141
Cuprous cyanide, reaction with diazo-
        nium salts, 1141
Cuprous oxide, reaction with diazonium
        salts, 1141
Curtius rearrangement
    of acyl azides, 1148–1150
    stereochemistry, 1152
    mechanism, 1149, SGL 23.4
Curved-arrow formalism, SGL 3.1
    for electron-pair displacement reac-
        tions, 91–94
    for free-radical reactions, 196
    for Lewis acid-base association reac-
        tions, 89–90
    for resonance structures, 95–96, 709
    review, 94–96
Cyanide, *see also* Hydrogen cyanide
    basicity, 99*t*
    reaction with alkyl halides, 1019
    use in Strecker synthesis, 1279–1280
Cyanoethylation, **1087**
Cyanogen bromide, reaction with pep-
        tides, 1286–1288
Cyano group, 977*t*
    activating and directing effects in elec-
        trophilic aromatic substitution,
        762*t*
    priority in *E,Z* system, 135
    polar effect, 1124
Cyanohydrins, **884**
    formation, relationship to aldol addi-
        tion, 1056
    in Kiliani-Fischer synthesis,
        1353–1355
    synthesis, 884–886
1,3-Cyclopentadiene
    acidity, 722
    in synthesis of bicyclic compounds,
        691–694
Cyclic compounds, 271–302; *see also* spe-
        cific compounds or classes, *e.g.,*
        Cycloalkanes
Cycloaddition reactions, **189,** 691, **1219,**
        1235–1239, SGL 15.1

Cycloaddition reactions (*continued*)
  and DNA damage, 1379–1380
  classification, 1235–1236
  selection rules, 1238*t*
Cycloalkanes, 271–298
  bicyclic and polycyclic, 294–299
  heats of formation, 272*t*
  nomenclature, 65–67
  planar structures, 287–289
  physical properties, 65*t*
  relative stability, 271–272
trans-Cycloalkenes, 299–300
Cycloalkynes, 649*p*
1,3-Cyclobutadiene
  antiaromaticity, 725
  complex with Fe(0), 724
Cyclobutane
  conformation, 293–294
  heat of formation, 272*t*
  physical properties, 65*t*
Cyclobutene derivatives, synthesis by
  cycloaddition reactions, 1237
Cyclodecane, heat of formation, 272*t*
Cyclododecane, heat of formation, 272*t*
Cycloheptane
  heat of formation, 272*t*
  physical properties, 65*t*
(*E*)-Cycloheptene, 171*p*
1,3-Cyclohexadiene, heat of hydrogenation, 776
Cyclohexane
  and derivatives, synthesis by catalytic
    hydrogenation, 775–776
  chair flip,
    effect on NMR spectrum, 619–621
  conformations, 272–279
  heat of formation, 272*t*
  mass spectrum, 560*f*
  physical properties, , 65*t*, 740
  planar structures, 287–289
  relative energies of conformations, 279*f*
  stability, 271–272
  structure, 64
Cyclohexane-d_{11}, NMR spectrum, temperature dependence, 621
Cyclohexanes, substituted, 279–287
  manipulating rings, SGL 7.1
Cyclohexanone, 869
  enolization, 1044
Cyclohexene, heat of hydrogenation, 776
1-Cyclohexenyl-1-ethanone, *see*
  1-Acetylcyclohexene
Cyclohexylbenzene, conformational analysis, 333*p*
Cyclononane, heat of formation, 272*t*
Cyclooctane
  heat of formation, 272*t*
  physical properties, 65*t*
1,3,5,7-Cyclooctatetraene (COT)

bromine addition, 716
  heat of formation, 719
  π-orbitals, 718
  planar, antiaromaticity, 725
  stability, comparison to benzene, 719
  structure, 717–718, 718*f*
trans-Cyclooctene, 299
Cyclopentadienyl anion, aromaticity, 722
Cyclopentane
  conformation, 292–293
  heat of formation, 272*t*
  physical properties, 65*t*, 346
Cyclopropane, heat of formation, 272*t*
Cyclopropane
  alkenelike behavior, SGL 7.2
  derivatives
    NMR spectra, 615
    ring opening, comparison to epox-
      ides, 527*p*
    synthesis by Simmons-Smith reac-
      tion, 430–433
  physical properties, 65*t*
  structure and bonding, 294
Cyclopropenyl cation, aromaticity, 722
Cyclopropyl halides, S_N2 reactions, 436*p*
Cyclotetradecane, heat of formation, 272*t*
Cyclotridecane, heat of formation, 272*t*
Cycloundecane, heat of formation, 272*t*
m-Cymene, structure, 814
p-Cymene, structure, 715
Cysteine
  L-isomer, absolute configuration,
    1269*p*
  structure and properties, 1267*t*
Cystine, 1267*t*
Cytidine, 1369*t*
Cytidylic acid, 1369*t*
Cytosine, 1368

D

D, stereochemical configuration,
  1268–1269
2,4-D, herbicide, 1053
Dansyl chloride, 1316*p*
DCC, *see* Dicyclohexylcarbodiimide
DDT
  structure and use, 374
  synthesis, 923*p*
Deactivating group, in electrophilic aro-
  matic substitution, **767**, 767–771,
  1176
Debye, unit of dipole moment, 11,
  SGL 1.2
Decalin, *cis*- and *trans*-, 297–298
Decane, 558*f*
  physical properties, 47*t*, 347

Decarboxylation, **958**
  of carbamic acid derivatives, 1149,
    1152
  of 2-furancarboxylic acid, 1188
  of heterocyclic carboxylic acids, 1188,
    1193
  role in acetoacetic ester synthesis,
    1080–1082
  role in malonic ester synthesis, 1075
Decoupled spectra, in CMR spectroscopy,
  624
Degree of unsaturation, *see* Unsaturation
  number
Dehydration, **443**
  of alcohols, 443–446
  of aldol addition product, 1057
  of β-hydroxy carbonyl compounds,
    acid-catalyzed, mechanism,
    SGL 22.3
7-Dehydrocholesterol, precursor of vitamin D, 1252
Delocalization of charge, 711
Demerol, 821*p*
Denaturants, for ethanol, 375
Denaturation, of proteins, 1306
Deoxyribonucleic acid
  alkylation, 1378–1379
  covalent structure, 1370–1374
  double-helical structure, 1372–1373,
    1373*f*
  replication, 1373–1375
  role in protein synthesis, 1375–1378
Deoxyribonucleoside, 1367–1368
Deoxyribonucleotide, **1368**
2'-Deoxyribose, component of deoxyribo-
  nucleic acids, 1368
DEPT, technique of CMR spectroscopy,
  625, 627*f*
Deshielded, **588**
Detergent, **942**
Deuterium, 217*p*
  exact isotopic mass and natural abun-
    dance, 559*t*
  introduction by Grignard protonolysis
    in D_2O, 371
  isotope effect, *see* Isotope effect
  use in proton NMR, 611
  D_2O shake, 618
Dewar benzene, and orbital symmetry,
  1233–1234, 1262*p*
Dextropimaricacid, 735*p*
Dextrorotatory, **235**
Dextrose, 1363
Diastereoisomers, *see* Diastereomers
Diastereomers, 241–244, **242**
  conformational, **253**
  formation in reactions, 307–308
  physical properties, 243
  relative reactivities, 304

role in enantiomeric resolution, 249–250

Diastereotopic, **478**

Diastereotopic groups
and chemical equivalence, 479–480
nonequivalence in NMR spectra, 585–586

1,3-Diaxial interactions, **280**

Diazomethane, in esterification of carboxylic acids, 950

Diazonium salts
aliphatic, decomposition, 1140
electrophilic aromatic substitution reactions, 1143–1144
Sandmeyer and related reactions, 1140–1141
synthesis, 1139–1143

Diazotization, **1139**, 1139–1143
mechanism, SGL 23.3
of acyl hydrazides, 1150
of aminopyridines, 1200

Dibenzo-[18]-crown-6
Rb$^+$ complex, structure, 359f
structure, 358

Diborane, structure, 184

Dibutyl dicarbonate, 1292

Di-*tert*-butyl peroxide, as free-radical initiator, 196–197

β-Dicarbonyl compounds, **1045**, *see also* specific types, *e.g.*, Esters, β-keto

Dichlorocarbene, *see* Dichloromethylene

Dichloromethane, *see* Methylene chloride

Dichloromethylene, 427
addition to alkenes, stereochemistry, 429
structure, 428

2,4-Dichlorophenoxyacetic acid (2,4-D), as selective herbicide, 373, 1053

1,3-Dichloropropane, NMR spectrum, 598f

Dichromate, oxidizing agent for alcohols, 467–469

Dicyclohexylcarbodiimide, 1294

Dicyclohexylurea, by-product of peptide synthesis, 1295

Dieckmann condensation, 1068

Dielectric constant, **352**
relationship to solvent polarity, 352
role in separating ions, 357

Diels, Otto, 690

Diels-Alder reaction, 690–699, 1235–1236
mechanism, 691
of furan, 1188
stereochemistry, 696–699
transition state, 695f
use in organic synthesis, 692–693

Dienes, **679**
conformation

effect on Diels-Alder reaction, 694–696
effect on UV absorption, 689–690
conjugated, **679**
addition of hydrogen halides, kinetic control, 703–707
Diels-Alder reaction, 690–699
structure and stability, 680–682
UV absorption, 688–690
cumulated, structure and stability, 682–683
heats of formation, 680t
polymerization, 707–708
structure and stability, 680–684

Dienophile, **691**

Diethyl acetamidomalonate, 1278

Diethylamine
boiling point and dipole moment, 1117
pK_a of conjugate acid, 1121t

Diethyl carbonate, in crossed Claisen condensations, 1070

Diethylene glycol dimethyl ether (diglyme), solvent in hydroboration, 184

Diethyl ether
cleavage, 507
complex with boron trifluoride, 367
flammability, 378
synthesis by dehydration of ethanol, 500
solvent in hydroboration, 184
solvent in synthesis of Grignard reagents, 368
solvent properties, 354t

Diethyl malonate
acidity, 1042p, 1074
in malonic ester synthesis, 1074–1076

Diethyl oxalate, in Reissert indole synthesis, 1192

Difluoroacetic acid, pK_a, 108, 939

Diglyme, *see* Diethylene glycol dimethyl ether

Dihedral angle, **18**

Dihydropyridines, Hantzsch synthesis, 1218p

Dihydroquinoline, intermediate in Skraup quinoline synthesis, 1205

Diiodomethane, *see* Methylene iodide

Diketopiperazine, 1321–1322

Dimethoxymethane (Formaldehyde dimethyl acetal), NMR spectrum, 583f

γ,γ-Dimethylallyl pyrophosphate, 809–811

Dimethylamine, pK_a of conjugate acid, 1121t

4-Dimethylaminopyridine, basicity, 1184p

N,N-Dimethylaniline, pK_a of conjugate acid, 1121t

7,12-Dimethylbenz[*a*]anthracene, carcinogenicity, 777–778

(2R,5R)-Dimethylborolane, chiral hydroborating reagent, 303–304, 331p

2,3-Dimethyl-2,3-butanediol, *see* Pinacol

3,3-Dimethyl-2-butanone, *see* Pinacolone

2,3-Dimethyl-2-butene, heat of formation, 143t

Dimethyl carbonate, 977
structure, 959

Dimethyl ether
boiling point and dipole moment, 346
physical properties, 350
structure, 345f, 934f

N,N-Dimethylformamide, solvent properties, 355t

2,2-Dimethylpropanoic acid, pK_a, 939t

Dimethyl sulfate
reaction with carbohydrates, 1346
structure and synthesis, 455
use as alkylating agent, 455

Dimethyl sulfide
reducing agent in ozonolysis, 189–190
structure, 345

Dimethylsulfoxide
solvent properties, 355t
synthesis, 522

2,4-Dinitrophenol, synthesis, 848

2,4-Dinitrophenylhydrazine, reaction with aldehydes and ketones, 905–906

2,4-Dinitrophenylhydrazones, 905–906

Dioscorea, source of steroids, 302

Diosgenin, 302

1,4-Dioxane
dipole moment, 384p
solvent properties, 354t
structure, 344

Dioxin, 861p

Dipeptide, **1267**

Dipeptidylaminopeptidase, 1317p

Diphenylisobenzofuran, use for trapping reactive dienes, 733–734p

Diphenylsulfone, by-product in sulfonation of benzene, 788p

Dipole, bond, **11**

Dipole moment, **11**
effect of electron delocalization, 1179–1180
effect on IR absorptions, 548

Dipoles, induced, 68

Directing effects, of substituents in electrophilic aromatic substitution, 761–767, 1176–1177

Disaccharides, **1326**, 1361–1364

Disiamylborane, in hydroboration of 1-alkynes, 658

Disparlure, 675p

Dispersion forces, 69; *see also* van der Waals forces, attractive

Displacement, *see* specific types, *e.g.*, Electron-pair displacements, Substitutions
Disrotatory, **1229**
Dissociation, free energy of, 103
Dissociation constant, **98**
Dissociation reactions, *see* Lewis acid-base dissociation
Disulfide, 483*f*
Disulfide bonds, *see* Disulfides
Disulfides
  equilibration, 485
  in proteins, 1299–1300
  in vulcanized rubber, 708
  reduction, 1299–1300
  synthesis from thiols, 484
Diterpenes, **807**
Dithiothreitol (DTT), reducing agent for disulfides, 1299
1,4-Divinylbenzene, as crosslinker in styrene polymerization, 220*p*
D,L-system, 1268–1269
  applied to monosaccharides, 1327–1331
DMAP, *see* γ,γ-Dimethylallyl pyrophosphate
DMF, *see* N,N-Dimethylformamide
DMSO, *see* Dimethylsulfoxide
DNA, *see* Deoxyribonucleic acid
2,4-DNP, *see* 2,4-Dinitrophenyl-hydrazones
D₂O shake, in NMR spectroscopy, 618
Dodecahedrane, structure, 296
(3E,5E,7E,9E)-1,3,5,7,9,11-Dodecahexaene, UV absorption, 688*t*
Dodecane, physical properties, 47*t*
Donor
  hydrogen-bond, **348**
  interactions, role in solvating cations, 357
  solvent classification, **352**
Double bond (carbon-carbon), 7
  conjugated, **679**
  cumulated, **679**
  orbital description, 124–125
  polar effect, 799, 1103–1104*p*, SGL 17.2
  treatment in E,Z-system, 135
Double helix, structure of DNA, 1372–1373, 1373*f*
  melting behavior, 1388*p*
Doublet, splitting pattern in NMR spectra, **596**
Downfield, 582
Doxorubicin, 843
DTT, *see* Dithiothreitol, 1299
Dulcitol, 1353
Dumas, Jean-Baptiste Andre, 374

E

E, stereochemical nomenclature, **132**
E1 reaction, 417–423, **419**
  contrast with S_N2 and E2 reactions, 421
  in alcohol dehydration, 444–445
  leaving-group effect on rate, 421
  of aryl and vinylic halides, 826–829
  prediction, 423–427
  rate law and mechanism, 417–419
  rearrangements, 422
  regioselectivity, 421–422
  solvent effect on rate, 417, 421
  summary, 423
E2 reaction, 404–416, **404**
  competition with S_N2 reaction, 411–415
  competition with S_N2 reaction, 803
  contrast with S_N1-E1 reaction, 420
  in chromate oxidation of alcohols, 468–469
  involving allylic and benzylic hydrogens, 802–803, 1136
  involving α-hydrogens in carbonyl compounds, 1136
  leaving-group effect on rate, 405
  of quaternary ammonium hydroxides, 1133–1137
    regioselectivity, 1134–1135
  of sulfonate esters, 454
  of vinylic halides, 826
  prediction, 423–427
  primary deuterium isotope effects, 405–406
  rate law and mechanism, 404
  regioselectivity, 410–411
  stereochemistry, 407–410, 1134
  summary, 416
Eaton, Philip, 296
Eclipsed conformation, **50**
Edman reagent, *see* Phenyl isothiocyanate
EI, *see* Electron-impact
Eicosane, physical properties, 47*t*
Electrocyclic reactions, **1219**
  of excited states, 1231–1232
  photochemical, 1232
  selection rules, 1232*t*
  thermal, 1229–1231
Electromagnetic radiation, **535**
Electromagnetic spectrum, 537*f*
Electron affinity, and acid strength, 106–107
Electron-deficient compounds, **88**
  as Lewis acids, 87–90
Electron-donating, *see* Polar effect
Electronegativity, **10**
  and Brønsted acidity, 107
  and chemical shift, 587–588

  table, 10*t*
Electronic configuration, **28**
Electron-impact mass spectra, **567**
Electron-pair displacement reactions, 90–94, **91**
  Brønsted acid-base reaction, 97
Electrons
  counting
    for formal charge, 7–8
    for 4n + 2 rule, 721–723
    for octet, 6–7
  energy and quantum number, 24
  loss or gain in oxidation-reduction, 462–463
  magnetic resonance (ESR), 644*p*
  position and probability, 22
  solvated, 660
  wave character, 21–22
  unshared pairs, **6**
    and molecular geometry, 17
  valence, 3
Electron spin resonance spectroscopy, 644*p*
Electron-withdrawing, *see* Polar effect
Electrophile, **88**
Electrophilic aromatic substitution, *see* Aromatic substitution
Electrophoresis, 1318*p*
Electrospray mass spectrometry, 569
Electrostatic attraction, **4**
Electrostatic interactions, in proteins, 1304–1305
Electrostatic law, **110**, 352
  role in Brønsted acidity, 109–110
β-Elemenone, 1260*p*
Elemental analysis, **72**
  by combustion, 72–75
  by mass spectrometry, 568
Element effect, on strength of Brønsted acids, 105–108
  role in alcohol and thiol acidity, 361
  trends in acidity, 107–108
β-Elimination, **387**, 387–389; *see also* specific types, *e.g.*, E2 reaction
  competition with nucleophilic substitution, 388–389, 423–427
α-Elimination, 427–429, **428**
Empirical formula, *see* Formula, empirical
Enamines, formation from aldehydes and ketones, 907–909
Enantiomerically pure, **238**
Enantiomeric resolution **239**, 249–250
  in nature, 309–310
  of α-amino acids, 1280–1281
Enantiomers, 225–227, **226**, 243
  conformational, **253**
  formation as reaction products, 305–307

nomenclature, 231–233
physical properties, 234–240
relationship of optical activities, 238
relative reactivities, 302–304
Enantiotopic, **477**
Enantiotopic groups
and chemical equivalence, 479–480
in NMR spectra, 585–586
Endopeptidase, 1289
β-Endorphin, 1313
Endo stereoisomer, 698
Endothermic reaction, **139**
Enediol, intermediate in base-catalyzed
monosaccharide isomerization,
1341–1342
Energy barrier, 154
analogy, 155
Enkephalins, 1313
Enolate ions, 1039–1042
alkylation, 1074–1075, 1079–1082,
SGL 22.7
conjugate-addition reactions,
1090–1093
electronic structure, 1040*f*
intermediates
in acetoacetic ester synthesis,
1079–1082
in aldol reactions, 1055
in Claisen and related condensations,
1065–1073
in base-catalyzed enolizations, 1047
in base-catalyzed monosaccharide
isomerizations, 1341–1342
in conjugate-addition reactions, 1085
in haloform reaction, 1050–1051
in malonic ester synthesis,
1074–1075
reactions, introduction, 1042–1044
Enol ethers, *see* Vinylic ethers
Enolization, mechanisms, 1047–1048
Enols, **654**, 1044–1048
conversion into aldehydes and ketones,
655
intermediates
in α-bromination of carboxylic acids,
1052
in acid-catalyzed aldol condensation,
1059
in acid-catalyzed conjugate addition,
1087
in acid-catalyzed enolization,
1047–1048
in acid-catalyzed halogenation of
aldehydes and ketones, 1048–1049
in alkyne hydration, 654–657
in decarboxylation of β-ketoacids,
958
in hydroboration-oxidation of
alkynes, 657–659

relationship to enamines, 908
Enthalpy of formation, *see* Heat of
formation
Entropy, SGL 4.4
and reaction probability, SGL 11.6
Envelope, conformation of cyclopentane,
292
Enzymes, **165**
as catalysts, 165–166, 1308–1311
as chiral reagents, 306–307
stereoselectivity, 1310–1311
use in enantiomeric resolutions,
1280
Enzyme-substrate complex, **1308**
Epimers, **1330**
Epinephrine, 1155
Episulfonium salt, reactive intermediate, 519
Epoxides
basicity, importance in ring-opening
reactions, 512
nomenclature, 344
reactions
Grignard reactions, *see* Ethylene
oxide
hydrolysis (conversion into glycols),
513–514
polymerization in base, 531*p*
ring-opening reactions under acidic
conditions, 511
ring-opening reactions under basic
conditions, 509–511
with amines, 1129
synthesis
halohydrin cyclization, 505–507
reaction of alkenes with peroxy acids,
503–505
summary, A-8
role in carcinogenicity of aromatic
hydrocarbons, 778
EPR spectroscopy, *see* Electron spin
resonance
Equatorial, substituent positions in cyclohexanes, **274**, 274–275
Equilibrium
and free energy, 102–105
constants
and catalysts, 166
and reaction rate, 155–156
relationship to free energies, 103,
104*t*
Equivalence, chemical, *see* Chemical
equivalence
Ergocalciferol (Vitamin D_2), 1253–1254
Ergosterol, 1253–1254
Erlich's reagent, 1189*p*
Ernst, Richard, 633
D-(−)-Erythrose
structure, 1329*f*

synthesis from D-glyceraldehyde, 1360
ESR spectroscopy, *see* Electron spin resonance spectroscopy
Essential oils, **806**, SGL 17.4
Esterification
by acid-catalyzed reaction of carboxylic
acids and alcohols, 946–949
by alkylation, 950–951
Esters, *see also* Sulfonate esters
acidity, 1040–1042
basicity, 987
tert-butyl, in peptide synthesis, 1292
IR spectroscopy, 982–983
β-keto, **1065**
alkylation, 1079–1082
synthesis by Claisen condensation,
1065–1073
NMR and CMR spectroscopy, 982–986
nomenclature, 971–973, 976–977
of carbamic acids, synthesis from isocyanates, 1149
of carbohydrates, 1346–1347
of strong inorganic acids. 455–456
physical properties, 980
reactions
acid-catalyzed hydrolysis, 990–992
alkylation, 1076–1078
Claisen condensations, 1065–1073
Dieckmann condensation, 1068
dienophiles in Diels-Alder reactions,
691
enolization, 1045
Grignard reactions, 1015–1016
reduction to primary alcohols,
1007–1008
saponification, 989–990
transesterification, 1005
with ammonia and amines, 1005
relative carbonyl reactivity, 1014
relative reactivity in nucleophilic acyl
substitution, 996–1000
synthesis
conjugate addition to α,β-unsatu-
rated esters, 1085–1093
from acid chlorides, 1002–1003
from carboxylic acids, 946–951
from other esters, 1005
malonic ester synthesis, 1074–1076
summary, A-9
Estrone, synthesis, 1261*p*
Et, abbreviation for ethyl group, 1064*t*
Ethane
bonding, 45, 46*f*
conformation, 49–52
models, 45
physical properties, 47*t*
structure, 14*f*, 122*f*
compared to other methyl derivatives, 345*t*

1,2-Ethanediol, *see* Ethylene glycol
Ethanethiol
   hydrogen bonding, 382*p*
   pK_a, 105
Ethanol
   absence of chirality, 225–226
   absolute, **375**
   as a drug, 375
   as a fuel, 375–376
   biological oxidation, 472–475
      stereochemistry, 480–482
   commercial synthesis, 164
   dehydration to diethyl ether, 500
   denatured, **375**
   from fermentation, 375, 474
   hydrogen bonding, 382*p*
   industrial, **375**
   NMR spectrum, effect of moisture, 617,
      618*f*
   physical properties, 350
   pK_a, 105, 365*t*
   production and use, 213*f*, 374–375
   solvent properties, 355*t*
Ethers, **336**; *see also* Acetals, Diethyl ether
   aryl
      cleavage, 850
      S_N2 reactivity, 849–850
      synthesis, 841
   basicity, 366–367
      role in ether cleavage, 507
   boiling points, 346–347
   complexes with borane, 184
   crown ethers, 358–359
   heterocyclic, 344
   IR spectroscopy, 553–554
   NMR spectroscopy, 616
   nomenclature, 343–344
      of carbohydrates, 1346–1347
   reactions
      autoxidation, 377–378
      Claisen rearrangement, 1247–1248
      cleavage, 507–509
         comparison to alcohol dehydra-
            tion, 508
      safety hazards, 377–378
   solvents, for synthesis of Grignard
      reagents, 368
   synthesis
      addition of alcohols to alkenes, 502
      alcohol dehydration, 496*p*
      alkoxymercuration-reduction of
         alkenes, 499–500
      dehydration of alcohols, 500–503
      reaction of alcohols with epoxides,
         509, 511
      summary, A-7–A-8
      Williamson synthesis, 497–499
Ethoxycarbonyl group, 977*t*

Ethoxyethane, *see* Diethyl ether
Ethoxyethylene, hydrolysis, 1109*p*
2-Ethoxy-2-methylpropane, synthesis by
      alcohol dehydration, 501
Ethyl acetate
   acidity, 1040
   Claisen condensation, 1065
   in crossed Claisen condensations, 1070
   solvent properties, 354*t*
Ethyl acetoacetate
   acidity, 1042*p*, 1079
   alkylation, 1079
   synthesis by Claisen condensation,
      1065
Ethylamine, pK_a of conjugate acid, 1121*t*
Ethylbenzene
   production from petroleum, 213*f*
   UV spectrum, 747*f*
2-Ethyl-1-butene, heat of formation, 141*p*,
   143*t*
Ethyldimethylamine, boiling point and
      dipole moment, 1117
Ethylene
   as fruit ripener, 212
   bonding, 124–125
   commercial synthesis, 210–212
   conformation, 18–19, 19*f*, 41–42
   in ethanol production, 164, 374–375
   molecular orbitals, 1223*f*
   polymerization, 208
   structure, 14*f*, 122*f*
   UV absorption, 688*t*
      physical basis, 687*f*
Ethylene glycol
   commercial synthesis, 212–213
   in formation of cyclic acetals, 899
   in polyester production, 1021
   production and use, 377
   production from petroleum, 213*f*
   structure, 338
Ethylene oxide
   Grignard reaction, 515–516
   production and use, 377
   structure, 344
Ethyl formate, crossed Claisen condensa-
      tions, 1069–1070
Ethyl group
   abbreviation, 1064*t*
   structure, 57
1-Ethyl-4-methoxybenzene, UV spectrum,
   747*f*
Ethyl 3-oxobutanoate, *see* Ethyl
   acetoacetate
Ethyl radical, heat of formation, 203*t*
Ethyl sulfate, in commercial synthesis of
      ethanol, 164
Ethynylmagnesium bromide, synthesis,
   664

Eudesmol, biosynthesis, 817–818*p*
Even-electron ions, **565**
Excited state, and molecular orbital
      theory, 1227–1229
Exclusion principle, **28**
Exhaustive methylation, **1130**
   in Hofmann elimination, 1134
Exopeptidase, 1289
Exo stereoisomer, 698
Exothermic reaction, **139**
Extinction coefficient, **686**

F

FAB, *see* Fast-atom bombardment
α- and β-Faces, of steroids, 301
Farnesol, biosynthesis, 812*p*
Fast-atom bombardment, mass spectrome-
      try ionization method, 569
Fats, **1023**
   saturated and unsaturated, 1023
Fatty acids, **942**
   biosynthesis, 1082–1085
FD & C #6, 1144
Feedstocks, 76
Fermentation, anaerobic, 474
Ferric bromide, catalyst for bromination
      of benzene, 748
Ferrocene, structure and aromaticity,
      723–724
Fiberglass, 851
Fingerprint region, of IR spectrum, 544
First-order
   NMR spectra, **607**
   reaction kinetics, 392
"First point of difference" rule
   in alkene nomenclature, 130–131
   in *E,Z* nomenclature, 133
Fischer, Emil, 256, 270*p*, 1359
Fischer indole synthesis, 1190–1191,
      SGL 24.1
Fischer projections, 256–260
Fishhook formalism, 196
   use in writing resonance structures, 709
Flame retardants, 373
Flash point, **378**
Fluoradene, acidity, 820*p*
(9-Fluorenyl)methyloxycarbonyl group,
      see Fmoc group
Fluoride, basicity, 99*t*
Fluorine
   activating and directing effects in elec-
      trophilic aromatic substitution,
      762*t*
   exact mass, 559*t*
Fluoroacetic acid, pK_a, 108, 939

Fluoroborate ion, 517
Fluorocyclohexane, 283p
Fluoroethane, physical properties, 350
Fluoromethane
    dipole moment, 41p
    structure, comparison to other methyl
        derivatives, 345t
1-Fluoropropane, 283p
Fluxional molecules, 1251–1252
Fmoc group, 1297
Forbidden
    pericyclic reactions, **1231**
    UV absorptions, 877
Formal charge, 7–8, **7**, SGL 1.1
Formaldehyde
    from periodate oxidation of carbohy-
        drates, 1352
    Grignard reactions, 896
    manufacture and use, 917
    polymerization, 901
    production from petroleum, 213f
    structure, 866f
Formamide, solvent properties, 355t
Formic acid
    from periodate oxidation of carbohy-
        drates, 1352
    pK_a, 939t
    solvent properties, 352, 355t
Formula
    condensed structural, **48**
    empirical, **73**
        determination, 73–74
    molecular, **47**, **74**
        determination by combustion, 74
        determination by mass spectrometry,
            577p
    relation to unsaturation number,
        136–137
    structural, **47**
Formyl group, 868
Four-helix bundle, in protein structure,
    1303
Fourier transform, NMR spectroscopy
    technique, 633, SGL 13.6
Fractional distillation, **75**
Fragmentation, in mass spectrometry,
    556–557
    mechanisms, 561–565
Franklin, Rosalind, 1372
Free energy
    and chemical equilibrium, 102–105
        relationship to equilibrium constant,
            103, 104t
    of acid dissociation,103
    of activation, **154**
        relationship to rate constant, 392
    of formation, 173p
    relationship to enthalpy, SGL 4.4

relationship to acid dissociation of
    ammonium ions, 1122
Free-radical chain reactions, 197–200
    initiation, 197–199, **199**
    inhibition by quinones, 844
    mechanisms, how to write, SGL 5.4
    propagation, 198–199
    termination, 199–200
Free radicals, **196**, 196–204
    allylic
        as reactive intermediates, 794–799
        relative stabilities, 795
    benzylic, as reactive intermediates,
        794–799
    intermediates
        in alkane halogenation, 372
        in alkyne reduction with sodium and
            liquid ammonia, 660–662
        in allylic and benzylic brominations,
            794–799
        in peroxide-promoted HBr addition
            to alkenes, 194–207
        in phenol oxidation to quinones,
            841–845
        in synthesis of Grignard reagents,
            369
    relative stabilities, 202–203
    structure and geometry, 203
    summary of fundamental reactions, 211
Free-radical substitution, 372
Freons, see Chlorofluorocarbons
Frequency, of a wave, **536**
Freud, Sigmund, endorsement of organic
    chemistry, 1156
Friedel, Charles, 756
Friedel-Crafts acylation, 754–757
    in synthesis of alkylbenzenes, 910–911
    intramolecular, 755
    of benzene, 754–757
    of furan, 1185
    substituent effects, 774
Friedel-Crafts alkylation
    of benzene, 757–759
    substituent effects, 774–775
Friedlander synthesis, of quinolines, 1217p
Fries rearrangement, SGL 18.4
Frontier orbitals, **1225**
Frontier-orbital theory, 1221
α-D-Fructofuranose, 1339
β-D-Fructofuranose, 1339
α-D-Fructopyranose, 1339
β-D-Fructopyranose, 1339
D-Fructose
    equilibrium composition, 1340t
    isomerization to D-glucose and
        D-mannose, 1341
    mutarotation, 1339
    structure, 1331

Fructose-6-phosphate, isomerization to
    glucose-6-phosphate, 1342
Fukui, Kenichi, 1221
Fumarase, 165, 495
    catalysis of fumarate hydration, 165,
        306
        stereochemistry, 306–307, 328p, 495p
Fumaric acid
    conversion into malic acid, see
        Fumarase
    structure, 931t
Fuming sulfuric acid, 753
Functional-group region, of IR spectrum,
    544
Functional groups, **78**, 78–80
Functional-group transformation, 523
Furan
    addition reactions, 1187–1188
    aromaticity, 724p
    derivatives, electrophilic aromatic substi-
        tution, 1186–1187
    Diels-Alder reactions, 1188
    dipole moment, 1179
    electronic structure, 1179–1180, 1180f
    electrophilic aromatic substitution,
        1185–1186
    Friedel-Crafts acylation, 1185
    in nylon production, 1022p
    resonance energy, 1181t
    structure, 344, 1178f
Furanose, **1332**
Furfural, 868
    in nylon production, 1022p
Fused bicyclic compound, **295**
Fusion, of rings, 297–299

G

$\Delta G^{\circ\ddagger}$, (Standard free energy of activa-
    tion), **154**
Gabriel synthesis, of amines, 1146–1147
Galactitol, synthesis from galactose, 1353
β-D-Galactopyranose, 1336
D-(+)-Galactose
    equilibrium composition, 1340t
    structure, 1329f
Gasoline, octane rating system,77–78
Gastrin, 1311
Gauche, see also Butane, conformations;
        Conformation; 1,3-Diaxial
        interactions
    interactions in methyl cyclohexane, 282
Genetic code, 1376–1377, 1377t
Geometry
    linear, **16**

Hydrogen fluoride, use in solid-phase peptide synthesis, 1296
Hydrogen halides
    addition to alkenes, 145–153
        application of Hammond's postulate, 158–160
        rearrangements, 151–153
    addition to conjugated dienes, 699–700, 703–706
    reactions with alcohols, 447–450
Hydrogen iodide, see also Hydrogen halides
    addition to alkenes, 145–153
        absence of peroxide effect, 200
Hydrogen peroxide
    conformation, 42p
    dipole moment, 42p
    oxidation of organoboranes, 185
    oxidation of pyridine, 1196
    oxidation of sulfides, 522
    oxidizing agent in ozonolysis, 190
    Ruff degradation of aldoses, 1360
α-Hydrogens, in carbonyl compounds, 868
    acid-catalyzed exchange, 1047
    acidity, 1039–1042
    base-catalyzed exchange, 1042–1043
Hydrogen sulfide, structure, 14f
Hydroiodic acid, pK_a, 99t
Hydrolysis, 989; see also specific reactions, e.g., Nitriles, hydrolysis
Hydronium ion, pK_a, 99t
Hydroperoxides, in ether autoxidation, 378
Hydroperoxy group, 503
Hydrophilic residues, in proteins, 1305
Hydrophobic bonds, 1305
Hydrophobic residues, in proteins, 1305
Hydroquinone, 739
    free-radical inhibitor 843–844
    oxidation, 841
    photographic developer, 845
Hydrosulfide, basicity, 99t
Hydrosulfuric acid, pK_a, 99t
Hydroxamate test, 1005
Hydroxamic acid, 1005
Hydroxide, base strength, 99t
Hydroxy
    electron affinity, 107
    group, 335
        activating and directing effects in electrophilic aromatic substitution, 762t
Hydroxy acids, lactonization, 991–992
α-Hydroxy carbocations, 881
    intermediates
        in acetal formation and hydrolysis, 899–900

in pinacol rearrangement, 882–883
Hydroxylamine, reactions
    with aldehydes and ketones, 906t
    with esters, 1005
2-Hydroxypyridine, see 2-Pyridone
Hyperconjugation, 149
Hypochlorous acid, 177
Hypohalous acid, and halohydrin formation, 177
Hypophosphorous acid, reaction with diazonium salts, 1142
Hz, see Hertz

I

Ibuprofen, 266p
D-(−)-Idose
    equilibrium composition, 1340t
    structure, 1329f
Imidazole
    basicity and pK_a of conjugate acid, 1183
    structure, 1178f
Imides
    acidity, 1042p
    nomenclature, 976
    synthesis, 1005
Imidic acids, intermediates in nitrile hydrolysis, 994
Imines, 904
    formation from aldehydes and ketones, 904–907
    intermediates
        in hydrogenation of nitriles, 1011
        in reduction of amides by LiAlH$_4$, 1009
        in reductive amination, 1130–1132
        in Strecker synthesis, 1279
Imminium ions, intermediates in reductive amination, 1131
Indole
    acidity, 1183
    reaction with organometallic reagents, 1183
    structure, 1178f
2-Indolecarboxylic acid, decarboxylation, 1193
Indolenine, 1213p
Indoles, Fischer indole synthesis, 1190–1191
Induced dipoles, 68
Inductive cleavage, 564, 879
Inductive effect, 100; see also Polar effect
Infrared active, 548
Infrared inactive, 548
Infrared spectrophotometer, 540

description, 555
Infrared spectroscopy, 540–555; see also specific compounds or classes, e.g., Alkenes, IR spectroscopy; see also Infrared spectrum
    Fourier-transform (FT), SGL 12.2
    physical principles, 541–543
    relation to chemical structure, 543–549
    summary of important absorptions, A-2–A-5
Infrared spectrum
    absorption intensities, 547–549
    absorption positions, 544–547
    regions, 544
Initiation, of free-radical reactions, 197–199
Initiators, free-radical, 197
Insects, control with pheromones, 669
Insulin, 1311
Integral, in NMR spectroscopy, 591–593
Integration, see Integral
Intermediates, reactive, 147
Intermolecular reaction, 520
Internal mirror plane, 229
Internal rotations, 50
    in butane, 53–54
    in ethane, 51
Intramolecular reaction, 520
Inversion
    of amines, 254–255, 255
    of configuration, in S$_N$2 reaction, 396–397
    stereochemistry of substitution, 312
Invertase, 1363
Iodide, basicity, 99t
Iodine
    activating and directing effects in electrophilic aromatic substitution, 762t
    addition to alkenes, 176
    atom
        electron affinity, 107
        exact mass, 559t
    in oxidation of thiols to disulfides, 484
Iodoform
    product of base-catalyzed iodination of methyl ketones, 1051
    structure, 337
Iodoform test, 1051
Iodohydrin, 177
Iodomethane (Methyl iodide)
    physical properties, 347
    reaction with carbohydrates, 1346–1347
    reaction with pyridine, 1202
    solvent effects on S$_N$2 reactions, 400t
    structure, compared to other methyl derivatives, 345t
Ion-dipole attraction, 358

Ion-exchange chromatography,
    1275–1277, 1283*f*
Ion-exchange resin, 1275–1276
Ionic bond, 4–5
Ionic compound, 4
    solvents for, 358
Ionization potential, 107
Ionophore, **358**
Ion pair, **356**
Ions
    aromaticity, 722
    dissociated, **356**
    solvation, 357
IPP, *see* Isopentenyl pyrophosphate
Ipsdienol, structure, 814
IR, *see* Infrared
Iron, catalyst in bromination of benzene,
    748
Isobutyl cation, 148
    heat of formation, 149*t*
Isobutyl group
    abbreviation, 1064*t*
    structure, 58*t*
Isobutyric acid, structure, 931*t*
Isocyanates, **1148**
    hydration to carbamic acids, 1149
    intermediates in Hofmann
        rearrangement, 1151
    synthesis by Curtius rearrangement,
        1148–1150
Isoelectric pH, *see* Isoelectric point
Isoelectric point, **1271**
    as a function of pK_a, 1272
    of amino acids, 1267–1268*t*
Isoleucine
    D- and L-, absolute configuration,
        1269*p*
    physical properties, 244*t*
    structure and properties, 1266*t*
Isomerism
    cis-trans, in alkenes, 126–129
    summary and analysis, 243–245, 245*f*
Isomers, **55**, 55–56, 243; *see also*
        Stereoisomers
    constitutional, 243, **55**
    stereoisomers, **126**, **255**
    structural, **55**
Isopentane, *see* 2-Methylbutane
Isopentenyl pyrophosphate, 809–811
    biosynthesis, 1084
Isoprene, 131, 806
    UV spectrum, 685
Isoprene rule, 806–808, **806**
Isoprenoids, *see* Terpenes
Isopropenyl group, **131**
2-Isopropoxybutane, mass spectrum, 563*f*
Isopropyl alcohol, *see* 2-Propanol
Isopropylbenzene, *see* Cumene

Isopropyl group
    abbreviation, 1064*t*
    structure, 58*t*
    treatment in *E,Z* system, 135
Isopropyl radical, heat of formation, 203*t*
Isoquinoline, structure, 1178*f*
Isotope effect, on reaction rate, 405–406
Isotopes; *see also* Isotope effect
    effect on infrared absorption positions,
        545, 547*p*, 575*p*
    exact masses, 559*t*
    in mass spectrometry, 558–561, 577*p*
    natural abundances, 559*t*
Isovaleric acid, structure, 931*t*
Isovanillin, 878*p*

J

Juglone, 843

K

K_a, **98**
Katz, Thomas J., 716
K_b, 100–101
Kekulé, August, 806
    benzene structure, 715
    structure theory, 46
Kel-F, 209*t*
α-Keratins, 1303
β-Keto esters, *see* Esters, β-keto
Ketones, *see also* α,β-Unsaturated car-
        bonyl compounds
    acidity, 1039–1042
    basicity, 881–884
        in gas phase, 883
    bonding, 866
    CMR spectroscopy, 875
    cyclic, carbonyl stretching frequencies,
        874
    formation in ozonolysis, 190–191
    formation in periodate cleavage of gly-
        cols, 470–472
    α-halo
        reactivity in nucleophilic substitu-
            tion, 1053–1054
        synthesis, 1048–1049
    intermediates in Grignard reactions of
        esters, 1015
    IR spectroscopy, 872–874
    mass spectrometry, 878–879
    NMR spectroscopy, 874
    nomenclature, 867–871
    physical properties, 872
    reactions

    acetal formation, 898–899
    aldol reactions, 1055–1063
    base-catalyzed halogenation,
        1050–1051
    Clemmensen reduction, 910
    enolization, 1044–1048
    exchange of α-hydrogens, acid-
        catalyzed, 1047
    exchange of α-hydrogens, base-
        catalyzed, 1042
    Grignard reactions, 895–897
    hydration, 884–886
    introduction, 880–881
    iodoform reaction, 1050–1051
    racemization, acid-catalyzed, 1047
    racemization, base-catalyzed, 1043
    reduction to alcohols, 890–894
    reductive amination, 1130–1132
    relative carbonyl reactivity, 1014
    with primary amines, 904–907
    with secondary amines, 907–909
    with sodium bisulfite, 919*p*
    Wittig reaction, 911–915
    synthesis
    acetoacetic ester synthesis,
        1080–1081
    aldol reactions, 1055–1063
    Claisen and Dieckmann condensa-
        tions, 1065–1073
    conjugate additions to α,β-unsatu-
        rated ketones, 1085–1093
    Friedel-Crafts acylation, 754–757
    hydration of alkynes, 654–657
    manufacture and use, 851–852, 917
    oxidation of secondary alcohols,
        467–469
    oxidation of phenols to quinones,
        841–845
    reaction of lithium dialkylcuprates
        with acid halides, 1017–1018
    hydration of alkynes, 654–657
    summary, 880, A-8–A-9
    stability, 889
    structure, 865
    α,β-unsaturated, synthesis by aldol con-
        densation, 1057–1063
    UV spectroscopy, 876–877
    Wolff-Kishner reduction, 909–910
Ketopentose, **1326**
Ketoses, **1326**
    isomerization in base, 1341–1342
Kiliani, Heinrich, 1354
Kiliani-Fischer synthesis, 1353–1355
Kinetic control, 703–707, 1089
    in naphthalene sulfonation, 1174–1175
Kinetic isotope effect, *see* Isotope effect
Kinetic order, **392**
Kolbe, Hermann, 263

Körner, Wilhelm, 785p
Kroto, Harold W., 1169

L

l, quantum number, **23**
L, stereochemical configuration, 1268–1269
Labetalol, 1158p
Lactams, nomenclature, 976
Lactobionicacid, 1362
Lactones
  equilibrium with carboxylic acid derivatives of monosaccharides, 1350–1351
  formation and hydrolysis, 991
  nomenclature, 973
Lactose, 1361–1363
  hydrolysis to glucose and galactose, 1362
  oxidation with bromine water, 1362
Ladenburg benzene, 716
Lauric acid, structure, 931t
Lavoisier, Antoine Laurent, 1
LCHIA, *see* Lithium cyclohexylisopropylamide
LDA, *see* Lithium diisopropylamide
Leaning, in NMR spectroscopy, **596**
  effect of chemical shift difference, 597
Leaving groups, **385**
  effect on reaction rate
    in E2 reactions, 404–405
    in nucleophilic acyl substitution, 949, 998–999
    in nucleophilic aromatic substitution, 830
    in S$_N$2 reactions, 403
  in nature, 810
  Lewis-acid assistance, 749, SGL 16.2
LeBel, Achille, 263
LeChatelier's principle, **390**
  in acetal formation, 899
  in alcohol dehydration, 444
  in Claisen condensation, 1066
  in esterification of carboxylic acids, 947
  in ester saponification, 990
  in nucleophilic substitution reactions, 390
  in reactions of alcohols with hydrogen halides, 447
  in storage of acetylene, 671
Lecithin, structure, 1025
Ledderhose, Georg, 1367
Lehn, Jean-Marie, 360
Leucine, structure and properties, 1266t
Levorotatory, **236**

Levulinic acid, 1098p
Levulose, 1363
Lewis acid, **88**
Lewis acid-base reactions
  associations, **88**
  classification, 94
  dissociations **90**
    in S$_N$1-E1 reactions, 417
  curved-arrow formalism, 89–94
  relation to Brønsted acid-base reactions, 97
Lewis acids
  assistance for leaving groups, 749, SGL 16.2
  association reactions with Lewis bases, 87–90, **88**
  catalysis of reactions, SGL 19.5
  electron-deficient compounds, 87–90
Lewis bases, **88**
  association reactions with Lewis acids, 87-90, 88
Lewis structures, 6–7
  and molecular orbital theory, 34
  rules for writing, 8
Light, velocity, 536
Lignin, 1364
Limonene, structure, 809f
Lindlar catalyst, 659
Lines, in NMR spectrum, 582
Literature, chemical, **62**
Lithium, in synthesis of organolithium reagents, 369
Lithium aluminum hydride
  in purification of ethers, 378
  in synthesis of lithium tri-*tert*-butoxyaluminum hydride, 1013
  reaction with aromatic nitro compounds, 1147–1148
  reaction with water, 891
  reduction
    of aldehydes and ketones, 890–894
    of amides, 1008–1010
    of carboxylic acids, 956–957
    mechanism, SGL 20.9
    of esters, 1007–1008
    of nitriles, 1010
    of α,β-unsaturated carbonyl compounds, 1093–1094
Lithium cyclohexylisopropylamide, 1077
Lithium dialkylcuprates
  reaction with acid chlorides, 1017
  synthesis, 1017
Lithium diisopropylamide, 1077
  synthesis, 1127
Lithium dimethylcuprate, 1017
Lithium reagents, *see* Organolithium reagents
Lithium tri-*tert*-butoxyaluminum hydride

reduction of acid chlorides, 1012
  synthesis, 1013
Lobry de Bruyn, Cornelius Adriaan van Troostenbery, 1341
Lobry de Bruyn-Alberda van Ekenstein reaction, 1341
Lucite, 209t
Lumisterol, 1254p
LUMO (Lowest unoccupied molecular orbital), **1225**
Lysine
  isoelectric point, 1273–1274
  structure and properties, 1267t
Lysozyme
  primary structure, 1300f
  structure, 1304f
D-(−)-Lyxose, structure, 1329f

M

m, quantum number, **23**
Magnesium, in synthesis of Grignard reagents, 368
Magnesium monoperoxyphthalate, in conversion of alkenes into epoxides, 504
Magnetic field
  effect of increase on NMR spectra, 609–610
  induced, *see* Induced field, 613
  role in NMR experiment, 579–581
Magnetic quantum number, **23**
Magnetic resonance imaging, 633
Malate, conjugate base of malic acid
  synthesis from fumarate, 165
  stereochemistry, 306–307, 328p
Maleic acid, structure, 931t
Maleic anhydride, dienophile in Diels-Alder reaction, 697, 1188
Malonic acid
  and derivatives, decarboxylation, 958–959, 1075
  pK_a values, 939t
  structure, 931t
Malonic ester synthesis, 1074–1076
  in synthesis of α-amino acids, 1278–1279
Malonyl CoA, 1082
Maltose, structure, 1388p
Manganese dioxide, in Baeyer test for unsaturation, 193
D-Mannopyranose, 1332
D-(+)-Mannose
  equilibrium composition, 1340t
  Fischer proof of stereochemistry, 1356–1359

isomerization to D-glucose and
    D-fructose, 1341
structure, 1329f
Marker, Russell, 302
Markownikoff, Vladimir, 146
Markownikoff's rule, **146**
Mass
    exact, and elemental analysis, 568
        table, 559t
    nominal, **568**
Mass spectrometer, **556**
    description, 567–569, SGL 12.3
Mass spectrometry, 74, 556–569; see also
        Mass spectrum
    fragmentation mechanisms, 561–565
    isotopic peaks, 558–561
    McLafferty rearrangement, 879
    molecular ion, identifying, 566–567
    sensitivity, 567
Mass spectrum, **557**
    base peak, **558**
    mass-to-charge ratio (m/z), **557**
    molecular ion (M), **557**
McLafferty, Fred, 879
McLafferty rearrangement, in mass spec-
        trometry, 879
mCPBA, see m-Chloroperoxybenzoic acid
Me, abbreviation for methyl group, 1064t
Mechanism, of a reaction, **147**
    evolution, 395
    how to write, SGL 4.7
    role of rate law, 395
Mechloramine, 532p
Megahertz, 581
Meisenheimer complex, 830
Melting points
    principles, 70–71
    trends with structure, 71
Membranes, of cells, constitution, 1025
2-Menthene, formation in E2 reaction,
        438–439p
3-Menthene, formation in E2 reaction,
        438–439p
Menthol, structure, 809f
Menthyl chloride, E2 reaction, 438–439p
Meperidine, 821
Mephenesin, 862p
Mercaptan, **336**; see also Thiol
Mercaptides, **361**; see also Thiolates
2-Mercaptoethanol, reduction of disulfide
        bonds, 1299
Mercapto group, see Sulfhydryl group
Mercuric acetate
    in alkoxymercuration of alkenes,
        499–500
    in oxymercuration of alkenes, 180
Mercuric ion, catalyst for alkyne hydra-
        tion, 654–657
Mercurinium ion, 180–181

Merrifield, R. Bruce, 1292
Merrifield resin, 1293
Mescaline, 1154f
Mesitoic acid, 1037p
Mesitylene, UV absorption, 748p
Mesityl oxide, synthesis, 1058
Meso-compounds, 244–248, **247**
    analysis of structures, 248
    in cyclic structures, 290–291
Messenger RNA (mRNA), 1376
Mesylate, 451
Meta, position on substituted benzene
        ring, 738
Meta-directing group, **761**
Methane
    bond dissociation energy, 106
    by-product of commercial ethylene pro-
        duction, 211
    chlorination, 371–372
    combustion, 72
    mass spectrum, 556–557, 557f
    models, 16f
    natural gas, 76
    physical properties, 47t
    production from petroleum, 213f
    structure, 14f, 14–15, 15f
Methanesulfonate esters, see Mesylates
Methanesulfonic acid, structure, 450
Methanesulfonyl chloride, synthesis, 953
Methanethiol, structure, 345f
    compared to other methyl derivatives,
        345t
Methanol
    pK_a, 365t
    production and use, 376
    production from petroleum, 213f
    solvent properties, 355t
    structure, 345f
        compared to other methyl deriva-
            tives, 345t
Methine group, 590
    diagnosing by DEPT technique,
        625–627
Methionine
    residue, reaction with cyanogen bro-
        mide, 1286–1288
    structure and properties, 1267t
4-Methoxybenzoic acid (Anisic acid), pK_a,
        939t
Methoxycarbonyl group, 977t
2-Methoxyethanol, dielectric constant,
        353p
Methoxyflurane, 373, 380p
Methoxy group
    activating and directing effect in electro-
        philic aromatic substitution, 762t,
        762–764
    resonance effect in S_N1 reactions, 792
Methoxymethane, see Dimethyl ether

Methoxymethyl cation, resonance struc-
        tures, 95
2-Methoxy-2-methylpropane (*tert*-Butyl
        methyl ether)
    commercial synthesis, 173p
    production and use, 377
    synthesis by dehydration of alcohols,
        502
    Williamson ether synthesis, 498
1-Methoxypentane, boiling point, 346
Methyl acrylate, conjugate-addition reac-
        tions, 1086–1087
Methylamine
    pK_a of conjugate acid, 1121t
    structure, compared to other methyl
        derivatives, 345t
2-Methylaniline, (*o*-Toluidine), pK_a of
        conjugate acid, 1121t
3-Methylaniline, (*m*-Toluidine), pK_a of
        conjugate acid, 1121t
4-Methylaniline, (*p*-Toluidine), pK_a of
        conjugate acid, 1121t
N-Methylaniline, pK_a of conjugate acid,
        1121t
Methylation, exhaustive, **1130**
Methylation, with *S*-adenosylmethionine,
        517–518
4-Methylbenzoic acid, (*p*-Toluic acid),
        pK_a, 939t
2-Methyl-1,3-butadiene, see Isoprene
2-Methylbutane (Isopentane)
    boiling point and dipole moment, 1117
    conformations, 55p
2-Methyl-1-butene, heat of formation,
        142
3-Methyl-1-butene
    heat of combustion, 171p
    heat of formation, 142
Methyl carbamate, 977
Methyl carbonate, decarboxylation, 959
Methylcyclohexane, conformational analy-
        sis, 280–282
Methylene chloride (Dichloromethane)
    solvent properties, 354t
    structure, 15, 336
Methylenecyclohexane, synthesis
    by E2 reaction, 414–415
    by Hofmann elimination, 1133
    by Wittig reaction, 911–912
Methylene group, **48**
    diagnosing by DEPT technique,
        625–627
Methylene iodide, in Simmons-Smith reac-
        tion, 430
Methyl group
    abbreviation, 1064t
    activating and directing effects in elec-
        trophilic aromatic substitution,
        762t, 764–765

Methyl group (*continued*)
  diagnosing by DEPT technique, 625–627
  structure, 57
5-Methyl-1-hexanone, mass spectrum, 878*f*
2-Methyl-1-hexene, IR spectrum, 553*f*
Methyl iodide, *see* Iodomethane
Methyl orange, 1143–1144
2-Methyl-1-pentene, heat of formation, 141*p*
4-Methyl-1-pentene, heat of formation, 143*t*
(*E*)-3-Methyl-2-pentene, heat of formation, 143*t*
2-Methyl-2-propanol (*tert*-Butyl alcohol)
  commercial synthesis, 165*p*
  pK_a, 365*t*
2-Methylpropene (Isobutylene),
  heat of formation, 141*p*
  mechanism of hydrogen bromide addition, 148
  use in *tert*-butyl methyl ether production, 377
2-Methylpyridine (2-Picoline), acidity, 1202
1-Methylpyridinium iodide, synthesis, 1202
1-Methyl-2-pyridone, synthesis, 1202
Methyl radical
  electron affinity, 107
  heat of formation, 203*t*
Methyl salicylate, structure, 715
Micelle, 943
Michael addition, 1090–1092
Micrometer, **540**
Microscopic reversibility, **445**
Millimicron, **684**
Miscible, **353**
Mixed anhydride, **974**
MMPP, *see* Magnesium monoperoxyphthalate
Models
  ball-and-stick, 15, 16*f*
  molecular, 15, 16*f*
  space-filling, 15, 16*f*
Modhephene, structure, 814
Molar absorptivity, *see* Extinction coefficient
Molar extinction coefficient, *see* Extinction coefficient
Molecular ion, in mass spectrometry, **557**
  identification, 566–567
Molecular orbitals, **30**
  and covalent bonds, 30–34
  antibonding, **32**, **1222**
  bonding, **32**, **1222**
  formation, 31–32

frontier orbitals, **1225**
  HOMO, **1225**
  LUMO, **1225**
  nodes, 1223–1224
  nonbonding, **1226**
  of conjugated alkenes, 1222–1226
  of conjugated ions and radicals, 1226–1227
  of hydrogen molecule, 31–32
  phase, 1224
  pi (π), 124–125
  pi-star (π^*), 124–125
  role in UV absorption, 687
  symmetry
    classification, 1224
    role in pericyclic reactions, 1229–1255
Molecular recognition, 360
Molozonide, 188–189
Monochromator, 555*f*
Monocyclic, **271**
Monomer, **207**
Monosaccharides, **1326**
  cyclic structures, 1331–1337
  equilibrium compositions, 1340*t*
  stereochemistry and configuration, 1327–1331
Monoterpenes, **807**
Morphine, 1115, 1154*f*
  biological effect, 1313
MPTP, 821*p*
MRI, *see* Magnetic resonance imaging
MTBE, *see* 2-Methoxy-2-methylpropane
Muscalure, 675*p*
Muscles, contraction and ATP hydrolysis, 1370
Mutarotation, of carbohydrates, 1337–1340, **1338**
Mycomycin, structure, 730
Myristic acid, structure, 931*t*
m/z, mass-to-charge ratio, 557

N

n, quantum number, **23**
n → π^* absorption, 876
n + 1 rule, and splitting in NMR spectra, 595–598
  breakdown, 607–610
4*n* + 2 rule, 719–726
NAD$^+$, coenzyme in biological oxidation reactions, 472–475
  structure, 473*f*
NADH, coproduct of biological oxidation of ethanol, 474
Nanometer, **684**

Naphthalene, 1169, 1171–1177
  and derivatives, electrophilic aromatic substitution reactions, 1173–1177
  aromaticity, 723
  derivatives, nomenclature, 1171–1172
  physical properties, 1171
  resonance energy, 1181*t*
  resonance structures, 1171
  structure, 1171
1-Naphthol, 1172
4-Naphthoquinone, 842
Naphthyl group, 1172
Natural gas, 76
NBS, *see* N-bromosuccinimide
Neighboring-group participation, 518–521, **520**
Neomenthyl chloride, E2 reaction, 438–439*p*
Neopentyl bromide, *see* 1-Bromo-2,2-dimethylpropane
Neopentyl group, structure, 58*t*
Neoprene, 671
Neurotransmitter, 1155
Newman projection, **18**, SGL 2.1
  of butane, 52*f*
  of ethane,4 9*f*
Niacin, 473
Nickel, hydrogenation catalyst, 162
Nicol prism, 234
Nicotine, 1154*f*
  oxidation, 1194
Nicotine adenine dinucleotide, *see* NAD$^+$
Nicotinic acid, synthesis, 1194
NIH shift, 864*p*
Ninhydrin, 890*p*
  in amino acid analysis, 1282–1284
Nitration
  of benzene, 752–753
  of naphthalene, 1173
  of pyridine-*N*-oxide, 1196
  of pyrrole, 1185
  of thiophene, 1185
  of toluene, 774
Nitric acid
  in nitration of benzene, 752; *see also* Nitration
  in oxidation of aldehydes and primary alcohols, 1350–1351
  in oxidation of nicotine, 1194
  in oxidation of sulfides, 523
  in oxidation of thiols, 484
Nitriles
  α-amino, synthesis, 1279–1280
  basicity, 987
  IR spectroscopy, 982–983
  NMR and CMR spectroscopy, 982–986
  nomenclature, 974–975, 976–977
  physical properties, 981

reactions
relative carbonyl reactivity, 1014
relative reactivity in nucleophilic acyl
  substitution, 996–1000
Diels-Alder reactions, 691
hydrolysis, 994–995
α-hydroxy, *see* Cyanohydrins
reduction to primary amines,
  1010–1012
synthesis
  conjugate addition of cyanide to car-
    bonyl compounds, 1085–1093
  conjugate addition to α,β-unsatu-
    rated nitriles, 1086–1087
  reaction of alkyl halides with cyanide
    ion, 1019
  summary, A-10
3-Nitroaniline, pK_a of conjugate acid,
  1121t
4-Nitroaniline, pK_a of conjugate acid,
  1121t
Nitrobenzene
  and derivatives, in Reissert indole syn-
    thesis, 1192–1193
  oxidizing agent in Skraup quinoline syn-
    thesis, 1205
  reaction with LiAlH$_4$, 1148
  synthesis from benzene, 752
4-Nitrobenzoicacid, pK_a, 939t
Nitro compounds, reduction to amines,
  1147–1148
Nitrogen, exact isotopic masses and natural abundances, 559t
Nitro group, activating and directing
  effects in electrophilic aromatic
  substitution, 762t, 765–766
  structure, 737
Nitromethane
  resonance structures, 19–20, 95
  solvent properties, 355t
Nitronium ion, intermediate in aromatic
  nitration, 752
3-Nitrophenol, pK_a, 838
4-Nitrophenol
  acidity and pK_a, 838–839
  UV spectrum, 857f
2-Nitropropane, ionization and enolization, 1104p
Nitrosamines, 1144
  and cancer, 1145
Nitrosation, of amines, 1144–1145
Nitrosyl cation, intermediate in nitrosation reactions, 1145
Nitrous acid, in amine diazotization,
  1139–1143
Nodes, of orbitals, 25
  of molecular orbitals, 1223–1224
  relationship to quantum number, 26

Nomenclature, *see also* specific compound
    classes, *e.g.*, Alkenes, nomenclature
  and chemical indexing, 62
  common, 336
  "first point of difference" rule, 60
  of E and Z stereoisomers, 132–136
  of enantiomers, 231–233
  priority of principal groups, 1115
  radicofunctional, 336
  substitutive, 56
    general rules, 339–340
    summary of rules, A-1–A-2
Nonactin, structure, 360, 361f
Nonane
  IR spectrum, 539f
  physical properties, 47t
Nonbonding molecular orbital, 1226
Nondonor, solvent classification, 352
Nonreducing sugar, 1363
Norepinephrine, 1155
Normal alkanes, *see* Alkanes, unbranched
Normal vibrational modes, 546
  of a methylene group, 547f
Nuclear magnetic resonance (NMR), 581
Nuclear magnetic resonance spectrometer,
  581, 632–633
Nuclear magnetic resonance spectroscopy,
  579–644, 581; *see also* specific com-
    pounds or classes
  carbon NMR (CMR) spectroscopy,
    622–628
  chemical shift, 582–590
  complex spectra, 603–610
  conformational equilibria, 619–621
  FT-NMR technique, 633, SGL 13.6
  integral, 591–593
  NMR tomography, 633
  oxygen (^{17}O) NMR, 643p
  phosphorus (^{31}P) NMR, 633
  solid-state NMR, 633
  spectrometer description, 632–633
  splitting, 594–610
  summary of chemical shifts, A-4–A-5
  use of deuterium, 611
Nuclear magnetic resonance spectrum,
  582–584
Nucleic acids, 1326
Nucleophile, 88
  in nucleophilic substitution reactions,
    385
Nucleophilic acyl substitution, 991
Nucleophilic addition, to carbonyl groups,
  885
  orbital diagram, 886f
Nucleophilic aromatic substitution, *see*
    Aromatic substitution, nucleophilic
Nucleophilic displacement, *see* Nucleo-
    philic substitution

Nucleophilicity, 399
  solvent effects, 399–402
Nucleophilic substitution, 385; *see also*
    Aromatic substitution, nucleo-
    philic; Nucleophilic acyl substitu-
    tion; and specific types, *e.g.*, S$_N$2
    reaction
  competition with β-elimination reac-
    tions, 388–389
    predicting, 423–427
  equilibria, 389–390
  reactions, 385–387
Nucleosides, 1367–1370
Nucleotides, 1368, 1367–1370
Nylon, 1020–1021

O

Octane
  isomers, 81–82p
  physical properties, 47t, 347
Octane number, 78
1-Octene, IR spectrum, 552f
Octet rule
  in covalent compounds, 6
  in ionic compounds, 4
  in Lewis acid-base reactions, 88
  violation in oxidized sulfur com-
    pounds, 484
1-Octyne, IR spectrum, 650f
Odd-electron ions, 565, 879
Olefins, 46, 121; *see also* Alkenes
Oleum, 753
Oligopeptide, 1267
Oligosaccharides, 1327
Operating frequency, in NMR spectroscopy, 582
OPP, abbreviation for pyrophosphate
  group, 810
Optical activity, 235, 234–240
  in nature, 309–310
  of enantiomers, 238
Optical density, 684; *see also* Absorbance
Optically pure, 238
Optical rotation, 235–236, 235
  and absolute configuration, 238
Orbitals, atomic, 22
  d, in oxidized sulfur compounds, 484
  directionality, 26
  energies in many-electron atom, 28,
    29f
    5f, 41p
    2p, 26, 26f
    3p, 27, 27f, 41p
    4p, 41p
    1s, 24, 24f

Orbitals (*continued*)
2*s*, 25, 25*f*
shape, 26
spatial characteristics, 24–27
hybrid, 34–37, **35**
sp, 648
*sp*², 123
*sp*³, 34–35, 35*f*
molecular, *see* Molecular orbitals
overlap
and resonance structures, 712–713
role in substituent effects in electrophilic aromatic substitution, 769–770
Orbital symmetry, role in pericyclic reactions, 1229–1255
Organic chemistry, **1**
reasons for studying, 2
Organic synthesis, **485**, 523–525
methods for carbon-carbon bond formation, A-11
planning, 486–488
with acetoacetic ester synthesis, 1080–1081
with aldol condensations, 1062–1063
with alkynes, 666–668
with Claisen and related condensations, 1071–1073
with conjugate-addition reactions, 1097–1098
with electrophilic aromatic substitution, 771–775
with malonic ester synthesis, 1075–1076
summary of preparative methods, A-5–A-11
three fundamental operations, 523–525
Organoboranes
oxidation, 185–186
stereochemistry, 317–318
synthesis from alkenes, 183–185
Organolithium reagents, **368**
in synthesis of lithium dialkylcuprates, 1017
synthesis, 369
reactions
protonolysis, 371
with aldehydes and ketones, 895–897
with carboxylic acids, 969*p*
with esters, 895–897
with indole and pyrrole, 1183
with pyridine, 1199
with α,β-unsaturated carbonyl compounds, 1094–1095
Organometallic compounds, **367**; *see also* specific types, *e.g.*, Grignard reagents
aromaticity, 723–724
Orlon, 209*t*

Ortho, position on substituted benzene ring, 738
Ortho esters, SGL 20.5
Ortho,para-directing group, **760**
Ortho,para ratio, in electrophilic aromatic substitution, 766–767
Osazones, 1385*p*, SGL 27.4
Osmate ester, **193**
stereochemistry of formation, 320
Osmium tetroxide, in conversion of alkenes into glycols, 193
Osmium tetroxide, stereochemistry of reaction with alkenes, 320
Osteomalacia, 1252
Overtones, bands in IR spectra, 742
Oxalic acid
pK_a values, 939*t*
structure, 931*t*
Oxaphosphetanes, 912–913
Oxazole, structure, 1178*f*
Oxepin, structure, 729*p*
Oxidation, **459**, *see also* specific compounds or classes, *e.g.*, Aldehydes, oxidation
benzylic, 804–805
Oxidation level, **459**
Oxidation number, 459–463, **459**
Oxidation states, of functional groups, 466*t*
Oxidizing agent, **463**
Oxime, 906*t*
Oxirane, structure, 344; *see also* Ethylene oxide
Oxonium salts, 516–517
OxyCope rearrangement, 1247
Oxygen
atomic
electronic configuration, 30
exact isotopic masses and natural abundances, 559*t*
molecular, reaction with ethers, 377–378
Oxymercuration, of alkenes, **180**, 180–181
stereochemistry, 319–320
Ozone
in alkene ozonolysis, 188
stratospheric, destruction by chlorofluorocarbons, 200, 373
synthesis, 189
Ozonides, 188–189
reduction, 189–190
Ozonolysis, of alkenes, 188–192

P

Palladium, hydrogenation catalyst, 162
Palmitic acid, structure, 931*t*
Paquette, Leo, 296

Para, position on substituted benzene ring, 738
Paracelsus, 1
Paraffins, **46**
Paraformaldehyde, 901
Paraldehyde, 901
Parent ion, *see* Molecular ion
Parkinson's disease, 821*p*
Parts per million, chemical shift scale, 584
Pasteur, Louis, 261–262
Pauli exclusion principle, **28**
Pedersen, C. J., 360
Pellagra, 473
Penicillamine, use in treatment of Wilson's disease, 363–364
Penicillin G, 964*p*
1,2-Pentadiene, heat of formation, 680*t*
(*E*)-1,3-Pentadiene, heat of formation, 171*p*, 680*t*
1,4-Pentadiene, heat of formation, 171*p*, 680*t*
2,3-Pentadiene, heat of formation, 680*t*
Pentane, physical properties, 47*t*
Pentanedioic acid, *see* Glutaric acid
2,3-Pentanediol, analysis of stereoisomers, 242–243
1,2-Pentanedione, 925*p*
2,4-Pentanedione (Acetylacetone), enolization, 1045
Pentose, **1326**
Pentothal, 1108*p*
Pentulose, **1326**
1-Pentyne, heat of formation, 680*t*
2-Pentyne, heat of formation, 680*t*
Peptidases, *see* Proteases
Peptide bond, **1263**
Peptides, 1263–1324, **1263**
acid-base properties, 1269–1278
backbone, **1267**
hydrolysis, 1282–1284
enzyme-catalyzed, 1289–1291
isoelectric points, 1271–1275
nomenclature, 1267–1268
occurrence, 1311, 1313
primary structure, 1299–1300
reaction with cyanogen bromide, 1286–88
separations using acid-base properties, 1275–1277
solid-phase synthesis, 1291–1298
specific cleavage, 1286–1291
structure determination, 1282–1291
zwitterionic structures, 1271
Peracid, **503**
Perchlorate, basicity, 99*t*
Perchloric acid, pK_a, 99*t*
Pericyclic reactions
Pericyclic reactions, 691, SGL 15.1, 1219–1262; *see also* specific types, *e.g.*, Electrocyclic reactions

application of microscopic reversibility, 1232–1233
role in vitamin D biosynthesis, 1252–1254
selection rules, summary, 1250–1251
Peri-interaction, 1175
Periodate, *see* Periodic acid
Periodate ester, intermediate in periodate cleavage of glycols, 471
Periodic acid
in oxidative cleavage of glycols, 470–472
properties, 470
reaction with carbohydrates, 1352
Periplanone B, structure, 814
Permanganate, *see* Potassium permanganate
Peroxide effect, in HBr addition to alkenes, 194–197
explanation, 200–204
Peroxides
effect on HBr addition to alkenes, 195
in ether autoxidation, 377–378
in ethers, test, 378
Peroxy acids, **503**
in epoxidation of alkenes, 503–505
in oxidation of sulfides, 523
Peroxycarboxylic acid, **503**
Petroleum, **75**
as foundation of chemical industry, 213f
components, 75t
importance in polymer industry, 1022
pH, **98**
Phase, of molecular orbitals, 1224
α-Phellandrene, 735p, 1170f
α-Phenethylamine, enantiomeric resolution, 249–250
α-Phenethyl chloride, 791t
Phenol, **823** *see also* Phenols
bromination, 773, 846–847
Friedel-Crafts acylation, 848–849
industrial synthesis and use, 851–852
ionization and pK_a, 837
IR spectrum, 741
ketone forms, 1045
nitration, 846
production from petroleum, 213f
structure, 738
Phenolate ion, *see* Phenoxide ion
Phenol-formaldehyde resins, 917
Phenols, *see also* Phenol
acidity, 837–840
chemical shift of OH protons, 745
reactions
electrophilic aromatic substitution, 846–847
ester formation with acid chlorides, 1002–1003

ester formation with anhydrides, 1004–1005
oxidation to quinones, 841–845
with aluminum trichloride, 848
synthesis
hydrolysis of diazonium salts, 1141
Claisen rearrangement, 1248
from aryl halides, 848
separation
from carboxylic acids, 940–941
from mixtures using acidity, 840–841
Phenoxide ion
as nucleophiles in S_N2 reactions, 841
formation and use, 837, 840–841
Phenylacetic acid, pK_a, 118p, 939t
Phenylalanine, structure and properties, 1266t
Phenyl cation, 828
Phenylcyclohexane, *see* Cyclohexylbenzene
Phenyl group, **79, 337, 739**
conformational preference in cyclohexylbenzene, 333p
directing effect in electrophilic aromatic substitution, 766p
polar effect, 799, SGL 17.2
Phenylhydrazine
in Fischer indole synthesis, 1190
in osazone formation, 1385p
reaction with aldehydes and ketones, 906t
Phenylhydrazones, 906t
intermediates in Fischer indole synthesis, 1190
Phenyl isothiocyanate (Edman reagent), in analysis of peptides, 1284–1285
Phenyllithium, synthesis, 860p
Phenylmagnesium bromide, synthesis, 836
Phenylthiohydantoins, formation in Edman degradation, 1285
Pheromones, 668–671
Phosgene, 959, 977
Phosphatidylcholine, structure, 1025
Phosphatidylethanolamine, structure, 1024f, 1025
Phospholipids, 1024–1025
Phosphoric acid, esters, 456p
Phosphorus
exact mass, 559t
^{31}P NMR, 633
reagent in α-bromination of carboxylic acids, 1052
Phosphorus pentachloride, in synthesis of acid chlorides, 952–953
Phosphorus tribromide
reaction with alcohols, 457p
reagent in α-bromination of carboxylic acids, 1052
Photochemical reactions, 1232

Photography, role of phenol oxidation, 845
Photon, **536**
Phthalamic acid, hydrolysis, 1038p
Phthalic acid, 805p
pK_a values, 939t
structure, 930
Phthalic anhydride, 955
Phthalimide, 976
in Gabriel amine synthesis, 1146–1147
Pi (π) bond, 124–125
Picolines, 1194; *see also* 2-, 3-, or 4-Methylpyridine
Picric acid, acidity, 839
Pimelic acid, structure, 931t
Pi (π) molecular orbitals, 124–125
Pinacol rearrangement, 882–883
Pinacolone, 882
α-Pinene, structure, 809f
biosynthesis, 812p
Piperidine, 1115
removal of Fmoc group, 1297
Pivalic acid
pK_a, 939t
structure, 931t
pK_a, **98**
of Brønsted acids, 99t
pK_b, 100–101
Planck's constant, 42p, **537**
Plane of symmetry, **229**
Plane-polarized light, *see* Polarized light
Planteose, structure, 1388p
Platinum, hydrogenation catalyst, 162
Pleated sheet, in structure of peptides, 1302–1303
Plexiglas, 209f
Plunkett, Roy J., 209–210
Poisons, catalyst, 161, 659
Polar
double usage, 352
solvent classification, **352**
Polar bond, 10–11, **10**
Polar effects, **110**
electron-donating, **110**
electron-withdrawing, **110**
of substituents in electrophilic aromatic substitution, 768
on acidity of phenols, 837
on amine basicity, 1124
on alcohol acidity, 364
on Brønsted acidity, 108–111
Polarimeter, **235**
Polarity, *see* Dipole moment
Polarization
mechanism for alkoxide stabilization, 365
mechanism for ammonium ion stabilization, 1123

Polarized light, 234–235
Polar molecules, **11**
Polyacrylonitrile, 209*t*
Polybutadiene, 707
Polychlorotrifluoroethylene, 209*t*
Polycyclic compounds, aromaticity, 723
Polyesters, 1021–1022
Polyethylene, 208
    production from petroleum, 213*f*
    synthesis from ethylene, 208
Poly(ethylene terephthalate), 1021–1022
Polyisoprene, *see* Rubber, natural
Polymer, **207**
    addition, **208**
    condensation, **1021**
Polymerization, **207**
    free-radical, 208
    of alkenes, 207–210
    of dienes, 707–708
Poly(methyl methacrylate), 209*t*
Polyphosphoric acid, 1190
Polypropylene, production from petroleum, 213*f*
Polysaccharides, **1327**, 1364–1367
Polystyrene, 209*t*, 220*p*
    production from petroleum, 213*f*
Polytetrafluoroethylene, 209*t*
Poly(vinyl chloride), 209*t*
Porphyrin, 1206
Potassium amide, reaction with aryl halides, 832
Potassium bromide, solubilization by crown ethers, 359
Potassium *tert*-butoxide
    as base in elimination reactions, 387
    as base in generating dichloromethylene, 428
Potassium chloride
    solubilization by crown ethers, 359
    structure, 4, 5*f*
Potassium ferricyanide, in synthesis of 1-methyl-2-pyridone, 1202
Potassium iodide, reaction with diazonium salts, 1141
Potassium permanganate
    in benzylic oxidations, 804
    in conversion of alkenes into glycols, 193
        stereochemistry, 320
    in oxidation of primary alcohols, 469
    in oxidation of sulfides, 523
    in thiol oxidation, 484
    solubilization by crown ethers, 359
Pott, Percivall, 777
ppm, *see* Parts per million
Pr, abbreviation for propyl group, 1064*t*
i-Pr, abbreviation for isopropyl group, 1064*t*

Previtamin D_3, 1252
Primary, classification
    of alcohols, 335
    of alkyl halides, 335
    of amides, 975
    of amines, 1113
    of carbocations, 149
    of carbon, 64
    of free radicals, 201
    of hydrogens, 64
Primary deuterium isotope effect, *see* Isotope effect
Primary structure, of peptides and proteins, 1299–1300
Principal chain, **56**, **339**; *see also* Nomenclature, substitutive
    in alkene nomenclature, 129
Principal group, **339**
    citation priority, 932
Principal quantum number, **23**
Principle of microscopic reversibility, *see* Microscopic reversibility
Prismane, 716
    and orbital symmetry 1233–1234
Probability
    and electron position, 22
    of a reaction, 520
        and entropy, SGL 11.6
Problems,
    involving reactions, solving, SGL 4.5
    involving unknown structures, solving, SGL 4.6
    mechanistic, solving, SGL 4.7
Product-determining step, **419**
    analogy, 419–420
Progesterone, structure, 301
Proline, structure and properties, 1267*t*
Propadiene (Allene), 679; *see also* Allenes
    structure, 682*f*, 683*f*
Propagation, in free-radical reactions, 198–199
Propane
    physical properties, 47*t*, 346, 350
    structure, 122*f*
Propanedioic acid, *see* Malonic acid
1,2-Propanediol, *see* Propylene glycol
1,2,3-Propanetriol, *see* Glycerol
Propanoic acid
    IR spectrum, 936*f*
    pK_a, 939*t*
    NMR spectrum, 936*f*
2-Propanol (Isopropyl alcohol)
    commercial synthesis, 165*p*
    pK_a, 365*t*
Propargyl group, **645**
2-Propenal, *see* Acrolein
Propene
    acidity of allylic hydrogens, 800

in synthesis of cumene, 776–777
    production, 212, 213*f*
    structure, 122*f*, 866*f*
Propionyl group, 868
Propiophenone, CMR spectrum, 876
Propylene, *see* Propene
Propylene glycol, structure, 338
Propyl group
    abbreviation, 1064*t*
    structure, 57
Propyl radical, heat of formation, 203*t*
Pro-*R*-hydrogen, 481
Pro-*S*-hydrogen, 481
Proteases, in peptide hydrolysis, 1289–1291
Protecting groups, **902**
    for carbohydrates, 1348
    in electrophilic substitution of aniline derivatives, 1138
Proteins, **1266**; *see also* Peptides
    occurrence, 1311, 1313
Proteolytic enzymes, *see* Proteases
Protic, solvent classification, 352
Protonated, **147**
Protonolysis, **371**
Protons, spin, 579–581
PTH, *see* Phenylthiohydantoins
Pulegone, 1108*p*
Purine, structure, 1178*f*
Purines, components of nucleic acids, 1368
Purple benzene, 359
PVC, *see* Poly(vinyl chloride)
Pyramidal geometry, **17**
    of carbon (hypothetical), consequences, 263–264, 270*p*
Pyranose, **1332**
Pyranoses, conformations, 1334–1337
Pyrene, 1170*f*
Pyridine
    and derivatives, electrophilic aromatic substitution, 1194–1197
    aromaticity, 721
    as a base
        in amide synthesis, 1001–1002
        in reactions of alcohols with thionyl chloride, 456
        in sulfonate ester formation, 452
    basicity and pK_a of conjugate acid, 1182
    catalyst poison, 659
    electronic structure, 721*f*, 1179–1180, 1180*f*
    NMR spectrum, 1181*p*
    reactions
        Chichibabin and related reactions, 1197–1199
        oxidation to pyridine-*N*-oxides, 1196

with methyl iodide, 1202
resonance energy, 1181*t*
structure, 1178*f*
Pyridine-*N*-oxide
  and derivatives, deoxygenation, 1197
  Grignard reactions, 1202
  nitration, 1196–1197
  synthesis, 1196
Pyridinium salts
  methyl derivatives, acidity, 1203
  reactions, 1201–1202, 1203
2-Pyridone
  conversion into 2-chloropyridine, 1200
  equilibrium with 2-hydroxypyridine, 1200
  hydrolysis, 1217*p*
  synthesis, 1200
Pyridoxal, 1207*f*
Pyridoxine, synthesis, 1218*p*
Pyrimidine, structure, 1178*f*
Pyrimidines, components of nucleic acids, 1368
γ-Pyrone, 1103*p*
Pyrophosphate group, 810
Pyrophosphoric acid, 810
Pyrrole
  acidity, 1183
  aromaticity, 721
  basicity and pK_a of conjugate acid, 1182
  dipole moment, 1181*p*
  electronic structure, 721*f*, 1180*f*
  electrophilic aromatic substitution, 1184–1185
  nitration, 1185
  NMR spectrum, 1181*p*
  reaction with organometallic reagents, 1183
  resonance energy, 1181*t*
  structure, 1178*f*
Pyrrolidine, 1115
  dipole moment, 1181*p*
  NMR spectrum, 1181*p*

Q

Quantum mechanics, 21–22
Quantum numbers, **23**, 23*t*, 23–28
  angular momentum (*l*), **23**
  magnetic (*m*), **23**
  principal (*n*), **23**
  spin (*s*), **23**
Quartet, splitting pattern in NMR spectra, **596**
Quartz, optical activity, 261

Quaternary ammonium salts, *see* Ammonium salts, quaternary
Quaternary carbon, **64**
  diagnosing by DEPT technique, 625–627
Quaternary structure, of proteins, 1307
Quaternization, of amines, 1129–1130
Quinine, 1154*f*
Quinoline, and derivatives
  aromaticity, 723
  basicity and pK_a of conjugate acid, 1182
  catalyst poison, 659
  nitration, 1197*p*
  structure, 1178*f*
  synthesis
    Combes synthesis, 1218*p*
    Friedlander synthesis, 1217*p*
    Skraup synthesis, 1204–1205
Quinones, **842**
  conjugate-addition reactions, 1087
  synthesis from phenols, 841–845
Quintet, splitting pattern in NMR spectra, **596**

R

R, configuration of asymmetric atom, **231**
Racemates, **239**
  formation in reactions of achiral substances, 305
Racemic mixture, **239**
Racemic modification, **239**
Racemization, **239**, 254
Radiation, electromagnetic, *see* Electromagnetic radiation
Radical, *see* Free radical
Radical anion, **660**
Radical cation, **556**
Radicofunctional nomenclature, 336
Radiofrequency, in NMR experiment, 581
Raffinose, 1384*p*
Random coil, in peptides and proteins, 1302
Rate, of reaction, **154**
  and equilibrium constant, 155–156
  effect of temperature, 155
Rate constant, **392**
Rate-determining, *see* Rate-limiting
Rate law, 391–392
  for acid-catalyzed halogenation of aldehydes and ketones, 1049
  for S$_N$1-E1 reactions, 417–419
  for S$_N$2 reactions, 394–395
  relationship to mechanism, 395
Rate-limiting step, 156–158, **157**

analogy, 158
and principle of microscopic reversibility, 445
relationship to product-determining step, 419–421
Rates, of reactions, *see* Reaction rates
Reaction-free energy diagram, **154**
Reaction mechanism, *see* Mechanism, reaction
Reaction rate, 154–160, 390–392, **391**; *see also* Rate law
  and neighboring-group participation, 519–520
  and rate law, 391–392
  and reaction conditions, 773–774; SGL 16.4
  and standard free energy of activation, 392
Reactions, conventions for writing, 179–180
Reactive intermediates, and Hammond's postulate, 158–160
Rearrangements
  allylic, **801**
  of carbocations, 151–153
    in acid-catalyzed alkene hydration, 164
    in acid-catalyzed alcohol dehydration, 446
    in Friedel-Crafts alkylation, 758
    in pinacol rearrangement, 882–883
    in reaction of alcohols with hydrogen halides, 448
    in S$_N$1-E1 reactions, 422
  of cumene hydroperoxide, 851
  sigmatropic, 1239–1250
Recombination, reactions of free radicals, 199
Reducing agent, **463**
Reducing sugar, **1363**
Reduction, **459**; *see also* specific compound classes, *e.g.*, Aldehydes, reduction
Reductive amination, 1130–1132
Reforming, of alkanes, 76
Regioselective, **146**
Reissert indole synthesis, 1192–1194
Relative abundance, of ions in mass spectrometry, **557**
Repulsions, *see* van der Waals repulsions
Residue, in peptides, **1267**
Resin, **917**
Resolution, *see* Enantiomeric resolution
Resolving agent, **249**
Resonance, 19–21, 708–715
  and benzene structure, 717
  destabilization, 725
  stabilization, 702, 713

Resonance effect, of substituents
    and dipole moments, 862p
    in electrophilic aromatic substitution,
        768
    on amine basicity, 1124
    on phenol acidity, 837–839, SGL 18.3
Resonance energy, empirical, **719**; *see also*
    specific compounds
Resonance hybrid, **20**
Resonance structures,
        SGL 15.2
    derivation by curved-arrow formalism,
        95–96
    drawing, 709–712
    relative importance, 710–712
    use, 712–715
Resorcinol, 739
Respiration, similarity to combustion, 1
Retention, stereochemistry of substitution,
    312
Rf, *see* Radiofrequency
R groups, 79
L-Rhamnose, 1387p
Rhodopsin, 689
β-D-Ribofuranose, 1336
Ribonuclease
    denaturation, 1306
    solid-phase synthesis, 1292
Ribonucleic acid
    role in protein synthesis,
        1375–1378
    structure, 1372
Ribonucleoside, **1367**
Ribonucleotide, **1368**
D-(−)-Ribose
    component of ribonucleic acids,
        1368
    equilibrium composition, 1340t
    structure, 1329f
Rickets, 1252
Ring current, **742**
Ring strain, **293**
    importance in epoxide ring opening,
        511
RNA, *see* Ribonucleic acid
Robinson annulation, **1092**
Rocking vibration, 547f
Rosanoff, M. A., 1327
Rosenmund reduction, of acid chlorides,
        1012
Rotation, internal, **50**; *see also* Internal
    rotations
Roth, Wolfgang, 1243–1244
Rubber
    natural, 219p, 708, 809f
    styrene-butadiene (SBR), 707
Ruff degradation, of aldoses, 1360
Ruhemann's purple, 1282

S

S, configuration of asymmetric atom, **231**
s, quantum number, **23**
s-cis conformation, of dienes, 681
Saffrole, structure, 715, 814p
Salicin, 1345f
Salicylic acid, 930, 1004
SAM, *see* S-adenosylmethionine
Saponification, of esters, 989–990
    in malonic ester synthesis, 1075
    relationship to Claisen condensation,
        1067
Saturated hydrocarbons, **121**
Saytzeff, Alexandr, 411
Saytzeff's rule, 411
SBR, 707
Schiff base, **904**; *see* Imines
Schotten-Baumann technique, for amide
        formation, 1002
β-Scission, **211**
    in mass spectrometry, **564**
Scissoring vibration, 547f
Secondary, classification
    of alcohols, 335
    of alkyl halides, 335
    of amides, 975
    of amines, 1113
    of carbocations, 149
    of carbon, 64
    of free radicals, 201
    of hydrogens, 64
Secondary structure, of peptides and pro-
        teins, 1301–1303, **1303**
Second-order
    reaction kinetics, 392
Selection rules,
    for cycloaddition reactions, 1238t
    for electrocyclic reactions, 1232t
    for pericyclic reactions, summary,
        1250–1251
    for sigmatropic reactions, 1248–1249
Semicarbazide, reaction with aldehydes
        and ketones, 906t
Semicarbazone, 906t
Semiquinone radical, 843–844
Septet, splitting pattern in NMR spectra,
    596
Sequence, amino acid, **1284**
Serine, structure and properties, 1267t
Sesquiterpenes, **807**
Sextet, splitting pattern in NMR spectra,
    596
β-Sheet, *see* β-Structure
Sigma (σ) bonds, **33**
Sigmatropic reactions, **1219**, 1239–1250
    classification, 1239

selection rules, 1248–1249
    stereochemistry, 1240–1250
Silicon, exact isotopic masses and natural
        abundances, 559t
Silk, 1303
Silver bromide, in photography, 845
Silver(I) oxide
    catalyst for methylation of carbohy-
        drates, 1347
    in aldehyde oxidation, 916
    in synthesis of quaternary ammonium
        hydroxides, 1133–1134
Simmons, Howard E., 430
Simmons-Smith reaction, 430–433
Singlet, in NMR spectroscopy, **596**
Skraup synthesis, of quinoline and deriva-
        tives, 1204
Skunk, essence of, 380p
Smalley, Richard E., 1169
Smith, Ronald D., 430
S_N1 reactions, 417–423, **418**
    as alkylation reactions, 455
    contrast with S_N2 and E2 reactions, 421
    effect of leaving groups on reaction
        rate, 421
    inertness of aryl and vinylic halides,
        826–829
    inertness of α-halo carbonyl com-
        pounds, 1054
    prediction, 423–427
    rate law and mechanism, 417–419
    rearrangements, 422
    solvent effect on rate, 417, 421
    summary, 423
S_N2 reactions, 394–404, **394**
    allylic, 803–804
        orbital diagram of transition state,
            804f
    and epoxide ring opening, 509
    as alkylation reactions, 455
    benzylic, 803–804
    competition with E2 reaction, 411–415,
        803
    contrast with S_N1-E1 reaction, 420
    effect of alkyl halide structure on rate,
        397–399, 398t
    effect of leaving group on rate, 403
    inertness of vinylic and aryl halides,
        824–825
    of cyclopropyl halides, 436p
    prediction, 423–427
    rate law and mechanism, 394–395
    solvent effects, 399–402
    stereochemistry, 396–397
    steric effects, 398
    summary, 403–404
S_N2' reaction, 821–822p
Soaps, **942**

Sodamide, *see* Sodium amide

Sodium, in liquid ammonia, 660–662

Sodium acetylide, addition to ketones, 896

Sodium amide, 663, 1127
  base in carbohydrate alkylation, 1347
  in Chichibabin reaction, 1197–1198

Sodium azide, in synthesis of acyl azides, 1150

Sodium bisulfite
  addition to aldehydes and ketones, 919*p*
  reducing agent for osmium, 193

Sodium borohydride
  discovery as reducing agent, 894
  inertness of nitro compounds, 1147–1148
  reactions
    reduction of aldehydes and ketones, 890–894
    reduction of aldoses, 1353
    reduction of alkoxymercuration adducts, 499–500
    reduction of oxymercuration adducts, 181
    reductive amination, 1130
  synthesis, 893
  with α,β-unsaturated carbonyl compounds, 1094

Sodium cyanoborohydride, in reductive amination, 1131

Sodium dichromate
  in benzylic oxidations, 804
  in synthesis of quinones, 841

Sodium D-line, use in measurement of optical activity, 235

Sodium dodecylsulfate, 965*p*

Sodium ethoxide, base in elimination reactions, 387

Sodium hydride
  base in Claisen and related condensations, 1068
  product of Chichibabin reaction, 1198
  use in synthesis of alkoxides from alcohols, 362

Sodium nitrite
  as food preservative, 1145
  in amine diazotization, 1139

Sodium pentothal, 1108*p*

Solid-phase peptide synthesis, 1291–1298

Solubility, 351, 353–358
  of covalent compounds, 353–356
  of ionic compounds, 356–358

Solvate, 356

Solvated electron, *see* Electron, solvated

Solvation, 356
  of Grignard reagents, 369

Solvation shell, 356

mimicked by crown ethers, 359

Solvent cage, **356**

Solvents, **351**, 351–358; *see also* specific reactions, *e.g.*, S_N2 reactions, solvent effects
  classification, 352–353
  dry-cleaning, 373
  properties, 354–355*t*

Solvolysis, **417**

Specificity, in enzyme catalysis, 1308

Specific rotation, **236**, 236–237

Spectrometer, *see* Spectrophotometer

Spectrometer, mass, *see* Mass spectrometer

Spectrophotometer, **538**

Spectroscopy, **535**
  absorption, **538**
  in solving structure problems, 628–632
  introduction, 535–540
  structural information available by spectroscopy type, 539

Spectroscopy, NMR, *see* Nuclear magnetic resonance spectroscopy

Spectrum, **538**; *see also* Electromagnetic spectrum
  ultraviolet-visible, 684–686

Spin
  of electrons, 23
  of nuclei, 579–581, 622*t*

Spin-spin splitting, *see* Splitting

Spirocyclic, **294**

Spiropentadiene, 734*p*

Splitting, in NMR spectroscopy, 594–610
  complex, 603–610
  diagram, 604–605, 605*f*
  effect of chemical exchange, 617–618, 620–621
  multiplicative, 603–607
  $n + 1$ rule, 595–598
  of alkene protons, 614*t*
  physical basis, 599–600
  theoretical line intensities, 596*t*
  use in determining unknown structures, 600–603

Squaric acid, 967*p*

Squiggly bond, in denoting uncertain configuration, 1336

Stability, kinetic vs. thermodynamic, SGL 22.1

Staggered conformation, **50**

Standard free energy, *see* Free energy

Starch, 1365–1366

Stearic acid, structure, 931*t*

Stereocenters, **229**, 128
  and reaction stereochemistry, 311–312

Stereochemistry, **225**, 225–269; *see also* specific reactions
  correlation, 240–241, **240**

D,L system, 1268–1269

E,Z (Cahn-Ingold-Prelog) system, 132
  group equivalance, 475–480
  in organic synthesis, 523
  of reactions, 310–320
  R,S (Cahn-Ingold-Prelog) system, 231–233

Stereogenic, 128; *see also* Stereocenter

Stereogenic atom, *see* Stereocenter

Stereoisomers, **126**, 225
  conformational, 253–255
  in cyclic compounds, 289–292
  nomenclature, 132–136
  in reactions, 302–310
  relationahip to other types of isomers, 243

Stereoselective reactions, 313

Stereospecific reactions, SGL 7.6, SGL 11.4

Steric effect, **202**
  in free-radical additions to alkenes, 202
  in S_N2 reactions, 398, 399*f*

Steroids, 300–302
  biosynthesis, 1084
  sources, 302

Stilbene, 775

Strain, ring and angle, **293**

s-trans conformation, of dienes, 681

Strecker synthesis, 1279–1280

Stretching vibration, **546**

β-Structure, in peptides and proteins, 1302–1303

Structures, **12**
  drawing, 15, SGL 1.3, SGL 4.1, SGL 4.2
  highly condensed, 61
  Lewis, *see* Lewis structures
  methods for determining, 12–13
  of covalent compounds, predicting, 13–19
  of ionic compounds, 4
  relationship to hybridization, 123
  skeletal, 64–67

Study Guide Links, flagged (checked), 10

Study Guide Links, unflagged, 11

Styrene, 131
  bromine addition, 716
  catalytic hydrogenation, 162
  copolymerization with butadiene, 707
  hydration, solvent isotope effect, 407*p*
  production from petroleum, 213*f*
  source and industrial use, 777
  structure, 738

Styrene-butadiene rubber, 707

Styrene oxide, structure, 344

Substituents, 57

Substitution, by free-radical mechanism, 372

Substitution, *cine*, 835

Substitution, aromatic, *see* Aromatic substitution

Substitution, nucleophilic, *see* Nucleophilic substitution

Substitution test, for stereochemical group relationships, 476–477

Substitutive nomenclature, *see* Nomenclature, substitutive

Substrate, **1308**

Subunits, in proteins, 1307

Succinic acid
 pK_a values, 939*t*
 structure, 931*t*

Succinimide, 797, 976
 acidity, 1042*p*

Sucrose, 1363
 hydrolysis to glucose and fructose, 1363

Sugars, *see* Carbohydrates

Sulfa drug, 1139*p*

Sulfanilamide, 1139*p*

Sulfathiazole, 1139*p*

Sulfenic acid, 483*f*

Sulfhydryl group, **336**

Sulfides
 nomenclature, 343–344
 oxidation, 522–523
 Williamson synthesis, 497–499

Sulfinic acid, 483*f*

Sulfolane,
 synthesis, 522
 solvent properties, 355*t*

Sulfonate anions
 as leaving groups in reactions of sulfonate esters, 452
 resonance structures, 939

Sulfonate esters, **450**
 E2 reactions, 454
 mechanism of formation, SGL 10.3
 reactivity in substitution reactions, 452–454
 structures, 450–451
 synthesis, 451–452, 1003

Sulfonation
 of benzene, 753–754
 of naphthalene, 1174
 reversibility, 753, 1174

Sulfones, synthesis, 522

Sulfonic acid group, activating and directing effects in electrophilic aromatic substitution, 762*t*

Sulfonic acids, **450**
 acidity, 939
 conversion into sulfonyl chlorides, 953
 structure, 929
 synthesis
  sulfonation of aromatic compounds, 753–754, 1174
  thiol oxidation, 484

Sulfonium salts, 516–518

Sulfonyl chlorides
 in synthesis of sulfonate esters, 451–452
 synthesis, 953
 reaction with alcohols, 1003

Sulfoxides, synthesis, 522

Sulfur, exact isotopic masses and natural abundances, 559*t*

Sulfuric acid
 in sulfonation of benzene, 753
 pK_a, 99*t*

Sulfur trioxide, in benzene sulfonation, 753

Supports, for hydrogenation catalysts, 162

Suprafacial, **1235**, 1240

Surfactants, **942**, SGL 20.4

Sweeteners, peptide, 1313

Symmetric molecular orbital, **1224**

Symmetry
 and chirality, 229–230
 cylindrical, **33**
 elements, **229**
 plane, **229**

Syn, addition stereochemistry, **311**

Syn addition, 696

Syn elimination, **407**, 407–409

Synthesis, *see* Organic synthesis

Synthetic equivalence, SGL 17.3

T

2,4,5-T, synthesis, 861*p*

D-(+)-Talose
 equilibrium composition, 1340*t*
 structure, 1329*f*

Target molecule, **486**

Tartaric acid,
 as a resolving agent for enantiomers, 250
 D- and L-, stereochemical correlation with D-glucose, 1360–1361
 in history of stereochemistry, 261
 sodium ammonium salt, crystals, 262*f*
 structure, 930

Tautomer, **1045**

Teflon, 209*t*
 discovery, 209–210

Temperature, effect on reaction rate, 155

Terephthalic acid
 industrial synthesis and use, 777
 in polyester production, 1021–1022

Termination, in free-radical reactions, 199–200

Terpenes, 806–812
 biosynthesis, 808–812

Tertiary, classification
 of alcohols, 335
 of alkyl halides, 335
 of amides, 975
 of amines, 1113
 of carbocations, 149
 of carbon, 64
 of free radicals, 201
 of hydrogens, 64

Tertiary structure, of proteins, 1303–1308

Testosterone, structure, 301

Tetrachloroethylene, industrial use, 373

Tetracyanoethylene, 1257*p*
 dienophile in Diels-Alder reaction, 695

Tetrafluoroborate ion, 7–8

Tetrahedral addition intermediate, **948**
 in Chichibabin reaction, 1198
 in ester hydrolysis, 991
 in ester saponification, 990
 in nucleophilic acyl substitution, 996–1000
 in nucleophilic aromatic substitution (Meisenheimer complex), 830
 in trypsin-catalyzed peptide hydrolysis, 1312*f*

Tetrahedral geometry, 15
 of carbon, postulation, 260–264

Tetrahedrane, structure, 296

Tetrahedron, **15**, 15*f*

Tetrahydrofuran
 as reaction cosolvent with water, 356
 boiling point and dipole moment, 346
 contamination with peroxides, 378
 dipole moment, 1179
 in nylon production, 1022*p*
 in synthesis of Grignard reagents, 368
 solvent in hydroboration, 184
 solvent in oxymercuration, 180
 solvent properties, 354*t*
 structure, 344

α-Tetralone, synthesis by intramolecular Friedel-Crafts reaction, 755

Tetramethylsilane, internal standard for NMR, 582

Thermodynamic control, **704**, 1089
 in naphthalene sulfonation, 1174–1175

THF, *see* Tetrahydrofuran

Thiamin, 1207*f*

Thiazole, structure, 1178*f*

Thiazolinones, intermediates in Edman degradation, 1285

Thioether, **336**; *see also* Sulfide

Thiolates, **361**; *see also* Mercaptides

Thiolates
 basicity, 99*t*
 heavy-metal derivatives, 363
  and Wilson's disease, 363–364
 in Williamson synthesis of sulfides, 497–498

nucleophiles in thiol oxidation, 485
synthesis from thiols, 363
Thiol esters, in fatty acid biosynthesis, 1082–1083
Thiol esters, reactivity, 1029p
Thiols, **336**
acidity, 360–364
amphoterism, 366
conversion into disulfides, 484
conversion into sulfonic acids, 484
IR spectroscopy, 574p
nomenclature, 338–342
oxidation, 482–485
pK_a, 99t
Thionyl chloride
in synthesis of acid chlorides, 952–953
in synthesis of alkyl chlorides, 456–457
properties, 456
Thiophene
and derivatives, electrophilic aromatic substitution, 1186–1187
Hinsberg synthesis, 1217p
nitratrion, 1185
resonance energy, 1181t
structure, 344, 1178f
Thioureas, intermediates in Edman degradation, 1284
Threonine, structure and properties, 1267t
D- and L-, absolute configuration, 1269
D-(−)-Threose, synthesis from D-glyceraldehyde, 1360
structure, 1329f
β-Thujone, structure, 814
Thymidine, 1369t
Thymidylic acid, 1369t
Thymine, 1368
Thymine dimer, in DNA, 1380
TMP, 1369t
TMS, see Tetramethylsilane
α-Tocopherol, 844
Tollens' test, 916
with monosaccharides, 1350
Toluene
acidity of benzylic hydrogens, 800
IR spectrum, 741f
nitration, 775
physical properties, 740
source and industrial use, 776
structure, 738
p-Toluenesulfonate anion, basicity, 99t
leaving groups in reactions of tosylate esters, 452–453
p-Toluenesulfonate esters, see Tosylates
p-Toluenesulfonic acid
acidity and pK_a, 99t, 939
structure, 450
p-Toluenesulfonyl chloride
in synthesis of tosylate esters, 451–452

reaction with alcohols, 1003
Toluidine, see Methylaniline
Tosylate, 451; see also p-Toluenesulfonate anion
Tosyl chloride, see p-Toluenesulfonyl chloride
Trans
alkene stereochemistry, 129
cycloalkane substitution, 283–287
ring fusion in polycyclic compound, 297–299
Transcription, of DNA to RNA, 1376
Transesterification, **1005**
of carbohydrate esters, 1347
$\pi \rightarrow \pi^*$ Transition, 688
Transition state, **154**
Translation, of messenger RNA, 1376
Transmittance, **541**
Transoid conformation, of dienes, 681
1,3,5-Tribromobenzene, synthesis, 1142
2,4,6-Tribromophenol, synthesis from phenol, 846–847
Tributyltin hydride, 439p
Tricarballylic acid, 1092
Trichloroethylene, industrial use, 373
Trichlorofluoroethane, industrial use, 373
Trichloromethyl anion, intermediate in dichloromethylene formation, 427
Triethylamine, pK_a of conjugate acid, 1121t
Triflate, 492p
Trifluoroacetic acid, pK_a, 108, 939
4,4,4-Trifluoro-1-butanol, pK_a, 364
2,2,2-Trifluoroethanol
dielectric constant, 353p
pK_a, 364
3,3,3-Trifluoro-1-propanol, pK_a, 364
Trifluralin, 1159p
Trihalomethylanion, as leaving group, 1051
Trimethylamine, pK_a of conjugate acid, 1121t
2,4,6-Trimethylbenzoic acid, see Mesitoic acid
Trimethyloxonium fluoroborate, 516–517
2,2,4-Trimethylpentane
dielectric constant, 353p
standard for octane number, 78
2,4,6-Trinitrobenzoic acid, pK_a, 939t
2,4,6-Trinitrophenol, see Picric acid
Tripeptide, **1267**
Triphenylmethane, acidity, 820p
Triphenylmethyl, see Trityl
Triphenylmethyl chloride, see Trityl chloride
Triphenylphosphine, in Wittig alkene synthesis, 913
Triphenylphosphine oxide, by-product of Wittig alkene synthesis, 911

Triple bond, **7**
treatment in E,Z system, 135
Triplet, splitting pattern in NMR spectra, **596**
Triptycene, synthesis, 863p
Trisaccharides, **1327**
Tritium, introduction by Grignard protonolysis in HTO, 371
Triton B, 1127
Trityl chloride, 791t
Trityl radical, 819–820p
Tropone, 1103p
Tropylium bromide, 732p
Trypsin
catalysis, 1308–1310, 1312
enzyme catalyst for peptide hydrolysis, 1289–1291
structure of enzyme-substrate complex, 1310
Tryptophan, structure and properties, 1266t
Twist-boat, conformation of cyclohexane, 277–279, 279
Twisting vibration, 547f
Tyrosine, structure and properties, 1266t

U

Ultraviolet spectrophotometer, 684
Ultraviolet spectroscopy, 684–690
and color, 688–689
physical basis, 686–688
forbidden absorptions, 877
of benzene derivatives, 746–747
UMP, 1369t
Uncertainty principle, **22**
Undecane, physical properties, 47t
α,β-Unsaturated carbonyl compounds, catalytic hydrogenation, 894
conjugate addition reactions, 1085–1093
Diels-Alder reactions, 690–699
equilibration with unconjugated isomers, 1102p
in Skraup quinoline synthesis, 1204
reactions with amines, 1128–1129
reaction with lithium dialkylcuprates, 1095–1096
reaction with organolithium reagents, 1094–1098
reduction, 1093–1094
synthesis by aldol condensation, 1062–1063
Unsaturated hydrocarbons, **121**
Unsaturation number, **137**, 136–137
Unshared pairs, **6**

Upfield, 582
Uracil, 1368
Urea, 977
  as a protein denaturant, 1306
  derivatives, synthesis by aminolysis of
      isocyanates, 1149
  structure, 959
  synthesis from ammonium cyanate, 2
  use in agriculture, 1154
Uridine, 1369t
Uridylic acid, 1369t
Uronic acids, structure, 1349t
UV spectroscopy, see Ultraviolet
      spectroscopy

V

Valence electrons, 3
Valence shell, 3
Valeric acid, structure, 931t
Valine, structure and properties, 1266t
Valproic acid, 963p, 1078
van der Waals force,
  attractive, 69
    effect on boiling point, 347
    in proteins, 1304–1305
van der Waals radius, 53
van der Waals repulsions, 54
van Ekenstein, Willem Alberda, 1341
Vanillin, 857, 878p
  as a glycoside, 1346
  structure, 715
Van't Hoff, Jacobus Hendricus, 263
Vector, 11
Veronal, 1108p
Vibrations, molecular
  and IR spectroscopy, 541
  classification, 546
Vicinal, 175
Vicinal glycol, 192, 336
Vinylacetylene, industrial synthesis and
      use, 671
Vinyl alcohol, 1044
Vinyl chloride
  industrial synthesis from acetylene, 670
Vinyl chloride, structure, 337
Vinyl 2,2-dimethylpropanoate, NMR spec-
      trum and analysis of splitting,
      603–605

Vinyl group, 131
  treatment in E,Z system, 135
Vinylic anion, 663
Vinylic cation, 827, 828f, SGL 18.1
Vinylic ethers, hydrolysis, 1109p
Vinylic halides, 823
  elimination reactions, 826
  inertness to S_N1 conditions, 826–829
  inertness to S_N2 conditions, 824–825
Vinylic protons, chemical shifts in NMR,
      612
Vinylic radical
  intermediate in alkyne reduction, 661
  inversion, 661
Visible spectroscopy, see Ultraviolet
      spectroscopy
Visual purple, 689
Vital force, 1
Vitamin A, structure, 809f
Vitamin B$_1$, 1207f
Vitamin B$_6$, 1207f
Vitamin C, see Ascorbic acid
Vitamin D, 1253
  biosynthesis, 1252–1254
Vitamin E, 844
Vulcanization, 708

W

Wagging vibration, 547f
Wallach, Otto, 806
Water
  as solvent for covalent compounds,
      353–355
  as solvent for ionic compounds, 358
  basicity, 99t
  orbital hybridization, 37p
  pK_a, 99t
  solvent properties, 355t
  structure, 14f, 42p
Water softeners, 1276
Watson, James D., 1372
Wave
  electrons, 21–22
  nodes, 25
  peaks, 25
  troughs, 25
Wavelength, 536, 536f
Wavenumber, 540
  as a frequency, 540

Wave-particle duality, 22
Wax, 1023
Wedges, and dashed wedges, 15
Weerman degradation, of aldoses, 1386p
Whitmore, Frank C., 153
Wilkins, Maurice, 1372
Williamson, Alexander William, 498
Williamson synthesis, 435p, 497–499
  intramolecular, 505
  of aryl ethers, 841
  of ethers and sulfides, 497–499
Wilson's disease, 363–364
Winstein, Saul, 520
Wittig, Georg, 915
Wittig alkene synthesis, 911–915
Wohl degradation, of aldoses, 1386p
Wöhler, Friedrich, 2, 1367
Wolff-Kishner reduction, 909–911
Woodward, Robert B., 1221
Woodward-Hoffmann-Fukui theory, 1221

X

Xeroderma pigmentosum, connection to
      DNA damage, 1380
Xylenes, 738
  source and industrial use, 777
D-(+)-Xylose
  equilibrium composition, 1340t
  structure, 1329f

Y

Yeast, role in fermentation, 474
Ylids, 912

Z

Z, stereochemical nomenclature, 132
Ziegler process, for ethylene polymeriza-
      tion, 208
Zinc-copper couple, use in Simmons-
      Smith reaction, 430–431
Zingiberene, biosynthesis, 817–818p
Zoapatanol, structure, 815
Zwitterions, 1263, 1269–1271

Organic Functional Groups

R = alkyl or aryl unless indicated otherwise.

	⟵	Aliphatic Hydrocarbons	⟶	
	Alkane	**Alkene**	**Alkyne**	**Aromatic Hydrocarbons**
General formula	R—H (R = alkyl)	$R_2C{=}CR_2$ (R = alkyl, aryl, H)	$R{-}C{\equiv}C{-}R$ (R = alkyl, aryl, H)	
Functional group	$-\overset{\|}{\underset{\|}{C}}-$	$\overset{\backslash}{\underset{/}{C}}{=}\overset{/}{\underset{\backslash}{C}}$	$-C{\equiv}C-$	benzene or other aromatic ring
Specific example	$CH_3{-}CH_3$	$CH_2{=}CH_2$	$HC{\equiv}CH$	
Common name	ethane	ethylene	acetylene	benzene
Substitutive name	ethane	ethene	ethyne	benzene

	Alkyl Halide	**Alcohol**	**Phenol**	**Ether**
General formula	R—X (R = alkyl X = halogen)	R—OH (R = alkyl)	R—OH (R = aryl)	R—O—R
Functional group	$-\overset{\|}{\underset{\|}{C}}-X$	$-\overset{\|}{\underset{\|}{C}}-OH$		$-\overset{\|}{\underset{\|}{C}}-O-\overset{\|}{\underset{\|}{C}}-$
Specific example	$CH_3{-}CH_2{-}Cl$	$CH_3{-}CH_2{-}OH$		$CH_3{-}O{-}CH_2{-}CH_3$
Common name	ethyl chloride	ethyl alcohol	phenol	ethyl methyl ether
Substitutive name	chloroethane	ethanol	phenol	methoxyethane